Advances in Intelligent Systems and Computing

Volume 1074

The series "Advances in Intelligent Systems and Computing" contains publications on theory, applications, and design methods of Intelligent Systems and Intelligent Computing. Virtually all disciplines such as engineering, natural sciences, computer and information science, ICT, economics, business, e-commerce, environment, healthcare, life science are covered. The list of topics spans all the areas of modern intelligent systems and computing such as: computational intelligence, soft computing including neural networks, fuzzy systems, evolutionary computing and the fusion of these paradigms, social intelligence, ambient intelligence, computational neuroscience, artificial life, virtual worlds and society, cognitive science and systems, Perception and Vision, DNA and immune based systems, self-organizing and adaptive systems, e-Learning and teaching, human-centered and human-centric computing, recommender systems, intelligent control, robotics and mechatronics including human-machine teaming, knowledge-based paradigms, learning paradigms, machine ethics, intelligent data analysis, knowledge management, intelligent agents, intelligent decision making and support, intelligent network security, trust management, interactive entertainment, Web intelligence and multimedia.

The publications within "Advances in Intelligent Systems and Computing" are primarily proceedings of important conferences, symposia and congresses. They cover significant recent developments in the field, both of a foundational and applicable character. An important characteristic feature of the series is the short publication time and world-wide distribution. This permits a rapid and broad dissemination of research results.

**** Indexing: The books of this series are submitted to ISI Proceedings, EI-Compendex, DBLP, SCOPUS, Google Scholar and Springerlink ****

More information about this series at http://www.springer.com/series/11156

Yong Liu · Lipo Wang ·
Liang Zhao · Zhengtao Yu
Editors

Advances in Natural Computation, Fuzzy Systems and Knowledge Discovery

Volume 1

Springer

Editors
Yong Liu
School of Computer Science
and Engineering
The University of Aizu
Aizu-Wakamatsu, Fukushima, Japan

Lipo Wang
School of Electrical
and Electronic Engineering
Nanyang Technological University
Singapore, Singapore

Liang Zhao
Computer Science and Mathematics
University of Sao Paulo
Ribeirao Preto, Brazil

Zhengtao Yu
School of Information Engineering
and Automation
Kunming University of Science
and Technology
Kunming, China

ISSN 2194-5357 ISSN 2194-5365 (electronic)
Advances in Intelligent Systems and Computing
ISBN 978-3-030-32455-1 ISBN 978-3-030-32456-8 (eBook)
https://doi.org/10.1007/978-3-030-32456-8

This Springer imprint is published by the registered company Springer Nature Switzerland AG
The registered company address is: Gewerbestrasse 11, 6330 Cham, Switzerland

Preface

The 2019 15th International Conference on Natural Computation, Fuzzy Systems and Knowledge Discovery (ICNC-FSKD 2019) was held from July 20 to 22, 2019, in Kunming, China, colocated with the 5th International Conference on Harmony Search, Soft Computing and Applications (ICHSA 2019).

ICNC-FSKD is a premier international forum for scientists and researchers to present the state of the art of machine learning, data mining, and intelligent methods inspired from nature, particularly biological, linguistic, and physical systems, with applications to computers, systems, control, communications, and more. We are delighted to receive many submissions from around the globe. After a rigorous review process, the accepted papers are included in this proceedings.

Kunming is the capital and largest city of Yunnan Province in southwest China. Kunming is also called the Spring City due to its mild sunny weather all year round. Kunming is located at the northern edge of the large Lake Dian, surrounded by famous temples. Attractions near Kunming include Stone Forest (South China Karst, a UNESCO Natural World Heritage Site), Village of Ethnic Culture, Western Hill, and Dragon Gate. Other UNESCO sites in Yunan are Old Town of Lijiang, Three Parallel Rivers of Yunnan Protected Areas, Chengjiang Fossil Site/Maotianshan, and Cultural Landscape of Honghe Hani Rice Terraces/Yuanyang.

We would like to sincerely thank all organizing committee members, program committee members, invited session organizers, and reviewers for their hard work and valuable contribution. Without your help, this conference would not have been possible. Special thanks go to Kunming University of Science and Technology for hosting this event. We thank Springer for publishing the proceedings. We are very grateful to the keynote and invited speakers for their authoritative speeches. We thank all the authors and conference participants for using this platform to communicate their excellent work.

August 2019

Yong Liu
Liang Zhao
Lipo Wang
Zhengtao Yu

Organization

Organizing Committee

General Chairs

Zhengtao Yu	Kunming University of Science and Technology, China
Joong Hoon Kim	Korea University, South Korea

Program Chairs

Yong Liu	University of Aizu, Japan
Lipo Wang	Nanyang Technological University, Singapore
Liang Zhao	University of São Paulo, Brazil
Anupam Yadav	NIT Jalandhar, India

Organizing Chairs

Zheng Xiao	Hunan University, Changsha, China
Asim Roy	Arizona State University, USA

Finance Chair

Jiwu Peng	Hunan University, China

Publication Chairs

Leszek Rutkowski	Technical University of Czestochowa, Poland
Guoqing Xiao	University of Waterloo, Canada

Publicity Chairs

Jiang Wang	Kunming University of Science and Technology, China
Huiyu Zhou	Queen's University Belfast, UK

Program Committee

Shigeo Abe	Kobe University, Japan
Henry N. Adorna	University of the Philippines, the Philippines
Davide Anguita	University of Genoa, Italy
Sabri Arik	Istanbul University, Turkey
Krassimir Atanassov	Bulgarian Academy of Sciences, Bulgaria
Sansanee Auephanwiriyakul	Chiang Mai University, Thailand
Philip Azariadis	University of the Aegean, Greece
Vladan Babovic	Singapore National University, Singapore
Thomas Bäck	Leiden Institute of Advanced Computer Science, the Netherlands
Emili Balaguer-Ballester	Bournemouth University, UK
Valentina Balas	Aurel Vlaicu University of Arad, Romania
Yaxin Bi	University of Ulster, UK
Federico Bizzarri	Politecnico di Milano, Italy
Tossapon Boongoen	Mae Fah Luang University, Thailand
Pierre Borne	Ecole Centrale de Lille, France
Hamid Bouchachia	Bournemouth University, UK
Ivo Bukovsky	Czech Technical University in Prague, Czech
Sujin Bureerat	Khon Kaen University, Thailand
Godwin Caruana	Harvest Technology, Malta
Michele Ceccarelli	University of Sannio, Italy
Kit Yan Chan	Curtin University, Australia
Chen-Tung Chen	National United University, Taiwan
David Daqing Chen	London South Bank University, UK
Jianxia Chen	Washington University in St. Louis, USA
Syuan-Yi Chen	National Taiwan Normal University, Taiwan
Chi Tsun (Ben) Cheng	RMIT University, Australia
Jao Hong Cheng	National Yunlin University of Science and Technology, Taiwan
France Cheong	RMIT University, Australia
Jen-Shiun Chiang	Tamkang University, Taiwan
Panagiotis Chountas	University of Westminster, UK
Huey-Der Chu	Takming University of Science and Technology, Taiwan
Hung-Yuan Chung	National Central University, Taiwan
Alessandro Colombo	Politecnico di Milano, Italy

José Alfredo F. Costa	Universidade Federal do Rio Grande do Norte, Brazil
Keeley Crockett	Manchester Metropolitan University, UK
Zoltán Ernö Csajbók	University of Debrecen, Hungary
Darryl N. Davis	University of Hull, UK
Andre C. P. L. F. de Carvalho	University of Sao Paulo, Brazil
Marc de Kamps	University of Leeds, UK
Mingcong Deng	Tokyo University of Agriculture and Technology, Japan
Minghua Deng	Peking University, China
Milena Djukanovic	University of Montenegro, Montenegro
Mustafa Dogan	Baskent University, Turkey
Prabu Dorairaj	Broadcom Inc., India
Giorgos Dounias	University of the Aegean, Greece
António Dourado	University of Coimbra, Portugal
Abdelali El Aroudi	Universitat Rovira i Virgili, Spain
Mohammed ElAbd	American University of Kuwait, Kuwait
Zuhal Erden	ATILIM University, Turkey
Geoffrey Falzon	STMicroelectronics (Malta) Ltd, Malta
Xiannian Fan	City University of New York, USA
Saeed Panahian Fard	Universiti Sains Malaysia, Malaysia
Elisabetta Fersini	University of Milan-Bicocca, Italy
Zbigniew Galias	AGH University of Science and Technology, Poland
Peter Geczy	AIST, Japan
Damian Giaouris	Newcastle University, UK
Onofrio Gigliotta	University of Naples Federico II, Italy
David Glass	University of Ulster, UK
Antonio Gonzalez	University of Granada, Spain
Giuseppe Grassi	University of Salento, Italy
Perry Groot	Radboud University Nijmegen, the Netherlands
Yuzhu Guo	University of Sheffield, UK
Jianchao (Jack) Han	California State University, USA
Thomas Hanne	University of Applied Sciences Northwestern Switzerland, Switzerland
Pitoyo Hartono	Chukyo University, Japan
Enrique Herrera-Viedma	University of Granada, Spain
Mhand Hifi	Université de Picardie, France
Ladislav Hluchy	Institute of Informatics, Slovak Academy of Sciences, Slovakia
Sean Holden	University of Cambridge, UK
Jun Hong	University of the West of England, Bristol, UK
Tzung-Pei Hong	National University of Kaohsiung, Taiwan
Wei-Chiang Samuelson Hong	Oriental Institute of Technology, Taiwan
Wen-xing Hong	Xiamen University, China

Rym MHallah	Kuwait University, Kuwait
Hongwei Mo	Harbin Engineering University, China
Dusmanta Kumar Mohanta	MVGR College of Engineering, India
Robert Newcomb	University of Maryland, USA
Yoshifumi Nishio	Tokushima University, Japan
Yusuke Nojima	Osaka Prefecture University, Japan
Dimitri Ognibene	CNR-ISTC, Italy
Maciej Ogorzalek	Jagiellonian University, Poland
Kok-Leong Ong	La Trobe University, Australia
Milos Oravec	Slovak University of Technology in Bratislava, Slovakia
Vasile Palade	Coventry University, UK
Linqiang Pan	Huazhong University of Science and Technology, China
Shaoning Pang	Auckland University of Technology, New Zealand
George Panoutsos	University of Sheffield, UK
Dong-Chul Park	Myongji University, Korea
Jessie Ju H. Park	Yeungnam University, South Korea
Petra Perner	Institute of Computer Vision and applied Computer Sciences, Germany
Valentina Plekhanova	University of Sunderland, UK
Petrica Pop	North University of Baia Mare, Romania
Man Qi	University of Canterbury, UK
Guangzhi Qu	Oakland University, USA
Rajesh Reghunadhan	Bharathiar University, India
Pedro Manuel Pinto Ribeiro	University of Porto, Portugal
Asim Roy	University of Arizona, USA
Álvaro Rubio-Largo	University of Extremadura, Spain
Alireza Sadeghian	Ryerson University, Canada
Indrajit Saha	National Institute of Technical Teachers Training & Research, India
Evangelos Sakkopoulos	University of Western Greece, Greece
Antonio Sala	Universitat Politecnica de Valencia, Spain
Christoph Schommer	University of Luxembourg, Luxembourg
Huseyin Seker	University of Northumbria at Newcastle, UK
Hirosato Seki	Osaka University, Japan
Neslihan Serap Sengör	Istanbul Technical University, Turkey
Subarna Shakya	Tribhuvan University, Nepal
Changjing Shang	Aberystwyth University, UK
Yain Whar Lawrence Si	University of Macau, Macau
Humberto Sossa	Instituto Politécnico Nacional, Mexico
João Miguel Sousa	Technical University of Lisbon, Portugal
Marco Storace	Telecommunications Engineering and Naval Architecture, University of Genoa, Italy

Mu-Chun Su	National Central University, Taiwan
Muhammad Sulaiman	Abdul Wali Khan University Mardan, Pakistan
Wen-Tsai Sung	National Chin-Yi University of Technology, Taiwan
Johan Suykens	KU Leuven University, Belgium
Eulalia Szmidt	Polish Academy of Sciences, Poland
Norikazu Takahashi	Okayama University, Japan
Vicenc Torra	University of Skövde, Sweden
Ljiljana Trajkovic	Simon Fraser University, Canada
Isis Truck	University Paris 8, France
Brijesh Verma	Central Queensland University, Australia
John Vlachogiannis	Industrial and Energy Informatics Laboratory (IEI-Lab), Greece
Michael N. Vrahatis	University of Patras, Greece
Feng Wan	University of Macau, China
Di Wang	Khalifa University, UAE
Lingfeng Wang	University of Wisconsin-Milwaukee, USA
Lipo Wang	Nanyang Technological University, Singapore
Xiaofan Wang	Shanghai Jiao Tong University, China
Hua-Liang Wei	University of Sheffield, UK
Santoso Wibowo	Central Queensland University, Australia
Slawomir Wierzchon	Polish Academy of Sciences, Poland
Ka-Chun Wong	City University of Hong Kong, China
Rolf Würtz	Ruhr-Universität Bochum, German
Jing Xiao	South China Normal University, China
Fan Xiong	Bio-Rad Laboratories, USA
Ning Xiong	Mälardalen University, Sweden
Yue Xu	Queensland University of Technology, Australia
Chan-Yun Yang	National Taipei University, Taiwan
Yingjie Yang	De Montfort University, UK
Zhijun Yang	Middlesex University London, UK
Yiyu Yao	University of Regina, Canada
Chung-Hsing Yeh	Monash University, Australia
Jian Yin	Sun Yat-sen University, China
Wen Yu	CINVESTAV-IPN (National Polytechnic Institute), Mexico
Yuqing Zhai	Southeast University, China
Jie Zhang	Newcastle University, UK
Jinglan Zhang	Queensland University of Technology, Australia
Liming Zhang	University of Macau, China
Liqing Zhang	Shanghai Jiao Tong University, China
Min-Ling Zhang	Southeast University, China
Zhongwei Zhang	University of Southern Queensland, Australia
Liang Zhao	University of Sao Paulo, Brazil
Wei Zheng	Xiamen University, China

Huiyu Zhou	Queen's University Belfast, UK
Ligang Zhou	Macau University of Science and Technology, Macau
Shangming Zhou	Swansea University, UK
Wenxing Zhu	Fuzhou University, China
William Zhu	Minnan Normal University, China
Jeffrey Zou	University of Western Sydney, Australia

Reviewers

Abobakr Khalil Al-Shamiri
Vangalur Alagar
Josep Arnal
Leqiang Bai
Luyi Bai
Dongming Chen
Feiqiang Chen
Ji Chen
Jianxia Chen
Weiyang Chen
Wenjuan Chen
Xiaogang Chen
Xuegang Chen
Yanping Chen
Hsien-Hsin Chou
Jianzhong Cui
Yingan Cui
Yingbao Cui
Shaobo Deng
Jozsef Dombi
Xiaomei Dong
Jishe Feng
Tak-chung Fu
Jian Gao
Yang Gao
Xianya Geng
Srimannarayana Grandhi
Wanrong Gu
Huaping Guo
Galib Hamidov
Bai Han
Song Han
Ming He
Ladislav Hluchy
Ling-Yuan Hsu
Yinfu Huang
Lisi Jia
Hong Jiang
Yongchen Jiang
Yung-Tsan Jou
Wengkin Lai
Kittichai Lavangnananda
Tao Lei
Haohao Li
Hui Li
Weigang Li
Xiaobin Li
Xue Li
Yafeng Li
Zhenxing Li
Hsing-Hung Lin
Haitao Lin
Ao Liu
Chunhui Liu
Fanghua Liu
Genggeng Liu
Jing Liu
Jingjing Liu
Kun Liu
Lanfen Liu
Lei Liu
Qingsheng Liu
Ruifang Liu
Wei Liu
Xiaoyan Liu
Yubao Liu
Wojciech Lorkiewicz
Hu Lu

Jin Lu
Ruhua Lu
Hongwei Ma
Hui Ma
Liangyu Ma
Shiwei Ma
Wei Mei
Hongying Meng
Alfredo Milani
Georgina Mirceva
Tianliang Peng
Manop Phankokkruad
Zhenhong Rao
Ghamgeen Izat Rashed
Hendrik Richter
Dingcai Shen
Yonghong Shen
Xiaoyu Shi
Chen-Chi Shing
Fang Su
Zhengru Tao
Liye Tian
Qiujuan Tong
Milan Tuba
Guixiang Wang
Jiesheng Wang
Shuching Wang
Xiao Wang
Xing Wang
Yujie Wang
Zhengfang Wang
Zhenhai Wang
Zhijun Wang
Santoso Wibowo
Xia Wu
Youxi Wu
Hong Xia
Sidong Xian
Gang Xie
Bo Xu
Saijuan Xu
Zhe Xu
Ye Xue
Chun Yan
Chunman Yan
Senlin Yan
Changsheng Yang
Jinfu Yang
Xiaojun Yang
Xinfeng Yang
Xiyang Yang
Yong Yao
Makoto Yasuda
Peng Yin
Xu Ying
Do Guen Yoo
Fusheng Yu
Xiujiu Yuan
Yinggao Yue
Zhihao Yun
Yuriy Zaychenko
Bin Zhang
Chijian Zhang
Huoming Zhang
Jialu Zhang
Jianke Zhang
Jin Zhang
Juxiao Zhang
Xiaojun Zhang
Xing Zhang
Yonghe Zhang
Yu-an Zhang
Yunong Zhang
Xinchao Zhao
Yunping Zheng
Xiaobin Zhi
Fujin Zhong
Aiping Zhou
Pucheng Zhou
Zhiwen Zhou
Xiaolan Zhu
Xianxia Zou

Contents

Natural Computation Applications: Natural Computation in Pattern Recognition and Diagnostics

Natural Computation Applications: Natural Computation in Signal Processing and Multimedia

Natural Computation: Theory and Algorithms - Deep Learning and Feedforward Neural Networks

Optimizing Convolutional Neural Network Architecture Using a Self-adaptive Harmony Search Algorithm

Yin-Fu Huang(✉) and Jung-Sheng Liu

National Yunlin University of Science and Technology, 123 University Road, Section 3, Touliu 640, Yunlin, Taiwan, R.O.C.
huangyf@yuntech.edu.tw

Abstract. In recent years, the advance of GPUs led to the development in neural networks and deep learning. However, it is difficult to find a good CNN architecture people desire. In the past, people had to manually find the CNN architecture, and this is quite time-consuming and labor-intensive. In this paper, we use a self-adaptive harmony search algorithm to find the optimized convolutional neural network architecture for image recognition. The system architecture is divided into two phases. In the first phase, we search for the most suitable layer length of a CNN. In the second phase, we fine-tune the architecture found in the first phase or a pre-trained architecture. In the experiments, three popular and well-known datasets are used to evaluate the proposed methods and the state-of-the-art CNN search methods. The experimental results show that our methods achieve competitive performances compared with the other methods on CIFAR-10, and have the best performance among all the methods on MNIST and Caltech-101.

Keywords: Convolutional Neural Networks · Deep learning · Hyper-parameter optimization · Self-adaptive harmony search algorithms · Transfer learning

1 Introduction

In recent years, the advance of GPUs led to the development in neural networks and deep learning. Because of more memory equipped with GPUs, neural networks can be trained deeper and wider, thereby achieving higher accuracy.

However, it is difficult to find a good CNN architecture people desire. In the past, people had to manually adjust the architecture hyper-parameters and then trained the architecture iteratively. It is not only time-consuming but labor-intensive. In recent years, some CNN architecture search methods have been proposed, and they can be divided into two categories. One is the methods proposed by Rosa et al. [9], Young et al. [15], and Khan [5] where the lengths of the searched architectures are fixed. Another is the methods proposed by Albelwi et al. [1], Al-Hyari et al. [2], Singh et al. [11], Pham et al. [8], and Tanno et al. [13] where the lengths of the searched architectures are dynamic or not fixed. Inspired by that an evolutionary algorithm can also iterate over a large number of solutions and finds out better architecture hyper-parameters, the

Y. Liu et al. (Eds.): ICNC-FSKD 2019, AISC 1074, pp. 3–12, 2020.
https://doi.org/10.1007/978-3-030-32456-8_1

self-adaptive harmony search (i.e., SAHS) algorithm [14] is used to optimize a CNN architecture.

In this paper, we use the SAHS algorithm to optimize a CNN architecture. The system architecture is divided into two phases. In the first phase, we search for the most suitable layer length of a CNN. In the second phase, based on the CNN found in the first phase, the hyper-parameters used inside the CNN are fine-tuned to promote the accuracy of image recognition.

The remainder of this paper is organized as follows. In Sect. 2, the system architecture consisting of two phases is proposed. In Sect. 3, Phase 1 process describes how to search new architectures based on Vgg16 as widely as possible. In Sect. 4, Phase 2 process focuses on how to fine-tune the hyper-parameters used inside the CNN. In Sect. 5, three popular and well-known datasets are used in the experiments to evaluate the CNNs built by the proposed methods and those done by the state-of-the-art CNN search methods. Finally, we make conclusions in Sect. 6.

2 System Overviews

In this section, we propose an approach to optimize a convolutional neural network (i.e., CNN) used in image recognition, using a self-adaptive harmony search (i.e., SAHS) algorithm [14]. As shown in Fig. 1, the system architecture is divided into two phases. In the first phase, we search for the most suitable layer length of a CNN. However, Phase 1 is optional based on training datasets. In general, if datasets are large or complex, the datasets are not suitable to train a CNN from scratch due to requiring a large number of parameters. Instead, we choose CNNs such as the original Vgg16 [10] and Inception-v3 [12], and start to train them from the second phase. Then, based on the CNN found in the first phase, the hyper-parameters used inside the CNN are fine-tuned to promote the accuracy of image recognition in the second phase.

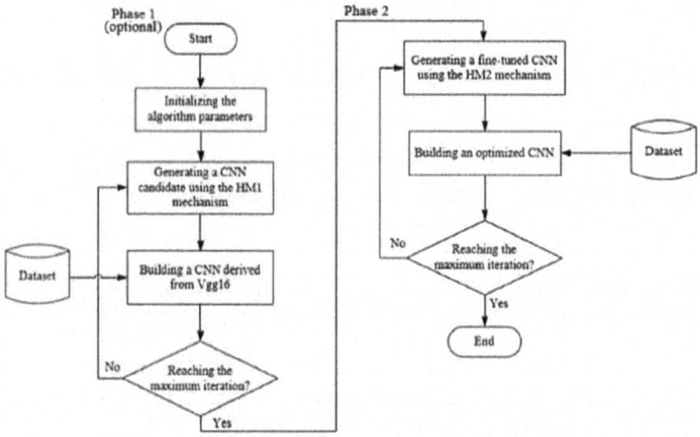

Fig. 1. System architecture to optimize CNNs.

In Phase 1, we use the SAHS algorithm to generate CNN candidates. These generated CNNs are all derived from the Vgg16 architecture. As usual in the SAHS algorithm, the maximum iteration is set as the termination condition. When reaching the maximum iteration, the system switches to Phase 2 for fine-tuning the CNN found in Phase 1. In Phase 2, to get the optimal hyper-parameter values used inside the CNN found in Phase 1, we generate different hyper-parameter values using a modified harmony search algorithm. Similar to Phase 1, when reaching the maximum iteration, the system eventually generates an optimized CNN used in image recognition.

3 Phase-1 Process

In the Phase-1 process, we aim to search new architectures based on Vgg16 as widely as possible. Here, we use the SAHS algorithm to generate CNN candidates.

3.1 Phase-1 Algorithm

In the Phase-1 algorithm as shown in Fig. 2, we initialize the algorithm parameters first, including *HMCR*, *PAR*, *hms*, *MI* (i.e., maximum iteration), and then randomly generate the initial harmony memory *HM*. The hyper-parameters used inside CNN architectures can be realized as the solutions in *HM*, and one solution in *HM* corresponds to the hyper-parameters inside a CNN architecture. A new CNN architecture is generated and built with its accuracy for each iteration. Here, accuracy is the criterion to evaluate the performance of a CNN architecture. The HM1 mechanism (i.e., the SAHS algorithm) is used to generate a new solution *NS* (or a new CNN architecture). Next, the new solution *NS* is compared with all current solutions in *HM* based on accuracy, and then we remove the worst solution. Finally, we use the remaining solutions (i.e., the *hms* solutions in *HM*) for the next iteration. When reaching the maximum iteration, the system moves forward to the second phase for fine-tuning the best CNN candidate found in Phase 1.

Phase-1 Algorithm
Input : *HMCR, PAR, hms, MI*
Output :CNN_{best}
1. Initialize *HMCR, PAR, hms, MI, and HM*
2. for *t* = 1 to *MI* do
3. *NS* = HM1-Mechanism(*HMCR, PAR, hms*)
4. Building a CNN candidate based on *NS*, comparing with the other CNN candidates in *HM*, and removing the worst CNN candidate
5. end for
6. Choosing the best CNN candidate CNN_{best} in *HM*

Fig. 2. Phase-1 algorithm.

3.2 Generating CNN Candidates Using the HM1 Mechanism

In the HM1 mechanism as shown in Fig. 3, we use the SAHS algorithm to generate new solutions. This intention of the SAHS algorithm is to achieve the self-adaptive effect when each generated solution is fine-tuned using a dynamic range, but not a fixed range.

```
HM1-Mechanism
Input : HMCR, PAR, hms
Output : x'
for j = 1 to n do
    if rand() < HMCR then
        x'_{i,j} = HM(int(rand()*hms) + 1, j)
        if rand() < PAR then
            if rand() < 0.5 then
                x'_{i,j} = x'_{i,j}−int([x'_{i,j} − min(HM)] * ran[0,1))
            else
                x'_{i,j} = x'_{i,j}+int([max(HM) − x'_{i,j}] * ran[0,1))
            end if
        end if
    else
        x'_{i,j} = int(rand()*(UB - LB))
    end if
end for
```

Fig. 3. HM1 mechanism.

3.3 Building a CNN Derived from Vgg16

We modify the original Vgg16 architecture by inserting more layers into the architecture. All the information of the inserted layers is realized as a solution in *HM*. As shown in Table 1, a solution in *HM* consists of the following information such as (1) layers: the total number of convolution layers to be added in Vgg16, (2) slot (i.e., space between layers): the position of a new layer to be added in Vgg16, (3) filter size: the filter size of a convolution layer, (4) filter number: the number of output filters used in a convolution layer, and (5) dropout: the percentage of dropout for reducing overfitting. How many groups of slot, filter size, filter number, and dropout in a solution are used is dependent on the number of convolution layers to be added in Vgg16.

Table 1. Information of the inserted layers in Vgg16.

Layers	Slot	Filter Size	Filter Number	Dropout	Slot	Filter Size	Filter Number	Dropout	Slot	Filter Size	Filter Number	Dropout
1~3	1~14 (MNIST) 1~18 (CIFAR10)	1~3	1~5	1~5	1~14 (MNIST) 1~18 (CIFAR10)	1~3	1~5	1~5	1~14 (MNIST) 1~18 (CIFAR10)	1~3	1~5	1~5
		1*1 3*3 5*5	32 64 128 256 512	10% 20% 30% 40% 50%		1*1 3*3 5*5	32 64 128 256 512	10% 20% 30% 40% 50%		1*1 3*3 5*5	32 64 128 256 512	10% 20% 30% 40% 50%

Besides, different from the original Vgg16 architecture, the last two fully-connected layers in Block 6 are changed into one convolution layer and one global average pooling (GAP) [7] in the new CNN architecture. As the fully-connected layers in Block 6 have a lot of parameters to be trained and cost much training time, they are replaced to reduce the training time. However, the new CNN architecture still achieves almost the same accuracy.

Furthermore, we have different basic CNN architectures for datasets MNIST and CIFAR-10 in the experiments due to their image sizes, as shown in Fig. 4.

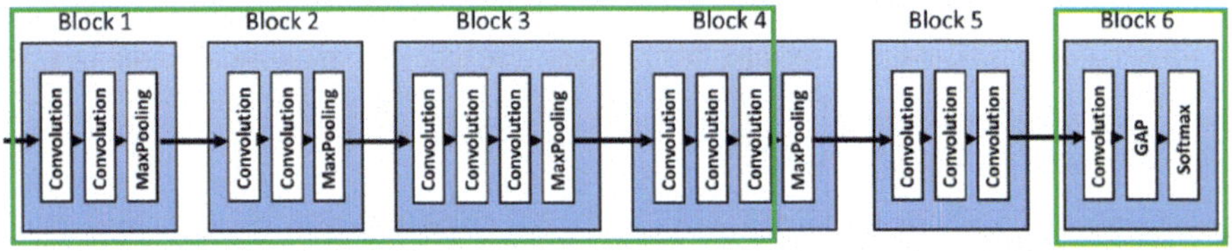

Fig. 4. Different basic architectures for MNIST (green parts) and CIFAR-10 (all).

4 Phase-2 Process

In the second phase, the hyper-parameters used inside the CNN are fine-tuned to find the optimized CNN. The second phase can be proceeded for two cases. For the case of MNIST or CIFAR-10, we use a modified harmony search algorithm to fine-tune the CNN found in the first phase. For another case of Caltech-101 more large and complex than the former two, we use the original Vgg16 and Inception-v3 pre-trained on ImageNet instead, and fine-tune them using the same modified harmony search algorithm.

4.1 Phase-2 Algorithm

In the Phase-2 algorithm as shown in Fig. 5, we use the same parameters as the Phase-1 algorithm, including *HMCR*, *PAR*, *hms*, *MI*, and randomly generate the initial harmony memory *HM*. The Phase-2 algorithm is very similar to the Phase-1 algorithm where we use a modified harmony search algorithm (i.e., the HM2 mechanism) to generate a new solution *NS*. Then, the CNN found in the first phase (or the pre-trained Vgg16 and Inception-v3) is integrated with *NS*, and a CNN candidate is built through the similar steps as done in the Phase-1 algorithm. Finally, when reaching the maximum iteration, the system generates an optimized CNN used in image recognition.

Phase-2 Algorithm
Input : *HMCR, PAR, hms, MI,* CNN_{best}(or Vgg16 or Inception-v3)
Output : $CNN_{optimal}$
1. Initialize *HMCR, PAR, hms, MI, and HM*
2. for $t = 1$ to *MI* do
3. *NS* = HM2-Mechanism(*HMCR, PAR, hms*)
4. Building a CNN candidate based on CNN_{best}(or Vgg16 or Inception-v3) and *NS*, comparing with the other CNN candidates in *HM*, and removing the worst CNN candidate
5. end for
6. Choosing the best CNN candidate $CNN_{optimal}$ in *HM*

Fig. 5. Phase-2 algorithm.

4.2 Generating a Fine-Tuned CNN Using the HM2 Mechanism

In the HM2 mechanism, we modify the HM1 mechanism using a fixed range, instead of using a dynamic range. If the current value should be adjusted downwards, it decreases by 1 unless it reaches the minimum value in *HM*. If the current value should be adjusted upwards, it increases also by 1 unless it reaches the maximum value in *HM*.

4.3 Building an Optimized CNN

An optimized CNN can be built based on (1) the CNN found in the first phase and (2) the pre-trained models Vgg16 and Inception-v3.

Based on the CNN Found in the First Phase. For the CNN found in the first phase, we fix Block 6 as shown in Fig. 4 (i.e., the last three layers including convolution layer, GAP, and softmax layer), and fine-tune the other convolution layers using the modified harmony search algorithm. The hyper-parameters to be fine-tuned include filter size, filter number, and dropout of each convolution layer. For the case, the new CNN architecture for CIFAR-10 found in Phase 1, with the 15 convolution layers including two inserted layers from Block 1 to Block 5, is fine-tuned as shown in Table 2 where the hyper-parameter values are more meticulous than Phase 1. In other words, we have 45 hyper-parameter values are fine-tuned in the modified harmony search algorithm.

Table 2. Information of the convolution layers from Block 1 to Block 5.

1			2				15		
Filter Size	Filter Number	Dropout	Filter Size	Filter Number	Dropout		Filter Size	Filter Number	Dropout
1~3	1~14	1~11	1~3	1~14	1~11	....	1~3	1~14	1~11
1*1 3*3 5*5	8 16 24 32 48 64 96 128 192 256 384 512 768 1024	0% 5% 10% 15% 20% 25% 30% 35% 40% 45% 50%	1*1 3*3 5*5	8 16 24 32 48 64 96 128 192 256 384 512 768 1024	0% 5% 10% 15% 20% 25% 30% 35% 40% 45% 50%		1*1 3*3 5*5	8 16 24 32 48 64 96 128 192 256 384 512 768 1024	0% 5% 10% 15% 20% 25% 30% 35% 40% 45% 50%

Based on the Pre-trained Models. For the pre-trained models, we use Vgg16 and Inception-v3 as bases to build optimized CNNs. Using a pre-trained model to classify other different datasets is an approach called transfer learning. It has the advantage of not having to retrain the entire model and reducing the time it takes for the model to converge.

For the Vgg16 architecture, we freeze the layers from Block 1 to Block 5 and fine-tune the layers in Block 6. In other words, only the weights within Block 6 are adjusted. Besides, the last softmax layer is specified with 101 classes for Caltech-101, not 1000 classes for ImageNet. The hyper-parameters to be fine-tuned include the neuron units and dropout of the fully-connected layers in Block 6 as shown in Table 3. The fully-connected unit 0 means that this fully-connected layer and dropout layer will not be built.

Table 3. Information of the fully-connected layers in Block 6.

1st Fully-connected Units	Dropout	2nd Fully-connected Units	Dropout
1~12 (Vgg16) 1~10 (Inception-v3)	1~5	1~12 (Vgg16) 1~10 (Inception-v3)	1~5
0 128 256 384 512 640 768 896 1024 2048 3072 4096	10% 20% 30% 40% 50%	0 128 256 384 512 640 768 896 1024 2048 3072 4096	10% 20% 30% 40% 50%

For the original Inception-v3 architecture, the last part is replaced with fully-connected layers and softmax layer (i.e. blue circle) in the new Inception-v3 architecture as shown in Fig. 6. We freeze the first 249 layers and fine-tune the rest part (i.e. red square as shown in Fig. 6). In other words, only the weights within the red square are adjusted. Like Vgg16, the hyper-parameters to be fine-tuned include the neuron units and dropout of the fully-connected layers in the blue circle, as shown in Table 3.

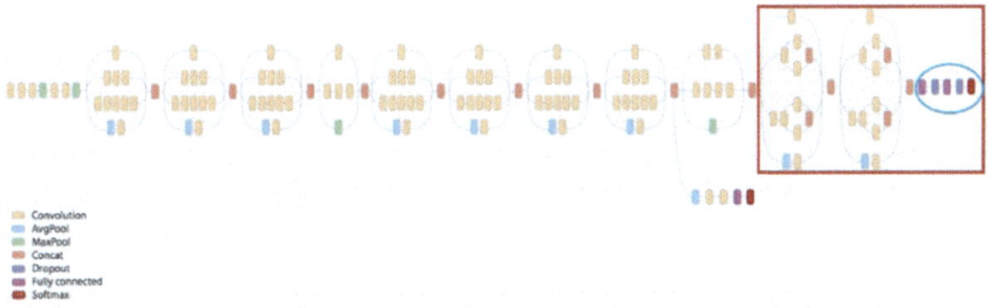

Fig. 6. New Inception-v3 architecture for Caltech-101.

5 Experiments

In this study, three popular and well-known datasets are used to evaluate the CNNs built by the proposed methods. The experiments are implemented in Python and use Keras to train the CNNs on GeForce GTX 1060 6 GB. First, the parameters of the experimental environment such as Keras and the SAHS algorithm are specified. Then, the CNNs built by our methods are compared with those done by the state-of-the-art CNN search methods. The experimental results show that our methods achieve competitive performances compared with the other methods on CIFAR-10, and have the best performance among all the methods on MNIST and Caltech-101.

5.1 Experimental Environments

In the experiments, suitable optimizers are adopted for different datasets; i.e., the Adadelta optimizer [16] for MNIST, and the SGD (i.e., stochastic gradient descent) optimizer [3] for CIFAR-10 and Caltech-101.

For MNIST, the weights ω are trained through the Adadelta optimizer with learning rate $lr = 1.0$, $rho = 0.95$, and batch size = 128. Each candidate architecture is run for 10 epoches. Besides, we also apply ℓ_2 weight decay [6] of $5 * 10^{-4}$ and batch normalization [4] after the activation function. The weight decay is to adjust the effect of model complexity on the loss function, which can be added to the loss function as a penalty to prevent overfitting. Furthermore, the batch normalization is to prevent gradient vanishing problems, which can speed up model convergence and helps training faster. For the SAHS algorithm, parameters *HMCR*, *PAR*, *hms*, and *MI* are set to 0.99, 0.2, 10, and 40, respectively.

For CIFAR-10, the weights ω are trained through the SGD optimizer with learning rate $lr = 0.1$, $lr\_decay = 10^{-6}$, momentum = 0.9, nesterov = True, and batch size = 128. Each candidate architecture is run for 250 epoches and with data augmentation for each epoch. Besides, we also apply ℓ_2 weight decay of $5 * 10^{-4}$ and batch normalization after the activation function. For the SAHS algorithm, parameters *HMCR*, *PAR*, *hms*, and *MI* are set to 0.99, 0.2, 10, and 30, respectively.

For Caltech-101, the weights ω are trained through SGD optimizer with learning rate $lr = 0.01$, $lr\_decay = 10^{-6}$, momentum = 0.9, nesterov = True, and batch size = 256. Each candidate architecture is run for 30 epoches. For the modified harmony search algorithm, parameters *HMCR*, *PAR*, *hms*, and *MI* are set to 0.99, 0.2, 10, and 90, respectively.

5.2 Experimental Results

In the section, the CNNs built by our methods are compared with those done by the state-of-the-art CNN search methods on (1) MNIST and CIFAR-10, and (2) Caltech-101.

The experimental results on MNIST and CIFAR-10 as shown in Table 4 can be divided into three parts. On the top part, the lengths of the CNN architectures built by Rosa et al. [9] and Young et al. [15] are fixed. On the middle part, the lengths of the CNN architectures built by Albelwi and Mahmood [1], Al-Hyari and Areibi [2], Singh and Kingsbury [11], Pham et al. [8], and Tanno et al. [13] are dynamic or not fixed. On the bottom part, the lengths of the CNN architectures built by our methods are also dynamic. For MNIST, our method achieves the best results among all the methods. For CIFAR-10, our method is better than most of them, except the methods proposed by Singh and Kingsbury and Pham et al. However, the best DTSCNN proposed by Singh and Kingsbury is the DTS-WResNet with 96.40%, but not the DTS-VGG with 92.40%. Actually, the CNN with 95.03% derived from the same Vgg16 architecture by our method has shown better than the DTS-VGG proposed by Singh and Kingsbury. Furthermore, Pham et al. proposed Efficient Neural Architecture Search (ENAS) to discover neural network architectures by searching for an optimal subgraph within a large computational graph. In other words, ENAS takes a lot of time to search the entire solution space for optimal convolutional architectures. Nevertheless, our method only generates CNNs derived from the Vgg16 architecture, and then fine-tunes these CNNs in the local solution space. In general, ENAS is able to find better solutions than our method. Besides, we also observe that the accuracy of the CNN architecture built by our method can be improved through the fine-tuning in Phase 2, if possible.

Table 4. Accuracy of all the methods on MNIST and CIFAR-10.

Methods	MNIST	CIFAR-10
Harmony Search (Rosa et al., 2015)	99.29%	72.43%
Genetic Algorithms (Young et al., 2015)	-	74.60%
Deconvnet + NMA (Albelwi & Mahmood, 2016)	-	83.63%
Genetic Algorithms (Al-Hyari & Areibi, 2017)	96.99%	-
DTSCNN (Singh & Kingsbury, 2017)	-	96.40%
ENAS + Micro Search Space + CutOut (Pham et al., 2018)	-	97.11%
Adaptive Neural Trees (Tanno et al.,2018)	99.36%	93.28%
Our Method Phase1	99.37%	94.85%
Our Method Phase2	99.37%	95.03%

The experimental results on Caltech-101 as shown in Table 5 can also be divided into three parts, like those as shown in Table 4. The original Vgg16 and Inception-v3 can be improved using our method. Although our method cannot get the satisfactory results using the pre-trained model Vgg16, it achieves the best results using the pre-trained model Inception-v3.

Table 5. Accuracy of all the methods on Caltech-101.

Methods	Caltech-101
Classifier Ensemble (Khan, 2018)	91.35%
Deconvnet + NMA (Albelwi & Mahmood, 2016)	55.75%
DTSCNN (Singh & Kingsbury, 2017)	91.22%
Original Vgg16	85.90%
Our Method (Vgg16)	88.01%
Original Inception-v3	91.31%
Our Method (Inception-v3)	92.88%

6 Conclusions

The performance of image recognition using CNN is deeply influenced by number of layers and inside hyper-parameters. In this paper, we use a self-adaptive harmony search algorithm to find an optimized convolutional neural network architecture for image recognition. The system is divided into two phases. In Phase 1, we use the self-adaptive harmony search algorithm to search for the most suitable layer length of a CNN. Then, the system switches to Phase 2 for fine-tuning the hyper-parameters inside the CNN found in Phase 1. However, for Caltech-101, it is not suitable to train a CNN from scratch due to requiring a large number of parameters. Instead, we choose two pre-trained CNN models Vgg16 and Inception-v3, and train them from Phase 2. Different from Phase 1, we use the modified harmony search algorithm in Phase 2 to fine-tune the CNN, and finally find out the optimized CNN. In the experiments, three popular and well-known datasets are used to evaluate the proposed methods and the state-of-the-art CNN search methods. The experimental results show that our methods achieve competitive performances compared with the other methods on CIFAR-10, and have the best performance among all the methods on MNIST and Caltech-101.

References

1. Albelwi, S., Mahmood, A.: Automated optimal architecture of deep convolutional neural networks for image recognition. In: Proceedings of the 15th IEEE International Conference on Machine Learning and Applications, pp. 53–60 (2016)
2. Al-Hyari, A., Areibi, S.: Design space exploration of convolutional neural networks based on evolutionary algorithms. J. Comput. Vis. Imaging Syst. **3**(1) (2017)
3. Bottou, L.: Large-scale machine learning with stochastic gradient descent. In: Proceedings of the 19th International Conference on Computational Statistics, pp. 177–186 (2010)
4. Ioffe, S., Szegedy, C.: Batch normalization: accelerating deep network training by reducing internal covariate shift. arXiv:1502.03167v3, pp. 1–11 (2015)
5. Khan, H.A.: DM-L based feature extraction and classifier ensemble for object recognition. J. Sig. Inf. Process. **9**(2), 92–110 (2018)
6. Krogh, A., Hertz, J.A.: A simple weight decay can improve generalization. Adv. Neural. Inf. Process. Syst. **4**, 950–957 (1992)
7. Lin, M., Chen, Q., Yan, S.: Network in network. arXiv:1312.1400v3, pp. 1–10 (2014)
8. Pham, H., Guan, M.Y., Zoph, B., Le, Q.V., Dean, J.: Efficient neural architecture search via parameter sharing. arXiv:1802.03268v2 (2018)
9. Rosa, G., Papa, J., Marana, A., Scheirer, W., Cox, D.: Fine-tuning convolutional neural networks using harmony search. In: Proceedings the 20th Iberoamerican Congress on Pattern Recognition, pp. 683–690 (2015)
10. Simonyan, K., Zisserman, A.: Very deep convolutional networks for large-scale image recognition. arXiv:1409.1556v6, pp. 1–14 (2015)
11. Singh, A., Kingsbury, N.: Efficient convolutional network learning using parametric log based dual-tree wavelet scatternet. In: Proceedings of the IEEE International Conference on Computer Vision Workshops, pp. 1140–1147 (2017)
12. Szegedy, C., Vanhoucke, V., Ioffe, S., Shlens, J., Wojna, Z.: Rethinking the inception architecture for computer vision. In: Proceedings of the IEEE Conference on Computer Vision and Pattern Recognition, pp. 2818–2826 (2016)
13. Tanno, R., Arulkumaran, K., Alexander, D.C., Criminisi, A., Nori, A.: Adaptive neural trees. arXiv:1807.06699v2 (2018)
14. Wang, C.M., Huang, Y.F.: Self-adaptive harmony search algorithm for optimization. Expert Syst. Appl. **37**(4), 2826–2837 (2010)
15. Young, S.R., Rose, D.C., Karnowski, T.P., Lim, S.H., Patton, R.M.: Optimizing deep learning hyper-parameters through an evolutionary algorithm. In: Proceedings of the Workshop on Machine Learning in High-Performance Computing Environments, Article No. 4 (2015)
16. Zeiler, M.D.: ADADELTA: an adaptive learning rate method. arXiv:1212.5701v1 (2012)

Global Features of Fused Frame Relationships Help Video Classification

MengYao Kong[1,2](✉) and Pin Lv[2]

[1] University of Chinese Academy of Sciences, UCAS,
80# ZhongGuanCun East Road, Beijing 100190, People's Republic of China
kmysbox@gmail.com
[2] Institute of Software, Chinese Academy of Sciences, ISCAS,
4# South Fourth Street, ZhongGuanCun, Beijing 100190, People's Republic of China

Abstract. ECO [20] is a newly proposed efficient video classification network that combines 2D and 3D networks. On this basis, we try to extract the global features composed of inter-frame relationships and fuse the global features with the initial frame feature, which can enhance the learning of video features to achieve better classification results. In this way, the new frame feature not only contains the information of itself but also contains the global feature which is related to other frames., so the module can more easily learn about distinguishing features. The main contribution of this paper is proposing a module for extracting global features to enhance video feature learning. It is very lightweight and basically does not add computation to the model. At the same time, it is flexible and easy to generalize to various network structures. Finally, based on the baseline the classification accuracy is increased by 0.8% on the HMDB51 dataset.

Keywords: Global feature · Inter-frame relation · Action recognition

1 Introduction

Recent years have witnessed significant progress in video action recognition using deep convolutional neural networks. Based on 2D convolution and 3D convolution, two branches based on Temporal Segment Network [17] and C3D [14] are developed respectively. In the recent work, by visualizing convolution kernel parameters, [18] found that the parameters in temporal dimension approach 0 which means low-level 3D convolution still tends to learn spatial features. Similarly, some other work [15,18,20] also mix 2D convolution and 3D convolution to build networks to achieve state of the art classification accuracy in an efficient way. ECO is one of them.

Video contains more complicated temporal information compared to pictures. The difficulty in video action recognition lies in the effective use of temporal information. Temporal Relation Network [19] uses a simple multi-layer perceptron for relational reasoning in video. [3] identify the objects in the frame first,

Y. Liu et al. (Eds.): ICNC-FSKD 2019, AISC 1074, pp. 13–20, 2020.
https://doi.org/10.1007/978-3-030-32456-8_2

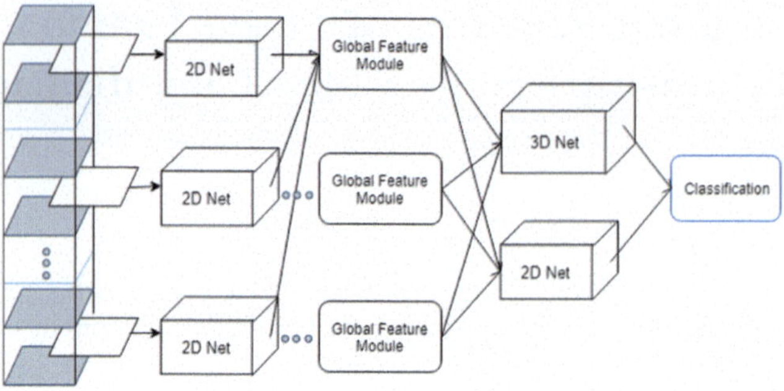

Fig. 1. ECO-full with Global Feature Module: the initial frame features extracted with low-layer 2D Net are fed to Global Feature Module to get the new enhanced feature.

then perform reasoning at the object level through the integration of state of the art object detection networks. Similar to these work, we want to be able to model the relation between video frames to make more efficient use of temporal information (Fig. 1).

Our approach is also motivated by the success of relation networks for object detection [3], which through interacting between appearance feature and geometry feature of a set of objects to model objects relations to help object recognition. Relation module in relation network was inspired by the scaled dot-product attention [16]. An attention module can affect an individual element by aggregating information from a set of elements, the aggregation weights are automatically learned, driven by the task goal. We believe that the scaled dot-product attention can also be used to extract relations between frames to help action recognition.

Our work is based on ECO-full. We extract each frame feature with 2D Net, then construct inter-frame relation feature based on each frame feature, and fuse inter-frame relation feature to obtain global feature. Finally, global features and original features are concatenated to obtain new frame feature, and then fed into the model for learning and classification. In the process of extracting the relation between frames, frame features are pooled and reshaped to one-dimensional vectors to compute the scaled dot-product to obtain weight. In this way, we hope that the frame which is more similar to the current frame or has a larger response value can get greater weight. And when concatenating the global feature and the initial frame feature, we pool the global feature first so that the global feature and the initial frame feature have the same number of channels.

2 Related Work

Convolution Neural Network for Action Recognition. Action recognition in video is an important problem in computer vision. 2D CNN and 3D CNN are two convolution structures widely used in action recognition. By constructing a two-stream CNN for RGB input [17] and optical flow input respectively, 2D CNN can achieve state of the art results. 2D CNN is computationally cheap but the calculation of optical flow is complicated. At the same time, 2D CNN cannot capture long-term temporal relationships.

3D CNN based methods can achieve better performance generally in view of its ability to model temporal information. Unfortunately, 3D CNN is computationally expensive, so there has been recent interest in more efficient variants. One way [2,8,15,18] to save computation is to replace 3D convolutions with separable convolutions, that is to say, first convolve spatially in 2D, and then convolve temporally in 1D. Some other work [15,18,20] proposed a variety of models based on top of the ResNet architecture, which use 3D convolution in either the bottom or top layers, and 2D in the rest; they call these "mixed convolutional" models and ECO is one of these models.

Relational Reasoning. Relational reasoning is a well-studied field for many applications ranging from visual reasoning to question answering. Santoro [9] introduced Relation Network, a plug-in module for reasoning in deep networks. RN shows human-level performance to semantically meaningful entities, but a recent study [6] has shown that some of these results are subject to dataset bias and do not generalize well to small changes in the settings of the dataset. [10,11] make use of hierarchical dynamic analysis and coding with time series information for motion recognition. [12,13] use a new timing alignment method for motion recognition. Following the similar perspective, Temporal Relation Network (TRN) model the multi-scale temporal relations in video with multilayer perceptions, [3] chose to perform the reasoning at the object level through the integration of state of the art object detection networks.

3 Global Feature Module

The global feature module can be implemented with three steps. Firstly, calculate inter-frame relation f_R by the scaled dot-product attention. Secondly, use relation features f_R to construct the global feature f_G. Lastly, concatenate global feature f_G and initial frame feature f extracted from 2D Net to obtain new frame feature f, which not only contains the information of itself but also other frames. The Global Feature Module can be easily implemented using basic operators, as illustrated in Fig. 2.

3.1 Inter-frame Relation

Inspired by Relation Network for object recognition, we believe that the scaled dot-product attention can be used to extract relations between frames to help action recognition in video. The input of the scaled dot-product attention consists of queries and keys of dimension d_k, and values of dimension d_v. The dot product is performed between the query and all keys to obtain their similarity. Then a softmax function is applied to obtain the weights on the values. Given a query q, all keys which are packed into matrices K, and values which are packed into V, the scaled dot-product attention is described as follow.

$$V^{out} = softmax(\frac{qK^t}{\sqrt{d}})V \tag{1}$$

In our work, we apply the scaled dot-product attention [16] to extract relations between frames. But the initial frame features extracted from low-layer 2D Net are three-dimensional vectors whose size is channel × height × width and cannot be directly used to calculate the dot product. So we pool the initial frame feature f to [1, 7, 7] first and reshape to one-dimensional vector f_P. Then we apply the scaled dot-product attention described as Eq. (1) to obtain relations between frames. Here the initial features f extracted from low-layer 2D Net are packed into matrices V, and the q is the current frame feature f_P which has been pooled and reshaped from f, rest of frame features are also pooled and reshaped and then packed into matrices K. Relations of the current frame and the rest each frame are packed into matrices V^{out}.

Algorithm 1. Global Feature Module. Input is N initial frame feature f extracted from low-layer 2D Net. And N is the number of sample frames.

```
pooling and reshape initial frame features to get f_p
for every (f^i, f_p^i, f_p^i) do
    compute relation feature f_R^i using Eq. (1)
    compute global feature f_G^i using Eq. (2)
    compute new enhanced frame feature f_new^i using Eq. (3)
end for
Output new features (f_new^n)_{n=1}^N
```

3.2 Global Feature Fusion

We will take one frame as an example to illustrate the calculation process. After getting a series of relation features of the current frame and the rest each frame $(f_R^1, \dots, f_R^{N_r})$, we want to use relation features to construct the global feature f_G. And then fuse the global feature f_G and the current frame initial feature f to get new enhanced feature f_{new}. The new feature f_{new} will be fed to the network to continue learning. Here we test two different fusion methods to construct the global feature.

Concatenate and Pooling. There are N_r relation features to construct global feature f_G for the current frame, N_r is the number of all frames except itself. The first method is to concatenate the relation features directly and different relation features will not influence each other in this way. The operation process can be represented as follows.

$$f_G = concat[f_R^1, ..., f_R^{N_r}] \tag{2}$$

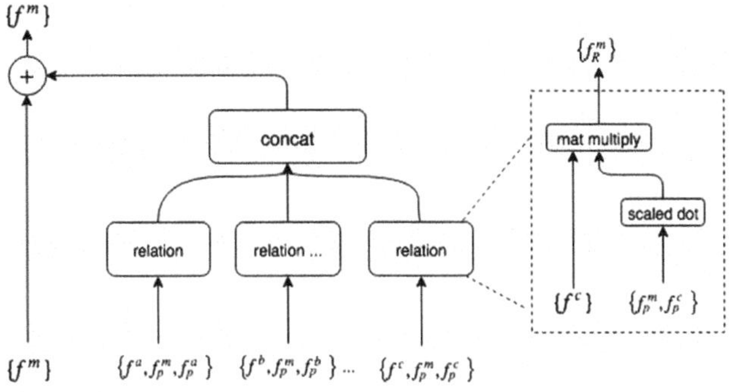

Fig. 2. The illustration of Global Feature Module.

Concat(.) is used to aggregate multiple relation features. But in this way, the number of channels of the global feature f_G obtained will be much larger than the initial frame feature f. And the new frame feature f_{new} is composed of the initial frame feature f and the global feature f_G. So the global feature f_G will take up too much weight in the new frame feature f_{new}. To match the channel dimension, pool global features f_G to make the number of channels becomes $\frac{1}{N_r}$ of the dimension of the initial feature f, the operation process to get f_{new} can be represented as follow. The complete calculation process is summarized in Algorithm 1.

$$f_{new} = concat[f, pooling(f_G)] \tag{3}$$

Add Directly. The other way to construct the global feature f_G is to add relation features directly. Adding equals to first concatenate the two feature maps and then apply the same convolution operation at the corresponding channel. In this way, the corresponding channel can strengthen each other, and the global feature f_G is also known as context here.

$$f_{new} = concat(f, f_R^1 + f_R^2 + ... + f_R^{N_r}) \tag{4}$$

Addition does not change the channel dimension of the global feature f_G, so there is no need to pool to reduce channel number.

4 Experiments

4.1 Dataset and Implementation Details

Datasets. We evaluate the ECO-full with global feature module on HMDB51 [5]. The HMDB51 dataset is a large collection of realistic videos from various sources, such as movies and web videos. The dataset is composed of 6,766 video clips from 51 action categories. Our experiments follows the original evaluation scheme using three training/testing splits and report average accuracy over these splits.

Training. The module only takes the raw RGB videos as input without further pre-processing, for instance, by providing optical flow or human pose, which is the same as ECO. We use the mini-batch stochastic gradient descent algorithm to learn the network parameters, where the batch size is set to 27 because of the limit of memory and momentum is set to 0.9. We initialize network weights with pretrained model for 2D Net which is provided by Temporal Segment Network and pre-trained model for 3D provided by C3DResNet18 [4]. Since many of the action recognition datasets are not large enough and are prone to over-fitting [7], we pretrianed the models on Kinetics [1] for 10 epoch with learning rate 0.001. We then fine-tuned the model on HMDB51. The fine-tuning is conducted for 40 epochs with an initial learning rate of 0.001 and decayed by a factor of 0.1 every 10 epochs.

We sampled 16 frames every video and compared the result with ECO-full in the same configuration, which is our baseline. We applied similar data augmentation strategy as [7] for pre-processing: first resizing the raw images to make the short side is 256, and then employing a fixed corner+center cropping and scale-jittering. The images were resized to 224×224 before fed to the network.

Test. Many states of the art work [14,19] use a different post-processing technique during the test stage. A common strategy is to sample more frames from the video and give the averaged output as the prediction. We only apply the center crop to the same number of frames as training (e.g. 16) to give the prediction directly.

4.2 Improving ECO-full Baseline

We use the same training protocol with ECO-full and the only difference is whether the Global Feature Module is used or not. We test two different fusion ways to calculate the global feature on HMDB51 and the results are listed in Table 1. The results show that the Concat and Pooling method to calculate the global feature is more effective than the Add method. Adding relation features directly even harms the classification accuracy of baseline module. But the Concat and Pooling method can produce better performance and the classification accuracy on three splits is higher than ECO-full, which is our baseline. The final performance of ECO-full with the Global Feature Module on HMDB51 dataset is 0.8% higher.

Table 1. The results of three networks.

Network	Split1	Split2	Split3	Average
ECO-full (baseline)	69.40	67.00	69.10	68.5
Concat and Pooling	70.34	67.30	70.27	69.3
Add Directly	66.56	64.28	66.40	65.7

The accuracy and loss curves of the ECO-full and ECO-full with Global Feature Module using Concat and Pooling method on the test dataset are shown as follows. The blue line is the ECO-full with Global Feature Module and the red line is ECO-full, which is our baseline. The accuracy curve shows that ECO-full will perform better and converge faster when applied with the Global Feature Module, and the loss also converges faster to a very low level. Although the loss of module with Global Feature Module is higher than ECO-full, we notice that the unit of the vertical axis is small so the gap is actually very small (Fig. 3).

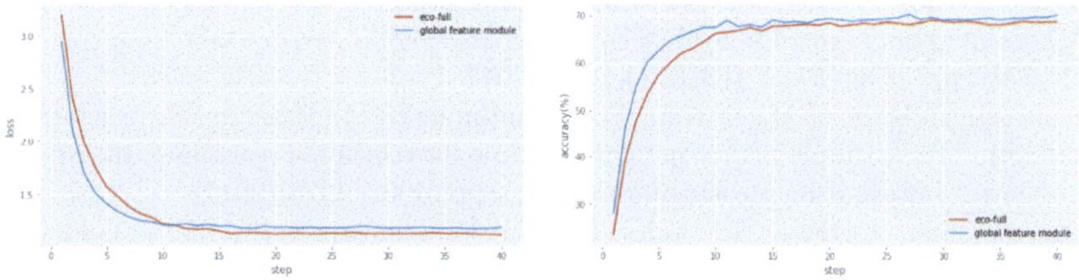

(a) loss (b) accuracy

Fig. 3. The loss and accuracy curves

5 Conclusion

Videos have more abundant information than images. This paper has investigated a method to extract the global feature to enhance the representation ability for each frame. The method calculates the relation feature of target frame with the rest each frame, then concat all relation features to get the global feature of the target frame. In this way, new frame feature not only contains the information of itself but also contains the global feature which is related to other frames. It increases the accuracy by 0.8% compared with baseline module ECO-full. We hope this work will foster further research in video recognition.

References

1. Carreira, J., Zisserman, A.: Quo vadis, action recognition? A new model and the kinetics dataset. In: 2017 IEEE Conference on Computer Vision and Pattern Recognition (CVPR), pp. 4724–4733. IEEE (2017)
2. Chen, Y., Kalantidis, Y., Li, J., Yan, S., Feng, J.: Multi-fiber networks for video recognition. In: Proceedings of the European Conference on Computer Vision (ECCV), pp. 352–367 (2018)
3. Hu, H., Gu, J., Zhang, Z., Dai, J., Wei, Y.: Relation networks for object detection. In: Proceedings of the IEEE Conference on Computer Vision and Pattern Recognition, pp. 3588–3597 (2018)

4. Hara, K., Kataoka, H., Satoh, Y.: Can spatiotemporal 3D CNNs retrace the history of 2D CNNs and ImageNet? In: Proceedings of the IEEE Conference on Computer Vision and Pattern Recognition, pp. 6546–6555 (2018)
5. Kuehne, H., Jhuang, H., Garrote, E., Poggio, T., Serre, T.: HMDB: a large video database for human motion recognition. In: 2011 IEEE International Conference on Computer Vision (ICCV), pp. 2556–2563. IEEE (2011)
6. Kim, J., Ricci, M., Serre, T.: Not-So-CLEVR: visual relations strain feedforward neural networks (2018)
7. Lin, J., Gan, C., Han, S.: Temporal shift module for efficient video understanding. arXiv preprint arXiv:1811.08383 (2018)
8. Qiu, Z., Yao, T., Mei, T.: Learning spatio-temporal representation with pseudo-3D residual networks. In: 2017 IEEE International Conference on Computer Vision (ICCV), pp. 5534–5542. IEEE, October 2017
9. Santoro, A., Raposo, D., Barrett, D.G., Malinowski, M., Pascanu, R., Battaglia, P., Lillicrap, T.: A simple neural network module for relational reasoning. In: Advances in Neural Information Processing Systems, pp. 4967–4976 (2017)
10. Su, B., Zhou, J., Ding, X., Wang, H., Wu, Y.: Hierarchical dynamic parsing and encoding for action recognition. In: European Conference on Computer Vision (ECCV), pp. 202–217 (2016)
11. Su, B., Zhou, J., Ding, X., Wu, Y.: Unsupervised hierarchical dynamic parsing and encoding for action recognition. IEEE Trans. Image Process. **26**, 5784–5799 (2017)
12. Su, B., Hua, G.: Order-preserving Wasserstein distance for sequence matching. In: Computer Vision and Pattern Recognition (CVPR), pp. 1049–1057 (2017)
13. Su, B., Hua, G.: Order-preserving optimal transport for distances between sequences. IEEE Trans. Pattern Anal. Mach. Intell. (2018)
14. Tran, D., Bourdev, L., Fergus, R., Torresani, L., Paluri, M.: Learning spatiotemporal features with 3D convolutional networks. In: Proceedings of the IEEE International Conference on Computer Vision, pp. 4489–4497 (2015)
15. Tran, D., Wang, H., Torresani, L., Ray, J., LeCun, Y., Paluri, M.: A closer look at spatiotemporal convolutions for action recognition. In: Proceedings of the IEEE Conference on Computer Vision and Pattern Recognition, pp. 6450–6459 (2018)
16. Vaswani, A., Shazeer, N., Parmar, N., Uszkoreit, J., Jones, L., Gomez, A.N., Kaiser, Ł., Polosukhin, I.: Attention is all you need. In: Advances in Neural Information Processing Systems, pp. 5998–6008 (2017)
17. Wang, L., Xiong, Y., Wang, Z., Qiao, Y., Lin, D., Tang, X., Van Gool, L.: Temporal segment networks: towards good practices for deep action recognition. In: European Conference on Computer Vision, pp. 20–36. Springer, Cham (2016)
18. Xie, S., Sun, C., Huang, J., Tu, Z., Murphy, K.: Rethinking spatiotemporal feature learning: speed-accuracy trade-offs in video classification. In: European Conference on Computer Vision, pp. 318–335. Springer, Cham (2018)
19. Zhou, B., Andonian, A., Oliva, A., Torralba, A.: Temporal relational reasoning in videos. In: European Conference on Computer Vision, pp. 831–846. Springer, Cham (2018)
20. Zolfaghari, M., Singh, K., Brox, T.: ECO: efficient convolutional network for online video understanding. arXiv preprint arXiv:1804.09066 (2018)

A Comparison and Strategy of Semantic Segmentation on Remote Sensing Images

Junxing Hu[1,2(✉)], Ling Li[1,2], Yijun Lin[1,2], Fengge Wu[2], and Junsuo Zhao[2]

[1] University of Chinese Academy of Sciences, Beijing, China
[2] Institute of Software Chinese Academy of Sciences (ISCAS), Beijing, China
junxing2017@iscas.ac.cn

Abstract. In recent years, with the development of aerospace technology, we use more and more images captured by satellites to obtain information. But a large number of useless raw images, limited data storage resource and poor transmission capability on satellites hinder our use of valuable images. Therefore, it is necessary to deploy an on-orbit semantic segmentation model to filter out useless images before data transmission. In this paper, we present a detailed comparison on the recent deep learning models. Considering the computing environment of satellites, we compare methods from accuracy, parameters and resource consumption on the same public dataset. And we also analyze the relation between them. Based on experimental results, we further propose a viable on-orbit semantic segmentation strategy. It will be deployed on the TianZhi-2 satellite which supports deep learning methods and will be lunched soon.

Keywords: Semantic segmentation · Deep learning · Remote sensing images

1 Introduction

Compared with natural images, remote sensing images always have more kinds of targets with the lower resolution, and irregular-shaped targets impact each other. These bring difficulties to object detection and recognition in remote sensing images.

As one basic method of image understanding, semantic segmentation conducts pixel-level classification of the image. Different from other methods like image classification and object detection, semantic segmentation can produce not only the category, size and quantity of the target, but also accurate boundary and position. Therefore, it is more suitable for processing remote sensing images.

Some traditional semantic segmentation methods are early proposed, such as active contour model [1], watershed algorithm [2] and graph cuts [3]. They usually require artificially setting thresholds or interaction controls, but they are not accurate enough.

In recent years, deep learning method has greatly promoted the development of semantic segmentation. Some deep learning based methods are applied to remote sensing images and achieve good performance. Kampffmeyer et al. [4]

Y. Liu et al. (Eds.): ICNC-FSKD 2019, AISC 1074, pp. 21–29, 2020.
https://doi.org/10.1007/978-3-030-32456-8_3

propose a deep Convolutional Neural Network (CNN) for small objects segmentation in remote sensing images of urban areas. Volpi et al. [5] use the Fully Convolutional Networks (FCN) [6] with 50 layers deep residual networks [7] for image segmentation. There have been some works focusing on the comparison of deep learning based semantic segmentation methods. Garcia-Garcia et al. [8] introduce and compare structures and evaluation results on several natural image datasets for most of the existing semantic segmentation methods. Ball et al. [9] provide applications and challenges for deep learning theories and related tools in remote sensing images. At present, a comparative research of semantic segmentation models about accuracy, parameters and resource consumption is still lacking for remote sensing images.

TianZhi satellites are designed to verify the feasibility of using satellite-based intelligent applications, and finally build the space-based intelligent system. With the successful launch of TianZhi-1, the feasibility has been initially verified. Unlike TianZhi-1, TianZhi-2 is equipped with a powerful computing system which supports deep learning algorithms, and it will be launched soon. Based on this platform, we can deploy deep learning models to perform on-orbit semantic segmentation of remote sensing images.

Motivated by such prior studies, we conduct a comparison of recent popular semantic segmentation methods of remote sensing images. The main contributions of this paper can be summarized as follows:

- First, we use a new image cropping method for the publicly available remote sensing image dataset. By using the training set of the same size, we fairly compare five efficient semantic segmentation methods in terms of structures, accuracy, parameters and resource consumption. These deep learning based methods are FCN, U-Net [10], SegNet [11], Pyramid Scene Parsing Network (PSPNet) [12] and DeepLab [13].
- Second, we propose an on-orbit semantic segmentation strategy for TianZhi-2 based on experimental results.

The remainder of this paper is organized as follows. In Sect. 2, we introduce the comparison of five methods in detail. We describe experiments and discusses results in Sect. 3. Section 4 talks about the on-orbit semantic segmentation strategy for TianZhi-2. Finally, the paper is concluded in Sect. 5.

2 Methods Comparison

Many deep learning based semantic segmentation methods are proposed, and some of them are applied to remote sensing images. We select five most representative methods: FCN, U-Net, SegNet, PSPNet and DeepLab. They and their variants achieve excellent results on some public competitions like the PASCAL VOC-2012 semantic segmentation task [14]. According to the structure, the main deep learning methods for semantic segmentation can be divided into two classes: (1) Encoder-Decoder structure; (2) Multi-Scale representation structure. In this section, we introduce the five methods based on their relations, respectively. As shown in Fig. 1, we arrange these methods according to the time when they are first proposed.

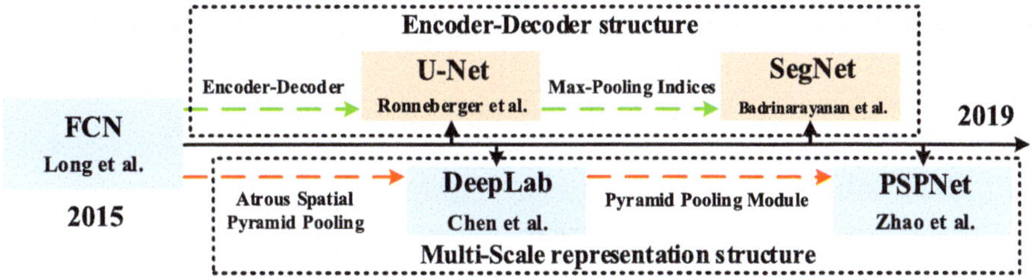

Fig. 1. The timeline of five methods. Note that DeepLab was first proposed in 2016 (arXiv), although its official publication was in 2018 (TPAMI).

2.1 FCN

FCN [6] leads to a rapid increase in the number of semantic segmentation networks. The model transforms all of the fully connected layers to convolutional layers and allows the input image with arbitrary size. In addition, it combines image semantic information from deep layers with shallow layers to produce the final segmentation by using a skip architecture. FCN has been used for remote sensing in [4]. The method mainly has three structures as follows:

FCN-32s. It is a basic FCN structure with a single stream, which produces the final segmentation with upsampling at stride 32.

FCN-16s. The method combines predictions from both FCN-32s and the pool4 layer which uses stride 16. It could predict the finer result.

FCN-8s. This model adds predictions from the pool3 layer which is stride 8, and provides further precision. Because remote sensing images contain both coarse, large objects and small, important details, we apply this structure to our experiments in this paper.

2.2 U-Net

Built upon FCN, U-Net [10] adopts an Encoder-Decoder architecture which consists of a contracting path to capture context and a symmetric expanding path to enable accurate localization. It is originally designed to segment medical images and achieve good results with the fewer training set. Recent years, some studies have shown that U-Net is also suitable for remote sensing images [16], and it has a great potential to be improved. In [17], Lu et al. propose a dual-resolution U-Net which uses pairs of images as inputs to capture both high and low resolution features.

2.3 SegNet

Similar to U-Net, SegNet [11] is also built on an Encoder-Decoder structure. Its encoder network is topologically same as the 13 convolutional layers of the VGG-16 [15]. And the decoder network first use max-pooling indices generated from the corresponding encoder to enhance the location information. In [18], Bischke et al. use SegNet with a new cascaded multi-task loss to further preserve semantic segmentation boundaries in high resolution satellite images.

2.4 DeepLab

DeepLab [13] exploits a powerful FCN architecture which mainly uses three components as follows:

Atrous Convolution. The algorithm is originally proposed for computing the undecimated wavelet transform in [19]. By using the atrous convolution with dilation d, the kernel size of a $k \times k$ filter is enlarged to $k_e \times k_e$ as the following formula:

$$k_e = k + (k-1)(d-1) \tag{1}$$

The algorithm fills consecutive filter values with $d-1$ zeros to avoid increasing parameters and computation of the model.

Atrous Spatial Pyramid Pooling (ASPP). ASPP exploits the atrous convolution with different dilation to capture the multiscale information of objects.

Fully-Connected Conditional Random Fields (CRFs). The fully connected CRFs is integrated into DeepLab to avoid reducing the spatial resolution of feature maps as proposed in [20]. And it greatly improves the localization performance.

2.5 PSPNet

Based on DeepLab, PSPNet [12] exploits the pyramid pooling module to aggregate the image global context information with an auxiliary loss. Since DeepLab provides two versions of the model adapted from VGG-16 and ResNet-101, PSPNet can also be applied to VGG and ResNet based network structures, respectively.

Table 1 indicates detailed comparisons about methods as above.

Table 1. Detailed comparisons of methods. The '-' indicates that the method does not belong to any classes. The 'LRP' indicates learning rate policy.

Method	Structure	Backbone	LRP	Loss
FCN-32s	-	VGG-16	Fixed	Cross Entropy
FCN-16s	Multi-Scale	VGG-16	Fixed	Cross Entropy
FCN-8s	Multi-Scale	VGG-16	Fixed	Cross Entropy
U-Net	Encoder-Decoder	VGG-16	Step	Cross Entropy
SegNet	Encoder-Decoder	VGG-16	Step	Cross Entropy
DeepLab	Multi-Scale	VGG-16	Poly	Cross Entropy
DeepLab	Multi-Scale	ResNet-101	Poly	Cross Entropy
PSPNet	Multi-Scale	VGG-16	Poly	Cross Entropy
PSPNet	Multi-Scale	ResNet-101	Poly	Auxiliary Loss

3 Experiments and Results

In this section, we first introduce the dataset and the image cropping method. Then, we compare five deep learning models in semantic segmentation with multiple indicators.

3.1 Datasets

We evaluate the methods on a public subset of the Inria aerial image labeling benchmark [21]. The available dataset contains 180 images of size 5000×5000 at 0.3 m resolution. It is collected from five cities: Austin, Chicago, Kitsap, Vienna and West Tyrol, and each of them has 36 images. As described in [21], the first 5 images of each city are used for testing and the other 155 images for training. These pixel-level labeled images contains two classes: building and non-building.

For all images, we extract 512×512 patches with 12 pixels of overlap between adjacent patches. The size of patch is more suitable for existing semantic segmentation methods and hardware. And the overlap of patches effectively prevent most small objects from being destroyed by image cropping. After extracting, we have 15500 images for training and 2500 images for testing. We use the same training set and test set for all models.

3.2 Experimental Results

We compare segmentation performance, parameters, resource consumption and edge prediction of five methods: FCN-8s, U-Net, SegNet, PSPNet and DeepLab v2. All of them are based on VGG structure. The base learning rate is 0.0001, and it is updated by the "poly" policy. All models are trained for 200 epochs and the best results are reported during the training process.

Segmentation Performance. To evaluate the segmentation performance of methods, we use the accuracy and the Intersection over Union (IoU) which is more suitable for evaluating the unbalanced dataset. We evaluate the methods on the overall test set and for each city separately. Table 2 shows that U-Net and DeepLab perform better than other models. More specifically, U-Net achieves the best result in the former three cities, and DeepLab is better in the latter two cities. For the overall test set, the IoU of DeepLab is better. It shows that DeepLab may be suitable for the unbalanced datasets which is more common in the real world environment.

Table 2. Comparisons of evaluation results of methods on Inria aerial image labeling dataset. 'Acc' indicates the accuracy.

Method	Austin		Chicago		Kitsap		West Tyrol		Vienna		Overall	
	IoU	Acc	IoU	Acc	IoU	Acc	IoU	Acc	IoU	Acc	IoU	Acc
FCN-8s	50.28	92.30	53.89	87.24	32.09	98.52	56.40	95.84	62.75	88.30	56.19	92.44
U-Net	**78.62**	**96.89**	**70.39**	**92.89**	**66.26**	**99.27**	70.93	97.71	78.28	93.85	74.79	**96.12**
SegNet	70.60	94.74	64.81	89.72	60.55	98.89	71.41	97.28	74.97	91.79	70.10	94.48
DeepLab	76.65	96.54	69.39	92.56	65.78	99.24	**75.01**	**97.98**	**79.24**	**94.06**	**74.86**	96.08
PSPNet	71.69	95.73	66.67	91.62	63.08	99.18	72.07	97.72	76.49	93.12	71.67	95.47

Parameters. Figure 2 (left) shows parameters of each model. We find that FCN is the largest model and SegNet is the smallest one. Note that this indicator is closely related to storage resource and transmission capability of satellites.

Resource Consumption. Figure 2 (right) illustrates floating point operations (FLOP) which can display the resource consumption of the model. SegNet and U-Net have smaller FLOP for Encoder-Decoder structure. And other models have larger FLOP for Multi-Scale representation structure.

Edge Prediction. Figure 3 shows several image segmentation results of the five methods. We observe that U-net and DeepLab do well in predicting edge which helps to restore the shape of targets. The edge predicted by SegNet is also sharp, but there are many points which are misclassified.

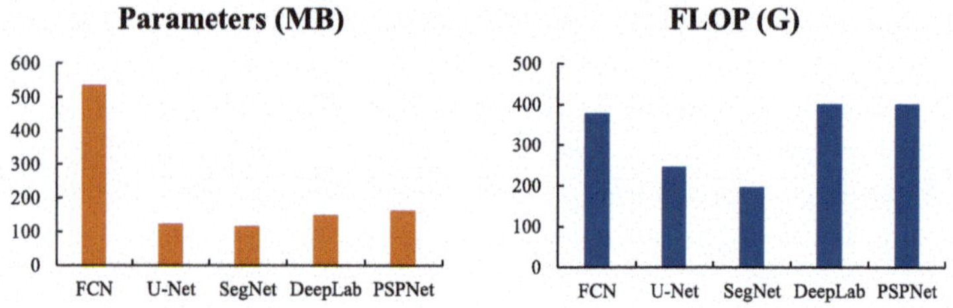

Fig. 2. The parameters (left) are calculated on VGG based models. The FLOP (right) is measured on 512 × 512 inputs, except that DeepLab is 513 × 513 because of the atrous convolution.

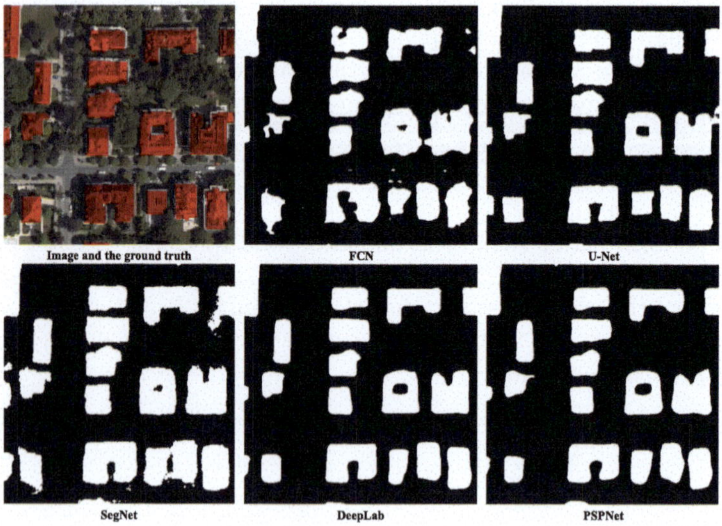

Fig. 3. Segmentation results of Inria aerial image labeling dataset by using five methods.

4 On-Orbit Semantic Segmentation Strategy for TianZhi-2

Due to the complexity of remote sensing image processing and limited energy on satellites, it is very necessary for us to consider segmentation performance and resource consumption simultaneously. Based on experimental results in Sect. 3, we propose that U-Net is more suitable for on-orbit semantic segmentation. We propose a strategy to segment remote sensing images on the satellite as follows:

Improve the Semantic Segmentation Performance for Remote Sensing Images. First, because the target distribution in remote sensing images is uneven and irregular, the object region extraction before segmentation will guide it and improve its performance. Second, due to there are many irregular-shaped targets in remote sensing images, optimizing the target edge prediction can further improve the semantic segmentation performance.

Reduce Resource Consumption by Model Pruning. In order to reduce the resource occupation on the satellite, we can use methods like l_1-norm to prune the model while keeping the performance levels almost intact and even better.

By using methods above, on-orbit semantic segmentation for remote sensing images can be achieved. After satellite capturing images, our image cropping method splits the images into smaller patches. The model takes these patches as input and outputs corresponding semantic segmentation results. Based on these results, there are many useful production such as the object image tile, target location and language description. They will greatly reduce the data transmission cost of satellite, which are powerful solutions for the limitation of satellite bandwidth.

5 Conclusions

In this paper, we make a research of deep learning based semantic segmentation methods on remote sensing images. We conduct a detailed comparative analysis of their structures and evaluate their performances using the same public dataset. The experimental result shows that U-Net does well in accuracy, parameters and resource consumption. By using this model, we propose a strategy for on-orbit semantic segmentation in TianZhi-2. We will further improve the method for the satellite environment, and strive to apply it to Tianzhi-2 which will be launched soon.

References

1. Kass, M., Witkin, A., Terzopoulos, D.: Snakes: active contour models. Int. J. Comput. Vis. **4**, 321–331 (1988)
2. Meyer, F., Beucher, S.: Morphological segmentation. J. Vis. Commun. Image Represent. **1**, 21–46 (1990)

3. Boykov, Y.Y., Jolly, M.P.: Interactive graph cuts for optimal boundary and region segmentation of objects in ND images. In: 2001 Proceedings of the Eighth IEEE International Conference on Computer Vision, ICCV 2001, vol. 1, pp. 105–112 (2001)
4. Kampffmeyer, M., Salberg, A. B., Jenssen, R.: Semantic segmentation of small objects and modeling of uncertainty in urban remote sensing images using deep convolutional neural networks. In: Proceedings of the IEEE Conference on Computer Vision and Pattern Recognition Workshops, pp. 1–9 (2016)
5. Volpi, M., Tuia, D.: Dense semantic labeling of subdecimeter resolution images with convolutional neural networks. IEEE Trans. Geosci. Remote Sens. **55**(2), 881–893 (2017)
6. Long, J., Shelhamer, E., Darrell, T.: Fully convolutional networks for semantic segmentation. In: Proceedings of the IEEE Conference on Computer Vision and Pattern Recognition, pp. 3431–3440 (2015)
7. He, K., Zhang, X., Ren, S., Sun, J.: Deep residual learning for image recognition. In: Proceedings of the IEEE Conference on Computer Vision and Pattern Recognition, pp. 770–778 (2016)
8. Garcia-Garcia, A., Orts-Escolano, S., Oprea, S., Villena-Martinez, V., Garcia-Rodriguez, J.: A review on deep learning techniques applied to semantic segmentation. arXiv preprint arXiv:1704.06857 (2017)
9. Ball, J.E., Anderson, D.T., Chan, C.S.: Comprehensive survey of deep learning in remote sensing: theories, tools, and challenges for the community. J. Appl. Remote Sens. **11**(4), 042609 (2017)
10. Ronneberger, O., Fischer, P., Brox, T.: U-Net: convolutional networks for biomedical image segmentation. In: International Conference on Medical Image Computing and Computer-Assisted Intervention, pp. 234–241. Springer, Cham (2015)
11. Badrinarayanan, V., Kendall, A., Cipolla, R.: SegNet: a deep convolutional encoder-decoder architecture for image segmentation. IEEE Trans. Pattern Anal. Mach. Intell. **39**, 2481–2495 (2017)
12. Zhao, H., Shi, J., Qi, X., Wang, X., Jia, J.: Pyramid scene parsing network. In: IEEE Conference on Computer Vision and Pattern Recognition (CVPR), pp. 2881–2890 (2017)
13. Chen, L.C., Papandreou, G., Kokkinos, I., Murphy, K., Yuille, A.L.: DeepLab: semantic image segmentation with deep convolutional nets, atrous convolution, and fully connected CRFs. IEEE Trans. Pattern Anal. Mach. Intell. **40**(4), 834–848 (2018)
14. Everingham, M., Eslami, S.A., Van Gool, L., Williams, C.K., Winn, J., Zisserman, A.: The pascal visual object classes challenge: a retrospective. Int. J. Comput. Vision **111**(1), 98–136 (2015)
15. Simonyan, K., Zisserman, A.: Very deep convolutional networks for large-scale image recognition. arXiv preprint arXiv:1409.1556 (2014)
16. Huang, B., Lu, K., Audebert, N., Khalel, A., Tarabalka, Y., Malof, J., Lefèvre, S.: Large-scale semantic classification: outcome of the first year of Inria aerial image labeling benchmark. In: IEEE International Geoscience and Remote Sensing Symposium-CIGARSS 2018 (2018)
17. Lu, K., Sun, Y., Ong, S.H.: Dual-resolution U-Net: building extraction from aerial images. In: 2018 24th International Conference on Pattern Recognition (ICPR), pp. 489–494. IEEE (2018)
18. Bischke, B., Helber, P., Folz, J., Borth, D., Dengel, A., Waterman, M.S.: Multi-task learning for segmentation of building footprints with deep neural networks. arXiv preprint arXiv:1709.05932 (2017)

19. Holschneider, M., Kronland-Martinet, R., Morlet, J., Tchamitchian, P.: A real-time algorithm for signal analysis with the help of the wavelet transform. In: Wavelets, pp. 286–297. Springer, Heidelberg (1990)
20. Krähenbühl, P., Koltun, V.: Efficient inference in fully connected CRFs with Gaussian edge potentials. In: Advances in Neural Information Processing Systems, pp. 109–117 (2011)
21. Maggiori, E., Tarabalka, Y., Charpiat, G., Alliez, P.: Can semantic labeling methods generalize to any city? The Inria aerial image labeling benchmark. In: IEEE International Symposium on Geoscience and Remote Sensing (IGARSS) (2017)

Att-ConvLSTM: $PM_{2.5}$ Prediction Model and Application

Zhe Xu and Yi Lv(✉)

Faculty of Information Technology, Beijing University of Technology, Beijing, China
781862308@qq.com

Abstract. Environment protection department need to grasp the concentration of $PM_{2.5}$ in a future moment when monitoring. However, the existing $PM_{2.5}$ prediction studies only forecast short-term time points, and cannot accurately give the trend of the next period of time. In this paper, a $PM_{2.5}$ prediction model based on Att-ConvLSTM model integrated training method is established by the advantage of ConvLSTM to obtain spatiotemporal information. Then, Experiments were performed using DNN, ARIMA and LSTM as control model with Att-ConvLSTM model and used it for application test. The result demonstrated that prediction model can extract spatiotemporal features with attention mechanism and ConvLSTM. The model can reduce the generalization error of the model when predicting other observation points.

Keywords: Attention mechanism · Spatiotemporal information · Att-ConvLSTM · Prediction · Generalization error

1 Introduction

$PM_{2.5}$ contains pollutants such as sulfates, nitrates and black carbon that can penetrate deep into the lungs and cardiovascular system and pose a significant risk to human health. About 3 million deaths a year are linked to exposure to outdoor air pollution. The environmental protection department needs to accurately identify the $PM_{2.5}$ change trend in the target area, so as to guide citizens to travel, regulate motor vehicle restrictions and supervise industrial production.

At present, there are two major methods for air quality prediction models: deterministic method and statistical method [1, 3].

Deterministic methods mainly use meteorological and statistical methods to simulate pollutant emission, transmission and diffusion processes based on atmospheric physical and chemical processes. When the actual conditions do not meet the model assumptions, the prediction results will produce large errors [4]. Chen et al. proposed a multi-scale air quality model CMAQ to predict the concentration of $PM_{2.5}$ in California SAN joaquin valley [2].

Statistical methods driven by data and do not pay attention to the internal physical and chemical changes of pollutants. It is currently the most widely used class of methods because it directly mine change patterns from observation data. The most famous is the u-Air from Microsoft research. This system use big data: Meteorological,

Y. Liu et al. (Eds.): ICNC-FSKD 2019, AISC 1074, pp. 30–40, 2020.
https://doi.org/10.1007/978-3-030-32456-8_4

human activities and other data to classify air pollution levels by ANN and CRF [5]. Li et al. designed a stacked self-encoder (SAE) to predict the concentration of $PM_{2.5}$ in Beijing, and found that the prediction performance would decline with the input time series becoming longer, but the test results were significantly better than ARIMA, SVR and other methods [6]. However, the prediction accuracy of the above research is generally low. In the research, only one-hour prediction accuracy is tested, and there are few studies with a long prediction time point.

In practical application, the predicted results of short-term time points are far from enough to guide people to travel and environmental protection departments to formulate emission reduction measures. Yan et al. used Encoder-Decoder model to try to predict the $PM_{2.5}$ concentration of a long time series, and found that the encoding and decoding structure can not only be used in the field of machine translation, but also effective in sequence prediction [7]. Wen et al. used 3D convolution and LSTM to build the C-LSTME model, and attempted to extract the correlation of the observation station in spatial position, and achieved good performance in 6-hour prediction. However this method needed to collect a large number of auxiliary data, such as meteorological and AOD data to ensure the accuracy [8]. Shi et al. proposed ConvLSTM structure and applied it to rainfall prediction, which directly integrated the convolution operation into the LSTM neuron, further improving the prediction performance [9]. The attention mechanism was first applied to the image classification task by Mnih et al., which can make the input information at different locations more accurately related. The classification effect similar to CNN network was obtained in the case of using only Attention and RNN [10]. In this paper, a $PM_{2.5}$ prediction model based on Attention, volume and cyclic layer is proposed. The Attention mechanism is utilized to weight the spatiotemporal information processing and the convolutional cyclic layer is used to extract spatial features. Main contents include:

- A spatiotemporal attention mechanism is designed to improve the weight of effective prediction by weighting spatiotemporal information.
- The $PM_{2.5}$ prediction model with spatial-temporal attentional mechanism and convolutional cyclic layer ConvLSTM was established. Combined with insensitive loss function, the model is trained and tested.
- The improvement of model prediction performance by attention mechanism is verified and compared with other models.
- The data other than the training data set were used to carry out application experiments to verify the actual generalization performance of the model.

2 Data and Methods

2.1 Data Description

$PM_{2.5}$ and pollutant concentration data were obtained from the data center of the ministry of environmental protection of China. The data of $PM_{2.5}$ monitoring station in Beijing are 35 observation stations. The study period was from January 1, 2015 to December 31, 2016. The data format is hourly measurement data.

Because the actual measurement data is affected by network transmission, human error and other reasons, the acquired data will be missing and abnormal. The missing data is usually the continuous missing of an observation station within a period of several hours, and the abnormal data shows that the value of an isolated observation station or a single observation station for several hours is seriously inconsistent with the reality, such as negative value or overflow value 999. According to the environmental air quality standard (GB 3095-2012), the data validity of particles with particle size less than or equal to 2.5 microns should be at least 27 days per month and at least 20 h of concentration sampling value per day. For a small number of random missing values, simply adding zero to the missing data will affect the prediction performance of the model. In order not to affect the consistency of time series data, the missing values are supplemented by linear interpolation method, and the abnormal points that are inconsistent with the actual value are deleted.

The sample size after treatment was 17410 records, each record including the hourly concentration of $PM_{2.5}$ of each ground station. 80% of the samples were randomly selected as the training data, and the remaining 20% were used for the index evaluation after the model training.

In order to ensure the comparability between the experiments, all the experiments chose the east fourth observation station near the city center as the target observation station.

The data are normalized, and the range of variables is transformed into the range of 0–1. The position information of the observation station is the longitude and latitude coordinates. According to the positions of each observation station, the observation station is put into a three-dimensional matrix. The latitude corresponds to the first dimension of the matrix, the longitude to the second dimension, and the time to the third dimension (see Fig. 1). The dimension of the spatial matrix is 57×48, where there are a large number of zero values. The input model will cause serious over-fitting problems, so the row and column of all zero are removed. Time dimension (len) represents that the input model is the length of time series and is a super parameter that needs to be adjusted, which has a great impact on the performance of the prediction model and needs to be determined in subsequent experiments.

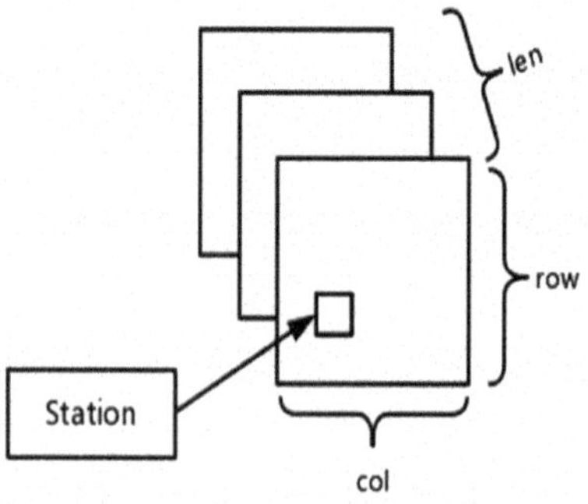

Fig. 1. Input data format

2.2 ConvLSTM

ConvLSTM is a neuron structure that extracts spatial and temporal information and processes space-time sequences [9]. Shi et al. proposed ConvLSTM and successfully applied it to the prediction of rainfall and the prediction of dynamic handwriting. Its basic structure is the most common LSTM neuron in the recurrent neural network [11]. LSTM is different from ConvLSTM in that dot product in each gate structure is replaced by convolution operation. The operation inside the neuron can be expressed as the following formula:

$$i_t = \text{sigmoid}(W_i * [h_{t-1}, x_t] + W_{ci} \cdot C_{t-1} + b_i) \tag{1}$$

$$C_t = f_t \cdot C_{t-1} + i_t \cdot \tanh(W_c * [h_{t-1}, x_t] + b_c) \tag{2}$$

$$f_t = sigmoid(W_f * [h_{t-1}, x_t] + W_{cf} \cdot C_{t-1} + b_f) \tag{3}$$

$$o_t = sigmoid(W_o * [h_{t-1}, x_t] + W_{co} \cdot C_{t-1} + b_o) \tag{4}$$

$$h_t = o_t \cdot \tanh(C_t) \tag{5}$$

As can be seen from the above expressions in (1)–(5), the input 3d variable x_t and the hidden state h_{t-1} of the neuron are both convolved, while the neuron state C_{t-1} is not convolved, but just the dot product. Input gate i_t, cell state update gate C_t, forgetting gate f, and output gate o_t all undergo similar convolution. The calculation of the convolution part replaces the operation process of the general two-dimensional convolution layer, and the update process of the cell state retains the memory and forgetting process of the time series.

2.3 Spatiotemporal Attention Mechanism

The attention mechanism imitates the way humans pay attention to images. Human eyes will focus on some parts of the image, which has the greatest influence on the classification of the image. In this way, the model only focuses on the regions of interest and ignores other unimportant regions [10, 17]. Vaswani et al. proposed several application backgrounds and basic forms of the Attention mechanism, and found that if the Attention mechanism is properly set, it can not only improve the prediction accuracy of the model, but even directly replace CNN or RNN [12]. The Attention mechanism used in this paper focuses on space and time respectively. The formula is as follows:

$$\text{Attention} = \text{softmax}\left(\frac{\tanh\left(XW^T_{\frac{time}{spatial}}\right)}{\sqrt{l}}\right)\text{X} \tag{6}$$

The formula of (6) is written as the calculation figure, as shown in Fig. 2. X is the input of the model, and the dimensions are: time, height and width. The input is first deformed, and only the dimensions that require attention are dotted. After deformation,

X is dotted with learnable parameter matrices W_{time} and $W_{spatial}$, respectively. After the activation by tanh function, the activated matrix needs to be divided by the square root of l for normalization. L is the size of the dimension in Attention for X. Softmax is used to get the weights for each dimension, expressed as a sum as a probability. Then after Reshape its deformation reduction for the original module is three dimensions, finally by page elements and weighted values of X multiplied to get attention.

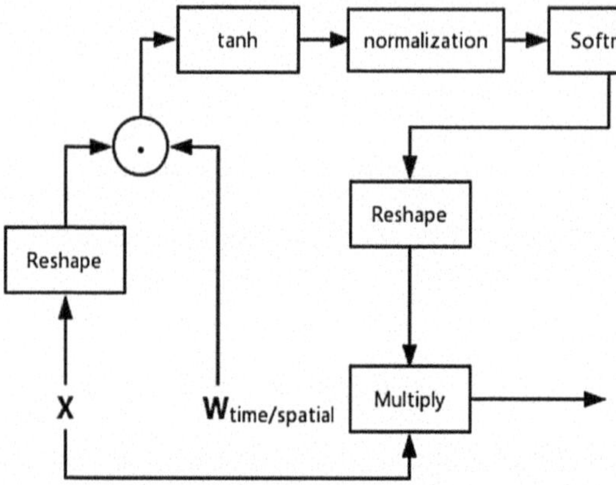

Fig. 2. Spatiotemporal attention module

2.4 ε-Insensitive Loss Function

The - insensitive loss function is a common loss function for SVR. The idea is to increase the tolerance of the regression function for the calculation of the deviation, not calculate the loss when the deviation is less than the tolerance, and calculate the loss function when the deviation is greater than the tolerance.

$$L(y,f(x,w)) = \mathrm{Max}(0, |y - f(x)| - \varepsilon) \tag{7}$$

In formula (7) above, $f(x, w)$ is the predicted value of the model with trainable parameter w. y is the true value; It has been found that when the model is set on a large scale, the over learning problem is solved well, while when the model is set on a small scale, the accuracy is only achieved [13].

3 Development of Att-ConvLSTM Model

The structure of the model is shown in Fig. 3. There is only one input data, namely, spatial data of pollutants organized in matrix. Input tensor includes: time length, rows, cols, channels. The time length is a parameter that has a great influence on the prediction, with the number of rows and columns being 8 × 7. Because the two-dimensional matrix composed of pollutant data has no RGB color channel, the channel parameter is set as one. The input dimension after adding the channel dimension is four dimensions.

Firstly, the input data are weighted by two cascading Attention modules for time and space. Secondly, the attention operation is carried out on the time dimension. At this time, the dimension of the attention weight used is the length of the input sequence. To perform attention operation on space, 8×7 dimensions need to be expanded into a one-dimensional vector, so the number of attention weights is 56; Conv-LSTM receives weighted spatiotemporal data. The convolution kernel size of this layer is 3×3, the step size is 1, and the spatial dimension is convolved according to the way of padding 0; Features on space-time are extracted in this layer to obtain the feature space-time sequence with the same dimension and output. As there are many parameters to be trained in this layer, it is easy to cause overfitting, so the number of layers is set as one layer; After the BN layer was batched and normalized, the Conv3D layer was input to adjust the size of the features. After the feature is processed into one dimension, the predicted time series value is obtained through the non-activated neurons in the layer, and the output value depends on the time series length to be predicted by the prediction model.

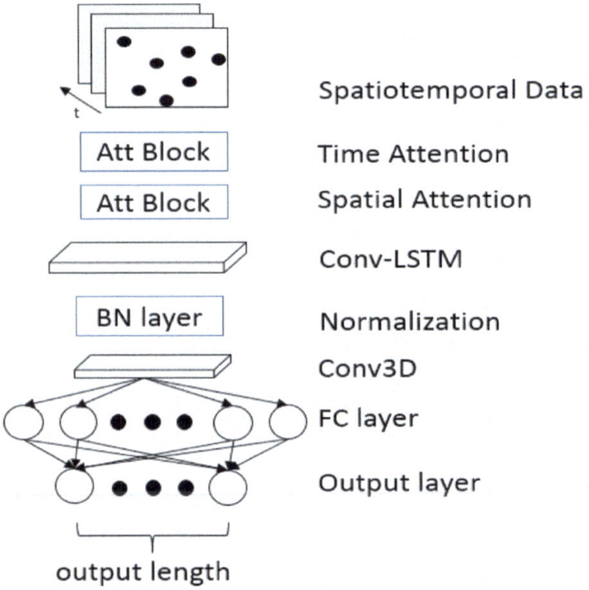

Fig. 3. Architecture of Att-ConvLSTM

In the training stage of the model, the insensitive loss function mentioned in 2.4 is used for training. In essence, Eq. (7) adds a relaxation interval to the original fitting curve, and the calculation method of error, $|y - f(x)|$, is to calculate the absolute average error of the batch in each round of training. The tolerance is controlled by the width of the relaxation interval, and the first part of the experiment is to adjust the tolerance because it is a super parameter on the performance.

4 Result and Analysis

4.1 Time Series Prediction

After repeated experiments to adjust the optimal hyper-parameter, the input sequence length was determined to be 8 h and the tolerance ε was 0.04.

The predicted results are compared with the actual values. The vertical axis is the actual pollutant concentration, and the horizontal axis is the predicted pollutant concentration. When the actual value is equal to the predicted value, the scatter will approach the oblique line in the figure (see Fig. 4). Most pollution is concentrated in the range of 0–400 ug/m^3, more than 500 ug/m^3 days is few, accord with $PM_{2.5}$ numerical distribution. As can be seen from the figure, the prediction performance of the model is relatively high. Most of the scattered points are near the diagonal line, and only a few of them are outliers.

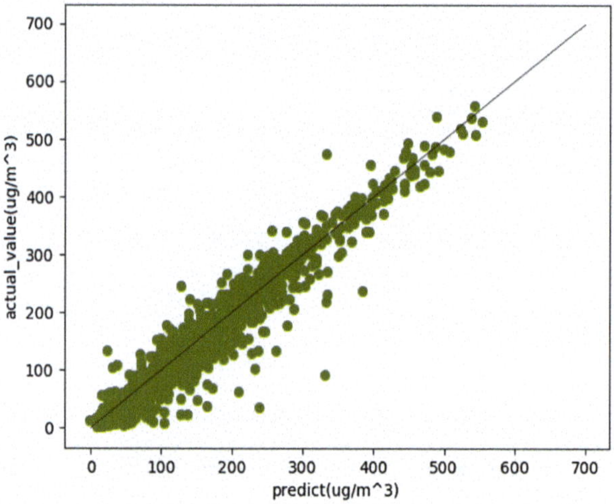

Fig. 4. Comparison between predicted results and actual values

$PM_{2.5}$ prediction method was used to construct the prediction comparison model, including ARIMA, LSTM, DNN, Bi-GRU [16]. The ARIMA model only requires the input variable to be a stationary signal, so the $PM_{2.5}$ sequence is differentially transformed [14]. Observe the results of the comparison experiment (see Table 1). The performance of DNN and ARIMA is poor, and the performance of bi-directional Bi-GRU is not as good as that of one-way LSTM in the 10-hour prediction performance even better than other neural network methods. The best performance in each set was Att-ConvLSTM, suggesting that the model was still good at predicting distant time points.

Table 1. Multiple time point prediction comparisons

Model	Output hour	R^2	RMSE (ug/m^3)	MAE (ug/m^3)
DNN	1	0.83	39.06	30.61
ARIMA		0.924	25.18	19.95
LSTM		0.934	24.56	20.15
Bi-GRU		0.936	23.88	17.95
Att-ConvLSTM		0.964	17.06	10.22
DNN	5	0.66	57.93	37.74
ARIMA		0.768	43.64	26.16
LSTM		0.842	45.31	26.9
Bi-GRU		0.795	46.83	30.171
Att-ConvLSTM		0.887	30.37	25.66
DNN	10	0.648	65.79	44.55
ARIMA		0.403	69.881	43.99
LSTM		0.723	58.27	36.89
Bi-GRU		0.684	62.31	39.6
Att-ConvLSTM		0.746	53.47	33.58

4.2 Comparison of Attention Mechanisms Models

In order to verify the improvement ability of attention mechanism to the model, this chapter also adds attention mechanism to other comparative models.

Att-ConvLSTM achieves the best performance (see Table 2). The attention mechanism also improved other models, and the RMSE of GRU type circular neural network decreased from 48.61 to 44.42. Since there is only a time attention mechanism, the performance improvement is not significant. DNN has little improvement in performance because DNN can only train attention on the input dimension rather than the time and space dimension. The ConvLSTM network showed a significant performance improvement when adding attentional mechanism, because the input dimension was three-dimensional, and the attentional weight could be increased according to the two dimensions of time and space. The experiment showed that the improvement effect of Attention was different in different structure models, and Att-ConvLSTM model achieved the highest performance improvement through the Attention mechanism in the 5-hour prediction.

Table 2. Comparison of attention mechanisms models for 5 h prediction

Model	R^2	RMSE (ug/m^3)	MAE (ug/m^3)
Bi-GRU	0.813	48.61	29.97
Att-Bi-GRU	0.831	44.42	28.35
DNN	0.7	51.61	34.74
Att-DNN	0.706	51.63	35.69
ConvLSTM	0.846	42.67	26.97
Att-ConvLSTM	0.887	30.37	25.66

4.3 Application Experiment of the Model

The deep learning model has excellent results in training and testing, but its fitting ability is often found to be poor in practical application. Therefore, it is necessary to test the trained model with the actual data, which refers to the data in the non-data set. In this paper, pollutant data were collected again in Beijing and the Att-ConvLSTM model was tested to evaluate the generalization ability of the model.

The test was conducted outside the data set in 2019, using data from 22 February to predict pollutant concentrations on 23 February. The comparison model was Att-ConvLSTM and LSTM which it is was second performance in 4.2 experiment. The input data were all from February 22 to predict February 23. According to the input form of the respective models, the Att-ConvLSTM input data was adjusted to a space-time 3d matrix, and the LSTM was a time series.

The dotted line is the mean daily change of pollutant concentration in February of 2015 and 2016 in the training data set (see Fig. 5). It can be clearly seen from the figure that Att-ConvLSTM's prediction at 10 am was obviously closer to the reality than LSTM, and other predicted results were basically the same. The similarity is that the predicted values are all low. Combined with the analysis of the mean dotted line of historical changes in February, it is speculated that the reason for the overall low is that the dataset used for training is generally low, so the model is conservatively predicted according to the lower numerical value. From 6:00–15:00, Att-ConvLSTM achieves more accurate prediction. LSTM model obviously overestimated the sudden drop of concentration.

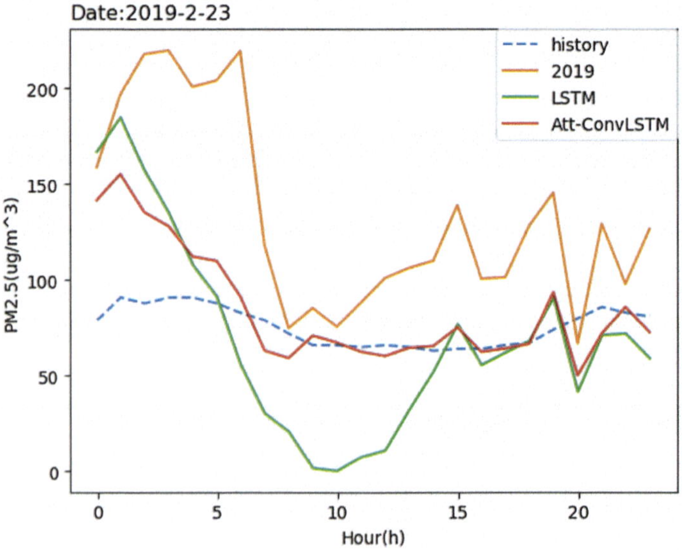

Fig. 5. All-day test results

5 Conclusion

In this paper, a spatiotemporal attention mechanism is designed to enhance the weight of elements with high correlation by weighting each input element, aiming at the low prediction accuracy and insufficient generalization of $PM_{2.5}$ at long time points; a model Att-ConvLSTM was used to extract temporal and spatial information, and $PM_{2.5}$ was predicted and tested.

Experimental results show that the Att-ConvLSTM model can get the characteristics of space-time data, and is significantly superior to other models in predicting time series tasks. Attention mechanism can be combined with any neural network-based prediction model, with Att-ConvLSTM being the most effective. In the applied experiments, the Att-ConvLSTM using spatiotemporal data input is obviously superior to the LSTM model using only time series input. The model also has certain errors in the concentration mutation, which is in line with the research conclusion of Zheng et al. that the neural network model has good prediction effect on slow $PM_{2.5}$ concentration change and is not easy to deal with the mutation concentration [15]. More studies are needed to address this shortcoming.

Project Support Fund

1. Supported by National Natural Science Foundation of China (61873009).
2. Supported by Beijing Natural Science Foundation (4192009).

References

1. Zhou, S., Li, W., Qiao, J.: Prediction of $PM_{2.5}$ concentration based on self-organizing recurrent fuzzy neural network. CAAI Trans. Intell. Syst. **13**(04), 509–516 (2018)
2. Chen, J., Lu, J., Avise, J.C., et al.: Seasonal modeling of $PM_{2.5}$ in California's San Joaquin Valley. Atmos. Environ. **92**, 182–190 (2014)
3. Fan, J., Li, Q., Zhu, Y., Hou, J., Feng, X.: Aspatio-temporal prediction framework for air pollution based on deep RNN. Sci. Surv. Mapp. **42**(7), 76–83
4. Li, X., Peng, L., Yao, X., et al.: Long short-term memory neural network for air pollutant concentration predictions: method development and evaluation. Environ. Pollut. **231**(Pt 1), 997–1004 (1987)
5. Zheng, Y., Liu, F., Hsieh, H.-P.: U-air: when urban air quality inference meets big data. In: Proceedings of the 19th ACM SIGKDD International Conference on Knowledge Discovery and Data Mining, pp. 1436–1444. ACM, New York (2013)
6. Li, X., Peng, L., Hu, Y., et al.: Deep learning architecture for air quality predictions. Environ. Sci. Pollut. Res. **23**(22), 22408–22417 (2016)
7. Yan, L., Wu, Y., Yan, L., et al.: Encoder-decoder model for forecast of $PM_{2.5}$ concentration per hour. In: 2018 1st International Cognitive Cities Conference (IC3), pp. 45–50 (2018)
8. Wen, C., Liu, S., Yao, X., et al.: A novel spatiotemporal convolutional long short-term neural network for air pollution prediction. Sci. Total Environ. **654**, 1091–1099 (2019)
9. Shi, X., Chen, Z., Wang, H., et al.: Convolutional LSTM network: a machine learning approach for precipitation nowcasting. arXiv:1506.04214 (2015)

10. Mnih, V., Heess, N., Graves, A., et al.: Recurrent models of visual attention. In: Advances in Neural Information Processing Systems, pp. 2204–2212 (2014)
11. Hochreiter, S., Schmidhuber, J.: Long short-term memory. Neural Comput. **9**(8), 1735–1780 (1997)
12. Vaswani, A., Shazeer, N., Parmar, N., et al.: Attention is all you need. In: 31st Annual Conference on Neural Information Processing Systems, NIPS 2017, 4–9 December 2017, pp. 5999–6009. Neural Information Processing Systems Foundation, December 2017
13. Yang, J., Zhang, Y., Zhu, Y.: Classification performance of support vector machine with ε-insensitive loss function. J. Xi'an Jiaotong Univ. **41**(11), 1315–1320 (2007)
14. Lang, Y., Xiao, L.: Forecasting concentrations of $PM_{2.5}$ in main urban area of Hangzhou and mapping using SARIMA model and ordinary Kringing method. Acta Sci. Circumstantiae **38**, 62–70 (2018)
15. Zheng, G., Wang, T., Yang Y., Zhang, X.: $PM_{2.5}$ concentration prediction model based on genetic wavelet neural network. Geomat. Spat. Inf. Technol. **41**(09), 248–250+256 (2018)
16. Schuster, M., Paliwal, K.K.: Bidirectional recurrent neural networks. IEEE Trans. Signal Process. **45**(11), 2673–2681 (1997)
17. Wang, X., Girshick, R., Gupta, A., et al.: Non-local neural networks. In: Proceedings of the IEEE Conference on Computer Vision and Pattern Recognition, pp. 7794–7803 (2018)

SAR Ship Detection Method Based on Convolutional Neural Network and Multi-layer Feature Fusion

Bangzheng Yue[1,2], Wenda Zhao[1,2(✉)], and Song Han[2]

[1] University of Chinese Academy of Sciences, Beijing 100049, China
{yuebangzheng16, zhaowenda17}@mails.ucas.ac.cn
[2] Institute of Electronics, Chinese Academy of Sciences, Beijing 100190, China
hans@mail.ie.ac.cn

Abstract. SAR ship detection plays an important role in marine traffic monitoring. Traditional SAR target detection methods are mostly based on intensity differences between target and clutter, which is limited especially in complex scenes, for instance coastal areas. In order to improve the detection performance in complex scenes, a SAR ship detection method based on convolutional neural network named LCMF is proposed in this paper. Firstly, a base network with low complexity is employed to extract features. Secondly, the 'top-down' approach is adopted to gradually fuse the semantically strong features, which is helpful for reducing false alarms, with the low-level high-resolution features to improve the detection performance on small targets. Finally, small-scaled anchor is designed to obtain region proposals, and these proposals are further fed to classification and regression network, which outputs the final detection results. Experiments on the sentinel-1A dataset demonstrate that the proposed method can detect ship targets in SAR images of complex scenes with high speed and accuracy.

Keywords: Synthetic Aperture Radar · Ship detection · Convolutional neural network

1 Introduction

Synthetic Aperture Radar (SAR) is active remote imaging system, which is widely used in remote sensing, reconnaissance, marine traffic monitoring and etc. With the increase of airborne SAR and space-borne SAR, a great amount of SAR images has been obtained. Developing methods which could detect artificial targets such as ships from SAR images automatically and quickly is of great importance and urgency [1].

Traditional SAR target detection methods [2, 3] have good performance in simple scenarios by focusing on the intensity difference between target and background clutter. However, in some complex scenes where the intensity difference is relatively small,

This work was supported in part by the national key research and development program of china under grant 2017YFB0503001.

Y. Liu et al. (Eds.): ICNC-FSKD 2019, AISC 1074, pp. 41–53, 2020.
https://doi.org/10.1007/978-3-030-32456-8_5

such as inshore regions with small islands, those methods could only get unsatisfactory detection results with a large number of false alarms. In order to improve detection performance in complex scenarios, Zhao *et al.* [4] proposed a method through feature recognition and adaptive background window to detect inshore ships in SAR images. Liang *et al.* [5] presented an approach via saliency and context information to deal with inshore SAR ship detection. However, these methods neglect the texture, shape priors, and etc., which signify the difference between the target and the clutter, restrict the further improvement of detection performance.

The Convolutional Neural Network (CNN) [6] can extract semantically strong features from the data layer by layer, and provide a feasible method for making full use of the information in SAR images. Presently, CNN has made breakthroughs in the field of optical image object detection [7]. R-CNN [8] proposed by Girshick et al. performs target detection through three modules. First, the selective search algorithm is used on the input image to obtain region proposals that might be targets; second, each region proposal is separately fed into the CNN to extract feature vector with fixed length; third, the feature vector is classified using Support Vector Machine (SVM). There are many redundant calculations in the R-CNN due to large overlap between region proposals. In order to solve this problem, Fast R-CNN [9] directly extracts feature of the entire image, and the feature of each region proposal is then obtained through the Region of Interest (RoI) Pooling layer. Faster R-CNN [10] proposed by Ren et al. replaces the traditional method with the Region Proposal Network (RPN) to generate region proposals, which shares feature with Fast R-CNN, further improves the detection speed and accuracy rate.

Although CNNs significantly improve the performance in target detection of optical images, these methods cannot be directly used with SAR images [12]. First, in order to be robust to changes of illumination, weather, incident angle and environment, CNN-based optical image object detection algorithms require a large amount of data in training process. In contrast, SAR images are not affected by the illumination, environment factors. Moreover, it is difficult to obtain large and comprehensive dataset for SAR target detection. Second, the optical image is acquired from the "first person" perspective, and the target usually covers a large area of the image. On the contrary, SAR image is usually of the perspective of aerospace observation, and targets such as ship are very small. Third, object detection algorithms used in optical images such as Faster R-CNN [10], Single Shot MultiBox Detector (SSD) [11] use only high-level features for detection. High-level features have strong semantics and feature resolution is compromised, which is not suitable for small target detection. Therefore, the CNN-based object detection algorithm used on optical images cannot be directly used on SAR images.

By taking the characteristics of SAR ship detection into consideration, we improve the Faster R-CNN and propose a SAR ship detection method named Low Complexity CNN and Multi-layer feature Fusion (LCMF). The main contribution of our work is two-fold:

- Inspired by Visual Geometry Group (VGG) [13], a low-complexity base network is designed by stacking small-scale convolution layers, whose parameters are only 1/32 compared to the convolutional layers weights of VGG16. The proposed network is able to extract effective features in the SAR image while needs less training data and reduces the risk of network over-fitting.
- To deal with the small size of the ship targets in SAR images, a 'top-down' fusion method is adopted to fuse the high-level features which have low resolution but strong semantics with the low-level high-resolution features.

2 Method

The overall flow of the LCMF algorithm is shown in Fig. 1. It can be divided into three stages: feature extraction, feature fusion, and predictive output. In the feature extraction stage, features of the input SAR image are extracted layer by layer using a base network with low complexity. In the feature fusion stage, a top-down approach is used to gradually integrate high-level features that help reduce false alarms with low-level high-resolution features. In the prediction output stage, two parallel branches are firstly used, and anchors with different scales and aspect ratios are combined to predict candidate regions. Then, the feature of each region proposal is extracted by ROI pooling layer, and the bounding box of region proposal is further classified and regressed.

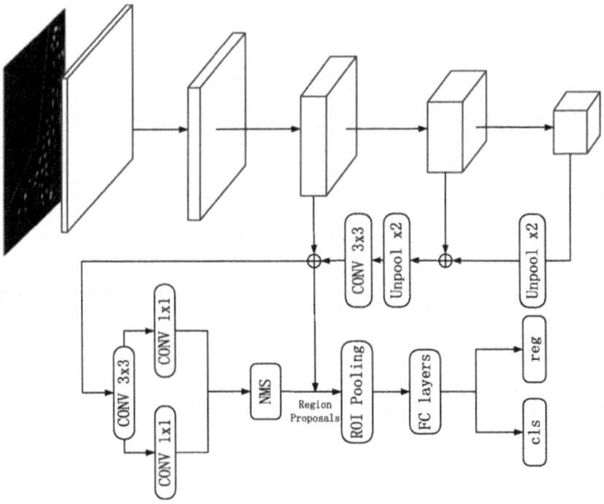

Fig. 1. Flow chart of the algorithm

2.1 Base Network

Two factors should be taken into consideration when designing a base network for SAR ship target detection tasks. Firstly, compared with optical images obtained in Natural scene. There is a huge gap in remote sensing SAR images between scene richness and image features. Secondly, there is currently no large public dataset in the field of SAR target detection, and the available data is quite limited. These factors prompted us to design a low-complexity base network to reduce the risk of over-fitting. VGG [13] stacks 3 × 3 convolution layers to replace the large-scale convolution layer, which not only result in fewer parameters of the convolution block with equal receptive field, but also more nonlinear operation. It enhances the representation while reduces the complexity of the network.

Inspired by VGG [13], the proposed base network of SAR images is built by stacking small-scale convolution kernels. In addition, we added a 1 × 1 convolution layer between each two 3 × 3 convolution layers to further reduce the number of parameters to be learned. As shown in Fig. 2, the receptive fields for the three modules are 5 × 5. When the number of channels of both input and output is C, the ‘a module’ has $25C^2$ parameters, the ‘b module’ has $18C^2$ parameters, and the ‘c module’ only has $14C^2$ parameters.

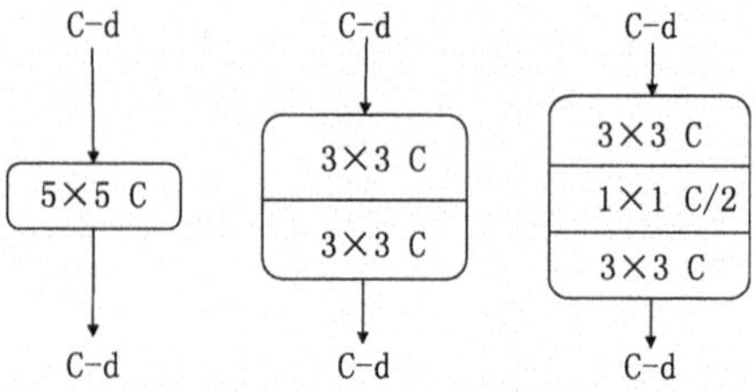

Fig. 2. Three convolution modules

The details of base network are shown in Table 1. It consists of 12 convolutional layers and 4 max-pooling layers. The stride of the convolutional layer is 1 and Rectified Linear Units (ReLU) is adopted as the nonlinear activation function. Padding operation is used before convolution to keep the input and output of the convolutional layer with the same size. In addition, we added Batch Normalization (BN) after the convolutional layer. Adding BN could not only speed up the convergence of the network, but also play a certain regularization role. The number of feature channels output by the five convolution modules is 16, 32, 64, 128, and 128, respectively. In addition, a max-pooling layer is applied after each convolution modules, and it is performed over a 2 × 2 pixel window, with stride 2. The parameters of the entire base network are 0.46M (Millions), which are only 1/32 compared to the 13 convolutional layers weights of VGG16 and 1/15 compared to the 21 convolutional layers weights of GoogleNet.

2.2 Feature Fusion

In the base network, high layers have larger receptive fields and utilize more context information. For example, ships are always on the sea, therefore the high layers can learn sea as the context. The high layer of the network can use this information to learn

Table 1. Base network configurations

Type	Size	Stride	Output
conv1_1	3 × 3	1	16
conv1_2	3 × 3	1	16
maxpool	2 × 2	2	16
conv2_1	3 × 3	1	32
conv2_2	3 × 3	1	32
maxpool	2 × 2	2	32
conv3_1	3 × 3	1	64
conv3_2	3 × 3	1	64
maxpool	2 × 2	2	64
conv4_1	3 × 3	1	128
conv4_2	3 × 3	1	64
conv4_3	3 × 3	1	128
maxpool	2 × 2	2	128
conv5_1	3 × 3	1	128
conv5_2	3 × 3	1	64
conv5_3	3 × 3	1	128

some abstract features that help reduce false alarms on the shore. However, as more pooling layers are operated in the high layers, the resolution of the feature map gradually decreases. In Faster R-CNN [10], only the low-resolution feature map of the last convolutional layer is used for detection, which directly causes the miss of some small targets. In SAR images, the ship of interest usually covers only a small area of the image. Therefore, it is necessary to comprehensively utilize both high-level semantically strong features and low-level high-resolution features to detect ships in SAR images.

In this paper, a high-level semantically strong feature is merged with low-level high-resolution features in a 'top-down' approach to obtain high-resolution features with strong semantics. As shown in Fig. 1, the process of feature extraction is divided into five stages (split by stride). We retain the feature maps output of the third, fourth and fifth stages, which are represented by C3, C4 and C5. The resolutions are 1/4, 1/8, 1/16 of the input SAR image. After the features of different stages are extracted, we gradually merge the features of the higher layers with the features of the lower layers. In order to obtain the fusion feature F3, two fusions are performed, and each consists of the following three steps:

Step 1: Perform the nearest neighbor sampling on the high-level features, and expand the feature map by 2 times.

Step 2: Concatenating the up-sampled high-level features with the low-level features.

Step 3: A 3 × 3 convolution is applied to the concatenated feature map to reduce the influence of aliasing effect, and output the fused feature.

$$g_i = unpool(F_i) \tag{1}$$

$$F_i = \begin{cases} C_i & i = 5 \\ conv_{3\times3}[concat(C_i, g_i)] & other \end{cases} \tag{2}$$

2.3 Output Network

In the predictive output section, we first use two parallel 1 × 1 convolution branches, combined with 9 small-sized anchors to obtain region proposals that may contain ship targets. One branch is used as a classification layer to map the features to 2 × 9 channels, and outputs the probability that the corresponding anchor being the target of interest. The other branch, as a regression layer, maps features to 4 × 9 channels, and outputs an offset between the corresponding anchor and the target area. Specifically, we use the two-class softmax function to output the probability that the corresponding anchor being a ship target, as shown in Eq. (3). The offset between the anchor and the target area is represented by a vector t (t_{cx}, t_{cy}, t_w, t_h), as shown in Eq. (4).

$$p_{ship} = \frac{\exp(s_{pos})}{\exp(s_{pos}) + \exp(s_{neg})} \tag{3}$$

$$\begin{aligned} t_{cx} &= (cx - cx_a)/w_a, t_w = \log(w/w_a) \\ t_{cy} &= (cy - cy_a)/h_a, t_h = \log(h/h_a) \end{aligned} \tag{4}$$

In order to detect targets of various size, the anchors of three scales of 128^2, 256^2, and 512^2 were selected in [10]. We selected 9 small anchors (3 scales with box areas of 8^2, 16^2, 32^2, and 3 aspect ratios of 1:2, 1:1, 2:1) based on the scale distribution of the ships on the dataset (Table 2). At each position on the feature map, 9 prediction results are output. For the feature map of size m × n after the fusion, the entire network outputs a total of 9 × m × n region proposals. To reduce redundancy, we adopt non-maximum suppression on the proposal regions based on their cls scores. After that we use the top-300 ranked proposal regions for detection.

Table 2. The distribution table of ship scale in sentinel-1

Scale	$(0, 8^2]$	$(8^2, 16^2]$	$(16^2, 32^2]$	$(32^2, 48^2]$	$(48^2, \infty]$
Number	39	2461	3976	237	0

2.4 Training

During training, we need to specify which anchor corresponds to the ground-truth and train the network accordingly. In this paper, each anchor is assigned a label according to the Intersection-over-Union (IoU). First, the anchor with the highest IoU overlap with a ground-truth box is designated as a positive sample. Second, an anchor that has an IoU overlap with any ground-truth box higher than 0.7 is also designated as a positive sample. We assign a negative label to a no-positive anchor if its IoU ratio is lower than 0.3 for all ground-truth boxes. Anchor of IoU between 0.3 and 0.7 are discarded.

$$IoU = \frac{area(A) \cap area(B)}{area(A) \cup area(B)} \tag{5}$$

With these definitions, we minimize a multi-task loss functions as shown in Eq. (6)

$$L = \frac{1}{N}(\alpha L_{cls} + \beta L_{reg}) \tag{6}$$

Where N is the number of anchor in a mini-batch. L_{cls} is the classification loss of the network, L_{reg} is the regression loss to the target area. α, β are the weights of the classification loss and the regression loss, in our experiment $\alpha = 1$, $\beta = 2$.

The classification loss L_{cls} is the softmax loss of two classes (object or non-object) as shown in Eq. (7).

$$L_{cls} = -\sum_{i} p_i^* \log(p_i) \tag{7}$$

Where i is the i-th anchor in a mini-batch. $p_i^* = 1$, if the anchor is positive, and $p_i^* = 0$ if the anchor is negative. p_i is the predicted probability of anchor i being an object.

The regression loss is the loss of Smooth L1 as shown in Eq. (8). Where i and pi* are defined the same as in the classification loss function. t is a four-dimensional vector representing the offset between the ship region predicted by the network and the corresponding anchor, and t* represents the offset between the real target region and the corresponding anchor.

$$L_{reg}(t_i, t_i^*) = \sum_{i} \sum_{j \in \{cx,cy,w,h\}} p_i^* smooth_{L_1}(t_j^i - t_j^{i*}) \tag{8}$$

$$smooth_{L_1}(x) = \begin{cases} 0.5x^2 & if|x| < 1 \\ |x| - 0.5 & other \end{cases} \tag{9}$$

The loss function for further classification and regression of candidate regions is consistent with Eq. (6). In addition, the region proposals with the ground–truth IoU greater than 0.5 is specified as a positive sample, and the rest are negative samples.

The entire network is trained end-to-end by gradient back propagation and Adam optimization method. The initial learning rate is 0.001 and the iterations of training are 15000. In region proposal stage, 256 anchors are randomly selected from one image to form a mini-batch according to a ratio of 1:1 to positive and negative samples. When the positive sample is less than 128, the negative sample is used for filling. In the regression and classification part of the region proposals, 32 candidate regions are extracted in the same way to form a mini-batch. The base network was pre-trained on the MSTAR (Moving and Stationary Target Acquisition and Recognition) dataset. Other parameters, initialized with a Gaussian distribution with a mean of 0 and a standard deviation of 0.1.

3 Experiment Results and Analysis

3.1 Sentinel-1A Dataset and Performance Evaluation Metrics

In this paper, data for experiment is the full resolution level-1 Ground Range Detected data obtained by sentinel-1A satellite, which works in IW mode. It is obtained from sentinel website for free.

Sentinel-1A satellite re-visits a same district each 12 days, the collected SAR images include multi-size, rotated ship targets. In this experiment, we use 22 pieces of offshore area SAR image in sum, the size of every image is about 25000 $\times$ 16000 pixels. The image was not preprocessed, we experimented directly with the original image. We tailor 17 of them into 1150 sub-images of size 512 $\times$ 512 pixels without overlapping, and the other 5 large-scene images are used as test data. When labeling these data, we refer to OpenSARShip [14], which provides ship ordination information and using manually interpretation to label data.

All of the experiments are achieved in Linux server with NVIDIA GTX1070 GPU, using deep learning frame TensorFlow.

As for evaluation metrics, we choose F1-score to evaluate the proposed algorithm. It is defined in Eq. (10).

$$F1{-}score = \frac{2 \times \text{precision} \times \text{recall}}{\text{precision} + \text{recall}} \tag{10}$$

$$recall = \frac{TP}{NP} \tag{11}$$

$$\text{precision} = \frac{TP}{TP + FP} \tag{12}$$

Where recall is the ratio of items that were correctly detected among all the ships that should have been detected. Precision is the ration of detected items that are correct. TP is the number of correct detected objects FP is the number of false-alarm, NP is the number of ground-truth.

3.2 Fusion Feature Comparison

In this section, experiment on the feature maps output at different stages is conducted to validate the influence of fusion feature. Specifically, we choose four groups C3, C4, C5 and F3 for compare, The results are shown in Table 3. It can be concluded that with the decrease of feature resolution, recall ratio gradually decreases. Among these, when low-level feature C3 is used for detection, false-alarm raises and precision decreases even we could get higher recall ratio. When fusion feature F3 is chosen, there is trade-off between precision of detection and recall-rate. F1-score is the highest at this scenario, with the value of 0.883. Figure 3 shows part of the experiment results, due to space limitations we only give experiment F3 in 4 scenes tailored image, each size of the tailored picture is 1500 × 1000.

Table 3. Comparison of detection results by different feature layers

Layer	Precision	Recall	F1-score
C3	0.787	0.932	0.854
C4	0.820	0.878	0.848
C5	0.789	0.749	0.769
F3	0.905	0.862	0.883

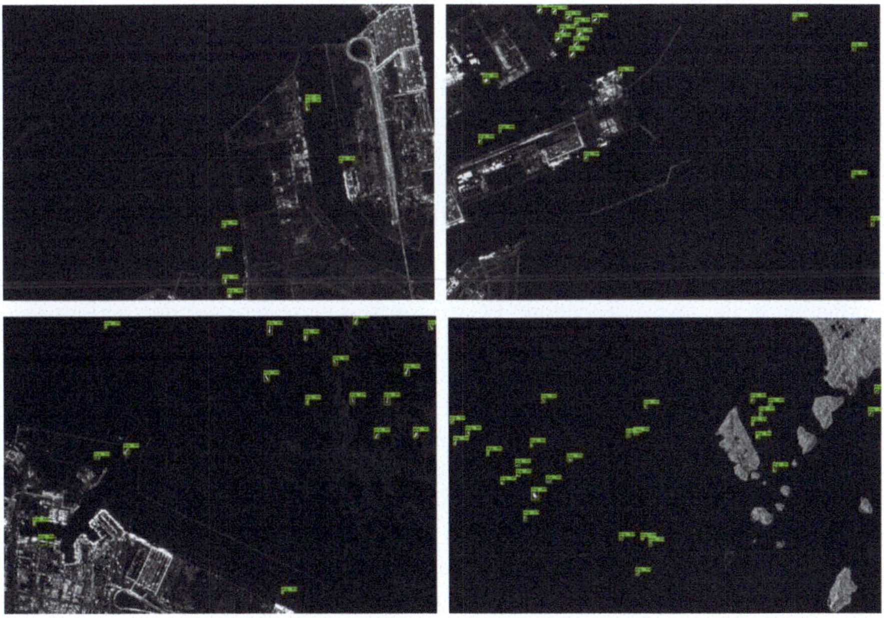

Fig. 3. The detection result with fusion features

3.3 Different Base Network

In order to validate the influence of low-complexity base network towards SAR ship target detection performance, we use 3 kind base networks for experiment: low-complexity network which is designed in this paper, VGG16 and GoogleNet. For fairness, we use the same feature fusion method in base network VGG16 and GoogleNet. From Table 4 we are able to conclude that while using VGG or GoogleNet as the base network, we could reach high precision of detection, and low false-alarm ratio. But the recall of these two networks is lower than the proposed method, several targets are not detected. Through the analysis of the missed target, we found many of the false negative targets are large size ship objects. This is because VGG16 and GoogleNet have too many parameters to learn and the models are too complex, thus the model gets serious over-fitting and causes detection failures. When using the low-complexity base network designed in this paper, it can keep higher precision and recall ratio, and effectively reduces the risk of net over-fitting issue.

Table 4. Comparison of detection results by different base network

Network	Precision	Recall	F1
LCMF	0.905	0.862	0.883
VGG16	0.920	0.735	0.817
GoogleNet	0.912	0.776	0.839

3.4 Comparison with Other Methods

In this part we compare our methods LCMF with other different detection methods to validate the efficiency.

Comparison with two-parameter Constant False Alarm Rate (CFAR): as shown in Fig. 4, in simple scene, two-parameter CFAR can detect ship target precisely, but in complex scene, it will regard small islands and near shore artificial clutter as ship targets. It is because the two-parameter CFAR uses only intensity differences for detection. In contrast, LCMF uses CNN to extract features and make better use of the information in the image. Therefore, LCMF can sustain better detection performance both in simple scene and complex scenario.

Comparison with CNN-based detection methods: in this part, we use LCMF, Faster R-CNN [10], SSD [11] and You Only Look Once (YOLO)v3 [15] for comparison. All experiments are conducted on the same training dataset. Table 5 shows the results of experiments in testing set, and Fig. 5 shows part of the comparison results. Among these experiments, Faster-RCNN causes lower recall ratio because of its complex base network and its scales of anchor that are different from the scales of ship target, implementing low-resolution high-level features to detect. SSD also has a complex base network, so it causes model over-fitting and many large-size targets are failed to be detected, and therefore its F1-score only reaches 0.763. In contrast to Faster R-CNN and SSD methods, LCMF considers the characteristics of SAR ship target. It extracts features with low-complexity base network, and fuses the multi-layer features, so

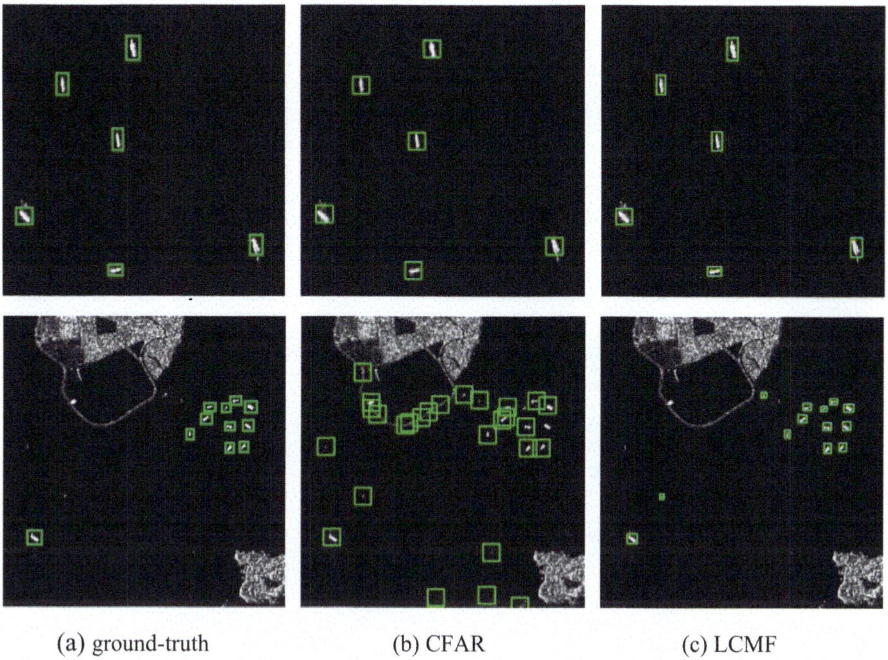

Fig. 4. Comparison of detection results with two-parameter CFAR

Table 5. Comparison of detection results by different methods

Methods	Precision	Recall	F1
LCMF	0.905	0.862	0.883
Faster R-CNN	0.914	0.683	0.782
SSD	0.835	0.703	0.763
YOLOv3	0.881	0.872	0.876

LCMF can detect ship target effectively. Besides, we compare LCMF with YOLOv3. As the preliminary experiments on Setntinel-1A show, YOLOv3 can detect all the targets in Fig. 5, however the performance of LCMF on the entire test dataset is slightly better than YOLOv3.

Experiments of Faster R-CNN, SSD, YOLOv3, and LCMF are all conducted on GPU (NVIDIA GTX1070). It takes 0.169 s with Faster R-CNN to detect a SAR image of 512 × 512 pixels, 0.042 s with SSD, 0.036 s with YOLOv3 and 0.059 s with LCMF.

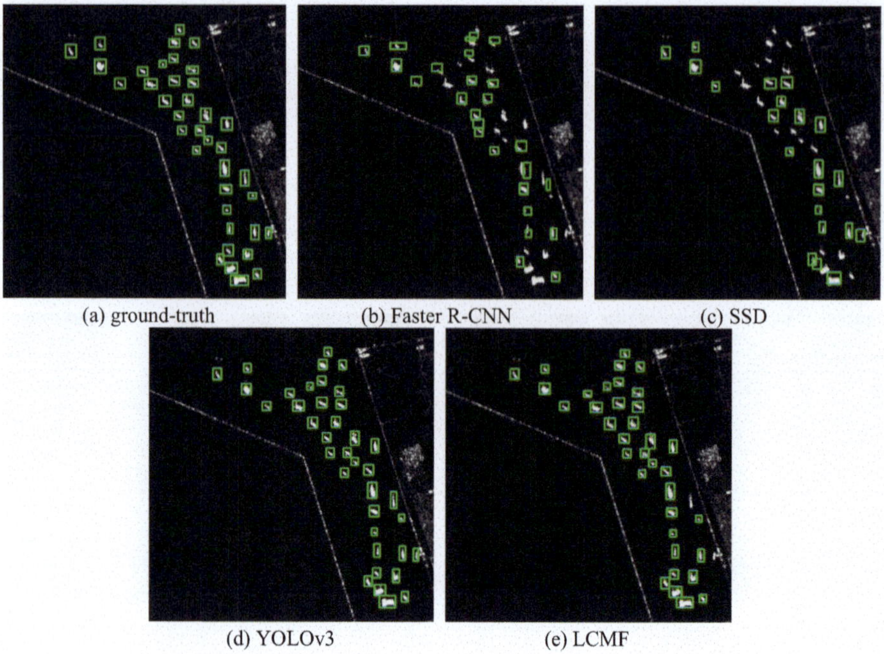

Fig. 5. Comparison of detection results with other methods

4 Conclusion

This paper proposes a SAR ship target detection algorithm LCMF based on low complexity CNN and multi-layer feature fusion. First, a low complexity base network is designed to extract the features of the input image layer by layer. Compared with common networks such as VGG and GoogleNet, the low complexity base network designed in this paper can not only extracts effective features from SAR images, but also requires less learning parameters and reduces the risk of network over-fitting. Second, using the 'top-down' approach, the high-level feature helps to reduce the abstract features of false alarms and the features of low-level high-resolution, and improve the detection performance of small targets. Experiments on the sentinel-1A satellite data prove that the proposed method can detect ship targets quickly and effectively in SAR images of complex scene.

References

1. Eldarymli, K., Mcguire, P., Power, D., et al.: Target detection in synthetic aperture radar imagery: a state-of-the-art survey. J. Appl. Remote Sens. **7**(7), 071598 (2013)
2. Goldstein, G.B.: False-alarm regulation in log-normal and Weibull clutter. IEEE Trans. Aerosp. Electron. Syst. **AES-9**(1), 84–92 (1973)

3. Gao, G., Liu, L., Zhao, L., et al.: An adaptive and fast CFAR algorithm based on automatic censoring for target detection in high-resolution SAR images. IEEE Trans. Geosci. Remote Sens. **47**(6), 1685–1697 (2009)
4. Zhao, H., Wang, Q., Huang, J., et al.: Method for inshore ship detection based on feature recognition and adaptive background window. J. Appl. Remote Sens. **8**(1), 083608 (2014)
5. Liang, Z., Yu, L., Yi, S.: Inshore ship detection via saliency and context information in high-resolution sar images. IEEE Geosci. Remote Sens. Lett. **13**(12), 1870–1874 (2016)
6. Lecun, Y., Bengio, Y., Hinton, G.: Deep learning. Nature **521**(7553), 436 (2015)
7. Huang, J., Rathod, V., Sun, C., et al.: Speed/accuracy trade-offs for modern convolutional object detectors, pp. 3296–3297 (2016)
8. Girshick, R., Donahue, J., Darrell, T., Malik, J.: Rich feature hierarchies for accurate object detection and semantic segmentation. In: Proceedings of the 2014 IEEE Conference on Computer Vision and Pattern Recognition (CVPR), pp. 580–587. IEEE, Columbus (2014)
9. Girshick, R.: Fast R-CNN. In: Proceedings of the 2015 IEEE International Conference on Computer Vision (ICCV), pp. 1440–1448. IEEE, Santiago (2015)
10. Ren, S.Q., He, K.M., Girshick, R., Sun, J.: Faster R-CNN: towards real-time object detection with region proposal networks. IEEE Trans. Pattern Anal. Mach. Intell. **39**(6), 1137–1149 (2017)
11. Liu, W., Anguelov, D., Erhan, D., Szegedy, C., Reed, S.E., Fu, C.Y., Berg, A.C.: SSD: single shot multibox detector. In: Proceedings of the 14th European Conference on Computer Vision (ECCV) 21–37. Springer, Amsterdam (2016)
12. Xu, F., Wang, H., Jin, Y.: Deep learning as applied in SAR target recognition and terrain classification. J. Radars **6**(2), 136–148 (2017). https://doi.org/10.12000/jr16130
13. Simonyan, K., Zisserman, A.: Very deep convolutional networks for large-scale image recognition. Computer Science (2014)
14. Huang, L., Liu, B., Li, B., Guo, W., Yu, W., Zhang, Z., Yu, W.: OpenSARShip, a dataset dedicated to Sentinel-1 ship interpretation. IEEE J. Sel. Top. Appl. Earth Obs. Remote. Sens. **11**(1), 195–208 (2018)
15. Redmon, J., Farhadi, A.: YOLOv3: an incremental improvement (2018)

Deep Learning Training Management Platform Based on Distributed Technologies in Resource-Constrained Scenarios

Jie Li, Guoteng Wang, Changsheng Zhang(✉), and Bin Zhang

Northeastern University, Shenyang 110819, People's Republic of China
zhangchangsheng@mail.neu.edu.cn, paper820@sohu.com

Abstract. Deep learning has attracted a lot of research attention in the past few years for its efficiency and accuracy. However, there exist two problems of their study. Firstly, the computing power of one single machine is limited and not suitable for handling with training deep learning models with massive cells. Secondly, it costs much to train models on different deep learning frameworks. Motivated by these problems, this paper proposed a deep learning training management platform based on distributed technologies, which integrates different kinds of deep learning frameworks through virtualization technologies and coordinates machines through distributed technologies. Specially, specific algorithms are proposed to solve the multi-task scheduling problem, the computing resources allocation problem and the fault tolerance problem in resource limited scenarios. It turns out that the platform can be widely used in small and medium-sized research teams.

Keywords: Deep learning · Distributed technologies · Virtualization technologies · Automation

1 Introduction

Deep learning has become particularly prevalent research topics in the artificial intelligent application. However, in practice deep learning technique is usually hard to be applied due to following reasons. Firstly, deep learning frameworks are usually hard to learn. Furthermore, since luxury computational resources are needed for deep learning training tasks to process, it's usually not easy for researchers without enough financial support to evaluate their work in relative normal and scientific steps. Although there are many different deep learning frameworks, the implementation ideas and core technologies are similar to some extent. [1–4] and some ongoing projects such as Kera [5] and DIGIST [6] have successfully achieved the unification of logic and interface by making a layer of encapsulation on the framework. And along with the appearance of nvidia-docker framework [7], it has been possible to use virtualization technologies to automate the configuration of deep learning processes. Most recently, a series

Y. Liu et al. (Eds.): ICNC-FSKD 2019, AISC 1074, pp. 54–62, 2020.
https://doi.org/10.1007/978-3-030-32456-8_6

of open source frameworks like zookeeper [8], mongodb [9] and kubernetes [10] have emerged, which provide fantastic support for the building and management of computing clusters. Besides, the basic theory of distributed deep learning has been initially established with the effort of scholars in relevant fields [11–15].

There are already some solutions based on Kubernetes and Docker [16], but the configuration and deployment processes of such solutions are complex, and the operation and maintenance costs are high. Besides, these solutions are not suitable for application in resource-constrained scenarios. Moreover, in resource limited situations, the limited computing resources, large differences in computing device performance and unstable machine states must be considered, which requires specially designed scheduling, distribution, and fault-tolerant schemes.

Based on the above, this paper proposes a distributed based deep learning training platform in resource-constrained scenarios, which provided an easy-to-maintain cluster architecture based on zookeeper and mongodb framework and a series of scheduling, allocation and fault tolerance algorithms to solve problems in resource-constrained scenarios.

2 Platform Architecture

The distributed deep learning training management platform proposed in this paper adopts a three-tier architecture, as shown Fig. 1.

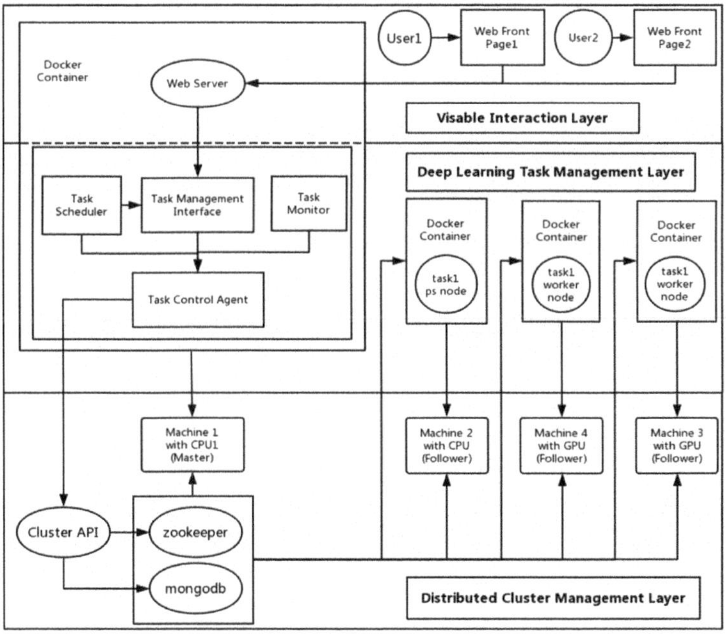

Fig. 1. Platform architecture

2.1 Distributed Cluster Management Layer

The distributed cluster management has to achieve three goals. The first is to use cluster resources simply and efficiently. The second is to ensure the stability of the software running on the cluster. The third is to meet the different environmental needs of different deep learning frameworks. To achieve these goals, this layer is based on zookeeper and mongodb. The implementation achieves efficient management of cluster resources with the help of zookeeper's distributed synchronization mechanism, and ensures the program stability through zookeeper's automatic selection mechanism and mongodb's replica set mechanism. Meanwhile, this layer achieves environmental isolation by packaging every program running on the cluster into a Docker container.

2.2 Deep Learning Management Layer

The main function of the deep learning management layer is to manage the deep learning distributed training tasks. Its core components include Task Scheduler, Task Monitor, Task Control Agent and Task Management Interface. Task Management Interface provides an interface to the visual interaction layer for it to manage deep learning distributed training tasks. Task Control Agent operates the cluster resources and resolves cluster operating conflicts by invoking various management operations through the interface provided by the Distributed cluster management layer. Task Scheduler implements scheduling algorithms and allocation algorithms to schedule and allocate resources for deep learning distributed training tasks. Task Monitor monitors the states of running tasks and execute a fault tolerance algorithm for error tasks. The scheduling algorithm, the allocation algorithm, and the fault-tolerant algorithm proposed by this layer are described in detail in the next section.

2.3 Visual Interaction Layer

The visual interaction layer is based on the BS architecture and supports the interaction between users and the platform front end according to the requirements of users and the characteristics of the deep learning technology.

3 Core Algorithms

A data parallel strategy based on the parameter server architecture is applied in the platform to achieve distributed training. This strategy divides the nodes of a training task into parameter server nodes (ps) and computing nodes (worker). For the convenience of discussion below, job is redefined to be a running ps or worker, and task is redefined to be a collection of jobs.

3.1 Multi-task Scheduling Algorithm

In a resource-constrained scenario, when the platform performs multiple deep learning training tasks, some tasks may have to wait because of insufficient computing resources. For tasks with a large amount of computation, users tend to expect a longer training time, while for tasks with a smaller amount of computation, users always expect a relatively smaller training time. So it's obviously unreasonable to schedule tasks only according to the request time order regardless of the user's expectations. And a multi-task scheduling algorithm is proposed to solve the problem. Now define the task priority as follow:

$$R = \begin{cases} \dfrac{C + S\alpha}{C} & S < \delta \\ \dfrac{C + S^2\alpha}{C} & S \geq \delta \end{cases} \tag{1}$$

$$C = SC * f * I \tag{2}$$

Where R is the scheduling priority of the task. C is the amount of computation of the task and is used to measure the total training time of a training task. S is the waiting time of task p in the waiting queue. α is a preset coefficient used to balance the magnitude of C and S. At the same time, a threshold δ is set. If the waiting time of a task exceeds the threshold, the proportion of the waiting time is increased. SC represents the total number of floating-point multiplications of one iteration of the deep network, fe denotes the time of a single floating-point operation, and I denotes the total number of iterations of the task. The optimization goal formula for defining the problem is as follow:

$$\begin{array}{ll} \max & R_1 \cdot x_1 + R_2 \cdot x_2 + \cdots + R_n \cdot x_n \\ s.t. & \sum_{i=1}^{m} u'_i \cdot x_i \leq Z, \\ & x_i \in \{0, 1\} \\ & i \in [1, n] \end{array} \tag{3}$$

Where R_i denotes the priority of the current i-th task, and u'_i represents the number of request machines for the i-th task. The goal of the scheduling algorithm is to select a task set that has the highest total priority. When only the first i tasks are processed, let S denotes the maximum value that can be obtained when less than x machines can are allocated. Its state transition equation is as Eq. (4):

$$\begin{array}{l} S_i(x) = max(S_{i-1}(x), S_{i-1}(x - u')_i + R_i) \\ S_0(x) = 0 \quad , x \in [0, Z] \\ S_i(0) = 0 \quad , i \in [0, k] \end{array} \tag{4}$$

3.2 Allocation Algorithm

According to the parameter server architecture, the Computational cost of PS is much smaller than that of woker. Therefore, for a job of type PS, an idle machine will be randomly assigned. But a job of type worker must be assigned to a machine with GPU. In resource-constrained scenario, in case of the difference in hardware device performance, if training is performed in a synchronous manner, then the training speed is limited to the slowest node. Meanwhile, for tasks using asynchronous method, the performance difference will lead to an unbalanced usage of the dataset. The experiment below illustrates the effect of machine performance differences on the training results in the asynchronous mode. The Nvidia 1080Ti and Nvidia 720M are used to train the MNIST dataset, and results are shown in the Figs. 2 and 3:

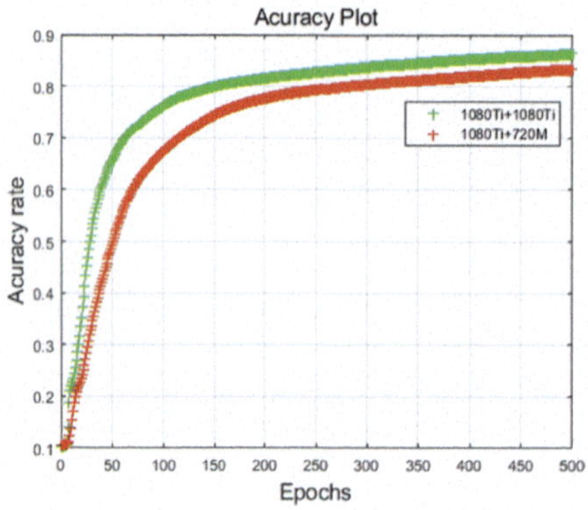

Fig. 2. Training accuracy trend in EX1

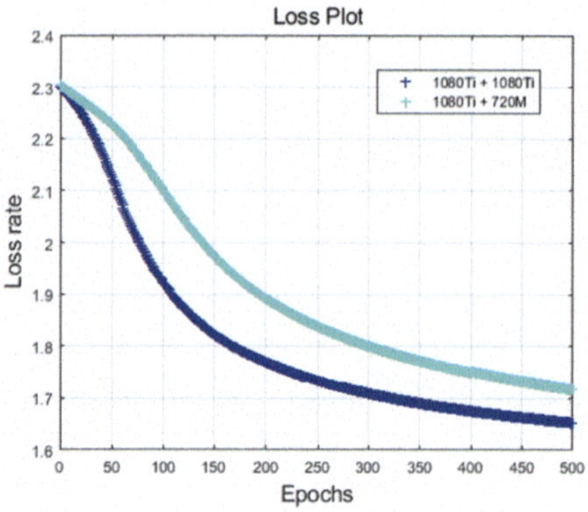

Fig. 3. Training loss trend in EX1

Therefore, it is necessary to design an algorithm to balance the training pace. Suppose that m stations are selected from n available machines to execute the job of type worker. Firstly, the algorithm is designed to select m devices with the smallest variance.

The optimization objectives of the definition algorithm are as follow:

$$v = \frac{1}{m}\sum_{i=1}^{m}(t_i - e)^2 \tag{5}$$

where $T = \{t_1, t_2, \cdots, t_n\}$, is the set of times ($t$ is an integer) that the available GPU trains a certain amount of data.

$e = \frac{1}{m}\sum_{i=1}^{m} t_i$, e is the mean of the time at which m GPUs train a certain amount of data.

$d = \frac{1}{m}\sum_{i=1}^{m}(t_i - e)^2$, d is the variance of the time at which m GPUs train a certain amount of data.

The time complexity of solving this problem without any optimization is $O(n!)$. In order to optimize the time complexity, consider introducing the parameter u, where $u = m * e$ and the range of u is $[u_{min}, u_{max}]$, specifically:

$$u_{min} = \min(\sum_{i}^{n} t_i x_i), \quad x_i \in \{0,1\}, \quad \sum_{n}^{i=1} x_i = m$$
$$u_{max} = \max(\sum_{i}^{n} t_i x_i), \quad x_i \in \{0,1\}, \quad \sum_{n}^{i=1} x_i = m \tag{6}$$

The optimized objective function becomes:

$$v = \frac{1}{m} \min(\sum_{i=1}^{n} (t_i - c)^2 \cdot x_i) \tag{7}$$

where $\sum_{i=1}^{n} x_i = m, x_i \in \{0,1\}$, $c \in [\frac{u_{min}}{m}, \frac{u_{max}}{m}]$.

In the formula (7), v is sum of squares of several terms. First, $(t_i - c)^2$ is calculated for all n items t_i, and then the first m minimum element is selected as the primary allocation result of the enumeration. After traversing $(u_{max} - u_{min} + 1)$ times, the group with the smallest variance is taken as the final allocation result. It can be proved that the time complexity of quickly selecting the first m-term minimum element algorithm is $O(n)$. The time complexity of this algorithm is $O(nm)$, where n is the number of machines currently available in the cluster, m equals $(u_{max} - u_{min} + 1)$, and m can be made smaller by reasonably setting the value of t.

The performance difference between the m machine machines selected by the above method may still be large. In order to further balance the training pace, smaller batch size is set for job on less powerful machines, and larger batch size is set for job on more powerful machine, which could balance the frequency of gradient submission. To prove this idea, Mnist was trained based on Nvidia 1080Ti and Nvidia 720M, and the batch size was changed to obtain the results in Figs. 4 and 5.

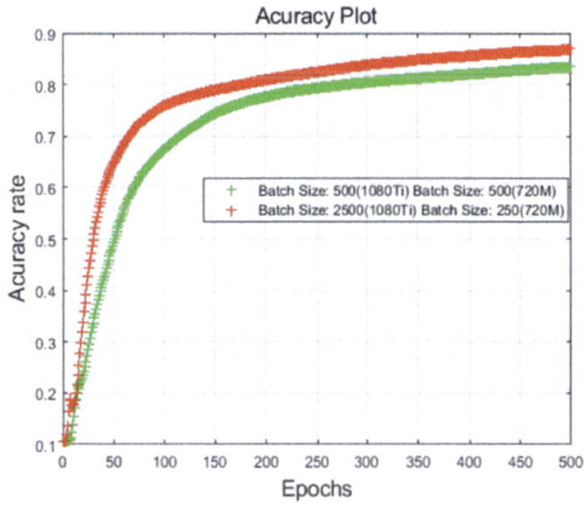

Fig. 4. Training accuracy trend in EX2

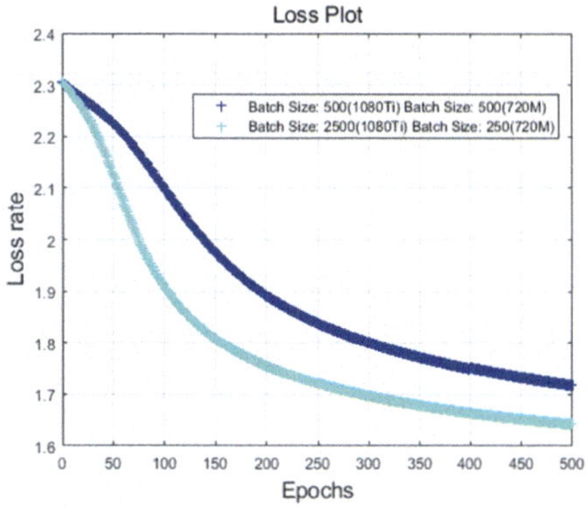

Fig. 5. Training loss trend in EX2

Define the batch allocation sequence $B: \{b_1, b_2, \cdots, b_m\}$, where $\sum_{i=1}^{m} b_i = C$, $b \in [b_{min}, b_{max}]$, b denotes the batch size allocated by the machine, and b_{min} and b_{max} are the batch size range set by users. Define The time sequence in which the machine trains a certain amount of data is $T = \{t_1, t_2, \cdots, t_m\}$, and t is an integer.

The optimization objective function is defined as follow:

$$v = \sum_{i=1}^{n} (b_i t_i - d)^2 \tag{8}$$

In formula (8), v denotes the sum of squared deviations of m machine training time. d denotes the mean value of m machines training data time, and $d \in [b_{min} \cdot \min(T), b_{max} \cdot \max(T)]$, for every value of d, the optimal sequence B needs to be determined to minimize v. Therefore, an allocation algorithm based on dynamic programming is proposed, and the state transition equation $F_k(x)$ is defined as follows:

$$F_k(x) = \begin{cases} 0, \quad x = 0 \quad \text{or} \quad k = 0 \\ INF, \quad x < kb_{min} \quad \text{or} \quad x > kb_{max} \\ \min((y \cdot t_k - d)^2 + F_{k-1}(x - y)), kb_{min} \leq x \leq kb_{max} \end{cases} \tag{9}$$

The Eq. (9) denotes the minimize v when x samples are assigned to the first k machines. Where y represents the batch size allocated by the kth machine. The time complexity of the algorithm is $O(mCR)$. C is the amount of data that can be processed in one iteration. R equals $(b_{max} \cdot \max(T) - b_{min} \cdot \min(T))$, and R can be small by reasonably setting the value of t.

3.3 Fault-Tolerant Algorithm

Due to the unstable state of each node in the cluster, the running Task is likely to go wrong. Based on the Mnist data set, experiments were carried out with fault-tolerant processing and no fault-tolerant processing. The training results are as Figs. 6 and 7:

Since only machines with GPUs can be assigned as workers, and GPU resources are usually scarce in resource-constrained scenarios. So a special fault-tolerant algorithm is needed to make full use of the available GPUs in the platform. Suppose that there are k tasks with node failures at some point. For these k error tasks, defining $M = \{m_1, m_2, \cdots, m_k\}$, in which mi denotes number of workers required by error task i, defining $M' = \{m'_1, m'_2, \cdots, m'_k\}$, in which mi denotes the number of workers that doesn't go wrong in task i, defining $X = \{x_1, x_2, \cdots, x_k\}$, in which xi denotes the number of workers newly assigned to error task i. A denotes the number of currently available GPU machines in the platform. Obviously, when A is greater than or equal to $\sum_{i=1}^{k}(m_i - m'_i)$, the allocation can complete without algorithm. When A is smaller than $\sum_{i=1}^{k}(m_i - m'_i)$, it is necessary to define the error task proper rate r as $r_i = \frac{m'_i}{m_i}$, $i \in [0, k]$. After

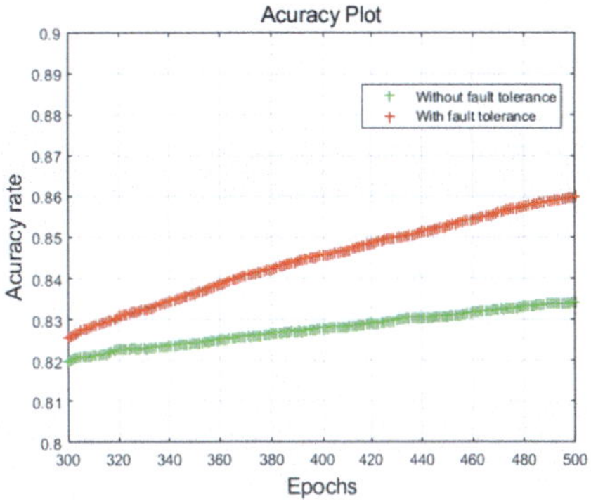

Fig. 6. Training accuracy trend in EX3

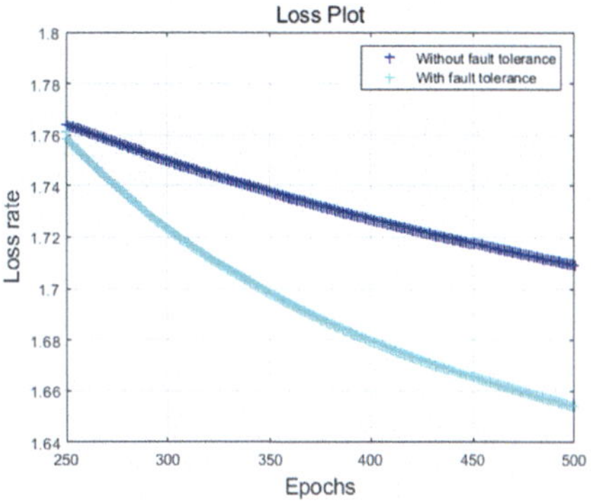

Fig. 7. Training loss trend in EX3

completing an error processing, the error task proper rates of error tasks are updated to: $r_i = \frac{m_i'+x_i}{m_i}$, $i \in [0,k]$. The sum of proper rates for defining all erroneous tasks is: $R = \sum_{i=1}^{k} r_i = \sum_{i=1}^{k} (\frac{m_i'+x_i}{m_i})$.

The goal of the fault-tolerant module is to maximize the R after fault-tolerant processing. This problem can be solved using a dynamic programming algorithm whose state transition equation $G(x)$ is:

$$G_i(x) = \begin{cases} 0, & x = 0 \quad \text{or} \quad i = 0 \\ \max(G_{i-1}(x-y) + \frac{y}{m}), & x \in [1, A] \quad , i \in [1, k] \end{cases} \tag{10}$$

where $y \in [0, \min((m_i - m_i'), x)]$, and $Gi(x)$ indicates the sum of the maximum appropriate rates of ith error tasks before processing when the number of assignable machines does not exceed x. The largest result of $G(x)$ is the optimal allocation result. And y indicates that the current kth task that is assigned to the y machine. It is obvious that the above state transition equation has the optimal substructure property, and the time complexity of the algorithm is $O(mn)$.

4 Conclusion

This paper introduces a design scheme for the distributed deep learning training platform in resource-constrained scenarios. For the limited computing resources, the differences in computing power and the unstable state of each machine in this scenario, a scheduling algorithm, an allocation algorithm, a fault-tolerant algorithm and a specific architecture are proposed. The platform can be applied to the automatically operate and manage distributed deep learning tasks in resource-constrained scenarios such as small and medium-sized laboratories.

References

1. Abadi, M., Barham, P., Chen, J., et al.: TensorFlow: a system for large-scale machine learning. In: 12th USENIX Symposium on Operating Systems Design and Implementation (OSDI 2016), pp. 265–283 (2016)

2. Paszke, A., Gross, S., Chintala, S., et al.: PyTorch: tensors and dynamic neural networks in Python with strong GPU acceleration (2017). https://github.com/pytorch/pytorch
3. Jia, Y., Shelhamer, E., Donahue, J., et al.: Caffe: convolutional architecture for fast feature embedding. In: Proceedings of the 22nd ACM International Conference on Multimedia, pp. 675–678. ACM (2014)
4. Chen, T., Li, M., Li, Y., et al.: MXNet: a flexible and efficient machine learning library for heterogeneous distributed systems. arXiv preprint arXiv:1512.01274 (2015)
5. Chollet, F.: Keras: the python deep learning library. Astrophysics Source Code Library (2018)
6. Yeager, L., Bernauer, J., Gray, A., et al.: DIGITS: the deep learning GPU training system. In: ICML 2015 AutoML Workshop (2015)
7. NVIDIA Docker. https://github.com/NVIDIA/nvidia-docker. Accessed 09 July 2017
8. Hunt, P., Konar, M., Junqueira, F.P., et al.: ZooKeeper: wait-free coordination for internet-scale systems. In: USENIX Annual Technical Conference, vol. 8, no. 9 (2010)
9. Chodorow, K.: MongoDB: The Definitive Guide: Powerful and Scalable Data Storage. O'Reilly Media, Inc., Sebastopol (2013)
10. Hightower, K., Burns, B., Beda, J.: Kubernetes: Up and Running: Dive Into the Future of Infrastructure. O'Reilly Media, Inc., Sebastopol (2017)
11. Dean, J., Corrado, G., Monga, R., et al.: Large scale distributed deep networks. In: Advances in Neural Information Processing Systems, pp. 1223–1231 (2012)
12. Sun, S., Chen, W., Bian, J., et al.: Slim-DP: a multi-agent system for communication-efficient distributed deep learning. In: Proceedings of the 17th International Conference on Autonomous Agents and Multiagent Systems, pp. 721–729. International Foundation for Autonomous Agents and Multiagent Systems (2018)
13. Ooi, B.C., Tan, K.L., Wang, S., et al.: SINGA: a distributed deep learning platform. In: Proceedings of the 23rd ACM International Conference on Multimedia, pp. 685–688. ACM (2015)
14. Gupta, S., Zhang, W., Wang, F.: Model accuracy and runtime tradeoff in distributed deep learning: a systematic study. In: 2016 IEEE 16th International Conference on Data Mining (ICDM), pp. 171–180. IEEE (2016)
15. Cui, H., Zhang, H., Ganger, G.R., et al.: GeePS: scalable deep learning on distributed GPUs with a GPU-specialized parameter server. In: Proceedings of the Eleventh European Conference on Computer Systems, p. 4. ACM (2016)
16. Frederic, J.: How to automate deep learning training with Kubernetes GPU-cluster [EB/OL], 07 September 2018. https://github.com/Langhalsdino/Kubernetes-GPU-Guide

A Deep Learning Based Reasoner for Global Consistency in Named Entity Recognition

Xiaoxiao Yin[1], Ruifang Liu[1](✉), Daqi Zheng[2], and Zhengdong Lu[2]

[1] Beijing University of Posts and Telecommunications, Beijing, China
{ygyyxx110,lrf}@bupt.edu.cn
[2] DeeplyCurious.ai, Beijing, China
{da,luz}@deeplycurious.ai

Abstract. Named Entity Recognition (NER) is a basic task of Natural Language Processing (NLP), it's a challenging task in a variety of special applications. This paper aims to solve the global consistency of NER, and to improve the performance. Inspired by human reading process, we propose a NE-Reasoner model, which combine deep neural networks and memory artificial neural network to identify named entities with global consistency. The advantages of the model are: (1) The multi-layer deep architecture, allowing it to bootstrap the recognized entity set from coarse to fine. (2) The candidate pool memory mechanism, allowing it to exchange identified entity information between layers. (3) The reasoner, combing encoder-decoder and cached information to infer to get global entities. The experimental results show that the NE-Reasoner can identity ambiguous words and named entities that rarely or never met before.

Keywords: Named Entity Recognition · NE-Reasoner · Global consistency · Deep neural network · Memory artificial neural network

1 Introduction

As a task to identify the named entities (NEs) such as person, location etc. in a text, Named Entity Recognition (NER) is considered as a basic problem of text understanding. NER is often solved as a sequence labeling problem [1–6].

Current methods make decisions through the calculation results of word representations in the input sequence, so they actually rely on local linguistic features learned from the training dataset to recognize NEs. Most recent researches focus on introducing more information to leverage these linguistic features better [7–9].

In our human reading process, we read a word not just based on the knowledge we had before, we can look at the entire article, remember all local information while reading, and then connect and integrate local knowledge to understand the meaning of words by combining the entire article. Even in some extreme

Y. Liu et al. (Eds.): ICNC-FSKD 2019, AISC 1074, pp. 63–71, 2020.
https://doi.org/10.1007/978-3-030-32456-8_7

cases, we may need to read until the end of the article to understand what a word really means.

Inspired by the human readings, we introduce a high-level "inference" mechanism to NER task and proposed NE-Reasoner. NE-Reasoner has a multi-layered architecture, each layer completes NER independently, and the NER results of each layer will be stored through a candidate pool as the reference for the next layer. With this architecture, when making decisions, the model can "see" and refer to relevant decisions elsewhere in the same article, so as to make global decisions. We implemented our model on CoNLL-2003 NER datasets, the results show that our model can make better decisions in many cases, compared to the basic 1-layer NER model.

2 Related Works

As a basic task in natural language processing, NER has received a lot of attention and research, and the neural networks work pretty well on this problem. [1] proposed a CNN-CRF structure, it could be viewed as an encoder-decoder model, which used CNNs as encoder and CRF as decoder. [2] used BiLSTM as encoder, and the BiLSTM-CRF model achieved state-of-art on many datasets. [3,4] proposed a hierarchical structure which used an additional CNN to represent character-level features. Not only the encoder, there are also lots of work on the decoder, [10] used LSTM as tag decoder, it turned out to outperform CRF decoder on some NER datasets.

[7] introduced pre-trained language model in sequence labeling and leverage external information resource. [8] presented a task-aware neural language model to extract abundant knowledge hidden in raw texts. [9] used lattice LSTM to get better leveraging explicit word information in Chinese.

[11] proposed a framework that use an external memory to augment neural networks. The memory can help to get better ability to remember facts from the past, [12] presented end-to-end memory networks that can be used to solve question answering problem. Such external memory neural network could store the embedding vector representations for named entities, so we design the candidate pool in our model.

3 The NE-Reasoner Model

In this section, we will introduce the overall architecture of our model and then explain the details of individual components.

3.1 Overall Architecture

As shown in Fig. 1, the NE-Reasoner model has a multi-layered architecture, each layer is an encoder-decoder structure NER model that can complete NER independently.

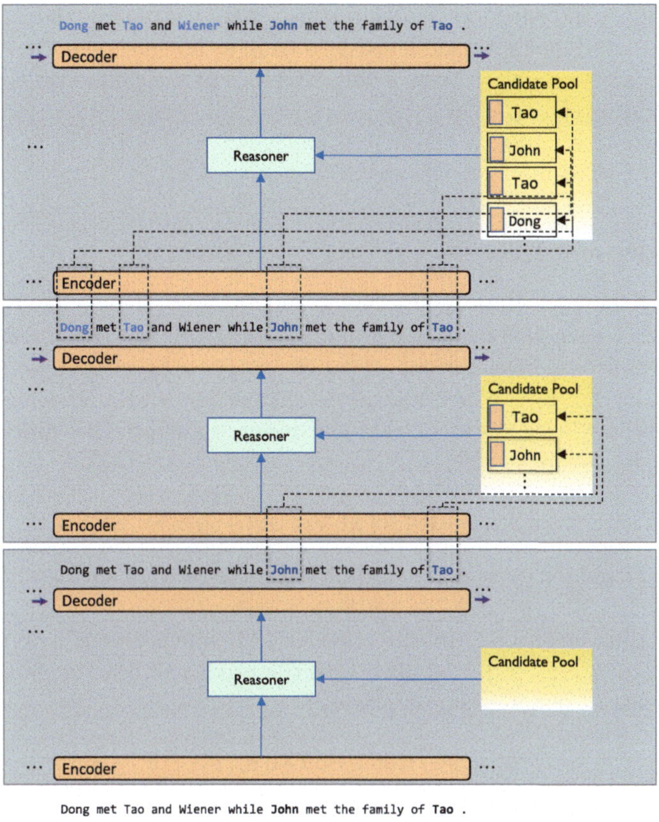

Fig. 1. The overall architecture of our model.

For each layer, the input sequence is encoded into a sequence of vector representations which are generated from the word representations and their context information, then the decoder can rely on this to give predict results independently. The predict results label out which words are entities, so the encoding representations of these words should contain the information that could be relied on to make these decisions. I other word, the model training could make the model to learn the entities vector representations.

Meanwhile, our model keeps a candidate pool throughout the entire NER process, which consists of entities information recognized already. The model completes NER layer by layer, so it can "see" all decisions made past, then each layer can take reference from it through the reasoner, and update this candidate pool from the predict results to help the next layer to keep global consistency and get better results.

Between encoder and decoder in each layer, the reasoner could combine the identified entity information from candidate pool, to improve the identify effects for next layer.

The decoders and encoders in all layers can share parameters to avoid growing of parameters, and make the model easy to train as an end-to-end model. So the only difference between each layer is the difference of the candidate pool and the entities they can refer to.

In this work, we took a character level CNN and BiLSTM model as the encoder, and LSTM model as the decoder.

3.2 Encoder and Decoder

Without the reference from the reasoner and the candidate pool, the encoder and decoder of each layer are a regular BiLSTM-LSTM NER model. But at first, we get word representations of input sequence by character level CNN, $X = \{\mathbf{x}_1, \mathbf{x}_2, \mathbf{x}_3, \ldots, \mathbf{x}_n\}$, which is fed into the encoder of the first layer.

The encoding layer could encode the input sequence into hidden vectors $H = \{\mathbf{h}_1, \mathbf{h}_2, \mathbf{h}_3, \ldots, \mathbf{h}_n\}$.

$$\mathbf{h}_t = \text{BiLSTM}(\mathbf{x}_t) = [\mathbf{h}_t^f, \mathbf{h}_t^b] \tag{1}$$

where h_t^f is the hidden state of forward LSTM, h_t^b is the hidden state of backward LSTM.

The decoding layer is an uni-direction LSTM, which uses H to predict tags $\mathbf{y}^i = \{y_1^i, y_2^i, y_3^i, \ldots, y_n^i\}$, and also takes the predict of last time step as input as follows, where i denotes the i-th layer.

$$\begin{aligned} \mathbf{p}_t^i &= Softmax(\text{LSTM}(\mathbf{h}_t, \mathbf{p}_{t-1}^i)) \\ y_t^i &= argmax(\mathbf{p}_t^i) \end{aligned} \tag{2}$$

3.3 Candidate Pool

With the encode-decode base model, we use a memory network to improve the performance of entity identification. We adopt the BMEOS (Begin, Middle, End, Other, Single) tagging scheme, so we can tell where is the beginning or end of every entity from $\mathbf{y}^i$, and then cache this boundary information of entities.

Since the baseline model relies on the local features to make decisions, we can consider how to store entities information more reasonable and effective from this. The encoding information of the beginning and end of an entity can totally represent this entity, which contains the information about what this entity looks like and how we understand it from the context. We can locate the boundary of entities easily through the predicted result, so we store an entity in these four aspects.

[Forward Context] [Entity Forward] [Entity Backward] [Backward Context]

As shown in Fig. 2 right side, for every entity in $\mathbf{y}^i$, we can get four embedding vector representations $\{\mathbf{h}_b^f, \mathbf{h}_b^b, \mathbf{h}_e^f, \mathbf{h}_e^b\}$, they represent the beginning or end of the entity learned by hidden layer. $\mathbf{h}_b^f$ has the context information before this entity, so we use it to represent forward context, while $\mathbf{h}_e^b$ is backward context, $\mathbf{h}_b^b$ and $\mathbf{h}_e^f$ are both entity information but one is from the entity beginning and the other one is from the end. Then we easily concatenate these representations from all entities in the document to generate four matrices $\{\mathbf{r}^{fc}, \mathbf{r}^{eb}, \mathbf{r}^{ee}, \mathbf{r}^{bc}\}$, where the k means k-th line of the matrix and also the k-th entity:

$$\{\mathbf{r}_k^{fc}, \mathbf{r}_k^{eb}, \mathbf{r}_k^{ee}, \mathbf{r}_k^{bc}\} = \{\mathbf{h}_{kb}^{f}, \mathbf{h}_{kb}^{b}, \mathbf{h}_{ke}^{f}, \mathbf{h}_{ke}^{b}\} \tag{3}$$

3.4 Reasoner and the Next Layer

Based on the candidate pool, we can make decisions refer to an entity recognized elsewhere because their representation is same. So for every word to predict, we can use the similarity between the current word and the candidate pool to make a better decision.

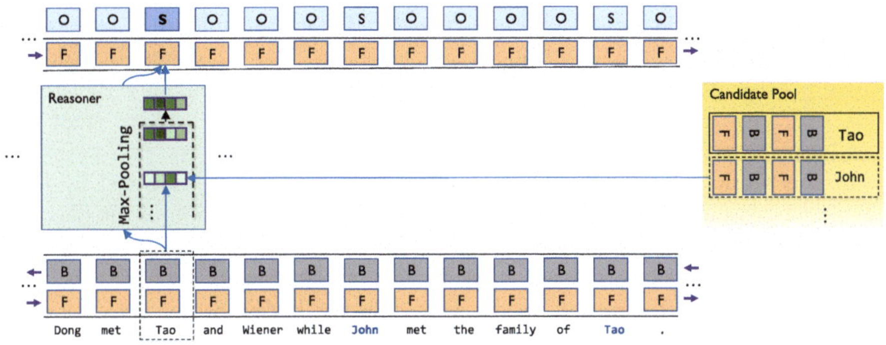

Fig. 2. The structure of reasoner which is also a multi-facts inference model.

As shown in Fig. 2, the reasoner is actually a multi-facts inference model, in this model, the current word information is query, and the entities information in candidate pool is fact, we use a kernel $K(\mathbf{q}uery, \mathbf{f}act)$ to calculate the relations between the current word information and every entity information in candidate pool, the calculate results $\mathbf{s} = \{s_1, s_2, s_3, \ldots, s_n\}$ represent the suggestions from each entity recognized in the last layer. Then we can get a final reference s from these suggestions through the reasoner.

In this work, we use similarity score to represent suggestions from entities and choose the dot product of vectors as the kernel. Since we use the relations between words and entities in candidate pool to as an inference, the only important information is these relations themselves, but not which entity to give out this relation, so we don't need to take any content from the candidate pool to be involved in after calculation. Furthermore, every calculate result in $\mathbf{s}$ represents the same type of relation, so we can use a max-pooling mechanism to get the most useful and typical value as the reference.

$$s_t^{fc} = max(sigmoid(\mathbf{h}_t^{f} \cdot \mathbf{r}_i^{fc}))\forall i \in [N_e] \tag{4}$$

$$s_t^{eb} = max(sigmoid(\mathbf{h}_t^b \cdot \mathbf{r}_i^{eb}))\forall i \in [N_e] \quad (5)$$

$$s_t^{ee} = max(sigmoid(\mathbf{h}_t^f \cdot \mathbf{r}_i^{ee}))\forall i \in [N_e] \quad (6)$$

$$s_t^{bc} = max(sigmoid(\mathbf{h}_t^b \cdot \mathbf{r}_i^{bc}))\forall i \in [N_e] \quad (7)$$

where the N_e means the number of entities.

Then the vector $\mathbf{s}_t = [s_t^{fc}, s_t^{eb}, s_t^{ee}, s_t^{bc}]$ can be viewed as a suggestion from the candidate pool, we feed this information to the decoder with H together to make better decisions.

4 Experimental Results

4.1 Dateset and Parameters Setting

The CoNLL-2003 dataset is a widely used NER dataset which has four types of named entities: person, location, organization and miscellaneous. In order to be comparable with other works, we didn't preserve the document structure of sentences, but shuffle them before training like other works, and we treat the mini-batch as a document.

In this work, we use the 50-dimensional word embedding released by [13]. The hyper-parameters setting is shown in Table 1. The dimension of the decoder is 289, because there are four types of entities and each of them has a four dimension reasoning information, and we take the output of last time step as input, which is a 17-dimension vector according the BMEOS tag system.

4.2 Result

Since we used the CNN-BiLSTM-LSTM model as NER model of each layer, for a fair comparison, we compared with works that based on standard LSTM or CNN encoder without any external information resource.

We implemented experiments with two settings for a better comparison: with and without the character level CNN. As shown in Table 2, our model can achieve

Table 1. Hyper-parameters of NE-Reasoner for experiments

Layer	Parameter	Size
CNN	Character embedding	30
	Window size	3
BiLSTM	State size	256
LSTM	State size	289
NE-Reasoner Layers	Depth	2
Others	Batch size	16
	Learning rate	0.01
	Tag scheme	BMEOS

Table 2. Evaluations on the test set of CoNLL-2003 English NER dataset

Char-encoder	Model	F1
None	Collobert et al. [1]	88.67
	Nguyen et al. [10]	89.86
	Huang, Xu, and Yu [2]	90.10
	Strubell et al. [6]	90.54(±0.18)
	NE-Reasoner	**90.78**
LSTM	Lample et al. [5]	90.94
CNN	Chiu and Nichols [4]	90.91(±0.20)
	Ma and Hovy [3]	91.21
	NE-Reasoner	**91.44**

better performance with both settings, which shows NE-Reasoner can work on different representations of words.

Some methods get high F1 score by introducing in extra features [4], extra information from external datasets and pre-trained language model like [7], co-training language model like [8], or extra encoder on a different level of words like [9]. NE-Reasoner could be implemented on these methods only if they are encoder-decoder based model. Therefore, how NE-Reasoner achieve better performance than the baseline model in each layer by reasoning is more important.

Table 3. Difference of each layer in NE-Reasoner on the test set of CoNLL-2003 English NER dataset

Character encoder	Layer	Precision	Recall	F1
None	First	90.08	90.69	90.38
	Final	**90.33**	**91.23**	**90.78**
CNN	First	90.24	91.95	91.09
	Final	**90.45**	**92.45**	**91.44**

Table 3 shows the F1 score gets an obvious improvement after refinement from the first layer, and the most important reason is the second decoding layer results can get a better recall rate, which benefit from the multi-layer reasoning.

The two results of the first layer are comparable with other works, that means the training of our model won't hurt the first decoding layer results, and the improvement is totally from the correction of NE-Reasoner.

4.3 Analysis

We choose some typical cases from the test dataset to explain how our model works. As shown in Table 4, the model can refer to the entities recognized elsewhere. In the first example, the model can realize that "Sturridge" is also a part of a name cause it saw the another recognized name "Darryl Powell" in the second layer. In the second example, the model predicted "Salem Bitar" as an organization at first but then changed it to a person with the help of reasoner.

Table 4. Typical cases of CoNLL-2003 English dataset to show how the second decoding layer refine the results

Decoder	Result
First layer result	The London club had been rocked by a two-goal burst from forwards Dean(PER) Sturridge and Darryl Powell(PER)
Final layer result	The London club had been rocked by a two-goal burst from forwards Dean Sturridge(PER) and Darryl Powell(PER)
First layer result	Hiroshige Yanagimoto(PER) cross towards the goal which Salem Bitar(ORG) appeared to have covered
Final layer result	Hiroshige Yanagimoto(PER) cross towards the goal which Salem Bitar(PER) appeared to have covered

5 Conclusions and Future Work

In this work, we proposed a NE-Reasoner to mimic the human reading process and keep the global consistency of entities in the NER task. NE-Reasoner can learn to integrate the local decisions in a document and then make inference to get the global decisions, experiments show that our model can learn to infer from the recognized entities information to achieve better performance than the baseline models.

NE-Reasoner is not only useful in NER task, the fusion of local decisions and inference from reasoner could be used on many other tasks, we could apply our model to other NLP tasks like POS tagging or chunking etc., and even on some sequence generating tasks.

References

1. Collobert, R., Weston, J., Bottou, L., Karlen, M., Kavukcuoglu, K., Kuksa, P.: Natural language processing (almost) from scratch. J. Mach. Learn. Res. **12**(Aug), 2493–2537 (2011)
2. Huang, Z., Xu, W., Yu, K.: Bidirectional LSTM-CRF models for sequence tagging. arXiv preprint arXiv:1508.01991 (2015)
3. Ma, X., Hovy, E.: End-to-end sequence labeling via bi-directional LSTM-CNNs-CRF. In: Proceedings of the 54th Annual Meeting of the Association for Computational Linguistics (Volume 1: Long Papers), vol. 1, pp. 1064–1074 (2016)
4. Chiu, J.P., Nichols, E.: Named entity recognition with bidirectional LSTM-CNNs. Trans. Assoc. Comput. Linguist. **4**, 357–370 (2016)

5. Lample, G., Ballesteros, M., Subramanian, S., Kawakami, K., Dyer, C.: Neural architectures for named entity recognition. In: Proceedings of NAACL-HLT, pp. 260–270 (2016)
6. Strubell, E., Verga, P., Belanger, D., McCallum, A.: Fast and accurate entity recognition with iterated dilated convolutions. In: Proceedings of the 2017 Conference on Empirical Methods in Natural Language Processing, pp. 2670–2680 (2017)
7. Peters, M., Ammar, W., Bhagavatula, C., Power, R.: Semi-supervised sequence tagging with bidirectional language models. In: Proceedings of the 55th Annual Meeting of the Association for Computational Linguistics (Volume 1: Long Papers), vol. 1, pp. 1756–1765 (2017)
8. Liu, L., Shang, J., Xu, F.F., Ren, X., Gui, H., Peng, J., Han, J.: Empower sequence labeling with task-aware neural language model. In: International Conference on Artificial Intelligence (2018)
9. Zhang, Y., Yang, J.: Chinese NER using lattice LSTM. In: Proceedings of the 56th Annual Meeting of the Association for Computational Linguistics (Volume 1: Long Papers), pp. 1554–1564. Association for Computational Linguistics (2018). http://aclweb.org/anthology/P18-1144
10. Nguyen, T.H., Sil, A., Dinu, G., Florian, R.: Toward mention detection robustness with recurrent neural networks. arXiv preprint arXiv:1602.07749 (2016)
11. Graves, A., Wayne, G., Danihelka, I.: Neural turing machines. arXiv preprint arXiv:1410.5401 (2014)
12. Sukhbaatar, S., Weston, J., Fergus, R., et al.: End-to-end memory networks. In: Advances in Neural Information Processing Systems, pp. 2440–2448 (2015)
13. Pennington, J., Socher, R., Manning, C.D.: Glove: global vectors for word representation. In: Empirical Methods in Natural Language Processing (EMNLP), pp. 1532–1543 (2014). http://www.aclweb.org/anthology/D14-1162

Multi-level Feature Combination in Dialogue State Tracking

Yang Zheng(✉), Ruifang Liu, and Sheng Gao

Beijing University of Posts and Telecommunications, Beijing, China
zy406625808@126.com, lrf@bupt.edu.cn

Abstract. Dialogue State Tracking is one of the most important component in task-oriented dialog system, which can update the dialogue state and accurately estimate the compact state representation. This paper introduces a multi-level feature combination to capture correlation on the basis of the entire dialog and slots. Based on the Memory-Network [1], which only consider the relevance of current user utterance with historical dialogue context under turn level, we both consider the correlation at the sentence-level and word-level, and give the slot-value pair as input replacing of candidate state pool to estimate the rare slot-value pair. We use Bi-LSTM and self-attention to capture the correlation between internal words from the utterance, the particular slot-value pair and previous unsettled actions, and finally give a confidence degree score to judge whether the particular slot-value pair is the right answer for the current utterance. Our model achieves a good result on the DSTC-2 dataset about 0.74, which is 3% higher than Memory-Network's 0.71.

Keywords: Task-oriented dialog system · Dialogue state tracking · Memory-Network · Multi-level feature combination

1 Introduction

In the recent years, task-oriented dialog systems help users to achieve specific goals with natural language such as movie booking, restaurant reservation etc. Currently, there are two ways of settling task-oriented dialog systems, one based on the traditional pipeline models, the other one based on end-to-end models. Traditional models have several pipeline modules, including natural language understanding (NLU), dialogue state tracking (DST), policy learning (PL) and natural language generation (NLG). Due to the error made in the upper stream modules propagate to downstream components and get amplified, so in the recent years, the most researches focus on predicting the current dialogue state in dialogue state tracking by using the current user utterance and historical dialogue context. Then the dialog system uses this estimated state to plan the next action and response to the user. Therefore, a precise dialogue state getting from dialogue state tracking module is an important research for a vivid and accurate task-oriented dialog system.

Y. Liu et al. (Eds.): ICNC-FSKD 2019, AISC 1074, pp. 72–80, 2020.
https://doi.org/10.1007/978-3-030-32456-8_8

In task-oriented dialog system, for example, in the case of booking a restaurant, the user's inquiry usually contains two types of actions: inform and request, inform is notifying the dialog system that they have already determined (eg. inform(food=Thai)), request is inquiring the dialog system for answers they unknown now (eg. request(address)). Figure 1 gives an example of ordering restaurant dialogue context to indicate dialogue state and the action representation in each turn.

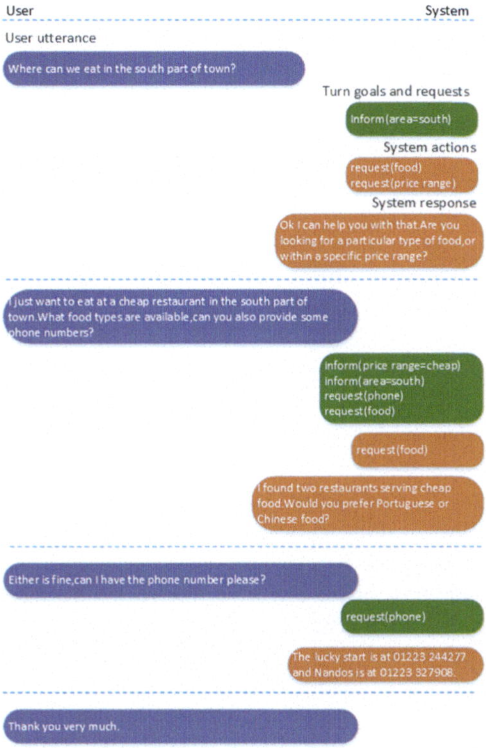

Fig. 1. Example dialogue about restaurant reserve

At present, the main problems of solving DST including several aspects. First is the single turn dialogue context only consider the correlation at the sentence-level or make a large number of hand-crafted feature engineering. Second is that previous researches get dialogue state in each turn always judging from the candidate state pool, which can not effectively track some rare slot-value pairs.

Yoshino et al. use neural network to solve DST in 2014 [2], which can not effectively capture the feature in historical dialogue information. In 2015, the Memory-Network framework has effectively improved the shortcomings of small memory capacity in Long Short-Term Memory (LSTM) and Perez and Liu [1] apply the Memory-Network framework into the task-oriented dialog system to store historical dialogue context. However, it still does not fully solve the main problems mentioned above.

In this paper, we make a multi-level feature combination to capture the correlation at word-level and sentences-level. We use Memory-Network and Bi-directional Long Short-Term Memory (Bi-LSTM), self-attention to solve feature capture from internal dialog and slots, decreasing hand-crafted feature engineering. Besides, we add dialogue state feature as input to precisely estimate the rare slot-value pair. We use our model in the DSTC-2 dataset and get a certain improvement.

2 Related Works

In 2014, the earliest using neural network to solve DST problem, Yoshino, Hiraoka, et al. [2] use LSTM, combining sentence embedding and two characteristics proposed by them to do classification. Henderson [3] considers a word-based tracker using n-gram features and recurrent neural network (RNN). In 2017, NBT-CNN [4] model was proposed using convolutional neural networks(CNN) and convolution kernels automatically extract features. In 2017, Memory-Network [1] used in task-oriented dialog system to improve DST accuracy by saving the historical dialogue context in a memory-matrix, and reading repeatedly through a multi-hop framework, similarity as using a reading comprehension idea. End-to-end methods use encoder-decoder neural network [5] modeling the utterances and interactive structure of dialogue that can automatically extract features. Besides, Madotto et al. [6], Li et al. [7], they all corporate external knowledge bases information to improve accuracy.

But most previous researches only considered the correlation at sentence level and can't have a good solution to estimate the rare slot-value pair. The model we proposed is a multi-level feature combination based on four parts below.

1. the correlation of the current user utterance with the historical dialogue context
2. the correlation of current session to the specific slot-value pair
3. the correlation of previous action to the slot-value pair
4. adding dialogue state feature as input replace of getting answers from the candidate state pool, which can track rare slot-value pairs in the state space.

3 Model Framework

The model framework is shown in Fig. 2. The structure consists with three parts: The first part is based on Memory-Network, settling the long-term dependence of the current user utterance with the historical dialogue context, and finally obtain sentence-level attention representation I. The second is contrary to the unfilling slot (action-request A) from the previous historical dialogue, the specific given slot-value pair V, and the currently obtained vector representation I from Memory-Network, they are all encoded into a Bi-LSTM+self-attention framework. After encoding, we get each part representations $t_{act}, t_{utt}, H_{utt}, t_{slot}$. Finally we score the encoded information in two parts. On the one hand we give a score to represent the confidence degree of the current session with the specific slot-value pair; On the other hand we give a score to represent the confidence degree of matching between the slot-value pair and the previous request action. The two-part scores normalized by a sigmoid layer to get a final score. We set a credibility threshold, if the score is higher than the threshold, it indicates that the given slot-value pair is the right dialogue state corresponding to the current user utterance.

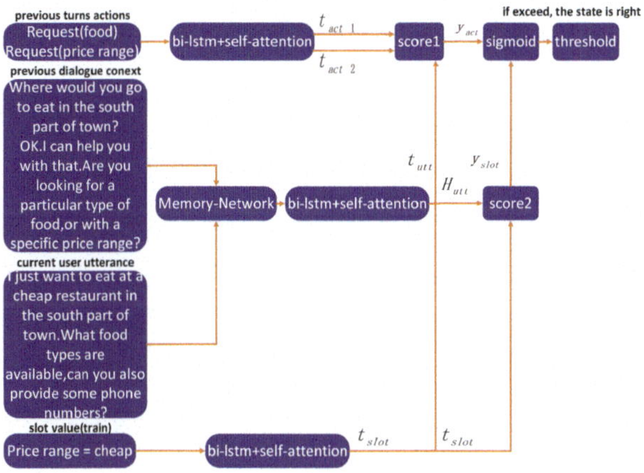

Fig. 2. Model framework

3.1 Memory-Network Module

The Memory-network framework was first applied in reading comprehension problems: when given a document d, and a question q, we need to get the correct answer a with the predicting conditional probability is $p(a|d,q)$. In the task-oriented dialog system, we think the historical dialogue context is d, the current user utterance is q, and the current correspondence dialogue state is a.

In our research we use a multi-hop Memory-Network module as one part of model input, to make current user utterance can be repeat calculated with the historical dialogue context, and finally capture sentence-level correlation. We give a single-hop Memory-Network module and a multi-hop Memory-Network module in Figs. 3 and 4 to explain how it works.

For the single-hop module in Fig. 3, the current user utterance is embedded into a vector u; u is calculated with the embedding historical dialogue context matrix using dot-operation to obtain different weight values p_i:

$$p_i = Softmax(u^T m_i) \tag{1}$$

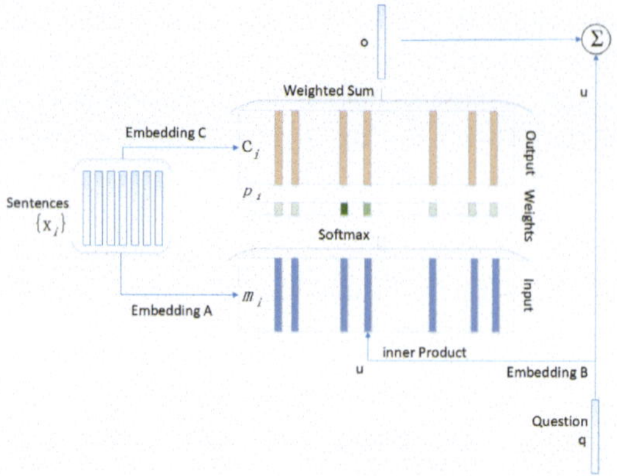

Fig. 3. Single-hop module

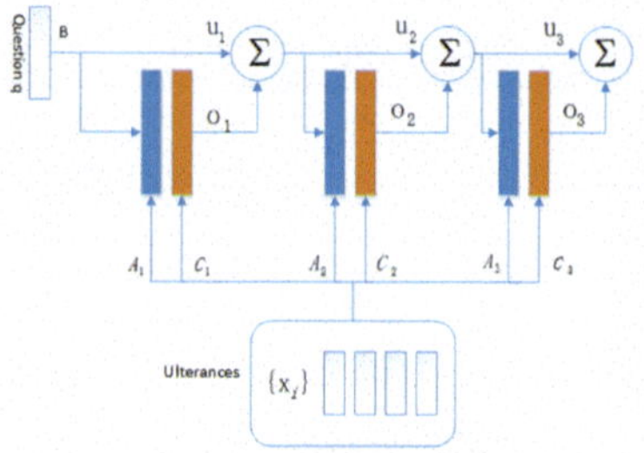

Fig. 4. Multi-hop module

And then we get the output vector o using another embedding matrix c_i:

$$o = \sum(p_i c_i) \tag{2}$$

In historical dialogue information, vector o represent the correlation to the current utterance. Finally we get the vector representation I which combining current utterance information and historical dialogue information using an add operation:

$$I = o + u \tag{3}$$

The method we calculate in multi-hop module is similar to the single-hop module. But it can fully consider the historical dialogue context.

3.2 Bi-LSTM+Self-attention Module

In the encoding part, we use Bi-LSTM to extract the information of the input sentence, followed by a self-attention layer to capture correlation at word-level.

The word-level in our framework includes several parts: The first one is the correlation between words in the internal sentence representation we get from the Memory-network's output. The second one is the correlation in the specific given slot-value pair. The third one is the correlation in the previous request-action pairs. Using self-attention, we get automatically extracting the word-level feature in the internal sentence representation, reducing hand-crafted feature engineering or adding external feature.

After encoding by the Bi-LSTM layer, we get the information from input, and represent it with the hidden state H:

$$H = BiLSTM(I) \in R^{n \times d_{lstm}} \tag{4}$$

where n means the length of input, d_{lstm} means the dimension of state in LSTM.

For the i_{th} element H_i in H, we use self-attention layer to calculate the correlation of the i_{th} element with other position elements H_j to get weights, and sum together to obtain the final self-attention representation of i_{th} element:

$$\sum_j Softmax(\frac{H_i H_j^T}{\sqrt{d_k}}) H_j \tag{5}$$

Here, d_k is the dimension of state H_j.

Finally, we get the three part encoding state $t_{act}, t_{utt}, t_{slot}$ and the hidden state H_{utt}.

3.3 Score-Module

After we encode the state, we do a confidence degree score including two parts. On the one hand we give a confidence degree score between the current sentence session and the specific slot-value pair, combining H_{utt}, t_{slot}:

$$a_i^{utt} = (H_i^{utt})^T t_{slot} \tag{6}$$

$$p^{utt} = Softmax(a^{utt}) \tag{7}$$

$$q^{utt} = \sum_i p_i^{utt} H_i^{utt} \tag{8}$$

$$y^{utt} = W q^{utt} + b \tag{9}$$

The answer y^{utt} is the first part score.

On the other hand we give a confidence degree score between previous request-action and the specific slot-value pair, using $H_{utt}, t_{slot}, t_{act}$. The idea is similar to the above and formula is omission, y^{act} is the second part score.

The final y is a normalized score between y^{act} and y^{utt} by using a sigmoid layer:

$$y = sigmoid(w y^{act} + y^{utt}) \tag{10}$$

We use the final score y compare to the threshold y_{base}. If the score is higher than y_{base}, we think the specific slot-value pair is the right dialogue state for the current user utterance.

4 Experiments

4.1 Data Set

The DSTC-2 data set is a traditional data set for DST tasks about restaurant ordering. The dialogue states include: cuisine, price, number of people, location (city), and some additional information that may be asked such as restaurant address and telephone number. In each turn of dialog, the content of the action is also divided into two types, one is the user has asked but no answer given we call it a request; the other is the type of answer that the user indicates we call it an inform.

4.2 The Implementation of Model Framework

In the process of model training, we evaluate the dialogue state accuracy based on the turn-level. During the training process, for the user's current utterance u, corresponding to a correct slot-value pair, we also have the historical dialogue context and its corresponding unresolved user's request-action; The four parts are used as input into the model framework to get a final confidence score.

In the process of model training, we embedded our four parts input and go through the 3-hop Memory-Network module, obtaining the user's utterance and historical dialogue context representation. In the encoding layer, the Adam optimizer is used to optimize the module, and the dropout is set to 0.7. Finally, we score the output to give the final confidence degree score. Because of the result we judge is yes or no, so in the score module, we set the threshold is 0.5. We use the idea of the N-best list as the final accuracy estimate.

4.3 Comparison of Experimental Results

We compare the results getting from our model with previous researches mentioned in the related works in the Table 1. And we give a case study predicted by our model showed in the Table 2.

Table 1. Comparison of experimental results

Model (DSTC-2)	Precision
LSTM [2]	0.37
Word-based with RNN [3]	0.46
NBT-CNN [4]	0.72
NBT-DNN [4]	0.72
Memory-Network [1]	0.71
Mem-Net+Bi-LSTM+self-attention	0.74

Table 2. Example predictions by our model

Actions in previous context	User utterance	Predict dialogue state
N/A	I would like an expensive restaurant in south part of town	inform(area=south) (price=expensive)
request(N/A) What kind of food would you like?	What food types are available, can you provide the phone numbers?	request(phone) inform(food=nocare)
request(phone) Ok, and how many people please?	4 people	request(phone) inform(people=4)

4.4 Analysis

Comparing with the results of the previous researches, our model considers the several parts is effective. The first part considers correlation at sentence-level, which is the correlation between current user utterance and historical dialogue context. Using a multi-hop module can fully calculate with historical information. The second part is the internal correlation of sentence to the specific slot-value pair. The third part is the previous request-action correlation to the specific slot-value pair. The fourth part add the slot-value pair as input replace of getting answers from the candidate dialogue state pool can improve extracting the slot-value pair which have a small proportion of dialogue state space. The model combines these four parts to get the correct dialogue state, indicating that the dialogue state is highly correlated with these parts.

5 Summary

In this paper, we propose a new model to improve DST accuracy in task-oriented dialog system, contrary to Memory-Network's single consider correlation at sentence-level and can not precisely estimate the rare slot-value pair. Without making hand-crafted features engineering, we add the slot-value pair as input feature and use Bi-LSTM and self-Attention layers to capture word-level features, including internal sentence word-to-word, word to slot-value pair, and the action slot to the slot-value pair. Finally, using a score module, we get the confidence degree of the sentence information to the given slot-value pair and the unsettled request action slot to the slot-value pair, then determine whether the given slot-value pair is the correct dialogue state corresponding to the current user utterance. We use this model improve 3% accuracy over Memory-Network on DSTC-2.

References

1. Perez, J., Liu, F.: Dialog state tracking, a machine reading approach using memory network. arxiv (2017)
2. Yoshino, K., Hiraoka, T., Neubig, G., Nakamura, S.: Dialogue state tracking using long short term memory neural networks. arxiv (2014)
3. Henderson, M., Thomson, B., Young, S.J.: Word-based dialog state tracking with recurrent neural networks. In: Proceedings of SIGDIAL (2014)
4. Mrksic, N., Seaghdha, D.O.S., Wen, T.H.., Thomson, B., Young, S.J.: Neural belief tracker: data-driven dialogue state tracking. In: Proceedings of ACL (2017)
5. Serban, I.V., Sordoni, A., Bengio, Y., Courville, A., Pineau, J.: Building end-to-end dialogue systems using generative hierarchical neural network models. arxiv (2015)
6. Madotto, A., Wu, C., Fung, P.: Mem2Seq: effectively incorporating knowledge bases into end-to-end task-oriented dialog systems. arxiv (2018)
7. Li, X., Chen, Y., Li, L., Gao, J., Celikyilmaz, A.: End-to-end task-completion neural dialogue systems. arxiv (2018)

Unmanned Aerial Vehicles Path Planning Based on Deep Reinforcement Learning

Guoqiu Wang, Xuanyu Zheng, Haitong Zhao, Qidong Zhao, Changsheng Zhang(✉), and Bin Zhang

Northeastern University, Shenyang 110819, People's Republic of China
wgq1220366878@gmail.com, zhangchangsheng@mail.neu.edu.cn

Abstract. Obstacle avoidance and path planning of unmanned aerial vehicles (UAVs) is an essential and challenging task, especially in the unknown environment with dynamic obstacles. To address this problem, a method of UAV path planning based on Deep Q-Learning is proposed. The experience replay mechanism is introduced in the deep reinforcement learning (DRL) process, and a value network is established to calculate the optimal value for the action of the UAV. The optimal flight policy of the UAV is determined through the ϵ-greed algorithm. The experimental results show that the UAV with well-trained model can avoid the obstacles in motion perfectly, and the cruise time is reduced by half compared with the untrained UAV.

Keywords: UAV path planning · Obstacle avoidance · Deep reinforcement learning · DQN

1 Introduction

Nowadays, UAVs are widely used in cargo delivery [1], urgency communication, reconnaissance, fire-fighting [2], aerial photography, disaster relief [3], et al. In the application of UAVs, the most important issue is path planning, especially in the complex environment with obstacles, including static obstacles and moving obstacles.

In recent years, many researchers have done a lot of work on path planning and obstacle avoidance, also proposed many algorithms. These methods fall into three categories. (1) geometric method [4]. (2) graph search method, such as Voronoi Graph Search method [5]. (3) methods based on machine learning [6–8].

For a person, he can follow a simple navigation rule, automatically avoiding obstacles that are already existing or are about to appear on the road, but it is difficult for a UAV. As an important way to implement artificial intelligence, the power of reinforcement learning (RL) has been shown in many fields. The famous application is AlphaGo developed by DeepMind team with the Approach of deep reinforcement learning. Besides, RL has excellent performance in video game [9], robot navigation [10,11], strategy optimization [12,13], detection and

Y. Liu et al. (Eds.): ICNC-FSKD 2019, AISC 1074, pp. 81–88, 2020.
https://doi.org/10.1007/978-3-030-32456-8_9

search [14,15], et al. Reinforcement learning is very suitable for solving sequential decision problems. Inspired by their work [9–11,16], this work proposes a method of UAV path planning based on DRL in the environment with moving obstacles. Meanwhile, this method can be applied to the path planning of the UAV group. For multiple cooperative UAV task, each UAV only needs to its own path planning while regarding others as obstacles.

2 Problem Formulation

Collision avoidance and path planning of UAV can be described as partially observable sequential decision making problems. Inspired by their research on robot navigation on the ground [10], this work proposes a three-dimensional problem formulation. Let s_t, a_t denote the state and action of the UAV (agent) at time step t, the state of agent is divided into two parts, that is, $s_t = [s_t^o, s_t^h]$, s_t^o denotes the part which can be observed by other agents, and s_t^h denotes the hidden part that is only known by oneself. Let $\boldsymbol{P}$, $\boldsymbol{V}$ denote the three-dimensional vector of position and velocity. Let $\boldsymbol{A}$ denotes the action. The action of the agent is the velocity, that is $\boldsymbol{A} = \boldsymbol{V}$. The observable part includes the position, velocity, and the size (radius) of the agent, denoted by $s^o = [p_x, p_y, p_z, v_x, v_y, v_z, r]$. The hidden part includes the target position, preferred velocity and bias angle of the agent, denoted by $s^h = [p_{gx}, p_{gy}, p_{gz}, v_{pref}, \psi_{xy}, \psi_{xz}]$, where ψ_{xy} is the angle between the vector pointing to the target position and the XOY plane, and ψ_{xz} is the angle between the vector pointing to the target position and the XOZ plane.

The following are the collision avoidance formulas of two agents. The state of one agent is s and the state of the other agent is $\hat{s}$. The policy, $q : (s : t, \hat{s} : t) \mapsto a_t$, minimizes the expected time to reach the target position, $E[t_g]$.

$$\underset{q}{\arg\min}\, E[t_g | s, \hat{s}, q, \hat{q}]. \tag{1}$$

$$\| \boldsymbol{P}_t - \hat{\boldsymbol{P}}_t \|_2 \geq r + \hat{r}. \tag{2}$$

$$\boldsymbol{P}_{tg} = \boldsymbol{P}_g. \tag{3}$$

$$\boldsymbol{P}_t = \boldsymbol{P}_{t-1} + \Delta t \cdot q(s_{t-1}, \hat{s}_{t-1}). \quad \hat{\boldsymbol{P}}_t = \hat{\boldsymbol{P}}_{t-1} + \Delta t \cdot q(\hat{s}_{t-1}, s_{t-1}). \tag{4}$$

Where (2) is the constraint of collision avoidance. The distance between two agents must be greater than their sizes, otherwise they will collide. (3) is the constraint of target position and detects whether the target position is reached. (4) is a kinematics update, which keeps updating the position of the next moment according to the policy. (1) is the time expectation, the smaller the better.

3 Approach

Generally, sequential decision making problems are expressed as Markov Decision Process (MDP), defined as $M = (S, A, P, R, \gamma)$, where S is the state, A is the action space, P is the state-transition model, R is the reward function, and γ is the discount factor.

State: Since collision relates to the state of two agents and cannot be determined by either one, the state is represented by the joint state of two agents, $s^{jn} = [s, \hat{s}^o]$, let s denotes the state of the UAV, and $\hat{s}_o$ denotes the observable state of the obstacle.

Action space: In the process of UAV flight, the action is the vector of velocity, and the action space is the set of it, $\boldsymbol{A} = \boldsymbol{V}$ $(\|\boldsymbol{V}\|_2 < v_{pref})$.

Reward function: The reward function is to evaluate whether the agent has reached the target position or the distance from other obstacle agents. When the agent reaches the target position, the reward will be the largest. The smaller the distance of two agents is, the smaller the reward will be. If the two agents are too close to each other or collide, the reward will be negative.

$$\boldsymbol{R}(s^{jn}, \boldsymbol{u}) = \begin{cases} -0.3 & if \quad d_{min} < 0 \\ -0.2 - d_{min}/2 & if \quad d_{min} < 0.2 \\ 1.0 & if \quad \boldsymbol{P} = \boldsymbol{P_g} \\ 0 & otherwise \end{cases} \tag{5}$$

Where d_{min} is the minimum distance between the two agents within a duration of Δt, assuming that the agent flying at velocity $\boldsymbol{V} = \boldsymbol{a}$, and the other agent flying at its observable velocity $\hat{\boldsymbol{V}}$.

State-transition model: In the process of reinforcement learning, it is impossible to known which policy the obstacle agent uses to navigate, so the state transition probability cannot be determined. Based on previous study [17], it is assumed that the obstacle agent and the UAV navigate with the same policy, $\hat{q} = q$.

Value function: The goal of reinforcement learning is to find the optimal value function.

$$\boldsymbol{V}(s^{jn}) = \sum_{t=0}^{T} \gamma^{t \cdot v_{pref}} R(s_t^{jn}, q(s_t^{jn})) \tag{6}$$

$\gamma \in [0, 1)$ is the discount factor, γ is close to 0, indicating that immediate reward is important; γ is close to 1, indicating that long-term reward is important. v_{pref} is agent's preferred speed and it is a constant, which is used here as a normalization factor.

$$q(s_{t+1}^{jn}) = \arg\max_{\boldsymbol{a}} R(s_t, \boldsymbol{a}) + \gamma^{\Delta \cdot v_{pref}} P(s_t^{jn}, s_{t+1}^{jn} | \boldsymbol{a}) \boldsymbol{V}(s_{t+1}^{jn}) \tag{7}$$

(7) is the optional policy function. According to the state and the optimal value function, the optimal policy in this state is calculated to find the optimal action to most rewards and achieve better result.

Inspired by Deep Q-Network (DQN) algorithm [9] and Deep V-Learning [11] algorithm, this work proposed the Deep UAV Path Planning Network algorithm combined with the characteristics of UAV path planning.

The algorithm consists of two parts. First, the value network and target value network are initialized with random parameters. Then the parameters of

Algorithm 1. Deep UAV Path Planning Network

initialize the value network $\boldsymbol{Q}(\cdot,\theta)$ with random weights θ
initialize the target value network $\hat{\boldsymbol{Q}}(\cdot,\hat{\theta}) \leftarrow \boldsymbol{Q}$
initialize replay memory $\boldsymbol{M}$ to capacity $\boldsymbol{D}$
for episode = 1,N ***do***
    initialize the random sequence s_0^{jn}
    for $t = 1, T$ ***do***
        with probability ϵ select a random action a_t
        otherwise select a_t =arg max $\boldsymbol{Q}(s_t^{jn}, a, \theta)$
        execute a_t and calculate r_t
        store tuple $(s_t^{jn}, a_t, r_t, s_{t+\Delta t}^{jn})$ in $\boldsymbol{M}$
        sample random minibatch tuples $(s_i.a_i, r_i, s_{i+1})$ from $\boldsymbol{M}$
        set target $y_i = r_i + \gamma^{\Delta t \cdot v_{pref}} max_{\hat{a}} \hat{\boldsymbol{Q}}(s_{i+1}, \hat{a}, \hat{\theta})$
        update value network $\boldsymbol{Q}$ by gradient decent
        until terminal state s_t or $t \geq t_{max}$
        update target value network $\hat{\boldsymbol{Q}} \leftarrow \boldsymbol{Q}$
    end for
end for
return Q

the value network are constantly updated in each exploratory training by using the ϵ-greed algorithm to select the best action. In each episode, when the agent reaches the target position or the time exceeds the maximum time limit, the target value network will be updated. Finally, the neural network for calculating the value function is obtained. As for the specific implementation details, the experiment part below will specify them.

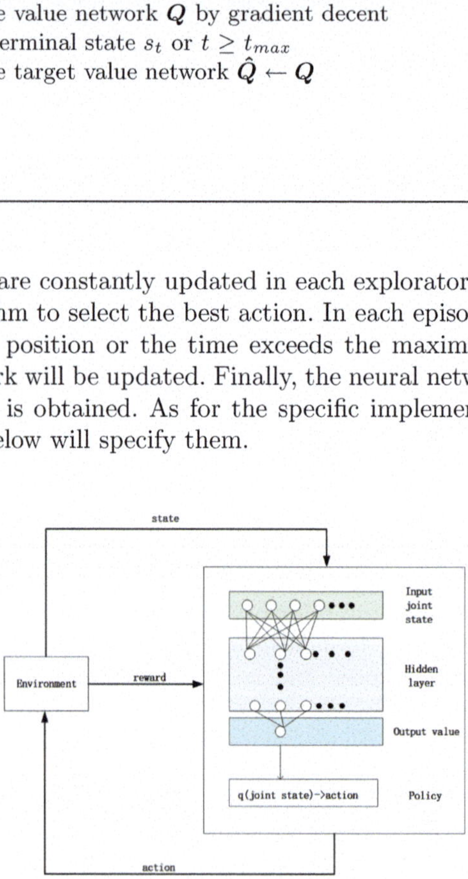

Fig. 1. The structure of the model.

The UAV uses the neural network to calculate the value according to the feedback state of environment. The policy is obtained through the value, and then generates action. The UAV performs the action, and the environment will give a certain reward. After many episodes' training, the neural network can calculate the optimal value, and plan a better action to get a bigger reward. Figure 1 shows the structure of the model.

4 Experiments

This experiment is implemented with the PyTorch and the training took about 30 h to converge on a 1080Ti GPU.

Simulation environment: This experiment uses Python-RVO2-3D to build simulation environment. RVO2-3D is an open source 3D space multi-agent real-time collision avoidance simulation environment developed by researchers from the University of North Carolina. In this experiment, the size of environment is set as $10 * 10 * 10$.

Action space: In this experiment, the action space is composed of 100 scattered spherical uniform action directions and 5 random velocities ($v_i < v_{pref}$).

Deep UAV Path Planning Network: The neural network inputs are 14 dimensions, they are integrated joint states and actions. The output dimension is 1 and that is the value of the input. Three hidden layers are included in this network. After many experiments, it is found that when the hidden layer dimension of the neural network is (150, 150, 100), the model works well and trained easily. Specific super parameters: the learning rate is 0.001, the discount factor γ is 0.9, the exploration rate of the ϵ-greedy policy is 0.3, and the number of episodes is 10000. When the UAV flies, the interval of movement change is 0.25 s, the UAV flies at a constant speed within the period of 0.25 s, which is consistent with the actual situation.

Take agent as the center: Considering that the value network should have good adaptability to the changing coordinate situation and the lightweight of neural network, there should not be too many input dimensions, which will reduce the complexity of calculation. Based on the above two considerations and inspired by [4], an agent-centric frame is defined. The agent is located at the origin of the coordinate and the X-axis points to the target position of the agent. The joint state is processed into a new state with agent as the center of the coordinate.

$$s_{new} = rotate(dg, v_{pref}, radius, vx, vy, vz, \hat{px}, \hat{py}, \hat{pz}, \hat{vx}, \hat{vy}, \hat{vz}, \hat{radius}, d) \quad (8)$$

Where dg is the distance between agent and target position, v_{pref} is the preferred speed, $radius$ is the size of the agent, (vx, vy, vz) is the velocity of the agent. $(\hat{px}, \hat{py}, \hat{pz})$ is the position of the obstacle agent, $(\hat{vx}, \hat{vy}, \hat{vz})$ is the speed of the obstacle agent. $\hat{radius}$ is the size of the obstacle agent and d is the distance between two agents.

The flight time to the destination and the reward were selected to measure the effect of the model. The starting position and the target position of the UAV are set, and the maximum flight time is set to 24 s. There are five obstacles in motion in the flight space. At first, the flight time for the UAV to reach the

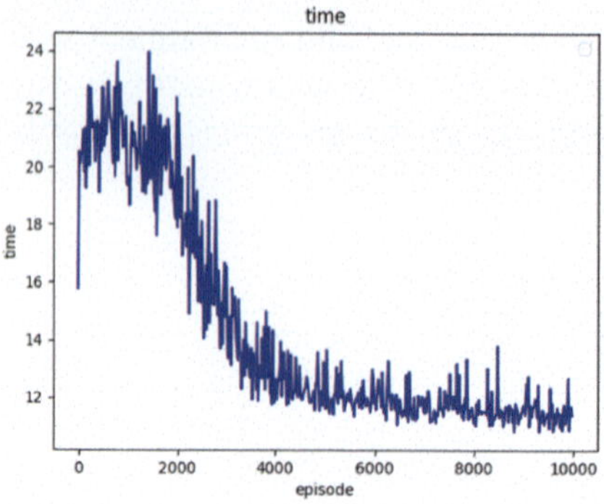

Fig. 2. The time of the UAV to reach its destination is decreasing.

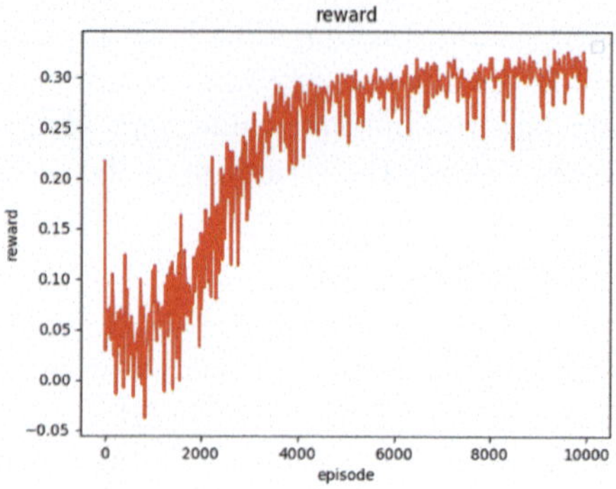

Fig. 3. The reward is increasing.

target position is mostly above 22 s. After training of about 2,000 episodes, the flight time starts to decrease. When it reaches 5,000 episodes, the flight time gradually converges and finally converges to about 11 s, which is about half of the beginning. The reward curve is consistent with the flight time curve, and the reward keeps improving, indicating that the UAV has gradually learned a better flight policy. Figure 2 shows the flight time curve and Fig. 3 shows the reward curve.

We also compared the flight effect of the UAV after increasing the number of obstacles in the same space (10 * 10 * 10), as shown in Table 1. Through comparison, it can be seen that with the increase of obstacles, the reward of UAV decreases a little, which is consistent with our experience. In the same space, the increase of obstacles is bound to make the increase of possible collisions and the difficulty of path planning. At the same time, it is also found that although the flight time and reward are changed, they are all changed within a certain small range, indicating that this method has a good adaptability.

Table 1. Flight time and rewards vary with the number of obstacles

Number of obstacles	Converges to time	Converges to reward
5	11.00 s	0.3123
8	11.75 s	0.2957
10	12.00 s	0.2889

5 Conclusion

This study proposes a method of path planning based on deep reinforcement learning of UAV. After many experiments, the most suitable neural network structure for this method was determined, which can effectively calculate the optimal value of UAV's current state and action. The trained model has been tested in the simulation environment and work well in obstacle avoidance and path planning.

References

1. Dorling, K., Heinrichs, J., Messier, G.G.: Vehicle routing problems for drone delivery. IEEE Trans. Syst. Man Cybern. **47**(1), 70–85 (2017). https://doi.org/10.1109/TSMC.2016.2582745
2. Yin, C., Xiao, Z., Cao, X., et al.: Offline and online search: UAV multiobjective path planning under dynamic urban environment. IEEE Internet Things J. **5**(2), 546–558 (2018). https://doi.org/10.1109/JIOT.2017.2717078
3. Bevacqua, G., Cacace, J., Finzi, A., Lippiello, V.: Mixed-initiative planning and execution for multiple drones in search and rescue missions. In: Twenty-Fifth International Conference on Automated Planning and Scheduling. AAAI Publications (2015)
4. Chen, Y.B., Luo, G.C., Mei, Y.S., et al.: UAV path planning using artificial potential field method updated by optimal control theory. Int. J. Syst. Sci. **47**(6), 1407–1420 (2016). https://doi.org/10.1080/00207721.2014.929191
5. Bortoff, S.A.: Path Planning for UAV. In: Proceedings of the 2000 American Control Conference. ACC (IEEE Cat. No. 00CH36334) (2000)
6. Peng, X., Xu, D., Zhang, F.: UAV online path planning based on dynamic multiobjective evolutionary algorithm. In: Proceedings of the 30th Chinese Control Conference, pp. 5424–5429. IEEE (2011)
7. Ragi, S., Chong, E.K.P.: UAV path planning in a dynamic environment via partially observable Markov decision process. IEEE Trans. Aerosp. Electron. Syst. **49**(4), 2397–2412 (2013). https://doi.org/10.1109/TAES.2013.6621824
8. Roberge, V., Tarbouchi, M., Labonte, G.: Comparison of parallel genetic algorithm and particle swarm optimization for real-time UAV path planning. IEEE Trans. Industr. Inf. **9**(1), 132–141 (2013). https://doi.org/10.1109/TII.2012.2198665
9. Mnih, V., Kavukcuoglu, K., Silver, D., Rusu, A.A., Veness, J., Bellemare, M.G., Graves, A., Riedmiller, M., Fidjeland, A.K., Ostrovski, G., et al.: Human-level control through deep reinforcement learning. Nature **518**(7540), 529–533 (2015). https://doi.org/10.1038/nature14236
10. Chen, Y.F., Liu, M., Everett, M., et al.: Decentralized non-communicating multiagent collision avoidance with deep reinforcement learning. In: 2017 IEEE International Conference on Robotics and Automation (ICRA), pp. 285–292. IEEE (2017). https://doi.org/10.1109/ICRA.2017.7989037
11. Chen, C., Liu, Y., Kreiss, S., et al.: Crowd-robot interaction: crowd-aware robot navigation with attention-based deep reinforcement learning. arXiv preprint arXiv:1809.08835 (2018)
12. Wan, Q., Liu, W., Xu, L., et al.: Research on optimal strategy planning of uncertain environment BDI agent based on Q-learning. Comput. Eng. Sci. **41**(01), 166–172 (2019)
13. Lillicrap, T.P., Hunt, J.J., Pritzel, A., Heess, N., Erez, T., Tassa, Y., Silver, D., Wierstra, D.: Continuous control with deep reinforcement learning. arXiv preprint arXiv:1509.02971 (2015)

14. Nair, A., McGrew, B., Andrychowicz, M., et al.: Overcoming exploration in reinforcement learning with demonstrations. In: 2018 IEEE International Conference on Robotics and Automation (ICRA), pp. 6292–6299. IEEE (2018)
15. Hu, Y., Da, Q., Zeng, A., et al.: Reinforcement learning to rank in e-commerce search engine: formalization, analysis, and application. In: Proceedings of the 24th ACM SIGKDD International Conference on Knowledge Discovery and Data Mining, pp. 368–377. ACM (2018)
16. Luo, W., Tang, Q., Fu, C., Eberhard, P.: Deep-Sarsa based multi-UAV path planning and obstacle avoidance in a dynamic environment. In: International Conference on Sensing and Imaging. Advances in Swarm Intelligence, pp. 102–111 (2018). https://doi.org/10.1007/978-3-319-93818-9_10
17. van den Berg, J., Guy, S.J., Lin, M., Manocha, D.: Reciprocal n-body collision avoidance. In: Robotics Research. Springer Tracts in Advanced Robotics, vol. 70, pp. 3–19 (2011). https://doi.org/10.1007/978-3-642-19457-3_1

Apple Freshness Recognition Technology Based on Gas Sensors

Wei Wang(✉), Zhihui Guo, Maozhen Li, and Yungang Liu

School of Information Science and Engineering, North University of China, Taiyuan 030051, Shanxi, China
41695559@qq.com

Abstract. At present, China is the world's largest apple producer. After being picked, apples rely on its own nutrients to maintain life activities, and with the consumption of energy, the freshness will gradually decrease. Nowadays, the storage and preservation awareness, facilities and methods of the apple processing industry and households are weak and lacking. Apples are prone to decay and deterioration, which seriously affects the flavor, nutritional value and food safety. This paper studies the artificial intelligence recognition technology based on sensors array apple freshness characteristics. The main research contents include: artificial intelligence odor recognition theory research and construction of sensor array information acquisition system. Combined with theoretical experiments, the research establishes apple odor characteristic parameter evaluation model, uses dielectric features, odor recognition technology and node location information to achieve apple freshness artificial intelligence recognition. This project has important significance in both theoretical research and practical application.

Keywords: Gas sensor · Odor recognition · Apple freshness · Freshness evaluation

1 Introduction

At present, apple has become one of the pillar industries of the rural economy in China's main producing areas and also plays an important role in increasing farmers' income and promoting exports. After harvest, apples lose the nutrient supply from the tree and need to rely on its own nutrients to maintain life activities. The living environment conditions after apples picking have changed radically, and if they are kept under good storage conditions, they can maintain their vitality and prolong their freshness. At present, China's apples processing industry and households are lack of storage and preservation awareness, facilities and methods. Many of them do not take any fresh-keeping measures and stack them at roadsides and yards. In the case of poor

Main lab open fund of North university of China (DXMBJJ2018-08) is funding.

Y. Liu et al. (Eds.): ICNC-FSKD 2019, AISC 1074, pp. 89–104, 2020.
https://doi.org/10.1007/978-3-030-32456-8_10

storage conditions, apples are more prone to be rotten, seriously affecting the flavor, nutritional value and appearance of the product, so keeping fresh has a very important significance for apples.

Fruit quality testing technology has a large number of relative researches at home and abroad. Longuetaud Fleur and Zhang Jingping mainly carried out research on agricultural product detection technology based on X-ray CT, scanned apples in different storage periods, obtained CT images, and measured the main internal parts of apples, established the model relationship between the composition content and the CT value, predicted the relationship between the main components of the fruit, and further predicted the internal quality of the apple. This detection method has great limitations: (1) it is necessary to use CT for image acquisition and analysis; (2) the detection efficiency is low; (3) the hardware requirements are relatively high and (4) it is not suitable for large-scale promotion. Nelson [3] uses the bio-impedance spectral characteristics to detect the freshness of apples. The value of resistive and capacitive components of biological tissues will obviously vary with the frequency of the loaded electrical signals and the time of placement of the organisms. However, this method is mainly for single apple detection, and has low efficiency and is not suitable for large-scale field applications.

Aiomascan's "Digital Odour Analysis System" in the UK, FOX2000 system from APha MOSSA in France, and NOSE system from Neotronics Scientific Ltd in the UK all have been widely adapted in a large number of tests with good results, such as: whisky and tobacco testing in the UK, quality of Japanese sake, quality of Italian olive oil judging. Fatih uses electronic odor recognition technology to identify several fruit types. Ma [6] established an Internet-based gas identification system, including twelve monitoring nodes, five control nodes and an environmental monitoring node distributed in seven different areas respectively.

Therefore, Apples have in-depth research values in the freshness and position information detection in fresh-keeping storage or general storage state, and it is worthwhile to have an in-depth study on gas sensors array and gas pattern recognition algorithm to establish a smart apple quality evaluation system to predict the preservation period and optimize storage conditions.

2 Apple Freshness Detection Technology

2.1 Dielectric Property Freshness Detection

The electrical characteristics of crops mainly include two types: one is the potential difference inside the material caused by the biological or chemical energy inside the agricultural material, like the potential difference in the living body; the other is like resistance, capacitance and the like, which affects the internal magnetic field and current distribution of materials. Compared with the magnetic field and current, the dielectric parameters such as resistance, capacitance and inductance inside the material are easier to measure.

The apple is a kind of crop, and its organizational structure is also a dielectric. Therefore, apples are also bound by electric fields. However, in the process of growing up, apples are accompanied by biochemical reactions and energy conversion, which will cause changes in the internal electric field strength of the fruit. Because the apple is a biological organism, from a microscopic point of view, there are many charged particles inside the apple to form a bioelectric field. The distribution and intensity of the bioelectric field also change, which basically determines the physical and chemical properties of apples, and cannot directly measure electrical performance. From a macro perspective, the internal field strength of apples has changed its dielectric properties.

2.2 Sensory Quality Evaluation

Sensory evaluation is to use the human sense of taste, touching, temperature and other intuitive senses to identify the quality and type of food, and then through a certain scientific analysis method to integrate the sensory evaluation of food to achieve food accurate identification. At the same time, the sensory assessment can first understand the specific indicators of food, and the evaluator can obtain important information about the sensory characteristics of the food through special indicators.

The apple fuzzy sensory evaluation uses human senses to identify specific indicators of apples, such as color, taste, texture, temperature and humidity, etc. Using a reasonable matrix multiplication algorithm, a comprehensive score is obtained. Sensory evaluation is an accurate measurement of the measured object, finding people's bias information on the measured object, minimizing the impact of brand and other information on people's perception, and reducing the error rate. Fruit sensory evaluation mainly consists of four parts: (1) Organization: It mainly refers to the composition of the assessors and establishes a reasonable process and method for food evaluation. (2) Measurement: according to the evaluator's sensory feelings: mainly visual, olfactory, gustatory, auditory and tactile, etc., the food is analyzed to collect sensory data. (3) Analysis: Analyze the sensory data of the food by the evaluator. (4) Conclusion: The data after analysis and the experimental results are judged.

2.3 Odor Recognition

Odor recognition has the characteristics of artificial intelligence and can be implanted into the robot so that it has the ability to judge different odors like humans and it also can make corresponding judgments on different environments. The principle is shown in Fig. 1. As an identification sensing unit, a gas sensor often uses an array of a

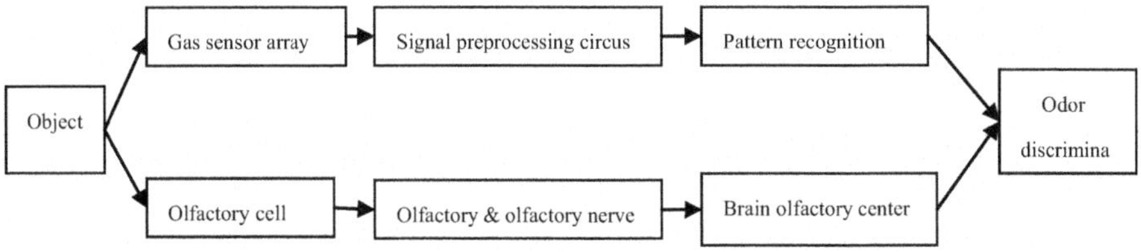

Fig. 1. Schematic of the sensor array odour recognition system

plurality of different types of gas sensors to obtain an identification pattern, that is, an odor fingerprint.

According to the characteristics of apple gas, an artificial intelligence odor pattern recognition algorithm is established. The research scheme is divided into three steps:

(1) Data preprocessing: Since the sensor is affected by circuit noise and reference voltage ripple during signal acquisition, the response curve is drawn according to the sensor response value which has a poor smoothness, so filtering is required. It is reliable to use gas sensor response curve steady state data as an identification feature.
(2) During the training process, there is generally a nonlinear relationship between the gas sensor response and the measured gas concentration. There are multiple overlaps and intersections in the measurement space. The general algorithm has poor anti-interference ability, and is limited to occasions where the environment does not change much. Training different categories can make the data characteristics more accurate, thus improving the recognition accuracy.
(3) Identifying methods, searching for a set of new bases, multiplying the multi-dimensional data by the matrix formed by the base, and transforming the representation into the space formed by the new base. If the number of new bases is smaller than the original data dimension, multiply the resulting data dimension is reduced, and the number of new bases is the dimension after dimension reduction.

Apple odor pattern recognition is achieved through training, and feature parameters are saved in the training data set or updated training model parameters. When there is a new unknown odor input, the feature is extracted according to the sensor response curve, and the recognized result is obtained by the pattern recognition method.

3 Information Collection System

3.1 Experimental Devices and Processes

The non-destructive testing system for the dielectric properties of apple freshness is mainly composed of the host, TH2822A intelligent LCR measuring instrument, electrode, 150 mm * 100 mm * 300 mm shielding box and data transmission line. A schematic diagram of the dielectric parameter detection system is shown in Fig. 2.

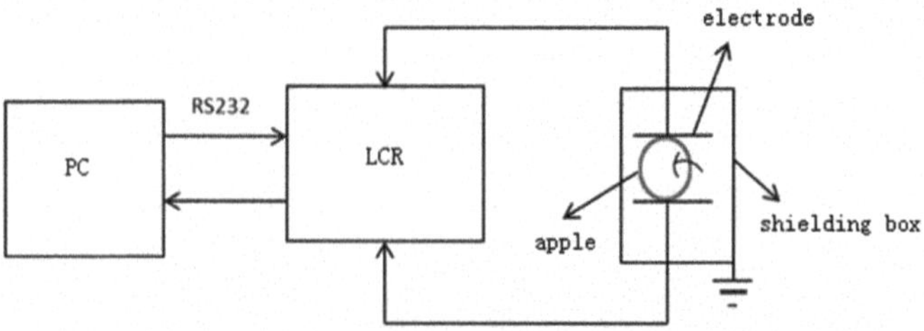

Fig. 2. Schematic diagram of non-destructive testing system for apple dielectric properties

This article uses TH2822A intelligent LCR measuring instrument, which is an instrument for automatic measurement of inductance, capacitance and impedance controlled by microcomputer, equipped with computer interface and test fixture. Throughout the test, apple samples and electrode plates were placed in a 150 mm 100 mm × 300 mm shielded box (made of galvanized fine iron with a thickness of 0.6 mm). The wires of the shield box are grounded, and the plates and apples are placed in the measurement.

The steps are as follows:

(1) After preheating the LCR meter for 10 min, start the LCR software;
(2) Set the measurement frequency of the LCR meter to 10 kHz, select the main parameter capacitor "c", and perform open circuit and short circuit calibration;
(3) When measuring, an apple is placed horizontally in the shielding box to reduce the interference of external electromagnetic waves, and the capacitance "c" and the loss factor "D" of the apple are measured. Finally, the equivalent capacitance formula is used to calculate the corresponding apple dielectric parameter "ε".

The relative dielectric constant is obtained by the known capacitance and loss factor, as shown in Fig. 3.

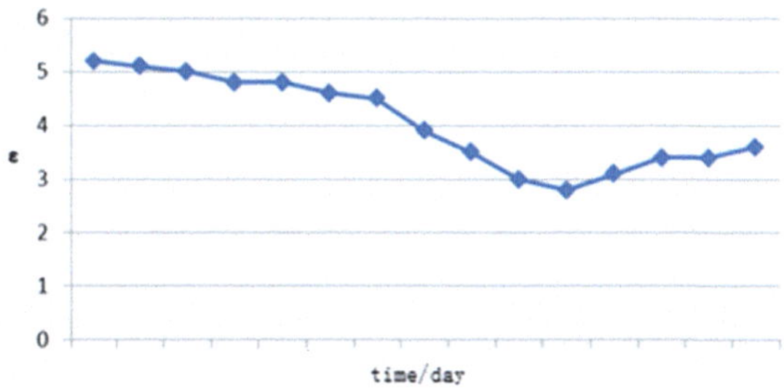

Fig. 3. Changes in the dielectric constant of the stored procedure

As can be seen from Fig. 3, during the change of apple from fresh to severe decay, the dielectric constant of apple decreased first and then increased. The main reason is that during the change of apple from fresh to decay, the apple's breathing intensity is enhanced, resulting in faster consumption of water and nutrients, and a decrease in dielectric constant. In the case of being severe rotten, the cell membrane of apple divides, resulting in an increase in the permeability of ions, and the colloid-bound water becomes free water, and the dielectric constant increases.

According to the detection requirements, establish sensor array odor recognition system and use gas sensor, wireless transmission node and related information processing circuit to form a hardware system, as shown in Fig. 4, complete A/D conversion, data storage, wireless sensor network transmission, serial communication, display, human-computer interaction and other related functional structures.

Main control chip based on the Zigbee protocol-based hardware node is CC2530F256RHAR, the working frequency band is 2.4 GHz, the working voltage is

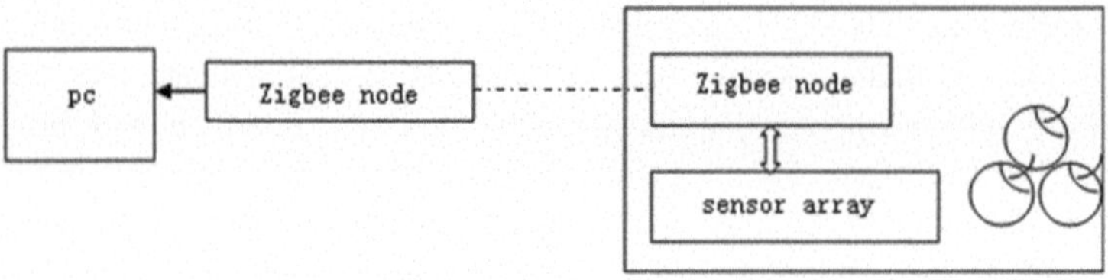

Fig. 4. A schematic diagram of the odor recognition system

2.0v–3.6v, the temperature range is −40 °C to 85 °C, and the hardware program is debugged and programmed by IAR software. The sensor array is composed of two sensors: MQ3 and CCS811. The relevant description and parameters are shown in Table 1. Since the performance of the sensor is affected by certain factors such as temperature and humidity, the indoor temperature is not much different, so it is performed indoors.

Table 1. Apple sensory evaluation criteria

Attribute	Definition	Score (10 points system)
Surface	Direct observation of the smoothness of the fruit surface and the degree of pest damage	Smooth fruit surface, no pest damage (9 points or more) Poor surface smoothness (6–8 points) Peel wrinkles and rot (3–5 points) The surface wrinkles and becomes dark, the rotten area is more than 1/2, and the deformation (0–2 points)
Color	Directly observe apple color	Fresh and clean color (9 points or more) The color is not too fresh and clean (6–8 points) The color is not fresh (3–5 points) No color and unclear (0–2 points)
Odor	Apple external odor intensity	Flavored, aromatic and odorless (9 points or more) No inherent fragrance and apple aroma, slight odor (6–8 points) No inherent fragrance and apple aroma, with alcohol odor (3–5 points) Have a strong alcoholic or bitter taste, no apple should have a clear fragrance (0–2 points)
Touching texture	Hardness of the apple, Smoothness, elasticity, etc.	Moderate soft and hard, full and firm flesh, smooth and tough, no sag (9 points or more) Soft and hard, moderate, firm flesh, smooth but not tough, with depression (6–8 points) The skin is soft, the hardness is uneven, and the smoothness is poor (3–5 points) The skin is shrinking and the flesh is thin and soft (0–2 points)

3.2 Data Processing

The sensor is subject to large noise caused by system and external interference during the acquisition process. Therefore, the raw data obtained after the sensor collected cannot be directly used for training and recognition. Data preprocessing is to preprocess the sensor array response mode to complete feature extraction. Four different apples, namely fresh apple (AAAA), less fresh apple (AAA), mildly rotted apple (AA) and severely rotted apple (A), were used to realize effective classification and identification by BP neural network. In order to avoid errors, each type of apple uses 5 apples as samples to detect the gas content, and finally conduct a pretreatment on data.

Gas sensors collect different data from apples with different freshness. In the experiment, fresh apples, less fresh apples, light apples and severely rotted apples were collected by a carbon dioxide sensor. The concentration characteristics are shown in Fig. 5.

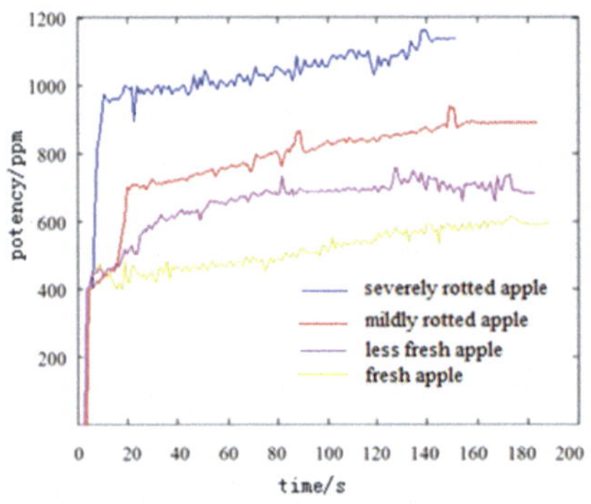

Fig. 5. Curve of sensor response to different freshness

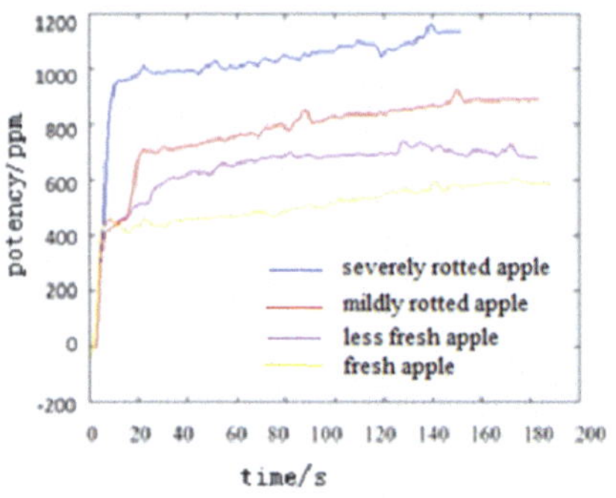

Fig. 6. Five point mean filtering

It can be seen from Fig. 5, the longer the apple is placed, the higher the concentration of carbon dioxide gas is, and it reaches a steady state after reaching a certain value; the freshness of the apple is directly related to the gas concentration: fresh, not fresh, rotten to severely rot, the carbon dioxide gas concentration value is getting higher and higher. Due to the influence of interference, the signal collected by the sensor fluctuates greatly, and misjudgment is easy to occur in information recognition. The signal is filtered and preprocessed, and from the output effect of the 8-point mean filtering output (as shown in Fig. 6), it can be seen that the 8-point mean filtering curve is smoother and the noise is removed. For situations where there is large noise in the system, the filtering process is indispensable. The value of the sensor response signal 40 s and the value of 60 s are extracted as the acquisition rate of the sensor, and secondly, the maximum value, the average value, and the stable value of the response signal are extracted.

4 Freshness Evaluation

4.1 Fuzzy Comprehensive Evaluation Scheme for Sensory Quality of Apple

The evaluators are students who are particularly sensitive to apple scent and color, and are asked to ban irritating foods such as tobacco, alcohol and spicy within 2 h before the evaluation. Before the evaluation, every student is explained with the evaluation content, evaluation criteria and evaluation method, and then the four apples to be evaluated which have been numbered are sent to the evaluation staff for evaluation by the same container.

4.2 Establishment of Factor Sets, Comment Sets, and Weighted Sets

A comment set is a collection of feedback information from an evaluator on an evaluation index, and determining a comment set is determining the level of the object to be tested. The evaluators concluded that apples' rating was fresh (V_1), not fresh (V_2), rotten (V_3), and heavily rotten (V_4), resulting in the review set V:

$$V = [V_1, V_2, V_3, V_4] \tag{1}$$

Scored by the system of ten points, the scored 9–10 is fresh, the scored 6–8 is not fresh, the scored 3–5 is rotten and the scored 0–2 is heavily rotten. The four grades of fresh, not fresh, rotten and severely rot in the test correspond to the four scores of the intermediate scores 9, 7, 4, and 1 respectively, and the evaluation grade set k:

$$k = [k_1, k_2, k_3, k_4] = [9, 7, 4, 1] \tag{2}$$

The weight set refers to the proportion of each indicator in the overall sensory quality of the evaluation set. Determining weight is essential and it has a large impact on the final assessment. Each comment set corresponds to a weighting factor. According to the different influence degree of various indicators on the sensory quality of apple, the weight of each index is determined by mathematical method. The factors affecting the sensory quality of apples are mainly the four indicators of fruit surface, fruit color, odor and texture. Through a large number of experimental comparisons, the proportions of the four indicators in the comprehensive evaluation are 0.2, 0.2, 0.35 and 0.25 respectively. Therefore, its weight set X can be expressed as:

$$X = [0.2, 0.2, 0.35, 0.25] \tag{3}$$

After the apple's sensory evaluation, the evaluation staff will evaluate the scores of various indicators of apples, and divide the number of votes of different levels of freshness by 10 to get the fuzzy relationship matrix R we need, and then we have the quality factor. The weight set X and R are synthesized by the matrix multiplication formula to obtain the fuzzy relation evaluation set Y of apple freshness:

$$Y = X \times R \tag{4}$$

Finally, the comprehensive scoring matrix T is introduced to process the fuzzy relation evaluation set Y. According to the particularity of the sensory evaluation, the evaluation grade $k = [9, 7, 4, 1]$ is obtained, and the fuzzy comprehensive evaluation score T of the apple sample is obtained:

$$T = Y \times k \tag{5}$$

The fuzzy comprehensive evaluation method is a method of comprehensive evaluation of the evaluated objects by considering the fuzzy transformation principle and the principle of maximum membership degree, considering various factors or main factors related to the things being evaluated.

4.3 Results and Analysis

Ten evaluators evaluated four indicators of 16 apple samples in turn. Get the No. 1 apple fruit surface fuzzy matrix U_1, color fuzzy matrix U_2, scent fuzzy matrix U_3, texture matrix U_4.

The calculated evaluation results of the four single-factor fuzzy matrices U_1, U_2, U_3, U_4 are written as a matrix, that is, the number 1 apple fuzzy matrix R1. 2-16 apple fuzzy matrix also can be obtained in the same way. It is known that the weight vector X of Apple's four indicators is:

$$X = [0.2, 0.2, 0.35, 0.25] \tag{6}$$

The evaluators concluded that the apple's rating was fresh (V_1), not fresh (V_2), rotten (V_3), and heavily rotten (V_4), resulting in the apple's review set V.

According to the standard fuzzy evaluation model (according to the algorithm of taking large and small), the evaluation result has a large error. Therefore, the ordinary matrix multiplication is used instead of the large and small algorithm to calculate the comprehensive membership degree. According to the formula, the fuzzy sensory evaluation set of Apple Number 1 is:

$$Y_1 = [0.14, 0.645, 0.125, 0.09] \tag{7}$$

Similarly, the fuzzy sensory evaluation set of apples No. 2-16.

According to the fuzzy comprehensive scoring formula, the comprehensive score T_1 of the apple sample No. 1 was calculated.

$$T_1 = Y_1 \times k = [0.14, 0.645, 0.125, 0.09] \times \begin{bmatrix} 9 \\ 7 \\ 4 \\ 1 \end{bmatrix} = 6.365 \tag{8}$$

Scored by the decimal system, the scored 9–10 is fresh, 6–8 is not fresh, the scored 6–8 is rotten, 3–5 is heavily rotten. The apple's comprehensive score of No. 1 is 6.635, and it is not fresh apple within 6–8 points. Similarly, according to the comprehensive score of 16 apple samples, the freshness results of apples are shown in Table 2.

Table 2. Apple freshness results

No.	1	2	3	4	5	6	7	8	9	10	11	12	13	14	15	16
Degree	V_2	V_3	V_1	V_4	V_4	V_4	V_3	V_3	V_3	V_3	V_1	V_4	V_2	V_3	V_1	V_4

According to the results, in the process of identifying 16 groups of apples, there are 15 groups of correctly identified apples, only one group has a recognition error, and the apple recognition rate can reach 93.75%, which indicates the use of fuzzy senses comprehensive evaluation of the apple freshness is highly accurate and can accurately identify the freshness and category of apples. However, the recognition result of No. 13 in the experimental results is inconsistent with the actual freshness of apple. This occurs because the fuzzy sensory evaluation cannot recognize the internal decay of the apple, and therefore the category of apple freshness is not completely recognized.

4.4 Gas Freshness Identification

The BP neural network's feature signal classification algorithm modeling includes three steps: BP neural network construction, training and BP neural network classification.

BP neural network construction determines the structure of BP neural network according to the characteristics of the input and output data of the system. Since the characteristic input signal of apple odor has 15 dimensions, there are 4 types of apple freshness characteristics to be classified, according to the formula:

$$N = \sqrt{n + m} + a \tag{9}$$

In the formula (9), N is the number of hidden layer units; n is the number of input units; m is the number of output units; a is between [1, 10];

Through many experiments, the optimal number of hidden layers N is 9. Therefore, the structure of the BP neural network is 15-9-4, that is, the input layer is 15, the hidden layer is 9, and the output layer is 4. Sixteen apples were measured and the training data was normalized. According to the fresh characteristics of apples, set the expectations of various apples, such as the expected output of fresh apples is [1, 0, 0, 0]; the expected output vector of no fresh apples is [0, 1, 0, 0]; the expected output vector of rotten apples is [0, 0, 1, 0]; the expected the output vector of heavily decaying apples is [0, 0, 0, 1].

(1) From the smoothness of fruit surface and the degree of pest damage, it can be divided into four grades: smooth fruit surface, no pest damage (9 points or more),

poor fruit smoothness (6–8 points), peel wrinkle and the phenomenon of decay (3–5 points), the wrinkles of the surface become dark, the area of decay is more than 1/2, and appears deformation (0–2 points).

The construction of the BP neural network is based on the characteristics of the input and output data of the system to determine the structure of the BP neural network. Since the input signal of the gas characteristic value has 15 dimensions, and 4 types of apple fruit surface features are classified, the structure of the BP neural network is 15-9-4. The training error curve of the fruit BP neural network is shown in Fig. 7:

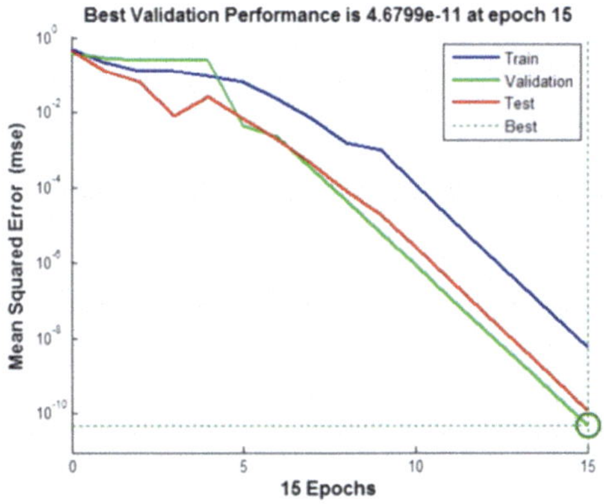

Fig. 7. Apple fruit surface BP neural network training error curve

According to the analysis of Fig. 7, after 15 iterations, the minimum error is $4.6799e^{-11}$, the training results are in line with expectations. The 16 sets of data to be tested are brought into the trained BP neural network model for verification. Finally, the actual output value of the test data can be obtained, and the fruit freshness classification result is shown in Table 3 below:

Table 3. Fruit surface freshness classification result

No.	1	2	3	4	5	6	7	8	9	10	11	12	13	14	15	16
Fresh degree	V_2	V_3	V_1	V_4	V_4	V_3	V_3	V_2	V_3	V_3	V_1	V_4	V_3	V_3	V_1	V_4

(2) The color degree of apple can be divided into four grades: fresh and clean color (9 points or more), color is not fresh and clean (6–8 points), color is not fresh (3–5 points), no color (0–2 points). Since the input signal of the gas characteristic value has 15 dimensions, there are 4 types of apple fruit color characteristics to be classified, so the structure of the BP neural network is 15-9-4. The apple fruit color BP neural network training error curve is shown in Fig. 8:

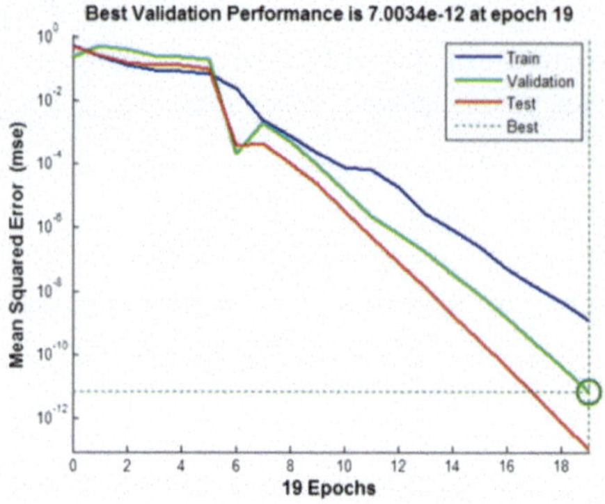

Fig. 8. Apple fruit color BP neural network training error curve

According to the analysis of Fig. 8, the fruit color input layer is 15, the hidden layer is 9, and the output layer is 4. After 19 iterations, the system reaches the minimum error $7.0034e^{-12}$, and the training result is in line with expectations. The 16 sets of data to be tested are brought into the trained BP neural network model for verification. Finally, the actual output value of the test data can be obtained, and the classification result of the fruit color freshness is as shown in Table 4 below:

Table 4. Fruit color classification results

No	1	2	3	4	5	6	7	8	9	10	11	12	13	14	15	16
Fresh degree	V_2	V_3	V_4	V_3	V_4	V_3	V_3	V_3	V_3	V_3	V_1	V_4	V_3	V_3	V_1	V_4

(3) From the external odor intensity of apples can be divided into four grades: flavor fragrance, aromatic no smell (9 points or more), no inherent fragrance flavor and apple aroma, slight odor (6–8 points), no inherent fragrance flavor and apple aroma, alcoholic odor (3–5 points), richer alcohol or bitterness, having no a clear aroma (0–2 points). Since the input signal of the gas characteristic value has 15 dimensions, the apple fruit color characteristics to be classified have 4 categories, so the structure of the BP neural network is 15-9-4. The odor BP neural network training error curve is shown in Fig. 9:

According to the analysis of Fig. 9, after 12 iterations, the system reaches the minimum error, and the training results are in line with expectations. The 16 sets of data to be tested are brought into the trained BP neural network model for verification. Finally, the actual output value of the test data can be obtained, and the classification result of the odor freshness is shown in Table 5 below:

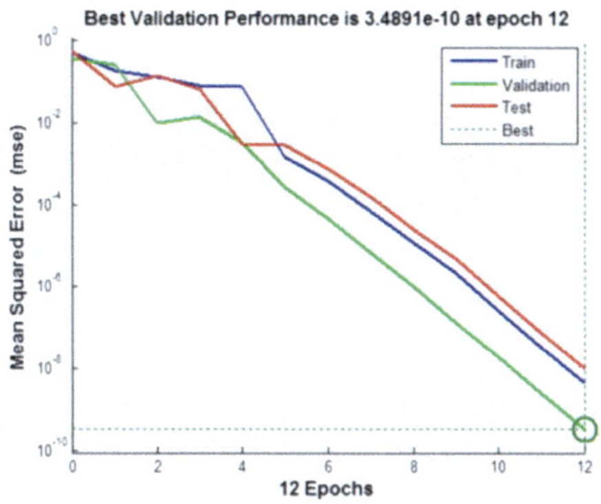

Fig. 9. Apple odor BP neural network training error curve

Table 5. Odor freshness classification results

No.	1	2	3	4	5	6	7	8	9	10	11	12	13	14	15	16
Fresh degree	V_2	V_2	V_1	V_4	V_4	V_4	V_2	V_3	V_3	V_2	V_1	V_4	V_3	V_3	V_1	V_4

(4) From the touching texture of apple (softness, smoothness, elasticity, etc.) can be divided into four grades: soft and hard, the flesh is full and firm, touch feeling smooth and tough and does not sink (9 points or more), soft and hard moderately, the flesh is firmer, the touch feeling smooth but not tough with sag (6–8 points), the skin is soft, the hardness is uneven, the smoothness is poor (3–5 points), the skin is shrinking, and the flesh is thin and outflows (0–2 points).

The construction of the BP neural network is based on the characteristics of the input and output data of the system to determine the structure of the BP neural network. Since the input signal of the gas characteristic value has 15 dimensions, the apple fruit color characteristics to be classified have 4 categories, so the structure of the BP neural network is 15-9-4. The apple hand-feeling texture BP neural network training error curve is shown in Fig. 10:

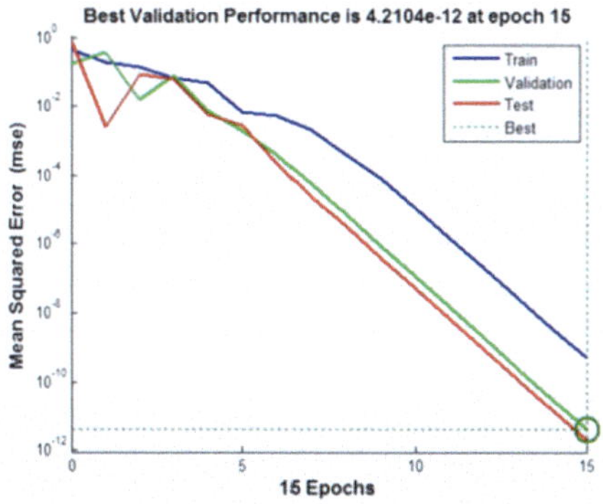

Fig. 10. Hand-feeling texture training results

By analyzing Fig. 10, the input layer of the texture is 15, the hidden layer is 9, and the output layer is 4. After 15 iterations, the system achieves the minimum error $4.2104e^{-12}$, and the training results are in line with expectations.

The 16 sets of data to be tested are brought into the trained BP neural network model for verification. Finally, the actual output value of the test data can be obtained, and the results of partial hand-feeling texture freshness classification are shown in Table 6 below:

Table 6. The hand-feeling texture classification results

No.	1	2	3	4	5	6	7	8	9	10	11	12	13	14	15	16
Freshness degree	V_2	V_3	V_1	V_3	V_3	V_4	V_3	V_3	V_2	V_3	V_1	V_3	V_2	V_2	V_1	V_3

According to the fuzzy comprehensive evaluation mathematical model principle (Sects. 4, 4.2, 4.3 has been introduced), combined with the above ordinary matrix multiplication, we calculate apple's comprehensive membership. The freshness results of 16 apple samples are shown in Table 7.

Table 7. Odor recognition apple freshness

No.	1	2	3	4	5	6	7	8	9	10	11	12	13	14	15	16
Freshness degree	V_2	V_3	V_1	V_4	V_3	V_4	V_3	V_3	V_3	V_3	V_1	V_4	V_3	V_3	V_1	V_4

Using the odor recognition combined with the fuzzy sensory comprehensive evaluation algorithm and the dielectric property identification to measure the freshness of 16 apples in multiple times yields the same results as shown in Table 8 below:

Table 8. Apple freshness degree

No.	1	2	3	4	5	6	7	8	9	10	11	12	13	14	15	16
Freshness degree	V_2	V_3	V_1	V_4	V_4	V_4	V_3	V_3	V_3	V_3	V_1	V_4	V_3	V_3	V_1	V_4

By comparing Tables 7 and 8 with the actual apple freshness degree, the higher the comprehensive score of the apple sample, the better the freshness of the odor-recognizing apple. In the process of identifying 16 groups of apples, there are 15 groups of correctly identified apples, only one group has recognition errors, and the recognition rate of apples can reach 93.75%, which indicates that apples are freshly identified by ordinary matrix multiplication in odor recognition combined with fuzzy comprehensive evaluation. The degree has high accuracy and can effectively identify the freshness and category of apples. The odor recognition incorporates a fuzzy sensory comprehensive evaluation algorithm (common matrix multiplication), which combines the dielectric properties to accurately measure apple freshness. This method has the dielectric properties to identify the advantages of internal decay of apples, the efficiency of odor recognition and the comprehensiveness of fuzzy sensory recognition. The four indicators can be judged to more fully identify the freshness of apples, thus the accuracy rate of freshness degree is improved.

5 Conclusion and Future Work

The sensor array is used to detect apple fruit surface, fruit color, odor and texture. The BP neural network was adapted to identify the apple index. Finally, the fuzzy sensory recognition algorithm was used to judge apple freshness. The odor recognition combined with the common matrix multiplication in fuzzy comprehensive evaluation has high accuracy in distinguishing apple freshness, and can accurately identify apple freshness and its category. Moreover, apples that are normal in appearance but whose interior begins to rot are also identified, indicating that odor recognition is also applicable to the identification of internally decaying apples. However, in the process of collecting data, it is affected by external noise, which causes certain errors in the data, and some apple freshness cannot be correctly recognized. Odor recognition is a non-contact measurement with high accuracy, so it is more suitable for apple freshness detection.

The traditional BP neural network algorithm is a simple and easy-to-follow static optimization method. It only corrects the weights and thresholds according to the error gradient direction of a certain step, and ignores the accumulation of previous learning experience, causing that learning rate is low, the convergence speed is slow, and it is easy to fall into a local minimum state, resulting in oscillation. Thus, further optimization algorithms are needed to reduce data processing time and improve the efficiency of testing and data processing.

References

1. Zhang, H., Wang, J., Ye, S., et al.: Application of electronic nose and statistical analysis to predict quality indices of peach. Food Bioprocess Technol. **5**, 65–72 (2012)
2. Shi, B., Zhao, L., Zhi, R., et al.: Quality recognition of Xihu Longjing tea based on intelligent olfactory. Trans. Chin. Soc. Agric. Mach. **43**(12), 130–135 (2012)
3. Nelson, S., Forbus, W., Lawrence, K.: Permittivities of fresh fruits and vegetables at 0.2 to 20 GHz. J. Microw. Power Electromagn. Energy: Publ. Int. Microw. Power Inst. **29**(2), 81–93 (1994)
4. Kuang, W., Nelson, S.: Dielectric relaxation characteristics of fresh fruits and vegetables from 3 to 20 GHz. J. Microw. Power **32**(2), 114–122 (1997)
5. Ma, D., Ding, Q., Li, Z., et al.: Prototype of an aquacultural information system based on internet of things E-Nose. Intell. Autom. Soft Comput. **18**(5), 569–579 (2012)
6. Byoung-Kwon, H., Hyochin, K., Atul, K.S.: Pathogen enrichment device (PED) enables one-step growth, enrichment and separation of pathogen from food matrices for detection using bioanalytical platforms. J. Microbiol. Methods **117**, 64–73 (2015)

7. Castro, J., Ramanathan, A., Hennubhotla, C.: Categorical dimensions of human odor descriptor space revealed by non-negative matrix factorization. PLoS One **8**(9), 78–89 (2013)
8. Hong, X., Wang, J., Hai, Z.: Discrimination and prediction of multiple beef freshness indexes based on electronic nose. Sens. Actuators: B Chem. **161**(1), 381–389 (2012)
9. Adak, M., Nejat, Y.: Classification of E-Nose aroma data of four fruit types by ABC-based neural network. Sensors **16**(3), 304–310 (2016)
10. Brezmes, J., Fructuoso, L., Llobet, E., et al.: Evaluation of an electronic nose to assess fruit ripeness. IEEE Sens. J. **5**(1), 97–108 (2005)
11. Baietto, M., Wilson, D.: Electronic-nose applications for fruit identification, ripeness and quality grading. Sensors **15**(1), 899–931 (2015)
12. Guohua, H., Yuling, W., Dandan, Y., et al.: Study of peach freshness predictive method based on electronic nose. Food Control **28**(1), 260–271 (2012)
13. Guohua, H., Wu, Y., Ye, D.: Study of peach freshness predictive method based on electronic nose. Food Control **28**, 1848–1852 (2012)
14. Barba, A., Lamberti, G.: Dielectric properties of pineapple as function of temperature and water content. Int. J. Food Sci. Technol. **48**(6), 1334–1338 (2013)
15. Birla, S.L., Wang, S., Tang, J., et al.: Characterization of radio frequency heating of fresh fruits influenced by dielectric properties. J. Food Eng. **89**(4), 390–398 (2008)

A Review of the Theory and Method for New Developed Feedforward Neural Networks

Daiyuan Zhang(✉)

College of Computer, Nanjing University of Posts and Telecommunications, Nanjing, People's Republic of China
dyzhang@njupt.edu.cn

Abstract. Two new theories and methods of feedforward neural network developed in recent years are systematically reviewed. They are Algebraic Algorithm of feedforward neural network and Spline Weight Function Algorithm of feedforward neural network. The network structure, basic theory and algorithm, and basic performance of the two networks are described. The algebraic algorithm of feedforward neural network can get the global optimum, but BP algorithm is difficult to get the global optimum. Algebraic algorithm has high accuracy and fast training speed. In engineering application, the algebraic algorithm gives the calculation formula of accurately determining the number of hidden layer neurons, but the BP algorithm does not give such an accurate formula. However, algebraic algorithm, like traditional BP algorithm, obtains constant weights, which is difficult to reflect the information of training samples. The spline weight function algorithm of feedforward neural network not only retains the advantages of algebraic algorithm, but also overcomes its shortcomings, so that the weight of trained neural network is a function of input sample (called weight function), which is composed of spline function, not a constant of traditional method. The trained weight function can extract useful information features from the trained samples. The topological structure of the spline weighted neural network is very simple, and the number of its neurons is independent of the number of samples, which is equal to the number of input and output nodes. The calculation speed is fast. Moreover, with the increase of the number of samples, the generalization ability of the network is also improving.

Keywords: Feedforward neural network · Algebraic algorithm · Spline Weight Function Algorithm

1 Introduction

The traditional feedforward neural network structure usually layers the nodes of neurons into input layer, hidden layer and output layer. The neurons in each layer only receive the output signal of the front layer neurons. For feedforward

Y. Liu et al. (Eds.): ICNC-FSKD 2019, AISC 1074, pp. 105–112, 2020.
https://doi.org/10.1007/978-3-030-32456-8_11

neural networks composed of non-linear neurons, traditional algorithms include BP algorithm [1,2], radial basis function algorithm and so on. However, due to some well-known shortcomings of these algorithms, such as local minimization, slow speed and so on. Many scholars have done a lot of research to overcome these shortcomings. For example, the hierarchical optimization learning algorithm [3] can be regarded as a penalty function method, and the connection parameters (weights and thresholds) after training are still constant. The second-order algorithm given in reference [4] improves the convergence of feedforward neural network algorithm to a certain extent. It can be said that these improvements basically focus on accelerating convergence speed, improving operation accuracy, selecting appropriate initial values and improving convergence. They do not overcome the inherent shortcomings of gradient descent method in finding global optimal solutions, nor do they pay attention to the interpretation of the physical meaning of weights stored in neural networks.

The Algebraic Algorithm [5] of feedforward neural network developed in recent years can get the global optimal point (when the cost function is equal to 0), but the BP algorithm is difficult to get the global optimal point.

Algebraic algorithms [5] train neural networks with quasi-support functions, which do not require function derivability. By introducing this kind of support function neurons into the hidden layer, the free weights can be easily adjusted so that the linear equations containing the undetermined weights are well-conditioned and easy to solve. If the traditional Sigmoid function is used as the hidden neuron, the algebraic algorithm can also get good results.

In engineering applications, it is hoped that the number of hidden layer neurons can be given accurately before network training begins. The algebraic algorithm gives a formula to accurately determine the number of hidden layer neurons when the precise mapping (i.e. the cost function is equal to 0), but the BP algorithm does not give such an accurate formula.

Algebraic algorithm has high accuracy and fast operation speed. Algebraic algorithm is much higher in accuracy and time efficiency than BP algorithm.

In order to overcome the shortcomings of traditional neural network training algorithms, a new structure of neural network and its training algorithm, namely Spline Weight Function Learning (Training) Algorithm of Feedforward Neural Network [5], is proposed, which is abbreviated as Spline Weight Function Algorithm. For traditional feedforward neural networks, the weights of the trained neural networks are constant. These constant weights can be arranged in a certain way to form a weight matrix. However, because the weights in the matrix are only discrete data, it is difficult to establish an intrinsic correspondence between these weights and the trained samples. That is to say, it is difficult for people to extract useful rules from the weights stored in the neural networks. Because the constant weights can hardly reflect the characteristics of training samples, the trained neural network is usually a nonlinear input-output system, and the weights stored in the neural network are only used as parameters of the nonlinear system. Obviously, these weights do not seem to have any "meaning" for getting information or rules from samples. The spline weight function learning algorithm makes the weight of the trained neural network be the function of input sample (called weight function), which is composed of cubic spline function instead of the constant of the traditional methods. The trained weight function can extract the information features of the trained samples well in a certain local space.

For the training algorithm of Spline Weight Function of Feedforward Network, the topology of the neural network is very simple. There are only two layers. The weights of the input layer are fully interconnected with the neurons, and the weights of the output layer are not. The traditional method of neural network has at least three layers, usually the weights of each layer are fully interconnected, so the number of weights that need to be trained is significantly reduced. As long as the dimensions of input and output sample vectors are known, the topological structure of the neural network can be determined. For a given problem, the dimensions of input and output sample vectors are known, so it is important to determine the topological structure of the neural network before training. Some traditional algorithms (such as BP algorithm or RBF algorithm) usually fail to do this. They need to be tested and tried again and again. It takes a lot of network training time, and the effect is not good. Obviously, a simple topology is conducive to hardware implementation.

Because of the simple topological structure of the Spline Weighted Neural network, the number of neurons required by the algorithm is independent of the number of samples (algebraic algorithm and traditional method are related to the number of samples), which is equal to m n, i.e. only depends on the number of input and output nodes (i.e. dimension m of input sample vector and dimension n of target sample vector).

Based on the error analysis of the spline weight function training algorithm of feedforward network, the quantitative description of the generalization ability of the network is given. The results show that the generalization ability of the network improves with the increase of the number of samples.

2 Algebraic Algorithm of Feedforward Neural Networks

This section is divided into three subsections: network structure, basic theory and algorithm, computational complexity of the algorithm. Figure 1 illustrates the network structure of algebraic algorithms. The core of the algebraic algorithm is to select a set of free weights between the input layer and the hidden layer, and then calculate a set of linear equations to obtain the weights between the hidden layer and the output layer. As long as the solution of the system of linear equations exists, it means that the global optimal solution is obtained. This shows that if the solution of the linear equation system exists, the algebraic algorithm has no error in theory, that is to say, it achieves accurate mapping. The calculation accuracy of the algebraic algorithm depends on the selected algorithm and the data accuracy of the computer, so the algebraic algorithm has high accuracy. The computational complexity of solving linear equations is polynomial order, so the algebraic algorithm has fast computational speed. In practical application, the algebraic algorithm gives a formula for accurately determining the number of hidden layer neurons when mapping accurately.

2.1 Network Structure

The algebraic theory and algorithm of feedforward neural network (algebraic algorithm for short) are basically the same as that of traditional feedforward neural network (such as BP or RBF). Neuron nodes are arranged hierarchically to form input layer, hidden layer and output layer. The input layer transmits the information to the hidden layer, and the hidden layer transmits the processed information to the output layer. There is no information feedback between the layers. Figure 1 shows a topological structure.

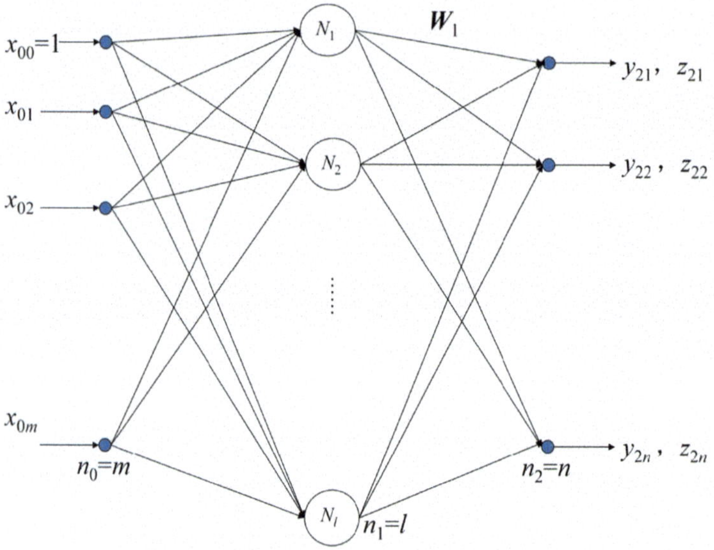

Fig. 1. Topological structure diagram of three-layer neural networks

2.2 Basic Theories and Algorithms

The most frequently used neural network is the three-layer feedforward network. The reason why we are interested in three-layer neural network is that three-layer neural network can realize accurate mapping of cost function 0. The structure of the neural network is simpler. The number of neurons in the input and output layers is equal to the dimension of the input and output sample vectors, and the transformation relation of neurons is identical transformation, that is to say, the number of neurons in the input and output layers is equal to the dimension of the input and output sample vectors.

$$\begin{cases} \mathbf{Y}_0 = f_0(\mathbf{X}_0) = \mathbf{X}_0 \\ \mathbf{Y}_2 = f_2(\mathbf{X}_2) = \mathbf{X}_2 \end{cases} \tag{1}$$

The input-output relationship of this neural network is as follows:

$$\begin{cases} \mathbf{Y}_0 = \mathbf{X}_0 \\ \mathbf{X}_1 = \tilde{\mathbf{W}}_0^T \tilde{\mathbf{Y}}_0 \\ \mathbf{Y}_1 = f(\mathbf{X}_1) \\ \mathbf{X}_2 = \mathbf{W}_1^T \mathbf{Y}_1 \\ \mathbf{Y}_2 = \mathbf{X}_2 \end{cases} \tag{2}$$

Theorem 1 (Key theorem for algebraic algorithm of feedforward neural networks). *For any given samples, if the number of neurons in the input layer is equal to the dimension of the input sample vector, the number of neurons in the output layer is equal to the dimension of the output sample vector, and the neuron transformation relations in the input layer and the output layer are both identical transformations, the number of neurons in the hidden layer equals the total number of samples, and the transformation relationship of neurons in the hidden layer is quasi-support function, there must be a three-layer feedforward network, its cost function is an accurate mapping without error.* □

The above theorem shows that the algebraic algorithm of neural network training is an accurate algorithm without theoretical error.

2.3 Time Complexity of Algebraic Algorithm

The main computational process of algebraic algorithm is to solve linear equations, so the computational complexity of algebraic algorithm is equivalent to that of solving linear equations. Since there are a total of weights between the hidden layer and the output layer to be solved, according to the $O\left(N^2n^2\right)$ and $O\left(N^3n^3\right)$, which is a polynomial order algorithm and belongs to a fast computationally acceptable algorithm.

The computational complexity of traditional BP algorithm is difficult to determine, because in BP algorithm, the computational complexity depends on the choice of initial weights. Different initial weights make the computational complexity of BP algorithm very different, sometimes limited to local minimum. So for BP algorithm, it is no longer meaningful to talk about computational complexity. A large number of computational practices show that BP algorithm consumes a lot of computational time. For example, an example in document [6] shows that the computational workload of BP algorithm is very large.

3 Weight Function Neural Network Theories and Methods

Like the traditional BP algorithm, the weight obtained by algebraic algorithm is constant, and it is difficult to extract hidden information from training samples. In order to retain the advantages of algebraic algorithm and overcome its shortcomings, the weights of trained neural networks are spline functions with input samples as interpolation points rather than constants. So as long as the trained spline function is extracted, the intrinsic information hidden in the samples can be obtained at least in the trained spatial dimension. Obviously, the more the number of samples, the more dense the interpolation nodes and the closer the calculated spline function is to the real curve, which shows that the generalization ability of the network is also enhanced with the increase of the number of samples. Moreover, the topological structure of the spline weight function neural network is simple, and the number of its neurons is independent of the number of samples, equal to the number of input and output nodes. The calculation complexity of solving spline function is also polynomial order, which is faster than algebraic algorithm. This section is divided into three subsections: network structure, basic theory and algorithm, error analysis and computational complexity.

3.1 Network Structure

The structure of the neural network is shown in Fig. 2. The input layer has m nodes, and the number of nodes is the same as the dimension of input sample vectors; the output layer has n nodes, and the number of nodes is the same as the dimension of output sample vectors, so the structure of spline weight function neural network is very simple.

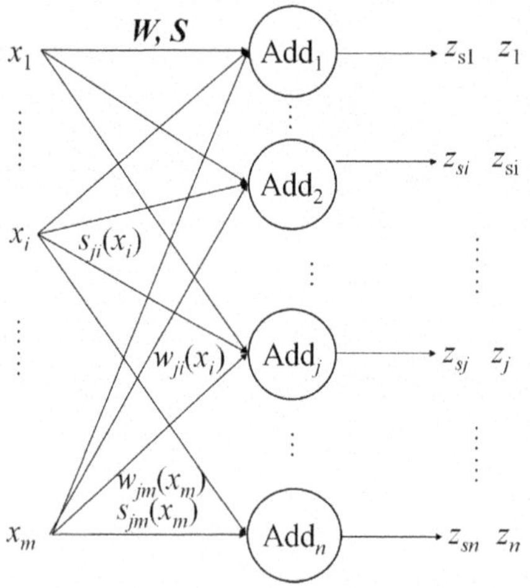

Fig. 2. Spline weight function neural network structure with m nodes in input layer and n nodes in output layer

3.2 Basic Theories and Algorithms

The expression of the interpolation point is

$$\mathbf{Ip}_{ji} = \left\{ (x_{i0},\ \eta_{ji}z_{j0}),\ (x_{i1},\ \eta_{ji}z_{j1}),\ \cdots,\ (x_{i(N+1)},\ \eta_{ji}z_{j(N+1)}) \right\} \tag{3}$$

The spline function is

$$\begin{aligned} s_{jip}(x_i) &= y_{jip}F_0(\xi) + y_{ji(p+1)}F_1(\xi) + h_{ip}\left(m_{jip}G_0(\xi) + m_{ji(p+1)}G_1(\xi)\right) \\ &= \eta_{ji}z_{jp}F_0(\xi) + \eta_{ji}z_{j(p+1)}F_1(\xi) + h_{ip}\left(m_{jip}G_0(\xi) + m_{ji(p+1)}G_1(\xi)\right) \end{aligned} \tag{4}$$

The parameter m_{jip} in above formula can be obtained by solving the following equations.

$$\mathbf{H}_{ji}\mathbf{m}_{ji} = \mathbf{C}_{ji} \tag{5}$$

where

$$\mathbf{H}_{ji} = \begin{pmatrix} q_{ji0} & q_{ji1} & & & & \\ h_{i1} & 2(h_{i0}+h_{i1}) & h_{i0} & & & \\ & h_{i2} & 2(h_{i1}+h_{i2}) & h_{i1} & & \\ & & \ddots & \ddots & \ddots & \\ & & & h_{iN} & 2(h_{i(N-1)}+h_{iN}) & h_{i(N-1)} \\ & & & & q_{jiN} & q_{ji(N+1)} \end{pmatrix} \tag{6}$$

$$\mathbf{m}_{ji} = \left(m_{ji0}\; m_{ji1} \cdots m_{ji(N+1)}\right)^T \tag{7}$$

$$\mathbf{C}_{ji} = \left(C_{ji0}\; C_{ji1} \cdots C_{ji(N+1)}\right)^T \tag{8}$$

3.3 Error Analysis

The error analysis results of the neural network can be obtained as follows.

Theorem 2 (Theorem for error analysis of weight function neural networks). *The error between the actual output of the network and the sample is*

$$\begin{aligned} J &= \max_j J_j = \max_j \left\| z_j^{(\alpha)} - z_j^{(\alpha)} \right\|_\infty = \max_j \left\| \sum_{i=1}^m w_{ji}^{(\alpha)}(x_i) - \sum_{i=1}^m s_{ji}^{(\alpha)}(x_i) \right\|_\infty \\ &\le \max_j \left(\sum_{i=1}^m \left\| w_{ji}^{(\alpha)}(x_i) - s_{ji}^{(\alpha)}(x_i) \right\|_\infty \right) \le m \cdot \max_j \left(\max_i \left(\left\| w_{ji}^{(\alpha)}(x_i) - s_{ji}^{(\alpha)}(x_i) \right\|_\infty \right) \right) \\ &\le m \cdot \max_j \left(\max_i \left(B_{i\alpha} \cdot h_i^{4-\alpha} \cdot \left\| w_{ji}^{(4)}(x_i) \right\|_\infty \right) \right) \end{aligned} \tag{9}$$

where $h_i = \max\{h_{i0}, h_{i1}, \cdots, h_{iN}\}$. □

The above inequality shows that as long as $w_{ji}^{(4)}(x_i)$ is bounded, the more samples, the more dense the interpolation nodes, and the closer the calculated spline function is to the real curve. This shows that the generalization ability of the network increases with the increase of the number of samples.

3.4 Computational Complexity

For m-dimensional input and n-dimensional output neural networks, if the total number of samples to be trained is N, the total computational time required is

$$T(m, n, N) = ((7N + 3)\, t_1 + (10N + 7)\, t_2)\, mn = O\,(mnN) \tag{10}$$

The required time t_1 for performing one addition and subtraction on a computer can be regarded as a constant, and the time t_2 for calculating one multiplication and division can also be regarded as a constant. Therefore, the asymptotic time complexity of the spline weight function neural network with m-dimension input and n-dimension output is $O\,(mnN)$, which is related to the dimension of the input sample, the dimension of the output sample and the number of samples to be trained. For a given problem, m and n are usually constant, so the calculation speed of spline weight function neural network is faster than algebraic algorithm, of course, much faster than BP algorithm!

4 Summary and Prospect

The new development of feedforward neural networks in recent years is systematically discussed. Of course, there are many other new developments that cannot be fully covered in one paper. For algebraic methods, the selection and optimization of free weights deserve further study. For weighted function neural network, there are many choices of weighted function. In some cases, it is feasible to choose rational functions. Its mathematical theory is based on rational expansion [7], rational spline function and so on. Of course, there are many other developments, and I won't do more discussion in this paper due to length limitations.

References

1. Hagan, M.T., Demuth, H.B., Beale, M.: Neural Network Design. China Machine Press, Beijing (2002)
2. Haykin, S.: Neural Networks. Macmillan College Publishing Company, New York (1994)
3. Ergezinger, S., Tomsen, E.: An accelerated learning algorithm for multilayer perceptrons: optimization layer by layer. Neural Netw. **6**, 31–42 (1995)
4. Ampazis, N., Perantonis, S.J.: Two highly efficient second-order algorithms for training feedforward networks. Neural Netw. **13**, 1064–1073 (2002)
5. Zhang, D.: New Theories and Methods on Neural Networks. Tsinghua University Press, Beijing (2006)
6. Tesauro, G., Janssens, R.: Scaling relationships in back-propagation learning. Complex Syst. **2**, 39–44 (1989)
7. Zhang, D.: New expansion and infinite series. Int. Math. Forum. **9**, 1061–1073 (2014). https://doi.org/10.12988/imf.2014.45102. http://www.m-hikari.com/imf/imf-2014/21-24-2014/zhangIMF21-24-2014.pdf

Awareness Learning for Balancing Performance and Diversity in Neural Network Ensembles

Yong Liu(✉)

School of Computer Science and Engineering, The University of Aizu,
Aizu-Wakamatsu, Fukushima 965-8580, Japan
yliu@u-aizu.ac.jp

Abstract. An ensemble learning system requires a set of cooperative modules that solve a task together. It is believed that awareness modules being aware of other modules in an ensemble could build strong cooperations each other. In this paper, two levels of awareness are introduced into negative correlation learning in order to control the differences among the individual modules in an ensemble. At the ensemble level, whether individual modules would learn to be more different to the rest of modules on a given data point would depend on how well the data point had been learned by the ensemble. At the individual level, each individual module would try to learn to be more different to the rest of modules on a given data point only if it had not been far away from the other individuals based on the distances measured by their outputs on the data point. Negative correlation learning with such awareness learning was tested on training different structures of neural network ensembles where the relations between differences and cooperation were analyzed in the learning process.

1 Introduction

Although some successful applications have been achieved by deep learning and neural network learning, both industry and researchers are starting to realize some limits in such machine learning [1–4]. One difficulty comes from the architecture design for deep neural networks. In practice, such architecture design is usually set by the human experts who have tested a number of different architectures, and finally chosen the best one among them. Currently most of deep networks set by experts in real applications are monolithic. They have become so immensely complex that it would be impossible to interpret a large number of rules created by such deep networks.

Different to a monolithic neural network, human brain has been known to be a modular architecture with different specialized modules for vision, language, facial recognition, and others. If artificial neural networks could also be constructed by a number of specialized modules, it might be possible to interpret

Y. Liu et al. (Eds.): ICNC-FSKD 2019, AISC 1074, pp. 113–120, 2020.
https://doi.org/10.1007/978-3-030-32456-8_12

these modules when their functions could be clearly determined during learning. Certainly, not only should their functions be different, but also cooperative. Some ensemble learning methods have been developed in training a set of modules rather than a single network. Different to the learning algorithms for designing a single model, ensemble learning methods mainly dealt with how to create a set of modules with different functions.

If some kinds of feedback could be introduced among all the individual modules in their learning process, cooperation could be enhanced. Negative correlation learning [5] introduced a correlation between each module and the rest of modules in an ensemble in the learning targets for each module. This correlation term serves a feedback to each module from all other modules in learning each data. This correlation term would let one module learn more from a data point if the rest of modules had not learned the data point well. Meanwhile, if the rest of modules had learned the data point well, the module would be left not to learn the data point. The correlation terms among all the individuals could be regarded feedbacks while these feedbacks would keep in changing on learning data points, and be used through the whole learning process.

Although such feedback based on correlation term is helpful in creating cooperative modules, each individual module has to use the feedback in the same way without its own choices in the original negative correlation learning. It appeared that individual modules created by negative correlation learning could be too different or too similar to cooperate each other. It is important for each module in an ensemble to be aware of how different should be between itself and the rest of modules [6–12]. In this paper, two levels of awareness are introduced into negative correlation learning in order to control the difference among the individual modules in an ensemble. At the ensemble level, whether individual modules would learn to be more different to the rest of modules on a given data point would depend on how well the data point had been learned by the ensemble. At the individual level, each individual module would try to learn to be more different to the rest of modules on a given data point only if it had not been far away from the other individuals based on the distances measured by their outputs on the data point. Negative correlation learning with awareness learning was tested on training different structures of neural network ensembles where the relations between difference and cooperation were analyzed in the learning process.

The next three sections describe both the implementation of negative correlation learning with awareness learning and simulation results. Section 2 describes how awareness was able to be built at both the level of neural network ensemble and the level of individual neural networks in the negative correlation learning process. Section 3 analyzes how the difference would cope with cooperation by examining the performance of both the ensembles and the individual neural networks in the ensembles. Finally, Sect. 4 summarizes the test results, and provides suggestions on how to build effective awareness in machine learning.

2 Awareness in Negative Correlation Learning

In an ensemble system with M individual neural networks, one way to define the output $F(n)$ of the ensemble on the n-th data is to have a simple average of the outputs $F_i(n)$ of all the individual neural networks on the n-th data:

$$F(n) = \frac{1}{M} \Sigma_{i=1}^{M} F_i(n) \tag{1}$$

Suppose that there are N data points $((\mathbf{x}(n), y(n))$, $n = 1, \cdots, N$, in a given training data set D, negative correlation learning could use the whole training set D to train all M individual modules in the ensemble without sampling, re-sampling or filtering data although such re-sampling data techniques could be applied. Negative correlation learning would be a good choice if the available data would be few. In negative correlation learning, the learning error function $E_i(n)$ for the i-th neural network on learning the n-th data is defined as [5]:

$$E_i(n) = \frac{1}{2} \left[(F_i(n) - y(n))^2 - \lambda (F_i(n) - F(n))^2 \right] \tag{2}$$

By minimizing $E_i(n)$, the i-th neural network would not only learn the target $y(n)$, but also be different to the ensemble output at the same time. λ is a small positive weight that could scale up and down the difference between output of the i-th neural network and the ensemble's output.

From the error function in Eq. (2), its partial derivative for learning the weights of the i-th neural network can be specified as:

$$\frac{\partial E_i(n)}{\partial F_i(n)} = (1 - \lambda)(F_i(n) - y(n)) + \lambda (F(n) - y(n)) \tag{3}$$

At $\lambda = 1$, the derivative of E_i in Eq. (3) becomes

$$\frac{\partial E_i(n)}{\partial F_i(n)} = F(n) - y(n) \tag{4}$$

If λ would be set to no more than 1, the importance for the two terms in Eq. (2) would be the same at $\lambda = 1$. If λ would be allowed to be bigger than 1, the ensemble might not be able to learn the given data points although each individual neural networks could be more different. How to let individual neural networks be more different while the ensemble would be able to learn the data points at the same time? This paper introduces awareness learning for letting individual neural networks learn to be different each other whenever the data points have been learned by the whole ensemble. For the two-class classification problem with the targets 0 and 1, the i-th data point would be learned if the absolute difference $|F(n) - y(n)|$ is less than T with its value no more than 0.5. Therefore, T could tell how well a data point has been learned by the ensemble. The smaller T is, the better the data point has been learned. If $|F(n) - y(n)|$ would be less than T, all the individual neural networks would switch to minimizing the following difference function in learning instead of the error function in Eq. (2):

$$E_i(n) = (F_i(n) - 0.5)(F(n) - 0.5) \quad (5)$$

In the next section, negative correlation learning with five values of T from 0 to 0.4 were tested in learning. The smaller T is, the less number of switching learning would happen. Meanwhile, the larger T is, the more difference among all the individual neural networks would be.

3 Relations Between Diversity and Cooperations

3.1 Descriptions of Experimental Setup

$\lambda = 1$ and five values of T from 0 to 0.4 with the step size 0.1 were used in negative correlation learning with awareness (NCLA). The data set used for testing NCLA is the Australian credit card assessment data set from the UCI machine learning bench-mark repository. The card data set consists of total 690 examples with 14 attributes. Although many applications provide large number of data examples, some applications only have few data examples. The ensemble systems might be more important for such applications with few data examples where noises exist. All the results were averaged from 10 runs of 10-fold cross-validation on the card data.

Two ensemble structures were tested, including one structure with 50 individual neural networks and the other with 200 individual neural networks. Each individual neural network has one hidden layer with 5 hidden nodes. Both hidden nodes and output nodes used sigmoid activation function. Large number of individual neural networks was tested to see whether the size of ensembles would affect the performance. The performance was observed in 2000 learning epochs on both the training and testing sets. Although less number of learning epochs was enough for the tested data set, it is to see whether the longer training would worsen the performance. In the practice, the longer training might not be avoided if the optimal learning epochs could not be determined.

3.2 Results of NCLA on the Card Data

Table 1 showed how the performance of the ensembles by NCLA changed for the different values of T in the whole learning process. When the threshold value T was raised from 0 to 0.4, the ensemble could learn the training data better, and reached the lower training error rates after the learning epochs were over 500. In the first 100 learning epochs, NCLA with the higher T learned the training data a litte slower. NCLA with $T = 0.2$ reached the best performance of the testing error rate 0.1329 at the learning epoch 1000. However, when the value of T was set to 0.4, NCLA's performance on the testing data turned worse although its training error rates were the lowest.

Table 1. Average error rates of the ensembles by 100-run NCLA on the card data at the learning epoch 100, 500, 1000, and 2000, respectively. The outputs of each measured ensemble were decided by a simple average of 50 individual neural networks with 5-hidden-node in the ensemble.

T	100		500		1000		2000	
	Training	Testing	Training	Testing	Training	Testing	Training	Testing
0	0.0924	0.1377	0.0597	0.1359	0.0525	0.1352	0.0481	0.1359
0.1	0.0933	0.1365	0.0564	0.1330	0.0508	0.1358	0.0460	0.1357
0.2	0.0944	0.1352	0.0552	0.1352	0.0479	0.1329	0.0426	0.1339
0.3	0.0993	0.1388	0.0546	0.1345	0.0406	0.1364	0.0322	0.1374
0.4	0.1024	0.1420	0.0483	0.1381	0.0273	0.1436	0.0145	0.1480

Table 2. Average error rates of the ensembles by 100-run NCLA on the card data at the learning epoch 100, 500, 1000, and 2000, respectively. The outputs of each measured ensemble were defined by a majority voting from 50 individual neural networks with 5-hidden-node in the ensemble.

T	100		500		1000		2000	
	Training	Testing	Training	Testing	Training	Testing	Training	Testing
0	0.0952	0.1430	0.0606	0.1388	0.0532	0.1384	0.0486	0.1394
0.1	0.0965	0.1414	0.0573	0.1374	0.0513	0.1399	0.0465	0.1393
0.2	0.0975	0.1390	0.0560	0.1383	0.0487	0.1375	0.0430	0.1371
0.3	0.1028	0.1425	0.0560	0.1401	0.0418	0.1419	0.0330	0.1448
0.4	0.1103	0.1516	0.0518	0.1517	0.0286	0.1620	0.0157	0.1694

When the outputs of the ensembles were decided by the majority voting, the training error rates became lower with the increased T as being showed in Table 2. The training performance was rather similar between the averaging outputs and the voting outputs. However, the testing performance by voting showed poor results at $T = 0.4$.

By examining the individual learning behaviors in Table 3, it could be seen how the individual neural networks adapted during learning different data. At $T = 0$, all the individual neural networks were forced to learn better and better in NCLA although the individual performance was much lower than the ensemble performance on both the training and testing sets. From the results in Table 4, NCLA with the bigger T could let the individual neural networks have more chances to learn to be different to the ensemble. Therefore, the individual neural networks were able to be more different each other. It could be seen that the same outputs on the measured data set between every pair of individual neural neworks in an ensemble became less and less with the increased T. Certain level of difference among the individual modules in an ensemble led to their

Table 3. Average error rates of individual neural networks by 100-run NCLA on the card data at the learning epoch 100, 500, 1000, and 2000, respectively. There are 50 individual neural networks with 5-hidden-node in each measured ensemble.

T	100		500		1000		2000	
	Training	Testing	Training	Testing	Training	Testing	Training	Testing
0	0.1431	0.1774	0.1106	0.1796	0.1033	0.1822	0.0978	0.1848
0.1	0.1841	0.2150	0.1672	0.2268	0.1600	0.2279	0.1551	0.2290
0.2	0.2509	0.2739	0.2414	0.2895	0.2359	0.2909	0.2318	0.2926
0.3	0.3328	0.3472	0.3237	0.3588	0.3179	0.3610	0.3136	0.3624
0.4	0.4228	0.4307	0.4109	0.4322	0.4050	0.4336	0.4012	0.4349

Table 4. Average overlapping rates among individual neural networks by 100-run NCLA on the card data at the learning epoch 100, 500, 1000, and 2000, respectively. There are 50 individual neural networks with 5-hidden-node in each measured ensemble.

T	100		500		1000		2000	
	Training	Testing	Training	Testing	Training	Testing	Training	Testing
0	0.8450	0.8377	0.8655	0.8345	0.8706	0.8294	0.8743	0.8252
0.1	0.7714	0.7649	0.7633	0.7419	0.7678	0.7411	0.7704	0.7395
0.2	0.6683	0.6650	0.6545	0.6413	0.6561	0.6388	0.6573	0.6375
0.3	0.5726	0.5714	0.5660	0.5595	0.5670	0.5574	0.5680	0.5562
0.4	0.5090	0.5087	0.5098	0.5077	0.5107	0.5074	0.5113	0.5074

Table 5. Average error rates of the ensembles by 100-run NCLA on the card data at the learning epoch 100, 500, 1000, and 2000, respectively. The outputs of each measured ensemble were decided by a simple average of 200 individual neural networks with 5-hidden-node in the ensemble.

T	100		500		1000		2000	
	Training	Testing	Training	Testing	Training	Testing	Training	Testing
0	0.0920	0.1388	0.0592	0.1375	0.0521	0.1364	0.0477	0.1362
0.1	0.0939	0.1365	0.0563	0.1346	0.0489	0.1352	0.0431	0.1361
0.2	0.0945	0.1354	0.0548	0.1336	0.0444	0.1345	0.0342	0.1359
0.3	0.0989	0.1393	0.0537	0.1355	0.0342	0.1370	0.0229	0.1393
0.4	0.1032	0.1416	0.0425	0.1394	0.0212	0.1443	0.0096	0.1461

cooperation. However, it should be noted that too much difference could destroy such cooperation. It happened when T was set to 0.4 in NCLA.

It would be interesting to see how the size of the ensembles would affect their performance from Tables 5 and 6. The same trend of the difference among the individual neural networks was observed when the number of individual neural networks in an ensemble was increased from 50 to 200. The larger the ensembles

Table 6. Average error rates of the ensembles by 100-run NCLA on the card data at the learning epoch 100, 500, 1000, and 2000, respectively. The outputs of each measured ensemble were defined by a majority voting from 200 individual neural networks with 5-hidden-node in the ensemble.

T	100		500		1000		2000	
	Training	Testing	Training	Testing	Training	Testing	Training	Testing
0	0.0918	0.1394	0.0589	0.1386	0.0522	0.1370	0.0478	0.1370
0.1	0.0936	0.1380	0.0562	0.1348	0.0491	0.1355	0.0434	0.1371
0.2	0.0948	0.1362	0.0547	0.1341	0.0445	0.1361	0.0345	0.1367
0.3	0.0990	0.1400	0.0536	0.1372	0.0344	0.1378	0.0231	0.1410
0.4	0.1037	0.1428	0.0436	0.1439	0.0212	0.1470	0.0101	0.1510

were, the lower their training error rates were at the end of the learning epoch 2000. With the lower training error rates, the large ensembles by NCLA were shown to be more stable by having the similar testing error rates to those small ensembles.

4 Conclusions

In the original version of negative correlation learning method, individual neural networks in an ensemble might become similar to each other if the learning would be conducted for long. One way to prevent these individual neural networks from being similar is to stop the learning at early stage. Such early stopping might run a risk of underfitting the given data. The other way discussed in this paper was to allow individual neural networks to have more chances to learn to be different each other. In this way, each individual neural network could gain its awareness in learning each data so that all the individual neural networks could cooperate each other better. Bear in mind that cooperation among all the individual neural networks in an ensemble should be able to be adapted through their whole learning process. Therefore, neither independent ensemble learning methods nor the sequential ensemble learning methods are capable of establishing such good cooperation in an ensemble.

References

1. Hinton, G.E., Vinyals, O., Dean, J.: Dark knowledge. Invited talk at the BayLearn Bay Area Machine Learning Symposium (2014)
2. Bengio, Y., Courville, A., Vincent, P.: Representation learning: a review and new perspectives. IEEE Trans. Pattern Anal. Mach. Intell. (PAMI) **35**(8), 1798–1828 (2013)
3. Chilimbi, T., Suzue, Y., Apacible, J., Kalyanaraman, K.: Project adam: building an efficient and scalable deep learning training system. In: 11th USENIX Symposium on Operating Systems Design and Implementation (2014)

4. Cohen, N., Sharir, O., Shashua, A.: On the expressive power of deep learning: a tensor analysis. arXiv:1509.05009 (2015)
5. Liu, Y., Yao, X.: Simultaneous training of negatively correlated neural networks in an ensemble. IEEE Trans. Syst. Man Cybern. Part B: Cybern. **29**(6), 716–725 (1999)
6. Liu, Y.: Error awareness by lower and upper bounds in ensemble learning. Int. J. Pattern Recognit. Artif. **30**(9) (2016)
7. Liu, Y.: Negative selection in negative correlation learning. In: The 12th International Conference on Natural Computation, Fuzzy Systems and Knowledge Discovery (2016)
8. Liu, Y.: Build correlation awareness in negative correlation learning. In: The 13th International Conference on Natural Computation, Fuzzy Systems and Knowledge Discovery (2017)
9. Liu, Y.: Negative correlation learning with multiple target values. In: The 14th International Conference on Natural Computation, Fuzzy Systems and Knowledge Discovery (2018)
10. Liu, Y.: Computational awareness for learning neural network ensembles. In: The 2017 IEEE International Conference on Information and Automation (2017)
11. Liu, Y., Zhao, Q., Pei, Y.: Error awareness by lower and upper bounds in ensemble learning. In: The 11th International Conference on Natural Computation (2015)
12. Liu, Y.: Control of the error signals by self-awareness in committee machines. In: The 9th International Congress on Image and Signal Processing, BioMedical Engineering and Informatics (2016)

Natural Computation: Theory and Algorithms - Genetic and Evolutionary Algorithms

A Model and an Algorithm for Empty Car Distribution in Railway Transportation

Lanfen Liu and Xinfeng Yang(✉)

School of Traffic and Transportation Engineering, Lanzhou Jiaotong University, Gansu 730070, Lanzhou, China
{liulanfen,xinfengyang}@mail.lzjtu.cn

Abstract. The empty car distribution is an important part of technical plan for railway transportation and can deal with the unbalanced distribution of empty car reasonably. So, taken the uncertainty of freight supply and demand into account, a fuzzy empty car distribution model is established in the aim of minimizing the total empty car running kilometers. Furthermore, according to the characteristic of this problem, a two-phase genetic algorithm (GA) is put on emphases to propose based on simulation. Since different objective values can be obtained in different conditions rather than a crisp value, more information is provided for making adjustment measures. Therefore, the model and algorithm have theoretical value and practical value.

Keywords: Empty car distribution · Genetic algorithm · Transportation problem · Fuzzy sets

1 Introduction

The empty car distribution is an important part of technical plan for railway transportation and can deal with the unbalanced distribution of empty car and arrange the efficient route between different stations reasonably. Along with the increasing of railway transportation demand, the contradiction between supply and demand of railway freight car is becoming increasingly conspicuous. If this contradiction is not solved suitably, the detention time waiting for loading in some stations is postpone because of lack of empty car, while empty car is overstocked in some other stations. Therefore, the reasonable empty car distribution is an important work for improving the service quality of railway transportation and speeding up the car turn round, and related directly to the car economic efficiency and the accomplishment of transportation index.

Actually, the empty car distribution problem is an application of transportation problem. A lot of researches have been carried out in recent years. Liang et al. [1] described a mathematical model for multi-type empty car distribution which availed to optimize further empty car distribution and combined. Li and

Y. Liu et al. (Eds.): ICNC-FSKD 2019, AISC 1074, pp. 123–131, 2020.
https://doi.org/10.1007/978-3-030-32456-8_13

Xia [2] proposed a time restraint-based model and algorithm which calculated the distribution of both loaded and empty wagons. Lei et al. [3] described a class of special stochastic chance-constrained programming models and designed a GA resolution. Zhang and Zhang [4] put forward an intelligent optimized method-real-coded GA, but this algorithm needed to find all the reasonable routes. Wang et al. [5] constructed an optimization model for refrigerated wagons adjustment and cooperation and designed a corresponding ant colony optimization (ACO) algorithm. Cao et al. [6] presented a collaborative optimization model based on the existing railway network, car flow organization and operation characteristics for the empty car distribution. Wu et al. [7] established an empty car allocation model to minimize the total traveling distance and used the adaptive mutation particle swarm algorithm to solve this problem. Jing et al. [8] built a cloud preference model of dynamic empty car scheduling considering cloud preference objective function and empty car delay time constraint. Liu and Zhu [9] established a distribution optimizing model of empty cars resource based on customer satisfaction and maximum enterprise benefits considering the enterprise benefits, customer satisfaction and enterprise benefits. Zhao et al. [10] formulated a stage-based optimization model for allocating freight empty cars in railway networks taking the storage cost, unit transportation cost, and demand into consideration. Zhang et al. [11] studied the problem of empty wagons adjustment of Chinese heavy-haul railway and set up an optimization model to solve the empty wagons adjustment of heavy-haul railway. Wang et al. [12] analyzed the dynamic characteristics of operational empty car distribution and proposed a dynamic multi-period optimization thought as well as a method to calculate the next day's empty car flow discharged over railway network boundaries, and presented a chaos-based particle swarm optimization algorithm based on the characteristics of model.

However, in the empty car distribution problem, there are cases that the supplies and demands of stations may be uncertain due to some uncontrollable factors. For example, the supplies and demands are affected by the delay time of a local car, information of car detaining repair and freight car arrived, the demand of a consignor, numbers of different type car for loading and so on. Therefore, it is a natural way to be considered as fuzzy numbers. Obviously, when the supply and demand quantities are fuzzy numbers, the total cost will be fuzzy as well. But it is few that the empty car distribution problem is studied by employing the fuzzy set theory. Consequently, it is difficult to guarantee the applicability and economy when uncertain factors are neglected. So studying on uncertain empty car distribution problem, there are both application value and practical value. The remainder of the paper is organized as follows: In Sect. 2, a fuzzy empty car distribution problem model is presented. Section 3.1 introduces the efficient-paths-based stochastic traffic assignment algorithm. Section 3.2 analyses the impact area of a link, and Sect. 3.3 provides a partial network scan algorithm for identifying the critical link based on the impact area. A numerical experiment is tested in Sect. 4. Finally, the conclusions are given in Sect. 5.

2 Fuzzy Empty Car Distribution Problem

2.1 Some Symbols

The empty car distribution problem is also formulated as the transportation problem. Transportation problem is a linear programming problem stemmed from a network structure consisting of a finite number of nodes and arcs attached to them. Efficient algorithms have been developed for solving the transportation problem when the cost coefficients and the supply and demand quantities are known exactly. In this study, we develop a solution procedure that is able to calculate the fuzzy objective value of the fuzzy transportation problem, where at least one of the parameters is a fuzzy number.

In order to express more conveniently, we firstly define some variables. Let $G = (V, E, W, L)$ be a directed railway network with stations set $V = \{v_i|i = 1, 2, ..., n\}$, railway lines set $E = \{e_{ij}\,|i, j = 1, 2, ..., n\}$, carrying capacity set $W = \{\tilde{w}_{ij}\,|(v_i, v_j) \in E\}$, where $\tilde{w}_{ij}$ is a triangular fuzzy number with $\tilde{w}_{ij} = (a_{ij}, b_{ij}, c_{ij})$ and lines length set $L = \{l_{ij}\,|(v_i, v_j) \in E\}$. Let $t = 1, 2, \cdots, k$ be the empty car types and f^t_{ij} be the car number of t type through line e_{ij}. Let $\tilde{S}^t_i$ be the empty car number of t type that the station i can supply and $\tilde{D}^t_i$ be the empty car number of t type that the station i demand, where $\tilde{S}^t_i$ and $\tilde{D}^t_i$ are the triangular fuzzy numbers with $\tilde{S}^t_i = (a^t_i, b^t_i, c^t_i)$ and $\tilde{D}^t_i = (a'^t_i, b'^t_i, c'^t_i)$.

2.2 Car Distribution Model

In many real-word situations, the total empty car supply of t type is not always equal to the total empty car demand of t type. It is well known that an unbalanced transportation problem is equivalent to an ordinary balanced transportation problem with one (dummy) column with zero costs added. In this case, a dummy supply origin is created if demand exceeds supply, or a dummy demand origin is created if supply exceeds demand. Then the unbalanced transportation problem can be converted to a balanced transportation problem. If we hope that the empty car number of t type that the station i send out is equal to the empty car number of t type that the station i can supply with a confidence level α^t_i. Then, we have the following chance constraint,

$$Pos\{\sum_{(v_i,v_j)\in E} f^t_{ij} - \sum_{(v_j,v_i)\in E} f^t_{ji} \leq \tilde{S}^t_i\} \geq \alpha^t_i(i, j = 1, 2, \cdots, n; t = 1, 2, \cdots, k). \quad (1)$$

Similarly, if we hope that the empty car number of t type that the station i receive is equal to the empty car number of t type that the station i demand with a confidence level β^t_i. Then, we have the following chance constraint,

$$Pos\{\sum_{(v_j,v_i)\in E} f^t_{ji} - \sum_{(v_i,v_j)\in E} f^t_{ij} \geq \tilde{D}^t_i\} \geq \beta^t_i(i, j = 1, 2, \cdots, n; t = 1, 2, \cdots, k). \quad (2)$$

In addition, the total freight quantity through edge e_{ij} must not exceed its carrying capacity with a confidence level γ_{ij}. Then, we have the following capacity chance constraint,

$$Pos\{f_{ij}^t \le \tilde{w}_{ij}\} \ge \gamma_{ij}, ((v_j, v_i) \in E). \tag{3}$$

If we want to minimize the total transportation cost, the mathematical description of the fuzzy empty car distribution problem is:

$$\min \sum_{t=1}^{m}\sum_{j=1}^{n}\sum_{i=1}^{n} f_{ij}^t l_{ij}$$
$$s.t. \begin{cases} Pos\{\sum\limits_{(v_i,v_j)\in E} f_{ij}^t - \sum\limits_{(v_j,v_i)\in E} f_{ji}^t \le \tilde{S}_i^t\} \ge \alpha_i^t, i,j = 1,2,\cdots,n, t = 1,2,\cdots,k \\ Pos\{\sum\limits_{(v_j,v_i)\in E} f_{ji}^t - \sum\limits_{(v_i,v_j)\in E} f_{ij}^t \ge \tilde{D}_i^t\} \ge \beta_i^t, i,j = 1,2,\cdots,n, t = 1,2,\cdots,k \\ Pos\{\sum\limits_{t=1}^{k} f_{ij}^t \le \tilde{w}_{ij}\} \ge \gamma_{ij}, i,j = 1,2,\cdots,n \\ f_{ij}^t \ge 0, \text{integer}, i,j = 1,2,\cdots,n, t = 1,2,\cdots,k. \end{cases}$$

2.3 Equivalent Crisp Model

One of the methods to solve fuzzy chance constrained programming is to convert the chance constraints into its crisp equivalents and then solve the equivalent crisp model by the traditional solution process.

Theorem 1 ([13]). *Assume that the fuzzy vector* $\boldsymbol{\xi}$ *degenerates to a fuzzy variable* ξ *with continuous membership function* μ, *and the function* $g(x,\xi)$ *has the form* $g(x,\xi) = h(x) - \xi$. *Then* $pos\{g(x,\xi) \le 0\} \ge \alpha$ *if and only if* $h(x) \le K_\alpha$, *where*

$$K_\alpha = sup\{K | K = \mu^{-1}(\alpha)\}. \tag{4}$$

Theorem 2 ([13]). *Assume that* ξ *is a fuzzy variable with continuous membership function* μ. *Then* $pos\{z \ge \xi\} \ge \alpha$ *if and only if* $z \ge K_\alpha$, *where*

$$K_\alpha = inf\{K | K = \mu^{-1}(\alpha)\}. \tag{5}$$

Hence the fuzzy chance constrained programming can be converted into the following equivalent crisp model by using Theorems 1 and 2,

$$\min \sum_{t=1}^{m}\sum_{j=1}^{n}\sum_{i=1}^{n} f_{ij}^t l_{ij}$$
$$s.t. \begin{cases} \sum\limits_{(v_i,v_j)\in E} f_{ij}^t - \sum\limits_{(v_j,v_i)\in E} f_{ji}^t \le (1-\alpha_i^t)c_i^t + \alpha_i^t b_i^t, i,j = 1,2,\cdots,n, t = 1,2,\cdots,k \\ \sum\limits_{(v_j,v_i)\in E} f_{ji}^t - \sum\limits_{(v_i,v_j)\in E} f_{ij}^t \ge (1-\beta_i^t)a_i'^t + \beta_i^t b_i'^t, i,j = 1,2,\cdots,n, t = 1,2,\cdots,k \\ \sum\limits_{t=1}^{k} f_{ij}^t \le (1-\gamma_{ij})c_{ij} + \gamma_{ij} b_{ij}, i,j = 1,2,\cdots,n \\ f_{ij}^t \ge 0, \text{integer}, i,j = 1,2,\cdots,n, t = 1,2,\cdots,k. \end{cases} \tag{6}$$

3 Design of GA

According to the characteristic of this problem, a two-stage GA is put forward. Herein, the first stage of this GA is determining the supply and demand relationship between different points and the second stage is used for confirming the quantity and route between the supply and the demand.

3.1 The First Stage

The first stage of this GA is adopted a priority-based encoding method for determining the supply and demand relationship between different points. The relationship orders of supply and demand are represented by the chromosome $CH = (ch_1, ch_2, \cdots, ch_m)$, where m is the product of the supply point number and the demand point number.

Assume that there are 3 supply points (point 1, 2, 3) and 4 demand points (point 11, 12, 13, 14). Figure 1 is a chromosome. From this chromosome, we can know that the order of supply and demand relationship is $3 \to 13, 2 \to 11, 3 \to 12, 2 \to 12, 1 \to 13, 1 \to 11, 3 \to 11, 1 \to 12, 3 \to 14, 2 \to 14, 1 \to 14, 2 \to 13$.

3	5	6	2	7	1	9	10	12	8	4	11

Fig. 1. Example for a chromosome

Initialization Process. An integer *pop_size* is defined as the number of chromosomes. By using the procedure below, we can generate *pop_size* chromosomes $CH_1, CH_2, \cdots, CH_{pop\_size}$ which compose the initial population.

The generating process of a chromosome is stated as follows:

Step1. For $i = 1$ to m, let $y_i = i$. Set $i = 1$, $j = m$.
Step2. If $i > m$, then stop. Otherwise generate a random integer k between 1 and j, let $ch_i = y_k$. Then go to Step3.
Step3. For $l = k$ to $m - 1$, let $y_l = y_{l+1}$. Set $j = j - 1$, $i = i + 1$ and go to Step2.

Selection Process. The selection process that we design integrates the elitist model and the fitness-proportional model. The selection process is based on spinning the roulette wheel $pop\_size - 1$ times. Each time a single chromosome is selected for a new population. The roulette wheel is a fitness-proportional selection. Simultaneously, the best one is always selected. Then we arrange the chromosomes by the objective function value in descending order. The selection process of the roulette wheel is stated as follows:

Step1. Compute $P_i = \sum_{j=1}^{i} f_j$ for $i = 1, 2, \cdots, pop\_size - 1$, where f_j is the probability that chromosome CH_j is selected. Set $i = 1$.

Step2. If $i > pop\_size - 1$, then stop. Otherwise go to Step3.

Step3. Generate a random number p_s from $u(0,1)$. If $p_s < P_1$, then CH_1 is selected. If $p_s \in [P_{i-1}, P_i)$, then CH_i is selected. Let $i = i + 1$ and go to Step2.

Crossover Operator. A parameter $P_c \in (0,1)$ of the genetic system is defined as the probability of crossover. Let us do the following process repeatedly from $i = 1$ to $pop\_size - 1$: generating a random number r_c from the interval $[0,1]$, the chromosome CH_i is selected as the parent if $r_c < P_c$. Let us illustrate the crossover operator on the chromosome CH_i. First, we generate two random numbers $n_1, n_2 \in [1, n)$ and $n_1 \neq n_2$, and then we exchange the values of ch_{n_1} and ch_{n_2}. So the new chromosome is got.

Mutation Operator. A parameter $P_m \in (0,1)$ is defined as the probability of mutation. In a similar manner to the process of selecting parent for crossover operation, we repeat the following steps from $i = 1$ to $pop\_size - 1$: generating a random number r_m from the interval $[0,1]$, the chromosome CH_i is selected as a parent for the mutation if $r_m < P_m$. We mutate the chromosome CH_i in the following way.

We randomly generate two mutation positions $n_1, n_2 \in [1, n]$ and $n_1 < n_2$, and rearrange the sequence $\{ch_{n_1}, ch_{n_1+1}, \cdots, ch_{n_2}\}$ at random to form a new sequence $\{ch'_{n_1}, ch'_{n_1+1}, \cdots, ch'_{n_2}\}$. We thus obtain a new chromosome

$$CH'_i = \{ch_1, \cdots, ch_{n_1-1}, ch'_{n_1}, \cdots, ch'_{n_2}, ch_{n_2+1}, \cdots, ch_n\}.$$

3.2 The Second Stage

The second stage is employed for confirming the quantity and route between the supply and the demand. Consequently, it is a multi-commodity flow problem (MCF) problem. The algorithm of this stage is also adopted a priority-based encoding method and the processes and operators are similar to those of the first stage. The algorithm of this stage is just utilizable for the relationship which the supply and the demand are both greater than zero.

The decoding process of solving MCF problem is stated as follows:

Step1. Let consumption $cost = 0$, path weight $length = 0$, i, O be the supply point and supply quantity, n, D be the demand point and demand quantity and $P = \phi$ be augmenting path.

Step2. If $i = n$, then go to Step5.

Step3. If the set of nodes S_i with all nodes adjacent to node i is ϕ or all the capacity of S_i is equal or lesser than zero, then output *cost* and return; otherwise, continue.

Step4. Select l from S_i with the highest priority, set $P = P \cup \{e_{il}\}, length = length + l_{il}$, $i = l$, and go back to Step2.

Step5. Let $cost = cost + flength$, where $f = min\{O, D, min\{w_{jk} | e_{jk} \in P\}\}$, and set the capacity $w_{ij} = w_{ij} - f, e_{jk} \in P$, $O = O - f$ and $D = D - f$. Output P, continue.

Step6. If $O > 0$ and $D > 0$, set i be the number of source point, n be the number of destination point and $P = \phi$, $length = 0$, then go to Step2. Otherwise, output $cost$ and return.

3.3 Global Algorithm Procedure

The global algorithm procedure of this problem is stated as follows:

Step1. Let $count = 1$, M be a large enough number, input the evolution generation G.

Step2. For $i = 1$ to pop_siz, generate the supply and demand relationship chromosomes $CH^i = (ch_1^i, ch_2^i, \cdots, ch_m^i)$, set the object value $objectvalue[i] = 0$. Set $i = 1$.

Step3. For chromosome $CH^i = (ch_1^i, ch_2^i, \cdots, ch_m^i)$, if the supply and demand are both greater than zero, call the second stage algorithm, let $objectvalue[i] = \sum_{j=1}^{m} cost_j$.

Step4. If there are both supply and demand greater than zero, set $objectvalue[i] = objectvalue[i] + M$. Let $i = i + 1$.

Step5. If $i \leq pop\_siz$, go to Step3. Otherwise, For $i = 1$ to pop_siz, arrange the chromosomes $CH^i = (ch_1^i, ch_2^i, \cdots, ch_m^i)$ by $objectvalue[i]$ in descending order. Call selection process. Let $count = count + 1$, go to Step6.

Step6. If $count > G$, return. Otherwise, for $i = 1$ to pop_siz, call crossover operator and mutation operator. Set $i = 1$, go to Step3.

4 Numerical Example

Figure 2 shows a transportation network. The quantities of supply and demand are given in Table 1. The capacities and weights of edges are shown in Table 2.

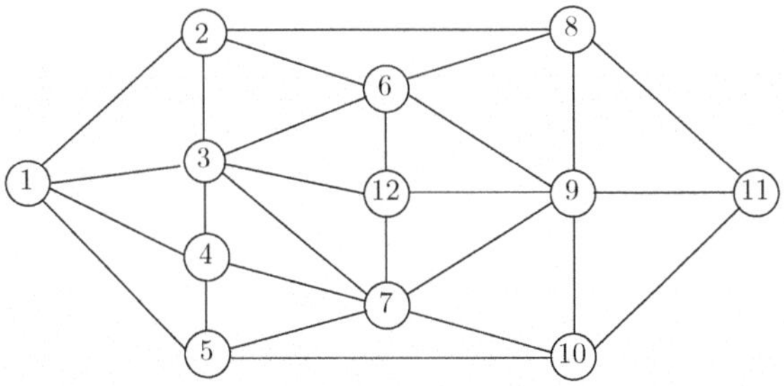

Fig. 2. Example network

We suppose that $pop\_size = 20$, $P_c = 0.6$ and $P_m = 0.2$. Let the confidence levels $\alpha_i, \beta_i, \gamma_{ji}$ be 0.9, and the evolution generation G be 1000. A run of G generations shows that the satisfactory result is given in Table 3.

Table 1. Quantities of supply and demand

Transfer direction	Quantities of supply	Demand Point	Quantities of demand
1	75,80,85	6	60,65,70
8	38,40,42	7	60,65,70
10	28,30,32	12	65,70,75
11	45,50,55		

Table 2. Edges capacity and weight

i, j	l_{ij}	$\tilde{w_{ij}}$	i, j	l_{ij}	$\tilde{w_{ij}}$	i, j	l_{ij}	$\tilde{w_{ij}}$	i, j	l_{ij}	$\tilde{w_{ij}}$	i, j	l_{ij}	$\tilde{w_{ij}}$
1, 2	10	55, 60, 65	3, 6	12	60, 70, 80	6, 2	10	50, 55, 60	8, 2	15	30, 35, 40	10, 7	15	55, 60, 65
1, 3	8	60, 70, 80	3, 7	15	55, 60, 65	6, 3	12	50, 55, 60	8, 6	8	40, 45, 50	10, 9	22	60, 70, 80
1, 4	15	50, 55, 60	3, 12	10	65, 70, 75	6, 8	8	55, 60, 65	8, 9	9	40, 45, 50	10, 11	28	60, 70, 80
1, 5	20	50, 55, 60	4, 1	15	50, 55, 60	6, 9	12	40, 45, 50	8, 11	17	55, 60, 65	11, 8	17	50, 55, 60
2, 1	10	55, 60, 65	4, 3	6	40, 45, 50	6, 12	6	30, 35, 40	9, 6	12	55, 60, 65	11, 9	10	40, 45, 50
2, 3	5	70, 80, 90	4, 5	8	55, 60, 65	7, 3	15	40, 45, 50	9, 7	15	55, 60, 65	11, 10	28	60, 70, 80
2, 6	10	55, 60, 65	4, 7	10	45, 50, 55	7, 4	10	50, 55, 60	9, 8	9	30, 35, 40	12, 3	10	40, 45, 50
2, 8	15	50, 55, 60	5, 1	20	55, 60, 65	7, 5	12	50, 55, 60	9, 10	22	60, 70, 80	12, 6	6	30, 35, 40
3, 1	8	50, 55, 60	5, 4	8	40, 45, 50	7, 9	15	40, 45, 50	9, 11	10	40, 45, 50	12, 7	10	50, 55, 60
3, 2	5	55, 60, 65	5, 7	12	60, 65, 70	7, 10	15	30, 35, 40	9, 12	8	45, 50, 55	12, 9	8	30, 35, 40
3, 4	6	55, 60, 65	5, 10	23	60, 70, 80	7, 12	10	30, 35, 40	10, 5	23	60, 70, 80			

Table 3. The satisfactory result

Path	8, 6	10, 7	1, 2, 6	1, 3, 7	1, 3, 12	11, 9, 12	11, 8, 6, 12	Total
Quantity	40.2	30.2	24.3	34.3	21.9	45.5	2.1	198.5
Cost	321.6	453	486	788.9	394.2	819	65.1	3327.8

5 Conclusions

Taken the uncertainty of freight supply and demand into account, a collaborative optimization model is formulated with transportation capacity constraint. In addition, a two-stage GA is put forward and the simulation results show that the algorithm is efficient. Moreover, this algorithm is suitable for balance and unbalance transportation on a directed network or undirected network. However, this algorithm needs to improve the operation speed when the network is large.

Acknowledgement. This work is supported by the National Natural Science Foundation of China (No. 71761024 and No. 71671079), and the Science and Technology Development Project Plan for 2019 of Lanzhou Bureau Group Company (No. KY201978). The authors wish to thank anonymous referees and the editor for their comments and suggestions.

References

1. Liang, D., Lin, B.L., Yan, H.X., Li, J.W.: Study of railway empty car distribution with substitution of empty car types. J. China Railw. Soc. **27**(4), 1–5 (2005)
2. Li, Z.P., Xia, J.F.: Time restraint-based model and algorithm for railway empty wagon distribution. J. Southwest Jiaotong Univ. **40**(3), 361–365 (2005)
3. Lei, Z.L., He, S.W., Song, R., Cai, J.: Stochastic chance-constrained model and genetic algorithm for empty car distribution in railway transportation. J. China Railw. Soc. **27**(5), 1–5 (2005)
4. Zhang, X., Zhang, Q.S.: Study on the real-coded genetic algorithm of the empty wagons flow assignment. Syst. Eng.-Theory Pract. **26**(11), 130–138 (2006)
5. Wang, D., Yan, H.F., Tan, Y.J.: Ant Colony Algorithm for the adjustment of empty refrigerated wagons between railway network nodes. China Railw. Sci. **29**(2), 131–135 (2008)
6. Cao, X.M., Wang, X.F., Lin, B.L.: Collaborative optimization model of the loaded & empty car flow routing and the empty car distribution of multiple car types. China Railw. Sci. **30**(6), 114–118 (2009)
7. Wu, K.J., Lu, H.W., Wang, T.J.: Railway empty car allocation based on adaptive mutation particle swarm algorithm. J. Lanzhou Univ. Technol. **37**(2), 102–105 (2011)
8. Jing, Y., He, S.W., Song, R., Li, H.D.: Dynamic empty car scheduling optimization based on immune clonal with cloud preference. J. Jilin Univ. **42**(1), 261–265 (2012)
9. Liu, X., Zhu, X.N.: Study on distribution of railway empty cars resource. Railw. Transp. Econ. **36**(10), 48–51, 69 (2012)
10. Zhao, C., Yang, L.X., Li, S.K.: Allocating freight empty cars in railway networks with dynamic demands. Discrete Dyn. Nat. Soc. **2014**(3), 1–12 (2014)
11. Zhang, J., Hao, Y., Wei, Y., Wang, Y.G.: The empty wagons adjustment algorithm of Chinese heavy-haul railway. Chaos, Solitons Fractals **89**, 91–99 (2016)
12. Wang, L., Ma, J.J., Lin, B.L., Li, J., Zhang, J., Ni, S.Q.: Dynamic empty car distribution for the whole rail system and calculation for empty car flow discharged over network boundaries. J. China Railw. Soc. **37**(6), 1–9 (2015)
13. Liu, S.T., Kao, C.: Solving fuzzy transportation problems based on extension principle. Eur. J. Oper. Res. **153**(3), 661–674 (2004)

Levelized Cost of Energy Optimization Method for the Dish Solar Thermal Power Generation System

Genye Dang[1], Hongsheng Su[1(✉)], and Biao Yue[2]

[1] School of Automation and Electrical Engineering, Lanzhou Jiaotong University, Lanzhou, China
2696443878@qq.com

[2] Experimental Teaching Center on Computer Science, Lanzhou Jiaotong University, Lanzhou, China

Abstract. In view of the high cost of power generation and the shortcomings of scale and industrialization of dish-Stirling optical thermal power station, the NSGA-II algorithm is proposed to optimize and analyze levelized cost of energy for dish solar thermal power generation system. The capacity and cost model of dish-Stirling thermal power generation system is established, and the influence of power station capacity, thermal storage medium and thermal storage time on the levelized cost of energy is analyzed by using NSGA-II genetic algorithm and particle swarm optimization. The results show that the larger the scale of power generation, the lower the levelized cost of energy, and the thermal storage medium and the thermal storage time have a greater impact on the levelized cost of energy.

Keywords: Dish-stirling optical thermal power generation system · Non-dominant sorting genetic algorithm II (NSGA-II) · Levelized cost of energy (LCOE) · Optimization

1 Introduction

At present, the development of fossil energy is limited and the problem of environmental pollution is prominent. Solar energy, as an inexhaustible renewable energy, is one of the important ways to solve the energy crisis. Solar thermal power generation is considered as one of the core technologies of "most likely to replace coal-fired power stations to achieve high-power generation".

Dish-Stirling optical thermal power generation system is a kind of power generation form with the highest photoelectric conversion efficiency and complete cleanliness, which uses the focusing effect of disc mirror to gather the heat of solar energy and provides high temperature and high pressure steam to drive Stirling engine through heat exchanger to generate electricity. Arora et al. [1] applied multi-objective genetic algorithm to optimize the output power and thermal efficiency of dish concentrating solar thermal power generation system. Li et al. [2] chose the optimal output power, thermal efficiency and economic performance to optimize the dish solar collector using non-dominated sorting genetic algorithm with elite strategy. Nithyanandam [3] has

Y. Liu et al. (Eds.): ICNC-FSKD 2019, AISC 1074, pp. 132–140, 2020.
https://doi.org/10.1007/978-3-030-32456-8_14

studied the cost and performance of concentrating solar thermal power generation system. Most of the above studies focus on the output power, thermal efficiency and economy of dish-Stirling optical thermal power generation system. However, there are relatively few studies on the influencing factors of the Levelized cost of energy of dish-Stirling optical thermal power generation system.

For this reason, the focus of this paper is to establish the capacity and cost model of dish-Stirling optical thermal power generation system, and to compare and analyze the influence of the capacity of power plant, the thermal storage medium and the thermal storage time on the leveling cost of power generation.

2 Mathematical Model of Dish-Stirling Optical Thermal Power Generation System

2.1 Capacity Modeling of Dish-Stirling Optical Thermal Power Generation System

Dish-stirling optical thermal power generation system mainly includes concentrator, receiver, Stirling machine, generator and other components. Among them, the concentrator concentrates the solar energy at the focal point of the rotating paraboloid; the receiver converts the focused solar energy into heat energy and transfers it to the Stirling machine, thereby converting the heat energy into electric energy.

The solar radiation heat obtained by the receiver can be expressed as:

$$Q_R = I_b A_0 E(\cos\theta_i)\rho_{ref}\tau\alpha_0\phi \tag{1}$$

Among them, I_b is the direct radiation intensity of the sun, W m^2. A_0 is the aperture area of the concentrator, m^2. E is the ratio factor of effective concentrator aperture area. ρ_{ref} is the specular reflectance of the concentrator. τ is the medium penetration between the mirror and the receiver. α_0 is the absorptivity of the receiver. ϕ is the truncation efficiency of the concentrator, which can be expressed as:

$$\phi = 1 - \exp(-1/2C)\sigma_f^2 \tag{2}$$

Among them, C is the concentrating ratio of the concentrator, $C = A_0/A_1$. A_0 is the open area of the concentrator, m^2. A_1 is the lighting port area of the receiver, m^2. σ_f^2 is the concentrator error.

There is heat loss in the heat transfer process of the receiver. The main heat loss is radiation heat loss and convection heat loss. Radiation heat loss can be expressed as:

$$Q_E = \varepsilon\sigma(T_R^4 - T_a^4)A_1 \tag{3}$$

Among them, ε is the effective emissivity of the absorber, σ is the Boltzmann constant, T_R is the inner chamber temperature of the absorber, T_a is the ambient temperature. Convective heat loss can be expressed as:

$$
\begin{aligned}
Q_C &= h(T_w - T_a)A_1 = N_u(\lambda/D)(T_w - T_a)A_1 \\
&= 0.088G_{rD}^{\frac{1}{3}}(\frac{T_w}{T_a})^{0.18}(\cos\theta)^{2.47}(\frac{d_{ap}}{D})^s \cdot (\frac{\lambda}{D})(T_w - T_a)A_1
\end{aligned}
\tag{4}
$$

Among them, G_{rD} is Grashov number based on length D, T_W is the inner cavity wall temperature of the absorber, $s = 1.12 - 0.98d_{ap}/D, d_{ap}$ is the diameter of the lighting outlet, D is the average intracavity diameter, h is the convective heat transfer coefficient of the heat absorber.

To sum up, the effective heat output of the heat absorber is as follows:

$$
\begin{aligned}
Q_T = {} & I_b A_0 E(\cos\theta_i)\rho_{ref}\tau\alpha_0\phi - \varepsilon\sigma(T_R^4 - T_a^4)A_1 \\
& - h(T_w - T_a)A_1
\end{aligned}
\tag{5}
$$

The working fluid circulation of Stirling machine is to heat the heat head of Stirling machine by the circulating heat obtained from the heat absorber, and by compressing and expanding the working fluid, the piston works, and the heat energy is converted into mechanical energy output.

An ideal analysis of Stirling machine is made. It is assumed that the working fluid is uniformly distributed in the process of circulation. By the ideal gas theorem:

$$
P = \frac{MR}{\frac{V_H}{T_H} + \frac{V_R}{T_R} + \frac{V_C}{T_C}} \tag{6}
$$

Among them, M is the quality of working fluid, R is the ideal gas constant, V_H is the volume of the expansion chamber, V_R is the regenerator volume, V_C is the volume of the compression chamber, T_H is the average temperature of the hot zone, T_C is the average temperature in the cold zone, T_R is the average temperature of the regenerator. Therefore, the cyclic work and power are respectively expressed as:

$$
W_t = \int PdV \tag{7}
$$

$$
P_t = W_t \cdot \omega \tag{8}
$$

After considering the influence of various heat losses on power loss in the process of circulation, such as heater, cooler and regenerator, and other power losses not taken into account are $0.05P_t$, the actual output power should be as follows:

$$
P_{out} = P_t - P_{HF} - P_{CF} - P_{RF} - 0.05P_t \tag{9}
$$

Among them, P_{HF} is the power loss of heater, P_{CF} is power loss of cooler, P_{RF} is power loss of regenerator.

2.2 Generation Cost Modeling of Dish-Stirling Optical Thermal Power Generation System

At present, China's dish-Stirling optical thermal power generation industry is in the demonstration stage, and relatively few power stations have been put into production.

Levelized Cost of Energy (LCOE) has been widely recognized and applied in the study of the economic characteristics of new energy power generation, and it is a common international index for evaluating power generation cost [4, 5]. Therefore, this paper uses LCOE to predict and measure the cost of dish solar power generation system.

$$LCOE = \frac{\sum_{t=1}^{n} \frac{C_t + M_t}{(1+r)^t}}{\sum_{t=1}^{n} \frac{E_t}{(1+r)^t}} \tag{10}$$

Among them, C_t is the cost of system construction. M_t is the operating cost of the system in year t. E_t is the generation capacity of the system in year t, kW/h. $E_t = P \times 8760 \times X_{function}$, 8760 is the number of hours per year, $X_{function}$ is the average generation time-length ratio, r is the discount rate, n is the system operation cycle.

$$C_t = C_{col} + C_{pow} + C_{stor} + C_{other} \tag{11}$$

Among them, C_{col} is the cost of the collector system, C_{pow} is the cost of power generation system, C_{stor} is the cost of heat storage system, C_{other} is the cost of other expenditures. The variables and their boundaries of the problem are as follows:

$$\begin{cases} 10000 \le S \le 400000 \\ 6 \le h \le 10 \\ 0 \le C_{tol} \le 80 \\ 0 \le E_{tol} \le 20 \end{cases} \tag{12}$$

3 Optimization Algorithms

NSGA-II is a multi-objective evolutionary algorithm improved on the basis of NSGA. The most prominent feature of NSGA-II is that it adopts fast non-optimal sorting and crowding mechanism [6]. NSGA-II mainly includes: crossover, mutation, fast non-dominated sorting, crowding degree sorting, truncation selection and other processes. The evolutionary process of NSGA-II algorithm is shown in Fig. 1.

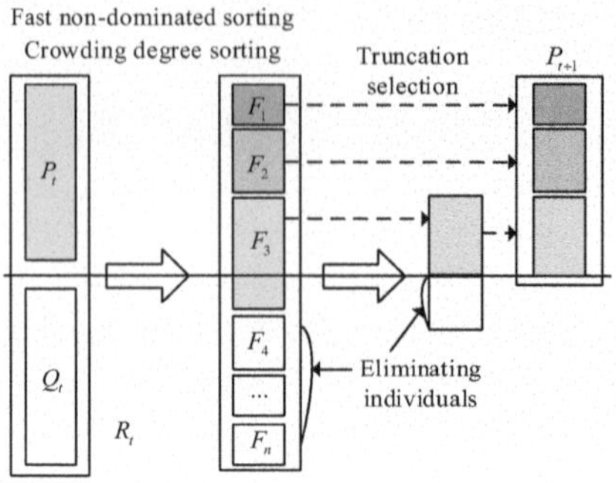

Fig. 1. The evolution process of NSGA-II algorithm

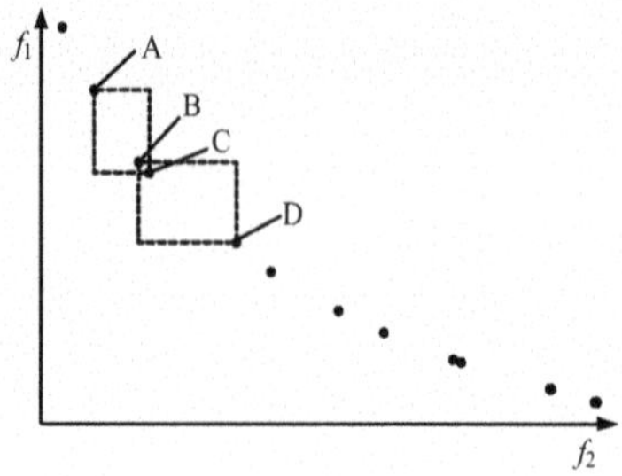

Fig. 2. The crowding distance diagram of NSGA-II

The selection process shown in the figure is as follows: Firstly, the individuals in P_t are crossed and mutated to produce offspring Q_t, and then combined with P_t and Q_t to produce R_t. Then, the non-dominant layer F is generated by fast non-dominant ranking of individuals in R_t. Finally, the non-dominant layer F is truncated and the first N individuals are selected to form P_{t+1} to enter the next generation of evolution. Since only N individuals are needed in the next generation of evolutionary pools, it is necessary to rank the crowding degree of individuals in the non-dominant layer, and those with high crowding degree rank ahead.

In NSGA-II, crowding degree is the degree of individual crowding. It is expressed by crowding distance. The crowding distance of NSGA-II is shown in Fig. 2. Points in Fig. 2 represent individuals in the population. Take point B as an example to calculate the crowding distance. Let point A and C be two points adjacent to point B, and the crowding distance at point B can be expressed as:

$$Q_{\mathrm{B}} = 2\left(\frac{\left|f_1^{\mathrm{A}} - f_1^{\mathrm{C}}\right|}{f_{1\max} - f_{1\min}} + \frac{\left|f_2^{\mathrm{A}} - f_2^{\mathrm{C}}\right|}{f_{2\max} - f_{2\min}}\right) \tag{13}$$

Among them, Q_{B} is the crowding distance at point B. f_i^X represents the objective value of objective function f_i corresponding to X point. $f_{i\max}$ and $f_{i\min}$ represent the maximum and minimum of objective function f_i, respectively.

The specific steps of NSGA-II are as follows:

(1) t = 0, taking N individuals generated randomly as initial population P_0.
(2) Crossing and mutating P_t to produce offspring population Q_t
(3) Merging P_t and Q_t to form R_t of 2 N individuals;
(4) Quick non-dominated ranking and congestion ranking are performed for P_t. The first N individuals in non-dominated layer F are truncated and selected as P_{t+1}.
(5) Repeat steps (2) to (4) with P_{t+1} instead of P_t position.
(6) If the maximum algebra is reached, stop searching, otherwise $t = t + 1$, go to step (2).

4 Simulation and Analysis

4.1 Model Parameter

Among the existing dish-Stirling optical thermal power generation systems, taking SES company's 25 kW dish-Stirling optical thermal power generation system as an example, the system parameters are shown in Table 1.

Table 1. Parameters of SES 25 kW dish-Stirling solar thermal power generation system

Parameter	Value	Parameter	Value	Parameter	Value
$A_0(m^2)$	87.7	$\tau(\%)$	90	$T_W(K)$	1056.5
$\rho_{ref}(\%)$	94	$A_1(m^2)$	0.6	$\theta(°)$	45
$I_b(Wm^{-2})$	780	$T_R(K)$	1190	$D(m)$	0.075
E	0.95	$T_a(K)$	293	$h[W \cdot (m^2 \cdot K)^{-1}]$	0.06
$\phi(\%)$	0.95	$V_H(cm^2)$	124.4	$T_C(K)$	300
ε	0.9	$V_R(cm^2)$	147.45	$T_H(K)$	1300
$d_{ap}(m)$	0.18	$V_C(cm^2)$	44.88	$\omega(r \cdot \min^{-1})$	1800
$\alpha_0(\%)$	90	$T_R(K)$	288	$M(kg)$	1.135×10^{-3}

In the simulation process, the maximum iterations number of NSGA-II algorithm is 200, the population size is 100, the crossover probability is 0.7, and the mutation probability is 0.01. The population size of PSO algorithm is 100, and the maximum number of iterations is 200.

4.2 Impact of Generation Capacity on LCOE

The total area of the concentrator mirror is selected as 10000–400000 m^2, the heat storage time is 6–10 h, the heat storage medium is molten salt, the solar multiple is 2.5, and the discount rate of funds is 8%/year. At this time, the relationship between the capacity and LCOE is shown in Fig. 3.

As can be seen from the figure, the optimization result of NSGA-II algorithm is better than that of PSO algorithm. Table 2 shows the analysis results of the relationship between LCOE and system capacity of dish-Stirling optical thermal power generation system using NSGA-II and PSO under the same conditions.

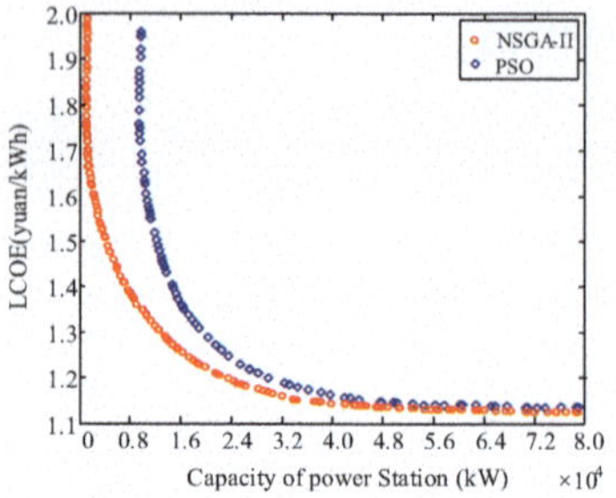

Fig. 3. The relationship between capacity of solar thermal power generation system and LCOE

Table 2. The relationship between system capacity and LCOE

Design variables				LCOE (yuan)	
Concentrator mirror area (m^2)	Heat storage time (h)	System capacity (MW)	System cost (RMB 100 million)	NSGA-II	PSO
100000	10	20	5	1.206	1.217
250000	10	50	12.5	0.964	0.983
400000	10	80	20	**0.785**	0.804

The results show that the capacity of dish-Stirling optical thermal power generation system has scale effect under the same conditions. The larger the system capacity, the lower the LCOE. When the system capacity reaches the maximum set, the LCOE optimized by NSGA-II algorithm is 0.785 yuan/kWh.

4.3 Effect of Different Heat Storage Fluids on LCOE

The concentrator mirror area of the power plant is selected as 10000–400000 m^2, the heat storage time is 6 h–10 h, the heat storage medium is heat conducting oil and molten salt, the solar multiple is 2.5, and the discount rate is 8%/year. At this time, the relationship between different heat storage medium and LCOE in dish-Stirling optical thermal power generation system is shown in Fig. 4.

In Fig. 4, the LCOE optimization results of the two algorithms under different heat storage media are compared. It can be seen that compared with heat conducting oil, molten salt can reduce the LCOE of the system. Table 3 gives the data analysis of the influence of different heat storage medium on LCOE under NSGA-II algorithm and PSO algorithm.

The results show that when the scale of dish-Stirling optical thermal power generation system is large enough, molten salt is used as heat storage medium, LCOE can be significantly reduced.

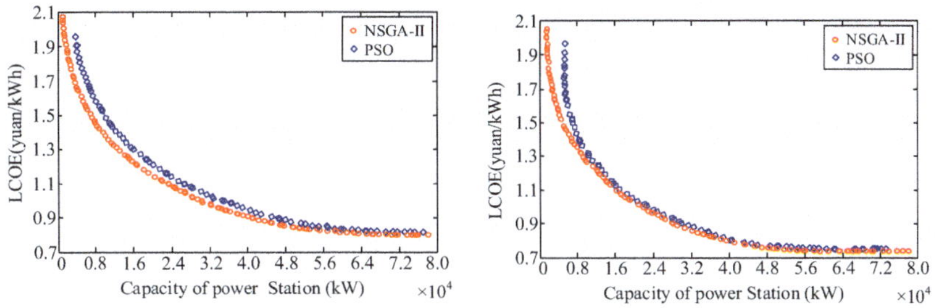

(a) Heat conducting oil is selected as the heat storage medium

(b) Molten salt is selected as the heat storage medium

Fig. 4. Effect of heat storage medium on LCOE

Table 3. Effect of different heat storage medium on LCOE

Design variables					LCOE (yuan)	
Heat storage medium	Concentrator mirror area (m^2)	Heat storage time (h)	System capacity (MW)	System cost (RMB 100 million)	NSGA-II	PSO
Heat conducting oil	375000	10	75	18.75	0.894	0.902
	275000	10	55	13.75	0.961	0.957
	125000	10	25	6.25	1.485	1.465
Molten salt	375000	10	75	18.75	**0.816**	0.884
	275000	10	55	13.75	0.923	0.938
	125000	10	25	6.25	1.440	1.429

4.4 Effect of Different Heat Storage Time on LCOE

When the installed capacity is 50 MW, the relationship between the heat storage time and LCOE of the dish-Stirling optical thermal power generation system is shown in Fig. 5.

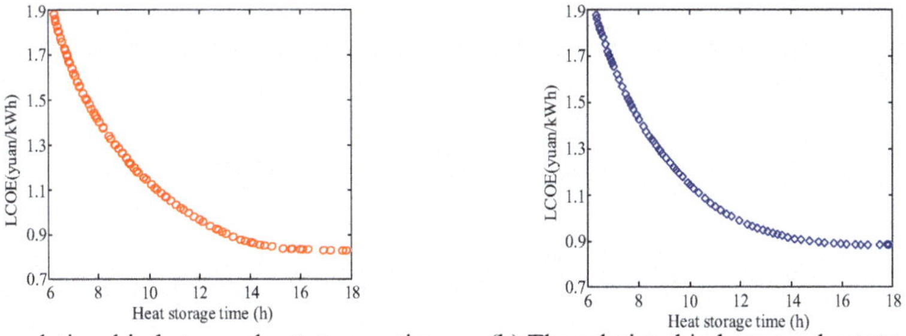

(a) The relationship between heat storage time and LCOE under NSGA-II algorithm

(b) The relationship between heat storage time and LCOE under PSO algorithm

Fig. 5. Effect of heat storage time on LCOE

As can be seen from the figure, the trend of LCOE change is the same under the two algorithms. The LCOE of dish-Stirling optical thermal power generation system decreases with the increase of heat storage time.

5 Conclusion

In this paper, by introducing the composition and energy transfer process of dish-Stirling optical thermal power generation system, the influence on LCOE is analyzed from different angles. The following conclusions can be drawn: (1) Dish-Stirling optical thermal power generation system has scale effect. The larger the scale, the lower the LCOE. When the system capacity reaches the maximum set, the LCOE optimized by NSGA-II algorithm is 0.785 yuan/kWh. (2) The LCOE of dish-Stirling optical thermal power generation system can be significantly reduced when the thermal storage medium is molten salt. (3) When the installation capacity is 50 MW, the LCOE will decrease with the increase of heat storage time.

Acknowledgment. The authors would like to thank the National Natural Science Foundation of China (61867003, 61263004 and 51867012) for financial support.

References

1. Arora, R., Kaushik, S.C., Kumar, R.: Multi-objective optimization of solar powered ericsson cycle using genetic algorithm and fuzzy decision making. In: 2015 International Conference on Advances in Computer Engineering and Applications, Ghaziabad, pp. 553–558. IEEE (2015)
2. Li, Y., Liao, S., Liu, G.: Thermo-economic multi-objective optimization for a solar-dish Brayton system using NSGA-II and decision making. Int. J. Electr. Power Energy Syst. **64** (64), 167–175 (2015)
3. Nithyanandam, K., Pitchumani, R.: Cost and performance analysis of concentrating solar power systems with integrated latent thermal energy storage. Energy **64**(1), 793–810 (2014)
4. Su, J.: Economic analysis and development policies research on concentrating solar power industry, pp. 17–26. North China Electric Power University, Beijing (2017)
5. Liang, Y., Zhu, Y., Wang, X.: Optimal configuration of micro-grid power supply based on levelized cost of electricity analysis. South. Power Syst. Technol. **10**(2), 56–61 (2016)
6. Wang, R., Xu, H., Guo, J.: Adaptive non-dominated sorting genetic algorithm. Control Decis. **33**(12), 82–87 (2018)

A Novel Ant Colony Optimization Algorithm with Dynamic Control Population for Community Detecting

Jianjun Chen[1], Shupeng Gao[2], Zhen Su[1], Siqi Chen[3], and Xianghua Li[1](✉)

[1] College of Computer and Information Science, Southwest University, Chongqing 400715, China
li_xianghua@163.com

[2] School of Mechanical Engineering, Northwestern Polytechnical University, Shanxi 710072, China

[3] College of Intelligence and Computing, Tianjin University, Tianjin 300072, China

Abstract. Nowadays, complex networks have driven great interests of scholars. As a special characteristic of a network, the community structure has wide research prospects. Many current algorithms are adopted for detecting the potential community structure, in which the ant colony algorithm is a typical one. However, the computational cost of the ant colony is too high which limits its performance. In this paper, we propose a novel ant colony optimization algorithm with dynamic control population. In the proposed algorithm, when a certain condition is reached, the number of ants starts to decrease based on the proposed rules. The efficiency of the proposed algorithm is estimated through comparing with the classical ant colony algorithm in real-world networks. Experiments show that the proposed algorithm has apparently lower computational cost, while the quality of the division is reserved relatively.

Keywords: Community mining · Ant colony optimization · Dynamic control population

1 Introduction

Nowadays, complex system comes into the eyes of many researchers, and many of the complex systems in the real world are in the form of complex networks or can be transformed into complex networks [1]. One of the intrinsic properties of the real complex network is the community structure, which shows the inherent organization of individuals [2–4]. The community structure has different connotations in different applications. For example, communities in social networks represent people with similar characteristics and functional groups in biological networks reveal biological organization modules with similar functions [5–7].

S. Gao—Contributed equally to this work and should be considered as co-first authors.

Y. Liu et al. (Eds.): ICNC-FSKD 2019, AISC 1074, pp. 141–148, 2020.
https://doi.org/10.1007/978-3-030-32456-8_15

Community mining aims to detect and reveal the inherent community structure of these different types of complex networks, which helps people understand the functions of complex networks, discover hidden rules in complex networks, and predict complex network behavior [3,4].

A series of algorithms have been proposed for mining the communities, such as Hierarchical algorithms, clustering-based algorithms, density-based, and grid-based clustering algorithms [8,9]. The modularity measure Q has also been proposed to estimate the qualities of community divisions [10]. The optimization method of the function Q as the objective function has also become one of the mainstream methods. In the field of evolutionary algorithms, the ant colony optimization (ACO) algorithm is effective in mining the community structure [9,11,12,14–16] and solving NP-hard problems [13,17]. However, the high time complexity limits its performance. The population size of ant has a great influence on the global search ability and computational cost of the algorithm. When the population size is large, the search space grows up, the probability of finding the optimal solution rise, but the running time increases simultaneously.

For the purpose of reducing the computational cost of the algorithm, a novel optimization algorithm that dynamically adjusts the population size of ant colony is proposed in this paper. The main contributions of this paper are as follows. The framework of dynamic control of population size is given, and we propose a method of deleting the number based on the ACOC algorithm (D-ACOC). The results of experiments in real-world networks verify that the proposed algorithm has a better performance in the computational cost.

The structure of this paper is organized as follows. Section 2 introduces the ant colony optimization algorithm for detecting the community structure. Then a novel ant colony optimization algorithm with dynamic control population is proposed in Sect. 3. Section 4 presents the results of experiments based on the four real-world networks. In the end, this paper comes to a conclusion in Sect. 5.

2 Related Work

2.1 Community Mining

The aim of the community mining is to divide the vertices in networks into communities, in which the vertices in different communities are in sparse connection, and the vertices in the same community are connected densely [2,18]. The modularity (Q) and the normalized mutual information index (NMI) are employed to estimate the quality of the detected community structure.

A network can be represented as $G = (V, E)$, where V stands for the set of vertices and E represents the edge set. And a community is a collection of vertices that have a common certain characteristic. The number of community is denoted as NC, and the result of the community mining is a set of communities, denoted as $C = \{C_1, C_2, C_3, ..., C_{NC}\}$.

Modularity Measure Q: A symmetric matrix $E = (e_{ij})$ is KK-dimension, where an element e_{ij} represents the proportion of edges of nodes connecting two different communities in a network in all edges. Two nodes are located in the ith community and the jth community. Let the sum of elements of diagonal in the matrix be $Tre = \sum e_{ij}$. The sum of elements in each row (or column) is defined as $a_i = \sum e_{ij}$. It represents the proportion of edges connected to nodes in the ith community in all edges. Therefore, the modularity of a network is given by:

$$Q = \sum e_{ij} - {a_i}^2 = Tre - ||e^2|| \tag{1}$$

Normalized Mutual Information (NMI): In some real-world networks, the structure is known. And *NMI* [19] measures the similarity between the real division and the detected division resulted by our algorithm. Suppose there are two divisions S and D, where S is the standard division and D is the detected division. The confusion matrix of S and D is denoted as Co, in which Co_{ij} represents the number of the vertices that not only belongs to the d_i of D but also belongs to the s_i of S. The $NMI(S,D)$ is defined in Eq. (2), where N indicates the number of nodes in the network, and Co_i and Co_j represent the sums of elements in row i and column j, respectively.

$$NMI(S,D) = \frac{-2\sum_{i=1}^{P_S}\sum_{j=1}^{P_D} Co_{ij} * \log(\frac{Co_{ij}*N}{C_i*C_j})}{\sum_{i=1}^{P_S} Co_i * \log(\frac{Co_i}{N}) + \sum_{i=1}^{P_D} Co_j * \log(\frac{Co_j}{N})} \tag{2}$$

2.2 Optimization Algorithms of Ant Colony Algorithms for Community Mining

The ant colony algorithm was proposed by Marco Dorigo, inspired by the behavior of the ants in the process of seeking food. It is a simulation evolutionary algorithm which can also be applied to solve the community mining problem [18]. The main parts of the optimization of the ant colony algorithm are as follows.

State Transfer Function: In order to help each ant find a community division, the transfer probability matrix is defined as:

$$P(i, c_j) = \frac{\eta(i, c_j)^{\beta} Tau(i, c_j)^{\alpha}}{\sum_{k=1}^{NC} \eta(i, c_k) Tau(i, c_k)^{\alpha}} \tag{3}$$

where $P(i, c_j)$ represents the probability of vertex i belonging to a community C_j, and c_j is the label of community C_j. On the basis of the qualities of the solutions, $Tau(i, c_j)^{\alpha}$ is the pheromone matrix which is updated in each iteration by the ants. $\eta(i, c_j)$ is the heuristic factor in which n_{i,c_j} indicates the number of edges connecting the vertex with the vertices in community C_j.

When the number of edges connecting a certain node in a certain community is large, the coupling degree will get strong. Therefore, that node is more likely to belong to the community. The heuristic factor is defined as Eq. (4):

$$\eta(i, c_j) = \frac{n_{i,c_j}}{\sum_{k=1}^{NC} n_{i,c_k}} \tag{4}$$

Pheromone Matrix Updating Strategy: When an ant constructs a solution, the ant updates the pheromone matrix locally according to its solution, and the local update rule is shown in the Eq. (5) where ρ indicates the volatile coefficient of pheromone, and Q is the modularity measure found by the ant.

$$Tau(i, c_j) = (1 - \rho) * Tau(i, c_j) + 2\rho * Q * \varphi(i, c_j) \tag{5}$$

When all ants finish their searching, in order to make the excellent division have positive feedback to the pheromone matrix, a global update strategy has been proposed which is shown in Eq. (6).

$$Tau(i, c_j) = (1 - \rho) * Tau(i, c_j) + Q_{top} * \varphi(i, c_j) \tag{6}$$

where Q_{top} is the maximum value of all the solutions. In Eqs. (5) and (6), if the vertex i belongs to the community C_j, $\varphi(i, c_j) = 1$; and if not, $\varphi(i, c_j) = 0$. After the whole process of searching the structure of a network, we will select the division with the highest Q value as the best solution.

Mutation Strategy: For finding the global optimal solution faster, the mutation strategy is proposed which avoids the uniqueness of the solution. The probability of mutation is defined as P_m which is set to be a constant in this paper. All solutions should be sorted in descending order based on the corresponding value of Q, and we construct the mutation to the solutions with top $Local'$ values. In the mutation process of a solution, every node will have a random floating point number ran between 0 and 1. If $ran < P_m$, then the corresponding node will mutate. The result of mutation will be accepted if and only if the mutation improves the value of Q. An example is shown in Fig. 1:

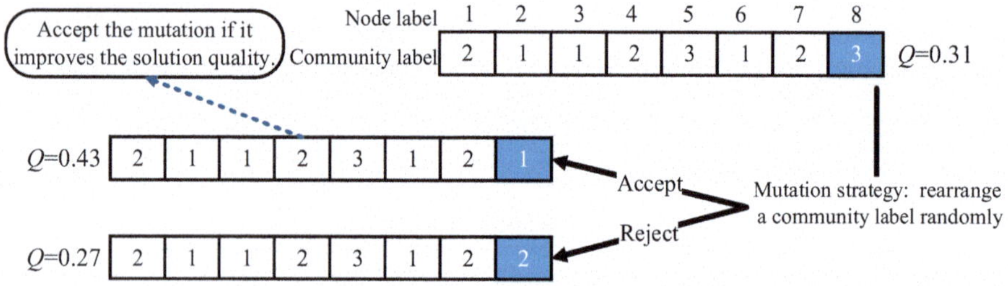

Fig. 1. An example of the mutation process

3 Dynamic Population Control for Ant Colony Optimization

3.1 The Rule Trigging the Delete Operation

We can dynamically adjust the population size of the ant colony to reduce the cost of time, but what is the time point can we construct the delete operation? In the process of iterations, the *Q(or NMI)* value increases continuously. When the *Q(or NMI)* grows slowly, the individual continuous K generation is not updated, and the number of ants $M_m > C_{\min}$ (the minimum value of the number of ants). It means that we do not need so many ants to search for the optimal solution, and it is the time we can construct the delete operation.

The value of K is adjusted according to the actual problem to solve. The delete operation will be difficult to carry out if the value of K is too large, but when the value of K is too small, the number of ants will reduce rapidly. In this paper, we let K be 10.

3.2 Deleting Rule

If the optimal solution is not updated in continuous K generations and $M_m > C_{\min}$, the delete operation should be constructed. At the first step, we should calculate how many ants in this iteration have the same solution as the ants in the last iteration. Then we will calculate the slope of the growth of *NMI(Q)*.

The specific delete operation is shown in Fig. 2(a), where M_m indicates the current number of ants. $C_{\min}$ represents the minimum value of the number of ants. And *kill* denotes the slope of the growth of *NMI*. k_1 and k_2 are the boundaries of the value of *kill*. An example is shown in Fig. 2(b) to illustrate the calculation of the slope.

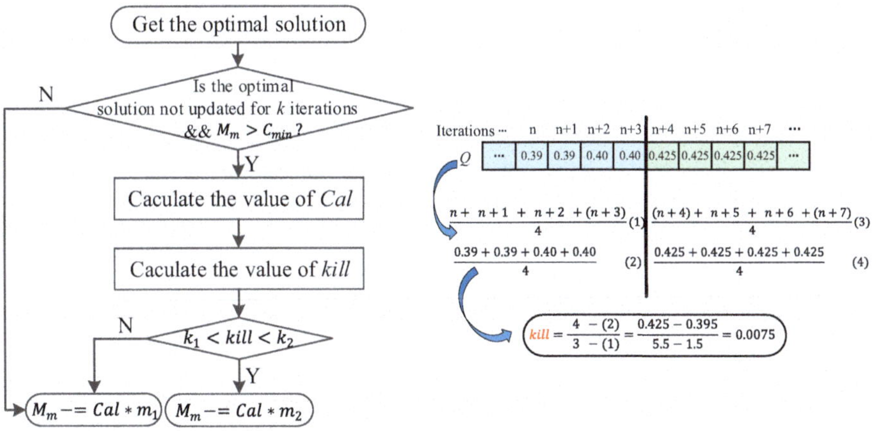

(a) The process of the delete operation. (b) The process to calculate the slope.

Fig. 2. An illustration of the delete rule.

4 Experiments

4.1 Datasets

The classical ant colony clustering algorithms and four real-world networks constructed by Newman [2] are used to verify the superiority of the proposed algorithm. The basic information of four networks is shown in Table 1. And the real divisions of four real-world networks are also adopted to compute the value of *NMI*. All experiments are constructed in the same environment and are averaged over 20 independent runs.

Table 1. The basic information of four real-world networks. k denotes the average degree of networks, C indicates the clustering coefficient of the network, and NC represents the number of communities in real networks.

Dataset	V	E	k	C	NC
Dolphins	62	160	5.129	0.303	2
Karate	34	78	4.588	0.588	2
Polbooks	105	441	8.400	0.488	3
Football	115	613	10.660	0.403	12

4.2 Experiment Results

All experiments are repeated 20 times and constructed in the same environment. The results of four real-world networks based on objective function *NMI* are shown in Fig. 3. With the growing of the number of ants, the differences come out between the original and proposed algorithms. D-ACOC has a lower average

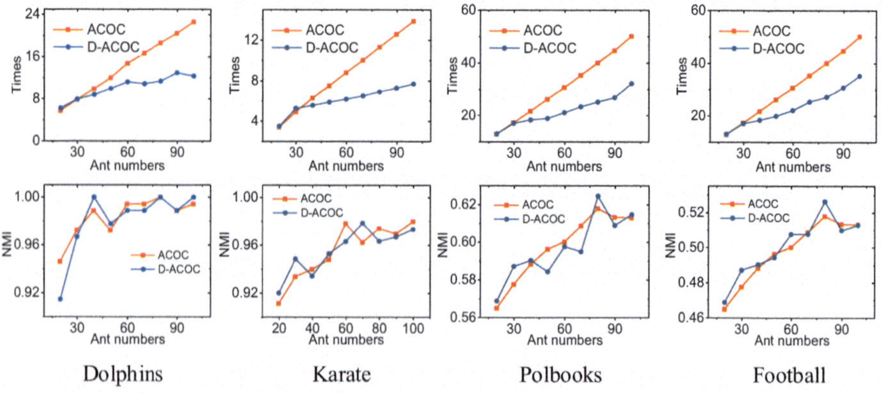

Fig. 3. Experiment results of four real-world networks in terms of *NMI* function.

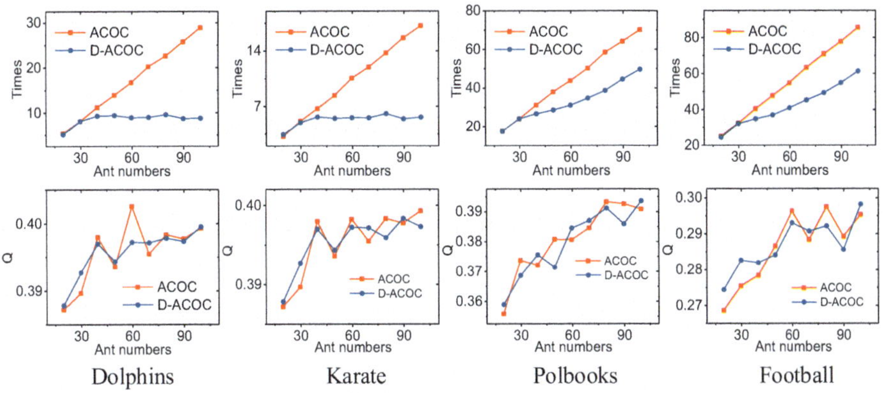

Fig. 4. Experiment results of four real-world networks in terms of Q function.

running time and the *NMI* of ACOC and D-ACOC are almost the same, which means the proposed algorithm can reduce the computational cost and does not affect the quality of result noticeably.

In order to further verifying the novelty of the proposed algorithm, another objective function modularity measure Q is also applied. The experiment results of four real-world networks based on objective function Q are shown in Fig. 4. In terms of the running time, D-ACOC has obvious superiority, and the values of Q of ACOC and D-ACOC are almost the same.

Experiments have verified that, in terms of Q and *NMI*, the proposed algorithm both have intelligent performance that the proposed algorithm can reduce the computational cost apparently and do not affect the quality of the solution.

5 Conclusion

A novel ant colony optimization algorithm with dynamic control population has been proposed in this paper. In the proposed algorithm, when the optimal solution is not updated for K iterations, the delete operation will be implemented by the specific rules. This algorithm can delete unnecessary ants to reduce the running time. Experiments on four real-world networks show the better performance of the proposed algorithm in terms of computational cost. More specifically, it can reduce the computational cost apparently, and the analysis of the solution shows that the proposed algorithm does not noticeably affect the quality of the division of the community.

Acknowledgment. This work was supported by National Natural Science Foundation of China (No. 61602391), Natural Science Foundation of Chongqing (No. cstc2018jcyjAX0274), and in part by the National Training Programs of Innovation and Entrepreneurship for Undergraduates.

References

1. Newman, M.E.J.: The structure and function of complex networks. Soc. Ind. Appl. Math. **45**(2), 167–256 (2003)
2. Fortunato, S.: Community detection in graphs. Phys. Rep. **486**(3–5), 75–174 (2010)
3. Weng, L.L., Menczer, F., Ahn, Y.Y.: Virality prediction and community structure in social networks. Sci. Rep. **3**, 2522 (2013)
4. Karrer, B., Newman, M.E.J.: Stochastic blockmodels and community structure in networks. Phys. Rev. E **83**(1), 016107 (2011)
5. Ruan, J.H., Zhang, W.X.: An efficient spectral algorithm for network community discovery and its applications to biological and social networks. In: 7th IEEE International Conference on Data Mining, pp. 643–648 (2007)
6. Nakagaki, T., Yamada, H., Tóth, Á.: Intelligence: maze-solving by an amoeboid organism. Nature **407**(6803), 470–470 (2000)
7. Tero, A., Takagi, S., Saigusa, T., Ito, K., Bebber, D.P., Fricker, M.D., Yumiki, K., Kobayashi, R., Nakagaki, T.: Rules for biologically inspired adaptive network design. Science **327**(5964), 439–442 (2010)
8. Gong, M.G., Cai, Q., Chen, X.W., Ma, L.J.: Complex network clustering by multi objective discrete particle swarm optimization based on decomposition. IEEE Trans. Evol. Comput. **18**(1), 82–97 (2014)
9. Mandala, S.R., Kumara, S.R., Rao, C.R., Albert, R.: Clustering social networks using ant colony optimization. Oper. Res. Int. J. **13**(1), 47–65 (2013)
10. Newman, M.E.J.: Modularity and community structure in networks. Proc. Natl. Acad. Sci. **103**(23), 8577–8582 (2006)
11. Gao, C., Liang, M.X., Li, X.H., Zhang, Z.L., Wang, Z., Zhou, Z.L.: Network community detection based on the Physarum-inspired computational framework. IEEE/ACM Trans. Comput. Biol. Bioinf. **15**(6), 1916–1928 (2018)
12. Mohan, B.C., Baskaran, R.: A survey: ant colony optimization based recent research and implementation on several engineering domain. Expert Syst. Appl. **39**(4), 4618–4627 (2012)
13. Liu, Y.X., Gao, C., Zhang, Z.L., Lu, Y.Q., Chen, S., Liang, M.X., Tao, L.: Solving NP-hard problems with physarum-based ant colony system. IEEE/ACM Trans. Comput. Biol. Bioinf. **14**(1), 108–120 (2017)
14. Jin, D., Liu, D.Y., Yang, B., Liu, J., He, D.X.: Ant colony optimization with a new random walk model for community detection in complex networks. Adv. Complex Syst. **14**(05), 795–815 (2011)
15. Shelokar, P.S., Jayaraman, V.K., Kulkarni, B.D.: An ant colony approach for clustering. Anal. Chim. Acta **509**(2), 187–195 (2004)
16. Mu, C.H., Zhang, J., Jiao, L.C.: An intelligent ant colony optimization for community detection in complex networks. In: 2014 IEEE Congress on Evolutionary Computation, pp. 700–706 (2014)
17. Gao, C., Chen, S., Li, X.H., Huang, J.J., Zhang, Z.L.: A physarum-inspired optimization algorithm for load-shedding problem. Appl. Soft Comput. **61**, 239–255 (2017)
18. Newman, M.E.J.: Fast algorithm for detecting community structure in networks. Phys. Rev. E **69**(6), 066133 (2004)
19. Jin, D., Chen, Z., He, D.X., Zhang, W.X.: Modeling with node degree preservation can accurately find communities. In: The 29th AAAI Conference on Artificial Intelligence, pp. 160–167 (2015)

A Hybrid Bat Algorithm Based on Combined Semantic Measures for Word Sense Disambiguation

Xu Ying[1(✉)] and Aws Hamed Hamad[1,2]

[1] College of Computer Science and Electronic Engineering, Hunan University, Changsha 410082, Hunan, China
{hnxy,mario}@hnu.edu.cn
[2] Ministry of Higher Education and Scientific Research, Baghdad, Iraq

Abstract. The task of assigning an appropriate sense for an ambiguous word depending on its context is referred to as Words Sense Disambiguation (WSD). The objective of WSD is to attain an improved accuracy for real world applications, such as in information extraction, automatic summarisation, or machine translation. The WSD is solved through the implementation of a computational intelligence approach known as the Bat Algorithm (BA). The BA has the potential to explore an expansive area of the search space as it is a population-based algorithm, which makes it considerably efficient in the diversification process. To further improve the search, a local search algorithm referred to as Hill Climbing (HC) is applied that balances the exploration as well as the exploitation aspects. The suggested algorithm has the ability to optimise the semantic value of the words from the inputted text. In this study, the semantic measure depends on the Leacock and Chodorow (LCH) algorithms and the extended Lesk (eLesk). The recommended algorithm is tested based on certain benchmark datasets. According to the experimental results, it is found that our algorithm can derive better quality performance than that of other relevant algorithms. It is, therefore, concluded that the method suggested in this study provides an effectual solution for the WSD problem.

Keywords: Word sense disambiguation · Meta-heuristics · Hybrid methods · Bat algorithm · Local search · WordNet

1 Introduction

In Natural Language Processing (NLP), semantic resemblance is a subject of importance due to its extensive applications. Also, it is an important area of research in the field of Artificial Intelligence and Cognitive Science. The different applications with respect to semantic similarity include Malapropism Detection, Information Retrieval, and Word Sense Disambiguation (WSD).

Y. Liu et al. (Eds.): ICNC-FSKD 2019, AISC 1074, pp. 149–157, 2020.
https://doi.org/10.1007/978-3-030-32456-8_16

A word in natural language is generally attributed with more than one sense. WSD has already undergone a significant amount of research. WSD is an open problem in ontology and natural language processing. When a word has different meanings, WSD ascertains the sense (that is, the meaning) of a word based on its use in a sentence. It is regarded as an AI-complete issue, in other words, it is a task whose difficulty level of resolution is at par with the most complex problems in artificial intelligence for example, a problem of symbolising dialectics about arbitrary real-world knowledge [10] or a problem which can be separated only by first resolving all the complex problems in AI, such as representing encyclopaedic knowledge or denoting prevalent sense [9].

In WSD, when bag words are processed for disambiguation, search methods are employed for knowledge-based approach. The heuristic information is employed by guided search method to accurately look out for problem space. For solving WSD, genetic algorithm (GA) had been frequently invested [2,3,6,17]. In these methods, the relatedness among the words in sentences can be maximised with GA. In the method of [8], the Lesk algorithm defines the relatedness measure, which has been derived by considering the gloss overlap concept. In the same manner, to maximise the relatedness as measured with the Lesk algorithm, exploitation of the ant colony system was done [12]. In [1], WSD is solved by employing the Harmony search algorithm. As a parsing approach, the study in [1] has employed Stanford Parser to gain additional information pertaining to grammatical relation. In [15], for solving WSD, the particle swarm optimisation (PSO) has been employed. The proposed algorithm in this study was examined based on a three-benchmark dataset (SensEval-2, SemCor 3.0 and SensEval-3).

In this paper, a precise evaluator has been put forward to estimate the semantic value pertaining to the target senses. This evaluator signifies the essence part pertaining to the put forward Bat Algorithm (BA), which is employed as the algorithms fitness function. Also, the local search algorithm has been included to enrich the search process pertaining to BA. The problem space can be searched effectively by integrating the two meta-heuristic approaches, BA and Hill Climbing (HC) algorithms, by balancing of the search in a wider area pertaining to the problem space through BA as well as around local area via HC. For all parts of speech (POSs), we also put forward a novel combination pertaining to the LCH method and extended Lesk (e-Lesk) algorithm.

The rest of the paper is structured as follows. Section 2 is devoted to introducing our proposed WSD system in detail. Experimental environments and outcomes are outlined in Sect. 3. The conclusion and recommendations are discussed in Sect. 4.

2 Proposed Method

This section describes the proposed the hybrid BA and HC algorithms that include three main components. The first one is reading the data and assign the sense count for each word using the WordNet database. The second component is to signify the semantic similarity as well as relatedness for each pair of words.

The last component is to search best combination of senses pertaining to the processed text. Figure 1 presents the framework pertaining to our put forward hybrid HC and BA algorithm.

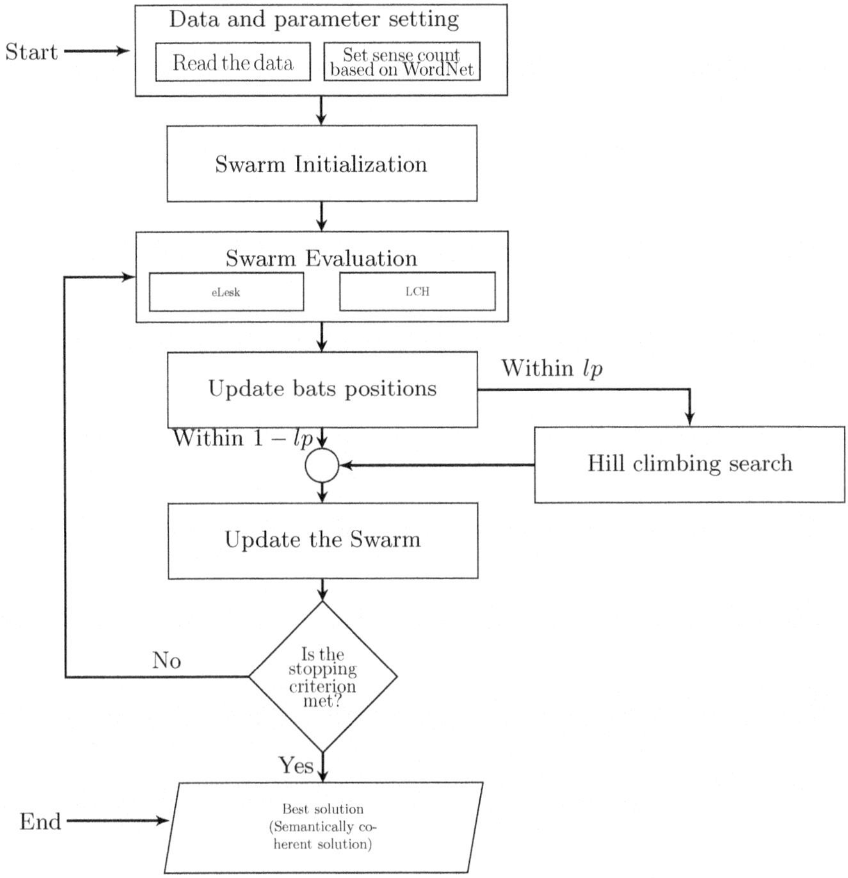

Fig. 1. Hybrid BA and HC algorithm.

Two kinds of meta-heuristic search algorithms are employed in the put forward method, wherein each type is defined with a special search capacity. BA [16] has the ability to explore the problem space of WSD effectively. Recently, BA has received intensive attention due to its successful application in different domains [7,13,14]. This algorithm represents the words in the given text as numbers which corresponds to their sense number. Thereafter, the search process is controlled by the frequency and the velocity which are the parameters of the BA.

In BA, every bat is associated with a velocity denoted by v_i^t and a location denoted by x_i^t, at iteration t. Amongst all the bats, there is a current best solution denoted by x^*. Thus, the abovementioned principles can be converted into updating equations that produce the best solution for the search procedure. These equations can be mapped as shown below:

$$f_i = f_{min} + (f_{max} - f_{min})\beta \tag{1}$$

$$v_i^t = v_i^{t-1} + (x_i^t - x^*) f_i \tag{2}$$

$$x_i^t = x_i^{t-1} + v_i^t \tag{3}$$

Where β is a uniformly distributed random number in the interval [0, 1], and x^* is the current global best solution.

The hybrid bat algorithm begins the search process by creating a set of solutions in a random manner, and initialising its parameters as depicted in Fig. 1. The set of solutions is assessed in the swarm evaluation step by gauging their relatedness and semantic similarity, based on Eq. 4. There are several iterations in the search and the solutions are shifted to a new problem space state by using Eq. 3 in every iteration. The new BA solution is generated by adding a special vector known as velocity to the prior solution. Subsequently, the ranking of the solutions is done on the basis of their fitness value, from the better (higher semantic value) to the worst (lowest). In BA, the best solution is tagged as the global best, and it is utilised for updating the value of the velocity of every solution as depicted in Eq. 2. The best solution is at all times in the initial swarm position. The global best is forwarded to the procedure of local search within a particular rate. In our study, hill climbing (HC) functions as the local search process and it attempts to better the universal best where it works on adjusting the solution repeatedly to find the optimum neighbour for the given solution. The hill climbing is used in every iteration within predefined probability called local search probability lp. This practice would be recurred for several iterations, and the ultimate output would be the global best solution.

$$f(x) = \sum_{i=1}^{n} Relatedness(C_i, C_j) + Similarity(C_i, C_j) \tag{4}$$

Where n is the number of variables of the solution x, and w is the size of the window of words. This window is considered the context of the ambiguous word. The relatedness is represented by the extended lesk algorithm [5], and the similarity measure is done by an improved Leacock and Chodorow (LCH) measure [4].

3 Experimental Results

The proposed method in this study is evaluated based on three available English WSD datasets, namely, SemCor, SensEval-2, SensEval-3. The evaluation is based on the following metrics:

1. **Coverage(C):** The ratio of all answered senses to the total number of senses.

2. **Precision (P):** The ratio of correctly answered senses to the total possible answered senses.
3. **Recall(R):** The ratio of correctly answered senses to the total number of senses.
4. **F-measure:** The harmonic average between the precision and the recall calculated:

In this research, a couple of semantic measures have been employed in the suggested hybrid technique. First of all, we employed LCH that served as a fitness function for hybrid BA, whereas LCH quantifies the resemblance for verbs and nouns only. Secondly, e-Lesk technique was used for assessing solutions. Lastly, e-Lesk and enhanced LCH are combined and serve as a fitness function for the suggested technique. The size of window is set as 11 because of its success in the earlier studies [1,15]. Also, selecting a bigger size of window may have an adverse effect by adding noisy attributes [11]. Thus, each one of the experiments carried out in this research has an 11 words window size. Table 1 shows the outcomes of POS where the similarity is quantified by the LCH technique. In our recommended technique, the LCH procedure of measuring semantic similarity acquired F-measures of (43.05%) and (72.39%) for verbs and nouns, respec-tively. As the LCH procedure is unable to quantify similarity for each POS as given in Table 1 (only verbs and nouns were detected), the proposed approach of this re-search used the adapted Lesk technique along with the LCH method. In Table 2 we present the result of the suggested disambiguation technique by using the e-Lesk technique to quantify the similarity among each conceptual pair. As the LCH technique only quantifies the similarity for verbs and nouns, these measures are integrated to carry out hybrid BAs fitness function. In Fig. 2, we include a comparison among hybrid BA and traditional BA that is based on an integrated measure which is a combination of LCH and e-Lesk techniques. Figure 2 demonstrates that hybrid BA technique got improved results in comparison to the conventional BA technique on the three used datasets.

Table 1. Hybrid bat algorithm based on improved LCH on SEMCOR dataset

POS	Coverage	Precision	Recall	F-measure
Noun	100%	72.39%	72.39%	72.39%
Verb	79.92%	48.45%	38.72%	43.05%

To demonstrate the efficiency of our suggested technique, we make a comparison of the outcomes of hybrid BA against other meta-heuristic techniques for the WSD issue in Figs. 3 and 4, together with a related discussion and assessment according to all POSs.

Figures 3 and 2a compare the suggested technique with GAs, GA new version, that is, self-adaptive GA (SAGA) [3], PSOs meta-heuristic approach [15], WSD as a traveling sales man problem [12], as well as the suggested techniques

Table 2. Hybrid bat algorithm based on e-Lesk relatedness method on SEMCOR dataset

POS	Coverage	Precision	Recall	F-measure
Noun	100%	71.57%	71.57%	71.57%
Verb	79.92%	47.56%	38.02%	42.25%
Adjective	100%	73.81%	73.81%	73.81%
Adverb	100%	63.15%	63.15%	63.15%
All	94.33%	66.14%	62.31%	64.17%

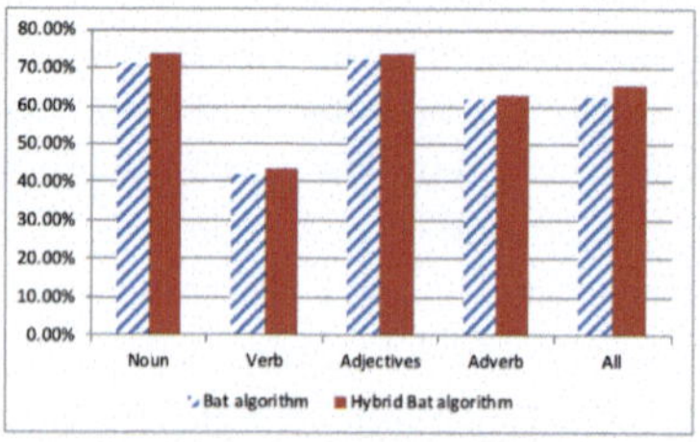

(a) Comparison of F-measure values on SEMCOR dataset

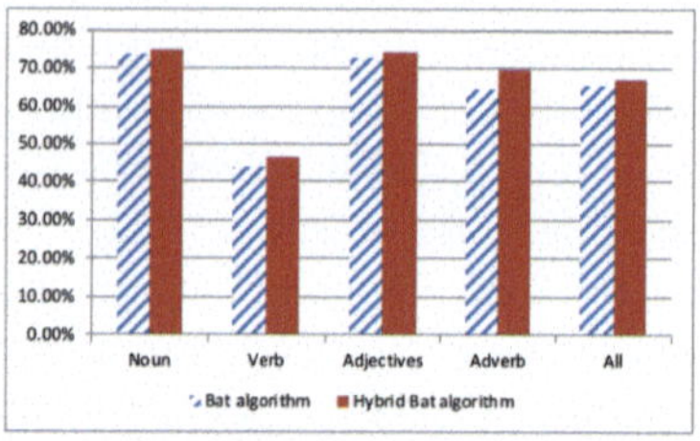

(b) Comparison of F-measure values on SensEval2 dataset

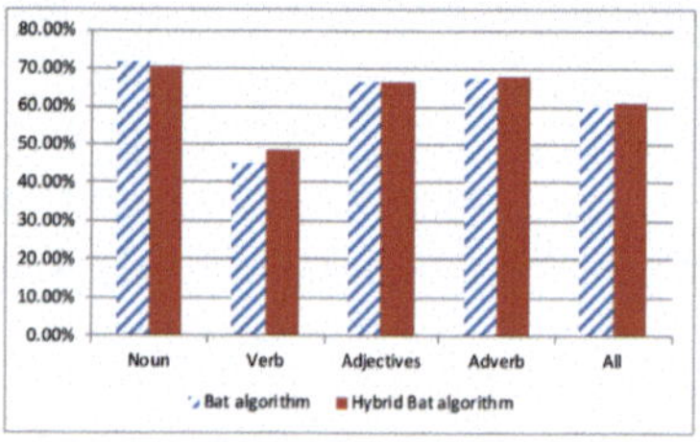

(c) Comparison of F-measure values on SensEval3 dataset

Fig. 2. The experimental comparison between the hybrid bat algorithm and the basic bat algorithm.

performance in relation to the HSA [2] (harmony search algorithm) on SensEval2 and SensEval3. In the case of SensE-val2, F-measure values acquired by using self-adaptive GA and HSA are 51.49% and 60.69%, respectively. The hybrid PSO obtained 65.83% for F-measure of verbs, whereas the hybrid bat algorithm obtained 67.39% for F-measure, which is the highest. The hybrid BA outshines all other techniques when each of the POSs is considered. Figure 4 compares hybrid BA with other techniques for each of the POSs.

In case of the SensEval3 data set, the values of F-measure are 52.95% and 46.15% for GA and self-adaptive GA, in that order. The hybrid PSO obtained 60.84% for the verbs F-measure. The hybrid BA obtained 61.27% for F-measure, which outshines the other techniques when each of the POSs is considered.

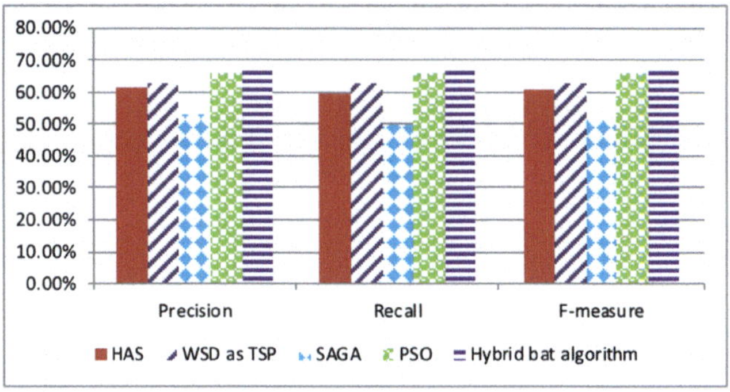

Fig. 3. Comparison of meta-heuristic algorithms based on SensEval2 dataset

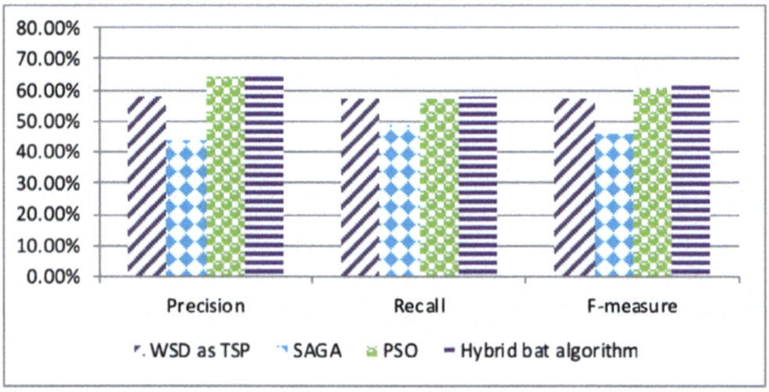

Fig. 4. Comparison of meta-heuristic algorithms based on SensEval3 dataset

Experimental outcomes show that the hybrid BA presented in this study acquired the best outcomes with respect to solution quality in comparison to the other 3 techniques given in the literature.

4 Conclusion

This study offered a computational technique to solve the issue of word ambiguity. The objective of WSD (Word Sense Disambiguation) is to search for the appropriate meaning of the word (the ambiguous word) based on context. To solve this use, we suggested a hybrid meta-heuristic technique which combines BA with hill climbing for the first time. Firstly, this study applies the BA to solve issue of WSD, then it applies the HC (hill climbing) which functions as a local search technique to enrich the search abilities. Thirdly, it combines both semantic resemblance and relatedness techniques to assess the fitness of the solutions. Several experiments have been conducted to confirm the efficiency of the HC local search and the integrated solution measurement for performance of hybrid BA. The outcomes of the experiments confirm that the hybrid BA technique suggested by us outperforms all other techniques with regard to the WSD issue.

Therefore, we can suppose that the hybrid BA technique proposed in this study can be viewed as an efficient solution for the ambiguity of the words. In further works, we propose to expand the local search to involve acceptance condition to look at more domains of the problem space. Moreover, examining more semantic techniques to perform the independent function of the hybrid BA may provide assistance in the evaluation of the produced solutions.

References

1. Abed, S.A., Tiun, S., Omar, N.: Harmony search algorithm for word sense disambiguation. PLoS ONE **10**(9), e0136614 (2015)
2. Abed, S.A., Tiun, S., Omar, N.: Word sense disambiguation in evolutionary manner. Connect. Sci. **28**(3), 226–241 (2016)
3. Alsaeedan, W., Menai, M.E.B.: A self-adaptive genetic algorithm for the word sense disambiguation problem. In: Ali, M., Kwon, Y.S., Lee, C.H., Kim, J., Kim, Y. (eds.) Current Approaches in Applied Artificial Intelligence, pp. 581–590. Springer, Cham (2015)
4. Altintas, E., Karsligil, E., Coskun, V.: A new semantic similarity measure evaluated in word sense disambiguation. In: Proceedings of the 15th Nordic Conference of Computational Linguistics (NODALIDA 2005), pp. 8–11 (2006)
5. Banerjee, S., Pedersen, T.: An adapted lesk algorithm for word sense disambiguation using wordnet. In: International Conference on Intelligent Text Processing and Computational Linguistics, pp. 136–145. Springer, Heidelberg (2002)
6. Gelbukh, A., Sidorov, G., Han, S.Y.: Evolutionary approach to natural language word sense disambiguation through global coherence optimization. WSEAS Trans. Comput. **2**(1), 257–265 (2003)
7. Hong, W.C., Li, M.W., Geng, J., Zhang, Y.: Novel chaotic bat algorithm for forecasting complex motion of floating platforms. Appl. Math. Model. **72**, 425–443 (2019)
8. Hung, C., Chen, S.J.: Word sense disambiguation based sentiment lexicons for sentiment classification. Knowl.-Based Syst. **110**, 224–232 (2016)
9. Ide, N., Véronis, J.: Introduction to the special issue on word sense disambiguation: the state of the art. Comput. Linguist. **24**(1), 2–40 (1998)
10. Kilgarriff, A.: What is word sense disambiguation good for? CoRR cmp-lg/9712008 (1997). http://arxiv.org/abs/cmp-lg/9712008
11. Lu, W., Huang, H., Zhu, C.: Feature words selection for knowledge-based word sense disambiguation with syntactic parsing. Przegl. Elektrotechniczny **88**(1b), 82–87 (2012)
12. Nguyen, K.H., Ock, C.Y.: Word sense disambiguation as a traveling salesman problem. Artif. Intell. Rev. **40**(4), 405–427 (2013)
13. Osaba, E., Yang, X.S., Fister Jr., I., Del Ser, J., Lopez-Garcia, P., Vazquez-Pardavila, A.J.: A discrete and improved bat algorithm for solving a medical goods distribution problem with pharmacological waste collection. Swarm Evol. Comput. **44**, 273–286 (2019)
14. Pattanaik, A., Sagnika, S., Das, M., Mishra, B.S.P.: Extractive summary: an optimization approach using bat algorithm. In: Hu, Y.C., Tiwari, S., Mishra, K.K., Trivedi, M.C. (eds.) Ambient Communications and Computer Systems, pp. 175–186. Springer, Singapore (2019)

15. Wafaa, A.S., Tiun, S., Ahmed, A.S., Awang, S., Al-Khaleefa, A.: Word sense disambiguation using hybrid swarm intelligence approach. PLoS ONE **13**(12), e0208695 (2018)
16. Yang, X.S.: A new metaheuristic bat-inspired algorithm. In: Nature Inspired Cooperative Strategies for Optimization (NICSO 2010), pp. 65–74. Springer, Heidelberg (2010)
17. Zhang, C., Zhou, Y., Martin, T.: Genetic word sense disambiguation algorithm. In: 2008 Second International Symposium on Intelligent Information Technology Application, vol. 1, pp. 123–127, December 2008

A Multi-objective Optimization Algorithm Based on Monarch Butterfly Optimization

Rui Hu, Jian Gao(✉), Rong Chen, and Jiahao Jiang

Dalian Maritime University, Dalian 116023, China
gaojian@dlmu.edu.cn

Abstract. Swarm intelligence optimization algorithm is an important technology that solves the complex optimization problem by simulating the behavior of biological groups in nature. Monarch butterfly optimization (MBO) algorithm is such a swarm intelligence algorithm that simulates the migration behavior of the monarch butterflies in nature. It has received great success on solving single-objective optimization problems, but few contributions on multi-objective problems. In this paper, we modify MBO to solve multi-objective problems, and then propose a new multi-objective optimization algorithm based by combining effective strategies from other swarm-based algorithms. A series of benchmark functions are employed to evaluate the performance of this algorithm. We compare the experimental results with three basic algorithms and state-of-the-art algorithms. It is shown that the proposed algorithm performs better than the compared algorithms on most of the benchmark functions.

Keywords: Hybrid swarm intelligence · Monarch butterfly optimization · Multi-objective optimization

1 Introduction

Optimization is an important research tool in the areas of scientific research, engineering technology and economic management. In fact, almost all scientific and engineering optimization problems involve non-linearity and complex constraints. To deal with these problems and find the optimal solution, efficient algorithms are crucial. In the past, many intelligent optimization methods have been proposed, such as Differential Evolution (DE) [1], Particle Swarm Optimization (PSO) [2, 3], Artificial Bee Colony (ABC) [4] and Monarch Butterfly Optimization (MBO) [5]. In recent years, they have shown promising performance in terms of the algorithm performance improvements and the applications in real-world problems. Taking MBO as an example, Wang *et al.* [6] proposed supplementing the original MBO algorithm with greedy strategy and self-adaptive crossover operator (GCMBO). In GCMBO, the self-adaptive crossover operator can significantly improve the population diversity during the later run phase of the search, the greedy strategy is used to select only those monarch butterfly individuals with improved fitness, and hence helping to accelerate convergence. Hu *et al.* [7] introduced a self-adaptive strategy to dynamically adjust the butterflies in subpopulation 1 and 2, and proposed an improved MBO algorithm, called monarch butterfly optimization with self-adaptive population (SPMBO).

Y. Liu et al. (Eds.): ICNC-FSKD 2019, AISC 1074, pp. 158–167, 2020.
https://doi.org/10.1007/978-3-030-32456-8_17

Simultaneously, as optimization problems become more and more complicated, traditional single optimization methods have been difficult to find a satisfactory solution within a limited time. Therefore, many researchers have proposed hybrid intelligent algorithms that combine different algorithms organically, according to optimization mechanisms, processes, search behaviors and operations, and make them complement each other for better optimization efficiency. Combining MBO with Harmony Search (HS), Ghetas *et al.* [8] proposed a hybrid metaheuristic algorithm, which can explore the search space very well and exploit the population. By modifying the butterfly adjusting operator in MBO algorithm and using it as a mutation operator to replace the employee phase of ABC algorithm, Ghanem and Jantan [9] proposed a new hybrid metaheuristic algorithm called Hybrid ABC/MBO.

In addition, since most of the real world optimization problems, especially engineering design problems, involve the simultaneous optimization of two or more conflicting objectives, many multi-objective optimization algorithms have been proposed. However, the research and application of the original MBO algorithm is still in the primary stage. It mainly deals with single-objective optimization problems, but has few contributions on multi-objective problems. In our proposed algorithm, we first transform the basic MBO algorithm into a multi-objective algorithm, and then introduce the information of individual extremum and neighbor in migration operator to guide the search of solutions. Lastly, we combine the multi-objective MBO algorithm with some operations in NSGA-II algorithm [10]. That is, we combine the fast non-dominated-sort, and then use the arithmetic crossover operator to further update the solution to achieve better overall performance of the algorithm. In order to validate the performance of our algorithm, we conduct intensive experiments on commonly-used multi-objective benchmark problems. The results show that our algorithm is superior to the compared algorithms on most of the benchmark functions.

2 Monarch Butterfly Optimization

Here, we only introduce the migration operator and the adjusting operator in MBO. The descriptions of the parameters and variables used below refer to [5].

Migration operator simulates the process that monarch butterflies move from their current position to a new position. In this process, a random number *rand* is first generated for each element of each butterfly in Subpopulation 1 and then, r can be calculated as

$$r = rand * peri \tag{1}$$

When $r \leq p$, the kth element of the newly generated butterfly is generated by

$$x_{i,k}^{t+1} = x_{r_1,k}^{t} \tag{2}$$

When $r > p$, the kth element of the newly generated butterfly is generated by

$$x_{i,k}^{t+1} = x_{r_2,k}^{t} \tag{3}$$

The adjusting operator is specifically described as follows: For all elements of butterfly j, a random number *rand* is generated.

If $rand \leq p$, it can be updated as

$$x_{j,k}^{t+1} = x_{best,k}^{t} \tag{4}$$

If $rand > p$, it can be updated as

$$x_{j,k}^{t+1} = x_{r_3,k}^{t} \tag{5}$$

Under this condition, if $rand >$ BAR, it can be further updated as follows

$$x_{j,k}^{t+1} = x_{j,k}^{t+1} + \alpha \times (dx_k - 0.5) \tag{6}$$

3 The Hybrid Multi-objective MBO

The research and application of the original MBO algorithm have received great success on solving single-objective optimization problems, but few contributions on multi-objective problems. In order to apply it to solve multi-objective optimization problems, we modified it.

First, we use an external archive *Rep* to preserve the non-dominated solutions found during the search process. When *Rep* is empty, we perform the judgment process that determines whether a solution is a non-dominated solution. If so, we add it to *Rep*; On the contrary, we first perform the judgment process. If it is a non-dominated solution, we further judge whether it is dominated by a solution in *Rep*. If not, we add it to *Rep*, and then delete the solution, which is dominated by the new solution, from *Rep*. Then, we use the grid method proposed in [3] to make the Pareto front well-distributed. In addition, the selection of the global optimal particle bestAgent and the retention strategy we adopted when *Rep* gets full are both according to [3]. Here, we call the modified MBO algorithm a basic multi-objective MBO algorithm named BMOMBO.

Moreover, by analyzing several classic multi-objective evolutionary algorithms and combining the effective strategies, a new hybrid multi-objective optimization algorithm named HMOMBO is proposed.

3.1 The New Migration Operator

In the migration operator of the original MBO algorithm, when updating a solution, the corresponding elements of other solutions in the population are directly used for value assignment. This makes the diversity of the solution limited, and thus is not conducive to the evolution of the population. Inspired by the idea of individual extremum in PSO and differential strategies in DE, we use a new mutation operator in the migration operator. That is, we introduce the information of individual extremum and neighbor to guide the search of solutions when $r > p$. Clearly, the mutation operator is described as follows.

$$x_{i,k}^{t+1} = x_{pbest,k}^{t} + (x_{neighbor1,k}^{t} - x_{neighbor2,k}^{t}) \times R \tag{7}$$

where x_{pbest} is the position of individual extremum, *neighbor1* and *neighbor2* are two different individuals randomly selected from Subpopulation 2. $R = (rand - 0.5) \times 2$ is a random number between $[-1, 1]$ that affects the differential variation between two individuals.

In the new mutation operator, a differential vector is generated by the subtraction of two randomly selected different individual vectors, after which a weight is assigned to it and then add a third randomly selected individual vector to generate a mutation vector. This increases the diversity of the solution; and at the same time, the introduction of individual extremum accelerates the convergence rate to some extent.

3.2 The Combination with Operations in NSGA-II

We know that NSGA-II adopts a fast non-dominated sorting approach, which greatly reduces the computational complexity; and uses the crowded-comparison operator as a winning criterion in the comparison between two solutions after fast non-dominated sorting, so that the individual can be evenly distributed to the entire Pareto domain to maintain the diversity of the population. Therefore, we combine the operations in NSGA-II to achieve better overall performance of our algorithms.

In detail, we combine the two newly-generated subpopulations processed by the migration operator and butterfly adjusting operator respectively into a whole population, and then the population is sorted according to nondomination. This returns two columns for each individual that are the rank and the crowding distance. After that, we select the parents based on the crowded-comparison operator and use the arithmetic crossover operator to further update the solution.

Combined with the above sections, we propose a multi-objective optimization algorithm based on MBO in this section. The detailed description is listed as follows.

Algorithm1 Multi-objective optimization algorithm based on MBO (HMOMBO)

Set the population size nPop and the Pareto archive size nRep, and initialize the Pareto archive $Rep = \emptyset$
Initialize the population Pop and evaluate each individual in Pop, and record $pbest$
Store the non-dominated solution in Rep
Generate hypercubes of the search space explored so far, and locate the particles
Select the global optimum $bestAgent$
while the termination condition is not satisfied **do**
  Divide monarch butterfly individuals into two subpopulations: Subpopulation 1 and 2
  for $i = 1$ to NP_1 (for each monarch butterfly in Subpopulation 1) **do**
    Generate new Subpopulation 1 according to the new migration operator
  end for i
  for $j = 1$ to NP_2 (for each monarch butterfly in Subpopulation 2) **do**
    Generate new Subpopulation 2 according to butterfly adjusting operator
  end for j
  Combine the two newly-generated subpopulations as Pop
  Perform non-dominated-sort on Pop
  Select the parents and perform arithmetic crossover operator to generate offspring
  Replace Pop with the offspring
  Evaluate each individual in Pop and update $pbest$, Rep and $bestAgent$
  Increment the loop counter
end while

4 Experiments

We select 11 widely used benchmark problems SCH, FON, POL, KUR, ZDT1-3, ZDT6, DTLZ1-2 and DTLZ4 [10, 11] to test the algorithm. Besides, we choose the generational distance (GD) [12] and spacing (SP) [13] as performance metrics to verify convergence and diversity respectively, and the inverted generational distance (IGD) [14] to reflect both convergence and distribution. For all the three metrics, the smaller their values are, the better the results will be.

4.1 Comparisons with Basic Algorithms

First, we use the three basic algorithms, namely basic multi-objective MBO (BMOMBO), multi-objective PSO (MOPSO) and NSGA-II, as comparisons. In order to have a fair comparison, we select 10 of the used benchmark problems and set the number of function evaluations (NFEs) to 50000 for ZDTN and DTLZN, and 25000 for other problems; set the population size and the Pareto archive size to 150 for DTLZN, and 100 for other functions. In the following experiments, we perform 30 independent runs on each benchmark problem, and take the maximum NFEs as the stopping condition unless the algorithm reaches convergence earlier. For the combination with operations in NSGA-II, it is called once every 10 iterations for ZDTN and DTLZN, and once per iteration for other test functions. Subsequently, three performance metrics are calculated. The average value, optimal value, worst value and the standard deviation are recorded with the best results marked in bold. Due to the limited length of the article, we only list the mean and standard deviation values here. Since MOPSO sometimes converges to a point on ZDT2, and NSGA-II sometimes converges poorly on SCH, their results cannot be used for comparison, and are represented by "-".

It can be noticed from Table 1 that HMOMBO performs relatively better than the three basic algorithms with regard to all considered problems, except that NSGA-II has better performance on POL, while BMOMBO performs better on DTLZ2. From Table 2, we found that the performance of NSGA-II is relatively optimal, and the proposed HMOMBO ranks second. From Table 3, it can be seen that HMOMBO obtains more optimal results than other algorithms.

In order to further demonstrate the convergence and diversity of the optimal solution set obtained by various algorithms, we use ZDT1, ZDT3 and DTLZ2 as examples to give a comparison of the Pareto-optimal front obtained by these four algorithms, and illustrate the results in Fig. 1. As can be seen, the PF found by HMOMBO can better cover the true PF whether for two or three objective problems. Compared to other algorithms, HMOMBO's PF has no missing parts and converges to the true PF relatively well. Therefore, we can also conclude that the algorithm obtains a solution set that is evenly distributed and close to the real PF for most multi-objective problems.

Table 1. The mean and standard deviation GD values obtained by NSGA-II, MOPSO, BMOMBO and HMOMBO on benchmark problems.

Problem	Criteria	NSGA-II	MOPSO	BMOMBO	HMOMBO
ZDT1	Mean	0.0089	0.0018	**3.2583e−04**	3.8397e−04
	Std	0.0013	0.0018	1.1400e−04	**6.5351e−05**
ZDT2	Mean	0.0104	–	3.1152e−04	**3.0300e−04**
	Std	0.0022		8.9276e−05	**4.5384e−05**
ZDT3	Mean	0.0087	0.0050	**2.8241e−04**	3.0556e−04
	Std	0.0015	0.0047	1.2581e−04	**5.3350e−05**
ZDT6	Mean	0.0398	0.0416	0.0015	**9.4211e−04**
	Std	0.0139	0.0502	0.0060	**0.0032**
FON	Mean	0.0012	0.0011	8.7738e−04	**4.5741e−04**
	Std	**3.9818e−05**	4.4589e−05	1.5755e−04	4.8731e−05
SCH	Mean	–	9.5898e−04	1.4717e−04	**1.4073e−04**
	Std		**6.3564e−05**	1.7435e−04	1.2350e−04
POL	Mean	**0.0035**	0.0118	0.0137	0.0084
	Std	**0.0077**	0.0207	0.0196	0.0167
KUR	Mean	0.0677	0.0039	0.0035	**0.0030**
	Std	0.0535	5.7437e−04	0.0014	**4.4269e−04**
DTLZ2	Mean	0.0046	0.0039	**8.4035e−04**	0.0024
	Std	5.4009e−04	6.8146e−04	4.5307e−04	**3.6112e−04**
DTLZ4	Mean	0.0026	0.0109	**0.0019**	**0.0019**
	Std	9.4657e−04	0.0062	4.7364e−04	**1.5319e−04**

Table 2. The mean and standard deviation SP values obtained by NSGA-II, MOPSO, BMOMBO and HMOMBO on benchmark problems.

Problem	Criteria	NSGA-II	MOPSO	BMOMBO	HMOMBO
ZDT1	Mean	**6.1960e−04**	9.1854e−04	9.3838e−04	7.6753e−04
	Std	1.1720e−04	1.9811e−04	1.1530e−04	**7.8103e−05**
ZDT2	Mean	**6.1905e−04**	–	0.0010	7.9319e−04
	Std	3.3862e−04		1.4977e−04	**9.0388e−05**
ZDT3	Mean	**7.7593e−04**	0.0011	0.0014	9.8861e−04
	Std	2.2560e−04	6.7398e−04	2.2714e−04	**1.2732e−04**
ZDT6	Mean	0.0111	0.0049	0.0021	**0.0014**
	Std	0.0132	0.0041	0.0059	**0.0023**
FON	Mean	**0.0012**	0.0013	0.0022	0.0018
	Std	**1.7642e−04**	1.5417e−04	3.9259e−04	1.9125e−04
SCH	Mean	–	0.0034	0.0043	**0.0033**
	Std		**2.9978e−04**	0.0010	4.1867e−04
POL	Mean	**0.0099**	0.0127	0.0142	0.0107
	Std	0.0066	0.0025	0.0047	**0.0016**
KUR	Mean	0.0114	**0.0094**	0.0107	0.0117
	Std	0.0069	0.0029	0.0028	**0.0020**
DTLZ2	Mean	0.0029	**0.0025**	0.0032	0.0029
	Std	4.5494e−04	2.6396e−04	3.7485e−04	**2.3778e−04**
DTLZ4	Mean	**0.0014**	0.0027	0.0030	0.0029
	Std	0.0010	0.0010	3.0383e−04	**2.2029e−04**

Table 3. The mean and standard deviation IGD values obtained by NSGA-II, MOPSO, BMOMBO and HMOMBO on benchmark problems.

Problem	Criteria	NSGA-II	MOPSO	BMOMBO	HMOMBO
ZDT1	Mean	0.0859	0.0132	0.0131	**0.0092**
	Std	0.0153	0.0091	0.0022	**0.0011**
ZDT2	Mean	0.1750	–	0.0140	**0.0084**
	Std	0.1471		0.0018	**7.7075e−04**
ZDT3	Mean	**0.0998**	0.1821	0.1978	0.1958
	Std	0.0169	0.0635	0.0039	**0.0033**
ZDT6	Mean	0.0567	**0.0045**	0.0095	0.0062
	Std	0.1997	**0.0011**	0.0024	**0.0011**
FON	Mean	**0.0052**	0.0065	0.0144	0.0077
	Std	5.9106e−04	7.0786e−04	0.0012	**5.3524e−04**
SCH	Mean	–	**0.0316**	0.0974	0.0326
	Std		**0.0023**	0.0125	0.0029
POL	Mean	**0.0709**	0.1179	0.2673	0.1348
	Std	**0.0047**	0.0223	0.0566	0.0344
KUR	Mean	0.6949	**0.0622**	0.1301	0.0629
	Std	0.4123	0.0073	0.0250	**0.0035**
DTLZ2	Mean	0.0977	0.2611	0.0708	**0.0633**
	Std	0.0055	0.0267	0.0090	**0.0021**
DTLZ4	Mean	0.2830	0.3529	**0.0645**	0.0666
	Std	0.1320	0.1143	0.0053	**0.0040**

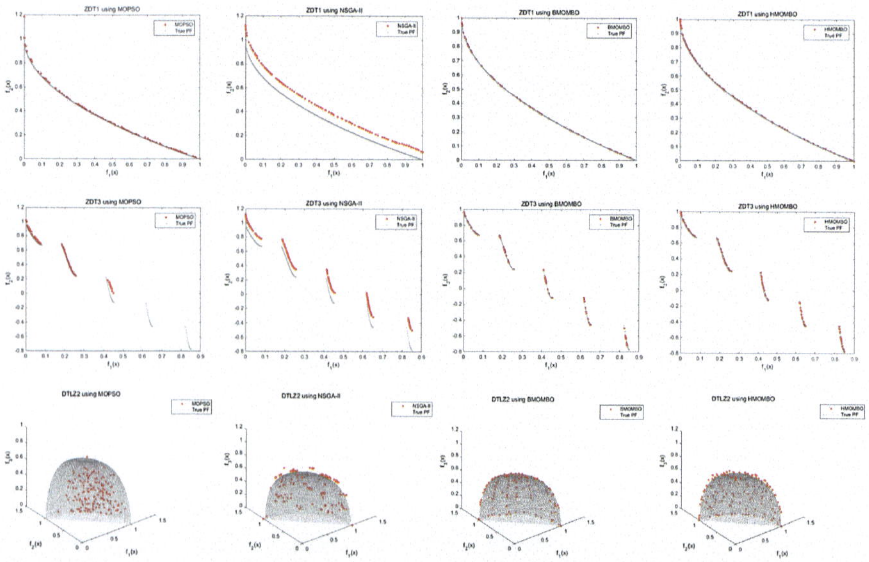

Fig. 1. PF comparison of test functions.

4.2 Comparisons with Other Algorithms

Moreover, we compare our algorithm with three other prominent algorithms BBMO [15], RM-MEDA [16] and MOTLBO [17] that employed in previous studies. We use the three ZDT and three DTLZ test functions to evaluate the performance of the proposed HMOMBO algorithm. Here, the parameter settings of our algorithm and the results of the three algorithms used for comparison refer to [18]. In terms of convergence, as shown in Table 4, HMOMBO gets the best result on four out of six benchmarks for the mean value and five benchmarks for the standard deviation. In terms of distribution uniformity, it can be seen from Table 5 that HMOMBO yields five out of six best results for the mean values and six for the standard deviation values.

Table 4. The mean and standard deviation GD values obtained by BBMO, RM-MEDA, MOTLBO and HMOMBO on benchmark problems.

Problem	Criteria	BBMO	RM-MEDA	MOTLBO	HMOMBO
ZDT1	Mean	2.53e−02	2.87e−03	1.59e−03	**1.40e−03**
	Std	2.98e−03	8.47e−03	3.27e−03	**2.72e−04**
ZDT2	Mean	3.26e−02	4.76e−03	**9.34e−04**	1.10e−03
	Std	2.72e−02	1.58e−03	1.06e−03	**3.65e−04**
ZDT3	Mean	7.23e−02	2.17e−03	2.04e−03	**4.45e−04**
	Std	1.41e−02	9.01e−03	8.47e−04	**2.01e−04**
DTLZ1	Mean	6.34e−02	**3.76e−02**	7.83e−02	1.68e−00
	Std	1.17e−01	**9.45e−02**	5.97e−01	5.82e−01
DTLZ2	Mean	4.81e−02	3.54e−02	1.06e−02	**4.70e−03**
	Std	5.22e−02	9.20e−02	4.92e−03	**8.16e−04**
DTLZ4	Mean	2.45e−01	7.73e−02	3.74e−02	**3.40e−03**
	Std	3.75e−02	5.76e−02	4.99e−02	**7.87e−04**

Table 5. The mean and standard deviation SP values obtained by BBMO, RM-MEDA, MOTLBO and HMOMBO on benchmark problems.

Problem	Criteria	BBMO	RM-MEDA	MOTLBO	HMOMBO
ZDT1	Mean	2.74e−02	1.10e−02	4.01e−03	**1.20e−03**
	Std	5.53e−03	1.18e−03	3.68e−03	**3.72e−04**
ZDT2	Mean	7.88e−01	3.88e−03	7.38e−03	**1.40e−03**
	Std	3.83e−01	1.56e−03	6.96e−03	**6.71e−04**
ZDT3	Mean	4.93e−03	1.19e−02	3.37e−03	**1.30e−03**
	Std	1.65e−03	2.80e−03	3.08e−03	**2.38e−04**
DTLZ1	Mean	3.85e−01	**7.65e−02**	8.34e−02	1.68e−01
	Std	2.28e−01	4.53e−01	8.13e−02	**6.77e−02**
DTLZ2	Mean	1.12e−02	3.98e−02	2.46e−02	**4.80e−03**
	Std	3.95e−02	5.92e−02	3.66e−03	**5.50e−04**
DTLZ4	Mean	1.75e−01	9.13e−02	3.30e−02	**5.50e−03**
	Std	2.75e−02	2.53e−02	3.06e−02	**8.66e−04**

5 Conclusion

In conclusion, a multi-objective optimization algorithm based on MBO is proposed. First, we modify the MBO algorithm so that it can be applied to solve multi-objective optimization problems. Second, inspired by PSO and DE, we use a mutation operator in the migration operator to increase the diversity of solutions and promote the evolution of the population. Finally, we combine the fast non-dominated-sort in NSGA-II, and then use the arithmetic crossover operator to further update the solution. Comparing the experimental results with other algorithms on several commonly used test problems, it demonstrates the effectiveness of our algorithm.

Acknowledgement. This work is supported by the National Natural Science Foundation of China (No. 61672122, No. 61602077), the Public Welfare Funds for Scientific Research of Liaoning Province of China (No. 20170005), the Natural Science Foundation of Liaoning Province of China (No. 20170540097), and the Fundamental Research Funds for the Central Universities (No. 3132016348, No. 3132018194).

References

1. Storn, R., Price, K.: Differential evolution–a simple and efficient heuristic for global optimization over continuous spaces. J. Global Optim. **11**(4), 341–359 (1997)
2. Kennedy, J., Eberhart, R.: Particle swarm optimization. In: Proceedings of IEEE International Conference on Neural Networks, pp. 1942–1948 (1995)
3. Coello, C.A.C., Pulido, G.T., Lechuga, M.S.: Handling multiple objectives with particle swarm optimization. IEEE Trans. Evol. Comput. **8**(3), 256–279 (2004)
4. Karaboga, D., Basturk, B.: A powerful and efficient algorithm for numerical function optimization: artificial bee colony (ABC) algorithm. J. Global Optim. **39**(3), 459–471 (2007)
5. Wang, G.G., Deb, S., Cui, Z.: Monarch butterfly optimization. Neural Comput. Appl. **31**, 1–20 (2015)
6. Wang, G.G., Deb, S., Zhao, X., Cui, Z.: A new monarch butterfly optimization with an improved crossover operator. Oper. Res. Int. J. **18**(3), 731–755 (2018)
7. Hu, H., Cai, Z., Hu, S., Cai, Y., Chen, J., Huang, S.: Improving monarch butterfly optimization algorithm with self-adaptive population. Algorithms **11**(5), 71 (2018)
8. Ghetas, M., Yong, C.H., Sumari, P.: Harmony-based monarch butterfly optimization algorithm. In: 2015 IEEE International Conference on Control System, Computing and Engineering (ICCSCE), Penang, Malaysia, pp. 156–161. IEEE (2015)
9. Ghanem, W.A., Jantan, A.: Hybridizing artificial bee colony with monarch butterfly optimization for numerical optimization problems. Neural Comput. Appl. **30**(1), 163–181 (2018)
10. Deb, K., Pratap, A., Agarwal, S., Meyarivan, T.A.M.T.: A fast and elitist multiobjective genetic algorithm: NSGA-II. IEEE Trans. Evol. Comput. **6**(2), 182–197 (2002)
11. Deb, K., Thiele, L., Laumanns, M., Zitzler, E.: Scalable test problems for evolutionary multiobjective optimization. In: Evolutionary Multiobjective Optimization, pp. 105–145. Springer, London (2005)
12. Van Veldhuizen, D.A., Lamont, G.B.: Evolutionary computation and convergence to a pareto front. In: Late Breaking Papers at the Genetic Programming 1998 Conference, pp. 221–228 (1998)

13. Schott, J.R.: Fault tolerant design using single and multicriteria genetic algorithm optimization. Cell. Immunol. **37**(1), 1–13 (1995)
14. Czyżak, P., Jaszkiewicz, A.: Pareto simulated annealing—a metaheuristic technique for multiple−objective combinatorial optimization. J. Multi-Criteria Decis. Anal. **7**(1), 34–47 (1998)
15. Ma, H.P., Ruan, X.Y., Pan, Z.X.: Handling multiple objectives with biogeography-based optimization. Int. J. Autom. Comput. **9**(1), 30–36 (2012)
16. Zhang, Q., Zhou, A., Jin, Y.: RM-MEDA: a regularity model-based multiobjective estimation of distribution algorithm. IEEE Trans. Evol. Comput. **12**(1), 41–63 (2008)
17. Zou, F., Wang, L., Hei, X., Chen, D., Wang, B.: Multi-objective optimization using teaching-learning-based optimization algorithm. Eng. Appl. Artif. Intell. **26**(4), 1291–1300 (2013)
18. Siling, F., Ziqiang, Y., Mengxing, H.: Hybridizing adaptive biogeography-based optimization with differential evolution for multi-objective optimization problems. Information **8**(3), 83 (2017)

An Improved Quantum Genetic Algorithm Based on Population Partition and Dynamic Probability Amplitude

Cheng Yao Shi(✉), Zhao Cheng Xuan, Chao Yang, and Yong Fei Yang

Department of Computer Science, Tianjin University of Technology and Education, Tianjin 300222, China
scy1994@126.com

Abstract. Aiming at the problem that quantum genetic algorithm is easy to fall into local optimization in the process of function optimization under the original framework, an improved quantum genetic algorithm using population partition method and dynamic inverse probability amplitude strategy is proposed. In the process of optimization, the improved algorithm utilizes the superior individuals in each generation of populations to construct a population genotype suitable for evolution through the individual binary form. On this basis, replace the inferior individuals in the population with a certain proportion, and then construct a number of similar individuals that tend to the genotype of the superior individual. Therefore, the algorithm can continuously improve the adaptability of the problem and gradually converge to the optimal solution. At the same time, for the premature or local optimization problem that the algorithm may be trapped, in the period of algorithm stagnation, the population diversity is enriched by resetting the genotype of some individuals and narrowing the search space by using the current optimal solution.

Keywords: Quantum genetic algorithm · Population partition · Dynamic probability amplitude

1 Introduction

Quantum genetic algorithm has attracted wide attention in recent years, thanks to its fusion of quantum computing theory [1] and genetic algorithm theory, which improves the search performance of genetic algorithms. After the introduction of quantum superposition state, quantum entanglement and quantum collapse in quantum computing theory, the algorithm has a good global optimization ability. In terms of convergence speed, the algorithm also has high computational efficiency.

The idea of quantum-inspired genetic algorithm was first proposed by Narayanan in 1996 [2], Han et al. actually used quantum bit coding and applied it to the combinatorial optimization problem [3]. Optimization problems are often considered to be NP-hard, and if general mathematical methods are applied, it is difficult to solve them in polynomial time. Therefore, for the optimization of such problems, various evolutionary algorithms are usually used to find the optimal solution or its approximate

Y. Liu et al. (Eds.): ICNC-FSKD 2019, AISC 1074, pp. 168–177, 2020.
https://doi.org/10.1007/978-3-030-32456-8_18

solution in the solution space. For example, quantum genetic algorithm based on real number coding and objective function gradient is used to optimize the weight of neural network [4]; K-means algorithm based on improved quantum genetic algorithm is used to improve the performance of traditional K-means algorithm in cluster analysis [5]; Using quantum genetic algorithm to solve TSP problem [6], knapsack problem [7], etc.

2 Quantum Genetic Algorithm (QGA)

Compared with traditional genetic algorithms, quantum bit coding used in quantum genetic algorithms [8, 9] can express each possible quantum state in a probabilistic manner. A rich population can be obtained by a random observation method that simulates quantum collapse. The evolutionary mechanism of the traditional quantum genetic algorithm mainly relies on the quantum revolving gate to adjust the probability amplitude expression of each qubit. After several iterations of algebra, the probabilistic amplitude form that is most suitable for the problem is obtained. However, the adjustment of the revolving gate rotation has a high degree of flexibility, both in the direction and in the magnitude of the rotation angle. If the rotation angle is too large, the algorithm is very easy to miss the optimal probability amplitude expression, and the probability amplitude changes more severely. On the other hand, if the rotation angle is too small, the number of iterations of the algorithm is large and the convergence speed is slow. Therefore, for the determination of the rotation angle and the rotation direction, it is necessary to adjust several times according to the actual situation, and gradually determine the most suitable scheme.

In order to improve the efficiency of quantum genetic algorithm and the reliability of the algorithm, this paper improves the binary decoding method after simulating quantum collapse. By means of Bidirectional decoding, the binary state after observation is used as much as possible to select better individuals. Moreover, according to the binary form of the optimal individual in each generation of population, the probability expression and probability amplitude of the individual are dynamically generated, so that both of them are replicated in the population in a certain proportion. This ensures that the superior individuals of each generation of the algorithm can be preserved, and the algorithm can perform effective local search in the vicinity of excellent individuals. Moreover, the algorithm applies quantum NOT-gate [10] to make the chromosomes in the population mutate with a certain probability, thereby enhancing its population diversity. In the later stage of operation, in order to cope with possible local optimization, the improved quantum genetic algorithm can reset the probability amplitude expression of some chromosomes in the population when evolution is stagnant, and then help the algorithm get rid of the above dilemma.

3 Dynamic Inverse Probability Amplitude

Dynamic inverse probability amplitude mainly includes two directions to provide a stable evolutionary possibility for the algorithm. Compared with the quantum revolving gate, each generation gradually adjusts the probability amplitude expression, so that the population has sufficient time to explore the search space. The dynamic inverse probability amplitude operator adjusts the probability amplitude very quickly, so that the algorithm can closely grasp the parameter setting of the superior individuals in each generation. According to the optimal individual of each generation, the dynamic inverse probability amplitude randomly constructs two types of population chromosomes with high similarity and similar trend, so that both of them replicate in a certain proportion. The construction method is as shown in the formula (1, 2). In the formula, $\alpha_i$and β_i respectively represent the probability that the quantum bit is in state 0 and state 1, and x_i represents the actual value in the case of optimal individual convergence. Equation (1) represents the range of probability amplitude of each quantum bit in the case of excellent individual high similarity. Equation (2) indicates the general range of the probability amplitude of each quantum bit in the case where the trend is guided. It is not difficult to find that the probability amplitude with high similarity can guarantee the phenomenon that the generation of the algorithm does not degenerate, and the excellent individual is used for small-scale local search. And with the probability of guiding the trend, you can make the algorithm close to each generation of the best individual, and carry out a large range of local search.

$$\left\{ \begin{array}{l} \left\{ \begin{array}{l} \alpha_i \in \mathrm{normrnd}[0.75,\ 0.95) \\ \beta_i \in \mathrm{normrnd}(0.00,\ 0.20] \end{array} \right., x_i = 0 \\ \left\{ \begin{array}{l} \alpha_i \in \mathrm{normrnd}(0.00,\ 0.20] \\ \beta_i \in \mathrm{normrnd}[0.75,\ 0.95) \end{array} \right., x_i = 1 \end{array} \right. \tag{1}$$

$$\left\{ \begin{array}{l} \left\{ \begin{array}{l} \alpha_i \in \mathrm{normrnd}[0.50,\ 1.00) \\ \beta_i \in \mathrm{normrnd}(0.00,\ 0.50] \end{array} \right., x_i = 0 \\ \left\{ \begin{array}{l} \alpha_i \in \mathrm{normrnd}(0.00,\ 0.50] \\ \beta_i \in \mathrm{normrnd}[0.50,\ 1.00) \end{array} \right., x_i = 1 \end{array} \right. \tag{2}$$

4 Bidirectional Decoding Operator

The Bidirectional decoding operator is an operator used by the quantum genetic algorithm to decode the quantum bits whose diversity disappears. The operator provides two different decoding values of the quantum bits in the same binary form. On this basis, the mutually different decoding values participate in the subsequent fitness calculation and evaluation operations. The existence of the bidirectional decoding mechanism extends the diversity of the population decoding and establishes a dual mapping of the population chromosome decoding values. After the fitness evaluation, the algorithm can filter out the decoding value more suitable for the problem as the final decoded value of the chromosome. As the number of qubits used in the optimization problem increases, the performance loss of the bidirectional decoding operator becomes smaller and smaller. When the number of qubits tends to infinity, the bidirectional decoding operator can use only half of the binary form. Exploring the entire search space reduces the number of observations used by the algorithm to simulate the quantum collapse process, saving resources. The performance loss of the bidirectional decoding operator is shown in Fig. 1. The decoding process is shown in Eq. (3).

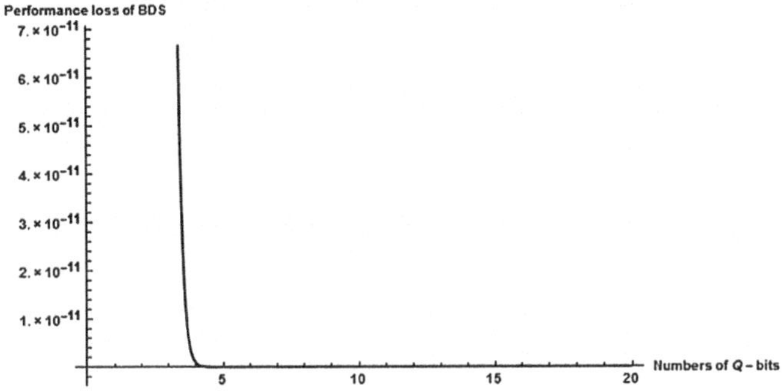

Fig. 1. The performance loss of BD.

$$101111\begin{matrix}\nearrow 47\\ \searrow 61\end{matrix}\qquad\qquad 110010\begin{matrix}\nearrow 50\\ \searrow 19\end{matrix}\tag{3}$$

5 Population Partition Strategy

Traditional quantum genetic algorithms do not divide chromosome populations, and chromosome update operations rely on quantum revolving gates. In the improved quantum genetic algorithm, we divide the population to a certain extent, which is mainly divided into three parts. The first segment of the population mainly performs the probability amplitude replication of excellent individuals with high similarity, so as to ensure that the algorithm can still present the optimal individuals of the previous generation population in the next generation. The second segment of the population is used to replicate the probability amplitudes that lead to elite individual trends, allowing the algorithm to perform a larger range of guided local searches. Finally, the rest of the chromosomes are updated by a quantum revolving gate. The population partition strategy is a method that takes into account both deep search and breadth search. The search for excellence near and within the elite belongs to the category of deep search. The gradual update of the quantum revolving gate makes the algorithm guarantee the breadth of the solution space. Later in the run, when the algorithm falls into a possible stagnation, the strategy can also reset some of the chromosomes to increase the population diversity. The population partition strategy of improved quantum genetic algorithm is shown in Fig. 2. Among them, U represents the whole population, and each divided part will give information to the whole population during the evolution of the algorithm, and strengthen the communication of each part.

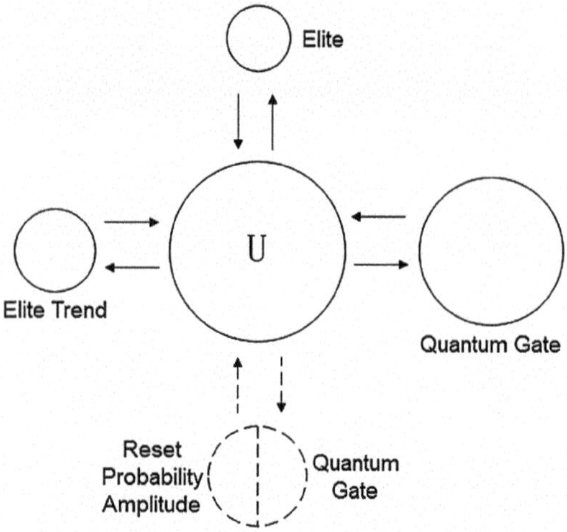

Fig. 2. Population partition strategy.

6 Quantum Variation

Quantum variation is a method to improve the performance of algorithm, accelerate the convergence speed of the algorithm, and improve the ability of the algorithm to jump out from local optimization. It is usually implemented in the form of a quantum NOT gate, which is a Pauli matrix. The quantum NOT-gate is mainly implemented in three steps. First, based on the probability of mutation, it is determined whether the corresponding individual has mutated. Then, according to the mutation probability, it is determined whether a certain position of the chromosome is mutated. Finally, the quantum NOT-gate is used to mutate the selected quantum bits to complete the mutation operation. The representation of the quantum NOT gate is shown in Eq. (4). The qubits after the quantum NOT gate are as shown in Eq. (5).

$$\text{NOT Gate} : \begin{bmatrix} 0 & 1 \\ 1 & 0 \end{bmatrix} \tag{4}$$

$$\begin{bmatrix} 0 & 1 \\ 1 & 0 \end{bmatrix} \begin{bmatrix} \alpha_i \\ \beta_i \end{bmatrix} = \begin{bmatrix} \beta_i \\ \alpha_I \end{bmatrix} \tag{5}$$

It is not difficult to find from Eq. (5) that the quantum NOT-gate changes the superposition state of the qubits, and the probability of getting 0 and the probability of getting 1 are interchanged, thus changing the direction of evolution.

7 Algorithm Description

The improved algorithm based on population partition and dynamic inverse probability amplitude mainly includes three parts: population partition mechanism, dynamic inverse probability amplitude mechanism, quantum NOT-gate and quantum revolving gate. During the evolution of the algorithm, most of the individuals still adjust the probability amplitude by quantum revolving gate. The rotation angle and rotation direction are obtained by looking up the Table 1. In the table, $f(\mathrm{x})$ shows the fitness value of the individual x, x_i represents the i-th binary value of a certain individual, b_i represents the i-th binary value of the optimal individual in the current population, and $\mathrm{S} = (\alpha_i, \beta_i)$ represents the direction of the rotation angle corresponding to the probability amplitude, θ_i is the rotation angle step size finally used by the i-th position of the individual. The formula for calculating the rotation angle step can be calculated by Eq. (6).

Table 1. Rotation angle step size.

b_i	x_i	$f(x) > f(b)$	$\delta\theta$	$\mathrm{S} = (\alpha_i, \beta_i)$			
				$\alpha_i\beta_i > 0$	$\alpha_i\beta_i < 0$	$\alpha_i = 0$	$\beta_i = 0$
0	0	True/False	0	–	–	–	–
1	1	True/False	0	–	–	–	–
1	0	True	0.01π	1	-1	0	± 1
1	0	False	0.1π	-1	1	± 1	0
0	1	True	0.01π	1	-1	0	± 1
0	1	False	0.1π	-1	1	± 1	0

Table 2. Procedure of IQGA

```
Procedure of IQGA
  Begin
    Initialize Q(t)
    Make P(t) by observing Q(t)
    Evaluate P(t) find the best P_best
    While(G<MAXG) do
      G←G+1
      Generate inverse probability amplitude by observing P_best
      Update Q(t) by population partition strategy
      Update Q(t) by quantum gates
      Make P(t+2) by observing Q(t)
      Evaluate P(t+2) with bidirectional decoding and find the best P_C_best
      Choose the better one between P_best and P_C_best
      If(P_C_best > P_best && P_C_best-P_best < 1E-7)
        Reset Part of Q(t)
        //Narrow the domain
  End
```

$$\theta_i = \delta\theta * S(\alpha_i, \beta_i) \tag{6}$$

The operation process of the improved quantum genetic algorithm based on population partition and dynamic inverse probability amplitude is shown in Table 2.

8 Experimental Design and Results

In order to test the performance of IQGA, it is experimentally analyzed and compared with the traditional genetic algorithm based on adaptive mechanism in the literature, traditional quantum genetic algorithm and parallel quantum genetic algorithm based on adaptive mechanism. Several optimization problems are selected in the experiment. The performance of the algorithm is judged by the result of function optimization. The algorithm runs 10 times in each complex function. The experimental results are partially excerpted from the results in the literature and compared with the algorithm in a table.

The experimental hardware platform run this algorithm is my laptop: CPU: Intel(R) Core(TM) i7-3630QM mobile quad-core 2.4 GHz; memory: 8G; software platform: Microsoft Visual Studio 2010 version number 10.0.40219.1 SP1Rel.

The main parameter settings of the algorithm are shown in Table 3.

Table 3. Parameter settings of the IQGA

Algorithm	Parameter settings
IQGA	Mutation rate: $p_m = 0.04$
	Biggest rotation angle: $r = 0.1\pi$
	Elite ratio: $p_e = 0.1$

In terms of testing functions, we have selected some functions, as shown in Table 4.

The above table gives the expression of the test function used, where $f_1(x, y)$ has a value range of $[-100, 100]$, and $f_2(x, y)$ has a value range of $[-10, 10]$, The value range of $f_3(x, y)$ is $\{x| -3 \le x \le 3\}, \{y| -2 \le y \le 2\}$, and the range of $f_4(x, y)$ is $[-2.048, 2.048][-2.048, 2.048]$, $f_5(x, y)$ has a value range of $\{x, y| -32.768 \le x, y \le 32.768\}\{x, y| -32.768 \le x, y \le 32.768\}$, and $f_6(x, y)$ has a value range of $\{x, y| -5 \le x, y \le 5\}\{x, y| -5 \le x, y \le 5\}$. In the above function, *f1* is Schaffer's F6 [11] function, *f2* is the Shubert function, *f3* is the Camel function, *f4* is the Rosenbrock [12] function, *f5* is the Ackley [13] function, and *f6* is the Sphere [14] function. In the experimental results, the reference data of *f1*–*f3* was derived from the literature (7), and the results are shown in Table 5. The reference data of *f4* and *f6* are derived from the literature (14), and the results are shown in Table 6. The reference data of *f5* is from the literature (13), and the results are also shown in Table 5.

Table 4. Test function.

Number	Function expression
$f_1(x,y)$	$f(x,y) = 0.5 - \frac{\sin^2\left(\sqrt{x^2+y^2}\right) - 0.5}{(1+0.001(x^2+y^2))^2}$
$f_2(x,y)$	$f(x,y) = \left(\sum_{i=1}^{5} i\cos((i+1)x+i)\right)\left(\sum_{i=1}^{5} i\cos((i+1)y+i)\right)$
$f_3(x,y)$	$f(x,y) = 4y^2(1-y^2) - \left(4 - 2.1x^2 + \frac{x^4}{3}\right)x^2 - xy$
$f_4(x,y)$	$f(x,y) = 100(x^2-y)^2 + (1-x)^2$
$f_5(x,y)$	$f(x,y) = 20\exp\left(-0.2\sqrt{\frac{1}{5}\left(\sum_{i=1}^{5} x_i^2\right)}\right) + \exp\left(\frac{1}{5}\left(\sum_{i=1}^{5}\cos 2\pi x_i\right)\right) - 20 - \exp(1)$
$f_6(x,y)$	$f(x,y) = x^2 + y^2$

Table 5. Algorithm comparison data (*f*1–*f*3, *f*5).

f(*x*)	Algorithm	Algorithm performance index						
		Best solution	Worst solution	Average solution	Min iteration	Max iteration	Average iteration	CT
f1	AGA	1.0	1.0	1.0	146	1967	894	10
	QEA	1.0	0.99014	0.99416	35	1808	483.25	4
	MPQEAAM	1.0	0.99058	0.99998	5	4851	702.8	9
	MOPAQGA	1.0	1.0	1.0	9	1866	222.7	10
	IQGA	1.0	1.0	1.0	15	483	99.4	10
f2	AGA	186.731	186.729	186.730	3926	4826	4384	4
	QEA	186.724	174.608	183.443	N	N	N	0
	MPQEAAM	186.730	185.099	186.199	251	4653	2452	2
	MOPAQGA	186.731	186.730	186.7307	80	359	190.86	7
	IQGA	186.731	186.730	186.7309	13	596	102.8	8
f3	AGA	1.03162	1.03145	1.031589	2149	2149	2149	1
	QEA	1.03162	0.99086	1.027344	4944	4944	4944	1
	MPQEAAM	1.031628	0.99701	1.023764	51	151	102	4
	MOPAQGA	1.031628	1.03157	1.036103	16	501	176	8
	IQGA	1.031628	1.03161	1.036200	8	288	75	7
f5	SPGA	2.390E-3	1.81E-2	7.05E-3	100	100	100	10
	MPGA	1.800E-5	4.89E-4	2.050E-4	–	–	–	10
	IQGA	0	1.6605E-3	7.631E-5	181	295	255	10

Table 6. Algorithm comparison data (*f*4, *f*6).

f(*x*)	Algorithm	Algorithm performance index			
		Best solution	Worst solution	average solution	Average iteration
*f*4	GA	3.3554E−2	–	3.19215E−1	83.7
	QGA	1.0490E−6	–	1.26100E−3	95
	HQGA	9.150E−10	–	7.69000E−5	150.8
	IQGA	2.73954E−8	7.40142E−6	1.02293E−6	71.4
*f*6	GA	6.274E−5	–	1.64600E−3	133
	QGA	4.777E−5	–	4.77700E−5	73.3
	HQGA	4.547E−11	–	4.5470E−11	43.4
	IQGA	3.63798E−10	4.37512E−4	1.66298E−5	112.5

9 Analysis of Results

It can be seen from the results that the improved quantum genetic algorithm based on population partition and dynamic inverse probability amplitude has a good performance when it comes to the function with multiple local minima such as *f*1, *f*2 and *f*5, whether it is the optimal solution or the worst solution or the average. The number of iterations also has a good result. The maximum number of iterations in the *f*2 function optimization process is worse than the algorithm in the reference literature. In the valley-shaped functions such as *f*3 and *f*4, *f*3 has a good performance on all data, but in the number of convergence, it is slightly inferior to the algorithm in the reference. *f*5 also has an order of magnitude difference with other algorithms in the worst solution, and slightly better than other algorithms in the mean solution, and still requires more algebra in the average number of iterations. The *f*6 function is a bowl-shaped function, and the global minimum is obtained in the middle of the interval. The improved algorithm is still inferior to other algorithms, and only surpasses the quantum genetic algorithm and the genetic algorithm under the native framework.

10 Conclusion

The improved quantum genetic algorithm based on population partition and dynamic inverse probability amplitude introduces the population partition mechanism into the quantum population, which makes the diversity of the population effectively improved, supplemented by the dynamic inverse probability structure to construct the elite trend, and gradually guides the evolution of the algorithm. With the help of quantum revolving gate, we constantly explore the solution space to find the global optimal solution to the problem. The application in the function optimization problem has verified its good performance.

References

1. Wang, A.M.: A quantum leap in computing. Physics **6**(29), 352–357 (2000)
2. Narayanan, A., Moore, M.: Quantum-inspired genetic algorithm. In: Proceedings of IEEE International Conference on Evolutionary Computation, pp. 61–66 (1996)
3. Han, K.H., Kim, J.H.: Genetic quantum algorithm and its application to combinatorial optimization problem. In: Proceedings of the 2000 Congress on Evolutionary Computing, vol. 2, pp. 1345–1360. IEEE Press, Piscataway (2000)
4. Li, S.Y., Li, P.C.: Quantum genetic algorithm based on real encoding and gradient information of object function. J. Harbin Inst. Technol. **38**(8), 1216–1218 (2006)
5. Guan, J.: Research of quantum genetic algorithm based on a combination of multi-operator. Harbin Engineering University, Heilongjiang (2014)
6. Wang, Y.P., Li, Y.H.: A novel quantum genetic algorithm for TSP. Chin. J. Comput. **30**(5), 748–755 (2007)
7. Qu, Z.J., Chen, Y.H., Li, P.J., Liu, X.H., Li, C.H.: Cooperative evolution of multiple operators based adaptive parallel quantum genetic algorithm. Acta Electronica Sinica **47**(2), 266–273 (2019)
8. Yang, J.A., Zhuang, Z.Q.: Actuality of research on quantum genetic algorithm. Comput. Sci. **30**(11), 13–15 (2003)
9. Yang, S.Y., Jiao, L.C., Liu, F.: The quantum evolutionary algorithm. Chin. J. Eng. Math. **23** (2), 235–246 (2006)
10. Nielsen, A.M., Chuang, L.I.: Quantum Computation and Quantum information (10th Anniversary Edition). Tsinghua University Press, Beijing (2015)
11. Ma, Y., Wang, H.X., Liu, H., Chen, Z.L.: Research on self-adaptive quantum genetic algorithm. Comput. Appl. **54**(20), 99–103 (2018)
12. Shao, P., Wu, Z.J.: Rosenbrock function optimization based on improved particle swarm optimization algorithm. Comput. Sci. **40**(9), 194–197 (2013)
13. Qin, B.Y., Qin, W.D., Lin, X.K.: Application of multi-population real number coding genetic algorithm in multimodal function optimization. J. Guangxi Univ. Sci. Technol. **26** (3), 13–17 (2015)
14. Gai, J.N., Zhou, J., Yang, F., Liu, C.X.: Quantum genetic algorithm for Hadamard gate mutation. Control Eng. China **25**(1), 143–148 (2018)

Natural Computation: Theory and Algorithms - Nonlinear Phenomena, Chaos, Complex Networks and Systems

Sink Location Privacy Protection Algorithm Based on Tangential Path in WSN

Leqiang Bai(✉) and Hongshu Zhu

Shenyang Jianzhu University, Shenyang, Liaoning, China
baileqiang@sjzu.edu.cn

Abstract. In order to improve the issue that the existing sink location privacy protection algorithms have lower safety period, sink location privacy protection algorithm based on tangential path in WSN is proposed. Sink determines circular security area, source nodes transmit real data packets through tangential path. Tangent point retransmits them to the sink and generates fake data packets through the original tangent path. In addition, fake sinks are randomly selected outside the security area by normal distribution model, and intersection nodes send fake packets to these fake sinks. Theoretical analysis and simulation results show that this algorithm can improve security period and achieve balance between sink location privacy protection and network energy consumption compared with the existing sink location privacy protection algorithm.

Keywords: Wireless sensor network · Sink location · Tangential path

1 Introduction

Wireless Sensor Network (WSN) is consisting of massive self-organizing sensor nodes, which is widely used in military. Deng et al. [1] first considered the location privacy of sink node in WSN, proposed antitraffic analysis strategies of sink location protection for traffic analysis attackers. Chen and Lou et al. [2] proposed four end-to-end location privacy protection technologies for both sink node and source node. Yao et al. [3] proposed a sink location privacy protection algorithm focus on hop-by-hop tracking attackers. Edith Ngai [4] proposed the research of location privacy of mobile sink in wireless sensor networks. Wang et al. [5] proposed a sink location privacy protection strategy for directional attackers. Jian et al. [6] proposed a new scheme to protect sink location privacy, called LPR. Long et al. [8] proposed an energy-friendly sink location privacy protection scheme. Baroutis et al. [9] proposed a privacy protection strategy for sink location. Rios et al. [10] proposed a probabilistic sink location protection algorithm. Model Definition

1.1 Attacker Model

Direction attacker: Direction attacker [5] refers to a kind of attacker who can calculate the transmission direction of data packets. They track data packets until get enough information, then fit out the transmission direction of the data packets, capture location of sensitive nodes along this direction.

Y. Liu et al. (Eds.): ICNC-FSKD 2019, AISC 1074, pp. 181–188, 2020.
https://doi.org/10.1007/978-3-030-32456-8_19

1.2 Concept Definition

Security Area of sink: A circle area with real sink as the center and as the radius is called safety area of sink. We choose fake sinks outside the security area of sink.

Target node T: Make a tangent from any source node to the security area of sink, called the tangent point as expected target node. In the process of sending data When a source node send packets to expected target node, each node that receives data packet judges the distance between itself and the expected target node. If the distance is less than the communication radius, consider the node as target node T.

Intersection node I: When two or more source nodes send data packets to the target node, it may pass through the same node, consider these nodes as Intersection node I.

Destination node D: The target node sends fake packets with lifetime Tf = 5 and forwards them along original tangential path, when a node receive the fake packet, it reduce Tf by one until reach zero. We call the node received a data packet with Tf = 0 as the destination node.

2 ABTP Algorithm

2.1 ABTP Algorithm Description

As shown in Fig. 1, B represent the sink, Bf_1 and Bf_2 represent fake sinks, I represent intersection node, S_1 and S_2 represent source node, T_1 and T_2 represent target node, solid line represents real packet transmission path and dotted line represents fake packet transmission path.

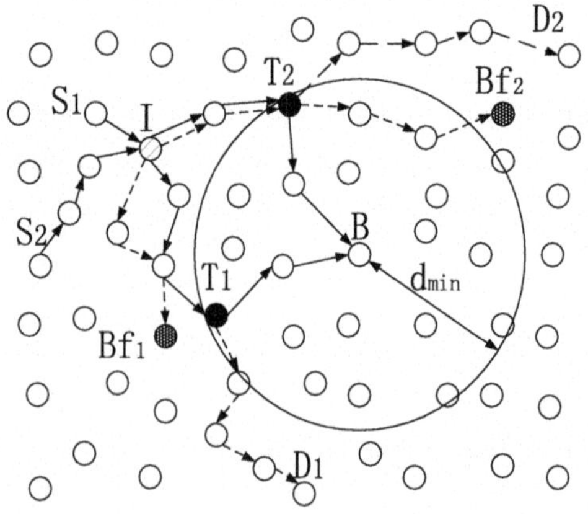

Fig. 1. Algorithm schematic diagram

2.2 Algorithm Proceeding

Calculate Coordinates of Fake Sinks Stage. Sink B determines a minimum distance d_{min}, distance between fake sinks and the real sink must be greater than or equal to d_{min}.

Assuming the distance between random selected fake sinks and the real sink is d_{rand}, the coordinates of fake sinks are (x_{Bf}, x_{Bf}), then $d_{rand} = \sqrt{x_{Bf}^2 + y_{Bf}^2}$.

In order to ensure that fake sinks only appears outside the security area, and nodes closer to real sink have more chance to be fake sinks. Before d_{rand} is determined, a random variable X should be generated first, X coincides with normal distribution with mean value of 0 and the variance of σ^2. Calculate d_{rand} according to $d_{rand} = d_{\min}(|X| + 1)$. Network used in this paper is a square area with side length l. In order to ensure d_{rand} does not overstep the bound, in other words, d_{rand} < l/2, value range of X can be obtained as follows:

$$1 - \frac{l}{2d_{\min}} < X < \frac{l}{2d_{\min}} - 1 \quad (1)$$

After determining the value of d_{rand}, the sink selects one of all coordinates randomly to be one fake sink. If we need more fake sinks, the above process needs to be repeated n times to get n fake sinks, their coordinates are (x_{Bf_i}, y_{Bf_i}), $i \in (1, n)$.

Source Nodes Send Data Packets to Target Nodes Stage. Nodes in this stage can be divided into two categories: source nodes and nodes receiving data packets.

Source nodes: Step1. Source node makes two tangents to a circle which has sink as its center and d_{min} as its radius, call the tangent point as T_1 and T_2, and selects one of them randomly as target node. Source node chooses a node which is closest to the target node from its neighbor to send packet, and the node process ends.

Nodes receiving data packets:

Step1. Determine whether the received packet is real. If it's a real packet, execute Step2; otherwise, execute Step6.
Step2. Node records the origin of each packet received, that is, ID of source node that sent the packet. Determine whether it is a target node, if it is target node, node process ends; otherwise, execute Step3.
Step3. Determine whether it is the first time to receive a packet, if it is the first time, execute Step4; otherwise, execute Step5.
Step4. Forward packets to a node in its neighbor list, which has the smallest distance to target node, process of node ends.
Step5. Determine whether ID of source node in the current received packet is the same as all previous received packets. If it is all the same, execute Step4. Otherwise, forward real packet to a node with the smallest distance to target node in its neighbor list, and generate n fake packets with different coordinates of fake sinks, forwarded them to a neighbor with the smallest distance from each fake sink respectively, end of node processing.
Step6. Determine whether the distance between itself and fake sink to be reached in the packet is more than communication radius r. If it is more than r, forwarding data packets to a node with the smallest distance from fake sink to be reached, end of node processing; otherwise, node processing ends.

Target Nodes Send Data Packets to the Sink Stage. Nodes in this stage can be divided into two categories: target nodes and nodes receiving data packets.

Target nodes: Step1. Target node forward real packets a neighbor with the smallest hops to the sink, and generates fake packets with lifetime Tf = 5, transmit along l_{S_T} tangent direction. The linear equation is as shown in formula (2):

$$y = \frac{y_S - y_T}{x_S - x_T} x + \frac{x_S y_T - y_S x_T}{x_S - x_T} \tag{2}$$

Let coordinate of random node to be (x_Q, y_Q), node receiving data packets calculates vertical distance d from all nodes within its communication range to line $l_{S_T}$is as shown in formula (3):

$$d = \left| \frac{(y_S - y_T)x_Q + (x_T - x_S)y_Q + (x_S y_T - y_S x_T)}{\sqrt{(y_S - y_T)^2 + (x_T - x_S)^2}} \right| \tag{3}$$

Select a node with the smallest value of d as next hop node to forward packets.

Nodes receiving data packets:

Step1. Determine whether the received packet is a fake one. If it is a fake packet, execute Step2; otherwise, execute Step3.

Step2. Determine whether the lifetime Tf is 0. If it is 0, abandon the packet, node process ends; otherwise, make lifetime Tf-1, calculate the vertical distance *d* from all nodes within the communication range to line l_{S_T}, and forward data packets to a node with the smallest value of d.

Step3. Node determines whether it is the sink or not. If it is the sink, node process ends; otherwise, forward data packets to a neighbor with the smallest hops to the sink, and node process ends.

3 Theoretical Analysis of Selecting Fake Sinks

In ABTP algorithm, sensor nodes with distance more than or equal d_{min} to the sink have a chance to be a fake sink. If distance between a fake sink and real base sink is too short, it may unable to induce attackers away from the real sink. If distance between a fake sink and the real sink is too far, communication overhead will increase. Therefore, this paper uses normal distribution model [11] to determine the location of fake sinks in a suitable range.

Normal distribution model is applied to data set *X*. Assumed X obeys normal distribution with expectation of 0 and standard deviation of σ, that is $X \sim (0, \sigma^2)$, its probability density is as shown in formula (4):

$$f(x) = \frac{1}{\sigma\sqrt{2\pi}} e^{-\frac{x^2}{2\sigma^2}} \; x \in R \tag{4}$$

The ratio of a certain interval area to the total area reflects the probability that a variable falls in that interval. In order to calculate the probability, probability density function is obtained as shown in formula (5):

$$F(x) = \int_{-\infty}^{x} \frac{1}{\sigma\sqrt{2\pi}} e^{-\frac{-x^2}{2\sigma^2}} dx \tag{5}$$

When σ tends to 1, data set X obeys standard normal distribution. Set $\rho > 1$, then the probability that d_{rand} falls in interval $[d_{\min}, (\rho+1)d_{\min})$ is as shown in formula (6):

$$\begin{aligned} P &= P(-\rho \le X \le \rho) = F(\rho) - F(-\rho) \\ &= F(\rho) - (1 - F(\rho)) = 2F(\rho) - 1 \end{aligned} \tag{6}$$

When $\sigma = 1$, take $x = \rho$ into formula (6), we can get the probability of d_{rand} falling in interval $[d_{\min}, (\rho+1)d_{\min})$ is as shown in formula (7):

$$2\int_{-\infty}^{\rho} \frac{1}{\sqrt{2\pi}} e^{-\frac{-\rho^2}{2}} d\rho \tag{7}$$

Select fake sinks through above method, we can get the actual probability of d_{rand} falling in interval $[d_{\min}, 2d_{\min})$ is 0.6826, probability of d_{rand} falling in interval $[d_{\min}, 3d_{\min})$ is 0.9544, probability of d_{rand} falling in interval $[d_{\min}, 3d_{\min})$ is 0.9744. That is to say, nodes whose distance to the real sink is more than or equal to d_{min} have a chance to be fake sinks, and the farther from the sink, the smaller the probability of becoming fake sinks, which can not only protect the location privacy of sink, but also control the communication overhead.

4 Simulation Experiment

In this paper, MATLAB simulation platform is used to test Lin Yao algorithm [3], LPR [6] algorithm, and ABTP algorithm. We analyze three indexes: Transmission delay, protection power and communication overhead. Protection power is divided into two aspects: security period and attack time. The network size is 6000 m $\times$ 6000 m, which is evenly divided into 100 $\times$ 100 grids. Communication radius of node is 100 m, and the detection radius of attacker is also 100 m. Each node has an average of 8.72 neighbor nodes. In the experiment, sink node is fixed in the center of network. Attacker's initial position is 20 hops from sink, and it has memory of 5 hops of his own trajectory.

4.1 Transmission Delay

In this paper, transmission delay refers to the number of hops required for a packet (real data packet) to be transmitted from the source node to the base station according to the algorithm flow. As shown in Fig. 2, transmission delay of all three algorithms increases with increase of hops from source to sink. In Lin Yao algorithm, real data packets is transmitted to sink through the shortest path, so the transmission delay is always minimum relatively; In LPR algorithm, real data packet is transmitted to distant or near neighbors randomly, path length increases greatly, and transmission delay increases with that fastest; In ABTP algorithm, real data packet is transmitted from source node to target node, and then from target node to sink. Path length of each transmission is approximately equal, so transmission delay of ABTP algorithm increases steadily and keeps in a relatively low range.

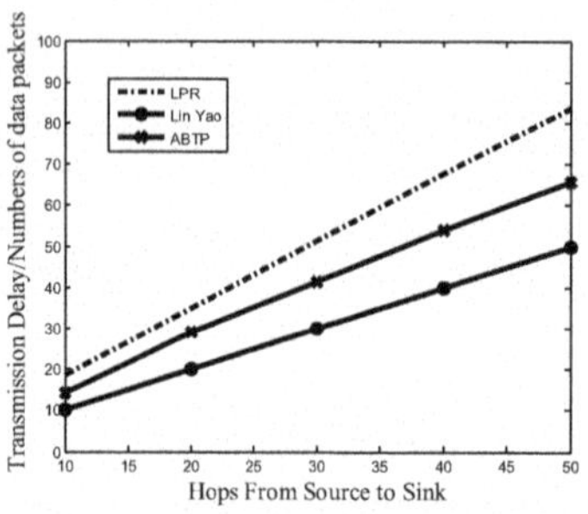

Fig. 2. Change of transmission delay

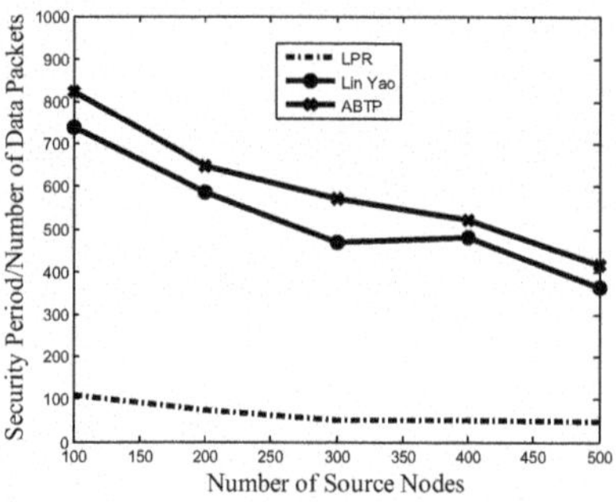

Fig. 3. Change of security period

4.2 Protection Strength

In this paper, two indexes are used to measure protection strength of the algorithm: security period and attack time. Security period refers to the number of packets received by sink before being captured; attack time refers to the number of steps that attacker moves before arriving at the sink.

Security Period. As shown in Fig. 3, when the number of source nodes increases, security period of all the three algorithm decreases, but ABTP algorithm can still maintain a high level. The increasing of source nodes will also lead to the increase of real paths directly, the probability of attackers appearing on real path will also increase. If the transmission path of an algorithm never changed and the initial location of attacker is not on real transmission path, attacker will not move at all, and security cycle reaches the maximum value.

LPR algorithm has no fake packets injection, so it has low security period. Both ABTP algorithm and Lin Yao algorithm use fake packets, security period of ABTP algorithm is still higher than that of Lin Yao algorithm under the same number of fake sinks. Lin Yao algorithm sends fake packets by intersection nodes; ABTP algorithm sends fake packets by intersection nodes and target nodes, fake packets sent by target nodes can induce direction attackers to deviate from real path effectively.

Attack Time. Figure 4 shows attack time varies with the number of source nodes. Compared with Lin Yao algorithm and LPR algorithm, attack time in ABTP algorithm is always at the highest level. As shown in Fig. 4, unlike the security period, correlation between attack time and the number of source nodes is weak.

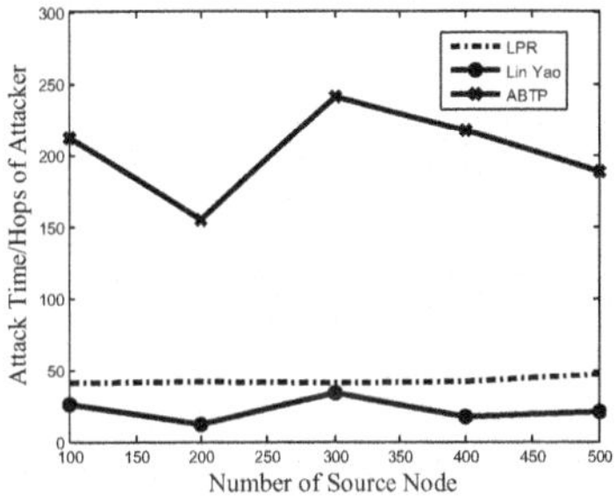

Fig. 4. Change of attack time

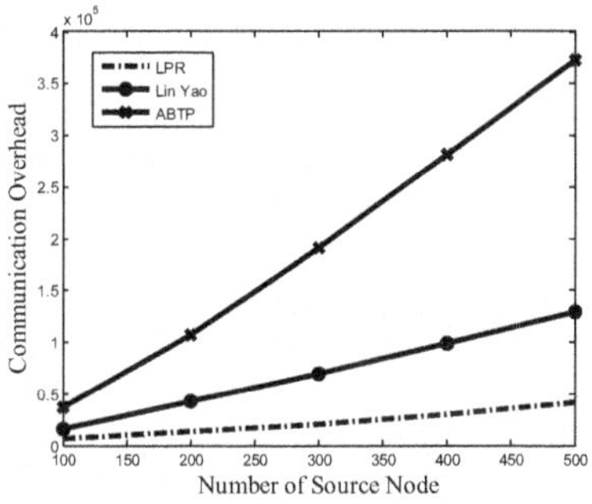

Fig. 5. Change of communication overhead

Combining with Figs. 3 and 4, we can see that the security cycle and attack time of ABTP algorithm are at the highest level. Security period of Lin Yao algorithm is slightly lower than that in ABTP algorithm, but attack time of Lin Yao algorithm is much lower than that in ABTP algorithm and even lower than that in LPR algorithm. It shows that the high security period of Lin Yao algorithm is due to the fact that the attacker has hardly moved.

Because if the attacker did not move, attack time will be zero but security period will hit the maximum value. In Lin Yao algorithm, real data packets are transmitted in shortest path, it is centralized and fixed. The probability of attackers appearing on real paths is small, and the security cycle will be high. But attackers' not moving makes attack time reduce.

4.3 Communication Overhead

In this paper, communication overhead refers to the total number of hops that all packets (true and false) need to be transmitted. As shown in Fig. 5, with the increase of the number of source nodes, communication overhead of all the three algorithms is increasing. LPR algorithm has the slowest increase in communication overhead, and ABTP algorithm has the fastest increase in communication overhead. Compared with Figs. 3 and 5, ABTP algorithm achieves a higher security period at the expense of communication overhead.

5 Concluding Remarks

This paper proposes a sink location privacy protection algorithm based on tangential path. In the algorithm, real data packets are transmitted along tangential path, and fake data packets are injected by tangential nodes and intersection nodes to increase the diversity of transmission paths. Theoretical analysis and experimental results show that ABTP algorithm can induce direction attacker to deviate from real path, achieve balance of security period and communication overhead, and improve security of sink.

References

1. Deng, J., Han, R., Mishra, S.: Intrusion tolerance and antitraffic analysis strategies for wireless sensor networks. In: DSN 2004: Proceedings of the 2004 International Conference on Dependable Systems and Networks, pp. 637–646. IEEE Computer Society, Washington, DC (2004)
2. Chen, H., Lou, W.: From nowhere to somewhere: protecting end-to-end location privacy in wireless sensor networks (2010)
3. Yao, L., Kang, L., Shang, P., et al.: Protecting the sink location privacy in wireless sensor networks. Pers. Ubiquit. Comput. **17**(5), 883–893 (2013)
4. Ngai, C.H., Rodhe, I.: On providing location privacy for mobile sinks in wireless sensor networks. In: ACM International Conference on Modeling, Analysis and Simulation of Wireless and Mobile Systems, pp. 116–123. ACM (2009)
5. Wang, J., Wang, F., Cao, Z., et al.: Sink location privacy protection under direction attack in wireless sensor networks. Wireless Netw. **23**(2), 1–13 (2017)
6. Jian, Y., Chen, S., Zhang, Z., et al.: A novel scheme for protecting receiver's location privacy in wireless sensor networks. IEEE Trans. Wireless Commun. **7**(10), 3769–3779 (2008)
7. Liu, C.X., Liu, Y., Zhang, Z.J., et al.: High energy-efficient and privacy-preserving secure data aggregation for wireless sensor networks. Int. J. Commun. Syst. **26**(3), 380–394 (2013)
8. Long, J., Liu, A., Dong, M., et al.: An energy-efficient and sink-location privacy enhanced scheme for WSNs through ring based routing. J. Parallel Distrib. Comput. **81–82**(C), 47–65 (2015)
9. Baroutis, N., Younis, M.: Load-conscious maximization of base-station location privacy in wireless sensor networks. Comput. Netw. **124**, 126–139 (2017)
10. Rios, R., Cuellar, J., Lopez, J.: Probabilistic receiver-location privacy protection in wireless sensor networks. Elsevier Science Inc. (2015)
11. Li, Y., Ren, J.: Mixing ring-based source-location privacy in wireless sensor networks. In: IEEE Communications Society Conference on Sensor, Mesh and Ad Hoc Communications and Networks, SECON 2009, pp. 1–9. IEEE (2009)

Study on the Energy Dissipation of Two Lane Traffic Flow with Lane Reduction

Li-Si Jia[1(✉)] and Wei Pan[2]

[1] Guangxi University XingJian College of Science and Liberal Arts, Nanning 530005, China
356587388@qq.com

[2] Guangxi Medical University, 530021 Nanning, China

Abstract. In this paper, a two-lane model with partial reduced lane was simulated based on the Kerner-Klenov-Wolf (KKW) cellular automaton model. In the traffic system with periodic boundary condition, the synchronized flow was investigated by local measurements of the density and flow in the KKW model near the entry of the reduced lane. And the length of the reduced lane and the maximum speed and the vehicle length are studied, respectively. The results show that the partial lane reduction strongly reduces the road capacity, and increases the energy dissipation. Moreover, the open boundary condition is given to analyze. The results implied that the injection rate as well as the removal rate has obvious influence on the energy dissipation.

Keywords: KKW model · Traffic bottleneck · Energy dissipation

1 Introduction

In modern society, the problem of traffic congestion, environmental pollution and energy dissipation caused by traffic have become more and more serious, and affect people's daily life. In accordance with previously reported results in refs [1], more than 20% additional fuel consumption and air pollution are caused by impeded traffic and "go and stop" traffic. How to solve the traffic problems comes to be a great challenge for us. The problems of energy dissipation in traffic system have attracted lots of scholars. Many traffic flow models such as car following model, city traffic model and CA model, hydrodynamic models are investigated in succession, respectively [2, 3]. The relation between the stability and the energy consumption by analyzing and comparing for many typical car-following models has been discussed [4]. The additional energy consumption of traffic flow under the action of an external disturbance was studied by using the optimal velocity model [2]. The effects of speed limits, stochastic noise and boundary conditions on energy consumptions were discussed in NaSch model [5, 6]. The influences of the maximum velocity and the length of vehicle in mixed traffic flow have been investigated in the NaSch model and the FI model [3, 7].

Traffic bottlenecks can be seen everywhere in daily life, which not only affect traffic capacity but also get rise to many complex traffic phenomenon. In this paper, we attempt to study the energy dissipation two-lane model with partial reduced lane based

Y. Liu et al. (Eds.): ICNC-FSKD 2019, AISC 1074, pp. 189–196, 2020.
https://doi.org/10.1007/978-3-030-32456-8_20

on KKW model. In Sect. 2, we give the definition of the energy dissipation, evolution rules of the KKW model, respectively. In Sect. 3 simulations and analysis are given. We discuss how the reduced lane and the maximum speed and the vehicle length and the injection rate of the road and the length of the reduced lane affect traffic capacity and energy consumption, respectively. Finally, some conclusion is given in Sect. 4.

2 Model of Reduced Lane and Energy Dissipation

The model is defined on a two-lane road with partial reduced lane (see Fig. 1). The traffic bottleneck is appeared on the right lane with the location of x1. The reduced lane (Lr) starts from the location of x1 and ends from x2.

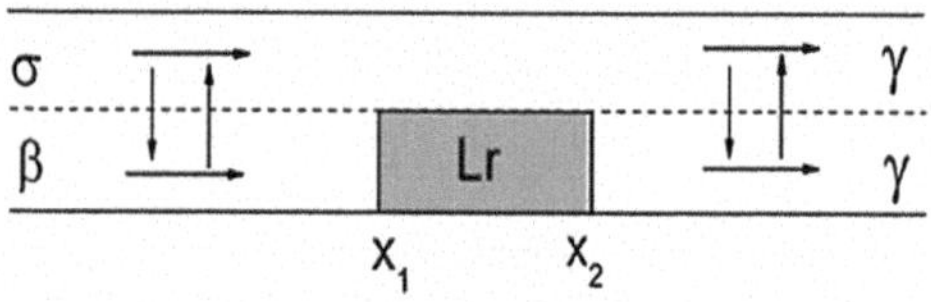

Fig. 1. Sketch of two-lane model with partial reduced lane and illustration of the lane changing. The length Lr is located on the right lane.

In order to mimic the synchronized flow near the entry of the reduced lane, the KKW model is used here.

The evolution rules of KKW-1 model [8] are as follows.

A. *Deterministic rule* ($t < t_1 < t+1$):
RAa. determine synchronization distance:

$$D_n \leftarrow D_0 + kv_n(t) \tag{1}$$

RAb. determine acceleration and deceleration:

$$\Delta = \begin{cases} -a & if \quad v_n(t) > v_{n+1}(t) \\ a & if \quad v_n(t) < v_{n+1}(t) \\ 0 & if \quad v_n(t) = v_{n+1}(t) \end{cases} \tag{2}$$

RAc. determine desired speed:

$$v_{des}(t+1) = \begin{cases} v_n(t) + a & if \quad v_{s,n}(t) > D_n(t+1) - d_n \\ v_n(t) + \Delta & if \quad v_{s,n}(t) \le D_n(t+1) - d_n \end{cases} \tag{3}$$

RAd. determine deterministic speed:

$$v_n(t_1) = \max\{0, \min\{v_{\max}, v_{s,n}(t), v_{des}(t)\}\} \tag{4}$$

B. *Stochastic rule:*

$$p_a(v_n) = \begin{cases} p_{a1} & if \quad v_n < v_p \\ p_{a2} & if \quad v_n \geq v_p \end{cases} \tag{5}$$

RBa. determine acceleration probability
RBb. determine braking probability:

$$p_b(v_n) = \begin{cases} p_0 & if \quad v_n = 0 \\ p & if \quad v_n > 0 \end{cases} \tag{6}$$

RBc. determine stochastic noise:

$$\eta_n = \begin{cases} -1 & if \quad rand < p_b \\ 1 & if \quad p_b < rand < p_b + p_a \\ 0 & otherwise \end{cases} \tag{7}$$

RBd. determine stochastic speed:

$$v_{n+1} = \max\{0, \min\{v_n(t_1) + \alpha\tau\eta_n, v_n(t_1) + \alpha\tau, v_{\max}, v_{s,n}(t)\}\} \tag{8}$$

C. *Vehicle movement:*

$$x_n(t+1) = x_n(t) + v_n(t) \tag{9}$$

The Lane changing rules are as follows:

D. *Incentive criterion*:

$$d_n < \min(v_n + 1, v_{\max}) \tag{10}$$

E. *Safety criterion*:

$$d_n < d_{n,pred}, d_{n,succ} > d_{safe} \tag{11}$$

Here $d_{n,pred}$, $d_{n,succ}$ denote the gap to the predecessor and the succeeding vehicle on the destination lane respectively. d_{safe} denotes safe distance ($d_{safe} = v_{max}$). Thus, a vehicle is allowed to change lane only if the lane change rules are fulfilled.

The kinetic energy of a vehicle with velocity v is $mv^2/2$, and m is the mass of the vehicle. In this paper, the mass of the vehicle is proportional to its length. When the vehicle decelerates, the kinetic energy is lost which is defined as the energy dissipation. For simplicity, we neglect frictional resistance, air drag and other dissipation in our analysis. The energy dissipation of the *n-th* vehicle from time $t-1$ to t is defined as

$$e(n,t) = \begin{cases} \frac{m}{2}[v^2(n,t) - v^2(n,t)] & for \quad v(n,t) < v(n,t-1) \\ 0 & for \quad v(n,t) \geq v(n,t-1) \end{cases} \tag{12}$$

Thus, the energy dissipation per time step per vehicle is

$$E_d = \frac{1}{T}\frac{1}{N}\sum_{t=t_0+1}^{t_0+T}\sum_{n=1}^{N} e(n,t) \tag{13}$$

Where N is the number of vehicles in the system and t_0 is the relaxation time.

The simulation of the traffic behavior in the traffic system with the partial reduced lane is respectively performed in open boundary condition and periodic boundary condition. The parameters are chosen as follows. The length of each lane is assumed as L = 60000 cells (30 km). Each vehicle is assumed to occupy 15 cells. Each cell is assumed to be 0.5 m. The maximum velocity is v_{max} = 108 km/h = 60 cells/s. The delay probability are P_{a1} = 0.2, P_{a2} = 0.052, P_0 = 0.425 and the other parameters are set as D_0 = 60, k = 2055, a = 1, p = 0.04, $d_{\text{safe}} = v_{\max}$ = 60. The relaxation time is $t_0 = 4\times10^4$ and the total simulation time is $T = 2\times10^4$. The length of the reduced lane Lr is adjustable parameters. In the open boundary systems, the injection rate σ and β are adjustable parameters and the extinction rate $\gamma(\gamma = 1)$ is determine.

3 Simulation and Analysis

3.1 Influence of the Length of the Reduced Lane on the Road Capacity in the Periodic Boundary Condition

First, we study the influence of the length of the reduced lane Lr on the traffic flow. We choose the parameters: Lr = 0, 1800, 15000, 36000. Figure 2 Show the relationship between the global flow and occupancy with various value of Lr. When Lr = 0, there are two critical value of occupancy $\rho 1$ = 0.19 and $\rho 2$ = 0.26. When the $\rho 1$–$\rho 2$, traffic flow is synchronized flow. When Lr larger than 0, the value of global flow increases with increasing the value of occupancy until reaches a maximum value and becomes a constant value, but abruptly decreases at critical point. With increasing length of the reduced lane Lr, the less than the value of occupancy will be.

Moreover, Fig. 3 given the fitting result of the road capacity which used liner least square method. The relationship between the road capacity and length of the reduced lane is Y = 0.40542 − 3.39835E − 6 * X, where Y is the road capacity and X is length of the reduced lane. The results implied that the road capacity decreases with increasing length of the reduced lane Lr.

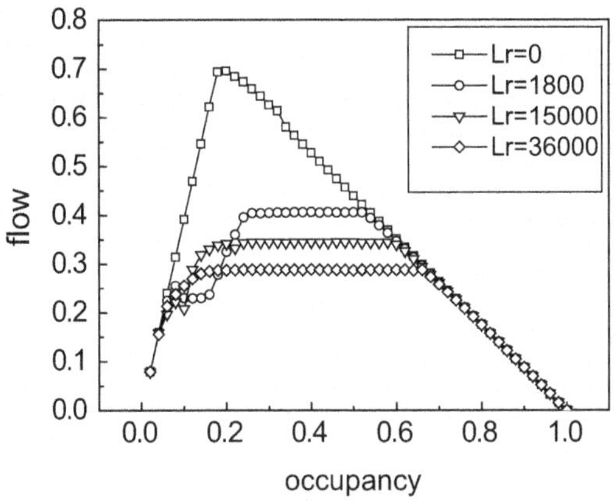

Fig. 2. The fundamental diagram of global flow and occupancy.

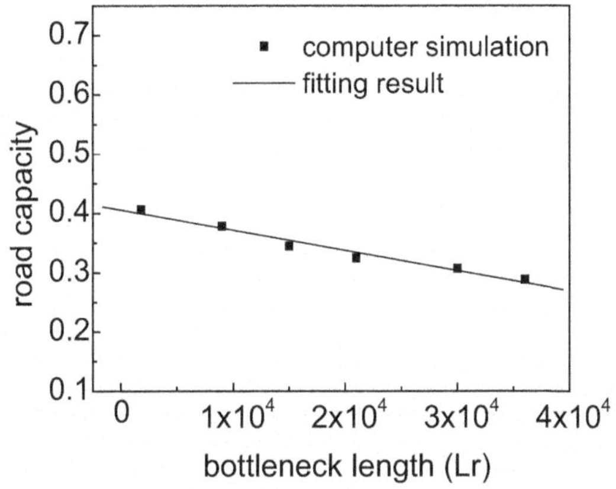

Fig. 3. The road capacity for different occupancy for different bottleneck length (Lr).

3.2 Influence of the Length of the Reduced Lane and the Maximum Speed and the Vehicle Length on the Energy Dissipation in the Periodic Boundary Condition

In the Fig. 4, the relationship between the energy dissipation (Ed) and the occupancy with various value of Lr is shown. The energy dissipation increases and reaches a maximum value at the critical point ρ1. When ρ1–ρ2, the Ed decreases reaches a minimum value. When $\rho > \rho 2$, the Ed decreases reaches. Here, the value of ρ corresponds to Fig. 2. The results implied that when the $Lr > 0$, the energy dissipation Ed increases with increasing length of the reduced lane Lr. Figures 5 and 6 are similar to results of the Fig. 4.

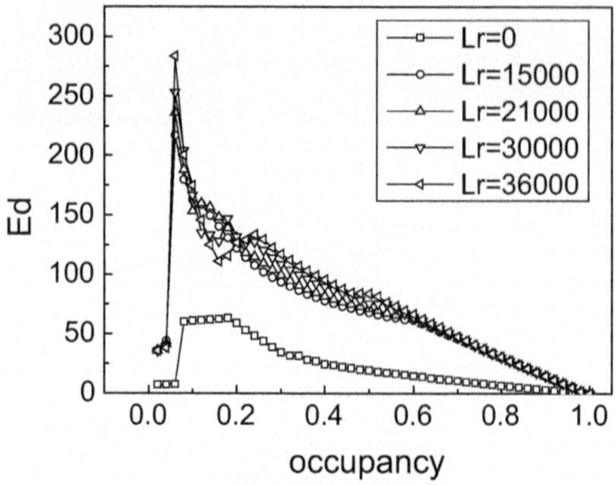

Fig. 4. The average energy dissipation for different Lr where $v_{max} = 60$, $d = 15$.

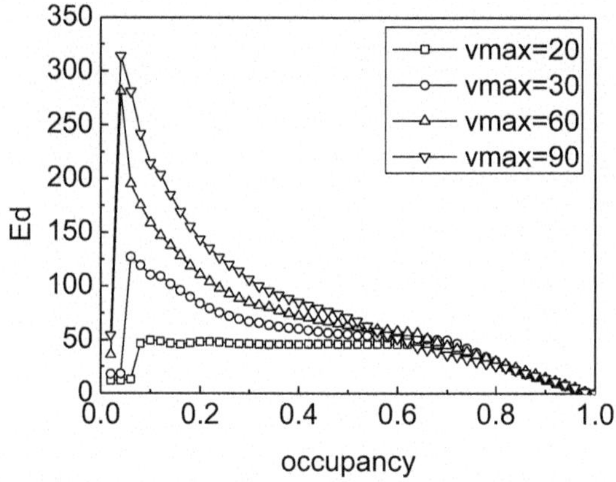

Fig. 5. The average energy dissipation for different maxing velocity where Lr = 6000, $d = 15$.

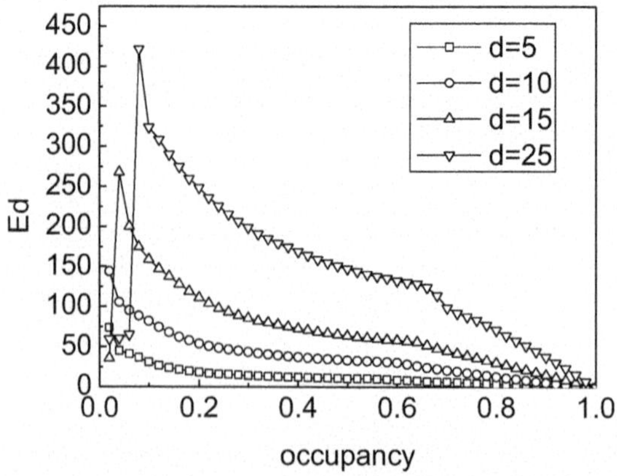

Fig. 6. The average energy dissipation for different vehicle length where Lr = 6000, Vmax = 60

3.3 Influence of the Injection Rate σ and β on the Energy Dissipation in the Periodic Boundary Condition

The simulation is performed on a highway of length L = 60000 cells and the reduced lane Lr = 15000 cells in the periodic boundary condition. Here is removal rate $\gamma = 1$. In the Figs. 7 and 8, the relationship between the energy dissipation Ed and the injection rate with various value of Lr is shown. The Simulation show the results of the Figs. 7 and 8 are similar. In the Figs. 7 and 8, the energy dissipation increases and reaches a maximum value at the critical point of injection rate, but abruptly decreases at critical point and becomes zero that traffic flow is wide moving jam. The Simulation show the energy dissipation (Ed) increases with increasing injection rate.

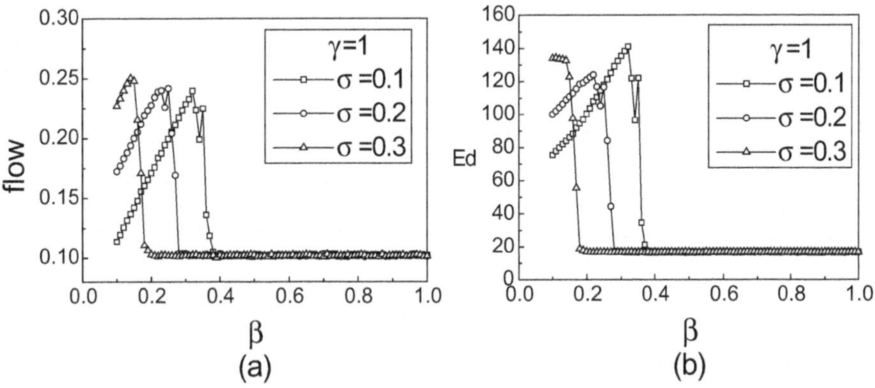

Fig. 7. (a) The injection rate β and flow with different σ. (b) The injection rate β and energy dissipation with different σ.

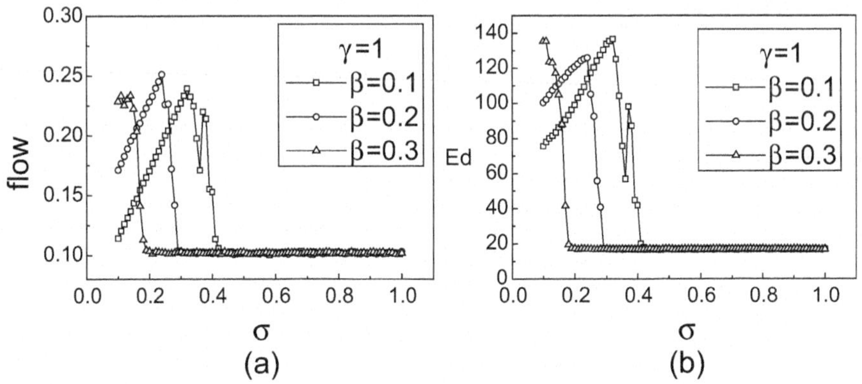

Fig. 8. (a) The injection rate σ and flow with different β. (b) The injection rate σ and energy dissipation with different β.

4 Summary and Conclusion

In this paper, based on KKW model, a two-lane model with a partial reduced lane and maximum speed limitation and at the entry of the reduced lane was simulated. The simulation results indicate that the length of the reduced lane can affect the capacity of the road and the energy dissipation. The energy dissipation (Ed) increases with increasing length of the reduced lane (Lr) and and the maximum speed and the vehicle length in the periodic boundary condition. Moreover, in the open boundary condition, the injection rate were found to have obvious influence on the change of the energy dissipation while the location of the reduced lane was not. The Simulation show the energy dissipation (Ed) increases with increasing injection rate.

Acknowledgment. This work is supported by Guangxi Higher Education and Young Teachers' Basic Ability Improvement Project (No. 2017KY1300).

References

1. Helbing, D.: Traffic and related self-driven many-particle systems. Rev. Mod. Phys. **73**(4), 1067–1141 (2001)
2. Nakayama, A.: Effect of looking at the car that follows in an optimal velocity model of traffic flow. Phys. Rev. E **65**(1), 016112 (2002)
3. Tian, H.H.: Study of the traffic flow on the various sloping way. Acta Phys. Sinica **59**, 7693 (2010)
4. Shi, W.: Study on stability and energy consumption in typical car-following models. Phys. A **381**(15), 399–406 (2007)
5. Zhang, W.: Energy dissipation in the deterministic and nondeterministic Nagel-Schreckenberg models. Phys. A **387**(18), 4657–4664 (2008)
6. Zhang, W.: Boundary effects on energy dissipation in a cellular automaton model. Physics and Society, arXiv:0904.3727 (2009)
7. Tian, H.H.: Study on the energy consumption using the cellular automaton mixed traffic model. Acta Phys Sinica **58**, 4506 (2009). (in Chinese)
8. Kerner, B.S.: Cellular automaton approach to three-phase traffic theory. J. Phys. A Math. Gen. **35**, 9971 (2002)

Chaos and Quasi-period in Erbium-Doped Fiber Laser

Senlin Yan(✉)

Department of Electronic Engineering, Nanjing Xiaozhuang University, Nanjing 21171, China
senlinyan@163.com

Abstract. Nonlinear dynamics in a short cavity Erbium (Er)-doped fiber laser is discussed while a complex dynamical model is presented by a two-level mode containing a periodic absorption of photon. The laser can produce a periodic state and a chaotic state when a periodic square-wave signal is used to modulate absorption of photon. A path to chaos is roughly presented by varying the modulation depth while a periodic pulsation, a quasi-periodic pulsation and a chaotic pulsation are produced and are analyzed. Another path to chaos is presented by varying the modulation frequency while a periodic pulsation, a quasi-periodic pulsation and a chaotic pulsation are produced and are discussed in detail. And some periodic regions, quasi-periodic regions and chaotic regions are introduced via the modulation depth and frequency. It can be found that the modulation depth and frequency determine dynamical behavior of the laser while the parameters of modulation depth and frequency can affect dynamics of the laser and can change behavior evolution of the laser. The effect of Er-doped concentration on dynamics of the laser is analyzed to find to change behavior of the laser. The obtained result is helpful to study on fiber laser technology and chaotic laser application.

Keywords: Fiber laser · Chaos · Dynamics · Quasi-period · Erbium-doped fiber

1 Introduction

Under certain conditions, a nonlinear dynamic system can be prone to show chaos. And chaotic characteristic has sensitivity to its starting condition while its dynamics results in an attractor of infinite dense trajectory in phase-space. A chaotic signal shows to vary in disorder, complication and pseudorandom. At present, a lot of chaotic lasers were reported to apply in many areas [1–4]. And an Er-doped fiber laser is found to show chaos and other nonlinear dynamics using an additional operation on the laser [5–9]. An Er-doped fiber laser has been widely used in optical communications and fiber sensors. An Er-doped fiber laser shows some attractive influences to researchers for its unique dynamical behavior. It is found that lifetime of erbium ion at the metastable state can exist for 1–10 ms. Er-doped fiber laser can produce some continuous transitions to show periodic pulses or chaos [5]. And many researcher are attracted to a short cavity single-mode Er-doped fiber laser [8]. We will focus on the

Y. Liu et al. (Eds.): ICNC-FSKD 2019, AISC 1074, pp. 197–205, 2020.
https://doi.org/10.1007/978-3-030-32456-8_21

short cavity single-mode Er-doped fiber laser and study its cycle pulse and chaos. And a path to chaos is discussed while periodic region and chaotic region are analyzed.

2 Dynamical Model

When the pump operates on the laser, erbium ions can transfer fastly from the level 3 to level 2, which results in activation level 3 at an almost vacancy state and the level 2 at a excited state. So a physical model of the Er-doped fiber laser can be described by a two level. The variable N is used to describe the normalized ion number density and the variable P is used to describe the photon number density. A physical model of the laser can is presented by the rate equation as following

$$\frac{dN}{dt} = \frac{P_p}{h\nu_p A_p}\sigma_1(1-N) - \frac{c}{n_c}\sigma_2(2N-1)P - \frac{N}{\tau} \tag{1}$$

$$\frac{dP}{dt} = \frac{c}{n_c}Q\sigma_2(2N-1)P - \frac{P}{\tau_c} \tag{2}$$

Where P_p is the pump of the laser. A_p is the effective mode field area of the laser at the pump wavelength. σ_1 is the absorption cross section of the laser. σ_2 is the stimulated emission and absorption cross section of the laser. $h\nu_p$ is the pump energy into the laser. c is the light velocity in vacuum. n_c is the medium refractive index of the laser. τ is the laser upper level lifetime. τ_c is the lifetime of photon in laser cavity where we can define $\gamma_c = 1/\tau_c$ as photon attenuation rate. Q is the total ion population density. Periodic states and chaos of the laser are generated by introducing an additional operation on the photon attenuation rate. This additional operation can be realized by introducing an photon attenuator or photon absorber on the laser to shift photon attenuation rate while the photon attenuation rate γ_c parameter will be rewritten as $\gamma_c[1 + \eta \times \text{square}(2\pi ft)]$, where "square" is a periodic signal of square wave, η is the depth and f is the frequency. When this additional freedom operation is operated on the laser to shift the photon attenuation rate, the laser can make some chaotic pulses and periodic pulses.

3 Result and Discussion

In numerical simulation, the laser parameters are taken as $\sigma_1 = 2 \times 10^{-21}$ cm^2, $\sigma_2 = 7 \times 10^{-21}$ cm^2, $n_c = 1.46$, $\tau = 10$ ms, $\tau_c = 4$ ns, $Q = 5 \times 10^{19}$/cm^3, $A_p = 2 \times 10^{-7}$ cm^2 and $P_p = 0.1$ W [8].

3.1 A Path to Chaos via Varying the Modulation Depths

Effect of the modulation depth is analyzed on dynamic and a path to chaos is presented when the modulation is operated on the photon attenuation rate of the laser. Figure 1 shows a periodic pulse state where the modulation parameters are taken as $\eta = 0.1$ and f = 0.67 MHz. When the modulation depth is taken as 0.09, the laser produces a period-3 state shown in Fig. 2. When the modulation depth is taken as 0.1, the laser produces chaos shown in Fig. 3. A chaotic region is found between 0.1 and 0.3.

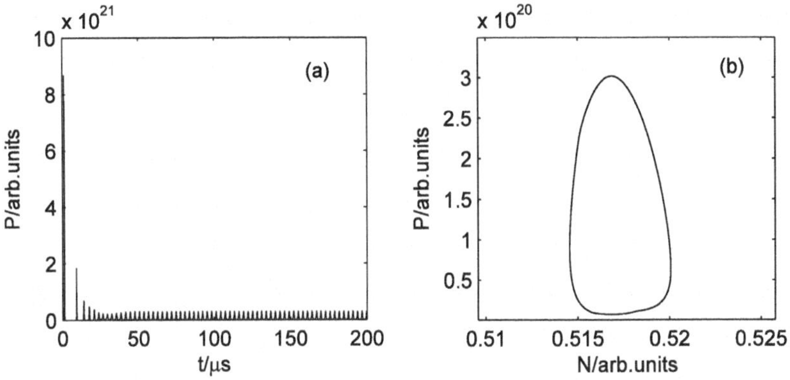

Fig. 1. A periodic state. (a) waveform and (b) a cycle ring.

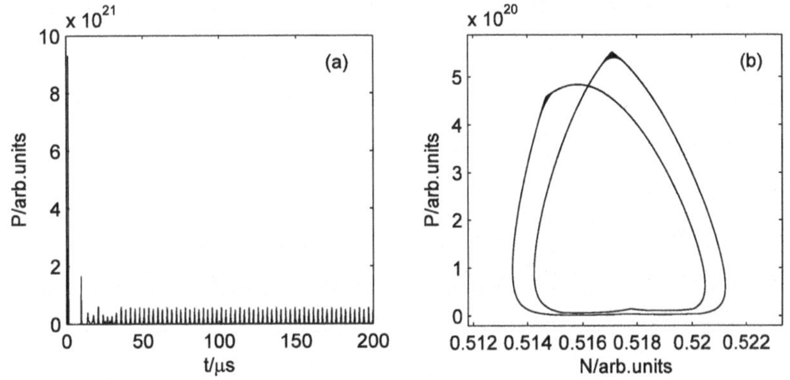

Fig. 2. A period-3 state. (a) waveform and (b) a cycle-3 orbit.

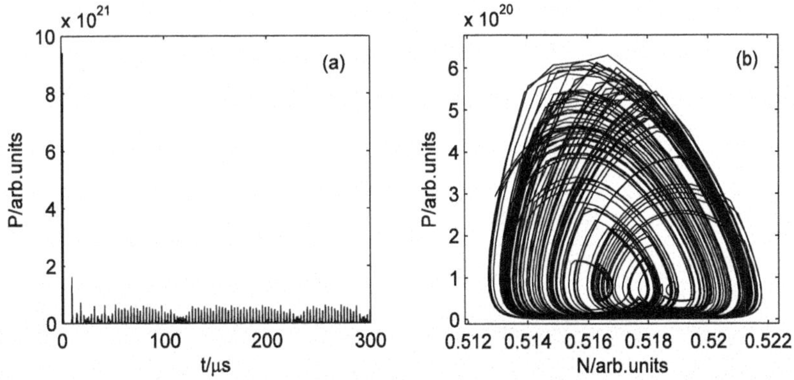

Fig. 3. Chaos. (a) waveform and (b) a chaotic attractor.

A full developed chaos is found to show in Fig. 4 where η = 0.25. By varying the depths, the obtained results show roughly a path to chaos from a cycle state.

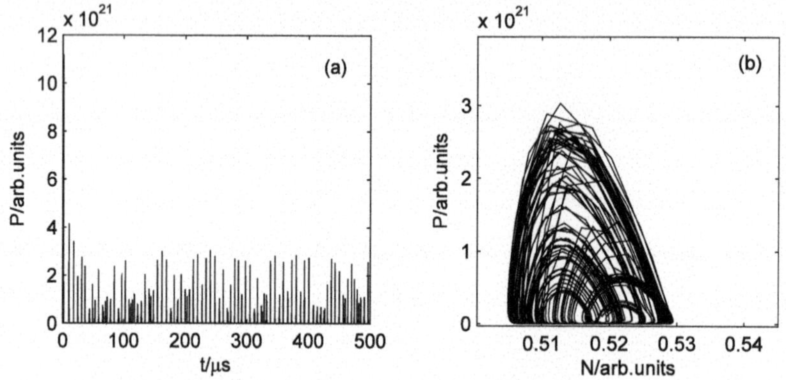

Fig. 4. Full developed chaos. (a) a chaotic wave and (b) another chaotic attractor.

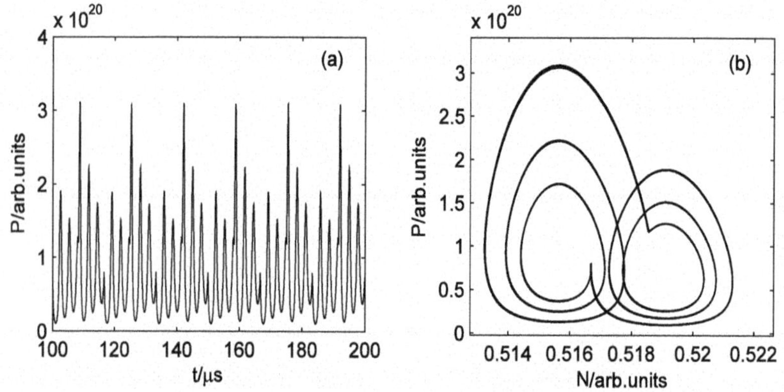

Fig. 5. A period-7 state. (a) a period-7 wave and (b) a period-7 orbit.

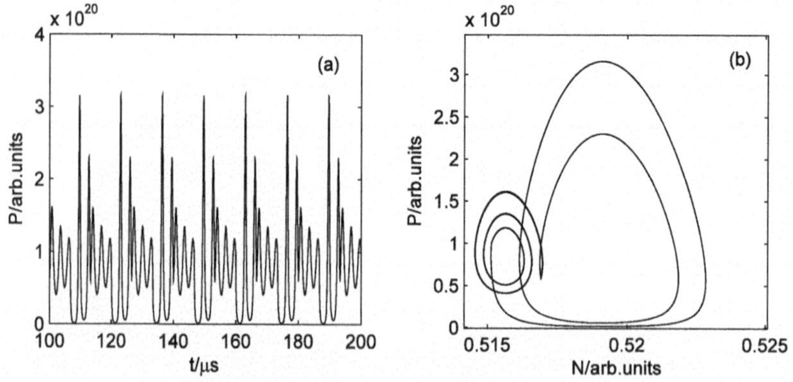

Fig. 6. A period-5 state. (a) a period-5 wave and (b) a period-5 orbit.

3.2 Another Path to Chaos via Varying the Modulation Frequencies

Effect of the modulation frequency is analyzed on dynamic and another path to chaos is presented by varying the modulation frequencies. When the parameter is taken as $\eta = 0.1$ and the frequency is taken as f = 0.06 MHz, a period-7 state results from the laser shown in Fig. 5. When the frequency is taken as f = 0.075 MHz, a period-5 state is produced to show in Fig. 6. When f = 0.075 MHz, a quasi-periodic state is produced to show in Fig. 7. When f = 0.09 MHz, a period-3 state is produced to show in Fig. 8. A period-3 region is found between 0.09 MHz and 0.13 MHz. A unstable region is found between 0.14 MHz and 0.16 MHz. When f = 0.17 MHz, another period-3 state is produced to show in Fig. 9. And when f = 0.18 MHz, a period-3 state shows in laser.

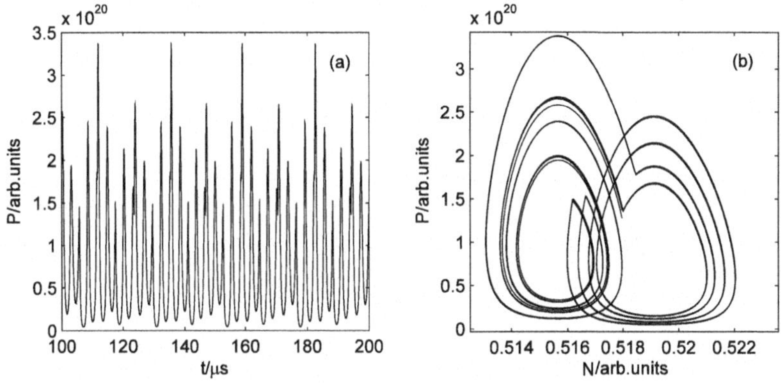

Fig. 7. A quasi-periodic state. (a) a quasi-periodic wave and (b) a quasi-periodic orbit.

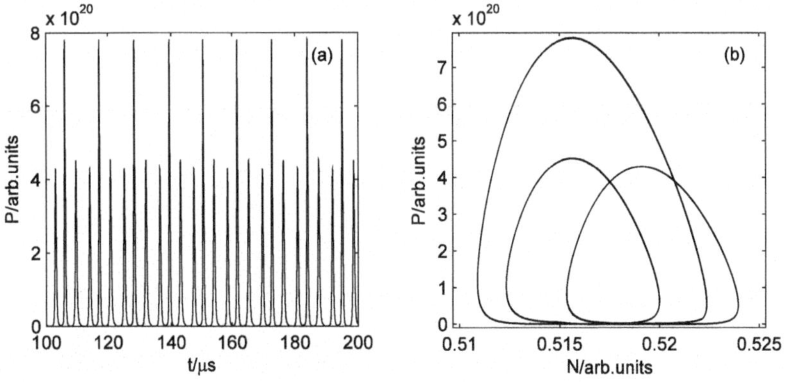

Fig. 8. A period-3 state. (a) a period-3 wave and (b) a period-3 orbit.

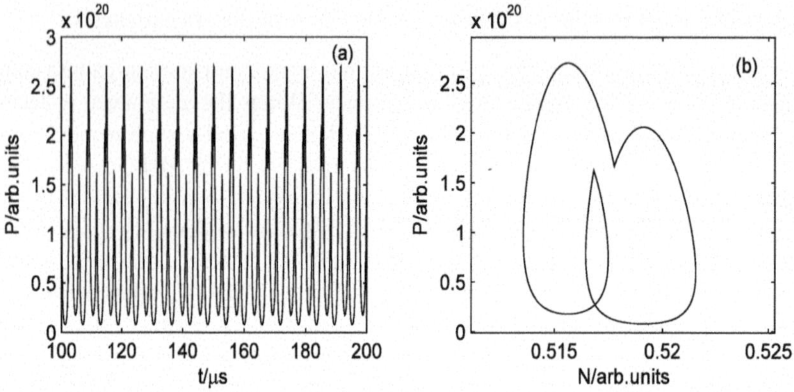

Fig. 9. Another period-3 state. (a) another period-3 wave and (b) another period-3 orbit.

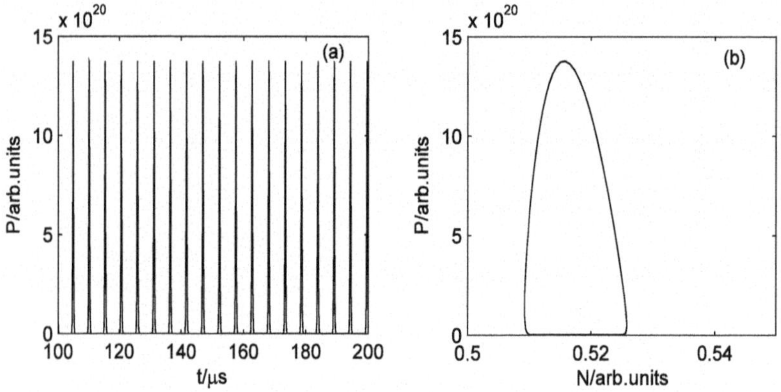

Fig. 10. A period-one state. (a) a periodic wave and (b) a periodic orbit.

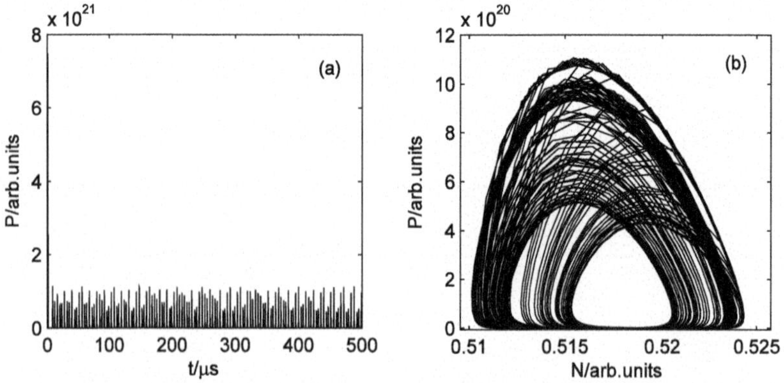

Fig. 11. A chaotic state. (a) a chaotic wave and (b) a chaotic orbit.

When f = 0.19 MHz, a period-one state is produced to show in Fig. 10 where the oscillation frequency locks at 0.19 MHz. And a periodic region is found between 0.19 MHz and 0.21 MHz. A unstable region is found between 0.22 MHz and 0.24 MHz. And a chaotic region is found between 0.25 MHz and 0.26 MHz. Figure 11 shows chaos in the laser when f = 0.25 MHz.

A period-3 region is found between 0.27 MHz and 0.31 MHz. Figure 12 shows a period-3 state in laser when f = 0.28 MHz.

A period-6 region is found between 0.32 MHz and 0.34 MHz. A period-3 region is found between 0.35 MHz and 0.36 MHz.

And a chaos region is found between 0.37 MHz and 0.4 MHz. Figure 13 shows a chaotic state in laser when f = 0.40 MHz.

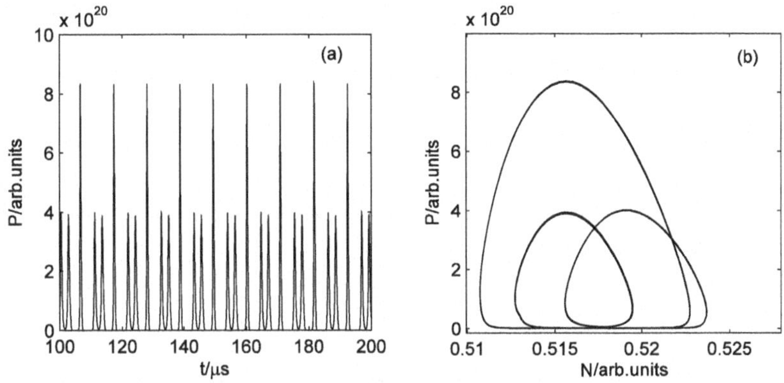

Fig. 12. A period-3 state. (a) a period-3 wave and (b) a period-3 orbit.

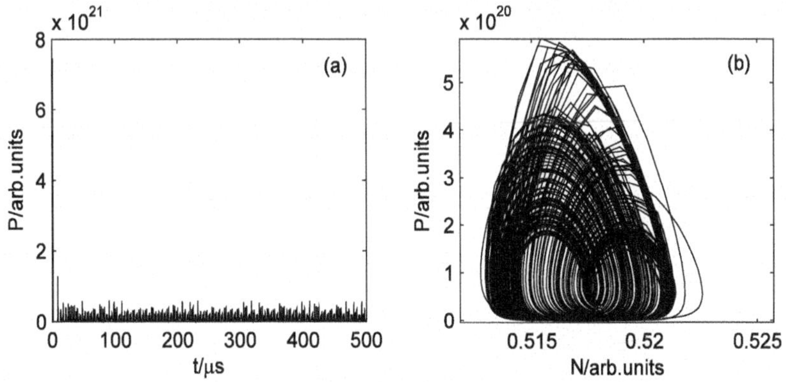

Fig. 13. Chaos. (a) a chaotic wave and (b) a chaotic orbit.

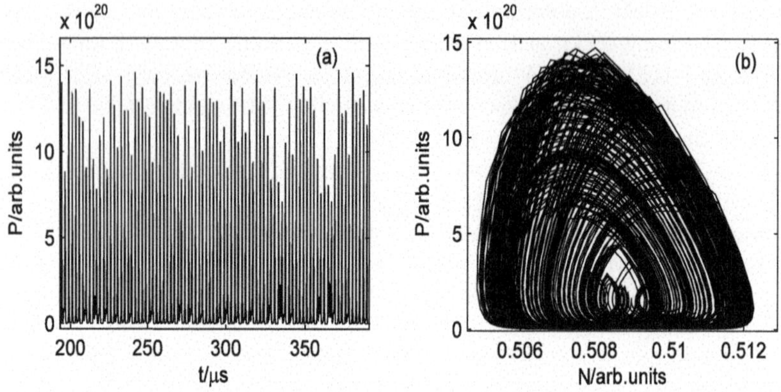

Fig. 14. Chaos. (a) a chaotic wave and (b) a chaotic attractor.

3.3 Effect of Er-Doped Concentration on the Dynamics

Effect of Er-doped concentration is analyzed on dynamic and another path to chaos is presented. When the Er-doped concentration is taken as $Q = 10 \times 10^{25}$, and let $\eta = 0.15$ and f = 0.1 MHz, the laser is reduced to show a period-5 state. A period-3 region is found between 0.15 MHz and 0.3 MHz. A period-one state, a period-2 state and other states appear alternately between 0.325 MHz and 0.7 MHz. And chaos shows between 0.82 MHz and 0.88 MHz. Figure 14 shows chaos in laser when f = 0.88 MHz.

4 Conclusion

Nonlinear dynamics in a short cavity Er-doped fiber laser is discussed. Two paths to chaos are presented by varying the depth and frequency. The effect of Er-doped concentration on the dynamics of the laser is analyzed. The results are helpful to study on fiber laser technology and chaotic laser application.

References

1. Luo, L.G., Chu, P.L., Liu, H.F.: 1-GHz optical communication system using chaos in erbium-doped fiber lasers. IEEE J. Quantum Electron. **12**(3), 269–271 (2000)
2. Wang, R., Shen, K.: Synchronization of chaotic systems modulated by another chaotic system in an erbium-doped fiber dual-ring laser system. IEEE J. Quantum Electron. **37**(8), 960–963 (2001)
3. Yan, S.L.: Dynamic characteristics in an external-cavity multi-quantum-well laser. Chin. Phys. B **27**(6), 060501-1-10 (2018)
4. Du, Y., Zheng, J., Wang, L., Wang, H., Zhu, N., Liu, J.: Widely-tunable and background-free ultra-wideband signals generation utilizing polarization modulation-based optical switch. IEEE Photon. Technol. Lett. **25**(4), 335–337 (2013)

5. Bayati, B.M., Ahmad, A.K., Naimee, K.A.M.A.: Influence of optical feedback strength and semiconductor laser coherence on chaos communications. J. Opt. Soc. Am. B **35**(4), 918–925 (2018)
6. Williams, O.L., Ojalvo, J.G., Roy, R.: Fast intercavity polarization dynamics of an erbium-doped fiber ring laser: inclusion of stochastic effects. Phys. Rev. A **55**(3), 2376–2385 (1997)
7. Sancheez, F., Stephan, G.: General analysis of instabilities in erbium-doped fiber laser. Phys. Rev. E **53**(3), 2110–2122 (1996)
8. Ming, D., Peter, K.C.: Analysis of Er-doped fiber laser stability by suppressing relaxation oscillation. IEEE J. Photonic Technol. Lett. **8**(9), 1151–1153 (1996)
9. Yan, S.L.: Controlling chaos to produce periodic-pulses and multi-periodic states phenomena in dual-ring Er-doped fiber laser. Acta Photonic Sinica **37**(4), 636–643 (2008)

Global Stability for a Coupled System of Fractional-Order Differential Equations with Discontinuous Terms on Network

Yang Gao(✉)

Daqing Normal University, Daqing 163712, People's Republic of China
gy19790607@163.com

Abstract. In this paper, a coupled system of fractional-order differential equations with discontinuous terms is investigated. The new model is constructed to realize the discontinuous control aim via the threshold policy (TP). Firstly, equilibrium's existence theorems for the new model are obtained by using the algebra theory and Filippov theory. Secondly, sufficient conditions to global Mittag-Leffler stability of the equilibrium are obtained based on the new model by applying graph theoretical approach of coupled systems and Laplace transform method.

Keywords: Filippov · Caputo · Mittag-Leffler · Global stability · Threshold policy · Differential inclusion

1 Introduction

The global stability problem of equilibria is investigated for coupled systems of differential equations on network for many years [1–6]. Recently, Li [7] investigated the global Mittag-Leffler stability of the following coupled system of fractional-order differential equations on network (CSFDEN):

$$\begin{cases} {}_{t_0}D_t^{\alpha} x_i = -\alpha_i x_i(t) + f_i(x_i(t)) + \sum_{j=1}^{n} \beta_{ij}^{x}(x_j(t) - x_i(t)), \\ x_i(t_0) = x_{it_0}, i = 1, \cdots, n. \end{cases} \tag{1}$$

where D denoted Caputo fractional derivative, $\alpha \in (0, 1)$. t_0 was the inial time, $n(n \geq 2)$ denoted the number of vertices in the network. $(x(t))^T = (x_1(t), x_2(t), \cdots, x_n(t))^T$ denoted the state variable of the system where $x_i(t) \in R$. α_i was a positive constant. Constant β_{ij}^{x} represented the influence of vertex j on vertex i with $\beta_{ii}^{x} = 0, \beta_{ij}^{x} = -\beta_{ji}^{x}$, if $i \neq j$. Function f_i was Lipschitz continuous. Several sufficient conditions were obtained to ensure the Mittag-Leffler stability of CSFDEN by using graph theory and the Lyapunov method.

Furthermore, Li [8] investigated a coupled system of fractional-order differential equations on network with feedback controls (CSFDENFCs). By using the contraction mapping principle, Lyapunov method, graph theoretic approach and inequality technique, some sufficient conditions were derived to ensure the

Y. Liu et al. (Eds.): ICNC-FSKD 2019, AISC 1074, pp. 206–213, 2020.
https://doi.org/10.1007/978-3-030-32456-8_22

existence, uniqueness and global Mittag-Leffler stability of equilibria of CSF-DENFCs.

Although well-improved results have been obtained in the above work on the model (1), the model is not well studied when the discontinuous term f_i is considered. In the sequel, we will use a threshold policy (TP) to denote the f_i for system (1).

In this paper, a coupled system of fractional-order differential equations with discontinuous terms is studied. The specific TP is used to stabilize the system. We will generalize the model (1) into the new model (2).

The Threshold Policy Method. Assume f_i can be denoted via the TP

$$f_i(x_i) = -\theta_i I(x_i)$$

$$I(x_i) = \begin{cases} 0, x_i < ET, \\ 1, x_i > ET, \end{cases}$$

Here, $\theta_i > 0$, and ET is a constant.

Remark 1. The generalized f_i is different from the f_i in model (1). New f_i is discontinuous without the Lipschitz continuous character.

Remark 2. The Threshold Policy method is used in the generalized work, which can be seen as discontinuous control to system (1). What's more, the discontinuous term seems as the activation function of one neuron of Hopfield neural networks. The term of θ_k comes from Zhao [10].

Therefore, the following coupled system of fractional-order differential equations with discontinuous terms is constructed.

$$\begin{cases} {}_{t_0}D_t^{\alpha} x_i = -\alpha_i x_i(t) - \theta_i I(x_i) + \sum_{j=1}^{n} \beta_{ij}^{x}(x_j(t) - x_i(t)), \\ x_i(t_0) = x_{it_0}, i = 1, \cdots, n. \end{cases} \quad (2)$$

The network method is applied widely in recently years [1–8]. In this paper, we interpret system (2) as a coupled system on a network. Using the method of Li and Shuai [6], we prove global Mittag-Leffler stability theorems for the equilibrium.

Let $\beta_{ij} = \beta_{ij}^{x}$ represent the influence of vertex j on vertex i. Let $A = (|\beta_{ij}|)_{n\times n}, A^{x} = (|\beta_{ij}^{x}|)_{n\times n}$. A digraph (G, A) with n vertices for system (2) can be constructed as follows. Each vertex represents a patch and $(j, i) \in E(G)$ if and only if $\beta_{ij}^{x} \neq 0$. Here $E(G)$ denotes the set of arcs (i, j) leading from inial vertex i to terminal vertex j. At each vertex of G, the vertex dynamics are described by the following system (3).

$$\begin{cases} {}_{t_0}D_t^{\alpha} x_i = -\alpha_i x_i(t) - \theta_i I(x_i). \\ x_i(t_0) = x_{it_0}. \end{cases} \quad (3)$$

The coupling among system (2) is provided by the network. The G is strongly connected if and only if the matrix $A = (|\beta_{ij}|)_{n\times n}$ is irreducible.

Our contribution is listed as follows:

(1) Existence conditions of equilibria for system (2) are obtained by the algebra theory and Filippov theory.
(2) Sufficient conditions that equilibria of the coupling model is globally Mittag-Leffler stable are derived by using the method of constructing Lyapunov functions based on graph-theoretical approach for coupled systems.

This paper is organized as follows. In Sect. 2, we obtain main results. Finally, the conclusions and outlooks are drawn in Sect. 3.

2 Main Results

Definition 1. *The solution of the system*

$$ {}_{t_0}D_t^{\alpha}x(t) = f(t,x) $$

is said to be Mittag-Leffler stable if

$$ ||x(t)|| \leq \{m[x(t_0)]E_{\alpha}(-\lambda(t-t_0)^{\alpha})\}^{b}. $$

Here t_0 is the initial time, $\alpha \in (0,1), \lambda > 0, b > 0, m(0) = 0, m(x) \geq 0$. $m(x)$ is locally lipchitz on $x \in B \subseteq R^n$ with Lipschitz constant m_0. $E_{\alpha}(t)$ is a Mittag-Leffler function. Moreover, the domain of the function $f(t,x)$ is $[t_0, +\infty) \times \Omega$, and the function $f(t,x)$ is piecewise continuous in t and locally Lipschitz in x [7].

Filippov solutions will be used for the discontinuous system (2). Consider the differential inclusion as follows:

$$ \begin{cases} {}_{t_0}D_t^{\alpha}x_i \in -\alpha_i x_i(t) - \theta_i \bar{co}[I(x_i)] \\ \qquad\qquad + \sum_{j=1}^{n} \beta_{ij}^{x}(x_j(t) - x_i(t)), \\ x_i(t_0) = x_{it_0}, i = 1, \cdots, n. \end{cases} \tag{4} $$

Here, $\bar{co}[I(x_i)]$ denotes the convex closure of $I(x_i)$.

Definition 2. *$x(t) = (x_1(t), x_2(t), \cdots, x_n(t))$ is the solution of system (2) with initial value $x_1(0) = x_{1_0}, x_2(0) = x_{2_0}, \cdots, x_n(0) = x_{n_0}$, if*

(1) $x(t)$ is defined in the interval $[0,T)$ with $T \in (0, +\infty]$.
(2) $x(t)$ is absolutely continuous in any subinterval of $[0,T)$.
(3) $x(t)$ is the solution of system (4) for a.e. $t \in [0,T)$.

Remark 3. Because the set-valued map is upper-semi-continuous with nonempty compact convex values, the local existence of a solution $x(t)$ of system (4) can be guaranteed. Growth condition is satisfied for system (4), therefore, the solution for system (2) exists.

We assume that $x^* = (x_1^*, x_2^*, \cdots, x_n^*)$ is an equilibrium of system (2). By system (4) and measurable selection theorem [9], there is $\eta_i^* \in \bar{co}[I(x_i^*)]$ such that

$$\begin{cases} -\alpha_i x_i^* - \theta_i \eta_i^* + \sum_{j=1}^n \beta_{ij}^x (x_j^* - x_i^*) = 0, \\ i, j = 1, \cdots, n. \end{cases} \tag{5}$$

for a.e. $t \in [0, T)$.

In this section, first of all, the existence of the equilibrium for system (2) is shown. Secondly, sufficient conditions that the coexistence equilibrium of the coupling model (2) is globally M-L stable in R_+^n are derived by using the method of constructing Lyapunov functions based on graph-theoretical approach for coupled systems.

2.1 The Existence Conditions of the Equilibrium for System (2)

Define

$$A_1(x) = diag[-\alpha_1, -\alpha_2, \cdots, -\alpha_n]$$

and

$$A_2(x) = (\beta_{ij}^x)_{n\times n} - diag[\sum_{j=1}^n \beta_{1j}^x, \sum_{j=1}^n \beta_{2j}^x, \cdots, \sum_{j=1}^n \beta_{nj}^x],$$

$$A_3(x) = [\theta_1, \theta_2, \cdots, \theta_n]^T.$$

Now we will consider existence conditions for the equilibrium of system (2). The theorem is listed as follows:

Theorem 1. *Assume that the rank of matrix $A_1(x) + A_2(x)$ is equivalent to the rank of augmented matix $(A_1(x) + A_2(x), A_3(x))$. Then system (2) has one equilibrium at least, when the suitable ET is chosen.*

Naturally, consider the matrix equation

$$(A_1(x) + A_2(x))X = A_3(x).$$

The theorem's condition demonstrates that matrix equation is solved. Then $X = x^*$, $\eta^* = 1$ is one equilibrium. Here, we choose ET less than $min_i\{x_i\}$.

Let

$$A_4(x) = [\eta_1^*\theta_1, \eta_2^*\theta_2, \cdots, \eta_n^*\theta_n]^T.$$

One corollary is obtained.

Corollary 1. *If we choose $\eta_i^* = 0, or, 1$, and assume that the rank of matrix $A_1(x) + A_2(x)$ is equivalent to the rank of augmented matix $(A_1(x) + A_2(x), A_4(x))$. Then system (2) has one equilibrium at least, when the suitable ET is chosen.*

Factually, suitable ET is very important to equilibrium existence for system (2). If $ET > 0$ is chosen, next corollary is obtained as follows.

Corollary 2. *Given $ET > 0$, then system (2) has one equilibrium at least with $x^* = 0, \eta^* = 0$.*

2.2 The Stability Analysis of the Equilibrium for System (2)

Similar to system (4), when

$$x(t) = (x_1(t), x_2(t), \cdots, x_n(t))$$

is the solution of system (2), then for any $v \in \bar{co}[I(x_i(t))]$, there is a measurable function $\eta_i(t) \in \bar{co}[I(x_i(t))]$ (Measurable Selection Theorem [9]) such that

$$\begin{cases} {}_{t_0}D_t^{\alpha} x_i = -\alpha_i x_i(t) - \theta_i \eta_i(t) + \sum_{j=1}^{n} \beta_{ij}^{x}(x_j(t) - x_i(t)), \\ x_i(t_0) = x_{it_0}, i = 1, \cdots, n. \end{cases} \tag{6}$$

for a.e. $t \in [0, T)$.

The main result is listed as follows:

Theorem 2. *Assume that the following assumptions hold.*

1. *Diagraph (G, A) is balanced;*
2. *$A = (|\beta_{ij}|)_{n \times n}$ is irreducible.*
3. *There exists an equilibrium $x^* = (x_1^*, x_2^*, \cdots, x_n^*)$ for system (2).*

Then the equilibrium x^* of system (2) is globally Mittag-Leffler stable in R_+^n.

Proof. Assume that $e_i(t) = x_i(t) - x_i^*(i = 1, 2, \cdots, n)$. Let $V_i(e_i(t)) = \frac{1}{2}e_i^2(t)$. We will prove that

$${}_{t_0}D_t^{\alpha} V_i(e_i(t)) \leq e_i(t) {}_{t_0}D_t^{\alpha} e_i(t),$$

for a.e. $t \in [0, T)$.

In fact, the i-th component of the Filippov solution $x_i(t)$ is differentiable except at the set $\Phi = \{t| \quad x_i(t) = ET, x_{i+}^{'}(t) \neq x_{i-}^{'}(t)\} \cup \Phi^{'}$ (derived set of Φ). Then we prove that the measure of Φ is 0.

From $x_{i+}^{'}(t) \neq x_{i-}^{'}(t)$, we obtain that $x_{i+}^{'}(t) \neq 0$ or $x_{i-}^{'}(t) \neq 0$. In the sequel, we assume that $x_{i+}^{'}(t) > 0$. The feature of guarantee code can be used. Therefore, t can be seen as end point of an open interval $I_i(t)$. When any $t^* \in I_i(t)$, $x_i(t^*)$ is continuously differentiable. By the fact that the set of open intervals that don't intersect each other on the straight line is at most countable, we obtain that the measure of Φ is 0. Furthermore, the measure of $\Phi \cup \Phi'$ is 0. So $x_i{}'(t)$ is piecewise continuous.

Therefore, L'Hopital rule can be applied for $t \notin \Phi \cup \Phi'$ in the proof similar to Lemma 1 in [11]. What's more, the ${}_{t_0}D_t^{\alpha} V_i(e_i(t)) \leq e_i(t) {}_{t_0}D_t^{\alpha} e_i(t)$(for a.e. $t \in [0, T)$) can be proved.

After calculating, we obtain that

$$\begin{aligned} {}_{t_0}D_t^{\alpha} V_i(e_i(t)) &\leq e_i(t)[-\alpha_i e_i(t) + \theta_i(\eta_i^* - \eta_i(t)) \\ &\quad + \textstyle\sum_{j=1}^{n} \beta_{ij}^{x}(e_j(t) - e_i(t))] \\ &\leq -\alpha_i e_i^2(t) + \theta_i e_i(t)(\eta_i^* - \eta_i(t)) \\ &\quad + \textstyle\sum_{j=1}^{n} \beta_{ij}^{x}(e_i(t)e_j(t) - e_i^2(t)). \end{aligned}$$

for a.e. $t \in [0, T)$.

From the condition of theorem, we obtain A^x is irreducible. Furthermore, (G, A) is strongly connected. Let c_i denote the cofactor of the ith diagonal element of Laplacian matrix of (G, A). Then we have $c_i > 0$. Let

$$V(t, e(t)) = \sum_{i=1}^{n} c_i V_i(e_i(t)).$$

Calculating the fractional-order derivative of $V(t, e(t))$ along the solution of system (2), we have

$$\begin{aligned}
&{}_{t_0}D_t^{\alpha} V(t, e(t)) = \tfrac{1}{2}\textstyle\sum_{i=1}^{n} c_i \; {}_{t_0}D_t^{\alpha}(e_i(t))^2 \\
&\leq \textstyle\sum_{i=1}^{n} c_i e_i(t)_{t_0} D_t^{\alpha} e_i(t) \\
&\leq \textstyle\sum_{i=1}^{n} c_i e_i(t)[-\alpha_i e_i(t) + \theta_i(\eta_i^* - \eta_i(t)) \\
&+ \textstyle\sum_{j=1}^{n} \beta_{ij}^{x}(e_j(t) - e_i(t))] \\
&\leq \textstyle\sum_{i=1}^{n}[-\alpha_i c_i e_i^2(t) + \theta_i c_i e_i(t)(\eta_i^* - \eta_i(t)) \\
&+ \textstyle\sum_{i=1}^{n} c_i a_{ij} F_{ij}(t, e_i, e_j)].
\end{aligned}$$

for a.e. $t \in [0, T)$.

Here, $a_{ij} = |\beta_{ij}| = |\beta_{ij}^{x}|$ and $F_{ij}(t, x, y) = sgn(\beta_{ij})xy$. Using the (G, A)'s balanced and strongly connected character, we obtain that

$$\begin{aligned}
&\textstyle\sum_{i=1}^{n} c_i a_{ij} F_{ij}(t, e_i, e_j) \\
&= \tfrac{1}{2}\textstyle\sum_{Q\in \mathbb{Q}} w(Q) \sum_{(j,i)\in E(C_Q)}[F_{ij}(t, e_i, e_j) + F_{ji}(t, e_i, e_j)] \\
&= \tfrac{1}{2}\textstyle\sum_{Q\in \mathbb{Q}} w(Q) \sum_{(j,i)\in E(C_Q)}[sgn(\beta_{ij})e_i e_j + sgn(\beta_{ji})e_j e_i] \\
&= \tfrac{1}{2}\textstyle\sum_{Q\in \mathbb{Q}} w(Q) \sum_{(j,i)\in E(C_Q)}[sgn(\beta_{ij})e_i e_j - sgn(\beta_{ij})e_i e_j] \\
&= 0.
\end{aligned}$$

Now the sign of $(x_i - x_i^*)(\eta_i^* - \eta_i(t))$ will be discussed. By applying the monotonic property of $I(x_i)$, we obtain that

When $x_i \geq x_i^*$,

$$\eta_i(t) \geq \eta_i^*$$

holds.

When $x_i \leq x_i^*$,

$$\eta_i(t) \leq \eta_i^*$$

holds. It means that

$$c_i\theta_i(x_i - x_i^*)(\eta_i^* - \eta_i(t)) \leq 0.$$

In the sequel,

$${}_{t_0}D_t^{\alpha} V(t, e(t)) \leq \sum_{i=1}^{n} -\alpha_i c_i e_i^2(t) + \sum_{j=1}^{n} a_{ij} F_{ij}(e_i, e_j),$$

for a.e. $t \in [0, T)$.

Furthermore, let $\lambda = 2min_i\{\alpha_i\}$, we have

$${}_{t_0}D_t^{\alpha} V(t, e(t)) \leq -\lambda V(t, e(t)),$$

for a.e. $t \in [0, T)$.

Let

$$_{t_0}D_t^\alpha V(t, e(t)) + M(t) = -\lambda V(t, e(t)),$$

for a.e. $t \in [0, T)$.

Using Laplace transform for the equation above, we obtain that

$$s^\alpha w(s) - w(0)s^{\alpha-1} + M(s) = -\lambda w(s),$$

where $w(s)$, $M(s)$ are the Laplace transform of $V(t, e(t))$ and $M(t)$, respectively. Using the inverse Laplace transform for the formula above, we have

$$V(t, e(t)) \leq V(0, e(0))E_\alpha(-\lambda t^\alpha).$$

By the definition of $V(t, e(t))$, we obtain that system (2) is globally Mittag-Leffler stable. This completes the proof.

3 Conclusions and Outlooks

In this paper, a coupled system of Fractional-Order differential equations with discontinuous terms is investigated. The new model is constructed to realize the discontinuous control aim via the threshold policy (TP). Our Theorem 2 has generalized main theorems in [7] and [8]. Further studies on this subject are being carried out by the presenting authors in the two aspects: one is to study the TP with time delay; the other is to discuss the method with hybrid coupled relationship.

This work is supported by Natural Science Foundation of HeiLongJiang Province (No. QC2009C99), and Science and Technology Planning Project of Daqing City under Grant No. zd-2017-49.

References

1. Freedman, H.I., Takeuchi, Y.: Global stability and predator dynamics in a model of prey dispersal in a patchy environment. Nonlinear Anal. Theory Methods Appl. **13**(8), 993–1002 (1989)
2. Kuang, Y., Takeuchi, Y.: Predator-prey dynamics in models of prey dispersal in 2-patch environments. Math. Biosci. **120**(1), 77–98 (1994)
3. Cui, J.G.: The effect of dispersal on permanence in a predator-prey population growth model. Comput. Math. Appl. **44**(8–9), 1085–1097 (2002)
4. Xu, R., Chaplain, M.A.J., Davidson, F.A.: Periodic solutions for a delayed predator-prey model of prey dispersal in two-patch environments. Nonlinear Anal. Real World Appl. **5**(1), 183–206 (2004)
5. Zhang, L., Teng, Z.D.: Permanence for a delayed periodic predator-prey model with prey dispersal in multi-patches and predator density-independent. J. Math. Anal. Appl. **338**(1), 175–193 (2008)
6. Li, M.Y., Shuai, Z.S.: Global-stability problem for coupled systems of differential equations on networks. J. Differ. Equ. **248**(1), 1–20 (2010)

7. Li, H.L., Jiang, Y.L., Wang, Z.L., Zhang, L., Teng, Z.D.: Globlal Mittag-Leffler stability of coupled system of fractional-order differential equations on network. Appl. Math. Comput. **270**, 269–277 (2015)
8. Li, H.L., Hu, C., Jiang, Y.L., Zhang, L., Teng, Z.D.: Global Mittag-Leffler stability for a coupled system of fractional-order differential equations on network with feedback controls. Neurocomputing **214**, 233–241 (2016)
9. Huang, L.H., Guo, Z.Y., Wang, F.J.: The theory and application for the differential equations with discontinuous right side. Science Press, Beijing (2013). (in Chinese)
10. Zhao, T.T., Xiao, Y.N., Smith, R.J.: Non-smooth plant disease models with economic thresholds. Math. Biosci. **241**, 34–48 (2013)
11. Aguila-Camacho, N., Duarte-Mermoud, M.A., Gallegos, J.A.: A Lyapunov functions for fractional order systems. Commun. Nonlinear Sci. Numer. Simulat. **19**, 2951–2957 (2014)

Optimal Identification of Multiple Diffusion Sources in Complex Networks with Partial Observations

Xiang Li[1], Xiaojie Wang[1], Chengli Zhao[1](✉), Xue Zhang[1], and Dongyun Yi[1,2]

[1] College of Liberal Arts and Sciences, National University of Defense Technology, Changsha 410073, China
chenglizhao@nudt.edu.cn

[2] State Key Laboratory of High Performance Computing, National University of Defense Technology, Changsha 410073, China

Abstract. Source localization is a typical inverse problem in complex networks, which is widely used in disease outbreak, rumor propagation and pollutants spread. In this paper, we propose that, based on network topology and the times at which the diffusion reached partial nodes, it is easy to identify the source. The results show that in six different networks, although the number of observers is small, the precision of source localization can be high. Precision increases with network size increasing and source number decreasing. Furthermore, our method makes the sources localization precision very robust, not only with the condition of three different given observers selection strategies, but with three various intensity noise on the diffusion path.

Keywords: Multiple sources localization · Precision · Robust

1 Introduction

In real life, there are many important diffusion processes, almost all of which start with several starting nodes. The outbreak of a disease, the spread of a rumor and the diffusion of pollutants all generally propagate from a few simple nodes. These phenomena are often harmful to human society and severely affect the normal activities of life; therefore, we urgently need to control and address these issues. In this regard, effectively and precisely locating the sources of diffusion is a critical issue.

In 2011, Shah introduced a new topological quantity - the rumor centrality to locate the source based on the SIR [1] model. As a traditional method, the distance centrality is the node which is closed to all other nodes. To locate the rumor source, the localization precision of rumor centrality is equivalent to that of distance centrality in the tree network and the localization precision of rumor centrality is superior in a general network [2]. In 2012, Pinto selected partial nodes as observers and located the source by the maximum likelihood

Y. Liu et al. (Eds.): ICNC-FSKD 2019, AISC 1074, pp. 214–223, 2020.
https://doi.org/10.1007/978-3-030-32456-8_23

estimation. For any tree network, this method can optimal the precision of source localization [3]. In addition, few previous work studied the source localization problem in a sequential network. For example, in 2017, Huang used observers to diffuse reversely and found the node as source with minimum variance yields, resolving the single source localization problem in the temporal network [4].

The above methods are used for single source localization [5,6]. However, some diffusion processes tend to have more than one source. Therefore, multiple sources localization models [7,8] were proposed to trace multiple sources. The Ebola virus, for example, has simultaneously spread in one place, severely affecting the security of society. To reduce its influence, we need to find multiple diffusion sources.

In 2013, Luo proposed a heuristic method for inferring multiple diffusion sources under the SIR model with partial observations. However, when the number of sources $N_s \geq 2$, the source localization precision will be low [9]. In 2014, Zang proposed a reverse diffusion method for detecting the recovered and susceptible nodes in the network. Then they applied the community clustering algorithm to remodel the multiple sources localization problem to a series of single-source localization problems. Finally, they applied maximum likelihood estimation to identify the source nodes among the infected nodes [10]. In 2016, Zhu proposed a new approach, named Optimal-Jordan-Cover (OJC). Firstly, extract a subgraph as source candidates using a candidate selection algorithm. Then OJC finds a set of nodes that cover all observed infected nodes with minimum radius. This nodes called Jordan Cover, and is regarded as the set of diffusion source [11]. In 2016, Fu studied a backward diffusion-based source localization method. Based on the times at which the diffusion reached partial observers, calculate the maximum time when the diffusion goes reversely from partial observers to each node. Then pick up the node with minimum value, recognized as source [12].

In this paper, we propose a new approach which locates the sources optimally. Based on the times at which the diffusion reach partial observers and network topology, calculate the maximum times when the diffusion go reversely from partial observers to each node with time-reversal backward spreading. Pick up a set of nodes which have the same maximum value into a group. In each group, find the smallest set of nodes which cover all the observed informed nodes. The set of nodes are the sources.

2 Model and Method

2.1 Problem Formulation

Given an undirected graph G, $G = \{V, E\}$, where V represents the node set and E represents the edge set. The network topology is preset and thus the diffusion process is invariant. Assume diffusion sources as S, $S = \{s_1, s_2, \cdots, s_{N_s}\}$ and $s_1, s_2, \cdots, s_{N_s}$ is the diffusion source, where N_s represents the number of sources. All sources have the same initial diffusion time.

A node in the network can be in two states: informed and uninformed. Uninformed nodes receive information from informed neighbor, meanwhile informed nodes diffuse information to uninformed neighbor. Once uninformed node i receives information, its state will change to informed. If a node receives information from several neighbors at the same time, the node chooses randomly a neighbor to receive. The diffusion time between two nodes is called the time delay. The diffusion process continues until all nodes in this network are informed (Fig. 1).

At this point, $\mathrm{V} = \mathrm{S} + \Pi_{s_1} + \Pi_{s_2} + \cdots + \Pi_{s_{N_s}}$, where $\Pi_{s_i}, i = 1, 2, \cdots, N_s$ is the region diffused at the source s_i.

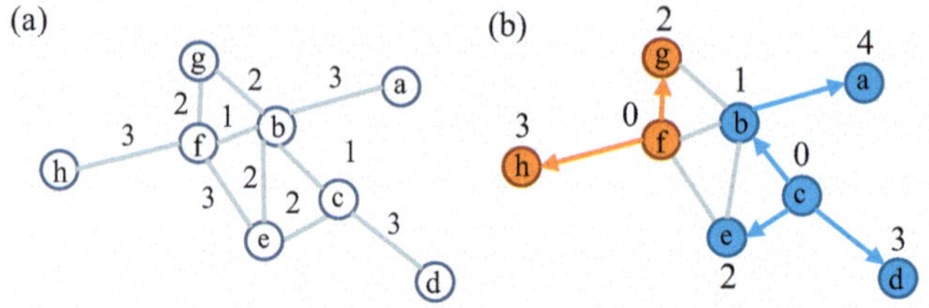

Fig. 1. (a) Network topology diagram. The topology keeps invariable, and the digital means time delay of each edge; (b) The diffusion process diagram, c and f are the sources. All the sources initial diffusion time is $t^0 = 0$, and the diffusion obey the shortest path algorithm. The digital means the time t_i when the diffusion reach node i, different colors represent different regions, $\mathrm{V} = \Pi_c + \Pi_f, \Pi_c = \{a, b, c, d, e\}, \Pi_f = \{f, g, h\}$.

2.2 Multiple Sources Localization

In reality, the diffusion usually comes up in a tremendous network. Usually, we can not acquire the times when diffusion reaches all the nodes. We can preset some nodes as observers in this network. Given the network topology and the times at which the diffusion reaches observers, we can locate the sources.

Maximum Principle. Due to unknown the initial diffusion time, based on the times $\xi = (t_i)$ at which the diffusion reaches partial observers, we firstly calculate the times T when the diffusion arrives at each node V.

$$\mathrm{T} = (t_{i,j}), t_{i,j} = t_i - \Delta_{i,j}. \tag{1}$$

where $t_{i,j}$ represents the times when the diffusion goes from node $i \in \mathrm{V}_0$ to node $j \in \mathrm{V}$, $\Delta_{i,j}$ ($\Delta_{i,j} = \Delta_{j,i}$) represents the shortest time delay between node i and node j.

Now, we don't know the sources and the initial diffusion time, and know the time information T of each node. Obviously, the sources are in the network and the initial diffusion time is in the time information T.

Considering j as a source, various i makes a difference to $t_{i,j}$:

1. For any diffusion source $j \in \mathrm{S}$, any $i \in \mathrm{V}_0$, if j and i satisfy that: $\Delta_{j,i} = \min\{\Delta_{s_1,i}, \Delta_{s_2,i}, \cdots, \Delta_{s_{N_s},i}\}$, i.e., $i \in \Pi_j$. Then the time when the information goes from observation i to node j:

$$t_{i,j} = t_i - \Delta_{j,i} = t^0 \tag{2}$$

2. For any diffusion source $j \in \mathrm{S}$, any $i \in \mathrm{V}_0$, if j and i satisfy that: $\Delta_{j,i} \neq \min\{\Delta_{s_1,i}, \Delta_{s_2,i}, \cdots, \Delta_{s_{N_s},i}\}$, i.e., $i \notin \Pi_j$. Then the time when the information goes from observation i to node j:

$$t_{i,j} = t_i - \Delta_{j,i} < t^0 \tag{3}$$

That is to say, if node j is a source, the maximum time when the diffusion gets to node j is nearest to initial diffusion time. Therefore, we choose the most possible initial diffusion times Γ from times T.

$$\Gamma = (\Gamma_j), \Gamma_j = \max_i \{t_{i,j}\} \tag{4}$$

where Γ_j is the maximum times when information goes from observations V_0 to node j.

Branching and Grouping. If the initial diffusion time of different sources is different, there will be a problem. For example, if the difference time between two sources is longer than the time delay. It is difficult to identify the node has information initially as a source or the node receives the information from the surrounding source. Therefore, we prefer to the same initial diffusion time.

Assume all the sources initial diffusion time is same, so we take the nodes of which the maximum times are the same into a group. Assume the nodes in V can be divided into k groups, which are expressed as $\mathrm{V}_1, \mathrm{V}_2, \cdots, \mathrm{V}_\mathrm{k}$. If $\Gamma_{j_1} = \Gamma_{j_2}, j_1, j_2 \in \mathrm{V}$, then node j_1 and node j_2 are divided into a group.

$$\mathrm{V} = \mathrm{V}_1 \bigcup \mathrm{V}_2 \bigcup \cdots \bigcup \mathrm{V}_\mathrm{k}, \mathrm{V}_\mathrm{i} \bigcap \mathrm{V}_\mathrm{j} = \varnothing, i \neq j \tag{5}$$

A Combinational Optimization Approach for Sources Localization. In the last two steps, we know the possible initial diffusion time of every node and group the same time into a group. Next, we need to judge the nodes we pick up, whether they are sources or not. Meanwhile, to avoid the false source, the sources need to be minimum. Therefore, we find the minimal node subset in each group to cover all informed observers by location set covering, called Group-Cover Model (GCM). That means the times when the diffusion reaches observers is equal to $\xi = (t_i)$, if these nodes are the sources. Meanwhile, we minimize the number of the sources.

Definition 1. The map g from V to V_0: $\forall j \in \mathrm{V}$, if $\mathrm{T}_f(i,j) \neq none$ and $i \in \mathrm{V}_0$, $g(j) = (i)$.

The location set covering method is as follows: For $\mathrm{V}^* \subset \mathrm{V}$, if $\bigcup_{j \in \mathrm{V}^*} s\,g(j) = \mathrm{V}_0$ and $|\mathrm{V}^*|$ is the smallest, then $\mathrm{S} = \mathrm{V}^*$ (Fig. 2).

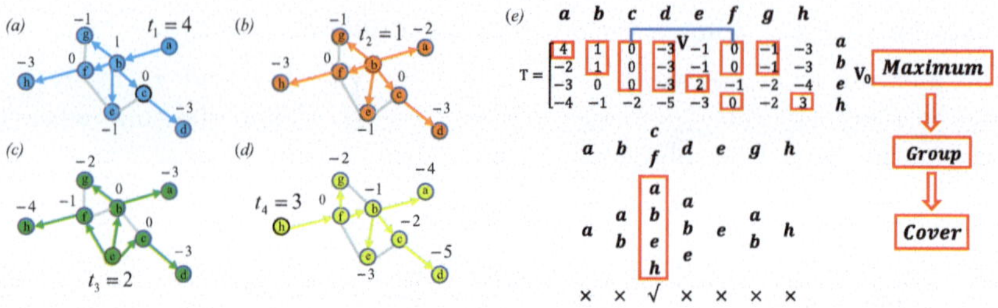

Fig. 2. (a)–(d) The backward diffusion diagram. It indicates that the times $\xi = [4, 1, 2, 3]$ when the diffusion reaches observers $V_O = \{a, b, e, h\}$ and the times T when the diffusion goes from observers to each node reversely. (e) Sources detection process diagram. It indicates that the key algorithm of the model in this paper. Firstly, pick up the maximum of each row in $\Gamma = [4, 1, 0, -3, 2, 0, -1, 3]$; then, group the same value into a group, $V = \{a\} \cup \{b\} \cup \{c, f\} \cup \{d\} \cup \{e\} \cup \{g\} \cup \{h\}$; finally, use the location set covering to judge the nodes, whether these are sources or not. Because $g(c) \bigcup g(f) = V_O$, we know $\{c, f\}$ is the sources.

3 Experiments and Analysis

When exploring the localization precision of GCM and MMM [12], we introduce ER, WS, BA, *dolphins*, *football*, and USAir networks as the experimental data, and we use the precision as the evaluation index. In the experiment, we assume that we have already known the number of sources (Table 1).

Table 1. Network parameter table. It is for *dolphins*, *football*, and USAir networks, where N represents the number of nodes, $|E|$ represents the number of edges, and $<k>$ represents the average degree.

Network type	N	$\|E\|$	$<k>$	Description
dolphins [14]	62	159	5.13	Sixty-two dolphins' group communication in New Zealand channel
football [14]	115	613	10.66	Games among 115 American college football teams
USAir [14]	332	2126	12.81	Flights in the USA

The formula of the evaluation index, precision, is expressed as

$$precision = \frac{1}{k} \sum_{k} \frac{N_k}{N_s} \tag{6}$$

where k represents the number of groups of the source set, N_k represents the correct number in group k, and N_s represents the number of sources in the real sources.

3.1 Source Localization in Synthetic Networks

We do some experiments in synthetic networks to research the relationship between source localization precision and network parameter.

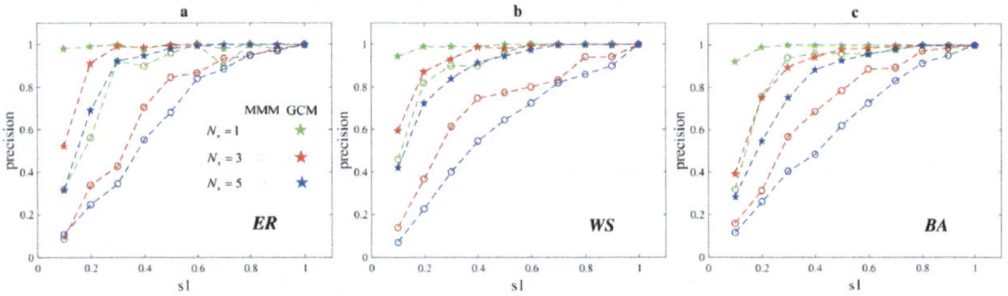

Fig. 3. Sources detection precision with the number of the sources diagram in synthetic networks. $s1$ represents the proportion of observers. (a–c) represent the ER, WS and BA networks. Number of sources $N_s = 1, 3, 5$, number of nodes $N = 100$, and average degree $\langle k \rangle = 6$. When the number of sources increases from 1 to 5, the precision of the source localization of GCM and MMM gradually decreases in the ER, WS and BA networks.

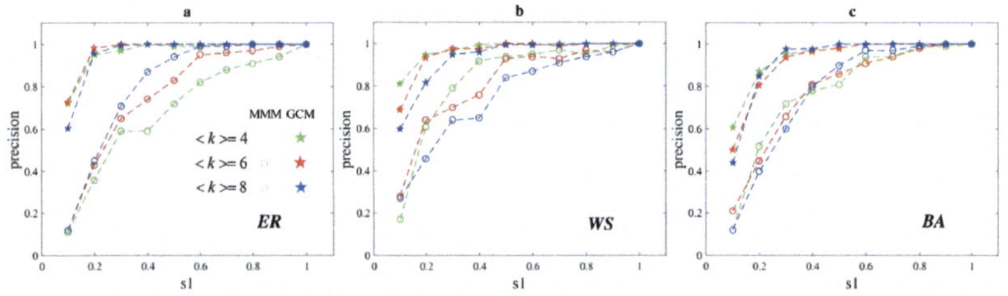

Fig. 4. Sources detection precision with the average degree diagram. (a–c) represent the ER, WS and BA networks. Average degree $\langle k \rangle = 4, 6, 10$, number of nodes $N = 100$, and number of sources $N_s = 2$. With the increase of the average degree, the source localization precision gradually increases, decreases and becomes irregular in the ER, WS and BA networks, respectively.

The situation in Fig. 3 is consistent with the fact that as the number of sources increases, the challenges of searching often increase, which is not conducive to the discovery of the source. When the number of sources N_s changes, the source localization precision of GCM is always higher than that of MMM.

Figure 4 shows that as the average degree increases, the nodes are more closely connected, which is conducive to finding the source. In contrast to the three phenomena of the networks, only the ER network is consistent with reality. When the network average degree $<k>$ changes, the source localization precision of GCM is always higher than that of MMM.

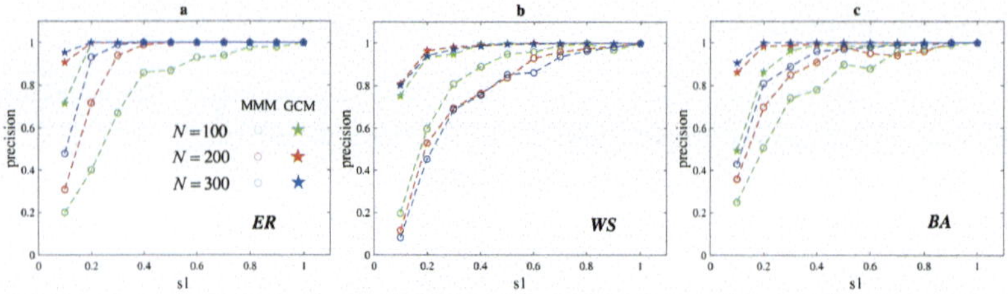

Fig. 5. Sources detection precision with the scale of network diagram. (*a*–*c*) represent ER, WS and BA networks. The number of nodes $N = 100, 200, 300$, the average degree $\langle k \rangle = 4$, and the number of sources $N_s = 2$. With the increase in network size, the source localization precision gradually increases in the ER and BA networks.

When the network size increases, the distance among the sources increases, reducing the mutual interference, so that it is easier to determine from which source the observation comes, showing in Fig. 5. In this way, the observation can be used to infer the diffusion source more accurately. When the network scale N changes, the source localization precision of GCM is always higher than that of MMM.

3.2 Source Localization in Real Networks

In the ER, WS and BA networks, precision is closely related to the number of sources and the average degree of the network. To prove the universality of this conclusion, corresponding tests were conducted in the *dolphins*, *football* and USAir networks. Since the real network topology is already given, the relationship between precision and $s1$ is various when the number of sources is different (Fig. 6).

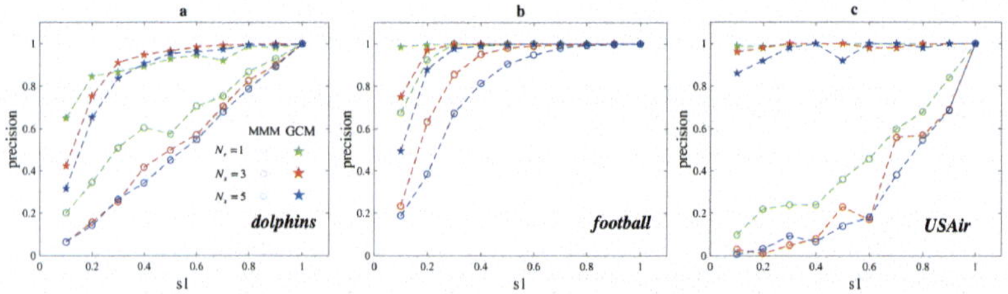

Fig. 6. Sources detection precision with the number of the sources diagram in real networks. (*a*–*c*) represent *dolphins*, *football* and USAir networks. Number of sources $N_s = 1, 3, 5$. With the increase of the number of sources, the source localization precision gradually decreases in GCM and MMM. In the *dolphins* and *football* networks, when $s1 < 20\%$, the source localization precision of GCM is at least 30% higher than that of MMM. In the USAir network, the source localization precision of GCM is much higher when $s1 < 80\%$. The network size and the average degree of the *dolphins*, *football* and USAir networks increase successively, and the source localization precision increases in GCM, but there is no detectable rule in MMM.

4 Discussions

4.1 Strategies for Observers Placements

By experiments, the source localization precision is related to the selection strategy of the observers. For example, in a general static network, nodes with a high degree of selection as observers can acquire more information to locate the diffusion sources accurately. In the dynamic network, nodes with a small degree of selection as observers can obtain sources easily [13]. For different networks, the optimal selection strategy will also be different. Therefore, our paper selects the maximum, minimum and random selection strategies to test the robustness of GCM and MMM, and the test results are shown in the figure below (Fig. 7).

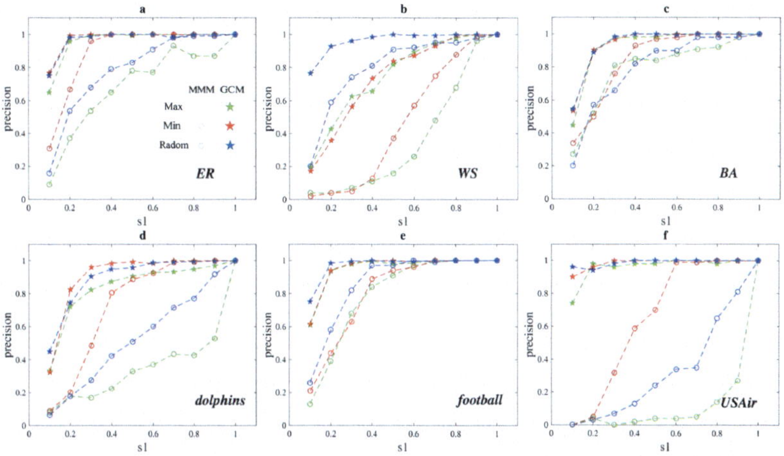

Fig. 7. Sources detection precision with observers selection strategy diagram. In synthetic networks, the number of nodes $N = 100$, the average degree $\langle k \rangle = 6$, and the number of sources $N_s = 3$. In real networks, the number of sources $N_s = 3$. The sources localization precision of GCM is more robust than that of MMM under the maximum, minimum and random selection strategies. Especially in the ER, *dolphins* and USAir networks, the source localization precision varies greatly with selection strategies. When the observation selection strategy changes, the source localization precision of GCM is always higher than that of MMM. In the WS, *dolphins* and USAir networks, precision increases exponentially in GCM, while precision increases linearly in MMM.

4.2 Uncertainty of Diffusion Delay Along Edges

The above test is carried out in a noise-free environment, and the results are relatively ideal. In the actual diffusion process, the time delay changes. To simulate the real situation better, Gaussian noise $t^{\sim}o(t, \sigma^2)$ is added to the diffusion path. After obtaining the $\xi = (t_i), i \in \mathrm{V}_0$, integer, next calculate the times at which each node receives information from observers, finally get the precision. Comparing the effects of σ on the stability of precision, the robustness of GCM and MMM against noise can be inferred (Fig. 8).

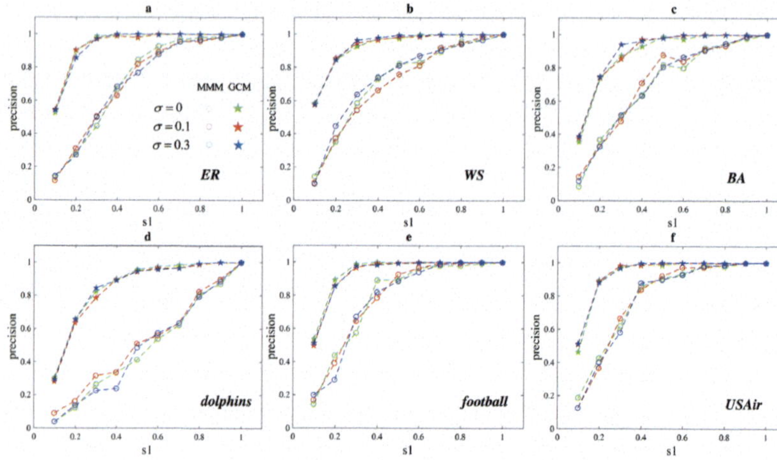

Fig. 8. Sources detection precision with noise diagram. In synthetic networks, the number of nodes $N = 100$, the average degree $\langle k \rangle = 6$, and the number of sources $N_s = 3$. In real networks, the number of sources $N_s = 3$. When $\sigma = 0, \sigma = 0.1, \sigma = 0.3$, the anti-noise ability of GCM is stronger than that of MMM. In above six networks, precision remains virtually unchanged with the change in σ in GCM, and precision fluctuates in MMM. When the noise σ changes, the source localization precision of GCM is always higher than that of MMM.

5 Conclusion

Disease outbreak, rumor propagation and spread of pollutants in nature pose great threats and challenges to human environment. The most direct and effective way to prevent these hazards is to find the source of propagation so that the problem can be solved from the source.

This paper proposes a multi-source localization model, called Group-Cover Model. Give the topology and the times when the diffusion reaches observers, we firstly use backward diffusion to receive the time information of each node, then group to guarantee the same initial diffusion time and judge the nodes whether they are the sources or not, by location set covering.

The numerical experiments show that: when network size increases and the number of sources decreases, sources localization precision has a upward trend. Observer placements strategies has little effect on GCM. The sources localization precision is robust when the noise is added to the diffusion path.

Acknowledgments. This work is partially supported by the National Key R&D Program of China (Grant No. 2017YCF1200301), the Postgraduate Innovation Fund of Hunan Province (Grant No. CX2015B010), and the Postgraduate Innovation Fund of the National University of Defense Technology (Grant No. B150203).

References

1. Altmann, M.: Susceptible-infected-removed epidemic models with dynamic partnerships. J. Math. Biol. **33**(6), 661–675 (1995). https://doi.org/10.1007/bf00298647
2. Shah, D., Zaman, T.R.: Rumors in a network: who's the culprit? IEEE Trans. Inf. Theory **57**(8), 5163–5181 (2011). https://doi.org/10.1109/TIT.2011.2158885
3. Pinto, P.C., Thiran, P., Vetterli, M.: Locating the source of diffusion in large-scale networks. Phys. Rev. Lett. **109**(6), 068702–068705 (2012). https://doi.org/10.1103/physrevlett.109.068702
4. Huang, Q., et al.: Locating the source of spreading in temporal networks. Physica A: Stat. Mech. Appl. **468**(C), 434–444 (2016). https://doi.org/10.1016/j.physa.2016.10.081
5. Li, X., Wang, X., Zhao, C.: Locating the source of diffusion in complex networks via Gaussian-based localization and deduction. Appl. Sci. **9**, 3758 (2019). https://doi.org/10.3390/app9183758
6. Li, X., Wang, X., Zhao, C.: Locating the epidemic source in complex networks with sparse observers. Appl. Sci. **9**, 3644 (2019). https://doi.org/10.3390/app9183644
7. Nguyen, H.T. , et al.: Multiple infection sources identification with provable guarantees. In: The 25th ACM International. ACM (2016). https://doi.org/10.1145/2983323.2983817
8. Altarelli, F., et al.: Bayesian inference of epidemics on networks via belief propagation. Phys. Rev. Lett. **112**(11), 118701 (2013). https://doi.org/10.1103/PhysRevLett.112.118701
9. Luo, W., Tay, W.P., Leng, M.: Identifying infection sources and regions in large networks. IEEE Trans. Sig. Process. **61**(11), 2850–2865 (2013). https://doi.org/10.1109/tsp.2013.2256902
10. Zang, W., et al.: Discovering multiple diffusion source nodes in social networks. Procedia Comput. Sci. **29**, 443–452 (2014). https://doi.org/10.1016/j.procs.2014.05.040
11. Zhu, K., Chen, Z., Ying, L.: Catch'Em all: locating multiple diffusion sources in networks with partial observations (2016)
12. Fu, L., et al.: Multi-source localization on complex networks with limited observers. EPL (Europhys. Lett.) **113**(1), 18006 (2016). https://doi.org/10.1209/0295-5075/113/18006
13. Hu, Z.L., Shen, Z., Cao, S.: Locating multiple diffusion sources in time varying networks from sparse observations. Sci. Rep. **8**(1) (2018). https://doi.org/10.1038/s41598-018-20033-9
14. Hu, Z.L., et al.: Localization of diffusion sources in complex networks with sparse observations. Phys. Lett. A **382**, 931–937 (2018). https://doi.org/10.1016/j.physleta.2018.01.037

Exact Solutions for (2 + 1)–Dimensional Nonlinear Schrödinger Schrodinger Equation Based on Modified Extended tanh Method

Mei Xiong, Longwei Chen(✉), Chaochao Li, and Juan Wang

College of Statistics and Mathematics, Yunnan University of Finance and Economics, Kunming 650221, Yunnan, People's Republic of China
clwxmff@163.com, tc715@sina.com

Abstract. A modified extended tanh method is used to construct more general exactsolutions of a (2+1)-dimensional nonlinear Schrödinger equation by traveling wave map, and the partial differential is changed into complex algebraic equations. With the aid of Maple and Matlab software, we obtain some exact explicit solutions, such as kink wave solutions, peakon wave solutions, periodic wave solutions and their images.

1 Introduction

It is well known that Schrödinger equation is one of the most basic equation of quantum mechanics. It reflects the state of micro particle changing with time. As it is a powerful tool for solving non relativistic problems in atomic physics, it was widely used in the field of atomic, molecular, solid state physics, nuclear physics, chemistry and so on [1,2]. Recently, Searching and constructing exact solutions of nonlinear partial differential (NLPD) equation is very meaningful for it can describe the problems of mechanics, control process, ecological and economic system, chemical recycling system and epidemiological [3–6]. In the past several decades, much efforts have been on this aspect and many useful methods have been proposed such as inverse scattering method, Jacobi elliptic function method, F-expansion method, Darboux transform, the sine-cosine method and the tanh method. The tanh method is widely used as it can find exact as well as approximate solutions in a systematic way. Subsequently, Fan has proposed an extended tanh method and obtained the travelling wave solutions that can not be obtained by the tanh method [7–9]. Based on this approach, we employed the modified extended tanh method to construct a series of exact travelling wave solutions of a (2+1)-dimensional nonlinear Schrödinger (NLS) equation as

$$iu_t + \alpha u_{xx} + \beta u_{yy} + r|u^2|u = 0. \tag{1}$$

Y. Liu et al. (Eds.): ICNC-FSKD 2019, AISC 1074, pp. 224–231, 2020.
https://doi.org/10.1007/978-3-030-32456-8_24

2 The tanh Method

Here we review the modified extended tanh method.

The modified extended tanh method is developed by Malflied in [10,11], and used in [12–14] among many others. Since all derivatives of a tanh can represented by tanh itself. consider the general NLPDE say in two variables

$$H(u, u_t, u_x, u_x x, u_t t, u_x t, ...) = 0. \tag{2}$$

Now we consider its travelling $u(x,t) = u(\xi)$, where $\xi = x - ct$ or $\xi = x + ct$ and the equation becomes an ordinary differential equation. We apply the following series expansion.

$$u(\xi) = \sum_{i=0}^{N} a_i \phi^i + \sum_{i=1}^{N} b_i \phi^{-i}, \phi' = b + \phi^2. \tag{3}$$

where b is a parameter to be determined, $\phi = \phi(\xi)$ and $\phi' = \frac{d\phi}{d\xi}$.

To determine the parameter N, we usually balance the linear terms of highest-order in the resulting equation with the highest-order nonlinear terms. Then we can get all coefficients of different powers of ϕ and determine a_i, b_i, b, c by make them equals to zeros.

The Riccati equation has the following general solutions:

(a) If $b < 0$
$\phi = -\sqrt{-b}\tanh(\sqrt{-b}\xi)$,
(b) If $b > 0$
$\phi = \sqrt{b}\tan(\sqrt{b}\xi)$,
(c) If $b = 0$
$\phi = -\frac{1}{\xi}$.

3 Exact Travelling Wave Solutions of of (2+1)-Dimensional Nonlinear Schrödinger (NLS) Equation

We consider the travelling wave solution $u(x, y, t) = u(\xi), \xi = x + ky - ct$ for (1), and also

$$u(\xi) = P(\xi) + iQ(\xi). \tag{4}$$

According to (1) and (4), we can get

$$i(P_t + iQ_t) + \alpha(P_{xx} + iQ_{xx}) + \beta(P_{yy} + iQ_{yy}) + r(P^2 + Q^2)(P + iQ) = 0. \tag{5}$$

From (5) can obtain two equation

$$-Q_t + \alpha P_{xx} + \beta P_{yy} + rP^3 + rQ^2 P = 0. \tag{6a}$$

$$P_t + \alpha Q_{xx} + \beta Q_{yy} + rP^2 Q + rQ^3 = 0. \tag{6b}$$

For $u(x, y, t) = u(\xi)$ and $\xi = x + ky - ct$, we can transmute (6a, 6b) into ordinary differential equations

$$cQ' + (\alpha + \beta k^2)P'' + rP^3 + rQ^2P = 0. \tag{7a}$$

$$-cP' + (\alpha + \beta k^2)Q'' + rP^2Q + rQ^3 = 0. \tag{7b}$$

The solution can be expressed as the following form

$$P(\xi) = \sum_{i=0}^{N} a_i\phi^i + \sum_{i=1}^{N} b_i\phi^{-i}. \tag{8a}$$

$$Q(\xi) = \sum_{i=0}^{N1} c_i\phi^i + \sum_{i=1}^{N1} d_i\phi^{-i}. \tag{8b}$$

Balancing the linear term of highest order with the nonlinear term in both equations, we find

$$N - 2 + 4 = 2N + N1 = 3N,$$

$$N1 - 2 + 4 = 2N1 + N = 3N.$$

Thus, $N = N1 = 1$, and

$$P(\xi) = a_0 + a_1\phi + b_1\phi^{-1}. \tag{9a}$$

$$Q(\xi) = c_0 + c_1\phi + d_1\phi^{-1}. \tag{9b}$$

With $\phi' = b + \phi^2$, we get

$$P'(\xi) = (a_1b - b_1) + a_1\phi^2 - b_1b\phi^{-2}, \tag{10a}$$

$$P''(\xi) = 2a_1\phi^3 + 2ba_1\phi + 2b^2b_1\phi^{-3} + 2bb_1\phi^{-1}, \tag{10b}$$

$$Q'(\xi) = (c_1b - d_1) + c_1\phi^2 - d_1b\phi^{-2}, \tag{10c}$$

$$Q''(\xi) = 2c_1\phi^3 + 2bc_1\phi + 2b^2d_1\phi^{-3} + 2bd_1\phi^{-1}. \tag{10d}$$

Substituting (8a, 8b), (9a, 9b), (10a, 10b, 10c and 10d) into two ordinary differential equations (7), and collecting the coefficients of ϕ gets two system of algebraic equations for $a_0, a_1, b_1, c_0, c_1, d_1, b, c, k$. ϕ^0 :$-c(a_1b - b_1) + r(a_0^2 + 2a_1b_1)c_0 + 2ra_0b_1c_1 + 2ra_0a_1d_1$
$+rc_0(c_0^2 + 2c_1d_1) + 4rc_0c_1d_1 = 0$,
ϕ :$2(\alpha + \beta k^2)c_1b + rc_1(a0^2 + 2a_1b_1) + 2rc_0a_0a_1 + ra_1^2d_1$
$+2rc_0^2c_1 + rc_1(c0^2 + 2c_1d_1) + rd_1c_1^2 = 0$,
ϕ^2 :$-ca_1 + rc_0a_1^2 + 2ra_0a_1c_1 + 3rc_0c_1^2 = 0$,
ϕ^3 :$2c_1(\alpha + \beta k^2) + rc_1a_1^2 + rc_1^3 = 0$,
ϕ^{-1} :$2d_1b(\alpha + \beta k^2) + 2ra_0c_0b_1 + rc_1b_1^2 + rd_1(a_0^2 + 2a_1b_1)$
$2rc_0^2d_1 + rc_1d_1^2 + rd_1(c_0^2 + 2c_1d_1) = 0$,
ϕ^{-2} :$cb_1b + rc_0b_1^2 + 2ra_0b_1d_1 + 3rc_0d_1^2 = 0$,
ϕ^{-3} :$2d_1b^2(\alpha + \beta k^2) + rd_1b_1^2 + rd_1^3 = 0$
and

ϕ^0 $:c(c_1b-d_1)+ra_0(a_0^2+2a_1b_1)+4ra_0a_1b_1+ra_0(c_0^2+2c_1d_1)$
$+2ra_1c_0d_1+2rc_0c_1b_1=0,$
ϕ $:2a_1b(\alpha+\beta k^2)+2ra_0^2a_1+ra_1(a_0^2+2a_1b_1)+ra_1^2b_1+2ra_0c_0c_1$
$+ra_1(c_0^2+2c_1d_1)+rb_1c_1^2=0,$
ϕ^2 $:cc_1+ra_1^2a_0+2ra_0a_1^2+2ra_1c_0c_1+ra_0c_1^2=0,$
ϕ^3 $:2a_1(\alpha+\beta k^2)+ra_1^3+ra_1c_1^2=0,$
ϕ^{-1} $:2b_1b(\alpha+\beta k^2)+2ra_0^2b_1+ra_1b_1^2+rb_1(a_0^2+2a_1b_1)+ra_1d_1^2$
$2ra_0c_0d_1+rb_1(c_0^2+2c_1d_1)=0,$
ϕ^{-2} $:-cbd_1+ra_0b_1^2+2ra_0b_1^2+ra_0d_1^2+2rc_0b_1d_1=0,$
ϕ^{-3} $:2b_1b^2(\alpha+\beta k^2)+rb_1^3+rb_1d_1^2=0.$

With the aid of Maple, we obtain $a_0, a_1, b_1, c_0, c_1, d_1, b, c, k$ as follows:

Case (1)

$a_0=-c_0c_1(-\frac{2\alpha+2\beta k^2+rc_1^2}{r})^{-\frac{1}{2}}, a_1=(-\frac{2\alpha+2\beta k^2+rc_1^2}{r})^{\frac{1}{2}}, b_1=-\frac{c_0^2}{4}(-\frac{2\alpha+2\beta k^2+rc_1^2}{r})^{-\frac{1}{2}}$

$b=-\frac{1}{4}\frac{c_0^2r}{2\alpha+2\beta k^2+rc_1^2}, c=-2c_0(\alpha+\beta k^2)(-\frac{2\alpha+2\beta k^2+rc_1^2}{r})^{-\frac{1}{2}}, c_0=c_0,$

$c_1=c_1, d_1=\frac{1}{4}\frac{c_0^2c_1r}{2\alpha+2\beta k^2+rc_1^2}, k=k,$

where $r, \beta, \alpha, k, c_0, c_1$ are arbitrary constants.

Case (2)

$a_0=-d_1(-\frac{2\alpha+2\beta k^2}{rd_1^2+rb_1^2})^{\frac{1}{4}}, a_1=0, b_1=b_1, b=(-\frac{2\alpha+2\beta k^2}{rd_1^2+rb_1^2})^{-\frac{1}{4}},$

$c=-r(d_1^2+b_1^2)(-\frac{2\alpha+2\beta k^2}{rd_1^2+rb_1^2})^{\frac{3}{4}}, c_0=b_1(-\frac{2\alpha+2\beta k^2}{rd_1^2+rb_1^2})^{\frac{1}{4}}, c_1=0, d_1=d_1, k=k,$

where $r, \beta, \alpha, k, b_1, d_1$ are arbitrary constants.

Case (3)

$a_0=(-\frac{2\alpha d_1^2+2d_1^2\beta k^2}{r})^{\frac{1}{4}}, a_1=0, b_1=0, b=d_1^2(-\frac{2\alpha d_1^2+2d_1^2\beta k^2}{r})^{-\frac{1}{2}},$

$c=\frac{r}{d_1}(-\frac{2\alpha d_1^2+2d_1^2\beta k^2}{r})^{\frac{3}{4}}, c_0=0, c_1=0, d_1=d_1, k=k,$

where r, β, α, k, d_1 are arbitrary constants.

Case (4)

$a_0=a_0, a_1=0, b_1=0, b=-\frac{1}{2}\frac{ra_0^2}{\alpha+\beta k^2},$

$c=-ra_0(-\frac{2\alpha+2\beta k^2}{r})^{\frac{1}{2}}, c_0=0, c_1=(-\frac{2\alpha+2\beta k^2}{r})^{\frac{1}{2}}, d_1=0, k=k,$

where r, β, α, k, a_0 are arbitrary constants.

Case (5)

$a_0=a_0, a_1=0, b_1=0, b=-\frac{1}{8}\frac{ra_0^2}{\alpha+\beta k^2}, c=-ra_0(-\frac{2\alpha+2\beta k^2}{r})^{\frac{1}{2}}, c=-ra_0(-\frac{4+2\beta k^2}{r})^{\frac{1}{2}},$

$c_0=0, c_1=(-\frac{2\alpha+2\beta k^2}{r})^{\frac{1}{2}}, d_1=-\frac{a_0^2}{4}(-\frac{2\alpha+2\beta k^2}{r})^{-\frac{1}{2}}, k=k,$

where r, β, α, k, a_0 are arbitrary constants.

Case (6)

$a_0=0, a_1=(-\frac{2\alpha+2\beta k^2}{r})^{\frac{1}{2}}, b_1=0, b=-\frac{1}{2}\frac{rc_0^2}{\alpha+\beta k^2}$

$c=rc_0(-\frac{2\alpha+2\beta k^2}{r})^{\frac{1}{2}}, c_0=c_0, c_1=0, d_1=0, k=k,$

where r, β, α, k, c_0 are arbitrary constants.

Case (7)

$a_0=-c_0c_1(-\frac{2\alpha+2\beta k^2+rc_1^2}{2})^{-\frac{1}{2}}, a_1=(-\frac{2\alpha+2\beta k^2+rc_1^2}{2})^{\frac{1}{2}}, b_1=0, b=\frac{-c_0^2r}{2\alpha+2\beta k^2+rc_1^2},$

$c = -2(\alpha + \beta k^2)c_0(-\frac{2\alpha+2\beta k^2+rc_1^2}{2})^{-\frac{1}{2}}, c_0 = c_0, c_1 = c_1, d_1 = 0, k = k,$
where $r, \beta, \alpha, k, c_0, c_1$ are arbitrary constants.
Case (8)

$a_0 = 0, a_1 = (-\frac{2\alpha+2\beta k^2}{r})^{\frac{1}{2}}, b_1 = \frac{1}{8}\frac{rc_0^2}{\alpha+\beta k^2}(-\frac{2\alpha+2\beta k^2}{r})^{\frac{1}{2}}, b = -\frac{1}{8}\frac{rc_0^2}{\alpha+\beta k^2}$
$c = rc_0(-\frac{2\alpha+2\beta k^2}{r})^{\frac{1}{2}}, c_0 = c_0, c_1 = 0, d_1 = 0, k = 0,$
where r, β, α, k, c_0 are arbitrary constants.
Case (9)

$a_0 = 0, a_1 = 0, b_1 = c_0^2\sqrt{-\frac{r}{2\alpha+2\beta k^2}}, b = -\frac{1}{2}\frac{c_0^2 r}{\alpha+\beta k^2},$
$c = -\frac{c_0 r}{\sqrt{-\frac{r}{2\alpha+2\beta k^2}}}, k = k, c_0 = c_0, c_1 = 0, d_1 = 0,$
where r, β, α, k, c_0 are arbitrary constants.

If $b > 0$, we get

$$u = a_0 + a_1 v(x,y,t) + b_1 v(x,y,t)^{-1} + i(c_0 + c_1 v(x,y,t) + d_1 v(x,y,t)^{-1}), \quad (11a)$$

Where $v(x,y,t) = \sqrt{b}\tan(\sqrt{b}(x + ky - ct))$.

If $b = 0$, we get

$$u = a_0 + a_1 v(x,y,t) + b_1 v(x,y,t)^{-1} + i(c_0 + c_1 v(x,y,t) + d_1 v(x,y,t)^{-1}), \quad (11b)$$

where $v(x,y,t) = \frac{1}{x+ky-ct}$.

If $b < 0$, we get

$$u = a_0 + a_1 v(x,y,t) + b_1 v(x,y,t)^{-1} + i(c_0 + c_1 v(x,y,t) + d_1 v(x,y,t)^{-1}), \quad (11c)$$

where $v(x,y,t) = -\sqrt{-b}\tanh(\sqrt{-b}(x + ky - ct))$.

Substituting all those situation into (11a, 11b and 11c) respectively, we can get all solutions of the derivative nonlinear Schrödinger equation.

4 Image Simulation

In order to grasp these exact travelling solutions,we choose several exact solutions and use the Matlab software to simulate images. In the process of image simulation, the figures and value of parameters we selected show as follows:

In the Fig. 1, we take $r = -1, \beta = 11, k = 0.1, c_0 = -0.25, c_1 = 0.1, \alpha = -0.2, t = 0.1$. In the Fig. 2, we take $r = -6, \beta = -3, k = 1, d_1 = 1, \alpha = 1, b_1 = -1t = 0.1$.

In the Fig. 3, we take $r = 0.1, \beta = 0.3, k = 0.1, d_1 = -2, \alpha = 0.2, t = 0.1$. In the Fig. 4, we take $r = -6, \beta = 2, k = 0.1, c_0 = 6, \alpha = -4, d_1 = 2, t = 0.1$.

In the Fig. 5, we take $r = -1, \beta = 3, k = 3, a_0 = 2, \alpha = -0.7, t = 0.1$. In the Fig. 6, we take $r = -1, \beta = -0.3, k = 0.1, a_0 = -0.25, \alpha = -1, t = 0.1$.

In the Fig. 7, we take $r = -6, \beta = 3, k = 2.5, c_0 = -0.5, \alpha = -2, t = 0.1$. In the Fig. 8, we take $r = -1, \beta = 3, k = 2.9, c_0 = 6, \alpha = -2, c_1 = 2, t = 0.1$.

In the Fig. 9, we take $r = -2, \beta = 2, k = 1.6, c_0 = 2.5, \alpha = 3.2, t = 0.1$. In the Fig. 10, we take $r = -0.1, \beta = 1, k = 1, c_0 = 2.5, \alpha = -1, c_1 = 1, t = 0.1$.

In the Fig. 11, we take $r = -6, \beta = 1, k = -6, c_0 = 1, \alpha = -1, t = 0.1$. In the Fig. 12, we take $r = 6, \beta = 1, k = -6, c_0 = 1, \alpha = -1, t = 0.01$.

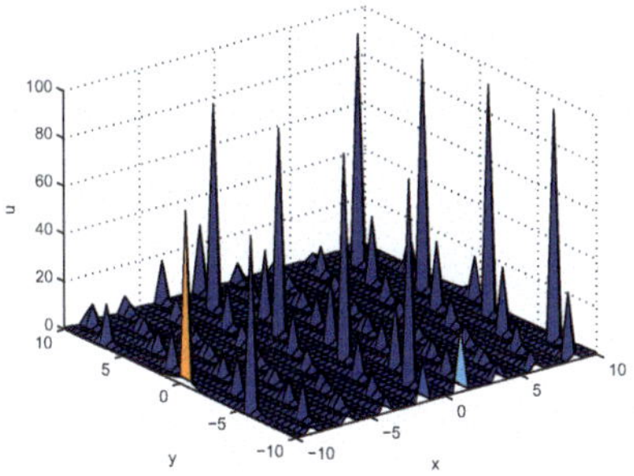

Fig. 1. 3-dimensional wave of Case (1).

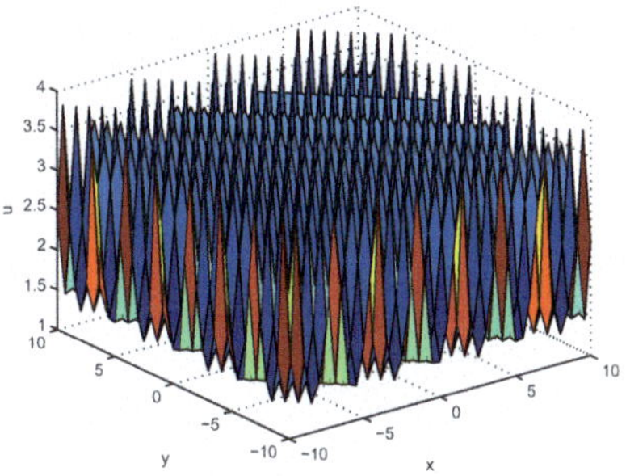

Fig. 2. 3-dimensional wave of Case (2).

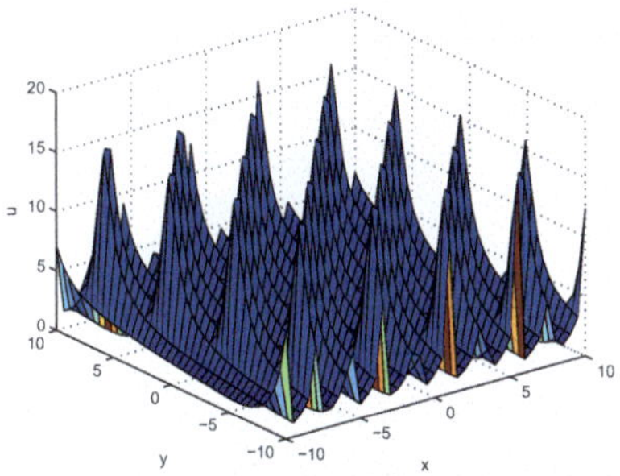

Fig. 3. 3-dimensional wave of Case (3).

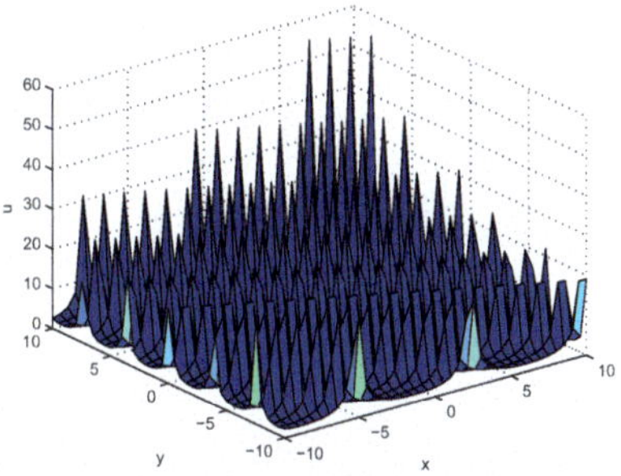

Fig. 4. 3-dimensional wave of Case (4).

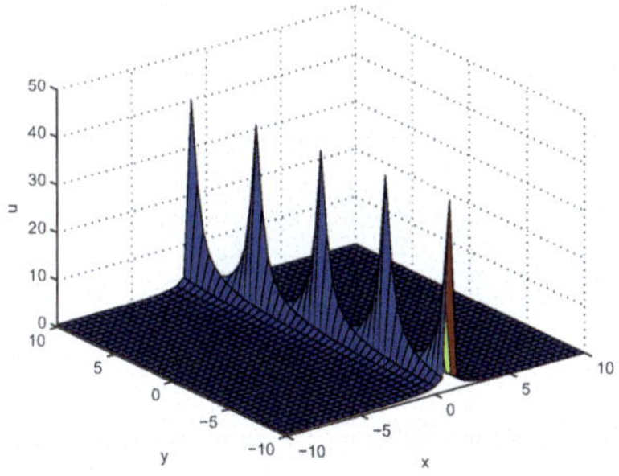

Fig. 5. 3-dimensional wave of Case (5).

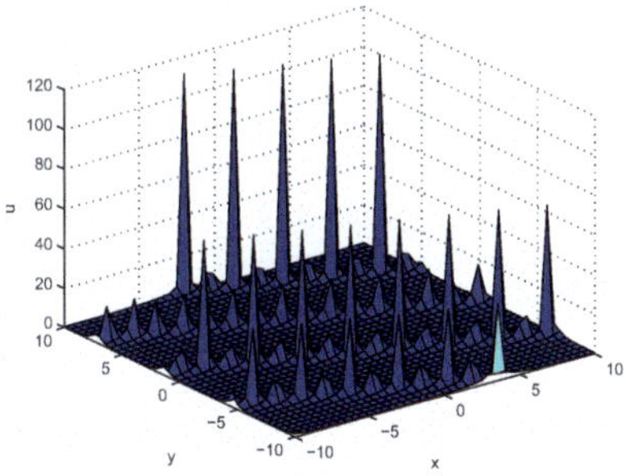

Fig. 6. 3-dimensional wave of Case (6).

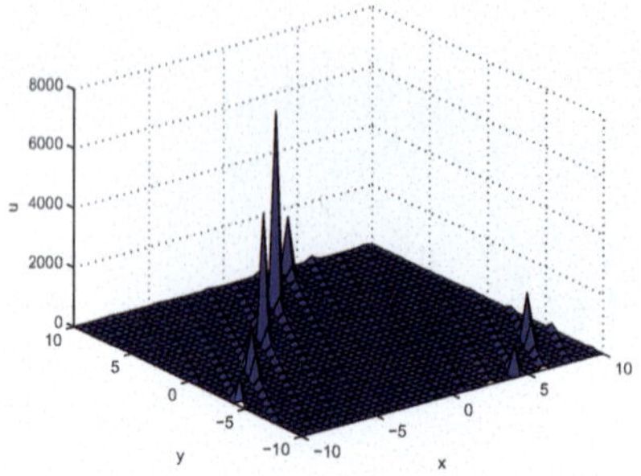

Fig. 7. 3-dimensional wave of Case (7).

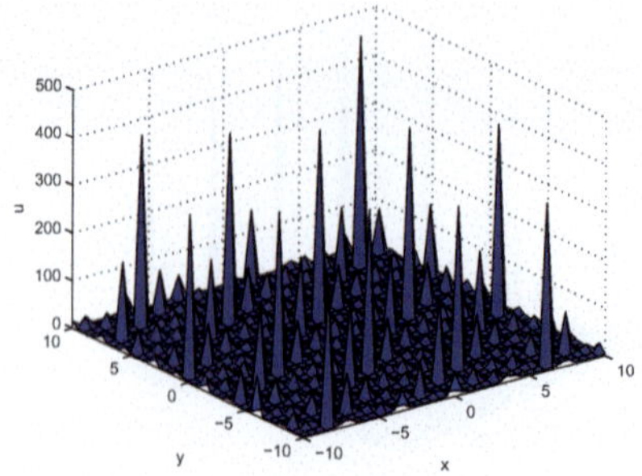

Fig. 8. 3-dimensional wave of Case (8).

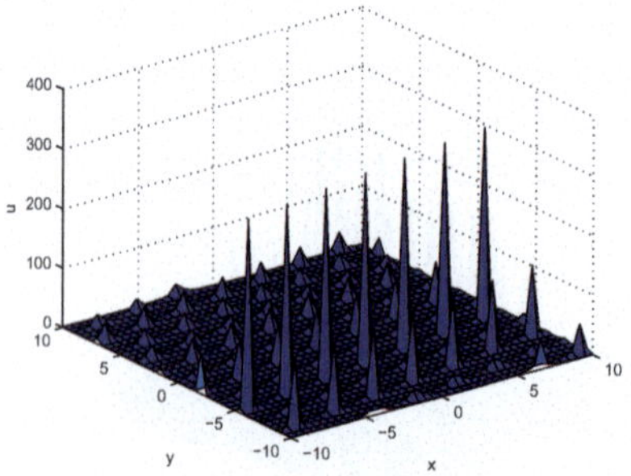

Fig. 9. 3-dimensional wave of Case (9).

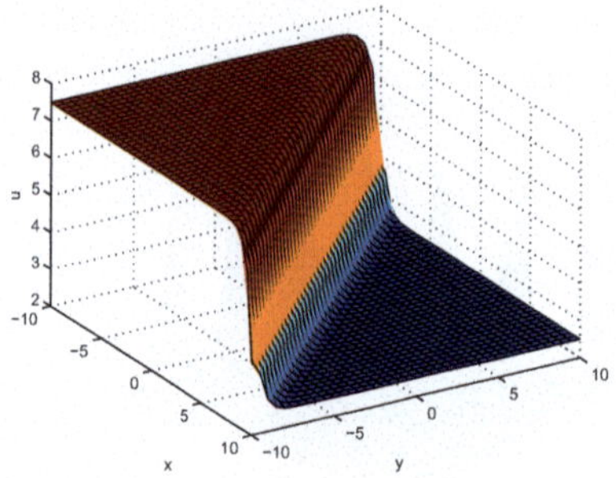

Fig. 10. 3-dimensional wave of Case (10).

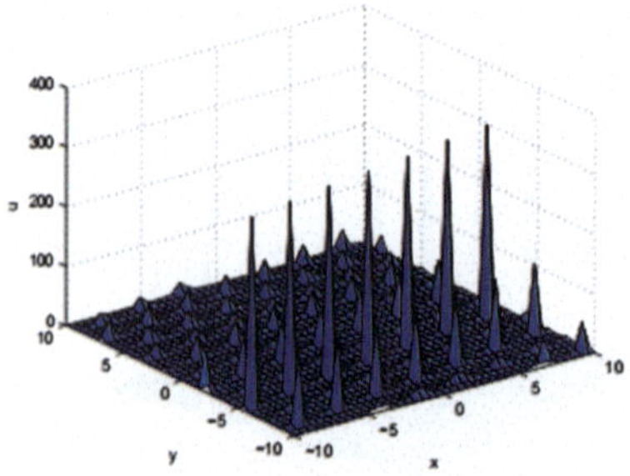

Fig. 11. 3-dimensional wave of Case (11).

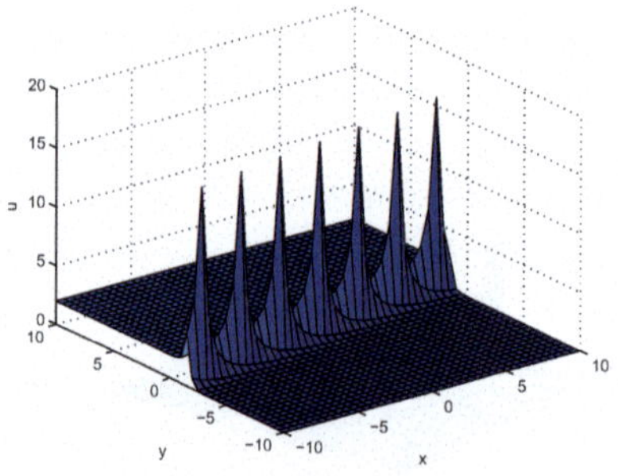

Fig. 12. 3-dimensional wave of Case (12).

5 Conclusion

In this paper we studied the nonlinear Schrödinger equation by finding its exact travelling wave solutions through the modified extended tanh methods. With the aid of waveform graphs of the solutions, we can obtain the related properties of the equation. However, we can also use the method of dynamical systems to obtain the bounded solutions.

Acknowledgements. The authors gratefully acknowledges the support of the Yunnan Tobacco Company's science and technology project "Research on Data Resource Mining Based on Multi-objective Decision Making".

References

1. Gray, P., Scott, S.: Chemical Oscillations and Instabilities. Clarendon Press, Oxford (1990)
2. Murray, J.: Mathematical Biology. Springer, Berlin (1989)
3. Dingjiang, H., Desheng, L., Hongqing, Z.: Explicit and exact travelling wave solutions for the generalized derivative Schrödinger equation. Chaos, Solitons Fractals **31**, 586–593 (2007)
4. Vakhnenko, V.O., Parkes, E.J., Morrison, A.J.: A Bäcklund transformation and the inverse scattering transform method for the generalized Vakhnenko equation. Chaos, Solitons Fractals **17**(4), 683–692 (2003)
5. Yomba, E.: The extended Fans sub-equation method and its application to KDV-mKDV BKK and variant Boussinesq equation. Phys. Lett. **A336**, 463–476 (2005)
6. Matveev, V.A., Salle, M.A.: Darboux Transformation and Solitons. Springer, Berlin (2001)
7. Khuri, S.A.: A complex tanh-function method applied to nonlinear equations of Schrödinger type. Chaos, Solitons Fractals **20**, 1037–1040 (2004)
8. Kavitha, L., Akila, N., Prabhu, A., Kuzmanovska-Barandovska, O., Gopi, D.: Exact solitary solutions of an inhomogeneneous modified nonlinear Schrödinger equation with competing nonlinearities. Math. Comput. Modell. **53**, 1095–1110 (2011)
9. Fan, E., Hon, Y.C.: Applications of extended tanh method to special types of nonlinear equations. Appl. Math. Comput. **141**, 351–358 (2003)
10. Malfliet, W.: Solitary wave solutions of nonlinear wave equations. Amer. J. Phys. **60**(7), 650–654 (1992)
11. Malfliet, W.: The tanh method: a tool for solving certain classes of nonlinear evolution and wave equations. J. Comput. Appl. Math. **164–165**, 529–541 (2004)
12. Wazwaz, A.-M.: The Camassa-Holm-KP equations with compact and noncompact travelling wave solutions. Appl. Math. Comput. **170**, 347–360 (2005)
13. Wazwaz, A.-M.: The tanh and the sine-cosine methods for the complex modified KdV and the generalized KdV equations. Comput. Math. Appl. **49**, 1101–1112 (2005)
14. Wazwaz, A.-M.: The extended tanh method for abundant solitary wave solutions of nonlinear wave equations. Appl. Math. Comput. **187**, 1131–1142 (2007)

Natural Computation: Theory and Algorithms - Other Topics in Natural Computation Theory and Algorithms

Extraction of Basic-Level Categories Using Dendrogram and Multidendrogram

Mariusz Mulka(✉), Wojciech Lorkiewicz, and Radosław P. Katarzyniak

Faculty of Computer Science and Management,
Wrocław University of Science and Technology, Wrocław, Poland
{mariusz.mulka,wojciech.lorkiewicz,radoslaw.katarzyniak}@pwr.edu.pl

Abstract. Cognitive agents should be equipped with computational methods enabling autonomous extraction of internal cognitive structures, i.e., supporting organization of large datasets and facilitating interaction with human users. In this research, we focus solely on the extraction of basic-level categories. We propose a computational approach that allows the agent to develop, through individual interaction with the external world, provisional clusters (using hierarchal clustering techniques) and filter them, using predefined measures of basic-levelness (inspired by psycholinguistic research). We focus on analysing the behaviour of two proposed computational approaches, namely dendrogram and multidendrogram, to establishing provisional clusters. Further, the approach is extensively studied through an array of simulations. Obtained results highlight that the proposed methods can be used to extract basic-level categories.

Keywords: Basic-level categories · Cognitive computing

1 Introduction

Categorization, a central part of human cognition, is the mental ability to structure and organise objects, established through the interaction with the external world, into meaningful categories. Each category groups individual objects that share some commonalities or specific purpose, i.e., kind [6]. One particular set of categories, i.e., basic-level categories, is of particular importance as they decompose the perception of the world into maximally informative categories [12] and provides fundamental building blocks of cognition.

As intentionality and autonomy should be embedded into artificial agents, they require mental-like processes in order to make decisions and take actions. Furthermore, as cognitive agents get smarter our expectations of their capabilities rise correspondingly. As such, basic-level categories are also of high importance to artificial systems, where they aid internal organization and communication with other agents.

In traditional approaches basic-level categories are predefined by the expert at the design time. Resultantly, incorporated assignments are static and require

Y. Liu et al. (Eds.): ICNC-FSKD 2019, AISC 1074, pp. 235–243, 2020.
https://doi.org/10.1007/978-3-030-32456-8_25

proper manual maintenance during agent operation. A natural extension is to employ an artificial agent with the ability to autonomously establish basic-level categories, i.e., through its individual interaction with the external world. Consequently, the agent is required to categorize its perceptions into basic-level categories.

The main goal of the paper is to study the ability of a cognitive agent to autonomously develop basic-level categories using computational methods and interaction with the external world. In this paper we continue our previous efforts [9] and focus on a computational solutions enabling an artificial agent to organise datasets into basic-level categories. Following psycholinguistic findings and cognitive computing paradigm we provide a novel approach, where the provisional clusters that result from hierarchical clustering techniques applied to the collected data are further filtered through basic-levelness measures. Additionally, we present an extensive simulation-based performance analysis of the proposed approach, i.e., using a broad array of annotated (by the experts) datasets utilised in psycholinguistic studies. As such, this paper fills in a current research gap, where different basic-levelness measures are utilised against different hierarchical methods, namely dendrogram and multidendrogram, to establish arrays of provisional clusters.

2 Approach

As aforementioned, in an idealised setting basic-level categories can be directly predefined by the expert, though limiting the autonomy and robustness of the system. In a more realistic setting, the expert could establish an array of possible category levels that need to be further analysed by the agent. Such a case is more complex, as it requires a computational approach to assess the basic-levelness of individual levels. However in a strictly practical scenario, an agent should be able to independently establish provisional clusters, i.e., perform preliminary partitioning of objects, and then assess their basic-levelness, i.e., using computational methods applied on collected perceptions without expert intervention.

In general, the task of extracting basic-level categories in a cognitive agent can be seen as a clustering problem, where specific properties of resulting clusters are sought. In particular, basic-level categories carry the most information and are the most differentiated from one another, so basic-level is the level at which most of our knowledge is organized [12]. In the proposed approach, this task is divided into two fundamental steps: extraction and filtering of provisional clusters.

The extraction of provisional clusters is aimed at delivering object categories [1] that refer to shared representations of different object kinds [6], i.e., objects that are alike but still are discriminable. The extraction is performed using hierarchical clustering techniques [13] that rely heavily on the notion of similarity between objects (similarity measure) and similarity between clusters of objects (linkage criteria). We rely on the hierarchical clustering approach as it follows the structural view of categorization [1] (as opposed to the processing view) and allows to capture the fact that natural tendency to group objects is performed on multiple levels of specificity, i.e., naturally formed categories are not flat and objects can be simultaneously categorized at multiple levels of generality. As such, provisional clusters are organised hierarchically, where each level represents a fundamentally different level of perception and ranges from very detailed, specific and distinct representations, e.g., a red Carrera GT, to a very broad and general representation, e.g., a vehicle.

Fernández and Gómez [4] proposed an agglomerative algorithm for creating hierarchy of clusters as a multidendrogram. This approach solved the common problem of ties, which typically introduced random characteristics into hierarchical clustering algorithms. In particular, in multidendrogram approach it is possible to merge more than two clusters in a single step, i.e., if the distance between clusters is tied (minimal for at least three clusters). Consequently, it is a deterministic algorithm and the resultant dendrogram has less levels.

Assuming that the hierarchy of clusters is defined, the agent can focus on filtering the basic-level. This step is performed to determine which level of provisional clusters is a basic-level. It is worth noting, that basic-level categories are typically characterised by the property that this particular categorisations maximizes the number of attributes shared by members of the category, and minimizes the number of attributes shared with other categories. In particular, four main approaches were found in the psycholinguistic literature for assessing basic-levelness of a given level of categories (described in paper [9]): category's cue validity [12] (*CCV* used to find categories with maximized informativeness and distinctiveness), feature-possession [8] (*FP* which assigns each attribute to specific category located in the hierarchy), category utility [2] (*CU* probabilistic measure, which thrives to capture following aspects: a category is only useful to the extent that it can improve ability to accurately predict attributes' values for members of that category and efficiently communicate information about attributes exhibited by members of that category), and category attentional slip [5] (*CAS* focused on cardinality of redundant tests and length of the optimal strategy needed to establish category). We further utilise two proposed measures (see paper [9]): cue validity with global threshold (*CVGT* uses a global predefined threshold in order to determine whether feature should be included in a specific category) and cue validity with feature-possession (*CVFP* uses *FP* measure to determine to which category a feature belongs to).

In order to extract basic-level categories, it is necessary to measure each level of categorization, and then select the one with the highest value of the measure (except attentional slip measure, where the value should be minimized). The selected level is regarded as a basic-level and contains basic-level categories.

3 Simulations

The underlying goal of simulations is to study the performance of dendrogram and multidendrogram approaches in the task of establishing provisional clusters in order to further extract basic-level categories.

All datasets consist of objects described by a set of binary attributes[1]. Moreover, all datasets are incorporated from psycholinguistic research papers, i.e., Gosselin and Schyns [5] (A), Corter and Gluck [2] (B), Murphy and Smith's [11] (C), Murphy's [10] (D, E), Tanaka and Taylor's [14] (F, G), Corter [3] (H), Tversky and Hemenway [16] (I, J, K) and Hoffmann and Ziessler's [7] (L). Each dataset has a predefined (by an expert) basic-level. Datasets differ in the number of objects (from 8 to 18, and dataset e 256) and attributes (from 10 to 30, and dataset f, g having 144), in number of overlapping attributes belonging to contrasting categories at different levels (low, middle or high), e.g., objects belonging to contrasting categories at higher level with no overlapping attributes (A, C, D, L), with some overlapping attributes (B, E, H), at middle level without overlapping attributes (D), with some overlapping attributes (A, B, C, E, F, G, H, I, J, K, L).

We test each approach, namely *dendrogram* (using agglomerative algorithm with Tversky similarity function [15] and complete linkage criteria) and *multidendrogram* (using multidendrogram approach [4] with Tvesky similarity function [15] and complete linkage criteria), against different basic-levelness measures and all assumed dataset. The simulation procedures is as follows: for a given approach and a given dataset we establish a hierarchy of provisional clusters (extraction step based on the data in the dataset) and calculate values of basic-levelness measures (assumed 5 measures) for each level. Then, we filter the categories at the level with the best value[2] as the basic-level. Finally, we compare this level with the predefined intended basic-level. We further repeat this procedure for all datasets, i.e., calculate precision and recall individually for each measure and approach. It should be noted that due to the inherent problem with handling ties in the dendrogram approach we repeat each individual simulation 1000 times using a randomly ordered data, as such the presented values of precision and recall are averages over 1000 repeats. For the sake of completeness, we also include results for the original dataset hierarchies as established by the experts – 3 *predefined levels* – for each dataset.

3.1 Results and Discussion

Average precision and recall for all measures for extracting basic-level categories and three versions of simulation are presented in the Fig. 1. Precision and recall of measures for version related to 3 predefined levels are presented as baseline. We further focus on the performance in three cases, namely comparison between 3 predefined levels and dendrogram approach, comparison between 3 predefined levels and multidendrogram approach, and comparison between dendrogram and multidendrogram approaches.

---

[1] Basic binarization process was performed for non-binary attributes.

[2] Value should be minimized for AS measure and maximized for remaining ones.

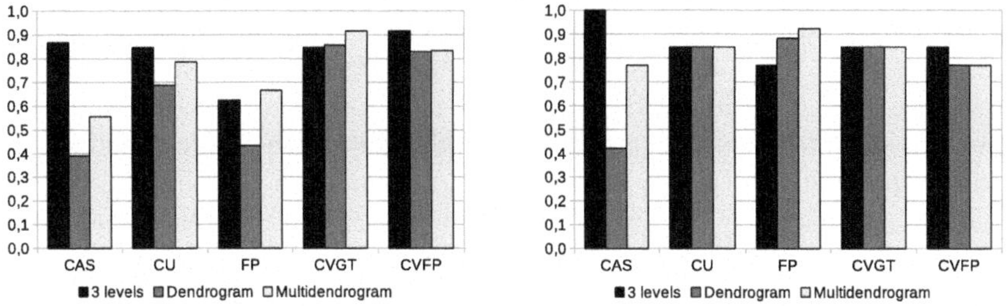

Fig. 1. General performance of measures: precision (left), recall (right).

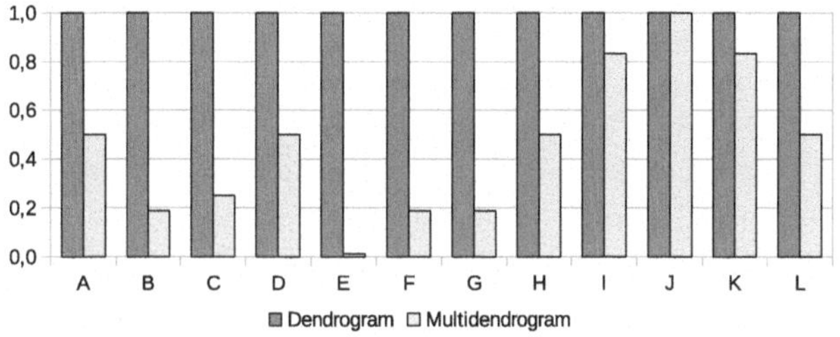

Fig. 2. Number of levels in hierarchy divided by the number of objects.

3.2 Comparison Between 3 Predefined Levels and Dendrogram

In general, in only 1 dataset (F) all measures correctly determined the basic-level if hierarchy of clusters was a dendrogram, as such we further focused on cases where individual measures resulted in erroneous determination of basic-level categories.

The performance of AS measure decreased highly in terms of both quality measures (precision nearly by 0.48 and recall nearly by 0.58). It resulted from the fact that the AS measure determined as basic-level one of additional levels which do not occur in predefined levels for A, D, I, K datasets. In addition, in case of B, E datasets, the measure selected levels containing 2 and 4 categories as basic one. Due to additional levels of categorization, AS selected level containing 2 categories as basic one, instead of correct one for C, H datasets which resulted from assigning attributes to additional levels. It is worth noting, that AS correctly determined basic-level without additional levels for such datasets. Only one level was still determined as basic one in case of G dataset.

In case of CU measure the decrease of precision (nearly by 0.16) resulted from the fact that for H, L datasets; the measure determined 3 levels as basic one, namely containing $2, 3, 4$ categories. While, in case of predefined levels only two levels were determined as basic one (containing 2 and 4 categories) in case of L dataset (for H dataset the measure correctly determined basic-level). The CU measure was still not able to determine correctly basic-level in case of C dataset and determined only one level as basic one in case of G dataset.

The increase of recall (nearly by 0.11), in case of measure *FP* resulted from determining more levels as basic one, which allowed to correctly determine basic-level in case of *B*, *G*, *J* and *L* datasets (while only for *B* dataset additional levels of categorization had no effect). Obviously, determining more levels as basic one decreased the precision (nearly by 0.19) of such a measure. The *FP* measure still was not able to correctly determine basic-level in case of *E* dataset but usage of dendrogram caused selecting root of hierarchy of clusters as basic-level.

The usage of dendrogram had almost none effect on *CVGT* measure. The *CVGT* measure still determined only one level as basic one in case of *G* dataset and was still not able to determine correctly basic-level in case of *L* dataset.

Finally, precision and recall of *CVFP* measure decreased (both nearly by 0.08) because root of hierarchy of clusters was selected as basic-level in case of *E* dataset (*FP* measure assigned many attributes to basic-level categories or superordinate clusters). The *CVFP* measure still determined only one level as basic one in case of *G* dataset and was still not able to determine correctly basic-level in case of *L* dataset.

Concluding, the precision slightly increased for *CVGT* measure and decreased for remaining ones, while recall decreased in case of *AS* and *CVFP* measures and increased in case of *FP* measure.

3.3 Comparison Between 3 Predefined Levels and Multidendrogram

In general, in only 3 datasets (*A*, *D*, *F*) all measures correctly determined the basic-level when hierarchy of clusters was a multidendrogram, as such we further focus on cases where individual measures resulted in erroneous determination of basic-level categories.

The performance of *AS* measure decreased in terms of both quality measures (precision nearly by 0.31 and recall nearly by 0.23). It resulted from the fact that the *AS* measure determined as basic-level one of additional levels of categorization which do not occur in predefined levels for *B*, *E*, *H*, *I*, *K*, *L* datasets (only in case of *B*, *H*, *L* datasets, the measure selected more levels of categorization but one of them was a basic-level).

In case of *CU* measure the decrease of precision (nearly by 0.06) resulted from the fact that for *H* datasets the measure determined 2 levels as basic one, namely containing 2, 4 categories (objects were differently assigned to categories in predefined hierarchy and by multidendrogram in case of level with 2 categories). The *CU* measure still determined only one level as basic one in case of *G* dataset and was still not able to determine correctly basic-level in case of *C*, *L* datasets.

The precision and recall of *FP* measure increased (precision nearly by 0.04, and recall nearly by 0.15), it resulted from the fact that for *B*, *J*, *H* datasets at least one determined level of categorization was a basic-level. In case of *E* dataset, the measure was still not able to determine basic-level but in that case the level containing 1 category was selected instead of that one containing 2 categories. However, the measure determined 3 levels as basic one in case of *G* dataset (including the correct one).

Precision of *CVGT* measure increased nearly by 0.07. Due to the change of hierarchy of clusters (it did not contain level with 8 categories as it was in case with predefined hierarchy) the *CVGT* measure correctly determined basic-level in case of *B* dataset. The measure still determined only one basic-level in case of *G* dataset and was still not able to determine correct basic-level in case of *L* dataset.

Finally, precision and recall of *CVFP* measure decreased (both nearly by 0.08) because root of hierarchy of clusters was selected as basic-level in case of *E* dataset (*FP* measure assigned many attributes to basic-level categories or superordinate clusters). The *CVGT* measure still determined only one level as basic one in case of *G* dataset and was still not able to determine correctly basic-level in case of *L* dataset.

Concluding, the precision increased only for *CVGT* measure and decreased for remaining measures, while recall increased only for *FP* measure and decreased for *CAS* and *CVFP* measures.

3.4 Comparison Between Dendrogram and Multidendrogram

Multidendrogram was a better choice than dendrogram because of two reasons, namely, better performance in terms of precision and recall, and lower number of levels of categorization.

Firstly, when multidendrogram was used the precision and recall of measures was higher or equal to precision and recall when dendrogram was used (see Fig. 1). The precision increased in case of *CAS* (nearly by 0.16), *CU* (nearly by 0.1), *FP* (nearly by 0.23), *CVGT* (nearly by 0.06), while the recall increased in case of *CAS* (nearly by 0.35), *FP* (nearly by 0.04).

Finally, multidendrogram coped with ties in proximity problem in contrast to dendrogram. Therefore, prepared hierarchy of clusters by multidendrogram contained less or the same number of levels of categorization (see Fig. 2). Only in case of *J* dataset the number of levels of categorization was the same. While, in case of *E* dataset, dendrogram contained 256 levels of categorization and multidendrogram contained only 3 levels of categorization. Therefore, containing less levels of categorization is related to performing less computations necessary to extract basic-level categories.

4 Conclusions

Basic-level categories, as core building blocks of cognition, are of high importance to artificial systems, especially in data-intensive environments, where human users are overwhelmed with information and require aid. We claim that a cognitive agent should be able to independently establish provisional clusters (preliminary partitioning of objects) and further be able to assess their basic-levelness, i.e., using computational methods applied on collected perceptions without need of an expert. As such, the main goal of this paper was to provide a concise and extensive analysis of the behaviour of two proposed computational approaches, namely *dendrogram* and *multidendrogram*, with 5 basic-levelness measures and using 12 well-studied datasets with predefined basic-levels.

Obtained results highlight that *multidendrogram* approach can be successfully utilised to extract basic-level categories, especially in combination with *CU*, *CVGT*, and *CCVFP* measures, i.e., all resulted in at least 80% of precision and recall. In general, precision and recall was higher when *multidendrogram* approach was used as compared to *dendrogram*. This seems intuitive, as the latter results in a creation of more levels in provisional clusters. Moreover, when provisional clusters were established computationally by an algorithm, the *CAS* measure resulted in the worst performance, i.e., it tended to select additional levels of categorization (other than 3 predefined ones).

Acknowledgment. This research was carried out at Wrocław University of Science and Technology (Poland) under Grant 0401/0190/18 titled Models and Methods of Semantic Communication in Cyber-Physical Systems.

References

1. Bornstein, M.H., Arterberry, M.E.: The development of object categorization in young children: hierarchical inclusiveness, age, perceptual attribute, and group versus individual analyses (2010)
2. Corter, J.E., Gluck, M.A.: Explaining basic categories: feature predictability and information. Psychol. Bull. **111**(2), 291–303 (1992)
3. Corter, J.E., Gluck, M.A., Bower, G.H.: Proceedings of the Tenth Annual Conference of the Cognitive Science Society (1988)
4. Fernández, A., Gómez, S.: Solving non-uniqueness in agglomerative hierarchical clustering using multidendrograms. J. Classif. **25**(1), 43–65 (2008)
5. Gosselin, F., Schyns, P.: Debunking the basic level. In: Cognitive Science Society (US) Conference/Proceedings, pp. 277–282 (1997)
6. Harnad, S.: To cognize is to categorize: cognition is categorization. In: Lefebvre, C., Cohen, H. (eds.) Handbook of Categorization. Elsevier, Amsterdam (2005)
7. Hoffmann, J., Ziessler, C.: Objektidentifikation in künstlichen Begriffshierarchien. Object Ident. Artif. Conceptual Hierarchies **191**(2), 135–167 (1983)
8. Jones, G.V.: Identifying basic categories. Psychol. Bull. **94**(3), 423–428 (1983)
9. Mulka, M., Lorkiewicz, W.: Measures for extracting basic-level categories. In: 2018 11th International Congress on Image and Signal Processing, BioMedical Engineering and Informatics (CISP-BMEI), pp. 1–6 (2018)
10. Murphy, G.L.: Parts in object concepts: experiments with artificial categories. Mem. Cogn. **19**(5), 423–438 (1991)
11. Murphy, G.L., Smith, E.E.: Basic-level superiority in picture categorization. J. Verbal Learn. Verbal Behav. **21**(1), 1–20 (1982)
12. Rosch, E., Mervis, C.B., Gray, W.D., Johnson, D.M., Boyes-Braem, P.: Basic objects in natural categories. Cogn. Psychol. **8**(3), 382–439 (1976)
13. Saxena, A., Prasad, M., Gupta, A., Bharill, N., Patel, O.P., Tiwari, A., Er, M.J., Ding, W., Lin, C.T.: A review of clustering techniques and developments. Neurocomputing **267**, 664–681 (2017)

14. Tanaka, J.W., Taylor, M.: Object categories and expertise: Is the basic-level in the eye of the beholder? Cogn. Psychol. **23**, 457–482 (1991)
15. Tversky, A.: Features of similarity. Psychol. Rev. **84**(4), 327–352 (1977)
16. Tversky, B., Hemenway, K.: Objects, parts, and categories. J. Exp. Psychol. Gen. **113**(2), 169–193 (1984)

Compressed Sensing of Complex Reflections Using Range-Azimuth Dictionary in a Bionic Sonar System

Changsheng Yang[1(✉)], Junxiong Wang[1], Hong Liang[1], and Herbert Peremans[2]

[1] School of Marine Science and Technology, Northwestern Polytechnical University, Xi'an 710072, China
yangchsh_nwpu@nwpu.edu.cn
[2] Department Engineering Management, University Antwerp, 2000 Antwerp, Belgium

Abstract. The FM-bats are known to be able to sense the environment by echolocation. In this paper, assuming the objects in the environment can be characterized by a sparse representation of the echoes in range and azimuth, a compressed sensing algorithm using a range-azimuth dictionary is proposed. The monaural and binaural range-azimuth dictionaries are constructed from measurements collected with a bionic sonar system consisting of one emitter and two receivers fitted with a 3-D printed replica of a real bat's external ears. To estimate the range and azimuth of a target, the L1-minimization method is used. Since the high coherence in templates could cause ambiguity, the non-uniform sampled dictionary derived from the coherence is investigated. The non-uniformly sampled monaural and binaural dictionaries are used to process the echoes collected from a real brick-wall. Results indicate that strong echoes can be correctly localized both in azimuth and range by all three dictionaries, but for weak, highly overlapping echoes, monaural dictionaries have problems interpreting these echo signals correctly while binaural dictionary could improve the result.

Keywords: Compressed sensing · Range-azimuth dictionary · Bionic sonar

1 Introduction

Bats can find and identify targets using FM signals even in quite complex environment such as a forest where many echoes from the surroundings are received [1]. The spectrogram correlation and transformation (SCAT) model of the sonar receiver in the big brown bat and its modified versions were proposed to explain how the bats could construct the acoustic image of the environment and targets [2, 3]. However, it was shown that these models break down when facing several (more than 3) closely spaced overlapping echoes [4]. However, an alternative method proposed by Fontaine and Peremans could distinguish such closely spaced objects based on a time-domain sparse representation of bat echolocation calls [5].

Y. Liu et al. (Eds.): ICNC-FSKD 2019, AISC 1074, pp. 244–251, 2020.
https://doi.org/10.1007/978-3-030-32456-8_26

Considering the azimuth information is also necessary to fully interpret the echoes, a method based on a range-azimuth dictionary (RAD) is proposed in this article. Using a bionic sonar system, the echoes from a real brick-wall are processed by this method.

2 Sparse Representation and Interpretation of the Echoes

2.1 Range-Azimuth Dictionary

For an active sonar system, reflections from the surroundings coming from different distances and directions combine into the echo arriving at the receiver. By a similar sparsity assumption as in [5], the received echo $E(t)$ in the sonar system can be represented as the linear superposition of reflections,

$$E(t) = \sum_{i,j} A_{ij} s(r_i, \theta_j) \tag{1}$$

Where $s(r_i, \theta_j)$ represents the normalized reflection of an object at the range r_i and the azimuth direction θ_j, A_{ij} indicating its amplitude.

Considering the range as the integer delay of the discretized reflection $s^j(n)$ which is the sampling of the reflection from azimuth θ_j, the range-azimuth dictionary (RAD) can be constructed, as shown in Eq. (2).

$$\Psi = \begin{bmatrix} s^1(1) & & & & s^p(1) & & & \\ s^1(2) & s^1(1) & & & s^p(2) & s^p(1) & & \\ s^1(3) & s^1(2) & & & s^p(3) & s^p(2) & & \\ \vdots & s^1(3) & & s^1(1) & \vdots & s^p(3) & & s^p(1) \\ s^1(N-2) & \vdots & & s^1(2) & \cdots\; s^p(N-2) & \vdots & & s^p(2) \\ s^1(N-1) & s^1(N-2) & \cdots & s^1(3) & s^p(N-1) & s^p(N-2) & \cdots & s^p(3) \\ s^1(N) & s^1(N-1) & & \vdots & s^p(N) & s^p(N-1) & & \vdots \\ & s^1(N) & & s^1(N-2) & & s^p(N) & & s^p(N-2) \\ & & & s^1(N-1) & & & & s^p(N-1) \\ & & & s^1(N) & & & & s^p(N) \end{bmatrix} \tag{2}$$

Every column in the dictionary represents the reflection for a specific time delay and azimuth. The sampled echo can be expressed as

$$\mathbf{E} = \mathbf{\Psi R} \tag{3}$$

where $\mathbf{R}$ is a sparse representation of the echo in the range-azimuth dictionary $\mathbf{\Psi}$. That is to say, if there are targets at multiple positions corresponding to specific columns in $\mathbf{\Psi}$, then $\mathbf{R}$ will have the same number of non-zero values at corresponding positions while all other values of $\mathbf{R}$ are zeros.

2.2 Parameter Estimation of the Targets

From the formula (3), it can be seen that the estimation of target azimuth and distance can be converted into finding the sparse representation of echoes using a range-azimuth waveform dictionary. Such problems can be solved by the L1 minimization method from convex optimization theory [6], also used in [5],

$$\min\left[\frac{1}{2}\|\mathbf{E}-\boldsymbol{\Psi}\mathbf{R}\|_2^2+\gamma\|\mathbf{R}\|_1\right],\mathbf{R}\geq 0 \tag{4}$$

where γ is a weighting coefficient quantifying the compromise between adherence to the data and sparsity. Its value depends on the input SNR and should increase with the level of noise.

So, the compressed sensing procedure proposed here have three steps:

(1) Construct of the Range-azimuth dictionary;
(2) Solve the sparse representation of echoes;
(3) Interpret the Range and azimuth information from the sparse representation.

3 Measurements and Processing

3.1 Dictionary Construction

A bionic sonar system was used to construct the RAD and measure the echo from a real wall. The experimental bionic sonar system is mainly composed of a sound transducer, a pair of 3D printed bat outer ears and receiving microphones, as shown in Fig. 1.

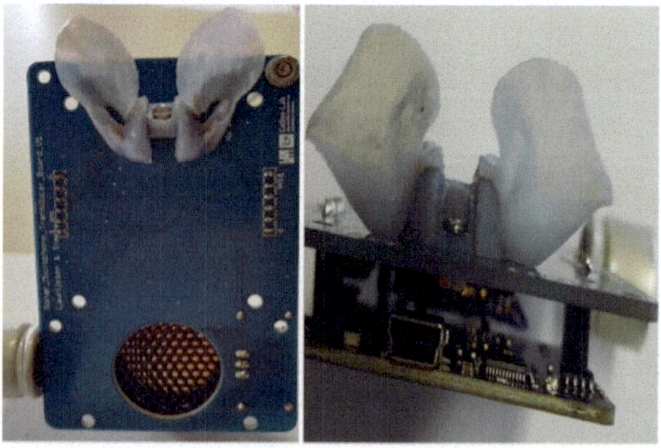

Fig. 1. The bionic sonar system

A cylindrical tube was used as a target to measure the reflected waveforms for azimuths ranging from −30° to 30°. A broadband harmonic signal with a fundamental component whose frequency sweeps down from 30 to 20 kHz, 2 ms in duration similar to an FM-bat was transmitted from the sonar, the target echoes for different azimuth values were collected by the right and left ear receivers. Every azimuth measurement was repeated 100 times and the average waveform was stored, as shown in Fig. 2.

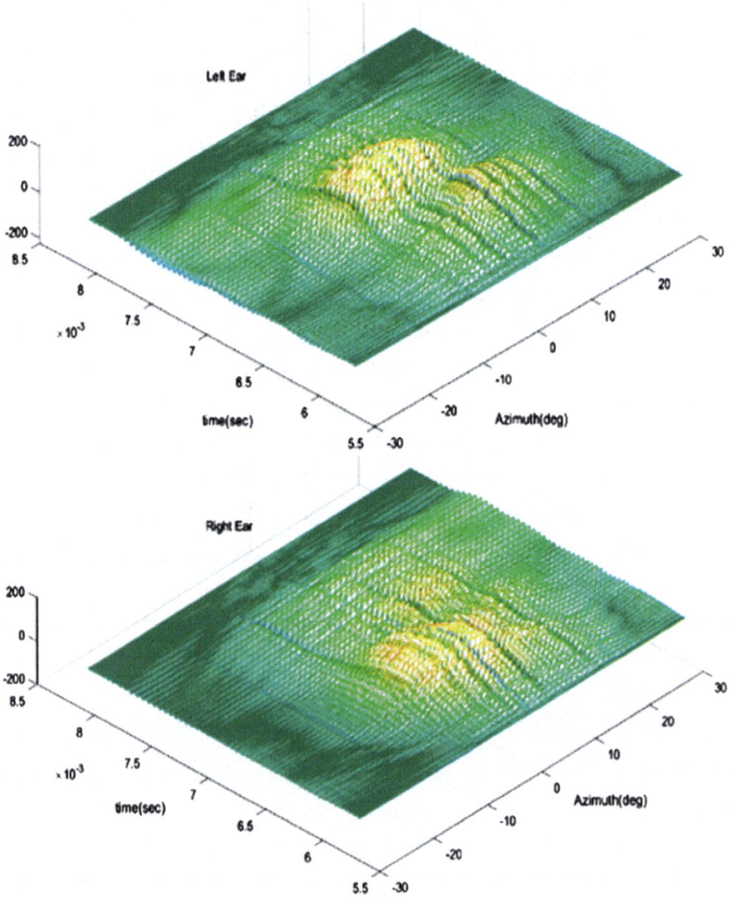

Fig. 2. Echoes received by left ear (top) and right ears (bottom)

According to the Eq. (2), these average waveforms were used to construct the two monaural dictionaries for the left and right ear respectively. Put the left and the right ear signals from the same azimuth in a vector noted as a binaural signal, then a binaural dictionary also was constructed.

3.2 Non-uniform Sampling of the RAD

L1 minimization method used here requires no redundancy in the dictionary, otherwise it will generate ambiguity in the estimation. In order to investigate the redundancy of the waveform dictionary constructed here, we use the coherence between dictionary column vectors

$$\boldsymbol{\rho} = \langle \boldsymbol{\Psi}(:,\mathbf{i}),\ \boldsymbol{\Psi}(:,\mathbf{j}) \rangle \quad (4)$$

where $\langle \bullet, \bullet \rangle$ denote the inner product.

In the most ideal case, there is no redundancy in the dictionary, that is, the coherence matrix is the unit diagonal matrix. In order to collect information from a larger field, the beamwidth of bat sonar is more than 20° [1]. Therefore, the dictionary constructed from uniformly sampled azimuth waveforms as received by the bionic sonar system would show obvious redundancy. Indeed, as shown in Fig. 3, coherence in the mainlobe direction for the left and right ear was relatively high. To deal with this, we non-uniformly selected azimuths and constructed a non-uniformly sampled dictionary. For the monaural right-eardictionary we selected azimuths: −30°, −28°, −26°, −23°, −20°, −17°, −13°, −12°, −10°, −5°, −1°, 0°, 6°, 8°, 10°, 12°, 16°, 22°, 28°. For the monaural left-eardictionary we selected azimuths: −30°, −24°, −16°, −11°, −6°, −2°, 0°, 4°, 6°, 8°, 10°, 14°, 16°, 18°, 21°, 24°, 26°, 28°, 30°. For the binaural dictionary we selected azimuths: −30°, −24°, −15°, −12°, −10°, −7°, −4°, −2°, 0°, 3°, 6°, 7°, 8°, 10°, 12°, 16°, 18°, 22°, 28°.

3.3 Wall Surface Echo Processing and Analysis

As shown in Fig. 4, the bionic sonar system was placed parallel to the side of a brick wall, the received signal of the left and right ears is shown in Fig. 5. The transmitted signal was 2 ms in duration, and the echo was measured after transmitting. It can be seen that as the right ear was closer to the wall, the echoes received in it were stronger than in the left ear. Although the doorframe was the reflector farthest away, its echo signal was the strongest due to its retroreflector properties, and its echoes received in left and right ear were quite close in signal intensity.

The monaural and binaural dictionaries are used to process the echo. Results are shown in Fig. 6.

The origin point of Fig. 6 is the location of the sonar system, 15 cm in the horizontal plane from the wall. Due to the small incident angle, the wall bricks mainly generated forward reflections, i.e. echoes reflected away from the sensor, whereas the backward scattering mainly came from the joints between the bricks and the farthest doorframe. The true locations of these points are marked with red crosses in Fig. 6. To judge the accuracy of the interpretation about the walls we compare the results from monaural and binaural dictionaries respectively with the red crosses.

Results (Fig. 6) show that all of the three dictionaries were able to locate the doorframe. Due to the strong echo there is no serious aliasing with the echoes from other locations. But for reconstructing the brick seams, the performance of the binaural dictionary is the best. More than three brick seams are successfully located (the positions of the pink colour stickers marked in Fig. 4, corresponding to the 4th–7th crosses from top to bottom), while the number of false reconstructed objects is minimal. On the other hand, both monaural dictionaries have problems interpreting these echo signals correctly. For the left ear, only the seam corresponding to the 4th cross was

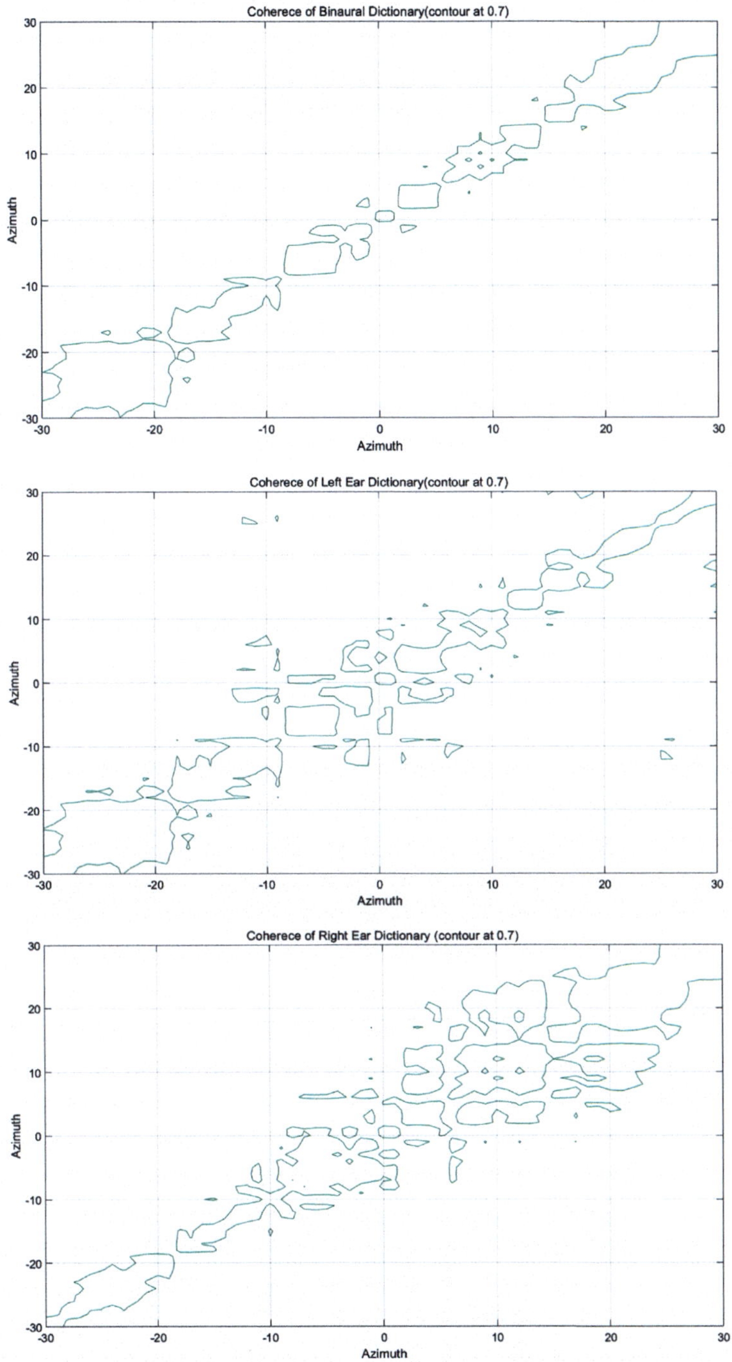

Fig. 3. Coherence of uniform Azimuth Dictionary

Fig. 4. Measurement of the wall

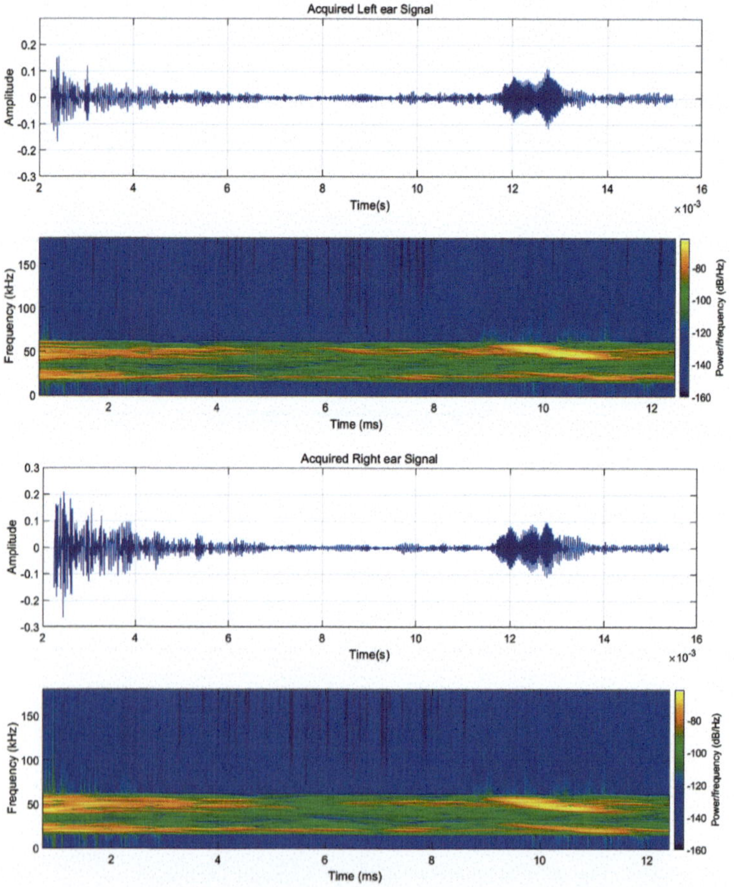

Fig. 5. Echoes from the wall received by left ear (top) and right ear (bottom)

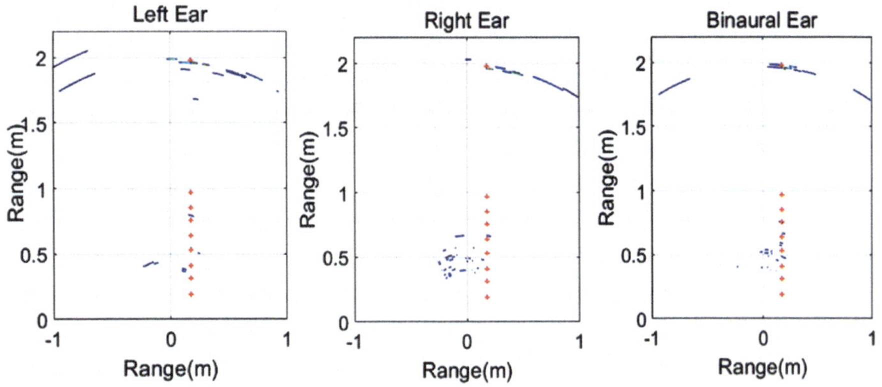

Fig. 6. Acoustic imaging of the wall using the monaural and binaural dictionaries

interpreted. For the right ear, only the seam corresponding to the 5th cross was interpreted. In addition to missing many of the real brick seams they also generate many false reconstructed objects.

The main reason for better performance of the binarual dictionary is the coherence of it is much better than the monaural dictionaries. By combining information from the both ears, the binaural dictionary is much robust in a challenging environment.

4 Conclusion

In this paper, a target location method based on the sparse representation of range-azimuth dictionary was proposed. A bionic sonar system is used to construct the waveform dictionary, and non-uniform sampling in azimuth was investigated. The real-world experimental results showed that a non-uniformly sampled binaural dictionary can distinguish multiple overlapping echoes.

References

1. Au, W.W.L., Simmons, J.A.: Echolocation in dolphins and bats. Phys. Today **60**(9), 40–45 (2007)
2. Saillant, P.A., Simmons, J.A., Dear, S.P., et al.: A computational model of echo processing and acoustic imaging in frequency-modulated echolocating bats: the spectrogram correlation and transformation receiver. J. Acoust. Soc. Am. **94**(5), 2691 (1993)
3. Boonman, A., Ostwald, J.: A modeling approach to explain pulse design in bats. Biol. Cybern. **97**(2), 159–172 (2007)
4. Peremans, H., Hallam, J.: The spectrogram correlation and transformation receiver, revisited. J. Acoust. Soc. Am. **104**(2 Pt 1), 1101–1110 (1998)
5. Fontaine, B., Peremans, H.: Determining biosonar images using sparse representations. J. Acoust. Soc. Am. **125**(5), 3052 (2009)
6. Kim, S.J., Koh, K., Lustig, M., et al.: An efficient method for compressed sensing. In: IEEE International Conference on Image Processing. IEEE (2007)

Relationships Between Dilemma Strength and Fixation Properties in Coevolutionary Games

Hendrik Richter(✉)

Faculty of Electrical Engineering and Information Technology,
HTWK Leipzig University of Applied Sciences,
Postfach 301166, 04251 Leipzig, Germany
hendrik.richter@htwk-leipzig.de

Abstract. Whether or not cooperation is favored over defection in evolutionary games can be assigned by structure coefficients for any arrangement of cooperators and defectors on any network modeled as a regular graph. We study how these structure coefficients relate to a scaling of dilemma strength in social dilemma games. It is shown that some graphs permit certain arrangements of cooperators and defectors to possess particularly large structure coefficients. Moreover, these large coefficients imply particularly large sections of a bounded parameter plane spanned by a scaling of gamble–intending and risk–averting dilemma strength.

Keywords: Coevolutionary games · Fixation properties · Structure coefficients · Dilemma strength

1 Introduction

A fundamental issue in understanding evolutionary dynamics of biological systems is the interplay between competition and cooperation. Evolutionary dynamics requires Darwinian selection in which biological entities compete in terms of survival and reproduction. But while we certainly observe competition in living entities, juxtaposed and intertwined with it, we also notice very frequently unselfish, altruistic and cooperative behavior. The paradox of competition and cooperation next to each other can be resolved by presuming that there must be situations where in evolutionary terms cooperation is more advantageous than competition. Mathematical models for discussing these questions are provided by evolutionary game theory, which gives a theoretical framework and a bio–inspired computational paradigm [3,7].

Studying the emergence of cooperation crystallizes into a well–defined form by considering so–called social dilemma games. In an evolutionary type of these games a population of players interacts in a predefined manner among themselves by each selecting one of two strategies, cooperate or defect, and receiving a payoff according to these selections. By converting the payoff into accumulable fitness, repeating the interaction and allowing players to change strategies

Y. Liu et al. (Eds.): ICNC-FSKD 2019, AISC 1074, pp. 252–259, 2020.
https://doi.org/10.1007/978-3-030-32456-8_27

depending on the accumulated fitness, the long–term effect of strategy selection becomes visible [2,4,6,17]. A frequently studied question of considerable biological relevance is whether or not one strategy is favored over another depending on the values of the payoff matrix *and* on the structure of the interaction network specifying who–plays–whom [5,6,11,12]. Recently, and independent from each other, two proposals have been made to formalize strategy selection and payoff allocation, on the one hand, and strategy distribution over interaction networks, on the other hand. Wang et al. [18] introduced an approach of universal scaling for payoff matrices that facilitates to study a continuum of social dilemmas including (but not restricted to) well–known examples such as the Prisoner's dilemma (PD), the stag hunt (SH) and the snow drift (SD) game. Chen et al. [5] presented a scheme to define structure coefficients for any arrangement of cooperators and defectors on interaction networks modelled as regular graphs. In this paper, these recent additions to evolutionary game theory are combined. There are simple algebraic relations between structure coefficients, the elements of any payoff matrix and whether or not cooperation is favored over defection [5,16]. Thus, we can study fixation properties across the universal scaling for payoff matrices and dilemma strength. The paper is structured as follows. In Sec. 2 we briefly review coevolutionary games, particularly highlighting the scaling of dilemma strength [18] as well as configurations and structure coefficients [5,11,12]. Numerical results are presented for different interaction networks modeled as regular graphs. It is shown how different networks may entail different combinations of dilemma strength where cooperation is favored over defection.

2 Game Description

2.1 Coevolutionary Games, Payoff Matrices and Scaling for the Dilemma Strength

We consider coevolutionary games of N players $\mathcal{I} = (\mathcal{I}_1, \mathcal{I}_2, \ldots, \mathcal{I}_N)$. The game has 2 strategies, cooperate (C_i) and defect (D_i) and each pairwise interaction between 2 players $\mathcal{I}_i$ and $\mathcal{I}_j$, $i \neq j$, (which thus are mutual coplayers) yields a payoff. A (possibly varying) interaction network describes which player interacts with whom, while the number of coplayers is the same for all players. We need three entities for describing such a game: (i) the network of interaction, (ii) the configuration describing the strategy of each player, and (iii) the payoff matrix [8–10].

The interaction network is given by an undirected graph $\mathcal{G} = (V, E)$ and specifies who–plays–whom. According to evolutionary graph theory [1,13,14], the set of vertices V equals the set of players $\mathcal{I}$ and the set of edges E shows which players are mutual coplayers, see Fig. 1 for examples. As each players has k coplayers, the interaction network is a k–regular graph. The configuration $\pi = (\pi_1 \pi_2 \ldots \pi_N)$ specifies the strategy $\pi_i \in \{C_i, D_i\}$ of each player $\mathcal{I}_i$, ($i = 1, 2, \ldots, N$). With 2 strategies (C_i and D_i) there are 2^N configurations. These configurations enumerate all possible arrangements of cooperators and defectors

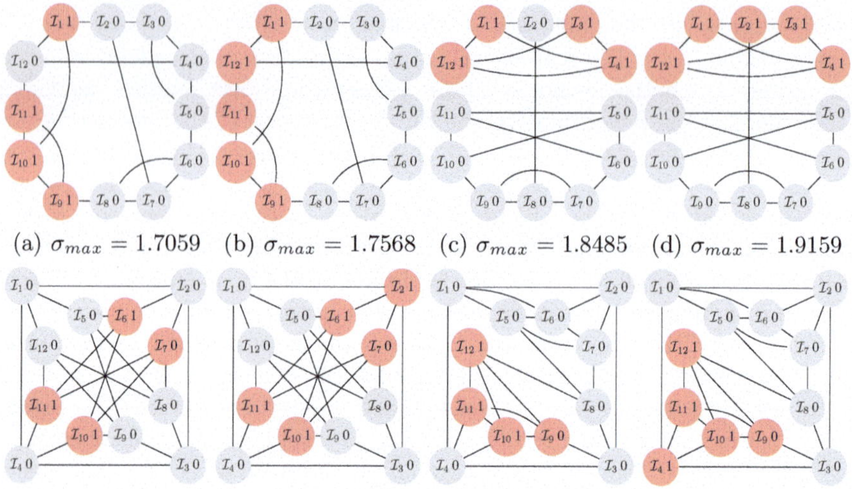

(a) $\sigma_{max} = 1.7059$ (b) $\sigma_{max} = 1.7568$ (c) $\sigma_{max} = 1.8485$ (d) $\sigma_{max} = 1.9159$

(e) $\sigma_{max} = 1.2857$ (f) $\sigma_{max} = 1.2957$ (g) $\sigma_{max} = 1.4433$ (h) $\sigma_{max} = 1.4727$

Fig. 1. Examples of interaction networks: (a)–(d) cubic (3–regular) and (e)–(h) quartic (4–regular) graphs. Two graphs are named: (a), (b) the Frucht graph, (e), (f) the Chvatal graph. The other two graphs have particularly large maximal structure coefficients $\sigma(\pi) = \sigma_{max}$ for some configurations π. A red vertex is a cooperator ($\pi_i = 1$), while a grey vertex is a defector ($\pi_i = 0$). The values of σ_{max} are shown for: (a), (c), (e), (f) $c(\pi) = 4$ cooperators; (b), (d), (f), (h) $c(\pi) = 5$ cooperators.

among the players. Additionally, the configurations describe any outcome of a player changing its strategy in a strategy updating process, for instance death–birth (DB) or birth–death (BD) updating [1,13,19]. It is convenient to binary code the strategies $\{C_i, D_i\} \rightarrow \{1, 0\}$, thus having a binary string to specify the strategies of all players [5,10]. As an example, see Fig. 1(a) with the configuration $\pi = (1000\,0000\,1110)$ showing players $\mathcal{I}_1$, $\mathcal{I}_9$, $\mathcal{I}_{10}$ and $\mathcal{I}_{11}$ cooperating, while the remaining 8 players defect.

The 2×2 payoff matrix is

$$\begin{array}{c} \\ C_i \\ D_i \end{array} \begin{array}{c} \begin{array}{cc} C_j & D_j \end{array} \\ \begin{pmatrix} R & S \\ T & P \end{pmatrix} \end{array} \quad (1)$$

where T is temptation to defect, R is reward for mutual cooperation, P is punishment for mutual defection, and S is sucker payoff for cooperating with a defector. According to the values and order of these 4 elements of the payoff matrix (1), we obtain different social dilemma games. Several suggestions have been made to rescale the payoff matrix (1) by freezing or linearly coupling its elements, which may reduce the 4–dimensional parameter space to a 2–dimensional plane [15,18,19], while preserving frequently–studied social dilemmas such as prisoner's dilemma (PD), snowdrift (SD), stag–hunt (SH), or harmony (H). Following Wang et al. [18], we consider two scaling parameters for the dilemma strength u and v to obtain a rescaled payoff matrix

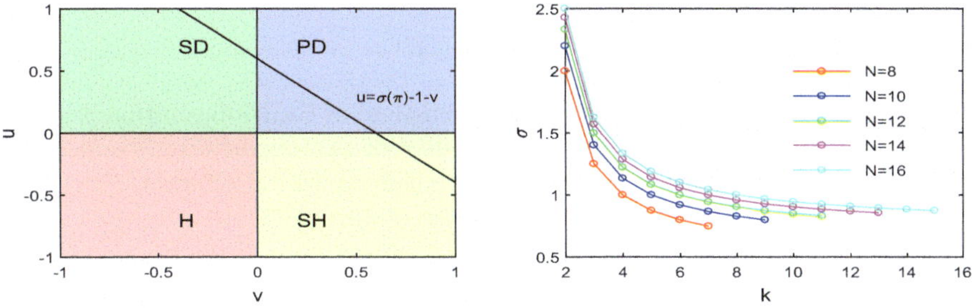

(a) (b)

Fig. 2. (a) Schematic description of social dilemmas in a uv–parameter plane, as defined by the rescaled payoff matrix (2). The plane can be divided into four sections (identified by different colors) which correspond to prisoner's dilemma (PD), snowdrift (SD), stag hunt (SH), and harmony (H) games. Condition (5) implies straight lines $u = \sigma(\pi)-1-v$ from north–west to south–east in the uv–plane for which cooperation is favored over defection below these lines. (b) The structure coefficient of a single cooperator (4) for different number of players N over the number of coplayers k. The structure coefficient σ falls rapidly for k increasing, which means an increasing number of coplayers reduces the sections in uv–plane where cooperation is favored over defection, see Fig. 2(a).

$$\begin{array}{c} \\ C_i \\ D_i \end{array} \begin{array}{cc} C_j & D_j \\ \left(\begin{array}{c} R \\ R+(R-P)u \end{array} \right. & \left. \begin{array}{c} P-(R-P)v \\ P \end{array} \right) \end{array} \tag{2}$$

with $u = \frac{T-R}{R-P}$ and $v = \frac{P-S}{R-P}$. We may interpret u as gamble–intending dilemma strength and v as risk–averting dilemma strength. The rescaling (2) requires $R > P$, while $T-R$ and $P-S$ may change sign to have different orders of (T, R, P, S), and thus different social dilemmas. Apparently, matrix (2) reduces to matrix (1) by inserting u and v. However, by varying u and v for $-1 \leq u \leq 1$ and $-1 \leq v \leq 1$, we may traverse a bounded two–dimensional uv–parameter plane encompassing all the social dilemmas given above, but also some intermediate forms, see Fig. 2(a). We obtain SD games for $0 \leq u \leq 1$ and $-1 \leq v \leq 0$, PD games for $0 \leq u \leq 1$ and $0 \leq v \leq 1$, and so on. Thus, a rescaling by matrix (2) significantly eases analyzing the games across social dilemmas. A square in the uv–plane generalizes the payoff matrix (1) and produces a multitude of dilemmas that are significant and interesting in evolutionary game theory. Moreover, Wang et al. [18] have shown that by the rescaling (2) fixation properties of the games over the uv–plane are fairly robust with respect to the choice of R and P.

2.2 Fixation Properties, Configurations and Structure Coefficients

Recently, Chen et al. [5] have shown that for 2×2 games with N players, payoff matrix (1), any configuration π of cooperators and defectors and for any interaction network modeled by a simple, connected, k–regular graph, in the case of weak selection strategy C_i is favored over D_i if

$$\sigma(\pi) > \frac{T-S}{R-P}. \tag{3}$$

The quantity $\sigma(\pi)$ is called the structure coefficient of the configuration π implying that it may have different values for different arrangements of cooperators and defectors described by π. It generalizes the structure coefficient

$$\sigma = \frac{(k+1)N - 4k}{(k-1)N} \tag{4}$$

applying to a single cooperator [5,16], see Fig. 2(b) with curves of σ for some N and k. For the rescaled payoff matrix (2) the condition (3) simplifies to

$$\sigma(\pi) > 1 + u + v. \tag{5}$$

Condition (5) depends linearly on the scaling parameters for dilemma strength $-1 \leq u \leq 1$ and $-1 \leq v \leq 1$, but not on the parameter of the payoff matrix (1) and particularly not on R and P. Obtaining such a simple algebraic expression as to whether or not cooperation is favored over defection over the whole uv–plane expressing all major social dilemmas is another advantage of the rescaling with payoff matrix (2), as compared to other types of rescaling [15,19].

The structure coefficient $\sigma(\pi)$ can be calculated with time complexity $\mathcal{O}(k^2N)$ for weak selection, the interaction graph $\mathcal{G} = (V, E)$ also describing the replacement structure, and DB updating:

$$\sigma(\pi) = \frac{N\,(1+1/k)\,\overline{\omega^1}\cdot\overline{\omega^0} - 2\overline{\omega^{10}} - \overline{\omega^1\omega^0}}{N\,(1-1/k)\,\overline{\omega^1}\cdot\overline{\omega^0} + \overline{\omega^1\omega^0}}. \tag{6}$$

The local frequencies $\overline{\omega^1}$, $\overline{\omega^0}$, $\overline{\omega^{10}}$, $\overline{\omega^1\omega^0}$ can be interpreted as follows [5,11,12]. Suppose a random walk is carried out with the starting vertex chosen uniformly–at–random on a given interaction network. The local frequency $\overline{\omega^1}$ (or $\overline{\omega^0} = 1 - \overline{\omega^1}$) is the probability that for a configuration π the player at the first step of the walk is a cooperator (or defector). The local frequency $\overline{\omega^{10}}$ is the probability that for a walk with 2 steps the player at the first step is a cooperator and at the second step it is a defector. The local frequency $\overline{\omega^1\omega^0}$ is the probability that for 2 random walks independent of each other the player at the first step on the first walk is a cooperator, but at the first step on the second walk is a defector.

3 Numerical Results and Discussion

We next present and discuss numerical results for the interaction networks given in Fig. 1. As shown above, condition (5) depends linearly on the scaling parameters for dilemma strength u and v yielding straight lines with $u = \sigma(\pi) - 1 - v$ from north–west to south–east in the uv–parameter plane, see Fig. 2(a). Such a straight line is the more towards the north–east corner, the larger the structure coefficient $\sigma(\pi)$ is. In other words, the largest $\sigma(\pi) = \sigma_{max}$ gives the largest section in the bounded uv–plane generalizing elements of the payoff matrix for which cooperation is favored over defection, while the smallest $\sigma(\pi) = \sigma_{min}$ gives the smallest section. We first briefly look at how the structure coefficients are distributed for each number of cooperators $c(\pi)$, see Fig. 3 for the Frucht ($k = 3$) and the Chvatal ($k = 4$) graph. We notice that for 1 cooperator ($c(\pi) = 1$) and 1 defector ($c(\pi) = N - 1 = 11$) we get the single values $\sigma = \{3/2, 11/9\}$ obtained for a single cooperator according to Eq. (4). For $2 \leq c(\pi) \leq 10$ we obtain a symmetric distribution with some values of $\sigma(\pi)$ larger and some smaller than σ, depending on the arrangement of cooperators and defectors on the evolutionary graph. This means there are for the same number of cooperators $c(\pi)$ some configurations π that are more prone to cooperation than others.

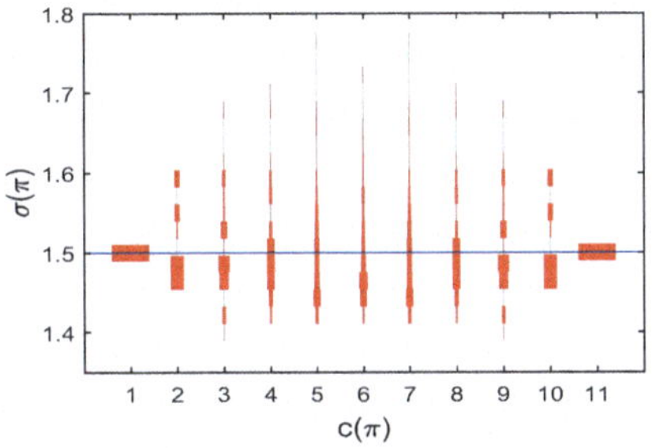

(a)

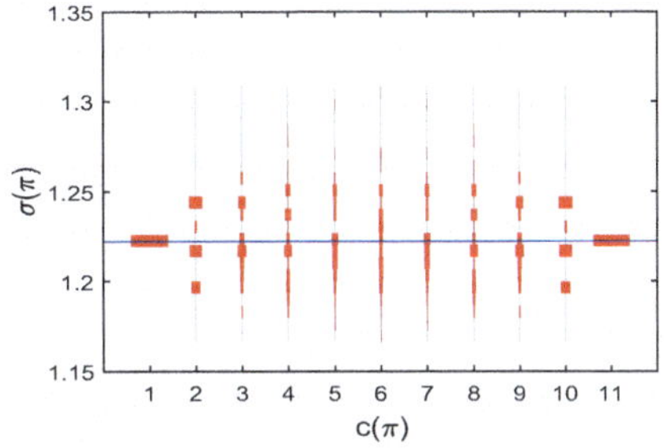

(b)

Fig. 3. Distributions of the structure coefficients $\sigma(\pi)$ for graphs: (a) Frucht, Fig. 1(a); (b) Chvatal, Fig. 1(e). The blue lines intersecting the distributions show the values $\sigma = \{3/2, 11/9\}$ obtained for a single cooperator with $N = 12$, $k = \{3, 4\}$ and Eq. (4).

The main result is given by Fig. 4, which shows for the interaction networks given in Fig. 1 the sections of condition $u < \sigma(\pi) - 1 - v$ between the maximal and minimal structure coefficients, $\sigma(\pi) = \sigma_{max}$ and $\sigma(\pi) = \sigma_{min}$ for the number of cooperators $2 \leq c(\pi) \leq 6$ (the results for $7 \leq c(\pi) \leq 10$ are omitted as they are symmetric, see Fig. 3). Different graphs yield different σ_{min} and σ_{max} for each $c(\pi)$, which in turn produce different ranges in the uv–plane indicating that cooperation is favored or not. For instance, the interaction graph given in Fig. 1(d) has for $c(\pi) = 5$ cooperators the largest value $\sigma_{max} = 1.9159$. For this configuration π cooperation prevails for almost all SD and SH games and a considerable fraction of PD games, see the green–yellow line in Fig. 4(b). Comparing the interaction graphs reveals that the lower line defined by σ_{min} is the same (or almost the same) for the examples considered, while the upper line is not and gives the largest range for the graph and configuration in Fig. 1(d). However, comparing $k = 3$ and $k = 4$ shows that the upper and lower lines are more in favor for cooperation for $k = 3$ than for $k = 4$, which is a consequence of the structure coefficient for a single cooperatorσ with $\sigma_{min} \leq \sigma \leq \sigma_{max}$ being smaller for $k = 4$ as for $k = 3$, see Fig. 2(b). Compare also to Fig. 3 showing that for all configurations the structure coefficients $\sigma(\pi)$ for the Frucht graph ($k = 3$)

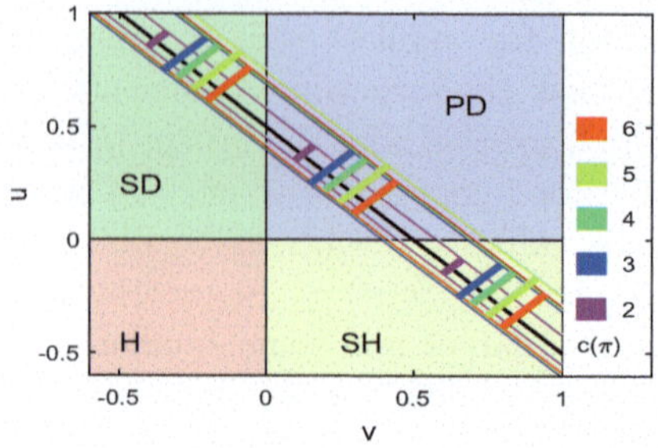

(a) Network in Fig. 1(a): Frucht graph

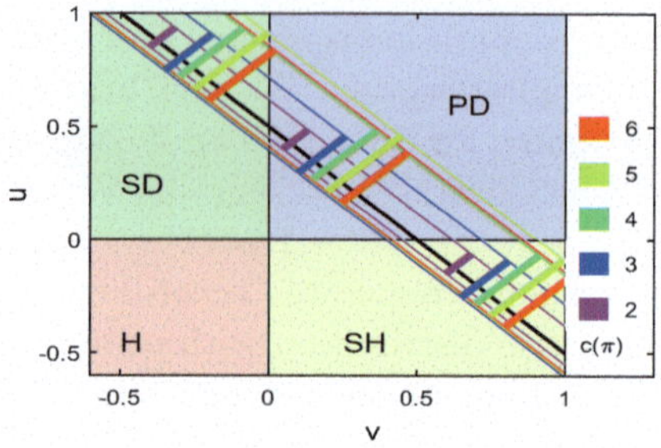

(b) Network in Fig. 1(c)

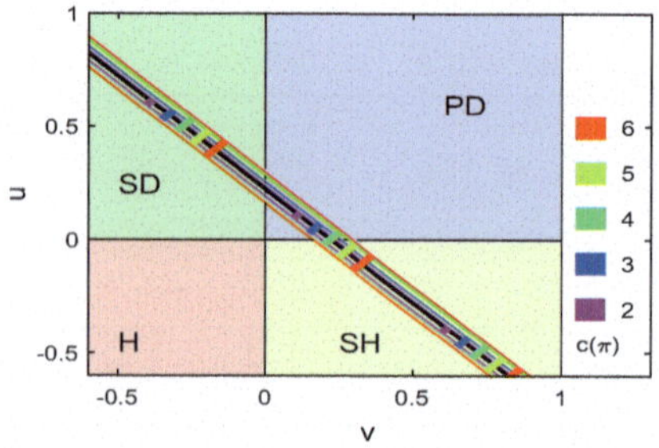

(c) Network in Fig. 1(e): Chvatal graph

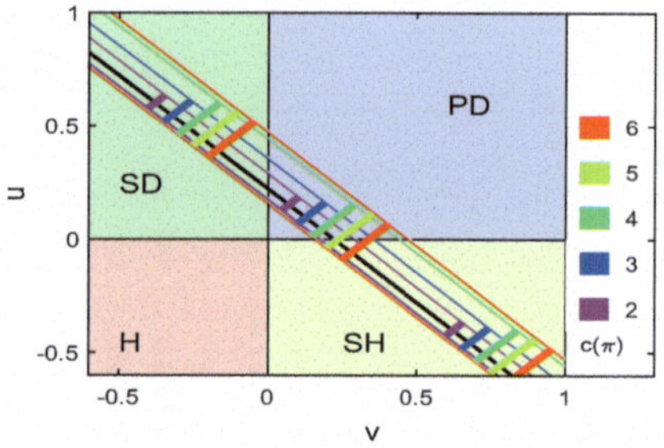

(d) Network in Fig. 1(g)

Fig. 4. Sections of condition $u < \sigma(\pi) - 1 - v$ between the maximal and minimal structure coefficients, $\sigma(\pi) = \sigma_{max}$ and $\sigma(\pi) = \sigma_{min}$ for the interaction networks given in Fig. 1 for the number of cooperators $2 \leq c(\pi) \leq 6$. The bars give the range of the condition $u < \sigma(\pi) - 1 - v$ between σ_{max} and σ_{min} according to the color code for each number of cooperators $c(\pi)$. The black line indicates the section for a single cooperator with σ according to Eq. (4).

are larger than for the Chvatal graph ($k = 4$). A recent work suggests that it is a general result for regular interaction graphs that structure coefficients fall for the degree k (= number of coplayers) of the graph getting larger [11, 12].

4 Conclusions

Different types of social dilemma games such as Prisoners dilemma, stag–hunt or snow–drift can be universally expressed by a scaling of dilemma strength. We have considered structure coefficients defined for each configuration describing any arrangement of cooperators and defectors on a regular evolutionary graphs. As these structure coefficients are linked to whether or not cooperation is favored over defection, we could study how the universal scaling of dilemma strength relates to specific favorable configurations of cooperators and defectors. The main findings are that some graphs permit certain arrangements of cooperators and defectors to possess particularly large structure coefficients. Moreover, these large coefficients imply particularly large sections of a bounded parameter plane spanned by a scaling of gamble–intending and risk–averting dilemma strength.

In addition, the sections can be described by linear inequalities depending only on the two scaling parameters.

References

1. Allen, B., Nowak, M.A.: Games on graphs. EMS Surv. Math. Sci. **1**, 113–151 (2014)
2. Allen, B., Lippner, G., Chen, Y.T., Fotouhi, B., Momeni, N., Yau, S.T., Nowak, M.A.: Evolutionary dynamics on any population structure. Nature **544**, 227–230 (2017)
3. Broom, M., Rychtar, J.: Game-Theoretical Models in Biology. Chapman and Hall, Boca Raton (2013)
4. Chen, Y.T.: Sharp benefit-to-cost rules for the evolution of cooperation on regular graphs. Ann. Appl. Probab. **3**, 637–664 (2013)
5. Chen, Y.T., McAvoy, A., Nowak, M.A.: Fixation probabilities for any configuration of two strategies on regular graphs. Sci. Rep. **6**, 39181 (2016)
6. Hindersin, L., Traulsen, A.: Most undirected random graphs are amplifiers of selection for birth-death dynamics, but suppressors of selection for death-birth dynamics. PLoS Comput. Biol. **11**, e1004437 (2015)
7. Nowak, M.A.: Evolutionary Dynamics: Exploring the Equations of Life. Harvard University Press, Cambridge (2006)
8. Perc, M., Szolnoki, A.: Coevolutionary games - a mini review. BioSystems **99**, 109–125 (2010)
9. Richter, H.: Analyzing coevolutionary games with dynamic fitness landscapes. In: Ong, Y.S. (ed.), Proceedings of IEEE Congress on Evolutionary Computation, IEEE CEC 2016, pp. 609–616. IEEE Press, Piscataway (2016)
10. Richter, H.: Dynamic landscape models of coevolutionary games. BioSystems **153–154**, 26–44 (2017)
11. Richter, H.: Properties of network structures, structure coefficients, and benefit-to-cost ratios. BioSystems **180**, 88–100 (2019)
12. Richter, H.: Fixation properties of multiple cooperator configurations on regular graphs. Theory Biosci. **138**(2), 261–275 (2019)
13. Shakarian, P., Roos, P., Johnson, A.: A review of evolutionary graph theory with applications to game theory. BioSystems **107**, 66–80 (2012)
14. Szabo, G., Fath, G.: Evolutionary games on graphs. Phys. Rep. **446**, 97–216 (2007)
15. Tanimoto, J.: A simple scaling of the effectiveness of supporting mutual cooperation in donor-recipient games by various reciprocity mechanisms. BioSystems **96**, 29–34 (2009)
16. Tarnita, C.E., Ohtsuki, H., Antal, T., Fu, F., Nowak, M.A.: Strategy selection in structured populations. J. Theor. Biol. **259**, 570–581 (2009)
17. Taylor, P.D., Day, T., Wild, G.: Evolution of cooperation in a finite homogeneous graph. Nature **447**, 469–472 (2007)
18. Wang, Z., Kokubo, S., Jusup, M., Tanimoto, J.: Universal scaling for the dilemma strength in evolutionary games. Phys. Life Rev. **14**, 1–30 (2015)
19. Zukewich, J., Kurella, V., Doebeli, M., Hauert, C.: Consolidating birth-death and death-birth processes in structured populations. PLoS One **8**, e54639 (2013)

Future Linear Matrix Equation of Generalized Sylvester Type Solved by Zeroing Neural Dynamics and 5-Instant ZeaD Formula

Liangjie Ming[1,2,3], Nini Shi[1,2,3], Yang Shi[1,2,3], and Yunong Zhang[1,2,3](✉)

[1] School of Information Science and Technology, Sun Yat-sen University, Guangzhou 510006, China
zhynong@mail.sysu.edu.cn, ynzhang@ieee.org, jallonzyn@sina.com

[2] Research Institute of Sun Yat-sen University in Shenzhen, Shenzhen 518057, China

[3] Key Laboratory of Machine Intelligence and Advanced Computing, Ministry of Education, Guangzhou 51006, China

Abstract. In this paper, a new discrete-time zeroing dynamics (or say, Zhang dynamics, ZD) model is proposed, analyzed and investigated for solving generalized-Sylvester-type future linear matrix equation (GS-type FLME). First of all, based on ZD design formula, a continuous-time ZD (CTZD) model is proposed for solving generalized-Sylvester-type continuous-time linear matrix equation (i.e., GS-type CTLME). Secondly, a novel one-step-ahead discretization formula (termed Zhang et al. discretization formula, ZeaD formula) is presented for the first-order derivative approximation with higher computational precision. Then, by exploiting the presented ZeaD formula to discretize the CTZD model, a novel discrete-time ZD (DTZD) model, i.e., ZeaD-type DTZD-E model, is proposed, analyzed and investigated for solving GS-type FLME. Theoretical analyses on the convergence and precision of the proposed DTZD models are presented. Comparative numerical experimental results further substantiate the efficacy and superiority of proposed ZeaD-type DTZD model for solving the GS-type FLME.

Keywords: Generalized Sylvester type · Zeroing neural dynamics · ZeaD formula · GS-type flme · Convergence

1 Introduction

Note that the generalized-Sylvester-type linear matrix problems are mainly divided into a fundamental research area, i.e., generalized-Sylvester-type linear matrix equations (GS-type LME). As fundamental issues, the problems of GS-type LME, e.g., Sylvester equations and Stein equation, have been widely studied [1,2]. For example, a sign-bi-power activation function has been proposed to accelerate the convergence of the Zhang neural network for solving Sylvester

Y. Liu et al. (Eds.): ICNC-FSKD 2019, AISC 1074, pp. 260–270, 2020.
https://doi.org/10.1007/978-3-030-32456-8_28

equation [1]. ZD has been formally generalized from Zhang neural network [3,4] since 2008 [4]. Note that it is different from the conventional dynamics methods in terms of the problem to be solved, indefinite error function, exponent-type design formula, dynamic equation and the utilization of time-derivative information [4]. In this paper, the ZD method is utilized to solve the GS-type future LME (GS-type FLME). First of all, by exploiting the ZD design formula, a CTZD model is developed (termed CTZD-E model). Secondly, a one-step-ahead discretization formula (termed ZeaD formula) is presented for the first-order derivative approximation with higher computational precision. Then, a discrete-time ZD model (termed ZeaD-type DTZD-E model) is designed and generalized for solving the GS-type FLME. Moreover, for comparative purposes, the recently developed Taylor&Zhang-type DTZD model and the widely used Euler-type DTZD model are also presented. The comparative numerical experiments, including the illustrative example, substantiate the efficacy and superiority of the proposed ZeaD-type DTZD model. The remainder of this paper is organized into four sections. In Sect. 2, the problem formulation is presented first, and the CTZD model (i.e., CTZD-E model) is proposed for solving the GS-type CTLME. In Sect. 3, we present a ZeaD formula, and further propose corresponding ZeaD-type DTZD model (i.e., ZeaD-type DTZD-E model). Corresponding theoretical analyses of such model are also proposed. Section 4 provides comparative numerical experiments to substantiate the efficacy and superiority of the proposed ZeaD-type DTZD model. Section 5 concludes this paper with final remarks. Before ending this section, it is worth pointing out the main contributions of this paper as follows.

(1) In this paper, based on ZD design formula, the CTZD model (i.e., CTZD-E model) is proposed for solving GS-type CTLME.
(2) For solving the GS-type FLME, by utilizing a ZeaD formula to discretize the CTZD model, the ZeaD-type DTZD model (i.e., ZeaD-type DTZD-E model) is proposed, analyzed and investigated.
(3) Theoretical analyses present the facts that the proposed ZeaD formula has a truncation error of $O(g^3)$, and the proposed ZeaD-type DTZD model is convergent with its maximal steady-state residual error having an $O(g^4)$ pattern.
(4) Comparative numerical experiments further substantiate the efficacy and superiority of the proposed ZeaD-type DTZD model (i.e., ZeaD-type DTZD-E model) for solving the GS-type FLME.

2 Problem Formulation and CTZD Model

In order to lay a basis for further investigation, the problem formulation is presented, and then, CTZD model is proposed in this section.

2.1 Problem Formulation

The GS-type FLME, to be solved at computational time interval $[t_k, t_{k+1}) \subseteq [t_{0,}t_{\mathrm{f}}) \subseteq [0, +\infty)$, can be described as

$$A_{k+1}X_{k+1}B_{k+1} + C_{k+1}X_{k+1}D_{k+1} = E_{k+1}, \tag{1}$$

where $A_{k+1} \in \mathbb{R}^{m\times m}$, $B_{k+1} \in \mathbb{R}^{n\times n}$, $C_{k+1} \in \mathbb{R}^{m\times m}$, $D_{k+1} \in \mathbb{R}^{n\times n}$ and $E_{k+1} \in \mathbb{R}^{m\times n}$ are respectively defined to be generated from the smoothly time-variant signals $A(t) \in \mathbb{R}^{m\times m}$, $B(t) \in \mathbb{R}^{n\times n}$, $C(t) \in \mathbb{R}^{m\times m}$, $D(t) \in \mathbb{R}^{n\times n}$ and $E(t) \in \mathbb{R}^{m\times n}$ by sampling at time instant $t = (k+1)g$ (termed t_{k+1}), and $X_{k+1} \in \mathbb{R}^{m\times n}$ is the unknown matrix. In addition, $g > 0$ denotes the sampling gap, and $k = 0, 1, 2, \cdots$ denotes the updating index. Note that the above generalized-Sylvester-type linear matrix equation includes the well-known Lyapunov matrix equation, Stein matrix equation and Sylvester matrix equation.

Remark 1: In this paper, we mainly focus on solving for future problem (or say, discrete time-variant problem), then the used methods should satisfy the requirement of real-time computation during the solution process. Specifically, based on the present and previous data (e.g., X_k, A_k, B_k and corresponding derivatives), we compute the result (e.g., X_{k+1}) for future use before the time instant t_{k+1}. For solving the future problem, if the computational time for each updating is smaller than the sampling gap, the efficacy of computational results (i.e., real-time computation requirement) can be guaranteed.

2.2 CTZD Model for Solving GS-Type CTLME

To obtain the solution of GS-type FLME (1), let us consider the following continuous time-variant problem as the continuation of (1):

$$A(t)X(t)B(t) + C(t)X(t)D(t) - E(t) = \mathbf{0} \in \mathbb{R}^{m\times n}. \tag{2}$$

Thus, to monitor and control (2), the time-variant zeroing error function of GS-type CTLME can be defined as

$$Z_{\mathrm{LME}}(t) = A(t)X(t)B(t) + C(t)X(t)D(t) - E(t). \tag{3}$$

Inspired by ZD design formula [3], the following design formula is employed:

$$\dot{Z}_{\mathrm{LME}}(t) = -\gamma\phi(Z_{\mathrm{LME}}(t)),$$

where $\dot{Z}_{\mathrm{LME}}(t)$ denotes the time-derivative of $Z_{\mathrm{LME}}(t)$; the design parameter $\gamma > 0$ is used to scale the convergence rate; and $\phi(\cdot) : \mathbb{R}^{m\times n} \to \mathbb{R}^{m\times n}$ denotes an activation-function mapping array. In general, parameter γ corresponds to the reciprocal of a capacitance parameter in potential electronic circuits. Its value should be set as large as the hardware would permit or it is selected appropriately for the simulation process [5]. For simplicity and clarity, in this paper, we select the linear activation function array. Thus, the above design formula reduces to $\dot{Z}_{\mathrm{LME}}(t) = -\gamma Z_{\mathrm{LME}}(t)$, and then we can obtain

$$\begin{aligned}&\mathrm{d}\left(A(t)X(t)B(t) + C(t)X(t)D(t) - E(t)\right)/\mathrm{d}t = \\ &-\gamma\left(A(t)X(t)B(t) + C(t)X(t)D(t) - E(t)\right).\end{aligned}$$

Inspired by Kronecker product and vectorization techniques [6,7], the above equation can be vectorized equivalently as

$$\begin{aligned}&\left(B^{\mathrm{T}}(t)\otimes A(t) + D^{\mathrm{T}}(t)\otimes C(t)\right)\mathrm{vec}(\dot{X}(t)) = \\ &-\gamma\left(\left(B^{\mathrm{T}}(t)\otimes A(t) + D^{\mathrm{T}}(t)\otimes C(t)\right)\mathrm{vec}(X(t))\right)\\ &-\left(B^{\mathrm{T}}(t)\otimes \dot{A}(t) + \dot{B}^{\mathrm{T}}(t)\otimes A(t)\right)\mathrm{vec}(X(t))\\ &-\left(D^{\mathrm{T}}(t)\otimes \dot{C}(t) + \dot{D}^{\mathrm{T}}(t)\otimes C(t)\right)\mathrm{vec}(X(t))\\ &+\mathrm{vec}(\dot{E}(t)) + \gamma\mathrm{vec}(E(t)),\end{aligned}$$

where symbol $\otimes$ denotes the Kronecker product, superscript ^{T} denotes the transpose of a matrix, and operator vec(·) generates a column vector, which can be obtained by stacking all column vectors of matrix together. Thus, we further have

$$\begin{aligned}M(t)\mathrm{vec}(\dot{X}(t)) = &-\gamma\left(M(t)\mathrm{vec}(X(t)) - \mathrm{vec}(E(t))\right)\\ &-\dot{M}(t)\mathrm{vec}(X(t)) + \mathrm{vec}(\dot{E}(t)),\end{aligned} \tag{4}$$

where time-variant matrixes $M(t) \in \mathbb{R}^{mn\times mn}$ and $\dot{M}(t) \in \mathbb{R}^{mn\times mn}$ are defined as

$$M(t) = B^{\mathrm{T}}(t)\otimes A(t) + D^{\mathrm{T}}(t)\otimes C(t),$$

and

$$\dot{M}(t) = B^{\mathrm{T}}(t)\otimes \dot{A}(t) + \dot{B}^{\mathrm{T}}(t)\otimes A(t) + D^{\mathrm{T}}(t)\otimes \dot{C}(t) + \dot{D}^{\mathrm{T}}(t)\otimes C(t).$$

In addition, in this paper, we use notations $\mathbf{x}(t)$, $\dot{\mathbf{x}}(t)$, $\mathbf{e}(t)$ and $\dot{\mathbf{e}}(t)$ to denote $\mathrm{vec}(X(t))$, $\mathrm{vec}(\dot{X}(t))$, $\mathrm{vec}(E(t))$ and $\mathrm{vec}(\dot{E}(t))$, respectively. Note that we would like to find a feasible solution matrix $X(t)$ such that (2) holds true for any time instant. Thus, it needs to guarantee the existence of the solution set first. To guarantee the existence of the solution set, and also for simplicity and clarity, guided by Horn and Johnson [6], the time-variant matrix $M(t)$ is full-rank in this paper, and Eq. (4) can be rewritten as

$$\dot{\mathbf{x}}(t) = -\gamma M^{-1}(t)\left(M(t)\mathbf{x}(t) - \mathbf{e}(t)\right) - M^{-1}(t)(\dot{M}(t)\mathbf{x}(t) - \dot{\mathbf{e}}(t)), \tag{5}$$

where $M^{-1}(t)$ denotes the inverse of $M(t)$. Note that (5) is a CTZD model for solving the GS-type CTLME. For presentation convenience, it is called the CTZD-E model in this paper.

3 DTZD Models

In this section, the DTZD model (i.e., ZeaD-type DTZD-E model) is proposed, developed, and investigated for solving GS-type FLME.

3.1 ZeaD Formula

In this subsection, we present a ZeaD formula, which can be obtained by exploiting Taylor expansion [8]. The ZeaD formula is formulated as follows:

$$\dot{u}_k \doteq \frac{9}{20g}u_{k+1} + \frac{1}{30g}u_k - \frac{3}{10g}u_{k-1} - \frac{3}{10g}u_{k-2} + \frac{7}{60g}u_{k-3}, \quad (6)$$

where $\doteq$ denotes the computational assignment operation, and the discretization formula has a truncation error of $O(g^3)$.

Theorem 1. *The optimal sampling gap of presented ZeaD formula (6) is*

$$g = \left(\frac{12\varepsilon_{\max}}{125\kappa_{\max}}\right)^{\frac{1}{4}}, \quad (7)$$

where $\varepsilon_{\max}$ denotes the maximal absolute value of round-off errors and $\kappa_{\max}$ denotes the maximal absolute value of $u^{(4)}(c_1)$, $u^{(4)}(c_2)$, $u^{(4)}(c_3)$ and $u^{(4)}(c_4)$.

Proof. The round-off error in the numerical computations can be simplified as

$$u_i = v_i + \varepsilon_i,$$

where u_i is approximated by numerical values v_i, and ε_i is the corresponding round-off error. In this paper, we present a 5-instant numerical discretization formula. Thus, i can be set as $k+1$, k, $k-1$, $k-2$ and $k-3$, correspondingly, and we further have

$$\dot{u}_k = \frac{9}{20g}v_{k+1} + \frac{1}{30g}v_k - \frac{3}{10g}v_{k-1} - \frac{3}{10g}v_{k-2} + \frac{7}{60g}v_{k-3} + e_{\mathrm{r}} + e_{\mathrm{t}},$$

in which the term e_{r} denotes the round-off error and the term e_{t} denotes the truncation error. Furthermore, we have

$$e_{\mathrm{r}} = \left|\frac{9}{20g}\varepsilon_{k+1} + \frac{1}{30g}\varepsilon_k - \frac{3}{10g}\varepsilon_{k-1} - \frac{3}{10g}\varepsilon_{k-2} + \frac{7}{60g}\varepsilon_{k-3}\right| \leqslant \frac{6}{5g}\varepsilon_{\max}$$

and

$$e_{\mathrm{t}} = \left|\frac{1}{8}g^3u^{(4)}(c_1) - \frac{1}{12}g^3u^{(4)}(c_2) - \frac{4}{3}g^3u^{(4)}(c_3) + \frac{21}{8}g^3u^{(4)}(c_4)\right| \leqslant \frac{25}{6}g^3\kappa_{\max}.$$

In view of $\varepsilon_{\max}$ denoting the maximal absolute value of round-off errors of u_{k+1}, u_k, u_{k-1}, u_{k-2} and u_{k-3} as well as $\kappa_{\max}$ being a positive constant, it is known that $(6/5g)\varepsilon_{\max}$ is the upper bound of round-off error e_{r}, and $(25/6)g^3\kappa_{\max}$ is the upper bound of truncation error e_{t}. It is worth noting that floating-point numbers have limited precision in computer; e.g., the precision of floating-point number in MATLAB environment is of order 10^{-16}, with its "eps" being 2.2×10^{-16} (i.e., 2^{-52}). In addition, we have

$$|e_{\rm r} + e_{\rm t}| \leqslant \frac{6}{5g}\varepsilon_{\max} + \frac{25}{6}g^3\kappa_{\max},$$

and the optimal sampling gap of presented ZeaD formula (6) can be obtained as

$$g = \left(\frac{12\varepsilon_{\max}}{125\kappa_{\max}}\right)^{\frac{1}{4}}.$$

Based on the above analyses, the proof is thus completed.

To substantiate the efficacy of the optimal sampling gap (7), we exploit the following function as the target function: $f(x) = \sin(x)$, and the first-order derivative of the target function is $f'(x) = \cos(x)$. Then, for the above target function, we have $\kappa_{\max} = 1$ and $\varepsilon_{\max} = 2.2 \times 10^{-16}$. In this paper, we use $x = 1.5$ as the target point in the following numerical experiment. At the target point, the errors between the practice value of first-order derivative and the approximate value of first-order derivative are presented in Table 1. As shown in Table 1, when the sampling gap g decreases, the values of error firstly decrease and then increase. Specifically, when the sampling gap $g = 7.5 \times 10^{-5}$, we obtain the minimal value of error in approximation to first-order derivative of the target function, which coincides with the theoretical results.

Table 1. Errors in approximation to $f(x) = \sin(x)$ using ZeaD formula (6) with different values of sampling gap g

Sampling gap g	Error	Sampling gap g	Error	Sampling gap g	Error
7.5×10^{-1}	6.18×10^{-2}	7.5×10^{-3}	8.41×10^{-8}	$\mathbf{7.5 \times 10^{-5}}$	$\mathbf{2.26 \times 10^{-14}}$
5.0×10^{-1}	2.16×10^{-2}	5.0×10^{-3}	2.49×10^{-8}	5.0×10^{-5}	4.32×10^{-13}
2.5×10^{-1}	2.99×10^{-3}	2.5×10^{-3}	3.12×10^{-9}	2.5×10^{-5}	2.71×10^{-12}
1.0×10^{-1}	1.98×10^{-4}	1.0×10^{-3}	2.00×10^{-10}	1.0×10^{-5}	5.43×10^{-12}
7.5×10^{-2}	8.36×10^{-5}	7.5×10^{-4}	8.42×10^{-11}	7.5×10^{-6}	1.09×10^{-11}
5.0×10^{-2}	2.48×10^{-5}	5.0×10^{-4}	2.50×10^{-11}	5.0×10^{-6}	5.48×10^{-12}
2.5×10^{-2}	3.11×10^{-6}	2.5×10^{-4}	3.09×10^{-12}	2.5×10^{-6}	1.27×10^{-11}
1.0×10^{-2}	1.99×10^{-7}	1.0×10^{-4}	4.77×10^{-13}	1.0×10^{-6}	1.26×10^{-10}

Remark 2: In this paper, we obtain the optimal sampling gap of presented ZeaD formula. As shown in Table 1, when the value of sampling gap g greater than 7.5×10^{-5}, e.g., $g = 1 \times 10^{-3}$, the value of error is increased. Besides, when the value of sampling gap g is smaller than 7.5×10^{-5}, e.g., $g = 1 \times 10^{-5}$, the value of error is not consistent of order 3. Because, at this stage, the influence of round-off error $e_{\rm r}$ can not be ignored, and such error is dominant relative to the truncation error $e_{\rm t}$.

3.2 DTZD Model for Solving GS-type FLME

In this subsection, based on the presented ZeaD formula (6), the ZeaD-type DTZD-E model is developed and investigated. Firstly, the presented ZeaD formula (6) is exploited to discretize CTZD-E model (2), and the ZeaD-type DTZD-E model can be further obtained as follows:

$$\begin{aligned}\mathbf{x}_{k+1} \doteq & -\frac{20}{9}hM_k^{-1}(M_k\mathbf{x}_k-\mathbf{e}_k)-\frac{20}{9}gM_k^{-1}\left(\dot{M}_k\mathbf{x}_k-\dot{\mathbf{e}}_k\right)\\ & -\frac{2}{27}\mathbf{x}_k+\frac{2}{3}\mathbf{x}_{k-1}+\frac{2}{3}\mathbf{x}_{k-2}-\frac{7}{27}\mathbf{x}_{k-3},\end{aligned} \tag{8}$$

where $h = g\gamma$. Based on the results about 0-stability and convergence presented in the Appendix, the convergence of ZeaD-type DTZD-E model (8) is guaranteed by the following theorems.

Theorem 2. *The proposed ZeaD-type DTZD-E model (8) is 0-stable.*

Proof. According to the Result 1 of the Appendix, we have the following characteristic polynomial of proposed ZeaD-type DTZD-E model (8) as

$$\varphi(\varsigma)=\varsigma^4+\frac{2}{27}\varsigma^3-\frac{2}{3}\varsigma^2-\frac{2}{3}\varsigma+\frac{7}{27}.$$

It can be obtained that the roots of characteristic polynomial are $-0.6920 + 0.5979\mathrm{i}$, $-0.6920 + 0.5979\mathrm{i}$, 1 and 0.3100 with i denoting imaginary unit. Evidently, they contain a real root and a couple of conjugate imaginary roots, and thus these roots are on or in the unit circle, then the proposed ZeaD-type DTZD-E model (8) is 0-stable. The proof is thus completed.

Theorem 3. *The proposed ZeaD-type DTZD-E model (8) is consistent and convergent, which converges with the order of its truncation error being* $\mathbf{O}(g^4)$.

Proof. As presented in Theorem 1, we have the following ZeaD formula as

$$\dot{x}_k=\frac{9}{20g}x_{k+1}+\frac{1}{30g}x_k-\frac{3}{10g}x_{k-1}-\frac{3}{10g}x_{k-2}+\frac{7}{60g}x_{k-3}+O(g^3).$$

By using above formula to discretize the proposed CTZD-E model (5), we have

$$\begin{aligned}\mathbf{x}_{k+1} = & -\frac{20}{9}hM_k^{-1}(M_k\mathbf{x}_k-\mathbf{e}_k)-\frac{20}{9}gM_k^{-1}(\dot{M}_k\mathbf{x}_k-\dot{\mathbf{e}}_k)\\ & -\frac{2}{27}\mathbf{x}_k+\frac{2}{3}\mathbf{x}_{k-1}+\frac{2}{3}\mathbf{x}_{k-2}-\frac{7}{27}\mathbf{x}_{k-3}+\mathbf{O}(g^4).\end{aligned}$$

Evidently, the truncation error of ZeaD-type DTZD-E model (8) is $\mathbf{O}(g^4)$. Therefore, the proposed ZeaD-type DTZD-E model (8) is said to consistent of order 4. As presented in Appendix, according to Result 2 and Result 3, the proposed ZeaD-type DTZD-E model (8) is 0-stable and consistent, which converges with the order of its truncation error being $\mathbf{O}(g^4)$. The proof is thus completed.

Theorem 4. *The maximal steady-state residual error* $\|M_{k+1}\mathbf{x}_{k+1} - \mathbf{e}_{k+1}\|_2$ *of the proposed ZeaD-type DTZD-E model (8) is of order* $O(g^4)$, *where* $\|\cdot\|_2$ *denotes the Euclidean norm of a vector.*

Proof. Let $\mathbf{x}^*_{k+1}$ denotes the vectorization of X^*_{k+1}, which is a theoretical solution of GS-type FLME (1). Thus, we know that $M_{k+1}\mathbf{x}^*_{k+1} - \mathbf{e}_{k+1} = \mathbf{0}$. In addition, we have $\mathbf{x}_{k+1} = \mathbf{x}^*_{k+1} + \mathbf{O}(g^4)$ with k large enough. Therefore, we have

$$\begin{aligned}\limsup_{k\to+\infty} \|M_{k+1}\mathbf{x}_{k+1} - \mathbf{e}_{k+1}\|_2 &= \limsup_{k\to+\infty} \left\|M_{k+1}\mathbf{x}^*_{k+1} - \mathbf{e}_{k+1} + M_{k+1}\mathbf{O}(g^4)\right\|_2 \\ &= \limsup_{k\to+\infty} \left\|M_{k+1}\mathbf{O}(g^4)\right\|_2 \\ &\leqslant \limsup_{k\to+\infty} \|M_{k+1}\|_{\mathrm{F}} O(g^4) = O(g^4),\end{aligned}$$

where $\|\cdot\|_{\mathrm{F}}$ denotes the Frobenius norm of a matrix. Thus we have the maximal steady-state residual error of the proposed ZeaD-type DTZD-E model (8) is of order $O(g^4)$. The proof is thus completed.

As mentioned above, the proposed ZeaD-type DTZD-E model (8) is convergent, and its maximal steady-state residual error is of order $O(g^4)$. Evidently, $\mathbf{x}_{k+1}$ and $\mathbf{e}_{k+1}$ denote the vectorization of X_{k+1} and E_{k+1}, respectively, and $M_{k+1} = B^{\mathrm{T}}_{k+1} \otimes A_{k+1} + D^{\mathrm{T}}_{k+1} \otimes C_{k+1}$. We further have

$$\|M_{k+1}\mathbf{x}_{k+1} - \mathbf{e}_{k+1}\|_2 = \|A_{k+1}X_{k+1}B_{k+1} + C_{k+1}X_{k+1}D_{k+1} - E_{k+1}\|_{\mathrm{F}}.$$

Thus, the feasibility and efficacy of the proposed ZeaD-type DTZD-E model (8) are presented for solving GS-type FLME (1).

4 Numerical Experiments and Comparisons

In this section, the comparative numerical experiments are performed.

4.1 GS-type FLME

For illustration and verification, let us consider the following GS-type FLME with X_{k+1} to be computed at each computational time interval $[kg, (k+1)g) \subseteq [0, 20)$ s:

$$A_{k+1}X_{k+1}B_{k+1} + C_{k+1}X_{k+1}D_{k+1} = E_{k+1} \tag{9}$$

with

$$A_k = \begin{bmatrix} 4+\cos((3/2)t_k) & (2/3)\sin((3/2)t_k) & (1/2)\sin(t_k) \\ (1/2)\sin(t_k) & 4+\cos((3/2)t_k) & (2/3)\sin((3/2)t_k) \\ (2/3)\sin((3/2)t_k) & (1/2)\sin(t_k) & 4+\cos((3/2)t_k) \end{bmatrix},$$

$$B_k = \begin{bmatrix} \sin(2t_k) & \cos(2t_k) \\ -\cos(2t_k) & \sin(2t_k) \end{bmatrix},$$

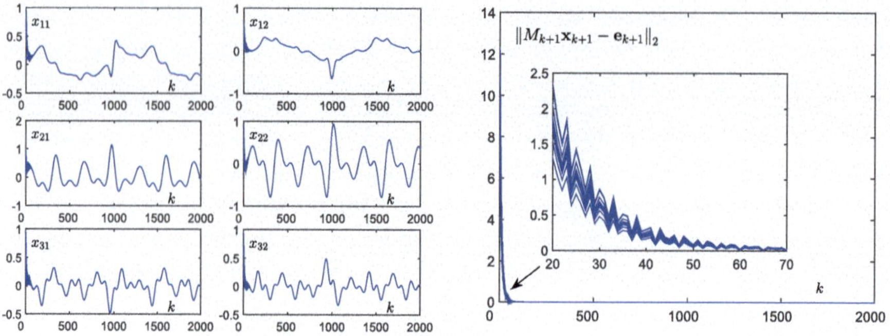

(a) (b)

Fig. 1. State trajectories and residual errors computed by proposed ZeaD-type DTZD-E model (8) with randomly generated initial states. (a) State trajectories of X_k. (b) Residual errors

$$C_k = \begin{bmatrix} 2+\sin(2t_k) & \cos(t_k)/3 & \sqrt{2}\cos(t_k)/2 \\ \sqrt{2}\cos(t_k)/2 & 2+\sin(2t_k) & \cos(t_k)/3 \\ \cos(t_k)/3 & \sqrt{2}\cos(t_k)/2 & 2+\sin(2t_k) \end{bmatrix},$$

$$D_k = \begin{bmatrix} (1/2)\sin(t_k) & (1/2)\cos(t_k) \\ -(1/2)\cos(t_k) & (1/2)\sin(t_k) \end{bmatrix},$$

and

$$E_k = \begin{bmatrix} \sin((3/2)t_k) & \cos((3/2)t_k) \\ \cos(2t_k)+1 & \sin(2t_k)+1 \\ \cos(2t_k)\sin((5/2)t_k) & \cos(2t_k)\sin((5/2)t_k) \end{bmatrix}.$$

To start the proposed ZeaD-type DTZD-E model (8), the initial states should be obtained. In these numerical experiments, based on the randomly generated initial states, the conventional Euler method is used to compute the other three initial states. Then, the proposed ZeaD-type DTZD-E model (8) is exploited to solve the above GS-type FLME depicted in (9), and the numerical experimental results are shown in Fig. 1. As seen from Fig. 1(a), starting from 10 randomly-generated initial states, the state trajectories of X_k are time-variant, which always converge to the theoretical solution accurately and rapidly. Besides, as illustrated in Fig. 1(b), the residual errors of the proposed ZeaD-type DTZD-E model (8) always converge to near zero as well.

For comparison purposes, in recent years, the Taylor&Zhang discretization formula and Euler discretization formula have been presented [9]. Then, we further have the Taylor&Zhang-type DTZD-E model as

$$\begin{aligned} \mathbf{x}_{k+1} \doteq & -hM_k^{-1}(M_k\mathbf{x}_k - \mathbf{e}_k) - gM_k^{-1}\left(\dot{M}_k\mathbf{x}_k - \dot{\mathbf{e}}_k\right) \\ & + \frac{3}{2}\mathbf{x}_k - \mathbf{x}_{k-1} + \frac{1}{2}\mathbf{x}_{k-2}, \end{aligned} \tag{10}$$

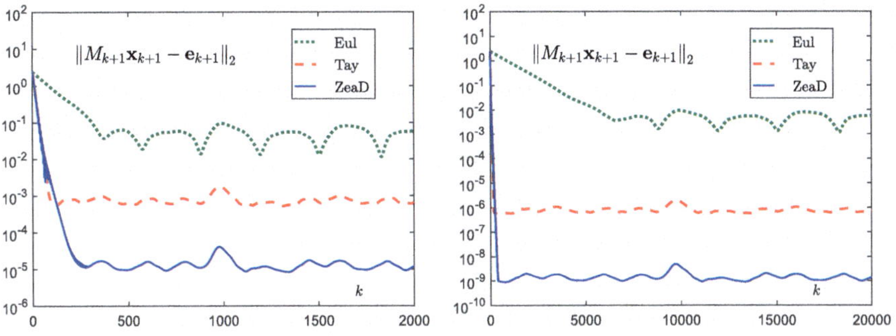

(a) (b)

Fig. 2. Residual errors of three discrete-time models with $h = 0.1$, where Eul, Tay and ZeaD represent Euler-type DTZD-E model (11), Taylor&Zhang-type DTZD-E model (10) and ZeaD-type DTZD-E model (8), respectively. (a) Semilog plot of the residual errors with $g = 0.01$ s. (b) Semilog plot of the residual errors with $g = 0.001$ s

and Euler-type DTZD-E model is presented as

$$\mathbf{x}_{k+1} \doteq -hM_k^{-1}\left(M_k\mathbf{x}_k - \mathbf{e}_k\right) - gM_k^{-1}\left(\dot{M}_k\mathbf{x}_k - \dot{\mathbf{e}}_k\right) + \mathbf{x}_k. \tag{11}$$

Furthermore, Fig. 2 shows the characteristics of residual error during the solving processes of the proposed ZeaD-type DTZD-E model (8). As shown in Fig. 2(a) and (b), starting from $\mathbf{x}_0 = [0, 0, 0, 0, 0, 0]^{\mathrm{T}}$, the maximal steady-state residual error of ZeaD-type DTZD-E model (8) changes in the manner of $O(g^4)$, and the maximal steady-state residual errors of Taylor&Zhang-type DTZD-E (10) and Euler-type DTZD-E model (11) change in the manner of $O(g^3)$ and $O(g^2)$, respectively.

5 Conclusion

In this paper, a new discrete-time zeroing dynamics (or say, Zhang dynamics, ZD) models has been proposed, analyzed and investigated for solving GS-type FLME. A novel ZeaD formula has been presented for the first-order derivative approximation with higher computational precision. Then, by exploiting the presented ZeaD formula to discretize the CTZD model, ZeaD-type DTZD-E model (8) has been proposed. Theoretical analyses have shown that the proposed DTZD model is convergent, and its steady-state residual errors have an $O(g^4)$ pattern. Comparative numerical experimental results have further substantiated the efficacy and superiority of the proposed ZeaD-type DTZD model for solving the GS-type FLME.

Acknowledgment. The work is supported by Shenzhen Science and Technology Plan Project (with number JCYJ20170818154936083). Kindly note that all authors of the paper are jointly of the first authorship.

Appendix

The following four basic results are presented, which are general concepts and theoretical results about linear multistep methods and difference equations [10].

Result 1: An N-step method $\sum_{j=0}^{N} \alpha_j \sigma_{k+j} = g \sum_{j=0}^{N} \beta_j \delta_{k+j}$ can be checked for 0-stability by determining the roots of its characteristic polynomial $\varphi(\varsigma) = \sum_{j=0}^{N} \alpha_j \varsigma^j$. If all roots of $\varphi(\varsigma) = 0$ satisfy $|\varsigma| \leq 1$ with $|\varsigma| = 1$ being simple, then the N-step method is 0-stable (i.e., has 0-stability).

Result 2: An N-step method is said to be consistent (i.e., has consistency) of order p if the truncation error for the exact solution is of order $O(g^{p+1})$ where $p > 0$.

Result 3: An N-step method is convergent, i.e., $\sigma_{[t/g]} \to \sigma^*(t)$, for all $t \in [0, t_{\mathrm{f}}]$, as $g \to 0$, if and only if the method is 0-stable and consistent. That is, 0-stability plus consistency means convergence, which is also known as Dahlquist equivalence theorem.

Result 4: A 0-stable consistent method converges with the order of its truncation error.

References

1. Li, S., Chen, S., Liu, B.: Accelerating a recurrent neural network to finite-time convergence for solving time-varying Sylvester equation by using a sign-bi-power activation function. Neural Process. Lett. **37**(2), 189–205 (2013)
2. Hajarian, M.: Finite algorithms for solving the coupled Sylvester-conjugate matrix equations over reflexive and Hermitian reflexive matrices. Int. J. Syst. Sci. **46**(3), 488–502 (2015)
3. Jin, L., Li, S., Liao, B., Zhang, Z.: Zeroing neural networks: a survey. Neurocomputing **267**, 597–604 (2017)
4. Guo, D., Nie, Z., Yan, L.: Theoretical analysis, numerical verification and geometrical representation of new three-step DTZD algorithm for time-varying nonlinear equations solving. Neurocomputing **214**, 516–526 (2016)
5. Jin, L., Zhang, Y., Qiu, B.: Neural network-based discrete-time Z-type model of high accuracy in noisy environments for solving dynamic system of linear equations. Neural Comput. Appl. **29**(11), 1217–1232 (2018). https://doi.org/10.1007/s00521-016-2640-x
6. Horn, R.A., Johnson, C.R.: Matrix Analysis. Cambridge University Press, Cambridge (2012)
7. Guo, D., Zhang, Y.: Zhang neural network for online solution of time-varying linear matrix inequality aided with an equality conversion. IEEE Trans. Neural Netw. Learn. Syst. **25**(2), 370–382 (2014)
8. Mathews, J.H., Fink, K.D.: Numerical Methods Using MATLAB. Prentice-Hall, Englewood Cliffs (2005)
9. Liao, B., Zhang, Y., Jin, L.: Taylor $O(h^3)$ discretization of ZNN models for dynamic equality-constrained quadratic programming with application to manipulators. IEEE Trans. Neural Netw. Learn. Syst. **27**(2), 225–237 (2016)
10. David, F.G., Desmond, J.H.: Numerical Methods for Ordinary Differential Equations: Initial Value Problems. Springer, London (2010)

A Heuristic Scout Search Mechanism for Artificial Bee Colony Algorithm

Ying Wu[1,2], Jian Xu[3], and Changsheng Zhang[3](✉)

[1] Key Laboratory of Fire Scene Investigation and Evidence Identification, Shenyang 110034, China
[2] Shenyang Fire Research Institute of Ministry of Public Security, Shenyang 110034, China
[3] School of Software, Northeastern University, Shenyang 110819, China
zhangchangsheng@mail.neu.edu

Abstract. Artificial bee colony (ABC) algorithm simulates the foraging behavior of honey bees, which has shown good performance in many application problems and large scale optimization problems. However, the scout foraging behavior in the ABC algorithm is completely random, which would sometimes make it consume more search efforts to discover some promising area and hamper the convergent speed of the ABC algorithm, especially for large scale optimization. To overcome this drawback, this paper proposes a heuristic scout search (HSS) mechanism based on the information obtained during running to guide the scout search. The ABC algorithm with HSS mechanism (HSSABC) has been tested on a set of test functions. Experimental results show that the HSS mechanism can greatly speed up the convergence of the ABC algorithm. After the use of HSS, the performance of the ABC algorithm is significantly improved for both rotated problems and large scale problems.

Keywords: Meta-heuristic · Swarm intelligence · Artificial bee colony algorithm · Continuous optimization

1 Introduction

The artificial bee colony (ABC) algorithm is a popular SI-based algorithm by simulating waggle dance and foraging behaviors of real honey bee colonies [1]. Due to the simple concept, easy implementation and fast convergence, ABC has attracted much attention and wide applications in numerical optimization domain and engineering applications [2].

The artificial employed bees and onlookers determine a new candidate solution by using the same formula in the most of the existing researches [3–5]. However, in real honey bee colonies, the onlookers choose their food sources in a different way from employed bees. To model the behavior of foragers more accurately, a new onlooker foraging strategy is defined and introduced by using the best neighbor to generate a candidate position [6]. Two new foraging strategies are presented to generate candidates for the employed bees and onlookers using the global best position information [7]. But different from the using way which generates a candidate around the current

Y. Liu et al. (Eds.): ICNC-FSKD 2019, AISC 1074, pp. 271–278, 2020.
https://doi.org/10.1007/978-3-030-32456-8_29

food source, a new candidate solution is generated around a randomly selected individual from the population in this study [8]. The "DE/best/1" mutation strategy of differential evolution algorithm and the search mechanism of particle swarm optimization are also used to generate candidate solutions by the employed bees and the onlookers respectively [9]. To efficiently solve optimization problems with different characteristics, Mustafa et al. propose the integration of multiple candidate generating strategy with ABC [10]. In this algorithm, a foraging strategy is selected by employed bees and onlookers based on a probability.

Furthermore, in the ABC algorithm, when the food sources are abandoned, the employed bee related to it becomes a scout. Then, a food source is produced for this scout, and the scout bee becomes an employed bee again to consume this food source. So, more search efforts can be saved if better food source can be found earlier by the scout. However, the scouts in these existing ABC variants forage in a completely random way, the same as in the initialization process. That is to say, any area of searching space can be made as the new food source for the scout in the same probability during the whole optimization process, including the food source just given up. Since each run of the ABC algorithm is a continuous process, more and more information related with the search space can be accumulated with the evolving of the bee colony. It has been proved that this kind of information can be used to make the optimizing process more effective [7].

To model the behavior of foragers in ABC more accurately, an improved artificial bee colony algorithm is presented, in which a heuristic search mechanism based on the information obtained during bee colony foraging is designed for the scout, and different foraging strategies are used by the employed bees and onlookers respectively. To study the performance of the proposed ABC algorithm with HSS mechanism (HSSABC), it has been tested on a set of different scale and type test problems and compared with the state of art related algorithms.

2 Standard Artificial Bee Colony Algorithm

In ABC algorithm, the colony of artificial bees consists of three groups of bees: employed bees, onlookers and scouts. A food source represents a possible solution to the problem to be optimized. The nectar amount of a food source corresponds to the quality of the solution represented by that food source. For every food source, there is only one employed bee. The employed bee whose food source has been abandoned becomes a scout. Then it finds a new food source randomly and starts to consume it as an employed bee. The ABC algorithm can be divided into the initialization stage, employed bee stage, onlooker bee stage and scout stage.

At the initialization stage, it is supposed that the initial population of the food sources is made up of SN number of n-dimensional real-valued vectors and the ith solution of the population can be represented as $\mathbf{x}_i = \{x_{i1}, \ldots, x_{in}\}$. Then the SN solutions are randomly generated by

$$x_{ij} = l_j + rand(0,1)(u_j - l_j) \quad (1)$$

where l_j and u_j are the lower and upper bound constraints of the jth variable of $\boldsymbol{x}_i$, respectively. At the employed bee stage, each employed bee flies to a food source and explores it by

$$v_{ij} = x_{ij} + \varphi_{ij}(x_{ij} - x_{kj}) \tag{2}$$

where φ_{ij} is uniformly distributed random real number in $[-1, 1]$, $k \in \{1, \ldots, i-1, i+1, \ldots, SN\}$ is randomly chosen and $j \in \{1, \ldots, n\}$ is a randomly chosen dimension. After generating the $\mathbf{v}_i$ by Eq. (2), there is a greedy selection between $\mathbf{x}_i$ and $\mathbf{v}_i$ by Eq. (3).

$$\mathrm{x}_i = \begin{cases} \mathrm{v}_i, & if \quad fit(\mathrm{v}_i) > fit(\mathrm{x}_i), \\ \mathrm{x}_i, & \text{otherwise}, \end{cases} \tag{3}$$

where $fit(\mathbf{x}_i)$ means the fitness value of $\mathbf{x}_i$. For a minimization problem, the fitness value of a solution can be defined as:

$$fit(\mathrm{x}_i) = \begin{cases} \frac{1}{1+f(\mathrm{X}_i)}, & if \quad f(\mathrm{x}_i) > 0 \\ 1 + |f(\mathrm{x}_i)|, & if \quad f(\mathrm{x}_i) \leq 0, \end{cases} \tag{4}$$

where $f(\mathbf{x}_i)$ is the value of the objective function at $\mathbf{x}_i$.

At the onlooker bee stage, every onlooker bee randomly selects a solution from the SN solutions with a probability p_i. The probability of a food source chosen by an onlooker bee can be calculated by:

$$p_i = fit(\mathrm{x}_i) \Big/ \sum\nolimits_{i=1}^{SN} fit(\mathrm{x}_i) \tag{5}$$

Once the onlooker bee has chosen her food source, exploitation is made for it by Eq. (2) to generate a new solution. Then a greedy selection is made by Eq. (3) between the new solution and the old one to remain a better solution.

After all employed bees and onlooker bees have explored their food sources, there is a check to see whether there is an exhausted food source need to be abandoned or not at the scout stage. Here the exhausted food means a food source that has not been improved over the last LIMIT cycles. The LIMIT is a predetermined control parameter of the ABC algorithm. If there is an exhausted food source $\mathbf{x}_i$, a scout bee discovers a new food source by Eq. (1) to replace it.

3 The Heuristic Scout Search Mechanism

We can see that the worth of scout does not only lie in that it can make the exhausted food source be abandoned timely to save search effort but also that it can discover promising search areas to guide the bees foraging in the ABC algorithm. More search efforts spent on exploration can be saved if the promising areas could be found earlier by the scout. Using these saved search efforts to make exploitation can further improve the algorithm performance when the termination is the same for an optimization problem. Moreover, during the optimization process, the best food source found so far is constantly changing. Its moving direction and the information related with the visited places can give us a clue to find better food source to some extent. Based on this, a heuristic scout foraging strategy is designed, in which a new food source v_i is generated as Eq. (1) when an employed bee i becomes a scout.

$$v_{i,j} = \begin{cases} l_j + r_1(u_j - l_j) & \text{if } \beta > I_c/I_{\max} \wedge T \\ x_{i,j} + r_2(cbest_j - x_{i,j}) + r_3(cbest_j - cworst_j) & \text{otherwise} \end{cases} \quad (6)$$

where r_1, r_2, r_3, and $\beta \in [0, 1]$ are all random real number. The I_c and I_{max} denote the current fitness evaluation number and the max fitness evaluation number respectively. The T is a Boolean, whose value is true if the best food source found so far has not been updated when "LIMIT*SN" fitness evaluations have past, otherwise its value is false. The *cbest* and *cworst* are the current best food source and the worst food source found so far. The implications of the heuristic scout foraging strategy can be explained by Fig. 1, in which the A, B and C denote the current position, the current best food source position and the current worst food source position. We can get the moving direction of to best food source B→C. Furthermore, we know that better food source can be found with a greater probability by moving away from the found worst food source, which can be denoted as A→C. So, we can get the promising foraging area shown as the dark area in Fig. 1, and better food source may exist in this area with a greater probability. The other parts of the HSSABC algorithm operate in the same way as ABC.

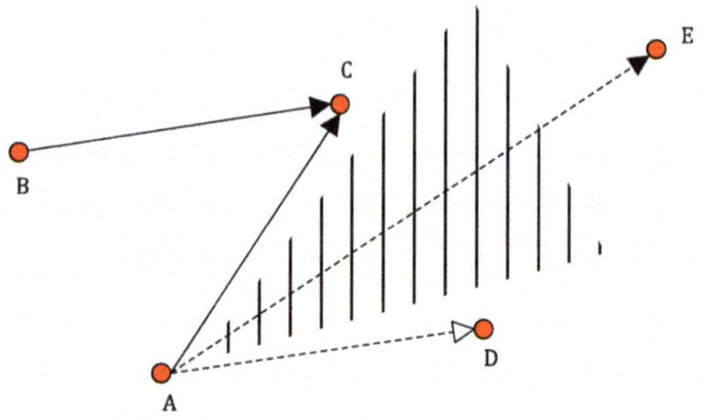

Fig. 1. The stressed foraging area

4 Experimental Results

To test the proposed HSSABC performance, 39 well known minimization benchmark functions coming from two test sets [11] with dimension D = 30, 100 and 1,000 are used in this part. Among these functions, $f_1, f_2, f_3, f_4, f_{26}$ and f_{27} are uni-modal. f_6 is a noisy multimodal function. The others are multi-modal functions. Moreover, the f_1, f_2, f_7, f_{10}, $f_{26}, f_{29}, f_{31} f_{33}, f_{36}, f_{38}$ are separable, while the others are non-separable functions. During experiments, five related algorithms are used to compare with the proposed HSSABC algorithm. They are ABC [2], OPIABC [11], MPIABC [11], qABC [14] and ABCVSS

[18]. The ABC is the standard artificial bee colony algorithm, and the other four are recently improved versions of ABC. All algorithms are implemented in Java. For each algorithm and each test function, 30 independent runs are executed with MFE = 5000*D as the termination criterion. The other parameters of HSSABC is set as SN = 50, LIMIT = 100. The parameters of the other compared algorithms are set as in their original paper.

Table 1 gives the mean value and standard deviations obtained by the compared algorithms on each test functions. It can be seen that the proposed HSSABC algorithm achieves the minimum mean values on 24 test functions. Moreover, The HSSABC obtains 14 minimum mean values out of 25 test functions with D = 30, 4 minimum mean values out of 7 test functions with D = 100, 6 minimum mean values out of 7 test functions with D = 1000. For the separable case, the HSSABC obtains 5 minimum mean values out of 10. But, for uni-modal, multi-modal, and un-separable cases, the HSSABC has achieved the minimum mean values on greatly more functions than the compared other algorithms.

Figure 2 illustrates the convergence graphs in terms of the best function value of the median run of each algorithm for f_3, f_{27}, f_{32} and f_{39}. These graphs are selected because they represent typical behavior. The convergence graphs of the other functions are not presented for saving space. It can be seen from Fig. 2 that HSSABC converges second fastest and deepest on test functions f_3, and fastest and deepest among the compared algorithms on the other test cases especially the high dimension test functions f_{27}, f_{32}, and f_{39}. So, through the experimental results, it is observed that the HSSABC works well on the most test functions especially the high dimension optimization problems.

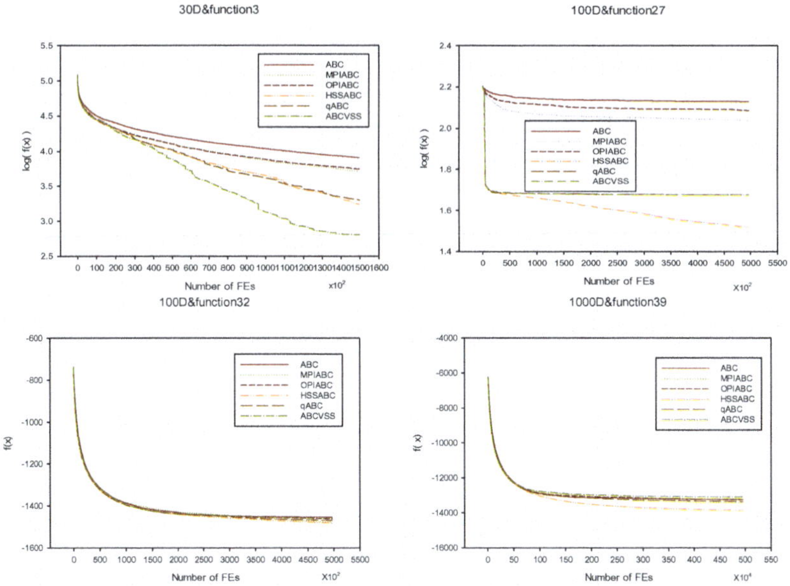

Fig. 2. The convergence graphs on parts of test functions

Table 1. The comparative results

Problems		HSSABC	OPIABC	ABC	MPIABC	qABC	ABCVSS
f1	mean	**2.99E−16**	3.49E−16	7.00E−16	3.05E−16	3.73E−16	3.40E−16
	std	7.71E−17	8.03E−17	1.04E−16	6.67E−17	8.41E−17	7.57E−17
f2	mean	9.48E−16	9.66E−16	1.27E−11	**7.05E−16**	9.77E−16	8.89E−16
	std	1.20E−16	1.22E−16	7.57E−12	9.82E−17	1.15E−16	1.51E−16
f3	mean	1.74E+03	5.52E+03	8.07E+03	5.12E+03	1.99E+03	**6.39E+02**
	std	1.63E+03	1.45E+03	1.20E+03	1.13E+03	2.14E+03	8.20E+02
f4	mean	**7.04E−02**	2.29E+01	3.86E+01	1.56E+01	8.49E−02	7.95E−02
	std	5.71E−02	4.05E+00	4.39E+00	2.08E+00	9.58E−02	6.31E−02
f5	mean	1.73E−01	1.91E−01	2.71E−01	5.36E−01	**1.51E−01**	3.98E−01
	std	2.55E−01	2.42E−01	3.27E−01	1.09E+00	2.43E−01	9.34E−01
f6	mean	3.67E−03	8.25E−02	1.87E−01	4.72E−02	**3.52E−03**	4.27E−03
	std	3.07E−03	1.66E−02	3.96E−02	1.01E−02	2.03E−03	2.70E−03
f7	mean	**0.00E+00**	**0.00E+00**	2.32E−14	**0.00E+00**	**0.00E+00**	**0.00E+00**
	std	0.00E+00	0.00E+00	3.07E−14	0.00E+00	0.00E+00	0.00E+00
f8	mean	5.55E−17	7.77E−17	8.14E−14	**1.11E−17**	1.29E−15	1.51E−15
	std	7.45E−17	1.15E−16	2.76E−13	4.39E−17	6.46E−15	5.68E−15
f9	mean	**−2.17E+04**	−1.26E+04	−1.25E+04	−1.26E+04	−1.26E+04	−1.26E+04
	std	5.67E+03	2.57E−12	6.42E+01	1.82E−12	2.44E−12	2.44E−12
f10	mean	3.00E−14	3.29E−14	5.70E−10	**2.23E−14**	3.24E−14	3.20E−14
	std	4.14E−15	3.31E−15	3.27E−10	2.82E−15	3.75E−15	2.54E−15
f11	mean	**0.00E+00**	**0.00E+00**	**0.00E+00**	8.55E−06	**0.00E+00**	1.01E−07
	std	0.00E+00	0.00E+00	0.00E+00	4.12E−05	0.00E+00	5.44E−07
f12	mean	**2.09E−16**	2.50E−16	6.99E−16	**2.09E−16**	2.59E−16	2.26E−16
	std	**3.44E−17**	4.79E−17	1.24E−16	3.89E−17	5.59E−17	7.46E−17
f13	mean	4.83E+00	1.43E+02	2.31E+02	8.19E+01	8.87E+00	**1.75E+00**
	std	6.50E+00	2.40E+01	3.32E+01	2.19E+01	9.56E+00	2.39E+00
f14	mean	**1.03E−07**	1.05E−07	3.37E−06	1.76E−05	2.75E−06	3.22E−07
	std	5.51E−07	5.06E−07	2.61E−06	8.04E−05	1.41E−05	1.32E−06
f15	mean	2.02E−02	4.34E+00	4.56E+00	2.34E+00	4.75E−03	**2.02E−03**
	std	4.16E−02	5.48E−01	2.64E−01	6.77E−01	7.39E−03	3.14E−03
f16	mean	−2.82E+01	−2.74E+01	−2.54E+01	−2.79E+01	−2.85E+01	**−2.88E+01**
	std	1.02E+00	9.45E−01	9.27E−01	1.04E+00	6.28E−01	3.43E−01
f17	mean	**0.00E+00**	**0.00E+00**	7.40E−17	**0.00E+00**	**0.00E+00**	**0.00E+00**
	std	0.00E+00	0.00E+00	2.01E−16	0.00E+00	0.00E+00	0.00E+00
f18	mean	**−2.28E+01**	−2.21E+01	−2.16E+01	−2.22E+01	−2.22E+01	−2.20E+01
	std	5.55E−01	4.71E−01	4.79E−01	7.14E−01	5.91E−01	7.46E−01
f19	mean	**1.24E−16**	1.73E−16	6.92E−16	1.26E−16	1.68E−16	1.40E−16
	std	4.05E−17	4.42E−17	1.12E−16	4.31E−17	3.77E−17	4.12E−17
f20	mean	**9.63E−01**	1.00E+00	1.00E+00	1.00E+00	9.68E−01	9.77E−01
	std	3.74E−02	2.33E−10	5.85E−05	9.95E−08	4.13E−02	3.74E−02

(*continued*)

Table 1. (*continued*)

Problems		HSSABC	OPIABC	ABC	MPIABC	qABC	ABCVSS
f21	mean	**4.68E−01**	1.01E+00	1.75E+00	6.34E−01	4.97E−01	4.83E−01
	std	3.11E−01	1.15E−01	2.26E−01	7.88E−02	2.86E−01	2.62E−01
f22	mean	−4.98E+34	−5.23E+34	−2.15E+34	**−6.25E+34**	−5.17E+34	−5.35E+34
	std	6.83E+33	5.63E+33	3.72E+33	1.81E+32	5.55E+33	7.96E+33
f23	mean	**−3.50E+00**	**−3.50E+00**	−1.90E+00	**−3.50E+00**	**−3.50E+00**	**−3.50E+00**
	std	1.61E−15	1.85E−15	9.31E−01	**2.15E−16**	8.39E−14	2.91E−15
f24	mean	**−2.85E+01**	−2.84E+01	−2.79E+01	−2.83E+01	−2.84E+01	−2.84E+01
	std	3.06E−01	4.48E−02	1.08E−01	7.98E−02	6.11E−02	7.70E−02
f25	mean	7.39E+00	5.03E+00	3.39E+02	2.64E+00	5.62E+00	**5.04E−02**
	std	1.30E+01	1.12E+01	1.22E+02	6.38E+00	1.11E+01	2.54E−01
f26	mean	**4.48E−13**	6.37E−13	1.28E−12	4.91E−13	6.37E−13	5.80E−13
	std	9.14E−14	1.30E−13	3.08E−13	8.49E−14	9.09E−14	9.91E−14
f27	mean	**3.30E+01**	1.22E+02	1.35E+02	1.09E+02	4.75E+01	4.75E+01
	std	7.24E+00	4.84E+00	3.95E+00	5.47E+00	2.98E−01	3.42E−01
f28	mean	1.03E+01	1.02E+01	**4.10E+00**	1.39E+01	6.41E+00	6.89E+00
	std	1.72E+01	1.72E+01	3.59E+00	1.91E+01	7.86E+00	8.68E+00
f29	mean	4.41E−13	5.59E−13	4.70E−01	**3.77E−13**	5.66E−13	5.37E−13
	std	1.12E−13	1.46E−13	6.34E−01	5.05E−14	1.69E−13	1.38E−13
f30	mean	**1.00E+00**	**1.00E+00**	**1.00E+00**	**1.00E+00**	**1.00E+00**	**1.00E+00**
	std	**3.55E−14**	5.38E−14	1.51E−13	4.02E−14	4.15E−14	4.22E−14
f31	mean	1.01E−12	1.21E−12	1.24E−08	**9.25E−13**	1.22E−12	1.18E−12
	std	1.01E−13	7.96E−14	9.54E−09	7.10E−14	5.87E−14	8.55E−14
f32	mean	**−1.48E+03**	−1.46E+03	−1.45E+03	−1.48E+03	−1.47E+03	−1.46E+03
	std	5.17E+00	8.03E+00	8.66E+00	9.00E+00	6.91E+00	9.87E+00
f33	mean	**1.19E−11**	1.82E−11	2.82E−10	1.27E−11	1.85E−11	1.80E−11
	std	3.26E−12	2.25E−12	1.84E−10	1.03E−12	2.22E−12	2.09E−12
f34	mean	**4.92E+01**	1.79E+02	1.80E+02	1.79E+02	4.99E+01	4.99E+01
	std	1.10E+00	9.54E−01	1.15E+00	1.04E+00	9.62E−03	1.05E−02
f35	mean	1.28E+00	1.53E+00	1.50E+01	2.84E+00	**1.20E+00**	1.91E+01
	std	2.57E+00	2.77E+00	8.67E+00	3.54E+00	1.67E+00	2.38E+01
f36	mean	**1.19E−11**	2.32E−11	8.35E+00	2.35E−01	3.82E−11	1.71E−03
	std	7.69E−12	8.99E−12	2.86E+00	3.96E−01	5.90E−11	8.36E−03
f37	mean	**1.00E+00**	**1.00E+00**	**1.00E+00**	**1.00E+00**	**1.00E+00**	**1.00E+00**
	stdMin	1.81E−12	**1.05E−12**	1.47E−10	3.80E−13	1.15E−12	1.14E−12
f38	mean	**1.58E−11**	2.19E−11	8.07E−08	1.36E−06	2.24E−11	2.14E−11
	std	2.11E−12	1.58E−12	1.33E−07	4.85E−06	9.68E−13	1.63E−12
f39	mean	**−1.38E+04**	−1.32E+04	−1.32E+04	−1.34E+04	−1.34E+04	−1.31E+04
	std	4.38E+01	6.80E+01	4.28E+01	7.47E+01	5.25E+01	6.74E+01

5 Conclusion

In this paper, a scout heuristic search mechanism for ABC is designed. Instead of presenting a new hybrid ABC algorithm or integrating an operator of an existing algorithm into ABC, our aim is to model the behavior of foragers in ABC more accurately by introducing a new definition for scout bees. By using the new definition proposed for scout bees, the formed HSSABC achieves a better performance than the compared recently proposed related algorithms in terms of solution quality and convergence characteristics. Through this work, we can see that using the best food source updating information during optimizing process can effectively improve the algorithm performance. So, how to use some obtained running information to improve the optimizing process is our further work.

References

1. Karaboga, D., Basturk, B.: A powerful and efficient algorithm for numerical function optimization: artificial bee colony (ABC) algorithm. J. Glob. Optim. **39**(3), 459–471 (2007)
2. Karaboga, D., Gorkemli, B., Ozturk, C., Karaboga, N.: A comprehensive survey: artificial bee colony (ABC) algorithm and applications. Artif. Intell. Rev. **42**(1), 21–57 (2014)
3. Zhang, X., Yuen, S.Y.: Improving artificial bee colony with one-position inheritance mechanism. Memetic Comput. **5**(3), 187–211 (2013)
4. Chen, J., Yu, W., Tian, J., Chen, L., Zhou, Z.: Image contrast enhancement using an artificial bee colony algorithm. Swarm Evol. Comput. **38**, 287–294 (2018)
5. Tansel, D., Ender, S., Ahmet, C.: Artificial bee colony optimization for the quadratic assignment problem. Appl. Soft Comput. **76**, 595–606 (2019)
6. Karaboga, D., Gorkemli, B.: A quick artificial bee colony (qABC) algorithm and its performance on optimization problems. Appl. Soft Comput. **23**, 227–238 (2014)
7. Gao, W., Liu, S., Huang, L.: Enhancing artificial bee colony algorithm using more information-based search equations. Inf. Sci. **270**, 112–133 (2014)
8. Zhu, G.P., Kwong, S.: Gbest-guided artificial bee colony algorithm for numerical function optimization. Appl. Math. Comput. **217**, 3166–3173 (2010)
9. Shan, H., Yasuda, T., Ohkura, K.: A self-adaptive hybrid enhanced artificial bee colony algorithm for continuous optimization problems. BioSystems **132–133**, 43–53 (2015)
10. Kiran, M.S., Hakli, H., Gunduz, M., Uguz, H.: Artificial bee colony algorithm with variable search strategy for continuous optimization. Inf. Sci. **300**, 140–157 (2015)

Correlation Filter Tracking Algorithm Based on Spatio-Temporal Context

Jin Die[1,2(✉)], Na Li[1,2], Ying Liu[2,3], and Yangyang Wu[1,2]

[1] School of Communication and Information Engineering, Xi'an University of Posts and Telecommunications, Xi'an 710121, China
1393101691@qq.com, 13873589@qq.com, 2745027040@qq.com
[2] Key Laboratory of Electronic Information Application Technology for Scene Investigation, Ministry of Public Security, Xi'an 710121, China
ly_ly_yolanda@sina.com
[3] International Joint Research Center for Wireless Communication and Information Processing Technology, Xi'an 710121, China

Abstract. Object tracking has been widely used in artificial intelligence, military reconnaissance, security monitoring and other fields. It has become a research hotspot of computer vision. To handle the drift problem in the presence of occlusions, a tracker combined with spatio-temporal context information and correlation filter is proposed in this paper. HOG (Histogram of Oriented Gradient), CN (Color Name) and gray features are extracted to learn the correlation filter. Meanwhile, the spatio-temporal context model is trained. The response map of correlation filter and spatio-temporal context model are normalized and fused. Experimental results show that the proposed algorithm can accurately track the object, and has better performance in terms of successful rate, center position error and distance precision.

Keywords: Object tracking · Spatio-temporal context · Correlation filter

1 Introduction

Object tracking is one of the most active research topics in the field of computer vision. It has a wide range of applications in motion analysis, activity recognition, human-computer interaction and traffic surveillance [1]. With the rapid development of information technology, scholars have proposed many new methods to visual tracking and continuously improve tracking performance. However, there were still some practical challenges, such as rapid motion of objects, changes in viewing angles, rapid appearance distortion, blurring and occlusion. Therefore, in order to solve the problem of object tracking failure better, designing an effective and stable tracking algorithm is still in great need in reality [2].

To handle the occlusion and blurring, a correlation filter tracking algorithm based on spatio-temporal context is proposed in this paper. The rest of this paper is organized as follows: Sect. 2 mainly introduces the related work. Section 3 focuses on the basic

Y. Liu et al. (Eds.): ICNC-FSKD 2019, AISC 1074, pp. 279–289, 2020.
https://doi.org/10.1007/978-3-030-32456-8_30

principles and flow charts of the algorithm. Section 4 introduces the experiment and result analysis based on the algorithm in this paper. Section 5 summarizes and forecasts the algorithm.

2 Related Work

Time information means that the object changes little between adjacent frames and the position will not be abrupt. Spatial information refers to the existence of a specific relationship between the object and the background around the object. When the appearance changes greatly, this relationship can help distinguish the object from the background. The combination of the two pieces of information is spatio-temporal context information, which can be used to track the object robust and quickly.

STC [3] is the use of spatio-temporal context information. By modeling the spatial-temporal relationship between the target and the local context region through the Bayesian framework, the statistical correlation of low-level features of the target and its surrounding region is obtained. The confidence map of the position of the target in the new frame is evaluated by integrating the spatial-temporal relationship and the focus characteristics of the biological vision system. The position corresponding to the maximum value of the confidence map is the position of the target in the predicted next frame.

CACF [4] solves the limitation of the cosine window and search area used by traditional correlation filter to suppress the boundary effect, which leads to the disadvantage that the CF template cannot obtain sufficient background information. When the object has a large deformation or is disturbed by a complex background, it is easy to lose the object. In response to the above problems, the literature [11] proposes a framework to explicitly learning the background information around the object, which can be widely used in CF-based tracking algorithms.

In order to alleviate the stability dilemma of online model updating in visual tracking, the TLD algorithm proposed by Kalel et al. [5] decomposes the tracking task into tracking, learning and detection. The tracking and detection mutually promote each other, that is, the result of the tracker provides training. The literature [6] proposes the definition of the interference term and the contributing region, and further refines the context structure. Literature [7] assigns different weights to the gray information around the object according to the similarity between different regions and the motion of the object region, and increases the anti-interference ability of the algorithm. In [8], an anti-occlusion mechanism is proposed, which divides the context area and determines whether the object is occluded based on the similarity of colors. After occlusion, block matching is used to locate the object. In [9, 10], the output frame of the nuclear particle filter KCF algorithm is proposed to perform conservative similarity calculation. The output frame with low similarity is detected by the detector, and a single cluster frame is detected and output. In [11], a long-term tracking algorithm based on STC is proposed. The image matching algorithm based on wavelet transform is used to determine the result of fast tracking.

The correlation filtering algorithm is from the least square error filter MOSSE [12] to the detection-based tracking algorithm CSK [13] using the nuclear cycle structure, to the classical kernel correlation filter KCF [14], in order to solve the variable scale tracking problem, Li et al. proposed the SAMF algorithm based on KCF [15]. Danelljan et al. proposed the DSST algorithm to improve the real-time performance of SAMF [16], and then improved the DSST and proposed the fDSST [17]. The robustness of the algorithm is improved and the speed is twice that of DSST. The SRDCF [18] suppresses the boundary effects caused by cyclic shifts in KCF and also improves the accuracy and speed of tracking.

3 Our Algorithm

CF trackers typically have very limited use of their background information and are prone to drift in the case of fast motion, occlusion, or background clutter. Literature [10] solves the traditional problem of correlation filtering. Without a stable appearance model when the object is deformed and occluded, the tracking robustness is poor and easy to lose. The algorithm proposed in [12, 13] does not have an anti-occlusion mechanism. The algorithm proposed in [14] adopts the color histogram similarity to make occlusion judgment, and the accuracy is not high. In some complex cases, the literature [15, 16] does not make comparison choices for multiple candidate object detected by the detector, which reduces the tracking success rate of the algorithm.

Our algorithm first initializes the tracking object and then tracks it, predicts the position of the next frame tracking point, and combines the spatial context model of the initialization object with the temporal relationship in the vision to obtain the spatio-temporal context model. The Fourier transform obtains a response value, the response value and another response value obtained based on the correlation filtering context are weighted and summed to determine object position of the next frame, then each frame spatio-temporal context model is updated one by one and the spatial model is utilized. Calculate the confidence, and finally the maximum value of the confidence value as the object output. Experiments show that this method can effectively improve the detection accuracy and reduce the number of occurrences of drift.

3.1 The Principle of the Algorithm

Training Spatial-Temporal Model

A traditional tracking problem can be described as a confidence map that calculates the likelihood of an estimated object position:

$$\begin{aligned} c(x) &= P(x/o) \\ &= \sum\nolimits_{c(z) \in X^c} P(x, c(z)/o) \\ &= \sum\nolimits_{c(z) \in X^c} P(x/c(z), o)P(c(z)/o) \end{aligned} \tag{1}$$

Where o indicates that the object exists in the scene, and the position corresponding to the maximum value of the confidence map $c(x)$ is the position of the object. The likelihood function can be broken down into two probability parts. The first is the conditional probability $P(x/c(z),o)$ of modeling the spatial relationship between the object and the surrounding context information, and the second is the context prior probability $P(c(z)/o)$ of modeling each point x of the local context. The conditional probability, the correlation between the object location and its spatial context, requires training learning.

The spatial context model can be described by a conditional probability function:

$$P(x/c(z),o) = h^{sc}(x - z) \tag{2}$$

Where $h^{sc}(x - z)$ is a function of the relative distance and direction of the object and local context location, representing the spatial relationship of the object and its context information. This function is not radially symmetric and helps to distinguish ambiguity. The model can be obtained through online learning and is continuously updated as the tracking progresses.

The context prior model can be modeled by prior probability as:

$$P(c(z)/o) = I(z)\omega_\sigma(z - x^*) \tag{3}$$

Where $I(z)$ is the grayscale of the point z.

$$c(x) = P(x/o) = be^{-\left|\frac{x-x^*}{\alpha}\right|^\beta} \tag{4}$$

The fast learning of spatial-temporal models is based on context-based prior models and confidence maps:

$$\begin{aligned} c(x) &= be^{-\left|\frac{x-x^*}{\alpha}\right|^\beta} \\ &= \sum\nolimits_{z\in\Omega_c(x^*)} h^{sc}(x - z)I(z)\omega_\sigma(z - x^*) \\ &= \mathrm{h}^{sc}(x) \otimes (I(x)\omega_\sigma(x - x^*)) \end{aligned} \tag{5}$$

Where $\otimes$ denotes the convolution operator. The time domain convolution can be calculated by FFT (the convolution of the time domain is equivalent to the product of the frequency domain), as follows:

The spatial context model is:

$$F\left(be^{-\left|\frac{x-x^*}{\alpha}\right|^\beta}\right) = F(\,\mathrm{h}^{sc}(x)) \odot F(I(x)\omega_\sigma(x - x^*)) \tag{6}$$

Where F denotes the FFT function and $\odot$ is the element-wise product.

$$\mathrm{h}^{sc}(x) = F^{-1}\left(\frac{F\left(be^{-\left|\frac{x-x^*}{\alpha}\right|^\beta}\right)}{F(I(x)\omega_\sigma(x - x^*))}\right) \tag{7}$$

Where F^{-1} denotes the inverse FFT function, update the spatio-temporal context model with this model:

$$H_{t+1}^{stc} = (1-\rho)H_t^{stc} + \rho h_t^{sc} \tag{8}$$

Where ρ is a parameter. After obtaining the spatio-temporal context model, the confidence map for the new one-frame object is:

$$c_{t+1}(x) = F^{-1}\left(F\left(H_{t+1}^{stc}(x)\right) \odot F\left(I_{t+1}(x)\omega_{\sigma_t}(x - x_t^*)\right)\right) \tag{9}$$

The value in the confidence map is the largest, which the position corresponding to the response value is most likely to be the object position.

$$R_2 = c_{t+1}(x) \tag{10}$$

Correlation Filter

The correlation filter-based tracking algorithm improves the traditional CF method. The traditional CF method is simple alternation between training and detection. The training phase intercepts a context area on the top, bottom, left, right sides of the object. During the training process, take the middle A_o as a positive sample, and the nearby context area A_i is taken as a negative sample to perform template training. The detection phase is the same as the traditional CF.

The objective function of the traditional CF:

$$\min_w \|A_o w - y\|_2^2 + \lambda_1 \|w\|_2^2 \tag{11}$$

Improved object function:

$$\min_w \|A_o w - y\|_2^2 + \lambda_1 \|w\|_2^2 + \lambda_2 \sum_{i=1}^{k} \|A_i w\|_2^2 \tag{12}$$

A penalty $\sum \|A_i w\|_2^2$ is added to make the response to the training template w and the background A_i as small as possible, and a closed solution can be obtained. Expanded A_o and y:

$$B = \begin{bmatrix} A_0 \\ \sqrt{\lambda_2 A_1} \\ \cdots \\ \sqrt{\lambda_2 A_k} \end{bmatrix} \quad \text{and} \quad \bar{y} = \begin{bmatrix} y \\ 0 \\ \cdots \\ 0 \end{bmatrix} \tag{13}$$

Where B is a new data matrix, the regression target $\bar{y}$ concatenates y with zeros, so that the third item can be merged into the first item:

$$f_p(w, B) = \|Bw - \bar{y}\|_2^2 + \lambda_1 \|w\|_2^2 \tag{14}$$

Where $f_p(w, B)$ has the same form as the traditional CF

$$w = \left(B^T B + \lambda_1 I\right)^{-1} B^T \bar{y} \tag{15}$$

Using the properties of the diagonalization of the circulant matrix, the solution in the Fourier domain is obtained under the single-channel original domain.

$$\hat{w} = \frac{\hat{a}_0^* \odot \hat{y}}{\hat{a}_0^* \odot \hat{a}_0 + \lambda_1 + \lambda_2 \sum_{i=1}^{k} \hat{a}_i^* \odot \hat{a}_i} \tag{16}$$

The learned filter w is convolved with image patch z (search window). The response can be rewritten as follows:

$$R_1 = F^{-1}(\hat{w} * \hat{z}) \tag{17}$$

Response Fusion

The response R_2 obtained based on the spatio-temporal context model in the same frame and the correlation filter based response value R_1 are normalized and weighted and summed to obtain a response value R.

$$R = \gamma R_1 + (1 - \gamma) R_2 = \gamma F^{-1}(\hat{w} * \hat{z}) + (1 - \gamma) c_{t+1}(x) \tag{18}$$

Where γ is the normalized parameter. Normalization can accelerate the convergence of the training network and eliminate the influence of other transform functions on image transformation. The two responses are normalized by using the method of linear function transformation.

The position corresponding to the maximum response value is the position of the tracked object. After updating the object location, the spatio-temporal context model of the frame can be retrained and filtered to detect the object that matches the next frame. In this paper, when γ is set to 0.7, the tracking performance is the best.

3.2 Flow Chart

As shown in Fig. 1, Reads the first frame and initializes the object. First, extract the background information twice the size of the object frame and obtain the correlation with it, calculate the confidence map and the prior probability of the local context, train the model and spatial context correlation, then add the time. The relationship obtains the spatial context model of the frame, the space-time context model is updated by the model to obtain a confidence map of the new frame. Secondly, four background frames are selected from the first frame and the features are extracted. The filter is quickly detected by the filter class, and a penalty is added to minimize the response of the template and the background to reduce background interference. Finally, the two responses are weighted together, the maximum response point is searched to update the

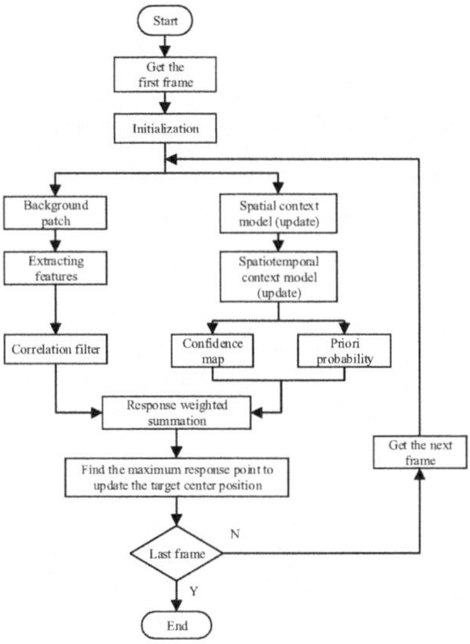

Fig. 1. Algorithm flow chart

object center position to detect whether the last frame is reached, otherwise, the next frame is continuously detected.

The normal flow of object tracking is initializing tracking object, training appearance model, setting the maximum response value point by matching the next frame image as the center location of forecast object. In this paper, the HOG, CN and gray features extracted from four background blocks chosen around the object are matched with the features extracted from the object. The appearance model by training is more robust. To weight with the response value got by matching with the spatio-temporal context model, will give better locating object and a higher accuracy rate.

4 Experiment

4.1 Experimental Platform and Database

The most widely used database is OTB-50 [19] at present, in which there are gray level images and colorful images, and they can be downloaded for free. It involves 11 attributes of object tracking, including illumination change, scale change, occlusion, deformation, motion blur, fast motion, in-plane rotation, out-of-plane rotation, out-of-view, background interference, low pixel, etc. Each image sequence corresponds to two or more attributes and one.txt file, which records the center position and size of the manually labeled object. The experiment was carried out on the computer of Intel(R) Core(TM) i3-4010u CPU @1.70ghz, 8GRAM, 64-bit operating system with the C language. Matlab mixed programming language and MATLAB2017a platform, by doing comparative analysis with several groups selected from published OTB video set, through CACF, CN, CACF+CN and our algorithm. The tracking performance is evaluated according to the three evaluation indexes of success rate SR (successful rate), central location error CLE (center position error) and DP (distance precision).

All baseline and adaptive object trackers operate according to the standard parameters provided by the author. The parameters of the mapping function α, β are set to 2.25, 1. Learning parameter ρ is set to 0.075. Experimental results show that the model frame weights got from other learning parameters are lower than this value. Size parameter initialization value $s1$ is set to 1, learning parameter λ is set to 0.25, and the number of frames used to update the scale is set to 5. To reduce the impact of illumination variations, each intensity value in the context region is normalized by subtracting the average intensity of that region. To be fair, run the context-aware tracker using the same parameters. Set the regularization operator to $\{2, 25, 0.4, 0.5\}$, the number of context patches k to 4, and evenly sample around the object. Increasing the k value above 4 will only result in minor improvements, but it will affect the running time.

4.2 Evaluation

In this paper, the performance comparison is made among the algorithm CACF, STC combined with spatio-temporal context information, the fusion of color feature CN into CACF and the improved algorithm in this paper in object tracking, through tracking accuracy and center location error. And the color naming tracking algorithm CN is also added. It can be seen from Table 1 that the average tracking accuracy of the algorithm on 20 test sets is 0.969, which is 23.28%, 29.02%, 6.25% and 25.03% higher than that of CACF, CN, CACF+CN and STC SR, respectively. The center location error CLE is significantly reduced. In the test of 20 video sequences, the success rate of the improved algorithm is obviously higher than that of the other three algorithms, and the CLE is also smaller, the tracking performance is more robust.

4.3 Result Analysis

According to the Fig. 2 we can see, our algorithm is able to obtain the optimal or suboptimal results in the accuracy, while the STC tracking algorithm is prone to drift when tracking the objects, the tracking accuracy of CN algorithm is lower, and its resistance force of object deformation and occlusion is not strong, the stability of the CACF algorithm is poorer, CACF+CN have improved in terms of accuracy and robustness, but the improved algorithm is more robust for occlusion, drift and deformation, etc. In the context algorithm based on correlation filtering, not only applied HOG feature but also extracted the gray-scale and color features, from 11-dimension to 42-dimension characteristics, which makes the algorithm is not sensitive to occlusion, deformation and the change of illumination. The weighted maximum obtains more accurate object location, eventually obviously improves the robustness and accuracy.

Table 1. Comparison of algorithm results. ('-' indicates tracking failure and object drift. Bold indicates the best tracking accuracy obtained by different sequences)

Data set	CACF		CN		CACF+CN		STC		Ours	
	SR	CLE	SR	CLE	SR	CLE	SR	CLE	SR	CLE
Basketball	0.924	6.092	0.676	14.59	0.971	4.482	0.526	75.41	**0.990**	3.950
Bird2	0.485	18.77	**0.780**	15.99	0.596	15.97	0.505	18.54	0.606	15.89
Bluecar2	0.923	7.990	0.234	86.87	0.990	4.475	–	–	**0.991**	1.680
Bolt	0.986	5.140	**1.000**	4.133	**1.000**	5.569	–	–	**1.000**	5.380
Bolt2	0.017	298.4	0.017	304.5	–	–	0.965	12.97	**0.993**	7.030
Car24	0.998	3.330	0.987	1.917	**1.000**	4.102	–	–	**1.000**	1.710
Car dark	0.999	5.120	0.990	2.827	0.999	3.137	–	–	**1.000**	1.510
David	**1.000**	8.880	0.240	35.82	0.998	7.535	0.737	12.16	**1.000**	5.760
David3	**1.000**	5.250	0.905	9.110	**1.000**	4.257	0.921	6.460	**1.000**	6.731
Deer	0.859	15.79	0.992	5.113	0.993	4.517	–	–	**1.000**	4.140
Dog	0.756	11.58	**1.000**	9.579	**1.000**	8.450	0.923	6.360	**1.000**	8.460
Girl	0.892	9.650	0.864	12.49	0.899	4.692	–	–	**0.998**	3.870
Kite surf	0.310	48.94	0.832	3.003	0.833	7.952	–	–	**0.897**	16.82
Man	**1.000**	2.230	**1.000**	1.766	**1.000**	2.390	**1.000**	1.640	**1.000**	1.580
Rubik	0.181	29.99	0.151	40.23	0.771	15.29	0.311	25.64	**1.000**	5.240
Singer1	0.843	12.65	0.966	8.880	**1.000**	7.626	–	–	**1.000**	7.252
Mountain bike	0.997	8.200	0.973	9.802	0.987	9.922	0.998	10.17	**1.000**	9.113
Red team	0.999	3.310	0.988	2.961	**1.000**	4.037	–	–	**1.000**	2.954
Walking	**1.000**	3.197	0.990	7.541	**1.000**	3.779	0.830	13.24	**1.000**	3.414
Walking2	0.564	22.33	0.424	47.73	0.484	23.03	0.812	16.71	**0.899**	11.02
Average value	0.786	26.34	0.751	31.24	0.912	7.432	0.775	18.12	**0.969**	6.174

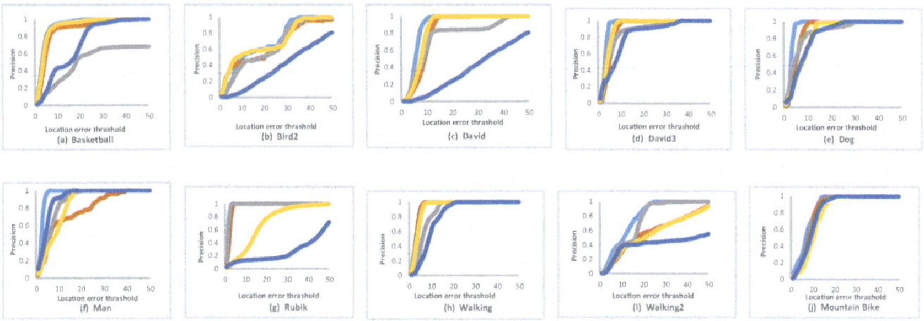

Fig. 2. Tracing accuracy curves of five algorithms on ten video sets (light blue represents the improved algorithm in this paper, dark blue represents CN, yellow represents CACF+CN, gray represents CACF, and orange represents STC.)

5 Conclusion

Contextual information about time and space is important for object tracking. This spatio-temporal context correlation can be used in object tracking, it can also help locate the object by extracting useful context information from scenes. This paper presents a general framework based on correlation filter, which incorporates context into filter training at a lower computational cost. A large number of experiments show that the framework improves the tracking performance of all CF trackers and has a high computational efficiency. Although it improves the performance of most attributes, it identifies specific scenes that benefit the most: fast motion, dramatic changes in appearance and cluttered backgrounds.

Deep learning [20] is a hot topic in the field of computer vision research at present. By constructing a deep nonlinear network structure, the essential features of image information can be learned from a small number of samples, and the accuracy of classifier classification can be improved eventually. In the following work, we can focus on how to apply the spatio-temporal context information into the training model and combine it with the neural convolution network to further improve the tracking performance.

Acknowledgments. This work is supported by the graduate innovation fund project of Xi'an university of posts and telecommunications. (Grant No. CXJJLY2018027). Shaanxi Science and Technology Innovation and Entrepreneurship Dual Tutor System Project (2019JM-604), National Science Foundation of China (61601362, 61571361).

References

1. Yilmaz, A., Javad, O., Shah, M.: Object tracking: a survey. ACM Comput. Surv. (CSUR) **38**(4), 1–45 (2006)
2. Kim, D.Y., Jeon, M.: Spatial-temporal auxiliary particle filtering with l1-norm based appearance model learning for robust visual tracking. IEEE Trans. Image Process. **22**(2), 511–522 (2013)
3. Zhang, K.H., Zhang, L., Liu, Q.S., et al.: Fast visual tracking via dense spatio-temporal context learning. In: ECCV, Zurich, pp, 127–141. Springer, Switzerland (2014)
4. Mueller, M., Smith, N., Ghanem, B.: Context-aware correlation filter tracking. In: CVPR, King Abdullah University of Science and Technology, pp, 1396–1404 (2017)
5. Kalal, Z., Mikolajczyk, K., Matas, J.: Tracking learning-detection. IEEE Trans. Pattern Anal. Mach. Intell. **34**(7), 1409–1422 (2011)
6. Dinh, T.B., Vo, N., Medioni, G.: Context tracker: exploring supporters and distracters in unconstrained environments. In: IEEE Computer Society Conference on Computer Vision and Pattern Recognition CVPR, pp, 1177–1184 (2011)
7. Xu, J.Q., Lu, Y., Liu, J.W.: Robust tracking via weighted spatio-temporal context learning. In: Proceedings of the IEEE International Conference on Image Processing, pp. 413–416 (2014)
8. Liu, W.J., Dong, S.H., Qu, H.C.: Anti-occlusion visual tracking algorithm based on spatio-temporal context learning. J. Image Graph. **21**(8), 1057–1067 (2016)

9. Yu, L.Y., Fan, C.X., Ming, Y.: Improved object tracking algorithm based on kernelized correlation filter. J. Comput. Appl. **35**(12), 3550–3554 (2015)
10. Qian, T.H., Luo, Z.Q., Li, G.J., et al.: Scale adaptive improvement of kernel correlation filter tracking algorithm. J. Comput. Appl. **37**(3), 811–816 (2017)
11. Lu, J.H., Chen, Y.M., Zou, Y.B., Zou, G.Z.: Long-term tracking based on spatio-temporal context. J. Shanghai Jiao Tong Univ. **22**(4), 504–512 (2017)
12. David, S.B.J., Ross, B., Bruce, A.: Visual object tracking using adaptive correlation filters. In: IEEE Computer Society Conference on Computer Vision and Pattern Recognition, pp. 2544–2550 (2010)
13. Joao, F.H., Rui, C., Pedro, M., Jorge, B.: Exploiting the circulant structure of tracking-by-detection with kernels. In: ECCV, pp. 702–715 (2012)
14. Henriques, J.F., Caseiro, R., Martins, P.: High-speed tracking with kernelized correlation filters. IEEE Trans. Pattern Anal. Mach. Intell. **37**(3), 583–596 (2015)
15. Li, Y., Zhu, J.: A scale adaptive kernel correlation filter tracker with feature integration. In: ECCV, pp. 254–265 (2014)
16. Danelljan, M., Hager, G., Khan, F.S., Felsberg, M.: Accurate scale estimation for robust visual tracking. In: BMVC (2014)
17. Danelljan, M., Hager, G., Khan, F.S., Felsberg, M.: Discriminative scale space tracking. IEEE Trans. Pattern Anal. Mach. Intell. **39**(8), 1561–1575 (2017)
18. Danelljan, M., Hager, G., Khan, F.S., Felsberg, M.: Learning spatially regularized correlation filters for visual tracking. In: ICCV, pp. 4310–4318 (2015)
19. Wu, Y., Lim, J., Yang, M.H.: Online object tracking: a benchmark, pp. 2411–2418. IEEE Computer Society, United States (2013)
20. Zheng, Y., Chen, Q.Q., Zhang, Y.J.: Deep learning and its new progress in object and behavior recognition. J. Image Graph. **19**(2), 175–184 (2014)

Anti-occlusion Video Target Tracking Based on Double Threshold Judgment

Ying Zhang(✉), Fang-cang Du, and Ke Xu

Shenyang Jianzhu University, Shenyang 110168, Liaoning, China
zy_1120@sina.com

Abstract. In recent years, great progress has been made in moving target tracking methods, but there is still no effective solution to the problem that affects the tracking accuracy, especially the problem that the target is occluded in the tracking process. To solve this problem, this paper proposes an anti-occlusion video target tracking method based on double threshold judgment, which combines Discriminative Scale Space Tracking (DSST) algorithm and Kalman filtering algorithm, according to the change of correlation coefficient P between adjacent templates. Firstly, the target is tracked by DSST algorithm, and the location information and scale information of the target are obtained. In the process of tracking the target, the occlusion threshold T_1 is set to determine whether occlusion occurs. If not, the tracking of DSST algorithm is continued. Otherwise, the Kalman tracking algorithm is invoked to track the target during occlusion. At the same time, according to the change of correlation coefficient, the threshold T_2 is set to judge whether the occlusion is over or not. If the occlusion is over, the DSST algorithm is returned. Otherwise, Kalman filter algorithm is used to track the occlusion until the end of the tracking. The simulation results show that this method can effectively solve the problem of tracking moving objects when they are occluded.

Keywords: Motion target tracking · Double threshold · Occlusion judgment

1 Introduction

In recent years, the rapid development of computer vision technology has brought tremendous impact to contemporary society. Among them, the research of moving target tracking has become one of the hotspots [1]. Motion target tracking is to find the tracking target from the original image containing the target accurately and real-time, and to track it accurately. It has been widely used in robot technology, medical image, traffic monitoring and other fields [2]. However, there are many disadvantageous factors in the whole tracking process, which lead to the decrease of tracking accuracy and even the loss of target. These include the uncertainties of the outside world and the target body. For example, illumination changes around the target, similar background interference, partial or total occlusion of the target, short-term occlusion and long-term occlusion, and when the target moves irregularly and so on. Many tracking algorithms are often difficult to achieve the desired results. Document [3] firstly introduces the correlation filtering method in the field of communication into the field of moving target tracking, which provides a new idea for many relevant researchers. Bolme et al.

Y. Liu et al. (Eds.): ICNC-FSKD 2019, AISC 1074, pp. 290–296, 2020.
https://doi.org/10.1007/978-3-030-32456-8_31

proposed the MOSSE [4] tracking algorithm, which has very low computational complexity, simple implementation and fast running speed. Wang et al. proposed the CSK [5] tracking algorithm, which uses the cyclic matrix to solve the weight of the classifier, greatly reduces the amount of calculation and improves the speed, but the algorithm is fixed target size, and is not suitable for scale transformation. Bradski applies MeanShift algorithm to face tracking [6, 7]. Henrique et al. proposed KCF [8] tracking algorithm, which uses discrete Fourier transform to diagonalize it, thus reducing storage and computation by several orders of magnitude, greatly improving efficiency. However, the algorithm relies on the cyclic matrix, which is not good for target tracking with variable scale. Zhang et al. proposed BACF [9] algorithm, and proved the effectiveness of substitution in visual detection by proposing a saliency model based on Boolean mapping (BMS). SAMF [10] algorithm is proposed by Itti et al. It effectively solves the problem of ground target identification in complex scenes. Kalal et al. proposed the TLD [11] tracking algorithm. The most important feature of this algorithm is that it can continuously learn the locked target to obtain the latest appearance features of the target, so as to improve the tracking in time to achieve the best state. Kalman [12, 13] filtering algorithm is based on the estimated background of the previous frame and the current frame data to get the foreground target of the current frame. Danelljan et al. proposed DSST algorithm [14], which is concise, excellent in performance and high in portability. This algorithm is improved on the basis of MOSSE algorithm. It mainly introduces multi-feature fusion mechanism. The features used are the combination of HOG feature, CN feature and gray feature and the improvement of scale. Recently, researchers have applied deep learning to target tracking, and achieved good results [15].

DSST tracking algorithm is an accurate scale estimation method in visual tracking. In addition to high tracking efficiency without interference, most of the scale change problems can be solved by DSST algorithm. However, the DSST algorithm can't solve the tracking problem when the target is occluded. Kalman filter tracking algorithm is an algorithm that can predict the target motion state. It can get the target state in the current frame according to the motion state of the previous frame. It's application range is very wide. The process noise and measurement noise of the system need not be removed. It is very suitable for real-time computer processing.

2 Target Tracking Based on DSST Algorithm

DSST algorithm is a correlation filter based ground tracking algorithm [16]. It mainly includes position filtering and scale filtering. The advantage is that the discriminant correlation filter is used to locate the target, and the principle and method of the two filters are identical. It can be combined with any other tracking algorithm without scale estimation to achieve better tracking effect. In this method, one of the inputs f is set as a d-dimensional eigenvector. In order to set the most suitable correlation filter h, the following function is used:

$$\epsilon = \left\| \sum_{l=1}^{d} h^l * f^l - g \right\|^2 + \lambda \sum_{l=1}^{d} \left\| h^l \right\|^2 \quad (1)$$

Where l represents the dimension number of features and λ is the regularization coefficient.

The upper formula is solved in frequency domain and the following formula is obtained:

$$H^l = \frac{\bar{G}F^l}{\sum_{k=1}^{d} \bar{F}F^k + \lambda} = \frac{A_t^l}{B_t} \tag{2}$$

Since any pixel in the image block needs to be computed and the amount of computation is large, both A_t^l and B_t in formula (2) need to be updated:

$$A_t^l = (1-\eta)A_{t-1}^l + \eta \bar{G}_t F_t^l \tag{3}$$

$$B_t = (1-\eta)B_{t-1} + \eta \sum_{k=1}^{d} \overline{F_t^k} F_t^k \tag{4}$$

Among them η is the learning rate.

In the next frame image, the position of the target can be obtained by solving the correlation filter response y:

$$y = F^{-1}\left\{\frac{\sum_{l=1}^{d} \overline{A^{\lambda}} Z^l}{B + \lambda}\right\} \tag{5}$$

The above is the estimation of target position by DSST algorithm.

Target scale estimation: Firstly, the latest position of the target is obtained by using two-dimensional correlation filter. At this time, the target scale is still consistent with the target scale of the previous frame. Then, the current target scale is accurately estimated by using one-dimensional scale filter. The selection principle of scale estimation is as follows:

$$a^n P \times a^n R, \quad n \in \left\{\left[-\frac{S-1}{2}\right], \cdots, \left[\frac{S-1}{2}\right]\right\} \tag{6}$$

R and Q are respectively the height and width of the previous frame.

(a) Tracking effect of DSST without occlusion (b) DSST tracking effect after occlusion

Fig. 1. DSST algorithm tracking target results

Figure 1 is the simulation result of tracking target with DSST algorithm. Figure 1(a) is the tracking effect when the target is not occluded. Figure 1(b) is the tracking result after the target is completely occluded. From Fig. 1a–b, it is obvious that the DSST tracking algorithm can track the target accurately before the target is not occluded. However, when the target is completely occluded, the DSST tracking algorithm immediately loses the target information, resulting in a serious decline in tracking accuracy and ultimately the tracking failure.

3 Anti-occlusion Video Target Tracking Based on Double Threshold Judgment

Kalman filtering algorithm is an optimal autoregressive data processing method. It has been widely used for more than 30 years, including radar systems in the military and directional tracking of missiles, etc. [17]. This algorithm is an optimal prediction algorithm for the system state by using the linear system state equation and the system input and output observation data. Then when the target is occluded, the DSST algorithm can't track effectively, we can consider the Kalman filter can accurately predict the speed and position of the target to solve the tracking problem when the target is occluded. Therefore, the author proposes an anti-occlusion video target tracking method based on double threshold judgment, which combines DSST algorithm and Kalman algorithm. The two adaptive thresholds, T1 and T2, are set on the basis of whether the correlation coefficients of the two templates can match or not in the tracking process. They are obtained adaptively along with the variation of the correlation coefficients throughout the tracking process, among them, $\rho = \sqrt{p * q}$, P is the target template parameter of the first frame, and q is the template parameter of each target in the next frame. After testing, the algorithm can effectively solve the problem of continuous tracking when the target is occluded in motion. The specific steps of this algorithm are as follows:

Step 1. Input video sequence.

Step 2. Tracking using DSST algorithm.

Input: image I_t, get the target position P_{t-1}, target scale S_{t-1}, position model A_{t-1}, scale model B_{t-1} of the previous frame.

Output: Predict target position P_t and scale S_t, update position A_t and scale model B_t.

The location estimates are as follows.

(1) According to the location of the target in the previous frame, the target sample Z is extracted in the current frame at a scale twice that of the target in the previous frame.
(2) Sample Z and A_{t-1} are used to obtain y according to formula (5).
(3) By calculating the maximum y_{max} of y, the new position P_t of the target is obtained.

The scale estimates are as follows:

(1) Thirty-three samples of different scales were extracted from the current frame with the new target location as the center.
(2) y is obtained from formula (5) by Z and A_{t-1}.
(3) The latest target scale S is obtained by calculating the maximum y_{max} of Y.

Through the above process, the latest position and scale of the target can be obtained. In this process, two thresholds T1 and T2 are set according to the correlation coefficient ρ between the front and back frame templates of the target. After processing a frame of image, occlusion judgment is made. If the correlation coefficient ρ is less than T1, occlusion occurs and the algorithm turns to step 3. Otherwise, the first step is returned and DSST algorithm is used to track.

Step 3. Give the information of the previous frame when the target is occluded to the Kalman filtering algorithm. Call the Kalman filtering algorithm to continue tracking the target. During this period, judge whether the correlation coefficient ρ between the templates satisfies the value of $\rho \geq T_2$. If it satisfies, explain the end of occlusion, terminate the Kalman filtering algorithm, and give the target information of the previous frame before the end of occlusion to DSST. The algorithm turns to the first step and begins to process the next frame of image. The DSST algorithm is used to track the next frame.

Step 4. Cycle 1–3 steps until the end of tracking.

4 Experimental Results and Analysis of Simulation

The author uses Windows 8.1 Chinese version operating system, Intel (R) Core (TM) i5-4258U CPU@2.40 GHz 4.00G memory configuration computer operation, using MATLAB language to write, in MATLAB R2017a to achieve simulation experiments.

This paper improves on the basis of DSST algorithm, so the basic parameters used are still the parameters of the original DSST, and they are consistent on all video sequences. The regularization coefficient is 0.0001, the learning rate is 0.02, the scale factor is a = 1.02, and the total scale series is S = 33.

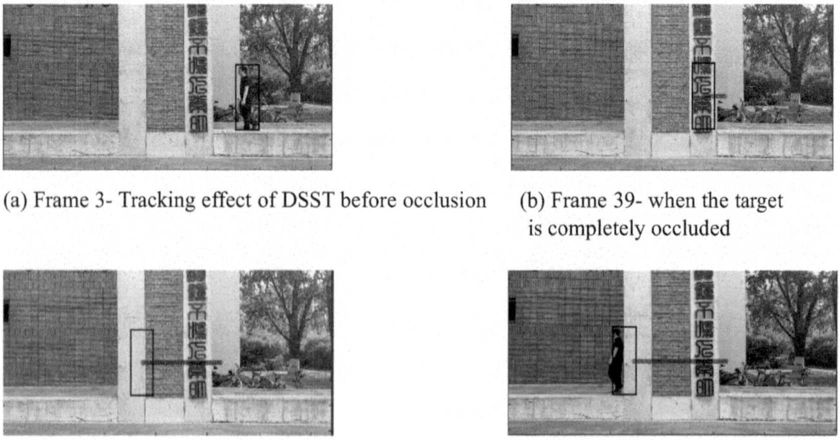

(a) Frame 3- Tracking effect of DSST before occlusion

(b) Frame 39- when the target is completely occluded

(c)Frame 78- Kalman tracking effect during occlusion

(d) Frame 95- Tracking effect at the end of occlusion

Fig. 2. The algorithm in this paper tracks the target result

In Fig. 2(a)–(d) is a tracking process using the proposed algorithm. Figure 2(a) is the effect of DSST tracking algorithm before the target is occluded. Figure 2(b) is the effect of calling Kalman tracking algorithm when the target is completely occluded for the first time. Figure 2(c) is the effect of tracking during occlusion. Figure 2(d) is the result of re-invoking DSST tracking algorithm after occlusion tracking. It can be seen from the above results that the DSST algorithm combined with Kalman algorithm has good tracking effect and effectively solves the continuous tracking of the target during the occlusion period.

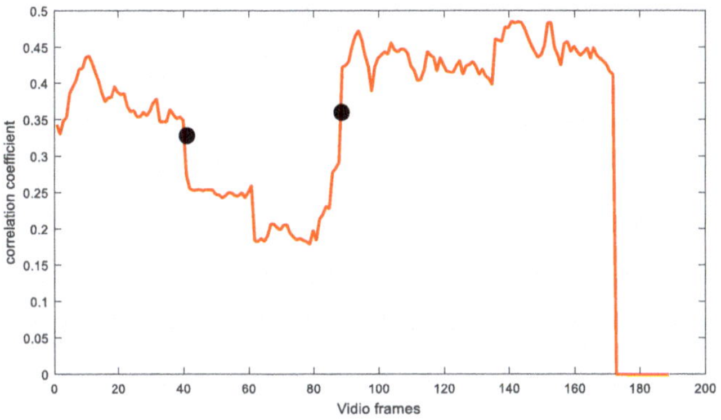

Fig. 3. The variation curve of correlation coefficients in the tracking process of this algorithm

Figure 3 is the correlation coefficient curve of this algorithm during tracking. The horizontal axis is the frame number of video, and the vertical axis is the correlation coefficient. It can be seen that the correlation coefficient P suddenly decreases at frame 39, which indicates that the target is occluded, and then Kalman filter algorithm is used to track the target. At frame 95, the correlation coefficient P suddenly increases again, indicating that the complete occlusion is over, and after the complete occlusion is over, the Kalman algorithm is also over, and the DSST algorithm is returned to track.

5 Conclusion

The classical DSST algorithm has poor tracking performance in the case of target scale change, background interference and occlusion. Therefore, on the basis of DSST algorithm, this paper uses the change of correlation coefficients between templates, sets adaptive threshold to judge whether the target has occlusion, and fuses Kalman algorithm to predict the motion state of the target, so as to solve the tracking problem when the target is completely occluded and before and after the target is occluded and the difficulty of tracking accuracy degradation when background is different. Finally, the simulation results show that the proposed algorithm has high robustness.

References

1. Yin, H., Chen, B., Chai, Y., et al.: A survey of vision-based target detection and tracking. J. Autom. **42**(10), 1466–1489 (2016)
2. Liu, Z., Tang, H., Feng, L.: Implementation of machine vision technology video tracking system based on FPGA. Electron. Technol. **31**(9), 41–44 (2018)
3. Hou, X., Zhang, L.: Saliency detection: a spectral residual approach. In: Proceedings of IEEE Conference on Computer Vision and Pattern Recognition, pp. 1–8. IEEE, Minneapolis (2007)
4. Bolme, D.S., Beveridge, J.R., Draper, B.A., et al.: Visual object tracking using adaptive correlation filters. In: 2010 IEEE Conference on Computer Vision and Pattern Recognition, San Francisco, pp. 2544–2550. IEEE Computer Society, New York (2010)
5. Wang, J., Lu, H., Li, X.: Saliency detection via background and foreground seed selection. Neurocomputing **152**, 359–368 (2015)
6. Borji, A., Itti, L.: Exploiting local and global patch rarities for saliency detection. In: Proceedings of the 2012 IEEE Conference on Computer Vision and Pattern Recognition, pp. 478–485. IEEE, Providence (2012)
7. Achanta, R., Hemami, S., Estrada, F., Susstrunk, S.: Frequency-tuned salient region detection. In: Proceedings of IEEE Conference on Computer Vision and Pattern Recognition, pp. 1597–1604. IEEE, Miami (2009)
8. Henriques, J.F., Caseiro, R., Martins, P., et al.: High-Speed tracking with kernelized correlation filters. IEEE Trans. Pattern Anal. Mach. Intell. **37**(3), 583 (2014)
9. Zhang, J., Sclaroff, S.: Exploiting surroundedness for saliency detection: a boolean map approach. IEEE Trans. Pattern Anal. Mach. Intell. **38**(5), 889–902 (2016)
10. Itti, L., Koch, C., Niebur, E.: A model of saliency-based visual attention for rapid scene analysis. IEEE Trans. Pattern Anal. Mach. Intell. **20**(11), 1254–1259 (1998)
11. Kalal, Z., Matas, J., Mikolajczyk, K.: Online learning of robust object detectors during unstable tracking. In: Proceedings of 2009 IEEE 12th International Conference on Computer Vision Workshops, pp. 1417–1424 (2009)
12. Chen, J., Liu, K., Zhan, Y.: Robust multi-person tracking algorithm based on Kalman particle filter framework. Comput. Eng. Des. **10**, 2759–2764 (2015)
13. Weng, S.K., Kuo, C.M., Tu, S.K.: Video object tracking using adaptive Kalman filter. J. Vis. Commun. Image Represent. JVCIR **17**(6), 1190–1208 (2006)
14. Danelljan, M., Hger, G., Khan, F.S., et al.: Accurate scale estimation for robust visual tracking. In: British Machine Vision Conference, pp. 65.1–65.11. British Machine Vision Association, Nottingham (2014)
15. Wu, G., Lu, W., Gao, G., et al.: Regional deep learning model for visual tracking. Neurocomputing **175**(PA), 310–323 (2016)
16. Valmadre, J., Bertinetto, L., Henriques, J., Vedaldi, A., Torr, P.H.S.: End-to-end representation learning for correlation filter based tracking. In: Conference on Computer Vision and Pattern Recognition, pp. 5000–5008. IEEE, Honolulu (2017)
17. Meng, X., Zhang, B., Yuan, Q.: Application of adaptive extended Kalman filter in mobile robot location. Appl. Comput. Syst. **24**(12), 176–181 (2015)

Natural Computation Applications: Natural Computation in Pattern Recognition and Diagnostics

Research on the Modern Power Grid Planning Method Based on the Nature and Characteristic of Power Network Planning

Ping Zhang[1(✉)], Jingbo Liu[1], and Zhijun Wang[2]

[1] Economic Technology Research Institute, State Grid Liaoning Electric Power Supply Co. Ltd., Shenyang, China
ping_zhang147@163.com

[2] Jinzhou Power Supply Branch, State Grid Liaoning Electric Power Supply Co. Ltd., Jinzhou, China

Abstract. City is the main power system load center, the city depends on the power grid is good or not urban power grid planning and construction is scientific, economic and reasonable, whether for fixed assets huge power supply enterprise, the city network planning work in the power supply enterprise's survival and development is always plays a decisive role. In this paper, on the basis of analyzing the essence and characteristics of power grid planning, focus on simulated evolution, swarm intelligence, artificial intelligence, uncertain systems and other modern power grid planning method, various methods are discussed in detail the basic principle, characteristics and the comprehensive evaluation. Looking forward to the new economic and technological environment on the influence of the various methods and the development trend of power grid planning method.

Keywords: City power grid · Planning method · Intelligence

1 Introduction

Power grid planning is a knowledge intensive research field, it involves a lot of research field of power system, such as load forecasting, the substation design, stability and flow control, over voltage and short circuit analysis, strain capacity evaluation, reliability evaluation, economic analysis, and the application of new technology. Power supply enterprises of city network planning goal mainly is to improve the urban power grid power supply capacity and the quality of power supply and power supply reliability to meet the social demand for electricity, at all levels of government in policies, investment and management support necessary, main consideration is the social benefits. But at present, the city network planning [1–3] when consider the value of enterprise assets.

Has been successfully applied in the power grid planning theory calculation parameters of parallel algorithms such as genetic algorithm, simulated annealing for route planning, and applied to the practice, has obtained the good effect, to ensure the optimal planning scheme choice, this will be the trend of the city network planning.

Y. Liu et al. (Eds.): ICNC-FSKD 2019, AISC 1074, pp. 299–306, 2020.
https://doi.org/10.1007/978-3-030-32456-8_32

City network planning work in the service and obey the municipal planning work at the same time, the power supply enterprise also should according to their own financial situation to come up with their own planning standards, timely communication with the municipal planning department, avoid the overemphasis on the improvement of science and technology, the municipal planning department in municipal planning for power supply enterprises put forward higher technical standard, load power supply enterprises with high investment. For a huge power supply enterprise fixed assets, to ensure good asset-liability ratio and profit margins, must strive for the realization of the diversification of the city grid. It affected the power supply enterprise in operating performance of the year, the more affected the sustainable development of the power supply enterprise [4].

Power grid planning method refers to the power planning personnel in the process of power grid planning by using a series of a specific set of logical relationship formed by the action of the whole. At home and abroad for application of power grid planning method, in the traditional planning method and modern heuristic optimization planning method and the status of and modern heuristic method has a tendency to replace the traditional method. In recent years, many new methods and new technology has been successfully used to solve the problem of power network planning, it is necessary for all kinds of modern power grid planning method are studied and compared.

2 Power Grid Planning

2.1 The Problem of Power Grid Planning

Is usually in the sense of power grid planning network programming, i.e., according to the existing grid structure, power planning, the basis of load forecasting, and then determine the need of the construction of the electric power line, in order to meet the load demand growth of planning period; At the same time also should satisfy all kinds of operation constraints, make the electric power network investment, operation, maintenance cost minimum, reliability is the best. Power grid planning goal is: to meet the load security under the premise of power supply and system operation constraints, through a series of measures to ensure that the establishment of the rack for the optimal solution [5, 6].

Power grid planning is essentially a process of dynamic multi-stage decision-making, through optimization and decision-making process to seek an optimal equipment investment plan. But during the planning period for each stage, the optimal scheme of the combination is not necessarily equal to the integrity of the optimal solution, so need to coordinate between each stage of the optimum scheme. In short, power grid planning is a large-scale combination optimization problem, in essence is to solve the problem of power equipment selection of location.

Please note that the first paragraph of a section or subsection is not indented. The first paragraphs that follows a table, figure, equation etc. does not have an indent, either.

Subsequent paragraphs, however, are indented (Fig. 1).

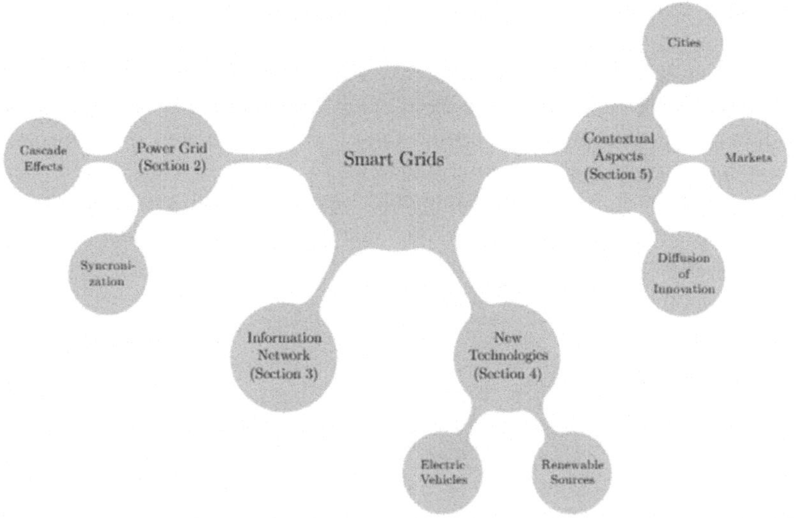

Fig. 1. A brief diagram of the grid-connected system

2.2 The Characteristics of the Power Grid Planning

(a) Multiple targets. Involved in the power grid planning goal is usually in the form of cost function in the model. Including high reliability and economy, the optimal operation and minimal environmental impact, etc.
(b) Many constraints. Through a lot of constraint conditions to achieve the goal of restrictions, reliable and economical operation of the system. Including substation capacity, voltage, power, track and corridor, the budget of the investment and other constraints.
(c) Multistage. For grid from a long-term point of view of overall layout, avoid the blindness of planning and short-sighted, power grid planning in every phases, and meter and influence between the front and back stage planning.
(d) Nonlinear. In addition to the objective function in the power system with nonlinear, most of the decision variables for the discrete values. If all the linearization, although can reduce the difficulty of the problem, can not reflect the actual problem.
(e) The uncertainty. Fundamental data such as load forecasting, power generation, investment unit price and the current and future policy, and marketization of power supply and utilization of associated, also include other unpredictable factors [4].
(f) Dynamics. Power grid planning is a kind of multistage planning, its dynamics is reflected in planning period to construction of power facilities are successive in stages, rather than a one-time completion in horizontal.
(g) Multidimensional nature. Dynamic grid planning problem involves more decision variables, the dimension refers to the state variables in the dimension of each stage. When the dimension increased, there was an exponential increase in the amount of calculation is also, namely the dimension disaster.

3 Modern Power Grid Planning Method

The modern power grid planning method is a kind of general optimization algorithm. It is an important feature of all these methods can realize the parallel computing. Due to the modern power grid planning method in solving combination optimum problem show excellent performance, in the past 20 years, it is under unprecedented attention. Many modern planning methods, commonly used method of power grid planning including simulated evolutionary method, method of swarm intelligence, artificial intelligence planning method, uncertain systems planning method, etc., the following these methods are discussed, respectively.

3.1 Simulated Evolutionary Methods

Simulated evolution method is used to imitate biological and human evolution method to solve the complicated optimization problems, mainly including genetic algorithm, evolutionary programming algorithm, evolutionary algorithm, etc.

Genetic Algorithm is the result of simulation of the laws of evolution in the nature. It is based on type theorem, introduce the evolution process of engineering problems, the degree of biological adaptation to the environment and engineering schemes of project technical and economic requirements of satisfaction with the degree of similarity, the optimization of project scheme selection and combination of (Fig. 2).

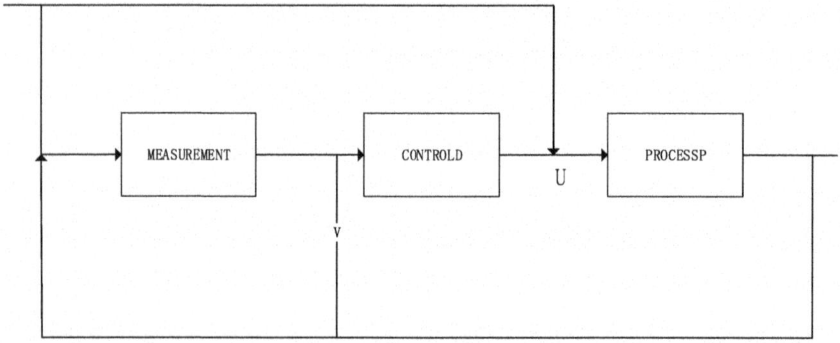

Fig. 2. An overview of the figure

Genetic algorithm in global optimization is not dependent on the gradient information, global optimization, no dimension limits, need not differentiated objective function, the advantages of implicit parallelism. Among them, the implicit parallelism, GA is on the selection and crossover mechanism. Since it has these advantages, is widely used in power grid planning [5].

3.2 Evolutionary Programming Algorithm

Evolutionary Programming algorithm with genetic algorithm [7, 8], is to simulate the evolution rule of species in the nature of a kind of method, can to larger probability search to the global optimal point.

Then it is possible to express the production rate x, in good as

$$\mathrm{x} = \gamma\tau\mathrm{p} \tag{1}$$

where

$$\gamma = \mathrm{diag}(\gamma_1, \gamma_2, \cdots \gamma_n) \tag{2}$$

and

$$\mathrm{p} = \mathrm{col}(p_1, p_2, \cdots p_n) \tag{3}$$

also

$$\mathrm{p(t)} = \mathrm{p(0)} + \int_0^t w(a)da \tag{4}$$

$$\mathrm{r(t)} = \mathrm{r(0)} + \int_0^t g(a)da \tag{5}$$

EP method in the process of operation, to determine the mutation rate is its difficulty, mutation rates determined too small or too leads to premature or trapped in local optimal algorithm, are unfavorable to EP optimization, which is also in the GA method. At present, the determination of mutation rates are generally from experience to set, not based on population size gives the theoretical value. Existing literature introduced species diversity strategy, to avoid the premature and fall into local optimum situation, understand the global optimization of space.

4 The Feasibility of Modern Power Grid Planning Method

Three issues in financial aspect should be considered: construction investment recovery, power grid economic operation situation and sustainable development.

(a) Construction investment to return, pay attention to recovery. Power grid construction investment is reasonable, whether to meet the economic operation of power grid. Grid construction investment recovery and economic operation is targeted at specific engineering projects, and sustainable development from the Angle of the power supply enterprise overall movement development to determine a certain period or a fiscal period of city network planning requirements, determine the scale and level of work in the city network planning.

Generally speaking, in the city network planning, first of all, according to the size of the sustainable development point of view to solve the long-term secondly according to the power grid economic operation situation to determine the specific planning project, according to the construction investment recovery rate to determine the specific planning of the project investment and equipment selection. It, according to the development of national economy and the people's bank interest rates to determine the theoretical calculation of the specific parameters, such as the discount rate. On this basis to research and determine the power grid, the optimal network connection mode, investment and the investment of time according to the row, the ability, power quality of power supply reliability of power supply can meet the user's demand, cost competitive with social marginal electricity price to provide customers high-quality electricity.

(b) City network planning in addition to meet the above a few goals, but also coordinated with urban planning, to meet the needs of the development of the whole city. For city network planning objects should be regional, industry, classification and analysis on time. According to the current economic development situation, should the city network planning object is divided into three categories: the old city, new town, city. Has the advantages of the old city is the molding area municipal planning, load, the municipal government of municipal planning work only for the adjustment of the locality, power supply enterprises can make full use of existing power grid and equipment at all levels to do a good job of power supply. Around the new town is the expansion of the old city area, it is in the process of promoting economic development to draw near the old villages and towns, the municipal government of municipal planning adjustment, the whole of the planning for power supply enterprises in some areas, subject to the municipal planning can be reference to the municipal planning and on the basis of load forecasting based on the city or region of the load characteristic of same type at home and abroad to do a good job of planning and power grid construction. Of villages and towns in urban area is the economy developed, can bring huge load and power consumption for power supply enterprises and the benefit, though probably not city in urban planning, this also must consider to the power supply enterprise, on the basis of load forecasting based on local features of the load or the same type at home and abroad work plan and the power grid construction.

(c) Load forecasting is the foundation of the city network planning, the planning of the long, medium and short-term scale, the quality of the specific project, planning play a key role.

Due to supply customers with good service, in recent years, for distribution network around the investment increase, thus become the main content of planning and distribution planning, stationing also should fully consider distribution substation planning, reasonable because of substation points can reduce the investment of the power supply enterprise distribution network, so as to reduce the power grid construction investment as a whole.

5 The Development Trend of Modern Power Grid Planning Method

City network planning work in the service and obey the municipal planning work at the same time, the power supply enterprise also should according to their own financial situation to come up with their own planning standards, timely communication with the municipal planning department, avoid the overemphasis on the improvement of science and technology, the municipal planning department in municipal planning for power supply enterprises put forward higher technical standard, load power supply enterprises with high investment.

For a huge power supply enterprise fixed assets, to ensure good asset-liability ratio and profit margins, must strive for the realization of the diversification of the city grid. It affected the power supply enterprise in operating performance of the year, the more affected the sustainable development of the power supply enterprise.

In the construction of urban and urban area in quasi investment especially take this into consideration. These areas on the new power supply projects often relative to the old city's power supply side wide, but load density is low, the investment benefit is not too good, so on the project of the construction of power supply enterprises in such areas like other public sectors, should strive for all levels of government investment, and to strive for all kinds of preferential policies, such as low-interest loans, tax, raise sell electricity prices, lower feed-in tariff, increase investment measures such as new users on the grid.

At present, the development of the economic potential of urban power grid greatly expansion pressure, this is completely different from western developed countries has high molding city power grid, so the power supply enterprises should face the reality, using the current state in the process of economic development for other public sector to support policy, especially in the urban area on urban power grid planning and quasi on city planning, strive for more diversified investment of urban power grid planning and construction is indispensable.

In the new economic and technological environment, under the influence of power grid planning method will be along the following direction:

(a) Hybrid or parallelism intelligent algorithm can solve more complex problems, is the main research object in the future.
(b) Multistage dynamic programming is still the difficult problems, should be combined with other technologies deal with periodic solutions of transition and the integrity of the final plan.
(c) Uncertain systems planning method can solve the uncertainty factors of the power market, the planning has flexibility and robustness, and is worth studying [6].

Under the condition of market economy, power supply enterprises in the planning and design of city network, on the premise of good social responsibility and benefit, we should plan to do a good job of economic analysis of various periods, level, make the economic point of view in a city network planning, only in this way, the benefit of power supply enterprise can be organic growth, in order to guarantee the sustainable development of the power supply enterprise, the role of planning and significance to fully emerge. Not only that, a power supply enterprise to use the inherent law of market economy, based on the actual, change passive to active, make full use of various factors, influence policy formulation department, strive for more policy is to the urban power grid planning and construction.

6 Summary

The development of industrial society, the construction of urban civilization growing demand for electricity, electricity has become the main power to promote the development of society, but the power as a kind of indispensable energy, often appear fault and dangerous situation, loss and even is the life of the people to harm. In order to reduce power's shortcomings in this aspect, the national power grid has joined the automation elements on the electric power increase power reliability, intelligence and security. China has made great progress in electric power automation, also has the very big development space in the future, China's intelligent power system would be to the next level. In the golden time of development, our related people, it is necessary to strengthen the research to make a certain contribution for the country's electric power industry development, the above this article is the application of electric power automation system in distribution network operation management research.

References

1. Cui, J., Yoon, H.: An omnidirectional biomechanical energy harvesting (OBEH) sidewalk block for a self-generative power grid in a smart city. Int. J. Precis. Eng. Manuf.-Green Technol. **5**(4), 507–517 (2018)
2. Volkov, A., Korovkin, N., Sokolova, O., Sorokin, E., Frolov, O.: Method for optimizing control actions following emergencies in large-city electric power systems. Power Technol. Eng. **45**(1), 50–52 (2011)
3. Odior, A., Omadudu, C.: Some factors responsible for erratic power supply in Benin City area of Edo state. Int. J. Syst. Manag. **4**(1), 48–56 (2013)
4. Zmieva, A.: Modeling of an industrial enterprise power supply system using direct current. Russ. Electr. Eng. **86**(5), 239–245 (2015)
5. Kuznetsov, E., Trokin, A.: Solution extensions: optimal parameter method in optimal control problems. J. Math. Sci. **199**(5), 564–570 (2014)
6. Benson, H.: On the optimal value function for certain linear programs with unbounded optimal solution sets. J. Optim. Theory Appl. **46**(1), 55–66 (1985)
7. Hayalioglu, M.: Optimum load and resistance factor design of steel space frames using genetic algorithm. Struct. Multidiscip. Optim. **23**(5), 404 (2002)
8. Lin, F., Yang, Q.: Improved genetic operator for genetic algorithm. J. Zhejiang Univ.-SCIENCE A **3**(4), 431–434 (2002)

Capsule Recurrent Neural Network with Weight Update Using Dynamic Routing by Agreement: A Unified Model for Action Recognition in Videos

Keyang Cheng[1,2](✉), Lubamba Kasangu Eric[1], Rabia Tahir[1], and Maozhen Li[3]

[1] Jiangsu University, Zhenjiang 212013, China
kycheng@ujs.edu.cn
[2] National Engineering Laboratory for Public Safety Risk Perception and Control by Big Data, Beijing 100041, China
[3] Brunei University, London UB8 3PH, UK

Abstract. The current state of the art for action recognition in computer vision field is still the Convolutional Neural Network (CNN) with its outstanding result and performances. But it lacks capability of resolving ambiguity on overlap action. A novel framework Capsule Recurrent Neural Network (Caps-RNN) is proposed in this paper that aim at achieving better accuracy in video action recognition. The proposed model is comprised of CNN, Primary Capsule and RNN. The CNN and Primary Capsule is in charge of extracting spatial feature information and the RNN with the dynamic routing is employed for temporal feature extraction and frames prediction. As the key component of the model Dynamic Routing by Agreement is used to update the weight during the training of RNN. Experiments were conducted on a subset of the UCF-101 dataset and the result reveals that our proposed model provides a competitive performance for action classification as compare to other methods.

Keywords: Capsule Network · Dynamic Routing by Agreement · Recurrent Neural Network · Action classification

1 Introduction

The technological advances are geared toward a time where systems will be able to perform better than human in almost all aspect of life, and the applications are limitless. Computer vision is among the field of Artificial Intelligence (AI) which increasingly getting more and more attention especially for actions recognition or action classification. Whether is in sport analysis, video surveillance or medical assistance and others, Human Action Recognition (HAR) is really getting more attractions. Nevertheless, it's still a challenging problem for researchers and practitioners to tackle properly. Peculiarly, in human action detection in a video which has both spatial and temporal attributes.

Many propositions have been made in the past years using deep learning techniques and methodologies, which lead us to outstanding result. Despite the achievements and

Y. Liu et al. (Eds.): ICNC-FSKD 2019, AISC 1074, pp. 307–316, 2020.
https://doi.org/10.1007/978-3-030-32456-8_33

state of the art solution obtain with using CNN, these models tend to be more complex and require more resources, time and testing before reaching an optimal solution [1–4]. Geoffrey H. in their paper [5] introduced a new method for features extraction which from the result obtain on the MNIST dataset show better performance called Capsule Network (CapNet). Contrary to CNN, capsule network enables the possibility of neuron to have more feature properties. And the learning of the features is viewpoint invariant under the routing by agreement algorithm, whereas CNN learning is viewpoint equivariant. With these capabilities, CapNet is able to achieve better state of the art result and performance for image classification, which is a requirement in case such as video surveillance or self-driving cars, where the need for accuracy is a principal requirement.

Our proposed method Caps-RNN, apply CNN and primary capsule for spatial feature extraction, followed by a RNN with dynamic routing so does the work in [6–8]. The first part of the model (CNN + Primary Capsule) captures spatial and pixel-wise localization features. Then following RNN + Dynamic Routing (DR) parts will capture temporal feature and update weight using DR. Furthermore, the architecture is being trained end-to-end by decreasing the complexity of the model and distinguish action at a frame level sequence. To the best of our knowledge, this is the first work to combine capsule network and recurrent neural network by dynamically routing the weight update for video action classification.

2 Related Works

In practice CNN models have issues with propagating the error or cost function through temporal range intervals, hence Long Short Term Memory [9] (LSTM) was created as an alternative that use non-linear mechanism of gating (input gate, forget gate, output gate). Many CNN architecture variant have been proposed in past years, researchers in [7] used two stream CNN with multiple optical flow, others have used spatio-temporal tubes [10, 11] and in [12] the authors have applied three dimensional CNN. In the case of 3D CNN, a more recent paper proposed in [13] also showed how we can apply 3D CNN for video action classification, where the authors leverage the use of a pretrained model on ImageNet dataset and convert the functionality of a 2D CNN to 3D CNN.

In fact, Geoffrey H. and his collaborators in their paper [14] introduced the capsule as vector and a new algorithm of routing by agreement that consist of passing capsule from one layer to another using a voting routine. Capsule network improves the feature extraction, as the vector contains more information than scalar neuron. In their work, the authors applied these capsules on 28*28*1 images. In our experiment the frame size is equal to 128*128*3, which makes the computation very expensive. To resolve the problem mean voting procedure is used as in [15].

As stated in previous paragraph, capsule network is a group of nested neurons instead of linear sequenced neurons, and, these neurons are characterized by various parameters such as pose or length of vector. In addition, Capsule Network replace pooling layers of CNN with a new technique called routing by agreement, where information is passed from lower layer to particular parent layer above meeting similar characteristics, and these operation between layer do not share the same coupling coefficient like for CNN. As in the original paper on capsule network, we consider a capsule i and an output u_i where the computation of its parent j is describe as:

$$\hat{u}_{j|i} = W_{ij}u_i \tag{1}$$

where the vector prediction of the capsule j^{th} is $\hat{u}_{j|i}$ computed by lower capsule i and the weight matrix describe as W_{ij}. And using a softmax function the coefficients is obtained by Eq. (2) as follow:

$$c_{ij} = \frac{\exp(b_{ij})}{\sum_k exp(b_{ik})} \tag{2}$$

where the probability of coupling the capsule i and j is describe by b_{ij} which is always initialize at zero.

The routing by agreement method replaced the Convolutional Neural Network scalar feature detectors output with vector feature detector output, as well as replacing the operation of max-pooling with the routing by agreement method.

$$S_j = \sum_i c_{ij}\hat{u}_{j|i} \tag{3}$$

With S_j to be equal to the summation of matrix of prediction vector $\hat{u}_{j|i}$ and the coefficient of coupling determined by the routing process c_{ij}.

As the output vector probability length vary from case to case, the Capsule Network introduce another operation called Squashing which purpose is to shrink the output vector to always be lower than 1. And the method is described below with S_j as the total input and v_j the capsule j vector output as:

$$v_j = \frac{\|S_j\|^2}{1 + \|S_j\|^2}\frac{S_j}{\|S_j\|} \tag{4}$$

Moreover, the agreement value is obtained by calculating the product of the output vector and the prediction of the output as in Eq. (11).

$$a_{ij} = v_j\hat{u}_{j|i} \tag{5}$$

3 Proposed Model

Below is the overall structure of our proposed model in Fig. 1, consisting of mainly two blocks where the first block comprise of the CNN and the Primary Capsule will be in charge of extracting Spatial Feature information; the second part an RNN with the dynamic routing properties for temporal feature extraction and frames prediction. All these block and part will be explained in detail in below paragraphs.

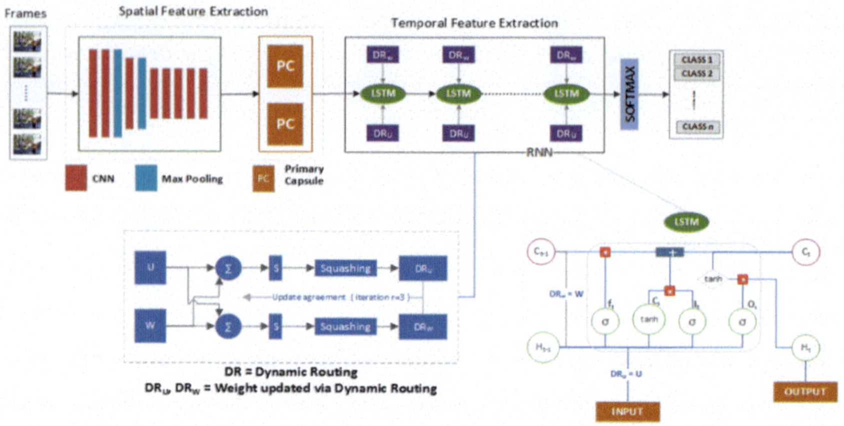

Fig. 1. Overview Structure of the model. Spatial Feature Extraction (CNN = Convolutional Neural Network, PC = Primary Capsule Model); Temporal Feature Extraction (LSTM = Long Short-Term Memory; Ct-1 = Previous Cell Memory; Ht-1 = Previous Cell Output; Ct = Current Cell Memory; Ht = vt = Current Cell Output; DRt = U = Current Weight for Element wise addition; DRw = W = Previous Weight for Element wise multiplication)

As in a matter summary, DR steps weights update can be summarized as: for the first block of LSTM both U and W are randomly determined and provide to the routing. Following process is described in paragraphs below. The value obtains for DR_U and DR_W are considered as the new value of weight for the LSTM block and these weights will be shared through the all RNN block. But, for the next epoch using previous technic the newly weight obtained are compared to the previous weights in memory. Let consider $(DR_U)'$ as saved weight. If $DR_U < (DR_U)'$, DR_U is considered as the new update, otherwise $(DR_U)'$ will be kept. The update process will continue unless the loss function is reduced close to zero.

3.1 Spatial Feature Extraction

At this stage, Convolutional Neural Network is employed as the spatial features extraction with input of size 128*128*3, which speed up the process of features extractions and enable us to get an output of size of 20*20*256 feature map. This output will be the input for the next step that is the primary capsule.

Just like in the original paper [14], a primary capsule (PC) block on the output from previous step (20*20*256) is applied and obtains new feature of size 6*6*8*32. The primary block will be comprised of 3 primary capsule block and each block is described as 32 sub-blocks with the width, height and depth by 6*6*8.

3.2 Temporal Feature Extraction

The output of the primary capsules $\hat{u}_{j|i}$ is passed to a Recurrent Neural Network as input. But instead of calculating directly the Softmax at the output of the RNN, the input is passed to v_j for dynamically prediction of the next frames via a cross-entropy function Eq. (6) first, then a Softmax is applied based on the output v_j.

The routing iteration r is equal to 3, because there is a chance of an overfitting, an additional squashing function is applied followed by a cross-entropy function to determine the next frames as in [16] described by the Eq. (6) as follow:

$$L = -(y \log(\hat{y}) + (1 - y) \log(1 - \hat{y})) \tag{6}$$

$$\hat{y} = \frac{1}{m} \sum_i softmax \left(\begin{bmatrix} v_1^T \\ v_2^T \\ \vdots \\ v_i^T \end{bmatrix} \right) \tag{7}$$

With the label ground truth y and the predicted frames label express by $\hat{y}$ in Eq. (7), m describe the output capsule v_j dimension number. Weights of RNN expressed in Eqs. (8–11) are determined or updated using the dynamic routing by agreement (Table 1).

$$f_t = \sigma\left(X_t * DR_{U_f} + H_{t-1} * DR_{W_f}\right) \tag{8}$$

$$C_t = \tanh(X_t * DR_{U_c} + H_{t-1} * DR_{W_c}) \tag{9}$$

$$I_t = \sigma(X_t * DR_{U_i} + H_{t-1} * DR_{W_i}) \tag{10}$$

$$O_t = \sigma(X_t * DR_{U_o} + H_{t-1} * DR_{W_o}) \tag{11}$$

where meanings of symbols in Eq. (8–11) are as follows:

σ: *sigmoid function*
tanh: *hyperbolic tangent*
i_t: *input gate*
f_t: *forget gate*
o_t: *output gate*
h_t: *hidden unit*
g_t: *input modulation gate*
c_t: *memory cell*
DR_W: *Saved weight (weight in memory)*
DR_U: *Current Weight*
X_t: *Capsule Features or input features*

Previous activation function a_{ij} enable to easily navigate through capsule parts and detect to which parts belongs each entity. Therefore, the model can detect distinctively whether the action is an open door or closing door for instance.

Table 1. Dynamic routing algorithm functionalities that enable the update of weight

DYNAMIC ROUTING ALGORITHM
procedure Routing $(u_{j\|i}, DR_U, r)$
$\widehat{DR}_U \leftarrow DR_W + rand(size(DR_U))$
$\hat{u}_{(j\|i)} \leftarrow \widehat{DR}_{Ui}\ squash(u_{j\|i})$
for all input i and all predicted output j **do**
$b_{ij} \leftarrow 0$
for r iterations **do**
for all input i **do** $c_i \leftarrow softmax(b_i)$
for all input j **do** $s_j \leftarrow \sum_i c_{ij}\hat{u}_{j\|i}$
for all input j **do** $v_j \leftarrow squash(s_i)$
for all input i and output j **do**
$b_{ij} \leftarrow b_{ij} + \hat{u}_{j\|i} . v_j$
return v_j

Initially for every predicted output a raw routing weight is set in the testing:

$$b_{ij} = 0 \tag{12}$$

or the LSTM block, where $DR_U = b_i = U$ and $DR_W = b_j = W$. And a Softmax function on each previous weight capsule or primary capsule b_i is used that give us:

$$c_i = softmax(b_i) \tag{13}$$

which gives the routing weight for each predicted output. Later a weighted sum s_j is computed from Eq. (3) of the prediction for each capsule in the next layer. As it is summation of weights, there is a higher chance of getting a value greater than 1. Hence, another squashing function is employed to reduce the dimensionality of the weighted sum by integrating Eq. (3) into Eq. (14):

$$v_j = squash(s_j) \tag{14}$$

This becomes the weight after 1 iteration and steps are repeat of Eq. (14) by applying a scalar product of the predicted output $\hat{u}_{j|i}$ and v_j that will be added to the initial weight b_{ij}.

$$b_{ij} += \hat{u}_{j|i} v_j \tag{15}$$

Good prediction will have higher weight, so that the predicted action or activity will have higher weight compare to other capsules. During back propagation the weight of the predicted output is kept and the other input weight are initialized again to zero from Eq. (12).

So, the weight DR_w and DR_U are initialized to be zero after routing the model update both weights. But on the next propagation, DR_w will memorize previous weight and DR_U will be initialized again to be zero and follow again the same process of dynamically being routed.

4 Experiments

Xifeng Guo[1] is our base for the implementation and python as our language programming. Some difficulties occur in Capsule Network such as the number of required parameters, as in the case of [14] the first layer has approximately 1,474,560 parameters corresponding to an input of size 28*28*1 for grayscale image whereas our input are of size 128*128*3. This indicate one of the reasons to reduce as much as possible our feature maps before supplying it to the primary capsule. Therefore, the expectation of the network is expensive network in terms of parameters for video as the size are often bigger, as for our work the image size is of 128*128*3. In fact, to resolve the memory problem in this work for each capsule receptive field only applying the mean transformation, also, routing lower layer to only defined higher spatially-local kernel. In addition, as the objective of our work is more or less modest the Capsule Network are not scale.

Moreover, capsule of same type will be sharing transformation matrix, their vote change will be disabled, and these capsules will model a variant position of same entity via capsule pooling as in [17] for each receptive field while averaging the capsule.

For the experiment 8 random classes from the UCF-101[2] dataset as our foundation were chosen (Fig. 2). The UCF-101 dataset consists of 101 different action or classes ranging from indoor to outdoor activities. Each of the class contain over 150 videos. With a total set of over 1200 videos for experimentation, 70% was considered for training and 30% for validation.

Fig. 2. Subset samples of the UCF-101 chosen for the experiments. From top-let to bottom-right: Drumming, Jumping Jack, Playing Guitar, Punch, Sky Diving, Soccer Penalty, Surfing, Tennis Swing

[1] https://github.com/XifengGuo/CapsNet-Keras.

[2] crcv.ucf.edu/data/UCF101.php.

Every class got split at around 120 videos for training and 30 videos for validation. The chosen activities are: Drumming, Jumping Jack, Playing Guitar, Punching, Sky Diving, Soccer Penalty, Surfing and Tennis Swinging. In addition, for each video 30 frames were extracted nearly 3 frames per second. Other settings consisted of a batch size of 32, Adam as the optimizer with a learning rate of 0.00001. All frames were cropped and resized to a set of 128*128 width and pixel size during the process of training.

The output of the RNN block is passed to the cross-entropy function in Eq. (6), with the predicted output class described by the Eq. (7). An important point to notice is H_t or the current cell output is equal to v_j. Although, the model still has room of improvement, the testing of feasibility outcome demonstrates that once it is fine tuned the performance will increase, which will be the next aim step in our research.

Another important key element of our model is the capability of resolving ambiguity on overlap action and distinctively consider the best output possible. As shown in below figure (Fig. 3), the model can distinguish a jumping jack action on from a tennis swing with no difficulties as the model applied capsule network methodology of splitting the input into parts and sending those part to the capsule that only routed each part of the input to the predicted output of higher probability in the next layer. Here, the RNN weight layer are updated using Dynamic Routing in accordance with the position and orientation of the actor hand.

Fig. 3. Similar actions which are easily misclassified: sequences frames in top line belongs to tennis swing and Sequences frames in bottom line belongs to jumping jack.

For each iteration, the weights are updated, the Dynamic routing and the squashing operation enable the model to get weight that are less than one and greater than 0. On the next iteration the model retrieves previous weights DR_U and DR_w and inserts them in the routing again hence they become U and W. In fact, for each iteration the weights are compared with saved memory weights and will be augmented or reduced depending on the loss function. The evaluation of the model over the testing set consisting of 30% of the dataset, resulted in an accuracy calculated using the probability metrics curve via: $\frac{TP+TN}{TP+TN+FP+FN}$ that lead to an average of nearly 95% accuracy (Table 2).

Table 2. Accuracy comparison based on the UCF-101 Dataset

Model	Accuracy on UCF-101
Wang et al. [17]	91.4%
Wu et al. [20]	92.6%
Diba et al. [18]	93.2%
Feichtenhofer et al. [21]	94.2%
Feichtenhofer et al. [19]	94.6%
Ours	**94.2%**

Based on the result shown in above table made us conclude that the proposed model performs better than some other methods. In fact, the proposed model not only has achieved a high performance of accuracy but also has good characteristics of interpretability and robustness. Nevertheless, the proposed solution still facing training challenges and errors. The results encourage us to further research and fine-tune some parameters and aim for even better outcome.

5 Conclusion

In this work, a comprised of CNN, Primary Capsule and RNN is proposed for video action classification using new algorithm of routing by agreement for weight update. To the best of our extent, this is the first time a proposition of applying Capsule Network Routing algorithm with an RNN is being made. Compared with other methods, the proposed model has capability of resolving ambiguity on overlap action. Besides, the model will be subject to further research, and it safe for us to acknowledge that more fine tuning is require to make the model better.

Acknowledgements. This research is supported by National Natural Science Foundation of China (61972183, 61602215) and the Director Foundation Project of National Engineering Laboratory for Public Safety Risk Perception and Control by Big Data(PSRPC).

References

1. Karpathy, A., Toderici, G., Shetty, S., Leung, T., Sukthankar, R., Fei-Fei, L.: Large-scale video classification with convolutional neural networks. In: 2014 IEEE Conference on Computer Vision and Pattern Recognition, pp. 1725–1732 (2014)
2. Simonyan, K., Zisserman, A.: Very deep convolutional networks for large-scale image recognition. CoRR, vol. abs/1409.1 (2014)
3. Kalogeiton, V., Weinzaepfel, P., Ferrari, V., Schmid, C.: Action tubelet detector for spatio-temporal action localization. In: Proceedings of the IEEE International Conference on Computer Vision (2017)
4. Herath, S., Harandi, M., Porikli, F.: Going deeper into action recognition: a survey. Image Vis. Comput. **60**, 4–21 (2017)
5. Sabour, S., Frosst, N., Hinton, G.: Matrix capsules with EM routing. In: ICLR 2018 (2018)

6. Donahue, J., et al.: Long-term recurrent convolutional networks for visual recognition and description. IEEE Trans. Pattern Anal. Mach. Intell. **39**, 677–691 (2017)
7. Girshick, R., et al.: Two-stream convolutional networks for action recognition in videos. arXiv Preprint arXiv:1406.2199 (2016)
8. Gkioxari, G., Girshick, R., Malik, J.: Contextual action recognition with R*CNN. In: 2015 IEEE International Conference on Computer Vision (ICCV), pp. 1080–1088 (2015)
9. Hochreiter, S., Urgen Schmidhuber, J.: Long short-term memory. Neural Comput. **9**, 1735–1780 (1997)
10. Peng, X., Schmid, C.: Multi-region two-stream R-CNN for action detection. In: Lecture Notes in Computer Science (including subseries Lecture Notes in Artificial Intelligence and Lecture Notes in Bioinformatics) (2016)
11. Yang, Z., Gao, J., Nevatia, R.: Spatio-temporal action detection with cascade proposal and location anticipation. CoRR, vol. abs/1708.0 (2017)
12. Tran, D., Bourdev, L., Fergus, R., Torresani, L., Paluri, M.: Learning spatiotemporal features with 3D convolutional networks. In: 2015 IEEE International Conference on Computer Vision (ICCV) (2015)
13. Carreira, J., Zisserman, A.: Quo Vadis, action recognition? A new model and the kinetics dataset. In: 2017 IEEE Conference on Computer Vision and Pattern Recognition (CVPR) (2017)
14. Sabour, S., Frosst, N., Hinton, G.E.: Dynamic routing between capsules. CoRR, vol. abs/1710.0 (2017)
15. Duarte, K., Rawat, Y.S., Shah, M.: VideoCapsuleNet: a simplified network for action detection. CoRR, vol. abs/1805.0 (2018)
16. Nguyen, H.H., Yamagishi, J., Echizen, I.: Capsule-forensics: using capsule networks to detect forged images and videos, October 2018
17. Wang, L., et al.: Temporal segment networks: towards good practices for deep action recognition, August 2016
18. Diba, A., et al.: Temporal 3D ConvNets: new architecture and transfer learning for video classification, November 2017
19. Feichtenhofer, C., Pinz, A., Wildes, R.P.: Spatiotemporal residual networks for video action recognition. In: Conference on Neural Information Processing Systems (2016)
20. Wu, Z., Jiang, Y.-G., Wang, X., Ye, H., Xue, X., Wang, J.: Fusing multi-stream deep networks for video classification, September 2015
21. Feichtenhofer, C., Pinz, A., Wildes, R.P.: Spatiotemporal multiplier networks for video action recognition. In: Proceedings - 30th IEEE Conference on Computer Vision and Pattern Recognition, CVPR 2017 (2017)

A High-Dimensional and Multi-granularity Feature Selection Method Based on CNN and RF

Yinghong Sun, Lei Liu(✉), Sheng Chen, and Liangwen Hou

College of Applied Sciences, Beijing University of Technology, Beijing, China
liuliu_leilei@bjut.edu.cn

Abstract. Feature engineering determines the upper limit of the performance of machine learning algorithm. And feature selection is the most critical step in feature engineering. However, the dimensional disasters are caused by high-dimensional and multi-granularity feature data, which makes effective feature selection very difficult. We propose a feature selection based on the Convolutional Neural Networks and Random Forest (FSCNNRF) for this issue. The model includes two parts, Feature Selection Convolutional Neural Networks (FSCNN) and Random Forest (RF). It can select more effective feature set by using FSCNN for dimensionality reduction and RF for feature selection. Firstly, the high-dimensional and multi-granularity feature data are subjected to dimensionality reduction processing by FSCNN, so that each feature becomes a single granularity feature. Then the RF is used to select valid features. Experiments show that the model has better effect on feature selection on high-dimensional and multi-granularity dataset and improves the performance of machine learning algorithms.

Keywords: Feature engineering · Feature selection · CNN · RF

1 Introduction

With the rapid development of the Internet, the various forms of data are exploding, such as video, image, audio, text and so on. An effective preprocessing method is urgently needed to convert these data into available data for machine learning. Feature engineering is a process of transforming raw data into the features that describe these data well. The experts in the field of machine learning believe that data and feature engineering determine the upper limit of machine learning algorithms, and the final model simply approximates this upper limit by a linear or non-linear way. Feature engineering mainly includes feature construction, feature extraction and feature selection.

Feature construction refers to manually identifying some meaningful features from the original data. Feature extraction is to automatically reduce the dimension of the original observation data, so that its feature set can be better modeled.

The feature selection is to select a subset of features with obvious physical or statistical significance from the feature set, and it can effectively eliminate irrelevant or

Y. Liu et al. (Eds.): ICNC-FSKD 2019, AISC 1074, pp. 317–325, 2020.
https://doi.org/10.1007/978-3-030-32456-8_34

redundant features, reduce the number of effective features and the time of model training, and improve the accuracy of the model. Each feature selection requires a model to verify, so that better features are selected to improve the performance of the model.

In this paper, we propose a feature selection method based on CNN and RF. This method combines feature extraction with feature selection, which can be divided into two stages. Firstly, the FSCNN model proposed reduces each feature from a multi-granularity dimension to a single granularity dimension, which is equivalent to feature extraction for each multi-granularity dimension feature through FSCNN. Secondly, we evaluate the importance of the feature by using RF and select the effective feature.

2 Related Work

The purpose of feature extraction and feature selection is to find the most effective feature from the original feature. Both of them can reduce the data dimension and redundancy and the training time, and improve the accuracy of the model. There are many methods for feature extraction and feature construction.

In feature extraction, Karl Pearson proposed (Principal Component Analysis) PCA, which is an unsupervised dimensionality reduction method [1–3]. The method transforms a set of features that may be correlated into a set of linear uncorrelated features by orthogonal transformation. Fisher's (Linear Discriminant Analysis) LDA is a supervised dimensionality reduction method [4, 5]. In this method, the data are projected in low-dimensional space.

Different from PCA and LDA, Josh Tenenbaum proposed a manifold learning, which assumes that data are uniformly sampled in a low-dimensional manifold in a high-dimensional Euclidean space [6]. The manifold learning is to recover the low-dimensional manifold structure from the high-dimensional sampled data. This method has good effect in dimensionality reduction.

In feature selection, it is mainly divided into Filter, Wrapper and Embedded. The Filter is to calculate the correlation between each feature and the label, and then select several features with greater correlation. In addition, Kira et al. proposed Relief, which designed a correlation statistic to measure the importance of features [7]. Different from the Filter, which does not take into account subsequent learners, the Wrapper directly takes the performance of the learner to be used as the evaluation criterion of feature subset. Liu et al. proposed Las Vegas Wrapper (LVW), which uses random strategies to search subsets within the framework of Las Vegas method and takes the error of the final classifier as the evaluation criterion of the subset [8, 9]. The method of Embedded combines the feature selection process with the learner training process, both in the same optimization process. Tibshirani et al. used L1 regularization to select features [10]. The method of feature selection based on tree model is also a commonly used Embedded [11]. In the feature selection method of tree model, the RF algorithm uses the error of out-of-bag to estimate the importance of the feature, and does not require an extra validation set.

With the mature application of deep learning in the field of image processing, deep learning also has some applications in feature extraction. Different from traditional pattern recognition methods, deep learning automatically learns features from large data rather than using manual construction. Li W J et al. proposed a double labels deep supervised hashing algorithm based on feature learning [12]. Wang H et al. used the AlexNet model to extract features from sogou image database and creating an index [13]. Huang et al. used multiple CNN models as feature extraction models to obtain multiple features and then integrated them for image retrieval [14]. Li and Liu used the convolutional neural network model to extract features of lower dimensions and optimize the retrieval process [15, 16].

In summary, The CNN is more similar to a process of feature extraction in the whole training. Therefore, aiming at the characteristics of high-dimensional and multi-granularity feature data, we first use the FSCNN model to make a dimensionality reduction for each feature, so that each feature becomes a single granularity dimension. Then the FSCNN combine the RF to select effective feature.

3 Our Method

Combining the advantages of deep learning in feature extraction and RF algorithm in feature selection, we describe our proposed FSCNNRF in this section. The whole model framework is shown in Fig. 1.

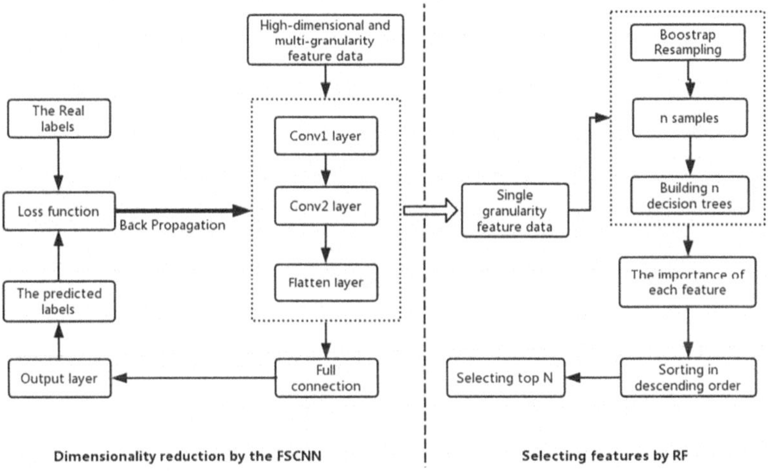

Fig. 1. The framework of the FSCNNRF

In Fig. 1, the left side of the dashed line is the flow chart of the FSCNN, and the right side is the flow chart of the RF. The Flatten layer is optimized by training the left FSCNN model and extracting the layer data as an input to the right RF. Then, RF is used to evaluate the importance of features, and the top *N* features with higher importance are selected.

The rest of the section is organized as follows. In Sect. 3.1, we introduce high-dimensional and multi-granularity feature data and corresponding representations. In

Sect. 3.2, we introduce our proposed FSCNN model. Finally, we introduce the FSCNNRF model and make feature selection in Sect. 3.3.

3.1 The Representation of High-Dimensional and Multi-Granularity Feature Data

Assuming that the original dataset can be denoted $D = \{X_1, X_2, \cdots, X_n\}$, where $X_i = (x_1, x_2, \cdots, x_m) \in R^m$ is a data that includes m features. Each feature $x_j \in R^l$ includes $l(l > 1)$ granularities, and can be denoted $x_j = (x_{j1}, x_{j2}, \cdots, x_{jl})$. Therefore, each data point $X \in D$ can be represented as matrix A. Each row of matrix A represents a feature of data point X, the number of columns of matrix A denote that the feature is distributed in l different dimensions. Data features in the form above are called multi-granularity features. This data A is called high-dimensional and multi-granularity data.

3.2 FSCNN Model

We propose a feature extraction model FSCNN based on the characteristics of convolution operations in the CNN. Its model structure is shown in Fig. 2 below.

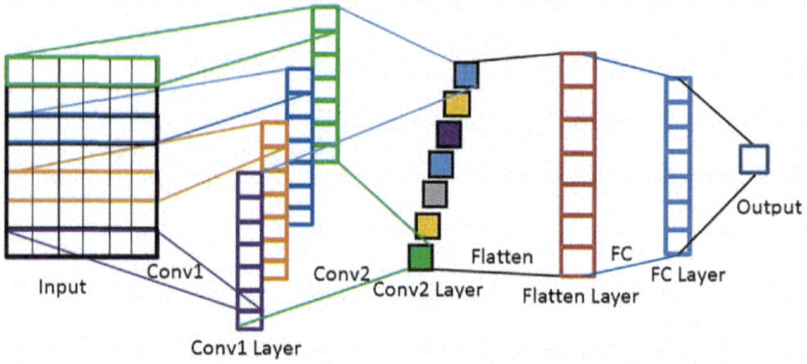

Fig. 2. The diagram of the FSCNN

In the construction of the FSCNN model, we mainly use the convolution operation of the CNN. Firstly, we make full use of the characteristics of 1D convolution to perform granularity reduction for each feature of the original data. It is significant that the 1D convolution can make the relative position of each feature unchanged. Secondly, the 1×1 convolution kernel we used not only effectively reduces the parameters of the model, but also fuses the features on the different channels of the previous layer. In the process of training, let X_i as input of the FSCNN.

3.3 FSCNNRF Model

Because the model of FSCNN cannot give the importance of each feature, we visualize the importance of each feature through a RF. And we introduce FSCNNRF model through the following three parts. The first section describes the process of RF generating. Finally, we present the assessment the of feature importance.

Building Random Forest According to New Dataset from the FSCNN. Firstly, each feature can be converted from a multi-granularity dimension to a single granularity dimension by training FSCNN. The dimensions of each of the original features $l = 1$ at this time.

Secondly, we extract the data D' of the Flatten layer of the trained FSCNN model, where each data $X'_i \in D'$ can be denoted:

$$X'_i = Flatten(\Phi^i_{21}, \Phi^i_{22}, \cdots, \Phi^i_{2m}) \in R^m. \tag{1}$$

Finally, using dataset D' as the input to the RF and building RF model.

Feature Importance Assessment. In the process of building a RF, firstly, the corresponding Out of Bag (OOB) data is used to calculate its out-of-bag data error errOOB1. The out-of-bag data refers to the data that has not been sampled by bootstrap sampling during the process of building RF. Secondly, the noise interference is randomly added to the feature X of all the OOB data, that is, randomly changing the value of the sample at feature X, then again calculating its out-of-bag data error errOOB2. Finally, the importance of feature X can be estimated by repeatedly updating the calculations. Assuming that there are n trees in the RF, and then the importance of the features can be calculating by the formula that is as follows:

$$importance(X) = \sum_{i=1}^{n} |OOB(i+1) - OOB(i)|, \tag{2}$$

where $OOB(i+1)$ is errOOB2, $OOB(i)$ is errOOB1. The out-of-bag data error by $\frac{|O_e|}{|O|}$, where $|O_e|$ is the number of wrongly categorized data, $|O|$ is the number of correctly classified data [17, 18]. The reason why formula (2) is used as a measure of the importance of the feature is that if the accuracy of the bag is greatly reduced after random addition of noise to a feature, it indicates that this feature has a great influence on the classification result of the sample. That is to say, its importance is relatively high.

Therefore, we can get the importance of m features by training RF and extracting the top N required features after sorting them in descending order, where N is a number of valid features that you want to select.

4 Experiments

4.1 Dataset and Experiment Preparation

The dataset selected is the data of the 8th APMCM[1], which is a dataset of the videos describing the physical balance of the elderly. Then we use whether the old man fell as a 0–1 label. Among them, each old person corresponds to 150 frames of data, and we use each frame as a data to construct a dataset with a size of 5000. Each of dataset has

[1] https://www.saikr.com/apmcm/2018

42 features and each feature has three granularity dimensions. We used 32 convolution kernels of size 1 × 3 in the first convolution layer. We used a convolution kernel of size 1 × 1 in the second convolution layer. The evaluation indicators we used during the experiment included Acc(Accuracy), F1(F1_score), P(Precision), and AUC values.

4.2 Comparative Analysis of Experiments

The FSCNN model is equivalent to reducing the dimension of each feature during training FSCNN model, so we compares it with PCA, LDA. However, since we want to evaluate the effect of dimensionality reduction, we use SVM and CNN as the classification learner model after PCA and LDA dimensionality reduction. The experimental results are shown in Table 1 below.

Table 1. Performance of the different methods

Model	Acc	P	F1	AUC
PCA+CNN	0.86	0.88	0.72	0.45
PCA+SVM	0.81	0.83	0.69	0.40
LDA+CNN	0.90	0.89	0.74	0.51
LDA+SVM	0.85	0.84	0.70	0.46
FSCNN	**0.94**	**0.92**	**0.81**	**0.57**

As can be seen from Table 1 above, the model proposed by us achieves a better effect (shown in bold in Table 1). The combination of PCA and LDA with CNN is significantly better than the combination of the two with SVM (Support Vector Machine). In addition, our model combines the feature extraction process with the learner training process, which is done in the same optimization process, fastly and simply. After the PCA and LDA models reduce the dimension of each feature, it is necessary to pass a classification model to verify the feature data after dimensionality reduction, which the process is too complex.

Since the importance of each feature is not known after each dimensionality reduction by FSCNN, we use RF based FSCNN processed data to visualize its features and output the importance of each feature. Then we can automatically select valid features based on the importance of each feature. The combination of PCA and LDA with RF as our comparison model to illustrate the effectiveness of our proposed FSCNNRF model. The experimental results are shown in Table 2 below.

Table 2. Performance of the different methods of feature selection

Model	Acc	P	F1	AUC
PCA+RF	0.83	0.80	0.68	0.39
LDA+RF	0.86	0.85	0.70	0.44
FSCNNRF	0.91	0.89	0.76	0.50

As can be seen from Table 2 above, the proposed model FSCNNRF has a better effect than PCA+RF and LDA+RF. Because the neural network itself has a good fitting effect, the classification result of the FSCNN model is slightly higher than that of the FSCNNRF model.

4.3 Feature Validity Analysis

We use FSCNNRF model to get the relative importance of each feature and arrange it by reverse order of the feature importance. The top 25 features we selected and their importance are shown in Fig. 3.

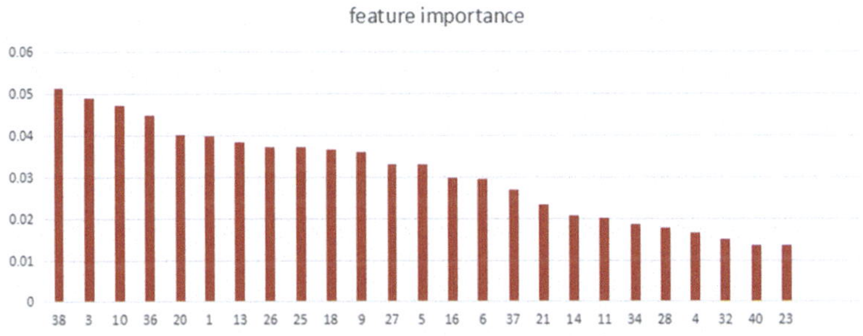

Fig. 3. The feature importance of the old body

In Fig. 3, the horizontal axis is the monitoring point of each part of the body of the elderly, and the vertical axis is the importance corresponding to these monitoring points. Based on the questions given, we selected the top $N = 25$ monitoring point characteristics from all the features as a part of balance indicator for assessing the elderly. Combined with the structure of elderly people, we can see that the points we selected are mainly located in the arms, knees, lower limbs, quadriceps and head. The amplitude of the swing of the arm when walking can determine the degree of walking of the elderly people, and the uneven swing will influence the steady state of walking. Secondly, injuries such as knee joints, lower limb muscle strength, quadriceps and head, or muscle loss can seriously affect the balance of the elderly people.

5 Conclusions and Future Work

In this paper, our proposed model effectively combines the advantages of CNN and RF for the feature selection of high-dimensional and multi-granularity feature data. According to the importance of the all features, we can select the effective features needed for the problem, and then remove those redundant features to avoid the dimensional disaster problem.

In future work, aiming at this type of dataset, firstly, we will exaggerate the amount of data to train the model and improve our model FSCNN. Secondly, we will improve the model with the model of AutoEncoder [19–21], thus improving the effectiveness of feature selection and the performance of machine learning. In addition, we will try to combine 3D convolution with RF to evaluate the validity of this feature.

Acknowledgment. This work is supported by the National Natural Science Foundation of China (Grant No. 61105040, 61203284), the Beijing Natural Science Foundation (Grant No 4133085), the general program of science and technology development project of Beijing Municipal Education Commission (Grant KM201810005005), the Beijing municipal commission of education young top-notch personnel plan and the Beijing University of Technology Science Foundation (Grant No. 006000543115502).

References

1. Zhou, Z.: Machine Learning, pp. 229–230. Tsinghua University Press, Beijing (2016)
2. Yu, L., et al.: Multi-response parameters optimization based on PCA and neural network. J. Syst. Simul. 176–183 (2018)
3. Yi, M.: Research on Infrared Feature Authentication and Counterfeiting Algorithm Based on PCA + SVM Paper Currency. University of Science and Technology Liaoling (2018)
4. Zhou, Z.: Machine Learning, pp. 60–63. Tsinghua University Press, Beijing (2016)
5. Han, Z., et al.: Gait recognition based on linear discriminant analysis and support vector machine. Pattern Recogn. Artif. Intell. **18**(2) (2005)
6. Zhou, Z.: Machine Learning, pp. 234–237. Tsinghua University Press, Beijing (2016)
7. Kira, K., Renddell, L.A.: Wrappers for feature selection problem: traditional methods and a new algorithm. In: Proceedings of the 10th National Conference on Artificial Intelligence (AAAI), pp. 129–134 (1992)
8. Zhou, Z.: Machine Learning, pp. 248–253. Tsinghua University Press, Beijing (2016)
9. Liu, H., Setiono, R.: Feature selection and classification a probabilistic wrapper approach. In: Proceedings of the 9th International Conference on Industrial and Engineering Applications of Artificial Intelligence and Expert Systems (IEA/AIE), pp. 419–424 (1996)
10. Tibshirani, R., et al.: Sparsity and smoothness via the fused LASSO. J. R. Stat. Soc.-Ser. B **67**(1), 91–108 (2005)
11. CSDN. https://blog.csdn.net/qq_26598445/article/details/8428790. Accessed 05 Sept 2018
12. Li, W.J., Wang, S., Kang, W.C.: Feature learning based deep supervised hashing with pairwise labels (2015)
13. Wang, H., Cai, Y., Zhang, Y., et al.: Deep learning for image retrieval: what works and what doesn't. In: IEEE International Conference on Data Mining Workshop, pp. 1576–1583. IEEE (2015)
14. Huang, H.-K., Chiu, C.-F., Kuo, C.-H., Wu, Y.-C., Chu, N.N.Y, Chang, P.-C.: Mixture of deep CNN-based ensemble model for image retrieval. In: IEEE 5th Global Conference on Consumer Electronics (2016)
15. Li, J.Y., Li, J.H.: Fast image search with deep convolutional neural networks and efficient hashing codes. In: International Conference on Fuzzy Systems and Knowledge Discovery, pp. 1285–1290. IEEE (2015)
16. Liu, H., Wang, R., Shan, S., et al.: Deep supervised hashing for fast image retrieval. In: IEEE Conference on Computer Vision and Pattern Recognition, pp. 2064–2072. IEEE (2016)
17. Breiman, L.: Bagging predictors. Mach. Learn. **24**(2), 123–140 (2001)
18. Genuer, R., et al.: Variable selection using random forests. Pattern Recogn. Lett. **31**(14), 2225–2236 (2010)

19. Deng, J., Zhang, Z., Marchi, E., et al.: Sparse autoencoder-based feature transfer learning for speech emotion recognition. In: Affective Computing & Intelligent Interaction (2013)
20. Liu, J., Li, C., Yang, W.: Supervised learning via unsupervised sparse autoencoder. IEEE Access **PP**(99), 1 (2018)
21. Wang, Y., et al.: Stacked sparse autoencoder with PCA and SVM for data-based line trip fault diagnosis in power systems. Neural Comput. Appl. (2018)

Structural Nonlinear Damage Detection Based on Time Series Model and Probability Theory

Huiyong Guo(✉), Feng Zhang, and Jinjun Cheng

School of Civil Engineering, Chongqing University, Chongqing 400045, People's Republic of China
ghy267@hotmail.com

Abstract. Damage of practical engineering structure usually possesses nonlinear feature and the results of damage diagnosis are also usually disturbed by different kinds of uncertainty, such as measurement noise and model error. Nonlinear and uncertain characteristics common in measurement signals bring challenges to damage detection. A novel method combined the probability theory with hybrid AR/ARCH model is proposed to detect structural nonlinear damage that couples with measurement uncertainty and structure uncertainty. The AR/ARCH model can fit structural acceleration response time series to effectively extract the nonlinear feature and the calculation probability of damage existence. Probability theory can deal with uncertainty caused by measurement noise or model errors. Therefore, the method based on AR/ARCH model and probability theory can improve the reliability of damage detection compared with deterministic method under different uncertainties. The results of a simulated 5-storey shear structure show the excellent performance to locate nonlinear damage and the potential to quantify the nonlinear damage degree.

Keywords: Nonlinear damage · Time series model · Probability theory · Uncertainty · Acceleration response

1 Introduction

Vibration-based damage detection has received considerable attention in recent thirty years. The feasibility of vibration-based damage detection is that structural characteristics such as stiffness will change due to damage and the characteristics will be reflected in measured vibration response [1]. Time domain methods are a kind of effective damage detection methods by constructing time series models and extracting damage features from the constructed models to realize damage diagnosis.

In general, most relevant time-domain studies are used to detect two kinds of damages: linear and nonlinear damages. The linear damage got more attention and AR, ARX and ARMA models have been widely used to fit the response data of linear structures [2]. In fact, most damages in practical civil engineering structures possess nonlinear feature, such as the damage cracks that open and close due to dynamic loading. Some methods and theories have been utilized for structural nonlinear damage detection. Pham and Yang [3] firstly employed a hybrid model, the combination of an ARMA model and a GARCH model (i.e., ARMA/GARCH), to diagnose the machine

Y. Liu et al. (Eds.): ICNC-FSKD 2019, AISC 1074, pp. 326–332, 2020.
https://doi.org/10.1007/978-3-030-32456-8_35

health state based on vibration signal. Chen et al. [4] introduced the hybrid model to diagnose the nonlinear damage of a three-storey shear structure and provided a comparison with the linear methods based on ARMA model. However, for practice engineering structure, the damage detection results may be unreliable due to uncertainties, which stems from model inputs, physical parameters, sensor measurements, model form and so on. These uncertainties may disturb the dynamic properties of a structure and render the results of diagnosis uncertainty [5]. For this reason, a probabilistic method based on the AR/ARCH model is proposed to detect structural nonlinear damage.

2 Probabilistic Damage Detection Based on AR/ARCH Model

2.1 AR/ARCH Model and Damage Index

AR models can be used for structural acceleration time series to describe conditional mean. The AR(p) model for conditional mean can be written as

$$y_t = c + \sum_{i=1}^{p} \varphi_i y_{t-i} + \varepsilon_t \tag{1}$$

in which y_t is the time domain response data, c is a constant, p is the number of autoregressive order, φ_i is autoregressive parameter, ε_t is the residual. In Eq. (1), the term $c + \sum_{i=1}^{p} \varphi_i y_{t-i}$ is the deterministic term that represents the forecast of the current state as the functions of past observations and errors. The residual is commonly assumed to be zero mean and constant variance. However, for some practical time series, the residuals do not satisfy the homoscedastic assumption of constant variance. The time-varying variance is named as conditional variance. The conditional variance of the residual ε_t is given by

$$\sigma_t^2 = \mathrm{Var}_{t-1}(\varepsilon_t) = \mathrm{E}_{t-1}\left(\varepsilon_t^2\right) \tag{2}$$

An ARCH (q) model for the conditional variance of residual ε_t is

$$\sigma_t^2 = k + \sum_{j=1}^{q} A_j \varepsilon_t^2 \tag{3}$$

where k and A_j (j = 1, …, q) are the parameters of ARCH model The values of the parameters are constrained by

$$k_0 > 0, A_j \geq 0, \quad j = 1, \cdots, q$$

$$\sum_{j=1}^{q} A_j < 1$$

The first two constraints ensure that the conditional variances are positive, and the third constraint ensures the covariance stationarity.

AR model is used to fit structural acceleration response time series and describe the conditional mean. ARCH model is used to fit the residual sequence obtained from AR parts and describe heteroscedastic of structural acceleration response time series. In this study, two states have been defined, which are undamaged state and damaged state. Thus, we can utilize AR models to obtain the residual sequences $\{\varepsilon_t^U\}\}$ and ε_t^D under undamaged and damaged states. Then, we can use ARCH model to obtain the conditional variance sequences $\left\{\sigma_t^{2^U}\right\}$ and $\left\{\sigma_t^{2^D}\right\}$ and further obtain the variance $std\left\{\sigma_t^{2^U}\right\}$ and $std\left\{\sigma_t^{2^D}\right\}$ under the undamaged and damaged states. Finally, second-order-variance (sov) is proposed as follows:

$$sov = std\{\sigma_t^{2D}\} - std\{\sigma_t^{2U}\} \tag{4}$$

2.2 Probability Theory

Given the uncertainties, the index sov can be calculated based on the AR/GARCH model, which obeys the probability distribution $f(sov)$ with mean value $\mu(sov)$ and standard deviation $\sigma(sov)$. Damage detection is based on comparison between the undamaged and the damaged structural response. The index sov is subjected to a hypothesis test at a chosen significance level α, as follows:

$$\begin{aligned} H_0 &: \mu = 0 \quad \text{(Undamaged)} \\ H_1 &: \mu \neq 0 \quad \text{(Damaged)} \end{aligned} \tag{5}$$

The probability of damage existence (*PDE*) can be calculated as the probability that the actual state of the system is H_1, having rejected the hypothesis H_0. In particular, the *PDE* can be estimated by calculating the probability distribution of variable sov_d in the damaged state with that of sov_u in the undamaged state. Usually we can assume that sov_u is normally distributed with mean $\mu(sov_u)$ and standard deviation $\sigma(sov_u)$ and sov_d is normally distributed with mean $\mu(sov_d)$ and standard deviation $\sigma(sov_d)$. The corresponding probability density functions (*PDF*) of the $f(sov_u)$ and $f(sov_d)$ can be obtained as illustrated in Fig. 1, where L_u is the lower bound of the undamaged index sov_u. With the confidence level α, the lower bound L_u is obtained by

$$L_u = \mu(sov_u) - \eta(1-\alpha)\sigma(sov_u) \tag{6}$$

where, $\eta(1-\alpha)$ indicates that there is a probability of α that the undamaged index falls into the range of $[L_u, \infty)$. Therefore, the *PDE* of the structure can be calculated by

$$PDE = \int_0^{L_u} f(sov_d)dsov_d \tag{7}$$

where *PDE* is a value between 0 and 1. The *PDE* close to 1 means the high probability of damage existence and the *PDE* close to 0 means the low probability of damage existence.

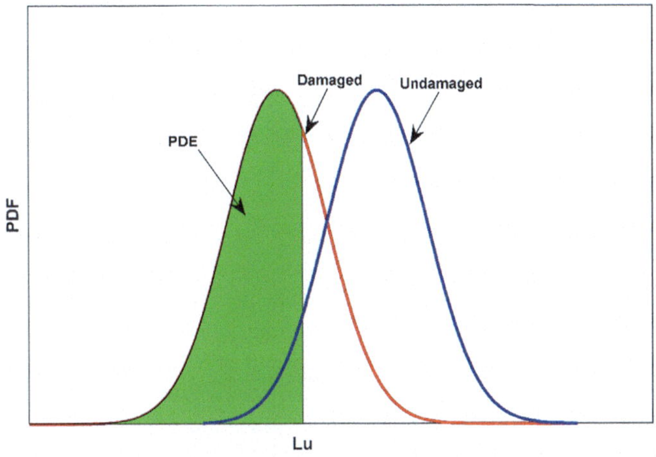

Fig. 1. Probability of damage existence

3 Numerical Example

A five degree-of-freedom shear structure (Fig. 2) is simulated and it is assumed that the mass of every storey is 100 kg and the initial lateral stiffness is 1 MN/m. Newmark – β algorithm is applied to calculate the structure acceleration response. The nonlinear damage in this system can be simply simulated by introducing a bilinear stiffness method proposed by Rébillat et al. [6]. The method can simulate a nonlinear breaking crack in columns between two adjacent storeys.

Six different damage scenarios are considered, which occurred in the 2th and 4th inter-storey stiffness with corresponding damage-parameter α being set as 10%, 30% and 50%, respectively. In this section, the structure uncertainty is assumed to be 10% reduction of the 3rd and 5th interlayer stiffness and the measurement uncertainty is assumed that the output acceleration time series are added to 10% measurement noise. In this example, 400 sets of acceleration response time series of each scenario are simulated to calculate mean value and standard deviation of the normal distribution of index sov. Akaike information criteria and partial correlation function are adopted for

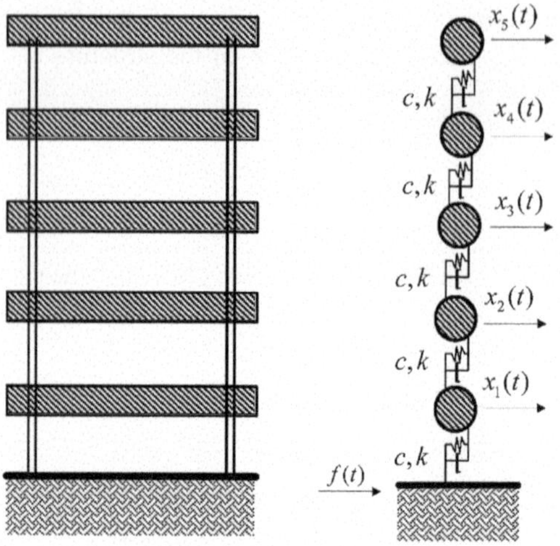

Fig. 2. Five-storey shear structure.

order determination of time series model. Here, the order of AR model is 28 and the order of ARCH model is 7.

The results of *PDE* are recorded in the Tables 1 and 2. The data in Table 1 is under the measurement uncertainty (MU) and the data in Table 2 is under both the measurement uncertainty (MU) and structural uncertainty (SU). They are also depicted in Fig. 3, where *L* represents the location of damaged inter-storey stiffness and α represents the nonlinear damage parameter. The underlined values in the Tables 1 and 2 are the *PDE* values of two storeys next to the nonlinear damaged inter-storey stiffness. From the results, it can be seen that all the *PDE* of targeted storeys with nonlinear damage are much larger than the *PDE* of other storeys without nonlinear damage, which indicates that the most probable existing location of the nonlinear damage accurately tallies with the targeted locations under two kinds of uncertainty. In addition, the *PDE* of storeys with nonlinear damage becomes larger along with the increasing degree of nonlinear damage, which also implies the method has the potential to quantify the degree of nonlinear damage.

Table 1. PDE (%) of 5-storeys under MU

L	α	1-storey	2-storey	3-storey	4-storey	5-storey
2	10%	19.25	38.16	0	0	0
	30%	42.20	98.72	0.14	0	0
	50%	67.38	99.70	2.61	0	0
4	10%	2.96	0	23.47	20.34	0
	30%	2.53	0.09	96.07	93.90	0.07
	50%	2.30	0.23	99.67	98.40	0.74

Table 2. PDE (%) of 5-storeys under MU and SU

L	α	1-storey	2-storey	3-storey	4-storey	5-storey
2	10%	17.58	38.19	0	0	0
	30%	40.17	98.73	0.28	0.01	0.02
	50%	65.71	99.70	3.97	0.01	0.01
4	10%	2.45	0.006	28.07	24.69	0.002
	30%	2.08	0.091	96.65	94.73	0.011
	50%	1.87	0.023	99.71	98.55	1.02

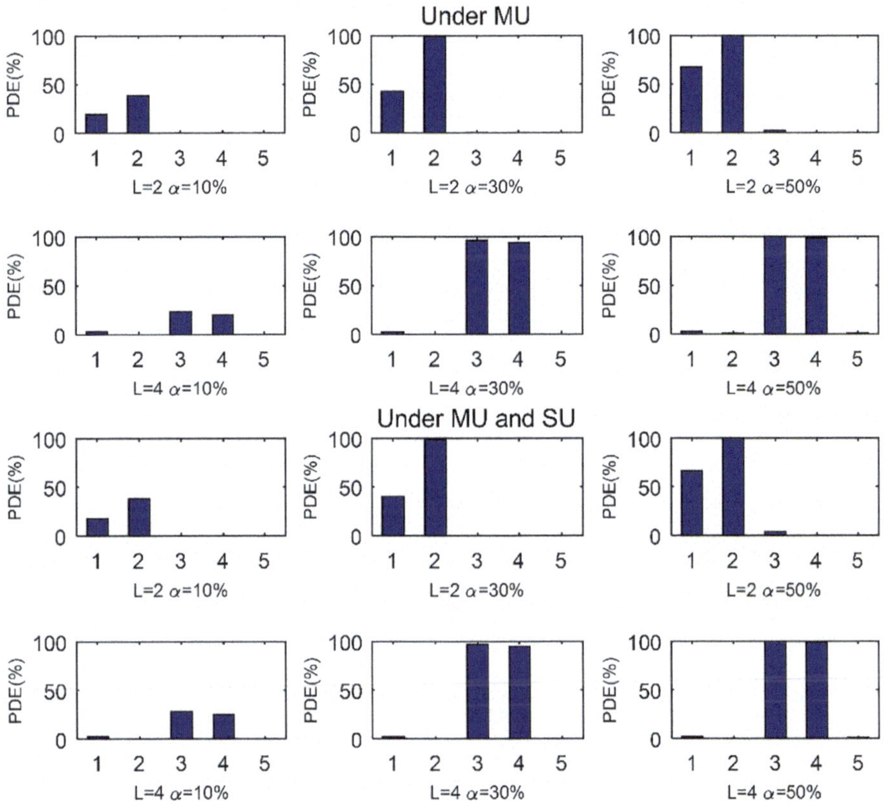

Fig. 3. The bar graph of the PDE under two kinds of uncertainty

4 Conclusions

This paper utilized probabilistic theory and AR/ARCH model to detect structural nonlinear damage. The basic theories and main procedure are described in this paper and a five-storey shear structure is employed to explore the effectiveness of the proposed method. The results showed that the method has excellent performance to locate the nonlinear damage and has the potential to reflect the changed degree of nonlinear damage under measurement noise and structure stiffness variation. However, the method is lack of the ability to quantify the degree of nonlinear damage and also further needed to be tested its adaptation under other kinds of uncertainty. In addition, since the method is only rely on the output acceleration response time series of a structure, it has the possibility to combine with continuous online monitoring to real time monitor the health condition of structure.

Acknowledgments. This work was supported by the National Natural Science Foundation of China (Grant No. 51578094).

References

1. Li, P.J., Xu, D.W., Zhang, J.: Probability-based structural health monitoring through Markov chain Monte Carlo sampling. Int. J. Struct. Stab. Dyn. **16**(7), 1550039-1-16 (2016)
2. Bao, C., Hao, H., Li, Z.X.: Integrated ARMA model method for damage detection of subsea pipeline system. Eng. Struct. **48**(3), 176–192 (2013)
3. Pham, H.T., Yang, B.S.: Estimation and forecasting of machine health condition using ARMA/GARCH model. Mech. Syst. Signal Process. **24**(2), 546–558 (2010)
4. Chen, L.J., Yu, L.: Structural nonlinear damage identification algorithm based on time series ARMA/GARCH model. Adv. Struct. Eng. **16**(9), 1597–1609 (2013)
5. Hong, M., Mao, Z., Todd, M.D., et al.: Uncertainty quantification for acoustic nonlinearity parameter in Lamb wave-based prediction of barely visible impact damage in composites. Mech. Syst. Signal Process. **82**(1), 448–460 (2017)
6. Rébillat, M., Hajrya, R., Mechbal, N.: Nonlinear structural damage detection based on cascade of Hammerstein models. Mech. Syst. Signal Process. **48**(1), 247–259 (2014)

Automatic Bug Priority Prediction Using DNN Based Regression

Wei Zhang(✉) and Chris Challis

Adobe Systems Inc., McLean, VA 22102, USA
wzhang@adobe.com

Abstract. Bugs are inevitable during software development. It is important to prioritize bugs and fix them based on their priorities. The priority assignment is usually done manually. Besides the cost of human effort, this process may also introduce bias since different people might have different opinions on the same issue. In this paper, we propose an approach to automate the process. It builds features from bug reports using Natural Language Processing, then trains a predictive model based on a deep Neural Network. The proposed approach was tested using a comprehensive data set containing more than 82 thousand bug reports. It runs in near real-time and its performance is significantly better than the previously reported results.

1 Introduction

In today's complicated software systems, bugs will emerge along with software development and evolution. In a software product that we are working with, nearly 1000 new bugs are reported each month in JIRA, which we used to track and monitor bug status. Usually we don't have enough resources to address all bugs. Therefore, a priority level is assigned to each bug, so that bugs can be prioritized and resolved accordingly. The typical list of priorities is: *Minor*, *Normal*, *Major*, *Critical* and *Blocker*, in ascending severity order.

Currently, bug priorities are assigned manually. This is not a trivial task as there are consequences if the priority is not set properly. We do not want to assign a high priority to a bug which is actually *Minor*, as this will cause unnecessary attention and waste resources. On the other hand, we need to avoid assigning a low priority to severe bugs, since they could have significant consequences if not addressed promptly. In addition, different bug reporters might have different criteria/scales. They could assign different priorities to the same bug. Thus a bug assigned with the *Major* priority might actually be more severe than a *Critical* bug assigned by a different reporter. Therefore, we would like to build a service that can assign bug priority automatically. In addition to the obvious benefit of alleviating human effort, it could provide a unified priority rating and eliminate individual bias.

The rest of the paper is organized as follows. Section 2 describes our approach for automatic bug priority prediction. We discuss experiments in Sect. 3 and related work in Sect. 4. We conclude the paper in Sect. 5.

Y. Liu et al. (Eds.): ICNC-FSKD 2019, AISC 1074, pp. 333–340, 2020.
https://doi.org/10.1007/978-3-030-32456-8_36

2 Our Approach

Each JIRA bug report comes with multiple fields. Among them, the field "Summary" gives a brief description about the bug, usually one sentence long. Here are two examples:

> "Segment based on Visit depth is not working as expected."
> "Data incomplete error returned in traffic report of VRS, Reports& analytics: Virtual Report Suites."

There is another field "Description" which presents a more comprehensive view of the bug. It can be as long as hundreds of sentences. Two examples are as follows (note the actual IDs and names are replaced with xxx):

> "We did a test IMS migration with the customer care login company 'xxx' on 5/17 Most of us got some email spam (I received 7 emails) but one user in particular received over 200 emails. Any ideas what's going on here? Who should take this?".
>
> "Any prop or evar value that contains the half-width kana Japanese character gets truncated unexpectedly. This might best be explained by viewing the attached document. Steps to reproduce: Company: xxx Data Center: San Jose RS: xxx RSID: xxx 1) Login to this Report Suite and load Custom Conversion→Custom Conversion 1-10→report (evar7, the bottom one in the list), selecting Oct. 2008 as time range. Use the Instances metric. 2) Notice line item 45, This item was truncated. xxx has confirmed this by sending test hits in to his own Report Suite and observed that the exact same truncation occurred (so the issue is not report suite specific and does not seem to be implementation based). There are more examples attached to this bug. Impact: Half width kana is sent often, especially when tracking Japanese mobile data. So 20+ xxx mobile customers including xxx are affected by this issue."

We can predict the bug priority based on either the "Summary" or "Description" field. To achieve our classification goal, we propose extracting features from these text fields and using a fixed length feature vector to represent each bug. Then we train a Deep Neural Network (DNN) [2] to predict the priority of a bug based on its feature vector.

2.1 Feature Extraction

We use topic modeling to construct features from text fields. First, a language model is built offline based on a large historical bug description corpus. This is done in the following steps: (1) preprocessing, e.g., tokenization and removing stop-words. (2) build a word frequency matrix based on the processed text, which is specific to the bug data set. Then we use Latent Dirichlet Allocation (LDA) [3] to extract a number of topics from this bug corpus. Each topic represents a distribution of words. Two example topics are as follows:

(1) 0.054 × "*classification*" + 0.047 × "*data*" + 0.035 × "*file*" + 0.032 × "*feed*" + 0.022 × "*percent*" + 0.019 × "*rule*" + 0.017 × "*error*" + 0.016 × "*campaign*" + 0.012 × "*processing*" + 0.010 × "*export*"
(2) 0.061 × "*error*" + 0.025 × "*admin*" + 0.023 × "*user*" + 0.016 × "*access*" + 0.015 × "*console*" + 0.014 × "*login*" + 0.014 × "*segment*" + 0.014 × "*message*" + 0.013 × "*page*" + 0.013 × "*table*"

Once the set of topics is extracted, the text of the chosen field ("Summary" or "Description") in any bug can be represented as a mixture of these topics. In other words, we can decompose any bug's "Summary" (or "Description") into a weighted sum of these topics. Since the number of topics is preset, the weights of all topics form a fixed size vector, which is used as the feature vector for the bug. Note that the number of topics is a hyper parameter to tune; we tried different numbers and empirically found that 500 topics to be the best number for our experiments. This discriminant feature vector enables us to build predictive models using various algorithms.

The task also involves imbalanced data: the bug priorities are not uniformly distributed, as shown in Fig. 1. In general, standard learning algorithms expect balanced class distributions. For imbalanced data, they tend to overfit training data and perform unfavorably on unseen data [6].

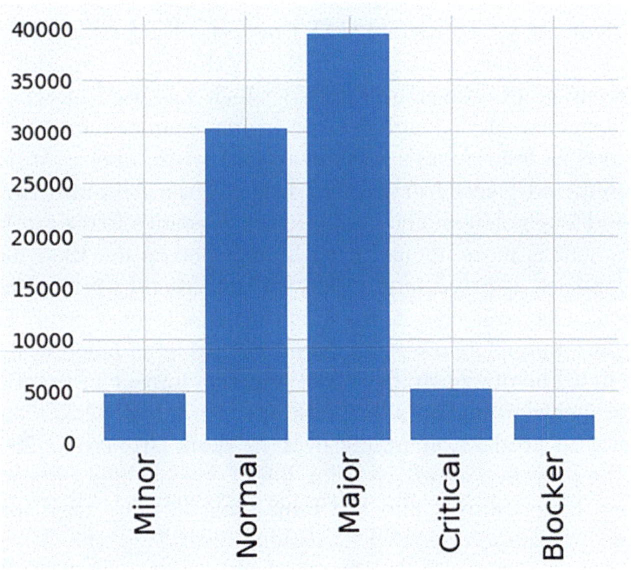

Fig. 1. Distribution of instances for all priority classes. The Y-axis is the number of instances per class. The X-axis is the priority label.

2.2 DNN Based Prediction

The bug priority has 5 levels, so it can be tackled as a multi-class classification problem. However, these 5 levels are ordered and the ordinal information will be lost if it is formulated as a classification problem. Therefore, we treat it as a regression problem, which outputs numerical values instead of class labels. We first map these priority levels to numerical values: *Minor*→0, *Normal*→1, ..., and *Blocker*→4. Then we train a machine learning module to map the relationship between a feature vector and its associated human labeled priority, based on a large set of training data. Here we use a Deep Neural Network (DNN) model. After trying different network structures, we found a 3 layer Neural Network, fully connected with one hidden layer of 30 nodes and Rectified Linear Unit (ReLU) [12] as the activation function, achieved the best trade-off between bias and variance for our problem. We also tried other algorithms, e.g., linear regression (as used by related work), but the performance is lower, as we will show in the experimental section.

The output from the DNN regression model is a floating point number rather than an integer. We then map each output value to its closest integer, so it can be converted back to the priority domain and assigned to the bug.

3 Experiments

We use the Mean Absolute Error (MAE) between the predicted level and the ground truth level to measure the prediction performance, in addition to accuracy (the percentage of correct prediction) which was used mostly by previous work. We believe that MAE is a better performance metric for this task, which can be explained as follows: let's say the ground truth label is *Major* for a bug report. For predicted levels *Normal* and *Critical*, the Absolute Error are both 1. For predicted levels *Minor* and *Blocker*, the Absolute Error are both 2. This error measurement is more aligned with human perception than metrics based on classification such as precision/recall, which are used by related works. Precision/recall treat every mis-classification as the same level of error. For example, labeling a *Minor* bug as *Blocker* is treated the same as labeling it as *Normal*, although the first one obviously has more negative impact in practice.

Our comprehensive bug data set contains a total of 82,299 bugs from an industrial software product, spanning over 12 years (from year 2004 to 2017). We evaluate the proposed approach using 5-fold cross-validation; each time 80% of data are used for training and the remaining 20% are used for testing. A useful baseline to compare against is assigning all bugs the *Normal* priority (constant priority assignment). In this case, the accuracy is 37% since 37% of the ground truth labels are *Normal*. The corresponding MAE is 0.76. The DNN model that we trained achieves much better performance, both in term of accuracy and MAE. We also trained a linear regression model for comparison. The DNN model clearly outperforms the regression model. Table 1 shows the performance achieved by using the "Summary" field. Although the field usually contains only one short sentence, we can predict the priority level fairly well, and the predicted level is usually close to the ground truth, with MAE equal to 0.51. In addition, we trained a model without considering the ordinal information; the results are much worse because the problem is ordinal by nature.

Table 1. Prediction performance based on the "Summary" field.

	Training data		Test data	
	Accuracy	MAE	Accuracy	MAE
Baseline	0.37	0.76	0.37	0.76
Regression	0.58	0.57	0.56	0.59
DNN	0.67	0.44	0.60	0.51

The "Description" field contains much more information. Therefore, we can attain even better prediction performance if the field is available, as shown in Table 2. Even on the unseen test set, almost 70% of the labels were assigned correctly. In addition, the misassigned labels are typically off by only one level.[1] This is quite important as we want to make sure our mistakes are small, since bug triage depends on it. A *Critical* bug mislabeled as *Major*, although undesirable, may not cause much of a trouble. On the other hand, if a *Blocker* bug is labeled as *Minor* (thus the mistake is big), there will be substantial consequences.

Table 2. Prediction performance based on the "Description" field.

	Training data		Test data	
	Accuracy	MAE	Accuracy	MAE
Baseline	0.37	0.76	0.37	0.76
Regression	0.66	0.50	0.64	0.51
DNN	0.84	0.22	0.69	0.39

3.1 Results Based on Precision/Recall

To be more comparable with the literature, we also measure the performance in terms of precision/recall and F-Measure which have been widely adopted in the literature. *Recall* for class C is defined as: $Recall = \frac{TP}{N}$, where TP represents the

[1] When the accuracy is at 69%, if all misassigned labels were off by 2 levels, MAE would be $0.69 \times 0 + 0.31 \times 2 = 0.62$. If all misassigned labels were off by just 1 level, MAE $= 0.69 \times 0 + 0.31 \times 2 = 0.31$. We achieved 0.39, which is very close to 0.31. With some simple calculation, this means that at least 75% of the wrong prediction results are off by only one level.

number of correct predicted results for the class C, N is the number of instances in class C. If all *Major* bugs are predicted as *Major*, the recall for the class *Major* is 100%. *Precision* measures the percentage of predicated labels that are correct labels. It is defined as: $Precision = \frac{TP}{N_c}$, where N_c is the number of instances predicted as class C. If all predicted *Major* bugs are truly *Major* bugs, the precision for the class *Major* is 100%. F-measure is a popular single metric for evaluating the classification performance. It combines *precision* and *recall* and is defined as:

$$F\text{-}measure = \frac{2 \times (precision \times recall)}{precision + recall} \tag{1}$$

The range of F-measure is [0, 1], and larger values mean better performance.

Table 3. Performance of the proposed approach, built using the "Summary" field. The last column "Average" is the simple average over all classes.

	Minor	*Normal*	*Major*	*Critical*	*Blocker*	Average
Precision	92.45%	54.20%	62.02%	23.40%	100%	66.17%
Recall	31.18%	59.10%	75.09%	1.04%	0.17%	33.32%
F-Measure	46.63%	56.54%	67.93%	2%	0.3%	34.69%

Table 4. Performance of the proposed approach, built using the "Description" field. The last column "Average" is the simple average over all classes.

	Minor	*Normal*	*Major*	*Critical*	*Blocker*	Average
Precision	70.52%	65.29%	73.51%	31.86%	42.60%	56.82%
Recall	39.58%	76.94%	77.67%	11.11%	9.42%	42.54%
F-Measure	50.70%	70.64%	75.53%	16.47%	15.43%	45.87%

The performance results on the test data are shown in Tables 3 and 4. The numbers are much better than those previously reported results. For example, the average precision/recall/F-Measure are: 28.30%, 30.97% and 29.16% respectively in the state-of-the-art paper [16]. We achieved average precision/recall/F-Measures of 56.82%, 42.54% and 45.87% respectively.

Note the averages reported in Tables 3 and 4 are simple averages over all classes. Each class is considered the same even though the *Blocke* class has significantly fewer instances than the *Normal* class. As can be seen from Tables 3 and 4, the performance numbers are much better for the *Normal* and *Major* classes, which have far more instances. If we take the size of each classes into consideration and take the weighted average (so we weigh each data instance the same in stead of weighing each class the same), the performance numbers are even better. For example, the weighted average F-measure overall all classes is 66.31% for the model that is learned based on the "Description" field.

Besides the promising performance numbers that we obtained, our solution is very efficient. Based on our experiments, the prediction is near real time with an average running time of 0.7 s, using a 2014 Macbook Pro laptop.

4 Related Work

There has been previous work [5,7,8,10,15,16] on the bug priority prediction problem. They mostly treat it as a classification problem, which does not account for the order between different priority levels. DRONE [16] is an state-of-the-art approach on automated bug priority assignment. It uses linear regression to address the problem. As we have shown, the deep learning model we adopted has much better performance than linear regression. Also, as opposed to [16] which uses manually designed feature, our approach is data driven and does not require any feature engineering. Another difference between our solution and the others is that we represent each description in the space of topics, rather than the typical word frequency. This leads to more powerful feature vectors. What's more, the performance metrics that they use are precision/recall, which are essentially only suitable for classification. We use the error in predicted levels, which is more suitable for the task. Nevertheless, we reported much better performance numbers in terms of precision/recall in this paper as well.

Bug priority prediction falls into the general problem of bug triage, which deals with the problem of prioritizing bugs and assigning them to appropriate developers. There has been considerable research on assigning bugs to developers [1,9,11,13,17–19]. It is essentially a classification problem rather than a regression problem. Prioritizing bugs and subsequently assigning to relevant developers for fixing are both crucial for software development and maintenance.

5 Conclusion

We have proposed and implemented an efficient solution for automatic bug priority assignment. It is based on topic modeling and a deep neural network. We have tested it using a large data set, with more than 82k JIRA bugs. Our experiments show that we can achieve 69% correct assignment on unseen data using the "Description" field. Moreover, the mis-assigned priorities are typically just off by one level, thus the impact of mis-assignment is moderate rather than catastrophic. Our solution runs in near real-time and can significantly speed up the bug triage process. In addition to alleviating human effort, it provides a unified priority rating and eliminates bias across users. To further improve the accuracy, we plan to use stacking [4] to combine the two prediction results (based on the "Summary" and "Description" fields respectively). We are also considering hard example mining [14] to further improve the prediction performance.

References

1. Anvik, J., Murphy, G.C.: Reducing the effort of bug report triage: recommenders for development-oriented decisions. ACM Trans. Softw. Eng. Methodol. (TOSEM) **20**(3), 10 (2011)
2. Arora, R., Basu, A., Mianjy, P., Mukherjee, A.: Understanding deep neural networks with rectified linear units. arXiv preprint arXiv:1611.01491 (2016)
3. Blei, D.M., Ng, A.Y., Jordan, M.I.: Latent Dirichlet allocation. J. Mach. Learn. Res. **3**(8), 993–1022 (2003)
4. Breiman, L.: Stacked regressions. Mach. Learn. **24**(1), 49–64 (1996)
5. Choudhary, P.A.: Neural network based bug priority prediction model using text classification techniques. Int. J. Adv. Res. Comput. Sci. **8**(5) (2017)
6. He, H., Garcia, E.A.: Learning from imbalanced data. IEEE Trans. Knowl. Data Eng. **21**(9), 1263–1284 (2009)
7. Lamkanfi, A., Demeyer, S., Giger, E., Goethals, B.: Predicting the severity of a reported bug. In: 2010 7th IEEE Working Conference on Mining Software Repositories (MSR), pp. 1–10. IEEE (2010)
8. Lamkanfi, A., Demeyer, S., Soetens, Q.D., Verdonck, T.: Comparing mining algorithms for predicting the severity of a reported bug. In: 15th European Conference on Software Maintenance and Reengineering (CSMR), pp. 249–258. IEEE (2011)
9. Linares-Vásquez, M., Hossen, K., Dang, H., Kagdi, H., Gethers, M., Poshyvanyk, D.: Triaging incoming change requests: bug or commit history, or code authorship? In: 28th IEEE International Conference on Software Maintenance (ICSM) (2012)
10. Menzies, T., Marcus, A.: Automated severity assessment of software defect reports. In: IEEE International Conference on Software Maintenance, ICSM 2008, pp. 346–355. IEEE (2008)
11. Murphy, G., Cubranic, D.: Automatic bug triage using text categorization. In: Proceedings of the Sixteenth International Conference on Software Engineering and Knowledge Engineering (2004)
12. Nair, V., Hinton, G.E.: Rectified linear units improve restricted Boltzmann machines. In: Proceedings of the 27th International Conference on Machine Learning (ICML 2010), pp. 807–814 (2010)
13. Shokripour, R., Anvik, J., Kasirun, Z.M., Zamani, S.: Why so complicated? Simple term filtering and weighting for location-based bug report assignment recommendation. In: Mining Software Repositories (MSR), pp. 2–11 (2013)
14. Shrivastava, A., Gupta, A., Girshick, R.: Training region-based object detectors with online hard example mining. In: Computer Vision and Pattern Recognition (CVPR), pp. 761–769 (2016)
15. Tian, Y., Lo, D., Sun, C.: Information retrieval based nearest neighbor classification for fine-grained bug severity prediction. In: 2012 19th Working Conference on Reverse Engineering (WCRE), pp. 215–224. IEEE (2012)
16. Tian, Y., Lo, D., Xia, X., Sun, C.: Automated prediction of bug report priority using multi-factor analysis. Empirical Softw. Eng. **20**, 1354–1383 (2015)
17. Xia, X., Lo, D., Wang, X., Zhou, B.: Dual analysis for recommending developers to resolve bugs. J. Softw.: Evol. Process **27**(3), 195–220 (2015)
18. Ye, X., Bunescu, R., Liu, C.: Learning to rank relevant files for bug reports using domain knowledge. In: Proceedings of the 22nd ACM SIGSOFT International Symposium on Foundations of Software Engineering, pp. 689–699. ACM (2014)
19. Zhou, J., Zhang, H., Lo, D.: Where should the bugs be fixed? More accurate information retrieval-based bug localization based on bug reports. In: 2012 34th International Conference on Software Engineering (ICSE), pp. 14–24. IEEE (2012)

Natural Computation Applications: Natural Computation in Signal Processing and Multimedia

Target Cropping: A New Data Augmentation Method of Fine-Grained Image Classification

JunFeng Lu[1,2](✉) and MingXue Liao[2]

[1] University of Chinese Academy of Sciences, UCAS, 80# ZhongGuanCun East Road, Beijing 100190, People's Republic of China
junfeng2017@iscas.ac.cn

[2] Institute of Software, Chinese Academy of Sciences, ISCAS, 4# South Fourth Street, ZhongGuanCun, Beijing 100190, People's Republic of China

Abstract. In this paper, we propose a novel data augmentation method of fine-grained image classification named target cropping. Previous work has demonstrated the effectiveness of data augmentation through simple technique, such as random cropping, image rotating and image flipping. But for fine-grained classification, due to its inter-class similarity and intra-class differences, traditional random cropping does not pay much attention to the discriminative regions and even may crop out the regions that have a critical impact on the classification results. To solve this problem, we propose target cropping which uses class activation maps to locate discriminative region. Compared with random cropping, our method significantly improves classification accuracy for all the tested datasets. For example, classification accuracy is improved from 71.1% to 73.9% for CUB200-2011 dataset with VGG-16 and from 77.2% to 79.0% in the FGVC-Aircraft dataset. It is a significant improvement in fine-grained image classification field.

Keywords: Data augmentation · Fine-grained image classification · Class activation maps · Target cropping

1 Introduction

For many machine learning tasks, such as image classification, object detection and object segmentation, the amount of samples available for training is critically important for achieving a good performance. Data augmentation is widely used to increase the amount of training data by applying a small change in the original training data and synthetically creating new samples [3]. There are a variety of methods for generating new samples from training data, such as flipping, rotating the input image, and cropping a patch from a random position [5].

In this paper, we introduce a simple but effective data augmentation technique of fine-grained classification. Due to the small differences between

Y. Liu et al. (Eds.): ICNC-FSKD 2019, AISC 1074, pp. 343–351, 2020.
https://doi.org/10.1007/978-3-030-32456-8_37

fine-grained classification images, additional attention needs to be paid to the discriminative regions, but traditional random cropping does not pay much attention to the discriminative regions and even may crop out the regions that have a critical impact on the classification results. To address this problem, we propose a method named Target Cropping. We learn from the idea of weakly-supervised object location to find the region with strong discrimination, select an anchor point in the region and crop the original image centered on the anchor. Our target cropping method draws on the idea of attention mechanism [4], in contrast, we shift the attention mechanism from the network training process to the data augmentation process, so that our method decouples from specific network architecture which makes it effective and easy to use.

Our data augmentation method makes significant accuracy improvement on fine-grained image classification datasets: CUB200-2011 and FGVC-Aircraft. For example, we compare our method with random cropping, the accuracy is improved from 71.1% to 73.9% for CUB200-2011 dataset with VGG-16 and from 77.2% to 79.0% in the FGVC-Aircraft dataset. Our main contributions are summarized as follows: (1) We propose a novel method to do data augmentation for fine-grained image classification by using weakly-supervised object location method to find the discriminative region and to crop purposefully. (2) Attention mechanism is used in our data augmentation method. We pay more attention to the discriminative regions, crop them and use them as training data instead of random cropping, which brings significant improvement in fine-grained image classification accuracy.

2 Related Work

2.1 Data Augmentation

Data augmentation has been widely used to avoid this overfitting problem by enlarging the size of the training set [10]. The two most popular data augmentation methods in the training of deep CNN are random flipping and random cropping. Random flipping randomly flips the input image horizontally, while random cropping extracts random sub-patch from the input image. As an analogous choice, our target cropping extracts sub-patch around the discriminative region and retain the local region information which is essential for fine-grained image classification.

2.2 Fine-Grained Image Classification

Fine-grained image classification is a challenging task in the field of computer vision, which aims to classify sub-categories such as different species of bird [13]. In recent years, deep learning, especially deep convolutional neural networks make great progress in the field of computer vision. Features extracted from deep convolutional neural networks have more powerful description capabilities than artificial features, and deep convolution features applied to fine-grained image classification have achieved better results [7,14,15]. The above mentioned methods focus more on the network structure level, and our approach enhances the final classification accuracy through data augmentation.

2.3 Weakly-Supervised Object Localization

There have been a number of recent works exploring weakly-supervised object localization using CNNs [1,2,9,16]. We use weighted activation maps generated for each image to localize object. The response value in the class activation map represents the contribution of this region to classifying the image into the corresponding label. Our target cropping method finds high response area, picks an anchor in the area and crops the image centered on the anchor point.

3 Method

In this section, we introduce our target cropping method in detail. We first introduce class activation maps which are used to find discriminative region. Then, the architecture of the model is analyzed in detail.

3.1 Discriminative Region Finding

In this section, we describe the procedure for generating class activation maps, and use class activation maps to find discriminative region. The procedure for generating these maps is illustrated in Fig. 1. Class activation maps highlight the class-specific discriminative regions. Conv means convolution operation and GAP means global average pooling operation.

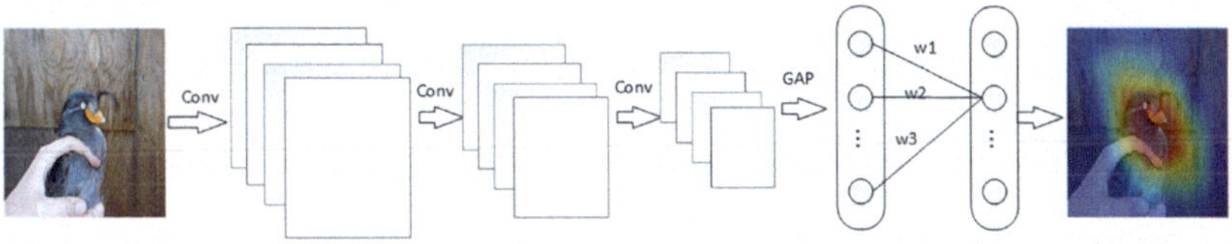

Fig. 1. Class Activation Maps: the predicted class score is mapped back to the previous convolutional layer to generate the class activation maps.

We formulate the process as follows: For a given image I, $f_k(x, y)$ represents the activation of unit k in the last convolutional layer at location (x, y).

$$F_k = \sum_{x,y} f_k(x, y) \tag{1}$$

which means the result of last convolutional layer after performing global average pooling. For a given category c, the input to the softmax layer is

$$S_c = \sum_k w_k^c F_k \tag{2}$$

where w_k^c is the fully-connected layer parameters of category c for unit k. Finally the output of the softmax for category c is

$$P_c = \frac{exp(S_c)}{\sum_c exp(S_c)} \tag{3}$$

Combining F_k and S_c, we can get

$$S_c = \sum_k w_k^c \sum_{x,y} f_k(x,y) = \sum_{x,y} \sum_k w_k^c f_k(x,y) \tag{4}$$

$$H_c(x,y) = \sum_k w_k^c f_k(x,y) \tag{5}$$

H_c directly indicates the importance of the activation at (x, y) resulting in the classification of an image to category c.

We use a network architecture which is similar to GoogLeNet [12] that mainly consists of convolutional layer, perform global average pooling of the convolution feature maps before the final output layer and project back the weights of the fully connected layer onto convolutional feature map to get $H_c(x, y)$. Once we get class activation maps, we can resize the class activation maps to the size of the input image, find the maximum response area and select an anchor in this area (in our experience, we select the maximum response value in the area), finally we crop the area around the anchor which means we pay more attention to the discriminative regions and use them as training data.

3.2 Model Architecture

In this section, we describe models used in our experience in detail.

CNN with Random Cropping (CRC): CRC is the baseline in our experience, we use both AlexNet [6] and VGG-16 [12] which take input image with random cropping and learn a multi-class classifier for fine-grained image classification.

CNN with Target Cropping (CTC): Compared with CRC, we propose a method named CTC, which takes the input image with target cropping described in Sect. 3.1, using convolutional layers and global average pooling to generate class activation mapps, and crop around the area which is important for fine-grained image classification.

CNN with Both Random Cropping and Target Cropping (CRCTC): In additional, we experiment with both random cropping and target cropping.

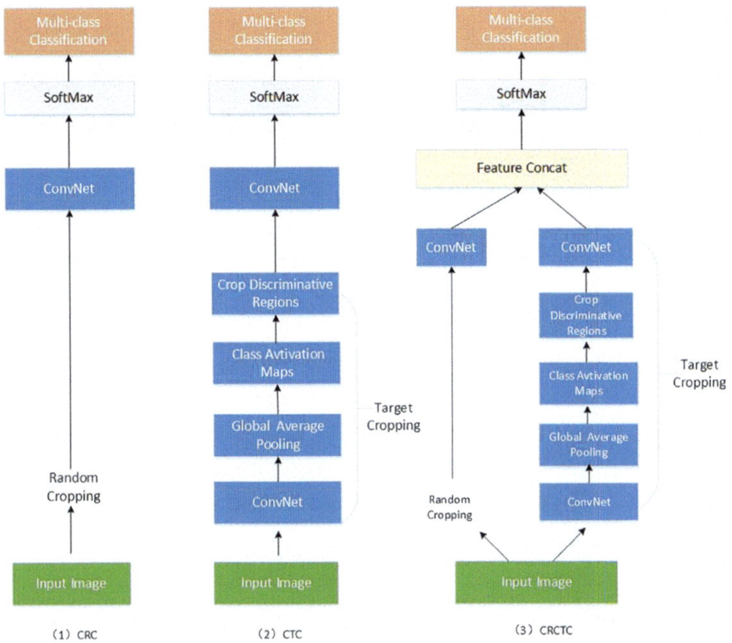

Fig. 2. Three different network architectures used in our experiments: (1) a standard convolutional neural network with random cropping for fine-grained image classification (2) a standard convolutional neural network with target cropping method described in Sect. 3.1 (3) a two-way convolutional neural network with both random cropping and target cropping.

We use a two-way convolutional neural network. For an input image, we use ConvNet with random cropping on one branch, on the other branch, we apply target cropping. Before the softmax layer, we concat the features extracted by two branches and decide which category it belongs to (Fig. 2).

4 Experiment

Our method is specifically designed for fine-grained image classification task, so we mainly evaluate our method on this task. In this section, we first introduce the datasets we used in the experiment including CUB200-2011 [13] and FGVC-Aircraft [8]. Then experimental settings of all models are introduced in detail. Finally, the result of our method is presented and analyzed in detail.

4.1 Dataset

CUB200-2011: CUB200-2011 is the most commonly used database in the field of fine-grained image classification. It contains 200 different categories and 11788

(a) CUB200-2011 (b) FGVC-Aircraft

Fig. 3. (a) images in CUB200-2011 database and (b) images in FGVC-Aircraft database

bird images totally. At the same time the database provides a variety of manual annotation data, but in our experiment, we only use images and labels.

FGVC-Aircraft: FGVC-Aircraft provides 102 different aircraft photos, each category contains 100 different photos. There are 10200 images in the database (Fig. 3).

4.2 Experimental Setting

For models CTC and CRC, we perform global average pooling of the convolution feature maps before the final output layer and create class activation maps for all training datas and test datas. We train the network using stochastic gradient descent with momentum as the optimization method with the batch size of 32 samples, and we use cross-entropy as loss function, set L2 regularization to be 1e-8, set the learning rate to be 0.02, when the training loss does not decrease for 3 epochs, learning rate will be divided by 10.

For all models CRC, CTC and CRCTC, we use both AlexNet and VGG-16 in the stage of image classification, and train the network using stochastic gradient descent with momentum as the optimization method with the batch size of 64 samples. Loss function, learning rate, and training strategy are the same with the target cropping stage mentioned above.

4.3 Result

We use CRC model as the baseline and use the multi-class classification accuracy as the criterion, which is the commonly used evaluation criteria in fine-grained classification [11,17]. Accuracies of image classification for CUB200-2011 and FGVC-Aircraft are described in Table 1 and the training processes are shown in Fig. 4.

For CRC model, we apply random cropping method for data augmentation. As a result, AlexNet converged after 10 epochs on CUB200-2011 dataset, the best accuracy is 53.0%, and converged after 13 epochs on the FGVC-Aircraft dataset,

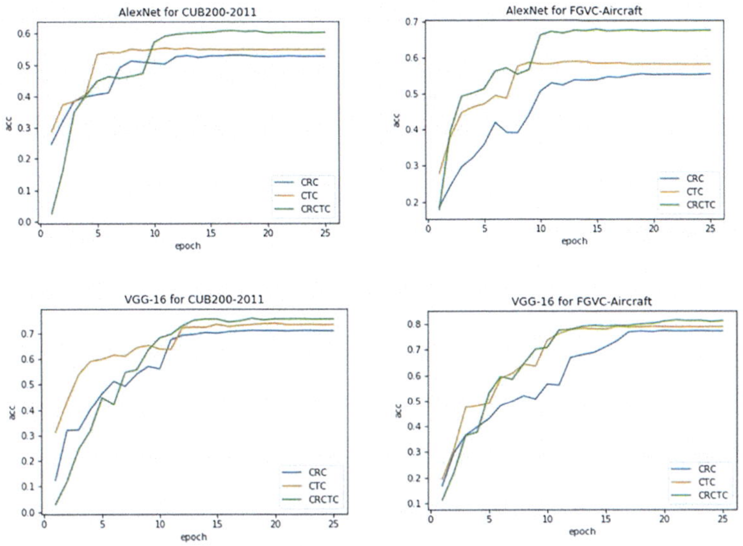

Fig. 4. Accuracy on test data for both CUB200-2011 and FGVC-Aircraft

Table 1. Accuracy on CUB200-2011 and FGVC-Aircraft

	CUB200-2011	FGVC-Aircraft
CRC (*AlexNet*)	0.530	0.555
CTC (*AlexNet*)	0.553	0.589
CRCTC (*AlexNet*)	0.608	0.679
CRC (*VGG*-16)	0.711	0.772
CTC (*VGG*-16)	0.739	0.790
CRCTC (*VGG*-16)	0.758	0.814

the best accuracy is 55.5%. VGG-16 converged after 12 epochs on CUB200-2011 dataset, the best accuracy is 71.1%, and converged after 17 epochs on the FGVC-Aircraft dataset, the best accuracy is 77.2%.

For CTC model, we apply target cropping method for data augmentation which is mentioned in Sect. 3.1. As a result, AlexNet converged after 7 epochs on the CUB200-2011 dataset, the best accuracy is 55.3%, and converged after 10 epochs on the FGVC-Aircraft dataset, the best accuracy is 58.9%. VGG-16 converged after 12 epochs on the CUB 200-2011 dataset, the best accuracy is 73.9%, and converged after 12 epochs on the FGVC-Aircraft dataset, the best accuracy is 79.0%.

Compared with random cropping, our target cropping data augmentation method achieves 2.3% improvement in accuracy on CUB200-2011, 3.4% improvement on FGVC-Aircraft for AlexNet, and 2.8% improvement in accuracy on CUB200-2011, 2.2% improvement on FGVC-Aircraft for VGG-16. In addition, we also use the CRCTC model in our experiment, which combines random cropping and target cropping. It is equivalent to adding additional discriminative region features to the original image features, and the classification accuracy is further improved. This shows that the attention mechanism used by our target cropping has a positive impact on the classification results.

5 Conclusions

The work is supported by both National scientific and Technological Innovation Zone (No.17-H863-01-ZT-005-005-01) and State Key Project of Research and Development Plan (No. 2016QY03D0505). In order to improve the accuracy of fine-grained image classification, we propose a novel data augmentation method in this work. It can be divided into two parts: data augmentation and image classification. In the process of data augmentation, compared with random cropping, class activation maps are used to find regions with strong discrimination and then to perform target cropping. In the process of image classification, we experiment CRC, CTC and CRCTC models with AlexNet and VGG-16 on both CUB200-2011 and FGVC-Aircraft. Compared with random cropping, our method achieve nearly 3% improvement in accuracy which is a great improvement in fine-greined image classification. Not limited to AlexNet and VGG-16, new model on this task can utilize our method to get a better result.

References

1. Bergamo, A., Bazzani, L., Anguelov, D., Torresani, L.: Self-taught object localization with deep networks (2016)
2. Cinbis, R.G., Verbeek, J., Schmid, C.: Weakly supervised object localization with multi-fold multiple instance learning (2017)
3. Fadaee, M., Bisazza, A., Monz, C.: Data augmentation for low-resource neural machine translation (2017)
4. Fu, J., Zheng, H., Tao, M.: Look closer to see better: recurrent attention convolutional neural network for fine-grained image recognition. In: Computer Vision & Pattern Recognition (2017)
5. Inoue, H.: Data augmentation by pairing samples for images classification (2018)
6. Krizhevsky, A., Sutskever, I., Hinton, G.E.: ImageNet classification with deep convolutional neural networks. In: Advances in Neural Information Processing Systems, pp. 1097–1105 (2012)
7. Lin, T.Y., Roychowdhury, A., Maji, S.: Bilinear CNN models for fine-grained visual recognition (2015)
8. Maji, S., Rahtu, E., Kannala, J., Blaschko, M., Vedaldi, A.: Fine-grained visual classification of aircraft. HAL - INRIA (2013)
9. Oquab, M., Bottou, L., Laptev, I., Sivic, J.: Learning and transferring mid-level image representations using convolutional neural networks. In: Computer Vision & Pattern Recognition (2014)
10. Perez, L., Wang, J.: The effectiveness of data augmentation in image classification using deep learning (2017)

11. Shu, X., Tang, J., Qi, G.J., Li, Z., Jiang, Y.G., Yan, S.: Image classification with tailored fine-grained dictionaries. IEEE Trans. Circ. Syst. Video Technol. **28**(2), 454–467 (2018)
12. Simonyan, K., Zisserman, A.: Very deep convolutional networks for large-scale image recognition. Computer Science (2014)
13. Wah, C., et al.: The caltech-ucsd birds-200-2011 dataset (2011)
14. Yu, Z., Wei, X.S., Wu, J., Cai, J., Lu, J., Nguyen, V.A., Do, M.N.: Weakly supervised fine-grained categorization with part-based image representation. IEEE Trans. Image Process. **25**(4), 1713–1725 (2016)
15. Zhang, X., Xiong, H., Zhou, W., Lin, W., Qi, T.: Picking deep filter responses for fine-grained image recognition. In: Computer Vision & Pattern Recognition (2016)
16. Zhou, B., Khosla, A., Lapedriza, A., Oliva, A., Torralba, A.: Learning deep features for discriminative localization (2015)
17. Zou, C., Luo, Y., Xiaolong, X.U.: Fine-grained image classification method based on multi-feature combination. J. Comput. Appl. (2018)

UAV Remote Sensing Image Stitching

Hui Wu[1(✉)] and Jun Chen[1,2]

[1] School of Automation, China University of Geosciences, Wuhan 430074, China
jwbgs@cug.edu.cn
[2] Hubei Key Laboratory of Advanced Control and Intelligent Automation for Complex Systems, Wuhan, China

Abstract. The majority of image stitching (include UAV remote sensing images stitching) models are homography matrix transformation function, which could effectively simulate the rigid transformation of 2D images in 3D coordinate system. In order to prevent the accumulation of alignment errors (drift) and the projection distortion in the multi-image stitching task, we proposes a novel UAV remote sensing image stitching access consisting of homography transformation and Helmert transformation. In the overlapping and non-overlapping region we utility homography transformation and Helmert transformation respectively. With proposed method the UAV remote sensing image stitching result has fewer alignment errors and less projection distortion.

Keywords: UAV remote sensing image stitching · VFC · Helmert · APAP

1 Introduction

UAV is widely used for image collection because of its convenient operation and low investment. However, the low flying altitude of UAV makes the images a limited field of view. In order to acquire more comprehensive regional information, its necessary to stitch lots of images into a wide range image. One of the important characteristics of UAV remote sensing images acquisition is sweeping captured, and the image scene can't be seen as an effective plane because of topographic relief and prominent stereo objects, in such condition, the images stitching using homography inevitably yield parallax errors. Moreover, in multi-image stitching task, the errors will accumulate and eventually lead serious distortion, which presents unnatural visual effect. And the homography model may cause projection distortion.

In order to addressing the deficiencies exposed by homography function in images stitching, several state-of-the-art approaches have been proposed and work really well. Homography function as a 2D model only works well in image alignment if the image scene is planar or the source images captured by camera's pure rotation. In UAV remote sensing image or other multiple perspectives image stitching task, the single homography model's effect is weak. Fortunately, if the

Y. Liu et al. (Eds.): ICNC-FSKD 2019, AISC 1074, pp. 352–360, 2020.
https://doi.org/10.1007/978-3-030-32456-8_38

influence of perspective change and scene depth difference on images' overlapping region is not particularly large, the difference between images could be considered as local deformation. And [1] uses mesh deformation (each unit corresponds to one homography function) to simulate the local deformation of image, which works well in reducing the parallax errors. In general, the homography function is estimated by sparse matching point set. The optimal plane corresponding to the point sets may differ from the optimal plane of the image scene. Using the homography function corresponding to the point sets may cause image projection distortion, especially in multi-image stitching task. For solving the problem of image distortion caused by homography function, [2,3] adopt global similarity transformation to prevent the distortion and make the result image look natural.

In this paper, we propose a new approach for multiple UAV remote sensing image stitching which can effectively prevent alignment errors accumulation and image distortion. We consider the image stitching as a transformation between the camera coordinate systems. And the Helmert transformation, which is widely used in geospatial information with 7 parameters (including 3 rotation parameters, 3 translation parameters and 1 scale parameter), is adopted to express the transformation between coordinate systems. Moreover, considering the approximate vertical shooting of UAV remote sensing image, we simplify the Helmert transformation to 5 parameters (including 1 rotation parameters, 3 translation parameters and 1 scale parameter), which not only preserves the shape of image scene but also prevents the accumulation of alignment errors effectively. The rest of the paper is organized as follows: Sect. 2 surveys important related work. Section 3 introduces the proposed method. Section 4 extends the method to stitch multiple UAV remote sensing images. And conclude in Sect. 5.

2 Related Works

In this part, we briefly review the previous work in image stitching. AutoStitch [4], which is the representative algorithm for previous image stitching task, gets almost perfect experimental results under the assumption of camera pure rotation. In this algorithm, the global homography model is used to simulate the transformation between images, the bundle adjustment model [5] is adopted to reduce the image alignment errors, and multi-band blending technology [6] is adopted to improve the visual effect. In most practical applications, it is difficult to meet the pure rotation assumption. If the camera moves and the image scene cannot be considered an effectively plane, there will be parallax errors or ghosting with single homography. DHW [7] divides the image scene into a distant plane and a ground plane, and uses two homography functions to align the distant and ground respectively which can stitch most real scenes well. SVA [8] estimates a global affine transformation and relaxed it to a affine filed in order to meet the local deformations with higher robustness in image stitching. APAP [1] divides the image into multiple local regions by adopting mesh grid, and a moving Direct Linear Transformation (DLT) [9] model is used to estimate the homography of each local region of the image. The homography corresponding to grid unit can simulate the images local transformation effectively. SPHP [2] proposed a warp method which gradually changes from the projective transformation in the overlapping region to similarity transformation in the non-overlapping region across the image, it draws on the Shape-Preserving method, corrects the shape of the stitched image and reduces projection distortion. And AANAP [3] reduces the perspective distortion by linearizing the homography function and gradually changes it to similarity transformation, which greatly improved the naturalness of stitching results.

3 Proposed Warp

3.1 Feature Point Sets Registration

Given two overlapping image i and image j representing the reference image and target image respectively. We use the SIFT [10] algorithm to extract features from image i, j and obtain the initial matching feature point sets X, Y. Considering the topographic relief in the UAV remote sensing image scene, which reflects the local deformation in the 2D image. We use Vector Field Consensus (VFC) [11] to refine the initial match. The VFC model is nonparametric model that is robust to local deformation. It considers the registration problem as a Gaussian mixture model (GMM) probability density estimation problem: A GMM is fitted to point set X, such that both the centers and local features of Gaussian densities are constrained to coincide with the other point set Y. The mathematical representation of GMM as follow:

$$p(\mathbf{Y}|\mathbf{X},\theta) = \prod_{n=1}^{N}\left(\frac{\gamma}{2\pi\sigma^2}e^{-\frac{\|\mathbf{y}_n - f(\mathbf{x}_n)\|^2}{2\sigma^2}} + \frac{1-\gamma}{a}\right) \tag{1}$$

where $\theta = \{f, \sigma^2, \gamma\}$ are the parameters to be solved, f is the transformation between X, Y, γ is the percentage of outliers, a is a constant. The model is solved under a unified maximum likelihood framework with an iterative expectation maximization (EM). The initial match is true could be expressed as posterior probability of the match:

$$p_n = \frac{\gamma e^{-\frac{\|\mathbf{y}_n - f(\mathbf{x}_n)\|^2}{2\sigma^2}}}{\gamma e^{-\frac{\|\mathbf{y}_n - f(\mathbf{x}_n)\|^2}{2\sigma^2}} + \frac{2\pi\sigma^2(1-\gamma)}{a}} \tag{2}$$

The update of the parameters $\theta* = \arg\max_{\theta} p(\mathbf{Y}|\mathbf{X},\theta)$ is expressed as follows:

$$\sigma^2 = \frac{(\mathbf{Y} - f(\mathbf{X}))^T P(\mathbf{Y} - f(\mathbf{X}))}{2 \cdot tr(P)} \tag{3}$$

$$\gamma = tr(P)/N \tag{4}$$

$$\varepsilon(f) = \frac{1}{2\sigma^2}\sum_{n=1}^{N} p_n \|\mathbf{y}_n - f(\mathbf{x}_n)\|^2 \tag{5}$$

The transformation between the point sets is expressed as a two-dimensional vector field. Each match corresponds to a two-dimensional vector. The closer the distance between matching pairs, the closer the corresponding vectors should be. In order to satisfy above assumption, add a regularization to Eq. (5), finally expressed as:

$$\varepsilon(f) = \frac{1}{2\sigma^2}\sum_{n=1}^{N} p_n \|\mathbf{y}_n - f(\mathbf{x}_n)\|^2 + \frac{\lambda}{2}\phi(f) \tag{6}$$

where λ is a constant, $\phi(f) = \|f\|_H^2$ is the smoothness function. The optimal solution is solved as:

$$(\Gamma + \lambda\sigma^2 P^{-1})C = Y \tag{7}$$

$$f(\mathbf{x}) = \sum_{n=1}^{N} \Gamma(\mathbf{x}, \mathbf{x}_n)c_n \tag{8}$$

where the Γ is an N*N block matrix with the (i,j)-th block $\Gamma(\mathbf{x}_i, \mathbf{x}_j)$ which is a matrix-valued kernels. After the EM algorithm converges, a correctly matched set of point pairs can be obtained according to the set threshold τ:

$$I = (n : p_n > \tau, n \in IN_N) \tag{9}$$

3.2 Images Stitching

We use DLT estimate the optimal global homography transformation:

$$p_*^{'} \sim Hp_* \tag{10}$$

which can be expression as follows:

$$0_{3\times1} = \begin{bmatrix} 0_{3\times1} & -\mathbf{p}^T & y^{'}\mathbf{p}^T \\ \mathbf{p}^T & 0_{3\times1} & -x^{'}\mathbf{p}^T \\ -y^{'}\mathbf{p}^T & x^{'}\mathbf{p}^T & 0_{3\times1} \end{bmatrix} \begin{bmatrix} \mathbf{h}_1 \\ \mathbf{h}_2 \\ \mathbf{h}_3 \end{bmatrix} \tag{11}$$

where $\mathbf{p} = [x \quad y \quad 1]^T$, $\mathbf{p}' = [x' \quad y' \quad 1]^T$, and $\mathbf{h}_1$, $\mathbf{h}_2$, $\mathbf{h}_3$ are given by the columns of H. The optimal $\mathbf{h}$ estimated by DLT:

$$\mathbf{h}_{optimal} = \arg\min_{\mathbf{h}} \sum_{i=1}^{N} \|a_i\mathbf{h}\|^2 \tag{12}$$

And estimate the local homography transformation for the images overlapping region using the Moving DLT model that mentioned in [1]. For each local region $\mathbf{x}_*$, using a location dependent homography to express the transformation:

$$\mathbf{x}_*^{'} \sim H_*\mathbf{x}_* \tag{13}$$

The $\mathbf{h}_*$ estimated by:

$$\mathbf{h}_* = \arg\min_{\mathbf{h}} \sum_{i=1}^{N} \|\omega_*^i a_i\mathbf{h}\|^2 \tag{14}$$

where $\omega_*^i = \exp(-\|\mathbf{x}_* - \mathbf{x}_i\|^2/\sigma^2)$, means the closer to $\mathbf{x}_*$ gives the more influence to estimate $\mathbf{h}_*$.

For the non-overlapping region, we use the global Helmert transformation to prevent the perspective distortion. Given two images i, j whose homogeneous coordinate systems are denoted as C_1, C_2 respectively. We regard the transformation between images as the transformation between the image homogeneous coordinate systems in 3D space:

$$C_2 = R(\phi) * C_1 + t \tag{15}$$

where $\phi = (\alpha, \beta, \gamma)^T$, and

$$R(\phi) = R_1(\alpha)R_2(\beta)R_3(\gamma) \tag{16}$$

$$R_1(\alpha) = \begin{bmatrix} \cos\alpha & -\sin\alpha & 0 \\ \sin\alpha & \cos\alpha & 0 \\ 0 & 0 & 1 \end{bmatrix} \tag{17}$$

$$R_2(\beta) = \begin{bmatrix} \cos\beta & 0 & -\sin\beta \\ 0 & 1 & 0 \\ \sin\beta & 0 & \cos\beta \end{bmatrix} \tag{18}$$

$$R_3(\gamma) = \begin{bmatrix} 1 & 0 & 0 \\ 0 & \cos\gamma & -\sin\gamma \\ 0 & \sin\gamma & \cos\gamma \end{bmatrix} \tag{19}$$

Convert t to a $3 * 3$ matrix:

$$C_2 = R(\phi)T * C_1, T = \begin{bmatrix} 1 & 0 & t_x \\ 0 & 1 & t_y \\ 0 & 0 & t_z \end{bmatrix} \tag{20}$$

Based on [12], we initialize the rotation matrix as follows:

$$R(\phi) = \begin{bmatrix} 1 & -\alpha & -\beta \\ \alpha & 1 & -\gamma \\ \beta & \gamma & 1 \end{bmatrix} \tag{21}$$

As the above $R(\phi)$ is no longer an orthognal matrix, we need to get a more accurate R. We use the Levenberg-Marquard (LM) method to optimize R and T, T is initialized to $T = R(\phi)^{-1}H$, and the errors function was defined as follows:

$$err = \sqrt{\sum_{n=1}^{N} \|\mathbf{y}_n - RT\mathbf{x}_n\|^2} \tag{22}$$

After the convergence of err in LM, we get the optimal global Helmert parameters corresponding to the global homography matrix. Based on the approximate vertical shooting of UAV remote sensing image, we ignore the two rotation parameters β and γ, and the 7 parameters Helmert converts to 5 parameters

Helmert. In multi-UAV remote sensing image stitching task, we assume that the average distortion corresponding to global homography function is d, stitching path is Breadth-First Search, the depth of the graph is n. So the total distortion could be expressed as d^n. But the distortion of the proposed method will always be d.

4 Experiments

The experiment consists of three parts. In the first part, we adopt the combination of VFC and RANSAC [13] algorithm to realize the registration operation of feature points set between images. In the second part, we compare the experimental results of the global homography model and the local homography model on the image alignment, and gives the intuitive result of image stitching. And the third part we gives the results of global homography model and the proposed method in this paper for multi-UAV remote sensing image stitching respectively.

Table 1. Feature points sets registration

	RANSAC	VFC	VFC+RANSAC
Matches of 65% overlapping	265	303	276
Correct matches of 65% overlapping	193	210	198
Matches of 50% overlapping	122	242	165
Correct matches of 50% overlapping	112	179	164
Matches of 35% overlapping	65	62	61
Correct matches of 35% overlapping	65	62	61
Matches of 20% overlapping	13	21	19
Correct matches of 20% overlapping	11	17	17
Matches of 5% overlapping	9	0	0
Correct matches of 5% overlapping	0	0	0

We select 5 pairs of images with about 65%, 50%, 35%, 20%, 5% overlapping rate respectively for feature matching, and use RANSAC, VFC, VFC+RANSAC to perform removing mismatches respectively. Then we manually label the matches follow the ground truth. Table 1 gives the statistical results of the matches after the mismatches removal operation. We need high-precision registration results because the mismatch have a big negative impact on the estimation of local homography matrices which is correspond to nearby region. The RANSAC and VFC+RANSAC method have higher precision than VFC. But RANSAC will definitely retain a partial match pair as long as there is an initial match pairs, such as the result of 5% overlapping rate's images pair, have potential threats to image stitching. But VFC could eliminate it. Figure 1 shows the visual results of points registration for a pair of images (from left to right are RANSAC, VFC, VFC+RANSAC). The green line indicates the correct matching pair preserved after the operation of outliers removing. It can be seen that the matches' distribution of RANSAC is relatively concentrated, which is unfavorable for local deformation estimation; the distributions of VFC and VFC+RANSAC are relatively scattered.

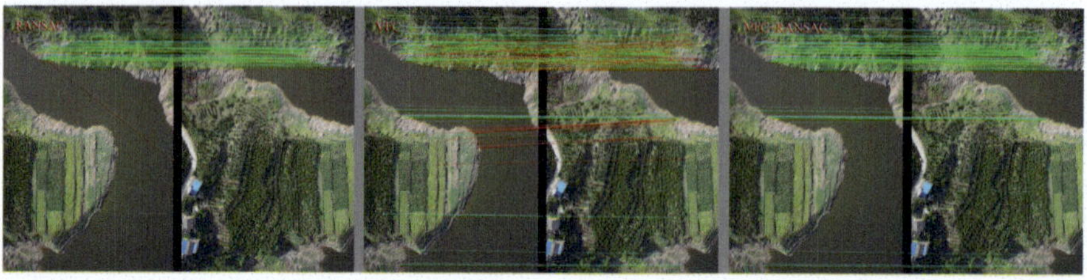

Fig. 1. Feature points sets registration

In order to show the stitching results more intuitively, we choose two pairs of images with extremely complex scene, and only use the most primitive linear weighted fusion strategy to fuse the image. Here are the experimental results of two pairs of images shown in Fig. 2, it can be seen that the parallex errors and ghosting of proposed method with local homography model are better than the global homography model.

Fig. 2. Image alignment comparison. The first row is two pairs of source images, and the second row is the stitching results of the corresponding source image with global homography model and proposed method respectively.

In the third part, we choose 8 source images. The global homography model+ bundle adjustment and the method proposed in this paper are used to realize multi-image stitching respectively. In addition, in order to improve the visual effect, we incorporate the best seam technology and multi-band blending for stitching. As can be seen from Fig. 3, in the experimental results based on the global homography matrix, the farther away from the central image, the greater the projection distortion of the image, and the visual effect of the method proposed this paper is much better.

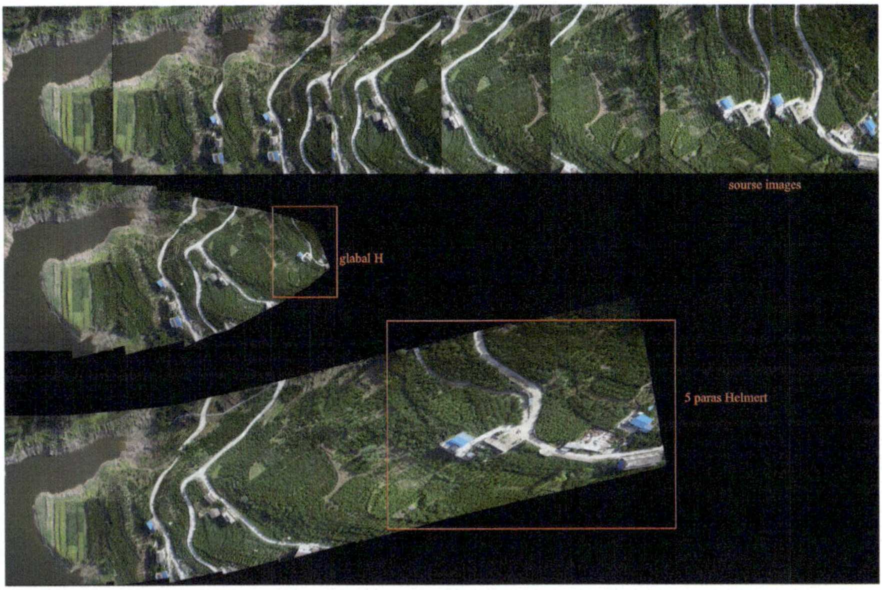

Fig. 3. Result of multi-image stitching

5 Conclusion

In this paper, a method is proposed to effectively prevent the projection distortion of multiple UAV remote sensing images stitching. Using APAP algorithm to realize accurate alignment for UAV remote sensing images with local deformation. But parallax errors still exist here, because the image alignment depends entirely on the sparse feature point set. The accuracy of the local deformation estimation depends to a large extent on the distribution of the feature points. Our future work will focus on the extraction of more evenly distributed sparse features. And the depth features may be a good solution to meet the even distribution.

References

1. Zaragoza, J., Chin, T.J., Brown, M.S., et al.: As-projective-as-possible image stitching with moving DLT. In: CVPR (2013)

2. Chang, C.H., Sato, Y., Chuang, Y.Y.: Shape-preserving half-projective warps for image stitching. In: CVPR (2014)
3. Lin, C.C., Pankanti, S.U., Ramamurthy, K.N., et al.: Adaptive as-natural-as-possible image stitching. In: CVPR (2015)
4. Brown, M., Lowe, D.G.: Automatic panoramic image stitching using invariant features. IJCV (2007)
5. Triggs, B.: Bundle Adjustment –A Modern Synthesis (1999)
6. Burt, P.J., Adelson, E.H.: A multi resolution spline with application to image mosaics. ACM TOG **2**(4), 217–236 (1983)
7. Gao, J., Kim, S.J., Brown, M.S.: Constructing image panoramas using dual-homography warping
8. Lin, W.Y., Liu, S., Matsushita, Y., et al.: Smoothly varying affine stitching. In: CVPR (2011)
9. Zhang, Z.: Parameter estimation techniques: a tutorial with application to conic fitting. Water Res. **39**(15), 3686–3696 (1997)
10. Lowe, D.G.: Distinctive image features from scale-invariant keypoints. Int. J. Comput. Vis. **60**(2), 91–110 (2004)
11. Ma, J., Zhao, J., Tian, J., et al.: Robust point matching via vector field consensus. IEEE Trans. Image Process. Publ. IEEE Sig. Process. Soc. **23**(4), 1706–1721 (2014)
12. Watson, A.: Computing Helmert transformations. J. Comput. Appl. Math. **197**(2), 387–394 (2006)
13. Fischler, M.A.: Random Sample consensus: a paradigm for model fitting with applications to image analysis and automated cartography. In: Readings in Computer Vision. pp. 726–740 (1987)

Towards Data-Driven Automatic Video Editing

Sergey Podlesnyy(✉)

Cinema and Photo Research Institute, 47 Leningradsky Ave., Moscow 126167, Russia
s.podlesnyy@nikfi.ru

Abstract. Automatic video editing involving at least the steps of selecting the most valuable footage from points of view of visual quality and the importance of action filmed; and cutting the footage into a brief and coherent visual story that would be interesting to watch is implemented in a purely data-driven manner. Visual semantic and aesthetic features are extracted by the ImageNet-trained convolutional neural network, and the editing controller is trained by an imitation learning algorithm. As a result, at test time the controller shows the signs of observing basic cinematography editing rules learned from the corpus of motion pictures masterpieces.

Keywords: Automatic video editing · Convolutional neural networks · Reinforcement learning

1 Introduction

Portable devices equipped with video cameras, like smartphones and action cameras, show rapid progress in imaging technology. 4 K sensor resolution and highly efficient video codecs like H265 provide firm ground for social video capturing and consumption. However, despite this impressive progress, users seem to prefer photos as the main medium to capture their daily events and to consume video clips produced by professionals or skilled bloggers. We argue that the main reason is the time needed to watch the video is orders of magnitude longer than photo browsing. Social video footage captured by ordinary gadget users are too long, not even speaking of quality. They should be edited before presenting them even to closest friends or relatives.

Video editing should involve at least the steps of selecting the most valuable footage from points of view of visual quality and the importance of action filmed. This is a time-consuming process on its own, but the next step is cutting the footage into a brief and coherent visual story that would be interesting to watch. This process in cinematographic and thus requires many artistic and technical skills, which makes it almost impossible for a broad range of users to success.

Since recently, the huge success of deep learning was shown in the visual data processing. We aim to apply the proven techniques of machine learning, convolutional neural networks and reinforcement learning to create automatic tools for social video editing that unifies compression/summarizing and cinematographic approaches for

Y. Liu et al. (Eds.): ICNC-FSKD 2019, AISC 1074, pp. 361–368, 2020.
https://doi.org/10.1007/978-3-030-32456-8_39

obtaining concise and aesthetically pleasant video clips of memorable moments for a broad audience of unprofessional video gadget users.

Contributions: We propose a system capable to learn the editing style from the samples extracted from the content created by professional editors including the motion pictures masterpieces, and to apply this data-driven style to cut unprofessional videos with the ability to mimic the individual style of selected reference samples.

2 Related Work

2.1 Video Summarizing

A broad literature exists for a video summarizing i.e. automatic rendering of video summary. Although this problem is less directly related to the topic of this paper, we refer to the method described in [1]. They cluster video shots using a measure of visual similarity such as color histograms or transform coefficients. Consecutive frames belonging to the same cluster a considered as video shot. The method of visual clustering is based on low-level graphical measurements and does not contain semantic information about frame content and geometry. In our work [2] we have shown that feature vectors obtained from frame image by a convolutional neural network trained to recognize wide nomenclature if classes, such as ImageNet contest [3] comprise semantic information suitable for visual examples-based information retrieval and for segmenting videos into distinct shots. Here we will show that the same feature vector can be used for automatic video editing.

2.2 Automated Editing of Video Footage from Multiple Cameras to Create a Coherent Narrative of an Event

Arev et al. [4] describe a system capable of automatic editing of video footage obtained from multiple cameras. Montage is performed by optimizing a path in the trellis graph constructed of frames of multiple sources, ordered in time. The system efficiently produces high-quality narratives by means of constructing the cost functions for nodes and edges of the trellis graph to closely correspond basic rules of cinematography. For example, they estimate 3D camera position and rotation for every source of video footage and further estimate the most important action location in a 3D scene as a joint focus of attention of multiple cameras. By estimating the distance between the camera and joint attention focus the system is capable of evaluating the scale of each shot. Cost functions for the graph edges penalize transitions between the shots more than two sizes apart. Lastly, the system promotes cuts-on-action transitions through actions estimation as local maxima of joint attention focus acceleration. The proposed method relies heavily on the availability of multiple video sources taken from different angles making it possible to derive a joint attention point in 3D space. Cinematography rules are hard-coded into the system.

To prevent including technically defective footage into the resulting film we propose to use the results shown in [5]. They used a crowd-sourced collection of rated photos to train a convolutional neural network to directly regress aesthetical score of an image. It appears that their system penalizes basic technical defects like blurred image, skyline slope, faces occlusions, etc.

2.3 Learning Editing Styles from Existing Films

As shown in [6], it is possible to model the editing process using a Hidden Markov Model (HMM). They hand-annotated existing film scenes to serve as training data. Concretely, they annotated shot sizes and action types as 'main character entering scene', 'both characters leave scene' etc. But manual annotation is tedious, and this approach still requires access to the underlying scene actions when applying the learned style to new animated scenes.

As can be seen from this review, existing methods of automatic film editing rely on broad meta-data either extracted from multiple footage sources in form of joint attention focus or parsing dialog scripts and even hand-annotating them. In particular, many methods described in the literature require the information about the shot size that is also a cornerstone for traditional cinematography editing rules.

The Rest of This Paper is Organized as Follows. In Sect. 3 we share practical details of the early prototype of the system for automatic video editing we are building. Section 4 describes the structure and methods of training of distinct system components, including the shot size classifier and visual quality assessor and reinforcement learning module for video editing control. Section 5 is devoted to the evaluation of results.

3 System Overview

Figure 1 shows the video footage features extraction pipeline. Frames are sampled from the video stream of possibly several video files. After simple preprocessing (downscaling to 227 × 227 pixels, mean color value subtraction) frame images are input into a GPU where a combination of three convolutional neural networks reside. A GoogLeNet [7] network is used to extract a semantic feature vector of length 1024. A network trained to regress aesthetical score on AVA dataset [5] produces a vector of length 2. A network trained to classify an image into three classes of shot sizes (close-up, medium shot, long shot) produces a vector of length 3.

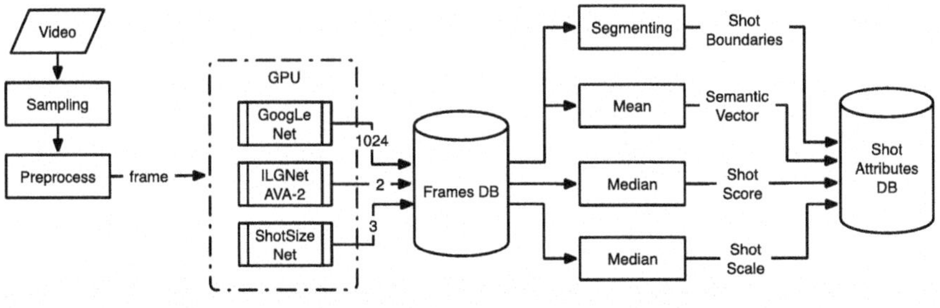

Fig. 1. Video footage features extraction pipeline.

The next step of the pipeline is segmenting video footage into coherent shots as described in [2]: to determine shot boundaries we analyze vector distance between semantic feature vectors of neighboring frames. If vector distance is large we place shot boundary there. For every shot, we calculate the attributes as the mean value of semantic feature vectors and median values of shot size vector and aesthetic score.

We perform the same pipeline for reference cinematography samples as well as for user-generated video content. For reference samples, we obtained DVD copies of 68 out of 100 best movies by The American Society of Cinematographers [8] and processed them excluding 2 min of content from the beginning and the end, thus eliminating captions, studio logos, etc. irrelevant material.

The process of automatic film editing works as follows. Features Preparation module reads data from the Shot Attributes database and feeds it into the learned model for automatic video editing. The Editing module produces a storyboard comprising complete shots or their segments from raw footage. Based on that, the Composing module composes the output video clip cutting from raw footage. We are using *ffmpeg* for composing.

4 Components and Training

4.1 Features Extraction Components

For semantic features extraction, we use GoogLeNet [7] trained by BVLC on the ImageNet dataset. We omit the final classification layer and use the output of layer 'pool5/7 × 7_s1' as a semantic feature vector of length 1024. After we have extracted feature vectors from over 1,670,000 frames of masterpieces mentioned in Sect. 3 (a), we performed an incremental PCA to reduce the dimensionality of feature vectors to 64. Residual error for the last batch on incremental PCA was 0.0029.

To automatically detect the shot size, we trained a classifier to distinguish images between 3 classes: close-up, medium shot, long shot. We created a dataset using detailed cinematography scripts of 72 episodes of various Russian TV series. The dataset contained 566,000 frames of nearly even distribution of frames belong to three classes. We trained the GoogLeNet-based network structure with softmax loss after 3-outputs fully connected the final layer. Top-1 testing accuracy was 0.938. We used a trained network published in [5] for the aesthetic scoring of the shot.

4.2 Automatic Editing Model Training

We use motion pictures masterpieces as reference samples of good editing. We model video editing as a process of making control decisions whether to include a shot into a final movie or skip it. Further, we model the editing rhythm by learning to make fine-grained decisions of the duration of a shot that was selected to be included in the final movie.

Concretely, we model the video editing process as a sequence learning problem [9] with the Hamming loss function. We selected the following labels for shots sequence labeling (see Table 1).

Table 1. Action labels for sequence learning.

Label	Action
1	Include shot, duration <1 s
2	Include shot, duration 1…3 s
3	Include shot, duration 3…9 s
4	Include shot, duration >9 s
5	Skip shot

We prepared the training data as follows. As described in 4.1 we collect a sequence of shots into a clip having a duration of around 2 min. Each shot in the clip sequence is labeled according to its duration with labels 1–4 (Table 1). This gives us a reference 'expert' movie cut. To give our model a concept of bad montage, we produce around 40 augmented clips from each reference ones. We do so by randomly inserting shots taken from other movies in masterpieces collection, thus breaking the author's idea of an edit. We assign such a shot a label 5. Additionally, we assign label 5 to all shots having an aesthetical score below some threshold (e.g. 0.1). As a result, we got the training set having 108,491 sample clips, and shuffled the samples. We used *vowpal wabbit* [10] for training in sequence learning task using DAGGER, an iterative algorithm that trains a deterministic policy in imitation learning setting where expert demonstrations of good behavior are used to learn a controller. We construct the enhanced state vector $\boldsymbol{s}$ as follows: we use action labels $\boldsymbol{a}$ in historical span 6, and add neighboring semantic feature vectors from -6th shot to $+3^{rd}$ shot from the current one. Held out loss after 32 epochs of training was 4.06 while the average length of a sequence was 21.

5 Evaluation

5.1 Qualitative Evaluation

In Fig. 2(a) the test footage is shown in the storyboard format. It contains an unmodified fragment of 'Cool Hand Luke', a 1967 film by Stuart Rosenberg. Figure 2(b) shows the result of automatic editing. We may observe that the algorithm has shortened the footage and deleted some shots that produced abrupt transitions. For example, the transition between medium I shot and the truck (very long distance shot) is quite abrupt, and since our model learned average rules of editing it removed the truck shot.

Figure 2(c) shows test footage constructed from the fragment of the movie not seen by the model during training. It is 'Gagarin the First in Space', 2013, editor Pavel Parhomenko. The fragment was augmented by random inserting shots taken from random places of the same movie. In Fig. 2(d) we see the result of auto-editing. It correctly removed the shot with a close-up on the radio but left a few other foreign shots. However overall cut looks smooth and shows a nice gradual change of tone throughout the length of the clip.

In Fig. 2(e) we used a fragment from 'Das Boot', 1981 by Wolfgang Petersen and inserted a few shots from 'Fanny and Alexander', 1982 by Ingmar Bergman, having

Fig. 2. Qualitative results: a, c, e, g – raw materials, b, d, f, h – edited clips. Shots automatically removed from raw materials are marked with a red frame.

the tone very much alike the close-up faces of ‘Das Boot’ but breaking the rule of 180° of action: in ‘Das Boot’ an officer speaks to a crew in front of him, so the correct cut would be to montage shots with facing directions. Our model correctly removed the wrong shots as shown in Fig. 2(f).

At last, Fig. 2(g) shows the typical unprofessional footage taken by a GoPro camera attached to a bike. Figure 2(h) demonstrates the result of automatic editing which has resulted in a brief clip showing some sights of dynamism. It contains the main character and beautiful surroundings. A professional editor would probably include shots with the sun shining and a bike wheel close-up. We should leave these artistic enhancements for future work.

5.2 Quantitative Evaluation

To estimate how well the proposed method learns basic video editing rules from unlabeled reference samples of motion pictures masterpieces, we have manually counted the number of transitions between the shot sizes in reference footage, in raw unprofessional footage, and in the automatically edited clips. The following shot sizes are distinguished: Detail - Close up - Medium 1 - Medium 2 - Long shot - Very long shot. It was advised, for example by famous Russian cinematographer Kuleshov in the early XX century, that transitions between the shots should occur at two size steps, e.g. between the medium 1 shot and the long shot. Transitions ‘Detail - Close up’ and ‘Long shot - Very long shot’ are also allowed.

In order to evaluate whether the trained DAGGER model has learned the very basic principles of video editing from the data features extracted from motion pictures masterpieces we have calculated the number of transitions between shots of different sizes in three corpuses of footage: clips sampled from masterpieces dataset, clips sampled from unprofessional video footage and clips automatically edited by the DAGGER model. Figure 3 shows histograms of the distribution of transitions between the shots. It is easy to see that unprofessional video has somewhat even distribution of transitions while motions pictures masterpieces and DAGGER-edited video clips demonstrate a distribution with distinct elevation in the part corresponding to ‘2 steps difference’ between the shot sizes. The standard deviation between masterpieces histogram and unprofessional video histogram is 0.004 while STDEV between masterpieces histogram and clips edited by our algorithm is 0.001 which is 4 times improvement.

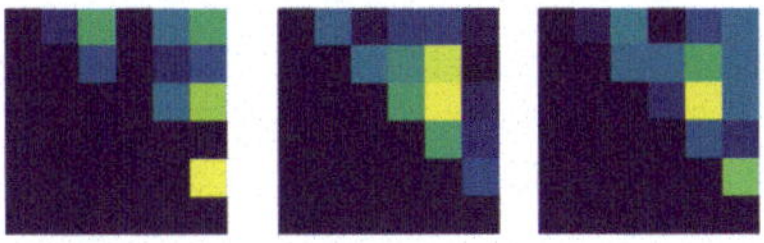

a) b) c)

Fig. 3. Histograms of the distribution of transitions between the shots. (a) Unprofessional video; (b) Motion Picture Masterpieces; (c) Automatically edited by our algorithm.

6 Conclusion and Future Work

These early results make us optimistic to pursue further development of an automatic video editing system capable to mimic the user-selected montage style. We plan to dramatically improve the quality of the resulting video clip by introducing new features into the state vector to give the model clues of the values the users expect from a video. For example, a measure of importance inspired by [1] but constructed from deep learning-based semantic features should promote unique fragments that are most important to preserve. Another line of research here is utilizing the advances in deep reinforcement learning with end-to-end training given visual input data.

References

1. Uchihachi, S., Foote, J.T., Wilcox, L.: US Patent 6,535,639 B1 (2003)
2. Podlesnaya, A., Podlesnyy, S.: Deep learning based semantic video indexing and retrieval. In: Proceedings of SAI Intelligent Systems Conference, pp. 359–372. Springer, Cham, 21 September 2016
3. Russakovsky, O., Deng, J., Su, H., Krause, J., Satheesh, S., Ma, S., Huang, Z., Karpathy, A., Khosla, A., Bernstein, M., Berg, A.C., Fei-Fei, L.: ImageNet large scale visual recognition challenge. CoRR, arXiv:1409.0575 (2014)
4. Arev, I., Park, H.S., Sheikh, Y., Hodgins, J., Shamir, A.: Automatic Editing of Footage from Multiple Social Cameras. ACM Trans. Graph. **33**(4), 1–11 (2014)
5. Xin, J., Chi, J., Peng, S., Tian, Y., Ye, C., Li, X.: Deep image aesthetics classification using inception modules and fine-tuning connected layer. In: 8th International Conference on Wireless Communications and Signal Processing (WCSP), Yangzhou, China, 13–15 October (2016)
6. Merabti, B., Christie, M., Bouatouch, K.: A virtual director using hidden markov models. In: Computer Graphics Forum, Wiley (2015). https://doi.org/10.1111/cgf.12775. Hal-01244643
7. Szegedy, C., Liu, W., Jia, Y., Sermanet, P., Reed, S., Anguelov, D., Erhan, D., Vanhoucke, V., Rabinovich, A.: Going deeper with convolutions. CoRR, arXiv:1409.4842 (2014)
8. ASC Unveils List of 100 Milestone Films in Cinematography of the 20th Century, 08 January 2019. https://theasc.com/news/asc-unveils-list-of-100-milestone-films-in-cinematography-of-the-20th-century
9. Ross, S., Gordon, G., Bagnell, J.A.: A reduction of imitation learning and structured prediction to no-regret online learning. In: International Conference on Artificial Intelligence and Statistics (AISTATS) (2011)
10. Langford, J., Li, L., Strehl, A.: Vowpal wabbit online learning project (2007). http://hunch.net/?p=309

A Deep Speaker Embedding Transfer Method for Speaker Verification

Kai Zhou, Qun Yang(✉), Xiusong Sun, and Shaohan Liu

College of Computer Science and Technology, Nanjing University of Aeronautics and Astronautics, Nanjing 210016, Jiangsu, China
1131966336@qq.com, Qun.Yang@nuaa.edu.cn

Abstract. Recently a kind of deep speaker embedding called x-vector has been proposed. It is extracted from deep neural network and considered as a strong contender for next-generation representation for speaker recognition. However, training such DNNs requires a lot of data, usually thousands of hours. If we want to apply x-vector to Mandarin task but we only have a small amount of data, we can train DNNs on another language and fine-tune the DNNs with these little data. Firstly, we proposed a pure data driven method to transfer DNNs across languages and tasks. Secondly, we investigated the question that how to choose between training DNNs from scratch and reusing a pre-trained model by the transfer method we proposed. To answer the question, in this paper, we present the results of adapting a x-vector based speaker verification system from English to Mandarin by fine-tuning the front-end DNNs. We investigate the trend of performance improvement in these two training strategies when data size increasing. Experiment results show that adapting a pre-trained English model with a small amount of Mandarin data can easily reduce the equal error rate (EER). They also demonstrate that system trained from scratch is able to achieve better performance only when feed enough data. Finally, we test the performance of the two systems in noisy environment and found that system trained from scratch outperforms system fine-tuned with a pre-trained model.

Keywords: Transfer learning · Speaker verification · Speaker recognition · Deep learning · Data augmentation

1 Introduction

Speaker verification (SV) is an important biometric recognition technology, which has been widely used in various fields. For example, in the fields of access control, transaction authentication, judicial forensics and personalization, automatic speaker verification system has become an indispensable technology [1]. After decades of research, speaker verification technology has achieved significant performance improvement. Nevertheless, there are still challenges to improve SV system performance [2], because the complex convolution of various speech factors, such as phone content and channel, makes it difficult to separate speaker information from them.

In previous work, David Snyder et al. proposed a robust speaker embedding extracted from DNN, namely x-vector [3–6]. They studied deep neural networks

Y. Liu et al. (Eds.): ICNC-FSKD 2019, AISC 1074, pp. 369–376, 2020.
https://doi.org/10.1007/978-3-030-32456-8_40

(DNNs) embeddings for speaker recognition and found that data augmentation is an effective strategy for improving their performance and also easily implemented. After including a large amount of augmented English speech, the x-vectors achieved much lower error-rates than their best baseline on Speakers in the Wild [18]. Their experiment showed that x-vector outperforms both acoustic i-vector and phonetic bottleneck i-vector.

In the present work, we will investigate the adaptability of x-vector to new language for the task of speaker verification in various cases. More concretely, we want to explore whether fine-tuning a pre-trained model to a new language is feasible and how much new data is needed. Some of the main research questions we aim to answer are: Does the system perform well for a language it has not been trained for? Is it worth to train from scratch or is it better to transfer a pre-trained x-vector DNN? How should we choose training methods for data of different scales and in noisy environment? To answer these transfer learning questions in the context of speaker verification and DNNs, we will carry out a series of comparative experiments.

Transfer learning [7] mainly refers to some techniques for adapting pre-trained models to new tasks. It has been used in speech recognition and improve the speed of training as well as performance [8]. Moreover, transfer learning techniques are applied in speech enhancement to adapt pre-trained English generative adversarial network to Korean and Catalan [9]. In the case of speaker verification and speaker recognition, most research on transfer learning focuses on end-to-end systems [10–12]. Baidu researchers carried out language transfer experiments on Deep Speaker in [12]. In their report, they present results that suggest adapting from a model trained with Mandarin can improve accuracy for English speaker recognition. And a series of experiment proof that Deep speaker can take full advantage of transfer learning to solve the speaker recognition problems on small data sets. This ability to migrate across languages stems from its special DNN structure. In the process of fine-tuning, DNN can automatically learn the feature of new data and adjust network parameters to the new task.

Our contributions are in the exploration of generalization capabilities of x-vector across languages, showing that (1) a pre-trained system does perform fine with new languages but easily progressing, (2) performance improved after pre-trained model fine-tuned with new training data and our pure data training method. And we observe that (3) system fine-tuned with pre-trained model perform better than trained from scratch when training data is not very large. However, (4) system trained from scratch obtain lower EER than fine-tuned with pre-trained model as long as feed enough data. In addition, (5) system trained from scratch perform better in noisy environment.

The rest of this paper is organized as follows. First, a kind of deep speaker embedding is introduced. Then, in Sect. 3 we proposed our pure data transfer method. In Sect. 4, the experimental details are covered, with the specifics of the architecture and data used to conduct the experiments. Results are discussed in Sect. 4, and some conclusions that wrap up our observations from the experiments are discussed in Sect. 5.

2 Deep Speaker Embedding

In this section, we will give a brief introduction to the deep speaker embedding. It is based on the DNN embeddings in [3–5] and described in detail in [6].

The features we used in DNN are 30 dimensional filterbanks (different from [6]) with a frame-length of 25 ms, mean-normalized over a sliding window of up to 3 s. We use energy SAD to filter out silent frames. The architecture of DNN is listed in Table 1. Firstly, we suppose an input segment has T frames. The first five layers do frame-level speech features calculations, with a small temporal context centered at the current frame t. Then the status pooling layer computes mean and standard deviation of all T frames-level outputs from layer frame5. After that, layer segment6 and layer segment7 operate on segment-level speech features to concatenate the mean and standard deviation together. Finally, layer softmax output the predicted speaker. The nonlinearities are all rectified linear units (ReLUs).

Table 1. The embedding DNN architecture. x-vectors are extracted at layer segment6, before the nonlinearity. The N in the softmax layer corresponds to the number of training speakers.

Layer	Layer context	Total context	Input × output
Frame1	$\{t-2, t+2\}$	5	120 × 512
Frame2	$\{t-2, t+2\}$	9	1536 × 512
Frame3	$\{t-3, t+3\}$	15	1536 × 512
Frame4	$\{t\}$	15	512 × 512
Frame5	$\{t\}$	15	512 × 1500
Status pooling	$[0, T)$	T	1500T × 3000
Segment6	$\{0\}$	T	3000 × 512
Segment7	$\{0\}$	T	512 × 512
Softmax	$\{0\}$	T	512 × N

We train this DNN to classify the N speakers in the training data. The input of DNN includes features of a segment speech and corresponding speaker label. After training, x-vectors are extracted from the affine component of layer segment6. Excluding the softmax output layer and segment7 (because they are not needed after training).

3 Pure Data Training Transfer Method

The DNNs we introduced in Sect. 2 originally trained for speaker classification, but we utilize the similarities between speaker classification and speaker verification tasks to reuse the first six layers in speaker verification. However, those layer of DNNs are jointly trained which means that if the output embeddings from layer six are used in different task, such as verification, the joint training will be suboptimal. The reason is that the last layer involves free parameters, so part of the discriminant information will be learned by the classifier, which, unfortunately, will be thrown away when performing verification tasks on other speakers.

So we drew lessons from [19] and discard the parametric classifier instead of using the speaker features to classify the speakers directly. Specifically, each speaker information of utterance u in the training set can be represented by 512 dimension embedding computed by the first six layers, so we represent speaker s with the average of all the utterances belonging to this speaker, given by:

$$x(s;\theta) = \frac{1}{|\varepsilon(s)|}\sum_{\varepsilon(s)} f(u;\theta), \tag{1}$$

where ε(s) is the set of utterance belonging to speaker s, and f(u; θ) is the speaker embedding of utterance u parameterized by θ. By these speaker vectors x(s; θ), each utterance u can be classified by a cosine distance based classifier as follows:

$$p(s|f(u;\theta)) = \frac{e^{\cos(f(u;\theta),x(s;\theta))}}{\sum_{s'} e^{\cos(f(u;\theta),x(s';\theta))}}, \tag{2}$$

where function cos(·,·) represents cosine distance. We use the cross entropy between the classification output p(s|f(u; θ)) and the ground truth label s as the cost function to train the system, formulated by:

$$L(\theta) = \sum_{n} \log \mathrm{s(n)}|f(u(n);\theta), \tag{3}$$

where u(n) and s(n) are the n-th utterance and the corresponding ground truth label. Note that, cost function involves only one free parameters θ, so all the discriminative information provided by the training data is solely learned by the first six layers of DNNs. For this reason, we call this approach pure data transfer method.

4 Experiments Setup

4.1 Database Details

The Open Source Mandarin Speech Corpus we used in experiments is AISHELL-ASR0009-OS1, and we call it AISHELL in this paper. The details of this corpus are described in [13]. The corpus is divided into several data sets as list in Table 2.

Table 2. Dataset details

Data Set	Speakers	Utts/spk	Duration
Train	340	8–60	140 h
Enroll_clean	20	3	5 min
Test_clean	20	45	1 h
Enroll_noisy	20	3	5 min
Test_noisy	20	45	1 h
Train_PLDA	340	30–40	10 h

Before the corpus divided, there are some preprocess, include removing utterances short than 4 s and speakers whose utterances less than 8. The enroll_noisy and test_noisy are test sets manually generated by add different kinds of noise, in order to evaluate performance of system in noisy environment. Our noise come from two databases called MUSAN and RIR in [14, 15], which are freely available from http://www.openslr.org. These two databases are used not only to generate noisy test set, but also to data augmentation before DNN training. The augmentation strategy in this work similar to that described in [6].

4.2 Experiment Steps

Our experiment evaluate two training strategies including training DNNs from scratch and fine-tuning DNNs from pre-trained model by the pure data transfer method. In both ways, we train the DNNs first and then use it to extract x-vectors. After that, the steps are the same. Firstly, mean normalization will be applied. Then we reduce dimension of x-vectors from 512 to 128 with LDA algorithm. Finally, we train a PLDA model as back-end to compute scores. In order to control experiment variable, we use the same dataset (Train_PLDA in Table 2) for PLDA training in all our experiments.

4.3 Pre-trained Model

The pre-trained model used in our experiment is available at http://www.kaldi-asr.org/models/m8. The DNNs in pre-trained x-vector system trained on augmented VoxCeleb1 and VoxCeleb2 in [16, 17]. The Speakers in the Wild dataset is used to evaluate the system. The results of the x-vector system in terms of equal error rate (EER%) is 3.5%, while 5.7% for i-vector. In the following experiments, we use this DNN model and its tuned model to extract x-vectors from Mandarin dataset.

4.4 Data Arrangement and Experiments

We train DNNs over a range of speech duration based on two different starting configurations, which are pre-trained model and random initialization. The duration of data we select to train DNNs is list as follow: 3.6, 5.6, 7.5, 9.5 and 11.3 h. And the number of speakers is the same 340, but the corresponding utterances per speaker are 10, 15, 20, 25, 30. In this way, we can observe the performance differences between the two initialization schemes, with varying speech duration in order to explain transfer learning and data requirements. Moreover, those systems are evaluated with noisy test data to observe their performance in noise environment.

A baseline is also setting to check out the performance of pre-trained model in both test_clean and test_noisy. Finally, we use all of data in Aishell to train those two systems.

5 Experiments Result

We report results in terms of equal error-rate (EER) and the minimum of the normalized detection cost function (DCF) at $P_{Target} = 10^{-2}$ and $P_{Target} = 10^{-3}$. The results is shown in Table 3. And the results of the previous metrics as a function training data time for both training methods retrain and transfer are shown in Fig. 1. We now elaborate on them.

First of all, we note that the pre-trained English system does not perform very bad with the new languages. However, fine-tuning the pre-trained model with 3.6 h data obtained a 6.6% EER relative reduction. It proves that our pure data transfer method is effective. Comparing the two red curves in Fig. 1, we can see that the dotted line is always above the solid line which means system transfer from pre-trained model perform better than that trained from scratch when training data is less than 12 h. However, when training data reaches 140 h, system trained from scratch obtain 0.23% in EER, which decreased by 52.1% compared with transfer system.

Table 3. Results of two training strategies using various size of data evaluated with two kind of test set. Y means system trained from pre-trained model, while N means system trained from scratch. The best results are **boldface**

Data size (h)	Pre-trained	Test_clean			Test_noisy		
		EER (%)	DCF (0.01)	DCF (0.001)	EER (%)	DCF (0.01)	DCF (0.001)
0	Y	3.01	0.472	0.699	9.12	0.889	0.981
3.6	N	3.62	0.562	0.731	8.38	0.846	0.960
	Y	2.82	0.421	0.682	8.70	0.830	0.951
5.6	N	3.01	0.403	0.623	8.06	0.791	0.926
	Y	2.41	0.359	0.522	8.72	0.814	0.931
7.5	N	2.45	0.365	0.576	7.88	0.770	0.924
	Y	2.11	0.301	0.496	7.96	0.762	0.920
9.5	N	2.12	0.312	0.503	6.93	0.726	0.885
	Y	1.87	0.289	0.426	6.96	0.681	0.878
11.3	N	1.98	0.259	0.442	5.33	0.647	0.848
	Y	1.76	0.226	0.398	5.55	0.642	0.833
140	N	**0.23**	**0.044**	**0.101**	**1.76**	**0.219**	**0.416**
	Y	0.35	0.052	0.116	1.93	0.212	0.356

Apart from the aforementioned experiment result, we also see that retrained system performs better than transfer system in noisy data which mostly due to that data augmentation plays a stronger role in retrain than transfer. More specifically, it is the strict restrictions on parameter changes in the fine-tuning process limit the effect of data augmentation.

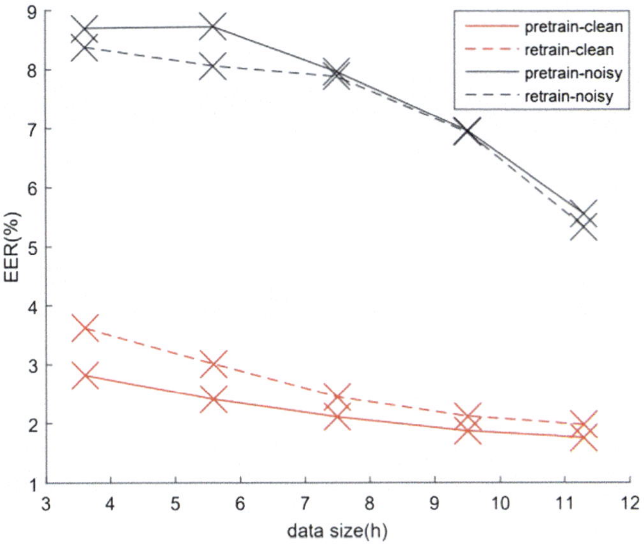

Fig. 1. EER curves with increasing data. Solid lines (transfer) represent model tuned with English pre-**trained** model. And dotted lines (retrain) represents model trained from scratch. Red liens represent results in clean data while black lines represent results in noisy data.

6 Conclusions

In this paper, we proposed a pure data driven transfer method. Then we show that this transfer method is effective for cross language speaker verification especially in low resource environment. But if there is enough data (140 h for 340 speakers) or application environment has noise interference, training model from scratch is a better choice than transfer a pre-trained model. In the future, we intend to apply attention mechanisms in front-end networks to suit different languages. We also plan to use attention model to separate speakers with different mother tongues.

References

1. Hansen, J.H.L., Hasan, T.: Speaker recognition by machines and humans: a tutorial review. IEEE Sig. Process. Mag. **32**(6), 74–99 (2015)
2. Zheng, T.F., Jin, Q., Li, L., et al.: An overview of robustness related issues in speaker recognition. In: Signal and Information Processing Association Annual Summit and Conference (APSIPA), 2014 Asia-Pacific, pp. 1–10. IEEE (2014)
3. Snyder, D., Garcia-Romero, D., Povey, D.: Time delay deep neural network-based universal background models for speaker recognition. In: 2015 IEEE Workshop on Automatic Speech Recognition and Understanding (ASRU), pp. 92–97. IEEE (2015)
4. Snyder, D., Ghahremani, P., Povey, D., et al.: Deep neural network-based speaker embeddings for end-to-end speaker verification. In: Spoken Language Technology Workshop (SLT) 2016 IEEE, pp. 165–170. IEEE (2016)

5. Snyder, D., Garcia-Romero, D., Povey, D., et al.: Deep neural network embeddings for text-independent speaker verification. In: Proceedings of Interspeech, pp. 999–1003 (2017)
6. Snyder, D., Garcia-Romero, D,, Sell, G., et al.: X-vectors: robust DNN embeddings for speaker recognition. In: 2018 IEEE International Conference on Acoustics, Speech and Signal Processing (ICASSP), pp. 5329–5333. IEEE (2018)
7. Pan, S.J., Yang, Q.: A survey on transfer learning. IEEE Trans. Knowl. Data Eng. **22**(10), 1345–1359 (2010)
8. Kunze, J., Kirsch, L., Kurenkov, I., et al.: Transfer learning for speech recognition on a budget. arXiv preprint arXiv:1706.00290 (2017)
9. Pascual, S., Park, M., Serrà, J., et al.: Language and noise transfer in speech enhancement generative adversarial network. In: 2018 IEEE International Conference on Acoustics, Speech and Signal Processing (ICASSP), pp. 5019–5023. IEEE (2018)
10. Heigold, G, Moreno I, Bengio S, et al. End-to-end text-dependent speaker verification. In: 2016 IEEE International Conference on Acoustics, Speech and Signal Processing (ICASSP), pp. 5115–5119. IEEE (2016)
11. Zhang, S.X., Chen, Z., Zhao, Y., et al.: End-to-end attention based text-dependent speaker verification. In: 2016 IEEE Spoken Language Technology Workshop (SLT), pp. 171–178. IEEE (2016)
12. Li, C., Ma, X., Jiang, B., et al.: Deep speaker: an end-to-end neural speaker embedding system. arXiv preprint arXiv:1705.02304 (2017)
13. Bu, H., Du, J., Na, X., et al.: AIShell-1: an open-source Mandarin speech corpus and a speech recognition baseline. IN: 2017 20th Conference of the Oriental Chapter of the International Coordinating Committee on Speech Databases and Speech I/O Systems and Assessment (O-COCOSDA), pp. 1–5. IEEE (2017)
14. Snyder, D., Chen, G., Povey, D.: Musan: a music, speech, and noise corpus. arXiv preprint arXiv:1510.08484 (2015)
15. Ko, T., Peddinti, V., Povey, D., et al.: A study on data augmentation of reverberant speech for robust speech recognition. In: 2017 IEEE International Conference on Acoustics, Speech and Signal Processing (ICASSP), pp. 5220–5224. IEEE (2017)
16. Nagrani, A., Chung, J.S., Zisserman, A.: VoxCeleb: a large-scale speaker identification dataset. arXiv preprint arXiv:1706.08612 (2017)
17. Chung, J.S., Nagrani, A., Zisserman, A.: VoxCeleb2: deep speaker recognition. arXiv preprint arXiv:1806.05622 (2018)
18. McLaren, M., Ferrer, L., Castan, D., et al.: The speakers in the wild (SITW) speaker recognition database. In: Interspeech, pp. 818–822 (2016)
19. Li, L., Tang, Z., Wang, D., et al.: Full-info training for deep speaker feature learning. In: 2018 IEEE International Conference on Acoustics, Speech and Signal Processing (ICASSP), pp. 5369–5373. IEEE (2018)

A Deep Neural Network Model for Music Genre Recognition

M. Suero, C. P. Gassen, D. Mitic, N. Xiong(✉), and M. Leon

Mälardalen University, Västerås, Sweden
ning.xiong@mdh.se

Abstract. Convolutional neural networks (CNNs) have become increasingly important to deal with many image processing and pattern recognition problems. In order to use CNNs in music genre recognition, spectrograms (visual representation of the spectrum of frequencies of a signal as it varies with time) are usually employed as inputs of the network. Yet some other approaches used music features for genre classification as well. In this paper we propose a new deep network model combining CNN with a simple multi-layer neural network for music genre classification. Since other features are taken into account in the multi-layer network, the combined deep neural network has shown better accuracy than each of the single models in the experiments (Code available at: https://github.com/risengnom/Music-Genre-Recognition.).

Keywords: Music genre classification · Multi-layer neural network · Convolutional neural network · Deep network · Mel frequency

1 Introduction

Music genre classification is a field of Music Information Retrieval (MIR), which, combined with other fields like: artist recognition, identification of a song's geographical origin and appliance of music emotion recognition, would result in better services to users based on their musical preferences. This is a requirement that many companies try to provide. Music genre recognition is a topic that has been under research for several years, so many different approaches for music classification according to genre have been developed [1–4]. The task of music classification according to genre is not easy, because of several reasons such as: audio signal representation and choice of subset of features from vast variety of possible ones [5], feature integration, [1,5], combination of different classifiers [6], problems with certain genres overlapping and many more. In the tests carried out by Perrot [7], humans could identify the correct genre 70% of the times.

One of the first works on music genre classification was published in 2002 [1] by G. Tzantakis and P. Cook. The authors described that different music genres have certain characteristics that differentiate themselves from each other. Multiple categories of features for classification were proposed in [1]. In general, features were split into three different feature sets: timbral, which is most

Y. Liu et al. (Eds.): ICNC-FSKD 2019, AISC 1074, pp. 377–384, 2020.
https://doi.org/10.1007/978-3-030-32456-8_41

commonly used in classification, rhythm and pitch. Later works, such as [5,8], apply the same set of features, among which the Mel-Frequency Cepstral Coefficients (MFCC) were shown to be the most important for classification. MFCC is derived from a Short Time Fourier transform (STFT) and then transformed into a logarithmic scale that mimics the accuracy with which humans perceive different frequencies. The most information about the audio signals is contained within the first few coefficients [9].

In recent works [4,10], it is shown that spectrograms obtained from music files can be successfully used to visually represent the audio signal, more specifically the changes in its frequency spectrum over time. These spectrograms are then used as inputs for the Convolutional Neural Network (CNN) that is proven to be a powerful technique for image classification. Additionally, many implementations for applying different types of neural networks (NN) as classifiers were found in the literature, where features extracted from music files were integrated and used as inputs in order to classify songs. In [11], authors used the combination of Long Short-Term memory Neural Network and Support Vector Machine (SVM) as a classifier, while the feature set is used as inputs for this combined network to learn. In [3], the authors used a Deep Belief Network (DBF) to identify which features from the set best describe the song, after which these features are passed to the SVM as inputs for classification.

This paper proposes a deep neural network model combining CNN with a Feed Forward Neural Network (FNN), stated as CNN-FNN, for music genre recognition. First the song is represented with a Mel-spectrogram which provides inputs to the CNN. Secondly the outputs of the CNN are adopted as inputs to the FNN together with a set of other features obtained from the same song. The idea behind this proposal is to exploit the power of CNN to extract information from images (Mel-spectrogram in this case), which is then fused with other features via FNN to enable more reliable classification. The proposed CNN-FNN model has been tested on the "GTZAN" dataset [1], demonstrating better classification accuracy than each of the single models.

The rest of the paper is organized as follows. In Sect. 2 the deep neural network CNN-FNN is described, together with the features that are fed to this combined model. The experiments and the results are presented in Sect. 3. Finally the conclusion is given in Sect. 4.

2 Methodology

In this section our approach will be explained in detail, this section will be divided as follows: In Sect. 2.1, the different features are described. In Sect. 2.2, our approach, CNN-FNN, is described.

2.1 Feature Extraction

Choosing the feature set that would give the most relevant information about a song and at the same time consists of a set of features that are not redundant, presented a challenging task. After some research, the features used in our combined model are: spectral centroid, spectral bandwith, spectral rolloff, zero crossing rate, chroma information and Mel-Frequency Cepstral Coefficients (MFCC). All of these features were calculated per song segment. Most of these features are categorized as timbral (with exception to chroma, which belongs to the pitch category), and are derived from STFT using short-time segmentation of the audio signal. After the values are successfully calculated for every frame, the mean value over all the frames are calculated. The different features are described below.

Spectral Centroid - Value that depicts the centroid of frequency amplitude spectrum. It is calculated as the weighted mean of the frequencies present in the sound [1].

Zero Crossing - Number that represents how many times the signal crossed the value 0 in the time domain. This feature is often used for classifying voiced and unvoiced audio signals [5].

Spectral Rolloff - Frequency bin R_t bellow which 85% of the magnitude distribution is contained for a specific frame [1].

Spectral Bandwidth - Represents the weighted average of difference between spectral components and the brightness (centroid). It is an indication of the frequency range in the frame [8, 12].

Mel Frequency Coefficient of Cepstrum (MFCC) - One of the regularly used timbre features in both speech and music genre recognition. It is obtained from STFT, by first splitting the audio signal in small frames (on which the signal is considered stationary), applying the hamming windowing function, calculating the Fast Fourier Transform (FFT) for each frame, applying log-amplitude on magnitude spectrum (which is used for mimicking the human hearing system) after which the frequency bins are then smoothed and grouped according to Mel-frequency scale. The last step is applying Discrete Cosine Transform (DCT) in order to decorrelate the components of the resulting Mel-spectrum vector [5].

Chroma Information - Used for representing the signal energy through 12 bins which represent 12 chroma classes (12 distinct semitones of musical octave). This feature is derived from pitch-based log-frequency spectogram, by taking bands of pitches that belong to the same [13].

2.2 CNN-FNN: A Deep Model Combining CNN and FNN

In this section, a deep model with combination of CNN [14] and FNN [15] is proposed. First a Mel-spectrogram [9] is extracted from a part of the song and used as inputs for the CNN. The CNN has already proven successful in the task of classifying conventional images [16] and has already been used in combination with Mel-spectrograms to classify music genres [4]. After the classification by the CNN, the results are then fed to the FNN as inputs together with different music features. The FNN then decides the final prediction for a song's genre. A general overview of the model can be found in Fig. 1.

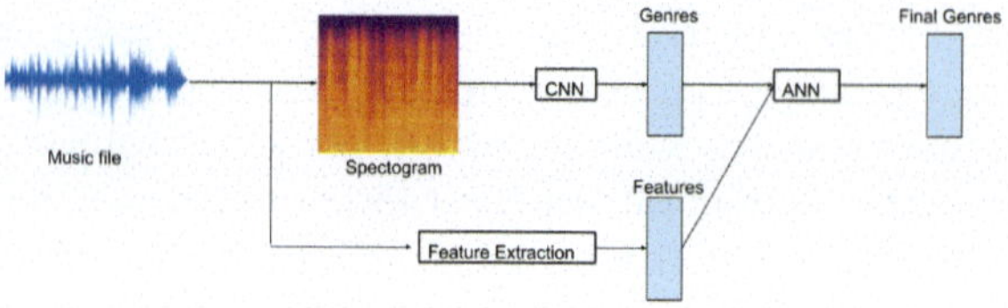

Fig. 1. Draft of the network architecture

Mel-spectrograms, and not normal frequency spectograms, are used as inputs to the CNN since they mimic better how humans hear. They have a greater emphasis on lower frequencies and smaller precision on higher frequencies [4,9]. The conversion of a frequency f from Hertz into Mel scale is done by the following equation [17]:

$$m = 2595 \log_{10}\left(1 + \frac{f}{700}\right) \quad (1)$$

Spectrograms extracted from the songs have 64 Mel scales and 256 time windows. The songs are split into multiple 3 second long snippets from which the Mel-spectrograms are calculated and then fed into the CNN. To extract the information, the CNN uses the following layout and parameters:

- Input layer: 64*256 neurons as defined by the size of the Mel-spectrogram
- 1st Convolution layer: 64 3*3 filters, stride of 1, ReLU activation function
- 1st Pooling layer: Size of 2*4
- 2nd Convolution layer: 64 3*5 filters, stride of 1, ReLU activation function
- 2nd Pooling layer: Size of 2*4
- Dropout Layer: Probability of 0.4
- Fully connected layer: 32 neurons, ReLU activation function
- Dropout Layer: Probability of 0.2
- Output layer: 10 neurons (one for each genre), Softmax activation function.

A small amount of dropout was used to help avoid overfitting to the training set. The layout of the CNN is based on the layout proposed by [4].

During training, the weights of the network are updated through backpropagation [18]. The original genre is used as a target, while the Mel-spectrogram of the sample is the input for the network. Once the loss of the validation set does not improve for several epochs the training is stopped and the network is tested for its accuracy with the remaining songs. Afterwards, the features described in Sect. 2.1 are calculated of the song snippet and appended to the output of the corresponding calculation from the CNN. All of this information is then put into the FNN for each song snippet separately. The layout of the FNN is given below:

- Input layer: 35 neurons as defined by the input data consisting of the following values:
  - Chroma STFT
  - Spectral Centroid

  - Spectral Bandwidth
  - Spectral Rolloff
  - Zero Crossing Rate
  - 20 MFCCs
  - 10 genre predictions from the CNN.
- 1st Fully connected layer: 35 neurons, ReLU activation function
- 2nd Fully connected layer: 20 neurons, ReLU activation function
- 3rd Fully connected layer: 15 neurons, ReLU activation function
- Output layer: 10 neurons (one for each genre), Softmax activation function.

Both networks used the same songs for training and testing. The CNN is always trained first and the final predictions for each sample are saved and used in the fully connected FNN to train and test. The networks are trained with the "Adams" optimizer [19]. At the end, the FNN can be tested for its accuracy. The predictions from the second network are the final predictions for the genre of the song sample.

3 Experiments

This section will be divided as follows. In Sect. 3.1 the different parameters and settings used in the experiment will be explained. In Sect. 3.2 the results and comparisons between algorithms will be shown and compared.

3.1 Experimental Settings

The final network was trained on 50%, validated on 20% and tested on 30% of the "GTZAN" data set containing 1000 songs from ten different genres. The songs used for training were randomly selected. Each song was split into 19 samples (3 s each) with 50% overlap. This results in 19.000 samples in total. In order to train the network, the following parameters were used:

- Cost function: Cross-entropy
- Training: Batch training
- Mini-batch-size: 512
- Optimizer: Adam.

3.2 Performance of CNN-FNN

In this section, a comparison between our proposed model and its individual components is presented. The three different models are tested on the "GTZAN" dataset [1]. The FNN uses all the features presented in Sect. 2.1, except spectrograms, the CNN is only using spectrograms and CNN-FNN is using both. The accuracies that could be achieved by the different neural networks are depicted in Table 1. The table shows the accuracy with which each algorithm predicted each genre of a song correctly. The higher percentages are marked in bold. It is apparent, that combining the two NNs resulted in a higher percentage for most

Table 1. Comparison of CNN-FNN against FNN and FNN in GTZAN dataset.

	Algorithm		
Genres	FNN	CNN	CNN-FNN
Blues	**82%**	62%	80%
Classical	87%	**96%**	94%
Country	43%	53%	**60%**
Disco	37%	55%	**60%**
Hip Hop	46%	54%	**61%**
Jazz	56%	66%	**70%**
Metal	75%	77%	**84%**
Pop	73%	62%	**81%**
Reggae	40%	63%	**71%**
Rock	32%	30%	**46%**
Overall	56%	64%	**72%**

of the genres. The combined CNN-FNN obtained an 8% improvement compared to CNN and a 16% increase with respect to only the FNN. Rock is the most difficult genre to classify in all three algorithms. This matches with most of the previous papers that show that these kind of genres have always been harder to classify [1,4]. On the other hand some other genres get good accuracy in other algorithms. We could say that if any genre has a high accuracy in the FNN algorithm, it is due to the fact that values extracted from the features are not similar to the ones extracted from other music files such as blues. Conversely, if a high accuracy is obtained in the CNN implementation, it can be because the patterns found in their spectrograms are distinct from the ones found in other genres.

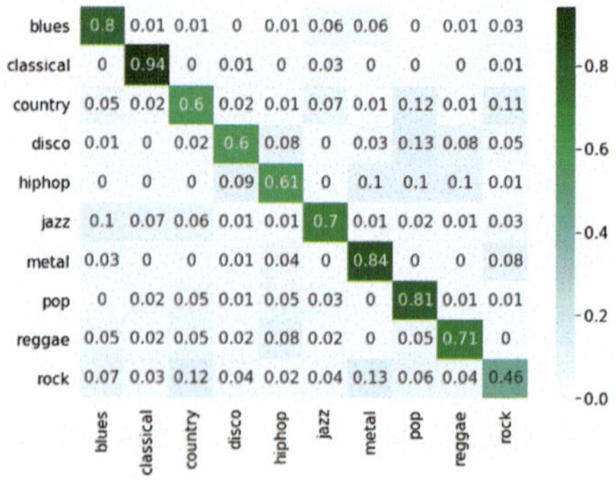

Fig. 2. Confusion matrix of the complete network for the testing data

In the final implementation, joining both NNs by using the output of the CNN as an input for the FNN together with some features extracted from the music files results in the confusion matrix Fig. 2. The matrix gives information about how the genre of a song is misinterpreted as a different one. As it can be seen, there is a considerable difference between genre accuracy, especially when rock is considered. A high grade of confusion also exists for the country and metal genre, however most of the confusion seems to be spread between the genres that were falsely identified. When songs of a particular genre get identified as one different genre often, this could indicate a similarity in features. We see in Fig. 2 that rock is often misinterpreted as country or metal. That could indicate, that the used features and the Mel-spectrogram do not separate these genres sufficiently. As it is represented in Table 1, the final implementation shows better performance in almost all genres from the database and consequently, in the average accuracy of the final implementation.

4 Conclusion

In this paper we proposed a new combined deep neural network model for music genre recognition. We evaluated spectrograms through a CNN and compared the results with standalone features of a music track in an FNN.

Our approach was tested and we could confirm that even the CNN performance could be improved by 8% when combining its outputs with additional features to the FNN. The results show that our solution is better in genre recognition than recognition done by a human [7] and it promises to improve on existing approaches to music genre recognition, by combining different neural networks.

As future work, different windowing functions will be tested, as well as the usage of statistical information about the features as is shown in [5]. Substantial improvements are also expected with additional samples and bigger spectrograms, as this increases the amount of data available to train the neural network. Furthermore different scenarios, where multiple networks could improve results, will be explored.

References

1. Tzanetakis, G., Cook, P.: Musical genre classification of audio signals. In: IEEE Transactions on Speech and Audio Processing, vol. 10.5, pp. 293–302, July 2002. ISSN: 1063-6676
2. Popescu, A., Gavat, I., Datcu, M.: Wavelet analysis for audio signals with music classification applications. In: 2009 Proceedings of the 5th Conference on Speech Technology and Human-Computer Dialogue, pp. 1–6. IEEE (2009)
3. Hamel, P., Eck, D.: Université De Montréal. Learning features from music audio with deep belief networks. In: 11th International Society for Music Information Retrieval Conference, ISMIR 2010 (2010)
4. Dong, M.: Convolutional neural network achieves human-level accuracy in music genre classification. *CoRR*, abs/1802.09697 (2018)

5. Lee, J., Baniya, B.K., Ghimire, D.: Automatic music genre classification using timbral texture and rhythmic content features (2014)
6. Kittler, J., Hater, M., Duin, R.P.W.: Combining classifiers. In: Proceedings of 13th International Conference on Pattern Recognition, vol. 2, pp. 897–901, August 1996
7. Perrot, D.: Scanning the dial: An exploration of factors in the identification of musical style. In: Proceedings of ICMPC 1999 (1999)
8. Elbir, A., Bilal am, H., Emre Iyican, M., Ztrk, B., Aydin, N.: Music genre classification and recommendation by using machine learning techniques. In: 2018 Innovations in Intelligent Systems and Applications Conference (ASYU), pp. 1–5, October 2018
9. Logan, B.: Mel frequency cepstral coefficients for music modeling. In: Proceedings 1st International Symposium Music Information Retrieval, November 2000
10. Ghosal, D., Kolekar, M.H.: Music genre recognition using deep neural networks and transfer learning, pp. 2087–2091, September 2018
11. Fulzele, P., Singh, R., Kaushik, N., Pandey, K.: A hybrid model for music genre classification using LSTM and SVM. In: 2018 Eleventh International Conference on Contemporary Computing (IC3), pp. 1–3, August 2018
12. Peltonen, V., Tuomi, J., Klapuri, A., Huopaniemi, J., Sorsa, T.: Computational auditory scene recognition. In: 2002 IEEE International Conference on Acoustics, Speech, and Signal Processing, vol. 2, pp. II-1941–II-1944, May 2002
13. Driedger, J.: Processing Music Signals Using Audio Decomposition Techniques. PhD thesis, January 2016
14. LeCun, Y., et al. Generalization and network design strategies. In: Connectionism in perspective, vol. 19. Citeseer (1989)
15. Nielsen, M.A.: Neural Networks and Deep Learning, vol. 25. Determination press, San Francisco (2015)
16. Krizhevsky, A., Sutskever, I., Hinton, G.E.: ImageNet classification with deep convolutional neural networks. In: Pereira, F., Burges, C.J.C., Bottou, L., Weinberger, K.Q., (eds.) Advances in Neural Information Processing Systems, vol. 25, pp. 1097–1105. Curran Associates, Inc. (2012)
17. O'Shaughnessy, D.: Speech Communication: Human and Machine, p. 150. Addison-Wesley, Boston (1987)
18. Hecht-Nielsen, R.: Theory of the back propagation neural network. In: Neural Networks for Perception, pp. 65–93. Elsevier (1992)
19. Kingma, D.P., Ba, J.: Adam: a method for stochastic optimization. *CoRR*, abs/1412.6980 (2014)

A Spatial Fusion Scheme of Multi-focus Image Combining SVM-Based Classification and PCA-Based Weight

Yikun Yang, Qian Jiang, Shaowen Yao, Gang Xue, Liwen Wu, and Xin Jin(✉)

School of Software, Yunnan University, Kunming, Yunnan, China
xinxin_jin@163.com

Abstract. The image fusion technique aims to generate a comprehensive image that can combine complementary information from different source images. Traditional signal processing-based image fusion methods are well-studied in recent years, and researchers try to explore new ideas for this task. Especially, the machine learning method likes support vector machine (SVM) has an apparent advantage in many situations. In this work, a sliding window technique is first used to extract the key features of source images based on several effective evaluation metrics which can reflect the detailed information. Second, the extracted image features are employed to distinguish the focused areas and unfocused areas of source images by a trained SVM model, so the decision results for each source image will be obtained. Third, consistency verification (CV) is utilized to assimilate a single singular point of a specific region in the decision results in order to correct possible errors. At last, a new weighted fusion method for the pixel is designed based on principal component analysis (PCA) to deal with the disputed areas which come from the same position of decision maps of different source images. Experimental results reveal the proposed method is effective and can achieve better performance than comparative methods.

Keywords: Feature extraction · Image fusion · Multi-sensor information fusion · Principal component analysis · Support vector machine

1 Introduction

The main purpose of image fusion is to fuse two or more images into a new synthetic image based on fusion algorithm, and widely-used in visual sensor network, such as military surveillance, monitoring system and image in painting [1]. The fused image provides more comprehensive information than any of the source images [4, 5]. In the field of digital photography, cameras like a digital single-lens reflex camera are difficult to take an image in which all the areas are captured in focus [2, 5]. In order to obtain an all-in-focus images, multiple images from the same scene with different focal settings will be fused in one image, which is known as multi-focus image fusion [6, 7]. In recent years, a variety of multi-focus image fusion methods have been proposed, and we consider the existing methods can be roughly classified into two main categories:

Y. Liu et al. (Eds.): ICNC-FSKD 2019, AISC 1074, pp. 385–398, 2020.
https://doi.org/10.1007/978-3-030-32456-8_42

conventional signal processing based fusion models and machine learning based fusion models.

The conventional signal processing based image fusion methods generally have following three steps: first, a specific transform method is implemented in the source images to get a set of sub-band coefficients (sub-images); second, fusion rules are carried out to combine the corresponding sub-band coefficients which come from different source images; at last, the fused image is generated by inverse transform [3]. However, the conventional signal processing based image fusion methods, such as PCA [4], wavelet transform [8, 9], nonsubsampled shearlet transform (NSST) [1], multi-resolution singular value decomposition (MSVD) [10], discrete cosine harmonic wavelet transform (DCHWT) [11] and multi-scale weighted gradient-based fusion (MWGF) [12], may produce unpredictable errors in the conversion between transform and spatial domain. These errors may cause the problem of distortion and artifacts.

Recently, researchers pay more attention to develop new ideas for image fusion, thus conventional machine learning (SVM and neural networks) and new deep learning algorithms are both applied in image fusion because the focused and unfocused areas can be regarded as a binary classification problem [13, 14]. The deep learning based image fusion algorithm can achieve satisfying performance in some situations; however, the deep learning based image fusion method has boundedness to some extent because a large number of image samples and computing resources are needed to train the good model [15]. By considering the balance of computational efficiency and performance, the shallow machine learning methods are introduced into this field as well. Shallow learning models such as SVM can be thought of as having a hidden layer node. These models are normally trained by extracted image features to identify focused and unfocused areas which can be further used as the decision map or mask for image fusion [13]. In general, the shallow machine learning models cannot extract image features (such as texture, structure, and edge) directly, thus a given feature extraction method must be used to accomplish this task which is significant for image fusion. The classification performance of shallow machine learning can be improved by a feature extraction method effectively; besides, the advantage of shallow machine learning is that they need limited computing resources and a small quantity of training samples.

A new image fusion method based on SVM, PCA and regional feature extraction is proposed to explore the possibility of shallow learning in image fusion. Sliding window technique is used to extract the regional features of source images. A trained RBF-SVM model is employed to distinguish the focused and unfocused areas. In the fusion strategy, the decision maps of different source images are integrated by logic operations, then CV is implemented to optimize the decision map, PCA based pixel weighted fusion method also is utilized to deal with disputed area mask which comes from the same position of different source image decision map. The contributions of this work are summarized as follow.

(a) The local feature extraction method is designed by five effective evaluation metrics, and the extracted features are used as the input of RBF-SVM to get a fusion decision model. This method can extract the detailed features of the source image and avoid inputting the whole image into the decision model.

(b) A SVM-based spatial fusion method of multi-focus image for visual sensors networks is proposed to mask the focused areas and unfocused areas to integrate the multi-sensor information. The proposed method only needs limited samples to train to distinguish the focused areas and un-focused areas.
(c) A sliding PCA based weighted pixel fusion method is designed to fuse the disputed areas which comes from the same position of different source image decision map. The proposed PCA based weighted fusion method can achieve better results than the conventional PCA based global image fusion method.

This paper is organized as follows. The feature extraction method and pixel classification of SVM is described in Sect. 2. Section 3 reports the fusion strategy. The experiments are described in Sect. 4. Section 5 concludes this work.

2 The Pixel Classification Based Image Fusion Method

In this section, the proposed pixel classification based image fusion method is introduced, including feature extraction, RBF-SVM and it's optimizing. The diagram of the proposed image fusion method is represented in Fig. 1.

2.1 Feature Extraction

In order to obtain the regional information of surrounding pixels in a given pixel $im(i,j)$ of source image, a $M \times N$ window is used to represent the regional features of source images [13], which means the given pixel $im(i,j)$ is regarded as the center of the window. In addition, we consider the boundary area of source images should be considered to represent the regional features of the boundary pixels, thus a mirroring optimizing is used to expand the boundary area according to the defined window. The size of the expanded area as:

$$s = (w - 1)/2, \tag{1}$$

where s is the size of expanding area, w is the size of the sliding window.

In order to extract the detailed information of source images and achieve the goal of dimensionality reduction, five indicators are commonly-used in image fusion and selected according to our repeated experiment. The employed indicators can be used to describe the regional features of a pixel $im(i, j)$ in the sliding window, and are used as the input features of the machine learning model (RBF-SVM). As a result, the sliding window can be represented as five vectors for model training, and the indicators are standard deviation, spatial frequency, energy of image gradient, average gradient and sum-modified Laplacian which are the often-used representation methods of image quality or features [13, 16]. In this section, $im(i, j)$ is the pixel value of an image at the position (i,j), M and N is the sizes of the given source image, respectively.

Standard deviation (STD) represents the statistical distribution and contrast features of an image by describing the dispersion degree between each pixel value and the mean value of a given window. Therefore, STD is selected as the input features to reflect image details and the definition as:

$$STD = \sqrt{\frac{1}{M \times N}\sum_{i=1}^{M}\sum_{j=1}^{N}(im(i,j) - \mu)^2}, \tag{2}$$

where is the mean value (MV) and is defined by:

$$\mu = \frac{1}{M \times N}\sum_{i=1}^{M}\sum_{j=1}^{N} im(i,j). \tag{3}$$

Spatial frequency (SF) can be used to represent the overall activity of an image in the spatial domain, and it is widely known as image definition. SF is a very important quality metric in image fusion and defined by:

$$SF = \sqrt{RF^2 + CF^2}, \tag{4}$$

$$RF = \sqrt{\frac{1}{M \times N}\sum_{i=1}^{M-1}\sum_{j=1}^{N-1}[im(i,j) - im(i,j-1)]^2}, \tag{5}$$

$$CF = \sqrt{\frac{1}{M \times N}\sum_{j=1}^{N-1}\sum_{i=1}^{M-1}[im(i,j) - im(i-1,j)]^2}. \tag{6}$$

where RF is row frequency, CF is column frequency.

Average gradient (AG) is used to measure image sharpness by the changing rate of density in multi-dimensional directions, and it represents the relative clarity of image details for analyzing the details and textures and is defined by:

$$AG = \frac{1}{M \times N}\sum_{i-1}^{M}\sum_{j-1}^{N}\sqrt{\frac{1}{2}(im(i,j) - im(i+1,j))^2 + (im(i,j) - im(i+j+1))^2}. \tag{7}$$

Energy of image gradient (EIG) can evaluate real-time image definition effectively. For an image window, the local energy of the focused area is greater than that of the unfocused area. The definition is represented by:

$$D(im) = \sum_j \sum_i \left(|im(i+1,j) - im(i,j)|^2 + |im(i,j+1) - im(i,j)|^2\right), \tag{8}$$

where the gradient value of a pixel is calculated by using the difference between the adjacent pixels.

The sum of modified Laplacian (SML) is a proposal to improve traditional energy to avoid the cancellation trend of the second derivate in the basic definition of energy of Laplacian [12]. The modified Laplacian is first defined as:

$$\nabla^2{}_{ML}im(i,j) = |2im(i,j) - im(i-\beta,j) - im(i+\beta,j)| \\ + \ldots\ldots|2im(i,j) - im(i,j-\beta) - im(i,j+\beta)|, \tag{9}$$

where β is a spacing parameter addressing the accommodation of extraction variation in an image and set as "1" in this work. Then, SML is represented as:

$$SML = \sum_{x=i-N}^{x=i+N} \sum_{y=i-N}^{y=i+N} \nabla^2_{ML} im(x,y) for \sum_{y=i-N}^{y=i+N} \nabla^2_{ML} im(x,y) \geq T, \tag{10}$$

where T is a discrimination threshold value and N is the window size.

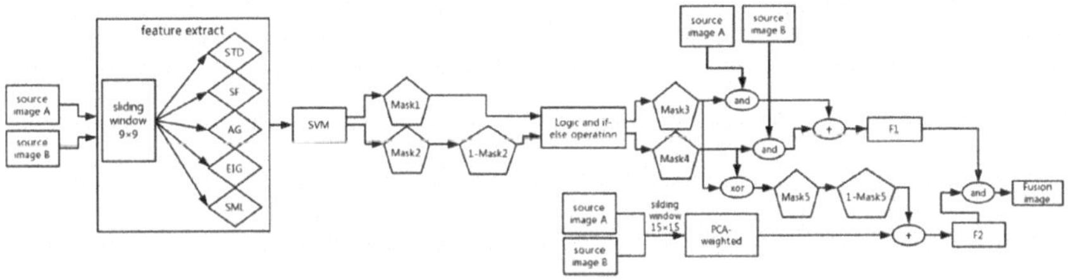

Fig. 1. The diagram of the proposed method

2.2 RBF-SVM Model

Cherkassky proposed SVM [17], which is a supervised learning algorithm that can be used for classification and regression problems. The pixel-based image fusion can be considered as a binary classification problem, and SVM is used in this work. The basic idea of classification learning of SVM is to find a dividing hyperplane based on the labelled trained set, and the hyperplane is further used to distinguish the different samples in the test set. The formula for dividing the hyperplanes is defined by:

$$\omega^T x + b = 0. \tag{11}$$

where $\omega = (\omega_1; \omega_2; \omega_3; \ldots; \omega_d)$ is a normal vector which defines the direction of the hyperplane; b is the displacement term which defines the distance between the hyperplane and the origin.

To solve the problem of linear indivisibility of samples, the kernel function which can map samples from the low dimensional feature space into the higher dimensional feature space, which can make samples divisible. In this paper, the RBF kernel function is used to solve this problem [18], as:

$$k(\chi_i, \chi_j) = \exp(-\frac{\|\chi_i, \chi_j\|^2}{2\sigma^2}), \tag{12}$$

where σ is the width of Gaussian kernel function.

In this work, cross-validation and grid parameter optimization are used to find optimized parameters which are penalty parameter c and RBF kernel parameter g to improve the classification performance of the SVM model. The grid parameter optimization is to divide c and g into grids within a given range and traverse the values of all points; then, k-fold cross-validation method is used to verify the classification accuracy, the parameters c and g with the highest classification accuracy are obtained, which means the optimal parameters are found. Thus, the source images are input the trained model to get two fusion decision masks.

3 The Image Fusion Strategy

In this section, a new fusion strategy is introduced based on the decision results of Sect. 2, and the strategy is composed of three main stages. In the first stage, the obtained masks of source images will generate some disputed and undisputed areas because different source images are input trained SVM to get the corresponding decision masks; then, the undisputed areas are fused by the decision masks of SVM directly, and the disputed areas of source images are extracted for the next stage. In the second stage, a pixel weighted image fusion method based on the sliding window and PCA is designed to deal with the disputed areas. At last, the results of the above two stages are integrated to obtain the final fused image. The image fusion strategy is represented in Fig. 1.

3.1 Disputed Area Extrication

In this stage, a pair of decision masks on each source images (im1 and im2) are generated by the trained SVM model, which means the focused and unfocused areas of the source images are marked as “1” and “0” on the decision results. The focused areas and unfocused areas in a pair of test images should be complementary, which means the decision masks of corresponding pixels are complementary in the same position of the image. But the decision masks generated by the SVM model cannot make the same position complementary completely because the decision results are inconsistent. Figure 2 shows a group of decision masks that come from image sets “hoed”. It can be found that the decision result from the corresponding pixels of the source images is the same as each other. Thus, it cannot be decided which pixel should be kept into the fused image. These pixels are denoted as disputed areas, such as the red arrows in Fig. 2.

In order to correct the misclassification decision of the model in some small areas, CV [20] is used to assimilate the single singular point in the specific region of decision masks, which can get optimized masks. The window size of the CV is set 5, which can get satisfying fusion performance. A pair of optimized masks are denoted as M1 and M2. The new matrixes M3 and M4 are set to record the integrated result of M1 and M2. *(x, y)* is the size of the decision masks. The process is defined as:

```
Input: M1 and M2.
Output: M3 and M4.
function JUDGEMENT(M1, M2)
for i = 1 → x do
    for j = 1 → y do
        if M1(i,j)&&(1−M2(i,j))=1
                M3(i,j)=1
        else
                M3(i,j)=0
        if M1(i,j)&&(1−M2(i,j))=0
                M4(i,j)=1
        else

    end for
end for
return M3, M4
end function
```

Then, the source images are fused according to the corresponding masks, as:

$$F_1(i,j) = M_3(i,j)im_1(i,j) + M_4(i,j)im_2(i,j), \tag{13}$$

where F_1 is a preliminary fused image

The decision mask of disputed area are calculated by the logic XOR operation, and defined as:

$$M_5(i,j) = M_3(i,j) \oplus M_4(i,j), \tag{14}$$

where $\oplus$ is logic XOR operation and M_5 is mask of disputed area.

Fig. 2. The decision mask of “hoed”

3.2 Pixel Weighted Fusion Method Based on PCA

PCA is a classical dimensionality reduction algorithm that can retain the critical information of the original variable [4]. Generally, the principal component of two source images is used to calculate the global weight of source images in PCA based fusion method. However, the weight obtained by the conventional PCA method cannot accurately reflect the regional features of an image due to the global weight for the source images.

In this study, the regional features of an image is considered as a significant factor for image fusion. Thus a sliding window is used to traverse the source image to calculate the corresponding weight of each pixel in the disputed areas by PCA based fusion method, which can reflect the regional feature of disputed areas and overcome the problem of conventional PCA image fusion method. In this work, the PCA based pixel weighted method of disputed areas is described as follows:

Input:	Window pairs: $D = \{imw_1\ ,\ imw_2\}$; The imw_1, imw_2 are the sliding window (blocks) of two source images, respectively.
Process:	
1:	Put window pairs D into PCA model to calculate eigenvector V.
2:	Select the eigenvector V which is corresponding to the largest eigenvalue.
3:	Generate weighted vector a which is calculated as: $a_1 = V_1 / sum(V)$ and $a_2 = V_2 / sum(V)$ · where a_1 and V_1 is the first value in vector a and V, respectively; a_2 and V_2 is the second value in vector a and V, respectively.
4:	Output weighted vector a; and record a_1 and a_2 as an element of weighted matrix wm_1 and wm_2, respectively.
5:	Fuse source images according to the weighted matrix, as: $F_2(i,j) = wm_1(i,j) \times im_1(i,j) + wm_2(i,j) \times im_2(i,j)$
Output:	Second fused image : F_2.

The complementary set of M5 and the second fused image F2 are employed to get the tertiary fused image F3, which is the complementary set of the preliminary fused image.

$$F_3(i,j) = (1 - M_5(i,j)) \times F_2(i,j). \tag{15}$$

Finally, F1 and F3 are added to get the final fused image F:

$$F = F_1 + F_3. \tag{16}$$

4 Experimental Results and Analysis

In this section, the libsvm software package is employed in the experiments. Four pairs of common multi-focus images from previous published-papers are utilized to verify the performance of the proposed method. The focused and unfocused areas of the experimental images are partly cut out as the trained set. Besides, some previous image fusion methods are used to compare with the proposed method by some widely evaluation metrics. The comparison methods are: PCA [19]; the proposed pixel weighted fusion method based on PCA (WPCA); discrete wavelet transform (DWT) [19]; gradient pyramid (GRP) [19]; multi-resolution singular value decomposition (MSVD [10]; convolutional sparse representation (CSR) [21]; gradient domain (GD) [22]; discrete cosine harmonic wavelet transform (DCHWT) [11]. multi-scale weighted gradient-based fusion (MWGF) [12]. The evaluation metrics are: mutual information (MI); entropy (EN); edge based on similarity measure (QAB/F); standard deviation (STD); spatial frequency (SF); feature mutual information (FMI); average gradient (AG) [23].

In the experiments, the size of the sliding window is 9 × 9. The fold of cross-validation is set as 5 for optimizing SVM, and the optimized parameters of SVM are g = 256, c = 0.0625.

The source images and fused images generated by different methods are displayed in Fig. 3, 4, 5, and 6. In Fig. 3, the fused image of Fig. 3(l) does not have obvious artifacts, and it is a little difficult for man to distinguish the difference of the fused images, except PCA and MSVD. In Fig. 4, the fused image generated by the proposed method is better than those of other methods in teams of edges and textures, and it is very close to the source images.

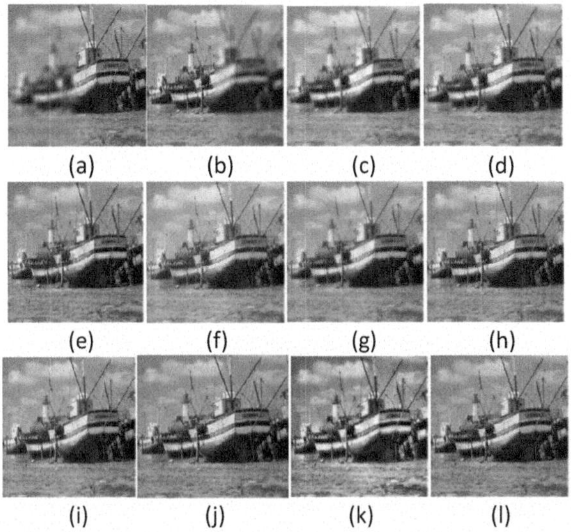

Fig. 3. Comparison of different image fusion algorithms with image sets "boat". (a) source image A (b) source image B (c) PCA [19] (d) WPCA (e) DWT [19] (f) GRP [19] (g) MSVD [10] (h) CSR [21] (i) DCHWT [11] (j) MWGF [12] (k) GD [22] (l) SVM-PCA.

In Fig. 5, the detailed features of source images are well represented by the fused image of Fig. 5(l) that is the most similar to the source images, which means the proposed method is more effective than other methods. In Fig. 6, the definition of the fused image of Fig. 6(l) is better than PCA, DWT, GRP, and MSVD obviously; but it is also difficult for people to distinguish the difference of other remaining fused images. Overall, the proposed method can generally present better visual effect when compared with these control methods visually. Besides, the proposed WPCA is much better than the conventional PCA method and it can achieve very competitive performance.

Some frequently-used evaluation metrics are further used to analyze the performance subjectively. In Tables 1, 2, 3 and 4, the best results for each assessment are highlighted in bold. Table 1 shows that the values of the proposed method teams of

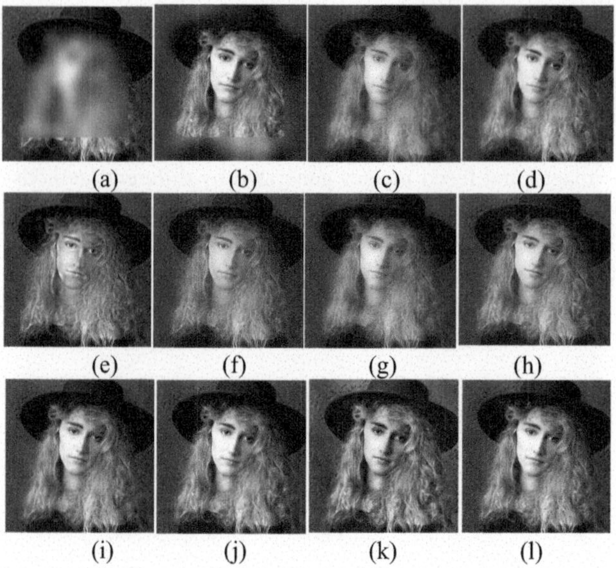

Fig. 4. Comparison of different image fusion algorithms with image sets "hoed". (a) source image A (b) source image B (c) PCA [19] (d) WPCA (e) DWT [19] (f) GRP [19] (g) MSVD [10] (h) CSR [21] (i) DCHWT [11] (j) MWGF [12] (k) GD [22] (l) SVM-PCA.

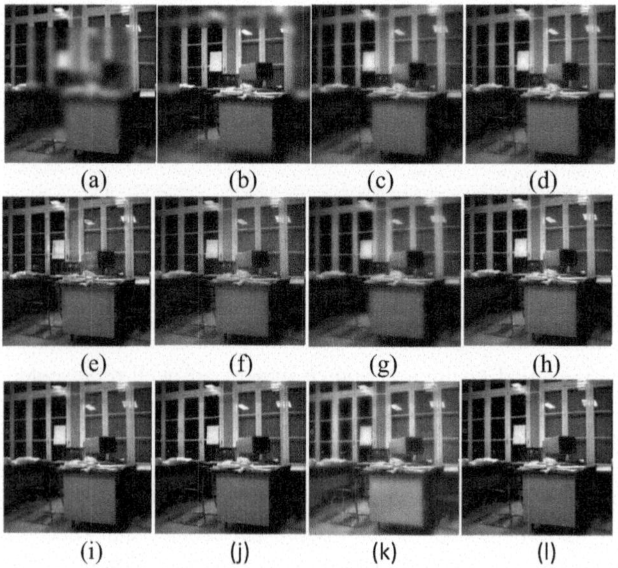

Fig. 5. Comparison of different image fusion algorithms with image sets "office256". (a) source image A (b) source image B (c) PCA [19] (d) WPCA (e) DWT [19] (f) GRP [19] (g) MSVD [10] (h) CSR [21] (i) DCHWT [11] (j) MWGF [12] (k) GD [22] (l) SVM-PCA.

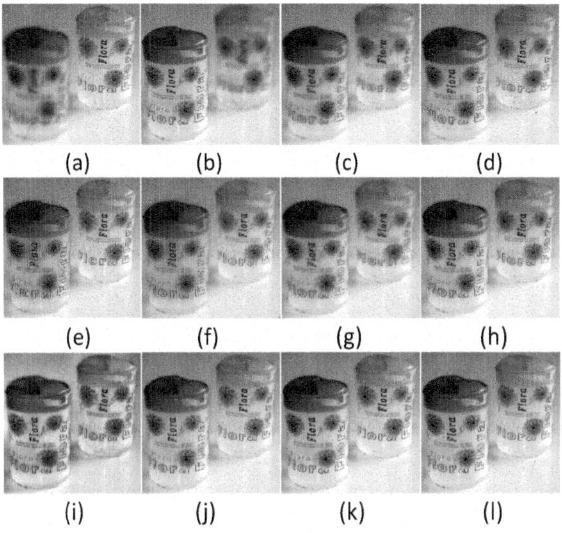

Fig. 6. Comparison of different image fusion algorithms with image sets "bottles". (a) source image A (b) source image B (c) PCA [19] (d) WPCA (e) DWT [19] (f) GRP [19] (g) MSVD [10] (h) CSR [21] (i) DCHWT [11] (j) MWGF [12] (k) GD [22] (l) SVM-PCA.

$Q^{AB/F}$, MI, ST, and FMI are the largest in the third pair of images. Table 2 shows that the values of the proposed method in teams of $Q^{AB/F}$, MI and STD are the largest in the first pair of images. Table 3 shows that the values of the proposed method in teams of $Q^{AB/F}$, MI, SF, and AG are the largest in the second pair of images. Table 4 shows that the proposed method obtains highest scores for three fused images in terms of $Q^{AB/F}$, MI, SF, and AG.

Table 1. Evaluation metrics of different image fusion algorithms with image sets "boat".

	QAB/F	MI	STD	SF	FMI	AG
PCA[19]	0.5431	5.1153	46.6333	12.5158	0.8620	5.7731
WPCA	0.6693	5.1889	47.5136	15.1021	0.8664	6.9757
DWT[19]	0.6845	5.0368	50.1138	20.7918	0.8639	9.7672
GRP[19]	0.6830	4.7941	45.1592	16.7996	0.8640	7.8353
MSVD[10]	0.5900	4.8608	47.3814	17.2383	0.8566	7.9434
CSR[21]	0.7227	5.3809	48.8383	18.7100	0.8655	8.4201
GD[22]	0.6999	3.1949	**53.4840**	**24.3752**	0.8585	**12.0170**
DCHWT[11]	0.6994	5.3117	49.1159	19.1425	0.8638	8.8718
MWGF[12]	0.7380	6.8722	50.3705	20.5385	0.8666	9.4865
SVM-PCA	**0.7419**	**6.9792**	50.1590	20.2634	**0.8681**	9.3049

Table 2. Evaluation metrics of different image fusion algorithms with image sets "hoed".

	QAB/F	MI	STD	SF	FMI	AG
PCA[19]	0.4706	4.8861	56.6482	17.6389	0.7880	9.2269
WPCA	0.7489	5.3014	59.8119	24.4070	**0.7910**	14.0758
DWT[19]	0.7037	5.5991	58.9731	34.3882	0.7847	17.9566
GRP[19]	0.7240	4.7167	57.1455	32.0316	0.7820	16.7318
MSVD[10]	0.5839	4.6095	58.1864	31.5158	0.7759	16.4240
CSR[21]	0.7763	5.0208	59.3009	31.4408	0.7867	16.0126
GD[22]	0.7707	3.5486	58.3318	**42.6904**	0.7830	**23.0448**
DCHWT[11]	0.7413	5.3390	61.9662	33.3022	0.7863	17.3192
MWGF[12]	0.8071	6.9006	63.2796	34.4156	0.7877	17.9884
SVM-PCA	**0.8082**	**8.0513**	**63.3636**	34.6847	0.7858	17.8993

Table 3. Evaluation metrics of different image fusion algorithms with image sets "office256.

	QAB/F	MI	STD	SF	FMI	AG
PCA[19]	0.5238	5.2868	59.3659	14.2466	0.8891	5.6931
WPCA	0.6364	5.0239	61.9976	18.0150	0.8898	6.8331
DWT[19]	0.6142	4.9734	64.0152	24.7496	0.8857	9.3293
GRP[19]	0.6261	4.7114	58.2125	20.0419	0.8839	7.5131
MSVD[10]	0.4490	4.7764	59.6761	19.0134	0.8768	6.9902
CSR[21]	0.6688	5.0989	63.3177	22.6544	0.8896	8.0580
GD[22]	0.6519	3.8418	58.3996	18.4592	0.8800	7.9927
DCHWT[11]	0.6446	5.0586	66.0015	23.9874	0.8873	8.9441
MWGF[12]	0.7017	6.0573	**68.6511**	25.3506	**0.8903**	9.3273
SVM-PCA	**0.7134**	**7.2144**	68.0386	**25.5431**	0.8897	**9.3847**

Table 4. Evaluation metrics of different image fusion algorithms with image sets "bottles".

	QAB/F	MI	STD	SF	FMI	AG
PCA[19]	0.5819	6.0339	53.5386	10.5907	0.9281	3.3497
WPCA	0.7552	6.2213	54.8585	17.5204	0.9293	4.7379
DWT[19]	0.7345	6.7702	**56.9598**	20.1160	0.9267	5.8923
GRP[19]	0.7231	5.8700	55.2217	17.3743	0.9261	4.9874
MSVD[10]	0.4997	5.5913	53.7290	13.8294	0.9138	3.9497
CSR[21]	0.7348	5.6427	55.1162	18.6646	0.9292	5.0100
GD[22]	0.7550	3.6948	53.4775	20.3445	0.9238	6.6982
DCHWT[11]	0.7621	6.4891	55.1291	20.0433	0.9299	5.6974
MWGF[12]	0.8031	8.8911	55.2504	20.4637	**0.9306**	5.8250
SVM-PCA	**0.8112**	**8.9506**	55.2261	**20.5184**	0.9303	**5.8343**

According to the experiments, the proposed method has the largest value of $Q^{AB/F}$ and MI, which is the two most crucial evaluation metrics in image fusion; thus, it can be found that the proposed method can achieve better results than other methods.

5 Conclusion

In this study, a regional feature based SVM pixel classification method and a PCA based pixel weighted method is proposed for image fusion; besides, the logic operations are also employed to optimize the fusion masks in the proposed method. Experiments show that the fused images produced by the proposed method are better than those of the contrast methods.

The designed regional features can reflect essential information of the focused and unfocused areas. The significant achievements of the proposed method are to cover the classification shortage of SVM in image fusion; besides, CV and PCA are employed to deal with disputed areas to optimize decision results. The next step of our research is to explore a new regional feature extraction method and the parameter optimization of SVM; besides, more conventional machine learning methods are also expected to introduce in different image fusion fields.

Acknowledgment. This study is supported by the National Natural Science Foundation of China (No. 61762089, 61863036), China Postdoctoral Science Foundation (2019M653507), and Doctoral Candidate Academic Award of Yunnan Province in China.

References

1. Zhaobin, W., Yide, M., Jason, G.: Multi-focus image fusion using PCNN. Pattern Recogn. **43**(6), 2003–2016 (2010)
2. Jia, Y., Rong, C., Wang, Y., Zhu, Y., Yu, Y.: A multi-focus image fusion algorithm using modified adaptive PCNN model. In: 12th International Conference on Natural Computation, Fuzzy Systems and Knowledge Discovery, pp. 612–617, October 2016
3. Jin, X., Chen, G., Hou, J.: Multimodal sensor medical image fusion based on nonsub sampled shearlet transform and S-PCNNs in HSV space. Sig. Process. **153**, 379–395 (2018)
4. María, G.A., José, L.S.: Fusion of multispectral and panchromatic images using improved HIS and PCA mergers based on wavelet decomposition. IEEE Trans. Geosci. Remote Sens. **42**(6), 1291–1299 (2004)
5. Kinoshita, Y., Kiya, H.: Scene segmentation-based luminance adjustment for multi-exposure image fusion. IEEE Trans. Image Process. (2019, In Press)
6. Yu, L., Chen, X., Peng, H., Wang, Z.: Multi-focus image fusion with a deep convolutional neural network. Inf. Fus. **26**, 191–207 (2017)
7. Liu, S., Chen, J., Rahardja, S.: A new multi-focus image fusion algorithm and its efficient implementation. IEEE Trans. Circ. Syst. Video Technol. (2019, In Press)
8. Ebenezer, D.: Optimum wavelet-based homomorphic medical image fusion using hybrid genetic-grey wolf optimization algorithm. IEEE Sens. J. **18**(16), 6804–6811 (2018)
9. Xiaorong, X., Fang, X., Hongfu, W., et al.: A parallel fusion algorithm of remote sensing images based on wavelet transform. In: Proceedings 10th IEEE International Conference of High Performance Computing and Communications (HPCC), pp. 13–15, November 2013

10. Naidu, V.P.S.: Image fusion technique using multi-resolution singular value decomposition. Defence Sci. J. **61**(5), 479–484 (2011)
11. Kumar, S.: Multifocus and multispectral image fusion based on pixel significance using discrete cosine harmonic wavelet transform signal. Sig. Image Video Process. **7**(6), 1125–1143 (2013)
12. Zhou, Z.Q., Sun, L.: Multi-scale weighted gradient-based fusion for multi-focus images. Inf. Fus. **20**, 60–72 (2014)
13. Atencio-Ortiz, P., Sanchez-Torres, G., Branch-Bedoya, J.W.: Evaluating supervised learning approaches for spatial-domain multi-focus image fusion. Dyna **84**(202), 137–146 (2017)
14. Yu, B.T.: Jia B, Lu D, Hybrid dual-tree complex wavelet transform and support vector machine for digital multi-focus image fusion. Neurocomputing **182**, 1–9 (2016)
15. Guo, X.P., Nie, R.C., Cao, J.D.: Fully convolutional network-based multifocus image fusion. Neural Comput. **30**(7), 1775–1800 (2018)
16. Eskicioglu, A.M., Fisher, P.S.: Image quality measures and their performance. IEEE Trans. Commun. **43**(12), 2959–2965 (1995)
17. Cherkassky, V.: The nature of statistical learning theory. IEEE Trans. Neural Netw. **8**(6), 1564 (1997)
18. Chen, S., Cowan, C., Grant, P.: Orthogonal least-squares learning algorithm for radial basis function networks. IEEE Trans. Neural Netw. **2**(2), 302–309 (1991)
19. Oliver, R.: Pixel-level image fusion and the image fusion toolbox, 30 December 1999. http://www.metapix/toolbox.htm
20. Jiang, Q., Jin, X., Lee, S.J., et al.: A novel multi-focus image method based on stationary wavelet transform and local features of fuzzy sets. IEEE Access. **5**, 20286–20302 (2017)
21. Liu, Y., Chen, X., Ward, R.K., et al.: Image fusion with convolutional sparse representation. IEEE Sig. Process. Letters **23**(12), 1882–1886 (2016)
22. Paul, S., Sevcenco, L.S., Agathoklis, P.: Multi-exposure and multi-focus image fusion in gradient domain. J. Circ. Syst. Comput. **25**(10), 1650123 (2016)
23. Petrovic, V., Xydeas, C.: Objective image fusion performance characterization. In: 10th IEEE International Conference on Computer Vision (ICCV), vol. 2, pp. 1866–1871 (2005)

An Improved K-SVD Algorithm and Its Application for Image Denoising

Chunman Yan, Di Zhang(✉), Youfei Hao, and Jiahui Chen

College of Physics and Electronic Engineering, Northwest Normal University, Lanzhou 730070, China
2812434011@qq.com

Abstract. The KSVD algorithm has been widely used in image processing fields due to its high efficiency for image sparse representation. The traditional KSVD alternately updates the dictionary atoms and sparse coefficients through iterative operations. The algorithm is with noise suppression ability, and the reconstructed image can maintain the internal structure information of the processed image. Thinking about the intrinsic structure information can be acquired by the dictionary update procedure, in this paper, the principal component analysis (PCA) process is adopted to replaces the singular value decomposition (SVD) performing on the error term, and the first principal component is extracted as update item for dictionary atom. Furthermore, the improved KSVD is applied to image denoising. The simulation results show that the improved algorithm can better reconstruct the essential features such as edge and texture of the denoising image, and obtain higher peak signal-to-noise ratio while significantly reducing the time consuming of the algorithm.

Keywords: Image denoising · K-SVD · Dictionary learning · PCA

1 Introduction

The traditional mean filtering, median filtering, wavelet threshold and other similar algorithms can be effectively used for image denoising [1–4]. Although these traditional methods are with simple model and high denoising efficiency, it is difficult to entirely maintain the details such as edges and textures of the image. Image sparse representation can effectively represent the essential features of images, which is one of the research hotspots in the field of image processing in recent years, and some denoising algorithms based on sparse representation are proposed. For example, Zhang proposed a double dictionary learning algorithm based on K-SVD [5]. He *et al.* proposed an image denoising algorithm by using double sparsity [6]. Gao *et al.* proposed another denoising algorithm based on collaborative sparse representation [7].

The K-SVD algorithm is one of the most representative dictionary learning algorithms, which is proposed by Elad *et al.* [8]. Instead of updating the entire dictionary, the K-SVD algorithm updates the dictionary atoms column by column [8, 9]. That is unlike other common dictionary learning algorithms, and can maintain the internal

Y. Liu et al. (Eds.): ICNC-FSKD 2019, AISC 1074, pp. 399–405, 2020.
https://doi.org/10.1007/978-3-030-32456-8_43

structure information of the image. In addition, the KSVD is with noise suppression ability. However, the fly in the ointment is that the K-SVD algorithm mostly uses random initialization to obtain the original dictionary. The inappropriate initialization will make it impossible to update all the columns of the dictionary, or some columns will fail. That will result in low efficiency of image information utilization. In addition, when the noise is relatively large, the dictionary will contain a large number of noise atoms, which will reduce the noise suppression ability and will take a long time to train a dictionary [10].

For the more effective applications, this paper selects principal component analysis method to improve the K-SVD, and the improved algorithm is used for image denoising. This one can improve the efficiency of feature utilization in the dictionary update stage and avoid the time-consuming problem of SVD decomposition. In the last, the denoising experiment is designed and the effectiveness of the improved algorithm is verified.

2 Principal Component Analysis

PCA is a multivariate statistical method that selects fewer important variables from multiple variables by means of K-L transformation [11]. The specific steps of the PCA algorithm are as follows.

There is an image of size $N \times N$. Each row of the pixel matrix of the image is connected together and constructed as a feature vector of length $D = N \times N$. There are n training samples, and x^i is the feature vector of the *i-th* sample. Then, the sample population scatter matrix is written as follows [12]:

$$s = \frac{1}{n}\sum\nolimits_{i=1}^{n} \left(x^i - \mu\right)\left(x^i - \mu\right)^T = \frac{1}{n}XX^T \tag{1}$$

where μ is the mean of the training sample, $\mu = \frac{1}{n}XX^T$, and $x = (x^1 - \mu, x^2 - \mu, \ldots\ldots x^n - \mu)$, $D \times D$ denotes the dimension.

Since D is usually greater than n, the matrix $Q = \frac{1}{\text{n}}X^TX$ is constructed in the first place, $n \times n$ denotes the dimension. According to the singular value decomposition, the former n eigenvalues $(\lambda_1 \geq \lambda_2 \geq \ldots \geq \lambda_n)$ of Q is identical with S. Firstly, the eigenvalues $(\lambda_1 \geq \lambda_2 \geq \ldots \geq \lambda_n)$ of matrix Q and its corresponding standard orthogonal eigenvectors $(\mu_1, \mu_2, \ldots \mu_n)$ are calculated. Then, according to (2), the orthogonal feature vector corresponding to the former d maximum eigenvalues of the matrix S is calculated.

$$v_i = \frac{1}{\sqrt{\lambda_1}}X * u_i, i = 1, 2, \ldots, d \tag{2}$$

For any sample x, its principal component eigenvector is showed as follows:

$$y = (v_1, v_2, \ldots, v_d)^T * x \tag{3}$$

3 Image Denoising Using the Improved K-SVD Algorithm

As one of the efficient sparse representation method, in the dictionary updating stage, the K-SVD follows the principle of error minimization and chooses SVD decomposition to update atom items. However, SVD decomposition is with high space complexity and long processing time. Thus, combined with [13], the PCA algorithm is introduced into the K-SVD algorithm, which is used to decompose the error term and extract the principal component for dictionary atom updating. Furthermore, the improved KSVD is used for image denoising, and the specific process based on K-SVD dictionary learning is as follows [14]:

Given the original image X of size $\sqrt{N} \times \sqrt{N}$, the noisy image Y, and the dictionary $D \in R^{n \times k}$, which contains k n-dimensional atoms. The denoising model based on image sparse representation is written by:

$$\left\{ \overset{\wedge}{\alpha}_{ij}, \hat{X} \right\} = \arg\min_{\alpha_{ij},X} \lambda\|X - Y\|_2^2 + \sum_{i,j} \mu_{ij}\|\alpha_{ij}\|_0 + \sum_{i,j} \|D\alpha_{ij} - R_{ij}X\|_2^2 \tag{4}$$

Where $\lambda\|X - Y\|_2^2$ represents the overall similarity between the original image X and the noisy image Y. The second and third items represent the sparsity constraint and the error constraint, respectively. The R_{ij} is a matrix for extracting image blocks. Then, $R_{ij}X$ expresses the corresponding image block and $D\alpha_{ij}$ represents the approximate value of $R_{ij}X$ after reconstruction.

For (4), it is assumed that both the dictionary D and the image X are given. And the Sparse representation coefficient $\overset{\wedge}{\alpha}$ is obtained by using the Orthogonal Matching Pursuit (OMP). Then, the sparse representation coefficient is used to update the dictionary with the K-SVD algorithm. Finally, if the updated dictionary and the sparse representation can meet the requirements, the denoised image can be outputted. The specific process of the image denoising method based on the improved K-SVD algorithm is as follows:

(1) In the initial stage of algorithm: Where $X = Y$, fixed over-complete dictionary D;
(2) Iterative operation (the number of iterations is J):
   (1) Sparse coding:

$$\forall ij \ \min_{\alpha_{ij}}\|\alpha_{ij}\|_0 s.t. \|R_{ij}X - D\alpha_{ij}\|_2^2 \leq (C\sigma)^2 \tag{5}$$

For each image $R_{ij}X$, formula (5) needs to solve the approximate solution by OMP algorithm.

(2) Dictionary update:

① Select all image blocks that satisfy the formula: $\omega_l = \{(i,j) | \alpha_{ij}(l) \neq 0\}$;
② For the index satisfying $(i,j) \in \omega_l$, calculate the residual value, $e_{ij}^l = R_{ij}X_{ij} - \sum_{m \neq 1} d_m \alpha_{ij}(m)$;
③ Set E_l to the residual matrix, corresponding column is written as $\{e_{ij}^l\}_{(i,j \in \omega_l)}$;
④ After PCA decomposition of E_l, the result is $\Omega \approx \frac{1}{N}\left(\overline{E}_l \overline{E}_l^T\right) = P \Lambda P^T$, where P is the main component matrix. The feature vector in P corresponding to the largest eigenvalue in Λ is used as the renewal of d_l;

(3) Get the denoised image:

$$\hat{X} = \left(\lambda I + \sum_{i,j} R_{ij}^T R_{ij}\right)^{-1} \left(\lambda Y + \sum_{i,j} R_{ij}^T D \hat{\alpha}_{ij}\right) \tag{6}$$

4 Experiment and Result Analysis

For the experiment, five standard images are selected, including Barbara, Boat, House, Lena and Peppers, and five algorithms are used for the denoising contrast. Simultaneously, the denoising PSNR and time-consuming of each algorithm are used as objective index to evaluate the performance of the algorithm.

4.1 Image Denoising

For this experiment, Gaussian noise ($\sigma = 20$) is added to the test image as noisy sample, and the noisy image is processed by five contrastive algorithms respectively. Figure 1 displays the subjective denoising effects of image Boat, House, and Lena by different algorithms.

It can be seen from Fig. 1 that the image denoised by Wavelet Threshold and Mean Filtering still contains a certain amount noise, and image blurring situation causes the loss of detail feature. By the DCT and the K-SVD dictionary learning algorithm, the image sharpness is greatly improved after denoising. Compared with these four algorithms, our improved algorithm not only effectively eliminates the noise but also largely preserves the edge and contour information of the image. Meanwhile, the obtained image has more preferable visual effects.

Table 1 lists the evaluation indices of the five groups about image denoising effects. It can be seen from the comparison of index "PSNRout", the improved algorithm outperforms the other four denoising methods to a certain extent. Compared with the DCT and the K-SVD dictionary learning algorithm, the time consuming of the improved algorithm has been greatly reduced, and the average denoising efficiency of

Fig. 1. Comparison of the denoising results of five algorithms ($\sigma = 20$).

the proposed algorithm is increased by 52.2% and 34.79%, respectively. It is mainly because of the error term is decomposed by using the PCA algorithm in the proposed algorithm. The procedure needs to arrange feature vectors in descending order and extracts the larger principal components in front to represent the original variables, which reduces the computational complexity.

4.2 Denoising Experiment with Noise Change

To further verify the performance of the improved K-SVD algorithm. Gaussian noise with variance change of 10 to 50 is added to image House image at every turn.

Figure 2 shows the comparison of the PSNR after denoising under different noise variance. It can be seen intuitively that the PSNR value of the proposed algorithm has certain surmounting compared with the other four algorithms with the noise variance increasing. The progress is ameliorated and the good denoising effect is also maintained when the noise variance is increased. The curve trend shows that the improved algorithm is with a certain extent robustness to noise change.

In terms of subjective effects and objective evaluation criterion, the proposed algorithm outperforms the other four methods to a certain extent. This one not only preserves image texture details, but also has the advantages of high denoising efficiency.

Table 1. Comparison of five methods for image denoising.

Image name/size	PSNRIn(db)	Denoising method	PSNROut(db)	Time consuming (s)
Barbara 512 × 512	22.11	Wavelet threshold	26.70	0.2656
		Mean filtering	22.83	0.1875
		DCT	29.94	138.8281
		KSVD	30.70	52.2813
		The proposed	**30.84**	**29.0781**
Boat 512 × 512	22.12	Wavelet threshold	27.61	0.5000
		Mean filtering	25.17	0.1563
		DCT	29.86	106.8906
		KSVD	30.29	46.5156
		The proposed	**30.42**	**26.7188**
House 256 × 256	22.08	Wavelet threshold	29.09	0.1876
		Mean filtering	25.37	0.1094
		DCT	32.09	17.9219
		KSVD	33.25	32.6719
		The proposed	**33.51**	**22.2344**
Lena 512 × 512	22.09	Wavelet threshold	29.59	0.2969
		Mean filtering	27.59	0.1250
		DCT	31.97	71.4063
		KSVD	32.39	35.5781
		The proposed	**32.45**	**25.0469**
Peppers 512 × 512	22.10	Wavelet threshold	29.27	0.2656
		Mean filtering	27.63	0.1719
		DCT	31.86	65.8750
		KSVD	32.28	32.3438
		The proposed	**32.42**	**24.0625**

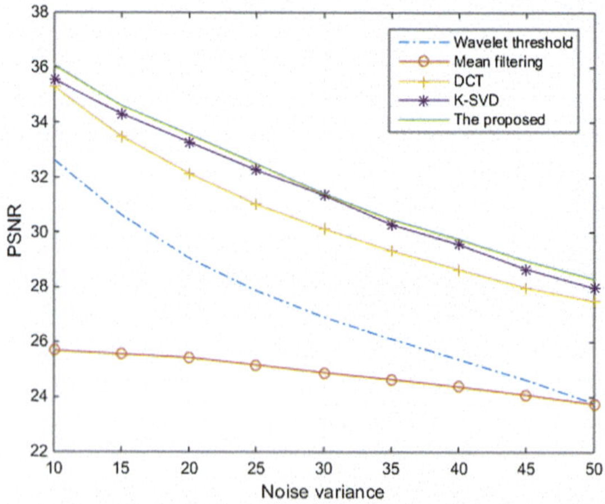

Fig. 2. PSNR curve under noise variance change of image house.

5 Conclusion

Combined with the noise suppression capability of K-SVD algorithm and the feature extraction ability of PCA algorithm, this paper proposed an image denoising method based on improved K-SVD algorithm. The experimental results shown that the improved algorithm outperformed other comparison methods in terms of subjective effects and objective evaluation indices. This algorithm could not only effectively filter out noise, but also better reconstruct image detail features. In the experimental results, it obtained a higher PSNR while significantly reducing the time consuming of the algorithm.

References

1. Ghimpeteanu, G.F., Batard, T.S., Bertalmio, M.T.: A decomposition framework for image denoising algorithms. IEEE Trans. Image Process. **25**(l), 388–399 (2016)
2. Zhu, F.Y.F., Chen, G.Y.S., Heng, H.A.T.: From noise modeling to blind image denoising. In: Proceedings of IEEE Computer Society Conference on Computer Vision and Pattern Recognition. Las Vegas, pp. 420–429. IEEE (2016)
3. Wu, Y.F., Tracey, B.S., Natarajan, P.T.: Probabilistic non-local means. IEEE Sig. Process. Letters **20**(8), 763–766 (2013)
4. Jiao Li-juan, F., Wang Wen-jian, S., Zhao Qing-shan, T.: MSK-SVD image denoising algorithm based on mean calculation. Comput. Eng. Des. **38**(12), 3380–3384 (2017)
5. Zhang Ming, F.: Double dictionary learning algorithm based on KSVD and its application in image processing. Beijing Industry University (2015)
6. He Jun-fei, F., Li Feng, S., Zhang Jian-ming, T.: Image denoising algorithm using double sparsity. J. Chin. Comput. Syst. **36**(5), 1109–1112 (2015)
7. Gao Mei-feng, F., Wang Chen, S.: Image denoising via clustering-based sparse representation over collaborative filter. Comput. Eng. Appl. **51**(24), 180–185 (2015)
8. Aharon, M.F., Elad, M.S., Bruckstein, A.M.T.: The K-SVD: an algorithm for designing overcomplete dictionaries for sparse representation. IEEE Trans. Sig. Process. **54**(11), 4311–4322 (2006)
9. Lian Qiu-sheng, F., Shi Bao-shun, S., Chen Shu-zhen, T.: Research Advances on Dictionary Learning Models, Algorithms and Application. J. Autom. **41**(2), 240–260 (2015)
10. Cheng Yi-feng, F., Liu Zeng-li, S.: Improved K-SVD image denoising algorithm. J. Metrol. **39**(03), 332–336 (2018)
11. Li, T.F., Jiang, S.S.: Adaptive threshold image denoising algorithm based on principal component analysis. Infrared Technol., **36**(04):311–314 + 319 (2014)
12. Yang Mei-fang, F., Shi Yi-long, S.: Face recognition based on 2DPCA + PCA and SVM. Inf. Technol. **02**, 32–36 (2018)
13. Yu Fu-ping, F., Feng You-qian, S., Fan Cheng-li, T.: Dictionary learning based on principle component analysis. Control Decis. **28**(07), 1109–1112 (2013)
14. Wang, Y.F.: Research on image denoising based on sparse representation. Anhui University, September 2014

Natural Computation Applications: Other Natural Computation Applications

Road Boundary Detection Using Ant Colony Optimization Algorithm

Tim Andersson, August Kihlberg, Anton Sundström, and Ning Xiong(✉)

Academy of Innovation, Design, and Engineering, Mälardalen University, Västerås, Sweden
{tan14007,akg15001}@student.mdh.se,
ning.xiong@mdh.se

Abstract. A common problem for autonomous vehicles is to define a coherent round boundary of unstructured roads. To solve this problem an evolutionary approach has been evaluated, by using a modified ant optimization algorithm to define a coherent road edge along the unstructured road in night conditions. The work presented in this paper involved pre-processing, perfecting the edges in an autonomous fashion and developing an algorithm to find the best candidates of starting positions for the ant colonies. All together these efforts enable ant colony optimization (**ACO**) to perform successfully in this application scenario. The experiment results show that the best paths well followed the edges and that the mid-points between the paths stayed centered on the road.

Keywords: Ant colony optimization · Lane detection · Autonomous vehicles

1 Introduction

The interest in autonomous vehicles has increased dramatically and almost all large car manufacturers are developing their own self-driving cars. Usually the natural environment for self-driving cars is highway where each car's main objective is to follow its own lane and stay away from swerving into another. This is the most common task for artificial intelligence research in this area, and how to develop stronger algorithms to function in worse conditions is an under-researched issue. There are a number of challenges associated with this, one of which is the problem to define a clear road boundary for unstructured, gravelled and snow-covered roads. The very common way to define road edges for lane detection is the use of an edge detection algorithm like the *canny edge* [3] algorithm in combination with different filters. But using these algorithms may not be sufficient, if the lane markings become discontinuous. To solve the problem with discontinuous roads, we employ ants to search and stitch together the

First three authors are alphabetically ordered.

Y. Liu et al. (Eds.): ICNC-FSKD 2019, AISC 1074, pp. 409–416, 2020.
https://doi.org/10.1007/978-3-030-32456-8_44

incoherent road boundaries [1] based on the principle of ant colony optimization (ACO).

ACO is an evolutionary algorithm that mimics the behaviour of biological ants. The ants form paths towards the food and leave traces of small amounts of pheromone for others to follow. The ants are therefore more likely to follow the path with the strongest pheromone scent, which is actually the most densely populated path [2]. This paper presents a new method based on a modified ACO algorithm that has been shown to be effective to detect road boundary in tough and night conditions.

2 Related Works

In recent years there have been several articles published on the topic of edge detection using an ant optimization algorithm. In article [4] the authors compared the result from the **ACO** approach with a more regular *canny edge detector* algorithm with promising results. In [4] they acknowledge the importance to experimentally test different weight factors for the evaporation, placement of the pheromone and the movement of the ant colony. They noticed that their **ACO** algorithm could not detect blurred contours like clouds which *canny* was able to do. Initially the **ACO** algorithm was used to detect edges but it was extended to a method to mark out road boundaries in [1,5].

In DARPA Grand Challenge 2004 there was a team that used an ant optimisation algorithm to effectively detect road boundaries of an unstructured road [1]. Their approach was a modified **ACO** algorithm to decrease the convergence time.

3 Proposed Approach and System

The **ACO** was implemented in two separate colonies, one for each side of the road. The movement of the ants were calculated using a mathematical function based on the information from the matrix that contained the pheromone traces, the pre-processed frame information and the weight for the given direction of the moves, rewarding those towards the *point of attraction.* Each move had a probability and the decision was randomized based on probability for each step of the ants. The point of attraction was defined by linear regression of all the paths along the road. The point of intersection from the regression became the new point of attraction for the next frame, initially it was set as the middle of the road. The implementation of a moving point of attraction was inspired by Alberto Broggi and Stefano Cattani [1] and their road boundary detection algorithm.

3.1 Pre-processing

First step was to convert the image into a gray scale, the hood of the car was cropped out and the image resized to a 226 by 456 pixels. A two-step **Sobel** filter was applied where the first filter removed vertical lines in the picture such as edges of road poles that didn't belong to the road edge, after that a regular **Sobel** filter was applied, the result from this can be seen in Fig. 2. The frame was divided into three sections of the same size where each part of the frame was individually filtered. The algorithm counted the number of identified edges in each section and adjusts its parameters to meet the amount of edges required. This process was repeated until the restrictions were met or the maximum number of iterations were exceeded (Fig. 1).

Fig. 1. The original gray scale image with hood of the car.

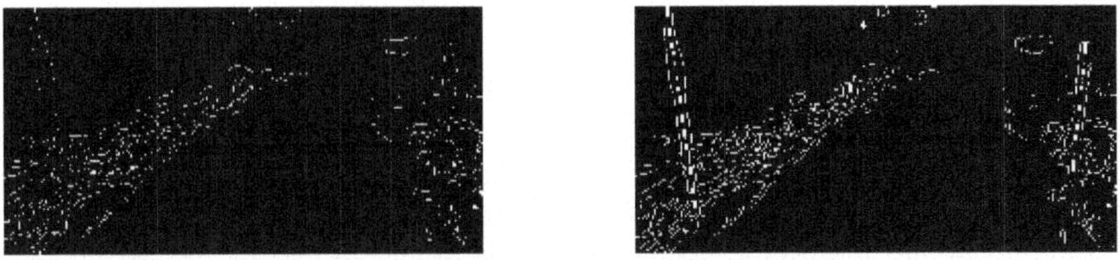

Fig. 2. Here is the pre-processing steps that is taking place on each frame from the camera. The left image at the bottom is the output image from the filtering process.

3.2 Finding the Start of the Road Lines

The position where the road lines travel past the camera's point of view is where the two ant colonies are placed respectively, the reason behind it was to hint the ants to traverse along the road line of the pre-processed image. The algorithm that finds the start position of the road lines was divided into two parts, the initialization phase and the iteration phase.

The Initialization Phase. is the first step where the algorithm searches for the initial starting positions along the bottom half and the vertical left and right side of the picture respectively, as seen in Fig. 4. The search is carried out by a local 21×21 pixel matrix also known as a search window and all the edge points inside the search window are used to fit a line relative to them. Each search window position and its corresponding line information in the image is saved and these are then evaluated towards the specified acceptable angular span of the road lines. Search window candidates with lines that did not have an angle inside the angular span are discarded. The remaining search window candidates are then evaluated and a final start position was chosen and passed on to the ant colonization algorithm. The intrinsic explanation of deciding the final start position can be seen in the section *final start position deduction* below (Fig. 3).

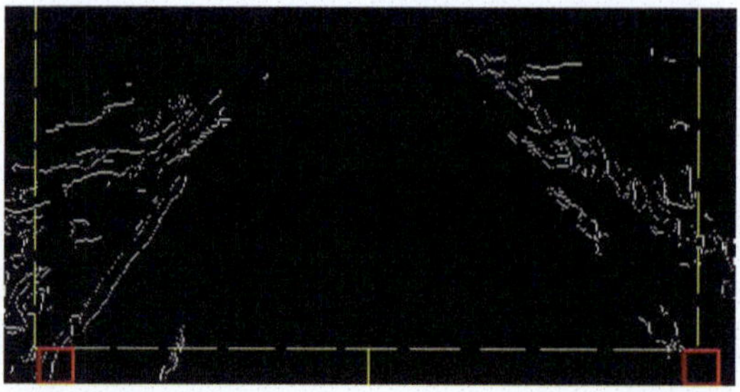

Fig. 3. The search intervals on the left and right side of the image respectively, the red squares are the search windows. The search performed on each side of the image only searches half of the image

3.2.1 Iteration Phase

The search area of the local 21×21 search window is limited to a local square placed around the previous start position. The purpose of using a local search rectangle instead of searching the edges of the image is due to the assumption that the start positions of the road lines will not change a lot between two frames given the recording is running at a speed of 30 frames per second as well as the assumption that the algorithm could be modified to run at such fast speeds. Since the road lines will not change their positions a lot between two iterations, there is no need of searching for them across the image edges, a local search in the proximity of the previous start position is believed to be enough. After the search window candidates are found, the ones with a line angle outside the angular span are discarded and the rest take part in a second evaluation that determines the final start position for the current iteration, see section below.

3.2.2 Final Start Position Deduction

The search window candidates were filtered by checking if the line fit to their local data points inside the acceptable angular span for the road line angle. If they were not inside the acceptable angular span, then they were discarded, and the remaining candidates took part in the final evaluation to find the start positions of the image. There were four parameters used to evaluate the final score of each search window candidate and those were the search window's:

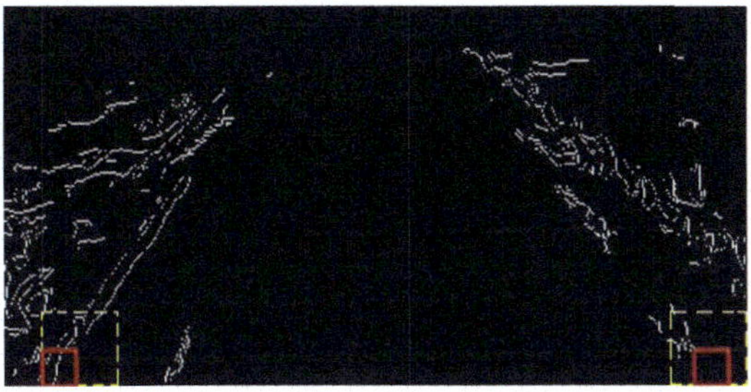

Fig. 4. The local search intervals placed in close proximity to the previous iteration's start positions. The red squares indicate the previous iteration's placement of start positions, the start position is in the pixel placed in the upper left corner of the red square.

- **Data size (D):** How many data points that are inside the local search window.
- **Root Mean Square error (E):** How good the search window's line is fit to the data points.
- **Continuity (C):** Is the angle conserved for data points placed further along the line. This value is determined by subtracting the current start position candidate's line angle by the mean angle of several local search window candidates' line angles evaluated along the angle of the current start position candidate's line angle.
- **Deviation (Dv):** The distance between the middle of the lowest horizontal edge in the image and the search window position.

Each of these parameters were combined in the equation below to determine the final score "F_s" of the search window candidate:

$$D_n = \frac{\mathbf{D}}{max(\mathbf{D})}$$
$$E_n = \frac{\mathbf{E}}{max(\mathbf{E})}$$
$$Dv_n = \frac{\mathbf{Dv}}{max(\mathbf{Dv})}$$
$$F_s = \frac{D_n * D_w}{E_n * E_w + C_n * C_w + Dv_n * Dv_w}$$

where $\mathbf{D}, \mathbf{Dv}, \mathbf{E}$ were vectors including the parameter value of each search window candidate and D_w, E_w, Dv_w were the user specified weights for each parameter, these weights allowed the user to weigh some parameters in favour of the others. The reason for the placement of the parameters in the numerator and denominator was to contribute to the final score by just using their own score multiplied with their corresponding weight. The parameters were normalized, hence their value ranges from 0 to 1 and therefor the parameters of which were better the larger value they had, were placed in the numerator and parameters of which were better the smaller value they had were placed in the denominator in order to contribute to the final score fairly. When the best start positions were determined they were traced back to the edge of the image along the winning search window's own line, given it was not already on the edge of the image. The final start positions of each side's search were then passed on to the **ACO** algorithm.

3.3 Ant Colony Optimization

The core of the algorithm was the **ACO** with the objective to connect the filtered edges of the road and to construct a path from the start position to the point of attraction along the road edges. This was a multi-objective **ACO** problem with the factors of direction and pheromone level affecting each transition between pixels.

3.3.1 Moving the Ants

Initially the two ant colonies were placed randomly in the two start positions defined earlier. The ants traversed one pixel at the time to any of the eight adjacent pixels. Each step from the current position (j) to the next (i) was assigned a probability ($p_{i,j}$) based on the direction ($\eta_{i,j}$) and pheromone level ($\tau_{i,j}$) where the factors were raised to the constants α and β for the n different possibilities to traverse according to the equation below.

$$p_{i,j} = \frac{\tau_{i,j}^{\alpha} * \eta_{i,j}^{\beta}}{\sum_{j=1}^{n} \tau_{i,j}^{\alpha} * \eta_{i,j}^{\beta}}$$

The direction chosen for the ants were based on the outcomes of a roulette where the probabilities ($p_{i,j}$) corresponded to the chance to be elected. The pheromone level was the value for a specific coordinate in the pheromone matrix at the given time instant. The pheromone level ($\tau_{i,j}$) of a black pixel was initially one hundredth of a white pixel. A move downwards in the picture had only a hundredth chance of occurring compared to the others available moves for the ants and a move in the direction towards the point of attraction was weighted the highest (in vector $\eta_{i,j}$). Whenever an ant moved to a pixel it left some pheromone on the pixel. The pheromone matrix (pm) was updated for each pixel ((i, j)) and the amount of pheromone (e) by the equation

$$pm_{i,j} = pm_{i,j} * (1 - e) + e$$

3.3.2 Reaching the Point of Attraction

When an ant reached a specified distance from the point of attraction and a certain height in the picture it was deemed to have reached its goal. The average pheromone level for the path the ant took is then calculated and evaluated. If the paths average pheromone level reached a specific threshold then the pheromone level of the pixels on the path was increased, like a boost. This threshold was proportional to the number of edges found in the specific section of the frame, a higher number of edges resulted in a higher threshold. The pheromone matrix was updated for all the pixels in a trace when the threshold for a boost is reached, similarly to the equation above for one step but at a greater pheromone amount.

When the ants also reached their goal or if they had wandered around for too long, they were randomly placed back in their starting square to start traversing all over again.

3.3.3 Aftermath of ACO

When a certain number of feasible paths were found in both the left and the right colony or if the maximum number of iterations was reached, then the path with the highest average pheromone level on each side was saved. The best paths were then used to calculate the mid-point of the road. To keep the mid-point from jumping to much between each frame a **PI**-controller was used, this was to make the movements of the mid-point more fluid. The approximated road boundaries from the last frame were inherited to the next frame and added with a weaker influence than the other edges in the next image. This helped the ant colony to converge, since the change in edges were not that different between two sequenced frames.

4 Experiment and Results

In Figs. 5 and 6 the image in the top left corner contains the pheromone traces from all the ants saved in the pheromone matrix on top of the pre-processed image. The top right image holds the two paths with the highest average pheromone level of all the traces. The circle in the top right image represents

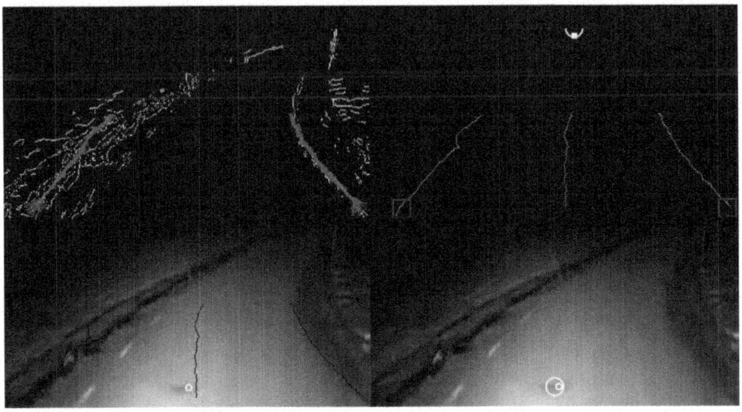

Fig. 5. This is the result from the second frame with inherited experience from other the previous frame.

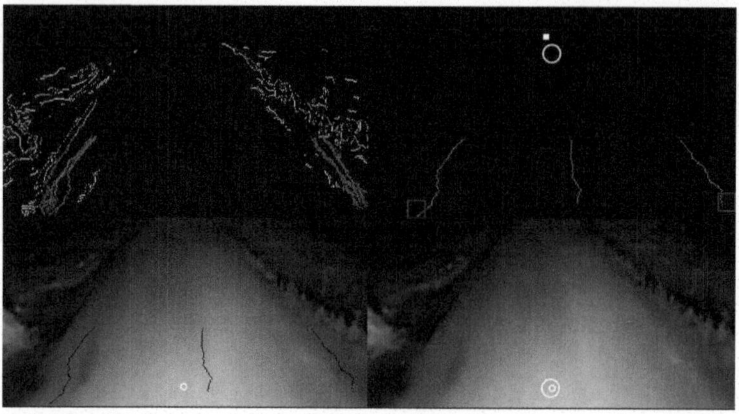

Fig. 6. A frame several second in to the video with inherited experience.

the point of attraction that was used for the ants in this image and the solid white square is the point of attraction calculated for the next frame based on the best paths. The two grey squares are the starting positions for the two colonies. The image in the bottom left corner is the original frame with the two best paths on top and the calculated mid-line of the road. The small white circle is the approximated mid-point of the road and the expected position of the cars mid-point is shown as the large white circle seen in the bottom right image.

5 Conclusion

From this project we found that the pre-processing in form of filtering and a structured edge detection is just as important as the intelligent system itself. The ant colony optimization requires a stable foundation in form of adaptive algorithms that can provide inputs such as start-, end-positions and edge detection in challenging environments. This made for a difficult task to manage in a time where autonomous vehicle has barely made it into the market and expertise of the subject require extensive research. The final algorithm in our project is far from applicable in a real-time car but it shows the extensive output you can achieve from a video using your phone.

References

1. Broggi, A., Cattani, S.: An agent based evolutionary approach to path detection for off-road vehicle guidance (2006)
2. Dorigo, M.: Ant colony optimization (2007). http://www.scholarpedia.org/article/Ant_colony_optimization
3. Mathworks. https://se.mathworks.com/discovery/edge-detection.html
4. Pinninghoff, M.A., Contreras, R., Ortega, J.: Using ant colony optimization for edge detection in gray scale images (2015)
5. Bertozzi, M., Broggi, A., Fascioli, A., Tibaldi, A.: An evolutionary approach to lane markings detection in road environments (2002)

Sharpening the Pan-Multispectral GF-1 Camera Imagery Using the Gram-Schmidt Approach: The Different Select Methods for Low Resolution Pan in Comparison

Qingsheng Liu(✉)

Institute of Geographic Sciences and Natural Resources Research, Chinese Academy of Sciences, Beijing 100101, China
liuqs@lreis.ac.cn

Abstract. The Gram-Schmidt spectral sharpening is one of the widely used image sharpening techniques with preserving the spectral information of images. It is a key step for selecting methods for simulating the low resolution pan image. In this paper, we compared the different select methods for low resolution pan applied to sharpening the GF-1 multispectral bands. The results indicated that the optimal GS sharpened GF-1 multispectral image was from the low resolution pan simulated from Lansat7 sensor. Although there was no significant difference between the ten sharpened images and the original low resolution multispectral images through visual inspection, the sharpening increase the correlation coefficients between the multispectral bands, which is not beneficial for land use classification. In the future, a comprehensive evaluation over more study areas with more images should be performed.

Keywords: Gram-Schmidt · Sharpening · GF-1 · Low resolution pan

1 Introduction

Because of a limited energy availability of the sensor, design considerations on satellite and sensors, satellite data transmission and satellite life [1], many satellites like Landsat7, GeoEye1, IKONOS, KOMPSAT2, QuickBird, RapidEye, SPOT5, World-View1, WorldView2, Sentinel2, China-Brazil Earth Resource Satellite (CBERS)-04, GaoFen-1 (GF-1) and GaoFen-2 (GF-2) commonly provide three or more relatively coarse spatial resolution multispectral bands along with a high spatial resolution panchromatic band. In order to utilize the information from the fine spatial resolution of a pan image and the high spectral resolution of multispectral images, several panchromatic sharpening methods have been developed to combine the fine spatial details of a pan image with the spectral information of multispectral images for compensating for the spatial/spectral trade-off in satellite imaging [2–11]. Some of them have been embedded in commonly used commercial remote sensing processing software packages such as ENVI, ERDAS, and PCI software [8, 12]. It is well known that a sharpening method may give different results for test images from different

Y. Liu et al. (Eds.): ICNC-FSKD 2019, AISC 1074, pp. 417–424, 2020.
https://doi.org/10.1007/978-3-030-32456-8_45

sensors and regions [7–9], so it is necessary to compare the performance of a sharpening method for a new sensor.

GF-1, the first satellite of the Gaofen series, was launched on April 26 2013 at the Jiuquan Satellite Launch Center of China. It is equipped with 1) two panchromatic and multispectral cameras with 2 m resolution panchromatic band (Band1: 450–900 nm) and 8 m resolution multispectral bands (Band2: 450–520 nm; Band3: 520–590 nm; Band4: 630–690 nm; Band5: 770–890 nm), a swath width of 60 km (two-camera stitching width), and repetition cycle of four days; and 2) four multispectral scanners with 16 m resolution (Band6: 450–520 nm; Band7: 520–590 nm; Band8: 630–690 nm; Band9: 770–890 nm), a swath width of 800 km (four-camera stitching width), and repetition cycle of two days [13]. The GF-1 imagery with a spatial resolution of 2 m should be enough to study the decametric-scale vegetation patch patterns [14, 15].

The Gram-Schmidt transform (GS) is one of the widely used image sharpening techniques with preserving the spectral information of images [16, 17], embedded in the commercial remote sensing processing software packages, and has been widely used to fusion the panchromatic, visible and near-infrared, shortwave, and thermal infrared bands from the same or different sensors [12, 16–19]. However, in ENVI software version 5.1, the different fusion results can obtain due to using the different select methods for low resolution pan during the GS image sharpening. There are four types of Select Method for Low Resolution Pan: (1) Average of Low Resolution Multispectral File; (2) Select Input File; (3) Create By Sensor Type; (4) User Defined Filter Function. The first and third methods for Select Method for Low Resolution Pan are simple and not needed to change and add the extra information by the users, so it is easy to evaluate and compare the sharpening performances for the GF-1 images covering the different regions in the future.

The visual inspection, several image quality similarity indexes and quantitative evaluation framework have been proposed to investigate the performance comparison of some widely used image sharpening methods [2–4, 7–10, 16, 17]. The comparison is ideally done through consistency measurements between the sharpened images and the available reference images. However, it is difficult to obtain a reference image at the same resolution of the sharpened one. Therefore, the performance comparison often is done through the comparisons between the sharpened imagery and the upscaled low spatial resolution multispectral imagery or between the downscaled sharpened imagery and the original low spatial resolution multispectral imagery, which may produce some errors because of the scale transformation methods. To overcome this problem, it may be a possible solution that the direct comparisons are done through statistical indexes between the sharpened imagery and the original low spatial resolution multispectral imagery without scale transformation.

The goal of this work is to compare the fusion results from the GS approaches with the first and third methods for Select Method for Low Resolution Pan based on the imagery of the panchromatic and multispectral camera of GF-1 using the Min, MAX, MEAN, STDEV, and coefficient of variance (CV) values of the spectral statistical indexes such as Band1 to Band 4, normalized difference vegetation index (NDVI), and normalized difference water index (NDWI), the global spatial statistical indexes such as the Moran's I (MoranI), Geary's C (GearyC), and Semivariance (SemiV), and the correlation coefficient (CC) between the four bands. Therefore, a total of fifty-four indexes were used in evaluating the sharpened image. All image processing and calculation of statistical indexes were done in ENVI software version 5.1, and statistical analyses were performed using SPSS 19.0 software.

2 Materials and Methods

2.1 Remote Sensing Data and Pre-processing

The previous study indicated that the spring and winter imagery was more suitable for detecting the decametric-scale vegetation patches [20]. One Scene of the Level 1A imagery (Samples 18192, Lines 18000 at 2 m spatial resolution pan image) from the first panchromatic and multispectral camera onboard the GF-1 satellite was acquired on March 11, 2017, which covers the northeast region of Dongying City, Shandong Province, China.

Twenty GCPs selected by the image-to image approach (an average root mean square error of 0.24) were used to wrap the four multispectral bands at 8 m spatial resolution using a quadratic polynomial cubic convolution method with the panchromatic band at 2 m spatial resolution as the base imagery. Because the radiometric calibration data are needed when the Select Method for Low Resolution Pan is the Create By Sensor Type, the digital number of co-registered panchromatic and multispectral imagery was converted to the spectral radiance used the parameters published the CRESDA website (http://218.247.138.119/CN/Downloads/dbcs/11307.shtml). To reduce the data processing time, a subset image (Samples 8314, Lines 6253 at 2 m spatial resolution pan image) was clipped to use in this study.

The two vegetation indexes were calculated using the following equations, the normalized difference vegetation index (NDVI, $NDVI = (\rho 4 - \rho 3)/(\rho 4 + \rho 3)$) and normalized difference water index (NDWI, $NDWI = (\rho 2 - \rho 4)/(\rho 2 + \rho 4)$) where $\rho 2$, $\rho 3$, and $\rho 4$ stand for the spectral radiance of Band 2, 3 and 4.

2.2 Select Method for Low Resolution Pan

The Average of Low Resolution Multispectral File for the low resolution pan means that an average of the original four multispectral bands is simulated as a low resolution panchromatic image. The Create By Sensor Type for the low resolution pan means that a low resolution panchromatic image is simulated according to the selected sensor including ten sensors such as IKONOS, IRS1, Landsat7, QuickBird, SPOT5, KOMPSAT2, RapidEye, GeoEye1, WorldView1 and WorldView2. The descriptions on how to apply GS spectral sharpening and choose one of selecting methods for the low resolution pan can be found in ENVI software version 5.1 help documents. In this paper, except IRS1 sensor, all of the other nine sensor types were applied in the GS spectral sharpening.

2.3 Global Spatial Statistics

Global spatial statistics provide a single value that describes the spatial autocorrelation of the dataset as a whole through looking for an overall pattern proximity and the similarity of pixel values. In ENVI software version 5.1, the spatial autocorrelation statistics known as Moran's I, Geary's C and semivariance are calculated.

The MoranI index measures the local homogeneity through comparing the differences between neighboring pixels and the mean. The value range is between +1 and −1, where +1, 0, and −1 stand for strong positive spatial autocorrelation, spatially uncorrelated data, and strong negative spatial autocorrelation, respectively. The GearyC index provides a measure of dissimilarity within a dataset through comparing the differences between neighboring pixels and the standard deviation. The value range is between 0 and 2, where 0, 1 and 2 stand for strong positive spatial autocorrelation, spatially uncorrelated data, and strong negative spatial autocorrelation, respectively. The SemiV index provides another measure of dissimilarity within a dataset using the squared difference between neighboring pixel values. The value range is constrained to be greater than or equal to 0.

2.4 Evaluations of the Sharpening Imagery

One-Sample T Test was used to determine whether the fifty-four index values were close to these values of the original multispectral bands, and imply whether this kind of the Select Method for Low Resolution Pan was good. One-Sample Nonparametric Test was used to measure the differences between the sharpened images from the ten different selected methods for low resolution pan image. The significant differences were determined at $p < 0.05$.

3 Results and Discussions

The visual inspections on the ten sharpened images (see Fig. 1), there no significant differences between the ten sharpened images and between the ten sharpened images and the original low spatial resolution multispectral image.

The results from One-Sample T Test presented that the fifty indexes were significant differences between the mean of the sharpened and original multispectral images except the minimum NDWI ($p = 0.129 > 0.05$), maximum NDVI ($p = 0.162 > 0.05$), the MoranI of Band2 ($p = 0.375 > 0.05$), and the GearyC of Band2 ($p = 0.364 > 0.05$). This indicated that the sharpened images derived from the ten kinds of the Select Method for Low Resolution Pan could not preserve the spectral information of the original four low spatial resolution multispectral bands. Except the Band2, the overall global spatial patterns of Band1, Band3, and Band4 had the significant differences between the sharpened images derived from the ten kinds of the Select Method for Low Resolution and the original low spatial resolution multispectral images, which indicated that the spatial patterns of the multispectral bands had been also changed after sharpening. These may be attributed to integrating the spatial and spectral information of the high spatial resolution panchromatic band into the sharpened multispectral bands.

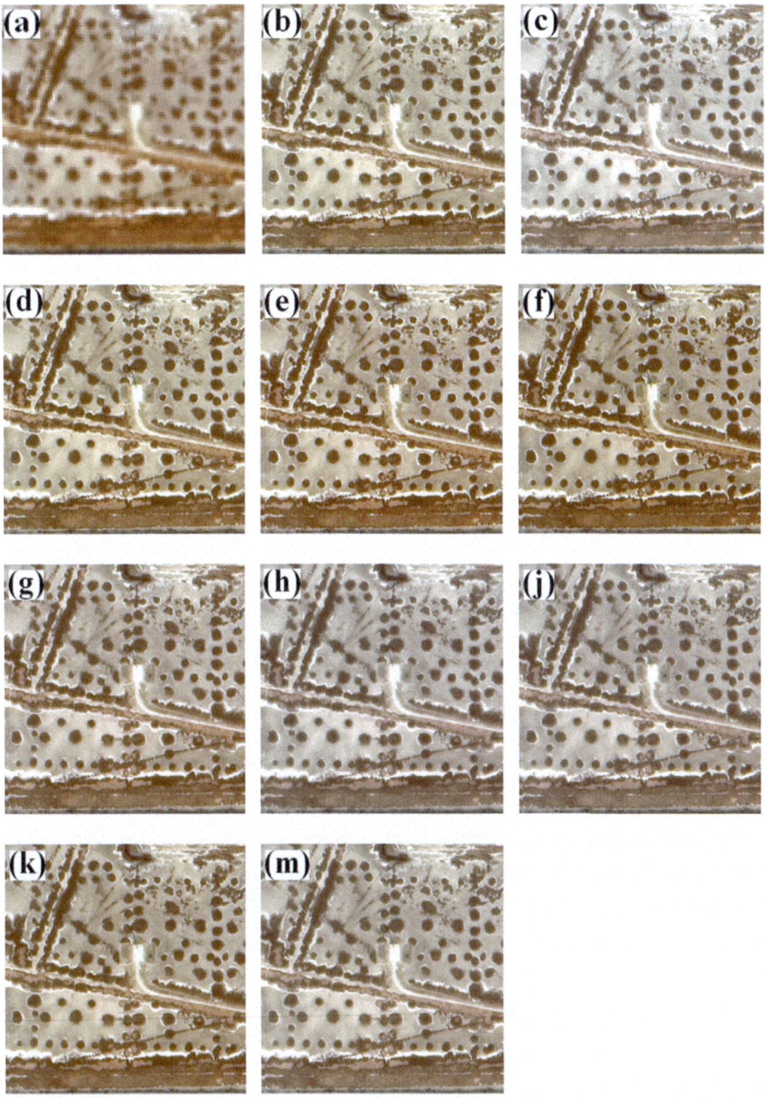

Fig. 1. The GF-1 false color composite images RGB432 (Fig. 1a is 40% of the original size, Fig. 1b to Fig. 1m is 20% size of the sharpened images using the Gram-Schmidt sharpening method). (a) the original multispectral image, (b) the sharpened image from Average of Low Resolution Multispectral File, (c) from GeoEye1, (d) from IKONOS, (e) from KOMPSAT2, (f) from Landsat7, (g) from QuickBird, (h) from RapidEye, (j) from SPOT5, (k) from WorldView1, and (m) from WorldView2.

The results from One-Sample Nonparametric Test presented that the forty-four indexes of the sharpened images from the ten selecting methods for low resolution pan had no significant differences with the original low spatial resolution multispectral image ($p > 0.05$), and only the global spatial statistical indexes including the MoranI and GearyC indexes of NDVI and NDWI, the SemiV of Band1 to Band4, NDVI and NDWI had significant differences with the original low spatial resolution multispectral image ($p < 0.05$). It indicated that the overall global spatial pattern of the NDVI and NDWI had no significant difference after image sharpening. The reasons need to be investigated in the future. The results from One-Sample Nonparametric Test presented that there were no significant differences among the fifty-four indexes of the sharpened images from the ten selecting methods for low resolution pan ($p > 0.05$). Although One-Sample Nonparametric Test detected no significant difference between the ten sharpened images, it could still analyze which sharpened image had the minimum difference with the original low spatial resolution multispectral image.

The difference value between the minimum values of Band1 to Band4, NDVI, and NDWI values of the sharpened image and the original low resolution multispectral image had the minimum values in the sharpened image derived from RapidEye, SPOT5, KOMPSAT2, Landsat7, Landsat7, and IKONOS sensors, respectively. Except the minimum difference of the maximum NDVI value in the sharpened image from the RapidEye sensor, the difference value between the maximum values of the sharpened image and the original low resolution multispectral image had the minimum values in the sharpened image derived from Landsat7 sensor. The difference value between the mean, standard deviation, coefficient of variance values of Band1 to Band4, NDVI, and NDWI values of the sharpened image and the original low resolution multispectral image had the minimum values in the sharpened image derived from Landsat7 sensor. The difference value between the correlation coefficient values of four bands of the sharpened and the original low resolution multispectral image had the minimum values in the sharpened image derived from Landsat7 sensor too.

Except the minimum difference of the MoranI and GearyC index values of Band1 and Band2 in the sharpened image from the RapidEye sensor and the Average of Low Resolution Multispectral File respectively, the sharpened image from Landsat7 sensor had the minimum differences of Band2, Band3, NDVI, and NDWI. The difference value between the SemiV values of Band1 to Band3 of the sharpened image and the original low resolution multispectral image had the minimum values in the sharpened image derived from Landsat7 sensor, and the minimum differences of Band4 and NDWI values in that of RapidEye sensor, and that of the NDVI in that of the Average of Low Resolution Multispectral File. In brief, forty-two of fifty-four indexes from the sharpened image from Landsat7 sensor had the minimum differences with the original low resolution multispectral image, and six indexes from the sharpened image from RapidEye sensor had the minimum difference, three indexes (Global spatial statistical index) for the Average of Low Resolution Multispectral File, and one index for IKONOS, SPOT5 and KOMPSAT2 sensor respectively. The findings indicated that the optimal GS sharpened GF-1 multispectral image was from the low resolution pan simulated from Lansat7 sensor. This may be partly attributed to the similarity of wavelength of bands covered the visible and near-infrared spectral ranges between GF-1 and Landsat7 sensor, which needs to be studied in the future.

After sharpening, the CC values between four multispectral bands increased. Moreover, except the CC value between Band1 and Band2, and Band3 and Band4, the other CC values increased by more than 0.20, and the CC value between Band2 and

Band4 increased by more than 0.33. It indicated that there was more redundant information between four multispectral bands, which is disadvantageous to land use classification. The above findings must be interpreted with caution because there was one scene of GF-1 image.

4 Conclusions

The Gram-Schmidt spectral sharpening is one of the widely used image sharpening techniques with preserving the spectral information of images. It is a key step for selecting methods for simulating the low resolution pan image. In this paper, the Average of Low Resolution Multispectral File and the Create By Sensor Type were applied to sharpen the four low resolution (8 m) multispectral bands with high resolution (2 m) panchromatic band. In general, the optimal GS sharpened GF-1 multispectral image was from the low resolution pan simulated from Lansat7 sensor. The sharpening increased the correlation coefficient between the four multispectral bands, which is not beneficial for land use classification. Of course, these findings must be investigated using more images over more study areas in the future.

Acknowledgment. This research work was jointly financially supported by the National Natural Science Foundation of China (Project No.41671422, 41661144030, 41561144012), the National Mountain Flood Disaster Investigation Project (SHZH-IWHR-57), the Innovation Project of LREIS (Project No.088RA20CYA, 08R8A010YA), and National Key Research and Development Project of China (2016YFC1402701). Thanks to China Center for Resources Satellite Data and Application for providing the GF-1 data.

References

1. Lanaras, C., Bioucas-Dias, J., Baltsavias, E., Schindler, K.: Super-resolution of multispectral multiresolution images from a single sensor. In: IEEE Conference on Computer Vision and Pattern Recognition Workshops, pp. 20–28 (2017)
2. Nikolakopoulos, K.G.: Comparison of six fusion techniques for SPOT 5 data. In: IEEE International Geoscience and Remote Sesning Symposium, pp. 2811–2814 (2005)
3. Du, Q., Younan, N.H., King, R., Shah, V.P.: On the performance evaluation of pan-sharpening techniques. IEEE Geosci. Remote Sens. Letters **14**, 518–522 (2007)
4. Nikolakopoulos, K.G.: Comparison of nine fusion techniques for very high resolution data. Photogramm. Eng. Remote Sens. **74**, 647–659 (2008)
5. Zhang, Y., Mishra, P.K.: A review and comparison of commercially available pan-sharpening techniques for high resolution satellite image fusion. In: IEEE International Geoscience and Remote Sensing Symposium, pp. 182–185 (2012)
6. Mandhare, R.A., Upadhyay, P., Gupta, S.: Pixel-level image fusion using Brovey transform and Wavelet transform. Int. J. Adv. Res. Electr. Electr. Instrum. Eng. **2**, 2690–2695 (2013)
7. Vivone, G., Alparone, L., Chanussot, J., Mura, M.D., Garzelli, A., Licciardi, G.A., Restaino, R., Wald, L.: A critical comparison among pansharpening algorithms. IEEE Trans. Geosci. Remote Sens. **33**, 2565–2586 (2015)
8. Nikolakopoulos, K.G., Oikonomidis, D.: Quality assessment of ten fusion techniques applied on worldview-2. Eur. J. Remote Sens. **48**, 141–167 (2015)

9. Maglione, P., Parente, C., Vallario, A.: Pan-sharpening worldview-2: IHS, Brovey and Zhang methos in comparison. Int. J. Eng. Technol. **8**, 673–679 (2016)
10. Li, H., Jing, L.H., Tang, Y.W.: Assessment of pansharpening methods applied to worldview-2 imagery fusion. Sensors **17**, 89 (2017)
11. Pak, H., Choi, J., Choi, S.: Sharpening the VNIR and SWIR bands of Sentinel-2A imagery through modified selected and synthesized band schemes. Remote Sens. **9**, 1080 (2017)
12. Liu, Q.S.: Sharpening the WBSI Imagery of Tiangong-II: gram-schmidt and principal components transform in comparison. In: 14th International Conference on Natural Computation, Fuzzy Systems and Knowledge Discovery, pp. 524–531 (2018)
13. Cresda, GF-1, Slate, http://www.cresda.com/EN/satellite/7155.shtml. Accessed 3 Jan 2019
14. Barbier, N., Couteron, P., Lejoly, J., Deblauwe, V., Lejeune, O.: Self-organized vegetation patterning as a fingerprint of climate and human impact on semi-arid ecosystems. J. Ecol. **94**, 537–547 (2006)
15. Liu, Q.S., Huang, C., Liu, G.H., Yu, B.W.: Comparison of CBERS-04, GF-1, and GF-2 satellite panchromatic images for mapping quasi-circular vegetation patches in the yellow river delta. China Sens. **18**, 2733 (2018)
16. Sarp, G.: Spectral and spatial quality analysis of pan sharpening algorithms: a case study in Istanbul. Eur. J. Remote Sens. **47**, 19–28 (2014)
17. Gao, L., Wang. Z.W., Jiang, M.: Assessment of fusion methods of ZY-3 satellite images. In: 2016 Academic Annual Conference of Jiansu Provincial Society for Surveying, Mapping and Geoinformation, pp. 44–46 (2016, In Chinese)
18. Liu, Q.S., Li, X.Y., Liu, G.H., Huang, C., Li, H., Guan, X.D.: Sharpening of the VNIR and SWIR bands of the wide band spectral imager onboard Tiangong-II imagery using the selected bands. In: The International Archives of the Photogrammetry, Remote Sensing and Spatial Information Sciences, Volume XLII-3, ISPRS TC III Mid-term Symposium “Developments, Technologies and Applications in Remote Sensing”, pp. 1085–1092 (2018)
19. Liu, Q.S.: Sharpening the VNIR-SWIR-TIR bands of the WIS of Tiangong-2 for mapping land use and land cover. In: Gu, Y., Gao., M., Zhao, G. (eds.) Proceedings of the Tiangong-2 Remote Sensing Application Conference-Technology, Method and Application 2018, Lecture Notes in Electrical Engineering, vol. 541, pp. 212–221. Springer, Singapore (2019)
20. Liu, Q.S.: Comparing the different seasonal CBERS 04 images to map the quasi-circular vegetation patches in the yellow river delta, China. In: 2018 International Conference on Sensor Networks and Signal Processing (SNSP 2018) (2018)

Detection of Vegetation Patch Growth by Absorption Feature Analysis on Tasseled Cap Brightness of Transects from Landsat 7 ETM+ Images

Qingsheng Liu(✉)

Institute of Geographic Sciences and Natural Resources Research, Chinese Academy of Sciences, Beijing 100101, China
liuqs@lreis.ac.cn

Abstract. Vegetation patches are worldwide distributed in arid and semi-arid ecosystems. Mapping vegetation patch dynamics provides valuable information for regional vegetation recovery and re-establishment. The high spatial resolution images may be powerful for the decametric-scale vegetation patch detection. However, the dense and long time series of fine spatial resolution (better than 2.5 m) imagery were not available for large regions until recently due to design considerations on satellite and sensors, satellite data transmission, satellite life and revisited period, and further effects like atmospheric absorption and cloud. For multispectral images (less than 10 m), it was often a challenge for detecting these vegetation patches through visual interpretation and the common image classifications. In this paper, our proposed method based on analysis of absorption features of tasseled cap brightness of Landsat 7 Enhanced Thematic Mapper Plus (ETM+) images along transect provided the acceptable results for vegetation patch recovery detection, which represented a efficient, economical and straightforward procedures for local vegetation management in the Yellow River Delta, China and other similar landscapes.

Keywords: Vegetation patch · Absorption feature · Landsat 7 ETM+ · Tasseled cap transformation · Transect

1 Introduction

In the Yellow River Delta (YRD), China, the quasi-circular vegetation patches (QVPs) have been identified within the grass land and bare land. These patches are often composed of *Suaeda salsa*, *Tamarix chinensis*, and *Phragmites australis*, forming three types of patches, ranging 2.9–31.5 m in diameter [1]. They are visible in high spatial resolution satellite imagery and on the ground level [2]. Petroleum exploration and production in the YRD were considered as one of the causes for the occurrence of these patches, but to date, their formation and succession remains a mystery. Oil wells and roads from petroleum exploration and production have considerably fragmented and damaged the protected wetlands and vegetation communities in the YRD [3]. The increased demand for petroleum and consequent exploration made government

Y. Liu et al. (Eds.): ICNC-FSKD 2019, AISC 1074, pp. 425–432, 2020.
https://doi.org/10.1007/978-3-030-32456-8_46

agencies, corporations, and local residents try to understand and reduce the effects of development on wetland ecosystems. So there is a strong need for scientific information on the long-term ecosystem response to disturbances and on the factors that affect the spatial and temporal patterns of vegetation recovery. Surveys on vegetation recovery following petroleum exploration will provide an indication of the persistence of the environmental and ecological impacts and an important opportunity to evaluate its long-term impacts on coastal marsh ecosystems, further benefiting wetland vegetation re-establishment and restoration in the YRD.

Vegetation recovery can often be detected using imagery of appropriate spatial resolution before and after an event in a cost-effective manner. One approach is to classify the multi-temporal data and to analyze vegetation changes. Aerial photographs may be powerful for the detection of vegetation dynamics [4, 5]. Barbier et al. showed that panchromatic aerial photographs with a pixel size of 2 m were sufficient to study decametric-scale vegetation patterns [6]. Liu et al. found that for the same detection method, the SPOT5 panchromatic imagery (2 m) had the higher detection accuracy, and the multispectral imagery (5 m) was well-suited for detecting the number of vegetation patches but not enough for area estimation [2, 7]. Liu et al. found that the GF-2 imagery with 0.8 m spatial resolution provided relatively high classification accuracy, independent of classifiers, however, the GF-1 imagery with the relatively high spatial resolution (2 m) could not performed better for identifying the vegetation patches than the CBERS-04 imagery with the relatively low spatial resolution (5 m) [8]. Liu et al. found a slight increase in vegetation patch number from 2000 to 2005. Another method involves the use of spectral indices of multispectral imagery [9]. So far, the numerous spectral indices have been used. Compared with the most common vegetation indices (e.g., normalized difference vegetation index (NDVI), soil adjusted vegetation index (SAVI), and normalized difference moisture index (NDMI)), the tasseled cap brightness (TCB), greenness (TCG), and wetness (TCW) obtain slightly better results [9–12].

In general, the decametric-scale vegetation patches were well-detected using high spatial resolution images (better than 10 m) based on various classifiers such as K-means unsupervised classifiers, Maximum likelihood (ML), object-oriented classification, and mathematical morphological method [2, 8, 13]. However, until recently dense and long time series of fine resolution (better than 2.5 m) imagery were not available for large regions because of high costs or no available archived images. Generally the detection results from multispectral images (less than 10 m spatial resolution) using the common classifiers such as K-means and ML classifier were poor. The main reason may be the small size of the vegetation patches for multispectral image and mixed pixels due to conglutination with other vegetation, which was a challenge for visual interpretation and sample selection.

Following the previous study [14], the aim of this study is to assess the effectiveness of the tasseled cap components (TCT) derived from 15 m fusion-ready Landsat 7 Enhanced Thematic Mapper (ETM+) images for mapping the vegetation patch recovery from 2000 to 2005 in greater detail. The study area and period were selected based on the observed vegetation patch growth (see Fig. 1). Landsat 7 ETM + imagery was used because it provides a free unique long-term 15 m spatial resolution archive after data fusion. This research will make more spatial resolution satellite images useful for mapping the decametric-scale vegetation patches.

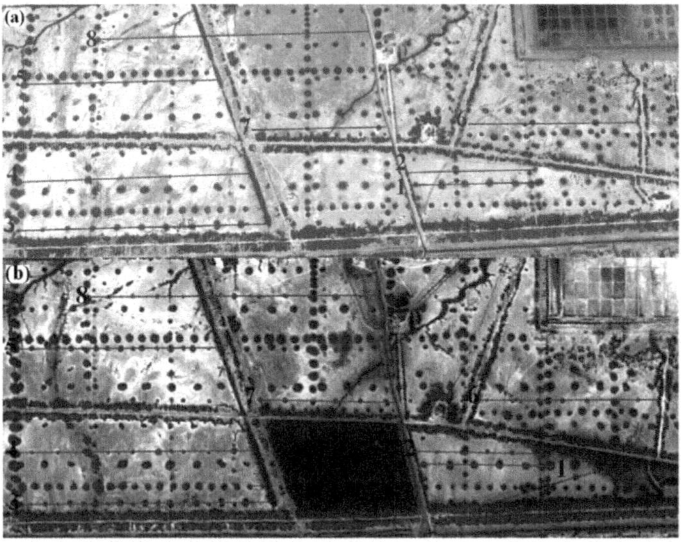

Fig. 1. (a) The eight transects on the image acquired on July 9 2000, (b) the eight transects on the image acquired on September 7 2005.

2 Materials and Methods

2.1 Remote Sensing Data

The unique long-term data record of the Landsat series sensors offers great possibilities to monitor environmental conditions and vegetation pattern changes. After the fusion of panchromatic with multispectral bands, six multispectral bands with spatial resolution of 15 m can be obtained from the Landsat7 ETM+ sensor (launched on April 15, 1999), making the data highly appropriate for mapping the vegetation patch dynamics. A near-annual time series of Landsat 7 ETM+ images (Path/Row 121/34, UTM, Zone 50 N, WGS84) were obtained with a low cloud cover from 2000–2005 (August 6 2000; September 10 2001; August 12 2002; August 31 2003; and October 23 2005) covered the YRD, freely downloaded from the USGS website (http://glovis.usgs.gov/). In addition, a high spatial resolution orthorectified fusion-ready SPOT5 (UTM, 50 N, WGS84; 2.5 m; from September 7, 2005) and a high spatial resolution fusion-ready image of July 9 2000 from Google Earth were used to help selecting the eight linear transects and assess the vegetation patch detection (see Fig. 1). All images were georeferenced to the orthorectified fusion-ready 2005 SPOT5 image with less than one pixel error. Huang et al. confirmed that a tasseled cap transformation (TCT) derived from the Landsat 7 ETM+ at-satellite reflectance was more suitable for applications when the atmospheric correction was not feasible [15]. Therefore, the raw digital number of each image was initially converted to at-satellite reflectance using the default values in the ENVI 5.1 for the Landsat7 ETM+ . Then, the Gram-Schmidt spectral sharpening in the ENVI 5.1 was applied to sharpen the images. Because the minimum and maximum TCB, TCG, and TCW values derived from the original multispectral bands and the pan-sharpened multispectral bands were slightly different, while the means and standard deviations were nearly the same, therefore, the tasseled cap method in the ENVI 5.1 was directly used to produce the TCB, TCG, and TCW in this paper.

2.2 Transect Selections

In general, the newly grown vegetation patches appeared along the west-east transects near the existing vegetation patches (see Fig. 1). Although there is no evidence, the reason may be attributed to the petroleum seismic exploration and drilling [1]. According to this phenomenon, it was possible to monitor the vegetation patch growth only along the west-east transects. To demonstrate it, the eight transects in the study area (see Fig. 1) were set up to detect the vegetation patch recovery. Each transect started in the edge of the canal or vegetation strip, crossed the centers of some vegetation patches, and ended in the other edge of the canal or vegetation strip. The range of the transect length was 525–1140 m (35–76 pixels).

2.3 Mapping the Position and Depth of the Absorption Features

Absorption features including the position, depth, width, and area can be used for identification and recognition of the targets, which are the local dips on the spectral, called as local valleys. The position of the absorption features is related to the type of material of the target, while the depth provides information related to the abundance or grain size of the absorbing materials. Therefore, absorption-based methods have been widely used for decades to detect the minerals and plant stress [16–18]. Because of the alternating vegetation patch and bare soil, there are the local peaks and valleys in the spectral profile of each Transect, which is like as the spectral curve of the target for a certain spectral range. So the absorption-based methods can be used to detect the absorption features of the transect spectrum for mapping the vegetation patch occurrences. In this paper, the position of absorption feature related to the position of the vegetation patch growth was detected by finding the local valleys of each TCT component for each transect. Then, the depth of absorption feature related to the growth status of the vegetation patch was determined through calculating the distance between the minimum valley point and the line of the two maximum values on both sides of the local valleys.

3 Results

3.1 Absorption Features in the Transects

For the quasi-circular vegetation patches, the TCB values were greater than zero and showed the deeper valley for each transect, and both the TCG and the TCW were below zero and showed the higher peaks for each transect. The absorption features derived from the TCB, TCG, and TCW of the eight transects indicated that the TCB was the better indicator to detect the vegetation patch growth because it had the greatest difference between the vegetation and non-vegetation patch pixels. The reason may be that the TCG mainly reflected the vegetation information, and the TCW mainly reflected soil and vegetation canopy moisture, the TCB included more information from the original six multispectral bands which was more suitable to monitor newly grown vegetation growth and environmental condition changes. So, the TCB were used to extract the absorption features of the spectral profiles of transect in this paper.

Through comparison of the absorption feature position and depth of TCB between vegetation patch and bare soil, it indicated that the absorption features could determine the location of the vegetation patch growth when the depth derived from the TCB was greater than 0.045, or the depth was greater than 0.04 with a standard deviation above or equal to 0.04, or the depth was greater than the difference between the mean depth and half of its standard deviation for a standard deviation of 0.02–0.04, or the depth was greater than the difference between the depth mean and its standard deviation for standard deviations below 0.02. According to these thresholds, Fig. 2 show the positions and depths of the detected absorption features for the eight transects in the different years. The deeper the absorption feature the more possibly it identified as the vegetation patch. The detection accuracy of the vegetation patches ranged from 28.57% to 100%. For 2000 and 2005, the detection accuracies of Transect 6, 7, and 8 were below 80%, which may be due to their smaller patches and the disturbances from other shaped vegetation (e.g., striped vegetation along the canals).

3.2 Temporal Analysis

According to visual interpretation (see Fig. 1), from 2000 to 20005, there were no newly grown vegetation patches or obvious area changes of the existing patches in Transect 1. There were three small newly grown patches in Transect 2. There were 7, 7, 6, 7, 4, and 10 more patches in Transects 3, 4, 5, 6, 7, and 8, respectively.

In fact, more absorption features were identified for each transect which often surpassed the number of actual quasi-circular patches but not analyzed because their absorption depths were relatively small (see Fig. 2). For instance, the absorption depth near position 32 in Transect 1 was 0.0390 in 2003 which was only slightly smaller than the given standard deviation of 0.04, so the vegetation patch near position 32 was omitted. Generally, the positions with smaller absorption depths in transects often became vegetation patches next year. For example, the three newly grown vegetation patches in transect 2 in 2005 (position 6, 20 and 25) had already shown absorption features in 2000 and 2001, respectively, indicating that the absorption features from the TCB were able to identify vegetation patch recovery. For the same year, the deeper the absorption depth, the more likely the vegetation patch would grow next year.

3.3 Analysis of the Errors

The main errors in the analysis of the absorption feature changes and in the accuracies for transects 6, 7, and 8, were because the channels and vegetation along the channels

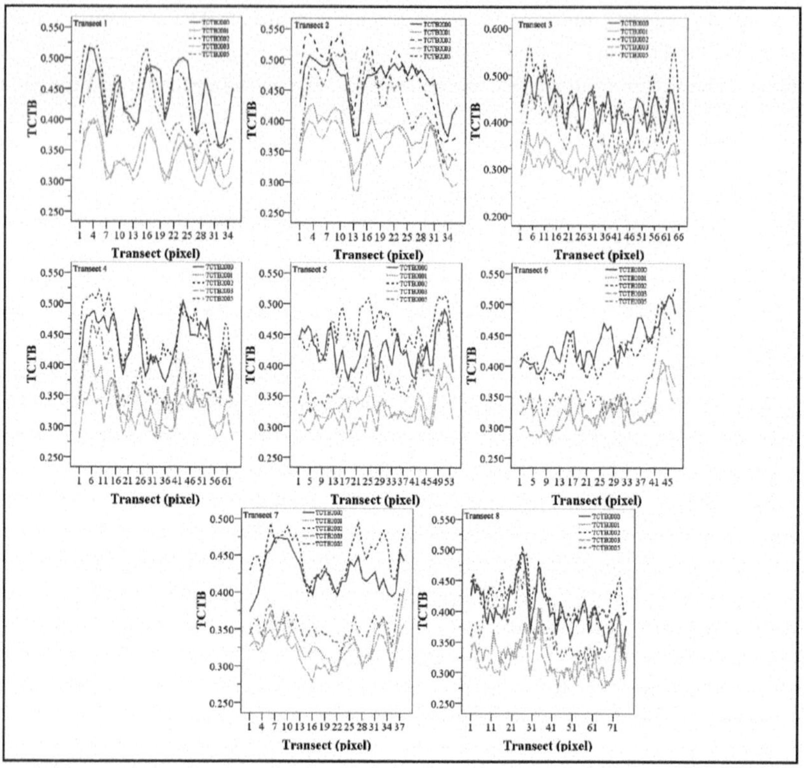

Fig. 2. Positions of the absorption features and TCTB curves of the eight transects from west to east in 2000–2005

had similar TCTB values to the vegetation patches or environmental conditions of the vegetation growth, and for small vegetation patch areas, the absorption depth was maybe reduced or not well-detected.

4 Discussion

Vegetation patches in arid and semi-arid ecosystems are attributed to their adaptation to local water dynamics and nutrient cycling. Regarding the shapes and patterns, the vegetation patches in the YRD are like the "fairy circles" from the Namib Desert [19]. However, their formation resembles the vegetation recovery of the Mackenzie Delta Region, Canada [20]. Although the formation is attributed to the seismic exploration of Shengli Oil Field, it requires a reasonable and further confirmation.

Comparing with the common classifiers for the whole image, our approach achieved better results since it only focused on transect, which reduced the disturbance of the background, decreasing calculation time and data storage space too. Both the TCG and the TCW can be used to map the vegetation patch growth [12], but the absorption position and depth from the TCB was the most effective for our study area.

The possible locations detected by the absorption features could still provide useful information for vegetation recovery and dynamics from 2000–2005.

The methods used in this research were feasible for regional vegetation growth mapping without sufficient funds where the vegetation patches distributed regularly, such as landscape tree health mapping along both sides of city roads. The radiometric calibration did not be applied, so it is difficult to compare the annual absorption depth changes on the same location. This may be resolved through using the pseudo-invariant feature based relatively radiometric corrections of multi-temporal images. In addition, it is necessary to analyze if the width of absorption feature can be applied to evaluate the area of vegetation patch, and how the growth status of vegetation patch can be quantitatively estimated based on the depth of absorption feature in the future.

5 Conclusions

Mapping the quasi-circular vegetation patch growth based on the absorption position and depth of transect in the YRD was implemented using a Landsat7 ETM + image time series. Although the 15 m spatial resolution image is comparatively inadequate in detecting the vegetation patch with less than 20 m in diameter through the common classifiers and visual interpretation approaches, we could evaluate where a new vegetation patch appeared through detecting the position and depth change on the valley of the TCB along transect each year because the newborn vegetation patch generally occurred around the existing patches regularly. Given the low data cost, our approach provides means to feasibly, efficiently and economically monitor vegetation patch recovery in the YRD and other similar landscapes.

Acknowledgment. This research work was jointly financially supported by the National Natural Science Foundation of China (Project No.41671422, 41661144030, 41561144012), the National Mountain Flood Disaster Investigation Project (SHZH-IWHR-57), the Innovation Project of LREIS (Project No.088RA20CYA, 08R8A010YA).

References

1. Liu, Q.S., Liu, G.H., Huang, C., Xie, C.J.: Vegetation patch structure and dynamics at gudong oil field of the yellow river delta, China. In: Bian, F., et al. (eds.) The 2013 International Conference on Geo-Informatics in Resource Management & Sustainable Ecosystem (GRMSE2013), Part I, CCIS, vol. 398, pp. 177–187. Springer, Heidelberg (2013)
2. Liu, Q.S., Liu, G.H., Huang, C., Xie, C.J.: Using SPOT5 fusion-ready imagery to detect Chinese Tamarisk (Saltcedar) with mathematical morphological method. Int. J. Digital Earth **7**(3), 217–228 (2014)
3. Bi, X.L., Wang, B., Lu, Q.S.: Fragmentation effects of oil wells and roads on the yellow river delta. North China. Ocean Coast Manage. **54**, 256–264 (2011)
4. Kadmon, R., Harari-Kremer, R.: Studying long-term vegetation dynamics using digital processing of historical aerial photographs. Remote Sens. Environ. **6**(8), 164–176 (1999)

5. Laliberte, A.S., Rango, A., Havstad, K.M., Paris, J.E., Beck, R.F., McNeely, R., Gonzalez, A.L.: Object-oriented image analysis for mapping shrub encroachment from 1937 to 2003 in southern New Mexico. Remote Sens. Environ. **93**, 198–210 (2004)
6. Barbier, N., Couteron, P., Lejoly, J., Deblauwe, V., Lejeune, O.: Self-organized vegetation patterning as a fingerprint of climate and human impact on semi-arid ecosystems. J. Ecol. **94**, 537–547 (2006)
7. Liu, Q.S., Liu, G.H., Chu, X.L.: Comparison of different spatial resolution bands of SPOT 5 to vegetation community patch detection. In: Proceedings of the 5th International Congress on Image and Signal Processing (CISP 2012), pp. 1190–1194 (2012)
8. Liu, Q.S., Huang, C., Liu, G.H., Yu, B.W.: Comparison of CBERS-04, GF-1, and GF-2 satellite panchromatic images for mapping quasi-circular vegetation patches in the Yellow River Delta. China Sens. **18**, 2733 (2018)
9. Sonnenschein, R., Kuemmerle, T., Udelhoven, T., Stellmes, M., Hostert, P.: Differences in landsat-based trend analyses in drylands due to the choice of vegetation estimate. Remote Sens. Environ. **115**, 1408–1420 (2011)
10. Dymond, C.C., Mladenoff, D.J., Radeloff, V.C.: Phenological differences in tasseled cap indices improve deciduous forest classification. Remote Sens. Environ. **80**, 460–472 (2002)
11. Jin, S.M., Sader, S.A.: Comparison of time series tasseled cap wetness and the normalized difference moisture index in detecting forest disturbances. Remote Sens. Environ. **94**, 364–372 (2005)
12. Czerwinski, C.J., King, D.J., Mitchell, S.W.: Mapping forest growth and decline in a temperate mixed forest using temporal trend analysis of Landsat imagery, 1987-2010. Remote Sens. Environ. **141**, 188–200 (2014)
13. Li, Y.Y., Liu, Q.S., Liu, G.H., Huang, C.: Detect quasi-circular vegetation community patches using images of different spatial resolutions. In: Proceedings of the 2013 6th International Congress on Image and Signal Processing (CISP 2013), pp. 824–829 (2013)
14. Liu, Q.S., Liu, G.H, Huang, C., Shi, L., Zhao, J.: Monitoring vegetation recovery at abandoned land. In: Proceedings of the 8th International congress on Image and Signal Processing (CISP 2015), pp. 88–92 (2015)
15. Huang, C.Q., Wylie, B., Yang, L.M., Zylstra, G.: A tasseled cap transformation for landsat 7 ETM+ at-satellite reflectance. Int. J. Remote Sens. **23**(8), 1741–1748 (2002)
16. Van der Meer, F.: Analysis of spectral absorption features in hyperspectral imagery. Int. J. Appl. Earth Observation Geoinf. **5**, 55–68 (2004)
17. Van Ruitenbeek, F.J.A., Bakker, W.H., Van der Werff, H.M.A., Zegers, T.E., Oosthoek, J.H.P., Omer, Z.A., Marsh, S.H., Van der Meer, F.D.: Mapping the wavelength position of deepest absorption features to explore mineral diversity in hyperspectral images. Planet. Space Sci. **101**, 108–117 (2014)
18. Sanches, L.D, Filho, C.R. S., Kokaly, R.F.: Spectroscopic remote sensing of plant stress at leaf and canopy levels using the chlorophyll 680 nm absorption feature with continuum removal. ISPRS J. Photogrammetry Remote Sens. **97**, 111–122 (2014)
19. Tschinkel, W.R.: The life cycle and life span of Namibian fairy circles. PLoS ONE **7**(6), e38056 (2012)
20. Johnstone, J.F., Kokelj, S.V.: Environmental conditions and vegetation recovery at abandoned drilling mud sumps in the Mackenzie Delta Region, Northwest Territories. Canada Arctic **61**(2), 199–211 (2008)

X-Architecture Steiner Minimal Tree Construction Based on Discrete Differential Evolution

Hailin Wu[1], Saijuan Xu[2], Zhen Zhuang[1,3], and Genggeng Liu[1,3(✉)]

[1] College of Mathematics and Computer Sciences, Fuzhou University, Fuzhou 350116, China
15505915786@163.com, zhuang_zhen@126.com, liugenggeng@fzu.edu.cn

[2] Department of Information Engineering, Fujian Business University, Fuzhou 350012, China
1044513210@qq.com

[3] Fujian Provincial Key Laboratory of Network Computing and Intelligent Information Processing, Fuzhou University, Fuzhou 350116, China

Abstract. As the best connection model for the multi-pin net of the non-Manhattan architecture global routing problem, the X-architecture Steiner Minimum Tree (XSMT) construction is a Non-deterministic Polynomial hard (NP-hard) problem. The Differential Evolution (DE) algorithm has shown good application effect in solving various NP-hard problems. For this reason, based on the idea of DE algorithm, this paper proposes an XSMT construction algorithm for solving this problem. First of all, because the traditional DE algorithm is designed for continuous problems, the optimization ability is limited in solving discrete problems. This paper proposes a novel crossover operator and mutation operator. At the same time, in order to maintain the effectiveness of the evolutionary algorithm, an Edge-to-Point coding strategy suitable for evolutionary algorithms is proposed to better preserve the optimal substructure of the population. Finally, in order to speed up the convergence speed and quality of the algorithm, this paper proposes an initial solution based on the minimum tree generation algorithm. Experiments show that the effectiveness of the proposed algorithm and related strategies can construct a high-quality XSMT solution.

Keywords: XSMT · Routing · VLSI · Steiner minimal tree · Different evolution

1 Introduction

Global routing is an important part of Very Large Scale Integration (VLSI) physical design. Steiner Minimal Tree (SMT) problem is to find a minimum connection tree for a given pin set by introducing some additional Steiner points. SMT problem is a Non-deterministic Polynomial hard (NP–hard) problem [1]. SMT model is the best connection model for VLSI global routing [2, 3]. How to construct a suitable SMT is a problem that needs to be solved in VLSI manufacturing process.

Y. Liu et al. (Eds.): ICNC-FSKD 2019, AISC 1074, pp. 433–442, 2020.
https://doi.org/10.1007/978-3-030-32456-8_47

At present, most of the existing algorithms for solving global routing problem are based on the Manhattan architecture [4–6]. With the development of VLSI fabrication technology, the optimization capacity of routing algorithms based on the Manhattan architecture is limited. It is gradually becoming the major challenges of chip performance. Based on this background, more and more researchers begin to study the non-Manhattan architecture which can use much more routing resources to better optimize wirelength.

The primary research work for non-Manhattan architecture routing is to construct a SMT based on non-Manhattan architecture. X-Architecture Steiner Minimal Tree (XSMT) is the representative architecture of non-Manhattan architecture routing [7–12]. SMT problem is a NP-hard problem. Meanwhile, evolutionary algorithm has shown good application effect in solving various NP-hard problems. So researchers began to study the application of evolutionary algorithm in SMT construction.

As a kind of swarm-based heuristic search algorithm, Differential Evolution (DE) algorithm was first proposed in 1995 [13]. DE is a kind of evolutionary algorithm and used to solve real number optimization problems. Compared with other optimization algorithms, DE algorithm has many characteristics such as undetermined parameters, simple structure, easy implementation, fast convergence and strong robustness. It is widely used in various fields such as pattern recognition, artificial neural network and so on [14–19].

Based on the above research work, this paper proposes an XSMT construction algorithm based on Discrete DE algorithm (XSMT-DDE). Firstly, in order to speed up the convergence speed and quality of the algorithm, this paper proposes an initial solution based on the minimum spanning tree algorithm. Secondly, according to the characteristics of XSMT, this paper proposes a discrete DE update strategy combined with the crossover operator and the mutation operator. And thus the proposed algorithm can be used for XSMT construction problem. Finally, the experimental testcases are carried out and they show the feasibility and effectiveness of the proposed algorithm.

2 Problem Description

The given set $P = \{P_1, P_2, \cdots, P_n\}$ in the XSMT problem is a set of n pins to be routed. The Cartesian coordinate system is introduced, and each P_i corresponds to one coordinate (x_i, y_i), For example, Table 1 shows that there are 5 pins in the net to be routed. The input information of the given pins is shown in the Table 1. The corresponding pin distribution is shown in Fig. 1. For example, the pin 1 is shown in the second column of Table 1, and the value of point 1 is (21, 2).

Table 1. The input information of the given pins

N	1	2	3	4	5
X	21	3	10	23	15
Y	2	4	5	12	23

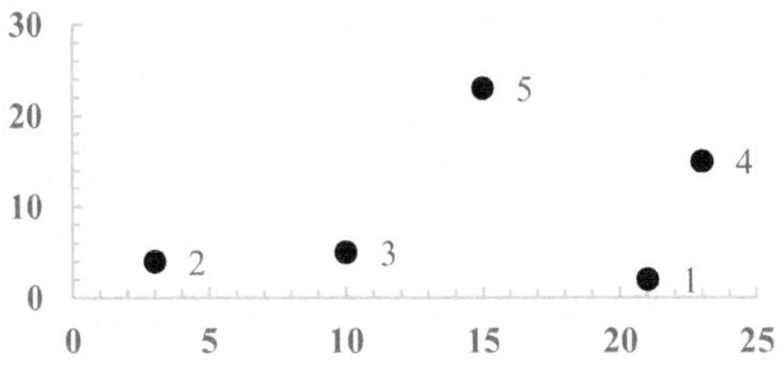

Fig. 1. The pin distribution for a given net in Table 1.

Definition 1 Pseudo-Steiner. For convenience, we assumes that the connection point except for the pins is called Pseudo-Steiner point, which is denoted as point *S*.

Definition 2 (Choice 0). As shown in Fig. 2a, let A = (x_1, y_1) and B = (x_2, y_2) be the two endpoints of segment $\overline{\mathrm{AB}}$, where $x_1 < x_2$. The Choice 0 of $\overline{\mathrm{AB}}$ leads rectilinear edge from A to S, and then leads octagonal edge from S to B.

Definition 3 (Choice 1). As shown in Fig. 2b, it leads octagonal edge from A to S, and then leads rectilinear egde from S to B,.

Definition 4 (Choice 2). As shown in Fig. 2c, it leads vertical edge from A to S, and then leads horizontal edge from S to B.

Definition 5 (Choice 3). As shown in Fig. 2d, it leads horizontal edge to S, and then leads vertical edge from S to B.

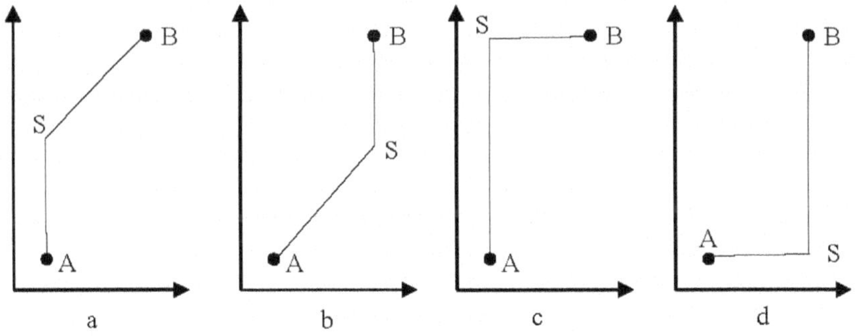

Fig. 2. Four choice of segment $\overline{\mathrm{AB}}$

3 DE Algorithm

The essence of DE is an efficient multi-objective global optimization algorithm. It is used to solve global optimal solution in multidimensional space. Through continuous differential evolution, it can get the optimal solution.

3.1 Process of Traditional DE Algorithm

According to Literature [13], the flow chart of traditional differential evolution algorithm is as follows.

Step1: Population initialization N particles are randomly generated in the solution space, each of which has an m-dimensional vector composition, and the j-dimensional vector of the *i-th* particle is denoted as follow.

$$X_i^j = L_{j\_min} + rand(0,1) * \left(L_{j\_max} - L_{j\_min}\right) \tag{1}$$

where L_{j_min} and L_{j_man} are the maximum value and the minimum value of dimension j respectively.
Step2: Mutation During the *g-th* iteration, the mutation strategy is denoted as follow.

$$V_i(g) = X_{p1}(g) + F * \left(X_{p2}(g) - X_{p3}(g)\right) \tag{2}$$

where $p_1 \neq p_2 \neq p_3 \neq i$, F is the scaling factor and is selected between [0, 2].
Step3: Crossover The crossover strategy of DE algorithm is denoted as follow.

$$v_i^j = \begin{cases} h_i^j(g) & rand(0,1) \leq cr \\ x_i^j(g) & else \end{cases} \tag{3}$$

where cr is a cross probabilistic, and $cr \in [0, 1]$.
Step4: Selection The selection strategy is a greedy strategy. It is denoted as follow.

$$X_i(g+1) = \begin{cases} V_i(g) & f(V_i(g)) < f(X_i(g)) \\ X_i(g) & else \end{cases} \tag{4}$$

Step5: If the algorithm termination condition is met, it ends. Otherwise, it returns **Step2**.

4 XSMT-DDE Algorithm

4.1 Coding Strategy

Generally, there are two encoding strategies of spanning trees, including Prufer number coding and Edge-to-Point coding. Because the Edge-to-Point coding strategy is more suitable for evolution algorithm, we use the Edge-to-Point coding strategy. The Edge-to-Point coding strategy is defined as below. If a net has n pins, a spanning tree would have *n–1* egdes, *n–1* Pseudo Steiner points and one extra bit which is the particle's fitness. For example, the spanning tree which is give in Fig. 1 can be expressed as the following numeric string.

1 3 0 2 3 1 3 5 2 4 5 2 23

where the number '23' is the fitness of the particle and it is also the length of the spanning tree.

4.2 The Fitness Function of the Particle

The length of an XSMT is the sum of the length of all the edge in the tree. The length of an XSMT is defined as below.

$$L(T_x) = \sum\nolimits_{e_i \in T_x} l(e_i) \tag{5}$$

where $l(e_i)$ represents the length of edge e_i in XSMT.

To measure the cost of an XSMT in our algorithm, the smaller the total length of the XSMT, the higher its cost is. In this paper, the fitness function which is used to measure the cost of an XSMT is defined as below.

$$f = \frac{1}{L(T_x) + 1} \tag{6}$$

When the pins of a routing tree are coincided, it causes that the length of the routing tree is 0 which means the denominator is meaningless. In order to avoid the happening of this problem, we add one to the denominator.

4.3 Particle Updating Formula

DE algorithm is a type of multi-objective optimization algorithms. It is used to get global optimal solution for continuous problem. However, XSMT problem is a discrete problem. It cannot be solved directly by DE algorithm. In order to solve this problem, this paper proposes a novel crossover operator and a novel mutation operator.

Novel Mutation Operator: Combining with the mutation strategy of formula (2), it is defined as below. Let F is equal to 1, the subtraction operation in the formula refers to the difference set operation, and the addition operation refers to the union operation of the set. For this improvement, the subtraction operation may produce two results as below.

A. As shown in Fig. 3, three selected particles $p1$, $p2$ and $p3$ are substituted into formula (2), and the result of $p2 - p3$ is empty set. Then the particle $p1$ selects an edge $m1$ which is randomly deleted to mutate. After deletion, $p1$ is divided into two sub trees. Combining with Disjoint Set Union (DSU), two points are randomly selected from the two sub trees respectively and reconnected as a mutation particle.
B. As shown in Fig. 4, according to formula (2), the result of $p2 - p3$ is not empty. Combining with DSU, the elements of the difference set are as part of the edge set of mutation result. Then we select the remaining from the edge set of the particle $p1$ to reconstruct a new tree as a mutation particle.

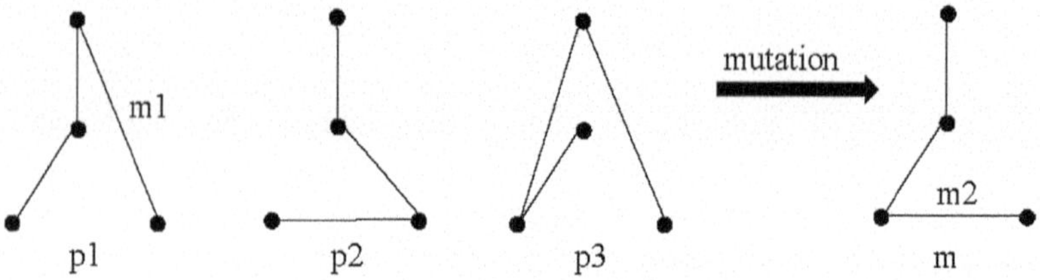

Fig. 3. Novel mutation operator 1

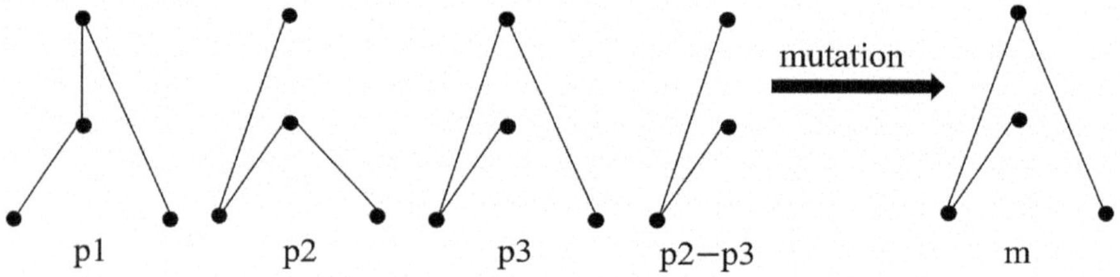

Fig. 4. Novel mutation operator 2

Novel Crossover Operator: The object to be crossed are two trees. If the formula (3) is used directly, the result of the crossover may be an illegal tree. To solve this problem, the following improvements are proposed. We cross the mutation particle with the current particle, and it is performed according to the probability *cr*. As shown in Fig. 5, after the mutation operation, the mutated particle *m* cross with the current particle *i*. The common edges of the two trees are used as the starting set, and the remaining edges are used as the candidate edges. Combining with DSU, we constantly extract the remaining edges from the candidate edges until the crossover result is a legal tree.

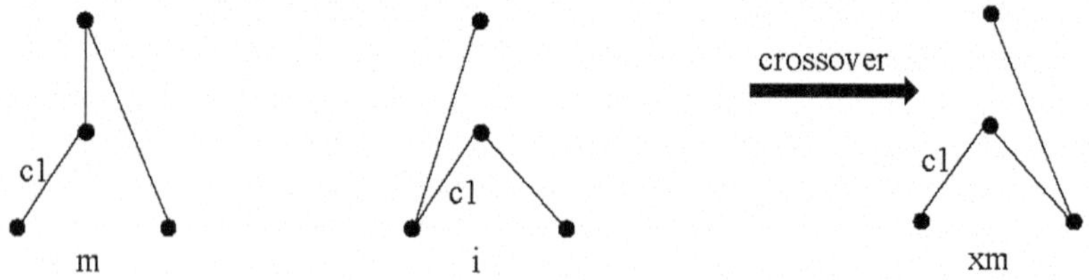

Fig. 5. Novel crossover operator

Selection Operator: The selection operator is based on formula (4).

4.4 Algorithm Flow of This Paper

The flow chart of the XSMT-DDE algorithm is shown in Fig. 6. The detailed steps are as below.

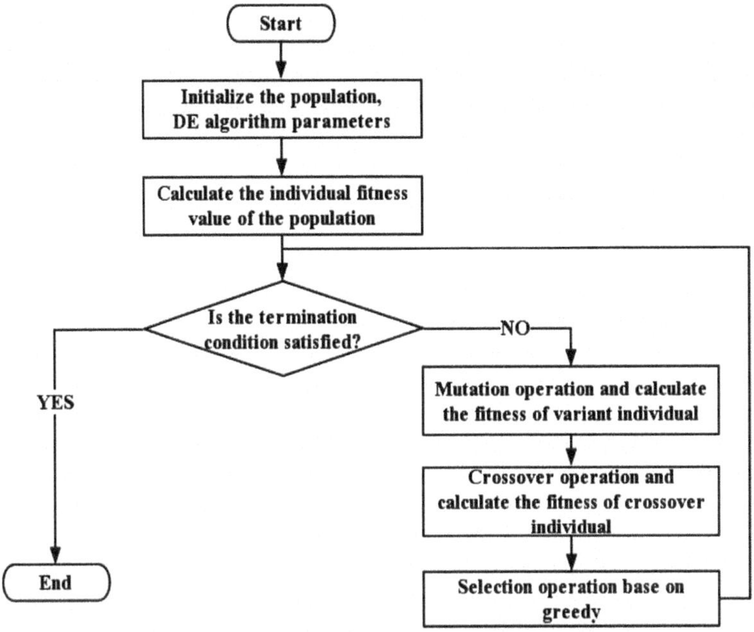

Fig. 6. The flow chart of the proposed algorithm

Firstly, read the pin information of the routing to be tested, and sort the pin coordinates in ascending order.

Secondly, the population is initialized by the minimum spanning tree method and the fitness value is calculated according to formula (6). Then set the number of iterations, the scaling factor F and the crossover probability cr.

Thirdly, in each iteration process, the particle population is updated by the mutation operator, crossover operator and selection strategy proposed in this paper, and the fitness value is calculated according to formula (6).

Finally, if the algorithm termination condition is met, it ends, otherwise, it returns to the third step.

5 Experimental Results

The algorithm in this paper is written in C/C++ language and implemented on Windows10 platform. The test routing data used to verify the validity of the algorithm comes from the literature [10]. In order to verify that the proposed algorithm can solve the XSMT problem, the algorithm is compared with the literature [4, 10]. The experimental results are given in the following sections.

5.1 Parameter Setting Relating to Algorithm

The main parameter setting of the XSMT-DDE algorithm includes the population size *NP*, the number of iterations, the scaling variable *F* and the crossover probability *cr*. In our implement, *F* is equal to 1 and *cr* adopts an adaptive strategy which is defined as below.

$$cr_i = \begin{cases} cr_l + (cr_u - cr_l)\frac{f_i - f_{min}}{f_{max} - f_{min}} & f_i > \bar{f} \\ cr_l & else \end{cases} \tag{7}$$

where cr_l is equal to 0.1 and cr_u is equal to 0.6.

5.2 Verify the Validity of the Minimum Tree Generation Method

In this paper, the population is generated by the minimum spanning tree method which avoids the situation that the algorithm solution space is too large to be converged due to random initialization. The specific experimental results are shown in Table 2. The last column is reduction rates. It can be seen from the last column in the Table 2 that, as the number of pins of basic circuits increases, the minimum spanning tree method can effectively optimize the situation to get over that the algorithm cannot converge due to the large solution space. It achieves an average optimization 35.72%.

Table 2. Optimization rate brought by minimum tree generation strategy

Test	Pins	Random initial solution	Minimum spanning tree	Reduction (%)
1	8	16900	16900	0.00%
2	9	18023	18023	0.00%
3	10	19397	19397	0.00%
4	20	32219	32198	0.07%
5	50	55751	47932	14.02%
6	70	80640	56347	30.13%
7	100	123767	68855	44.37%
8	410	1042081	141666	86.41%
9	500	1378707	154697	88.78%
10	1000	3388682	221739	93.46%
Average				35.72%

5.3 Comparison with Other Algorithms

In order to verify the effectiveness of the proposed algorithm, the results of our algorithm are compared with the results of the literature [4, 10]. The experimental results are shown in Table 3. The experimental results show that the XSMT-DDE algorithm achieves an optimization rate of 9.61% and 0.37% in solving the XSMT problem compared with the traditional strategies of Rectilinear Steiner Minimal Tree (RSMT)[4] and Octagonal Steiner Minimal Tree (OSMT) [10]. This algorithm can solve the XSMT construction problem better.

Table 3. Comparison between XSMT-DDE and other algorithms

Test	Pins	RSMT (R)	OSMT	XSMT-DDE	Reduction (%)	
					XSMT-DDE:RSMT	XSMT-DDE:OSMT
1	8	17931	17040	16900	5.75%	0.82%
2	9	20503	18163	18023	12.10%	0.77%
3	10	21910	19818	19397	11.47%	2.12%
4	20	35723	32199	32198	9.87%	0.00%
5	50	53383	47960	47932	10.21%	0.06%
6	70	61987	55980	56347	9.10%	−0.66%
7	100	76016	68743	68855	9.42%	−0.16%
8	410	156520	142880	141666	9.49%	0.85%
9	500	170273	154290	154697	9.15%	−0.26%
10	1000	245201	222050	221739	9.57%	0.14%
Average					9.61%	0.37%

6 Conclusion

In this paper, we apply an improved DE algorithm to solve VLSI routing problem. Compared to traditional DE algorithms which are used to solve continuous problems, our algorithm can be effectively used to solve discrete problems by adopting some improved strategies so that it can solve the problem of constructing XSMT. The experiment results show that our algorithm is effective to construct XSMT.

Acknowledgements. This work was partially supported by the National Natural Science Foundation of China (Nos. 61877010 and 11501114), and the Fujian Natural Science Funds (No.2019J01243).

References

1. Garey, M.R., Johnson, D.S.: The rectilinear Steiner tree problem is NP-complete. SIAM J. Appl. Math. **32**(4), 826–834 (1977)
2. Lavagno, L., Markov, I.L., Martin, G., et al.: Electronic Design Automation for IC Implementation, Circuit Design, and Process Technology: Circuit Design, and Process Technology. CRC Press, Boca Raton (2016)
3. Stroud, C.E., Wang, L.T., Chang, Y.W.: Chapter 1: Introduction. In: Ang, L.T., Chang, Y. W., Cheng, K.T., Electronic Design Automation: Synthesis, Verification, and Testing, pp. 1–38, Elsevier/Morgan Kaufmann (2009)
4. Liu, G.G., Chen, G.L., Guo, W.Z., et al.: DPSO-based rectilinear Steiner minimal tree construction considering bend reduction. In: Proceeding of the 7th International Conference on Natural Computation, pp. 1161–1165 (2011)
5. Held, S., Muller, D., Rotter, D., et al.: Global routing with timing constraints. IEEE Trans. Comput. Aided Des. Integrated Circuits Syst. **37**(2), 406–419 (2018)
6. Siddiqi, U.F., Sait, S.M.: a game theory based post-processing method to enhance VLSI global routers. IEEE Access **5**, 1328–1339 (2017)

7. Coulston, C.S.: Constructing exact octagonal steiner minimal trees. In: Proceedings of the 13th ACM Great Lakes symposium on VLSI, pp. 1–6. ACM (2003)
8. Chiang, C., Chiang, C.S.: Octilinear Steiner tree construction. In: The 2002 45th Midwest Symposium on Circuits and Systems, vol. 1, pp. 1–603 (2002)
9. Zhu, Q., Zhou, H., Jing, T., et al.: Spanning graph-based nonrectilinear steiner tree algorithms. IEEE Trans. Comput. Aided Des. Integrated Circuits Syst. **24**(7), 1066–1075 (2006)
10. Liu, G.G., Chen, G.L., Guo, W.Z.: DPSO based octagonal steiner tree algorithm for VLSI routing. In: Proceeding of the 5th International Conference on Advanced Computational Intelligence (ICACI2012), pp. 383–387 (2012)
11. Liu, G.G., Huang, X., Guo, W.Z., et al.: Multilayer obstacle-avoiding X-architecture Steiner minimal tree construction based on particle swarm optimization. IEEE Trans. Cybern. **45**(5), 989–1002 (2015)
12. Liu, G.G., Guo, W.Z., Niu, Y.Z., et al.: A PSO-based timing-driven Octilinear Steiner tree algorithm for VLSI routing considering bend reduction. Soft. Comput. **19**(5), 1153–1169 (2015)
13. Storn, R., Price, K.: Differential evolution – a simple and efficient heuristic for global optimization over continuous spaces. J. Global Optim. **11**(4), 341–359 (1997)
14. Zhou, Y.Z., Yi, W.C., Gao, L., et al.: Adaptive differential evolution with sorting crossover rate for continuous optimization problems. IEEE Trans. Cybern. **47**(9), 2742–2753 (2017)
15. Qiu, X., Xu, J.X., Xu, Y., et al.: A new differential evolution algorithm for minimax optimization in robust design. IEEE Trans. Cybern. **48**(5), 1355–1368 (2018)
16. Yong, W., Hao, L., Long, H., et al.: Differential evolution with a new encoding mechanism for optimizing wind farm layout. IEEE Trans. Ind. Inf. **14**(3), 1040–1054 (2018)
17. Ge, Y.F., Yu, W.J., Lin, Y., et al.: Distributed differential evolution based on adaptive mergence and split for large-scale optimization. IEEE Trans. Cybern. **48**(7), 2166–2180 (2018)
18. Zhou, X.G., Zhang, G.J.: Differential evolution with underestimation-based multimutation strategy. IEEE Trans. Cybern. **49**(4), 1353–1364 (2019)
19. Wang, J., Liang, G., Zhang, J.: Cooperative differential evolution framework for constrained multiobjective optimization. IEEE Trans. Cybern. **49**(6), 2060–2072 (2018)

Detection Models Study of Chlorophyll in Winter Wheat's Leaves by Reflectance Spectra and Artificial Neural Networks

Zhenhong Rao[1(✉)], Li Zhang[1], and Xue Liang[2]

[1] College of Science, China Agricultural University, Beijing 100083, China
ndrzh@sina.com
[2] College of Information and Electrical Engineering, China Agricultural University, Beijing 100083, China

Abstract. Reflectance spectra of living wheat's leaves were acquired by portable radiation spectrometer. Multivariate scatter correction (MSC) method was used to preprocess the spectra, and quantitative analysis detection models were built up by artificial neural networks (ANN). The models of chlorophyll, chlorophyll a, and chlorophyll b in leaves were built up and tested. In calibration set and predicted set, for chlorophyll in leaves, the correlation coefficient (R) were 0.927 and 0.818, the standard deviation (SD) were 0.252 and 0.470; for chlorophyll a in leaves, the correlation coefficient (R) were 0.780 and 0.627, the standard deviation (SD) were 0.441 and 0.592; for chlorophyll b in leaves, the correlation coefficient (R) were 0.876 and 0.871, the standard deviation (SD) were 0.205 and 0.259. The results show that, using portable radiation spectrometer to acquire the reflectance spectra, multivariate scatter correction to preprocess the spectra, and artificial neural networks to build up the models, the chlorophyll, chlorophyll a and chlorophyll b in winter wheat's leaves may be quantitative detected.

Keywords: Artificial neural networks · Spectral analysis · Winter wheat's leaves · Chlorophyll · Chlorophyll a · Chlorophyll b

1 Introduction

The measurement and diagnosis of chlorophyll, chlorophyll a and chlorophyll b in winter wheat plants are important links in the production process of winter wheat. The conventional technique for the diagnosis of winter wheat nutrients is destructive sampling and its analysis time is longer. This sampling analysis technique is difficult to meet the requirements of modern agriculture for water and nutrient management. Although portable instruments can measure the SPAD (Soil and Plant Analyzer development) value, which has relation with chlorophyll, but the content of chlorophyll a and chlorophyll b cannot be measured. Retrieving the biochemical components of crops with crop spectral information has the advantages of timely, rapid, non-invasive sampling, so spectral analysis techniques are ideal techniques for the diagnosis of crop nutrient conditions [1–3].

Y. Liu et al. (Eds.): ICNC-FSKD 2019, AISC 1074, pp. 443–451, 2020.
https://doi.org/10.1007/978-3-030-32456-8_48

Multivariate scatter correction (MSC) and artificial neural network (ANN) are methods that have been used more actively in crop spectra processing in recent years. It is mostly used for the detection of biochemical components such as oils, proteins, starches, fatty acids, etc. which in cereals, seeds, dried leaves (tobacco), powders, etc. MSC was used as preprocess method [4, 5] and ANN was used as to build quantitative analysis models [6–9]. ANN was also used to estimation of rice pigment content [10] and estimation of sugar to nitrogen ratio in wheat leaves [11].

In this paper, the reflectance spectra of living winter wheat leaves were acquired by portable radiation spectrometer. The acquired spectra were pretreated by multivariate scatter correction pretreatment method. Standard chlorophyll value and preprocessed spectral data were used by artificial neural network (ANN) to establish a quantitative analysis model, aiming at establishing a model with high prediction accuracy, studying the spectrum detection mechanism of winter wheat biochemical components, and achieving the goal of directly predicting the content of chlorophyll, chlorophyll a and chlorophyll b in winter wheat leaves by spectral information.

2 Experiment

2.1 Acquisition of Winter Wheat Leaf Reflection Spectra

Select 40 winter wheat plants, cut the second leave of each plant, wipe off the surface dirt for use. The reflectance spectra of the 40 winter wheat leaves were scanned by using the FieldSpec® Pro FR portable spectroradiometer from the US ASD and the 1800 integrating sphere produced by LI-COR. The spectral range is 400 nm to 1550 nm, the data are collected in each nanometer of the spectral region. The absorption of chlorophyll is in the visible region, so the spectral range of 400–750 nm was selected to study. The front of the wheat leave was placed in front of the aperture, and the reference plate and the wheat leaves were separately measured. Each leave was scanned 5 times and averaged to reduce the error. The measured reflectance spectrum is shown in Fig. 1.

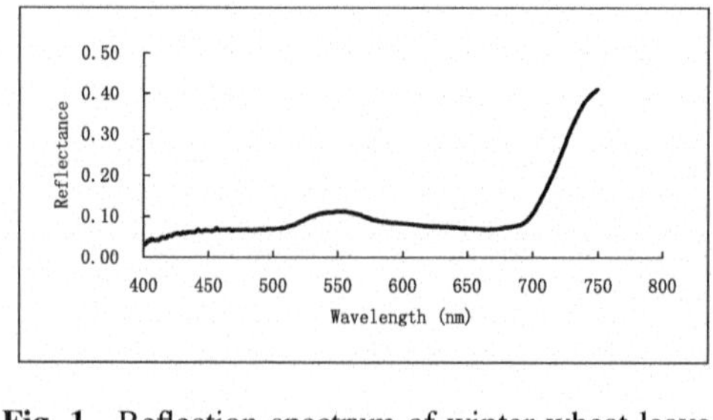

Fig. 1. Reflection spectrum of winter wheat leaves

2.2 Determination of Standard Value of Leaf Chlorophyll Content

After spectral scanning, the chlorophyll, chlorophyll a, and chlorophyll b contents in the leaves were measured by standard methods and used as the standard chemical values of the leaves to build quantitative models by artificial neural networks.

The standard methods is as following: After the leaves are ground and extracted, the absorbance values of the extracts at 663 nm and 645 nm were measured on a spectrophotometer, and the concentrations of chlorophyll, chlorophyll a and chlorophyll b in the solution were calculated by substituting A-M equation.

$$\text{Chlorophyll a concentration}: \quad C_{\mathrm{a}} = 12.7A_{663} - 2.69A_{645} \tag{1}$$

$$\text{Chlorophyll b concentration}: \quad C_{\mathrm{b}} = 22.9A_{663} - 4.68A_{645} \tag{2}$$

$$\text{Chlorophyll concentration}: \quad C_T = 8.02A_{663} + 20.12A_{645} \tag{3}$$

Among them, A_{645} and A_{663} are the absorbance values of chlorophyll extract at monochromatic light at wavelengths of 645 and 663 nm, respectively, and the unit of chlorophyll content is mg/g.

The chlorophyll a, chlorophyll b, and total chlorophyll content were calculated according to the following formula.

$$\text{Chlorophyll a content (mg/g)} = \frac{C_{\mathrm{a}} \times W_{\mathrm{t}}}{W_Y \times 1000} \tag{4}$$

$$\text{Chlorophyll b content (mg/g)} = \frac{C_{\mathrm{b}} \times W_{\mathrm{t}}}{W_Y \times 1000} \tag{5}$$

$$\text{Total chlorophyll content (mg/g)} = \frac{C_{\mathrm{T}} \times W_t}{W_Y \times 1000} \tag{6}$$

Where W_t is the total amount of extract (ml), W_Y is fresh weight (g) for the sample.

It was determined that the highest chlorophyll content in the 40 winter wheat leaves was 4.980 mg/g, the lowest value was 2.021 mg/g, and the average value was 2.873 mg/g. The highest chlorophyll a content was 2.809 mg/g, the lowest value was 0.826 mg/g, and the mean value was 1.364 mg/g; The highest chlorophyll b content was 2.476 mg/g, the lowest value was 0.606 mg/g, and the mean was 1.509 mg/g.

3 MSC Pretreatment of Winter Wheat Reflectance Spectra

The acquired reflection spectrum of winter wheat leaves was pretreated by multivariate scattering correction method. The Algorithm of MSC is as following: Usually calculate the average spectrum of all the samples as the standard spectrum, operate each sample spectrum with the method of unary linear regression to get their regression constant and regression coefficients, and then correct the original spectra of each sample to improve the signal-to-noise ratio.

Multivariate scattering correction eliminates the influence of environmental factors on spectral information, separates the effects of scattering from the original spectrum, and enhances the spectral information associated with the content of the component to be tested. Thereby increasing the signal to noise ratio of the spectrum. The acquired reflection spectrum of winter wheat leaves was pretreated by multivariate scattering correction method to eliminate the influence of scattering.

4 The Structure of ANN

A back-propagation neural network was constructed by using three layers of nodes: an input layer, a hidden layer, and an output layer, as shown in Fig. 2.

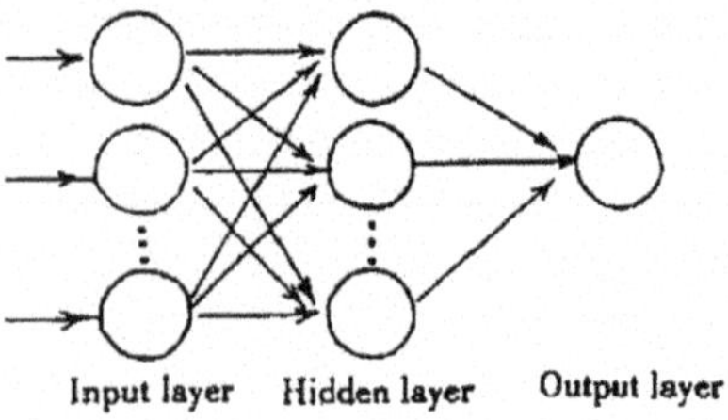

Fig. 2. Three layers B-P ANN for chlorophyll content in winter wheat leaves

Artificial neural network is a nonlinear modeling method that can establish a nonlinear model between input parameters and output parameters. The advantage is that the established model is stable and the relative error is small. The ANN consists of an input layer, a hidden layer, and an output layer. In the spectral analysis, the characteristic parameter of the spectrum is taken as the node value of the input layer, and the sample content value to be obtained is used as the node value of the output layer. In this paper, the node value of the input layer of ANN is the principal component of the spectrum, and the node value of the output layer is the value of chlorophyll (chlorophyll a, chlorophyll b) to be obtained.

5 Establishment of Spectral Detection Model for Chlorophyll Content in Winter Wheat Leaves

The reflection spectrum of winter wheat leaves was pretreated by multivariate scattering correction method, and then the principal components of the spectrum were calculated. The number of suitable principal components is determined by comprehensively comparing the prediction results. The extracted principal component is used as the input of the neural network to establish a quantitative analysis test model of the artificial neural network.

The sample is divided into two parts: the calibration set and the predicted set. The calibration set contains 30 samples for establishing a calibration model for quantitative analysis. There are 10 samples in the predicted set to test the applicability of the model. The optimal principal component number is calculated.

In Matlab, newff is used to create a network. For estimation chlorophyll, the parameters of neural network model are as follows:

net.trainParam.lr = 0.1; % the learning rate of training is set to 0.1.
net.trainParam.mc = 0.1; % the momentum coefficient is set to 0.1.
net.trainParam.epochs = 20; % the training epochs is set to 20.
net.trainParam.goal = 0.08; % the error of training objective is set to 0.08.

For estimation chlorophyll a and chlorophyll b, the four parameters of neural network model are set as 0.1, 0.1, 20, 0.3.

After much training, a model with better prediction results is obtained. The calibration set data was modeled. Table 1 shows the optimal principal component number, as well as correlation coefficient (R) and standard deviation (SD) between chemical values and predicted values in the calibration set and predicted set. Figures 3 and 4 is a plot of the predicted and chemical values of chlorophyll in winter wheat leaves in calibration set and predicted set respectively. Figures 5 and 6 is a plot of chlorophyll a. Figures 7 and 8 is a plot of chlorophyll b.

Table 1. The results of calibration and predicted set for ingredients

Ingredient	PC number	Calibration set	Predicted set
Chlorophyll	5	R = 0.927 SD = 0.252	R = 0.818 SD = 0.470
Chlorophyll a	3	R = 0.780 SD = 0.441	R = 0.627 SD = 0.592
Chlorophyll b	4	R = 0.876 SD = 0.205	R = 0.871 SD = 0.259

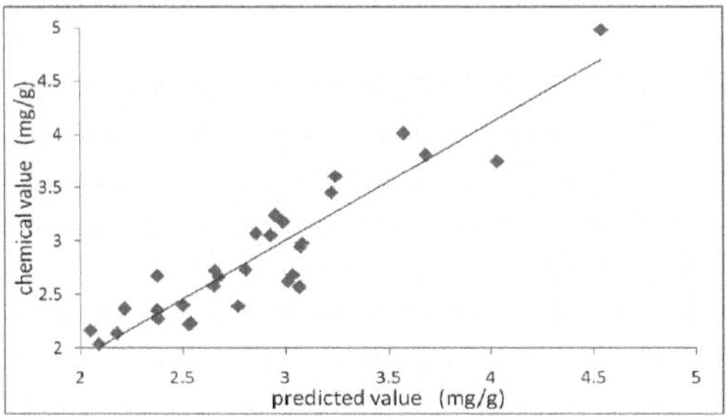

Fig. 3. Relationship between predicted and chemical values of chlorophyll content in winter wheat leaves in the calibration set

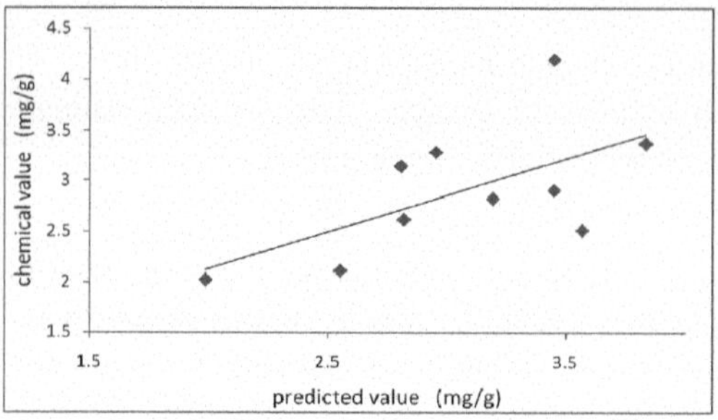

Fig. 4. Relationship between predicted and chemical values of chlorophyll content in winter wheat leaves in the predicted set

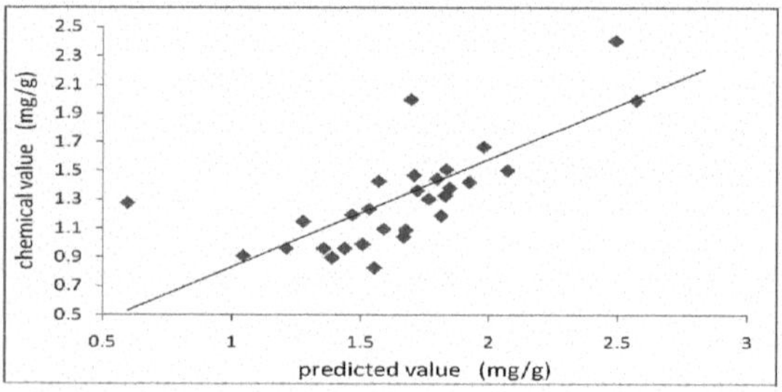

Fig. 5. Relationship between predicted and chemical values of chlorophyll a content in winter wheat leaves in the calibration set

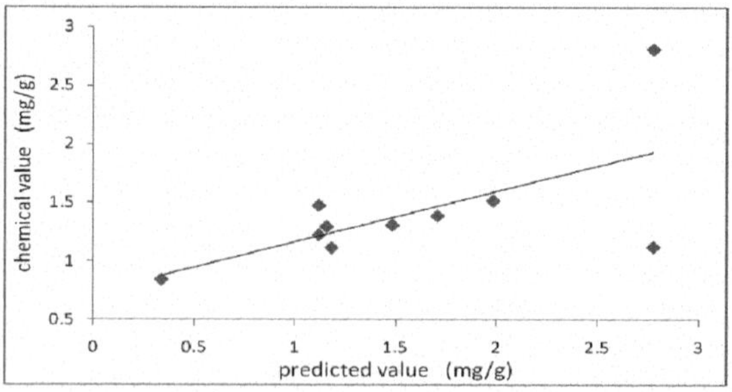

Fig. 6. Relationship between predicted and chemical values of chlorophyll a content in winter wheat leaves in the predicted set

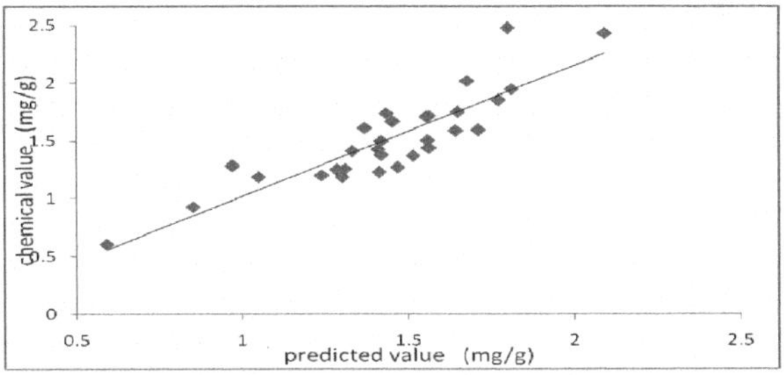

Fig. 7. Relationship between predicted and chemical values of chlorophyll b content in winter wheat leaves in the calibration set

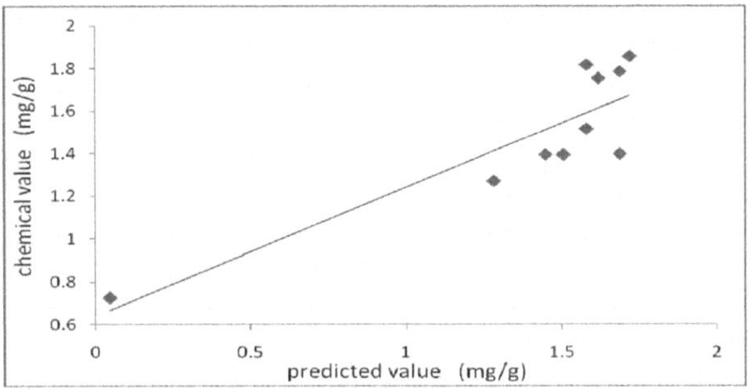

Fig. 8. Relationship between predicted and chemical values of chlorophyll b content in winter wheat leaves in the predicted set

6 Discussion

The spectral data of the wild winter wheat leaves are acquired. The scattering caused by the external environmental factors will cause the collected spectral data to contain invalid information. Therefore, the spectral data is preprocessed by the multivariate scatter correction method, and the scatter correction is obtained. Spectral data after scatter correction can effectively eliminate the effects of scattering and enhance the spectral information associated with the component content. The back propagation artificial neural network is a powerful learning system that can achieve a highly nonlinear mapping between input and output. After multiple learning and training, a predictive model with better predictability can be obtained.

The error mainly comes from two aspects. One is that when measuring the reflection spectrum, the placement position of the wheat leave has certain influence on the spectrum measurement. The other is the errors caused in the measurement of the chlorophyll, chlorophyll a and chlorophyll b by laboratory method, that is the degree of chlorophyll extraction in leaves has a certain effect on the measurement of chlorophyll content in leaves.

7 Results and Conclusion

In this paper, the reflectance spectra of winter wheat leaves were acquired by portable radiation spectrometer. The spectra were preprocessing by multivariate scatter correction. The quantitative analysis model of chlorophyll, chlorophyll a and chlorophyll b in leaves was established by artificial neural network method. The correlation coefficients of modeling and test for chlorophyll in leaves were 0.927 and 0.818 respectively; the standard deviations of modeling and test were 0.252 and 0.470 respectively. The correlation coefficients of modeling and test for chlorophyll a in leaves were 0.780 and 0.627 respectively; the standard deviations of modeling and test were 0.441 and 0.592 respectively. The correlation coefficients of modeling and test for chlorophyll b in leaves were 0.876 and 0.871 respectively; the standard deviations of modeling and test were 0.205 and 0.259 respectively.

In summary, the reflectance spectra of winter wheat leaves were acquired by portable radiation spectrometer. The multivariate scatter correction pretreatment of the spectra was carried out. The quantitative analysis model of artificial neural network was established to quantitatively analyze the chlorophyll, chlorophyll a and chlorophyll b in winter wheat leaves. The results are satisfactory. This method can be used in agricultural practice.

Acknowledgements. The authors acknowledge the National Key Research and Development Program (Grant No. 2016YFD0200602).

References

1. Fan, Z., Simone, G., Yuping, B., et al.: Estimation of wheat water status by reflectance measurements. Trans. CSAE **21**(S), 215–217 (2005)
2. Ji, H., Wang, P., Yan, T.: Estimations of chlorophyll and water contents in live leaf of winter wheat with reflectance spectroscopy. Spectrosc. Spectr. Anal. **27**(3), 514–516 (2007)
3. Li, F., Zhao, C., Wang, J., et al.: Diagnosis of nitrogen nutrition of flue-cured tobacco with chlorophyll meter. Plant Nutr. Fertil. Sci. **13**(1), 136–142 (2007)
4. Wang, D., Ji, J., Gao, H.: The effect of MSC spectral pretreatment regions on near infrared spectroscopy calibration results. Spectrosc. Spectr. Anal. **34**(9), 2387–2390 (2014)
5. Yu, D., Cheng, W., Wang, Q., et al.: Construction of analysis model of rice seed components based on near infrared reflectance spectroscopy. J. Light. Scatt. **27**(4), 384–389 (2015)
6. Li, L., Huang, H., Zhao, S., et al.: NIR spectra detection model of protein, fat, total sugar and moisture in rice. J. Chin. Cereals Oils Assoc. **32**(7), 121–126 (2017)
7. Cao, H., Li, D., Liu, L., et al.: Near infrared spectroscopy quantitative analysis model based on incremental neural network with partial least squares. Spectrosc. Spectr. Anal. **34**(10), 2799–2803 (2014)
8. Lai, L., Ma, W., Chen, H.: Wheat protein nondestructive analysis with near infrared reflectance spectroscopy combined with artificial neutral networks. J. China Univ. Metrol. **26** (1), 55–59 (2015)

9. Kuang, J., Guan, X., Liu, J.: Rapid determination of protein and fat contents in raw milk by near infrared spectroscopy analysis. J. Anal. Sci. **31**(6), 783–786 (2015)
10. Zheng, W., Ming, J., Yang, M., et al.: Hyperspectral estimation of rice pigment content based on band depth analysis and BP neural network. Chin. J. Eco-Agric. **25**(8), 1224–1235 (2017)
11. Yao, X., Wang, X., Huang, Y., et al.: Estimation of sugar to nitrogen ratio in wheat leaves with near infrared spectrometry. Chin. J. Appl. Ecol. **26**(8), 2371–2378 (2015)

Forecasting Tourist Arrivals in China Based on Seasonal Decomposition and LSSVR Model

Gang Xie[1(✉)], Jian Zhang[1,2], Boyu Yang[1,2], and Shouyang Wang[1]

[1] CFS, MDIS, Academy of Mathematics and Systems Science, Chinese Academy of Sciences, Beijing 100190, China
gxie@amss.ac.cn

[2] University of Chinese Academy of Sciences, Beijing 100049, China

Abstract. In this study, a novel decomposition-ensemble methodology is proposed based on seasonal decomposition and the least squares support vector regression (LSSVR) model for forecasting foreign tourist arrivals in China. Firstly, the original time series of foreign tourist arrivals in China are decomposed into several components. Then, an individual forecasting model is selected for the prediction of a component. Finally, these prediction results of the components are combined as an aggregated output of forecasts of foreign tourist arrivals. Using the time series of tourist arrivals from twelve foreign countries to China, an empirical study is implemented for the purposes of illustration and verification. The results suggest that our proposed hybrid model TS-SL-LSSVR can achieve better forecasting performance than other models such as seasonal autoregressive integrated moving average (SARIMA), back propagation neural network (BPNN), and single LSSVR model.

Keywords: Foreign tourist arrivals · Seasonal decomposition · Time series forecasting · Hybrid model

1 Introduction

In the past decades, tourism has experienced continued expansion and diversification, becoming one of the largest and fastest-growing economic sectors in the world [1]. As the largest developing country, China has made great progress in tourism. Since 2006, China has been ranked as the fourth largest tourist destination behind France, the United States and Spain, reflecting the country's popularity in the global tourism market [2]. As competition increases in the global tourism industry, the growth in the demand of Chinese inbound tourism in these source markets exhibits greater volatility and a growing discrepancy. Particularly after the financial crisis in 2008, the growth rates of the tourism markets have varied immensely from year to year [3].

For destination countries, the tourist industry benefits the transportation, entertainment, catering, accommodation, and retailing sectors, so there is a need for substantial investments in infrastructure and promotional activities [4]. As a consequence, accurate forecasts of tourist arrivals are necessary for the effort of safeguarding positive returns on investment [5]. Specifically, accurate forecasts are important to policy-makers as such forecasts can serve as tools for policy decisions, aimed at boosting

Y. Liu et al. (Eds.): ICNC-FSKD 2019, AISC 1074, pp. 452–459, 2020.
https://doi.org/10.1007/978-3-030-32456-8_49

economic development, well-being and employment, particularly for tourism destination countries [6]. In addition, accurate forecasts are also important at the industrial level (e.g. airlines, tour operators, hotels), as they help firms to produce more accurate budgets. Therefore, it is valuable to develop models with high forecasting accuracy for the prediction of foreign tourist arrivals.

It has been found that the monthly inbound tourist flow demonstrates a complex nonlinear characteristic [7] and seasonality is one of the most salient and significant characteristics of tourism [8]. Therefore, for accurate tourist arrival forecasts, it is necessary to describe the nonlinearity and seasonality of the time series.

In this study, we propose a novel decomposition-ensemble methodology for the prediction of foreign tourist arrivals in China. First, the sample data of tourist arrivals are decomposed into several components. Then, a single forecasting model is selected for the prediction of each component. Finally, the forecasting results (in terms of the decomposed subtasks) are combined as an aggregated output of tourist arrival forecasts. Using the time series of tourist arrivals from 12 foreign countries to China, an empirical study is implemented for the purposes of illustration and verification. Finally, some related issues are discussed, and our conclusions are drawn.

2 Methodology Formulation

In this section, firstly, the overall process of the proposed methodology is presented. Then, the methods used to perform the seasonal decomposition are introduced. Finally, our forecasting modeling is described.

2.1 Overall Process of the Proposed Methodology

To enhance the degree of the accuracy of the tourist arrival predictions, a novel decomposition-ensemble methodology is proposed. Our proposed approach actually improves the existing decomposition-ensemble techniques in the "divide and conquer" framework, by selecting a single model for the component that should be predicted. The overall process of the proposed methodology can be described in Fig. 1 as being comprised of the following three main steps:

Step 1: Data decomposition. The original time series x_t $(t = 1, 2, \ldots, T)$ is decomposed into n components via decomposition techniques.
Step 2: Individual prediction. A single forecasting model is selected for the prediction of each component. Accordingly, we can obtain the prediction results for the corresponding component.
Step 3: Ensemble prediction. The predicted results for the components are combined as an aggregated output.

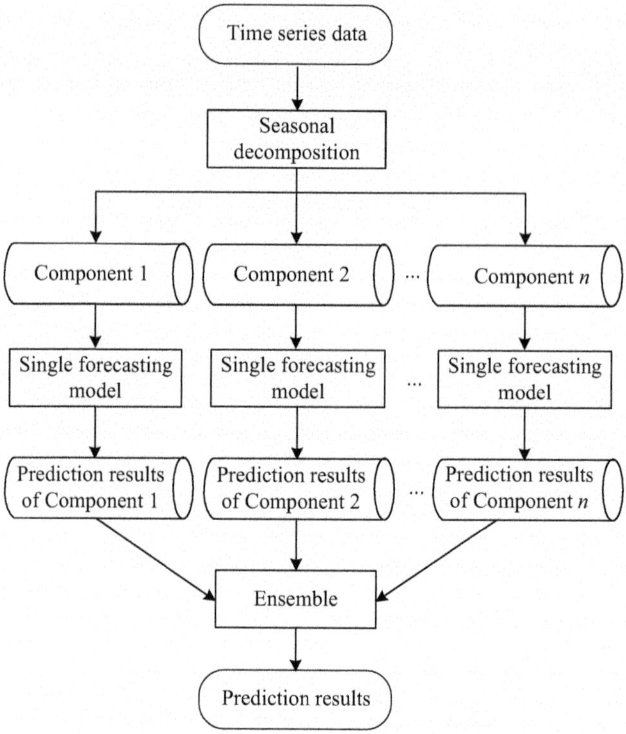

Fig. 1. Overall process of the proposed methodology

2.2 Seasonal Decomposition Methods

The TRAMO/SEATS method [9] decomposes time series x_t into the three components, namely the trend-cycle component, tc_t, seasonal factor component, sf_t, and irregular component, ir_t, with both multiplicative and additive forms. Here, we adopt the multiplicative form as follows:

$$x_t = tc_t \times sf_t \times ir_t, \tag{1}$$

In order to measure the seasonal variation of time series, sf_t, we use the Coefficient of Variance (CV) as follows:

$$CV = \sigma / |\mu|, \tag{2}$$

where the mean $\mu = \sum_{t=1}^{n} x_t / n$ and $\sigma = \sqrt{\sum_{t=1}^{n} (x_t - \mu)^2 / n}$. Here, CV is commonly used to measure the dispersion degree of a data set, and the seasonal variation of time series, sf_t, increases in its CV value. Obviously, if the CV value of time series sf_t is rather small (indicating that the seasonality characteristic is not significant), the TRAMO/SEATS method should not be adopted.

2.3 Forecasting Modeling

Two types of single forecasting models, the SARIMA model and the LSSVR model, are employed for the prediction of decomposed components.

Suykens and Vandewalle [10] proposed the least squares support vector regression (LSSVR) model that solves a system of equations as opposed to a quadratic programming (QP) problem. As such, the LSSVR model provides significantly improved speeds of calculation. The LSSVR first maps the original data, x_t, into a high-dimensional feature space via a nonlinear kernel function, and then performs a linear regression as follows:

$$\hat{x}_t = w^T \varphi(x_t) + b, \tag{3}$$

where $\varphi(x_t)$ is the nonlinear mapping function. The weight, w, and bias, b, are estimated according to structural risk minimization as follows:

$$\begin{aligned} &\text{Min } (w^T w)/2 + (\gamma \sum_{t=1}^{T} e_t^2)/2 \\ &\text{s.t. } \hat{x}_t = w^T \varphi(x_t) + b + e_t, (t = 1, 2, \ldots, T) \end{aligned} \tag{4}$$

where γ is the penalty parameter and e_t is the estimation error at time t.

After solving the function, we can obtain the solution to the problem in the following form:

$$\hat{x} = \sum_{t=1}^{T} \alpha_t K(x, x_t) + b, \tag{5}$$

where $K(\cdot)$ is the kernel function. Here, the usual Gaussian RBF $K(x, x_i) = \exp[||x - x_i||/(2\delta^2)]$ with a width of δ is employed. The LSSVR model has been used for container throughput forecasting [11–13] and air passenger forecasting [14].

For ensemble prediction, all multivariate regression models can be used to generate the final results for the original data.

$$\hat{x}_t = f(\hat{d_t(1)}, \hat{d_t(2)}, \ldots, \hat{d_t(K)}), \tag{6}$$

where $\hat{x}_t$ denotes the final prediction result at time t, and $\hat{d_t(j)}$ is the prediction value for the j th component $j = 1, 2, \ldots, K$).

For the TS-SL-LSSVR model, the prediction results of the components are combined with another LSSVR to formulate an ensemble result for the original foreign tourist arrival series. The aggregated outputs are:

$$\hat{x}_t = f(\hat{tc}_t, \hat{sf}_t, \hat{ir}_t). \tag{7}$$

Moreover, in order to verify the effectiveness of the proposed approach, an empirical study is conducted in the next section.

3 Empirical Study

3.1 Data Description and Experiment Design

The time series data employed in this study were obtained from the WIND Database (http://www.wind.com.cn/). The sample data are the monthly data of foreign tourist arrivals in China (covering the period from January 2006 to February 2016 with a total of 122 observations of each country), as shown in Fig. 2, where the foreign countries

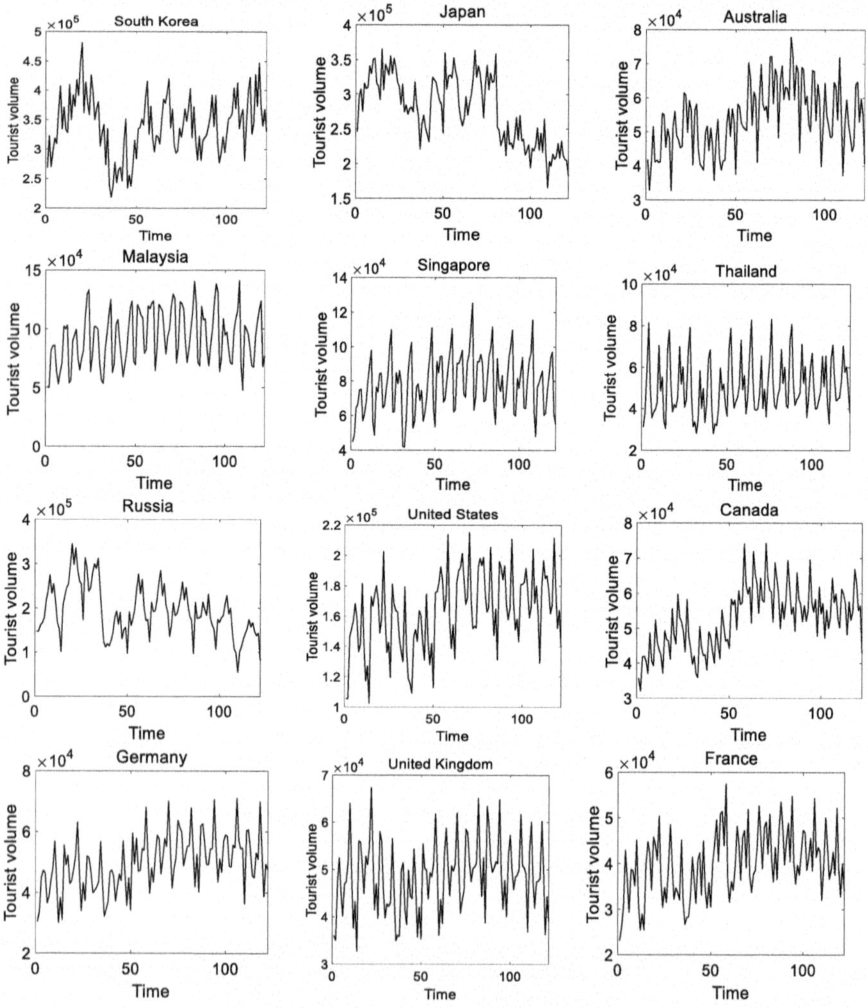

Fig. 2. Monthly foreign tourist arrivals in China

are South Korea, Japan, Australia, Malaysia, Singapore, Thailand, Russia, United States, Canada, Germany, United Kingdom and France, respectively.

Five models, including SARIMA, BPNN, LSSVR, empirical model decomposition (EMD) based LSSVR model EMD-LSSVR [14] and TS-SL-LSSVR, are used for the prediction of foreign tourist arrivals in China. In addition, the criterion MAPE is chosen as the measure of the level prediction accuracy as follows:

$$MAPE = 100\% \times \sum_{t=1}^{N} |1 - \hat{x}_t/x_t| \Big/ N, \tag{8}$$

where N is the size of the testing set, x_t is the real value of foreign tourist arrivals in China, and $\hat{x}_t$ is the corresponding forecast of foreign tourist arrivals in China in the t th month.

3.2 Prediction Results

When the TRAMO/SEATS method with the multiplicative form is implemented, the countries and the CV values (in the brackets) of the seasonal factors of the tourist arrival time series are: South Korea (0.084), Japan (0.056), Australia (0.129), Malaysia (0.217), Singapore (0.192), Russia (0.216), United States (0.127) and Canada (0.097). When the TRAMO/SEATS method with additive form is implemented, the countries and the range values of seasonal factors of the tourist arrival time series are: Thailand (73.9), Germany (64.1), United Kingdom (49.2) and France (67.6). Obviously, the seasonal variation in the time series of tourist arrivals from Japan is not significant.

In the LSSVR model, the values of the γ and δ^2 parameters are first determined via a five-fold cross-validation grid search method, in the range of [0.01, 10000]. The values are then adjusted using the trial-and-error approach to produce the smallest error in the training set. Apart from the single LSSVR, the BPNN in this study used 3 input neurons, 10 hidden nodes and 1 output neuron. The BPNN models are iteratively run 10,000 times to train the model using the training subset. In the SARIMA model, the best model for each training sample is determined through the minimization of the Schwarz Criterion (SC).

In our experiments, the training data sets are the first 100 observations of tourist arrival time series from each country (covering the period from January 2006 to April 2014), whilst the testing data sets are the latter 22 observations (covering the period from April 2014 to February 2016). Table 1 provides the criteria MAPE of models for different time series of sources when one-step-ahead prediction is implemented.

4 Discussion

Experimental forecasts of foreign tourist arrivals in China were made by using the experiment design and methodologies mentioned above. Forecasting performances were then evaluated by the criteria MAPE. The proposed hybrid model TS-SL-LSSVR

Table 1. MAPE of forecasting models for different time series.

Source	SARIMA	BPNN	LSSVR	EMD-LSSVR	TS-SL-LSSVR
Korea	8.81	8.22	10.27	7.29	**5.59**
Japan	5.78	23.87	11.6	**5.33**	6.88
Australia	5.18	13.24	13.57	13.69	**3.04**
Malaysia	9.11	23.21	22	22.69	**6.94**
Singapore	4.87	20.41	16.27	17.47	**4.37**
Thailand	10.58	21.44	19.93	16.59	**4.6**
Russia	10.65	52.19	30.66	14.38	**7.33**
USA	4.54	10.94	9.47	7.07	**2.33**
Canada	5.04	8.88	9.36	5.84	**2.68**
Germany	6.9	12.38	12.1	13.45	**4.62**
UK	5.19	10.49	10	8.61	**3.59**
France	7.58	10.46	11.1	10.14	**4.05**

can achieve a better performance than do the other methods in all countries, except for EMD-LSSVR with a time series of foreign tourist arrivals from Japan to China, as shown in Table 1.

From the comparisons and analysis of the performances as detailed above, deep insights regarding foreign tourist arrival forecasting can be derived.

The decomposition-ensemble methodology can enhance the level of prediction accuracy for forecasting foreign tourist arrivals in China. The hybrid models, TS-SL-LSSVR and EMD-LSSVR, can achieve all the best forecasting performances. The reason is that the principle of "divide and conquer" can be implemented to overcome the current difficulties experienced in forecasting foreign tourist arrivals.

Moreover, it is important to capture the seasonality characteristic of a time series to achieve a better forecasting performance of foreign tourist arrivals in China. The proposed hybrid models based on seasonal decomposition methods, i.e. TS-SL-LSSVR, usually outperform their counterparts, the SARIMA, BPNN, LSSVR and EMD-LSSVR models. However, the proposed hybrid models based on seasonal decomposition cannot achieve the best forecasting performance with the time series of foreign tourist arrivals from Japan to China. The reason is that the seasonal variation of the time series is not significant and the seasonal decomposition is not a good choice for the time series.

5 Conclusions

In this study, a new hybrid model based on the seasonal decomposition method and the LSSVR model is proposed for forecasting foreign tourist arrivals in China. The contribution of this study is in recommending the decomposition-ensemble methodology for the enhancement of prediction accuracy for forecasting foreign tourist arrivals in China. We also developed a new hybrid approach for comparison with benchmark methods. Our investigation suggests that seasonal decomposition is effective and important for capturing the seasonality characteristic of time series, especially those with significant seasonal variation, in order to achieve a better performance for forecasting foreign tourist arrivals in China.

Acknowledgements. This work was supported by the National Natural Science Foundation of China (No. 71771207, 71642006), and the National Center for Mathematics and Interdisciplinary Sciences, CAS.

References

1. Hassani, H., Silva, E.S., Antonakakis, N., Filis, G., Gupta, R.: Forecasting accuracy evaluation of tourist arrivals. Ann. Tour. Res. **63**, 112–127 (2017)
2. Yang, C.-H., Lin, H.-L., Han, C.-C.: Analysis of international tourist arrivals in China: the role of world heritage sites. Tour. Manag. **31**, 827–837 (2010)
3. Sun, X., Sun, W., Wang, J., Zhang, Y., Gao, Y.: Using a Grey-Markov model optimized by Cuckoo search algorithm to forecast the annual foreign tourist arrivals to China. Tour. Manag. **52**, 369–379 (2016)
4. Peng, B., Song, H., Crouch, G.I.: A meta-analysis of international tourism demand forecasting and implications for practice. Tour. Manag. **45**, 181–193 (2014)
5. Chatziantoniou, I., Degiannakis, S., Eeckels, B., Filis, G.: Forecasting tourist arrivals using origin country macroeconomics. Appl. Econ. **48**(27), 2571–2585 (2016)
6. Gounopoulos, D., Petmezas, D., Santamaria, D.: Forecasting tourist arrivals in Greece and the impact of macroeconomic shocks from the countries of tourists' origin. Ann. Tour. Res. **39**(2), 641–666 (2012)
7. Chen, C.-F., Lai, M.-C., Yeh, C.-C.: Forecasting tourism demand based on empirical mode decomposition and neural network. Knowl.-Based Syst. **26**, 281–287 (2012)
8. Wu, L., Cao, G.: Seasonal SVR with FOA algorithm for single-step and multi-step ahead forecasting in monthly inbound tourist flow. Knowl.-Based Syst. **110**, 157–166 (2016)
9. Maravall, A.: An application of the TRAMO-SEATS automatic procedure; direct versus indirect adjustment. Comput. Stat. Data Anal. **50**(9), 2167–2190 (2006)
10. Suykens, J.A.K., Vandewalle, J.: Least squares support vector machine classifiers. Neural Process. Lett. **9**, 293–300 (1999)
11. Xie, G., Zhang, N., Wang, S.: Data characteristic analysis and model selection for container throughput forecasting within a decomposition-ensemble methodology. Transp. Res. Part E: Logist. Transp. Rev. **108**, 160–178 (2017)
12. Xie, G., Wang, S., Zhao, Y., Lai, K.K.: Hybrid approaches based on LSSVR model for container throughput forecasting: A comparative study. Appl. Soft Comput. **13**, 2232–2241 (2013)
13. Xie, G., Qian, Y., Yang, H.: Forecasting container throughput based on wavelet transforms within a decomposition-ensemble methodology: a case study of China. Marit. Policy Manag. **46**(2), 178–200 (2019)
14. Xie, G., Wang, S., Lai, K.K.: Short-term forecasting of air passenger by using hybrid seasonal decomposition and least squares support vector regression approaches. J. Air Transp. Manag. **37**, 20–26 (2014)

Techno-Economic Design and Optimal Operation of Active Radial Active Distribution Network Using Online Optimizer: Real Case Study in Sohag, Egypt

M. B. Shafik[1,3(✉)], G. I. Rashed[1(✉)], Hongkun Chen[1], M. R. Elkadeem[2], and Shaorong Wang[2]

[1] School of Electrical Engineering and Automation, Wuhan University, Wuhan 430072, China
mohamed.shafeeq@eng.kfs.edu.eg, ghamgeen@whu.edu.cn
[2] School of Electrical and Electronic Engineering, Huazhong University of Science and Technology, Wuhan 430074, China
[3] Electrical Power System and Machines Department, Faculty of Engineering, Kafrelsheikh University, Kafrelsheikh 33516, Egypt

Abstract. Continuous development in renewable resources technologies and a lot of trials to overcome integration problems to main grid or even work as isolated grid. Depending on region nature; renewables with grid can form hybrid renewable energy resources (HRERs) that can introduce economical, technical and environment friendly solutions for load supply. This research produces a framework to HRERs system from design plan to optimal operation for a real case study grid connected in Sohag-Egypt. This paper proposes set of simulators working in parallel; HOMER PRO software for techno-economic system electrification and NEPLAN software for system simulation both of them connected to C++ online optimizer to control all parameters for optimal operation. The optimal size and configuration of the hybrid system are determined based on different scenarios of hybridization of energy resources are studied and assessed considering technical, economic and environmental perspectives. The results indicate that the proposed hybrid Grid/PV/WT/Diesel system is founded as the most feasible economic solution for supplying the investigated area because it reaches the minimum net present cost (NPC) and cost of energy (COE) with the lowest levels of gases emissions and fuel consumption compared to the grid only feeding system. In addition, this hybrid system with soft open points (SOPs) potentially gets the minimum system losses to beside enhancing system voltage profile and lines loading reserve where it is working as an Active Distribution network (ADN).

Keywords: Hybrid energy resources · Techno-economic analysis · HOMER PRO · NEPLAN · Online optimizer · Soft open points · Optimal power flow

Y. Liu et al. (Eds.): ICNC-FSKD 2019, AISC 1074, pp. 460–471, 2020.
https://doi.org/10.1007/978-3-030-32456-8_50

1 Introduction

By reaching 2040 it is expected that the total energy demand will increase by around 30% [1] however, nearly 75% of global electrical energy production still comes from fossil fuel-based (i.e. Coal and Natural gas) [2]. In 2017, around 167 GW (solar: 94 GW, wind: 47 GW, others: 26GW) produced from renewables are added to the total global capacity with an average increase by 10% per year since 2010 [3]. The renewable energy development to 2050 documented by the International Renewable Energy Agency (IRENA) has configured a goal of increasing the share of renewables in electrical energy generation to 85% to 25% in 2017 [2]. So, hybrid energy resources (HRERs) are initiated as a cost-effective solution and also for dealing with technical issues. To get the maximum benefits from HRERs, the system must be optimally planned and managed. After the techno-economic study system also must be tested for technical benefits via optimal power flow study.

Up to now, many studies have been considered in the literature studied hybridization of energy systems including design, planning, and energy management subjects [4, 5]. Most of these researches have been based on a software program developed by National Renewable Energy Laboratory (NREL), USA named Hybrid Optimization Model for Electric Renewables (HOMER) [4, 6].

Concerning with PV/WT/Diesel/Battery HRERs system, different scenarios of hybrid resources for rural electrification of an off-grid community region in Yemen has been studied in [7] with the help of HOMER. From the results, the combination of PV/WT/diesel has been awarded as the most feasible configuration. The system achieved around 45% and 70% reduction in the total cost of energy (COE) and CO_2 emission, respectively. The possibility of a hybridization of PV/WT/Battery with a diesel generator in a rural isolated area in Hurghada, Egypt has been studied in [8]. From the results, a significant reduction in COE from 0.444 $/kWh (using diesel) to 0.139 $/kWh, as well as decreasing the total emitted CO_2 from 461,667 kg/yr. to 78,500 kg/yr. have been obtained with hybrid PV/WT/Diesel/Battery system.

For feeding the electricity to an urban center having a group of residential buildings in Hargeisa, Somali [9] in the grid-connected mode, a reduction by 24% and 53% has been achieved. The same conclusion has been obtained with the stand-alone system at COE of 12.6 ₵/kWh and RF of 43.9%. The techno-economic feasibility of the hybridization of PV/WT/Diesel stand-alone system for rural area electrification in Colombia has been evaluated in [10].

A lot of studies exploited the wide area of distribution networks after the penetration of DG units that transfer it to ADNs. Some studies covered control strategies for penetration and high penetration of DGs challenges and problems solving [11, 12], and others considered the benefits of presence of DG units in the distribution system and its role of optimizing all distribution networks functions [13, 14]. Also, some studies focused on providing Distribution systems with new technologies like SOPs [15, 16] and automation systems and all of this can be in short term studies or for long term planning [12, 17]. In 2010 [18] the SOPs concept was proposed as active/reactive power management tool to modify network topology to mitigate the high penetration of DG units.

Various optimization techniques were investigated for solving engineering problems, especially swarm-based techniques such as particle swarm (PSO), Genetic Algorithms (GA) and Ant colony optimizer (ACO) [19, 20]. PSO is the simplest well-known structure, lowest number of controlled parameters, most powerful, and most frequently used optimizer that can be implemented to solve engineering problems and especially with FACTS devices installation problems [19].

2 Sohag Case Study Planning Data

Egypt possess outstanding wind profile especially along the Suez Gulf with an average wind speeds between 7 and 10 m/s, and Egypt is one of 38 countries in the world with a published National Wind Atlas. Egypt's Solar Atlas states that Egypt is considered a "sun belt" country with 2,000 to 3,000 kWh/m^2/year of direct solar radiation. However, the good potential of solar and wind resources, no much work has been done in the literature to evaluate the feasibility of using hybrid renewable energy resources in Egypt.

The selected site of studied area is located geographically at 26° 32.7' N latitude and 31° 42.2' E longitude in Sohag city in South portion, Egypt. According to NASA surface metrology and solar energy database [21], the annual average solar radiation and annual average wind speed are founded to be 6.08 kWh/m^2/day and 5.26 m/s, respectively. The global solar radiation, monthly averaged profile over a 22-year period (July 1983–June 2005) profile is presented in Fig. 1. Whereas, the monthly averaged wind speed profile over a 10-year period (July 1983–June 1993) is illustrated also in Fig. 1.

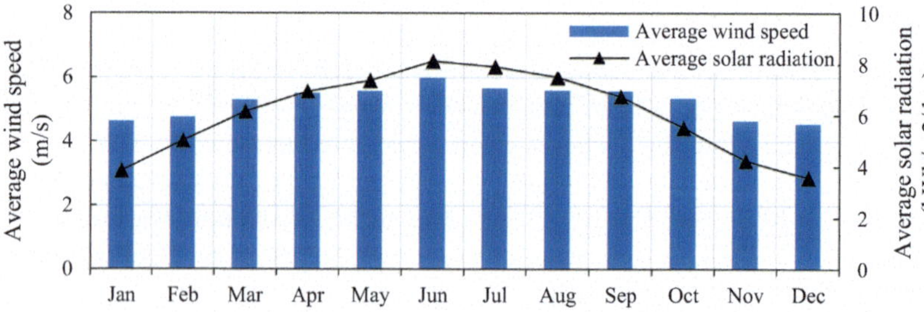

Fig. 1. Monthly average solar radiation and wind speed profiles in Sohag

3 Research Methodology

3.1 Online Optimizer

HOMER PRO is a powerful and robust software tool for simulation, optimization and sensitivity analysis of hybrid renewable energy systems in off-grid and on-grid mode [4]. Depending on the configuration of the lowest cost; text file of the number, Size and type of units is then sent to the optimizer to search for the best allocation.

NEPLAN is a software tool to analyze, plan, optimize and simulate networks. The graphical user interface allows the user to perform study cases very efficiently[add]. Capabilities of NEPLAN is used to build the system nodes, branches, loads, …etc. in medium and low voltage stages then, perform power flow analysis then uses NEPLAN Programming Library (NPL) to export all system-controlled variables to Optimizer for Optimal power flow framework.

Network optimizer as illustrated from Fig. 2 reads DDL files from NPL and text files from HOMER, then uses a C++ Particle swarm optimizer (PSO) [19] to perform Optimal power flow process by modifying the controlled variables with specific number of iterations to reach the best system performance.

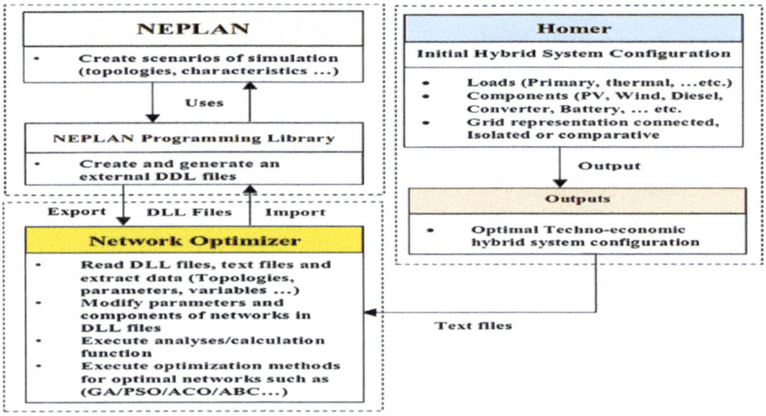

Fig. 2. Frame work of online optimizer

3.2 Techno-Economic Modelling

The detailed technical and economic data in addition to the solar and wind resources data of the presented PV/WT/Diesel/Battery/Converter hybrid system are defined for HOMER. Then, the system is simulated and optimization analysis is started for every hour in the first year subjected to the constraints of power balance.

In this paper, the first main objective is minimization of total system NPC and COE subject to technical and environmental constraints are the evaluation criteria to determine the optimal configuration plan of the proposed grid connected hybrid energy system. From Eq. (1), the total system NPC is a function of total system annualized cost (Ct_an) divided by capital discounting factor (DF). Equation (2), gives the total components n years of life time system annual cost which is the summation of capital, replacement and O&M costs of system components. DF is calculated from Eq. (3), where, d is the discount rate (interest rate) on capital investments of the installed equipment, and p is the project lifetime. The cost of energy is given as in Eq. (4), where Et_an is the total annual useful energy generated from the HDERs to supply the electrical load demand.

$$\mathrm{NPC} = \frac{\mathrm{C_{t\_an}}}{\mathrm{DF}} \tag{1}$$

$$\mathrm{C_{t\_an}} = \sum\nolimits_{k=1}^{n} \left(\mathrm{C_{capital,k}} + \mathrm{C_{repl,k}} + C_{O\&M,k}\right) \tag{2}$$

$$\mathrm{DF} = \frac{\mathrm{d(1+d)^p}}{\mathrm{[(1+d)^p - 1]}} \tag{3}$$

$$\mathrm{COE} = \frac{\mathrm{C_{t\_an}}}{\mathrm{E_{t\_an}}} \tag{4}$$

3.3 Optimal Power Flow Model

A multi-objective function is introduced for handling the main goals of minimizing losses and reducing total voltage Deviation. In Eq. (6), the main objective of the problem is formulized by minimizing power losses P_{Losses} and the total voltage deviancy of feeders' voltages V_i beyond nominal voltage V_n with number of feeders Nf and weighting factor W_1 to exclude worst solutions.

$$Obj = Min\left(P_{Losses} + W_1 * \sum\nolimits_{i \in Nf} |V_n - V_i|\right) \tag{5}$$

$$P_{Losses} = \sum\nolimits_{i.j \in Nf} Real\left(I_{ij}^2 * Zl_{ij}\right) \forall Zl \in Nl \tag{6}$$

Distribution network losses P_{Losses} is the sum of all lines Nl real power losses when line between nodes i and j has impedance Zl_{ij} and current through it is I_{ij}. The combined objective function in Eq. (6) is exposed to set of constraints are stated in Eqs. from 7:14.

$$P_{Gt} = \sum\nolimits_{i \in Nf} PD_i + P_{losses} + P_{Losses}^k \tag{7}$$

$$Q_{Gt} = \sum\nolimits_{i \in Nf} QD_i + Q_{losses} + Q_i^K + Q_j^K \forall Q_{Gt} = \sum\nolimits_{i \in NDG} Q_{DGi} + Q_{slack} \tag{8}$$

$$Q_{Losses} = \sum\nolimits_{i.j \in Nf} Imag\left(\mu_{ij} * I_{ij}^2 * Zl_{ij}\right) \forall Zl \in Nl \tag{9}$$

$$P_{DGi} \le P_{DGi}^{max} \forall i \in NDG \tag{10}$$

$$Q_{DGi} \le Q_{DGi}^{max} \forall i \in NDG \tag{11}$$

$$V_i^{min} \le V_i \le V_i^{max} . i \in Nf \tag{12}$$

$$S_{ij} \le S_{ij}^{max} \forall i \ne j \& i.j \in Nf \tag{13}$$

$$P_{ij}^{k} \leq P_{ij}^{max} \forall k \in N_{SOP} \tag{14}$$

$$I_{ij}^{n} \leq I_{ij}^{max} \tag{15}$$

Total active/reactive power balance stated in Eqs. 7 and 8 where PD_i, QD_i are load concentrated at each node and Q_{losses} is system reactive power losses that can be calculated in terms of expression (9); where Q_i^K, Q_j^K are kth SOP reactive power and P_{Losses}^{k} is its real power losses. Active/reactive limit of DG units P_{DGi} and Q_{DGi} should be selected according to the available ranges, so objective function is also constrained with Eqs. 10, 11 where NDG is number of DG unites defined from HOMER economic decision. Security constraints are also considered for feeders' voltage levels that must be kept within limits also lines loadings S_{ij} that can't exceed lines thermal loading S_{ij}^{max} as expressed in (12) and (13) respectively. Power flow direction P_{ij}^{k} through the kth SOP can be controlled according to Eq. 14 which is limited to the Kth SOP rating. Finally, n^{th} line flow current I_{ij}^{n} is restricted to the lines thermal limit I_{ij}^{max}.

3.4 Soft Open Points (SOPs) Modelling

SOP can be considered multi-function power electronic device installed into DNs to be used as active/reactive power flow control tool in place of tie line switches. Figure 3 shows the static model is initiated, as there are no dynamic considerations in this work focus. Back-to-back VSC series with unified power flow controller based SOP with P-Q control mode for VSC_i and V_{DC}-Q mode for VSC_j [22] for a smooth control of voltage and active/reactive power flow in steady state.

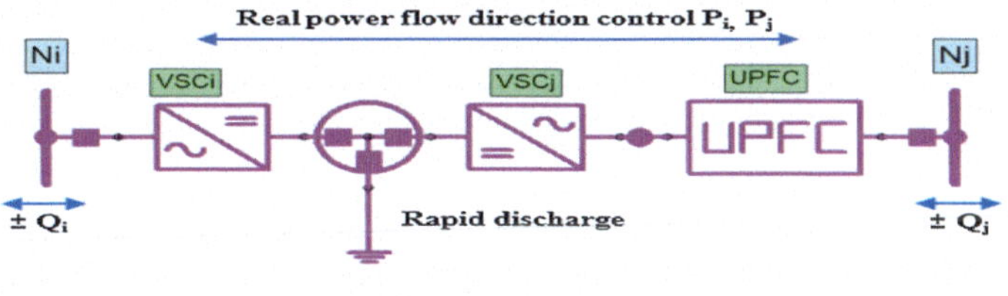

Fig. 3. SOP modelling using NEPLAN

3.5 Sohag Active Radial Distribution Network

After the optimal hybrid system for Sohag network electrification by HOMER; Online optimizer defines the best locations of committed DG units as will be explained in section of results IV. Node 11 is selected for PV system feeding where node 25, 40 for Diesel generator and wind turbine. System is 48 nodes with 47 branches balance system with 4 tie lines as in Fig. 4. But optimizer also finds the best place for SOP devices that replaces tie lines between nodes 25, 34 and 35, 40. Sohag system is balanced radial DN which is fed from one substation 22/11 kV at node 1 and all loads at the low voltage side of transformers 11 kV/400 V. Electrical loads with peak power of 6154 kW and 3271 kVar has been distributed over all system nodes except main bus connected to grid with average energy consumption of are 59882 kWh/day/year with load factor of 0.41.

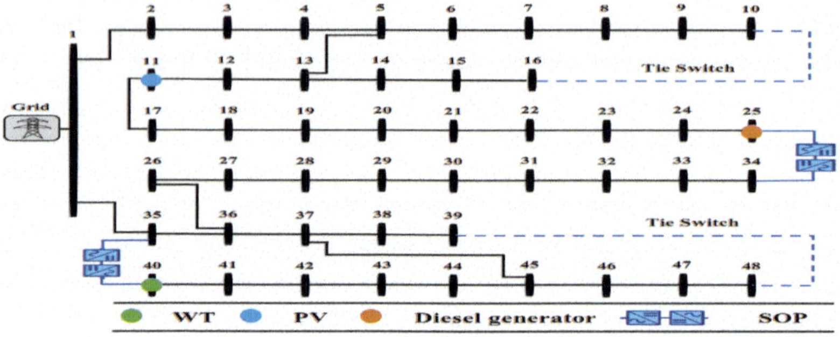

Fig. 4. Suhage-Egypt active radial distribution network

4 Simulation Results and Discussion

In this paper, for HOMER to cover all possible scenarios, eight different scenarios are studied and simulated. Each scenario is evaluated based on techno-economic analysis methodology previously in Sect. 3.2. Where the optimal one is obtained with minimum NPC and COE with a reasonable emission level. With respect to optimizer and NEPLAN two cases are studied for optimal system operation, the first is normal operation (case1) of system without any DG or SOP with basic configuration and second case is the optimal result (case2) with DG and SOP best sizing and sitting.

4.1 Optimal Configuration Plan of the Hybrid Grid Connected System of Sohag

With the help of Hybrid Optimization Model for Multiple Energy Resources (HOMER) software package; Table 1 gives the detailed simulation results of different simulation scenarios under study. The table shows the optimal size of the system component at each configuration. Also, the cost data including NPC ($) and COE ($/kWh) are calculated. Moreover, the table provides the value of the total system emission (kg/yr.). In addition, the table presents the value of renewable fraction (RF) which is defined as the percentage the energy delivered to the load that generated from renewable power sources and calculated from Eq. 16, where E_{g_nonren} and H_{g_nonren} are the total electrical and thermal energy produced from conventional i.e. non-renewable sources and E_{load} and H_{load} are the total electrical and thermal energy consumption served by the system.

$$\%\ \mathrm{RF} = \left(1 - \frac{E_{g\_nonren} + H_{g\_nonren}}{E_{load} + H_{load}}\right) \times 100 \tag{16}$$

Points to be highlighted from the optimized techno-economic results:

- The combination of PV/WT/G with system converter and grid (Scenario 1) gives the optimal configuration with a cost-effective solution of the hybrid system of Sohag. Where it has the minimum values of NPC and COE of 34.3 M$ and 0.117$/kWh, respectively.
- The optimal configuration has obtained at RF of 39.4%, which mean a moderate penetration of renewable especially in presence of main grid. The hybrid system consists from 2000 kW PV, a 25 WT with 100 kW rating each, 700 kW diesel generator in with converter rating of 1500 kW.
- It is observed that, with the proposed hybrid system, there are a weighty reduction by about 16% in NPC and 18.75% in COE using the hybrid system compared to the base case using Grid only.
- It is also noted that the total greenhouse gas emission of the optimal system has a value of 8,909,810 kg/yr. But, it worth to mention that the total gasses emitted by the optimal configuration are relatively the lowest compared to the zero sustainable configuration (Scenario 8) with grid fed only. Optimal configuration reduces the total emissions by 35.65% compared to base scenario 8.
- From the previous discussion, it is concluded that the hybrid renewable system with Grid/PV/WT/G inclusion with converter gives a satisfactory result regarding technical, economic and environmental aspects which make it the optimal configuration plan for Sohag.

Table 1. Techno-economic optimization results of HRES.

Scenario	System configuration	Components size				Cost				RF (%)	CO_2 Emissions (kg/yr.)	Fuel (L/yr.)
		PV (kW)	WT (Qty.)	G (kW)	C (kW)	NPC (M$)	COE ($//kWh)	O&M (M$/yr)	Cap. (M$)			
1	Grid-PV-WT-G-C	2000	35	700	1500	34.3	0.117	1.9	9.67	39.4	8,909,810	1,220,409
2	Grid-PV-WT-C	2000	40	–	1500	34.8	0.177	1.92	9.97	42.2	8,410,011	0
3	Grid-PV-G-C	2000	–	800	1500	35.2	0.124	2.51	2.77	16.3	11,887,659	1,719,788
4	Grid-PV-C	2000	–	–	1500	36.1	0.128	2.64	1.97	16.3	11,586,400	0
5	Grid-WT-G	–	45	1000	–	38.3	0.13	2.19	10	30.4	10,296,618	1,576,236
6	Grid-WT	–	50	–	–	38.9	0.13	2.24	10	38.9	9,730,151	0
7	Grid-G	–	–	1000	–	39.9	0.141	3.01	1	0	14,177,239	1,922,647
8	Grid	–	–	–	–	40.8	0.144	3.15	0	0	13,840,800	0

Table 2. Summary of electrical energy production/consumption for optimal Scenario 1.

Source	(kWh/yr.)	%	Load	(kWh/yr.)	%
PV	3,804,888	16.6	Load	21,900,000	96.4
			Grid sales	820,172	3.61
G	4,726,594	20.6	Excess electrical	50,030	0.218
WT	5,392,251	23.5	Unmet load	0	0
Grid	9,034,212	39.4	Capacity shortage	0	0
Total	22,957,945	100			

As shown in Table 2, PV and WT contributes by 12,839,100 kWh/yr. (40.1%) in total electricity production compared to main grid of 39.4% and rest comes from diesel (20.6%). As expected from the cheaper cost of PV besides the high value of average solar radiation in Sohag, PV has a high value of electricity generation but is limited with space for PV array. Because of this, diesel share by 20.6% only in the total energy production. It is noted also from the table that the optimal configuration plan succeeds to meet 24-hr load demand with zero capacity shortage, as well as a backup excess energy of 50,030 kWh/yr.

4.2 Optimal Operation of Sohag Active Radial Distribution Network

System under the new techno-economical design must be tested from technical perspectives. So maximum load demand case of 6154 kW and 3271 kVar is selected to be fed with the new hybrid system and see the effect on system losses, lines load flow and voltage profile. Table 3 compares between case 1 of normal operation where losses is 140.2 kW and 658.34 kVar with minimum voltage of 0.948 PU at medium bus 25 with large voltage deviation. When online optimizer perform optimal power flow; it is found that the best number of SOP units is two and best place is between nodes 25–34 and 35–40. In addition optimizer select best place and sharing of DG units where number and type is econmically defined for system by HOMER.

Table 3. OPF results of case 1 and 2 for Sohag AND with HRES.

Case	DG			SOP control			Power losses			Voltage (PU)				% lines loading Decline	% *Pl*
	Size (kVA)	Type	@ bus	P (kW)	Q (kVar)	Between buses	P (kW)	Q (kW)	Decline %	V_{min}	@ bus	Sum Div. to Vn	Sum Div. to V_{limits}		
1	…	…	…	…	…	…	140.2	658.34	…	0.948	25	2.885	0.485	…	0
2	700	PV	11	±800	±500	25-34	60.11	166.36	57.11	0.993	48	2.205	0.195	48.68	36.05
	900	Diesel	25	±700R	±600C	35-40									
	960	Wind	40												

Table 3 also, shows that in case of using DG and SOP losses decreased by 57.11% when DG units is placed at buses 11, 25 and 40 with penetration level (total energy generated by renewables to total energy generated) Pl 36.05% also lines loading decreased also by 48.68% beyond the maximum permissible limit.

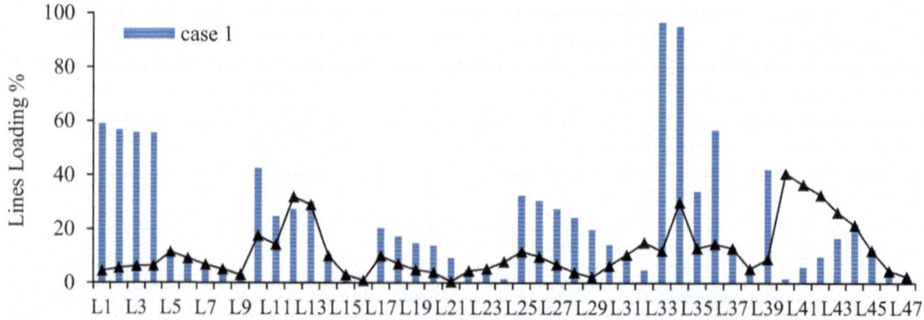

Fig. 5. System branches loading comparison between case1 and case 2

Figure 5 illustrates that presence of DG and SOP reduces most of lines loading moreover from table IV it reduces the total lines loading bu 48.68%. Voltage profile enhancement is very important especially in lower voltage directly connected to loads. It is shown from Fig. 6 that both medium voltage and consequently low voltage profile is enhanced within limits; moreover, table IV show that total voltage deviation to voltage permissible limits is reduced from 0.485 to 0.195. and minimum voltage at bus 25 is increased from 0.948 PU to 0.999 PU where minimum voltage moved to bus 48 in that case.

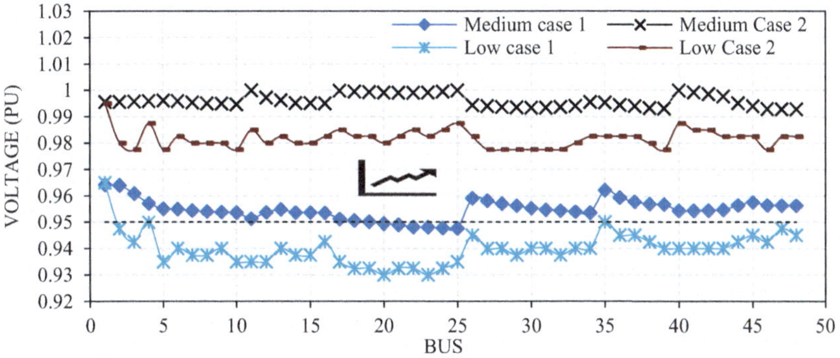

Fig. 6. Voltage profile improvement from case 1 to case 2 for low and medium voltage

5 Conclusions

In this paper, the potentials of the hybridization of renewable energy systems of PV/WT/Diesel/Converter connected to grid for electrical load supply in Sohag-Egypt have been studied. In addition, the hybrid system has been tested for total system losses and enhancing voltage profile for maximum load. The techno-economic analysis is conducted on eight different hybrid system configurations (considering different penetration level of renewables) with the help of HOMER PRO software. System is then simulated in NEPLAN and optimal power flow is established with online optimizer. The study emphasized technical, economic and environmental aspects through a comparative study between different scenarios. The simulation results indicate that the hybridization of 2000 kW PV, 35 × 100 kW WT together with 700 kW diesel generator as well as a and 1500 kW system converter shows the optimal performance among all scenarios. The proposed optimal plan has been obtained with minimum NPC, COE and minimum gas emissions as well as a reasonable value of RF compared to grid share. It is observed also that combination between grid and renewables beside using SOPs reduces total system losses by 57% and consequently lines loading decreased also. Also, enhances all nodes voltage to be within limits and near to nominal value. So, the paper demonstrates that the implementation of hybrid systems can achieve both technical and economical target of real case study especially in Egypt with good average wind speed and high solar irradiation.

References

1. Logesh, R.: Resources, configurations, and soft computing techniques for power management and control of PV/wind hybrid system. Renew. Sustain. Energy Rev. **69**, 129–143 (2017)
2. International Renewable Energy Agency: Global energy, a roadmap to 2050 by Irena. www.irena.org. Accessed 21 Oct 2018
3. Zia, M.F., Elbouchikhi, E., Benbouzid, M.: Microgrids energy management systems: a critical review on methods, solutions, and prospects. Appl. Energy **222**, 1033–1055 (2018)
4. Bahramara, S., Moghaddam, M.P., Haghifam, M.R.: Optimal planning of hybrid renewable energy systems using HOMER: a review. Renew. Sustain. Energy Rev. **62**, 609–620 (2016)
5. Dawoud, S.M., Lin, X., Okba, M.I.: Hybrid renewable microgrid optimization techniques: a review. Renew. Sustain. Energy Rev. **82**, 2039–2052 (2018)
6. HOMER Energy: Homer PRO. www.homerenergy.com. Accessed 21 Oct 2018
7. Ajlan, A., Tan, C.W., Abdilahi, A.M.: Assessment of environmental and economic perspectives for renewable-based hybrid power system in Yemen. Renew. Sustain. Energy Rev. **75**, 559–570 (2017)
8. Dawoud, S., Lin, X., Sun, J.W., Okba, M.I., Khalid, M.S., Waqar, A.: Feasibility study of isolated PV-wind hybrid system in Egypt, pp. 145–151 (2015)
9. Abdilahi, A.M., Yatim, A.H.M., Mustafa, M.W., Khalaf, O.T., Shumran, A.F., Mohamed Nor, F.: Feasibility study of renewable energy-based microgrid system in Somaliland's urban centers. Renew. Sustain. Energy Rev. **40**, 1048–1059 (2014)
10. Baneshi, M., Hadianfard, F.: Techno-economic feasibility of hybrid diesel/PV/wind/battery electricity generation systems for non-residential large electricity consumers under southern Iran climate conditions. Energy Convers. Manag. **127**, 233–244 (2016)
11. Shafik, M.B., Chen, H., Rashed, G.I., El-Sehiemy, R.A.: Adaptive multi objective parallel seeker optimization algorithm for incorporating TCSC devices into optimal power flow framework. IEEE Access **7**, 36934–36947 (2019)
12. Wang, X., Wang, C., Xu, T., Meng, H., Li, P., Yu, L.: Distributed voltage control for active distribution networks based on distribution phasor measurement units. Appl. Energy **229**, 804–813 (2018)
13. Jiang, T., Li, Z., Jin, X., Chen, H., Li, X., Mu, Y.: Flexible operation of active distribution network using integrated smart buildings with heating, ventilation and air-conditioning systems. Appl. Energy **226**, 181–196 (2018)
14. Usman, M., Coppo, M., Bignucolo, F., Turri, R.: Losses management strategies in active distribution networks: a review. Electr. Power Syst. Res. **163**, 116–132 (2018)
15. Cao, W., Wu, J., Jenkins, N., Wang, C., Green, T.: Benefits analysis of soft open points for electrical distribution network operation. Appl. Energy **165**, 36–47 (2016)
16. Qi, Q., Wu, J.: Increasing distributed generation penetration using network reconfiguration and soft open points. Energy Procedia **105**, 2169–2174 (2017)
17. Bai, L., Jiang, T., Li, F., Chen, H., Li, X.: Distributed energy storage planning in soft open point based active distribution networks incorporating network reconfiguration and DG reactive power capability. Appl. Energy **210**, 1082–1091 (2018)
18. Bloemink, J.M., Green, T.C.: Increasing distributed generation penetration using soft normally-open points. In: IEEE PES General Meeting, 25–29 July 2010, pp. 1–8 (2010)
19. Jordehi, A.R.: Particle swarm optimisation (PSO) for allocation of FACTS devices in electric transmission systems: a review. Renew. Sustain. Energy Rev. **52**, 1260–1267 (2015)

20. Lee, C.K.H.: A review of applications of genetic algorithms in operations management. Eng. Appl. Artif. Intell. **76**, 1–12 (2018)
21. NASA power: Power data Acess/Predict of the world wide Energy resource. power.larc.nasa.gov. Accessed 21 Oct 2018
22. Cao, W., Wu, J., Jenkins, N., Wang, C., Green, T.: Operating principle of soft open points for electrical distribution network operation. Appl. Energy **164**, 245–257 (2016)

Artificial Honey Bee Swarm Intelligence for the Autograding of EBN

Weng Kin Lai(✉), Jack En Gan, and Pui Mun Koh

Department of Electrical and Electronic Engineering,
Faculty of Engineering and Technology,
Tunku Abdul Rahman University College, Kuala Lumpur, Malaysia
laiwk@tarc.edu.my

Abstract. Swarm intelligence represents a group of computing paradigms inspired by the collective behaviour of animals and insects. While there are many swarm intelligence inspired algorithms there are those modeled after the intelligent behavior bees which have shown to be effective in doing many challenging tasks in recent years. Edible bird nest (EBN) is an expensive animal bioproduct produced by swiftlets which is deemed to have many health benefits. Currently, the common way of grading the EBN that is effective is by using trained human operators. In addition to being an unattractive job this approach is time consuming, cost ineffective with inconsistencies in the grading. Moreover, the whole time consuming training regime has to start all over again when the operator has to be replaced when they leave. In this paper we investigated the use of the artificial honey bee swarms on a set of unique features extracted by image processing to autograde EBN. An accuracy of nearly 86% was achieved on the datasets used with our proposed method.

Keywords: Swarm intelligence · Artificial bees colony (ABC) · Bat algorithm · Classification · Edible bird nest · Feature extraction

1 Introduction

The set of computing techniques inspired the behaviour of animals and insects are collectively known as Swarm intelligence (SI). SI exhibits collective behavior of decentralized, self-organized systems including both natural and artificial. The expression was introduced by Gerardo Beni and Jing Wang [1] in 1989 in the context of cellular robotic systems. While bees may have tiny brains, but they are surprisingly intelligent [2]. Research has showed that bees can learn from their environment to gain a reward, and then teach other bees to do the same [3]. More specifically, honey bee swarms have been observed and known for their foraging behaviour. This was quickly exploited by Dervis Baragoa as a meta-heuristic algorithm in 2005. Since then Artificial Bee Colony (ABC) has been successfully applied to solve various practical problems. A recent study on the use of bee-based swarm intelligence showed that the top four are ABC (Artificial Bee Colony) at 54%, BA (Bee Algorithm) at 18%, followed by HBMO (Honey Bees Mating Optimization)) at 15% and BCO (Bee Colony Optimization) at 13% [4]. In the ABC model, the colony consists of three groups of bees: employed

Y. Liu et al. (Eds.): ICNC-FSKD 2019, AISC 1074, pp. 472–480, 2020.
https://doi.org/10.1007/978-3-030-32456-8_51

bees, onlookers and scouts. There is division of labour where the employed bees go to their food source and come back to the hive and dance on this area. When the food source has been abandoned the employed bee then becomes a scout and starts to search for a new food source. Finally, the onlookers watch the dances of employed bees and choose food sources depending on dances. It had been used to discover the transcription factor binding sites (TFBS) in DNA sequences [5] as well as to detect the tumor and its size using real Brain MR images [6]. Zhang and Wu found that ABC was able to determine the correct solution faster than either genetic algorithm or particle swarm optimization for multilevel thresholding [7]. Lai et al. had also investigated ABC to characterise the latex particles to help identify the quality of natural rubber latex [8]. These are some of the many different applications of ABC in various domains.

Edible Bird's Nest (EBN) is one of the indulgences in Chinese cuisine. Nevertheless, EBN was perceived to have numerous health advantages. They are high-protein glutinous secretion produced by the salivary glands of certain species of swiftlets. Malaysia is the third largest EBN producing country in the world [9] having been in the business for more than 500 years [10]. However the raw EBN contains a lot of impurities such as swiftlet dropping, twigs and dirt. Therefore, the raw EBN needs to go through a series of cleaning processes to remove these impurities before further product development can proceed. Nevertheless, EBN can be sold either in their original form or cleaned or processed form. In its unprocessed form, it can be categorized into 4 grades. Grade AA has the highest quality followed by Grade A, Grade B, and Grade C. They are graded based on their size, shapes (or curvature), amount of impurities and colour of the nest [11, 12]. Traditionally, EBN producers grade the nests solely by observations of the nest colour, shape, size and the impurities. Even though this is done by trained operators, nevertheless the results may not be consistent as they were prone to the significant differences from different individuals [13]. Moreover, it is also very time-consuming to train a human operator who has zero experience and relevant knowledge before they can grade the EBN accurately. Due to technology advancement, new ways of grading the EBN were introduced [9, 14].

In this paper we are incorporating image processing with the Artificial Bee Colony to automatically grade EBNs. Details of some of the prior work are described in the next section and this is followed by the methodology of our proposed new approach. In addition, the performance of the ABC will be basedlined with the results from the immensely popular k-Means as well as that from another swarm intelligence, i.e. the Bat algorithm. The results of these and the analysis are in Sect. 4 with the conclusions and recommendations for further work in Sect. 5.

2 Prior Work

In general, an effective method for automated grading systems is usually integrated with machine vision. Some prior work has been reported using computer vision based system for automatic grading and sorting of agricultural products like mangoes according to their maturity level. Nandi et al. [15] showed that the Gaussian Mixture Model (GMM) performance of their system is as good as the average classification accuracy of the human experts. Furthermore, diseases on the fruits can also be detected using a combination of feature extraction and classification techniques. Fruits tested included apples, palm oil fresh fruits, mangoes, dates and strawberries. The images of fruits were segmented to identify the healthy and infected parts on the fruit [16]. The fruit was then graded according to the percentage of infection on the fruit. Various segmentation techniques were applied to identify the region of interest from the image. There does not seem to be much prior work done to autograde EBNs. It is only recently that Koay *et al.* [17] reported their work in combining machine vision with ANFIS to autograde EBN. Gan and Lai [18] followed this up where they investigated the effectiveness of the Bat to identify the correct EBN grades, achieving an accuracy of about 80%. Our motivation here is to explore if these results can be further improved, especially with ABC.

3 Proposed Methodology

The manual way to grade an EBN is by comparing the size of the EBN with the index finger, middle finger and the ring finger of an adult. A high-grade EBN should have the same size as the three fingers or larger. However, this will lead to inconsistencies of EBN grade as each individual would have different finger sizes. A scientific way to measure the size of EBN is by getting the length and width of the nest [11]. Another factor that affects the grading of the EBN is the number of impurities found in the nest. The nests with low impurities will have a higher market value. When the impurities were cleaned from the nest it will leave holes on the nest which will reduce its market value. Colour is another key factor in grading EBNs. Most unprocessed EBNs are white although some might have different colours due to environmental contamination. White nests are the highest grade while the others are a mixture of yellowish nests [11]. Lastly, the shape of the nests is another factor. Bird nests that can be placed evenly on a horizontal surface where they form a perfect half cup shape, for instance, would be graded as the highest grade, i.e. AA [19]. The lower grade, for example A would have similar shape as these AA nests, but these nests will not be even when placed on a horizontal surface like Grade AA.

3.1 Feature Extraction

Image processing techniques were used to extract the features from the unprocessed EBN images. The four features extracted were the total area of the EBN, the curvature of the EBN, the total amount of the impurities and colour of the EBN. Figure 1 shows the methodology used for extracting features from the EBN images.

Fig. 1. Flow of feature extraction

Area of EBN

In order to obtain the total area of the EBN, the EBN image was segmented out as the contrast of the object and the background differs significantly. The changes in contrast were detected by calculating the gradient of the image. The gradient of the image was set as a threshold value for the image using the *Sobel* operator. The threshold value was adjusted with a fudge factor to attain the binary mask that contains the object of the image. Lastly, the area of the object was computed from the total number of pixels that are white. Examples of the area for each of the 3 grades are shown in Fig. 2(b).

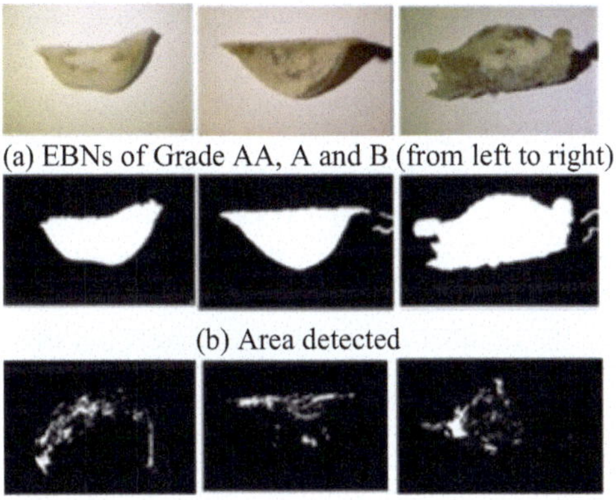

Fig. 2. EBNs – estimation of area and impurities

Impurities of EBN

The HSV colour model was used to detect the impurities on the EBN with the saturation layer of the image used for further processing. As the shadow has almost the same colour as the impurities, the *Saturation* layer is the best layer to distinguish between the shadow and the impurities. From the histogram of the saturation layer, it was observed that for saturation values between 0.1 to 0.5, there is a wide distribution of pixels. This means that there is noise at these values. From the histogram of the image, saturation values that are larger than 0.5 have a smaller number of pixels too. These correlate to the impurities in each piece of EBN. The number of pixels representing the impurities is much smaller compared to the number of pixels of the EBN and also the background. The more impurities detected, the higher the number of white pixels. Examples of the impurities detected for each of the 3 grades of EBN are shown in Fig. 2(c).

Curvature of EBN

Aperfect circle detected will give a maximum value of 1.0. Therefore, if the shape of the EBN is a perfect circle, the curvature value would then be assigned a 1.0. However, the curvature value would decrease if it deviates from a perfect circle. Firstly, the original EBN image would have to undergo the process to crop the main object and place it against a black background. Next, the image was converted to a black and white image before tracing the boundary of the EBN. The curvature of the EBN can be calculated by using Eq. (1), i.e.

$$\text{curvature} = 4 \times \pi \times \text{area} / [\![\text{perimeter}]\!]^2 \tag{1}$$

Both the perimeter and area of the EBN needs to be determined in order to generate this curvature value. The curvature of the EBN was calculated and then compared to determine if it is indeed the EBN or just noise from the image. We adopted a simple heuristic by choosing the larger metric as the curvature of the EBN would be significantly larger compared to the noise.

Colour of EBN
From the images of the EBN, Grade AA EBNs are clearer and white whereas Grade B EBNs are off-white or yellowish. Impurities like mud and sand will also affect the colour of the EBN. The *Value* layer in the HSV colour model best distinguishes the colour of the EBN from the impurities and the background. The median of the histogram of the *Value* layer was used to detect the colour of the EBN, as the number of EBN pixels is less than the number of background pixels but more than the impurities pixels.

Data Normalisation
The features extracted from the EBN color images have to be pre-processed with decimal scaling to normalise the different range of values obtained from each feature. i.e.

$$X' = \frac{X}{10^J} \tag{1}$$

where j is the largest value which can capture the range of data values.

3.2 Autograding

Artificial Bee Colony (ABC)
The Artificial Bee Colony (ABC) algorithm is modeled after the behaviour of honey bees. n this algorithm, the food source's actual position corresponds to a probable solution of the optimization problem. The quality (fitness) of the potential solution is represented by the amount of nectar found in a food source.

The bees were first separated into a group of *employed* bees and another as *onlooker* bees. The number of employed bees is equal to the number of onlooker bees and the number of food sources. This algorithm performs in an iterative manner and the employed bees will be initialized with randomly generated food source as it is required for each employed bee to be assigned to a food source. In each iteration, the employed bees will head to their current food source in the neighborhood to evaluate the amount of nectar which represents the fitness. $S_i = (S_{i1}, S_{i2}, \ldots, S_{iD})$ represents the position of ith food source.

$$P_i = \frac{fit_i}{\sum_{k=1}^{FS} fit_k} \tag{2}$$

After returning to the hive, the employed bees will share information to the onlooker bees and the onlooker bees will then head to the food source that the employed bees have explored at S_i based on the probability (P_i) which is shown in Eq. (3) where FS equals to the total number of Food Sources and fit_i is the fitness value which can be calculated from Eq. (3).

$$fit_i = \frac{1}{1 + f(S_i)} \tag{3}$$

$f(S_i)$ is the objective function considered. The onlooker bees will explore new food sources in the region of S_i. To produce a new candidate food position from the old one, the equation below is used:

$$S_{new} = S_{ij} + rx(S_{ij} - S_{kj}) \tag{4}$$

where

S_{new} = new food source,
k = solution in the neighborhood of i
r = random number in the range of −1 to +1
j = dimension of the problem considered

When a better fitness score is found, the bee will abandon the old food source and move to the new food source. Otherwise the bee will remain at the current food source searching until one with a better fitness is found. Each of the onlooker bees will select their food source based on the probability given in Eq. (4). The onlooker bees will accommodate those good food sources and every bee will keep searching for a better food source for a required number of cycles (limit). If the fitness value of the food source does not improve over time – defined by the stagnation threshold, that particular bee will become a *scout* bee instead.

Bat Algorithm

Gan and Lai [18] had investigated the use of the Bat Algorithm (BA) to autograde EBNs. The Bat Algorithm is one of the algorithms in the family of Swarm Intelligence. It is a metaheuristic optimization algorithm where the inspiration comes from natural biological behavior, inspired by the echolocation behavior of microbats, based on varying pulse rates of emission and loudness [20]. The Bats were modeled to fly randomly with velocity v_i at position x_i, and performed echolocation at a constant frequency f_{min}, variable loudness A_0 and wavelength λ to search for preys. They adjust their frequency (or wavelength) of the pulses emitted naturally as well as the rate of pulse emission *r* in the range of [0, 1] - depending on the distance of their target. Their loudness and emission rates will be updated only if the new solutions are improved. This basically signifies that these bats are moving towards the optimal solution. The number of bats was set to 20, while the number of generations was set to 10 in the simulations here. In addition, the results from these two swarm intelligence were also compared with that obtained from k-Means.

k-Means Clustering

The k-Means clustering algorithm is a popular clustering method [21] as it is fast and easy to implement. It first generates *k* objects as the initial centroids arbitrarily, where *k* is a user-specified parameter. Each point was then assigned to the cluster of the closest centroid. After that, the centroid of each cluster was updated by taking the means of the data points for each cluster. Various data points may move from one cluster to another cluster. Again, the new centroid was computed with the data points assigned to the suitable clusters. The assignment was repeated with updates to the centroids until the convergence criteria is met i.e., no point changes clusters, or equivalently until the centroid remains the same.

4 Results and Discussions

The ABC were trained with the three classes of EBN data (Grades AA, A and B) with four features (total of 63 samples). 70% of the data was used for training with the remainder for testing. This was repeated 1,000 times with different random selection of the data. The size of the bees colony was set to 50, number of food sources set at half of the colony – as recommended, stagnation threshold of 100 and finally, the stopping criteria at 250. The results are summarized in Table 1.

Table 1. Grading accuracies

Algorithm	Class			Overall accuracy
	AA	*A*	*B*	
ABC	78.35%	82.56%	98.60%	86.29%
Bat algorithm	71.53%	73.44%	94.14%	80.29%
k-Means	73.07%	77.20%	99.77%	83.02%

Significantly better results were obtained by the ABC compared to another SI algorithm – the Bat Algorithm as well as the popular k-Means. The position of swarm updates in ABC by two contradictory activities: first one is a process of adaptation, which empowers exploring the diverse search space, and the second one is a process of selection, which ensures the exploitation of the earlier experience. It is these two process mechanism that has proved to be effective in accurately grading the EBN.

5 Conclusions

Features such as the size, curvature, impurities and the colour of the EBN were extracted from the images of the EBN for classification. From the features extracted, the three classifiers were able to classify the different grades (AA, A and B) of EBNs but with different degree of accuracies. The Bat Algorithm Clustering was able to identify the different grades up to an accuracy of 80.29% using decimal scaled data. Meanwhile k-Means was able to correctly identify grade B almost all the time. Nevertheless, on the average across all the three grades, k-Means only managed 83% whereas ABC was able to achieve an average accuracy by 86.29%, and performing better than k-Means on grades AA and A.

The images tested were those obtained after the raw EBN had been meticulously graded by the trained operators. Manually grading of EBNs is a time consuming process and it is precisely because of this that provided the motivation to automate this process. However even though the samples used may not be large but as these had been manually graded we believe the results obtained here would be representative of the autograding results of EBNs obtained from this facility. Nevertheless our future work will focus on testing this technique on more data.

References

1. Beni, G., Wang, J.: Swarm intelligence in cellular robotic systems. In: Proceedings NATO Advanced Workshop on Robots and Biological Systems, Tuscany, Italy, 26–30 June, pp. 703–712 (1989)
2. Hugo, K.: Intelligence test shows bees can learn to solve tasks from other bees. Science, 23 Feburary 2017. https://www.pbs.org/newshour/science/intelligence-test-shows-bees-can-learn-to-solve-tasks-from-other-bees. Accessed 8 Mar 2019
3. Loukola, O.J., Perry, C.J., Coscos, L., Chittka, L.: Bumblebees show cognitive flexibility by improving on an observed complex behavior. Science **355**(6327), 833–836 (2017)
4. Karaboga, D., Gorkemli, B., Ozturk, C., Karaboga, N.: A comprehensive survey: artificial bee colony (ABC) algorithm and applications. Artif. Intell. Rev. **42**(1), 21–57 (2014)
5. González Álvarez, D., Vega-Rodŕiguez, M., Pulido, G., Sánchez-Pérez, J.: Finding motifs in DNA sequences applying a multi-objective artificial bee colony (MOABC) algorithm. In: Evolutionary Computation, Machine Learning and Data Mining in Bioinformatics, pp. 89–100 (2011)
6. Balasubramani, K., Marcus, K.: Artificial bee colony algorithm to improve brain MR image segmentation. Int. J. Comput. Sci. Eng. (IJCSE) **5**(01), 31 (2013)
7. Zhang, Y., Lenan, W.: Optimal multi-level thresholding based on maximum tsallis entropy via an artificial bee colony approach. Entropy **13**, 841–859 (2011)
8. Lai, W.K., Phoon, Z.H., Chew, K.H., Goh, K.M., Chua, P.Y.: Image multi-thresholding to characterise the mechanical stability of rubber latex concentrate. In: Proceedings of the International Conference on Digital Image Computing: Techniques and Applications (DICTA 2016), Gold Coast, Australia, pp. 372–379 (2016)
9. Hao, Q., Rahman, A.: Swiftlets and edible bird's nest industry in Asia. PJSRR Pertanika J. Sch. Res. Rev. **2**(1), 32–48 (2016)
10. Sankaran, R.: The status and conservation of the Edible-nest Swiftlet (*Collocalia fuciphaga*) in the Andaman and Nicobar Islands. Biol. Conserv. **97**(3), 283–294 (2001)

11. Tan, K.H., Chia, F.C., Alan, H.K.: Impact of Swiftlet's moult season on the value of edible bird nests. In: Proceedings of the 2014 International Conference on Intelligent Agriculture, vol. 63, pp. 17–21 (2014)
12. Syahir, F.A.S., Shakaff, A.Y.M., Zakaria, A., Abdullah, M.Z., Adom, A.H., Ezanuddin, A.A.M.: Edible bird nest shape quality assessment using machine vision system. In: Proceedings of 3rd International Conference Intelligence System Modellling Simulation, ISMS 2012, pp. 325–329 (2012)
13. Goh, K.M., Lai, W.K., Ting, P.H., Koh, D., Wong, J.K.R.: Size characterisation of edible bird nest impurities: a preliminary study. In: 21st International Conference on Knowledge Based and Intelligent Information and Engineering Systems (KES2017), Marseille, France, 6–8 September, pp 1072–1081 (2017)
14. Tsim, K.W.K., Chan, G.K.L., Wong, Z.C.F., Zhang, L.M.L., Cheng, L.K.W., Lin, H.Q., Dong, T.T.X.: The quality assurance of edible bird's nest. In: Edible Bird Nest Industry Conference 2014, pp. 25–26 (2014)
15. Nandi, C.S., Tudu, B., Koley, C.: An automated machine vision based system for fruit sorting and grading. In: 2012 Sixth International Conference on Sensing Technology (ICST), pp. 195–200 (2012)
16. Solanki, U., Jaliya, U.K., Thakore, D.G.: A survey on detection of disease and fruit grading. Int. J. Innov. Emerg. Res. Eng. **2**(2), 109–114 (2015)
17. Koay, M.Y., Loh, S.X.C., Goh, K.M., Lai, W.K.: Feature selection for automated grading of edible birds nest with ANFIS. In: Proceedings of the International Conference on Control, Robotics and Informatics (ICCRI 2018), Kuala Lumpur, Malaysia, 7–9 April 2018
18. Gan, J.E., Lai, W.K.: Automated grading of edible birds nest using a novel hybrid bat algorithm. In: 2019 IEEE International Conference on Automatic Control and Intelligent Systems (accepted), Kuala Lumpur, Malaysia, June 2019
19. Chua, Y.G., Chan, S.H., Bloodworth, B.C., Li, S.F.Y., Leong, L.P.: Identification of edible bird's nest with amino acid and monosaccharide analysis. J. Agric. Food Chem. **63**, 279–289 (2014)
20. Yang, X., Hossein Gandomi, A.: Bat algorithm: a novel approach for global engineering optimization. Eng. Comput. **29**(5), 464–483 (2012)
21. Trevino, A.: Introduction to K-means clustering. https://www.datascience.com/blog/k-means-clustering. Accessed 05 Jan 2019

Predicting Potential Years of Most Costly War Involving USA via ASF Approach

Yunong Zhang[1,2,3](✉), Guanqun Yang[1,2,3], Ruifeng Wang[1,2,3], and Liangjie Ming[1,2,3]

[1] School of Information Science and Technology, Sun Yat-sen University, Guangzhou 510006, China
zhynong@mail.sysu.edu.cn, ynzhang@ieee.org, jallonzyn@sina.com
[2] Research Institute of Sun Yat-sen University in Shenzhen, Shenzhen 518057, China
[3] Key Laboratory of Machine Intelligence and Advanced Computing, Ministry of Education, Guangzhou 510006, China

Abstract. In this paper, the addition-subtraction frequency (ASF) approach is used to predict the years of potentially costly war, which may erupt with the United States of America in the future. In the numerical experiments involved in this paper, we select the years of historic 10 most expensive wars as input data, using three-variable, full-traversal ASF approach to predict the potential years of most costly war that may occur in the future. By using different 9-input data (removing one each year datum) to test the robustness of the experimental results, the final conclusion is that there is a relatively high possibility of most costly war occurring in 2021 or 2036–2038.

Keywords: ASF (addition-subtraction frequency) approach · War year prediction · Consistency analysis · Numerical experiments

1 Introduction

In the time of rapid development of science and technology, the cost of war is enormous. Every large-scale war may have a huge impact on politics and economy. Therefore, the year prediction of most costly war based on the years of wars in the past is of practical significance.

As a superpower, the United States of America (i.e., the USA) has dominated many expensive wars in history [1]. The occurrence of the wars had or has huge impacts on economy and politics all around the world. In this paper, the wars that involved USA historically are considered and selected on the basis of the cost. Then, the investigations are carried out on the basis of 10 most expensive wars [2].

The ASF approach involved in this paper is used to predict the potential years of most costly war involving the USA. It is improved from the commensurability method presented by Wenbo Weng, which is often used to discover the intrinsic connections between events. The commensurability method can be used

Y. Liu et al. (Eds.): ICNC-FSKD 2019, AISC 1074, pp. 481–490, 2020.
https://doi.org/10.1007/978-3-030-32456-8_52

to predict the occurrence of natural disasters [3], but it has some shortcomings, such as the need of much subjective adjustment, hand calculations, and human intervention, which may influence the accuracy and objectiveness of the prediction results. Therefore, to improve the scientificalness of the commensurability method, the ASF approach has been proposed and developed by Zhang et al [4–7].

The rest of this paper is divided into five sections. In Sect. 2, the preliminaries and ASF approach are introduced. Then, in order to predict the years of most costly war involving the USA, numerical experiments are carried out on the basis of ASF approach in Sect. 3. In addition, we also focus on improving the robustness of the prediction results. Specifically, in Sect. 4, we remove one year datum in each experiment and use the remaining nine data for numerical experiments, which can effectively eliminate less possible years among the prediction results. The consistency analysis is presented in Sect. 5, and we comprehensively predict the years of most costly war which are potential occurring and involving the USA. In the last section, we come to a conclusion that, in the years around 2021 or 2036–2038, the USA has a relatively higher possibility of most costly war. Besides, in 2053–2057, 2069–2072 or 2105–2107, there is also a high possibility of most costly war.

2 Preliminaries

In this section, we introduce how to select the sample data set, and give a basic introduction to the principle of the ASF approach.

2.1 Ten Most Costly Wars in USA History

Wars have affected the tendency of development of mankind in history. As a superpower in the world, the USA and its related wars have research value. According to statistics, there have been more than 30 wars involving the USA in history. In this paper, the years of 10 most costly and also representative wars are selected as input data, which are shown in Table 1. For example, as for World War II, the USA consumed about $341 billion dollars and the number of casualties was as high as 400,000 [8]. This type of costly war can be regarded as turning points in the history of the USA and the world, and is of big research value. The wars which are selected to do numerics have the characteristics of high cost and large number of casualties. By using them we can predict the future year of war which the USA might participate in.

2.2 ASF Approach

The ASF approach is originated from the commensurability method, which was once used to predict the occurrence of natural disasters such as earthquakes. In Sect. 1, we briefly introduce the ASF approach. In this section, the details of ASF approach are introduced.

Table 1. Years of 10 most costly wars selected as input values [2]

Year	War name
1775	American Revolutionary War
1846	Mexican-American War
1861	American Civil War
1898	Spanish-American War
1917	First World War
1941	Second World War
1950	Korean War
1965	Vietnam War
1990	Gulf War
2001	War on Terror

In this paper, the three-variable, full-traversal ASF approach is used, and the related ASF formula is as follows: $p_{\text{pre}} = x_a + x_b - x_c$, where p_{pre} represents the predicted year of most costly war that might occur, and x_a, x_b and x_c indicate the years in which the wars have occurred.

With the above ASF formula, we can process the inputs by using the full traversal approach. The full traversal means using loops to compute the input data from the first to the last, which tolerates repetition [4–7,9], and finally obtaining the prediction results. The algorithm is presented as follows.

Algorithm 1. ASF Approach of 3 Inputs

Require: Previous years of most costly wars of the USA in x
Ensure: Predicting years in p_{pre}

function ASF_FULL_TRAVERSAL(input_data x)
  for $a = 1 \rightarrow n$ **do**
    for $b = 1 \rightarrow n$ **do**
      for $c = 1 \rightarrow n$ **do**
        $p_{\text{pre}} = x_a + x_b - x_c$
      end for
    end for
  end for
end function

3 Numerics with Ten War Data

In this section, we focus on how to analyze and predict the most costly war that might involve the USA in the future. First, 10 years shown in Sect. 3 are used as integer inputs, and the full traversal approach (that is, all input years are used for each layer) is conducted. Then, we use Matlab [10] to perform

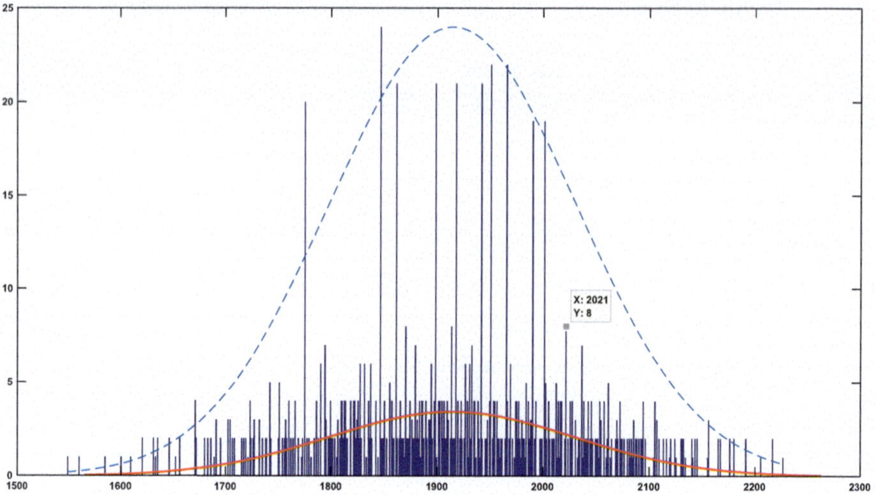

Fig. 1. Frequency of years computed from ten war data, in addition to peak year 2021 near 2036 (right)

statistics of the resultant data in order to get the frequency corresponding to each year. Correspondingly, Fig. 1 shows the numerical experimental results of the boundary years and the corresponding frequency distribution.

Observing that the entire frequency distribution shows a normal distribution, we draw a normal distribution curve for the convenience of observation. It is ready to find out that the frequency of the input years (the years in which most costly wars occurred) is much higher than the others, which is consistent with historical records. Therefore, we can roughly conclude that the most costly wars are more likely to occur in years with high frequency [3–7]. Thus, the years with high frequency in the experiments can be predicted as the future years in which a most costly war or wars may occur.

In order to facilitate the observation, the difference between the frequency and the corresponding year value of the normal distribution is subtracted and shown in Fig. 2. The existence of positive value or peak value in this figure indicates that there is a greater possibility of a war which may happen in that year.

Based on the numerics and figures, the years denoted by A1 through A9 in Fig. 2 are the years in which there are high possibilities of a most costly war occurring, and the corresponding detailed information is shown in Table 2.

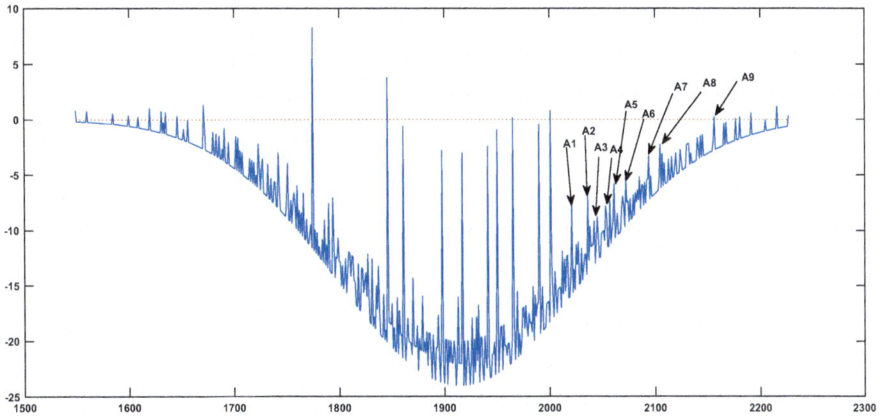

Fig. 2. Auxiliary figure of frequency minus normal-distribution value generated with ten war data

Table 2. Numerical results with ten war data

Years	Notation in Fig. 2	Peak	Frequency
2021	A1	High	8
2036–2038	A2	High	7
2042–2045	A3	High or Middle	4
2053–2057	A4	High or Middle	4
2061	A5	High	5
2069–2072	A6	High or Middle	4
2094	A7	High	4
2105–2107	A8	High	4
2156	A9	High or Middle	3

4 Numerics with Nine War Data

In the above section, we use ten input data for simple prediction and discussion. In this section, we show how to use nine input data to test the robustness of the prediction.

In order to test the numerics in Sect. 3, we remove one year each time, and use the remaining nine years to perform the three-variable, full-traversal ASF approach to compute and predict.

4.1 Numerics without 2001 War Datum

In this section, we select and use the most representative 9-input numerics (totally, ten group of 9-input numerics) to introduce and implement our thoughts. In the numerics where 2001 is removed, we use the remaining nine

Table 3. Numerical results without 2001 war datum

Years	Notation in Fig. 3	Peak	Frequency
2021	A1	High	8
2025–2027	A2	High or Middle	4
2036–2038	A3	High	7
2042–2045	A4	High	4
2061	A5	High	4
2069–2072	A6	High or Middle	3
2094	A7	High	4
2105–2107	A8	High	3
2130–2132	A9	High or Middle	2
2156	A10	High or Middle	2
2165	A11	High or Middle	2
2180	A12	High	2

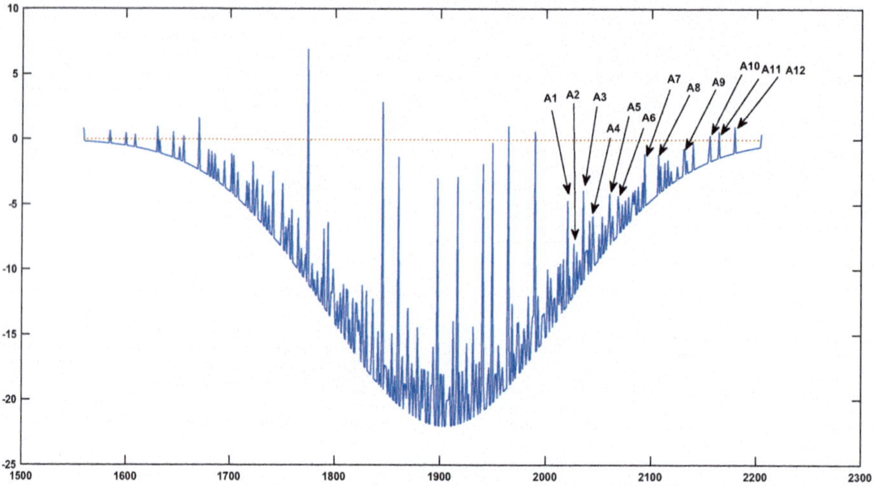

Fig. 3. Auxiliary figure of frequency minus normal-distribution value generated without 2001 war datum

years $\{1775, 1846, 1861, 1898, 1917, 1941, 1950, 1965, 1990\}$ as input to compute based on ASF approach. Then, we get the frequency results and make the discussion about the numeric results. The results are shown in Table 3 and the auxiliary figure is shown in Fig. 3.

4.2 Summary Results of Ten Groups of Numerics

In the above subsection, we show how to remove a datum for numerical experiments and get the results of prediction. In this subsection, we present the comparison between these results and the final conclusion. After doing ten numerics with different nine input data, ten results of prediction can be made. Table 4 shows the predicted results for all numerics.

Table 4. Original results of war years predicted by all numerics

All	No 2001	No 1990	No 1965	No 1950	No 1941	No 1917	No 1898	No 1861	No 1846	No 1775
2021	2021	2021	2021	2021	2021	2021	2021	2021	2021	2021
2036-2038	2025-2027	2025-2027	2025-2027	2025-2027	2036-2038	2025-2027	2025-2027	2025-2027	2025-2027	2025-2027
2042-2045	2036-2038	2036-2038	2030	2036-2038	2042-2045	2030	2030	2036-2038	2030	2030
2053-2057	2042-2045	2042-2045	2036-2038	2053-2057	2053-2057	2036-2038	2036-2038	2042-2045	2036-2038	2036-2038
2061	2061	2053-2057	2042-2045	2061	2061	2042-2045	2042-2045	2053-2057	2053-2057	2042-2045
2069-2072	2069-2072	2069-2072	2053-2057	2069-2072	2069-2072	2053-2057	2053-2057	2061	2107	2053-2057
2094	2094	2105-2107	2061	2085	2094	2061	2061	2069-2072	2156	2061
2105-2107	2105-2107	2140	2069-2072	2105-2107	2105-2107	2064	2069-2072	2085	2165-2167	2069-2072
2156	2130-2132	2167	2085	2156	2130-2132	2069-2072	2085	2105-2107	2176	2085
	2156	2176	2156	2165-2167	2165-2167	2094	2094	2156	2180	2094
	2165	2191	2165	2180	2176	2105-2107	2105-2107	2165-2167	2191	2105-2107
	2180		2167	2191	2180	2156	2156	2176	2216	2145
			2176	2216	2191	2165-2167	2165-2167	2180		
					2216	2176	2176	2191		
						2180	2180	2216		
						2191				
						2216				

Besides, we perform a statistic for all involved years and get the frequency of these years. The frequency of a certain year is obtained by counting the number of the numerics which have a same prediction result corresponding to that year.

Table 5. Possibilities of war years predicted by all numerics

	All	No 2001	No 1990	No 1965	No 1950	No 1941	No 1917	No 1898	No 1861	No 1846	No 1775	Possibility
2021	1	1	1	1	1	1	1	1	1	1	1	11
2036-2038	1	1	1	1	1	1	1	1	1	1	1	11
2053-2057	1	0	1	1	1	1	1	1	1	1	1	10
2069-2072	1	1	1	1	1	1	1	1	1	0	1	10
2105-2107	1	1	1	0	1	1	1	1	1	1	1	10
2025-2027	0	1	1	1	1	0	1	1	1	1	1	9
2042-2045	1	1	1	1	0	1	1	1	1	0	1	9
2061	1	1	0	1	1	1	1	1	1	0	1	9
2165-2167	0	1	1	1	1	1	1	1	1	1	0	9
2156	1	1	0	1	1	0	1	1	1	1	0	8
2176	0	0	1	1	0	1	1	1	1	1	0	7
2180	0	1	0	0	1	1	1	1	1	1	0	7
2094	1	1	0	0	0	1	1	1	0	0	1	6
2191	0	0	1	0	1	1	1	0	1	1	0	6
2085	0	0	0	1	1	0	0	1	1	0	1	5
2216	0	0	0	0	1	1	1	0	1	1	0	5
2030	0	0	0	1	0	0	1	1	0	1	0	4
2130-2132	0	1	0	0	0	1	0	0	0	0	0	2
2064	0	0	0	0	0	0	1	0	0	0	0	1
2140	0	0	1	0	0	0	0	0	0	0	0	1

A higher frequency indicates a greater possibility of a most costly war. After the statistics of the frequency, we can sort the results in a descending order of the frequency of each year. The ranking represents the year in which the most costly war is likely to occur. Based on these sorted results, we make the overall predictions, which are shown in Table 5.

Specifically, as illustrated in Table 5, there are some years that show higher frequency in many numerics. Thus, we can make a prediction that a most costly war are likely to occur in one of those years.

5 Consistency Analysis

In Sect. 3, the data of 10 war years are computed by three-variable, full-traversal ASF approach. After making statistics and analysis, the years that are prone to a most costly war occurring are 2021, 2036, 2038, 2042, 2045, 2053, 2057, 2061, 2072, 2094, 2105, 2107 and 2156. As for a war, it is usually the result of multiple ongoing events. Therefore, the occurrence of some errors of the predicted outbreak year is reasonable. With such characteristics, we can merge adjacent two years, indicating that a most costly war may occur during these two years.

In Sect. 4, we perform robustness tests on these predictions and adjust the predicted results. Based on the 10-group numerics predictions and previous predictions, it is not difficult to find that there are some years with very high frequency, occurrence or possibility, shown in Table 5. Sorting the years by frequency, we can see that years 2021 and 2036–2038 have a very high frequency in multiple numerical tests, and thus we can make a prediction that there is a higher possibility of a most costly war in that year. A high frequency can also be observed in 2053–2057, 2069–2072 and 2105–2107.

According to the consistency analysis above, we finally predict the years of most costly war to be 2021, 2036–2038, 2053–2057, 2069–2072 or 2105–2107.

It is worth noting that, in another paper, we use weights-and-structure-determination (WASD) neuronets and a sinusoidal model to fit and predict the possible major internal crisis (MIC) in the USA [11]. The results show that, in 2093, the sinusoid reaches the lowest value, which is consistent with the prediction result of 2094 in this paper; it is possible to speculate on the war caused by the worst situation of the MIC.

6 Conclusion

In this paper, 10 USA-involving most costly wars have been selected in terms of the scale and cost, and these years of wars have been extracted as input data for the three-variable, full-traversal ASF approach. By using ASF approach, combined with statistics and analysis, the conclusion can be drawn that there is a high possibility of most costly war in 2021 or 2036–2038. Additionally, in 2053–2057, 2069–2072 or 2105–2107, there is also a possibility of most costly war.

Acknowledgments. This work is supported by Shenzhen Science and Technology Plan Project (with number JCYJ20170818154936083). Kindly note that all authors of the paper are jointly of the first authorship.

References

1. American Involvement in Wars from Colonial Times to the Present. https://www.thoughtco.com/american-involvement-wars-colonial-times-present-4059761
2. The Top Ten Burning Money Wars in American History Announced in World War II. http://www.81.cn/jsdj/2015-07/17/content_6589879.htm

3. Wen, W.: Commensurability. Acta Geophys. Sin. (in Chinese) **24**(2), 151–154 (1981)
4. Zhang, Y., Wang, N., Wan, J., Ling, Y., Ye, C.: USA future war prediction using ASF method with 3 inputs and full traversal: no new war till 2030 or 2034 though 2021, 2023, 2027 and 2032 risky? In: Proceedings of Chinese Automation Congress (CAC), Jinan, China, pp. 116–121 (2017)
5. Zhang, Y., Wang, R., Yang, M., Zhu, M., Ye, C.: Using full-traversal addition-subtraction frequency (ASF) method to predict possible El Nino events in 2019, 2020 and so forth. In: Proceedings of Chinese Control and Decision Conference (CCDC), Shenyang, China, pp. 2653–2658 (2018)
6. Zhang, Y., Wang, N., Lai, F., Ling Y., Ye, C.: Predicting potential valley-point dates of stock market numerically based on ASF algorithm. In: Proceedings of the 4th International Conference on Information, Cybernetics and Computational Social Systems (ICCSS), Dalian, China, pp. 605–610 (2017)
7. Zhang, Y., Yang, S., Zou, M., Ye, C., Ling, Y.: Numerical prediction of short-term snowy weather in Guangzhou via addition-subtraction frequency (ASF) algorithm with unequally half traversal. In: Proceedings of the 13th International Conference on Natural Computation, Fuzzy Systems and Knowledge Discovery (ICNC-FSKD), Guilin, China, pp. 2557–2562 (2017)
8. Cost of World War II, Human Cost of World War II: Official Total Military & Civilian. https://caseagainstbush.blogspot.com/2005/04/financial-cost-of-world-war-ii1u.html?showComment=1414341097057
9. Ye, C., Wang, R., Xiao, M., Xue, Z., Zhang, Y.: Using different types of addition-subtraction frequency (ASF) approach with different numbers of input variables to predict years of financial events after fed tightening. In: Proceedings of the 5th International Conference on Systems and Informatics (ICSAI), Nanjing, China, pp. 382–387 (2018)
10. Jia, P.: Application of image processing technology based on MATLAB. In: Proceedings of International Conference on Data Processing, Artificial Intelligence, and Communications (DPAIC), Shenzhen, China, pp. 13–16 (2018)
11. Zhang, Y., He, P., Wen, J., Ding, Y., Ling, Y.: Potential MIC season of US since March 2017 or so predicted via WASD neuronets and sine wave with historic data and events. In: Proceedings of the 5th International Conference on Computer Science and Network Technology (ICCSNT), Changchun, China, pp. 218–223 (2016)

Guidance Instrumentation Systematic Error Separation Method Based on Particle Swarm Optimization

Zhen-xing Li(✉), Zhao-gang Wang, and Dong Li

PLA 92124 Unit, Dalian, Liaoning, China
lizhenxing@126.com

Abstract. A new separation method of guidance instrumentation systematic error of vehicle based on particle swarm optimization (PSO) was proposed. The telemetry environment function matrix is seriously ill-conditioned, which results in the performance degradation of guidance instrumentation systematic error separation. Hereby the problem of guidance instrumentation systematic separation is transformed into an optimization problem and PSO is used to estimate the systematic error coefficients. Furthermore, the guidance instrumentation systematic error is separated from the vehicle trajectory measurement data. The measured data processing results show that the accuracy of the separation of guidance instrumentation systematic error based on PSO is better than that of the traditional Bayesian estimation and principal component estimation methods. The proposed method has practical engineering application value in vehicle test.

Keywords: Vehicle test · Particle swarm optimization · Guidance instrumentation systematic error · Telemetry environment function matrix

1 Introduction

Guidance error is the main factor affecting the accuracy of vehicle, which can be divided into guidance method error and guidance instrumentation error [1]. The error of the guidance method has been controlled within a few meters to several tens of meters and can be ignored. The existence of platform systematic error has a great influence on the positioning accuracy of the inertial systematic, accounting for more than 80% of the guidance error. A large positioning error of vehicle may occur in the case of long term, long distance flight if no correction is made [2]. The guidance instrumentation error is further divided into the guidance instrumentation random error and the guidance instrumentation systematic error. The guidance instrumentation random error is generally much smaller than the guidance instrumentation systematic error, so the guidance instrumentation random error can be ignored and the guidance instrumentation systematic error must be considered [3]. Practice has proved that simply relying on hardware to improve the precision of the inertial device of the guidance systematic consumes a lot of money, the development cycle is long, and the accuracy is very limited. Therefore, from the software point of view, it is one of the key technologies to improve the precision engineering of the vehicle to establish a precise model that is

Y. Liu et al. (Eds.): ICNC-FSKD 2019, AISC 1074, pp. 491–497, 2020.
https://doi.org/10.1007/978-3-030-32456-8_53

easy to separate the guidance instrumentation systematic error [4]. The telemetry environment function matrix in the guidance instrumentation systematic error model has proved to be seriously ill-conditioned. Bayesian estimation and principal component estimation using numerical methods are difficult to obtain ideal estimation results. Even if the world-to-ground consistency is good, the Bayesian estimation with the ground test value as the prior information faces great difficulties in the choice of weights. Underwater launching vehicle is more complex than land-based because land-based launch position coordinates can be accurately measured and underwater platform positioning is limited by the accuracy of the measuring equipment. The underwater platform positioning error must be taken into account the guidance instrumentation systematic error, which makes it more difficult to separate the error of the underwater guided aircraft guidance instrumentation systematic. The number of the guidance instrumentation systematic error items is increased, which results in the coupling terms in the telemetry environment function are increased. Therefore, whether Bayesian estimation or principal component estimation is adopted, the correlation of coupling terms directly leads to the degradation of the accuracy of the separation results.

The problem of the guidance instrumentation systematic error separation can be transformed into an optimization solution problem. Therefore, the intelligent optimization algorithms can be used to estimate the error coefficients of the guidance instrumentation systematic error. Intelligent optimization algorithms that have been reported include genetic algorithms, particle swarm optimization, fish swarm algorithms, ant colony algorithms, etc. Each optimization algorithm has its advantages but also has certain defects. In contrast, the particle swarm optimization algorithm avoids more complicated genetic evolution operations by using the evolutionary model of two simple parameters of position and velocity while maintaining the global search strategy of the population. Therefore, it has a good advantage in terms of computational efficiency and global optimization performance. In this paper, the objective function is designed by analyzing the guidance instrumentation systematic error model. The constraint coding is performed based on the ground test value of the guidance instrumentation systematic error. Furthermore, the PSO is used to estimate the guidance instrumentation systematic error coefficients. The experimental data processing results show that the proposed method can effectively avoid the influence of the ill-conditioned telemetry environment function matrix on the error separation precision of the guidance instrumentation systematic, and it can obtain more effective separation items. At the same time, it can estimate the drop point with high precision after compensating the guidance instrumentation systematic error.

2 The Guidance Instrumentation Systematic Error Model

For a vehicle using platform inertial navigation, the platform is generally a three-frame structure. Three gyroscopes are mounted on the platform body, which are used as three sensitive components in the orthogonal direction to correct the angular deviation caused by the interference torque and maintain the stability of the three axes of the platform [5]. At the same time, three accelerometers are installed to measure the apparent acceleration in the corresponding three orthogonal directions. Ignoring the platform dynamic error, the guidance instrumentation systematic error can be seen as consisting of the platform systematic static error, gyroscope drift error, and accelerometer error. The platform systematic static error model can be expressed as follow.

$$\begin{cases} \Phi_x(t) = \Phi_{0x} + k'_{1x}\dot{W}_y(t) + k''_{1x}\dot{W}_z(t) + k_{xy}\ddot{W}_y(t) + k_{xz}\ddot{W}_z(t) \\ \qquad + k'_{2x}\dot{W}_y(t)\dot{W}_z(t) + k_{2x}\frac{d[\dot{W}_y(t)\dot{W}_z(t)]}{dt} \\ \Phi_y(t) = \Phi_{0y} + k'_{1y}\dot{W}_x(t) + k''_{1y}\dot{W}_z(t) + k_{yx}\ddot{W}_x(t) + k_{yz}\ddot{W}_z(t) \\ \qquad + k'_{2y}\dot{W}_x(t)\dot{W}_z(t) + k_{2y}\frac{d[\dot{W}_x(t)\dot{W}_z(t)]}{dt} \\ \Phi_z(t) = \Phi_{0z} + k'_{1z}\dot{W}_x(t) + k''_{1z}\dot{W}_y(t) + k_{zx}\ddot{W}_x(t) + k_{zy}\ddot{W}_y(t) \\ \qquad + k'_{2z}\dot{W}_x(t)\dot{W}_y(t) + k_{2z}\frac{d[\dot{W}_y(t)\dot{W}_x(t)]}{dt} \end{cases} \tag{1}$$

where $\Phi_x(t)$, $\Phi_y(t)$ and $\Phi_z(t)$ are the angle deviation for each axis of the platform. W_i, $\dot{W}_i$ and $\ddot{W}_i$, $i = x, y, z$ are the true speed, acceleration and additional acceleration of the vehicle in the i direction of the inertial coordinate systematic. Φ_{0x}, Φ_{0y} and Φ_{0z} are the platform angle deviations independent of overload. k'_{1i} and k''_{1i}, $i = x, y, z$, are the error coefficients related to overload. $k_{xy}, k_{xz}, k_{yx}, k_{yz}, k_{zx}, k_{zy}$ are the error coefficients related to the rate of change of overload. k_{2x}, k_{2y}, k_{2z} are the error coefficients related to the rate of change of the square of the overload. $k'_{2x}, k'_{2y}, k'_{2z}$ are the error coefficients related to the square of the overload.

The gyroscope error model can be expressed as follow.

$$\begin{cases} \alpha_x(t) = K_{0x}t + K_{11x}W_z(t) + K_{12x}W_y(t) + K_{13x}W_x(t) \\ \qquad + K_{2x}\int_0^t \dot{W}_y(\tau)\dot{W}_z(\tau)d\tau \\ \alpha_y(t) = K_{0y}t + K_{11y}W_z(t) + K_{12y}W_y(t) + K_{13y}W_x(t) \\ \qquad + K_{2y}\int_0^t \dot{W}_x(\tau)\dot{W}_z(\tau)d\tau \\ \alpha_z(t) = K_{0z}t + K_{11z}W_z(t) + K_{12z}W_y(t) + K_{13z}W_x(t) \\ \qquad + K_{2z}\int_0^t \dot{W}_x(\tau)\dot{W}_y(\tau)d\tau \end{cases} \tag{2}$$

where $\alpha_x(t)$, $\alpha_y(t)$ and $\alpha_z(t)$ are the angular deviation caused by drift of the gyroscope in the x, y, and z directions. K_{0x}, K_{0y} and K_{0z} are the drift coefficients for gyroscopes unrelated to overload. K_{11x}, K_{11y} and K_{11z} are the drift coefficients for gyroscopes related to overload, which depend on the centroid shift of the gyro sensitive axis direction. K_{12x}, K_{12y} and K_{12z} are the drift coefficients of the gyroscope related to the overload in the direction of the sensitive axis depends on the centroid offset along the direction of the spinning axis of the gyroscope. K_{2x}, K_{2y} and K_{2z} are the drift coefficients when there is an overload at the same time as the gyro's sensitive axis and the axis of rotation.

The accelerometer error model can be expressed as follow.

$$\begin{cases} \Delta\dot{W}_{xg}(t) = C_{0x} + C_{1x}\dot{W}_x(t) + C_{2x}\dot{W}_x^2(t) + \theta_{xz}w_y - \theta_{xy}w_z \\ \Delta\dot{W}_{yg}(t) = C_{0y} + C_{1y}\dot{W}_y(t) + C_{2y}\dot{W}_y^2(t) + \theta_{yz}w_x - \theta_{yx}w_z \\ \Delta\dot{W}_{zg}(t) = C_{0z} + C_{1z}\dot{W}_z(t) + C_{2z}\dot{W}_z^2(t) + \theta_{zx}w_y - \theta_{zy}w_x \end{cases} \tag{3}$$

where $\Delta\dot{W}_{xg}(t)$, $\Delta\dot{W}_{yg}(t)$ and $\Delta\dot{W}_{zg}(t)$ are the accelerometer errors. C_{0x}, C_{0y} and C_{0z} are the error coefficients of accelerometer unrelated to acceleration as the Partial stability. C_{1x}, C_{1y} and C_{1z} are the error coefficients of accelerometer related to acceleration as the scale factor stability. C_{2x}, C_{2y} and C_{2z} are the error coefficients of accelerometer related to the square of acceleration. $\theta_{xy}, \theta_{xz}, \theta_{yx}, \theta_{yz}, \theta_{zx}, \theta_{zy}$ are the install angular errors for the platform. Due to the angular deviation of the inertial platform, the platform coordinate systematic does not coincide with the inertial coordinate systematic. After three coordinate rotations, the error model under the nonlinear calibration of the platform can be unified into the inertial coordinate systematic. Since the systematic error of the guidance instrumentation in the inertial coordinate systematic is a nonlinear function of the error coefficient of the guidance instrumentation, the systematic error linearity of the guidance instrumentation in the velocity domain can be directly obtained by doing Taylor expansion and ignoring the high-order term, and integrating the telemetry visual acceleration as follow.

$$\Delta W = W_p(t) - W(t) = S(W(t), \dot{W}(t))C + \xi(t) \tag{4}$$

where $\xi(t)$ is the random error and ΔW is the difference between telemetry and trajectory measurement which can be given by

$$\Delta W = \int_0^t \Delta\dot{W}(s)ds = \left[\Delta W_x, \Delta W_y, \Delta W_z\right]^T \tag{5}$$

In actual calculation, it can be obtained by using the telemetry sight speed and the trajectory measurement fusion solution results. $S(W(t), \dot{W}(t))$ is the telemetry environment function matrix and C is the guidance instrumentation systematic error vector. The guidance instrumentation systematic error coefficients can be estimated by using Bayesian estimation or principal component estimation method based on Eq. 4. However, because the telemetry environmental function matrix has strong complex collinear characteristics, Regardless of the Bayesian method and the principal component estimation method, the accuracy of the estimation results is not ideal. Therefore, there is a need for an algorithm that can have higher estimation accuracy in the case of ambiguity of the environmental function matrix.

3 Guidance Instrumentation Systematic Error Separation Based on PSO

Based on the linear model of the guidance instrumentation systematic error, the problem of the error coefficient estimation of the guidance instrumentation systematic can be transformed into the optimization problem by using the telemetry sight speed and trajectory fusion measurement. Therefore, the intelligent optimization algorithm can be considered. In this paper, PSO is used to estimate the error coefficients of the guidance instrumentation systematic. The objective function is given by

$$f(t) = \sqrt{\sum_{i=1}^{N} \left\| S(W(t), \dot{W}(t))C - \Delta W(t) \right\| \Big/ N} + \frac{1}{K} + \lambda \sum_{i=1}^{m} |C - C_0| \tag{6}$$

where N is 3 times the number of data samples. K is the number of the estimated error coefficient of the guidance instrumentation systematic meets evaluation criteria. m is the number of terms of the error coefficient of the guidance instrumentation systematic to be estimated. C_0 is the ground test value of the guidance instrumentation systematic error coefficients. λ is the weight, which reflects the credibility of the harmony between heaven and earth. PSO realizes an optimization process toward the evolution of the objective function by changing the position and velocity [6], and finally completes the global optimal evolutionary solution by sharing the individual optimal information. The particle swarm is initialized as $P = \{p_i\}_{i=1}^{l}$, where l is the size of the particle swarm. Let the position of the i-th particle p_i is $x_i = [x_{i1}, x_{i2}, \cdots, x_{id}]^T$, where d is the dimension of the variable to be optimized. Let the optimal value of each individual during the evolutionary process is $pbest_i = [pbest_{i1}, pbest_{i2}, \cdots, pbest_{id}]^T$, and the overall optimal value of the particle swarm search is $gbest = [gbest_1, gbest_2, \cdots, gbest_d]^T$. In this way, each individual in the particle swarm updates its position and speed by tracking individual optimal and overall optimal values as follows.

$$v_i(k+1) = w(k)v_i(k) + c_1 r_1 [pbest_i(k) - x_i(k)] + c_2 r_2 [gbset(k) - x_i(k)] \tag{7}$$

$$x_i(k+1) = x_i(k) + v_i(k+1) \tag{8}$$

$$w(k) = w_{\max} - (w_{\max} - w_{\min})k/k_{\max} \tag{9}$$

where k is the current times of evolution iterations and $k_{\max}$ is the maximum times of evolution iterations. c_1 and c_2 are the acceleration factor. r_1 and r_2 are the random number between 0 to 1. $w(k)$ is the Inertia weight coefficient.

The steps of estimating the error coefficient of the guidance instrumentation systematic according to the particle swarm algorithm can be summarized as follows.

Step 1. Initializing the current number of iteration is $k = 1$ and the maximum number of iteration is $k_{\max}$. The particle swarm size is l and the acceleration factors are c_1 and c_2. The maximum inertia weight coefficient is $w_{\max}$ and the minimum inertia weight coefficient is $w_{\min}$. The position and speed of the i-th particle are $x_i(k)$ and $v_i(k)$, which are set according to the guidance instrumentation systematic error coefficients ground test value C_0.
Step 2. Calculating the value of the fitness function according to $x_i(k)$ and Eq. 6.
Step 3. Updating the individual optimal value $pbest_i(k)$ of the particle p_i according to the fitness function $f(n)$ and calculating the optimal value of the overall particle swarm $gbest(k)$. Updating the position and speed of the particle p_i according to Eqs. 7–9.
Step 4. Determining if the maximum number of evolution iterations has been reached, if yes, go to step 5, otherwise go back to step 2.
Step 5. The optimal individual particle position is obtained as an estimation result of the systematic error coefficients of the guidance instrumentation.

4 Data Processing and Analysis

Three vehicle test data is used to verify the guidance instrumentation systematic error separation method based on PSO. The telemetry data is telemetry sight speed in inertial coordinate systematic which collected from the acceleration gyro plus table output and pretreatment by equivalent reduction, wild value detection and filtering etc. the trajectory data is used the fusion result of multi-measurement equipment, which is considered to be the true vehicle trajectory parameter. The number of the guidance instrumentation systematic error coefficients which to be estimated is 29, where the initial uncertainty error is included. The initial parameters of PSO are $l = 200$, $k_{max} = 50$, $c_1 = c_2 = 1.5$, $w_{max} = 0.9$, $w_{min} = 0.3$. The processing results of Bayesian estimation, principal component estimation and PSO estimation are shown in Table 1.

Table 1. The guidance instrumentation systematic error estimation results.

Serial number	Effectively estimated number of items		
	PCA estimation	Bayesian estimation	PSO
1	9	12	26
2	12	16	28
3	5	7	17

Based on the aircraft telemetry trajectory, the error of the guidance instrumentation systematic is discounted to calculate the aircraft landing point. The results are shown in Table 2.

Table 2. Falling point deviation estimation result (km).

Serial number	PCA estimation		Bayesian estimation		PSO	
	Portrait	Landscape	Portrait	Landscape	Portrait	Landscape
1	1.40	0.72	1.24	0.41	1.11	0.26
2	1.35	0.48	0.92	0.25	0.52	0.23
3	2.46	0.94	2.20	0.70	1.21	0.20

It can be seen from the error separation results of the guidance instrumentation systematic that in the three-vehicle test mission, the number of effective separation terms of the error separation method using the particle swarm guidance instrumentation systematic is better than the principal component estimation and Bayesian estimation method. The calculation results of the drop point deviation after discounting the systematic error of the guidance instrumentation prove that the systematic error of the guidance instrumentation using particle swarm optimization has higher precision, and the deviation of the drop point obtained after separation is the smallest among the three methods.

5 Conclusion

The guidance instrumentation systematic error separation based on PSO was proposed to overcome the ill-conditioned problem of the telemetry environment function matrix for the vehicle test. The test data processing results show the effectiveness of the proposed method. The guidance method will become increasingly complex with the development of vehicle tests. Effectively improving the error separation accuracy of the guidance instrumentation systematic is of great significance in the future optimization of vehicle guidance component design and guidance equations. The research of the corresponding guidance instrumentation systematic error model and separation algorithm still needs to be continuously improved according to the guidance method changes.

Acknowledgement. This work was supported in part by The National Natural Science Foundation of China (61703408, 61801482).

References

1. Xie, Y.Z.: The research of combining ridge and principal components estimate in separating guidance instrument systematicatic error. J. Proj. Rocket. Missiles Guid. **33**(3), 189–191 (2013)
2. Yang, H.B., Zhang, S.F., Cai, H., Hu, Y.Z.: Modeling and parameters of guidance instrumentation systematicatic error and initial launched parameters error for marine-missile. J. Astronaut. **28**(6), 1638–1642 (2017)
3. Xie, Y.Z., Li, Z.X.: Influence factors of guidance instrumentation systematic error linear model estimation. Aerosp. Control **31**(5), 46–49 (2013)
4. Yao, J., Duan, X.J., Zhou, H.Y.: Modeling and parameters estimation of marine guidance instrumentation systematicatic error. J. Ballist. **17**(1), 33–39 (2005)
5. Luo, S.W., Wei, K.H.: Effect of guidance instrument error on reentry point's parameters. Inf. Electron. Eng. **3**(2), 101–104 (2005)
6. Chai, X.D., Liu, L.W., Chang, B.X., Jiang, T., Wang, Z.: On a batch matching systematic with impatient servers and boundedly rational customers. Appl. Math. Comput. **354**, 308–328 (2019)

Assessment of Global Left Ventricle Deformation Using Recursive Spherical Harmonics

Malika Jallouli[1(✉)], Wafa Belhadj Khalifa[1(✉)], Anouar Ben Mabrouk[2,3,4(✉)], and Mohamed Ali Mahjoub[1(✉)]

[1] Ecole Nationale d'Ingénieurs de Sousse, LATIS-Laboratory of Advanced Technology and Intelligent Systems, Université de Sousse, 4023 Sousse, Tunisie
jallouli.malika3@gmail.com, wafa.bhk@gmail.com, medalimahjoub@gmail.com

[2] Research Unit of Algebra, Number Theory and Nonlinear Analysis UR11ES50, Department of Mathematics, Faculty of Sciences, 5019 Monastir, Tunisia
anouar.benmabrouk@gmail.com

[3] Department of Mathematics, Higher Institute of Applied Mathematics and Informatics, Street of Assad Ibn Al-Fourat, Kairouan 3100, Tunisia

[4] Department of Mathematics, Faculty of Sciences, University of Tabuk, King Faisal Rd, Tabuk, Saudi Arabia

Abstract. In this paper, the Heart Left Ventricle is modelised using a set of recursive spherical harmonics. We model the left ventricle in a compact and relevant way and secondly, it permits to define a global deformation of the anatomical structure between systole and diastole phases by applying Recursive Spherical Harmonics are applied on three left ventricle meshes acquired from cardiac pet scan to analyze their deformation during the cardiac cycle, evaluate cardiac global function and diagnose ventricular diseases. Experimental results proved the efficiency and accuracy of the present method compared to classical spherical harmonics modeling in terms of necessary number of coefficients and running time.

Keywords: Spherical harmonics · Recursivity · Left ventricle · Global deformation · Cardiac function

1 Introduction

Cardiovascular disease is one of the leading causes of death. This makes the understanding of the heart structure, functions, deformations, analysis and modeling of great interest. It holds for example that functional and morphological analysis of Heart Left Ventricle (HLV) during the cardiac cycle provides insight into its functioning for early prevention and accurate treatment of associated diseases. HLV analysis occurs mainly between the blood ejection phase (systole) and the heart's dilatation phase, when its cavities are filled with blood (diastole).

Y. Liu et al. (Eds.): ICNC-FSKD 2019, AISC 1074, pp. 498–505, 2020.
https://doi.org/10.1007/978-3-030-32456-8_54

However, classical diagnostic tests are based essentially on well known parameters of cardiac global function such us End-Diastolic Volume (EDV), End-Systolic Volume (ESV) or Ejection Fraction (EF) [1], which are based especially on the ventricular volume and does not take into account his global deformation.

Furthermore, the heart, due to its dynamic nature and its non-rigid structure continues to be problematic since, the acquisition of data with good spatial and temporal precision requires a considerable period of review and storage capacity. So, we need to model LVH in order to extract quantitative parameters which describe more efficiently its function and its anatomy [2].

Superquadrics have been applied for modeling LVH. They are defined by a small number of parameters and can reconstruct a surface model even from incomplete data [4]. See also [5]. In [6] the Superquadric parameter model is used to describe the timewise deformation of the LV.

The present work aims to apply a different type of mathematical models of LVH based on Harmonic analysis on the sphere. SHs form a complete set for decomposing spherical shapes. They allow the reconstruction of a compact model of an initial complex surface. However, classical models based on SHs have rised some disadvantages in terms of complexity evaluation, since the SHs coefficients have to be computed on all the grid points of the initial surface. This induces a big time cost and biased results due to the big number of coefficients used in the reconstruction model.

In [9], 3-level recurrent SHs (RSHs) bases have been exploited allowing the decompositions of signals in eigen modes. Special filters have been constructed and proved to be more efficient and accurate than existing ones as they lead to faster and more accurate algorithms. In the present work, we continue to exploit these RSHs to provide 'good' models for LVH. The RSHs are characterized by the low number of necessary coefficients computed directly from integral forms for the reconstruction of the model. Higher reconstruction orders if needed are automatically deduced from the recurrence rule. Else, RSHs model is characterized by a reduced running time algorithms 'somehow' independent of the size and the complexity of the analyzed or intial surface. In the present paper, this technique is applied to LV surfaces acquired from cardiac pet scan. The findings here improved those in [9] by considerably reducing the size of the data to be processed, simplifying the calculation of the parameters of the global cardiac function and, mainly defining numerically the LV deformation during different instants of the cardiac cycle and more particularly between the systole and diastole phases.

2 Description of the Method

Cardiac function assessment has been widely studied by the technical development of medical acquisition tools, especially cardiac computed tomography. LV functional parameters provide direct quantitative information about the patients' current cardiovascular health status and may be useful as a prognostic marker. The most common parameters that qualify this function are essentially calculated between two specific instants of the cardiac cycle which are end systole phase and end diastole phase:

$$EV = EDV - ESV, \quad EF = EV/EDV \times 100, \quad LVF = EV \times CF$$

where: EV: Ejection Volume, EDV: End Diastolic Volume, ESV: End Systolic Volume, EF: Ejection fraction, LVF: LV Flow, EV: Ejection Volume and CF: Cardiac Frequency.

In the present paper, based on the last assesments on RSHs in [9] where the last aims have been attained, we proposed to exploit the RSHs for modeling the HLV. The Fig. 1 illustrates the global process modeling.

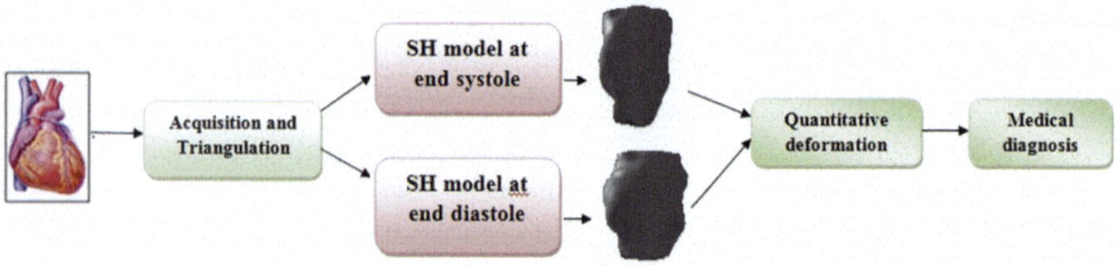

Fig. 1. Assessment of global deformation of the epicardium wall between ES and ED

The proposed approach, based on a set of RSHs, allows a compact and pertinent anatomical description of the epicedium wall of the LV and a quantification of the global deformation of this cardiac wall between end systole and end diastole of the cardiac cycle. We aim to extract anatomical parameters that describe the LV global deformation during the cardiac cycle or between two specific phases. The modeling approach is a revisit of the well known SHs based on a recurrence rule that aim to reduce considerably complexity and computing time comparing to classical SHs. We start by a spherical parameterization of the LV surface. Next, $S_{0,0}$ and $S_{k,1}$, $k = 0, + - 1$ levels of SHs coefficients are intialized as inputs (computed via the original/classical mathematical integral form). Next, higher order decomposition are deduced directly from the recursive formula which induced a significant gain on the complexity and the processing time.

Legendre SHs are expressed by means of the Legender polynomials $P_{l,m}$ as

$$Y_l^m(\theta, \varphi) = K_{l,m} P_{l,m}(cos\theta) e^{im\varphi}, \; l \in \mathbb{N}, \;\; m \in \mathbb{Z}; \; |m| \leq l, \tag{1}$$

where $K_{l,m}$ is a normalization constant in the sense of $L^2(S^2)$-norm,

$$K_{l,m} = \sqrt{\frac{(2l+1)(l-m)!}{4\pi(l+m)!}} \tag{2}$$

S^2 being the unit sphere in $\mathbb{R}^3$, and SHs form an orthonormal basis in the Hilbert space $L^2(S^2)$. A spherical supported signal or image $S(\theta, \varphi)$ is decomposed into a SHs series

$$S(\theta,\varphi)=\sum_{l=0}^{\infty}\sum_{m=-l}^{l} S_{l,m}Y_l^m(\theta,\varphi) \tag{3}$$

where $S_{l,m}$ are the SHs coefficients of S relatively to the basis $Y_{l,m}$.

It holds that the proposed method did not necessitate neither the explicit nor the exact computation of all the SHs coefficients $S_{l,m}$ above. Denote the filters-type sequences

$$a_l=\frac{2l+1}{l+1},\quad b_l=\frac{l}{l+1}, \tag{4}$$

$$\alpha_{l,m}=\sqrt{\frac{(2l+1)(l-m)}{(2l-1)(l+m)}},\quad \beta_{l,m}=\sqrt{\frac{(2l+1)}{(2l-1)(l+m)(l+m-1)}} \tag{5}$$

and

$$\gamma_{l,m}=\sqrt{\frac{(2l+1)(l-1)(l-m-1)}{(2l-3)(l+m)(l+m-1)}}. \tag{6}$$

Denote also

$$A_{l,m}=a_{l-1}\alpha_{l,m},\ B_{l,m}=ma_{l-1}\beta_{l,m}\text{ and }C_{l,m}=-b_{l-1}\gamma_{l,m}, \tag{7}$$

we showed in [9] the 3-level induction rule for the Legendre spherical harmonics basis,

$$Y_{l,m}=A_{l,m}\cos\theta Y_{l-1,m}+e^{i\varphi}B_{l,m}\sin\theta Y_{l-1,m-1}+C_{l,m}Y_{l-2,m} \tag{8}$$

which in turns has yielded a 3-level recurrence for the coefficients $S_{l,m}$ as

$$S_{l,m}=A_{l,m}S_{l-1,m}+B_{l,m}S_{l-1,m-1}C_{l,m}S_{l-2,m}. \tag{9}$$

Algorithm 1 : Recursive spherical harmonic

Input : $S_{0,0}$, $S_{1,0}$ *and l*
Output : $S_{l,m}$

Begin
  For i=1 **to** l
    $r_1=a_{i-1}$; (see equation 8)
    $r_2=b_{i-1}$; (see equation 8)
    For j=0 **to** i
      $S_{i,j}=r_1\,S_{i-1,j}+j\,r_1\,S_{i-1,j-1}-r_2\,S_{i-2,j}$; (see equation 16)
    End for
  End for
Return $S_{l,m}$;
End

The coefficients of the SHs are calculated recursively. Finally, the generalized shift theorem will make it possible to determine invariant descriptors under rotations of $SO(3)$ giving by:

$$I_l(f)=\sum_{|m|\le l}\overline{Y_{l,m}}Y_{l,m}. \tag{10}$$

A global deformation between shapes will be represented by the distance:

$$D^2 = \sum_{l \leq L} \sum_{m=-l}^{l} \mid I_l(f) - I_l(h) \mid^2, \tag{11}$$

where f and h are two functions defined on S^2.

3 Experimental Results

3.1 LV Modeling Using RSHs

In this section, we will illustrate the different steps of the LV surface modeling using recursive SHs. Figure 2 shows the evolution of the reconstruction of the spherical model of some surfaces of the LV when the reconstruction order L increases.

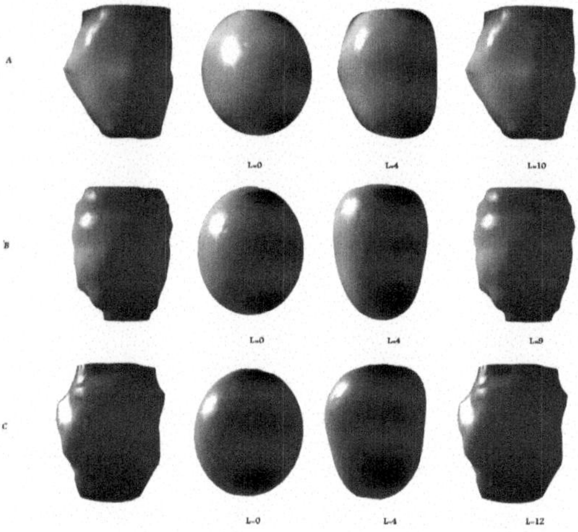

Fig. 2. SH modeling of some LV surfaces in different orders of reconstruction.

The most common problem in modeling with SHs is how to define the optimal reconstruction order that represents the closest reconstruction to the initial surface. For this purpose, we calculate the average square error applied on spherical harmonics coefficients using the formula

$$D_l^2 = \frac{1}{N} \sum_{i=1}^{N} (R_i(O_l) - R_i(O_{l-1}))^2, \tag{12}$$

where, R_i is the ith ray on the ith point of a SHs model, O_l is the SHs model reconstructed at order l and N is the number of points of the initial surface.

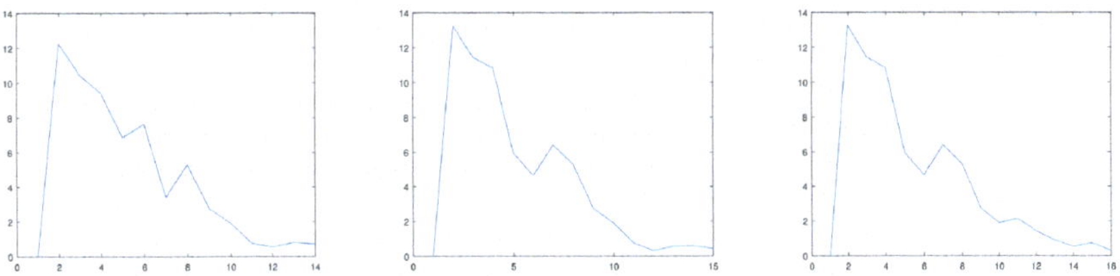

Fig. 3. The variation of D_l^2 for the surface A, B and C of Fig. 2.

This distance should be minimum when the optimal model is reached. Figure 3 illustrates the variation of D_l^2 due to Fig. 2.

Table 1 below illustrates for each surface the number of calculated SHs coefficients for each modeling method. It illustrates for each surface the processing time for the two modeling method.

Table 1. Numerical result for the reconstruction of different surface.

LV surface	Nb of point	Optimal reconstruction order	Nb of SH. coeff	Nb of RSH. coeff	Running time of classical method	Running time of recursive method
A	2601	10	120	2	6.65 s	2.10 s
B	2387	9	99	2	3.72 s	1.95 s
C	3150	12	168	2	9.98 s	3.09 s

Numerical results show a considerable gain of computational complexity. This value is estimated as $O(L(L+2))$ for the classical SHs modeling which depends on the resolution of the initial surface. Whereas, for the recursive process, the number of SHs coefficients to calculate is always two, independently of the initial resolution. This also affects the processing time.

3.2 Medical Application

Medical application is performed on a set of 50 patients and 10 volunteers from RABTA university hospital, Tunisia. A cardiac computed tomography images is obtained for the entire cardiac cycle (Fig. 4).

Table 2 illustrates the results obtained for some volunteers and patients. It is important to note that we aim to get a correlation between classical global functional LV parameters (EV, EF) and the proposed global LV deformation. This approach gives the opportunity to the practitioner to have both functional and morphological description of the LV of the heart which contributes to a better clinical diagnosis.

In cardio medecine, the normal left ventricular EF ranges ususally from 50% to 70%. In the case of a heart failure (myocarditis or ischemic heart disease),

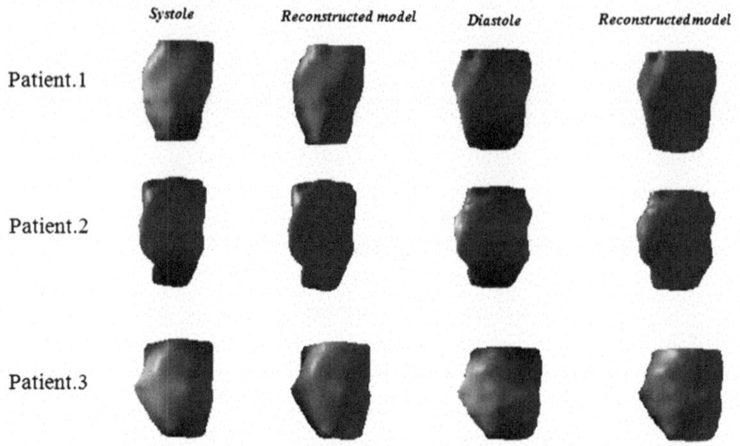

Fig. 4. Recursive SH modeling of the LV at systolic and diastolic states.

Table 2. Numerical result for the reconstruction of different surfaces.

	Ejection volume (EV (ml))	Ejection fraction (EF)	Global deformation
Volunteer 1	70	51%	1,4
Volunteer 2	65	48%	0,98
Patient 1	47	32%	0,15
Patient 2	36	26%	0,11
Patient 3	29	18%	0,09

EF value may decrease. It is important to add to the global function quantitative morphological parameters to give more precise diagnosis. For this purpose, the global deformation is defined. Experimentations show that a normal global deformation ranges from 0,5 and 1,5. In the case of myocarditis, deformation lies between 0,01 and 0,08 and for ischemic heart disease global deformation values are between 0,08 and 0,2.

It is important to note that experimentations are carried out on healthy volunteers, patients with myocarditis (morphological pathology) and patients with ischemic disease (functional pathology). The global deformation is a new quantitative parameter that helps the practitioner to conclude easily thanks to a simple exam the nature of the cardiac pathology. Finally, the global deformation gives the real nature of cardiac pathology (functional or morphological) in 95% of patients.

4 Conclusion

In this paper, RSH have been applied on medical cardiac images from computed tomography to define a global deformation of the LV surface of the heart between two specific states of the cardiac cycle (end systolic and end diastolic states). A correlation between this new morphological parameter and the classical global LV functional parameter is performed in order to help the practitioner to get a more precise diagnosis and to be able to define the nature of the heart disease. Experimental results proved that RSH are more efficient and accurate than classical SH in terms of number of coefficient and running time. Multiple perspectives should be considered for this work. The enrichment of the database will allow to better test our programs. It is important also to deploy a statistical tool that will better judge the results.

References

1. Schneider, J.E.: Assessment of global cardiac function. Methods Mol. Biol. 387–405 (2011)
2. Mihalef, V., Ionasec, R., Wang, Y., Zheng, Y., Georgescu, B., Comaniciu, D.: Patient-specific modeling of left heart anatomy, dynamics, and hemodynamics from high resolution 4D CT. In: IEEE International Symposium on Biomedical Imaging From Nano to Macro, Netherlands (2010)
3. Zhu, Y., Papademetris, X., Sinusas, A.J., Duncan, J.S.: Segmentation of the left ventricle from cardiac MR images using a subject-specific dynamical model. IEEE Trans. Med. Imaging **29**(3), 669–687 (2010). Author, F.: Article title. Journal **2**(5), 99–110 (2016)
4. Robert, A.: Etude de la forme et du mouvement du cœur á partir de données lacunaires. Thèse de doctorat, Ecole Nationale Supérieure des Télécommunications (1996)
5. Cohen, I., Cohen, D.: A hybrid hyperquadric model for 2D and 3D data fitting. Rapport Technique 2188 INRIA, (1994)
6. Bardinet, E., Cohen, L.D., Ayache, N.: A parametric deformable model to fit unstructured 3D data. Comput. Vis. Image Unders. **71**, 39–54 (1998)
7. Amini, A., Duncan, J.: Pointwise tracking of left-ventricular motion in 3D. In: Proceedings of the IEEE Workshop on Visual Motion, Princeton, New Jersey (1991)
8. Strichartz, R.S.: Local harmonic analysis on spheres. J. Funct. Anal. **77**, 403–433 (1988)
9. Jallouli, M., Zemni, M., Ben Mabrouk, A., Mahjoub, M.A. :Toward recursive spherical harmonics-issued bi-filters: part I: theoretical framework. Soft Computing (2018). https://doi.org/10.1007/s00500-018-3596-9
10. Khelifa, W. B., Ben Abdallah, A., Ghorbel, F.: Three dimensional modeling of the left ventricle of the heart using spherical harmonic analysis. In: ISBI, pp. 1275–1278 (2008)

A Hybrid Swarm Intelligence-Based Algorithm for Finding Minimum Positive Influence Dominating Sets

Geng Lin(✉), Jinyan Luo, Haiping Xu, and Meiqin Xu

College of Mathematics and Data Science, Minjiang University, Fuzhou 350108, China
lingeng413@163.com

Abstract. The minimum positive influence dominating set problem is one of the central problems in the study of online social networks. This paper presents a hybrid swarm intelligence-based algorithm to solve the minimum positive influence dominating set problem. The proposed swarm intelligence-based algorithm is based on genetic algorithm and particle swarm optimization. Firstly, a greedy randomized adaptive construction procedure is employed to generate initial swarm. Secondly, a crossover procedure is applied to obtain new solutions. Then, a mutation procedure is introduced to diversify the population. Finally, a repair procedure is used to ensure the feasibility of new solutions. Nine social networks from the literature are applied to test the proposed algorithm. The experimental results show that the proposed algorithm can achieve significant improvements over the existing greedy algorithms.

Keywords: Minimum positive influence dominating set · Swarm intelligence · Heuristic · Local search

1 Introduction

The minimum positive dominating set problem (MPDSP) has a lot of real word applications including college drinking, smoking problem, influence coverage in the market [1,2]. Let $G = (V, E)$ be an undirected graph with a node set $V = \{1, \cdots, n\}$ and an edge set E, and $d(i)$ be the degree of vertex $i \in V$. A positive influence dominating set (PIDS) of G is defined to be a subset D such that any vertex i in V is dominated by at least $\lceil \frac{d(i)}{2} \rceil$ nodes in D. The MPDSP seeks to find a PIDS of minimum cardinality.

Let $A = (a_{ij})_{n\times n}$ be the symmetric adjacency matrix of G with $a_{ij} = 1$ for $(i,j) \in E$ and $a_{ij} = 0$ for $(i,j) \notin E$. We introduce a binary vector $x = (x_1, \cdots, x_n)$, where $x_i = 1$ means node i is selected, and 0 otherwise. Then, the minimum positive dominating set problem can be formulated as following constrained 0-1 programming problem [3]:

$$(P)\begin{cases} \min\ f(x) = \sum_{i=1}^{n} x_i, \\ s.t.\ \sum_{j=1}^{n} a_{ij}x_j \geq \lceil \frac{d(i)}{2} \rceil, i = 1, \cdots, n, \\ \qquad x \in \{0,1\}^n, \end{cases}$$

Y. Liu et al. (Eds.): ICNC-FSKD 2019, AISC 1074, pp. 506–511, 2020.
https://doi.org/10.1007/978-3-030-32456-8_55

The objective function $f(x)$ minimizes the number of selected nodes, while the constraints ensure that each node i is dominated by at least $\lceil \frac{d(i)}{2} \rceil$ selected nodes. The MPDSP is known to be APX-hard [4]. A lot of approximation algorithms [4–6] have been proposed to the MPDSP. The research of Dinh et al. [7] implied that it is hard to improve the approximation rate of an approximation algorithm. Recently, several heuristic algorithms [3,8] and constructive methods [9–11] were designed to find positive influence dominating sets.

Swarm intelligence-based methods such as genetic algorithm (GA) [12], differential evolution algorithm [13], particle swarm optimization [14,15], have been successfully applied to solve many hard optimization problems. The successful applications of these swarm intelligence-based methods motivate us to design a swarm intelligence-based algorithm to solve the MPDSP.

In this paper, we propose a hybrid swarm intelligence-based algorithm (HSIA) for finding positive influence dominating sets in social networks. The HSIA is based on GA and PSO. It combines the genetic operators such as crossover and mutation in the notion of PSO.

The remainder of the paper is organized as follows: Sect. 2 introduces the proposed swarm intelligence-based algorithm HSIA. The computational experiments and results are reported in Sect. 3. Finally, conclusions are drawn in Sect. 4.

2 The Proposed Hybrid Algorithm

In this section, firstly, the general framework of the proposed hybrid swarm intelligence-based algorithm HSIA is introduced. Then, we describe the main components of HSIA in detail.

2.1 General Framework

GA and PSO are two popular swarm intelligence algorithms. Their variants have been applied to solve lots of hard optimization problems. Combing GA and PSO [15] is an effective way to overcome their shortcomings. The proposed HSIA is based on GA and PSO, and the general framework of HSIA is given in Algorithm 1. Let $pbest^i$ be the personal best solution found by the corresponding particle, and $gbest$ be the global best position found by the swarm so far.

HSIA starts with an initial swarm $P = \{x^1, \cdots, x^s\}$. In line 2, each solution x^i is generated by the greedy randomized adaptive construction method proposed by Lin et al. [3], and we initialize $pbest^i = x^i$ (line 3). In each generation, we randomly select a personal best solution or the global best solution. Then, a crossover procedure is applied on the selected solution and x^i to generate a new solution (line 9). In order to avoid the premature convergence in PSO, a mutation procedure is applied subsequent to each newly obtained solution (line 10). Since each node should be dominated by at least $\lceil \frac{d(i)}{2} \rceil$ selected nodes, a new solution can easily fall into infeasible region. In order to avoid this, we simply modify the greedy randomized adaptive construction method to repair infeasible solutions (line 11). Finally, the newly obtained solution is used to update $pbest^i$ and $gbest$. We repeat the process until a fixed number of generations G_{max} is reached, and return the best solution $gbest$ found so far.

Algorithm 1. HSIA

Require: $G = (V, E)$.
Ensure: The best solution $gbest$ found so far.
1: **for** $j = 1$ to s **do**
2: Generate x^i by the greedy randomized adaptive construction method.
3: Let $pbest^i = x^i$.
4: **end for**
5: Let $P = \{x^1, \cdots, x^s\}$ and $gbest = argmin\{f(x^i), i = 1, \cdots, s\}$.
6: **while** a predetermined maximum number of generations G_{max} is not reached **do**
7: **for** $j = 1$ to s **do**
8: Select a personal best solution or the global best solution, denoted by y.
9: Apply a crossover procedure to generate a new particle.
10: Employ a mutation procedure to the newly generated particle.
11: Use a repair procedure to repair the new particle.
12: Use the new particle to update $pbest^i$ and $gbest$.
13: **end for**
14: **end while**
15: output the best solution $gbest$ obtained during the search.

2.2 Greedy Randomized Adaptive Construction Procedure

The greedy randomized adaptive construction procedure [3] starts from a node set T. Let U be the set of unselected nodes. The greedy randomized adaptive construction procedure iteratively chooses an unselected node by the cover-degree $cd(i)$, and adds it into T. The cover-degree $cd(i)$ is the number of uncovered nodes in the set of neighbors of node i. Let RCL be the set of unselected nodes with cover-degree larger than a predefined value $Thre$. More formally, we define $Thre = cd_{min} + \alpha(cd_{max} - cd_{min})$, where $cd_{min} = min\{cd(i), i \in U\}$, $cd_{max} = max\{cd(i), i \in U\}$, and $\alpha \in (0, 1)$ is a parameter, and $RCL = \{i \in U : cd(i) \geq Thre\}$.

We randomly select an unselected node from RCL, and add it into T. The above process is repeated until T is a positive influence dominating set.

2.3 Crossover Procedure and Mutation Procedure

The crossover procedure is used to generate new solutions. Suppose that x and y are two selected solutions, and their corresponding sets of selected nodes are T_x and T_y, respectively. Let T_z be the selected node set of the newly generated solution z. We add the nodes, which are simultaneously selected by their parents, i.e., we set $T_z = T_x \cap T_y$.

To diversify the search, a mutation procedure is applied. Each node in T_z is removed from T_z with a probability $p_r = 0.15$. The newly generated solutions obtain good information from their parents, and guide the solution to more promising areas.

2.4 Repair Procedure

The newly solutions generated by the crossover procedure and the mutation procedure may be infeasible. Repair procedure directly uses the greedy randomized adaptive construction procedure to repair infeasible solutions. Specially, repair procedure starts from the corresponding selected node set T_z, and then employs the greedy randomized adaptive construction procedure to eliminate uncovered nodes.

3 Experimental Results

This section presents the experimental results to evaluate the performance of the proposed algorithm. In the experiments, we use seven real social networks from Stanford large network dataset collection, a scientific collaboration network, and an actor collaboration network. The values of the parameters are set as follows: $s = 10, G_{max} = 200, \alpha = 0.9, p_r = 0.15$.

The proposed algorithm was run 10 times on each social network. Table 1 displays the results of our proposed algorithm HISA on these tested instances. The results of the existing greedy algorithms including Wangs greedy algorithm [4], Raeis greedy algorithm [10], and Feis greedy algorithm [11], are also reported in Table 1.

Table 1. Computational results obtained by HISA, and existing greedy algorithms

Instances			Wang's algorithm	Raei's algorithm	Fei's algorithm	HSIA	
Name	n	m				Best	Average
ego-facebook	4089	85087	1827	1802	1815	1726	1726.6
CA-GrQc	5242	14485	2628	2620	2623	2597	2598.4
CA-HepTh	9877	25975	4592	4584	4576	4515	4516.2
CA-HepPh	12008	118491	4892	4888	4880	4791	4792.4
CA-AstroPh	18772	198051	7086	7078	7070	6905	6906.6
CA-CondMat	23133	93439	9877	9849	9846	9729	9734.0
Email-Enron	36692	183831	12009	12186	11968	11865	11873.4
Ncstrlwg2	6396	15872	3030	3027	3026	3004	3005.4
Actors-data	10042	145682	3200	3208	3189	3143	3144.5

From Table 1, one can see that the proposed algorithm HSIA is able to find better solutions than Wangs algorithm, Raeis algorithm, and Feis algorithm on all tested social networks. Moreover, the average solution obtained by HSIA is also better than those of Wangs algorithm, Raeis algorithm, and Feis algorithm. These results illustrate that the HSIA perform better than the existing greedy algorithms.

4 Conclusions

This work proposed a hybrid swarm intelligence-based approach to find minimum positive influence dominating sets in large scale social networks. The proposed algorithm HSIA is based on genetic algorithm and particle swarm optimization, and can directly search the binary space. The HSIA employed a greedy randomized adaptive construction procedure to generate initial swarm, and used a crossover procedure to generate new solutions. Then, we developed a simple mutation procedure to avoid premature convergence in HSIA. Finally, a repair procedure was introduced to repair infeasible solutions. Several social networks were used to test the proposed algorithm. The comparison between the proposed algorithm and the existing greedy algorithms showed that the proposed algorithm is efficient.

References

1. Zeng, Y., Chen, X., Cong, G., Qin, S., Tang, J., Xiang, Y.: Maximizing influence under influence constraint in social networks. Expert Syst. Appl. **55**, 255–267 (2016)
2. Zhang, W., Wu, W., Wang, F., Xu, K.: Positive influence dominating sets in power-law graphs. Soc. Netw. Anal. Min. **2**, 31–37 (2012)
3. Lin, G., Guan, J., Feng, H.: An ILP based memetic algorithm for finding minimum positive influence dominating sets in social networks. Phys. A **500**, 199–209 (2018)
4. Wang, F., Du, H., Camacho, E., Xu, K., Lee, W., Shi, Y., Shan, S.: On positive influence dominating set in social networks. Theor. Comput. Sci. **412**, 265–269 (2011)
5. Zhu, X., Yu, J., Lee, W., Kim, D., Shan, S., Du, D.Z.: New dominating sets in social networks. J. Global Optim. **48**, 633–642 (2010)
6. Ran, Y., Zhang, Z., Du, H., Zhu, Y.: Approximation algorithm for partial positive influence problem in social network. J. Comb. Optim. **33**, 791–802 (2017)
7. Dinh, T.N., Shen, Y., Nguyen, D.T., Thai, M.T.: On the approximability of positive influence dominating set in social networks. J. Comb. Optim. **27**(3), 487–503 (2014)
8. Khomami, M.M.D., Rezvanian, A., Bagherpour, N., Meybodi, M.R.: Minimum positive influence dominating set and its application in influence maximization: a learning automata approach. Appl. Intell. **48**(3), 570–593 (2018)
9. Wang, F., Camacho, E., Xu, K.: Positive influence dominating set in online social networks. In: Du, D.Z., Hu, X. (eds.) Combinatorial Optimization and Applications. Lecture Notes in Computer Science, vol. 5573, pp. 313–321. Springer, Heidelberg (2009)

10. Raei, H., Yazdani, N., Asadpour, M.: A new algorithm for positive influence dominating set in social network. In: Proceedings of the 2012 International Conference on Advances in Social Networks Analysis and Mining, pp. 253-257. IEEE Computer Society (2012)
11. Fei, M., Chen, D.: An improved algorithm for finding minimum positive influence dominating sets in social networks. J. South China Normal Univ. Nat. Sci. Ed. **48**(3), 59–63 (2016)
12. Holland, J.H.: Adaptation in Natural and Artificial Systems. The University of Michigan Press, Ann Arbor (1975)
13. Liang, J., Xu, W., Yue, C., Yu, K., Song, H., Crisalle, O.D., Qu, B.: Multimoda multiobjective optimization with differential evolution. Swarm Evol. Comput. **44**, 1028–1059 (2019)
14. Lin, G., Guan, J.: A hybrid binary particle swarm optimization for the obnoxious p-median problem. Inf. Sci. **425**, 1–17 (2018)
15. Ozsoydan, F.B., Baykasoglu, A.: A swarm intelligence-based algorithm for the set-union knapsack problem. Future Gen. Comput. Syst. **93**, 560–569 (2019)

Limit Cycles Analysis in a Fifth-Order Vector Field with Asymmetric Perturbation Terms

Yanjie Wang, Lijun Hong, and Xiaochun Hong(✉)

School of Statistics and Mathematics, Yunnan University of Finance and Economics, Kunming 650221, Yunnan, People's Republic of China
xchong@ynufe.edu.cn

Abstract. The limit cycle bifurcation of a plane fifth-order vector field with double homoclinic polycyclic rings is studied by qualitative analysis and numerical exploration. This study is based on a detection function that is particularly effective for perturbed planar polynomial system. The research shows that a class of five-order vector field has 5 limit cycles under asymmetric disturbance. The asymmetric perturbation here has 4 arbitrary parameters. Using the numerical simulation method, the distributed orderliness of the 5 limit cycles is observed. This will help to further study Hilbert's 16th question.

Keywords: Limit cycle · Five-order vector field · Detection function · Numerical exploration · Qualitative analysis

1 Introduction and Main Conclusions

For the following planar polynomial system

$$\begin{cases} \dfrac{dx}{dt} = y, \\ \dfrac{dy}{dt} = R_n(x) + \epsilon R_m(x)y, \end{cases} \tag{1}$$

where $R_n(x)$ and $R_m(x)$ are polynomials of degree n and m about x, and $0 < \epsilon \ll 1$. This type of (m, n) of Liénard system has received extensive research and attention. When $\epsilon = 0$, the Hamiltonian function corresponding to system (1) is

$$H(x, y) = 0.5y^2 + \int_0^x R_m(s)ds. \tag{2}$$

Obviously, $H(x, y)$ is a polynomial function of $m + 1$ order. When $m = 2$ or 3, we call $H(x, y)$ an elliptic Hamiltonian function. When $m \geq 4$, we call $H(x, y)$ a hyperelliptic Hamiltonian function. On the study of the number of limit cycles for elliptic Hamiltonian functions, reference article [1,2]. The study of the correspondence problem of hyperelliptic Hamiltonian functions has the following results. Literature [3] studied the system as follows:

Y. Liu et al. (Eds.): ICNC-FSKD 2019, AISC 1074, pp. 512–521, 2020.
https://doi.org/10.1007/978-3-030-32456-8_56

$$\begin{cases} \dfrac{dx}{dt} = y, \\ \dfrac{dy}{dt} = x(x-1)^3 + \epsilon(\alpha + \beta x + \gamma x^2 + x^3)y, \end{cases} \tag{3}$$

branch problem of the limit cycle. when $0 < \epsilon \ll 1$, it is proved that the system has three limit cycles at most. When $m = 5$, the vector field corresponding to the 6-symmetric hyperelliptic Hamiltonian function as follows:

$$\begin{cases} \dfrac{dx}{dt} = y, \\ \dfrac{dy}{dt} = -(ax + bx^3 + cx^5), \end{cases} \tag{4}$$

where $(a, b, c) \in R^3$, and $c \neq 0$.

Published a large number of articles about this disturbance of Liénard system. The article [4–7] studied the problem of the number of limits in (4) when (a, b, c) is equal to $(1, \frac{21}{10}, -1)$, $(1, -2, 1)$ and $(0, -1, 1)$ respectively under the disturbance of $\epsilon(\alpha + \beta x^2 + \gamma x^4)y$. When $(a, b, c) = (0, -1, 1)$, reference [8] studied the branching problem of the limit cycle under asymmetric perturbations. Concretely, the papaer [8] considered the system as follows:

$$\begin{cases} \dfrac{dx}{dt} = y, \\ \dfrac{dy}{dt} = -x^3(x^2 - 1) + \epsilon P(x)y, \end{cases} \tag{5}$$

where $P(x) = a_0 + a_1 x + a_2 x^2 + a_3 x^3 + a_4 x^4$, the number of isolated zeros of Abelian integrals in a bounded periodic ring domain under a combination of three parameters.

For the system identified by the 6-order asymmetric hyperelliptic Hamiltonian function, the following progress has been made. The literature [9] proves the system

$$\begin{cases} \dfrac{dx}{dt} = y, \\ \dfrac{dy}{dt} = -x(x-1)^3(x + \frac{1}{2}) + \epsilon Q(x)y, \end{cases} \tag{6}$$

where $Q(x) = a_0 + a_1 x + a_3 x^3 + a_4 x^4$. It branches up to 3 limit cycles.

Some applications of detection functions and numerical exploration method are as follows:

In 2011, Hong [12] considered the following disturbance system,

$$\begin{cases} \dfrac{dx}{dt} = y(x^2 - 9), \\ \dfrac{dy}{dt} = -(x^5 - 3^3 + x)(x^2 - 9) + \epsilon y R(x, y), \end{cases} \tag{7}$$

where $R(x, y) = -a + bx^2 + cy^2 + dx^2y^2 + ey^4$, $a, b, c, d, e \in R$, and $0 < \epsilon \ll 1$, Using the detection functions and numerical exploration method, the system has 8 limit cycles under disturbance.

In 2012, Hong [13] considered the following disturbance system,

$$\begin{cases} \dfrac{dx}{dt} = y(x-2), \\ \dfrac{dy}{dt} = -x(x^2+1)(x-2) + \mu y p(x,y,\lambda), \end{cases} \tag{8}$$

where $P(x,y,\lambda) = mx^2 + ny^2 - \lambda$, $m, n, \lambda, \mu \in R$, and $0 < \mu \ll 1$, He proved that this system has three limit cycles under three disturbances.

Document [10] studies the number of limit cycles for the following systems by using the extended complete Chebyshev system and geometric qualitative theory.

$$\begin{cases} \dfrac{dx}{dt} = y, \\ \dfrac{dy}{dt} = -x(x^2 - \frac{1}{2})(x^2+2) + \epsilon S(x) y, \end{cases} \tag{9}$$

where $0 < \epsilon \ll 1$, $S(x) = a_0 + a_1 x + a_2 x^2 + a_4 x^4$, $a_0, a_1, a_2, a_4 \in R$, and $a_4 \neq 0$.

Inspired by the above literature, this paper on the basis of [10] studies system (9) limit cycle number and distribution problem by qualitative analysis and numerical exploration. And further verified the results of [10] by numerical calculations.

2 Detection Functions and Numerical Exploration Method

Firstly, learned the preliminary detection function results on for the Hamiltonian system with disturbance terms as follows:

Lemma 1. [11] Result on the Hamiltonian system

$$\begin{cases} \dfrac{d\nu}{dt} = -\dfrac{\partial H}{\partial \omega}, \\ \dfrac{d\omega}{dt} = \dfrac{\partial H}{\partial \nu}, \end{cases} \tag{10}$$

and relevant to disturbanced Hamiltonian system

$$\begin{cases} \dfrac{d\nu}{dt} = -\dfrac{\partial H}{\partial \omega} + T(\nu,\omega,\delta), \\ \dfrac{d\omega}{dt} = \dfrac{\partial H}{\partial \nu} + B(\nu,\omega,\delta). \end{cases} \tag{11}$$

Clearly, system (11) turn into system (10) when $T(\nu,\omega,\delta) \doteq B(\nu,\omega,\delta) \neq 0$. Υ^h is the closed curve of the Hamiltonian integral of system (10) when $H(\nu,\omega) = h$. When h increases, the closed curve Υ^h is continuously expanding outward. And $\Upsilon^h(D)$ indicates the area that surrounded by Υ^h, Suppose there is h^* so that the following function satisfies the condition

$$A(h) = \iint\limits_{\Upsilon^h(D)} \left(T''_{\nu\delta}(\nu,\omega,0) + B''_{\omega\delta}(\nu,\omega,0) \right) d\nu d\omega, \tag{12}$$

$A(h^*) = 0$, $A^{'}(h^*) \neq 0$. In the case of $\delta A^{'}(h^*) < 0(> 0)$, if δ is a small variable, system (11) has a stable (or unstable) limit cycle around Υ^{h^*}; if closed curve Υ^h increases at h, it shrinks inward, resulting in a limit cycle that the stability is reversed from the situation described above, and when $\delta A^{'}(h^*) > 0(< 0)$, the limit cycle is unstable (stable). If $A(h) = 0$, then the system has no limit cycle. In [9], the authors studies were carried out on the following system,

$$\begin{cases} \dfrac{d\nu}{dt} = -\dfrac{\partial H}{\partial \omega} - \mu\nu[T(\nu,\omega) - \lambda)], \\ \dfrac{d\omega}{dt} = \dfrac{\partial H}{\partial \nu} - \mu\omega[B(\nu,\omega) - \lambda)], \end{cases} \tag{13}$$

when $T(0,0) = B(0,0) = 0$, which can be obtained by the idea of Lemma 1. We get that

$$A(h) = \iint_{\Upsilon^h(D)} (f)\, d\nu d\omega - 2\lambda(h) \iint_{\Upsilon^h(D)} d\nu d\omega, \tag{14}$$

where $f = f(\nu, \omega)$, and

$$f = \nu T_{\nu}^{'}(\nu,\omega) + \omega B_{\omega}^{'}(\nu,\omega) + T(\nu,\omega) + B(\nu,\omega), \tag{15}$$

let $A(h) = 0$, we obtain that

$$\lambda = \lambda(h) = \frac{\iint_{\Upsilon^h(D)} f(\nu,\omega) d\nu d\omega}{2\iint_{\Upsilon^h(D)} d\nu d\omega}. \tag{16}$$

We call $\lambda(h)$ as a system of (13) detection functions. Due to the idea of detection function and Lemma 1, and the following conclusions can established:

Proposition 1. For any given λ_0,

(1): If $(h_0, \lambda(h_0)$ is a focus of the line $\lambda = \lambda_0$ and the detection curve of $\lambda = \lambda(h)$, and satisfies $\lambda^{'}(h_0) > 0(< 0)$. So when $\lambda = \lambda_0$, the system (11) has a stable (unstable) limit cycle around Υ^{h^0}.

(2): If line $\lambda = \lambda_0$ has no intersection with detection curve $\lambda = \lambda(h)$, then when $\lambda = \lambda_0$, the system (13) has no limit cycle.

Proof: In Lemma 1, let $\delta = \mu$, $T(\nu,\omega,\delta) = \delta\nu[T(\nu,\omega) - \lambda)]$, $B(\nu,\omega,\delta) = \delta\omega[B(\nu,\omega) - \lambda)]$, we can know $T(\nu,\omega,0) = B(\nu,\omega,0) \equiv 0$, $T_{\nu\delta}^{''} + B_{\omega\delta}^{''} = f(\nu,\omega) - 2\lambda$. So in Lemma 1, $A(h)$ becomes as follows:

$$A(h) = \iint_{\Upsilon^h(D)} (f(\nu,\omega))\, d\nu d\omega - 2\lambda \iint_{\Upsilon^h(D)} d\nu d\omega, \tag{17}$$

let $\psi(h) = \iint_{\Upsilon^h(D)} (f(\nu,\omega))\, d\nu d\omega$, $\phi(h) = 2\iint_{\Upsilon^h(D)} d\nu d\omega$, here

$$A(h) = \psi(h) - \lambda\phi(h), \tag{18}$$

when $A(h_0) = 0$, we can obtain $\lambda = \frac{\psi(h_0)}{\phi(h_0)}$. Substituting $h = h_0$ into the (18) formula. We can simultaneously derivate the two sides of the equal and we can get the equation

$$A^{'}(h_0) = \psi^{'}(h_0) - \lambda\phi^{'}(h_0). \tag{19}$$

Put $\lambda = \frac{\psi(h_0)}{\phi(h_0)}$ into the equation of (19) which get

$$\begin{aligned} A^{'}(h_0) &= \psi^{'}(h_0) - \lambda\phi^{'}(h_0) \\ &= \frac{\psi^{'}(h_0)\phi(h_0) - \psi(h_0)\phi^{'}(h_0)}{\phi(h_0)}. \end{aligned} \tag{20}$$

On the other hand, form the (16) formula, we get that

$$\lambda = \lambda(h) = \frac{\iint_{\Upsilon^h(D)} (f(\nu, \omega))\, d\nu d\omega}{2\iint_{\Upsilon^h(D)} d\nu d\omega} = \frac{\psi(h)}{\phi(h)}. \tag{21}$$

We can simultaneously derivate the two side of the equation of (21) at the point of h_0.

$$\lambda^{'}(h_0) = \frac{\psi^{'}(h_0)\phi(h_0) - \psi(h_0)\phi^{'}(h_0)}{[\phi(h_0)]^2}. \tag{22}$$

Obviously, from the (16) formula that we know $\phi(h_0) > 0$. And through the (20) and (22) formulas. We can find that $\lambda^{'}(h_0)$, $A^{'}(h_0)$ have the same sign, so the symbols of $-\mu\lambda^{'}(h_0), -\mu A^{'}(h_0)$ are also the same. It can be known from Lemma 1 that Proposition 1 is true.

3 Analysis of the Unperturbed System

Consider the next unperturbed system of

$$\begin{cases} \dfrac{dx}{dt} = y, \\ \dfrac{dy}{dt} = -x(x^2 - \frac{1}{2})(x^2 + 2). \end{cases} \tag{23}$$

Obviously, the system (23) is a Hamiltonian system. And it has has three finite singular points. System (23) has a hyperbolic saddle point $O(0, 0)$ and two centers $(\frac{\sqrt{2}}{2}, 0)$, $(-\frac{\sqrt{2}}{2}, 0)$ (see Fig. 1).

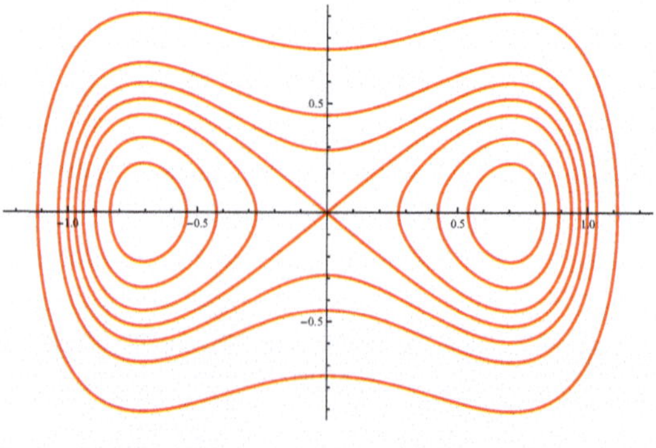

Fig. 1. Closed orbits of unperturbed system (23).

As the change of h, the closed curves defined by the system (23) as follows:

(**A**): When $h \in (-\frac{13}{96}, 0)$, the system has two left and right bounded periodic ring domains, wherein the closed trajectories are recorded as ℓ_1^h and ℓ_2^h respectively.

(**B**): There is also an unbounded periodic ring domain, in which the closed orbits is recorded as ℓ^h, corresponding to $h \in (0, +\infty)$ (see Fig. 1).

Note that as h increases, the curves ℓ_1^h, ℓ_2^h and ℓ^h extend outward.

The first integral of the system (23) given by bifurcation of limit cycles in given planar polynomial system and as follows:

$$H(x, y) = \frac{1}{2}y^2 - \frac{1}{2}x^2 + \frac{3}{8}x^4 + \frac{1}{6}x^6 = h. \tag{24}$$

And its integral factor is Fig. 1. From the Eq. (24), we can get $H(0, 0) = 0$, $H(\pm\frac{\sqrt{2}}{2}, 0) = -\frac{13}{96}$ and the following equation.

$$y_{\pm} = \pm\mu(h, x) = \pm\frac{1}{6}\sqrt{36x^2 - 27x^4 - 12x^6 + 72h},$$
$$(-\frac{13}{96} < h < +\infty). \tag{25}$$

Letting $y = 0$, we can obtain the Eqs. (26) from (25)

$$4x^6 + 9x^4 - 12x^2 - 24h = 0, \tag{26}$$

where $-\frac{13}{96} < h < +\infty$. From (26), for any number that belongs to of $h \in (-\frac{13}{96}, 0)$, we can define $\pm x_1$ and $\pm x_2$. For an arbitrary constant of $h \in (0, +\infty)$, we can get definite $\pm x_2$.

4 Limit Cycles Analysis of the Perturbed System

For the non-Hamiltonian perturbed integrable system (9), we get the Abelian integral as follows:

$$J(h, a_0) = \oint_{\Upsilon^h} (a_0 + a_1x + a_2x^2 + a_4x^4)ydx. \tag{27}$$

Now, we premeditate the perturbed system (9). Because some symmetry corresponds to three families of closed orbits in Fig. 1. We can get the following (28) for system (9) by using Lemma 1 of the Eq. (12).

$$\iint_{\Upsilon^h(D)} \left(\frac{\partial^2 T}{\partial x \partial \epsilon} + \frac{\partial^2 B}{\partial y \partial \epsilon}\right) dxdy = 0. \tag{28}$$

We can obtain the equation of (29) form the system (9)

$$\frac{\partial^2 T}{\partial x \partial \epsilon} + \frac{\partial^2 B}{\partial y \partial \epsilon} = a_0 + a_1 x + a_2 x^2 + a_4 x^4. \tag{29}$$

By referring to (28) and (29), we are able to determine the detection function a_0.

$$a_0(h) = -\frac{\iint_{\Upsilon^h(D)} \left(a_1 x + a_2 x^2 + a_4 x^4\right) dxdy}{\iint_{\Upsilon^h(D)} dxdy}. \tag{30}$$

Due to (30), we can get the detection functions of the system (9).

$$a_{0_{1L}}(h) = -\frac{\int_{-x_1}^{-x_2} x\mu(h,x)dx}{\int_{-x_1}^{-x_2} \mu(h,x)dx} a_1 - \frac{\int_{-x_1}^{-x_2} x^2\mu(h,x)dx}{\int_{-x_1}^{-x_2} \mu(h,x)dx} a_2 - \frac{\int_{-x_1}^{-x_2} x^4\mu(h,x)dx}{\int_{-x_1}^{-x_2} \mu(h,x)dx} a_4, \quad h \in (-\frac{13}{96}, 0). \tag{31}$$

$$a_{0_{1R}}(h) = -\frac{\int_{x_1}^{x_2} x\mu(h,x)dx}{\int_{x_1}^{x_2} \mu(h,x)dx} a_1 - \frac{\int_{x_1}^{x_2} x^2\mu(h,x)dx}{\int_{x_1}^{x_2} \mu(h,x)dx} a_2 - \frac{\int_{x_1}^{x_2} x^4\mu(h,x)dx}{\int_{x_1}^{x_2} \mu(h,x)dx} a_4, \quad h \in (-\frac{13}{96}, 0). \tag{32}$$

$$a_{0_2}(h) = -\frac{\int_{-x_2}^{x_2} x\mu(h,x)dx}{\int_{-x_2}^{x_2} \mu(h,x)dx} a_1 - \frac{\int_{-x_2}^{x_2} x^2\mu(h,x)dx}{\int_{-x_2}^{x_2} \mu(h,x)dx} a_2 - \frac{\int_{-x_2}^{x_2} x^4\mu(h,x)dx}{\int_{-x_2}^{x_2} \mu(h,x)dx} a_4, \quad h \in (0, +\infty). \tag{33}$$

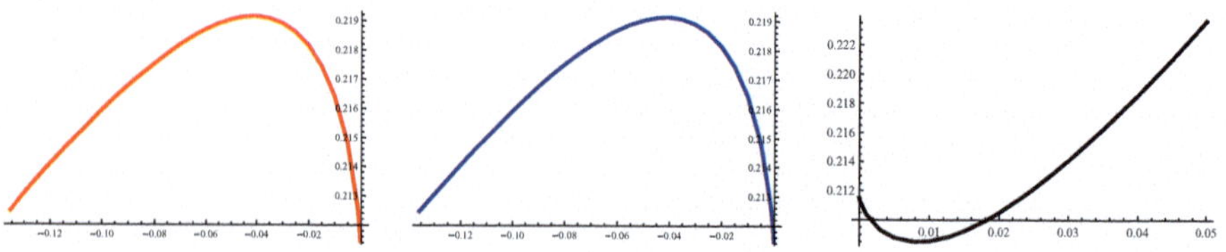

(a) $a_0 = a_{0_{1L}}(h)$ (b) $a_0 = a_{0_{1R}}(h)$ (c) $a_0 = a_{0_2}(h)$

Fig. 2. Detection curves of system (9) when $a_1 = 0$, $a_2 = 1$ and $a_4 = -2.85$.

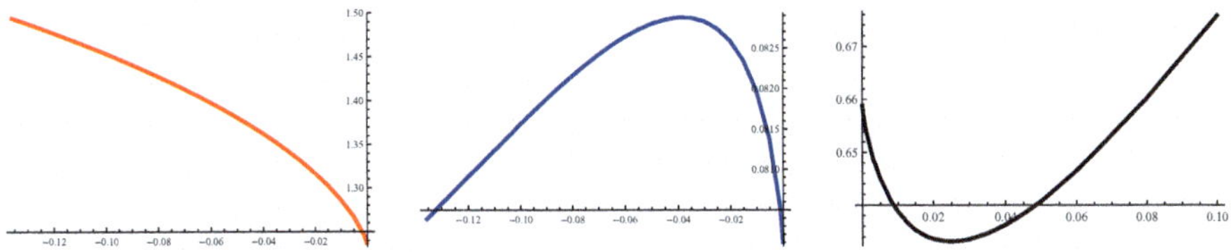

(a) $a_0 = a_{0_{1L}}(h)$ (b) $a_0 = a_{0_{1R}}(h)$ (c) $a_0 = a_{0_2}(h)$

Fig. 3. Detection curves of system (9) when $a_1 = 1$, $a_2 = 0$ and $a_4 = -3.15$.

From the (31)–(33), we can obtain Tables 1 and 2. From the Tables 1 and 2, when $a_1 = 0$, $a_2 = 1$, $a_4 = -2.85$, we can obtain detection curves $a_0(h)$ (see Fig. 2), when $a_1 = 1$, $a_2 = 0$, $a_4 = -3.15$, we can obtain detection curves $a_0(h)$ (see Fig. 3).

For $a_1 = 0$, $a_2 = 1$, $a_4 = -2.85$, and $0 < \epsilon \ll 1$, from detection curves of the Fig. 2 and the Tables 1 and 2, we can get the following results:

(1) The system (7) has a limit cycle when $a_0 > 0.219093$ or $a_0 = 0.210500$.

(2) The system (7) has 5 limit cycles when $0.212501 < a_0 < 0.219093$ (see Fig. 4).

(3) The system (7) has 3 limit cycles when $0.211488 \leq a_0 < 0.212501$ or $a_0 = 0.219093$.

(4) The system (7) has 2 limit cycles when $0.210500 < a_0 < 0.211488$.

From detection curves of the Fig. 3 and the Tables 1 and 2, we can get some results, and from Tables 1 and 2, we can obtain some results too. They are omitted for concision.

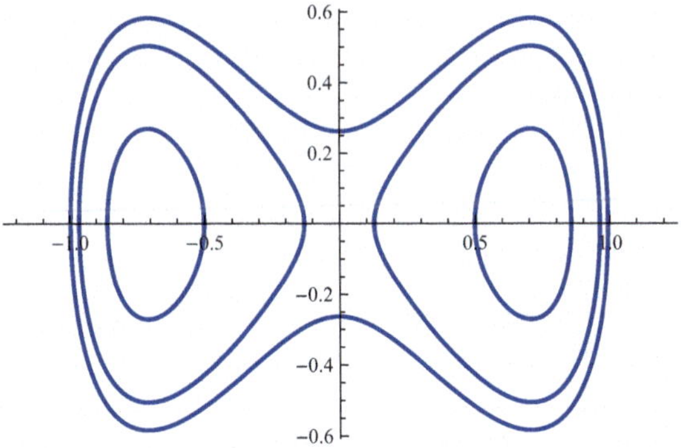

Fig. 4. The 5 limit cycles of system (9) when $a_0 = 0.216$, $a_1 = 0$, $a_2 = 1$, $a_4 = -2.85$ and $\epsilon = 0.001$. The big limit cycle l^b passes point $(0, 0.2625)$, it is stable, the two small limit cycles $l_i^s (i = 1, 2)$ pass points $(\mp 0.5001, 0)$ respectively, they are stable, and the two middle limit cycles $l_i^m (i = 1, 2)$ pass points $(\mp 0.1291, 0)$ respectively, they are unstable.

Table 1. Values of detection functions $a_{01_L}(h)$ and $a_{01_R}(h)$ of system (7)

h	$a_{01_L}(h)$	$a_{01_R}(h)$
-0.13541	0.7071031986 a_1 - 0.4999962667 a_2 - 0.2499989334 a_4	-0.7071031986 a_1 - 0.4999962667 a_2 - 0.2499989334 a_4
-0.13540	0.7070978244 a_1 - 0.4999906665 a_2 - 0.2499973333 a_4	-0.7070978244 a_1 - 0.4999906665 a_2 - 0.2499973333 a_4
-0.13539	0.7070924497 a_1 - 0.4999850660 a_2 - 0.2499957331 a_4	-0.7070924497 a_1 - 0.4999850660 a_2 - 0.2499957331 a_4
-0.13500	0.7068826791 a_1 - 0.4997665269 a_2 - 0.2499332748 a_4	-0.7068826791 a_1 - 0.4997665269 a_2 - 0.2499332748 a_4
-0.13300	0.7058017967 a_1 - 0.4986419344 a_2 - 0.2496113502 a_4	-0.7058017967 a_1 - 0.4986419344 a_2 - 0.2496113502 a_4
-0.13000	0.7041640494 a_1 - 0.4969426553 a_2 - 0.2491232524 a_4	-0.7041640494 a_1 - 0.4969426553 a_2 - 0.2491232524 a_4
-0.12500	0.7013890761 a_1 - 0.4940763510 a_2 - 0.2482952926 a_4	-0.7013890761 a_1 - 0.4940763510 a_2 - 0.2482952926 a_4
-0.11900	0.6979802362 a_1 - 0.4905775921 a_2 - 0.2472764647 a_4	-0.6979802362 a_1 - 0.4905775921 a_2 - 0.2472764647 a_4
-0.11000	0.6926926903 a_1 - 0.4851992302 a_2 - 0.2456919083 a_4	-0.6926926903 a_1 - 0.4851992302 a_2 - 0.2456919083 a_4
-0.10000	0.6865458243 a_1 - 0.4790213262 a_2 - 0.2438425653 a_4	-0.6865458243 a_1 - 0.4790213262 a_2 - 0.2438425653 a_4
-0.09000	0.6800736972 a_1 - 0.4726033893 a_2 - 0.2418858195 a_4	-0.6800736972 a_1 - 0.4726033893 a_2 - 0.2418858195 a_4
-0.08000	0.6732254198 a_1 - 0.4659098535 a_2 - 0.2398034095 a_4	-0.6732254198 a_1 - 0.4659098535 a_2 - 0.2398034095 a_4
-0.06000	0.6581143892 a_1 - 0.4514971388 a_2 - 0.2351586279 a_4	-0.6581143892 a_1 - 0.4514971388 a_2 - 0.2351586279 a_4
-0.04000	0.6403322658 a_1 - 0.4351748618 a_2 - 0.2295903368 a_4	-0.6403322658 a_1 - 0.4351748618 a_2 - 0.2295903368 a_4
-0.03000	0.6299008747 a_1 - 0.4259283659 a_2 - 0.2262648820 a_4	-0.6299008747 a_1 - 0.4259283659 a_2 - 0.2262648820 a_4
-0.02000	0.6178248359 a_1 - 0.4155378393 a_2 - 0.2223546423 a_4	-0.6178248359 a_1 - 0.4155378393 a_2 - 0.2223546423 a_4
-0.01500	0.6108570431 a_1 - 0.4097017117 a_2 - 0.2200667344 a_4	-0.6108570431 a_1 - 0.4097017117 a_2 - 0.2200667344 a_4
-0.01000	0.6029177931 a_1 - 0.4032014208 a_2 - 0.2174295147 a_4	-0.6029177931 a_1 - 0.4032014208 a_2 - 0.2174295147 a_4
-0.00500	0.5933481817 a_1 - 0.3955948549 a_2 - 0.2142042709 a_4	-0.5933481817 a_1 - 0.3955948549 a_2 - 0.2142042709 a_4
-0.00100	0.5829344790 a_1 - 0.3876545228 a_2 - 0.2106272688 a_4	-0.5829344790 a_1 - 0.3876545228 a_2 - 0.2106272688 a_4
-0.00050	0.5811227104 a_1 - 0.3863210296 a_2 - 0.2099957010 a_4	-0.5811227104 a_1 - 0.3863210296 a_2 - 0.2099957010 a_4
-0.00010	0.5793470393 a_1 - 0.3850369335 a_2 - 0.2093724822 a_4	-0.5793470393 a_1 - 0.3850369335 a_2 - 0.2093724822 a_4
-0.00005	0.5790689250 a_1 - 0.3848390912 a_2 - 0.2092742944 a_4	-0.5790689250 a_1 - 0.3848390912 a_2 - 0.2092742944 a_4
-0.00001	0.5788123837 a_1 - 0.3846581858 a_2 - 0.2091834397 a_4	-0.5788123837 a_1 - 0.3846581858 a_2 - 0.2091834397 a_4

Table 2. Values of detection function $a_{0_2}(h)$ of system (7)

h	$a_{0_2}(h)$	h	$a_{0_2}(h)$
0.00001	$-0.3845444747\ a_2 - 0.2091253651\ a_4$	0.45000	$-0.4182700649\ a_2 - 0.3120960893\ a_4$
0.00010	$-0.3841671924\ a_2 - 0.2089371378\ a_4$	0.50000	$-0.4264113639\ a_2 - 0.3253966404\ a_4$
0.00100	$-0.3816296512\ a_2 - 0.2077271964\ a_4$	0.60000	$-0.4419738925\ a_2 - 0.3513572495\ a_4$
0.00300	$-0.3777215790\ a_2 - 0.2059822639\ a_4$	0.70000	$-0.4566502852\ a_2 - 0.3765354541\ a_4$
0.00500	$-0.3747499907\ a_2 - 0.2047491243\ a_4$	0.80000	$-0.4705369859\ a_2 - 0.4010146688\ a_4$
0.01000	$-0.3693251597\ a_2 - 0.2027637083\ a_4$	0.90000	$-0.4837236350\ a_2 - 0.4248688334\ a_4$
0.02000	$-0.3627681746\ a_2 - 0.2011466261\ a_4$	1.00000	$-0.4962884996\ a_2 - 0.4481613926\ a_4$
0.04000	$-0.3570996353\ a_2 - 0.2019927398\ a_4$	1.10000	$-0.5082985787\ a_2 - 0.4709463711\ a_4$
0.05000	$-0.3561055788\ a_2 - 0.2034202931\ a_4$	1.20000	$-0.5198109720\ a_2 - 0.4932698415\ a_4$
0.06000	$-0.3558139252\ a_2 - 0.2052284787\ a_4$	1.40000	$-0.5415307861\ a_2 - 0.5366848135\ a_4$
0.08000	$-0.3566017336\ a_2 - 0.2095808151\ a_4$	1.60000	$-0.5617617082\ a_2 - 0.5786627115\ a_4$
0.10000	$-0.3585267485\ a_2 - 0.2145336328\ a_4$	1.80000	$-0.5807397768\ a_2 - 0.6193993282\ a_4$
0.15000	$-0.3657446791\ a_2 - 0.2281300277\ a_4$	2.00000	$-0.5986467686\ a_2 - 0.6590479632\ a_4$
0.20000	$-0.3743681664\ a_2 - 0.2423525287\ a_4$	2.20000	$-0.6156257059\ a_2 - 0.6977313029\ a_4$
0.23000	$-0.3614503176\ a_2 - 0.2010175647\ a_4$	2.40000	$-0.6317912675\ a_2 - 0.7355494026\ a_4$
0.25000	$-0.3833774171\ a_2 - 0.2566528031\ a_4$	2.60000	$-0.6472369331\ a_2 - 0.7725851906\ a_4$
0.28000	$-0.3596963928\ a_2 - 0.2010449338\ a_4$	2.80000	$-0.6620400073\ a_2 - 0.8089083680\ a_4$
0.30000	$-0.3923890759\ a_2 - 0.2708318919\ a_4$	3.00000	$-0.6762652282\ a_2 - 0.8445782323\ a_4$
0.35000	$-0.4012450625\ a_2 - 0.2848132318\ a_4$	3.50000	$-0.7096407682\ a_2 - 0.9312129704\ a_4$
0.40000	$-0.4098799652\ a_2 - 0.2985697600\ a_4$	4.00000	$-0.7403935902\ a_2 - 1.0147002140\ a_4$

5 Conclusion

In the above discussion, Using the qualitative analysis and numerical exploration, a plane fifth-order vector field with double Homoclinic polycyclic rings are studied. In particular, for the system (9), when $a_1 = 0$, $a_2 = 1$, $a_4 = -2.85$, and $0 < \epsilon \ll 1$, the study reveals that it has 5 limit cycles if $0.212501 < a_0 < 0.219093$. It is also found that each limit cycle passes a particular point and the position of these points is obtained by using the numerical exploration method (see Fig. 4).

Acknowledgment. This work was financially supported by the Natural Science Foundation of China (Grant No. 11761075).

References

1. Dumortier, F., Li, C.: Perturbations from an elliptic Hamiltonian of degree four, saddle loop and two saddle cycle. J. Differ. Equ. **176**, 114–157 (2001)
2. Dumortier, F., Li, C.: Perturbations from an elliptic Hamiltonian of degree four figure eight-loop. J. Differ. Equ. **188**, 512–554 (2003)
3. Wang, J., Xiao, D.: On the number of limit cycles in small perturbations of a class of hyper-elliptic Hamiltonian systems with one nilpotent saddle. J. Differ. Equ. **250**, 2227–2243 (2011)
4. Qi, M., Zhao, L.: Bifurcations of limit cycles from a quintic Hamiltonian system with a figure double-fish. Int. J. Bifurcat. Chaos Appl. Sci. Eng. **23**, 1350116 (2013)
5. Asheghi, R., Zangeneh, H.R.Z.: Bifurcations of limit cycles from quintic Hamiltonian systems with an eye-figure loop. Nonlinear Anal. **68**, 2957–2976 (2008)
6. Asheghi, R., Zangeneh, H.R.Z.: Bifurcations of limit cycles from quintic Hamiltonian systems with an eye-figure loop (II). Nonlinear Anal. **69**, 4143–4162 (2008)
7. Asheghi, R., Zangeneh, H.R.Z.: Bifurcations of limit cycles from quintic Hamiltonian system with a double cuspidal loop. Comput. Math. Appl. **59**, 1409–1418 (2010)
8. Zhao, L.: The perturbations of a class of hyper-elliptic Hamilton systems with a double homoclinic loop through a nilpotent saddle. Nonlinear Anal. **95**, 374–387 (2014)
9. Kazemi, R., Zangeneh, H.R.Z., Atabaigi, A.: On the number of limit cycles in small perturbations of a class of hyper-elliptic Hamiltonian systems. Nonlinear Anal. **75**, 574–587 (2012)
10. Zhao, L.: Zeros of Abelian integral for a class of quintic planar vector fields with a double homoclinic loop under non-symmetry perturbations. Acta Math. SINICA **47**, 227–240 (2017)
11. Li, J., Li, C.: Distribution of limit cycles for planar cubic Hamiltonian systems. Acta Math. SINICA **28**, 509–521 (1985)
12. Hong, X., Tan, B.: Bifurcation of limit cycles of a perturbed integrable non-Hamiltonian system. In: Proceedings of the 4th International Workshop on Chaos-Fractal Theories and Applications, pp. 22–26 (2011)
13. Hong, X.C.: Limit cycles analysis of a perturbed integal Abel non-Hamiltonian system. Ann. Differ. Equ. **3**, 263–268 (2012)

Exploring Optimal Model for Machine Learning by Differential Evolution

Yi-Chuan Chiu[1(✉)], Yung-Tsan Jou[1], and Hsing-Hung Lin[2]

[1] Department of Industrial and System Engineering, Chung Yuan Christian University, Taoyuan, Taiwan, R.O.C.
angel.chiu957@gmail.com

[2] Department of Industrial Engineering, Chung Hua University, Hsinchu, Taiwan, R.O.C.

Abstract. In the big data era, the data scientist or business analyst own business data can apply machine learning algorithm to train inference models easily since the application of big data attracts more and more attention and the technique of training models can be obtained easier and cheaper than ever. The focus of big data applications has gradually shifted from model training to the prediction and inference. For enterprise application scenarios, selecting the most precise model among many trained models has become a significant topic of research. Though ensemble methods have been proposed to discover best model by multiple training phase, studies of exploring best combination within multiple modes are still few. Finding the appropriate parameters to configure different machine learning models is an NP-hard problem that needs metaheuristic algorithm to solve. This study proposes a differential evolution algorithm to integrate multiple trained machine learning models into a hybrid model. For experiment, the regression model is taken as an example and the differential evolution algorithm is compared with the ant colony optimization algorithm in this paper. Three benchmark datasets are employed to examine, and the results discovered that the differential evolution algorithm outperforms ant colony optimization.

Keywords: Machine learning · Differential evolution · Ant colony optimization

1 Introduction

The application of big data and machine learning model usually consist with four major steps that includes: data acquisition, data preparation, model training and validation, model inference. Since more and more machine learning algorithms have developed for model training, the model selection and performance evaluation have become the significant role before model inference for production applications.

There are different methods to enhance the accuracy of machine learning models for production during model training phase. Ensemble learning was firstly introduced for strengthening precision of classification problems [1]. The basic concept of ensemble learning is that the integration of a number of experts to make decisions in some specific ways (such as voting method, weighting method), the results will be

Y. Liu et al. (Eds.): ICNC-FSKD 2019, AISC 1074, pp. 522–530, 2020.
https://doi.org/10.1007/978-3-030-32456-8_57

better than only a single expert. Due to the different expertise of each expert, the combined mechanism allows the experts to complement each other and get better results. Ensemble learning focus on training data replacement or not to obtained more efficient model. Sagi and Rokach introduced the concept of ensemble learning, reviews traditional, novel and state-of-the-art ensemble methods [2].

Since model selection is choosing the best performance model from different machine learning models, model selection is different thinking from ensemble methods that is choosing the best result from different approaches (ex. bagging, boosting) with one model. With the advent of big data, more and more machine learning models will be trained and applied to business operation, however the past literature mostly focuses on ensemble methods, and the literature of model selection is relatively rare. The goal of this paper is to explore the best use of the algorithm to find the optimal weights combination solution of multiple machine learning models.

This study applied different regression algorithms with same training dataset, then exploring the optimal parameters combination via differential evolution methods. Three different famous datasets were employed to verify the hybrid model and the outcomes performed better than any single machine learning model. Furthermore, the ACO is introduced to compare with DE to find the optimal model combination. The remain parts of this paper is arranged as follows. In Sect. 2, related works are reviewed. Experiment methods are described in Sect. 3. Section 4 reveals the results of the experiments. At last, conclusion is presented in Sect. 5.

2 Related Studies

There have been numerous studies in the literature dealing with regression including generalized linear model, rigid regression and robust regression. In addition, a lot of research works have been conducted in the fields of differential evolution and ant colony optimization. This section provides review of these works.

2.1 Regression Models

The Generalized Linear Model (GLM) is a flexible linear regression model, in statistic. This model allows the distribution of the deviation of the number of strains to have other distributions than the normal distribution. This model assumes that the distribution function of the random variables measured by the experimenter and the systematic effects (i.e., non-random effects) in the experiment can establish a function that can explain its correlation via a link function. The representative literature of the generalized linear model is an outline of the principles, calculations (such as the most approximate estimators) and practical applications of generalized linear models [3].

Ridge regression analysis is a technique for the presence of multiple collinear (automatically independent variables) data. In the case of multicollinearity, although the ordinary least squares (OLS) is fair to each variable; they vary widely, shifting the observations away from the true values. Ridge regression reduces the standard error by adding a degree of deviation to the regression. Ridge regression solves the multicollinearity problem by the contraction parameter λ (lambda). The assumption of this regression is similar to the least square regression, except for the constant term. It shrinks the value of the correlation coefficient but does not reach zero, which indicates that it has no feature selection function. This is a regularization method and uses L2 regularization.

The models for linear regression in the previous article were all based on the least square method. However, when there are many outliers in the data sample points, the impact of these anomalies on the regression model will be very large, and the traditional regression method based on least squares will not be applicable. Though, preprocessing the data and eliminating those abnormal points can be considered to conduct before doing regression analysis. However, in the actual data, there are two problems: Outliers are not well defined, and there is no good standard for determining which points are outliers. It is very likely that this seemingly anomalous point is the data of the original model. If this is the case, then the points of these anomalies will carry a large amount of information of the original model, and a large amount of information will be lost after the culling.

Robust regression is an algorithm used to replace the least squares method when the least squares method encounters the above-mentioned data sample points with abnormal points. In addition, robust regression can also be used for outlier detection, or to find those sample points that have the greatest impact on the model.

2.2 Differential Evolution

Differential evolution (DE) is a stochastic, population-based optimization algorithm that developed to optimize real parameter, real valued functions and was firstly introduced by Storn and Price in [4]. Figure 1 is the standard flow chart of differential evolution.

DE has the advantage of incorporating a relatively simple and efficient form of self-adapting mutation. The population size does not need to be overly high, and smaller populations can be considerably more efficient. With its ease of implementation and proven efficiency, DE is ideally suited to both novice and experienced users wishing to optimize their simulation models [5].

Through two decades development, DE has improved by different configuration including population, mutation and cross rate to enhance the performance of this state-of-the-art evolution algorithm [6]. An improved version of the differential evolution (DE) based on the orthogonal design (ODE) was proposed that makes the DE faster and robust [7]. Simulations result shoed that the ODE can find the near- optimal solution in all cases and outperforms other state-of-the-art evolutionary algorithms in terms of the quality, stability as well as computational cost. Another enhanced differential evolution optimization algorithm has been developed [8]. The enhancement lies in reducing the number of control parameters from three (NP, F and CR) to two (NP and F), thereby simplifying the tuning process. Comparison is made with the original DE and other famous DE algorithm, and the results demonstrate the superiority of the proposed approach for most of the functions considered. Elsayed and Sarker proposed an adaptive configuration of differential evolution algorithms for solving big data optimization competition problems [9]. The proposed algorithm automatically determines the best variant and shows the superiority compared with the baseline algorithm.

Piotrowski briefly reviews the opinions regarding DE population size setting and verifies the impact of the population size on the performance of DE algorithms [10]. Based on the extensive experimental results the use of adaptive population size is highly recommended, especially for higher-dimensional and real-world problems. The study of Das and Suganthan attempted to provide an overall picture of the DE [11]. It discussed the different schemes of parameter control and adaptation for DE and extended review of the modifications of DE for tackling constrained, multi-objective, uncertain, and large-scale optimization problems. In this study, DE is applied to search optimal combination of the three machine learning models (GLM, rigid and robust regression).

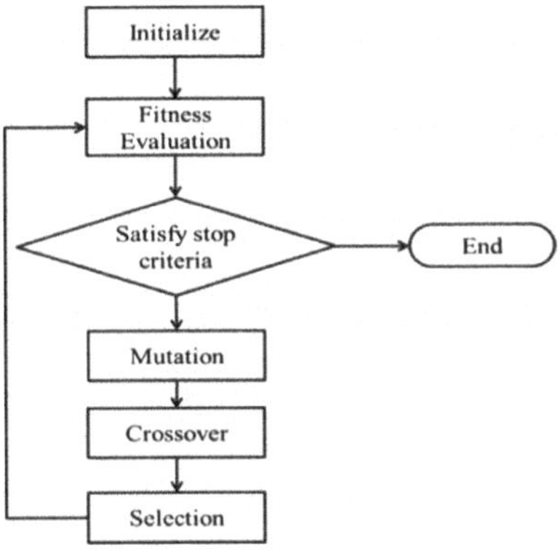

Fig. 1. Flow chart of standard differential evaluation

2.3 Ant Colony Optimization

Ant Colony Optimization (ACO), also known as ant colony system (ACS), is a kind of probabilistic algorithm used to find optimized paths in the graph. It was presented by Marco Dorigo in his doctoral thesis in 1992 [12], inspired by the behavior of ants finding paths during food search. Ant colony optimization is a kind of simulated evolution algorithm. Preliminary research shows that the algorithm has many excellent properties. At the beginning of the algorithm, a single artificial ant seeks the solution disorderly. After a period of evolution, the artificial ants gradually become more and more close to the optimal solution through the effects of pheromones.

Essentially, the ant colony optimization is a parallel algorithm. The searching process of each ant is independent to each other and communicates only through pheromones with information. Therefore, the ACO can be regarded as a decentralized multi-agent system. It starts independent search in multiple points of the problem space, which not only increases the reliability of the algorithm, but also makes the algorithm have a stronger global search ability [13].

Compared with other algorithms, the ant colony system has low requirements on the initial route, that is, the solution result of the ant colony algorithm does not depend on the selection of the initial route, and no manual adjustment is needed during the search process. Secondly, the number of parameters of the ACO is small, the setting is simple, and the ant colony algorithm is easy to apply to the solution of other combinatorial optimization problems.

The ant colony optimization was originally used to solve the TSP problem. After years of development, it has gradually penetrated other fields, such as graph coloring, large-scale integrated circuit design, routing problems in communication networks, load balancing problems, and vehicle scheduling problems. Niu et al. proposed a new feature selection mechanism based on ant colony optimization in attempt to combat the difficulties. The method is then applied to find optimal feature subsets in the fuzzy-rough data reduction process [14]. Alwan and Ku-Mahamud presents an algorithm that integrates Incremental Mixed Variable Ant Colony Optimization (IACOMV) and SVM. Experiment results showed that the proposed algorithm can enhance the classification accuracy with small number of features [15]. Ant colony optimization have been successfully applied in several fields, the most successful of which is the application in combinatorial optimization problems.

In this study, ACO is employed to compare with DE in explore the optimal combination of three regression models with three control parameters including pheromone evaporation ratio (ρ), pheromone concentration (α) and absolute distance parameter (β).

3 Method and Experiment

3.1 Benchmark Datasets

The benchmark dataset of this study are Boston housing, Concrete Compressive Strength and Combined Cycle Power Plant. The Boston data frame has 506 rows and 14 columns. The target field is median value of owner. The Concrete dataset contains 1030 instances and 9 attributes, while the output variable is the concrete compressive strength in MPa. The CCPP dataset contains 9568 data points collected from a Combined Cycle Power Plant over 6 years (2006–2011), when the power plant was set to work with full load. Features consist of hourly average ambient variables Temperature, Ambient Pressure, Relative Humidity and Exhaust Vacuum to predict the net hourly electrical energy output of the plant. The datasets summary is listed in Table 1.

Table 1. Summary of benchmark datasets

Dataset	Amounts of instances	Amounts of attributes
Boston	506	14
Concrete	1030	9
CCPP	9568	4

3.2 Model Training

Model Training of three different regression methods (Glm, Ridge and Robust) in the first step was conducted by RStudio. Figure 2 presented the example code of R to execute data preparation and model training.

3.3 DE Methods

Based on the outcomes of the first stage, the optimal algorithm is applied to exploring the optimal combination of three different regression models. Differential evolution was conducted to search weights combination for the actual value of each data in the datasets. Parameter setting of differential evolution is also considered in this study. There are three parameters should be set while conduct differential evolution including Population, F and CR. According to the former experience, the parameter setting of the experiments is as Table 2 and the pseudocode of differential evolution is as Fig. 2.

Table 2. Differential evolution parameter configuration

Parameter	Iteration	Population	F	CR
Value	20, 40, 60, 80	50, 100, 150, 200	0.2, 0.4, 0.6, 0.8	0.2, 0.4, 0.6, 0.8

Initialize the populating, setting F and CR
Do while *not stop criteria*
  For *each individual j in the population*
    Choose three numbers r_1, r_2, r_3 where $1 \leqq r_1, r_2, r_3 \leqq N$
    Generate random integer i_{rand} in (1, N)
    For *each parameter i*
      $y_{i,g} = x_{i,g} + F(x_{r2,g} - x_{r3,g})$
      if $rand() \leqq CR$ or $j = i_{rand}$ ***then*** $z^j_{i,g} = y^j_{i,g}$
      else $z^j_{i,g} = x^j_{i,g}$
    if *zi,g is better than xi,g* ***then*** *xi,g =zi,g*
Loop

Fig. 2. Pseudo code of differential evolution

3.4 ACO Methods

Ant Colony Optimization is used to compare with differential evolution for finding optimization parameter from the three selected datasets. Table 3 illustrates the parameter setting of ACO according to the former experience [12].

Table 3. Ant colony optimization parameter configuration

Parameter	Iteration	Ant population	α	β	ρ
Value	20, 40, 60, 80	50, 100, 150, 200	.2, .4, .6, .8	.2, .4, .6, .8	.2, .4, .6, .8

4 Results

In the comparison results of Boston dataset (Table 4), DE can reach the performance of general linear model in R-squared and RMSE. However, ACO can also perform as good as DE in all three metrics.

Table 4. Results of different optimization methods for Boston house dataset

Boston	Regression models			DE			ACO		
	GLM	Robust	Rigid	Avg.	Min.	Max.	Avg.	Min.	Max.
R^2	**0.7406**	0.4743	0.6451	0.7392	0.5957	**0.7406**	0.7405	0.7333	**0.7406**
MAE	3.2708	3.1690	3.2527	3.1322	**3.1313**	3.1342	3.1334	**3.1313**	3.1500
RMSE	**4.6791**	5.1108	4.6831	4.6792	**4.6791**	4.6802	4.6796	**4.6791**	4.6981

The convergence status of execution differential evolution through several iterations for CCPP dataset is illustrated in Fig. 3. DE presented excellent outcomes in the comparison of concrete dataset (Table 5) in R^2 and MAE. General linear model and ACO got the same result of DE in RMSE. Table 6 showed the comparison outcomes that DE is superior to ACO in two of three metrics, while GLM performs better than DE in R-squared.

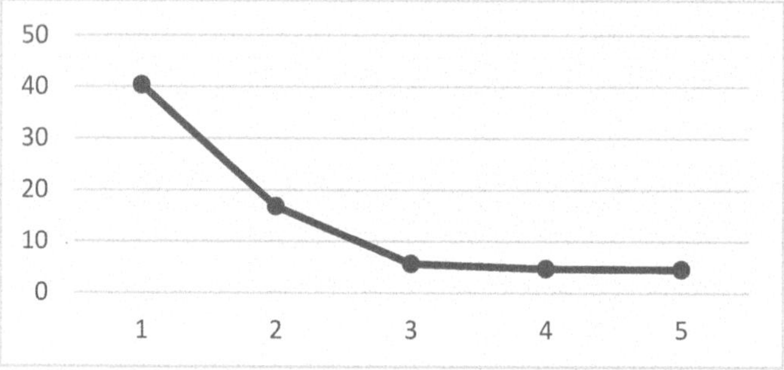

Fig. 3. The convergence of DE for CCPP dataset

Table 5. Results of different optimization methods for Concrete dataset

Concrete	Regression models			DE			ACO		
	GLM	Robust	Rigid	Avg.	Min.	Max.	Avg.	Min.	Max.
R^2	0.6724	0.7478	0.6530	0.7810	0.5633	**0.8372**	0.6148	0.5954	0.6155
MAE	8.2143	8.9418	10.011	8.0873	**8.0851**	8.1254	8.1129	8.0860	8.2804
RMSE	**10.353**	15.644	12.346	10.360	**10.353**	10.535	10.363	**10.353**	10.495

Table 6. Results of different optimization methods for CCPP dataset

CCPP	Regression models			DE			ACO		
	GLM	Robust	Rigid	Avg.	Min.	Max.	Avg.	Min.	Max.
R^2	**0.9606**	0.9512	0.9594	0.9255	0.2974	0.9286	0.9272	0.8094	0.9286
MAE	3.6252	3.6252	3.6179	3.6180	**3.6160**	3.6203	3.6195	3.6162	3.6219
RMSE	**4.5571**	**4.5571**	4.5607	**4.5571**	**4.5571**	**4.5571**	4.619	4.5571	10.228

5 Conclusion

In the big data era, more and more machine learning model will be trained and employed to inference for business and industrial application. Combining multiple models simultaneously will enhance the precision of prediction for decision support. The contribution of this study is proposing a method to explore the optimal combination of machine learning models by metaheuristic algorithms. There are two phases in the method that applied three different regression models (Glm, Ridge and Robust) to three benchmark datasets. At first, the estimated label value of each single data in the three dataset is inferenced by three regression algorithms and the metrics (r-squared, MAE, RMSE) are also calculated for performance evaluated. Secondly, the evolution algorithms are employed to explore the weight combination of three regression models to explore the closest value to the actual value of each data in benchmark datasets.

The comparison results reveal that DE performs better than ACO, especially in the datasets that consisted with more data. Some important directions for further research are comparison of more benchmark datasets and evolutional algorithms or finding the optimal combination of other machine learning algorithms (ex, classification or clustering).

References

1. Nilsson, N.J.: Learning Machines: Foundations of Trainable Pattern-Classifying Systems. McgrawHill, New York (1965)
2. Sagi, O., Rokach, L.: Ensemble learning: a survey. WIREs Data Min. Knowl. Discov. **8**(4), e1249 (2018)
3. McCullagh, P., Nelder, J.: Generalized Linear Models. Chapman and Hall, London (1989)
4. Storn, R., Price, K.: Differential evolution - a simple and efficient heuristic for global optimization over continuous spaces. J. Glob. Optim. **11**(4), 341–359 (1997)
5. Mayera, D.G., Kinghornb, B.P., Archer, A.A.: Differential evolution – an easy and efficient evolutionary algorithm for model optimization. Agric. Syst. **83**(3), 315–328 (2005)
6. Oparaa, K.R., Arabasb, J.: Differential Evolution: a survey of theoretical analyses. Swarm Evol. Comput. **44**, 546–558 (2018). in press
7. Gong, W., Cai, Z.: Differential evolution made faster and more robust. In: Proceedings of 2006 IEEE International Conference on Industrial Technology (2006)
8. Arafa, M., Elsayed, A.S., Fahmy, M.M.: An enhanced differential evolution optimization algorithm. In: Proceedings of 2014 Fourth International Conference on Digital Information and Communication Technology and its Applications (2014)

9. Elsayed, S., Sarker, R.: An adaptive configuration of differential evolution algorithms for big data. In: Proceedings of IEEE Congress on Evolutionary Computation, pp. 695–702 (2015)
10. Piotrowski, A.P.: Review of differential evolution population size. Swarm Evol. Comput. **32**, 1–24 (2017)
11. Das, S., Suganthan, P.N.: Differential evolution: a survey of the state-of-the-art. IEEE Trans. Evol. Comput. **15**(1), 4–31 (2011)
12. Dorigo, M.: Optimization, learning and natural algorithms, Ph.D. Thesis, Politecnico di Milano, Italia (1992)
13. Dorigo, M., Blum, C.: Ant colony optimization theory: a survey. Theor. Comput. Sci. **344** (2–3), 243–278 (2005)
14. Niu, D., Wang, Y., Wu, D.D.: Power load forecasting using support vector machine and ant colony optimization. Exp. Syst. Appl. **37**(3), 2531–2539 (2010)
15. Alwan, H.B., Ku-Mahamud, K.R.: Feature selection and model selection algorithm using incremental mixed variable ant colony optimization for support vector machine classifier. Int. J. Math. Comput. Simul. **7**(5), 406–414 (2013)
16. Kaggle dataset Homepage. https://www.kaggle.com/jhchiuh/boston-housing
17. UCI Machine Learning Repository Homepage. http://archive.ics.uci.edu/ml/datasets.html

Power Transformer Fault Diagnosis Based on Improved Bat Algorithms to Optimize RNN

Chun Yan[1], Meixuan Li[1], and Wei Liu[2(✉)]

[1] College of Mathematics and System Science, Shandong University of Science and Technology, Qingdao 266590, China
[2] College of Computer Science and Engineering, Shandong University of Science and Technology, Qingdao 266590, China
liuwei_doctor@yeah.net

Abstract. In order to improve the accuracy of transformer fault diagnosis, this paper presents a fault diagnosis method based on Bat algorithm to optimize randomized neural network. Because the initial connection weights and hidden neuron biases of the randomized neural network are generated randomly, the network results are easily unstable. Therefore, based on the excellent performance of at algorithm in function optimization, this paper uses the improved Bat algorithm to optimize the weight of randomized neural network. In order to overcome the shortcomings of early convergence and poor results of Bat algorithm, the inertia weight of particle swarm optimization (PSO) is introduced into Bat algorithm, and the velocity update formula is improved. In order to improve the diversity of population and the local search ability of the Bat algorithm, the chaotic local search method is added to balance the global search ability of the Bat algorithm in the early stage and the local search ability in the later stage. In order to control the range of bat's later position, the empirical factor is introduced into the position change formula to accelerate the convergence speed of the algorithm. Finally, the empirical results show that the proposed fault identification model can better diagnose various types of transformer faults.

Keywords: Bat algorithm · Randomized neural network · Transformer fault diagnosis

1 Introduction

Power transformer is an important part of power system. Its operation state is directly related to the stability and reliability of the whole power system. It is important to detect and diagnose transformer faults in time for maintaining the stable operation of power system. The composition, content and gas yield of dissolved gases in power transformer oil are closely related to the type and intensity of faults, which can effectively reflect the operation status of transformers.

Dissolved gas-in-oil analysis (DGA) is a relatively ideal monitoring and analysis method because it can monitor and analyze oil chromatography on-line and sample analysis at any time. At present, domestic and foreign scholars acknowledge that DGA technology is a powerful method to find latent faults in transformers [1]. Over the

Y. Liu et al. (Eds.): ICNC-FSKD 2019, AISC 1074, pp. 531–538, 2020.
https://doi.org/10.1007/978-3-030-32456-8_58

years, scholars at home and abroad have devoted themselves to the development of on-line monitoring devices and systems characterized by the content of dissolved gases in oil. The most basic method is the three-ratio method, but the three-ratio method exposes the problems of incomplete coding and too absolute coding [2]. With the development of research, more and more artificial intelligence methods are introduced into transformer fault diagnosis [3–7]. However, all kinds of artificial intelligence methods have some limitations, such as BP neural network is easy to fall into local minima, and when the parameters are not selected properly (such as learning rate and momentum term), it will cause the shortcomings of slow training process or non-convergence [8]. The Bayesian network requires a large number of sample data [9] etc.

This paper presents a fault diagnosis method for power transformer based on Bat algorithm to optimize Randomized neural network (RNN). RNN randomly generates the connection weights of input layer and hidden layer and the neuron threshold of hidden layer. It does not need to be adjusted in the training process. However, the parameters randomly generated by RNN will lead to poor generalization performance of the network. Bat algorithm (BA) is an intelligent optimization algorithm proposed by Yang (2010) by simulating bat echolocation. The algorithm has the advantages of simple model, easy to use and good optimization performance [10]. Therefore, this paper uses BA to optimize the initial parameters of RNN. As a novel swarm intelligence optimization algorithm, BA algorithm has many advantages, but like other algorithms, Bat algorithm also has some shortcomings. Aiming at the shortcomings of Bat algorithm, this paper puts forward corresponding improvement strategies. In order to improve the global search ability and local search ability of the bat algorithm, the inertia weight and chaotic local search of the PSO algorithm are introduced into the BA algorithm. At the same time, in order to control the range of the bat's later position, the empirical factor is introduced into the position change formula to accelerate the convergence speed of the algorithm. Then, by solving the test function, the excellent performance of the improved BA algorithm (IBA) is proved. Finally, the IBA algorithm is used to optimize RNN to realize transformer fault diagnosis.

2 Algorithm Description

2.1 Traditional Bat Algorithm

Bat algorithm is a global swarm intelligence optimization algorithm which simulates the bat sending and receiving its own ultrasound to prey and avoid obstacles.

The iteration steps of the bat algorithm are described as follows [11]:

(1) Random initialization of bat population $x_i (i = 1, 2, 3, \ldots, N)$. The minimum and maximum wave frequencies are $f_{\min}$ and $f_{\max}$, and the bat's speed is v_i. Bats emit ultrasound with a response of A_i and a pulse frequency of r_i. Set the dimension d of the search space, the maximum number of iterations T, the number of bats N.

(2) According to formulas (1), (2), (3), the velocity v_i and position x_i of individual bats are updated. When the number of iterations is t, the update formulas of bat's position x_i and velocity v_i in d-dimensional search space are as follows:

$$f_i = f_{\min} + (f_{\max} - f_{\min})\beta \tag{1}$$

$$v_i^t = v_i^{t-1} + (x_i^{t-1} - x_*)f_i \tag{2}$$

$$x_i^t = x_i^{t-1} + v_i^t \tag{3}$$

Formula (2) x_* is the global optimal position, $f_i \in [f_{\min}, f_{\max}]$ is the acoustic emission frequency of the ith bat, $\beta \in [0, 1]$ is a random number subject to uniform distribution.

In the process of optimization, when the bat population gradually approaches the global optimal solution, BA algorithm automatically turns to local search. Choose a solution from the current optimal solution set and make each bat walk randomly in its neighborhood to produce a local new solution. The local position update formula is as follows:

$$x_{new} = x_{old} + \varepsilon A^t \tag{4}$$

In formula (4) x_{new} is a local new solution near the optimal solution x_{old}, and $\varepsilon \in [-1, 1]$ is a random vector of d dimension. A^t is the average loudness at the time of iteration number t.

The variation formula of the loudness and the pulse rate is as follows:

$$A_i^{t+1} = \alpha A_i^t \tag{5}$$

$$r_i^{t+1} = r_i^0[1 - \exp(-\gamma t)] \tag{6}$$

In Formula (6), r_i^0 is the initial rate of pulse emission. $r > 0, 0 < \alpha < 1$ is a constant and α is determined according to the actual problem to be solved. Satisfaction under the condition that α remains unchanged: when $t \to \infty$ has $r_i^t \to r_i^0$, $A_i^t \to 0$.

2.2 Improvement of Bat Algorithm

2.2.1 Introducing Adaptive Inertial Weight

Formula (2) shows that each bat moves at a fixed speed, which leads to a single change in speed. Bat algorithm and particle swarm optimization (PSO) belong to swarm intelligence algorithm. The mathematical models of the two algorithms are similar. Therefore, the improved strategy of PSO can be applied to the improvement of bat algorithm [12]. Therefore, the inertia weight of PSO algorithm is introduced into BA algorithm, that is, adding a weight factor $w \in [w_{\min}, w_{\max}]$ to the bat's speed. In order to reflect the influence of individual optimal solution and global optimal solution on velocity, the formula of velocity change is changed into formula (7). In formula (7), x_i is the optimal solution of the i-th individual and $\delta,\ \gamma \in [0,1]$ is a random number with uniform distribution.

$$v_i^{t+1} = w_i v_i^t + f_i \delta (x_i^t - x_*) + (1 - f_i)\gamma (x_i - x_*) \tag{7}$$

When improving the PSO algorithm, Shi and Ebehart (1999) think that larger inertia weight factor is beneficial to jump out of local minimum and global search. When the inertia weight factor is small, it is easier for the particle to strengthen the local search, thus speeding up the convergence of the algorithm [13]. Therefore, in order to improve the bat algorithm, this paper proposes an inertia factor which can be reduced adaptively according to the number of iterations.

$$w_i = w_{\max} - (w_{\max} - w_{\min}) \times \frac{t}{T\_\max} \tag{8}$$

2.2.2 Introducing Experience Factor

In order to control the range of bat's position at the later stage of searching, the empirical factor is introduced into the formula of bat's position change, and the formula is changed to formula (9) [14].

$$x_i^t = x_i^{t-1} + (1 - \frac{t}{T\_\max})^b \times V_i^t \tag{9}$$

$(1 - \frac{t}{T\_\max})^b$ is an empirical factor, and $b(1 < b < n)$ is a constant, which can be used for different situations. In this paper, b is 3 and n is the number of bats.

2.2.3 Introducing Chaotic Local Search

Chaotic sequence optimization is widely concerned in engineering practice because the individual's behavior trajectory in the population is seldom repeated, which reduces the possibility of falling into local optimum and makes it easier to get the optimal solution. The chaotic maps used in this paper are cubic maps:

$$\varepsilon_{n+1} = 4\varepsilon_n^3 - 3\varepsilon_n\ ,\quad -1 < \varepsilon_n < 1 \tag{10}$$

In the local search formula of the original bat algorithm (4), ε is randomly generated between $[-1,1]$. In the improved bat algorithm, for local search, ε_0 is randomly generated between $[-1,1]$ at the beginning, and the value of ε_n is generated by cubic mapping on the basis of the previous generation, which indirectly enhances the diversity of population and speeds up local convergence [13].

2.3 Experimental Simulation and Result Analysis

In order to verify the performance of improved bat algorithm (IBA), this paper compares the performance of IBA algorithm with traditional genetic algorithm (GA), particle swarm optimization (PSO) and original bat algorithm (BA). Two typical benchmark function optimization problems are selected for verification.

The first test function is:

$$f_1 = (1 + (x + y + 1)^2 (19 - 14x + 3x^2 - 14y + 16xy + 3y^2))(30 + (2x - 3y)^2 (18 - 32x + 12x^2 + 48y - 36xy + 27y^2)) \quad -2 \leq x, \ y \leq 2 \tag{11}$$

The second test function is:

$$f_2 = \sum_{i=1}^{d} [x_i^2 - 10\cos(2\pi x_i) + 10], \quad -5.12 \leq x_i \leq 5.12, \ d = 2 \tag{12}$$

In order to make the algorithm comparable, this paper tries to make the parameter settings of each algorithm consistent. Specific parameters are set as follows: The population size of each algorithm is 100 and the maximum number of iterations is 300. Each test function runs 30 times independently. The experimental results obtained by optimizing the test are shown in Table 1.

Table 1. Comparison of experimental results of each algorithm

	Algorithm	Optimal value	Standard deviation	Average running time	Average convergence algebra
f_1	GA	3.000000000000064	1.148824112450234e-10	2.607779 s	286.07
	PSO	3.000001105606386	6.244076914686642e-05	1.774666 s	151.57
	BA	3.000012641473877	0.003262191745047	0.916938 s	208.13
	IBA	2.999999999999921	7.141704126996961e-16	1.085536 s	175.37
f_2	GA	0	3.350916691845618e-12	2.549448 s	278.3
	PSO	1.127066679984523e-06	7.775640856012103e-06	1.822066 s	207.33
	BA	2.675601427526431e-06	3.330326611455099e-05	0.922120 s	238.03
	IBA	0	0	1.085354 s	93.80

From the experimental comparison Table 1, we can see that the optimal value, standard deviation and average convergence algebra of the proposed algorithm IBA are much smaller than those of GA, BA and PSO. As far as function $f_1(x)$ is concerned, the optimal value of the function solved by IBA algorithm is 2.99999999921, which is much smaller than BA and PSO. Although the optimization results of GA algorithm are similar to those of IBA algorithm, the convergence algebra of GA is obviously higher than that of IBA algorithm, and the GA algorithm takes a long time. Which shows that the stability, accuracy and convergence speed of IBA algorithm are obviously better than the other four algorithms.

2.4 Randomized Neural Network

The structure of PNN is a single hidden layer feedforward neural network. The neural network consists of input layer, hidden layer and output layer [15–17].

Give a certain number of training samples (x_i, y_i), $x_i \in R^d$, $y_i \in R$. According to the given excitation function f, the predictive value y_i can be obtained from the feedforward layer.

$$y_i = \sum_{i=1}^{L} \beta_i f(w_i^T x_j + b_i), \quad j \in [1, N] \tag{13}$$

In formula (13), parameter (w_i, b_i) is randomly selected, and parameter β_i is the output weight, which can be calculated by the following steps:

First, suppose w^*, B^*, β^*. and then make the error range between the input y_j value and the real value y_j' in RNN infinitely close to 0, that is $y_i' = \sum_{i=1}^{L} \beta_i' f(w_i^{*T} x_j + b_i^*)$ $= y_j, j \in [1, N]$ where $Y = (y_1, y_2, \ldots, y_N)^T$, $\beta = (\beta_1, \beta_2, \ldots, \beta_z)$, Then the above relationship can be simplified as follows:

$$H\beta = Y \tag{14}$$

After determining the values of W and B, the solution is obtained by least square method:

$$\beta = H^+ Y \tag{15}$$

H^+ is the generalized inverse matrix of H.

3 Empirical Analysis

3.1 Data Selection

Because of the difference of transformer internal faults, the gas produced by each fault is not completely the same. Since the auto-volume and integral components of H_2, CH_4, C_2H_6, C_2H_4 and C_2H_2 are closely related to the fault types of transformers, all the five gases are taken as the characteristic parameters of transformer fault diagnosis in this paper [18]. Inspired by IEC three-ratio method, this paper also takes C_2H_2/C_2H_4, CH_4/H_2, C_2H_4/C_2H_6 as the characteristic parameters of transformer fault diagnosis. This paper selected 100 representative samples from the historical faults of several oil-immersed power transformers in a 220 V substation for empirical analysis. In Sect. 2.3, by solving the test function, the excellent performance of IBA algorithm is proved. In this paper, the IBA algorithm is used to optimize RNN(IBA-RNN) for transformer fault diagnosis.

Eight variables were selected as input vectors of the model, and then 20 samples were randomly selected as test data, and the remaining 80 samples were used as training data. In order to realize the accurate diagnosis of transformer faults by IBA-RNN, according to the proportion of the total sample of each fault type, the corresponding test samples are randomly selected from each fault type.

3.2 Comparative Analysis of Prediction Examples

In order to verify the diagnostic effect of the improved model, the improved model and RNN model were tested by 10 fold cross validation. Figure 1 is the diagnostic accuracy results of the testing set of the two models after 10 fold cross-checking. From Fig. 1, it can be seen that the diagnostic accuracy of the improved diagnostic model in this paper is significantly higher than that of RNN model almost every time.

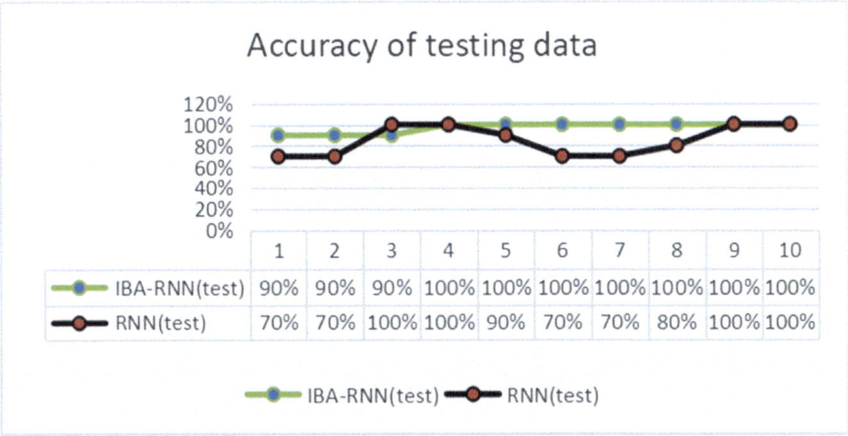

Fig. 1. 10 fold cross test results of testing data

4 Conclusions

Power transformer plays an important role in power transmission and distribution. The performance of power transformer directly affects the operation of the whole power system. Therefore, it is very important to discover the faults of the transformer in advance. Whether early fault of the transformer can be eliminated as soon as possible is the key to ensure the stability of the power supply for users.

In this paper, the inertia weight and chaotic local search of PSO algorithm are introduced into the bat algorithm to improve the global search ability in the early stage and the local search ability in the later stage. At the same time, in order to control the range of the bat's later position, the empirical factor is introduced into the position change formula to accelerate the convergence speed of the algorithm. Then, by solving the test function, the excellent performance of IBA algorithm is proved. Finally, the improved algorithm is used to optimize RNN for transformer fault diagnosis. Through the final diagnosis results, it can be seen that the diagnosis model proposed in this paper can well diagnose the transformer fault type. This provides an effective solution for the fault diagnosis of voltage transformer in the future.

References

1. Liang, Y., Li, K., Zhao, J., et al.: Study on dynamic adjustment strategy of transformer oil chromatography on-line monitoring cycle. Proc. Chin. J. Electr. Eng. **34**(09), 1446–1453 (2014)
2. Zheng, R., Zhao, J., Zhao, T., et al.: Fault diagnosis of power transformer based on genetic support vector machine and grey artificial immune algorithm. In: CSEE, vol. 31, no. 07, pp. 56–63 (2011)
3. Seifeddine, S., Khmais, B., Abdelkader, C.: Power transformer fault diagnosis based on dissolved gas analysis by artificial neural network. In: First International Conference on Renewable Energies & Vehicular Technology, no. 03, pp. 230–236. IEEE (2012)
4. Siddique, M.A.A., Mehfuz, S.: Artificial neural neworks based incipient fault diagnosis for power transformers. In: India Conference, pp. 1–6. IEEE (2016)
5. Long, Q., Guo, S., Li, Q., et al.: Research of converter transformer fault diagnosis based on improved PSO-BP algorithm, vol. 231, p. 012015 (2017)
6. Setiawan, N.A., Sarjiya, Adhiarga, Z.: Power transformer incipient faults diagnosis using dissolved gas analysis and rough set. In: International Conference on Condition Monitoring and Diagnosis, Bali, Indonesia, pp. 950–953. IEEE (2012)
7. Aizpurua, J.I., Catterson, V.M., Stewart, B.G., et al.: Power transformer dissolved gas analysis through Bayesian networks and hypothesis testing. IEEE Trans. Dielectr. Electr. Insul. **25**(2), 494–506 (2018)
8. Wang, L., Zhang, L., Pu, J., et al.: Application of improved BP neural network in object recognitio. Electron. Opt. Control **19**(4), 68–71 (2012)
9. Wang, K., Li, J., Zhang, S., et al.: New characteristic parameters of dissolved gas in oil for transformer fault diagnosis. Proc. Chin. J. Electr. Eng. **36**(23), 6570–6578 + 6625 (2016)
10. Yang, X.: A new metaheuristic bat-inspired algorithm. In: Computer Knowledge & Technology, pp. 65–74 (2010)
11. Song, Y.: Research on logistics center location problem based on improved bat algorithm, Henan University (2017)
12. Zhang, X., Hu, Z., Tang, T.: Application of improved bat algorithm in Fuzzy Analytic hierarchy process. Appl. Comput. Syst. **27**(03), 143–148 (2018)
13. He, Z.: Improvement and application of bat algorithm, Guangdong University of Technology (2016)
14. Wang, W.: Research on the improvement of cuckoo search algorithm and bat optimization algorithm, Guangxi University for Nationalities (2014)
15. Schmidt, W.F., Kraaijveld, M.A., Duin, R.P.W.: Feedforward neural networks with random weights. In: Proceedings of 11th IAPR International Conference on Pattern Recognition, Conference B: Pattern Recognition Methodology and Systems, vol. 2, pp. 1–4 (1992)
16. Pao, Y.H., Park, G.H., Sobajic, D.J.: Learning and generalization characteristics of random vector functional-link net. Neurocomputing **6**, 163–180 (1994)
17. Suganthan, P.N.: On non-iterative learning algorithms with closed-form solution. Appl. Soft Comput. **70**, 1078–1082 (2018)
18. Cheng, J.T., Duan, Z.M., Xiong, Y., et al.: Research on QPSO-BP model transformer fault diagnosis method based on DGA. High Volt. Appl. **52**(02), 57–61 (2016)

Architecture of Real-Time and Dynamic Audit for Network Behavior Security

Yan Zhang[1,2] and Caiming Liu[1,2(✉)]

[1] School of Computer Science, Leshan Normal University, Leshan 614000, Sichuan, China
caiming_liu@163.com
[2] Key Lab of Internet Natural Language Processing of Sichuan Provincial Education Department, Leshan Normal University, Leshan 614000, Sichuan, China

Abstract. Dynamics and complexity of the Internet security environment make that real-time and dynamic auditing of network behavior security is more difficult. For the resolution of difficulties on dynamics and complexity in the network behavior audit, an architecture of real-time and dynamic audit for network behavior security under the Internet environment is created. The proposed architecture consists of pretreating network behavior data, synchronizing network data, dynamically recognizing network behavior and comprehensively auditing network behavior security. Pretreating network behavior data solves the problems of the integrity of network behavior data to be treated and the scientific modeling of network behaviors. Synchronizing network data solves the problem of real-time treating of network behavior data. Dynamically recognizing network behavior uses and simulates immune mechanisms to realize the dynamic auditing for network behavior security. Comprehensively auditing network behavior security solves the problem of direct service of audit results of network behavior security for the network security management. The proposed architecture is expected to provide a new reference for the establishment of active network security management model.

Keywords: Network behavior security · Behavior recognition · Behavior audit · Real-time audit · Dynamic audit

1 Introduction

Network Behavior Security Audit (NBSA) tracks and monitors the whole process of network activities and network operation. It analyses the dynamic rules of network activities' changes, checks, examines and verifies the security of network behaviors, and provides an important supplement and improvement for network security management. NBSA is one of the important branches of Network Security Audit (NSA) [1]. It finds out the rules of network activities through analyzing the relationship among network access records, network user activities and system network activities, and conducts in-depth audit of network behavior security hidden behind network security events.

Y. Liu et al. (Eds.): ICNC-FSKD 2019, AISC 1074, pp. 539–546, 2020.
https://doi.org/10.1007/978-3-030-32456-8_59

NSA began with Computer Security Audit (CSA). Anderson put forward the concept of computer security audit in 1980, and put forward a preliminary conception of information system security audit [2]. The data source of CSA is mainly from the log information of local computer system, which plays a vital role in maintaining the security of computer information system.

Current research on NSA mainly focuses on system log audit. For example, Wei et al. [3] adopt log audit method based on rule library. Through matching log information with rule library and mining the relationship between security events, the researchers use log audit results to evaluate the theoretical security threats of nodes, and then calculate the security situation of nodes through performance correction algorithm. Haggerty et al. [4] proposed a visual analysis method for system log file to detect past network security events. The above research results of network security audit lack real-time performance. Its drawback is that they cannot be tracked and recorded in real time, and cannot reflect the network security situation in time.

Governments of many countries are gradually forcing Internet operators to strengthen the audit of network behavior security in order to lay the last line of defense for network security [5, 6].

Network Behavior [7] is manifested as the dynamic change rule of specific network activities and network operations [8]. The occurrence of network security events is accompanied by a large number of illegal or abnormal network behaviors. In order to establish an effective and active network security protection system, it is urgent to audit a large number of network behaviors in the Internet environment.

This paper constructs an architecture of real-time and dynamic audit for network behavior Security. The rest of the paper is organized as follows. Section 2 introduces the details of the proposed architecture. Section 3 concludes this paper.

2 Proposed Architecture

2.1 General Description

The proposed architecture is based on three key technologies of realizing real-time and dynamic network behavior security audit, which includes network data synchronization technology of real-time audit, network environment adaptation technology of dynamic audit and comprehensive audit technology of network behavior pattern. Pretreating network behavior data and synchronizing network data of real-time network behavior audit with multi-queue achieve the goal of real-time audit in network behavior security audit through high-speed and parallel treating of network behavior data. Dynamically recognizing network behavior based on immune principles refers to the good immune mechanisms to realize the dynamic audit goal of network behavior security audit. Comprehensively auditing network behavior security uses the audit results to provide data support for network security management, such as classified alarm, vulnerability and hidden danger analysis, etc.

2.2 Pretreating Network Behavior Data

Pretreating Network Behavior Data provides real-time and comprehensive data base for real-time and dynamic auditing of network behavior security through online capturing and treating of network behavior data sources. It is one of the key technologies for the proposed architecture to achieve real-time goals.

Parallel and Distributed High-Speed Capture of Network Data Traffic. According to the characteristics of network traffic transmission in the Internet, the proposed architecture adopts a high-speed data **capture** technology which combines parallel and distributed data capture to ensure the timeliness and integrity of network data capture. This is also the primary technical problem solved by the proposed architecture, as shown in Fig. 1.

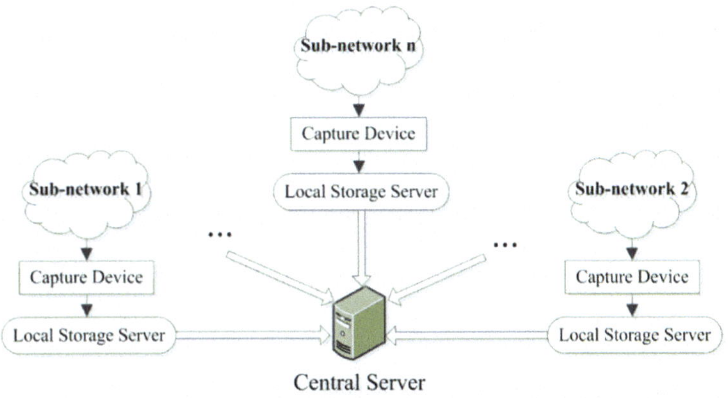

Fig. 1. Parallel and distributed high-speed capture of network data traffics.

In the aspect of parallelism, special data capture devices are deployed at each key network node. They access network devices. Each device runs independently, concurrently captures the network data traffics that are transmitted in the network devices, and stores them into the local network data storage server. Each independent network data capture device transmits the gathered network behavior data in the local storage server to the central server at a high speed.

Comprehensive Network Behavior Data Fusion. Comprehensive network behavior data fusion technology needs to solve the following key technologies: creation of network activity and network behavior data pattern based on network data traffics, creation of network behavior data pattern corresponding to network user's online behaviors, creation of network behavior data pattern corresponding to network user's system logs, creation of fusion criteria and process for security audit data patterns.

Creation of Network Activity and Network Behavior Data Pattern. A complete process of sending and receiving network data is called a network activity. A network activity corresponds to one or more network data packets. Let the network activity data set be

N_A. Through analyzing the network activity data set, the data patterns of network behavior are created. Network behavior can be completed by one or more network activities. They can be created by collecting similar network operations in the network activity data set. Let the network behavior data set be B_N.

Creation of Network Behavior Data Pattern Corresponding to Network User's Online Behaviors. User's online behaviors cover all kinds of activities and rules of network users accessing the Internet through local computers. The network behaviors formed by user's online operations can be created by surveying user's online operations. The data pattern expressing this kind of network *behavior* comes from the characteristic data of the user's online operation. Let this kind of network behavior data set be B_U.

Creation of Network Behavior Data Pattern Corresponding to Network User's System Logs. All kinds of application network access are stored in the system log. They are filtered. Log entries closely related to network activities are extracted to express the corresponding network behavior data patterns of log information. Let the data set of system log be *LOG*, and the data set of network behavior generated by system log be B_L.

Creation of Fusion Criteria and Process for Security Audit Data Patterns. Several security audit data patterns have their own data characteristics, which need to be fused to store together. The creation of data fusion criteria and processes for security audit data patterns can standardize the *comprehensive* analysis of several network behavior data sources, and realize the fusion storage of different security audit data patterns.

Creation of Network Behavior Patterns. The sources of network behavior data are diversified and need to be treated step by step in order to create a reasonable network behavior effectively. The specific steps are as the following: data analysis, data pre-treating, data feature extraction, definition method of network behavior, representation method of network behavior, network behavior creation.

2.3 Synchronizing Network Data

The proposed architecture uses the multi-queue technology to synchronize network data, so as to achieve multi-queue network data synchronization of real-time network behavior audit, solve the conflict problem of network data in different treating stages, and realize real-time synchronization of network behavior security audit.

The Multi-queue Creation Mechanism of Network Data. According to the needs of treating network data, the implementation mechanism of transactions of network data queue's storage, reading, updating, deleting, querying, etc., is created, so that each network data queue not only can run in parallel, but also can work together to realize pretreating the network behavior pattern data, which are captured in real time. The network data queues can be used as data source for **dynamic** recognition of network behavior in real time, so as to achieve the purpose of real-time and dynamic audit for network behaviors.

The Data Exchange Interface Mechanism Between Network Data Queues. In order to standardize the data conversion between queues, the data exchange interface between network data queues is created. A complete data exchange interface mechanism need to be established. There are two main **synchronization** problems in the proposed architecture: data synchronization between Pretreating Network Behavior Data and Dynamically Recognizing Network Behavior, and data synchronization interface between distributed network behavior recognition agents.

The Conflict Resolution Mechanism of Network Data Queues. The priority definition mechanism of network data queue needs to be constructed, and the priority of atomic operations such as writing, reading, updating, deleting and querying of network data needs to be defined in detail. When data conversion between network data queues has data treating conflict, a priority mechanism conforming to network behavior audit needs to be established.

2.4 Dynamically Recognizing Network Behavior

The proposed architecture adopts the excellent mechanisms of artificial immune system (AIS) [9] to construct the dynamic recognition technology of network behavior based on immune principles, so as to solve the dynamic audit problem in network behavior security audit.

Creating Libraries Related to Network Behaviors. The key technologies of the libraries related to network behaviors include: Network Behavior **Security** Audit Library, Network Behavior Pattern Library and Data Transmission Interfaces of Network Behavior Security Audit Data Pattern.

The database of network behavior security audit library stores the data patterns of network behavior security audit, which is set as *SAL*.

Network Behavior Pattern Library can identify network behavior data patterns of network behavior types. The initial data of this library comes from classical illegal network behavior. In the application of the proposed architecture to practical process, classical network behavior patterns need to be collected. Let Network Behavior Pattern Library be *BEL*.

Data Transmission Interfaces of Network Behavior Security Audit Data Pattern standardizes the format of the interface data mode, which is convenient for data transmission. Designing these interfaces can provide a unified specification for the information flow of security audit data patterns, so as to facilitate the treating of network behavior data.

Simulating Immune Environment. In order to make use of the dynamic [10] and adaptive [11] characteristics of immune principles, it is necessary to simulate the analysis environment of network behavior security audit data pattern to artificial immune environment. The key point of immune environment simulation technology is to determine the mapping relationship between network behavior security audit pattern data and its analysis method to artificial immune mechanisms. The key points of immune environment simulation in the proposed architecture are: network behavior data – antigens, normal network behavior – self, abnormal or illegal network behavior - non-self, network behavior detector - artificial immune antibody.

Simulating Immune Elements. Simulating Immune Elements mainly constructs the data conversion criterion and method from network behavior security audit data patterns to artificial immune elements. Its technical points include: antigen simulation, self simulation, non-self simulation, immature immune element simulation, mature immune element simulation, memory immune element simulation, detector gene library simulation.

Simulating Immune Mechanisms. The corresponding relationship of immune mechanism simulation is shown in Table 1. The main technical **points** are antigen matching method, self-tolerance, dynamic evolution of detectors and gene recombination.

Table 1. The corresponding relationship of immune mechanism simulation.

Artificial immune mechanisms	Network behavior recognition mechanisms
Antigen matching	Detector matches network behavior
Self-tolerance	Exclude detectors which match normal network behaviors
Dynamic evolution of detectors	Update illegal or abnormal network behavior patterns
Gene recombination	Use illegal or abnormal network behavior patterns to generate new network behavior patterns

Recognizing Network Behaviors. A network behavior recognition agent and its distributed operation mechanism and collaboration **mechanism** are constructed to recognize the network behavior in a distributed way. Its technical points include: creation of network behavior recognition agent, creation of cooperative operation mechanism between recognition agents and distributed network behavior recognition.

An immune-based network behavior recognition agent needs to be created. A single network behavior recognition agent runs independently to achieve the purpose of parallel analysis.

A cooperative mechanism between recognition agents needs to be established. It simulates the distributed treating characteristics of AIS. It is necessary to create a vaccine library, shared memory detector library and other key data to solve the key problems of cooperative mechanisms such as vaccine formation, vaccine packaging, vaccine transmitting, vaccine integrity testing, local tolerance, and etc.

2.5 Comprehensively Auditing Network Behavior Security

The proposed architecture uses the results of real-time and dynamic recognition of network behavior to conduct real-time and comprehensive audit.

Illegal Network Behavior Recording, Warning and Forensics. Illegal **network** behaviors are network behaviors that match the records in *BEL*. Let the data set of illegal network behaviors be *IAg*, as shown in Eq. (1).

$$IAg = \{iag | iag = ag \wedge f_{match}(ag, r) = true \wedge ag \in Ag \wedge r \in D_R\} \quad (1)$$

Where, Ag is the antigen data set, D_R is the memory detector data set, $f_{match}()$ is a match function.

The audit of illegal network behaviors adopts recording, alarming and forensic methods. First, the process of illegal network behavior is recorded in a special database. Secondly, the process of illegal network behavior is sent to the administrator in the form of summary information for the administrator to grasp the network security situation. Finally, the whole process of illegal network behavior is transmitted to the forensic server through a secure way.

Abnormal Network Behavior Recording and Warning. Abnormal network behavior does not match the records in *BEL*, which is recognized by **mature** detectors to be activated. Let the abnormal network behavior data set be *AAg*, as shown in Eq. (2).

$$AAg = \{aag | aag = ag \wedge f_{match}(ag, m) = true \wedge ag \in Ag \wedge m \in D_M\} \quad (2)$$

Where, D_M is the mature detector data set.

Abnormal network behaviors are potential illegal network behaviors, which are recognized. Therefore, the proposed architecture only records and alarms the abnormal network behaviors. Whether they are illegal or not, they need to be confirmed by the network administrator.

Analysis of Network Security Vulnerabilities and Hidden Dangers. Illegal or abnormal network behaviors often take advantage of network security vulnerabilities and causes hidden dangers to network security. Once **malicious** network behaviors cause network security events, it may correspond to network security vulnerabilities. The proposed architecture further identifies the consequences of network behavior, and establishes a database of network security vulnerabilities. According to the results of network behavior identification, the vulnerabilities of network security are scientifically mapped. According to the risk of network behavior, the hidden dangers of network security are analyzed.

3 Conclusion

According to the operation rules of network behavior, the proposed architecture treats real-time network activity data affecting the security situation of network behavior, adapts to the changes of network security, dynamically analyses the data patterns of network behavior, and realizes the purpose of real-time and dynamic audit of network behavior security. The proposed architecture lays important theoretical foundations for the construction of active management model for network security. Moreover, it has important application prospects and superior practical values to draw up reasonable protection strategies for network security.

Acknowledgement. This work is supported by Sichuan Science and Technology Program (No. 2018JY0523), the Scientific Research Fund of Sichuan Provincial Education Department (No. 18ZA0233), the Scientific Research Project of Leshan Normal University (No. ZZ201825).

References

1. Zhang, H.: Research of Network Security Auditing System. Thesis of Jilin University (2006)
2. Liu, Z., Wang, X., Li, J., Zhang, S.: Research on an integrated security audit system model. Police Technol. **2014**(B06), 32–35 (2014)
3. Wei, Y., Lian, Y.: A network security situational awareness model based on log audit and performance correction. Chin. J. Comput. **32**(4), 763–772 (2009)
4. Haggerty, J., Hughes-Roberts, T.: Visualization of system log files for post-incident analysis and response. In: Human Aspects of Information Security, Privacy, and Trust. HAS 2014. LNCS, vol. 8533, pp. 23-32 (2014)
5. Standing Committee of the National People's Congress of China. People's Republic of China Network Security Law. Issued on November 7 (2016)
6. Yang, Y.: The Enlightenment of Network Security Audit of American Audit Office. Audit Monthly (2018)
7. Zhang, H., Fang, B., Hu, M., Jiang, Y., Zhan, C., Zhang, S.: A survey on internet measurement and analysis. J. Softw. **14**(1), 110–116 (2003)
8. Wang, J., Wu, J.: Internet Network Ethology. http://wenku.baidu.com/link?url=6LXb8IWY7WaAs1vDc7jGAz0XRjqpjuV2diRnwz0MZEW45vcLHcbtCm6bJ62IDZtxJsloZRm0RkH8he_FkerBgEYAAdL0HmVyHbcQCwNlx1O
9. De Castro, L., Timmis, J.: Artificial Immune Systems: A Novel Computational Intelligence approach. Springer, London (2002)
10. Li, T.: An immune based dynamic intrusion detection model. Sci. Bull. **50**(22), 2650–2657 (2005)
11. Farmer, J., Packard, N., Perelson, A.: The immune system, adaptation, and machine learning. Physica **22D**, 187–204 (1986)

A Comparison of Extreme Gradient Boosting and Convolutional Neural Network-Long Short-Term Memory for Service Demand Forecasting

Manop Phankokkruad[1(✉)] and Sirirat Wacharawichanant[2]

[1] Faculty of Information Technology, King Mongkut's Institute of Technology Ladkrabang, Bangkok 10520, Thailand
manop@it.kmitl.ac.th

[2] Department of Chemical Engineering, Faculty of Engineering and Industrial Technology, Silpakorn University, Nakhon Pathom 73000, Thailand
wacharawichanan_s@su.ac.th

Abstract. The accurate demand forecasting gives powerful insights on amount of the available resources should be allocated for the operation and can improve the effectiveness of the management to assure the completeness of the services. This work studied to forecast the number of patients, that applied for service demand forecasting. We proposed the comparison between XGBoost model and CNN-LSTM model for service demand forecasting. We created the XGBoost model and configured the optimal parameters by using grid search optimization techniques. In the same way, the CNN-LSTM model was constructed and adjusted the layers in the network structure. Later, the model has configured the optimal parameters obtained from the grid search optimization techniques. Both models were applied to two different datasets for forecasting the number of patients in the future. The results indicated that two models had skillful, and made the reliable forecasting in two datasets. This work evaluated and measured the model performance by calculating RMSE, and sMAPE, which was acceptable in all case study. The CNN-LSTM model gave better efficiency forecasting than XGBoost model. In contrast, XGBoost model operated faster, took a lower performing time, and lower CPU consumption than the CNN-LSTM model.

1 Introduction

The medical services are an essential requirement for being healthy thus the most hospital must ensure the availability of the medicine needed at all times. A suitable allocation of medical resources deal with the quality of services, and it becomes a long-term challenges because these resources relates with an increasing of the operating cost and patient care quality declines [1]. Therefore, the accurate demand forecasting gives powerful insights on amount of the available resources should be allocated for the operation, and can improve the effectiveness

Y. Liu et al. (Eds.): ICNC-FSKD 2019, AISC 1074, pp. 547–556, 2020.
https://doi.org/10.1007/978-3-030-32456-8_60

of hospital management to assure the completeness of the services [2]. In general, most hospital make an estimation for a future serivce demand by forecast a number of patients to provide the medical services. Therefore, the accurately forecast a number of patients is an valuable knowledge for managing, and providing better and more efficient patient care while minimizing resource use.

A time series forecasting is used in a variety of applications such as sales forecasting, weather forecasting, inventory studies, etc. An autoregressive integrated moving average (ARIMA), is one of the most popular classical time series forecasting. However, ARIMA makes particularly serious problem, it has a difficult to create the model in the cast of the non-stationary dataset, and deal with seasonal variation data. Futhermore, there are several deep learning techniques to effectively forecast the time series data such as convolutional neural network (CNN), long short-term memory (LSTM), boosting decision trees, and more. Convolutional Neural Network (CNN) is a category of neural networks that has very effective and significant performances for solving many hard problems. Many kinds of literature reported the CNN model can be very effective at automatically extracting and learning features from sequence data. Long Short-Term Memory (LSTM) is a variant of artificial recurrent neural network (RNN) that has ability to learn and process the sequence property from a given data. LSTM is applied to time series forecasting such as a software reliability prediction, load forecasting, and rail transit passenger flow, etc. An extreme gradient boosting (XGBoost) [3] is the most effective techniques for building predictive models, that is used as a numerical optimization problem. XGBoost is a based formulation of boosting decision trees, that is designed for computing speed and performance [4]. A time series forecasting is a challenge task due to the unprecedented changes in the customer demand, scheduling decisions, and economic trends.

For this reason, the objectives of this work are forecasting a number of patients, and compare the performance of two models. The proposed method will combine the CNN model with LSTM backend. The CNN model is used to interpret the input data, then it provides this data to LSTM model for interpreting, and making the forecasting. This hybrid model is called a CNN-LSTM model. Concurrently, the XGBoost model will be created to make a forecast on the same dataset. Both XGBoost and CNN-LSTM model are applied to forecast the number of outpatients who visits to a hospital. This work choose the XGBoost model and CNN-LSTM model due to it based on the two different forecasting techniques, XGBoost is based the decision tree, whereas CNN-LSTM is a class of artificial neural networks. The make use of the two models will help the hospital can manage the suitable allocation of medical resources, quality of services, resource planning and scheduling decisions more significantly.

This paper is organized as follows. Section 2 describes the background of the time series forecasting, the theoretical basis of extreme gradient boosting, convolutional neural network, long short-term memory, and the details of proposed techniques. Section 3 describes the experiments, data preparation, and the model configurations. Section 4 reveals the results, evaluations, and discussion. Finally, Sect. 5 presents the conclusion.

2 Proposed Method

In this section, the state of the problem, the background of the time series forecasting, the theoretical basis of extreme gradient boosting, convolutional neural networks - long short-term memory, and the details of proposed techniques will be described.

2.1 Extreme Gradient Boosting

The extreme gradient boosting (XGBoost) [5] is an algorithm based on the gradient boosting algorithm [6], that merge a set of weak learner to a strong one in the iterative model in order to improve the better prediction accuracy [3].

$$\tilde{L}^{(t)}(q) = -\frac{1}{2}\sum_{j=1}^{T}\frac{(\sum_{i\in I_j} g_i)^2}{\sum_{i\in I_j} h_i + \lambda} + \gamma T \tag{1}$$

Equation (2) is the scoring function that is used to determine the quality of tree structure q.

$$Gain = \frac{1}{2}\left[\frac{(\sum_{i\in I_L} g_i)^2}{\sum_{i\in I_L} h_i + \lambda} + \frac{(\sum_{i\in I_R} g_i)^2}{\sum_{i\in I_R} h_i + \lambda} - \frac{(\sum_{i\in I} g_i)^2}{\sum_{i\in I} h_i + \lambda}\right] - \gamma \tag{2}$$

For creating the base learner, a gradient boosting based on the only first derivative γ_i, whereas XGBoost stand on the first derivative g_i and second derivative h_i. XGBoost is an optimization algorithm that perform on the right cost function, and can make a prediction model though the weak prediction models.

2.2 CNN-LSTM Model

A Convolutional Neural Network-Long Short-Term Memory (CNN-LSTM) is a combination model that Convolutional Neural Network (CNN) model is applied in a hybrid with Long Short-Term Memory (LSTM) backend. The CNN-LSTM model has complex network structure, that consists of two parts; CNN and LSTM. The CNN model is used to interpret and extract the features from the input data before this data is provided to LSTM model for interpreting and making a forecast. The summarizes of CNN-LSTM architecture includes input layer, convolutional layers, pooling layers, flatten lay LSTM layer, fully-connected layer, and output layer.

Convolutional Neural Network (CNN) is a class of neural networks that consists of the different kinds of layers includes input layer, convolutional layers, pooling layers, fully-connected layer, and output layer. The input layer feeds the data into the network. The convolutional layer comprises the set of fixed size filters, that it operates the convolution on the data. This operation is called the feature map, which has ability to extract some patterns. This operation can be formulated as shown in Eq. (3).

$$x_j^l = f\left(\sum_i w_{ij}^l * x_i^{l-1} + b_j^l\right) \tag{3}$$

where * depicts the convolution operation. x_i^{l-1} denotes the i_{th} feature map in the $l-1_{th}$ layer, x_l^j denotes the jth feature map in the lth layer. The trainable parameters b_j^l, and w_{ij}^l are the bias of the x_j^l feature map in the jth layer, and the weight connecting x_i^{l-1} to x_j^l, respectively.

This work uses scaled exponential linear units (SELU) as the activation function instead of the rectified linear unit (ReLU) because on the preliminary study, we found that SELU has self-normalising properties [7], better convergence, and can effectively avoid the problem of gradient vanishing [8].

$$SELU(x) = \lambda \begin{cases} x & \text{if } x > 0 \\ \alpha e^x - \alpha & \text{otherwise} \end{cases} \tag{4}$$

where λ is about 1.0507, α is about 1.6732, these numerical values are defined by Klambauer et al. [8].

Fully-connected layer is used to recognize the extracted patterns of data from previous layers. The most neurons in this layer have full connections to every neurons in the previous layer.

Long Short-Term Memory (LSTM) [9] is a kind of recurrent neural network (RNN), that has the capability of remembering the values from historical stages for predicted values. LSTM consists of three major gates: an input gate (i_t), an output gate (o_t), and a forget gate (f_t).

$$i_t = \sigma\left(W_i x_t + b_{ii} + U_i h_{t-1} + b_{hi}\right) \tag{5}$$

$$f_t = \sigma\left(W_f x_t + b_{if} + U_f h_{t-1} + b_{hf}\right) \tag{6}$$

where σ denotes the standard sigmoid activation function, x_t denotes the input to the unit, t denotes the current input at each step, c_t denotes the cell state time t, and c_{t-1} is the previous cell state at time $t-1$.

$$g_t = tanh\left(W_g x_t + b_{ig} + U_g h_{t-1} + b_{hg}\right) \tag{7}$$

$$o_t = \sigma\left(W_o x_t + b_{io} + U_o h_{t-1} + b_{ho}\right) \tag{8}$$

$$c_t = f_t \circ c_t + i_t \circ g_t \tag{9}$$

where $tanh$ is the hyperbolic tangent function that is used to scale the values into the range −1 to 1, $\circ$ denotes the Hadamard product [10], and g_t denotes the cell gate. The output gate o_t is responsible for generating the output and updating the hidden vector h_{t-1}. This process can be defined as Eq. (10).

$$h_t = o_t \circ tanh(c_t) \tag{10}$$

W and U and b are the learnable weights, b biase terms, i_t denotes the updated information, o_t denotes the information output of each state, and h_t denotes the hidden state time t. As described above, LSTM consists of input gate, output gate, and forget gate. The cooperation among three gates makes LSTM has capability to solve the problem of time dependencies.

This work constructs the CNN-LSTM structure by combining the CNN model and LSTM model together. Consequently, the CNN-LSTM model includes eight layers; input layer, two convolutional layers, max pooling layer, flatten layer, LSTM layer, fully connected layer, and output layer. The structure and details of the different types of layers are illustrated in Fig. 1. Since, this work defines the model as a sequential model of the different kinds of layers, the last layer is an output.

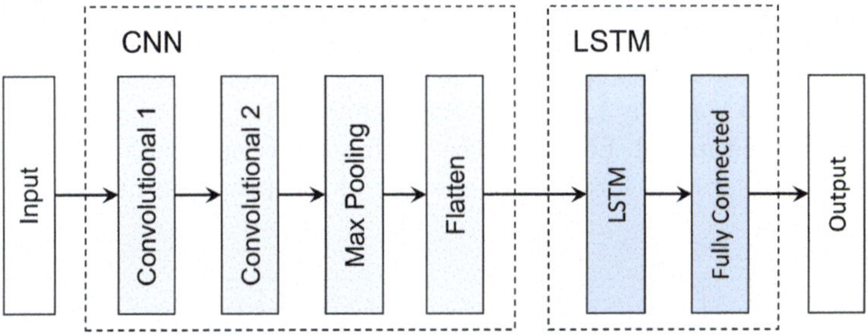

Fig. 1. The architecture of the proposed CNN-LSTM model

2.3 Model Evaluation Metric

This work evaluates the proposed model performance by using the Root Mean Square Error (RMSE), and Symmetric Mean Absolute Percentage Error (sMAPE), which was described by Hyndman and Koehler [11].

RMSE measures the differences between observed values and predicted values, and compares the errors of the different models for a specific data. The sMAPE is an accuracy measure based on percentage errors. The formula of RMSE, and SMAPE are defined in Eqs. (11), and (12) respectively.

$$RMSE = \sqrt{\frac{\sum_{i=1}^{N}(x_i - \hat{x}_i)^2}{N}} \tag{11}$$

$$sMAPE = \frac{1}{N}\sum_{i=1}^{N}\frac{|x_i - \hat{x}_i|}{|\hat{x}_i| + |x_i|} \tag{12}$$

where N is the total number of observations, x_i denotes the observed values, and, $\hat{x}_i$ denotes the predicted values. As described, a lower RMSE, and sMAPE mean better forecasting performance.

3 Experiments

In this section, the data preparation, creating and configuring CNN-LSTM model and XGBoost model will be described. The proposed model will be done using the deep learning framework such as XGBoost python package, Tensorflow, Keras, Pandas, and cuDNN, etc. The experiment use of Intel Core i7 with 16 GB of RAM, and NVIDIA GTX 1060 GPU.

3.1 Data Preparation and Description

The dataset is the daily outpatient of a hospital in Bangkok, from 2012 to 2016. This dataset is collected from two departments include the Medical (MED), and Obstetrics-Gynaecology (OB-GYN) department. The data structure includes two columns of data, which consists of a pair of date, and the daily number of patients. The dataset is split into two parts; training data and testing data. The training data contains 1,096 rows between 2012 to 2015, this data is used to train the proposed model. Likewise, the testing data contains 365 rows in 2016, it is used to evaluate the forecasting efficient of the proposed model.

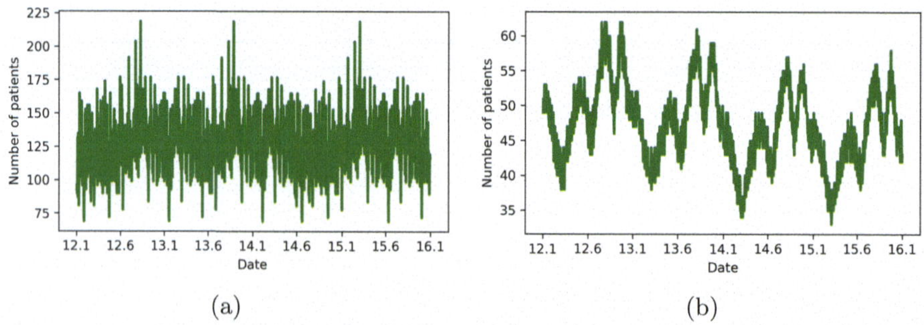

Fig. 2. Time series plot of patient visits during 2012 to 2016 of (a) MED and OB-GYN dataset

Figure 2 shows the plot of the number of patients and its density for each dataset, from 2012 to 2016 of (a) MED and (b) OB-GYN dataset. The MED dataset has a level of noise, whereas the OB-GYN dataset contains the seasonal variation.

3.2 XGBoost Modeling and Configuration

The XGBoost model requires careful parameter tuning to ensure optimal model performance. The essential parameters are the number of rounds for performing, number of trees, maximum depth, and learning rate. Technically, this work determines the optimal parameter values by using grid search optimization technique. We preferred a learning rate is 0.07, a number of rounds is 500, and the maximum tree depth is 7 for MED dataset. In the same way, we preferred a learning rate is 0.03, a number of rounds is 500, and the maximum tree depth is 5 for OB-GYN dataset. A benefit of using ensembles of decision tree methods like XGBoost, it can automatically provide the estimates of feature importance from a trained predictive model. We defines seven important features for the predictive model. These features are selected based on the hospital requirements includes the day of month, day of year, day of week, week of year, quarter, month, and year.

3.3 CNN-LSTM Modeling and Configuration

We construct the CNN-LSTM as a sequential model which is described in previous section. The architecture of the proposed CNN-LSTM is depicted in Fig. 1. Since the CNN-LSTM model has many significant parameters, it has a difficult to define the suitable parameters. We perform the grid search optimization technique to find the optimal parameters. Then, we parameterized the optimal parameters to each layer which was constructed in Fig. 1. We train the CNN-LSTM models on the 1,096 days of data for different epochs using Adam optimizer and test on the last 365 days of data. Concurrently, the RMSE and sMAPE values were calculated in these experiments.

4 Results and Evaluation

This section is organized in two parts include the XGboost, and CNN-LSTM forecasting results.

4.1 XGboost Forecasting Results

As described, the performance of XGBoost model commonly depends on the number of selected features. When the boosted trees were created, we had got the importance scores for each feature. This score represented the importance of each feature, that it was in creating of boosted trees. As depicted in Fig. 3, a feature "Day of Year" gives the highest importance with F score at 1858, and 1804, respectively. On the other hand, a feature "Quarter" gives the lowest importance score with F score 30, and 56, respectively. The higher F score indicate that the feature is more important. Using the most of important features allows us to increase the performance of models. Concurrently, the model accuracy is evaluated using RMSE, and sMAPE, which was described by

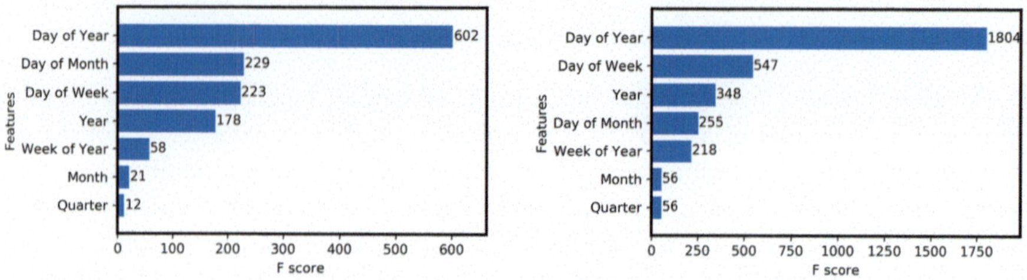

(a) (b)

Fig. 3. A bar chart of feature importance

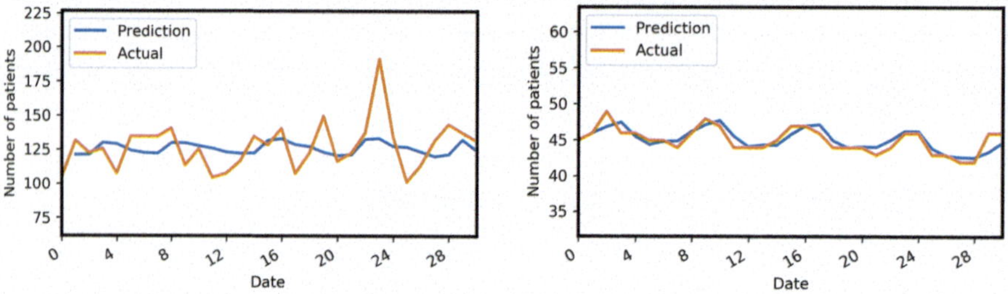

(a) (b)

Fig. 4. The plot of actual values compare to predicted values of 2016 using XGBoost model (a) MED dataset and (b) OB-GYN dataset

Granderson et al. [12]. The results shows the RMSE, and sMAPE of MED dataset are 21.812, and 13.389, respectively. In the same way, the RMSE, and sMAPE of OB-GYN dataset are 1.831, and 3.281, respectively. Figure 4 depicts the plot of observed values compare to predicted values of 2016 of two datasets. The plot shows the XGBoost model has skill to forecast the next values, that makes two lines move in the same direction.

4.2 CNN-LSTM Forecasting Results

Figure 5 shows the forecasting results from CNN-LSTM model. The observed values for the test dataset are plotted to compare with the predicted value. The results shows the RMSE, and sMAPE of MED dataset are 3.736, and 4.299, respectively. In the same way, the RMSE, and sMAPE of OB-GYN dataset are 1.366, and 2.199, respectively. Figure 5 shows the plot of observed values compare to predicted values of 2016 on the two datasets. The plot shows the CNN-LSTM model has skill to forecast the next values, that makes the two lines move in the same direction.

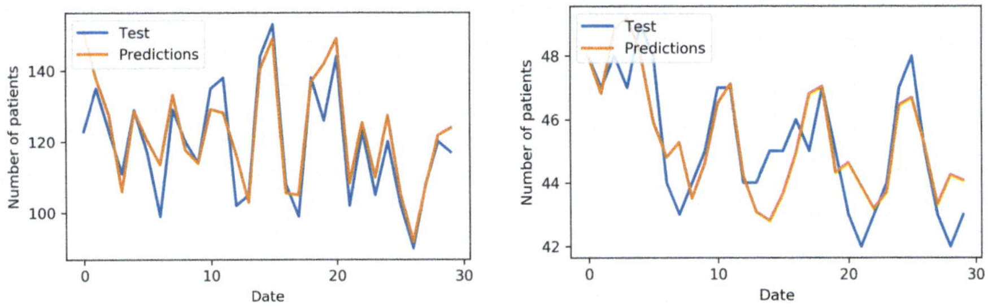

(a) (b)

Fig. 5. The plot of actual values compare to predicted values using CNN-LSTM model (a) MED dataset and (b) OB-GYN dataset

4.3 Evaluations

The results indicate that XGBoost model and CNN-LSTM model can forecast the number of outpatient accurately. The XGBoost and CNN-LSTM model make the reliable forecasting by giving the low RMSE, and sMAPE in two different datasets. Both models are evaluated with RMSE, and sMAPE function. The model accuracy of forecasting is compared as shown in Table 1. The CNN-LSTM makes the forecast more accurate than XGBoost model in two datasets. In the case of MED dataset, the XGBoost model give the lowest accuracy that XGBoost is based on the tree model thus it tend to select predictors with a higher number of distinct values that is caused of unstable, when faced with predictors with many categories, and have high variance [13].

Table 1. An average RMSE, an sMAPE of XGBoost and CNN-LSTM model

Datasets	XGBoost		CNN-LSTM	
	RMSE	sMAPE	RMSE	sMAPE
MED	21.812	13.389	3.736	4.299
OB-GYN	1.831	3.281	1.366	2.199

In the case of processing time, the XGBoost model had operated on MED and OBGYN dataset in 3.8770, and 4.3011 s, respectively. Whereas the CNN-LSTM model had operated on MED and OBGYN dataset in 111.5793, and 112.0442 s, respectively.

5 Conclusion

This work studied to forecast the number of patients, that applied for service demand forecasting. We proposed the comparision between XGBoost model, and

CNN-LSTM model for service demand forecasting. The results indicated that two models had skillful, and made the reliable forecasting in two datasets. This work evaluated and measured the model performance by calculating RMSE, and sMAPE, which was acceptable in all case study. The CNN-LSTM model gave better efficiency forecasting than XGBoost model. In contrast, XGBoost model operated faster, took a lower performing time, and lower CPU consumption than the CNN-LSTM model.

The forecasting the number of patients can help the hospital to estimate the service demand for managing resources the medicine demand in the future. In addition, this work can apply to manage medical resources and improve the efficiency of medical services. Although, the accuracy of forecasting are not good enough in some case study, it caused by insufficient the number of the data, and the collection of data is widely scattered.

References

1. Xinxiang, Z., Bo, Z., Huijuan, F.: A comparison study of outpatient visits forecasting effect between ARIMA with seasonal index and SARIMA. In: 2017 International Conference on Progress in Informatics and Computing (PIC), pp. 362–366, December 2017
2. Qingkui, C., Junhu, R.: Study on the demand forecasting of hospital stocks based on data mining and BP neural networks. In: 2009 International Conference on Electronic Commerce and Business Intelligence, pp. 284–289, June 2009
3. Friedman, J.H.: Greedy function approximation: a gradient boosting machine. Ann. Stat. **29**(5), 1189–1232 (2000)
4. Freund, Y., Schapire, R.E.: A decision-theoretic generalization of on-line learning and an application to boosting. J. Comput. Syst. Sci. **55**(1), 119–139 (1997)
5. Chen, T., Guestrin, C.: XGBoost: a scalable tree boosting system. In: Proceedings of the 22nd ACM SIGKDD International Conference on Knowledge Discovery and Data Mining, KDD 2016, pp. 785–794. ACM, New York (2016)
6. Breiman, L.: Arcing the edge. Statistics Department, University of California at Berkeley, Technical report (1997)
7. Pedamonti, D.: Comparison of non-linear activation functions for deep neural networks on MNIST classification task. CoRR abs/1804.02763 (2018)
8. Klambauer, G., Unterthiner, T., Mayr, A., Hochreiter, S.: Self-normalizing neural networks. CoRR abs/1706.02515 (2017)
9. Hochreiter, S., Schmidhuber, J.: Long short-term memory. Neural Comput. **9**(8), 1735–1780 (1997)
10. Styan, G.P.: Hadamard products and multivariate statistical analysis. Linear Algebra Appl. **6**, 217–240 (1973)
11. Hyndman, R.J., Koehler, A.B.: Another look at measures of forecast accuracy. Int. J. Forecast. **22**(4), 679–688 (2006)
12. Granderson, J., Price, P.N., Jump, D., Addy, N., Sohn, M.D.: Automated measurement and verification: performance of public domain whole-building electric baseline models. Appl. Energy **144**, 106–113 (2015)
13. Hastie, T., Montanari, A., Rosset, S., Tibshirani, R.J.: Surprises in high-dimensional ridgeless least squares interpolation. CoRR abs/1903.08560 (2019)

Bifurcation of Limit Cycles and Their Relations in Three Perturbed Integrable Systems

Lijun Hong, Wenyue Fu, and Xiaochun Hong(✉)

School of Statistics and Mathematics,
Yunnan University of Finance and Economics,
Kunming 650221, Yunnan, People's Republic of China
xchong@ynufe.edu.cn

Abstract. Bifurcation of limit cycles of three perturbed integrable systems is investigated by using both qualitative analysis and numerical exploration. The study reveals that the three systems has 2 limit cycles, 3 limit cycles, 4 limit cycles severally. By using method of numerical simulation, the distributed orderliness of these limit cycles is observed, and their nicety places are determined. The study also indicates that each of these limit cycles passes the corresponding nicety point.

Keywords: Limit cycle · Integrable system · Detection function · Numerical exploration

1 Introduction and Main Conclusions

During the past decades the bifurcation of limit cycles of the following planar polynomial system

$$\begin{cases} \dfrac{dx}{dt} = P_n(x, y), \\ \dfrac{dy}{dt} = Q_n(x, y), \end{cases} \tag{1}$$

has been considerably investigated, and it has now become a very popular topic in the area of applied mathematics, where $P_n(x, y)$ and $Q_n(x, y)$ are two polynomials of degree n. It should be mentioned that the system (1) is related to the celebrated Hibert's 16th problem.

In Ref. [1], Hong considered the two differential systems as follows:

$$\begin{cases} \dfrac{dx}{dt} = y, \\ \dfrac{dy}{dt} = -(x + x^2 + x^3) + \mu y R(x, y, \lambda), \end{cases} \tag{2}$$

Y. Liu et al. (Eds.): ICNC-FSKD 2019, AISC 1074, pp. 557–568, 2020.
https://doi.org/10.1007/978-3-030-32456-8_61

$$\begin{cases} \dfrac{dx}{dt} = y(x^2+1), \\ \dfrac{dy}{dt} = -(x+x^2+x^3)(x^2+1) + \mu y R(x,y,\lambda), \end{cases} \tag{3}$$

where $R(x, y, \lambda) = mx + nx^2 + ky^2 - \lambda$, $\mu (0 < \mu \ll 1)$, m, n, k and λ are all real parameters.

He discovered that the system (2) and the system (3) has 3 limit cycles.

For the perturbed quadratic system

$$\begin{cases} \dfrac{dx}{dt} = y, \\ \dfrac{dy}{dt} = x^2 - 1 + \mu R(x,y), \end{cases} \tag{4}$$

and the perturbed Bogdanov-Takens system

$$\begin{cases} \dfrac{dx}{dt} = y, \\ \dfrac{dy}{dt} = x^2 - x + \mu R(x,y), \end{cases} \tag{5}$$

where $\mu(0 < \mu \ll 1)$ is a real parameter, $R(x, y)$ is a real polynomial, and $\deg(P(x, y)) = n$.

Bogdanov [2] proved that the system (4) has only one limit cycle when $R(x, y) = (a + x)y$, Dumortier et al. [3] proved that the system (4) has two limit cycles when $R(x, y) = (a + bx + cx^3)y$, Li et al. [4] proved that the system (4) has three limit cycles when $R(x, y) = (a + bx + cx^3 + dx^4)y$; Petrov [5] proved that $Z(2, n) = n - 1$ for the system (5).

In this paper, for the following three perturbed integrable systems, we study the bifurcation of their limit cycles.

$$\begin{cases} \dfrac{dx}{dt} = y, \\ \dfrac{dy}{dt} = x^2 - 1 + \mu y R(x,y), \end{cases} \tag{6}$$

$$\begin{cases} \dfrac{dx}{dt} = y(x+2.5), \\ \dfrac{dy}{dt} = (x^2-1)(x+2.5) + \mu y R(x,y), \end{cases} \tag{7}$$

$$\begin{cases} \dfrac{dx}{dt} = y(x+2.5)(x-1.3), \\ \dfrac{dy}{dt} = (x^2-1)(x+2.5)(x-1.3) + \mu y R(x,y), \end{cases} \tag{8}$$

where $R(x, y) = a + bx + cx^2 + dy^2$, $\mu(0 < \mu \ll 1)$, a, b, c and d are all real parameters.

Their perturbed conditions are the same $\mu y(a+bx+cx^2+dy^2)$. Further, their corresponding unperturbed systems are all integrable system, and integrating factors are 1, $1/(x + 2.5)$, $1/[(x + 2.5)(x - 1.3)]$ severally. By using the method of detection function [6] and the method of numerical exploration [7], we will show that the system (6) has 2 limit cycles when $b = -1$, $c = 1$, $d = 0.16$, $-2.06860 < a < -2.01351$; the system (7) has 3 limit cycles when $b = -5$, $c = 3.6$, $d = -1$, $-8.59995 < a < -8.58035$; and the system (8) has 4 limit cycles when $b = -1$, $c = -9$, $d = 0.77$, $7.99172 < a < 7.99873$.

The rest of this paper is organized as follows: In Sect. 2, we introduce the concepts of detection functions and detection curves, obtain a proposition. In Sect. 3, we study the unperturbed systems accordingly. In Sect. 4, we analyze the bifurcation of limit cycles of the perturbed systems, and obtain three propositions. In Sect. 5, we give a short conclusion.

2 Detection Functions and Detection Curves

In this section some preliminary results on the detection functions for perturbed Hamiltonian system are briefly reviewed. Let us begin with considering the perturbed Hamiltonian system.

It is helpful to begin with introducing Ye's some useful results [8]. He obtained these results by introducing:

Lemma 1. Consider the perturbed Hamiltonian system

$$\begin{cases} \dfrac{dx}{dt} = -\dfrac{\partial H}{\partial y} + P\left(x, y, \alpha\right), \\ \dfrac{dy}{dt} = \dfrac{\partial H}{\partial x} + Q\left(x, y, \alpha\right), \end{cases} \tag{9}$$

and the corresponding unperturbed Hamiltonian system

$$\begin{cases} \dfrac{dx}{dt} = -\dfrac{\partial H}{\partial y}, \\ \dfrac{dy}{dt} = \dfrac{\partial H}{\partial x}. \end{cases} \tag{10}$$

Obviously system (9) reduces into system (10) when $P\left(x, y, 0\right) \equiv Q\left(x, y, 0\right) \equiv 0$. The curve Γ^h defined by Hamiltonian $H\left(x, y\right) = h$ of system (10) is, then, closed orbits and will extend to the outside of Γ^h as h increases, and $\Gamma^h\left(D\right)$ is the area inside of Γ^h. If there exists h^* such that the function

$$A\left(h\right) = \iint_{\Gamma^h(D)} [P''_{x\alpha}\left(x, y, 0\right) + Q''_{y\alpha}\left(x, y, 0\right)] dxdy, \tag{11}$$

satisfies $A\left(h^*\right) = 0$ and $A'\left(h^*\right) \neq 0$, $\alpha A'\left(h^*\right) < 0\ (> 0)$, then system (9) has only one stable (unstable) limit cycle near to Γ^{h^*} when α is very small. Conversely, if Γ^h is constrained inside as h increases, the stable properties of the limit cycle are opposite of that described above, i.e., when $\alpha A'(h^*) < 0\ (> 0)$ the limit cycle is unstable (stable). If $A(h) \neq 0$, then system (9) has no limit cycle.

Li and Li [9] considered the following system:

$$\begin{cases} \dfrac{dx}{dt} = -\dfrac{\partial H}{\partial y} - \mu x \left[p\left(x,y\right) - \lambda \right], \\ \dfrac{dy}{dt} = \dfrac{\partial H}{\partial x} - \mu y \left[q\left(x,y\right) - \lambda \right], \end{cases} \tag{12}$$

where $p\left(0,0\right) = q\left(0,0\right) = 0$. Using the results given in [8] above, it follows from $A\left(h\right) = 0$ that

$$\lambda = \lambda\left(h\right) = \frac{\iint_{\Gamma^h(D)} f\left(x,y\right) dxdy}{2\iint_{\Gamma^h(D)} dxdy}, \tag{13}$$

where

$$f(x,y) = xp'_x + yq'_y + p + q. \tag{14}$$

The function $\lambda\left(h\right)$ is usually known as the detection function of system (12). Using the detection function $\lambda\left(h\right)$ and Lemma 1 above, the following proposition regarding the limit cycle of system (12) can be obtained [10]:

Proposition 1. For any given λ_0: (i) If $(h_0, \lambda(h_0))$ is an intersecting point of the line $\lambda = \lambda_0$ and the detection curve $\lambda = \lambda(h)$, and $\lambda'(h_0) > 0$ (< 0), then the system (12) has only one stable (unstable) limit cycle near Γ^{h_0} when $\lambda = \lambda_0$. (ii) If the line $\lambda = \lambda_0$ and the detection curve $\lambda = \lambda(h)$ have no intersecting point, then the system (12) has no limit cycle when $\lambda = \lambda_0$. Conversely, If Γ^h is constrained inside as h increases, the stability of the limit cycle is opposite to the results above.

The proof of this proposition can be found elsewhere [11,12]. For the sake of completeness, we briefly present the proof as below:

Proof. In Lemma 1, let $\alpha = -\mu$, $P(x,y,\alpha) = \alpha x[p(x,y) - \lambda]$, $Q(x,y,\alpha) = \alpha y[q(x,y) - \lambda]$. We have $P(x,y,0) \equiv Q(x,y,0) \equiv 0$ and $P''_{x\alpha} + Q''_{y\alpha} = f(x,y) - 2\lambda$. Thus the $A(h)$ in Lemma 1 becomes

$$A(h) = \iint_{\Gamma^h(D)} f(x,y)dxdy - \lambda \iint_{\Gamma^h(D)} 2dxdy. \tag{15}$$

By denoting $\psi(h) = \iint_{\Gamma^h(D)} f(x,y)dxdy$ and $\phi(h) = \iint_{\Gamma^h(D)} 2dxdy$, we have $\lambda = \dfrac{\psi(h_0)}{\phi(h_0)}$, from $A(h_0) = 0$, which leads to

$$A'(h_0) = \psi'(h_0) - \lambda\phi'(h_0) = \frac{\psi'(h_0)\phi(h_0) - \psi(h_0)\phi'(h_0)}{\phi(h_0)}. \tag{16}$$

We also have from (15) that

$$\lambda'(h_0) = \frac{\psi'(h_0)\phi(h_0) - \psi(h_0)\phi'(h_0)}{[\phi(h_0)]^2}. \tag{17}$$

It is noted from (15) that $\phi(h_0) > 0$. Moreover, both $A'(h_0)$ and $\lambda'(h_0)$ have the same sign by comparing (16) with (17). Therefore, $-\mu A'(h_0)$ and $-\mu\lambda'(h_0)$ also have the same sign. From Lemma 1, the proof of Proposition 1 is thus completed.

3 Analysis of the Unperturbed System

For the three corresponding unperturbed systems, in this section, we study their essential property.

The three corresponding unperturbed systems as follows:

$$\begin{cases} \dfrac{dx}{dt} = y, \\ \dfrac{dy}{dt} = x^2 - 1. \end{cases} \tag{18}$$

$$\begin{cases} \dfrac{dx}{dt} = y(x + 2.5), \\ \dfrac{dy}{dt} = (x^2 - 1)(x + 2.5). \end{cases} \tag{19}$$

$$\begin{cases} \dfrac{dx}{dt} = y(x + 2.5)(x - 1.3), \\ \dfrac{dy}{dt} = (x^2 - 1)(x + 2.5)(x - 1.3). \end{cases} \tag{20}$$

For the systems (18)–(20), they are all integrable systems, they have all a center $(-1, 0)$, a hyperbolic saddle point $(1, 0)$, a family of closed orbits $\{\Gamma^h\}$ which enclose the finite singular point $(-1, 0)$. Furthermore, the system (19) has an integral curve $x = -2.5$, the system (20) has two integral curves $x = -2.5$ and $x = 1.3$ (see Figs. 1, 2 and 3).

Their first integral is all

$$H(x, y) = \frac{1}{2}y^2 - \frac{1}{3}x^3 + x = h, \tag{21}$$

with integrating factor 1, $1/(x + 2.5)$, $1/[(x + 2.5)(x - 1.3)]$ severally.

From (21), we can get $h(-1, 0) = -2/3$, $h(1, 0) = 2/3$, and

$$y_{\pm} = \pm u(h, x) = \pm\sqrt{\frac{2}{3}x^3 - 2x + 2h}. \tag{22}$$

Setting $y = 0$, from (21), we can get

$$x^3 - 3x + 3h = 0, h \in (-\frac{2}{3}, \frac{2}{3}). \tag{23}$$

From (23), for an arbitrary constant of $h \in (-2/3, 2/3)$, we can get three definite roots x_1, x_2, x_3, and they satisfy $x_1 < x_2 < x_3$.

Note that as h increases, the curves $\{\Gamma^h\}$ extend outward.

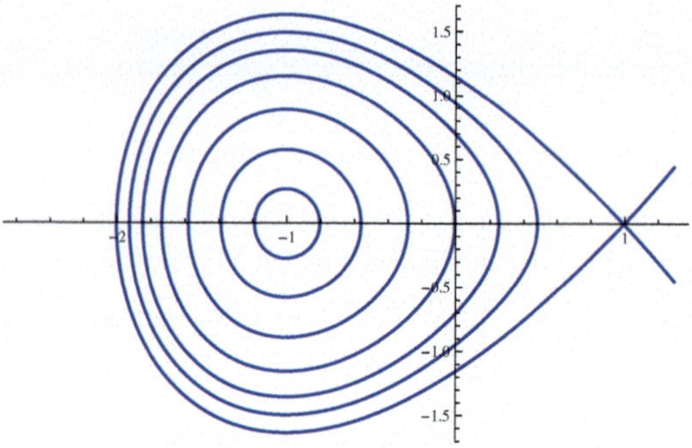

Fig. 1. Closed orbits of the unperturbed system (18).

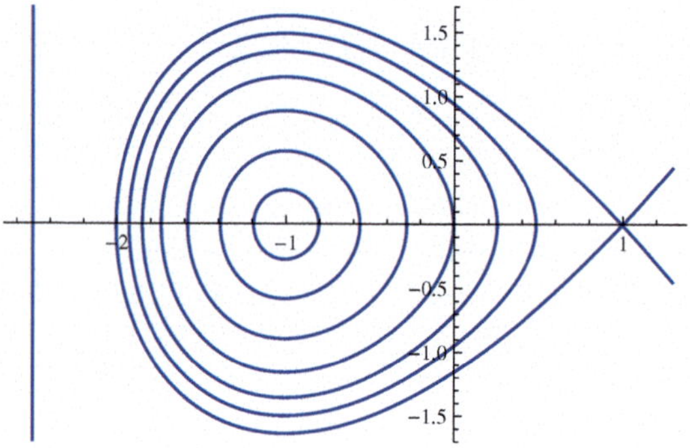

Fig. 2. Closed orbits of the unperturbed system (19).

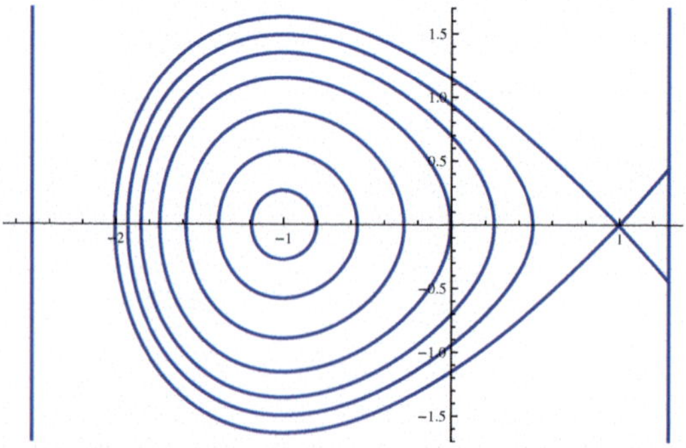

Fig. 3. Closed orbits of the unperturbed system (20).

4 Limit Cycles Analysis of the Perturbed System

Consider the three perturbed integrable systems (6)-(8).

For the systems (6)–(8), using the [7], we can get three Abelian integrals as follows:

$$I_1(h) = \oint_{\Gamma^h} (a + bx + cx^2 + dy^2)ydx, \quad h \in (-\frac{2}{3}, \frac{2}{3}). \tag{24}$$

$$I_2(h) = \oint_{\Gamma^h} \frac{1}{x + 2.5}(a + bx + cx^2 + dy^2)ydx, \quad h \in (-\frac{2}{3}, \frac{2}{3}). \tag{25}$$

$$I_3(h) = \oint_{\Gamma^h} \frac{1}{(x + 2.5)(x - 1.3)}(a + bx + cx^2 + dy^2)ydx, \quad h \in (-\frac{2}{3}, \frac{2}{3}). \tag{26}$$

Letting $I_j(h) = 0(j = 1 - 3)$, from (24)–(26), we have three detection functions [6] as follows:

$$a_1(h) = -\frac{\int_{x_1}^{x_2} xu(h,x)dx}{\int_{x_1}^{x_2} u(h,x)dx}b - \frac{\int_{x_1}^{x_2} x^2u(h,x)dx}{\int_{x_1}^{x_2} u(h,x)dx}c - \frac{\int_{x_1}^{x_2} u^3(h,x)dx}{\int_{x_1}^{x_2} u(h,x)dx}d, \quad h \in (-\frac{2}{3}, \frac{2}{3}). \tag{27}$$

$$a_2(h) = -\frac{\int_{x_1}^{x_2} \frac{xu(h,x)}{x+2.5}dx}{\int_{x_1}^{x_2} \frac{u(h,x)}{x+2.5}dx}b - \frac{\int_{x_1}^{x_2} \frac{x^2u(h,x)}{x+2.5}dx}{\int_{x_1}^{x_2} \frac{u(h,x)}{x+2.5}dx}c - \frac{\int_{x_1}^{x_2} \frac{u^3(h,x)}{x+2.5}dx}{\int_{x_1}^{x_2} \frac{u(h,x)}{x+2.5}dx}d, \quad h \in (-\frac{2}{3}, \frac{2}{3}). \tag{28}$$

$$a_3(h) = -\frac{\int_{x_1}^{x_2} \frac{xu(h,x)}{(x+2.5)(x-1.3)}dx}{\int_{x_1}^{x_2} \frac{u(h,x)}{(x+2.5)(x-1.3)}dx}b - \frac{\int_{x_1}^{x_2} \frac{x^2u(h,x)}{(x+2.5)(x-1.3)}dx}{\int_{x_1}^{x_2} \frac{u(h,x)}{(x+2.5)(x-1.3)}dx}c - \frac{\int_{x_1}^{x_2} \frac{u^3(h,x)}{(x+2.5)(x-1.3)}dx}{\int_{x_1}^{x_2} \frac{u(h,x)}{(x+2.5)(x-1.3)}dx}d, \quad h \in (-\frac{2}{3}, \frac{2}{3}). \tag{29}$$

For system (6), when $b = -1$, $c = 1$, $d = 0.16$, from detection function (27), we obtain Table 1. For system (7), when $b = -5$, $c = 3.6$, $d = -1$, from detection function (28), we obtain Table 2. For system (8), when $b = -1$, $c = -6.55$, $d = 0.528$, from detection function (29), we obtain Table 3.

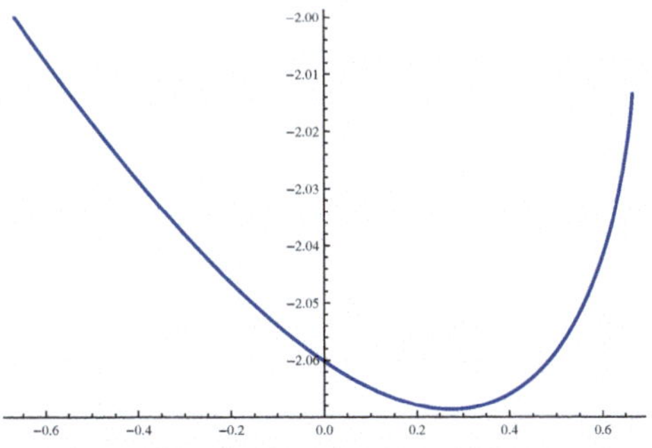

Fig. 4. Detection curve of system (6) when $b = -1$, $c = 1$, and $d = 0.16$.

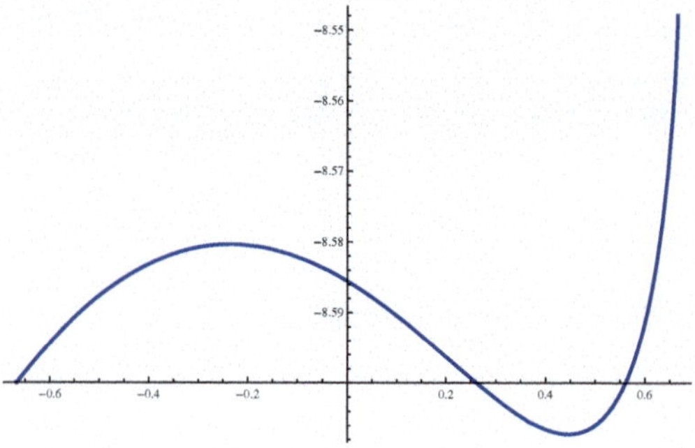

Fig. 5. Detection curve of system (7) when $b = -5$, $c = 3.6$ and $d = -1$.

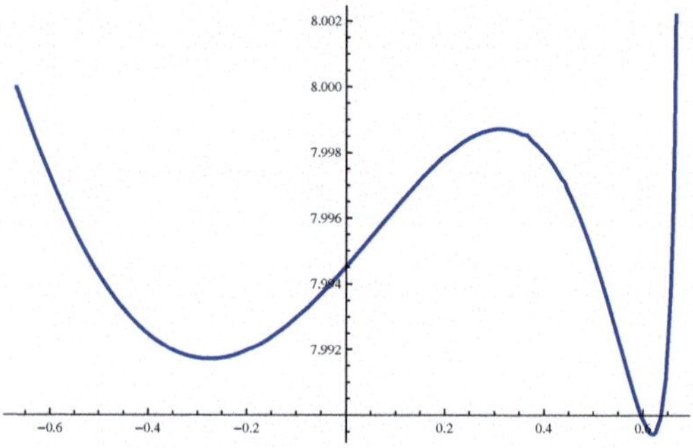

Fig. 6. Detection curve of system (8) when $b = -1$, $c = -9$, and $d = 0.77$.

Values of the three detection functions are listed in Tables 1, 2 and 3 are also illustrated in Figs. 4, 5 and 6.

Based on the results for the three detection curves shown in Figs. 4, 5 and 6, following conclusions are presented:

Proposition 2. For the system (6), when $b = -1$, $c = 1$, $d = 0.16$ and $0 < \mu \ll 1$, from detection curve Fig. 4, we get following results.

(i) It has 2 limit cycles if $-2.06860 < a < -2.01351$ (see Fig. 7).
(ii) It has 1 limit cycle if $a = -2.06860$ or $-2.01351 \leq a < -2$.
(iii) It has no limit cycle if $a < -2.06860$ or $a \geq -2$.

Proposition 3. For the system (7), when $b = -5$, $c = 3.6$, $d = -1$ and $0 < \mu \ll 1$, from detection curve Fig. 5, we get following results.

(i) It has 3 limit cycles if $-8.59995 < a < -8.58035$ (see Fig. 8).
(ii) It has 2 limit cycles if $a = -8.58035$ or $-8.60719 \leq a \leq -8.59995$.
(iii) It has 1 limit cycle if $a = -8.60719$ or $-8.58035 < a < -8.54785$.
(iiii) It has no limit cycle if $a < -8.60719$ or $a \geq -8.54785$.

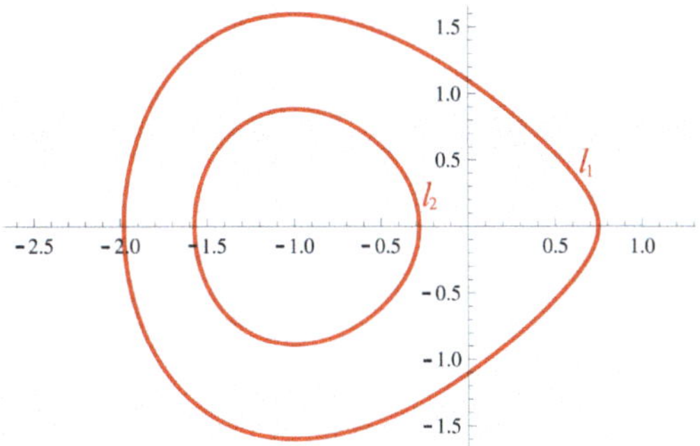

Fig. 7. The two limit cycles and their locations of system (6) when $a = -2.04$, $b = -1$, $c = 1$, $d = 0.16$, and $\mu = 0.01$. Big limit cycle l_1 passes $(0.7463, 0)$, it is stable, and small limit cycle l_2 passes $(-0.2826, 0)$, it is unstable.

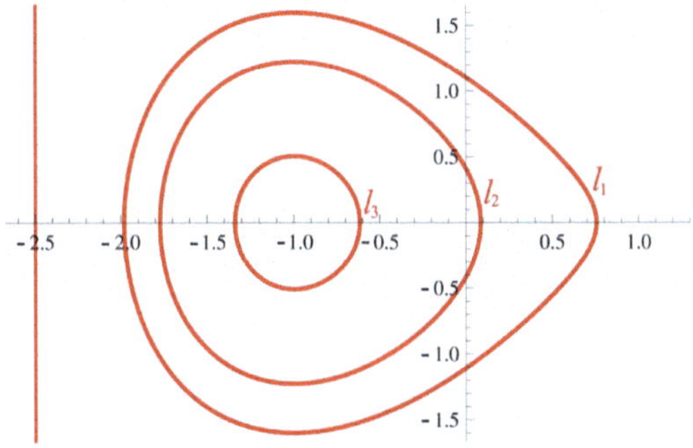

Fig. 8. The three limit cycles and their locations of system (7) when $a = -8.59$, $b = -5$, $c = 3.6$, $d = -1$, and $\mu = 0.01$. Limit cycles l_1 and l_3 passes $(0.7539, 0)$ and $(-0.6161, 0)$ severally, they are stable, limit cycle l_2 passes $(0.0843, 0)$, it is unstable.

Proposition 4. For the system (8), when $b = -1$, $c = -9$, $d = 0.77$ and $0 < \mu \ll 1$, from detection curve Fig. 6, we get following results.

(i) It has 4 limit cycles if $7.99172 < a < 7.99873$ (see Fig. 9).
(ii) It has 3 limit cycles if $a = 7.99172$ or $a = 7.99873$.
(iii) It has 2 limit cycles if $7.98943 < a < 7.99172$ or $7.99873 < a < 7.99996$.
(iiii) It has 1 limit cycle if $a = 7.98943$ or $7.99996 \leq a \leq 8.00219$.
(iiiii) It has no limit cycle if $a < 7.98943$ or $a \geq 8.00219$.

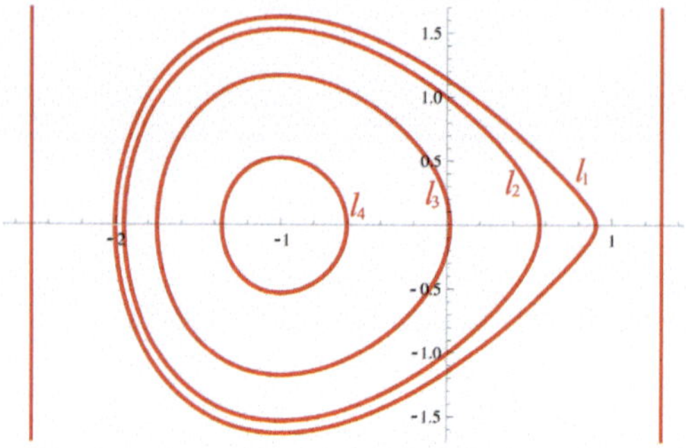

Fig. 9. The four limit cycles and their locations of system (8) when $a = 7.995$, $b = -1$, $c = -9$, $d = 0.77$, $\mu = 0.01$. Stable limit cycles l_1 and l_3 passes $(0.9077, 0)$ and $(0.0220, 0)$ severally, unstable limit cycles l_2 and l_4 passes $(0.5663, 0)$ and $(-0.5979, 0)$ severally.

Table 1. Values of detection function $a_1(h)$ when $b = -1$, $c = 1$, $d = 0.16$

h	$a_1(h)$	h	$a_1(h)$	h	$a_1(h)$	h	$a_1(h)$
−0.666666	−2.00000	−0.416666	−2.02683	0.278666	−2.06860	0.526666	−2.05537
−0.666566	−2.00001	−0.356666	−2.03262	0.286666	−2.06858	0.546666	−2.05249
−0.665666	−2.00012	−0.296666	−2.03812	0.306666	−2.06845	0.566666	−2.04911
−0.660666	−2.00069	−0.266666	−2.04075	0.316666	−2.06834	0.586666	−2.04511
−0.660566	−2.00070	−0.236666	−2.04330	0.336666	−2.06801	0.606666	−2.04031
−0.659666	−2.00080	−0.216666	−2.04495	0.356666	−2.06754	0.616666	−2.03751
−0.657666	−2.00103	−0.156666	−2.04967	0.376666	−2.06692	0.620666	−2.03630
−0.646666	−2.00229	−0.076666	−2.05534	0.396666	−2.06612	0.622666	−2.03567
−0.626666	−2.00455	−0.016666	−2.05907	0.416666	−2.06514	0.624666	−2.03503
−0.596666	−2.00791	0.036666	−2.06197	0.436666	−2.06396	0.636666	−2.03079
−0.566666	−2.01121	0.076666	−2.06386	0.443666	−2.06349	0.656666	−2.02142
−0.536666	−2.01445	0.156666	−2.06679	0.449666	−2.06307	0.663666	−2.01662
−0.506666	−2.01764	0.236666	−2.06839	0.466666	−2.06175	0.666566	−2.01368
−0.486666	−2.01973	0.266666	−2.06859	0.486666	−2.05994	0.666664	−2.01351
−0.456666	−2.02281	0.274666	−2.06860	0.506666	−2.05783	0.666666	−2.01351

For the system (6), taking $a = -2.04$, $b = -1$, $c = 1$, $d = 0.16$, $\mu = 0.01$. For the system (7), taking $a = -8.59$, $b = -5$, $c = 3.6$, $d = -1$, $\mu = 0.01$. For the system (8), taking $a = 7.995$, $b = -1$, $c = -9$, $d = 0.77$, $\mu = 0.01$. Using numerical exploration method [7], we find out the position of the each limit cycle of systems (6)–(8) is depicted in Figs. 7, 8 and 9 respectively.

Table 2. Values of detection function $a_2(h)$ when $b = -5$, $c = 3.6$, $d = -1$

h	$a_2(h)$	h	$a_2(h)$	h	$a_2(h)$	h	$a_2(h)$
−0.666666	−8.59995	−0.226666	−8.58035	0.296666	−8.60195	0.443666	−8.60716
−0.666662	−8.60002	−0.216666	−8.58038	0.306666	−8.60250	0.449666	−8.60718
−0.666566	−8.59999	−0.156666	−8.58094	0.316666	−8.60303	0.476666	−8.60684
−0.660606	−8.59945	−0.116666	−8.58170	0.326666	−8.60354	0.506666	−8.60563
−0.659666	−8.59936	−0.076666	−8.58275	0.336666	−8.60403	0.536666	−8.60322
−0.656666	−8.59909	−0.036666	−8.58407	0.356666	−8.60495	0.566666	−8.59913
−0.626666	−8.59650	−0.016666	−8.58483	0.366666	−8.60545	0.596666	−8.59254
−0.586666	−8.59335	0.016666	−8.58624	0.376666	−8.60575	0.616666	−8.58601
−0.546666	−8.59054	0.056666	−8.58813	0.386666	−8.60609	0.621666	−8.58401
−0.506666	−8.58808	0.076666	−8.58915	0.396666	−8.60640	0.624666	−8.58271
−0.456666	−8.58548	0.116666	−8.59132	0.406666	−8.60666	0.646666	−8.57036
−0.386666	−8.58274	0.156666	−8.59363	0.416666	−8.60688	0.663666	−8.55362
−0.296666	−8.58074	0.196666	−8.59603	0.426666	−8.60704	0.666566	−8.54819
−0.256666	−8.58040	0.256666	−8.59965	0.436666	−8.60715	0.666664	−8.54786
−0.231666	−8.58035	0.276666	−8.60082	0.440666	−8.60719	0.666666	−8.54785

Table 3. Values of detection function $a_3(h)$ when $b = -1$, $c = -9$, $d = 0.77$

h	$a_3(h)$	h	$a_3(h)$	h	$a_3(h)$	h	$a_3(h)$
−0.666666	7.99996	−0.266666	7.99172	0.311666	7.99873	0.621666	7.98943
−0.666662	8.00002	−0.256666	7.99173	0.326666	7.99871	0.622666	7.98944
−0.665666	7.99996	−0.236666	7.99178	0.356666	7.99856	0.623666	7.98945
−0.660566	7.99973	−0.226666	7.99182	0.386666	7.99822	0.624666	7.98946
−0.659966	7.99970	−0.176666	7.99213	0.416666	7.99768	0.626666	7.98950
−0.657666	7.99960	−0.116666	7.99275	0.440666	7.99710	0.636666	7.98995
−0.626666	7.99832	−0.036666	7.99390	0.449666	7.99681	0.646666	7.99111
−0.566666	7.99618	0.016666	7.99481	0.476666	7.99588	0.656666	7.99374
−0.506666	7.99448	0.056666	7.99552	0.506666	7.99460	0.660666	7.99567
−0.456666	7.99338	0.116666	7.99659	0.536666	7.99313	0.663666	7.99787
−0.406666	7.99258	0.196666	7.99785	0.566666	7.99148	0.665666	8.00015
−0.356666	7.99204	0.256666	7.99850	0.596666	7.99000	0.666566	8.00186
−0.316666	7.99180	0.276666	7.99863	0.614666	7.98947	0.666656	8.00215
−0.296666	7.99174	0.296666	7.99871	0.618666	7.98944	0.666664	8.00218
−0.276666	7.99172	0.306666	7.99873	0.620666	7.98943	0.666666	8.00219

5 Conclusion

In the above discussion, we use both qualitative and numerical methods to investigate the number and distribution of limit cycles for three perturbed integrable systems (6)–(8). In particular, for the system (6), when $b = -1$, $c = 1$, $d = 0.16$

and $0 < \mu \ll 1$, the study reveals that it has 2 limit cycles if $-2.06860 < a < -2.01351$; for the system (7), when $b = -5$, $c = 3.6$, $d = -1$ and $0 < \mu \ll 1$, the study reveals that it has 3 limit cycles if $-8.59995 < a < -8.58035$; for the system (8), when $b = -1$, $c = -9$, $d = 0.77$ and $0 < \mu \ll 1$, the study reveals that it has 4 limit cycles if $7.99172 < a < 7.99873$. It is also found that each limit cycle passes a particular point and the position of these points is obtained by using the numerical exploration method [7] when $\mu = 0.01$ for the particular value of parameter $a = -2.04$, $a = -8.59$, $a = 7.995$, severally.

Acknowledgment. This work was financially supported by the Natural Science Foundation of China (Grant No. 11761075).

References

1. Hong, X.C.: Bifurcation of limit cycles for two differential systems. In: Proceedings of the Fifth International Conference on Natural Computation, vol. 5, pp. 455–459 (2009)
2. Bogdanov, R.I.: Bifurcation of limit cycles of a family of planar vector fields. Sel. Math. Sov. **1**, 373–387 (1981)
3. Dumortier, F., Roussarie, R., Sotomayor, J.: Generic 3-parameter family of vector fields on the plance, unfolding a singularity with nilpotent linear part. The cusp case of codimension 3. Ergod. Theory Dyn. Syst. **7**, 375–413 (1987)
4. Li, C.Z., Rousseau, C.: A system with three limit cycles appearing in a Hopf bifurcation and dying in a homoclinic bifurcation, the cusp of order 4. J. Diff. Equ. **79**, 132–167 (1989)
5. Petrov, G.S.: Number of zeros of complete elliptic integrals. Funct. Anal. Appl. **18**, 148–149 (1988)
6. Li, J.B., Huang, Q.M.: Bifurcations of limit cycles forming compound eyes in the cubic system. Chin. Ann. Math. **8(sec. B)**(4), 391–403 (1987)
7. Nusse, H.E., Yorke, J.A.: Dynamics: Numerical Explorations (Accompanying Computer Program Dynamics Eric J. Kostelich). Springer, New York (1998)
8. Ye, Y.Q.: The theory of limit cycles. Trans. Math. Monogr. (Am. Math. Soc.) (1986)
9. Li, J.B., Li, C.F.: Distribution of limit cycles for planar cubic Hamiltonian systems. Acta Mathematica Sinica **28**, 509–521 (1985)
10. Liu, Z.R., Qian, T.F., Li, J.B.: Detection function method and its application to a perturbed quintic Hamiltonian system. Chaos Solitons Fractals **13**, 295–310 (2002)
11. Tang, M.Y., Hong, X.C.: Fourteen limit cycles in a cubic Hamiltonian system with nine-order perturbed term. Chaos Solitons Fractals **14**, 1361–1369 (2002)
12. Hong, X.C., Qin, Q.H.: Limit cycle analysis on a cubic Hamiltonian system with quintic perturbed terms. Int. Math. Forum **1**, 1805–1818 (2006)

Modeling of Information Diffusion in Sina Weibo Based on Random Forest Classifier and SIR Model

Zhang Jianyi, He Ping, Ken K. T. Tsang(✉), and Deng Yuhui

Beijing Normal University-Hong Kong Baptist University United International College, Zhuhai, Guangdong Province, China
kentsang@uic.edu.hk

Abstract. Recent developments in information diffusion model for social network have not taken into account its topological structures. Characteristics such as the degree of connections and clustering of nodes in a network are known to influence the speed of information propagation. Yet, existing models (such as SIR with an average probability to repost received message) are not sophisticate enough to reflect the fine-grain characteristics. Differences among nodes are often overlooked, leading to inaccurate description of the information dissemination process.

In this work, a new approach to predict the information diffusion probability in social network is studied. We combine the Random Forest classification and the SIR model together to analyze the dissemination of information in Weibo. Python crawlers are employed to obtain a total of 316,329 microblogs concerning major news events in 2018, together with related features of nodes from Sina Weibo. The unbalanced positive and negative repost behavior together with 15 features that characterize the nodes and edges data are rebalanced by SMOTE resampling, then used to train a Random Forest classifier to predict individual user's forwarding behavior. For comparison, we find the performance of the Random Forest classifier, judging from the AUC of receiver operating characteristic (ROC) curve, is higher than a comparable SVM model. Finally, a Susceptible Infected Recovered (SIR) information propagation model with the forwarding rates obtained from the Random Forest classifier as input parameter is used to simulate the information dissemination process of Weibo. The predicted time behaviors of the Susceptible, Infected, and Recovered populations are in good agreement with real-life data obtained from Sina Weibo.

Keywords: Social network · Information diffusion · Random Forest classifier · Machine learning · SIR model · SMOTE resampling

1 Introduction and Background

Information dissemination in social network is an important tool for marketing and public opinion monitoring. Recently, study of the topological characteristics of social network and their effects on information diffusion across the network become an active area of research. However, so far most researchers have focused their efforts either on

Y. Liu et al. (Eds.): ICNC-FSKD 2019, AISC 1074, pp. 569–576, 2020.
https://doi.org/10.1007/978-3-030-32456-8_62

one or the other and very few holistic approaches have been attempted. The pioneering work of Watts and Strogatz [1] showed that just by rewiring a few edges in the network randomly, the mean shortest path between any pair of nodes in a network can be reduced significantly. This result implies the profound effect of a small change in topology may have on the propagation of information across the network. In this work, we attempt to quantify how topological and other characters of network affect information diffusion based on a Random Forest Classifier combined with SIR model.

As the biggest online social network in China, Sina Weibo is an ideal marketing and public opinion monitoring tool as well as a laboratory for studying such behaviors. In recent years, models of information diffusion in Weibo have been studied actively by researchers [2]. However, existing models are not sufficiently accurate taking into account some of the relevant characteristics of Weibo's users. For example, the individual differences in message forwarding rate among users (nodes) are often overlooked, causing the information dissemination process inadequately modelled.

The relationship between users in Sina Weibo is a directed one. Users of Sina Weibo can follow anyone and browse his/her Microblog homepage without asking for permission. According to one figure [3] the registered accounts of Sina Weibo topped 240 million in June 2015, however it was also reported that there were already 324 million registered Sina Weibo users at the end of the first quarter of 2012, according to another [4]. Anyway, top celebrity on Sina Weibo in 2018 reportedly attracted over 117 million followers [5]. Sina Weibo has a 140 character limit to each post and users are connected in a unidirectional follower/fans-blogger/celebrity relationship. One can *follow* a celebrity and read his/her posts, like and repost (share) them with fans, without being *followed* back.

2 Methodology

In this study, we propose a new approach to model the information diffusion in social networks. For such large and complex network as Sina Weibo it is impossible to get the whole picture of the entire population. We started with a python crawler to collect data on repost activities from Sina Weibo through the fan-blogger relationship related to 316,329 microblogs. From these data a partial (sample of nodes and edges) network associated with these microblogs was recovered, together with characteristic details of the nodes. We analyzed from this those features (as shown in Table 1) that affected the message forwarding behavior. The unbalanced dataset of positive and negative behaviors with regard to reposting is corrected by SMOTE (Synthetic Minority Oversampling Technique) resampling before they are used as target data to train a random forest classification model to predict forwarding behavior of individual users and estimated the message forwarding rates according to the results of classifier.

Finally, a Susceptible Infected Removed (SIR) propagation model was constructed with the forwarding rates estimated in the step above as input parameter, to simulate the overall dissemination process of in Sina Weibo.

This new approach combines the Random Forest classifier with the SIR model together to model the information diffusion process of Weibo. The result shows that the

Table 1. Extracted Features as input to Random Forest classifier.

#	Feature name
1	Blogger's influence
2	Fan's forwarding tendency
3	Historical forwarding frequency between blogger and fan
4	Category of microblog contents
5	Similarity between blogger and fan's interests
6	Similarity between the microblog contents and fan's interest
7	Number of bloggers followed by fan
8	Number of followers of fan
9	Gender of fan
10	Gender of blogger
11	Activity level of fan
12	Activity level of blogger
13	Number of fans of blogger
14	Time of microblog publication
15	Length of microblog in number of words

Random Forest classification with SMOTE achieves a precision of more than 80%. Furthermore, the time history of diffusion process predicted by our SIR model deviated from the actual data with an error less than 15%.

3 Data Pre-processing

When the blogger followed by a fan posts a new microblog, the fan can choose to repost it so that his/her own followers can receive the same message, or ignore it and do nothing so that the spread of information stops. This repost (forwarding) behavior of a fan determines how fast the message of an original microblog diffuses across the network.

There are evidences that this behavior is determined by a number of factors which is summarized in Table 1. Some of these factors are related to the topology of the network, such as the incoming and outgoing edges from the nodes. Some are due to the intrinsic characters of the nodes, like the gender and interest of the fans. Still others are related to the nature of the microblog, like its length and the subject of discussion the microblog. In this work, all these factors are lumped together as the input factors to a supervised machine learning model from which the output is the '*repost*' (positive, 1) or '*not repost*' (negative, 0) response of the node to the microblog.

In reality, the fans would most likely to repost just a few microblogs that they read and really appreciated or excited by. The '*not repost*' is naturally the majority response. Hence the sample we got is disproportionately negative and this lopsided dataset will affect the training of the classification model because the model tends to follow the data to optimize accuracy. In particular, unbalanced dataset where some target factor values occur much more frequently than others may lead to biased trees since the frequently occurring target values are preferred over the less frequently occurring ones.

Using a python crawler, 188750 positive instances and 1916035 negative instances from 500 original microblogs published between 2018/9/13 and 2018/10/21 were obtained. These are the training data for the machine learning model. Later in November of 2018, we further obtained the forwarding data of another ten microblogs from ten bloggers for validation purpose. Specifically, the factors listed in Table 1 are grouped together as input training data, with the forwarding behavior (either positive or negative) as the target data to train our random forest model. The proportion of number of negative instances were roughly 90%, implying that even if the 'trained' model blindly output negative, it still would have an accuracy of at least 90%.

There are two general methods for adjusting the proportion of unbalanced data sets: oversampling and under-sampling. Under-sampling refers to eliminate redundant data in the majority class, while oversampling uses the dataset of minority class to generate pseudo samples that are similar to them. Since under-sampling can easily lead to the loss of valuable information, oversampling is often more popular. SMOTE is a popular heuristic oversampling method which adopts the neighbor interpolation strategy to generate synthetic instances to avoid random oversampling and over-fitting. A schematic illustration of the SMOTE operation is shown in Fig. 1. After data preprocessing with SMOTE, we increased the number of positive instances by a factor of 4, so that the numbers of positive and negative instances are closer to balance.

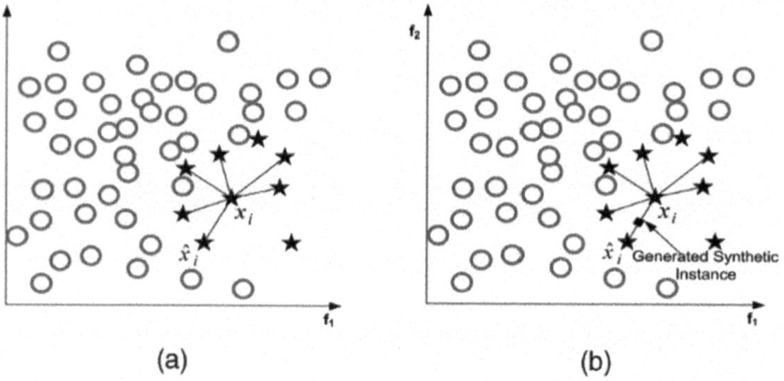

Fig. 1. A schematic illustration of the SMOTE process is shown in 2-dimensional space where f_1 and f_2 are two randomly chosen factors. In (a) the positive (stars) and negative (circles) instances in the original dataset are plotted with most of them negative. To generate a synthetic positive instance, a pair of random positive instances is selected, and a synthetic instance is generated by linear interpolation as shown in (b).

4 Random Forest Model to Predict Repost Behavior

Random Forest is one of the most popular and powerful machine learning algorithms, particularly for problems with categorical factors both in input and target data. It is an ensemble technique that trains a number of decision tree classifiers on various subsamples of the dataset generated by bootstrap with replacement, and uses averaging to improve the predictive accuracy and control over-fitting. In the following, we use 100 trees and out-of-bag samples to estimate the generalization accuracy. The model outputs the classification probabilities in positive or negative response.

For imbalance data, the *accuracy*, which is the percentage of true positive and true negative instance identified, cannot properly measure the performance of classifier. Another common standard, the *Area under ROC Curve* is used as the metric to compare the classifiers. The result is shown in Table 2. The training data is preprocessed by regular SMOTE described above and alternatively under-sample the negative data by one half with one-half SMOTE generated positive data. We have compared the performance of our Random Forest with a SVM Classifier [6] and find the Random Forest Classifier with mixed resampling has better result.

Table 2. AUC_ROC of different classifiers

Resample mode	Classifiers	AUC_ROC
Regular SMOTE	SVM	0.69
SMOTE with random under-sampling	SVM	0.73
Regular SMOTE	Random Forest	0.72
SMOTE with random under-sampling	Random Forest	0.79

As the bagged decision trees are constructed, we can calculate how much the error function (Gini score) drops for a variable at each split point. These drops in error can be averaged across all decision trees and output to provide an estimate of the importance of each input variable. The greater the drop when the variable was chosen, the greater the importance. The Random Forest classifier can assign a numerical value to quantify the relative importance of the input factors. There are six factors which have importance score larger than 0.1 in our study, which are shown in Table 3.

Table 3. Ranking of important factors influencing the repost behavior

Rank	Feature	Importance
1	Historical forwarding frequency between blogger and fan	0.26
2	Similarity between the microblog contents and fan's interest	0.17
3	Blogger's forwarding influence	0.17
4	Activity level of fan	0.13
5	Category of microblog contents	0.11
6	Time of microblog publication	0.10

5 SIR Model to Simulate Information Diffusion

The SIR model is a classical and simple diffusion model that describe the process of outbreak and propagation of epidemic across a population. It breaks down the entire population into three groups: S (Susceptible), I (Infected) and R (Recovered). The set of differential equations that governs the time development of these three groups: *S(t), I(t)* and *R(t)* are:

$$dS/dt = -\lambda IS, \tag{1}$$

$$dI/dt = \lambda IS - \gamma I, \tag{2}$$

$$dR/dt = \gamma I, \tag{3}$$

where λ is the infection rate that changes a Susceptible to an Infected, and γ is the recovery rate that moves an Infected to the Recovered. Note that the total sum ($S + I + R$) is time independent so that the total population is constant in time. In our case, S is the group of fans that is exposed to the microblogs. When the fans decide to repost the microblog they move from the Susceptible to the Infected. After a while, the Infected can be recovered from the excitement and counted as Recovered.

The classification probabilities in positive or negative response of a fan output from the Random Forrest model above is interpreted as the infection rate λ in the SIR model. The recovery rate is a parameter to be determined by model fitting minimizing the discrepancy between model and data. The results of a simulation based on 2018 Sina data is shown in Figs. 2 and 3 below.

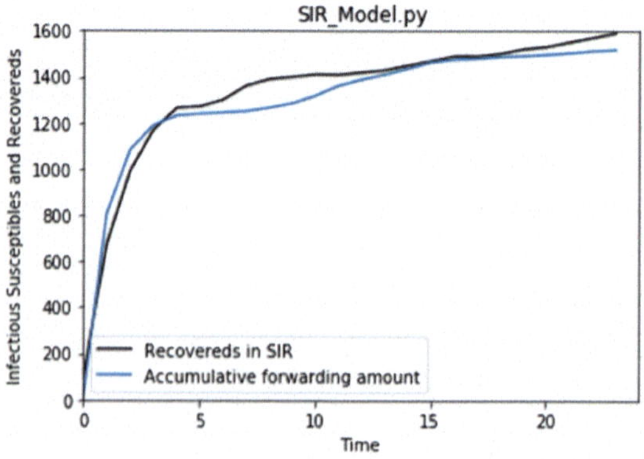

Fig. 2. Result of a SIR model simulation, based on an event in 2018 Sina data, is shown here where the black curve is the time (in unit of hour) history of recovered population, which can be compared with the actual accumulated forwarding repost microblogs amount (blue curve). The total accumulated forwarding repost at the end of the simulation is 1463.

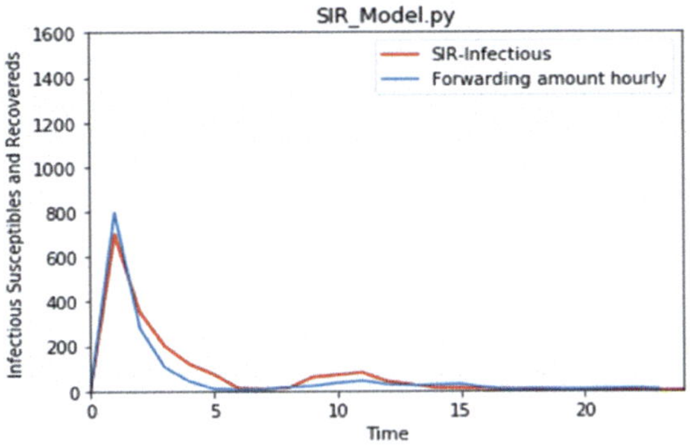

Fig. 3. Result of the same SIR model simulation (Fig. 2) is shown where the red curve is the time history of the newly infected population, which can be compared with the hourly forwarding repost microblogs amount (blue curve).

6 Conclusions and Discussions

In conclusion, the contribution of this work can be summarized in four aspects:

1. We have solved the unbalanced data problem by SMOTE (Synthetic Minority Over-sampling Technique) resampling technique.
2. We have introduced a set of factors from the blogger-fans relationship to predict reposting behavior on Microblog. Some of the new factors introduced by us are later shown to be very important in predicting the reposting behavior.
3. We have developed a Random Forest classifier, trained by Sina Weibo microblog data, to predict the repost probability after the fans received the microblogs.
4. We have employed a SIR model to simulate the information diffusion process in Sina Weibo using the repost probability calculated from the Random Forest classifier.

The results from the last step above are compared with data obtained from Python crawlers gathered from Sina Weibo microblogs during events in November of 2018. We have found that the model predictions and actual data are closely in agreements with each other, except some minor deviations likely due to statistical noises.

The approach adopted here may be extended for further improvement. For example, the most serious limitation of our model accuracy is due to the unbalanced dataset of the positive and negative behaviors with regard to reposting of microblogs. This is corrected by SMOTE resampling in our work before the training process. However, it may be possible or even more efficient to use semi-supervised learning [7, 8] to train the classification model for better results.

The Random Forest classifier can assign a numerical value to quantify the relative importance of the input factors. The most important six factors are shown in Table 3. This useful feature of the Random Forest classifier may be further exploited to reduce the dimension [9, 10] of the input training data. Such dimension reduction technique can lead to a simplified classification model which may have more predictive power.

Acknowledgement. This work is supported by internal research grant from Beijing Normal University-Hong Kong Baptist University-United International College (UIC) R201809.

References

1. Watts, D.J., Strogatz, S.H.: Collective dynamics of 'small-world' networks. Nature **393** (6684), 440–442 (1998)
2. Guo, Z., Li, Z.: Sina microblog: an information-driven online social network. In: International Conference on Cyberworlds, pp. 160–167 (2011)
3. The 36th China Internet network development state statistics report. CONNIC (2015)
4. Wang, X., Chen, Z., Liu, P., Gu, Y.: Edge balance ratio: Power law from vertices to edges in directed complex network. IEEE J. Sel. Top. Sig. Process. **7**(2), 189–194 (2013)
5. Top Weibo accounts of 2018. https://www.whatsonweibo.com/top-weibo-accounts-of-2018-most-popular-celebrities-on-sina-weibo/. Accessed 21 May 2019
6. Chu, F., Wang, L.P.: Applications of support vector machines to cancer classification with microarray data. Int. J. Neural Syst. **15**(6), 475–484 (2005)
7. Liu, B., Wan, C., Wang, L.P.: An efficient semi-unsupervised gene selection method via spectral bi-clustering. IEEE Trans. Nano-Biosci. **5**(2), 110–114 (2006)
8. Zhang, L., Wang, L.P., Lin, W.: Semi-supervised biased maximum margin analysis for interactive image retrieval. IEEE Trans. Image Process. **21**(4), 2294–2308 (2012)
9. Azar, A.T., Hassanien, A.E.: Dimensionality reduction of medical big data using neural-fuzzy classifier. Soft Comput. **19**, 1115–1127 (2015)
10. Wang, L.P., Chu, F., Xie, W.: Accurate cancer classification using expressions of very few genes. IEEE-ACM Trans. Bioinform. Comput. Biol. **4**, 40–53 (2007)

Automatic Segmentation of Visible Epicardium Using Deep Learning in CT Image

Ziyu Zhao[1], Yutaro Iwmoto[1], Yuji Tezuka[2], Hiroki Okada[2], Kiyosumi Maeda[2], Atsuyuki Wada[2], Atsunori Kashiwagi[2], and Yen-Wei Chen[1(✉)]

[1] Graduate School of Information Science and Engineering, Ritsumeikan University, Shiga, Japan
chen@is.ritsumei.ac.jp
[2] Kusatsu General Hospital, Shiga, Japan

Abstract. Recently, many studies indicate that epicardial adipose tissue (EAT) volume closely associates with progression of coronary atherosclerosis as well as increased prevalence of coronary artery disease. Segmentation of the epicardium is an important step for estimation of EAT volume. There are two main challenges facing automatic epicardium segmentation. One is that the epicardium exists as a very thin line and the contrast with the surrounding adipose tissue, blood vessels, and the heart is low. The other is that there are quite a few invisible parts. In this study, we propose a fully deep-learning automatic method for visible epicardium segmentation, which is developed using 2D DenseU-Net, demonstrated that it is possible to automatically segment epicardium, and we also find that the segmentation results of visible epicardium are affected by EAT volume.

Keywords: Epicardium segmentation · Computed Tomography (CT) · Deep learning · DenseU-Net

1 Introduction

Recently, many studies indicate that epicardial adipose tissue (EAT) volume closely associates with progression of coronary atherosclerosis as well as increased prevalence of coronary artery disease [1–3]. Epicardium is the outer layer of the heart wall. The epicardium functions to protect the inner heart layers and also assists in the production of pericardial fluid. Segmentation of the epicardium is an important step for estimation of epicardial adipose tissue (EAT) volume and. A typical cardiac Computed Tomography (CT) image is shown in Figs. 1. and 4(a). The fat inside of the epicardium is the epicardial adipose tissue (epicardial fat) and the fat outside of the epicardium is the paracardial fat. As shown in Fig. 1, the epicardium exists as a very thin line and the contrast with the surrounding adipose tissue, blood vessels, and the heart is low. There are also quite a few invisible parts of the epicardium. Because of the above two

Y. Liu et al. (Eds.): ICNC-FSKD 2019, AISC 1074, pp. 577–584, 2020.
https://doi.org/10.1007/978-3-030-32456-8_63

characteristics, the automatic segmentation of the epicardium is a challenge research topic. There has been some researches on automatic epicardium segmentation by convolutional neural network [4, 5]. Since these methods are patch-based methods, the structure of the entire cardiac region cannot be taken into consideration.

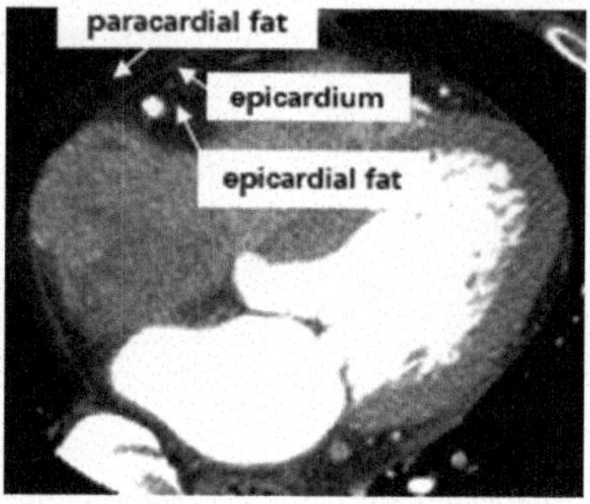

Fig. 1. Typical CT image of the cardiac.

In this paper, we proposed a fully deep-learning automatic method for visible epicardium segmentation in CT images, which is developed using fully convolutional network (DenseU-Net) [6]. In the proposed method, the whole image is used as input. The remainder of this paper is organized as follows. In Sect. 2, we describe the details of the proposed method; experiments and results are presented in Sect. 3. Section 4 presents the conclusions.

2 Proposed Method

We propose a fully deep-learning automatic method for visible epicardium segmentation, which is developed using 2D DenseU-Net [6], DenseU-net is based upon the fully convolutional network U-net. In this section, we describe the details of U-Net [6] and dense U-Net [7].

2.1 Fully Convolutional Network (U-Net)

In 2015, Ronneberger et al. proposed the U-Net, which is a state-of-the-art method to solve the biomedical-image-segmentation problem with very few annotated images [7], as shown in Fig. 2. The typical U-net contains encoding path (down-sampling) and decoding part (up-sampling). In our study, we also use U-net architecture for the visible epicardium segmentation.

2.2 Fully Dense Connected Convolutional Network (DenseU-Net)

In 2018, Li et al. presented a densely connected network (called H-DenseU-Net) for tumor and liver segmentation [6], which combines a 2D DenseU-Net and 3D DenseU-net. The H-DenseU-Net outperformed other state-of-the-arts on the segmentation results of tumors and livers.

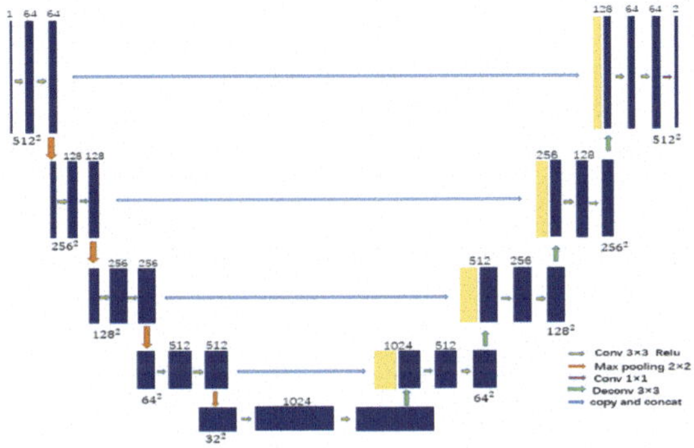

Fig. 2. The architecture of U-Net we used.

In our study, we proposed automatic visible epicardium segmentation method is based upon the 2D DenseU-net. The illustration and detailed structure are shown in Fig. 2. The DenseU-net encoder part follows DenseNet-161 [8], which designed for the object classification task. It has may dense block as shown in Fig. 3(a). For segmentation tasks, DenseU-net also apply UNet-like connections as encoding part. A simplified architecture of DenseU-Net is shown in Fig. 3(b). The loss function of origin 2D DenseU-Net is based on weighted standard cross-entropy [6]. But in our experiments, Dice loss is used on the visible epicardium segmentation. The Dice Coefficient (DICE), described in [9, 10], also called the overlap index, is a commonly used in medical segmentation. For the binary ground truth images of each class, Dice can be written as:

$$DICE_k = \frac{2\sum_{i=1}^{N} p_{ik} g_{ik}}{\sum_{i=1}^{N} p_{ik} + \sum_{i=1}^{N} g_{ik}} \tag{1}$$

Where $g_{ik} \in \{0, 1\}$ is the ground truth label of class k and pixel i, $p_{ik} \in [0, 1]$ specifies the output probability, and N denotes the total number of pixels. Therefore, the Dice Loss (DL) is:

$$\text{DL} = \sum_k \omega_k (1 - DICE_k) \tag{2}$$

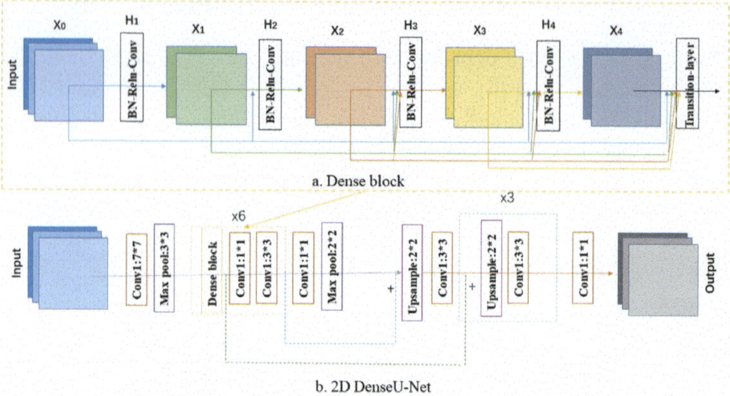

Fig. 3. The architecture of 2D DenseU-Net we used.

3 Experiments and Results

3.1 Dataset

In this study, we used 14 cases of contrast-enhanced cardiac CT images collected in Kusatsu General Hospital, Japan, which include cases with 4 cases of large EAT volume, 5 cases of middle EAT volume and 5 cases of small EAT volume. All images were acquired by a CT scanner (LightSppedVCT, GE healthcare Japan) with retrospective ECG gating scan. The thickness of the CT volumes is 0.625 mm. In total we have 2,726 slices, with the size of each image being 512 × 512 pixels. All ground truth images are annotated by experts. Pairs of CT images (Fig. 4(a)) and visible epicardium traces (Fig. 4(b)) are used as training data for the network.

Our experiments were run on a TITANX 12 GB memory GPU and the network was implemented using Keras. The training of 2D DenseU-Net model took about 26 h in our experiment, in test phase, one case took about 100 s.

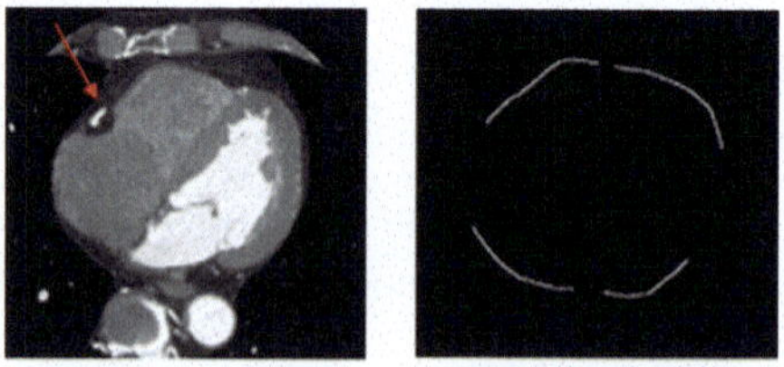

(a) CT image of the cardiac **(b) Trace result of the visible epicardium**

Fig. 4. The data we used in experiment.

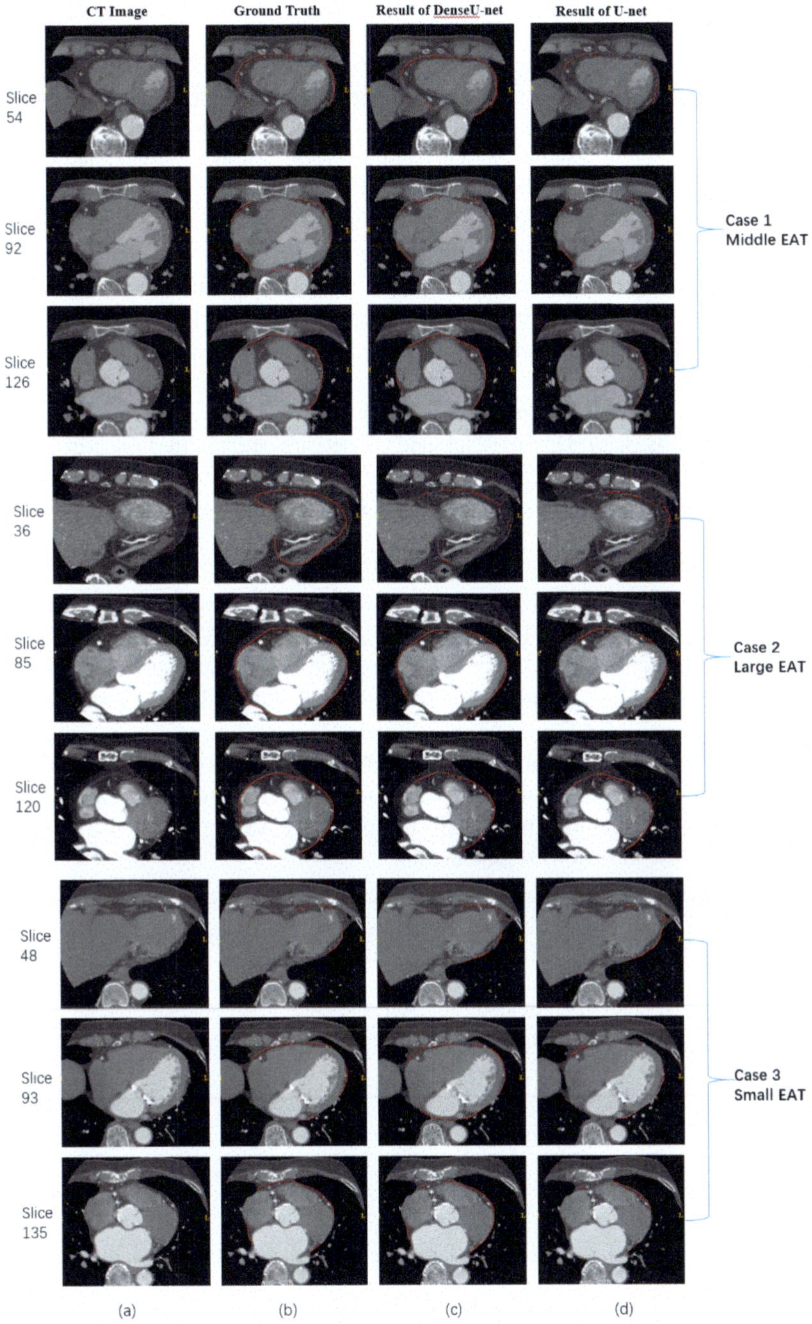

Fig. 5. Epicardium segmentation results of three test cases.

3.2 Experiments

In order to evaluate the performance of the epicardium segmentation, we conducted segmentation experiments using a leave-one-out method. 13 cases were used as training dataset and the remaining one was used as test. The experiments were repeated 14 times with a different test image. The segmentation results are evaluated quantitatively using the Dice coefficient shown in Eq. 1.

We also conducted comparative experiments as to verify the performance of the proposed method, which included U-net method, fully convolution network (FCN) method, and graph-cut method for epicardium segmentation [7, 11, 12]. The results are summarized in Table 1. Detailed segmentation results of three test cases (large EAT, middle EAT, small EAT) with three slices (upper, middle, lower) obtained by the proposed method are shown in Fig. 5. CT slice images (input images) and ground truth of the visible epicardium, are shown in Fig. 5(a), (b) respectively. The segmentation results with DenseU-Net and U-net are shown in Fig. 5(c) and (d), respectively. As shown in Fig. 5, the epicardium can be automatically segmented well by DenseU-Net than U-Net method. There are some bad segmentation results of DenseU-Net shown in Fig. 6.

We also conducted comparative experiments by large EAT volume, middle EAT volume and small EAT volume, we have 14 cases of contrast-enhanced cardiac CT images, with 4 cases large EAT volumes, 5 cases middle EAT volume and 5 cases small EAT volume. The average Dice results are summarized in Table 2. As shown in Fig. 5, there are three cases segmentation results which with middle EAT volume, large EAT volume and small EAT volume respectively, each case has three test slices from top to bottom. we can find that the segmentation results of visible epicardium are affected by EAT volume. Since the epicardium have lower contrast than fat, large EAT volumes have poor results than middle EAT and small EAT, the larger the fat volume, the worse result of epicardium segmentation.

Table 1.

Method	Average Dice
DenseU-Net	0.48
U-Net	0.26
FCN	0.21
Graph-cut	0.04

Table 2. Comparison of the visible epicardium-segmentation results with large EAT, middle EAT and small EAT volumes.

EAT volume	Number of cases	Dice
Large EAT	4 cases	0.46
Middle EAT	5 cases	0.51
Small EAT	5 cases	0.5

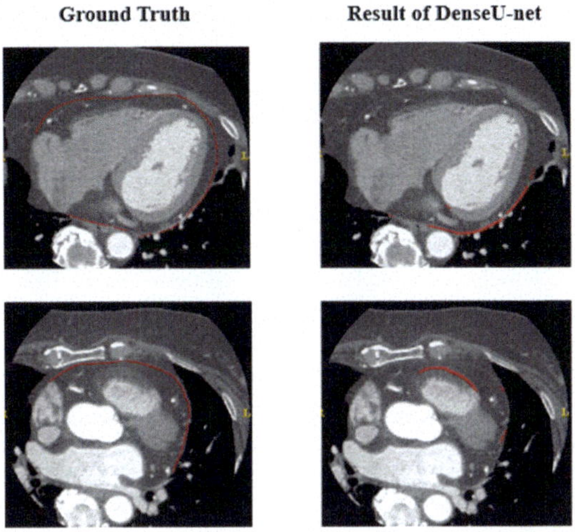

Fig. 6. Epicardium segmentation results of some bad cases.

4 Conclusions

In this study, we proposed a method to automatically segment the epicardium using 2D DenseU-net. Our proposed method is fully automated and used whole image as input. The segmentation result achieved mean Dice of 0.48 by using 14 cases CT images. we also find that the segmentation results of visible epicardium are affected by EAT volume. Since automatic segment the region where the epicardium cannot be seen is a fundamental task for segmentation of epicardial adipose tissue (EAT). For future work, we are going to segment invisible epicardium and epicardial adi-pose tissue (EAT).

Acknowledgements. This work was supported in part by the Grant-in Aid for Scientific Research from the Japanese Ministry for Education, Science, Culture and Sports (MEXT) under the Grant No. 18H03267 and No. 17K00420.

References

1. Shimabukuro, M., et al.: Epicardial adipose tissue volume and adipocytokine imbalance are strongly linked to human coronary atherosclerosis. Arterioscler. Thromb. Vasc. Biol. **33**, 1077–1084 (2013)
2. Lu, M.T., et al.: Epicardial and paracardial adipose tissue volume and attenuation. Atherosclerosis **251**, 47–54 (2016)
3. Mahabadi, A.A., et al.: Association of pericardial fat, intrathoracic fat, and visceral abdominal fat with cardiovascular disease burden: The Framingham heart study. Eur. Heart J. **30**(7), 850–856 (2009)
4. Commandeur, F., et al.: Deep learning for quantification of epicardial and thoracic adipose tissue from non-contrast CT. IEEE Trans. Med. Imaging **37**, 1835–1846 (2018)

5. Barbosa, J.G., Figueiredo, B., Bettencourt, N., et al.: Towards automatic quantification of the epicardial fat in non-contrasted CT images. Comput. Methods Biomech. Biomed. Eng. **14** (10), 905–914 (2011)
6. Li, X., et al.: H-DenseUNet: hybrid densely connected UNet for liver and tumor segmentation from CT volumes. IEEE Trans. Med. Imaging **37**, 2663–2674 (2018)
7. Ronneberger, O., Fischer, P., Brox, T.: U-Net: convolutional networks for biomedical image segmentation. In: Navab, N., Hornegger, J., Wells, W.M., Frangi, A.F. (eds.) MICCAI 2015. LNCS, vol. 9351, pp. 234–241. Springer, Cham (2015)
8. Huang, G., et al.: Densely connected convolutional networks (2016)
9. Milletari, F., Navab, N., Ahmadi, S.-A.: V-Net: fully convolutional neural networks for volumetric medical image segmentation, pp. 1–11. arXiv (2016)
10. Sudre, C.H., Li, W., Vercauteren, T., Ourselin, S., Jorge Cardoso, M.: Generalised dice overlap as a deep learning loss function for highly unbalanced segmentations. In: Cardoso, M.J., et al. (eds.) DLMIA/ML-CDS-2017. LNCS, vol. 10553, pp. 240–248. Springer, Cham (2017)
11. Long, J., Shelhamer, E., Darrell, T.: Fully convolutional networks for semantic segmentation. IEEE Trans. Pattern Anal. Mach. Intell. (2014)
12. Freedman, D., Zhang, T.: Interactive graph cut based segmentation with shape priors. In: IEEE Computer Society Conference on Computer Vision & Pattern Recognition (2005)

Semi-automatic Cephalometric Landmark Detection on X-ray Images Using Deep Learning Method

Yu Song[1], Xu Qiao[2], Yutaro Iwmoto[1], and Yen-Wei Chen[1](✉)

[1] Graduate School of Information Science and Engineering, Ritsumeikan University, Shiga, Japan
chen@is.ritsumei.ac.jp

[2] Department of Biomedical Engineering, Shandong University, Shandong, China

Abstract. Accurate quantitative cephalometry plays a significant role in both diagnosis and treatment. However, it is tedious work and time consuming to label cephalometric landmarks manually. It also requires professional doctors to implement. Deep learning has been a hot topic in recent years and has achieved great success in many image classification and image regression problems. In this paper, we propose a semi-automatic method for detection of cephalometric landmarks using deep learning. We first roughly extract an ROI region for each landmark manually. Then we utilize Resnet50, which is a state-of-the-art convolutional neural network, to detect the landmark in the ROI region. The network directly output the coordinates of the landmark. The dataset we used in this paper is a public dataset—ISBI 2015 grand challenge in dental x-ray image analysis. Experiments demonstrated that the proposed method achieved better results compared with state-of-the-art methods.

Keywords: Cephalometric landmark · X-ray · Deep learning · Resnet

1 Introduction

Cephalometric analysis is mandatory for doctors to make orthodontic diagnosis. It provides doctors necessary information like bone structure, soft tissue, dental structure and so on [1–3]. In order to perform such analysis, 2D x-ray images of head bones are needed. Cephalometric analysis usually needs experienced doctors to identify different landmarks' locations, then measure the angles and distances between these landmarks, so that they can classify patients' situation to different levels. This is really time consuming and tedious work, sometimes can even cause intra-observer variability issues since different doctors may vary a lot in identifying some landmarks [4]. Therefore, computer-aided judgement which can identify landmarks with high accuracy is highly desired.

Y. Song and X. Qiao—Contributed equally to this paper.

Y. Liu et al. (Eds.): ICNC-FSKD 2019, AISC 1074, pp. 585–592, 2020.
https://doi.org/10.1007/978-3-030-32456-8_64

There have been many studies for anatomical landmark detection. In particular, the international Symposium launched the challenge on Cephalometry landmark detection in 2014 and 2015 [5, 6], several benchmark methods were raised. Among them, Lindner and Cootes used random-forest based method which achieved the best result, with successful detection rate of 74.84% for 2 mm precision range [7] (Certain amount of error between prediction and ground-truth is acceptable, medically acceptable range is 2 mm). Ibragimov et al. used Harr-like feature extraction and random forest regression method, then used global-context shape model to modify the result, they achieved the second-best result, with 71.7% successful detection accuracy within 2 mm range [8]. In 2017, Ibragimov et al. improved their method by adding convolutional neural network to do binary classification, they achieved a little improvement compared with Lindner's result, with around 75.3% prediction accuracy within 2 mm range [9]. Despite the variety of techniques being used, the problem is still an open topic due to the limitation of the limited prediction accuracy.

In recent years, deep learning has been applied to many image processing problems and has achieved extremely great performance. Different types of convolutional neural network models are widely used in landmark detection [10, 11], image classification [12–14] and image segmentation [15, 16] problems. In many of these applications, their performance can already surpass human-beings. In this paper, we propose a semi-automatic method for detection of cephalometric landmarks using deep learning. We first roughly extract an ROI region (patch image) for each landmark manually. Then we utilize Resnet50, which is a state-of-the-art convolutional neural network, to detect the landmark in the ROI region. Experiments demonstrated that the proposed method achieved better results compared with state-of-the-art methods.

The paper is organized as follows, in Sect. 2, we will illustrate the details of our work. Then we will describe the dataset and show the experimental results in Sect. 3. At last, we will make a conclusion in Sect. 4.

2 Proposed Method

2.1 Preprocessing

Figures 1 and 4 are typical cephalometry x-ray images. There are 19 landmarks that need to be detected, which are listed in Table 2 and shown in Fig. 4. For each cephalometry x-ray image, the resolution is 1935 by 2400 in Tiff format. The scale is so huge that we cannot input the whole image into the neural network to predict all 19 landmark's location at one time. Therefore, we crop the image into patches. Each patch contains only one landmark as input, which can be considered as an ROI region. We trained the network individually for each landmark. That means we have 19 models (networks) for detecting 19 landmarks. For training samples, we randomly select 200 ROI patches for each landmark with a size of 128×128 pixels as shown in Fig. 1, which can be considered as a kind of data augmentation with translation. The procedure is described in Table 1. The extracted ROI patch is used as an input image of a deep convolutional neural network for landmark detection.

Table 1. Procedures to extract ROI patches from training data

Procedures to extract ROI patches from training data:
Step1: Find the ground-truth location of landmark A: (x, y). Create a list V, which is used to store the ROI vertices' coordinates
Step2: Randomly select a 128 × 128 patch in the region of [x−128: x + 128, y−128: y + 128]
Step3: Check if the four vertices are in the list V, if not, append four vertices into V, else, repeat step2
Step4: Repeat step 1 to step 3, until we have 200 sets of patch vertices in V
Step5: Crop 200 ROI patches using vertices in V for each landmark

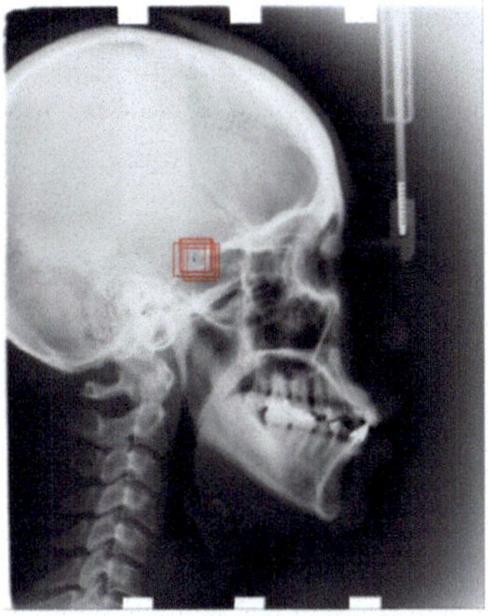

Fig. 1. The way we crop ROI patches for each landmark: the blue dot is a target landmark, the red boxes are extracted 128 × 128 ROI patches. We can see the landmark is included in all the ROI patches. We randomly extract 200 patches for one landmark in one x-ray image for training.

2.2 Resnet50 Model

In our work, we use Resnet50 model proposed by He et al. in [17] to extract features for landmark detection. Convolutional neural networks can automatically extract the useful features from the input image. Based on the features we can use methods like SVR, random-forests or fully-connected neural networks to estimate the landmark location. General neural networks will encounter gradient vanishing problems if the network goes too deep. However, Resnet provides a solution as shown in Fig. 2 by adding the former feature maps to the later feature maps. In this way, the feature information is passed to the later layer which works well even with deep architecture.

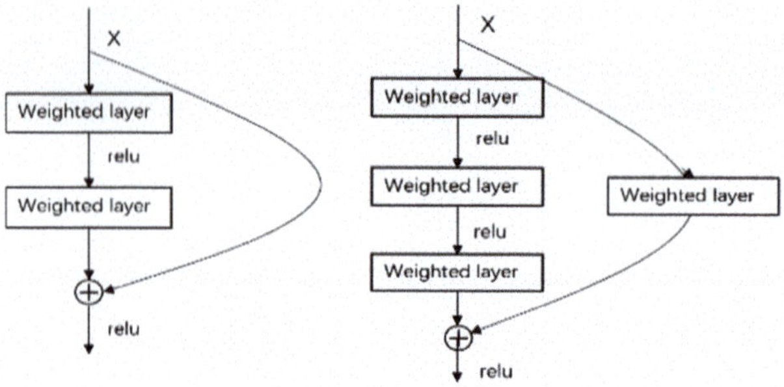

Fig. 2. Identity block on the left, add former input directly to the later output feature map. Convolutional block on the right, first apply convolutional and batch normalization on the input, then add to the later feature map.

The over-all architecture of our proposed method is shown in Fig. 3. The input is the ROI patch image. After feature extraction using Resnet50, we use fully connected (FC) layers to estimate the landmark location. The output is the coordinate (x and y) of the landmark as shown in Fig. 3.

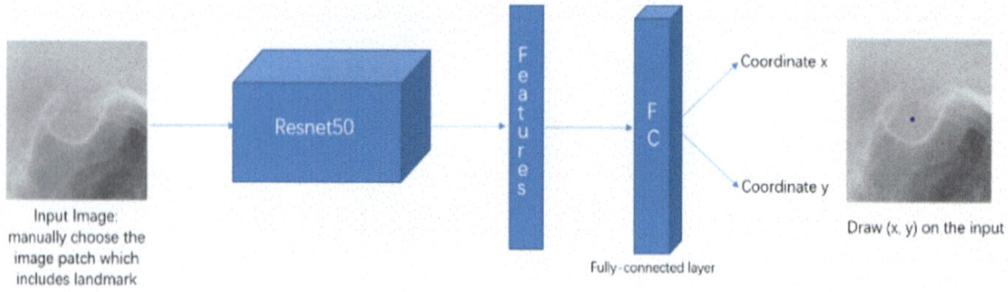

Fig. 3. The over-all architecture of proposed method, the output is x and y coordinate of landmark.

The loss function we used is MSE (Mean squared error). It is commonly used for regression problems, which is represented as:

$$\mathrm{MSE} = \frac{1}{n}\sum\nolimits_{i=1}^{n}\left((x_i - \hat{x}_i)^2 + (y_i - \hat{y}_i)^2\right) \quad (1)$$

where x_i and y_i represent the ground-truth predicted coordinate, $\hat{x}_i$ and $\hat{y}_i$ represent the predicted coordinate.

3 Experiment and Result

3.1 Dataset

The dataset we used is 2015 ISBI grand challenge data [6]. It contains 150 images for training, 150 images as testset1, 100 images as testset2. Each image is 1935 by 2400 in

Tiff format. Each pixel's size is 0.1 mm. The images are labeled by two experienced doctors, one with 5 years of experience, the other with 15 years of experience. In our experiment, we take the mean coordinate value of two doctors as the ground-truth value.

3.2 Experiments

In our experiments, we use Titan X GPU (12 Gb) to accelerate training procedure. The experiments are implemented using python with Keras. For training, the learning rate is set to 0.001. The optimizer we used is Adam optimizer. Besides. we spare 10% of our training data for validation. We set early-stopping if the validate MSE value remain unchanged for 5 epochs.

According to ISBI grand challenge [3]. we use the mean radial error to evaluate the performance. Radial error is described in the Eq. 2, where Δx Δy are the absolute differences of x-axis and y-axis, respectively, between the predicted landmark and ground truth landmark.

$$\mathrm{R} = \sqrt{\Delta x^2 + \Delta y^2} \tag{2}$$

And the mean radial error(MRE) is defined as follows:

$$\mathrm{MRE} = \frac{\sum_{i=1}^{n} R_i}{n} \tag{3}$$

If the radial error between the predicted landmark and the ground-truth value is no greater than z mm (where z = 2, 2.5, 3, 4), the detection is considered as accurate. The equation of the successful detection rate is defined as Eq. 4, where N_a means the number of accurate detection, N means the number of total detection:

$$P_z = \frac{N_a}{N} * 100\% \tag{4}$$

For testing, we first randomly crop a ROI patch image with 128 × 128 pixels for each landmark in one x-ray image based on the ground-truth value. Then we input all patches into the trained models. After we have prediction results, we calculate the mean radial error between predictions and ground-truth values. At last, we calculate the successful detection rate for four types of measurement (2 mm, 2.5 mm, 3 mm, 4 mm) and the overall mean radial error of all the landmarks. The results for test dataset 1 and test dataset 2 are shown in Table 2.

We also compared our results with state-of-the-art methods. Table 3 shows the comparison results. We can see that the proposed method achieved better results (higher accuracy) than the state-of-the-art methods.

Figure 4 shows one of the examples of detection results by the proposed method. The green dots represent the detection result, while the red dots represent the ground-truth location of the landmarks. All landmarks were detected well.

Table 2. Results on Testset1 and Testset2 of range 2 mm, 2.5 mm, 3 mm, 4 mm and mean radial error

Testset1/Testset2					
Anatomical Landmarks	2 mm (%)	2.5 mm (%)	3 mm (%)	4 mm (%)	Mean radial error (mm)
1. Sella turcica	99.3/96	99.3/96	100/96	100/99	0.6248/0.6936
2. Nasion	86.7/95	91.3/95	93.3/97	96.7/99	1.0602/0.7582
3. Orbitale	81.3/56	92/75	94.7/85	100/97	1.3536/1.9988
4. Porion	66.7/80	71.3/83	79.3/89	95.3/93	1.83756/1.4296
5. Subspinale	76/82	84/89	90.7/93	97.3/99	1.58698/1.4360
6. Supramentale	73/39	84/54	90.7/67	95.3/86	1.45519/2.5356
7. Pogonion	87.3/94	93.3/96	96/99	98/99	1.0288/1.4360
8. Menton	94/99	96/100	98/100	99.3/100	0.9198/0.6525
9. Gnathion	92/99	95.3/100	98/100	100/100	0.9524/0.6661
10. Gonion	59.3/80	74/86	86/91	96/97	1.83/1.3558
11. Lower incisal incision	96.7/93	98/94	98.7/97	100/98	0.5588/0.6744
12. Upper incisal incision	96.7/97	99.3/98	100/98	100/99	0.43381/0.4809
13. Upper lip	86/67	95.3/85	99.3/95	100/97	1.3199/1.7903
14. Lower lip	99.3/76	100/89	100/97	100/100	0.8295/1.7497
15. Subnasale	95.3/96	96.7/96	98/99	99.3/99	0.8095/0.7979
16. Soft tissue pogonion	77.3/54	87.3/71	92.7/91	98.7/100	1.4679/1.9559
17. Posterior nasal spine	90.7/80	94.7/90	97.3/94	99.3/99	1.0086/1.3573
18. Anterior nasal spine	87.3/84	92/93	95.3/96	98.7/100	1.1568/1.1054
19. Articulate	69.3/88	79.3/94	87.398	96/100	1.5676/1.0054
Average	85.0/81.8	90.7/88.06	94.5/93.8	98.4/97.95	1.14746/1.22347

Table 3. Success detection rates of the proposed method compared with the state-of-the-art methods on ISBI 2015 grand challenge Testset1 and Testset2.

Testset1/Testset2				
Method	2 mm (%)	2.5 mm (%)	3 mm (%)	4 mm (%)
Ibrgimov [8]	71.9/62.74	77.4/70.47	81.93/76.53	88.04/85.11
Lindner [7]	73.68/66.11	80.21/72	85.19/77.63	91.47/87.43
Ibragimov [9]	75.37/67.68	80.91/74.16	84.32/79.11	88.25/84.63
Proposed	**85.0/81.8**	**90.7/88.6**	**94.5/93.79**	**98.4/97.95**

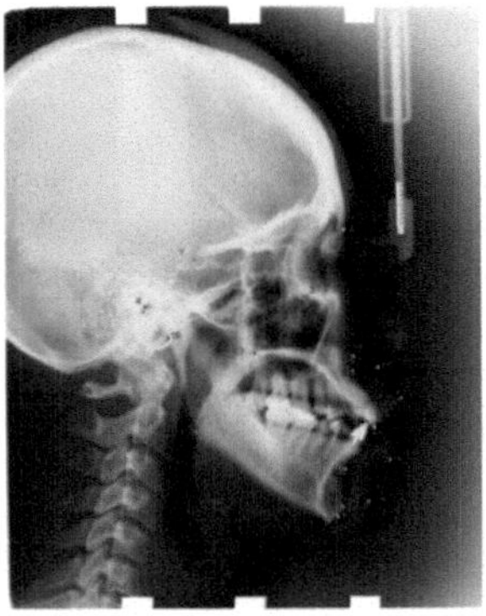

Fig. 4. The green dots represent the prediction result, while the red dots represent the ground-truth location of the landmarks.

4 Conclusion

The experiments demonstrated that the proposed method achieved better performance than state-of-the-art methods, which proves that the features extracted by convolutional neural network is essential to identify landmarks' locations. Some landmarks like the 6th, supramentale, its accuracy is relatively lower than other landmarks' accuracy. From the manual annotations, we can see that even between two experienced doctors, the judgement varies a lot. For the same picture, there are cases that the radial error between two doctor's labels is more than 2 mm. This proves that the 6th landmark itself is very hard to judgement. The intra-variety of the training labels explains the reason of the relatively low accuracy.

So far, in the test phase we used a manual way to find the ROI region. In the future, we are aimed to design a fully automatic landmark detection method which can decide the ROI region first, then input the patch into the neural network to do detection. We are also going to increase the training data in the future.

Acknowledgements. This work was supported in part by the National Natural Science Foundation of China under the Grant No. 61603218, and in part by the Grant-in Aid for Scientific Research from the Japanese Ministry for Education, Science, Culture and Sports (MEXT) under the Grant No. 18H03267 and No. 17K00420.

References

1. Kaur, A., Singh, C.: Automatic cephalometric landmark detection using Zernike moments and template matching. SIViP **9**(1), 117–132 (2015)
2. Grau, V., Alcaniz, M., Juan, M.C., Monserrat, C., Knoll, C.: Automatic localization of cephalometric landmarks. J. Biomed. Inform. **34**(3), 146–156 (2001)
3. Ferreira, J.T.L., de Souza Telles, C.: Evaluation of the reliability of computerized profile cephalometric analysis. Braz. Dent. J. **13**(3), 201–204 (2002)

4. Yue, W., Yin, D., Li, C., Wang, G., Tianmin, X.: Automated 2-D cephalometric analysis on X-ray images by a model-based approach. IEEE Trans. Biomed. Eng. **53**(8), 1615–1623 (2006)
5. Wang, C.-W., Huang, C.-T., Hsieh, M.-C., Li, C.-H., Chang, S.-W., Li, W.-C., Vandaele, R., et al.: Evaluation and comparison of anatomical landmark detection methods for cephalometric x-ray images: a grand challenge. IEEE Trans. Med. Imaging **34**(9), 1890–1900 (2015)
6. Wang, C.-W., Huang, C.-T., Lee, J.-H., Li, C.-H., Chang, S.-W., Siao, M.-J., Lai, T.-M., et al.: A benchmark for comparison of dental radiography analysis algorithms. Med. Image Anal. **31**, 63–76 (2016)
7. Lindner, C., Cootes, T.F.: Fully automatic cephalometric evaluation using random forest regression-voting. In: IEEE International Symposium on Biomedical Imaging (2015)
8. Ibragimov, B., Likar, B., Pernus, F., Vrtovec, T.: Computerized cephalometry by game theory with shape-and appearance-based landmark refinement. In: Proceedings of International Symposium on Biomedical imaging (ISBI) (2015)
9. Arik, S.Ö., Ibragimov, B., Xing, L.: Fully automated quantitative cephalometry using convolutional neural networks. J. Med. Imaging **4**(1), 014501 (2017)
10. Ramanan, D., Zhu, X.: Face detection, pose estimation, and landmark localization in the wild. In: Proceedings of the 2012 IEEE Conference on Computer Vision and Pattern Recognition (CVPR), pp. 2879–2886 (2012)
11. Belhumeur, P.N., Jacobs, D.W., Kriegman, D.J., Kumar, N.: Localizing parts of faces using a consensus of exemplars. IEEE Trans. Pattern Anal. Mach. Intell. **35**(12), 2930–2940 (2013)
12. Simonyan, K., Zisserman, A.: Very deep convolutional networks for large-scale image recognition. arXiv preprint arXiv:1409.1556 (2014)
13. Krizhevsky, A., Sutskever, I., Hinton, G.E.: ImageNet classification with deep convolutional neural networks. In: Advances in Neural Information Processing Systems, pp. 1097–1105 (2012)
14. Ciresan, D.C., Meier, U., Masci, J., Gambardella, L.M., Schmidhuber, J.: Flexible, high performance convolutional neural networks for image classification. In: Twenty-Second International Joint Conference on Artificial Intelligence (2011)
15. Long, J., Shelhamer, E., Darrell, T.: Fully convolutional networks for semantic segmentation. In: Proceedings of the IEEE Conference on Computer Vision and Pattern Recognition, pp. 3431–3440 (2015)
16. Ronneberger, O., Fischer, P., Brox, T.: U-Net: convolutional networks for biomedical image segmentation. In: International Conference on Medical Image Computing and Computer-Assisted Intervention, pp. 234–241. Springer, Cham (2015)
17. He, K., Zhang, X., Ren, S., Sun, J.: Deep residual learning for image recognition. In: Proceedings of the IEEE Conference on Computer Vision and Pattern Recognition, pp. 770–778 (2016)

Robust Hand Gesture Recognition Using Multimodal Deep Learning for Touchless Visualization of 3D Medical Images

Kotaro Furusawa[1], Jiaqing Liu[1], Seiju Tsujinaga[1], Tomoko Tateyama[2], Yutaro Iwamoto[1], and Yen-Wei Chen[1,3,4](✉)

[1] Ritsumeikan University, Kusatsu, Shiga 525-8577, Japan
chen@is.ritsumei.ac.jp
[2] Hiroshima Institute of Technology, Hiroshima, Hiroshima 731-5193, Japan
[3] Zhejiang Lab, Hangzhou, China
[4] College of Computer Science and Technology, Zhejiang University, Hangzhou, China

Abstract. Three-dimensional (3D) visualization of medical images is an important technology for efficiently conducting a surgery. However, efficient review of 3D anatomical models is required to maintain sterile field conditions. An operation using touchless interface for gesture recognition is one of the review methods. Real-time hand gesture application for supporting a surgery requires a robust recognition of various gestures. This study proposes a robust hand gesture recognition using multimodal deep learning to perform recognition using color and depth images. We evaluated the recognition accuracy of 25 different gestures and compared its recognition accuracy with conventional recognition methods. Resultantly, it was found that the proposed system achieves better real-time robust recognition than conventional methods.

Keywords: Hand gesture recognition · Multimodal deep learning · Surgery aid system

1 Introduction

High-definition medical images are important for performing advanced medical practices on a clinical site [1, 2]. In addition, it is important for the surgeon to review medical images while maintaining sterility during surgery [3]. Several touchless surgery support systems using Kinect have been developed for medical image visualization [4–6]. However, some limitations still persist in these systems. Gesture interface for medical interaction [7] is not sufficient for robust and real-time touchless visualization system.

In our previous studies, four versions have been developed [8–11]. In the fourth version [11], we built a new multiview RGB-D dataset, namely MaHG–RGBD. Moreover, we performed a recognition experiment to visualize depth images using deep learning. MaHG–RGBD comprises 25 classes of gestures, proposing 9 classes of gestures with high recognition accuracy. However, the average recognition accuracy of 9 and 25 classes was 96.51% and 91.87%, respectively. These showed that some

Y. Liu et al. (Eds.): ICNC-FSKD 2019, AISC 1074, pp. 593–600, 2020.
https://doi.org/10.1007/978-3-030-32456-8_65

gestures were difficult to recognize using only depth image. Accurate gesture recognition is important for performing medical operations within the image interaction system. The proposed system is an improved version because we have constructed a multimodal deep learning network that adds color information to depth information. This network uses the depth and color images for recognition. In the depth image, gesture recognition accuracy is improved. Thus, the proposed system outperforms our previous system from the viewpoint of recognition accuracy, rapidity, and flexible Kinect based touchless visualization.

The remainder of this paper is as follows. Section 2 describes the proposed system, including a detailed description of a hand preprocessing and hand gesture recognition using deep learning. Experimental results are discussed in Sect. 3, and conclusion is given in Sect. 4.

2 Proposed System

Figure 1 summarizes the system architecture of the proposed system, which comprises two modules, i.e., interaction and visualization modules.

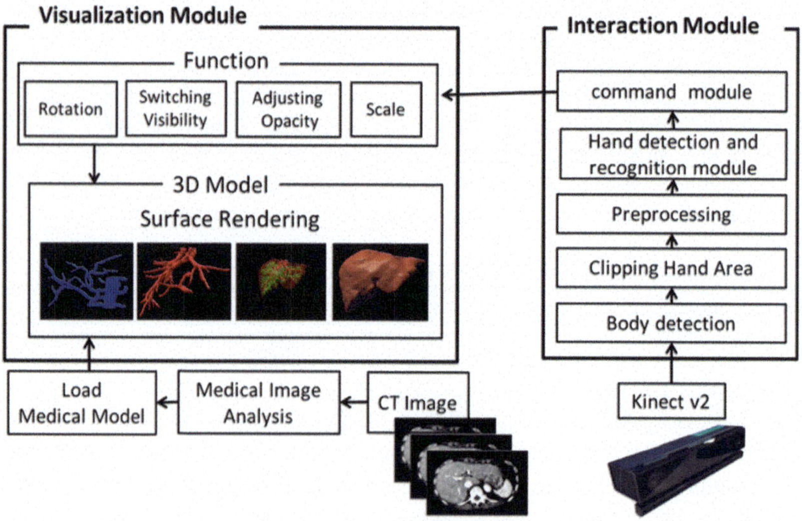

Fig. 1. Proposed system.

2.1 Hand Image Pre-processing

Hand region images are generated using skeleton tracking provided by Kinect. Figure 2 shows the procedure for generating a hand depth image. First, we acquire a depth image of the user (Fig. 2(a)). Then, we cut out a 300 $\times$ 300 square area centered on the coordinates of the right hand from the depth image (Fig. 2(b)). As can be seen in Fig. 2 (c), from the cut depth image, an area with a distance value of 5 and 30 cm at the front and back, respectively, from the distance of the right hand of the user was extracted. We performed the opening process to remove the noise (Fig. 2(d)) and removed small areas by shrinking and expanding the area. Depending on the distance between Kinect and user, the pixel value of the hand area changes. Corresponding to this change, the distance, d, of the right hand minus 30 cm is subtracted from the pixel value of the hand area (Fig. 3). To reduce the computation time in our analysis of the hand shape regions, we cropped and resized the image to 224 × 224 for an input of the deep convolutional neural network.

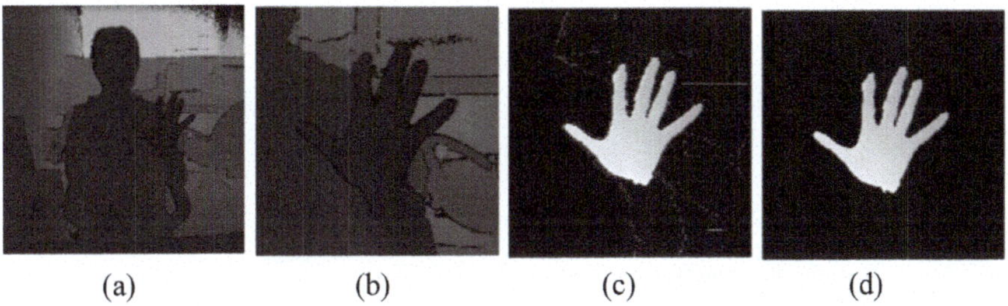

Fig. 2. The procedure for generating a hand depth image: (a) The depth image of the user, (b) Cut out a 300 × 300 square area of the hand, (c) Extracted hand depth image, including noise, and (d) Hand depth image, excluding noise.

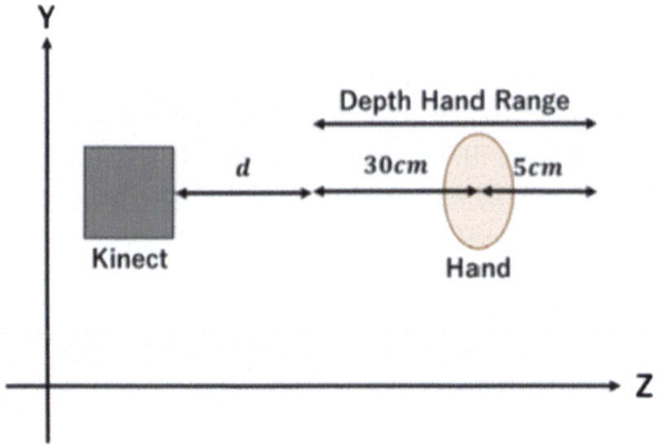

Fig. 3. Hand depth image normalization.

2.2 Recognition Using Multimodal Image

The proposed method employs multimodal deep learning gesture recognition consisting of two types of images, i.e., depth and color images. This method aims to improve gesture recognition accuracy using depth and color images. Figure 4 shows the network architecture of multimodal deep learning. The depth and color images are input into the two AlexNet [12] learned with the distance and color images, respectively. The average value of the network output is taken as the final output. The output is computed with the following Eq. (1):

$$y = \arg\max_k \left(\frac{yd_k + yc_k}{2}\right), \quad (1)$$

where yd_k and yc_k are the probability that the output of the depth and color image network, respectively, belongs to class k.

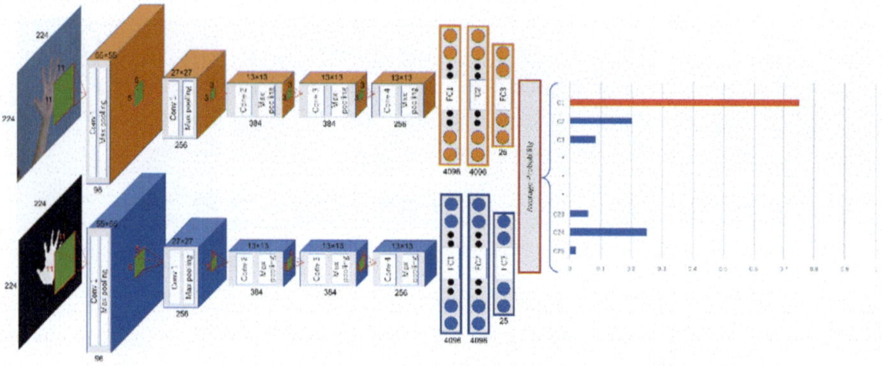

Fig. 4. Multimodal deep learning network for hand gesture recognition.

2.3 Visualization Module

Herein, the surface model of the hepatic artery, hepatic vein, hepatic portal vein, and liver parenchyma, which is segmented semiautomatically from computed tomography (CT) images under the guidance of a doctor, is generated and presented to the user [1, 2]. The respective visualized organ models are shown in Fig. 5. The visualization results obtained by overlaying the respective models are shown in Fig. 6 with the opacity of a reduced liver parenchyma. The detailed information about CT, segmented liver, and vessel data is presented in [1, 2]. This can be efficiently reviewed with four rotations, which are zoom in/out, adjustment of transparency, and fusion and selection of vessel.

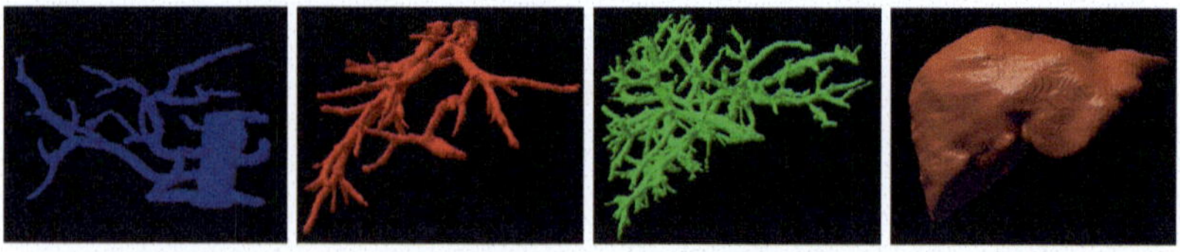

(a) Hepatic artery (b) Hepatic portal vein (c) Hepatic vein (d) Liver

Fig. 5. Surface model of the liver.

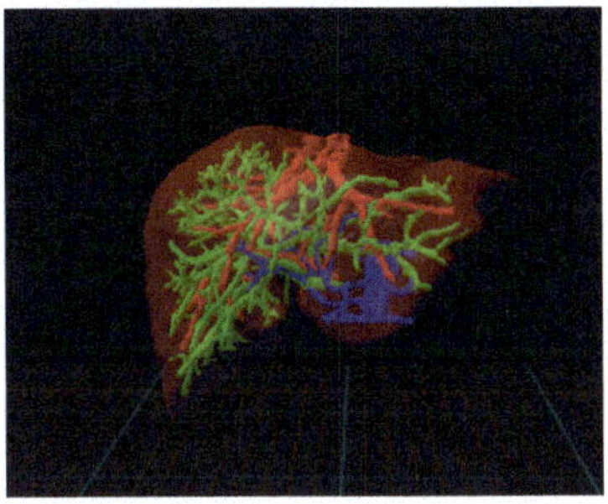

Fig. 6. Simultaneous visualization of models.

3 Experimental Results

3.1 Dataset

The experiment is performed using MaHG–RGBD; Fig. 7 shows an example of MaHG–RGBD dataset, which comprises of 25 gestures. The shooting environment is shown in Fig. 8. For each gesture, we shot images with Kinect at two different angles. A total of 100 images and 15 people were taken for each gesture. The size of the hand image is 300 × 300 pixels. The data were taken with Kinect1 within the surgical environment.

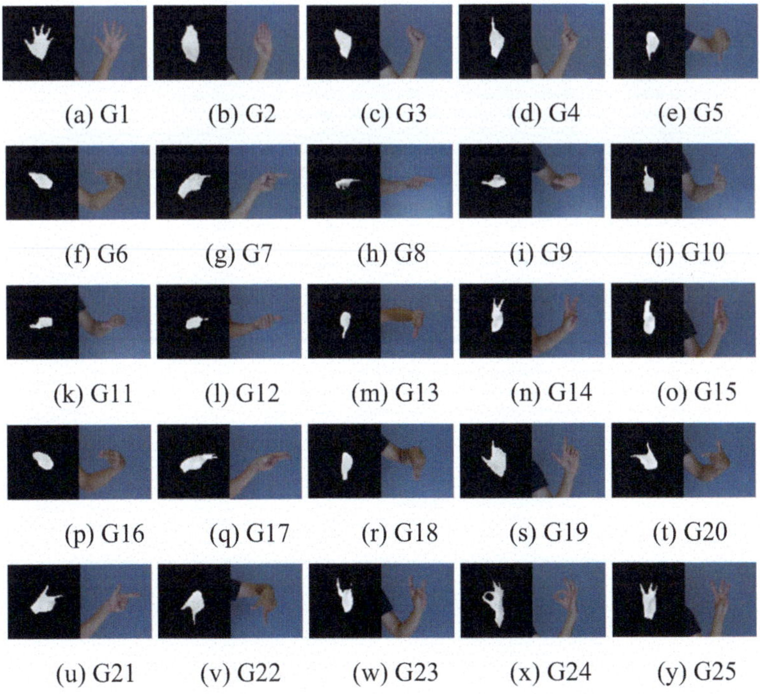

Fig. 7. Example of 25 classes in the MaHG–RGBD dataset.

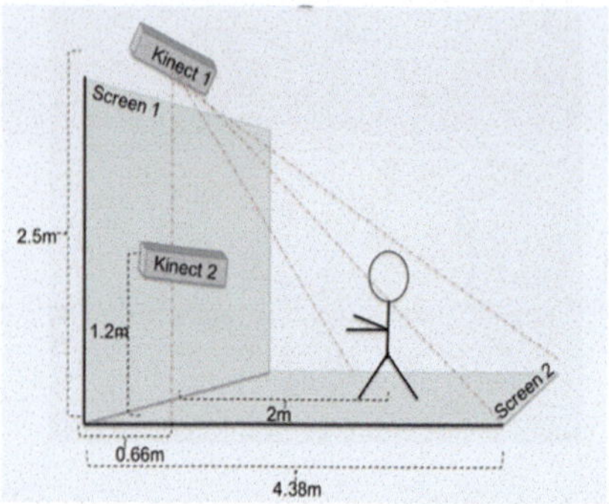

Fig. 8. The shooting environment of hand gesture images

3.2 Recognition Result

We use a 15-fold cross-validation method. A total of 14 persons were used as training data, and 1 person for testing. Twenty percent of the training data are selected randomly as the validation set. We repeat the process 15 times and verify the results of all the cases. Assuming that the verification result of one dose is $result_k$ and final recognition accuracy is *accuracy*, this is expressed as following Eq. (2):

$$accuracy = \frac{\sum_{k=1}^{15}(result_k)}{15}. \tag{2}$$

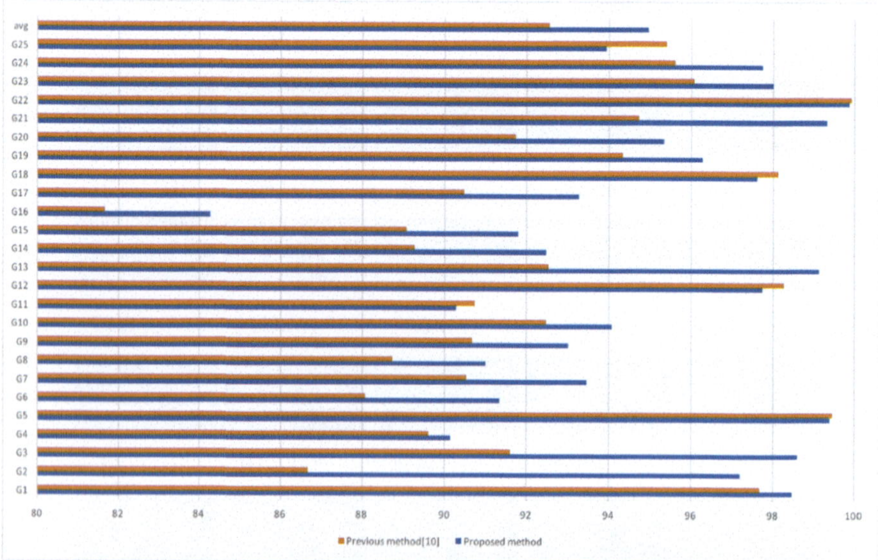

Fig. 9. Comparison of the recognition result of the proposed method and previous method [10]. The orange graph is the previous method and the blue graph is the proposed method.

Figure 9 presents a comparison between the results of recognition accuracy obtained from the previous research [11] and those obtained using the proposed method. In the graph, results obtained using the previous and proposed method are denoted in orange and blue, respectively. Most gesture recognition accuracy has been improved. In the recognition using only the depth image, the accuracy is 92.53%, whereas in the recognition using the depth and color images, the accuracy is 94.94%, showing a 2.41% improvement in the accuracy. Using a color image, the accuracy is significantly improved for the gesture recognition accuracy. However, poor accuracy is remarked for only the depth image. Since the time taken for recognition is 0.0146 s, the implementation in real-time is similar.

Table 1 lists the average confusion matrix of the proposed method. Misrecognition of a specific gesture is remarked when gesture recognition accuracy is less than 90%. Thus, the gesture recognition is more robust, if the gesture in the touchless interface does not use an easily misrecognized gesture.

Table 1. The average confusion matrix of the proposed method. Lager misclassification errors (>5) are indicated in red and zero values and not included.

Truth	G1	G2	G3	G4	G5	G6	G7	G8	G9	G10	G11	G12	G13	G14	G15	G16	G17	G18	G19	G20	G21	G22	G23	G24	G25
G1	98.47	0.07									0.20	0.07				0.60			0.27		0.13			0.13	0.07
G2		97.20	0.07	0.07	0.13	0.40			0.07		0.13	0.07				1.20		0.20	0.27			0.13		0.07	
G3			98.60	0.07				0.20								0.40		0.13						0.60	
G4			1.13	90.13						2.80					5.13	0.73							0.07		
G5			0.13		99.40				0.07									0.40							
G6	0.07					91.33	0.13		0.07		0.53					6.87	0.07		0.87	0.07					
G7						0.07	93.47					5.87				0.07	0.53								
G8			5.13					91.00	2.07			0.27													1.53
G9		0.07	4.47					0.13	93.00		1.80							0.53							
G10						0.27				94.07	0.07		1.07	0.67	3.13			0.20		0.40					0.13
G11			2.27	0.07		0.13				1.33	90.27		0.07			5.87									
G12	0.13	0.33					0.13	0.33				97.73	1.27					0.07							
G13								0.07	0.47				99.13					0.33							
G14	1.67			0.13						0.53				92.47	0.07								4.73	0.07	0.33
G15				2.80					0.07	3.07			0.07	0.80	91.80								1.33		0.07
G16						8.13	0.07				0.67		5.80		0.07	84.27				0.67		0.33			
G17	0.07						1.13	0.20			0.07	4.47					93.27				0.80				
G18					0.20								2.20					97.60							
G19						0.40			0.07		0.20					1.93			96.27	1.13					
G20						0.80					0.07								3.73	95.33					0.07
G21							0.60	0.07													99.33				
G22											0.07							0.07				99.87			
G23				0.20										0.80					0.40				98.00		0.60
G24	0.80	0.27	1.13																				0.07	97.73	
G25				0.27						2.60			1.87	0.67									0.67		93.93

4 Conclusion

Herein, we proposed a robust hand gesture recognition system with multimodal deep learning for touchless visualization for 3D medical images. Two features of this system are identified. First, surgeons can review 3D medical images without touching the device, which reduces the risk of infection. Second, gesture detection is more robust than the previous recognition systems. The previous system required both hands for operation; however, the proposed system can perform the previous operation with only one hand. In the future, we plan to respond to dynamic gestures, which can be manipulated more intuitively.

Acknowledgment. Authors would like to thank Dr. M. Kaibori of KANSAI Medical University for providing medical images and advice on surgical support systems. This work is supported in part by the Grant-in Aid for Scientific Research from the Japanese Ministry for Education, Science, Culture and Sports (MEXT) under the Grant Nos. 18H03267, 18K11454, 17H00754, 17K00420; and in part by Zhejiang Lab Program under the Grant No. 2018DG0ZX01.

References

1. Kaibori, M., Chen, Y.W., Matsui, K., Ishizaki, M., Tsuda, T., Nakatake, R., Sakaguchi, T., Matsushima, H., Miyawaki, K., Shindo, T., Tateyama, T., Kwon, A.H.: Novel liver visualization and surgical simulation system. J. Gastrointest. Surg. **17**, 1422–1428 (2013)
2. Tateyama, T., Kaibori, M., Chen, Y.W., et al.: Patient-specified 3D-visualization for liver and vascular structures and interactive surgical planning system. Med. Imaging Technol. **31**, 176–188 (2013). (in Japanese)
3. Nasr-Esfahani, E., et al.: Hand gesture recognition for contactless device control in operating rooms. arXiv Preprint arXiv1611.04138 (2016)
4. Gallo, L.: Controller-free exploration of medical image data: experiencing the Kinect, National Research Council of Italy Institute for High Performance Computing and Networking (2011)
5. Yoshimitsu, K., Muragaki, Y., Iseki, H., et al.: Development and initial clinical testing of "OPECT": an innovative device for fully intangible control of the intraoperative image-displaying monitor by the surgeon. Neurosurgery **10**(Suppl. 1), 46–50 (2014)
6. Ruppert, G.C., Coares, C., Lopes, V., et al.: Touchless gesture user interface for interactive image visualization in urological surgery. World J. Urol. **30**, 687–691 (2012)
7. Mewes, A., Hensen, B., Wacker, F., et al.: Touchless interaction with software in interventional radiology and surgery: a systematic literature review. Int. J. CARS **12**, 291 (2017)
8. Liu, J.Q., Fujii, R., Tateyama, T., Iwamoto, Y., Chen, Y.W.: Kinect-based gesture recognition for touchless visualization of medical images. Int. J. Comput. Electr. Eng. **9**(2), 421–429 (2017)
9. Liu, J.Q., Tateyama, T., Iwamoto, Y., Chen, Y.W.: A preliminary study of kinect-based real-time hand gesture interaction systems for touchless visualizations of hepatic structures in surgery. J. Med. Imaging Inf. Sci. (2019, in press)
10. Liu, J.Q., Tateyama, T., Iwamoto, Y., Chen, Y.W.: Kinect-based real-time gesture recognition using deep convolutional neural networks for touchless visualization of hepatic anatomical models in surgery. In: KES-IIMSS-18 2018. Smart Innovation, Systems and Technologies, vol. 98. Springer, Cham (2018)
11. Liu, J.Q., Furusawa, K., Tsujinaga, S., Tateyama, T., Iwamoto, Y., Chen, Y.W.: An improved kinect-based real-time gesture recognition using deep convolutional neural networks for touchless visualization of hepatic anatomical models in surgery. J. Image Graph. (2019, in press)
12. Krizhevsky, A., Sutsjever, I., Hinton, G.: ImageNet classification with deep convolutional neural networks. In: Advances in Neural Information Processing Systems, p. 9 (2012)

Tensor-Based Subspace Learning for Classification of Focal Liver Lesions in Multi-phase CT Images

Jian Song[1,2], Sihang Zhu[1], Lanfen Lin[3], Hongjie Hu[4], and Yen-Wei Chen[2,3,5](✉)

[1] School of Mathematical Sciences, Huaqiao University, Quanzhou, Fujian, China
[2] College of Information Science and Engineering, Ritsumeikan University, Shiga, Japan
chen@is.ritsumei.ac.jp
[3] Department of Computer Science and Technology, Zhejiang University, Hangzhou, China
[4] Department of Radiology, Sir Run Run Shaw Hospital, Hangzhou, China
[5] Zhejiang Lab, Hangzhou, China

Abstract. Medical images play an important role in clinics. Machine learning has been widely used in the fields of computer vision and pattern recognition and computer-aided diagnosis with medical images become an active research topic. Efficient representation of the medical images or effective extraction of discriminative features from CT images is one of crucial steps in computer-aided diagnosis. Principal component analysis (PCA) is a subspace learning method and is widely used for efficient representation of data. The limitation of PCA is that a multi-dimensional data (e.g. an image or a video image) should be unfolded into a vector resulting in loss of spatial and spatial-temporal relationship of the data. In this paper, we proposed an efficient representation of multi-phase CT images based on a tensor-based subspace learning method known as generalized N-dimensional principal component analysis (GND-PCA). In the proposed method, the multi-phase CT image is treated a tensor without vector-unfolding for subspace learning. The core tensor obtained by GND-PCA is used as temporal and spatial features for focal liver lesion classification. Experiments show that in the case of fewer samples, GND-PCA achieved better results than conventional PCA and 2D-PCA, which is an extension of PCA.

Keywords: Generalized N-dimensional principal component analysis (GND-PCA) · Multi-phase medical image · Focal liver lesion (FLL)

J. Song and S. Zhu—The first two authors contributed equally to this paper.

Y. Liu et al. (Eds.): ICNC-FSKD 2019, AISC 1074, pp. 601–608, 2020.
https://doi.org/10.1007/978-3-030-32456-8_66

1 Introduction

Liver cancer is one of the leading causes of death worldwide. However, diagnosis of focal liver lesions (FLLs) is a challenging task. Currently, multi-phase computer tomography (CT) images, which are also known as contrast-enhanced dynamic CT images, are widely used to detect, locate and diagnose focal liver lesions. Liver lesions have different visual characteristics at different time points before/after intravenous contrast injection. In common situations, four CT scans are captured to obtain the visual transition features of liver tumors over time, including noncontrast-enhanced (NC) phase images that scanned before contrast injection, arterial (ART) phase images that scanned 25–40 s after contrast injection, portal venous (PV) phase images scanned 60–75 s after contrast injection, and delayed (DL) phase images scanned 3–5 min after contrast injection. So the multi-phase CT image can be considered as a 4D image. Extraction of temporal and special features from the multi-phase image is an important issue for accurate classification of FLLs [1].

To date, several studies have reported on the characterization of FLLs using multiphase images. In the most previous work, spatial features are first extracted from each phase by the use of bag-of-visual-words (BoVW) methods [2–8] or deep learning methods [9–11]. Then the extracted features from each phase are concatenated to form a spatiotemporal feature for classification of FLLs. Though their results showed that multi-phase images generate better results than the use of single-phase ones, they ignore the temporal enhancement information and relationship among phases. Recently, Liang et al. [12] proposed a method by combining a convolutional neural network with a recurrent neural network to extract temporal enhancement information and achieved the state-of-the-art results.

On the other hand, principal component analysis (PCA) is a subspace learning method and is widely used for efficient representation of data. The limitation of PCA is that a multi-dimensional data (e.g. an image or a video image) should be unfolded into a vector resulting in loss of spatial and spatial-temporal relationship of the data. In order to overcome the limitation of conventional PCA, Xu et al. [13] proposed a method called generalized N-dimensional PCA (GND-PCA), which is a tensor-based subspace learning method. They used it to reconstruct a statistical model and the higher-dimensional medical data are treated as a series of higher-order tensors and the bases on each mode subspace are calculated in order to approximate the tensors accurately. Inspired from the work of GND-PCA, we propose to use it to classify the FLLs in multi-phase CT images. In order to contrast the effectiveness of the proposed method, from single-phase to multi-phase of medical images, PCA, 2D-PCA and GND-PCA are used to extract features for classification of FLLs, respectively. The nearest neighbour algorithm is used as a classifier.

This paper is organized as four sections. Section 2 gives a brief introduction to PCA, 2D-PCA and GND-PCA. The experiments and results are shown in Sect. 3. The last part is the conclusion.

2 Methods

2.1 PCA-Based Methods

Expand a series of data into a one-dimensional column vector a_i, $i = 1, 2, \ldots, n$ then obtain a matrix of m × n, and after removing mean by column, $b_i, i = 1, 2, \ldots, n(\sum_{i=1}^{n} b_i = 0)$ is obtained, where each row is characterized. Then, the bases of each subspace can be obtained by the covariance matrix, $Cov_1 = A \times A^T$, $A = [b_1, b_2, .., b_n]$, which takes the eigenvector corresponding to the largest j eigenvalues of the covariance matrix, can obtain the bases of the subspace $v_1, v_2, \ldots, v_j$. Each vector b_i can be represented by a more compact representation of these bases, $c_i = [v_1, v_2, \ldots, v_j]^T \times b_i$. Here the dimension of the covariance matrix is m × m, so when the dimension of the data is large, the computational cost is quite large.

Another PCA method is eigenface method [14]. The eigenvectors $u_1, u_2, \ldots, u_j$ can be calculated by calculating $Cov_2 = A^T \times A$, and the bases of the subspace can be calculated by $v_i = A^T \times u_i$. When the number of samples n is much smaller than the dimension of the sample m, the eigenface method can efficiently calculate the bases of the subspace. This is because the dimension of Cov_2 is n × n, which is much smaller than the dimension of Cov_1. Therefore, the cost of computational bases will be smaller. The eigenface method can only get n−1 bases, while the first PCA method can get more bases, but when n < m, it can only get (n−1) bases whose corresponding eigenvalues are greater than zero. Therefore, when the number of samples is much smaller than the dimension of the training feature vector, although the feature face method can effectively replace the first method, the two methods are essentially the same, and the bases estimated by a small number of samples cannot accurately represent the original data.

2.2 2D-PCA

For n two-dimensional data A_i, $i = 1, 2, \ldots, n, \sum_{i=1}^{n} A_i = 0$. In 2D-PCA [15], it can be directly compressed without the expansion processing. First calculate the covariance matrix $Cov = (1/n) \sum_{i=1}^{n} A_i^T * A_i$, get the first j eigenvectors $v_1, v_2, \ldots, v_j$, then the matrix A_i. The principal component on the base v_j can be obtained by $b_j = A_i * v_j$, whereby it can be seen that the principal component of the 2D-PCA is a vector on each bases, rather than a scalar. At the same time, the matrix A_i can be represented by its principal component B_i, where $B_i = [b_1, b_2, .., b_J] = A_i \times [v_1, v_2, .., v_j] = A_i \times V$. In 2D-PCA, the dimension of the covariance matrix is determined by the number of columns of A_i, the column dimension of B_i is j, and the row dimension is the same as A_i. It can be seen that 2D-PCA only compresses the column subspace, so 2D-PCA requires a large number of coefficients to represent the data.

2.3 GND-PCA

Given a series of the N th-order tensors with zero-mean, $\mathcal{A}_i \in \mathbb{R}^{I_1 \times I_2 \times .. \times I_N}$, $i = 1, 2, .., M, \sum_{i=1}^{M} \mathcal{A}_i = 0$, we hope to find another series of lower rank tensor $\hat{\mathcal{A}}_i \in \mathbb{R}^{J_1 \times J_2 \times .. \times J_N}, J_N < I_N$ which can most accurately approximate the original tensors. The new series of tensors can be decomposed by the same N matrices $U^{(n)} \in \mathbb{R}^{I_n \times J_n}$ with orthogonal columns according to Tucker model which can be shown by

$$\hat{\mathcal{A}}_i = \mathcal{B}_i \times {}_1U^{(1)}{}_1U^{(2)} \times \ldots \times {}_NU^{(N)} \tag{1}$$

Where $\mathcal{B}_i \in \mathbb{R}^{J_1 \times J_2 \times .. \times J_N}$ are the core tensors [13].

The orthogonal matrix group $U^{(n)}$ can be determined by minimizing the following cost function:

$$S = \sum_{i=1}^{M} \left\| \mathcal{A}_i - \hat{\mathcal{A}}_i \right\|^2 = \sum_{i=1}^{M} \left\| \mathcal{A}_i - \mathcal{B}_i \times {}_1U^{(1)}{}_1U^{(2)} \times \ldots \times {}_NU^{(N)} \right\|^2 \tag{2}$$

In Eq. (2), only the tensors $\mathcal{A}_i$ are known. However, supposing the N matrices $U^{(n)}$ are known, the answer of $\mathcal{B}_i$ to minimize Eq. (2) is merely the result of the traditional linear least-square problem. For the N optimal matrix groups $U_{opt}^{(1)}, U_{opt}^{(2)}, \ldots, U_{opt}^{(N)}$, we use the iterative algorithm[13] to find out, when the value of the cost function does not change significantly in two consecutive times, we terminate the algorithm.

Using the calculated matrices $U_{opt}^{(n)}, n = 1, 2, \ldots, N$, each of the volume $\mathcal{A}_i$ can be approximated by $\hat{\mathcal{A}}_i$ with least errors.

Where $\hat{\mathcal{A}}_i = \mathcal{B}_i \times {}_1U_{opt}^{(1)}{}_2U_{opt}^{(2)} \times \ldots \times {}_NU_{opt}^{(N)}$ and $\mathcal{B}_i = \mathcal{A}_i \times {}_1U_{opt}^{(1)^T}{}_2U_{opt}^{(2)^T} \times \ldots \times {}_NU_{opt}^{(N)^T}$. The approximation can be illustrated by Fig. 1 for the 3D case. In GND-PCA, the matrices $U_{opt}^{(n)}, n = 1, 2, \ldots, N$ construct the bases on the N mode subspaces; and the core tensor $\mathcal{B}_i$ are the principal components.

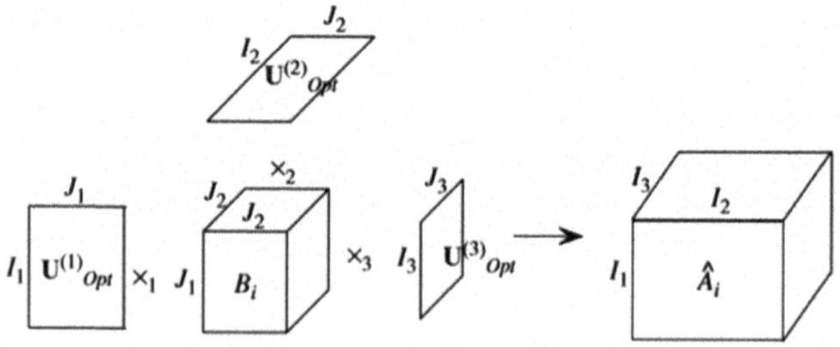

Fig. 1. Illustration of reconstructing a third-order tensor by the principal components $\mathcal{B}_i$ and the three orthogonal bases of mode subspaces $U_{opt}^{(1)}, U_{opt}^{(2)}, U_{opt}^{(3)}$

3 Experiments

3.1 Data

In this paper, a medical database consisting of multi-phase computer-tomography (CT) images is used. The dataset comprises five types of FLLs, which were collected from 256 medical cases. For each medical case, triple-phase (NC/ART/PV) CT images were collected including 37 lesions of the cyst class, 44 cases of focal nodular hyperplasia (FNH), 76 cases of hepatocellular carcinoma (HCC), 73 cases of hemangioma (HEM), and 26 cases of metastases (METs). The ROI of lesions are annotated by doctors manually. Each ROI is normalized to a size of 32 × 32. Typical 2D multi-phase slice images are shown in Fig. 2. We can see that different type of FLL will have different temporal enhancement pattern. Thus, it is important to extract both spatial and temporal features from the multi-phase CT images for classification of FLLs.

In real clinical applications, a single phase (usually PV phase) or multiple phases are commonly used for diagnosis. In this paper, we first introduce a feature extraction method based on a single phase usually the PV phase of an enhanced CT scan. Next, in the experiments of PCA and 2D-PCA, we spliced triple-phase CT images in the manner as shown in Fig. 3 to form multi-phase data. Finally, we combine the triple-phase as a volume to represent the 3D data with the size of 32 × 32 × 3 for the GND-PCA experiment.

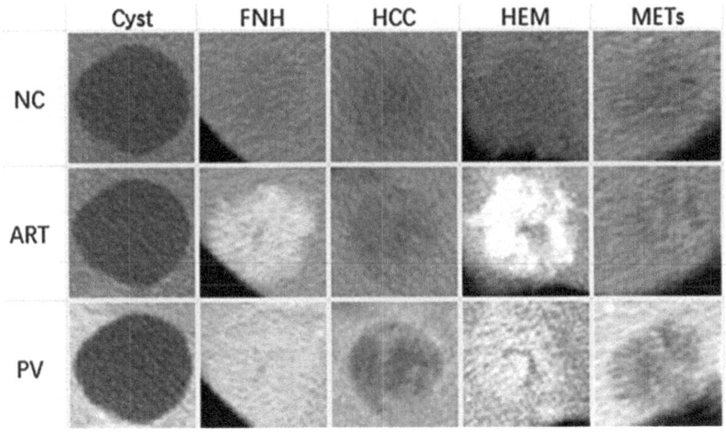

Fig. 2. Typical multi-phase CT images

3.2 Experimental Settings

We contacted following three experiments.

(1) The first experiment was focused on classification of FLLs with single-phase CT images. Typical slice image of PV phase with the largest cross section of tumor

are used for experiments. PCA and 2D PCA were used for feature extraction from the 2D slice image, respectively.

(2) The second experiment was focused on classification of FLLs with multi-phase CT images. PCA and 2D PCA were used to extract features from each phase image (2D image). The extracted features from NC, ART, and PV phases were concatenated for classification.

(3) In the third experiment, the multi-phase images (typical slice images of NC, ART and PV phases) are combined into a 3D image (3rd-order tensor) as shown in Fig. 2. The GND-PCA was used to extract features from the 3D image. The 3D data with the original size of $32 \times 32 \times 3$ was compressed into a core tensor with the new size of $4 \times 4 \times 3$ by GND-PCA. The core tensor was used as a feature vector for classification.

In all experiments, the nearest neighbor classifier (NN) was used for classification. In our experiments, we divided each type of data set into training samples and testing samples with a ratio of 17:3. Each distribution is listed in Table 1.

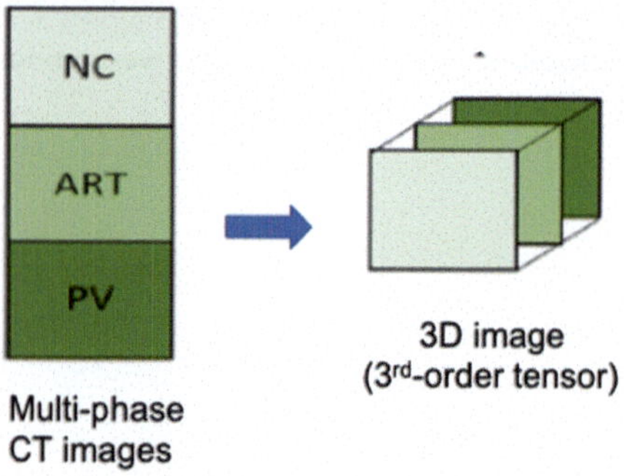

Fig. 3. 3rd-order tensor of multi-phase data

Table 1. Data division (15% division, downward rectification)

Category	Cyst	FNH	HCC	HEM	METS	Total
Training set	32	38	65	63	23	221
Test set	5	6	11	10	3	35
Total	37	44	76	73	26	256

3.3 Results

Experimental results are summarize in Table 2. We first compared the results between single phase and multi-phase. As shown in Table 2, the classification accuracy using multi-phase CT images is higher than that using single phase for both PCA and 2D PCA. We then compared the results with different feature extraction methods (PCA, 2D PCA and GND-PCA). We can see that 2D PCA achieved better performance than PCA. The proposed GND-PCA achieved the best performance among three methods.

Table 2. Experimental results

Method		Accuracy
Single phase	PCA	67.39%
Single phase	2D-PCA	71.74%
Multiphase	PCA	68.57%
Multiphase	2D-PCA	74.29%
Multiphase	**GND-PCA**	**80%**

4 Conclusion

With the multi-linear algebra has been used into the fields of computer vision and pattern recognition, tensor is popular used in the processing of high-dimensional data even in medical images. This paper used a tensor-based subspace learning method, GND-PCA, which processes the training samples into a series of tensors and calculates the bases of each model space to represent the data more concisely and efficiently. Experiments showed that the proposed method achieved much more significant improvement from single-phase to multi-phase images than conventional subspace learning methods.

Acknowledgement. This research was supported in part by the Grant-in Aid for Scientific Research from the Japanese Ministry for Education, Science, Culture and Sports (MEXT) under the Grant No. 18H03267, and No. 18H04747, and in part by Zhejiang Lab Program under the Grant No. 2018DG0ZX01.

References

1. Ryerson, A.B., et al.: Annual report to the nation on the status of cancer, 1975–2012, featuring the increasing incidence of liver cancer. Cancer **122**(9), 1312–1337 (2016)
2. Roy, S., et al.: Three-dimensional spatiotemporal features for fast content-based retrieval of focal liver lesions. IEEE Trans. Biomed. Eng. **61**(11), 2768–2778 (2014)
3. Yu, M., et al.: Extraction of lesion-partitioned features and retrieval of contrast-enhanced liver images. Comput. Math. Meth. Med. **2012**, 12 (2012)
4. Yang, W., et al.: Content-based retrieval of focal liver lesions using bag-of-visual-words representations of single-and multiphase contrast-enhanced CT images. J. Digit. Imaging **25**(6), 708–719 (2012)
5. Diamant, I., et al.: Improved patch-based automated liver lesion classification by separate analysis of the interior and boundary regions. IEEE J. Biomed. Health Inform. **20**(6), 1585–1594 (2016)
6. Xu, Y., et al.: Bag of temporal co-occurrence words for retrieval of focal liver lesions using 3D multiphase contrast-enhanced CT images. In: Proceedings of 23rd International Conference on Pattern Recognition, ICPR 2016, pp. 2283–2288 (2016)
7. Wang, J., et al.: Sparse codebook model of local structures for retrieval of focal liver lesions using multiphase medical images. Int. J. Biomed. Imaging **2017**, 13 (2017)

8. Xu, Y., et al.: Texture-specific bag of visual words model and spatial cone matching-based method for the retrieval of focal liver lesions using multiphase contrast-enhanced CT images. Int. J. Comput. Assist. Radiol. Surg. **13**(1), 151–164 (2018)
9. Frid-Adar, M., Diamant, I., Klang, E., Amitai, M., Goldberger, J., Greenspan, H.: Modeling the intra-class variability for liver lesion detection using a multi-class patch-based CNN. In: Wu, G., Munsell, B.C., Zhan, Y., Bai, W., Sanroma, G., Coupé, P. (eds.) Patch-MI 2017. LNCS, vol. 10530, pp. 129–137. Springer, Cham (2017). https://doi.org/10.1007/978-3-31967434-6_15
10. Yasaka, K., et al.: Deep learning with convolutional neural network for differentiation of liver masses at dynamic contrast-enhanced CT: a preliminary study. Radiology **286**(3), 887–896 (2017)
11. Liang, D., et al.: Residual convolutional neural networks with global and local pathways for classification of focal liver lesions. In: Geng, X., Kang, B.H. (eds.) PRICAI 2018: Trends in Artificial Intelligence, PRICAI 2018, Nanjin, China, 28–31 August 2018. Lecture Notes in Artificial Intelligence, vol. 11012, pp. 617–628. Springer (2018)
12. Liang, D., et al.: Combining Convolutional and recurrent neural networks for classification of focal liver lesions in multi-phase CT images. In: Frangi, A., Schnabel, J., Davatzikos, C., Alberola-López, C., Fichtinger, G. (eds.) Medical Image Computing and Computer Assisted Intervention – MICCAI 2018. Lecture Notes in Computer Science, LNCS, vol. 11071, pp. 666–675, Springer (2018)
13. Xu, R., Chen, Y.-W.: Generalized N-dimensional principal component analysis (GND-PCA) and its application on construction of statistical appearance models for medical volumes with fewer samples. Neurocomputing **72**, 2276–2287 (2009)
14. Turk, M., Pentland, A.: Eigenfaces for recognition. J. Cogn. Neurosci. **3**, 71–86 (1991)
15. Yang, J., et al.: Two-dimensional PCA: a new approach to appearance-based face representation and recognition. IEEE Trans. Pattern Anal. Mach. Intell. **26**, 131–137 (2004)

Analysis of Time Series Anomalies Using Causal InfoGAN and Its Application to Biological Data

Takaya Ueda, Masataka Seo, Yukako Tohsato, and Ikuko Nishikawa(✉)

College of Information Science and Engineering, Ritsumeikan University, Kusatsu, Shiga 525-8577, Japan
nishi@ci.ritsumei.ac.jp

Abstract. Data generation is an inverse function of the data recognition, which is a map from a low dimensional latent space to a high dimensional data. Generator of normal data acquires the data distribution in its latent space, and is expected to generate any normal data, but unable to generate abnormal data. Therefore, the generator can be used for the anomaly detection. We propose a model of Generative Adversarial Nets (GANs) with an encoder to characterize the data anomaly. GANs are trained by normal data, and the encoder is trained to infer the latent space of the trained generator. Then, by combining the encoder and the generator, anomaly of the data input to the encoder is characterized by the data reconstruction error and by its latent representation. If any interpretable representation is obtained for the latent space, we can characterize the abnormality by the latent variable. In this study, Causal InfoGAN is trained by normal time series data to acquire temporal state variables in the latent space. Then, a temporal abnormality can be characterized by an abnormal transition in the state space. Proposed method is applied to a simple toy data, which mimics a cell division in the biological system, and it is observed that the state transition which expresses the normal time development is obtained in the latent space.

Keywords: Generative model · Causal InfoGAN · Data encoder · Anomaly detection · Time series data · Disentangled representation · Biological data

1 Introduction

Statistical machine learning achieves a high performance in the pattern recognition through acquiring hierarchical levels of features from the training data. Data recognition is realized by mapping the high dimensional data onto the lower dimensional feature space, in general. The inverse process of the recognition is the data generation, which is a map from a lower dimensional latent space to a higher dimensional data space. To realize the data generation, Generative Adversarial Nets (GANs) [1] acquire a latent space of the generator and a probability distribution on the latent space, which is mapped to a high dimensional probability distribution of real-world data. Another generative model, variational auto-encoder (VAE) [2] obtains the data distribution as a low-dimensional simple distribution using an encoder that infers the latent representation of

Y. Liu et al. (Eds.): ICNC-FSKD 2019, AISC 1074, pp. 609–617, 2020.
https://doi.org/10.1007/978-3-030-32456-8_67

the data. VAE learns the distribution of high-dimensional data by maximizing the variational lower bound represented by the reconstruction rate and the constraint term on the latent space. As the data generator, VAE often tends to generate rather blur data compared to the generator in GAN.

Once a generator of the data of a certain category is successfully obtained, any data in that category has a corresponding point in the latent space, while a data outside the category does not have a corresponding point and be unable to be generated. Based on this, several approaches have been proposed to utilize the generator for the anomaly detection of the data. The data which can be generated is normal, otherwise it is abnormal. In other words, the data which is encoded to an appropriate point within a leant distribution on the latent space, and is successfully reconstructed by the generator, is normal. On the other hand, the data which is mapped to a point outside the given distribution, and cannot be well reconstructed, is not normal. In AnoGAN [3] proposed by Schlegl et al., and ADGAN [4] proposed by Deecke et al., anomalies in each data are characterized by the reconstruction result. In those GAN framework, the latent representation is inferred by the back propagation from the generated data. In contrast, auto encoder can use the encoder part to infer the latent representation. Especially, VAE assumes a certain distribution in the latent space and models statistical distribution of real data. DAGMM [5] using mixture of Gaussian model derives statistical anomaly criteria based on each cluster of Gaussian distribution.

The disentanglement in the latent space can be described as to obtain the latent representation where one latent variable corresponds to a quantitative expression of a certain character in original data, such as width, length, color and so on [6]. Acquisitions of some interpretable or quantitative latent variables as a disentangle representation enable us to generate a desired data. In that direction, several attempts have been reported including InfoGAN [7]. Among them, Causal InfoGAN [8] is proposed to obtain a latent variable which corresponds to a temporal order of a time series data, in other words, the causality of the dynamics in a real world. In Causal InfoGAN, a generator is trained to generate a pair of data at two consecutive time. Causal time development in a real world is expressed by a state transition rule in a latent state space of the generator, and the state space is expressed by a few latent variables in a disentangle representation.

In this study, time series anomalies are analyzed based on the framework of Causal InfoGAN with an additional encoder. The generator is trained in the original framework only by normal time series data to acquire some latent variables which express the temporal change of the normal data. Next, an encoder is attached to the generator and trained to map a data onto a latent space of the generator. The encoder searches a point in the latent space which minimizes the reconstruction error between original input data and a generated data. Then, if the input data is not normal, it can be characterized by a large reconstruction error and by a large deviation from the normal distribution in the latent space. Moreover, if there is an abnormality only in time development, the reconstruction error at each time point may not be notable, but the abnormality can be found in the temporal transition in the latent space. Thus, the anomaly in the temporal development of the time series data can be detected by the abnormal transition in disentangled state variables on the latent space in Causal InfoGAN with encoder.

As one of the most interesting applications, we will analyze the biological phenotype anomaly caused by a genetic manipulation for the gene function analysis. Cell division in embryonic developmental process and the related essential genes are well investigated on several model animals. Instead of real experimental biological data in this paper, simplified artificial data are used to mimic the cell division dynamics, and the proposed method is applied to acquire a latent variable corresponding to normal dynamics during the division.

2 Causal InfoGAN to Acquire the Dynamics of the Data

Let us consider a temporal sequence of data, and a data pair $\mathbf{x}$ and $\mathbf{x}'$ denote two successive data in time series, whose distribution is given by $p(\mathbf{x}, \mathbf{x}')$. Moreover, we assume a dynamical system which gives a transition from a current data $\mathbf{x}$ to the next data $\mathbf{x}'$. Generator $G(\mathbf{z})$ (referred to as G) in GAN generates a pair of data from a latent variable $\mathbf{z}$. Here, a discriminator (referred to as D) identifies whether each given data pair is sampled from real data or generated from a generator. In addition to latent variable $\mathbf{z}$, latent state variables $\mathbf{s}$ and $\mathbf{s}'$ are introduced to represent the time development of real data from $\mathbf{x}$ to $\mathbf{x}'$. The next state $\mathbf{s}'$ is obtained from the current state $\mathbf{s}$ by $\mathbf{s}' \sim T_M(\mathbf{s}'|\mathbf{s})$, where T_M is a stochastic transition function, given as a probability distribution of $\mathbf{s}'$ conditioned by $\mathbf{s}$. This abstract state $\mathbf{s}$ is introduced to represent the causality in data sequence $\mathbf{x}$, and transition T_M is introduced to express the dynamical system of $\mathbf{x}$. Same as in InfoGAN [3], abstract state $\mathbf{s}$ is inferred by calculating the mutual information between a generated data pair $(\mathbf{x}, \mathbf{x}') \sim G(\mathbf{z}, \mathbf{s}, \mathbf{s}')$ and the corresponding states $\mathbf{s}$ and $\mathbf{s}'$. The loss function for training the generator, discriminator and transition T_M in Causal InfoGAN is given by:

$$\min_{G,M} \max_{D} \{V(G, D) - \lambda I(\mathbf{s}, s'; G(\mathbf{z}, \mathbf{s}, \mathbf{s}'))\}, \tag{1}$$

where the first term $V(G, D)$ is same as the objective function in the original GAN. The second term I is the mutual information to obtain abstract state $\mathbf{s}$. It is further replaced by the lower bound using the auxiliary distribution Q as in [3], to lead to the following equation:

$$\min_{G,M,Q} \max_{D} \left\{ V(G, D) - \lambda \mathbb{E}_{(\mathbf{x},\mathbf{x}') \sim G(\mathbf{s},\mathbf{s}',\mathbf{z}), \mathbf{s}' \sim T_M(\mathbf{s}'|\mathbf{s}), \mathbf{s} \sim p(\mathbf{s}), \mathbf{z} \sim p(\mathbf{z})} [\log Q(\mathbf{s}, \mathbf{s}'; \mathbf{x}, \mathbf{x}')] \right\} \tag{2}$$

The transition rule T_M is assumed to derive the next state $\mathbf{s}'$ by adding a Gaussian perturbation to the current states $\mathbf{s}$. Namely, the next state is given as $\mathbf{s}' = \mathbf{s} + \boldsymbol{\delta}$, where $\boldsymbol{\delta} \sim \mathcal{N}(0, \Sigma_\Theta(\mathbf{s}))$, with a variance-covariance matrix Σ_Θ. For simplicity, diagonal matrix is assumed for Σ_Θ with zero covariance, and is acquired by MLP with parameters Θ. During the training, following L2 loss (3) is minimized, to obtain the small transition in state space, which corresponds to a short time development of the data:

$$\min_{\Theta} \mathbb{E}_{s \sim p(s)} ||\Sigma_\Theta(\mathbf{s})||_2. \tag{3}$$

Thus, T_M is acquired as a stochastic flow in the abstract state space $\mathbf{s}$.

3 Extraction of Causality by Data Encoder

In the framework of Causal InfoGAN, the latent space is composed of the state variable **s**, which describes the time development whose dynamics is given by T_M, and the other latent variable **z**, which is independent of time. Therefore, in order to detect temporal anomaly of the time series data, the proposed method introduces an encoder, which infers both **s** and **z** for a given data **x**.

First, Causal InfoGAN is trained with the normal data (Fig. 1(a)). Next, an encoder (referred to as *E*) is attached to the trained generator, and it is trained to infer abstract state **s** and latent variable **z** corresponding to a given data **x** (Fig. 1(b)). The generator is not trained any more at this stage. The loss function to be minimized for the encoder training is given by:

$$\mathbb{E}_{\mathbf{s}' \sim T_M(\mathbf{s}'|\mathbf{s}),(\mathbf{s},\mathbf{z}) \sim E(\mathbf{x}),\mathbf{x} \sim p(\mathbf{x})}\left[||\mathbf{x} - G(\mathbf{s},\mathbf{s}',\mathbf{z})||^2\right] + \alpha\,\mathbb{E}_{(\mathbf{x},\mathbf{x}') \sim G(\mathbf{s},\mathbf{s}',\mathbf{z})}\left[||\mathbf{z} - E(\mathbf{x}')||^2\right], \quad (4)$$

where α is a hyper parameter. Output of encoder $(\mathbf{s},\mathbf{z}) = E(\mathbf{x})$ is given by $\mathbf{s} = f_{\Phi}(\mathbf{x})$, and $\mathbf{z} \sim \mathcal{N}\left(\mathbf{z}; \boldsymbol{u}_{\Phi}(\mathbf{x}), \sigma_{\Phi}^2 I(\mathbf{x})\right)$, where $\boldsymbol{\Phi}$ is encoder's parameter, and **z** is derived from Gaussian with isotropic variance. The first term of Eq. (4) minimizes the L2 norm of the error between the input data **x** to *E* and the reconstructed data by $G \circ E$. Namely, $G \circ E$ should be the identity. The second term of Eq. (4) forces the same latent variable **z** to be shared by both the current state **s** and the next state $\mathbf{s}'$. Thus, an encoder is trained to reconstruct a pair of consecutive inputs with the same **z**.

Furthermore, to minimize reconstruction errors, we add the following loss function to make *E* be the inverse function of generator *G*. Namely, $E \circ G$ should be the identity. For the data generated from **s** and **z**, the alternative loss function to be minimized is given by:

$$\mathbb{E}_{(\mathbf{x},\mathbf{x}') \sim G(\mathbf{s},\mathbf{s}',\mathbf{z}),\mathbf{s}' \sim T_M(\mathbf{s}'|\mathbf{s}),\mathbf{s} \sim p(\mathbf{s}),\mathbf{z} \sim p(\mathbf{z})}\left[||(\mathbf{s},\mathbf{z}) - E(\mathbf{x})||^2\right]. \quad (5)$$

Encoder *E* is trained to minimize the loss functions Eqs. (4) and (5), alternately with a certain ratio. Through the training of *E* by Eq. (5), *E* covers the distribution of the generated data. Furthermore, real data **x** which is close to a generated data $G(\mathbf{s},\mathbf{s}',\mathbf{z})$ is expected to be encoded in the neighbor of $(\mathbf{s},\mathbf{z})$ because of the continuity of *G*, so Eq. (4) also helps to reduce reconstruction error of real data **x** (the first term of Eq. (4)) effectively in the early stage of the training.

The proposed method detects the anomalies in time series data in two ways using the trained encoder and generator. The first is to detect the data which deviates from the normal data distribution indicated by a large reconstruction error as ordinary detection method. It is measured by the first term in Eq. (4). The second and our original way is to detect the temporal anomalies of time series data. In such case, data at each time point is normal and cannot be detected by the reconstruction error or by the deviation in the latent space, but the time development or the trajectory in the state space **s** is not normal.

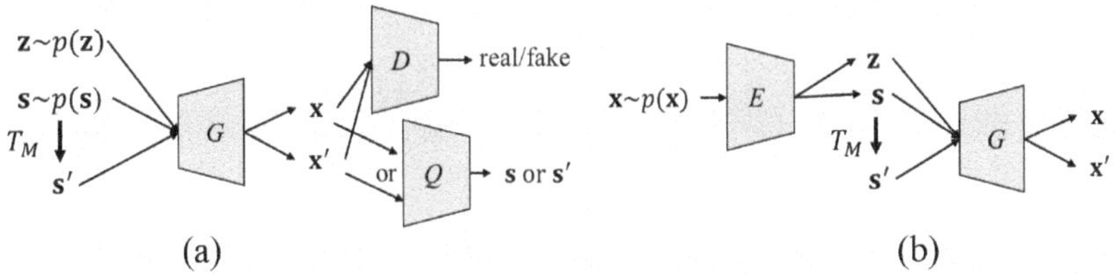

Fig. 1. (a) Causal InfoGAN (b) An encoder infers the abstract state **s** from the data **x**.

Next chapter shows numerical experiments, where the proposed method is applied to acquire the dynamics of given time series data in abstract state space **s**.

4 Numerical Experiments on a Toy Data of Division Process

Our final target is to detect morphological and temporal anomalies in the biological data. As three-dimensional time series data, we focus on the cell division dynamics in the early embryonic development process. If we are able to characterize any phenotype abnormality in an embryo with a certain gene knocked down in advance, then it can be linked to the function of the knock down gene. As a preliminary step before using experimental biological data, we use simplified artificial image as a toy data which mimics nuclear division dynamics in this section. Figure 2 is an example of artificial data set of nuclear division, where one cell is divided into two cells.

4.1 Artificial Datasets

The artificial image data of cell division was created by the following procedure. Each data is 64 × 64 pixels grayscale image. White region stands for a cell nucleus with black background, and one data sequence for the division process has 25 time points. At the initial time $t = 1$, the white region is a square of size 4 × 4 pixels, shown at the upper left in Fig. 2. Then it is vertically extended up to $t = 10$. Next, the region is divided into two separate parts during $t = 11 - 18$. Finally, each of two parts expands keeping a round shape until the last time $t = 25$. Each data at 25 time points is then augmented by the rotation by the angle $\Phi = 0, 6, 12, \ldots, 174$.

Then, the total number of images is $25 \times 30 = 750$, with 25 different time points and 30 different rotation angles. The image pair of the training dataset for Causal InfoGAN is $(\mathbf{x}_t, \mathbf{x}_{t+1})$, $(t = 1, 2, \ldots, 24)$ with the same rotation angle Φ. Thus, the generator is trained by 25 sequences with different rotation. The training data for the encoder is same as for the generator.

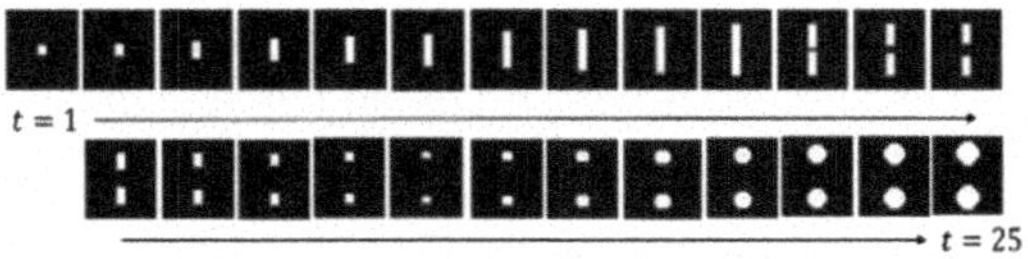

Fig. 2. One sequence of data without rotation.

4.2 Experimental Setting

Table 1 shows the structure of each neural network in the proposed model. The Discriminator (D) and the auxiliary network Q share the convolutional layers except the first layer and the last fully connected layer (fc). The minibatch-std [9] is applied to the second layer of D and Q to keep the data variation. And the feature vector normalization [9] is applied to G to maintain the stability. The transition rule T_M with parameter Θ is acquired by a two-layer network with 10 hidden units, and uses Relu as an activation function. Note that T_M is given by a variance matrix Σ_Θ for each element s_i of the state $\mathbf{s}$. The dimensions of the latent variable $\mathbf{z}$ is 16 and the state variable $\mathbf{s}$ is 2, respectively. The prior probability $p(\mathbf{z})$ for the latent variable $\mathbf{z}$ is $\mathcal{N}(\mathbf{0}, \mathbf{1})$, and $p(\mathbf{s})$ for the state variable $\mathbf{s}$ is uniform $U[-1, 1]^2$. The parameters in the loss functions [2] and [4] are $\lambda = 1$ and $\alpha = 1$. The parameter $\boldsymbol{\Phi}$ of the encoder is updated once by Eq. (5) and 50 times by Eq. (4), alternatively.

Table 1. The structure of each network used in the numerical experiment

D/Q	G
$\mathbf{x}, \mathbf{x}' \in R^{64\times64\times2}$ for D and $\mathbf{x}$ for Q	$\mathbf{z} \in R^{16} \sim \mathcal{N}(\mathbf{0}, \mathbf{1}), \mathbf{s} \in R^2 \sim U[-1,1]^2$
4×4 conv. 16, stride 2, lReLU	4×4 Deconv. 64, stride 2, lReLU, FVN
4×4 conv. 32, stride 2, lReLU Minibatch-std	4×4 Deconv. 32, stride 2, lReLU, FVN
4×4 conv. 64, stride 2, lReLU	4×4 Deconv. 16, stride 2, lReLU, FVN
fc. 4096→1 for D and fc. 4096→2 for Q	4×4 Deconv. 2, stride 2, sigmoid

E
$\mathbf{x} \in R^{64\times64\times1}$
4×4 conv. 16, stride 2, lReLU
4×4 conv. 32, stride 2, lReLU
4×4 conv. 64, stride 2, lReLU
fc. 4096→16 for $\boldsymbol{\mu}$, fc. 4096 → 16 for $\boldsymbol{\sigma}^2$ and fc. 4096 → 2 for $\mathbf{s}$

4.3 Acquired Dynamics in the Latent State Space

Let us investigate the obtained latent space of the generator, especially the two-dimensional state space $\mathbf{s}$, after the training of InfoGAN and the encoder.

Figure 3(a) is a visualization of state space $\mathbf{s}$, in the range s_1, $s_2 \in [-1.0, 0.8]$. Generated image $\mathbf{x}$ with $\mathbf{z}$ fixed to $\mathbf{0}$ is shown at each point of $\mathbf{s}$. Continuous change of the image is observed according to both s_1 and s_2. Temporal evolution of the artificial splitting process is coded along the negative direction of both s_1 and s_2, while no rotation is observed and only the horizontal elongation and splitting is observed in this plane. The whole sequence, which starts from the initial image of the smallest single white dot and ends up as two large circles, is observed.

Figure 3(b) is a visualization of the acquired dynamics in the state space $\mathbf{s}$ in the range s_1, $s_2 \in [-1.0, 0.8]$. As the visualization of the transition function $T_M(\mathbf{s})$, the standard deviation $\sigma_1(\mathbf{s})$ and $\sigma_2(\mathbf{s})$ are shown by the relative length of the arrows.

$\sigma_1(\mathbf{s})$ and $\sigma_2(\mathbf{s})$ are the square roots of the two diagonal components of the variance-covariance matrix $\Sigma_\Theta(\mathbf{s})$ of the Gaussian perturbation added to the state $\mathbf{s}$, which gives the stochastic transition to the next state $\mathbf{s}'$. It is observed the transition is mainly in the direction of variable s_1.

The expansion of two parts during $t = 19 - 25$ is expressed by s_1, while the initial stage of vertical extension and the splitting stage during $t = 1 - 18$ is expressed by both s_1 and s_2. This result indicates that it is hard to express the entire dynamics of the time series by one state variable, or by a single trajectory. But two dimensional state space is enough to express every time series with different rotation angles.

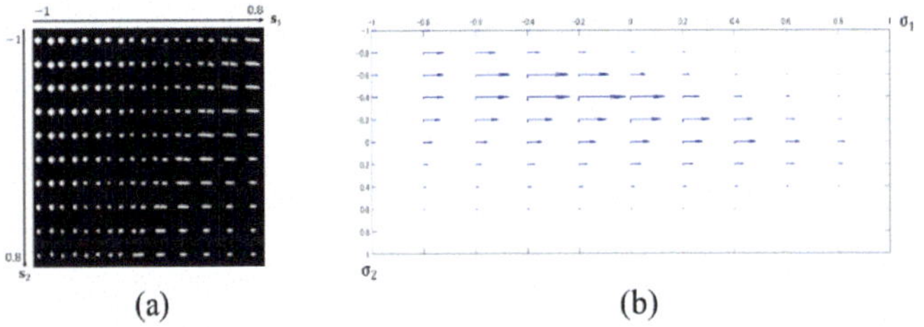

Fig. 3. Visualization of state space $\mathbf{s}$ and the transition rule obtained by the training. (a) Generated image $\mathbf{x}$ from each $\mathbf{s}$ by setting latent variable $\mathbf{z}$ to $\mathbf{0}$. Both s_1 and s_2 are changed from -1 to 0.8 by 0.2. (b) Visualization of the transition function $T_M(\mathbf{s})$ by showing the standard deviation $\sigma_1(\mathbf{s})$ and $\sigma_2(\mathbf{s})$ by the relative length of two arrows. $\sigma_1(\mathbf{s})$ and $\sigma_2(\mathbf{s})$ are obtained as the square root of the diagonal components of the variance-covariance matrix $\Sigma_\Theta(\mathbf{s})$, and correspond to the Gaussian perturbation added to the current state $\mathbf{s}$. Note that each length is multiplied by 10^3 in the figure.

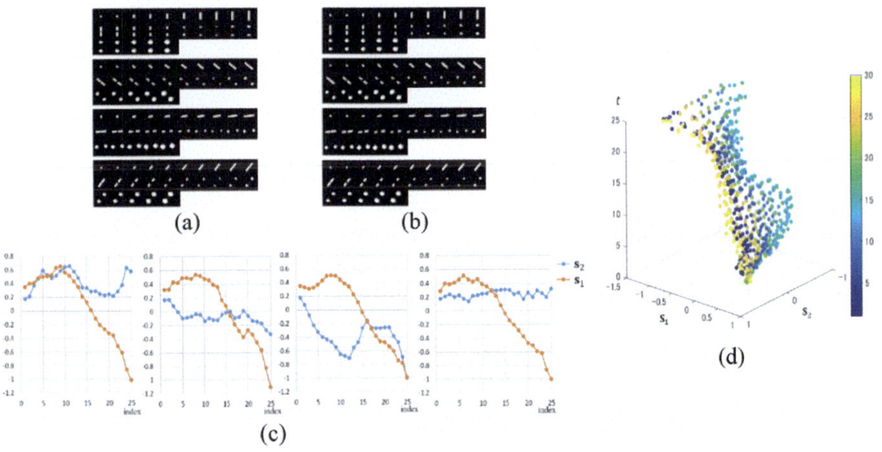

Fig. 4. (a) Four series of artificial data with the rotation angles 0, 48, 96 and 144. (b) Reconstruction data for each data shown in (a). (c) Temporal change of state variables s_1 and s_2 given by the encoder, for the four input series shown in (a). (d) The state value $\mathbf{s}$ ($25 \times 30 = 750$ data) of all 30 time series data with different rotation angles obtained by the encoder. The color indicates each angle from 0 to 176° in the order from 1 to 30 as shown by a color bar.

4.4 Encoding Result of the Time Development onto the Latent State Space

Next, let us study how the time series data is mapped to the state space **s** by the encoder. Figure 4 shows that the input data is well reconstructed by the encoder and the generator, where Fig. 4(a) are four temporal series of input data and (b) are the corresponding reconstruction results.

It is also observed that all data in a single time series take almost same value of the latent variable **z**, as the encoder output. The standard deviation of each component of **z** over the whole time series is calculated for all 30 series. The average over 16 components z_i is 0.153 at the maximum. Thus the time development is coded only by the state space **s**, and the angle information is coded by **z**.

Figure 4(c) indicates that s_1 expresses a morphological change according to time development, regardless of the rotation angle, while s_2 shows the different change. Figure 4(d) shows the trajectories in the state space **s**. The vertical axis is the time, and 30 different rotation angles are indicated by each color. As shown in Fig. 3, s_1 is more responsible for the dynamics.

5 Present Summary and Future Works

The model of Causal InfoGAN with encoder is proposed to characterize the temporal dynamics of the time series data. As the preliminary study for the biological developmental dynamics, simple artificial image data of splitting cell is used for the numerical experiment. As the results, a latent state variable is acquired which corresponding to the time development during the splitting process. The next step is the application of the proposed method to real three-dimensional temporal data obtained by the biological experiment, including the normal and abnormal dynamics in the cell division.

References

1. Goodfellow, I.J., Pouget-Abadie, J., Mirza, M., Xu, B., Warde-Farley, D.: Generative adversarial nets. In: Advances in Neural Information Processing Systems, pp. 2672–2680 (2014)
2. Kingma, D.P., Welling, M.: Auto-encoding variational Bayes. In: International Conference on Learning Representations (2014)
3. Schlegl, T., Seeböck, P., Waldstein, S.M., Schmidt-Erfurth, U., Langs, G.: Unsupervised anomaly detection with generative adversarial networks to guide marker discovery. In: Information Processing in Medical Imaging (2017)
4. Deecke, L., Vandermeulen, R., Ruff, L., Mandt, S., Kloft, M.: Anomaly detection with generative adversarial networks. In: International Conference on Learning Representations (2018)
5. Zong, B., Song, Q., Min, M.R., Cheng, W., Lumezanu, C., Cho, D., Chen, H.: Deep autoencoding gaussian mixture model for unsupervised anomaly detection. In: International Conference on Learning Representations (2018)

6. Higgins, I., Matthey, L., Pal, A., Burgess, C., Glorot, X., Botvinick, M., Mohamed, S., Lerchner, A.: β-VAE: learning basic visual concepts with a constrained variational framework. In: International Conference on Learning Representations (2017)
7. Chen, X., Duan, Y., Houthooft, R., Schulman, J., Sutskever, I., Abbeel, P.: InfoGAN: interpretable representation learning by information maximizing generative adversarial nets. In: Advances in Neural Information Processing Systems, pp. 2180–2188 (2016)
8. Kurutach, T., Tamar, A., Yang, G., Russell, S.J., Abbeel, P.: Learning plannable representations with causal InfoGAN. In: Advances in Neural Information Processing Systems, pp. 8747–8758 (2018)
9. Karras, T., Aila, T., Laine, S., Lehtinen, J.: Progressive growing of GANs for improved quality, stability, and variation. In: International Conference on Learning Representations (2018)

Characterizing Phenotype Abnormality by Variational Auto Encoder

Yuki Kimura(✉), Takaya Ueda, Seo Masataka, Yukako Tohsato, and Ikuko Nishikawa

Ritsumeikan University, Kusatsu, Japan
is0304pr@ed.ritsumei.ac.jp

Abstract. Variational auto-encoder (VAE) acquires not only a low dimensional representation of the data, but also the probability distribution of the data, in its latent space. Therefore, if an input data is not from the trained category, it fails to find an appropriate point in the latent space, which leads to a poor reconstruction. Thus, it can be used for data abnormality detection. We use VAE to detect a phenotype abnormality in a biological system. When each of lethal genes in early embryogenesis is knocked down by RNAi technique, an observed abnormality can be linked to the corresponding gene function. We train VAE by the wild type (WT) data without any gene manipulation, and use to characterize a phenotype of RNAi embryo in two cell stage. Abnormality is defined by a data reconstruction error, and several genes are found whose absence causes the abnormality, including some already known genes.

Keywords: Variational auto encoder · Anomaly detection · Latent space · Biological dynamics data

1 Introduction

Generative model in statistical pattern recognition considers data as a stochastic variable, and assumes a probability distribution as a data generator for each category. Variational auto-encoder (VAE) [1] is a convolutional neural network [2], which is trained to acquire a distribution of high dimensional data in its low dimensional latent space, while an auto-encoder [3] is trained to acquire a point-wise mapping between a high dimensional data and its low dimensional representation. Trained VAE maps a data derived from the trained distribution to a corresponding point in the latent space by an encoder part, and reconstructs the input by a decoder part. But, if an input to VAE is not a data derived from the trained distribution, then it fails to find an appropriate point in the latent space, which leads to a poor reconstruction. Therefore, if VAE is trained only by the data of a certain category, it can be used as a detector which discriminates whether an input sample belongs to the trained category or not, by the quality of reconstruction and by the mapped point in the latent space.

Y. Liu et al. (Eds.): ICNC-FSKD 2019, AISC 1074, pp. 618–626, 2020.
https://doi.org/10.1007/978-3-030-32456-8_68

In this paper, VAE is used for the detection of morphological abnormality in the biological data. We focus on a cell division process in early embryogenesis. Temporal change of the three dimensional shape of an embryo is coded in the gene. Knocking down of a lethal gene leads to any phenotype anomaly during the development. In other words, the role and function of each gene can be investigated by knocking down the gene and then observing the phenotype during the development. In that direction, extensive data has been obtained by the biological experiments on one of the model animals: *Caenorhabditis elegans*(*C. elegans*) [4]. All genes had been analyzed for *C. elegans*, and lethal genes had been listed up. Furthermore, one of those lethal genes is knocked down for each embryo by RNA interference (RNAi) technique. Then temporal series of three dimensional data, composed of multiple sliced images observed by differential interference contrast microscope (DIC), are obtained during the early development of each RNAi embryo. At the same time, a set of the same temporal data of wildtype embryo, which means no gene manipulation, is obtained as the reference.

Based on the above database, VAE is trained by three dimensional data of wildtype embryos to acquire the phenotype distribution of the wildtype. Then it is used to characterize a phenotype anomaly of each RNAi embryo. If an abnormality of RNAi is detected as a significant deviation from the wildtype distribution, then it may be caused by the knocking down of corresponding lethal gene.

Section 2 introduces the biological data used in this study, and the training of VAE is described in Sect. 3. Section 4 shows the obtained results, and Sect. 5 summarizes the present result.

2 Temporal Series of Three-Dimensional Biological Data

2.1 Worm Developmental Dynamics Database

Biological dynamics data are obtained from Worm Developmental Dynamics Database (WDDD), an public open database provided by Riken in Japan [4]. Three dimensional data of the embryo of *C. elegans* in early embryogenesis is composed of 66 sliced DIC images with different depth of focus. Data is taken with sampling time 40[s], and one time series corresponds to one worm embryo.

Fifty data of wildtype, and 136 data of RNAi are used in the following computer experiments. For each lethal gene, there are one to three RNAi data. 40 wildtype data are used for the training of VAE, and the others are used for the test and the detection of the abnormality.

We focus on the nuclear region in each cell. Binary segmentation for the nuclear region is also provided in the database. As the first step, we use the time series of two cell stage. That is, the data starts just after the division into two cells, named AB and P1, and ends up just before the division into four cells, which lasts about 20 time steps with the above sampling time.

2.2 Input Data to VAE

Sixty-six slices of DIC images of nuclear region at each time point are transformed into three dimensional $150 \times 150 \times 150$ voxel data, with adjusting a relatively coarse resolution in the depth direction. Then the position is normalized by the embryo center, which is also provided on the database. The angle should also be normalized, as the direction of embryo is arbitrarily set to the microscope. However in the two cell period, only the direction of anterior and posterior can be defined, while the direction of ventral and dorsal, or the direction of left and right is still not defined. Namely, we can identify only the axis of anterior and posterior (AP axis) to express the data. Therefore, the angle of embryo is normalized by AP axis according to the database, though it should be noticed that there still remains the ambiguity of the rotation around AP axis.

This rotation ambiguity of the volume data causes a difficulty, so we sliced the volume in a radial direction with AP axis as the center, instead of ordinary slices by the parallel planes, as sagittal and coronal. More detailed description is as follows. As is shown in Fig. 1(a), 12 slices are taken by $\frac{\pi}{6}$ rotation around AP axis for the volume data at each time point. Let us notice that the first slice is identical through all the time of one embryo, while it is not identical between the different embryos, because of the rotation ambiguity. Then, sliced image data is 150×150 gray scale data (Fig. 1(a)).

We prepare two types of dataset which are sliced image data for the input to VAE. First set is composed of one sliced image. Then 12 image data are obtained from one embryo at each time point, as a right most image in Fig. 1(a). Second one is composed of a set of 6 sliced images with $\frac{\pi}{3}$ rotation, as semi-3D data (Fig. 1(b)). Then 12 data are obtained from one embryo at each time point by rotating $\frac{\pi}{6}$. Right part of Fig. 1(b) shows this rotation process: a green slice swaps next to next for six times. No temporal information is given as the input.

3 Network Structure of VAE

The structure of VAE is shown in Fig. 2. The encoder and decoder parts are composed of three convolution and deconvolution [5] layers, respectively. The first and the second layers are followed by batch normalization [6]. Activation function is ReLU except for the last layer in the decoder. The final layer of the decoder takes sigmoid function to output a binary data.

The input size n and the dimension $d_{\mathbf{z}}$ of latent space $\mathbf{z}$ take different values for 2 datasets. For the first dataset, input size is $n = 1$ and latent space dimension is $d_{\mathbf{z}} = 2$. So, the obtained latent space can be simply visualized. For the second dataset, input size is $n = 6$, and latent space dimension is $d_{\mathbf{z}} = 64$. Higher dimension of the input data requires higher dimension of the latent space for the successful reconstruction.

Training data for VAE is 9,684 image data obtained from 40 wildtype embryos, and the remaining 2,508 image data from 10 wildtype embryos are used for the test.

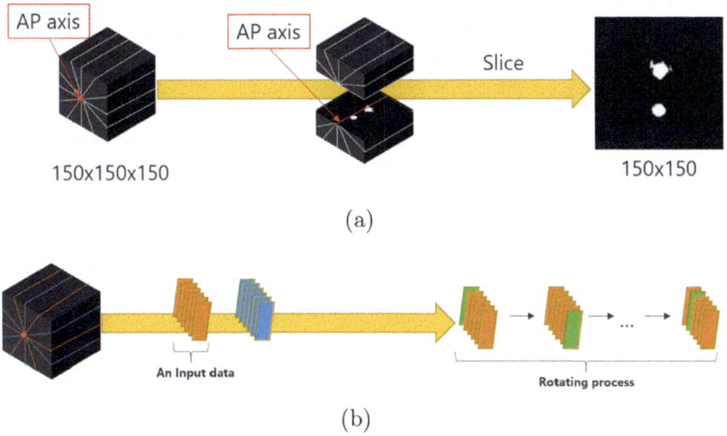

Fig. 1. Input data for VAE (a) First dataset: one data is one image (b) Second dataset: one data is a set of six images with $\frac{\pi}{3}$ rotation around AP axis. 12 data are obtained from one embryo at each time point in both (a) and (b).

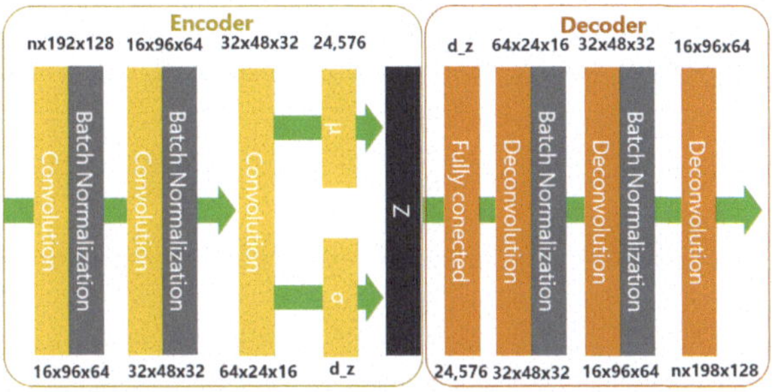

Fig. 2. Network structure of VAE

4 Experimental Results

VAE is trained by wildtype data, and an input data which deviates from the wildtype distribution is detected by the large reconstruction error. Two datasets described in Sect. 2 are used, and both VAE trained by each dataset show the similar results. Therefore, only the result for the first dataset is shown.

4.1 Abnormality Defined by the Reconstruction Error by VAE

Loss function to be minimized by VAE training is given by KL divergence for the latent space distribution and the reconstruction error. The reconstruction error L_r between input image data $\mathbf{x}_{it}$ of i-th slice at time point t and its output $\mathbf{y}_{it}$ is given by the following average of Bernoulli negative log-likelihood:

$$L_r = -\frac{1}{TN}\sum_{t=0}^{T-1}\sum_{i=1}^{N}\log B(\mathbf{x}_{it}, \mathbf{y}_{it}), \tag{1}$$

where $-\log B(\mathbf{x}, \mathbf{y})$ stands for Bernoulli negative log-likelihood function between two images $\mathbf{x}$ and $\mathbf{y}$ defined by:

$$\log B(\mathbf{x}, \mathbf{y}) = \sum_i \{x_i \log y_i + (1 - x_i)\log(1 - y_i)\}. \tag{2}$$

Input data is nearly binary valued image with white nucleus region and black background, though B is also applicable to continuous gray scale image. $N = 12$ is the number of slices at each time point and T is the time length of two cell stage.

Figure 3 shows the reconstruction error for 50 wildtype data. The error is given by time average over two cell period T for each data. 40 data indicated by the left side of a vertical blue line in the figure are used for the training, while the lest 10 data are used for the test. The average error for 40 training data is shown by a horizontal black line, and the average for 10 test data is shown by a red line. The larger error for the test data indicates the over-fitting, though VAE training is stopped at 50 epochs, after which the loss for the test data begins to increase.

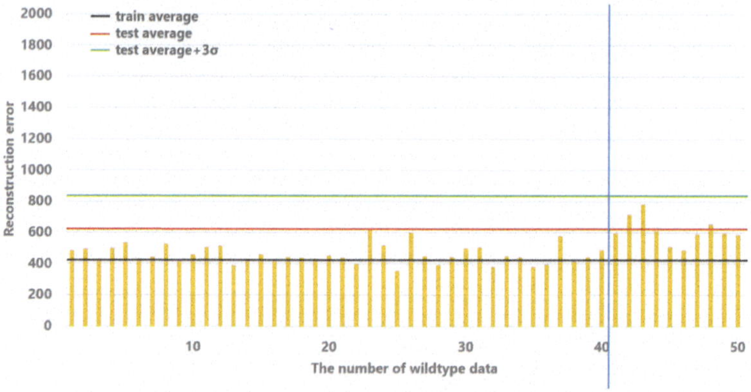

Fig. 3. Reconstruction error for 50 wildtype data. 40 data on the left side of a blue vertical line is used for VAE training. The average of 40 training data shown by a black line is smaller than the average of 10 test data shown by a red line. Threshold for the abnormality detection is set to a green line, defined by 3σ deviation from the average in test data distribution. No anomaly is found in 50 wildtype data.

The threshold of the error for abnormality detection is given based on the average and the standard deviation σ of 10 wildtype test data. Deviation of 3σ from the average is shown by a green line in the figure, and set as the threshold.

Though no wildtype data reaches the threshold, some RNAi data show the error over the threshold.

Table 1 is a list of RNAi data, which are found abnormal from the above definition. Each line in the table corresponds to one RNAi data, one of whose lethal gene is knocked down. The name of knocked down gene is given by Open Reading Frame (ORF), that is, a name of RNA sequence which will be translated to a gene, in the left most column. In the middle column, the gene name is given by three letter alphabets which is a gene class, followed by a number in the class. The vacancy in the middle column means the gene class is still not clarified. Right most column is the time span of the data in the two cell stage. The time span is mostly 20 in the wildtype data.

The list is divided into three groups according to the rate of abnormal data, and shown separately by a vertical line in the table. One of multiple data means that there are more than one RNAi data for the same knocked down gene, and only one data is found abnormal among them. In the same way, one of one data means there is only one data for the knocked down gene, and the unique data is found abnormal. All of multiple data means that there are more than one RNAi data for the same knocked down gene, and all data are found abnormal. Though the last group has the highest reproducibility, one realization of the abnormality should not be missed for the gene function study. It should be noticed that the expression level of the gene is suppressed by knocking down, but not extinguished as the knocking out.

Table 1. List of abnormal RNAi data detected from VAE reconstruction error

One of multiple data			One of one data			All of multiple data		
ORF	Gene name	Time span	ORF	Gene name	Time span	ORF	Gene name	Time span
C02F5.9	*pbs-6*	36	K06H7.1		31	F10E9.8	*sas-4*	54
F11H8.4	*cyk-1*	2	R07E5.14	*rnp-4*	21	F10E9.8	*sas-4*	45
F56A8.6	*cpf-2*	17	R08D7.2		21	F35G12.8	*smc-4*	10
R107.6	*cls-2*	21	C02F5.1	*knl-1*	11	F35G12.8	*smc-4*	14
R144.2	*pcf-11*	22	C03C10.3	*rnr-2*	1	F54E7.3	*par-3*	22
T20H4.5		28	T03F6.5	*lis-1*	26	F54E7.3	*par-3*	23
Y111B2D.18	*rps-3*	25	T23G5.1	*rnr-1*	1	F58B6.3	*par-2*	22
Y37D8A.14	*cco-2*	55	T26A5.9	*dlc-1*	81	F58B6.3	*par-2*	14
Y49E10.19	*ani-1*	19				H38K22.2	*dcn-1*	47
						H38K22.2	*dcn-1*	21
						Y43F4B.6	*klp-19*	34
						Y43F4B.6	*klp-19*	13

As the comparison with the already known gene function, genes *par-3*, *par-2* and *dcn-1* are known to play a key role in asymmetric division also in two cell period of AB and P1. On the other hand, we find no report on the function at this stage for other genes in the list: *sas-4*, *smc-4*, *klp-19*, *knl-1* and so on. Further discussion and study is expected to link to a new finding.

Figure 4 shows the temporal series of the reconstruction error during two cell stage. Figure 4(a) is an example of a wildtype test data. The temporal change is similar for 50 wildtype data, and the error becomes small in the middle during the series. This middle period corresponds to the *interphase* in the cell division. In interphase, both nuclei take round shape, and it takes relatively long span to give more data with the similar shape. Therefore in the wildtype data distribution, the data in interphase has higher density distribution also in the latent space of VAE, to give less error in the reconstruction. Figure 4(b) is the temporal change of RNAi data, whose *par-2* is knocked down. This data is listed in Table 1, and the error exceeds the threshold through the period.

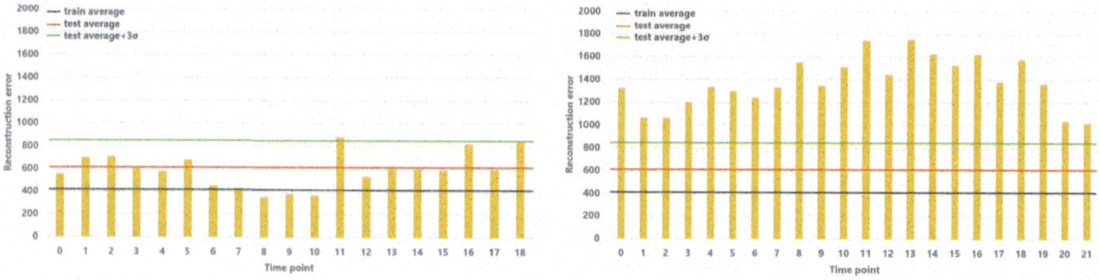

(a) (b)

Fig. 4. Temporal change of the error (a) Wildtype test data (b) *par-2* RNAi data

4.2 Time Development in Latent Space

Figure 5 is a visualization of 2 dimensional latent space obtained by VAE. Output image is drawn at each point in the plane raging from -5.0 to 4.5 by 0.5 in both axises. The origin is shown by a red square.

Morphological change during two cell period can be traced in the figure. At the beginning, both nuclei are vertically elongated after the first division. Then two nuclei become separated each other, and take spherical round shape in interphase. At the end, the division starts from AB cell, which is the upper white region in the image (A direction in AP axis), and nuclei become again extended for the next division. A trace of a wildtype data is shown in Fig. 6.

Figure 6(a) is a trace of a wildtype test data, following a single slice from $t = 0$ to 18. Figure 6(b) shows traces of 12 slices by 12 different colors, with horizontal axis for the time. 12 traces grow closely each other until the interphase, and then spread out in different directions at the final stage before the division. These traces are commonly observed for wildtype data. On the contrary, several RNAi data show irregular traces. Though we should not trust a trace of RNAi with large reconstruction error, we can also find RNAi data, which is not abnormal from the error, but the trace is different from the wildtype data. Thus, the temporal abnormality can be detected from the trace in the latent space.

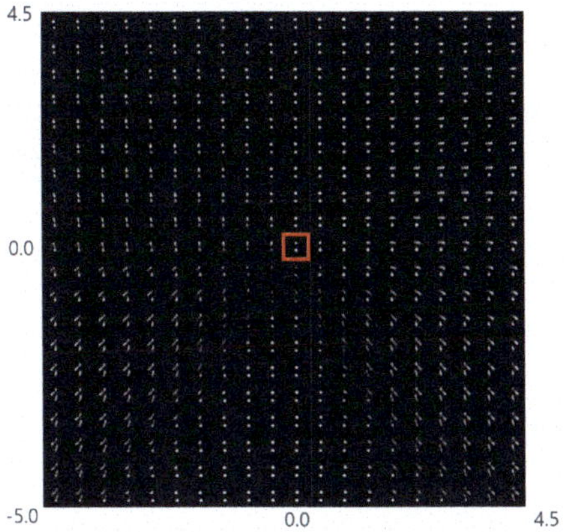

Fig. 5. Visualization of 2 dimensional latent space obtained by VAE

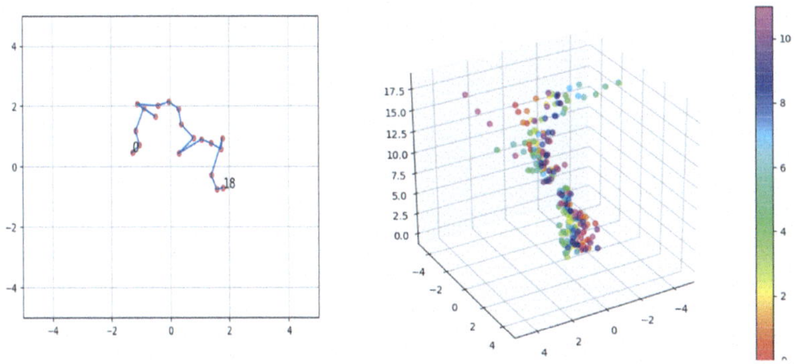

(a) (b)

Fig. 6. Temporal development of a wildtype data characterized by the trace in the latent space (a) Trace of one slice data on a plane (b) Trace of 12 slices by 3D visualization with horizontal time axis. 12 slices are indicated by 12 different colors.

5 Present Summary

VAE is applied to characterize the morphological and temporal abnormality of the phenotype of RNAi embryo. VAE is trained by wildtype data, and the morphological anomaly is defined by the reconstruction error. Several RNAi data are found abnormal, and corresponding genes are listed, including some well known genes for the function in cell division. Temporal abnormality can be characterized by a trajectory in the latent space, which deviates from the trajectory common to the wildtype. Our next plan includes a further study on the later period of four cells, where the dynamics becomes more challenging.

References

1. Kingma, D.P., Welling, M.: Auto-encoding variational bayes. In: International Conference on Learning Representations (2014)
2. LeCun, Y., Bottou, L., Bengio, Y., Haffner, P.: Gradient-based learning applied to document recognition. Proc. IEEE **86**, 2278–2324 (1998)
3. Hinton, G.E., Salakhutdinov, R.R.: Reducing the dimensionality of data with neural networks. Science **313**, 504–507 (2006)
4. Kyoda, K., et al.: WDDD: worm developmental dynamics database. Nucleic Acids Res. **41**, D732–D737 (2013)
5. Zeiler, M.D., Krishnan, D., Taylor, G.W., Fergus, R.: Deconvolutional networks. In: 2010 IEEE Computer Society Conference on Computer Vision and Pattern Recognition, pp. 2528–2535 (2010)
6. Ioffe, S., Szegedy, C.: Batch normalization: accelerating deep network training by reducing internal covariate shift. In: International Conference on Machine Learning, pp. 448–456 (2015)

Diversity Preservation in Genetic Algorithm by Lifespan Control

Yu Yamane(✉), Masataka Seo, and Ikuko Nishikawa

Ritsumeikan University, Kusatsu, Shiga 525-8577, Japan
is0335iv@ed.ritsumei.ac.jp

Abstract. Genetic algorithms (GA) application to a real world problem which possesses various complicated constraints often requires to design each problem dependent genetic operations to keep the feasibility of the individual, which could decrease the population diversity and cause the early convergence. We propose Life Control Genetic Algorithm (LCGA) to maintain the diversity under such biased operations, by setting the lifespan of each individual depending on the relative fitness. LCGA is first applied to a typical functional optimization with largely biased crossover, and compared with several types of GA including uniform lifespan. Next, it is applied to a building design optimization as an example of a real world combinatorial optimization with complicated constraints, and the effectiveness is studied by numerical experiments.

Keywords: Genetic algorithm · Diversity · Combinatorial optimization · Lifespan

1 Introduction

One of the difficulties in applying genetic algorithms (GA) to real-world problems with complicated constraints is the effective search in the solution space without losing the feasibility of newly generated individual. Simple GA (SGA) [1] with completely stochastic crossover or mutation often tends to generate an infeasible individual. Therefore, we design some biased genetic operations to keep the feasibility. Then, the generated individuals become similar, the population looses diversity and causes the early convergence.

In this report, we propose Life Control Genetic Algorithm (LCGA) to maintain the diversity under such biased operations, by setting the lifespan of each individual. The proposed lifespan is set depending on the relative fitness of each individual in the population, while GA with fixed lifespan proposed in the past often set the common lifespan as in [2].

Section 2 explains the effectiveness and difficulty of GA with the biased genetic operations, with an example optimization problem in building structure design [3]. Next, Sect. 3 describes the algorithms and the techniques of LCGA. In Sect. 4, it is first applied to a functional optimization of Rastrigin function with an intensionally biased operation, and the result is compared with other

Y. Liu et al. (Eds.): ICNC-FSKD 2019, AISC 1074, pp. 627–635, 2020.
https://doi.org/10.1007/978-3-030-32456-8_69

GAs. Then in Sect. 5, it is applied to the building structure design described in Sect. 2, and the appropriate setting of lifespan is discussed.

2 Building Structure Optimization

2.1 Building Member Placement as an Combinatorial Optimization

As an example of combinatorial optimization with multiple complex constraints in a real world, building structure optimization is described in the following. We consider the construction of steel housing as shown in Fig. 1, which is a two story ordinary house. Once the external form of the building is given, building members, such as pillar and beam, should be appropriately placed to attain the legally required resistance against the earthquake. The required earthquake resistance are defined by hundreds of inequality constraints, and verified by a physical structural calculation. Selection of multiple members with various type, size and the position can be described as a combinatorial problem in the industrialized construction. On the other hand, the total cost should be minimized by reducing the total amount of members. Thus the building member placement is considered as a combinatorial optimization.

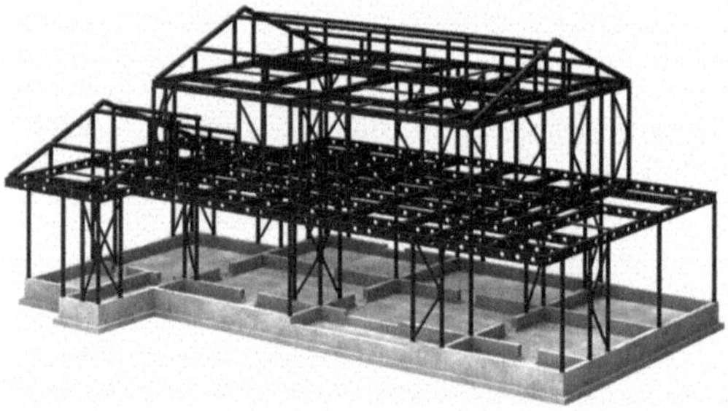

Fig. 1. An example of steel industrialized housing as an optimizing target (from [3])

Yoshitomi et al. [3] defined the objective function to be minimized by the total volume of all members. Violation of each physical constraint is added as a large penalty to the objective function.

2.2 Optimization by Genetic Algorithm

Then GA is applied for the optimization to search a practical solution in a reasonable calculation time. Crossover and mutation are rather biased to keep the resistance, which is easily lost by a small change of position, size or type of a single member. Namely, fully stochastic operation of crossover and mutation mostly generates an infeasible solution. In addition, all individual but except elite are replaced at each generation. As the result, once the feasible solution is generated in a population, the diversity stays low, and the obtained optimal solution during the search tends to vary for each trial, depending on the initial population. This low ability of search is found to be improved by increasing the population size and the number of generation, which requires large computational cost.

Therefore, GA adopted continuous generation model [4] is introduced to improve the search efficiency in the next section. Meanwhile, the above original GA proposed by Yoshitomi et al. [3] is based on discrete generation model [4] and we call this Discrete generation GA (DGA).

2.3 Continuous Generation Genetic Algorithm for Efficient Search

Continuous generation GA (CGA) is adopted continuous generation model [4] and generation gap [5]. Genetic operation in CGA is same as DGA except for the method of individual replacement. CGA is considered to accelerate the evolution, instead of increasing the population size and generation number in DGA. In stead of full replacement of population in the above DGA, CGA replaces $M(< N)$ individuals in each generation, where N is the population size. Generation gap g is defined as a fraction of replacement $\frac{M}{N}$. Figure 2 describes the individual replacement in the population in DGA and CGA. DGA corresponds to $M = N(g = 1)$. If generation gap g is smaller, newly generated individuals have a chance to become the parent in crossover and mutation earlier. Therefore, CGA is expected to accelerate the evolution especially in the early stage.

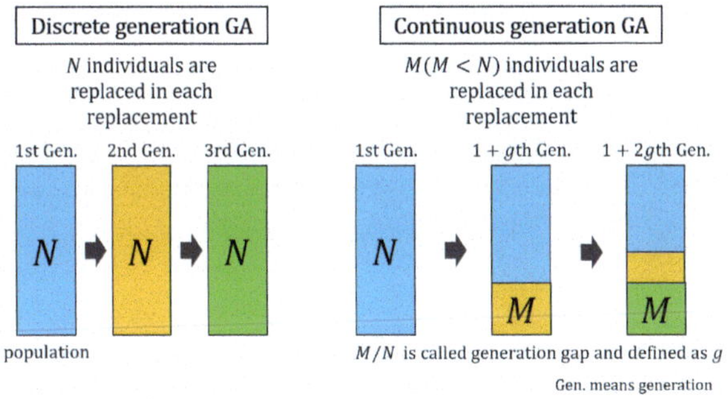

Fig. 2. Replacement of the individuals in DGA and CGA

CGA are compared with DGA with the same number of evaluation. Numerical experiments on the optimization of building member placement show the higher efficiency of CGA when the number of evaluations is 10,000. However, when the evaluation number is increased to 15,000, DGA outperforms. This reversal is considered as the earlier convergence of CGA, which is caused by high-rank individuals remaining for long generation and generating many similar offspring.

A method to control the lifespan of individuals is proposed to realize both the accelerated improvement observed in CGA and the diversity preservation in the next section.

3 Life Control Genetic Algorithm

We propose Life Control Genetic Algorithm (LCGA), where each individual has the lifespan depending on the relative fitness among the population at its birth. Followings are the flow of the individual replacement in LCGA:

1. Start n-th generation.
2. Increase the age by one to all individuals.
3. Count the number of individuals, which reach the lifespan and define as d.
4. Generate r individuals by the crossover and mutation, where $r = \max(d, M)$.
5. Evaluate the fitness of newly generated r individuals.
6. Replace the old individuals by new ones. If $r > d$, then choose $r-d$ individuals from any individuals except for the elite to be replaced.
7. Finish n-th generation, and start $(n + \frac{r}{N})$-th generation.

LCGA is applied to a typical functional optimization in the next section, and then to the building design optimization in Sect. 5, to evaluate the effectiveness especially in the diversity preservation.

4 Evaluation in Function Optimization

4.1 Experimental Settings for Rastrigin Function Optimization

Rastrigin Function:

$$f(x_1, x_2, \cdots, x_n) = 10n + \sum_{i=1}^{n}({x_i}^2 - 10\cos(2\pi x_i)) \quad (-5.12 \leq x_i \leq 5.12)$$

is known as a multimodal function which takes a minimum value at the origin. Parameter settings are given as follows for Rastrigin function optimization by GA in this section:

- The number of variables in Rastrigin function: $n = 5$
- Real value coding: 20 bits gray code for each variable, while first bit stands for sign
- Fitness of solutions which violate a range constraint is set 1000 as penalty
- Population size: 100
- The number of evaluations: 10,000
- Crossover: one-point crossover
- Selection rule: roulette
- Except for DGA, start over creation if there is same individual as a newly generated one

We compare LCGA with SGA, CGA, Steady State GA (SSGA) [6] and Uniform Lifespan GA (ULGA). We defined ULGA as the GA adopted the uniform lifespan regardless of the individual fitness. We can measure the effectiveness of setting the lifespan depending on the relative fitness by comparing LCGA with ULGA. LCGA with two lifespan settings are also considered. Table 1 shows all GA settings.

Table 1. GAs applied for Rastrigin function optimization

	SGA	SSGA	CGA	ULGA	LCGA setting 1	LCGA setting 2
Crossover	Crossover rate: 0.8	One times in each replacement	Generate 70% of all			
Mutation	Mutation rate: 0.05		Generate 30% of all			
Reserve the elite	1	Depending on probability	1	None		
Generation gap	None		0.2			
Lifespan	None			All individual: 5	Superior 20 individual: 9 Inferior 80 individual: 4	Superior 20 individual: 7 Inferior 80 individual: 3

Furthermore, in order to simulate the biased genetic operation, experiments with 5 different conditions in the crossover are compared. As the intentionally restricted crossover operations, Table 2 shows each crossover point and crossover rule. ALL in the table means that one crossover point is chosen from all 99 points. 5 POINTS means that one crossover point is chosen from 5 points, i.e. the middle of each variable. 1 POINT means that one crossover point is the rear of the second variable.

As a strong bias in the crossover, one offspring is generated from two parents, by choosing the better substring for each part. As each variable is independent in Rastrigin function, the evaluation for each variable is obtained independently. Figure 3 shows the outline of this crossover operation with superior substring selection. This strong bias leads to the rapid loss of the diversity, which is often observed during the evolution in Sect. 2. The effectiveness of LCGA is evaluated in such biased situation.

100 trials are executed for each GA shown in Table 1 and each crossover operation shown in Table 2, and the average best fitness are compared at the evaluation numbers of 2,500, 5,000 and 10,000.

4.2 Results

First, we compared LCGA settings 1 and 2. We found that setting 2 was superior to setting 1 at all time points of all 5 conditions. Therefore, we use setting 2 as LCGA in the following.

Table 2. 5 crossover operations for Rastrigin function optimization

	Crossover point	Superior substring selection
R-1	ALL	Off
R-2	5 POINTS	
R-3	1 POINT	
R-4	ALL	On
R-5	5 POINTS	

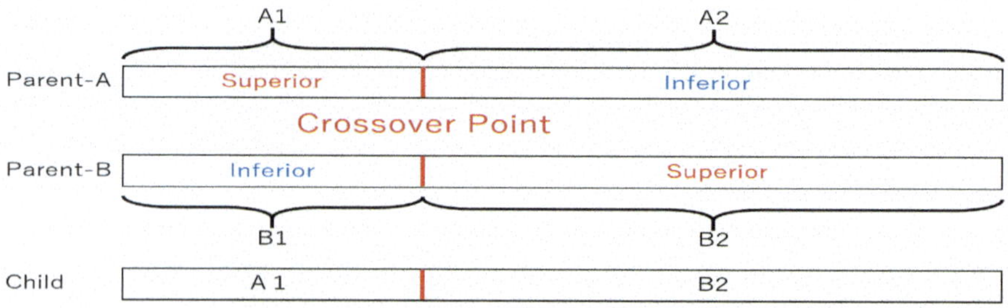

Fig. 3. Superior substring selection in crossover

Next, we evaluated LCGA against SGA and CGA. The result indicated that LCGA reached solution whose fitness is significantly lower than the other two in all situations except for 10,000th evaluation of CGA in R-2 and R-4.

Furthermore, we focused on ULGA whose individuals had uniform lifespan and LCGA. The result showed that LCGA is more excellent significantly than the other in R-1 and R-3, and LCGA is not inferior significantly in all settings. However, LCGA setting 1 did not perform similarly. Therefore, an appropriate setting of lifespan is important.

Finally, we compared LCGA and SSGA. LCGA is superior to SSGA in 10,000th evaluation of R-3, though it is reversed in 10,000th evaluation of R-1 and R-2. When we arranged R-1, R-2 and R-3 in the order of the tendency to lose the diversity, the order is R-1, R-2, R-3. In the case of R-1, R-4 and R-5, the order is R-1, R-4, R-5. Therefore, LCGA would be effective equivalently or better than SSGA in the difficult condition in reserving diversity. In terms of the number of searching the global minimum, SSGA was larger than LCGA in R-4, though it was inverted in R-5. This invert shows the same effect. Besides, we compared 5,000th generation and 10,000th evaluation in each GA. We found that the search efficiency of these is higher than the other four GAs because there were small differences in both GA.

5 Application to Building Structure Optimization

5.1 Experimental Settings

In this section, LCGA is applied to the building structure optimization, and compared with the other GAs. Followings are common setting for all GA, with the other settings same as in [3].

- Population size N: 300
- Selection rule: rank selection based on rank raised to the power of $-\frac{1}{2}$
- If there is same individual as a newly generated one, start over creation

Table 3 shows the list of GAs with each setting.

Table 3. Various GAs applied for building structure optimization

	DGA	SSGA	CGA	ULGA	LCGA
Termination condition	The number of evaluation 15000	The number of creation 15000	The number of evaluation 15000		
No. of preserved elite	1	—	1	None	
Generation gap g	None		1/6		
Lifespan	None			All individuals: 6	Superior 50 individuals: 8 Inferior 250 individuals: 4

5.2 Results

As can be seen from Table 4, SSGA is expressively inferior to the others. Thus, we made an only 5 attempts for SSGA while we conducted 30 trials for all GA excluding SSGA. Table 4 shows the trial average of the elite fitness at 10,000th and 15,000th evaluations for all GAs.

Table 4. Elite fitness for building structure optimization

(Thousand units)						
GA type	10,000th evaluation			15,000th evaluation		
	Average	Standard error	Standard deviation	Average	Standard error	Standard deviation
DGA	679	1	7	667	1	6
CGA	677	3	14	671	2	12
ULGA	676	2	10	671	2	9
LCGA	671	2	11	666	2	10
	15,000th evaluation			15,000th individual creation		
	Average	Standard error	Standard deviation	Average	Standard error	Standard deviation
SSGA	770	10	30	750	10	20

First, we compared between LCGA and CGA. The result shows that LCGA is significantly superior to CGA in every time point and indicates the effectiveness of adopting the lifespan.

Next, we evaluated LCGA against ULGA had the uniform lifespan. As a result, we found that LCGA showed higher performance significantly than the other as in Sect. 4. Those results prove that setting the lifespan depending the fitness is more efficient than setting the uniform lifespan. Moreover, we would see the tendency to obtain the excellent result when we make the lifespan average shorter than the reciprocal of the generation gap.

Finally, we compared LCGA and DGA. Table 4 shows that LCGA is significantly more effective than DGA at the 10,000th evaluation though the difference becomes tiny at the 15,000th evaluation. This means that we failed to reserve the diversity while we succeeded to improve the search efficiency.

Unlike the results in Sect. 4, SSGA did not perform effectively. The reason is that the speed of improvement is slower than the speed of convergence. Though we experimented Rastrigin function optimization under the small diversity, the optimization of building structure seemed to lose the diversity more easily.

In addition, DGA is inferior to CGA at the 10,000th evaluation while it is reversed at the 15,000th evaluation same as in Sect. 2.

5.3 Setting of Lifespan

Let us study the effect of setting the lifespan depending on the relative fitness. We equalized the average lifespan over the population for both LCGA and ULGA and compared the results. It is found that LCGA is superior to the other not only for Rastrigin function optimization in Sect. 4, but also for Knapsack problem.

When we conduct LCGA, we must decide the range of the superior individuals and the lifespans of both superior and inferior individuals. However, we still do not have the definite answer and need the further investigation.

In addition, Kubota et al. [2] recommend the setting of lifespan where 3 to 5% of population are replaced in each generation, while our setting replaces more rapidly. The less replacement like SSGA was not effective for the search in the building structure optimization.

6 Summary

We proposed LCGA which adopted the lifespan depending the fitness of each individual to reserve the feasibility and diversity. It is expected to be efficient when the biased genetic operation, which is often required for a real world optimization, leads to the loss of diversity. LCGA is first applied to a function optimization as an exemplar problem, and compared it to other GAs under the various conditions. Then, it is applied to building structure optimization as a complicated real world problem, and the improvement of the search efficiency is observed. Besides, followings are found through various experiments.

- In the case of adopting the lifespan to GA, it is better to set lifespan according to the fitness of each individuals
- LCGA has the advantage of the efficient search and disadvantage of the early convergence
- Under the difficulty in reserving the diversity, which is found in the building structure optimization, LCGA is effective to obtain the better solution

Appropriate parameter setting for preserving the diversity still needs the further study.

References

1. Goldberg, E.D.: Genetic Algorithms in Search, Optimization and Machine Learning. Addison-Wesley, Boston (1989)
2. Kubota, N., Date, T., Fukuda, T.: Introduction of age structure to genetic algorithm and its convergence. Trans. Soc. Instrum. Control Eng. **31**(5), 560–568 (1995)
3. Yoshitomi, S., Nakagawa, D., Sada, T.: Research on structural optimization for steel industrialized housing - practical method for optimal placement of structural members based on genetic algorithm. Archit. Inst. Japan J. Struct. Constr. Eng. **80**(714), 1347–1355 (2015)
4. Kitano, H.: Continuous generation genetic algorithms. J. Soc. Instrum. Control Eng. **32**(1), 31–38 (1993)
5. De Jong, K.A.: An analysis of the behavior of a class of genetic adaptive systems. Doctoral Dissertation, University of Michigan (1975)
6. Syswerda, G.: Unifrom crossover in genetic algorithms. In: Proceedings of the Third International Conference on Genetic Algorithms, pp. 2–9 (1989)

Detection of Toothbrush Hair Loss Based on Machine Vision

Nengsheng Bao[1,2(✉)] and Haitao Fang[1]

[1] Department of Mechanical and Electronic Engineering, Shantou University, Shantou 515063, China
nsbao@stu.edu.cn, 785735155@qq.com

[2] Key Laboratory of Intelligent Manufacturing, Shantou University, Ministry of Education, Shantou 515063, China

Abstract. In the process of toothbrush production, the hair implanter is usually used to implant the brush hair. It is easy to miss the implant in the process of wool implantation, resulting in defective toothbrushes. In the process of subsequent tensile stress testing, the brush that has not reached the standard will also have the defect of the tooth brush hair in this process. At present, most toothbrush manufacturers still use manual to detect toothbrush hair loss. Aiming at this situation, an online detection method of toothbrush hair loss based on machine vision is proposed. Firstly, an image acquisition system is designed. Then, image graying, image sharpening and image matching are used to detect toothbrush hair loss in turn, and online detection software is designed. The experimental results show that this method can effectively detect the absence of toothbrush hair, and the recognition accuracy can reach more than 95%.

Keywords: Machine vision · Toothbrush hair · Defect detection · On-line detection

1 Introduction

Toothbrush is closely related to people's daily life. With the increasing awareness of global oral hygiene, using toothbrush to clean teeth and protecting oral health will become an indispensable part of people's life. The demand for toothbrush is positively related to the number of people [1]. On the market side, the demand for toothbrushes in mature markets has not declined. Even in the economic downturn, no one will not brush their teeth because of the economic crisis. Therefore, there is a huge market for toothbrush production [2]. The toothbrush consists of a brush head, a handle and a toothbrush bristle. The production process of toothbrush is injection moulding, wool planting, grinding treatment, tensile stress detection and packaging. In the process of hair implantation of toothbrush, it is easy to leak or fail to implant, which results in some holes in the head of toothbrush being empty, and in the process of subsequent tensile stress testing, if the tensile stress does not meet the standard, it will pull out the holes, which will also lead to some holes being empty.

Y. Liu et al. (Eds.): ICNC-FSKD 2019, AISC 1074, pp. 636–642, 2020.
https://doi.org/10.1007/978-3-030-32456-8_70

At present, most enterprises in China still use traditional manual detection methods to detect toothbrush bristles. There are few reports on the automatic detection of toothbrush bristles using machine vision [3]. Shen [4] of Nanjing University of Aeronautics and Astronautics and others proposed a visual-based method for measuring the quality of gum injection on toothbrush surface. Artificial naked eye detection is inefficient and labor intensive, and the detection results are easily affected by the level of the inspector; moreover, long-term and high-intensity naked eye detection can easily lead to product misinspection and missed detection [5]. Using machine vision technology to detect the hair loss of toothbrush can realize the integrative online detection of toothbrush. The detection accuracy is high and the efficiency is fast. It is an effective method to solve the low efficiency and high intensity of manual detection of toothbrush hair.

Based on the above situation, a brush hair defect detection method based on machine vision is proposed. The brush image is collected by industrial camera, and then the obtained brush image is processed by image graying, image sharpening and image matching. Finally, the test results are uploaded by computer to the upper computer to display the test results. Experiments show that this method can realize online detection of toothbrush hair loss, and the detection accuracy can reach more than 95%.

2 Research Methods and Contents

2.1 Vision Detection System

Machine vision inspection system for toothbrush hair loss is mainly composed of image acquisition, image processing and result analysis. Hardware system includes computer, industrial camera, dome light source, worktable and so on. In machine vision and image processing system, the quality of light source and illumination scheme is often the key to the success of the whole vision processing system, and plays a very important role [6]. Therefore, in order to prevent the interference of external light sources, the hardware equipment of the whole image acquisition unit is placed in a closed dark box, and the dome light source is used as the system light source, which is installed directly above the toothbrush head. The camera uses MV-GE500C-T produced by Mindvision Company. Detailed camera parameters are shown in Table 1. The layout of the system is as follows: the focal length of the lens is 8 mm–12 mm, the distance between the lens and the surface of the toothbrush head is 120 mm, and the exposure time is 50 ms. The software is a self-programming program based on VS2013 platform. The whole image acquisition process is controlled by soft trigger signal. The missing brush hair images collected by image acquisition system are shown in Figs. 1 and 2:

Table 1. Mindvision camera parameters.

Serial number	Name	Parameters
1	Camera model	MV-GE500C-T
2	Sensor	1/2.5"CMOS
3	Pixel size	2.2 * 2.2 μm
4	Resolution	2592 * 1944

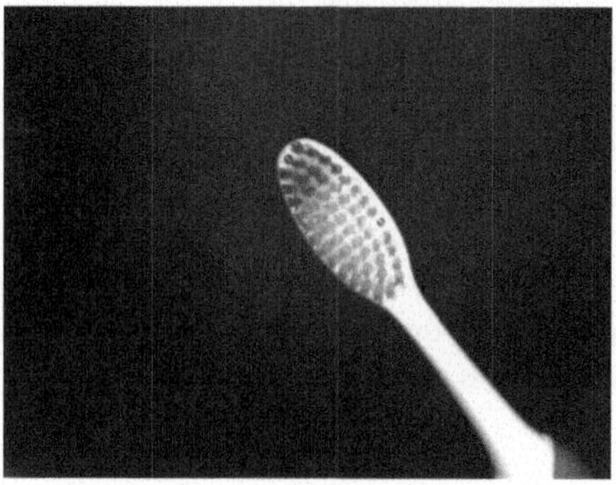

Fig. 1. The missing brush hair image1

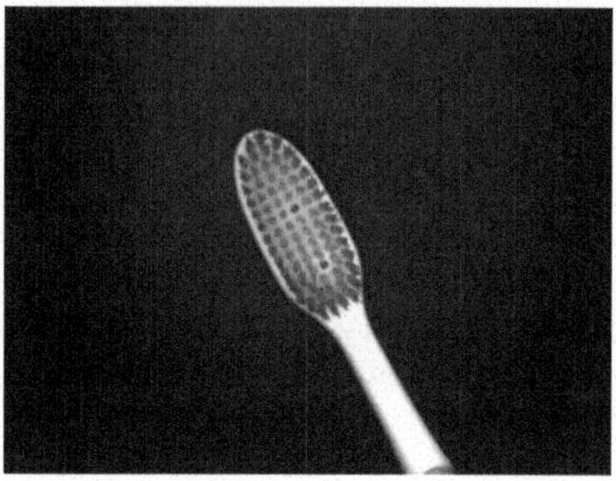

Fig. 2. The missing brush hair image2

2.2 Grayscale Image

The collected toothbrush image is a three-channel real-color (RGB) image with high resolution and large amount of image data. It takes a long time to process the toothbrush image directly to the computer, and the real-time processing is affected [7]. In order to reduce the processing time of computer, it is necessary to convert the acquired image into a single channel gray image with less data. In RGB model, if R = G = B, the color represents a gray level color. The value of R = G = B is called gray level value. Therefore, only one byte of gray level value (also called intensity value and brightness value) is needed for each pixel of gray level image. The gray level range is 0–255. The brightness of three components in a color image is taken as the gray value of three gray images, and a gray image can be selected according to the application needs. The formula is shown in Formula (1):

$$f_1(i,j) = R(i,j) f_2(i,j) = G(i,j) f_3(i,j) = B(i,j) \tag{1}$$

$f_k(i,j)(k = 1, 2, 3)$ is the gray value of the converted gray image at (i, j). By changing the range of gray value of the image, the processing time of the computer can be reduced, and the image quality can be improved, and the contrast of the image can be enhanced [8].

2.3 Image Sharpening

Because the defect of tooth brush is not obvious in the collected image, it is necessary to sharpen the image to improve the contrast. Because the sharp changes of edges and other gray levels are related to high frequency components, image sharpening can be achieved by high-pass filtering in frequency domain. High-pass filtering can attenuate the low-frequency components of Fourier transform without disturbing the high-frequency information. Fourier transform can be compared to a mathematical prism, which decomposes functions into different components based on frequency. The image as shown in Fig. 3. The formula is shown in Formula (2) and (3):

$$F(u,v) = \frac{1}{MN}\sum_{x=0}^{M-1}\sum_{y=0}^{N-1} f(x,y)e^{-j2\pi(ux/M+vy/N)} \tag{2}$$

$$f(x,y) = \frac{1}{MN}\sum_{x=0}^{M-1}\sum_{y=0}^{N-1} F(u,v)e^{j2\pi(ux/M+vy/N)} \tag{3}$$

F (x, y) image matrix, row of X and Y images, f (x, y) spectrum of F (u, v) in the formula.

In this paper, the ideal high-pass filter is used for high-pass filtering in frequency domain. In the formula X is the cut-off frequency. The filter zeroes all frequencies in the circle with radius and passes through all frequencies outside the circle without attenuation. The formula is shown in Formula (4):

$$H(u,v) = \begin{cases} 1, D(u,v) > D_0 \\ 0, D(u,v) \leq D_0 \end{cases} \tag{4}$$

The filter filters out the low-frequency part, so the background part of the image is not clear, only retains the details of the original image, that is, the high-frequency part, for further processing. Figure 4 is a spatial domain image filtered in frequency domain.

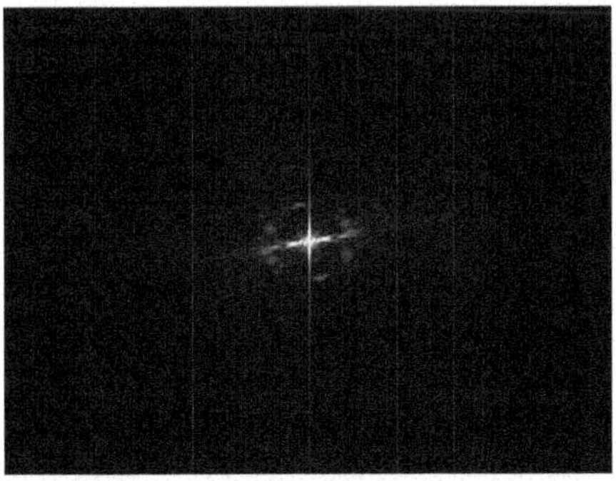

Fig. 3. Fourier transform image

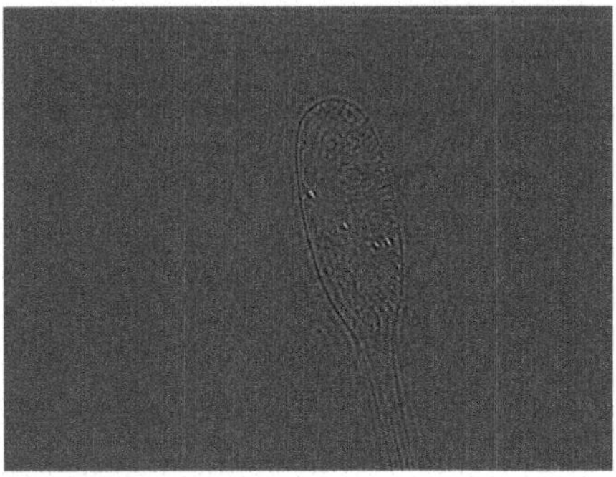

Fig. 4. Filtered image in spatial domain

2.4 Image Matching

The image matching algorithm adopted in this paper is based on the normalized product correlation algorithm NCC, which has high accuracy and adaptability, and has "immunity" to the linear transformation of image gray value [9]. NCC algorithm is a statistical algorithm for calculating the correlation between two sets of sample data. Its range is between [−1, 1]. For an image, each pixel can be regarded as a RGB value, so that the whole image can be regarded as a collection of sample data. If it has a subset that matches another sample data, its NCC value is 1, which indicates high correlation, and if it is −1, it means completely unrelated. The mathematical formula is as follow (5):

$$R(i,j) = \frac{\sum_{s-1}^{M}\sum_{t-1}^{N}|S^{i,j}(s,t) - E(S^{i,j})| * |T(s,t-(T)|}{\sqrt{\sum_{s-1}^{M}\sum_{t-1}^{N}[S^{i,j}(s,t) - E(S^{i,j})]^2 * \sum_{s-1}^{M}\sum_{t-1}^{M}[T(s,t) - E(T)]^2}} \quad (5)$$

Among them, E($S^{i,j}$) and E (T) are the average gray values of (i, j) subgraph and model respectively.

As shown in Fig. 5, any toothbrush bristle defect is selected as NCC template and saved in. NCM file. In the process of defect detection, only need to read the. NCM file of NCC template, and then match the template to realize defect detection. the final detection result is Fig. 6.

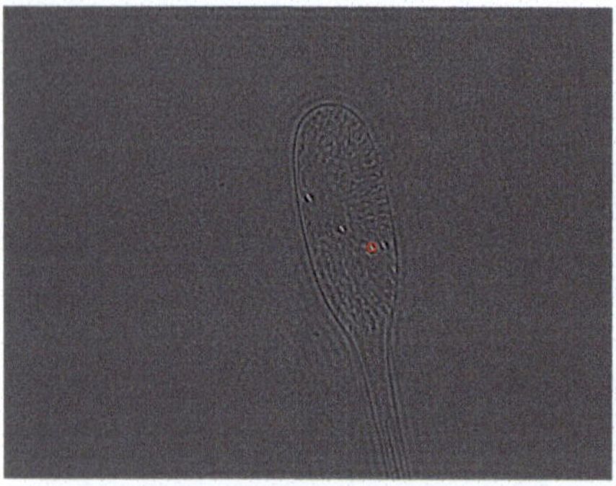

Fig. 5. Create NCC template image

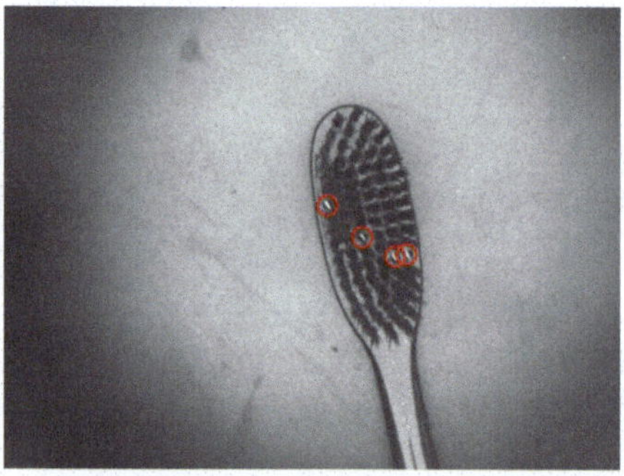

Fig. 6. Test result image

3 Experimental Results and Analysis

3.1 Experimental Scheme

The hardware of the detection system mainly includes detection platform, industrial camera, white dome light source, PC and so on. The structure sketch of the inspection platform is shown in Fig. 7. The industrial camera is above, the dome light is below, fixed on the coaxial line, and the toothbrush is placed on the inspection platform within the camera's field of vision.

The software is implemented by Visual Studio 2013 and Halcon. The upper computer software of machine vision inspection is written by Visual Studio, and Halcon operator is used to realize image acquisition, image processing and defect detection. At the same time, the test results and the processed images can be displayed. Detection platform image as shown Fig. 7

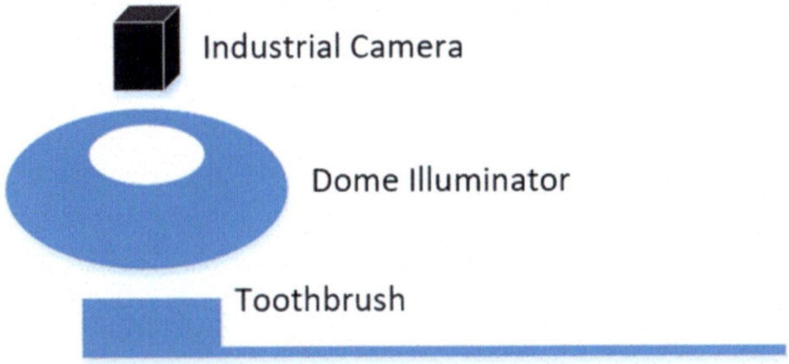

Fig. 7. Detection platform image

3.2 Experimental Results

Fifty defective toothbrushes were used in the experiment. Each toothbrush contained 90 pinches of bristles, totaling 4500 pinches of bristles. The test results are shown in Table 2.

Table 2. Test results

Name	Missing hair number	No missing hair number
Correct number	**132**	**4368**
Checked out number	**127**	**4322**
Number of missing inspections	**5**	\
Number of false checks	\	**46**

The result is calculated as shown below:
System missed detection rate is:

$$O = \frac{5}{4500} \times 100\% = 0.11\% \tag{6}$$

The false detection rate is:

$$N = \frac{46}{4500} \times 100\% = 1.02\% \tag{7}$$

Total recognition rate is:

$$R = \frac{127 + 4322}{4500} \times 100\% = 98.87\% \tag{8}$$

From the experimental results, it can be seen that this method can effectively detect the loss of toothbrush hair, and has high detection accuracy and efficiency. Compared with the simple naked eye detection, the detection efficiency is high, and it is not affected by human factors. It can meet the actual production needs.

4 Conclusion

On the premise that there are many shortcomings in manual detection of toothbrush hair defect, combined with the principle of machine vision, Halcon image processing operator is invoked through Visual Studio programming. A set of machine vision processing system is designed for toothbrush hair defect detection. Through the detection algorithm designed in this paper, template matching of toothbrush hair defect image can be completed quickly and accurately, and the missing area can be determined. Compared with the current artificial naked eye detection, it has high efficiency and stability, can effectively realize the detection of toothbrush hair loss, and has good theoretical reference and practical value.

References

1. Xie, D.Q.: Toothbrush and its quality control. Oral Care Ind. **21**(03), 33–37 (2011)
2. Gong, J.Y.: Research progress on toothbrush products in China. Oral Care Ind. **21**(06), 47–49 (2011)
3. Guo, M., Hu, L., Zhao, J.: Surface defect detection method of ceramic bowl based on Kirsch and Canny operator. Acta Opt. Sin. **36**(09), 27–33 (2016)
4. Shen, W., Li, K., Wang, B., et al.: Quality inspection of rubber injection of toothbrushes based on vision. Mach. Build. Autom. **47**(03), 198–200 + 215 (2018)
5. Peng, H.: The detection of ceramic product appearance defect based on digital image processing technology research. Jingdezhen Ceramic Institute (2015)
6. Li, Z.-P., Li, P., Guo, X.-Y., et al.: Three programming methods of gray processing based on GDI + . Comput. Technol. Dev. **19**(07), 73–75 + 79 (2009)
7. Zhu, F.Q.: Study on shape recognition and detection system for revolving parts based on image processing. Shandong University of Technology (2014)
8. Shi, P., Zhang, L.-P.: Machine vision for axial pitch detection based on spatial moment location. Mach. Des. Manuf. (12), 205–208 (2016)
9. Yang, T.-Y., Peng, G.-H.: Fast algorithm for image matching based on NCC. Modern Electron. Tech. **33**(22), 107–109 (2010)

Analysis of the Influencing Factors of the Artificial Intelligence Judicial Referee System

Deqiang Chen(✉)

East China University of Political Science and Law, Shanghai 201620, China
dqchen@ecupl.edu.cn

Abstract. This paper discusses the influencing factors of the artificial intelligence judicial referee system. Through the AHP model, the weights and relationship levels between the involved factors are fully examined. Through algorithm analysis and test, the experimental results show that the analytic hierarchy process can effectively analyze the weights of the judicial judgments and prevent the risk of different judgments.

Keywords: Analytic hierarchy process · Artificial intelligence · Judicial referee system

1 Introduction

With the development of artificial intelligence and the arrival of the era of big data, jurists and experts in other fields have begun to explore how to apply artificial intelligence to the field of legal ruling. In fact, the legal provisions themselves are a prediction of the consequences of behavior. An important milestone in the rule of law is the stability and predictability of judicial decisions.

In theory, the predictability of judicial decisions reflects the certainty of the application of the law. In this case, the real question should be the degree of legal certainty. Therefore, the judicial referee prediction is a scientific analysis of the relationship between laws and regulations. It can analyze the information and legal consequences of a large number of cases in the past, and probabilistic speculation on the legal consequences of pending cases.

With the advent of the era of artificial intelligence, there is a way to deal with such as large amounts of legal documents. But there are many problems with this kind of judgment based on artificial intelligence. For example, most of the main data sources for legal artificial intelligence research are open referendums. But the factors that have a significant impact on case handling are not necessarily written in the judgment. The important premise of artificial intelligence to identify data is the structural characteristics. The judgment structure of different courts in different regions has significant differences. Although the legal ruling document clearly stipulates the legal subject, the facts of the case and the application of the law, no unified regulations have been formulated to further improve various legal terms of the law. Many artificial intelligence technology companies are using knowledge maps to build legal artificial

Y. Liu et al. (Eds.): ICNC-FSKD 2019, AISC 1074, pp. 643–650, 2020.
https://doi.org/10.1007/978-3-030-32456-8_71

intelligence legal applications. They use them as mainstream algorithms, but their specific algorithms are generally not public. So it is difficult to be convinced.

2 Models and Review

2.1 Literature Review

In 1958, Lucien first proposed the concept of information processing in legal science. However, it was not until 1970 that artificial intelligence and law were officially born as research topics [1]. It was proposed to use computer programming to solve judgment analysis and analogy conclusions. In recent years, especially in the early 21st century, with the combination of artificial intelligence and big data technologies, more progress has been made. More and more scientists are investing in research in related fields. For example, in 2007, Strand used Bayesian methods to analyze legal evidence and compared the differences between observed and normative values. In 2011, Dung used the selection model to analyze conflicts between jurisdiction and law. In 2013, Wisconsin v. Eric Loomis used artificial intelligence sentencing assistance, which sparked heated debates. In 2016, the International Institute of Electrical and Electronics Engineers (IEEE) issued an ethical code for artificial intelligence and recommended implementation in judicial practice. In 2018, the General Data Protection Regulation (GDPR) was officially implemented. This law involves the intelligent assistant referee system that needs to explain and explain its decision-making process.

In China, Professor Qian Xuesen proposed a framework concept combining artificial intelligence with law in 1981. However, for various reasons, no one has continued to study. After 2000, Chinese experts began to pay attention to the research and development of the artificial intelligence legal system. In 2005, Professor Yang Yiquan proposed a quantitative analysis of Chinese legislation. In 2012, Dr. Zhang Ni tried to establish a judicial discretion model to analyze mental damage compensation. Beginning in 2013, China began to build a "smart court". In 2016, the State Council included the construction of "smart courts" in the national informationization development strategy. In 2018, the Supreme People's Court launched the "class case intelligent push system" and began to officially run online.

In short, the research on the intersection of artificial intelligence and law in China started late, and the related theoretical research results are not rich. But related artificial intelligence legal aid system is online faster. The evaluation of related technical models and algorithms is still opaque and needs further research.

2.2 Judicial Referee Intelligent Assistant Systems

The ASSYST System. The ASSYST system [2] is an expert sentencing system developed by the US Sentencing Commission in accordance with existing legal rules. In June 1988, the system was promoted in the United States, which allowed judges, prosecutors, defense attorneys, and other relatively easy calculations, records, preservation, and related sentencing laws. After a period of promotion, the police officers who responded to the case can effectively reduce the processing time of each case. For example, the system can input relevant information of the case at hand and conduct a search. It can obtain the processing result of similar cases and the sentencing standard. However, due to limitations of artificial intelligence technology at the time, this system was discontinued in 1996.

The Sentencing Information System. The Sentencing Information System [3] seeks to ensure that "the same case is judged" by classifying the information. The system retrieves key information entered by the judge, such as the age of the suspect, criminal record, and so on. The results of the search will classify statistically relevant information, such as the number of criminal records, the type of crime, and so on.

The Rui Judge System. In December 2016, the Beijing Court Intelligent Judgment System named "Rui Judge" was launched in the Beijing Higher People's Court. It can provide judges with accurate information. It can promote the application of laws and the uniformity of refereeing standards with big data. The "Rui Judge" learns through intelligent machines to form a complete knowledge system. It can also conduct autonomous, deep learning. This system has multi-dimensional data support, which automatically matches all kinds of relevant information according to the trial case.

The Shanghai Criminal Case Intelligent Aid Case Handling System. In May 2017, the "Shanghai Criminal Case Intelligent Aid Case Handling System" was born in Shanghai. This system uses big data, cloud computing and artificial intelligence as the technical core. After studying the files and documents of tens of thousands of criminal cases in Shanghai, it provides evidence information capture, verification and logic analysis. The system makes full use of artificial intelligence technologies such as graphic recognition, natural language understanding, intelligent speech recognition, judicial entity identification, entity relationship analysis. It provides guidance for case handlers to collect fixed evidence and verifies the evidence.

The legal system is a complex system. In addition to the rule factors, there are many hidden rules that cannot be expressed one by one. In addition to considering legal factors, artificial intelligence referees must also consider social and legal factors. The following article will analyze how to use the analytic hierarchy process (AHP) method to assist judges in sentencing.

3 Design and Analysis of Algorithms

About 140,000 judgments on all traffic accidents were published on the Internet [4], and the key six factors were extracted as representatives of the relationship between the criminal terms and the statutory circumstances. The analytic hierarchy process modeling can be carried out in the following four steps [5].

3.1 Establish a Hierarchical Structure Model

First of all, it is necessary to analyze the problem of decision-making, and organize it into a hierarchical structure. The hierarchical structure required generally consists of the following three levels.

- Target layer A
- Criteria layer B
- Solution layer C.

The analytic hierarchy process needs to clarify the factors at different levels and their locations, and connect the relationships between them to form a hierarchical structure. This article takes the relevant data of the traffic accident ruling as a case, and gets the following Fig. 1.

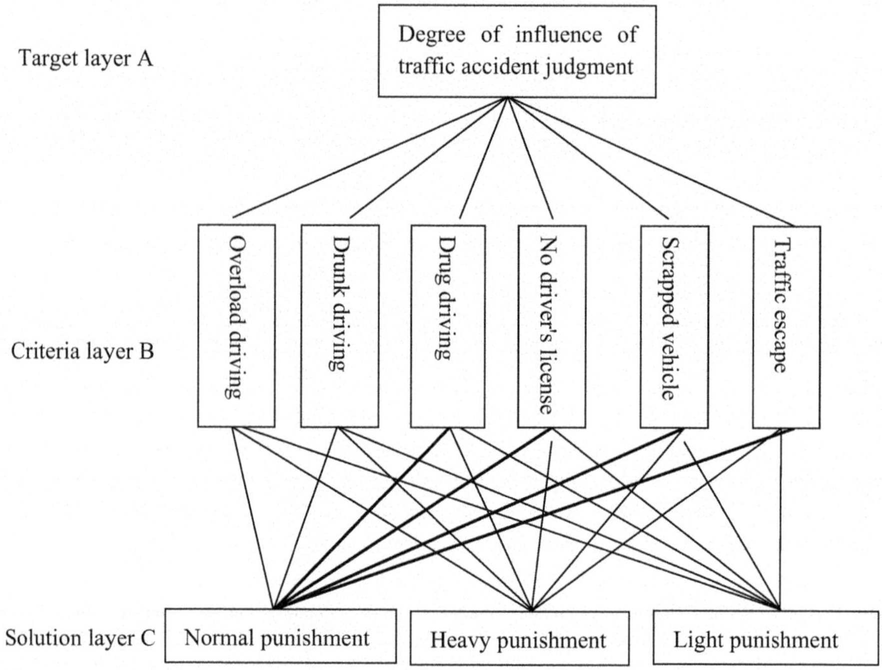

Fig. 1. The traffic accident referee AHP model.

In the criteria layer, the relevant factors are called B_1–B_6. In the Solution layer, the relevant results are called C_1–C_3.

3.2 Constructing All the Judgment Matrices in Each Level

The judgment matrix is constructed based on the analysis of the data of the traffic accident case. The method of constructing the judgment matrix is as follows. Each element with a downward membership (called a criterion) is used as the first element of the judgment matrix (in the upper left corner), the elements are arranged in turn in the first row and the first column.

The judgment matrix between the solution layer for C_1, C_2 and C_3 of B_1 is as follows,

$$P_1 = \begin{pmatrix} 1 & 1/5 & 1/2 \\ 5 & 1 & 3 \\ 2 & 1/3 & 1 \end{pmatrix}.$$

The judgment matrix between the solution layer for C_1, C_2 and C_3 of B_2 is as follows,

$$P_2 = \begin{pmatrix} 1 & 1/5 & 1/6 \\ 5 & 1 & 1/2 \\ 6 & 2 & 1 \end{pmatrix}.$$

The judgment matrix between the solution layer for C_1, C_2 and C_3 of B_3 is as follows,

$$P_3 = \begin{pmatrix} 1 & 3 & 1/3 \\ 1/3 & 1 & 7 \\ 3 & 1/7 & 1 \end{pmatrix}.$$

The judgment matrix between the solution layer for C_1, C_2 and C_3 of B_4 is as follows,

$$P_4 = \begin{pmatrix} 1 & 1/3 & 5 \\ 3 & 1 & 7 \\ 1/5 & 1/7 & 1 \end{pmatrix}.$$

The judgment matrix between the solution layer for C_1, C_2 and C_3 of B_5 is as follows,

$$P_5 = \begin{pmatrix} 1 & 1 & 7 \\ 1 & 1 & 7 \\ 1/7 & 1/7 & 1 \end{pmatrix}.$$

The judgment matrix between the solution layer for C_1, C_2 and C_3 of B_6 is as follows,

$$P_6 = \begin{pmatrix} 1 & 7 & 9 \\ 1/7 & 1 & 1 \\ 1/9 & 1 & 1 \end{pmatrix}.$$

3.3 Hierarchical Single Sorting and Consistency Check

Hierarchical single sorting refers to the relative weight of each criterion matrix for its criteria. So, it essentially calculates the associated weight vector. For the consistency judgment matrix, each column is normalized to the corresponding weight. For the non-uniformity judgment matrix, each column is normalized to approximate its corresponding weight. The arithmetic mean is obtained for the n column vectors as the final weight.

In practice, the judgment matrix is required to satisfy the general consistency, and a consistency check is required. Only by passing the test, the explanation matrix can be logically reasonable. The analysis of the results can continue.

The steps for the consistency check are as follows.

The first step is to calculate the consistency index (C.I.). $\lambda_{\max}$ is the average of the largest eigenvalues of the positive reciprocal matrices.

$$C.I. = \frac{\lambda_{\max} - n}{n - 1} \tag{1}$$

In the second step, we look up the table to determine the corresponding average random index (R.I.).

According to the different orders of the judgment matrix, we check the table below to get the average random index.

The third step is to calculate the consistency ratio (C.R.) and judge.

$$C.R. = \frac{C.I.}{R.I.} \tag{2}$$

When C.R. < 0.1, it is considered that the consistency of the judgment matrix is acceptable. When C.R. > 0.1, it is considered that the judgment matrix does not meet the consistency requirement, and the judgment matrix needs to be re-corrected.

3.4 Hierarchical Total Ordering and Consistency Check

The total ordering refers to the relative weight of each factor of the judgment matrix for the target layer (the top layer). This weight is calculated using a top-down approach, layer by layer.

Similarly, a consistency check on the overall ranking results is required. When $C.R.^{(k)} < 0.1$, it is considered that the overall consistency of the judgment matrix is acceptable.

4 Experiments and Results

The experimental data comes from Table 1. Through the calculation and analysis, we obtain the following Fig. 2.

From the results of the overall ranking of the program level, the weight of the heavy penalty (0.4050) is far greater than the weight of the lighter penalty (0.2749). Therefore, the final decision-making plan is to impose a heavier punishment on the case of the data satisfying Table 1. For the six factors of the criterion layer B, unlicensed driving is the least harmful. Traffic escaping is the most harmful. The second hazard is poison driving and drunk driving. It indicates that these factors are more important in judicial decisions.

Table 1. Judgment matrix.

A	B_1	B_2	B_3	B_4	B_5	B_6
B_1	1	1	1	5	1	1/2
B_2	1	1	2	5	1	1/2
B_3	1	1/2	1	6	3	1/2
B_4	1/5	1/5	1/6	1	1/3	1/3
B_5	1	1	1/3	3	1	1
B_6	2	2	2	3	3	1

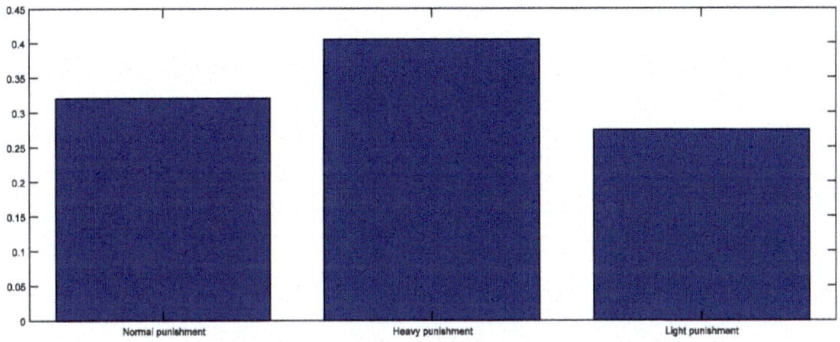

Fig. 2. Hierarchical total ordering.

5 Conclusion

Based on the analysis of the common artificial intelligence judicial ruling system, we find the correlation between the sentencing referee indicators through calculation. This method can help judges more accurately calculate the sentencing. This model can quantify the weighting indicators that achieve the legal factors of the trial and promote the number of trial decisions.

In this way, the arbitrariness in the application of laws and regulations can be effectively avoided. Further narrowing the discretion, based on the combination of artificial intelligence and judge inference, this method can effectively reduce the risk of different judgments in the same case.

Acknowledgements. Thanks for the funds support of Chinese Law Society 2018 Ministerial Law Research Project (CLS-2018-D108) and East China University of Political Science and Law's research projects (A-0333-18-139015) and thanks for East China University of Political Science and Law Judicial Identification Center.

References

1. Buchanan, B.G., Headrick, T.E.: Some speculation about artificial intelligence and legal reasoning. Stanford Law Rev. **23**, 40–62 (1970)
2. Simon, E., Gaes, G.: International Conference on Artificial Intelligence & Law (1989)

3. St Amand, M.D., Zamble, E.: Impact of information about sentencing decisions on public attitudes toward the criminal justice system. Law Hum Behav. **25**, 515–528 (2001)
4. Bai, J.: Sentence prediction research based on the collective experience of judges. Jurisprudential Res. **06**, 140–154 (2016)
5. Srdjevic, B., Pipan, M., Srdjevic, Z., Blagojevic, B., Zoranovic, T.: Virtually combining the analytical hierarchy process and voting methods in order to make group decisions. Univ. Access Inf. Soc. **14**, 231–245 (2015)

An Application of Artificial Intelligence Technique in Horizontal Crustal Stress Measurement

Shike Zhang(✉), Yuan Yuan, and Dewei Fu

Department of Civil Engineering, Anyang Normal University, Anyang 455000, China
shikezhang1021@163.com

Abstract. In petroleum engineering, the horizontal crustal stress magnitude and direction are necessary geomechanical parameters to conduct the wellbore stability analysis, geomechanics simulation and optimization of production well location and direction. For this purpose, this paper presents an artificial intelligence technique of multiple parameters identification. In this method, we apply the field measured wellbore deformation information to identify the field crustal stresses backward based on the artificial intelligence optimization algorithm. The algorithm is to combine back propagation neural network (BPNN) with genetic algorithm (GA) to implement the parameters determination. There is an obvious linear/nonlinear relationship between the horizontal crustal stresses and the wellbore deformation, which naturally applies the BPNN to map it. For the parameter optimization determination, the GA is used to search the optimal values in a large search space on the basis of the fitness function. Results from the learning samples and identification horizontal crustal stress state show that the hybrid BPNN-GA displacement back analysis model can effectively map the relationship between geomechanical parameters and mechanical behavior, and accurately determine the horizontal crustal stress magnitude and direction by the field measured wellbore deformation when drilling.

Keywords: Horizontal crustal stress · BP neural network · Genetic Algorithm · Displacement back analysis · Petroleum engineering

1 Introduction

In petroleum rock mechanics, the crustal stress state is seen as the necessary parameters to analyze the deformation state of the reservoir and surrounding rock, because it represents the original stress field state of formation without any engineering and/or human disturbances [1, 2]. The crustal stress state is usually reduced to four field principal stress components [3, 4]. The first is the vertical principal stress σ_v in the same direction as the vertical well. The second is the smallest horizontal principal stress σ_h in the perpendicular direction to the vertical well. The third is the largest horizontal principal stress σ_H is on the same level of the σ_h, but there is a 90° angle with the

Y. Liu et al. (Eds.): ICNC-FSKD 2019, AISC 1074, pp. 651–659, 2020.
https://doi.org/10.1007/978-3-030-32456-8_72

σ_h. The fourth is the direction of the horizontal principal stress, normally, it is defined as the azimuth of leveling to the largest horizontal principal stress. Notice that σ_H and σ_h exist in pairs, and each pair are symmetric and the same magnitude.

In practical engineering, in order to perform the wellbore construction and stability analysis, optimization of wellbore location and direction, and various simulation of petroleum rock mechanics, researchers try to adopt a series of approaches to determine the crustal stress state. These methods are overcoring gauge test, borehole analysis, acoustic emission, fault plane solution, stress-strain theoretical analysis, inelastic strain relaxation, core discing, observation of discontinuity states, hydraulic fracturing tests, inverse analysis and so forth [5–7]. These methods are mainly on the basis of the framework of elasticity except for inversion methodology [8]. So that results from the above methods are generally valid for shallow rock mass to analyze the mechanical behavior of formation.

For inversion methodology, it can be summarized as inverse solution method and direct quantification inversion method [9]. The inverse solution method is to employ the equations to back solve the input parameters, apparently, it is just suitable for one or two input parameters. With the development of artificial intelligence technique, the direction quantification inversion method based on the artificial intelligence technique is widely used to solve the nonlinear problems in petroleum rock mechanics [10, 11]. The advantages of the method is to (1) take more than one parameter at a time, (2) quantitatively obtain the optimal parameters in a large possible parameter space based on an objective function. Wang and Li [12] carried out a review analysis for the identification results by the inverse optimal algorithm, which showed that the inverse method can effectively improve the precision of identified parameters.

Therefore, this paper is to integrate the BP neural network (BPNN) and genetic algorithm (GA) to develop an artificial intelligence technique of multiple parameters. This needs to build an objective function that is defined as the minimum absolute error between the target value and the calculated value. The calculated values are the predictions from the BPNN model, and the learning specimens for BPNN model are obtained via The Universal Distinct Element Code (UDEC) [13, 14]. The target values are the practical field monitoring displacements when drilling [15]. The genetic algorithm (GA) is used to identify the crustal stress magnitude and direction based on the objective function. In addition, the identified crustal stress magnitude and direction are input the numerical computation model to achieve the mechanical behavior, and comparing numerical calculations, predictions and measurements is to verify the effectiveness that the proposed artificial intelligence technique back identifies geotechnical parameters in petroleum engineering.

2 Methodology of the Crustal Stress State Identification

2.1 Back Propagation Neural Network (BPNN)

In practical engineering, the most commonly used artificial neural network model is the multilayer feed-forward network trained by error back-propagation algorithm, because the back-propagation neural network (BPNN) model is more effective to approximatively map a nonlinear function relationship [16]. BPNN can conduct the arbitrary specific functions via changing the weights and thresholds of the input layer, hidden layer(s) and output layer. The forward propagation uses the input vector to produce the target vector, and perform a comparison for target outputs expected outputs to determine the effect of network learning. The backward propagation reduce the error between target outputs and expected outputs by adjusting the weights and thresholds for each layer repeatedly. Figure 1 is the topology of BP neural network model including input, hidden and output layers.

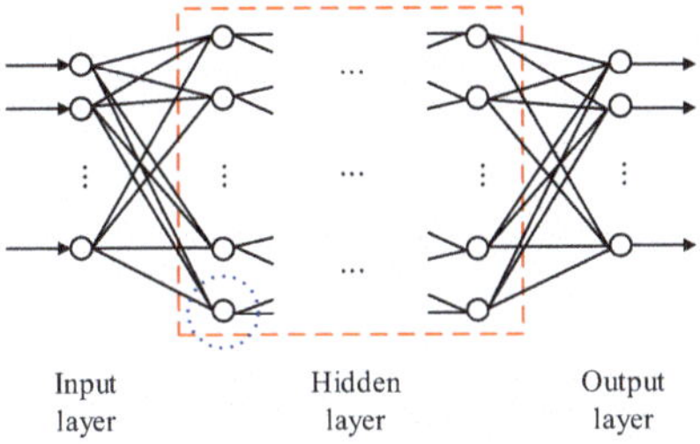

Fig. 1. A topology of BP neural network structure.

Input of the hidden layer in BPNN is expressed as

$$X_j^l = \sum_i w_{ij}^l Y_i^{l-1} - \theta_j^l \tag{1}$$

and the output of the hidden layer is written as

$$Y_j^l = f\left(X_j^l\right) = f\left(\sum_i w_{ij}^l Y_i^{l-1} - \theta_j^l\right) \tag{2}$$

where l (l = 1, 2, …, R), R the number of hidden layer, w_{ij} is the connection weight from the ith node to the jth node between two layers, θ_j is the threshold of node j, f represents the activate function of node.

In the back-propagation process, the error is expressed as

$$E_k = Y_k - O_k, \quad k = 1, 2, \ldots, n \tag{3}$$

where Y_k is target outputs, O_k is network outputs.

The weights and thresholds of hidden layers can be respectively updated as

$$w_{ij}^l(g+1) = w_{ij}^l(g) + \eta Y_j^l\left(1 - Y_j^l\right) Y_j^{l-1} \sum_i w_{ij}^{l+1}(g) E_k \tag{4}$$

and

$$\theta_j^l(g+1) = \theta_j^l(g) + \eta Y_j^l\left(1 - Y_j^l\right)\sum_j w_{ij}^{l+1}(g)E_k \tag{5}$$

The mean square error (MSE) is the one of evaluation indictor of BPNN model as

$$\mathrm{MSE} = \frac{1}{nN}\sum_{a=1}^{N}\sum_{b=1}^{n}(Y_{ab} - O_{ab})^2 \tag{6}$$

The correlation coefficient (R–value) is the other evaluation indictor of BPNN model as follows:

$$R = \left|\frac{\sum_{a=1}^{N}\sum_{b=1}^{n} y_{ab}o_{ab}}{\sqrt{\sum_{a=1}^{N}\sum_{b=1}^{n} y_{ab}^2}\sqrt{\sum_{a=1}^{N}\sum_{b=1}^{n} o_{ab}^2}}\right| \tag{7}$$

where N is the number of samples, n is the number of output node, Y and O are the target outputs and the predicted outputs, respectively. $y = Y - Y_{\text{ave}}, o = O - O_{\text{ave}}, Y_{\text{ave}}$ are O_{ave} are the corresponding average values.

2.2 Genetic Algorithm (GA)

The genetic algorithm (GA) is an optimal searching algorithm, and it can find the optimum in a large searching space based on the fitness function [17]. According to previous investigation, it can find that as long as the parameters of genetic operation can be reasonably selected, it can effectively figure out the complicated nonlinear problems [18]. This paper adopts the roulette wheel method to select gene parents to form population of GA. The criterion of selection is that a smaller fitness is better, the selected probability of individual is as follows:

$$P_k = \frac{f(G_k)}{\sum_{k=1}^{POP_{\max}} f(G_k)}, \quad f(G_k) = \frac{1}{\text{Fitness}} \tag{8}$$

where $POP_{\max}$ is the number of population individual. Fitness is defined as follows:

$$\text{Fitness} = \min\left(\frac{1}{n}\sum_{i=1}^{n}(|\mathrm{BPNN}_i(X) - U_i|)\right) \tag{9}$$

where n is displacement evaluation factors, $\mathrm{BPNN}_i(X)$ is the predicted displacement of the ith evaluation factor. U_i is the monitored information of the ith evaluation factor.

Once the selection operation is finished, GA will conduct successively the crossover operation and mutation operation for the selected individuals to form the new population. The detailed description of genetic operation and its probability selection criterion are introduced in reference [19].

2.3 Artificial Intelligence Model Combing BPNN and GA

An artificial intelligence model consists of BP neural network learning and prediction and GA optimization identification. The major procedure of crustal stress state prediction based on the monitoring wellbore displacements is illustrated as follows:

Step 1: Build BPNN model of setting proper initial network type and contructure.
Step 2: Initialize the parameters of BPNN model such as weights and thresholds.
Step 3: Initialize the parameters of GA such as population size, N_{pop_size}, maximum generation, N_{max_gen}, crossover probability, P_c, mutation probability, P_m, etc.
Step 4: Determine the searching range of identified parameters, and input the monitoring wellbore displacements to build the fitness function.
Step 5: Conduct the proposed artificial intelligence model to achieve the crustal stress state parameters. Meanwhile, check if termination criterion is satisfied?
Step 6: If termination criterion is not satisfied, back to Step 1 until we achieve the satisfactory consequence.

3 Verification of Artificial Intelligence Technique

3.1 Parameter Setting of Artificial Intelligence Technique

Table 1 is the 30 learning specimens calculated by UDEC, they are used to train and test BPNN model, and the random 20% of all samples are employed to validate the trained BPNN model. For BPNN model parameter setting, the node number of input layer and output layer are 3 and 5, respectively. The hidden layer number is 2 and the number of node for each hidden layer is respectively 6 and 8. Activate function of output layer is pure-linear. Transmission function of hidden layers is log–sigmoid.

Table 1. Learning specimens for artificial intelligence model.

No.	Inputs			Outputs (mm)				
	σ_H(MPa)	σ_h(MPa)	θ (°)	P_1	P_2	P_3	P_4	P_5
1	58.95	32.86	0	−2.23	−4.43	−5.235	−4.513	−2.529
2	56.86	48.12	0	−2.015	−3.894	−4.453	−3.773	−2.198
3	54.01	32.38	0	−2.046	−3.958	−4.655	−3.972	−2.248
4	59.42	33.12	20	−0.801	−3.435	−4.961	−5.703	−4.107
……								
30	59.72	38.32	60	2.266	0.0186	−2.271	−4.483	−5.244

The operation parameters of GA are as follows: maximum generation $N_{max\_gen} = 200$, population size $N_{pop\_size} = 80$, crossover probability $P_c = 0.4$ and mutation probability $P_m = 0.2$. The monitoring wellbore displacements are respectively −1.291 mm, −3.486 mm, −4.175 mm, −4.115 mm, −2.932 mm. The range of the largest horizontal crustal stress σ_H is 30.0–60.0 MPa, The range of the smallest horizontal crustal stress σ_h is 20.0–50.0 MPa, The range of the maximum horizontal crustal stress in east–north direction θ is 0–90°.

3.2 Result Analysis

The identification results are $\sigma_H = 53.0516$ MPa, $\sigma_h = 31.2957$ MPa, the direction angle, $\theta = 40.1786°$. MSE variation curve with iteration is shown in Fig. 2. The final MSE from training, validation and test specimens are 0.0002, 0.0008 and 0.0004, respectively. Results indicate that MSE is very close to the actual value 0, and there is no significant overfitting during training, so BPNN model training is successful.

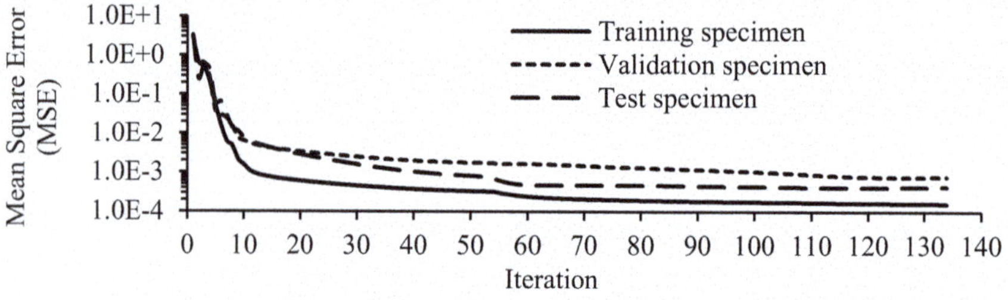

Fig. 2. MSE variation for the learning specimens with iteration

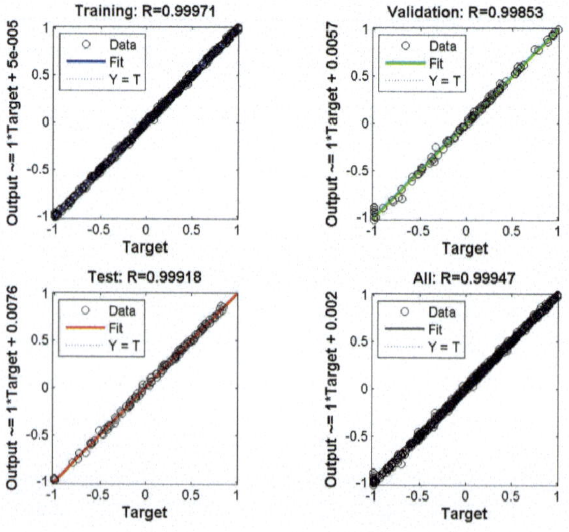

Fig. 3. Regression analysis for learning specimens and their R–values.

The regression analysis between expected outputs and targets for learning and *R*-values are shown in Fig. 3. The predictions are uniformly distributed in both sides of fitness line, R–values for learning specimens are more than 0.9. This demonstrates a good correlation between the expected outputs and targets.

The fitness variation values with 200 iterations is shown in Fig. 4. The average individual fitness values have an obvious fluctuation during optimum searching, which shows the searching values can effectively keep the global optimum. The final best individual fitness is 0.03, it is very close to the real value zero.

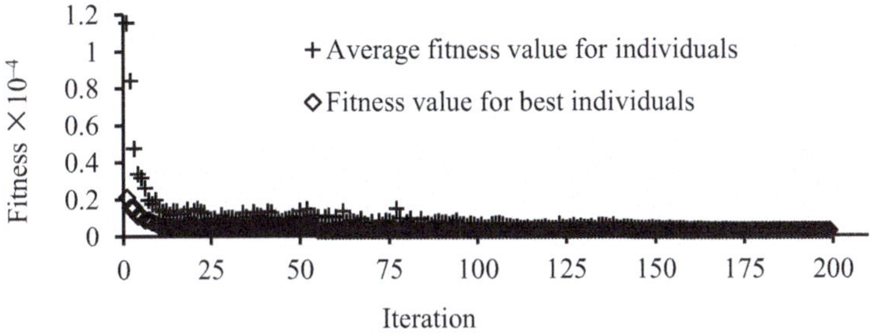

Fig. 4. Fitness values of GA with iterations.

To verify the identified parameters, the obtained crustal stresses are input the UDEC model to perform numerical computation. The comparison between UDEC calculations, field monitoring values and network predictions is shown in Fig. 5. All absolute errors are less than 0.00002 m, and all relative errors are within 10%. This illustrates that GA has found the optimal crustal stress state parameters.

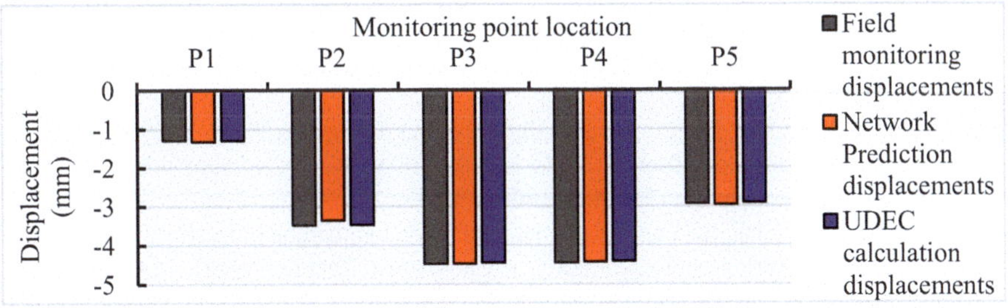

Fig. 5. The comparison between predictions, calculations and measurements.

4 Conclusions

This paper presents an artificial intelligence technique of integrating BPNN and GA and its application in the crustal stress state identification based on the wellbore monitoring displacements when drilling. Results from MSE and regression analysis of learning specimens illustrate that BPNN model can accurately map the nonlinear relationship between the crustal stresses and its mechanical behavior. Results from GA fitness and the comparison show that the identified geomechanical parameters are enough accurate. Thus, the propose technique is an excellent method for multi-parameter determination in petroleum engineering.

Acknowledgements. This study was supported by Henan Science and Technology Research Planning Project (182102310804, 182102310763), Natural Science Foundation of Henan Province (182300410160), Training Project for Young Backbone Teachers of Higher Education of Henan (2018GGSJ122), Key Research Project of Institution of Higher Learning in Henan Province (20B560002), Anyang Science and Technology Research Planning Project (Anke[2018] 66), College Student Innovation Fund Project (ASCX/2019-Z154).

References

1. Schmitt, D.R., Currie, C.A., Zhang, L.: Crustal stress determination from boreholes and rock cores: fundamental principles. Tectonophysics **580**(2012), 1–26 (2012)
2. Ljunggren, C., Chang, Y., Janson, T., Christiansson, R.: An overview of rock stress measurement methods. Int. J. Rock Mech. Min. Sci. **40**(2003), 975–989 (2003)
3. Bell, J.S., Gough, D.I.: Northeast-southwest compressive stress in Alberta evidence from oil wells. Earth Planet. Sci. Lett. **45**(2), 475–482 (1979)
4. Zhang, S., Yin, S., Yuan, Y.: Estimation of fracture stiffness, in situ stresses, and elastic parameters of naturally fractured geothermal reservoirs. Int. J. Geomech. **15**(3), 1–9 (2015)
5. Liu, Y., Liu, Y.: Study on 3-D geostress determining method with hydrofracturing technique. Crustal Deform. Earthq. **19**(3), 64–71 (1999)
6. Hudson, J.A., Cornet, F.H., Christiansson, R.: ISRM suggested methods for rock stress estimation – part 1: strategy for rock stress estimation. Int. J. Rock Mech. Min. Sci. **40** (2003), 991–998 (2003)
7. Li, P., Cai, M., Guo, Q., Miao, S.: In situ stress state of the northwest region of the Jiaodong Peninsula, China from overcoring stress measurements in three gold mines. Rock Mech. Rock Eng. **2019**, 1–11 (2019)
8. Zoback, M.D., Moos, D., Mastin, L.: Well bore breakouts and in situ stress. J. Geophys. Res. **90**(1985), 5523–5530 (1985)
9. Gioda, G., Maier, G.: Direct search solution of an inverse problem in elastoplasticity: identification of cohesion, friction angle and in situ stress by pressure tunnel tests. Int. J. Numer. Methods Eng. **15**(12), 1823–1848 (1980)
10. Yang, Z., Gao, Q., Zhai, S., Yang, X.: Intelligent inversion method of in-situ stress field for a complicated engineering geological body. J. Harbin Inst. Technol. **48**(4), 154–160 (2016)
11. Zhang, S., Yin, S.: An application of natural computation in geo-stress field analysis of hydropower engineering. In: 14th International Conference on Natural Computation, Fuzzy Systems and Knowledge Discovery, pp. 431–436. IEEE, Huangshan (2018)
12. Wang, J., Li, F.: Review of inverse optimal algorithm of in-situ stress field and new achievement. J. China Univ. Min. Technol. **44**(2), 229–233 (2015)
13. Cundall, P.A., Hart, R.D.: A discrete numerical model for granular assemblies. Geotechnique **29**, 47–65 (1979)
14. Itasca: UDEC user's manual, Itasca Consulting Group, Minneapolis, MM, USA (2011)

15. Borah, S.B., Chatterjee, R.S., Thapa, S.: Detection of underground mining induced land subsidence using Differential Interferometric SAR (D-InSAR) in Jharia coalfields. ADBU J. Eng. Technol. **6**(2), 72–77 (2017)
16. Wang, L., Fu, X.: Data Mining with Computational Intelligence. Springer, Berlin (2005)
17. Gen, M., Cheng, R.: Genetic Algorithms and Engineering Design. Wiley, New York (1997)
18. Zhang, S., Yin, S.: A hybrid ANN-GA method for analysis of geotechnical parameters. In: 12th International Conference on Natural Computation. Fuzzy Systems and Knowledge Discovery, pp. 1–6. IEEE, Changsha (2016)
19. Fu, X., Wang, L.: Rule extraction by genetic algorithms based on a simplified RBF neural network. In: Proceedings of the 2001 Congress on Evolutionary Computation, vol. 1, pp. 753–758 (2001)

Short-Term Bus Passenger Flow Forecast Based on the Multi-feature Gradient Boosting Decision Tree

Zixian Xu[1], Rui Zhu[1], Qiyue Yang[1], Lixia Wang[2], Runfang Wang[1], and Tong Li[1(✉)]

[1] School of Software, Yunnan University, Kunming 650500, Yunnan, China
tli@ynu.edu.cn
[2] School of Economics, Yunnan University, Kunming 650500, Yunnan, China

Abstract. Accurate prediction of bus passenger flow is an important basis for the dynamic scheduling of urban smart transportation system. In order to accurately predict short-term bus passenger flow and help managers achieve efficient dispatching operations and alleviate traffic pressure, a Multi-feature Gradient Boosting Decision Tree (GBDT) model is proposed. Using the flexibility of Gradient Boosting Decision Tree algorithm in complex data processing, a Gradient Boosting Decision Tree basic model is established. Through data analysis and processing, multiple features such as the week, time and environmental factors that related to passenger flow were mined to construct effective feature engineering, which can make the model's prediction results more accurate. Experiments running on the real data set of Guangzhou show that the multi-feature Gradient Boosting Decision Tree model can predict bus passenger flow accurately and effectively. The MAPE of the model is 1.152%, the RMSE is 4.58, under the same conditions, which are better than the prediction effect of traditionally used models such as the Linear Regression and the BP Neural Network, etc.

Keywords: Urban smart transportation · Bus passenger flow · Forecast · Gradient Boosting Decision Tree (GBDT) · Multi-features

1 Introduction

With the high development of the city, the problem of urban traffic congestion is getting more and more serious. In 2012, China established the priority development strategy of urban public transportation system at the national level to alleviate the traffic congestion problem. As a public transportation facility with a large passenger capacity and occupying a small number of road resources, buses can significantly share the pressure of urban public resources. Therefore, it is very important to establish an efficient and efficient intelligent public transportation system for urban transportation development. Bus route planning and dynamic dispatching are one of the main modules of the intelligent transportation system, and its important basis is the change of bus passenger flow. Only by grasping the law of passenger flow changes in advance, the enterprise operation manager can scientifically formulate the operation plan and

Y. Liu et al. (Eds.): ICNC-FSKD 2019, AISC 1074, pp. 660–673, 2020.
https://doi.org/10.1007/978-3-030-32456-8_73

reasonably use people, car resources to improve service levels while reducing management costs. Although many cities have established or are building intelligent public transportation systems now and many intelligent dispatching systems have also been widely used, but in fact, their scheduling schemes are still determined personnel based on experience and subjective judgments, making schedule management less timely, and capacity arrangements. Compared to the past, the efficiency has not been significantly improved, the urban traffic pressure cannot be effectively shared. The fundamental reason is that we lack effective means of accurately predicting short-term bus traffic. In addition, with the construction and development of intelligent public transportation systems in cities, a large amount of system data will also be generated, such as the GPS data for bus operation, the bus line data, the IC card data, etc. However, these data have not been validly mined, many useful data are wasted. Therefore, it is imperative to establish an efficient forecasting model, make full use of various public transport data information, and solve the problem of accurate forecasting of short-term bus passenger flow to get the most efficient arrangement of public transport vehicles.

2 Related Work

The research of short-term bus traffic forecasting problems has gradually become a hot issue in current transportation research. Nowadays, on this issue, domestic and foreign scholars have mainly studied from the aspects of the IC card data acquisition and processing and the passenger flow forecast.

In terms of the IC card data acquisition and processing, Li [1] proposed an information processing method for passenger flow, this calculation is mainly based on the indicators such as site passenger flow, line passenger flow, section passenger flow and so on to calculate passenger flow. Han [2] used grey relational analysis method to calculate the passenger flow's time distribution of total time and the correlation and distribution level of different passenger groups to provide data parameters for urban public transportation planning and scheduling. Zhao [3] extracted the space-time association rules of passengers on the bus by mining frequent stations, route models, transfer models in the IC card database, but did not give a further application. Qun-Yong [4] calculated the departure and destination of large-scale passengers from IC card data based on the Map-Reduce, in order to reflect the travel characteristics and needs of residents, thus providing a data basis for bus system scheduling and route optimization.

In terms of passenger flow forecast, Ma [5] used the BP Neural Network to model the historical passenger flow data of the Beijing urban rail transit. Their study of passenger flow through time-phase passenger flow distribution provides research inspiration for this paper, but it only targets the transfer traffic of two of the No. 2' sites were predicted. A method based on station attributes to predicting rail transit's passenger flow was proposed by He [6]. They founded the changing rules and influencing factors of passenger flow from the historical data to evaluate the new attributes of the station so that the passenger flow can be predicted more accurately. Bai [7] proposed a multi-mode deep fusion model to predict the passenger flow of the bus. The algorithm sets the dataset of the short-term bus passenger flow into different clusters and then identify the passenger flow's distribution patterns of these clusters. In each mode, the passenger flow represented by a deep belief network. Chen [8] proposed a predictive model based on support vector machine and genetic algorithm in 2011. By optimizing the use of parameters in the vector machine and enhancing the intersection and variation, they designed an improved genetic algorithm to predict the short-term passenger flow. Also, Jung [9] introduced the development of a computer vision system for bus passenger counting. The image of passengers getting on and off captured by the camera system that installed on the bus, to achieve the purpose of passenger counting, and provide demand forecast for the bus dispatching system. The car has to equip a camera system to capture the portrait accurately. In comparison, the information obtained from the IC card's passenger flow data is more accurate.

The above researches provided a good basis for the further study of short-term passenger flow forecasting for this paper. The common forecasting models has advantages and disadvantages. The traditional linear model regression, time series, and other models are simple to calculate, but their prediction accuracy is unstable in complex environments. The neural network, ant colony algorithm, support vector machine, and other models' prediction are more accurate, but their requirements of data are strict, thus, there are applicability problems on passenger flow prediction. The wavelet analysis, traffic simulation and other models can be applied to prediction in complex environments, but the parameter setting is troublesome and the computational complexity is high. In addition, in terms of traffic flow, researchers have applied various characteristics to the prediction model by analyzing the influencing factors of traffic flow [10] to make the prediction more accurate, but most of the current research on passenger flow prediction only use the single feature. In recent years, the Gradient Boosting Decision Tree has been widely used in various prediction fields and has achieved good effects. See the detail in [11–15], it is a machine learning algorithm that integrates multiple weak classifiers to improve the effect. The Gradient Boosting framework is used, and the base learner is a CART (Classification and Regression Tree). As a predictive model, the Gradient Boosting Decision Tree has the advantages of preventing over-fitting and generalization.

Therefore, in order to solve the above problems, this paper proposed a short-term bus passenger flow forecast method based on the multi-feature Gradient Boosting Decision Tree. This method is based on the Gradient Boosting Decision Tree model, and analyzes the factors affecting passenger flow and mines its information as a feature introduced into the model. This method can accurately predict the passenger flow of the bus, but also has good applicability and low computational complexity.

3 Passenger Flow Forecasting Method Based on the Gradient Boosting Decision Tree

3.1 Introduction to the Gradient Boosting Decision Tree

Leo Breiman first proposed the idea of the Gradient Boosting Decision Tree. In 1999, Friedman applied the Gradient Boosting Decision Tree algorithm to regression analysis and then formally proposed the algorithm [16]. The algorithm performs multiple rounds of iteration by using the additive model. The base learner is CART, which generates a weak regression tree (i.e. weak classifier) for each iteration, and uses the negative gradient of the loss function to fit the approximation of each round of loss. The weak regression tree is trained on the basis of the residual of the previous round of regression trees, and the purpose of data regression is achieved by continuously reducing the residual generated by the training process. Figure 1 shows the structure of the Gradient Boosting Decision Tree algorithm.

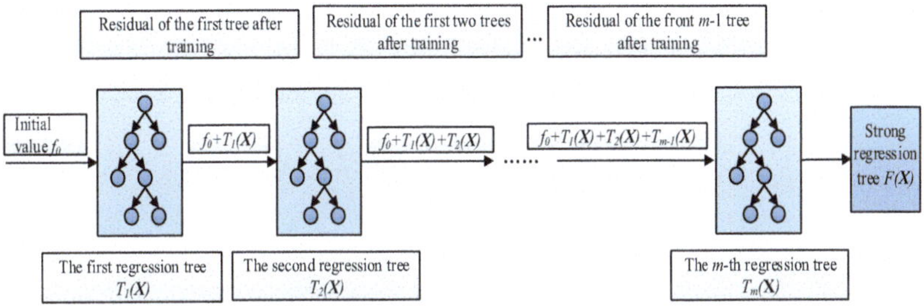

Fig. 1. Schematic diagram of GBDT training.

It can be seen from Fig. 1 that the model has a total of m weak regression trees, f0 is the initial value of the model input, and T_i represents the i-th regression tree, where i = 1, 2,..., m, and $\boldsymbol{X}$ represent attribute vectors. Then the final prediction result F($\boldsymbol{X}$) is expressed as the cumulative sum of m predicted values:

$$F(\boldsymbol{X}) = \mathrm{f}_0 + T_1(\boldsymbol{X}) + T_2(\boldsymbol{X}) + \ldots + T_m(\boldsymbol{X}) \tag{1}$$

3.2 Short-Term Passenger Flow Forecast Method

1 The Gradient Boosting Decision Tree Algorithm

The main steps are as follows: given training samples, set $T = \{(\boldsymbol{X}_1, y_1), (\boldsymbol{X}_2, y_2), \ldots, (\boldsymbol{X}_m, y_m)\}$, m is the total number of training samples, and $\boldsymbol{X}$ is the variable matrix. Set the maximum number of iterations to T and the loss function to L(y, $f(x)$).

(1) Initialize the weak regression tree, look for the constant c to minimize the loss function L:

$$f_0(\boldsymbol{X}) = \arg\min \sum_{i=1}^{m} L(y_i, c) \tag{2}$$

(2) Enter the iteration, for the iterations t = 1, 2,..., T:

a. For the samples i = 1, 2,...,m, with the negative gradient of the model as the search direction, the negative gradient value of the current model loss function L is obtained:

$$r_{ti} = -\left[\frac{\partial L(y_i, f(X_i))}{\partial (X_i)}\right]_{f(x)=f_{t-1}(x)} \tag{3}$$

The loss function used in this paper is the square loss function, namely:

$$L(y, f(X) = (y - f(X))^2 \tag{4}$$

b. Use (x_i, r_{ti}), where $i = 1, 2, \ldots, m$, to fit a CART regression tree, and get the t-th regression tree whose corresponding leaf node area is R_{tj}, $j = 1, 2, \ldots, J$. Where J is the number of leaf nodes of the t-th regression tree.

c. Calculate the best fit constant c for the leaf region R_{tj}, $j = 1, 2, \ldots, J$:

$$c_{tj} = \arg\min_{c} \sum_{X_i \in R_{tj}} L(y_i, f_{t-1}(X_i) + c) \tag{5}$$

Where I is the indicator function, and when X_i is in the R_{tj} region, the value is 1, otherwise, it is 0.

d. Update the strong regression model:

$$f_t(X) = f_{t-1}(X) + \sum_{j=1}^{J} c_{tj} I(X \in R_{tj}) \tag{6}$$

(3) After the iterative process finished, the final prediction model is obtained according to formula (1):

$$F(X) = f_T(X) = f_0(X) + \sum_{t=1}^{T} \sum_{j=1}^{J} c_{tj} I(X \in R_{tj}) \tag{7}$$

2 A Multi-feature Gradient Boosting Decision Tree

The process of establishing a multi-feature Gradient Boosting Decision Tree algorithm based on this paper is shown in Fig. 2.

(1) Let the bus passenger flow in the current time period be $g_i(t)$, $h(T)$ is the time data, which is a 16-dimensional vector (16 h of the daily forecast), and $r(T)$ is environmental data, which is a 7-dimensional vector. (The highest temperature, the lowest temperature, the average temperature, the highest wind, the lowest wind, the average wind, the weather conditions), $w(T)$ is the week data, which is the 7-dimensional vector (Monday to Sunday). The training samples of the passenger flow $g_i(t+1)$ in the next period are $x_i = (h(T), r(T), w(T), g_i(t), g_i(t-1), \ldots, g_i(t-n))$.

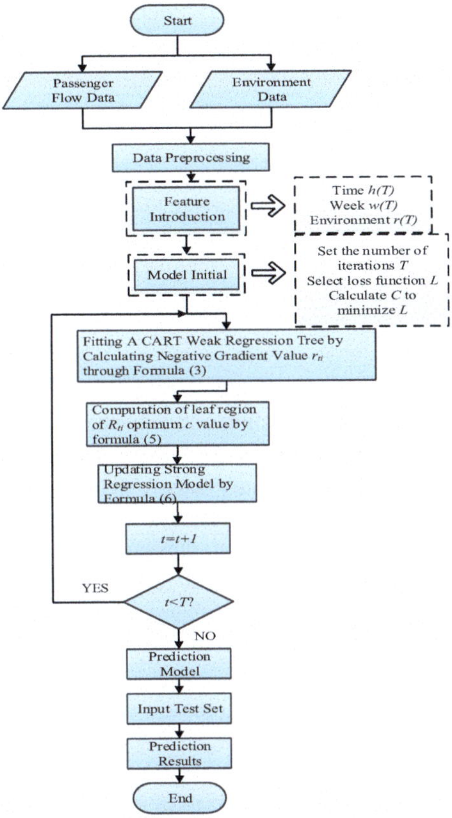

Fig. 2. Flow chart for the multi-feature model GBDT prediction.

(2) Set the main parameters, select the loss function and the evaluation index.
(3) Calculate r_{tj} by formula (3), and obtain a training data set to construct a weak regression tree.
(4) Searching the leaf node region R_{tj} to obtain a value of c that minimizes the loss function.
(5) Update the strong regression model, continue iterating until m is equal to sample value, and return the result.
(6) Enter the test set to get the final predicted result $F(X)$.

4 Experimental Results and Analysis

4.1 Experimental Data and Purpose

In order to verify the validity of the model prediction, based on the data of the IC card of the Guangzhou bus, this paper combines the environmental characteristic data of the corresponding date to carry out experiments. The IC card data includes passenger flow data from August 1, 2014, to January 7, 2015, including time, date, IC card number, and other attributes. The environmental characteristic data includes the temperature, wind, and rain of the corresponding date.

Due to the characteristics of public transport vehicles, there are fewer buses running before 6 o'clock and after 21 o'clock, and fewer passengers travel by bus. Therefore, this paper only predicts the total number of passengers in each hour from 6:00 to 21:00 on one line from January 1st to January 7th, so that the experimental results are more in line with the actual value. The data sets of this paper are divided as follows: the time of the training set is from August 1, 2014, to December 31, 2014, the forecast set is from January 1 to January 7, 2015.

4.2 Data Preprocessing

In order to ensure the accuracy of modeling and prevent the interference of abnormal data in the original data, this paper preprocesses the raw data of the bus IC card [17]. We need to clean, aggregate, and sort the raw data to get the data format we need. Data cleaning includes processing error data, such as removing data of two lines corresponding to the user's card information at the same time, incomplete data. Data aggregation is not a simple merger of data but refers to the classification of heterogeneous data, such as temperature, wind and other environmental data to normalize and unify, to obtain the visual situation of different types of data, to provide ideas for feature selection.

Since the passenger flow data is continuous, in order to reduce the range of numerical variation and accelerate the convergence speed of the model gradient, before the model training, the zero-mean normalization method is used to normalize the data and it also called standard deviation normalization, and the formula is as follows:

$$x^* = \frac{x - \bar{x}}{\sigma} \tag{8}$$

$\bar{x}$ is the mean of the sample data, ∂ is the standard deviation of the sample data.

Since the period, day of the week, temperature and wind data are discrete values, this paper uses the one-hot coding method [18] to abstract these data and extends the value of discrete features to the European space. A certain value of the discrete feature corresponds to a certain point in the European space, which can make the calculation of the feature distance more reasonable. The weather data (sunny, rain, thunderstorm, etc.) and the week data (Monday to Sunday) are digitally mapped and then abstracted, for example, $w = 1, 2, \ldots, 7$ means Monday to Sunday. Temperature data and wind data are still represented by the original data. Partially processed data are shown in Table 1. Among them, whether it is a weekend column, 1 means weekend, 0 means non-weekend, the middle part of the table is the highest temperature, the lowest temperature, the highest wind, the lowest wind, the average wind, the weather and so on.

Table 1. Partial sample data after preprocessing.

ID number	Date	Time	...	Average temperature	The day of the week
1639	2014-8-1	6	...	11	4
4108	2014-8-1	7	...	11	4
...	...	...	...	...	...
1914	2014-12-23	12	...	1	1
2211	2014-12-23	13	...	1	1
...	...		...	...	...

4.3 Feature Introduction

Through analyzing the data of the passenger flow and the current environment, such as temperature and weather conditions, the following attributes are selected as features.

(1) Week

The characteristic week refers to that the week is the unit and the different days of the week as features. As shown in Fig. 3(a) and (b) are randomly selected two-week passenger flow histograms. The different colors represent different moments of the day, and the colors are the same at the same time in different days.

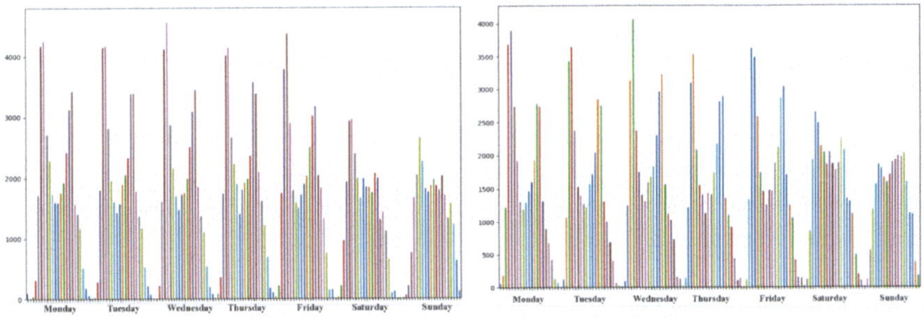

(a) September 7-13, 2014 (b) November 10-16, 2014

Fig. 3. The relationship between the week and the passenger flow.

As can be seen from Fig. 3, the peak passenger flow in the morning and evening from Monday to Friday and the trend in the same day are roughly the same, while the peak passenger flow on Saturday and Sunday is obviously lower than the working day. The passenger flow trend and working day in one day are also completely different, so the day of the week has a significant impact on the change of passenger flow, and should be selected as an important feature of the training of this model.

(2) Time

The time referred to the different moments of the day as a feature in days. As shown in Fig. 4(a) and (b) are randomly selected two-day passenger flow histograms, and the same time on different days are represented by the same color.

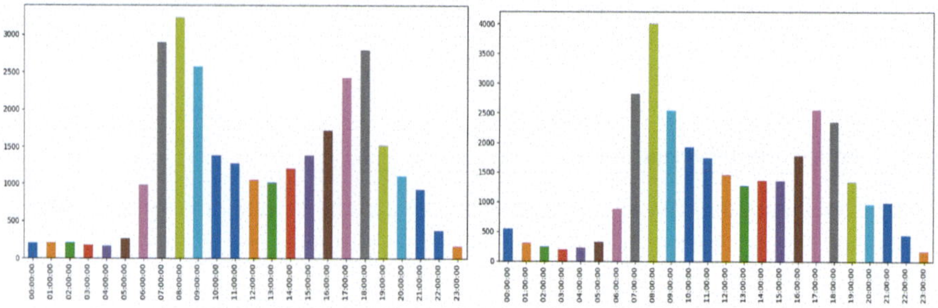

(a) August 19, 2014 (b) August 20, 2014

Fig. 4. The relationship between time and passenger flow.

It can be seen from Fig. 4 that there is a regular change in the daily passenger flow during the week. The passenger traffic at different times of the day varies greatly, and the morning and evening peaks are obvious (high passenger flow). Therefore, the time should be included in the features of the model.

(3) Environmental characteristics

In order to quantify show the relationship between environmental characteristics and passenger flow, this paper calculates the correlation between environmental characteristics and passenger flow by using the Pearson Correlation Coefficient [19]. The formula is as follows:

$$r = \frac{N\sum x_i y_i - \sum x_i \sum y_i}{\sqrt{N\sum x_i^2 - \left(\sum x_i\right)^2} - \sqrt{N\sum y_i^2 - \left(\sum y_i\right)^2}} \tag{9}$$

When there is a linear relationship between the two variables, the Pearson Correlation Coefficient can be used to represent the degree of correlation between the two variables. Positive and negative numbers represent positive correlation and negative correlation, respectively. The larger the absolute value of the coefficient, the stronger the correlation. Among them, the correlation coefficient 0.8–1.0 is an extremely strong correlation, 0.6–0.8 is a strong correlation, and 0.2–0.4 is a weak correlation. In this experiment, the calculation results of correlation coefficients between environmental characteristics and passenger flow are shown in Table 2.

Table 2. Pearson Correlation Coefficient values for each environmental feature.

Environmental characteristic	Pearson correlation coefficient	The judgment of relevance degree
Temperature	−0.466148	Negative medium intensity correlation
Wind power	0.658395	Positive strong correlation
Weather	−0.207843	Negative week correlation

From the calculation results, the temperature and wind and passenger flow are moderately strong and strongly correlated, and there is a weak correlation between weather and passenger flow. Therefore, the three attributes of the environment should be included as the characteristics of model training.

4.4 Comparative Analysis of Experimental Results

The main parameters of the Gradient Boosting Decision Tree model are set as follows: the number of model iterations is 90, and the learning rate is 0.2. In order to better analyze the model prediction effect, the evaluation indicators used in the paper are the MAPE (Average Absolute Percentage Error) and the RMSE (Root Mean Square Error). MAPE is a percentage value indicating the deviation of the predicted result from the real result; RMSE is also called the standard deviation of the regression system and is the mean of the square root of the error between the predicted value and the true value. It is often used to measure the deviation between the observed value and the true value. The calculation formulas are as follows:

$$MAPE = \sum_{t=1}^{m} \left| \frac{obseved_t - predicted_t}{observed_t} \right| \times \frac{100}{m} \tag{10}$$

$$RMSE = \sqrt{\frac{1}{M} \sum_{t=1}^{m} (observed_t - predicted_t)^2} \tag{11}$$

To further verify the performance of the model, this paper compared the three representative models—Linear Regression, Lasso, and BP Neural Network model with the same data sets. Figure 5 shows the comparison of the predicted passenger flow results with the actual passenger flow at each time from 1 to 7 day. Figure 6 shows the error of the models' prediction results.

It can be seen from Fig. 5 that the Gradient Boosting Decision Tree, the Linear Regression, BP Neural Network, and Lasso model predict the passenger flow trend is basically consistent with the actual value, but overall, the Gradient Boosting Decision Tree' results are closer to the actual passenger flow.

The error calculation in Fig. 6 is the MAPE. As can be seen from Fig. 6, the prediction error of the Lasso model is relatively stable, but the overall trend of errors is higher than other models; the predicted value of the Linear Regression model has a large error compared with the actual value on the 2nd day to the 4th day. Moreover,

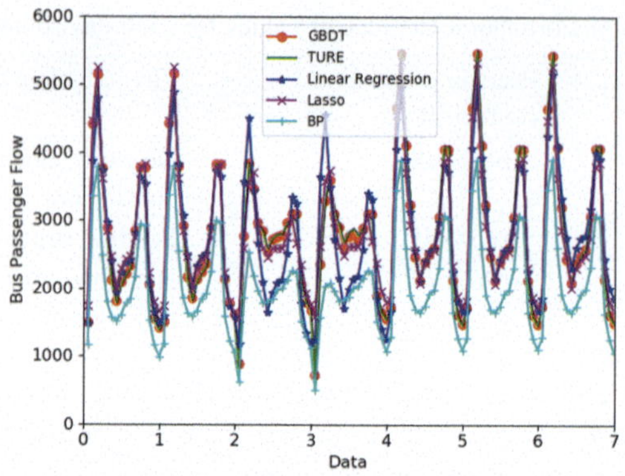

Fig. 5. Comparison of the forecasted and actual traffic flow.

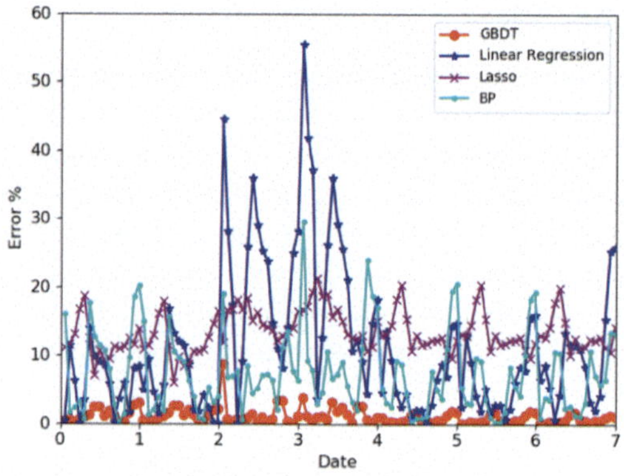

Fig. 6. Model error comparison.

compared with the Linear Regression model, the BP Neural Network model has a slightly lower error, but some partial predicted point error values are higher. The Gradient Boosting Decision Tree' error is the lowest compared to other models, its great majority prediction error points are under other models' points, indicating that the predicted value of the Gradient Boosting Decision Tree has the smallest deviation from the true value.

In order to quantify the experimental results, the MAPE and RMSE of the four models on all test data were calculated. The results are shown in Table 3.

Table 3. Comparison of prediction results of different models.

Algorithm	Evaluating index	
	MAPE (%)	RMSE
Linear Regression	11.25	5.91
Lasso	13.51	5.36
BP Neural Network	7.397	5.42
Gradient Boosting Decision Tree (GBDT)	1.152	4.58

Table 3 shows that although the MAPE value of the Linear Regression model is lower than that of the Lasso model, its RMSE value is higher than the Lasso model because of the high penalty of the RMSE for the outliers. Similarly, the MAPE value of BP Neural Network is lower than the Lasso model. However, its RMSE value is also slightly higher than that of the Lasso, which confirms that there is a large error in the predicted values of the Linear Regression model and the BP Neural Network in Fig. 5 at multiple times during the period from 2nd to 4th. The Gradient Boosting Decision Tree' errors calculation results are smaller, and the MAPE value is 1.152, which is much lower than other models. Since there is a relatively large error of prediction value on the 2nd day, its RMSE value is 4.58, but it is still lower than other models, therefore, it fully demonstrates that the multi-feature Gradient Boosting Decision Tree proposed in this paper can accurately predict the passenger flow.

In addition, in order to verify the validity of the introduced multi-features to the prediction results, this paper experiments the feature selection and calculates the mean square error under different features. The error between feature selection and model prediction results is shown in Table 4.

Table 4. Comparison of effects of multi-features on the model.

Algorithm	Feature selection	RMSE
Linear Regression	Four hours before the forecast period	28.75
	Including time, week, environment	5.91
Lasso	Four hours before the forecast period	25.57
	Including time, week, environment	5.36
BP Neural Network	Four hours before the forecast period	29.38
	Including time, week, environment	5.42
Gradient Boosting Decision Tree (GBDT)	Four hours before the forecast period	21.04
	Including time, week, environment	4.58

Tables 3 and 4 shows that the multi-feature Gradient Boosting Decision Tree model obtains lower MAPE and RMSE values than the Linear Regression, the Lasso, and the BP Neural Networks, indicating that Gradient Boosting Decision Tree model's prediction effect is more excellent than other traditional prediction models.

Furthermore, after adding multi-feature such as time, week and environmental characteristics, the RMSE of each model is significantly reduced, and the RMSE of the multi-feature Gradient Boosting Decision Tree model is the smallest, indicating that the selection of multiple features can effectively reduce the impact of outliers on the model and improve model prediction accuracy. The above fully proves that the Gradient Boosting Decision Tree algorithm adopted in this paper and the introduced multi-features are effective, which can improve the accuracy of the prediction.

5 Conclusion

Real-time bus scheduling is an important step in building smart traffic, and short-term bus passenger flow forecasting is an important part of operational decision-making. Compared with the traditional passenger flow prediction algorithm, this paper mines new features through data analysis, and proposes a multi-feature Gradient Boosting Decision Tree forecast algorithm for passenger flow prediction. In addition, experiments are carried out on real traffic data sets and verified the method's effectiveness: the comparison results of several models show that the multi-feature Gradient Boosting Decision Tree prediction model has higher prediction accuracy and higher applicability. In applications, this method can accurately predict the passenger flow of the bus, improve the theoretical basis for the operation and dispatch management of the bus enterprise, and alleviate the traffic pressure. This method also has some limitations, such as the complexity of the algorithm when processing data, which reduced the real-time performance of the algorithm. Therefore, further optimizing the model, speeding up the algorithm and improving the prediction accuracy and the real-time performance of the model are the focus of the subsequent research work.

Acknowledgment. This study is supported by the following projects: The National Natural Science Foundation of China (61662085, 61662065), The Science Research Fund of Yunnan Provincial Department of Education (2017ZZX227), and The Data-Driven Software Engineering Provincial Science and Technology Innovation Team Project of Yunnan University (2017HC012).

References

1. Li, X.: Temporal outlier detection in vehicle traffic data. In: Proceedings of the 25th International Conference on Data Engineering. IEEE (2009)
2. Han, X.: Research on data mining of public transit IC card and application. In: 2010 International Conference on Intelligent Computation Technology and Automation, vol. 2, pp. 1134–1137. IEEE (2010)
3. Zhao, X.: Spatio-temporal association rules in bus IC card databases. In: 2nd International Conference on Power Electronics and Intelligent Transportation System, vol. 1, pp. 125–128. IEEE (2009)
4. Qunyong, W.U.: A map reduce-based method for parallel calculation of bus passenger origin and destination from massive transit data. J. Geo-Inf. Sci. **2**(5), 99–110 (2018)
5. Ma, C.: Sensitivity analysis on urban rail transit passenger flow forecast. In: 2011 International Conference on Electric Technology and Civil Engineering, pp. 1537–1541. IEEE (2011)
6. He, Z.: Station passenger flow forecast for urban rail transit based on station attributes. In: 3rd International Conference on Cloud Computing and Intelligence Systems, pp. 410–414. IEEE (2014)
7. Bai, Y.: A multi-pattern deep fusion model for short-term bus passenger flow forecasting. Appl. Soft Comput. **58**, 669–680 (2017)
8. Chen, Q.: The use of LS-SVM for short-term passenger flow prediction. Transport **26**(1), 5–10 (2011)

9. Jung, J.: Deep-learning architecture to forecast destinations of bus passengers from entry-only smart-card data. J. Intell. Transp. Syst. **11**(6), 334–339 (2017)
10. Lin, P.-Q.: Short-term traffic flow forecast of toll station based on multi-feature GBDT model. J. Guangxi Univ. (Nat. Sci. Ed.) **43**(3) (2018)
11. Ma, Q., Xu, B., Sun, B., et al.: Terminal access data anomaly detection based on GBDT for power user electric energy data acquisition system. In: Nutzwertanalysen in Marketing und Vertrieb (2019)
12. Chen, F.: Prediction of luciferase inhibitors by the high-performance MIEC-GBDT approach based on interaction energetic patterns. Phys. Chem. Chem. Phys. (PCCP) **19**(15), 10163 (2017)
13. Ma, X.: Prioritizing influential factors for freeway incident clearance time prediction using the gradient boosting decision trees method. IEEE Trans. Intell. Transp. Syst. **18**(9), 2303–2310 (2017)
14. Zhu, Z.: GBDT based hierarchical model for commodity distribution prediction. J. Beijing Jiaotong Univ. **42**(2), 15–19 + 51 (2018)
15. Weng, X.: GBDT subway IC card commuter crowd identification based on GBDT algorithm. J. Chongqing Jiaotong Univ. (Nat. Sci.) **38**(05), 8–12 (2018)
16. Friedman, J.H.: Greedy function approximation: a gradient boosting machine. Ann. Stat. **29** (5), 1189–1232 (2011)
17. Schmidt, R.: Leveraging textual information for improving decision-making in the business process lifecycle. In: International Conference on Intelligent Decision Technologies, pp. 563–574. Springe (2017)
18. Chen, W.: One-hot residue coding for low delay-power product CMOS design. IEEE Trans. Circ. Syst. II Analog Digit. Sig. Process. **45**(3), 303–313 (1998)
19. Benesty, J., Chen, J., Huang, Y.: Pearson Correlation Coefficient//Noise Reduction in Speech Processing. Springer, Berlin (2009)

Fuzzy Theory and Algorithms: Fuzzy Theory and Models

One Pattern Recognition Algorithm Based on Centroids of Fuzzy Hyper-Pyramid Numbers

Yifeng Xu, Guixiang Wang(✉), and Chenjie Shen

Institute of Operations Research and Cybernetics, Hangzhou Dianzi University, Hangzhou 310018, China
g.x.wang@hdu.edu.cn

Abstract. In this paper, a pattern recognition algorithm is given based on centroids of fuzzy hyper-pyramid numbers which are special type fuzzy n-cell numbers. The specific calculation formula (which can be easy calculated by computer program in applications) is obtained for fuzzy hyper-pyramid numbers. Using the results obtained by us, we propose a algorithm of pattern recognition. At last, a practical example is given to demonstrate the application of the algorithm given by us in recognizing uncertain or imprecise multidimensional digital information.

Keywords: Fuzzy number · Fuzzy hyper-pyramid number · Centroid · Pattern recognition

1 Introduction

The concept of fuzzy numbers was introduced by Chang and Zadeh [1] in 1972 with the consideration of the properties of probability functions. Since then both the numbers and the problems in relation to them have been widely studied, see for example, [2,3] and the references therein. With the development of science and society, classifying and identifying information in digital information set is becoming more and more important, many experts and scholars have been studying on this field, see for example, [4–6]. In [8], we proposed the concept of centroid of n-dimensional fuzzy numbers and obtained the simplified form of centroid of fuzzy n-cell numbers. In order to achieve fast and real-time computing, we study the simplified form of centroid of fuzzy hyper-pyramid numbers which are special type fuzzy n-cell numbers, and propose a pattern recognition algorithm based on the specific calculation formula in this paper. We also give a practical example in pattern recognition.

This work was partially supported by the Nature Science Foundation of China (Nos. 61771174 and 61433001) and Graduate Innovation Foundation of Hangzhou Dianzi University (CXJJ2019034).

Y. Liu et al. (Eds.): ICNC-FSKD 2019, AISC 1074, pp. 677–685, 2020.
https://doi.org/10.1007/978-3-030-32456-8_74

2 Basic Definitions and Notations

Let n be a natural number, R be the real number set, and R^n be the n-dimensional Euclidean space. A fuzzy subset (in short, a fuzzy set) of R^n is a function u: $R^n \to [0,1]$. For each such fuzzy set u, we denote by $[u]^r = \{x \in R^n : u(x) \geq r\}$ for any $r \in (0,1]$, is its r-level set. We denote the support of u, i.e., the set $\{x \in R^n : u(x) > 0\}$ by supp$u$. By $[u]^0$, we denote the closure of the supp$u$, i.e., $[u]^0 = \overline{\{x \in R^n : u(x) > 0\}}$.

If u is a normal and fuzzy convex fuzzy set of R^n, $u(x)$ is upper semi-continuous, $[u]^0$ is compact, then we call u a n-dimensional fuzzy number, and denote the collection of all n-dimensional fuzzy numbers by E^n.

If $u \in E^n$, and for each $r \in [0,1]$, $[u]^r$ is a hyper-rectangle, i.e., $[u]^r$ can be represented by $\Pi_{i=1}^n[\underline{u_i}(r), \overline{u_i}(r)]$ (i.e. the Cartesian product $[\overline{u_1}(r), \underline{u_1}(r)] \times [\overline{u_2}(r), \underline{u_2}(r)] \times \ldots \times [\overline{u_n}(r), \underline{u_n}(r)]$), where $\underline{u_i}(r)$, $\overline{u_i}(r) \in R$ with $\underline{u_i}(r) \leq \overline{u_i}(r)$ when $r \in [0,1]$, $i = 1, 2, \ldots, n$, then we call u a fuzzy n-cell number. And, we denote the collection of all fuzzy n-cell numbers by $L(E^n)$.

Let $u_i \in E\,(= E^1)$, $i = 1, 2, \ldots, n$. We call the ordered 1-D fuzzy number class $u_1, u_2, \ldots, u_n$ (i.e. the Cartesian product of 1-D fuzzy numbers $u_1, u_2, \ldots, u_n$) a n-dimension fuzzy vector, denote it as $(u_1, u_2, \ldots, u_n)$, and call the collection of all n-dimensional fuzzy vectors (i.e. the Cartesian product $\overbrace{E \times E \times \ldots \times E}^{n}$) n-dimensional fuzzy vector space, and denote it as $(E)^n$.

By Theorem 3.1 of [7], we see fuzzy n-cell numbers and n-dimensional fuzzy vectors can represent each other, and the representation is unique, so $L(E^n)$ and $(E)^n$ may be regarded as identity. If $(u_1, u_2, \ldots, u_n)$ is the unique n-dimensional fuzzy vector represented by fuzzy n-cell number u, then we denote as $u = (u_1, u_2, \ldots, u_n)$.

Let $u \in E$, if there exist $a, b, c \in R$ with $a \leq b \leq c$ such that

$$u(x) = \begin{cases} \frac{x-a}{b-a}, & x \in [a,b] \\ \frac{c-x}{c-b}, & x \in (b,c] \\ 0, & x \notin [a,c] \end{cases}$$

then we call u a triangular fuzzy number, denote as $u = (a, b, c)$. We define the collection of all triangular fuzzy numbers by $Tri - E$.

Let $u = (u_1, u_2, \ldots, u_n) \in L(E^n)$. If $u_1, u_2, \ldots, u_n$ are all triangular type fuzzy numbers, we call u a n-dimensional fuzzy pyramid number, and denote as

$$u = \begin{pmatrix} a_1 & b_1 & c_1 \\ a_2 & b_2 & c_2 \\ \vdots & \vdots & \vdots \\ a_n & b_n & c_n \end{pmatrix}$$

$$(for\ short,\ u = (a_i, b_i, c_i)_{i=1}^n)$$

We define the collection of all n-dimensional fuzzy pyramid numbers by $Tri - E^n$.

Let $u = (u_1, u_2, \ldots, u_n) \in L(E^n)$, then for any $i = 1, 2, \ldots, n$, the ith component $C(u_i)$ of centroid $C(u) = (C(u_1), C(u_2), \ldots, C(u_n))$ is [8]

$$C(u_i) = \frac{\frac{1}{2}\int_0^1 (\overline{u_i}(r) + \underline{u_i}(r))\Pi_{j=1}^n(\overline{u_j}(r) - \underline{u_j}(r))dr}{\int_0^1 \Pi_{j=1}^n(\overline{u_j}(r) - \underline{u_j}(r))dr}$$

3 Centroid of Fuzzy Hyper-pyramid Numbers

Lemma 1. *Let n and k be positive integers, then the following equation holds.*

$$\sum_{k=0}^{n} C_n^k \frac{(-1)^k}{k+1} = \frac{1}{n+1} \tag{1}$$

Proof: Note that if n and k are positive integers, we have the following equations.

$$\frac{1}{k+1}C_n^k = \frac{1}{n+1}C_{n+1}^{k+1} \tag{2}$$

$$C_{n+1}^1(-1)^1 + C_{n+1}^2(-1)^2 + \ldots + C_{n+1}^{n+1}(-1)^{n+1} = -1 \tag{3}$$

In the following, by Eqs. (2) and (3), we can show the process of the proof of Lemma 1.

$$\begin{aligned}\sum_{k=0}^{n} C_n^k \frac{(-1)^k}{k+1} &= \sum_{k=0}^{n} C_{n+1}^{k+1}\frac{(-1)^k}{n+1} \\ &= \frac{1}{n+1}\sum_{k=0}^{n} C_{n+1}^{k+1}(-1)^k \\ &= \frac{-1}{n+1}[C_{n+1}^1(-1)^1 + C_{n+1}^2(-1)^2 + \ldots + C_{n+1}^{n+1}(-1)^{n+1}] \\ &= \frac{1}{n+1}\end{aligned}$$

Lemma 2. *Let i, n, k be positive integers, and $f(n,i) = \sum_{k=0}^{n} C_n^k \frac{(-1)^k}{k+i}$. Then we have the following recursion formula.*

$$f(n,i) = f(n,i-1) - f(n+1,i-1), \ \textit{for } i \geq 2$$

Proof:

$$\begin{aligned}
f(n+1,i-1) &= C_{n+1}^{0}\frac{(-1)^0}{i-1} + C_{n+1}^{1}\frac{(-1)^1}{i} + \ldots + C_{n+1}^{n+1}\frac{(-1)^{n+1}}{n+i} \\
&= C_{n+1}^{0}\frac{(-1)^0}{i-1} + (C_n^0 + C_n^1)\frac{(-1)^1}{i} + (C_n^1 + C_n^2)\frac{(-1)^2}{i+1} + \ldots \\
&+ (C_n^{n-1} + C_n^n)\frac{(-1)^n}{n+i-1} + C_{n+1}^{n+1}\frac{(-1)^{n+1}}{n+i} \\
&= (C_n^0\frac{(-1)^0}{i-1} + C_n^1\frac{(-1)^1}{i} + C_n^2\frac{(-1)^2}{i+1} + \ldots + C_n^n\frac{(-1)^n}{n+i-1}) \\
&- (C_n^0\frac{(-1)^0}{i} + C_n^1\frac{(-1)^1}{i+1} + \ldots + C_n^{n-1}\frac{(-1)^{n-1}}{n+i-1} + C_n^n\frac{(-1)^n}{n+i}) \\
&= \sum_{k=0}^{n} C_n^k\frac{(-1)^k}{k+i-1} - \sum_{k=0}^{n} C_n^k\frac{(-1)^k}{k+i} \\
&= f(n,i-1) - f(n,i)
\end{aligned}$$

Such that, we get the result

$$f(n,i) = f(n,i-1) - f(n+1,i-1),\ for\ i \geq 2$$

In particular, if $i = 2$, we can get the result of $f(n,2)$ as follows

$$\begin{aligned}
f(n,2) = \sum_{k=0}^{n} C_n^k\frac{(-1)^k}{k+2} &= f(n,1) - f(n+1,1) \\
&= \frac{1}{n+1} - \frac{1}{n+2}
\end{aligned}$$

For fuzzy hyper-pyramid numbers, as special type fuzzy n-cell numbers, by Lemmas 1 and 2, we can obtain the following calculation formula of centroid:

Theorem 1. *Let* $u = (a_i, b_i, c_i)_{i=1}^n \in Tri - E^n$. *Then for any* $i = 1, 2, \ldots, n$, *the i-th component* $C(u_i)$ *of centroid* $C(u) = (C(u_1), C(u_2), \ldots, C(u_n))$ *is*

$$C(u_i) = \frac{1}{2}(\frac{n+1}{n+2}a_i + \frac{2}{n+2}b_i + \frac{n+1}{n+2}c_i) \tag{4}$$

Proof: with using the formula (5) of the centroid of fuzzy n-cell numbers

$$C(u_i) = \frac{\frac{1}{2}\int_0^1 (\overline{u_i}(r) + \underline{u_i}(r))\Pi_{j=1}^n(\overline{u_j}(r) - \underline{u_j}(r))dr}{\int_0^1 \Pi_{j=1}^n(\overline{u_j}(r) - \underline{u_j}(r))dr} \tag{5}$$

And note that, $\underline{u_i}(r) = a_i + (b_i - a_i)r$, $\overline{u_i}(r) = c_i - (c_i - b_i)r$. We calculate the denominator of formula (5) firstly.

$$\int_0^1 \Pi_{j=1}^n(\overline{u_j}(r) - \underline{u_j}(r))dr = \int_0^1 \Pi_{j=1}^n[(c_j - a_j)(1 - r)]dr \tag{6}$$

$$= \Pi_{j=1}^n(c_j - a_j)\int_0^1 (1 - r)^n dr \tag{7}$$

$$= \Pi_{j=1}^n(c_j - a_j)\frac{1}{n+1} \tag{8}$$

Secondly, we compute the numerator of formula (5).

$$\int_0^1 (\overline{u_i}(r) + \underline{u_i}(r))\Pi_{j=1}^n(\overline{u_j}(r) - \underline{u_j}(r))dr \tag{9}$$

$$= \int_0^1 [a_i + c_i + (2b_i - a_i - c_i)r]\Pi_{j=1}^n[(c_j - a_j)(1 - r)]dr \tag{10}$$

$$= (a_i + c_i)\Pi_{j=1}^n(c_j - a_j)\frac{1}{1+n} \tag{11}$$

$$+ (2b_i - a_i - c_i)\Pi_{j=1}^n(c_j - a_j)\sum_{k=0}^n C_n^k(-1)^k \int_0^1 r^{k+1}dr \tag{12}$$

$$= (a_i + c_i)\Pi_{j=1}^n(c_j - a_j)\frac{1}{1+n} \tag{13}$$

$$+ (2b_i - a_i - c_i)\Pi_{j=1}^n(c_j - a_j)\sum_{k=0}^n C_n^k \frac{(-1)^k}{k+2} \tag{14}$$

Lastly, by Lemma 2 and the formulas (8) and (14) above, we have that

$$C(u_i) = \frac{\frac{1}{2}[(a_i + c_i)\Pi_{j=1}^n(c_j - a_j)\frac{1}{1+n} + (2b_i - a_i - c_i)\Pi_{j=1}^n(c_j - a_j)\sum_{k=0}^n C_n^k \frac{(-1)^k}{k+2}]}{\Pi_{j=1}^n(c_j - a_j)\frac{1}{n+1}}$$
$$= \frac{1}{2}(\frac{n+1}{n+2}a_i + \frac{2}{n+2}b_i + \frac{n+1}{n+2}c_i)$$

In particular, if $u \in Tri - E$ and denote as $u = (a, b, c)$, we get the result of centroid of triangular fuzzy number $C(u) = \frac{1}{3}(a + b + c)$.

4 Fuzzy Approximation Relation and Pattern Recognition

In this section, for fuzzy hyper-pyramid numbers, we are going to give the specific calculation formula of fuzzy approximation relations proposed in [8], and set up a pattern recognition algorithm based on the specific calculation formula.

4.1 Fuzzy Approximation Relation

In [8], we have given the definition of fuzzy approximation relation as follows.

Definition 1. *Let* $p = (p_1, p_2, \ldots, p_n) \in R^n$ *with* $p_i \geq 0$ *and* $\sum_{i=1}^{n} p_i = 1$, *and* α *be a positive real number. We define fuzzy binary relations* $S_{p,\alpha}$ *and* $R_{p,\alpha}$ *on* $L(E^n)$, *i.e., mappings* $S_{p,\alpha} : L(E^n) \times L(E^n) \to [0,1]$ *and* $R_{p,\alpha} : L(E^n) \times L(E^n) \to [0,1]$:

$$S_{p,\alpha}(u,v) = \frac{1}{1 + \alpha |\sum_{i=1}^{n} p_i (C(u_i) - C(v_i))|} \tag{15}$$

$$R_{p,\alpha}(u,v) = \frac{1}{1 + \alpha \sum_{i=1}^{n} p_i |C(u_i) - C(v_i)|} \tag{16}$$

for any $u = (u_1, u_2, \ldots, u_n)$ *and* $v = (v_1, v_2, \ldots, v_n) \in L(E^n)$, *respectively. We call the fuzzy binary relations* $S_{p,\alpha}$ *and* $R_{p,\alpha}$ *fuzzy approximation relations on fuzzy* n*-cell number space, respectively. And call the values* $S_{p,\alpha}(u,v)$ *and* $R_{p,\alpha}(u,v)$ *to be the approximation degrees of* u *and* v *with respect to fuzzy approximation relations* $S_{p,\alpha}$ *and* $R_{p,\alpha}$, *respectively. According to Theorem 1 and Definition 1, it is not hard to get the following proposition.*

Proposition 1. *Let* $p = (p_1, p_2, \ldots, p_n) \in R^n$ *with* $p_i \geq 0$ *and* $\sum_{i=1}^{n} p_i = 1$, *and* α *be a positive real number. The specific calculation formulas of fuzzy approximation relation are obtained as follows:*

$$R_{p,\alpha}(u,v) = \frac{1}{1 + \frac{\alpha}{2} \sum_{i=1}^{n} p_i |\frac{n+1}{n+2}(a_i^u - a_i^v + c_i^u - c_i^v) + \frac{2}{n+2}(b_i^u - b_i^v)|} \tag{17}$$

$$S_{p,\alpha}(u,v) = \frac{1}{1 + \frac{\alpha}{2} |\sum_{i=1}^{n} p_i (\frac{n+1}{n+2}(a_i^u - a_i^v + c_i^u - c_i^v) + \frac{2}{n+2}(b_i^u - b_i^v))|} \tag{18}$$

for any $u = (a_i^u, b_i^u, c_i^u)_{i=1}^n$ *and* $v = (a_i^v, b_i^v, c_i^v)_{i=1}^n \in Tri - E^n$, *respectively.*

4.2 Pattern Recognition

We introduce the algorithm of recognizing imprecise or uncertain multichannel digital information by using the specific calculation formula of fuzzy approximation relation.

Problem: Suppose that the following l sets of data respectively come from the l objects (denoted by $O_1, O_2, ..., O_l$), which have n characteristics with some imprecise or uncertain attributes.

$$O_k : \begin{array}{lcccc} \textit{Sample } 1 & o_{11}^k & o_{12}^k & \cdots & o_{1n}^k \\ \textit{Sample } 2 & o_{21}^k & o_{22}^k & \cdots & o_{2n}^k \\ \vdots & \vdots & \vdots & & \vdots \\ \textit{Sample } m_k & o_{m_k1}^k & o_{m_k2}^k & \cdots & o_{m_kn}^k \end{array}$$

$$(k = 1, 2, \ldots, l)$$

Our goal to be done is to recognize these imprecise or uncertain objects.

Pattern Recognition: We construct a suitable fuzzy hyper-pyramid number to represent the object which possesses some imprecise attributes, and compute the centroid of fuzzy number.

The first step: For all $k = 1, 2, \ldots, l$ and $j = 1, 2, \ldots, n$, we calculate the means μ_j^k and the standard deviation σ_j^k of each attribute of all objects

$$\mu_j^k = \frac{1}{m_k} \sum_{i=1}^{m_k} o_{ij}^k, \quad \sigma_j^k = \sqrt{\frac{1}{m_k - 1} \sum_{i=1}^{m_k} (o_{ij}^k - \mu_j^k)^2} \tag{19}$$

The second step: For each $k = 1, 2, \ldots, l$ and $i = 1, 2, \ldots, n$, we construct triangular fuzzy number u_j^k as

$$u_j^k = \begin{cases} \frac{x - (\mu_j^k - 2\sigma_j^k)}{2\sigma_j^k}, & x \in [u_j^k - 2\sigma_j^k, \mu_j^k] \cap D_j \\ \frac{(\mu_j^k + 2\sigma_j^k) - x}{2\sigma_j^k}, & x \in [\mu_j^k, u_j^k + 2\sigma_j^k] \cap D_j \\ 0, & x \notin [u_j^k - 2\sigma_j^k, u_j^k + 2\sigma_j^k] \cap D_j \end{cases}$$

$$(j = 1, 2, \ldots, n; k = 1, 2, \ldots, l)$$

where D_j is the proper domain according to practical case.Then fuzzy hyper-pyramid number $u^k = (u_1^k, u_2^k, \ldots, u_n^k)$, and we can use u^k to represent the object O_k. And we do the same for the target to be identified, and use w to represent the target to be identified.

The third step: Work out the centroid $C(u^k) = (C(u_1^k), C(u_2^k), \ldots, C(u_n^k))$ of fuzzy hyper-pyramid number u^k. According to the importance of each factor which characterized the objects, we take out proper weight $p = (p_1, p_2, \ldots, p_n)$ with $p_i \geq 0$ $(i = 1, 2, \ldots, n)$ and $\sum_{i=1}^n p_i = 1$. Take proper parameter $\alpha \in (0, +\infty)$. We calculate $S_{p,\alpha}(u^k, w)$ (or $R_{p,\alpha}(u^k, w)$).

The fourth step: According to the value of fuzzy approximation relation, we recognize the target which need to be identified to get the label whose value of fuzzy approximation relation is the largest.

Example 1: Let us suppose an area is covered by three different types of land-based cover: C_1: road; C_2: mixed broad-needle forest; C_3: the white birch forest. And suppose that we use four wavebands: MSS-4, MSS-5, MSS-6, MSS-7. We take ten samples for each type of land-based cover and acquire the following data from C_1:

C_1 :

	MSS-4	*MSS*-5	*MSS*-6	*MSS*-7
sample 1	18.62	20.71	58.20	26.72
sample 2	19.76	17.01	51.02	24.32
sample 3	18.24	19.46	48.12	26.33
sample 4	18.76	15.95	56.35	22.89
sample 5	18.96	18.78	45.32	28.32
sample 6	19.90	20.13	50.82	25.05
sample 7	19.16	18.58	52.30	22.21
sample 8	19.36	17.32	55.02	25.28
sample 9	19.36	17.98	48.26	23.86
sample 10	18.48	16.46	46.63	27.46

The first step: According to formula (19), for C_1, we work out:

	MSS-4	*MSS*-5	*MSS*-6	*MSS*-7
j	1	2	3	4
μ_j^1	19.06	18.24	51.20	25.24
σ_j^k	0.55	1.58	4.28	1.98

The second step: Consider the practical situation, let $D_j = [0, +\infty)(j = 1, 2, 3, 4)$, we construct a 4-dimensional fuzzy hyper-pyramid number $u^1 = (u_1^1, u_2^1, u_3^1, u_4^1)$ as

$$u^1 = \begin{pmatrix} 17.96 & 19.06 & 20.16 \\ 15.08 & 18.24 & 21.40 \\ 42.64 & 51.20 & 59.76 \\ 21.28 & 25.24 & 29.20 \end{pmatrix}$$

to represent C_1. With using the same method, we work out other 4-dimensional fuzzy hyper-pyramid numbers u^2, u^3 as

$$u^2 = \begin{pmatrix} 14.94 & 16.22 & 17.50 \\ 11.62 & 12.78 & 13.94 \\ 36.67 & 42.41 & 48.15 \\ 18.22 & 21.22 & 24.22 \end{pmatrix}, u^3 = \begin{pmatrix} 15.37 & 17.01 & 18.65 \\ 12.36 & 13.20 & 14.04 \\ 43.02 & 45.00 & 46.98 \\ 22.36 & 23.20 & 24.04 \end{pmatrix}$$

to represent C_2, C_3, respectively.

For the target to be identified, we also work out a 4-dimensional fuzzy hyper-pyramid number w as

$$w = \begin{pmatrix} 15.55 & 17.09 & 18.63 \\ 12.17 & 13.17 & 14.17 \\ 43.08 & 45.10 & 47.12 \\ 22.53 & 23.19 & 23.79 \end{pmatrix}$$

to represent.

The third step: For $k = 1, 2, 3$, by Eq. (4), we work out $C(u^k) = (C(u_1^k), C(u_2^k), C(u_3^k), C(u_4^k))$ and $C(w) = (C(w_1), C(w_2), C(w_3), C(w_4))$ as follows:

$$C(u^1) = (19.06, 18.24, 51.20, 25.24),\ C(u^2) = (16.22, 12.78, 42.41, 21.22)$$
$$C(u^3) = (17.01, 13.20, 45.00, 23.20),\ C(w) = (17.09, 13.17, 45.10, 23.16)$$

Taking weight $p = (\frac{1}{4}, \frac{1}{4}, \frac{1}{4}, \frac{1}{4})$ and $\alpha = 1$, by Eq. (17), we obtain that $S_{p,\alpha}(u^1, w) = 0.21$, $S_{p,\alpha}(u^2, w) = 0.40$, $S_{p,\alpha}(u^3, w) = 0.97$.

The fourth step: As $S_{p,\alpha}(u^3, w) = max\{S_{p,\alpha}(u^1, w), S_{p,\alpha}(u^2, w), S_{p,\alpha}(u^3, w)\} = 0.97$, so we can obtain a recognition result that the target to be identified belongs to the class C_3, i.e., the region is the white birch forests.

5 Conclusions

In this paper, we obtained a specific calculation formula of centroid which is easy to use in application for fuzzy hyper-pyramid number by Theorem 1. And then, we propose a pattern recognize algorithm and gave a practical example to demonstrate the application of the algorithm in recognizing uncertain or imprecise multichannel digital information.

References

1. Chang, S.S., Zadeh, L.A.: On fuzzy mapping and control. IEEE Trans. Syst. Man Cybern. **SMC-2**(1), 30–34 (1972)
2. Asady, B.: Revision of distance minimization method for ranking of fuzzy numbers. Appl. Math. Model. **35**(3), 1306–1313 (2011)
3. Kaufmann, A., Gupta, M.M.: Introduction to Fuzzy Arithmetic Theory and Applications. Reinhold, New York (1991)
4. Havens, T.C., Bezdek, J.C., Leckie, C., Hall, L.O., Palaniswami, M.: Fuzzy c-means algorithms for very large data. IEEE Trans. Fuzzy Syst. **20**(6), 1130–1146 (2012)
5. Ishibuchi, H., Mihara, S., Nojima, Y.: Parallel distributed hybrid fuzzy GBML models with rule set migration and training data rotation. IEEE Trans. Fuzzy Syst. **21**(2), 355–368 (2013)
6. Liang, D., Liu, D.: A novel risk decision making based on decision-theoretic rough sets under hesitant fuzzy information. IEEE Trans. Fuzzy Syst. **23**(2), 237–247 (2015)
7. Wang, G., Li, Y., Wen, C.: On fuzzy n-cell numbers and n-dimension fuzzy vectors. Fuzzy Sets Syst. **158**(1), 71–84 (2007)
8. Wang, G., Xu, Y.: Ranking and classifying uncertain multichannel digital informations based on centroids of high dimensional fuzzy numbers (submitted)

Approximation of Fuzzy Numbers by Using α-β-knots Piecewise Linear Fuzzy Numbers

Chenjie Shen, Guixiang Wang(✉), and Yifeng Xu

Institute of Operations Research and Cybernetics, Hangzhou Dianzi University, Hangzhou 310018, China
g.x.wang@hdu.edu.cn

Abstract. In this paper, the problems of approximating general fuzzy number by using α–β–knots piecewise linear fuzzy number are studied. Firstly, α–β–knots piecewise linear fuzzy number are defined, and the conceptions of nearest $\alpha - \beta$–knots piecewise linear approximation and nearest $\{0, \alpha, 1\}$–$\{0, \beta, 1\}$–knots piecewise linear approximation are introduced for a general fuzzy number. Then, it is also the most important work of this paper that for a general fuzzy number, we obtain a formula to get the nearest $\alpha - \beta$–knots piecewise linear approximation and the nearest $\{0, \alpha, 1\} - \{0, \beta, 1\}$–knots piecewise linear approximation using weighted metric as a criterion. And then, we give specific example to show more reasonable and effective of the methods proposed by us.

Keywords: Approximations · Membership functions · Fuzzy numbers · α–β–knots piecewise linear fuzzy numbers

1 Introduction

It is known that approximation theory is an important research field in mathematics, and has strong application background. In this research field, many important results have been made [2,4–7]. Recently, there is still a lot of work about function approximations. For example, In [1], Coroianu, Gagolewski and Grzegorzewski obtained approximations which are simple enough and flexible to reconstruct the input fuzzy concepts under study by using 1-knot fuzzy numbers; In [6], we proposed methods of approximating general fuzzy number by using step type fuzzy number (i.e., simple fuzzy number). In fact, for some practical applications, some problems that need more accurate approximation (using piecewise linear functions to approximate general fuzzy number) may be encountered, so it is meaningful work to study using 1-knots piecewise linear fuzzy number to approximate general fuzzy number. However, the approximation method given

This work is supported partially by the Nature Science Foundation of China (Nos. 61771174 and 61433001) and Graduate Innovation Foundation of Hangzhou Dianzi University (Nos. CXJJ2019034).

Y. Liu et al. (Eds.): ICNC-FSKD 2019, AISC 1074, pp. 686–694, 2020.
https://doi.org/10.1007/978-3-030-32456-8_75

in [1] failed to give the formulas to calculate 1–knots piecewise linear approximation. So in this paper, the weight metric d is defined, which takes into account the contribution of points corresponding to the membership function value of the fuzzy number to the fuzzy numbers. It is more reasonable and objective to describe the difference between two fuzzy numbers. Then, we are going to study the approximation problem of α–β–knots piecewise linear fuzzy number using weighted metric as a criterion.

2 Basic Definitions and Notations

A fuzzy subset (in short, a fuzzy set) of the real line $\mathbb{R}$ is a function $u : \mathbb{R} \to [0,1]$. For each such fuzzy set u, we denote by $[u]^r = \{x \in \mathbb{R} : u(x) \geq r\}$ for any $r \in (0,1]$, its r–level set. By supp$u$ we denote the support of u, i.e., the $\{x \in \mathbb{R} : u(x) > 0\}$. By $[u]^0$ we denote the closure of the supp$u$, i.e., $[u]^0 = \overline{\{x \in \mathbb{R} : u(x) > 0\}}$.

If u is a normal and fuzzy convex fuzzy set of $\mathbb{R}$, $u(x)$ is upper semi-continuous, and $[u]^0$ is compact, then we call u a fuzzy number, and denote the collection of all fuzzy numbers by E.

It is known that if $u \in E$, then for each $r \in [0,1]$, $[u]^r$ is a nonempty compact convex set in $\mathbb{R}$, i.e., a closed interval. For $u \in E$, we denote the closed interval as $[u]^r = [\underline{u}(r), \overline{u}(r)]$ for any $r \in [0,1]$.

In [3], Grzegorzewski defined a metric $\hat{d}$ on E as

$$\hat{d}(u,v) = \sqrt{\int_0^1 (\underline{u}(r) - \underline{v}(r))^2 dr + \int_0^1 (\overline{u}(r) - \overline{v}(r))^2 dr} \tag{1}$$

for any $u, v \in E$.

We defined the weight metric d on E as

$$d(u,v) = \sqrt{\int_0^1 2r(\underline{u}(r) - \underline{v}(r))^2 dr + \int_0^1 2r(\overline{u}(r) - \overline{v}(r))^2 dr} \tag{2}$$

for any $u, v \in E$.

3 Approximation by Using $\alpha - \beta$–nodes Piecewise Linear Fuzzy Number

In [1], Coroianu, Gagolewski and Grzegorzewski gave a method of approximating general fuzzy numbers by using piecewise linear 1-knot fuzzy numbers with respect to the metric $\hat{d}$. In the section, we are going to obtain a formula to get the nearest α–β–knots piecewise linear approximation by using the metric $\hat{d}$ and the weighted metric d. First, we give the following theorem and definition:

Theorem 1. *Let $\alpha, \beta \in (0,1)$ and $a_i, b_j \in \mathbb{R}$ $(i = 0,1,2, j = 0,1,2)$ with $0 < \alpha < 1 > \beta > 0$ and $a_0 \le a_1 \le a_2 \le b_2 \le b_1 \le b_0$. If the fuzzy set u of $\mathbb{R}$ is defined as*

$$u(x) = \begin{cases} 0, & x < a_1 \\ \alpha\frac{x-a_0}{a_1-a_0}, & x \in [a_0, a_1) \\ \alpha + (1-\alpha)\frac{x-a_1}{a_2-a_1}, & x \in [a_1, a_2) \\ 1, & x \in [a_2, b_2] \\ \beta + (1-\beta)\frac{b_1-x}{b_1-b_2}, & x \in (b_2, b_1] \\ \beta\frac{b_0-x}{b_0-b_1}, & x \in (b_1, b_0] \\ 0, & x > b_0 \end{cases} \tag{3}$$

then $u \in E$

Proof. It can be directly proved by the definition of fuzzy number.

Definition 1. *Let $\alpha, \beta \in (0,1)$ with $0 < \alpha < 1 > \beta > 0$, and any $a_i, b_j \in \mathbb{R}$ $(i = 0,1,2, j = 0,1,2)$ with $a_0 \le a_1 \le a_2 \le b_2 \le b_1 \le b_0$. We call the fuzzy number u defined in Theorem 1 a $\alpha - \beta$-knots piecewise linear fuzzy number, and denote it as*

$$u_{pl} = PL\left[\begin{pmatrix} a_0 & 0 \\ a_1 & \alpha \\ a_2 & 1 \end{pmatrix}; \begin{pmatrix} b_0 & 0 \\ b_1 & \beta \\ b_2 & 1 \end{pmatrix}\right] \tag{4}$$

and call α and β the left threshold value set and the right threshold value set of u_{pl}, respectively. Specially, as $\alpha = \beta$, we call the fuzzy number u_{pl} a α–knots (i.e., β–knots) piecewise linear fuzzy number, and call α (i.e., β) the threshold value set of u_{pl}.

For fixed α and $\beta \in (0,1)$ with $0 < \alpha < 1 > \beta > 0$, we denote the collection of all $\alpha - \beta$–knots piecewise linear fuzzy numbers by ${}^{\alpha}PL^{\beta}(E)$, and denote the collection of all α–knots (β–knots) piecewise linear fuzzy numbers by $PL^{\alpha}(E)$ (i.e. $PL^{\beta}(E)$).

In the following, we are going to establish a recursion formula to obtain the $\alpha - \beta$–knots piecewise linear fuzzy number which is the nearest (with respect to metric d) approximation with the left threshold value set α and the right threshold value set β. To do this, we first give the following definition.

Definition 2. *Let $u \in E$ and $\alpha, \beta \in (0,1)$ with $0 < \alpha < 1 > \beta > 0$. If there exists $u_{pl} \in {}^{\alpha}PL^{\beta}(E)$ with $\underline{u_{pl}}(0) = \underline{u}(0)$, $\overline{u_{pl}}(0) = \overline{u}(0)$, $\underline{u_{pl}}(1) = \underline{u}(1)$ and $\overline{u_{pl}}(1) = \overline{u}(1)$ (i.e., $[u_{pl}]^0 = [u]^0$ and $[u_{pl}]^1 = [u]^1$) such that*

$$d(u, u_{pl}) = \min\left\{d(u,v) : \ v \in {}^{\alpha}PL^{\beta}(E)\, with\, [v]^0 = [u]^0, [v]^1 = [u]^1\right\}$$

then we say u_{pl} to be the nearest (with respect to metric d) α–β–knots piecewise linear approximation (in short, KPLA) of fuzzy number u. Specially, as $\alpha = \beta$, we say u_{pl} to be the nearest (with respect to metric d) α–knots (i.e., β–knots) piecewise linear approximation (in short, KPLA) of fuzzy number u.

Definition 3. *Let $u \in E$ and $\alpha, \beta \in (0,1)$ with $0 < \alpha < 1 > \beta > 0$. If there exists $u_{pl} \in {}^{\alpha}PL^{\beta}(E)$ such that*

$$d(u, u_{pl}) = \min \{d(u, v) : v \in {}^{\alpha}PL^{\beta}(E)\}$$

then we say u_{pl} to be the nearest (with respect to metric d) $\{0, \alpha, 1\}$–$\{0, \beta, 1\}$–knots piecewise linear approximation (in short, KPLA) of fuzzy number u. Specially, as $\alpha = \beta$, we say u_{pl} to be the nearest (with respect to metric d) $\{0, \alpha, 1\}$–knots (i.e., $\{0, \beta, 1\}$–knots) piecewise linear approximation (in short, KPLA) of fuzzy number u.

By the definitions of the nearest α–β–knots piecewise linear approximation and the nearest $\{0, \alpha, 1\} - \{0, \beta, 1\}$–knots piecewise linear approximation of a fuzzy number, we can directly get the following result:

Proposition 1. *Let $u \in E$ and $\alpha, \beta \in (0,1)$ with $0 < \alpha < 1 > \beta > 0$, u_{pl} be the nearest $\alpha - \beta$–knots piecewise linear approximation of fuzzy number u, and v_{pl} be the nearest $\{0, \alpha, 1\} - \{0, \beta, 1\}$–knots piecewise linear approximation of fuzzy number u. Then*

$$d(u, v_{pl}) \leq d(u, u_{pl})$$

i.e., v_{pl} is a better approximation of u than u_{pl}.

Proof. From

$$\left\{d(u, v) : \ v \in {}^{\alpha}PL^{\beta}(E) \text{ with } [v]^0 = [u]^0, [v]^1 = [u]^1\right\} \subset \left\{d(u, v) : v \in {}^{\alpha}PL^{\beta}(E)\right\}$$

we see that

$$\begin{aligned} d(u, u_{pl}) &= \min \{d(u, v) : \ v \in {}^{\alpha}PL^{\beta}(E) \text{ with } [v]^0 = [u]^0, [v]^1 = [u]^1\} \\ &\geq \min \{d(u, v) : v \in {}^{\alpha}PL^{\beta}(E)\} \\ &= d(u, v_{pl}) \end{aligned}$$

The proof of the proposition can be completed.

Lemma 1. *Let $u = u_{pl} \in {}^{\alpha}PL^{\beta}(E)$. Then*

$$\underline{u}(r) = \begin{cases} a_0 + (a_1 - a_0)\frac{r}{\alpha}, & r \in [0, \alpha] \\ a_1 + (a_2 - a_1)\frac{r-\alpha}{1-\alpha}, & r \in (\alpha, 1] \end{cases} \tag{5}$$

and

$$\overline{u}(r) = \begin{cases} b_0 + (b_1 - b_0)\frac{r}{\beta}, & r \in [0, \beta] \\ b_1 + (b_2 - b_1)\frac{r-\beta}{1-\beta}, & r \in (\beta, 1] \end{cases} \tag{6}$$

for $r \in [0, 1]$.

Proof. From the definition of $u = u_{pl}$, we can directly obtain the conclusion by the definitions of $\underline{u}(r)$ and $\overline{u}(r)$.

Theorem 2. *Let $u \in E$ and $\alpha, \beta \in (0,1)$ with $0 < \alpha < 1 > \beta > 0$. If $a_1, b_1 \in \mathbb{R}$ are respectively determined by following formulas:*

$$\begin{cases} a_1 = \frac{1}{1+2\alpha}\left[\frac{12}{\alpha}\int_0^\alpha r^2\underline{u}(r)dr + \frac{12}{1-\alpha}\int_\alpha^1 r(1-r)\underline{u}(r)dr - \alpha^2\underline{u}(0)\right. \\ \quad \left. -(1-\alpha^2)\underline{u}(1)\right] \\ b_1 = \frac{1}{1+2\beta}\left[\frac{12}{\beta}\int_0^\beta r^2\overline{u}(r)dr + \frac{12}{1-\beta}\int_\beta^1 r(1-r)\overline{u}(r)dr - \beta^2\overline{u}(0)\right. \\ \quad \left. -(1-\beta^2)\overline{u}(1)\right] \end{cases} \tag{7}$$

and satisfy $\underline{u}(0) \le a_1 \le \underline{u}(1) \le \overline{u}(1) \le b_1 \le \overline{u}(0)$, then

$$u_{pl} = PL\left[\begin{pmatrix} \underline{u}(0) & 0 \\ a_1 & \alpha \\ \underline{u}(1) & 1 \end{pmatrix}; \begin{pmatrix} \overline{u}(0) & 0 \\ b_1 & \beta \\ \overline{u}(1) & 1 \end{pmatrix}\right]$$

is the nearest (with respect to matric d) $\alpha-\beta$-knots piecewise linear approximation of fuzzy number u.

Proof. For any $x_1, y_1 \in \mathbb{R}$ with $0 < \alpha < 1 > \beta > 0$, we denote

$$u_{x,y} = PL\left[\begin{pmatrix} \underline{u}(0) & 0 \\ x_1 & \alpha \\ \underline{u}(1) & 1 \end{pmatrix}; \begin{pmatrix} \overline{u}(0) & 0 \\ y_1 & \beta \\ \overline{u}(1) & 1 \end{pmatrix}\right]$$

and define $D : \mathbb{R}^2 \to \mathbb{R}$ as

$$D(x_1, y_1) = D(x, y) = (d(u, u_{x,y}))^2$$

for any $(x, y) = (x_1, y_1) \in \mathbb{R}^2$. By the Definition of metric d, we see that

$$D(x_1, y_1) = \int_0^1 (\underline{u}(r) - \underline{u_{x,y}}(r))^2 dr + \int_0^1 (\overline{u}(r) - \overline{u_{x,y}}(r))^2 dr$$

By Lemma 1, we have that

$$\begin{aligned} &\int_0^1 2r(\underline{u}(r) - \underline{u_{x,y}}(r))^2 dr \\ &= \int_0^\alpha 2r(\underline{u}(r) - \underline{u}(0) - (x_1 - \underline{u}(0))\tfrac{r}{\alpha})^2 dr + \int_\alpha^1 2r(\underline{u}(r) - x_1 - (\underline{u}(1) - x_1)\tfrac{r-\alpha}{1-\alpha})^2 dr \end{aligned}$$

and

$$\begin{aligned} &\int_0^1 2r(\overline{u}(r) - \overline{u_{x,y}}(r))^2 dr \\ &= \int_0^\beta 2r(\overline{u}(r) - \overline{u}(0) - (y_1 - \overline{u}(0))\tfrac{r}{\beta})^2 dr + \int_\beta^1 2r(\overline{u}(r) - y_1 - (\overline{u}(1) - y_1)\tfrac{r-\beta}{1-\beta})^2 dr \end{aligned}$$

Then we have that

$$\begin{aligned} &\frac{\partial D(x_1, y_1)}{\partial x_1} \\ &= \tfrac{1}{3}\alpha^2\underline{u}(0) + \tfrac{1}{3}(1+2\alpha)x_1 + \tfrac{1}{3}(1-\alpha^2)\underline{u}(1) - \tfrac{4}{\alpha}\int_0^\alpha r^2\underline{u}(r)dr \\ &\quad - \tfrac{4}{1-\alpha}\int_\alpha^1 r(1-r)\underline{u}(r)dr \end{aligned}$$

Likewise, we have that

$$\begin{aligned}&\frac{\partial D(x_1,y_1)}{\partial y_1}\\&=\tfrac{1}{3}\beta^2\overline{u}(0)+\tfrac{1}{3}(1+2\beta)y_1+\tfrac{1}{3}(1-\beta^2)\overline{u}(1)-\tfrac{4}{\beta}\int_0^\beta r^2\overline{u}(r)dr\\&-\tfrac{4}{1-\beta}\int_\beta^1 r(1-r)\overline{u}(r)dr\end{aligned}$$

Let $\frac{\partial D(x_1,y_1)}{\partial x_1}=0$ and $\frac{\partial D(x_1,y_1)}{\partial y_1}=0$. Then we can obtain that

$$\begin{cases}\alpha^2\underline{u}(0)+(1+2\alpha)x_1+(1-\alpha^2)\underline{u}(1)=\frac{12}{\alpha}\int_0^\alpha r^2\underline{u}(r)dr+\frac{12}{1-\alpha}\int_\alpha^1 r(1-r)\underline{u}(r)dr\\ \beta^2\overline{u}(0)+(1+2\beta)y_1+(1-\beta^2)\overline{u}(1)=\frac{12}{\beta}\int_0^\beta r^2\overline{u}(r)dr+\frac{12}{1-\beta}\int_\beta^1 r(1-r)\overline{u}(r)dr\end{cases}$$

We can obtain a unique solution of the equation set $(x_1,y_1)=(a_1,b_1)$, where a_1,b_1 are determined by formula (7).

On the other hand, from the definition of $D(x_1,y_1)$, we can obtain the Hessian matrix of the function $D(x_1,y_1)$ at (a_1,b_1) as follow:

$$C=\frac{1}{3}\begin{pmatrix}1+2\alpha & 0\\ 0 & 1+2\beta\end{pmatrix}$$

From $1+2\alpha>0, 1+2\beta>0$, we see that the all of order principal minor determinants of the Hessian matrix are positive, so C is a positive matrix. Therefore (a_1,b_1) is the minimum point of $D(x_1,y_1)$. From it, we see that

$$u_{pl}=PL\left[\begin{pmatrix}\underline{u}(0) & 0\\ a_1 & \alpha\\ \underline{u}(1) & 1\end{pmatrix};\begin{pmatrix}\overline{u}(0) & 0\\ b_1 & \beta\\ \overline{u}(1) & 1\end{pmatrix}\right]$$

is the nearest (with respect to matric d) $\alpha-\beta$-knots piecewise linear approximation of fuzzy number u.

Remark 1. *In Theorem 2, for a general fuzzy number, we obtain a formula to get the nearest $\alpha-\beta$-knots piecewise linear approximation with respect to weighted metric d, which is more reasonable and effective to approximate fuzzy number u. More importantly, Theorem 2 gave a specific formula of calculating the $\alpha-\beta$-knots piecewise linear approximation, but the approximation method given in* [1] *failed to do that.*

From Proposition 1, we see that for $u\in E$ and $\alpha,\beta\in(0,1)$ with $0<\alpha<1>\beta>0$, the nearest $\{0,\alpha,1\}-\{0,\beta,1\}$–knots piecewise linear approximation v_{pl} of fuzzy number u is better than the nearest $\alpha-\beta$–knots piecewise linear approximation u_{pl} of fuzzy number u (i.e., $d(u,v_{pl})\le d(u,u_{pl})$). In the following, we are going to give the recursion formulas to calculate $\{0,\alpha,1\}$–$\{0,\beta,1\}$–knots piecewise linear approximation.

Theorem 3. *Let $u \in E$ and $\alpha, \beta \in (0,1)$ with $0 < \alpha < 1 > \beta > 0$. If $a_i, b_j \in \mathbb{R}$, $i = 0,1,2$, $j = 0,1,2$ are respectively determined by following formulas:*

$$\begin{cases} a_0 = \frac{12}{\alpha^3}\int_0^\alpha r(\alpha - r)\underline{u}(r)dr - a_1 \\ a_1 = \frac{18+6\alpha}{1+3\alpha}\Big[\frac{1}{\alpha}\int_0^\alpha r(2r-\alpha)\underline{u}(r)dr + \frac{1}{1-\alpha}(\int_\alpha^1 r(1-r)\underline{u}(r)dr \\ \quad -\frac{1+\alpha}{3+\alpha}\int_\alpha^1 r(r-\alpha)\underline{u}(r)dr)\Big] \\ a_2 = \frac{12}{(3+\alpha)(1-\alpha)^2}\int_\alpha^1 r(r-\alpha)\underline{u}(r)dr - \frac{1+\alpha}{3+\alpha}a_1 \\ b_0 = \frac{12}{\beta^3}\int_0^\beta r(\beta - r)\overline{u}(r)dr - b_1 \\ b_1 = \frac{18+6\beta}{1+3\beta}\Big[\frac{1}{\beta}\int_0^\beta r(2r-\beta)\overline{u}(r)dr + \frac{1}{1-\beta}(\int_\beta^1 r(1-r)\overline{u}(r)dr \\ \quad -\frac{1+\beta}{3+\beta}\int_\beta^1 r(r-\beta)\overline{u}(r)dr)\Big] \\ b_2 = \frac{12}{(3+\beta)(1-\beta)^2}\int_\beta^1 r(r-\beta)\overline{u}(r)dr - \frac{1+\beta}{3+1\beta}b_1 \end{cases} \tag{8}$$

and satisfy $a_0 \le a_1 \le a_2 \le b_2 \le b_1 \le b_0$, then

$$u_{pl} = PL\left[\begin{pmatrix} a_0 & 0 \\ a_1 & \alpha \\ a_2 & 1 \end{pmatrix}; \begin{pmatrix} b_0 & 0 \\ b_1 & \beta \\ b_2 & 1 \end{pmatrix}\right]$$

is the nearest (with respect to matric d) $\{0,\alpha,1\}-\{0,\beta,1\}$–knots piecewise linear approximation of fuzzy number u.

Proof. The theorem can be proved by the similar method as proving Theorem 2. We omit the specific proof process.

Remark 2. *By Proposition 1, we know that the method gave in Theorem 3 is better than the method gave in Theorem 2 for approximating general fuzzy number by using piecewise linear fuzzy number.*

4 Specific Example of Solving the Nearest Approximates

Example 1. *Let fuzzy number u be defined as:*

$$u(x) = \begin{cases} \frac{2}{x^2+1} - 1, & x \in [-1,1] \\ 0, & x \notin [-1,1] \end{cases}$$

By Lemma 1, we have that $\underline{u}(r) = -\sqrt{\frac{2}{r+1} - 1}$ and $\overline{u}(r) = \sqrt{\frac{2}{r+1} - 1}$, and $\underline{u}(0) = -1$, $\underline{u}(1) = 0$, $\overline{u}(1) = 0$ and $\overline{u}(0) = 1$.

(1) If we take $\alpha = \beta = \{\frac{1}{2}\}$, then by Formula (7), we have that $a_1 = -0.64$ and $b_1 = 0.64$. Therefore, by Theorem 2, we know that the piecewise linear fuzzy number (see Fig. 1)

$$u_{pl} = PL\left[\begin{pmatrix} -1 & 0 \\ -0.64 & \frac{1}{2} \\ 0 & 1 \end{pmatrix}; \begin{pmatrix} 1 & 0 \\ 0.64 & \frac{1}{2} \\ 0 & 1 \end{pmatrix}\right]$$

is the nearest $\{\frac{1}{2}\}$*–knotes piecewise linear approximation of fuzzy number* u.

Likewise, by Formula (8), we have that $a_0 = -0.96$, $a_1 = -0.60$, $a_2 = -0.12$, $b_0 = 0.96$, $b_1 = 0.60$ *and* $b_2 = 0.12$. *Therefore, by Theorem 3, we know that the piecewise linear fuzzy number (see Fig. 2)*

$$u_{pl} = PL\left[\begin{pmatrix} -0.96 & 0 \\ -0.60 & \frac{1}{2} \\ -0.12 & 1 \end{pmatrix}; \begin{pmatrix} 0.96 & 0 \\ 0.60 & \frac{1}{2} \\ 0.12 & 1 \end{pmatrix}\right]$$

is the nearest $\{0, \frac{1}{2}, 1\}$*–knotes piecewise linear approximation of fuzzy number* u.

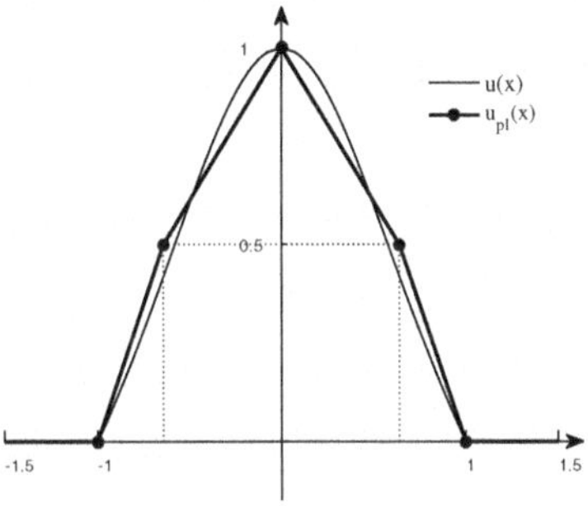

Fig. 1. u and $\{\frac{1}{2}\}$–KPLA u_{pl}

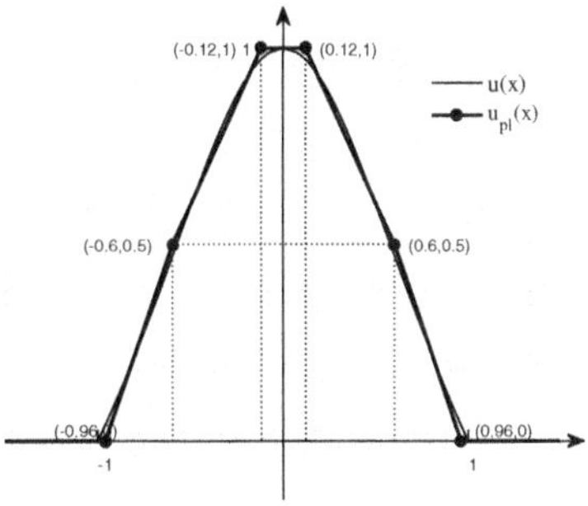

Fig. 2. u and $\{0, \frac{1}{2}, 1\}$–KPLA u_{pl}

Remark 3. *From the example, we can see that using weighted metric* d *in this paper is better than the metric* $\hat{d}$ *gave in* [1] *for approximating general fuzzy number by using piecewise linear fuzzy number.*

5 Conclusion

In this paper, we obtain the methods to get the nearest α–β–knots piecewise linear approximation (Theorem 2) and the nearest $\{0, \alpha, 1\} - \{0, \beta, 1\}$–knots piecewise linear approximation (Theorem 3) for a general fuzzy number, and gave some remarks (Remarks 1–3) to show that the methods proposed by us in this paper greatly extend and improve the methods given by predecessors. And then, we gave specific example (Example 1) to show more reasonable and effective of the methods proposed by us using weighted metric d as a criterion.

References

1. Coroianu, L., Gagolewski, M., Grzegorzewski, P.: Nearest piecewise linear approximation of fuzzy numbers. Fuzzy Sets Syst. **233**, 26–51 (2013). https://doi.org/10.1016/j.fss.2013.02.005

2. Drugowitsch, J., Barry, A.M.: A formal framework and extensions for function approximation in learning classifier systems. Mach. Learn. **70**, 45–88 (2008). https://doi.org/10.1007/s10994-007-5024-8
3. Grzegorzewski, P.: Metrics and orders in space of fuzzy numbers. Fuzzy Sets Syst. **97**(1), 83–94 (1998)
4. Nikolaev, G.N.: Renormalization group approach to function approximation and to improving subsequent approximations. Theoret. Math. Phys. **164**(2), 1035–1050 (2010). https://doi.org/10.1007/s11232-010-0083-6
5. Pekarskii, A.A.: Approximation to the function z^{α} by rational fractions in a domain with zero external angle. Math. Notes **91**(5), 714–724 (2012). https://doi.org/10.1134/s0001434612050136
6. Wang, G., Li, J.: Approximations of fuzzy numbers by step type fuzzy numbers. Fuzzy Sets Syst. **310**, 47–59 (2017). https://doi.org/10.1016/j.fss.2016.08.003
7. Yeh, C.T.: Weighted semi-trapezoidal approximations of fuzzy numbers. Fuzzy Sets Syst. **165**, 61–80 (2011). https://doi.org/10.1016/j.fss.2010.11.001

The Lattice of Interval Valued $(\in, \in\vee q)$-fuzzy Filters in a Given MTL-algebra

Chunhui Liu(✉)

School of Mathematics and Statistics, Chifeng University,
Chifeng 024000, Inner Mongolia, China
chunhuiliu1982@163.com

Abstract. In the present paper, the interval valued $(\in, \in\vee q)$-fuzzy filter theory in MTL-algebras is further studied. Some new properties of interval valued $(\in, \in\vee q)$-fuzzy filters are given. It is proved that the set consisting of all interval valued $(\in, \in\vee q)$-fuzzy filters in a given MTL-algebra, under the partial order $\sqsubseteq$, forms a complete distributive lattice.

Keywords: Many-valued logic · MTL-algebra · Interval-valued $(\in, \in\vee q)$ -fuzzy filter · Complete distributive lattice

1 Introduction

To formalize the many-valued logics induced by continuous t-norms on the real unit interval $[0, 1]$, in 1998, Hájek [1] introduced a very general many-valued logic, called Basic Logic (**BL** for short). It is a well-known result that a t-norm has a residuum if and only if the t-norm is left-continuous, so this shows that **BL** is not the most general t-norm-based logic. In fact, a logic weaker than **BL**, called Monoidal t-norm-based logic (**MTL** for short), was defined by Esteva and Godo in [2] and proved to be the logic of left-continuous t-norms and their residua. Thus, the **MTL** is indeed the logic of left-continuous t-norms. In connection with the logic **MTL**, Esteva and Godo also introduced a new kind of algebra, namely, the MTL-algebra and they studied several basic properties of such algebra. It is noteworthy that filter theory plays an important role in studying these algebras. We notice that Esteva and Godo introduced the concepts of filters and prime filters in MTL-algebras in [2]. Then, Borzooei et al. discussed some other types filters in MTL-algebras (cf., [3,4]). On the other hand, the theory of fuzzy sets was first introduced by Zadeh [5] in 1965 and has been applied to many branches in mathematics. By using fuzzy sets, many Scholars studied the fuzzy structure of filters in MTL-algebras (cf., [6–8]). Among many theories, Zadeh [9] also introduced the concept of interval valued fuzzy sets, IV-fuzzy set for short. Recently, Ma et al. [10] applied the ideas of IV-fuzzy sets to the study of filters in MTL-algebras, introduced the notions of interval valued $(\in, \in\vee q)$-fuzzy filters, IV-$(\in, \in\vee q)$-fuzzy filters for short and investigated some their properties.

Y. Liu et al. (Eds.): ICNC-FSKD 2019, AISC 1074, pp. 695–703, 2020.
https://doi.org/10.1007/978-3-030-32456-8_76

As a continuation of the work of [10], in this paper, we will investigate the lattice structural feature of the set containing all of IV-$(\in, \in\vee q)$-fuzzy filters in an MTL-algebra. Some interesting results are obtained.

2 Preliminaries

In this section, we review related basic knowledge of MTL-algebras and interval valued fuzzy sets.

Definition 1 [1,2,11,12]. *A residuated lattice is an algebra* $(L, \vee, \wedge, \otimes, \rightarrow, 0, 1)$ *of type* $(2, 2, 2, 2, 0, 0)$ *such that for all* $x, y, z \in L$*:*

(R1) $(L, \vee, \wedge, 0, 1)$ *is a bounded lattice with the greatest element* 1 *and the least element* 0*;*

(R2) $(L, \otimes, 1)$ *is a commutative monoid;*

(R3) $(\otimes, \rightarrow)$ *is an adjoint pair on* L*, i.e.,* $x \otimes y \leqslant z$ *if and only if* $x \leqslant y \rightarrow z$.

An MTL-algebra is a residuated lattice $(L, \vee, \wedge, \otimes, \rightarrow, 0, 1)$ *satisfying the following pre-linearity equation: for all* $x, y \in L$,

(MTL) $(x \rightarrow y) \vee (y \rightarrow x) = 1$.

An MTL-algebra $(L, \vee, \wedge, \otimes, \rightarrow, 0, 1)$ will be denoted by L in short.

Lemma 1 [1,2,11,12]. *Let* L *be an MTL-algebra, then for all* $x, y, z \in L$,

(M1) $x \leqslant y$ *if and only if* $x \rightarrow y = 1$,
(M2) $x \rightarrow x = 1$ *and* $x \rightarrow 1 = 1$ *and* $1 \rightarrow x = x$,
(M3) $x \otimes y \leqslant x \otimes (x \rightarrow y) \leqslant x \wedge y \leqslant x \wedge (x \rightarrow y) \leqslant x$;
(M4) $x \rightarrow (y \rightarrow z) = (x \otimes y) \rightarrow z = y \rightarrow (x \rightarrow z)$;
(M5) $x \leqslant y$ *implies* $x \otimes z \leqslant y \otimes z$ *and* $z \rightarrow x \leqslant z \rightarrow y$ *and* $y \rightarrow z \leqslant x \rightarrow z$;
(M6) $(y \vee z) \otimes x = (y \otimes x) \vee (z \otimes x)$ *and* $x \vee (y \otimes z) \geqslant (x \vee y) \otimes (x \vee z)$.

In the unit interval [0, 1] equipped with the natural order, we denote $\vee = \max$ and $\wedge = \min$. By an interval number $\overline{a}$, we mean an interval $[a^-, a^+]$, where $0 \leqslant a^- \leqslant a^+ \leqslant 1$. Denote the set of all interval numbers by $D[0, 1]$. The interval $[a, a]$ can be identified by the number $a \in [0, 1]$. For the interval numbers $\{\overline{a}_i\}_{i=1}^n = \{[a_i^-, a_i^+]\}_{i=1}^n \in D[0, 1]$, $n \in \mathbb{Z}_+$, we define:

(i) $\overline{a}_1 \sqcap \overline{a}_2 \sqcap \cdots \sqcap \overline{a}_n = [a_1^- \wedge a_2^- \wedge \cdots \wedge a_n^-, a_1^+ \wedge a_2^+ \wedge \cdots \wedge a_n^+]$;
(ii) $\overline{a}_1 \sqcup \overline{a}_2 \sqcup \cdots \sqcup \overline{a}_n = [a_1^- \vee a_2^- \vee \cdots \vee a_n^-, a_1^+ \vee a_2^+ \vee \cdots \vee a_n^+]$.

And for the interval numbers $\{\overline{a}_\lambda\}_{\lambda \in \Lambda} = \{[a_\lambda^-, a_\lambda^+]\}_{\lambda \in \Lambda} \in D[0, 1]$, we define:

(iii) $\bigsqcap_{\lambda \in \Lambda} \overline{a}_\lambda = \left[\bigwedge_{\lambda \in \Lambda} a_\lambda^-, \bigwedge_{\lambda \in \Lambda} a_\lambda^+\right]$ and $\bigsqcup_{\lambda \in \Lambda} \overline{a}_\lambda = \left[\bigvee_{\lambda \in \Lambda} a_\lambda^-, \bigvee_{\lambda \in \Lambda} a_\lambda^+\right]$.

For any interval numbers $\overline{a}_1, \overline{a}_2 \in D[0, 1]$, we define an order relation $\leqslant$ on $D[0, 1]$ such as $\overline{a}_1 \leqslant \overline{a}_2 \Longleftrightarrow a_1^- \leqslant a_2^-$ and $a_1^+ \leqslant a_2^+$, it easily observe that $(D[0, 1], \leqslant, \sqcap, \sqcup)$ forms a complete lattice with $0 = [0, 0]$ as its least element and $1 = [1, 1]$ as its greatest element. And $\overline{a}_1 = \overline{a}_2 \Longleftrightarrow a_1^- = a_2^-$ and $a_1^+ = a_2^+$.

Definition 2 [5]. *Let a set X be the fixed domain. An interval valued fuzzy set (IV-fuzzy set for short) on X is a mapping $\overline{F} : X \to D[0,1]$, we denote for each $x \in X$, $\overline{F}(x) = [F^-(x), F^+(x)] \in D[0,1]$, where F^- and F^+ are fuzzy sets on X such that $F^-(x) \leqslant F^+(x)$, for all $x \in X$.*

For the sake of simplicity, in the sequel, we shall use the symbol $\overline{F} = [F^-, F^+]$ denote an IV-fuzzy set on X. Let $\overline{F} = [F^-, F^+]$ and $\overline{G} = [G^-, G^+]$ be two IV-fuzzy sets on X, we define $\overline{F} \Cap \overline{G}$ and $\overline{F} \Cup \overline{G}$ as follows: for all $x \in X$,

(i) $(\overline{F} \Cap \overline{G})(x) = \overline{F}(x) \sqcap \overline{G}(x) = [F^-(x) \wedge G^-(x), F^+(x) \wedge G^+(x)]$;
(ii) $(\overline{F} \Cup \overline{G})(x) = \overline{F}(x) \sqcup \overline{G}(x) = [F^-(x) \vee G^-(x), F^+(x) \vee G^+(x)]$.

3 On Interval Valued $(\in, \in\vee q)$-fuzzy Filters

In the sequel, we emphasize that every interval valued fuzzy set $\overline{F} = [F^-, F^+]$ on L must satisfies the comparable condition and the following properties:

$$[F^-(x), F^+(x)] < [0.5, 0.5] \text{ or } [0.5, 0.5] \leqslant [F^-(x), F^+(x)], \text{ for all } x \in L. \quad (1)$$

Definition 3 [10]. *Let L be an MTL-algebra. An IV-fuzzy set $\overline{F} = [F^-, F^+]$ on L is called an interval valued $(\in, \in\vee q)$-fuzzy filter of L, IV-$(\in, \in\vee q)$-fuzzy filter of L for short, if it satisfies the following conditions for all $\overline{\alpha}, \overline{\beta} \in D[0,1]$ and $x, y \in L$,*

(IVF1) $x \leqslant y$ *implies* $\overline{F}(y) \geqslant \overline{F}(x) \sqcap [0.5, 0.5]$;
(IVF2) $\overline{F}(x \otimes y) \geqslant \overline{F}(x) \sqcap \overline{F}(y) \sqcap [0.5, 0.5]$.

The set of all IV-$(\in, \in\vee q)$-fuzzy filters of L is denoted by **IVFil**(L).

Lemma 2 [10]. *Let L be an MTL-algebra and F an IV-fuzzy set on L. Consider the following conditions for all $x, y, z \in L$,*

(IVF3) $\overline{F}(1) \geqslant \overline{F}(x) \sqcap [0.5, 0.5]$;
(IVF4) $\overline{F}(y) \geqslant \overline{F}(x) \sqcap \overline{F}(x \to y) \sqcap [0.5, 0.5]$;
(IVF5) $x \leqslant y \to z$ *implies* $\overline{F}(z) \geqslant \overline{F}(x) \sqcap \overline{F}(y) \sqcap [0.5, 0.5]$.

Then $F \in$ **IVFil**(L) *$\Longleftrightarrow$(IVF5)+(IVF6)$\Longleftrightarrow$(IVF7).*

Definition 4. *Let L be an MTL-algebra and $\overline{F} = [F^-, F^+]$ an IV-fuzzy set on L. An IV-fuzzy set $\overline{F}^{\overline{\alpha}}$ on L is defined such as $\overline{F}^{\overline{\alpha}}(x) = \begin{cases} \overline{F}(x), & x \neq 1, \\ \overline{F}(1) \sqcup \overline{\alpha}, & x = 1 \end{cases}$ for all $x \in L$, where $\overline{\alpha} \in D[0,1]$.*

Theorem 1. *Let L be an MTL-algebra and $\overline{F} = [F^-, F^+] \in$* **IVFil**$(L)$. *Then for all $\overline{\alpha} \in D[0,1]$, $\overline{F}^{\overline{\alpha}} \in$* **IVFil**$(L)$.

Proof. Firstly, for all $x, y \in L$, let $x \leqslant y$, we consider the following two cases:

(i) Assume that $y = 1$. If $x = 1$, we have that $\overline{F}^{\overline{\alpha}}(y) = \overline{F}^{\overline{\alpha}}(1) = \overline{F}(1) \sqcup \overline{\alpha} = \overline{F}^{\overline{\alpha}}(x) \geqslant \overline{F}^{\overline{\alpha}}(x) \sqcap [0.5, 0.5]$. If $x \neq 1$, by using $\overline{F} \in \mathbf{IVFil}(L)$ and (IVF3), we have that $\overline{F}^{\overline{\alpha}}(y) = \overline{F}^{\overline{\alpha}}(1) = \overline{F}(1) \sqcup \overline{\alpha} \geqslant \overline{F}(1) \geqslant \overline{F}(x) \sqcap [0.5, 0.5] = \overline{F}^{\overline{\alpha}}(x) \sqcap [0.5, 0.5]$.

(ii) Assume that $y \neq 1$, then $x \neq 1$. It follows that $\overline{F}^{\overline{\alpha}}(y) = \overline{F}(y) \geqslant \overline{F}(x) \sqcap [0.5, 0.5] = \overline{F}^{\overline{\alpha}}(x) \sqcap [0.5, 0.5]$ from $\overline{F} \in \mathbf{IVFil}(L)$ and (IVF3).

Summarize these two cases, we conclude that $x \leqslant y$ implies $\overline{F}^{\overline{\alpha}}(y) \geqslant \overline{F}^{\overline{\alpha}}(x) \sqcap [0.5, 0.5]$, for all $x, y \in L$. i.e., $\overline{F}^{\overline{\alpha}}$ satisfies (IVF1).

Secondly, for all $x, y \in L$, we consider the following two cases:

(i) Assume that $x \otimes y = 1$. If $x = y = 1$, it is obvious that $\overline{F}^{\overline{\alpha}}(x \otimes y) = \overline{F}(1) \sqcup \overline{\alpha} = \overline{F}^{\overline{\alpha}}(x) \sqcap \overline{F}^{\overline{\alpha}}(y) \geqslant \overline{F}^{\overline{\alpha}}(x) \sqcap \overline{F}^{\overline{\alpha}}(y) \sqcap [0.5, 0.5]$. If $x = 1, y \neq 1$ or $x \neq 1, y = 1$, then $x \oplus y \neq 1$, it is a contradiction. If $x \neq 1$ and $y \neq 1$, it follows that $\overline{F}^{\overline{\alpha}}(x) \sqcap \overline{F}^{\overline{\alpha}}(y) \sqcap [0.5, 0.5] = \overline{F}(x) \sqcap \overline{F}(y) \sqcap [0.5, 0.5] \leqslant \overline{F}(x \otimes y) = \overline{F}(1) \leqslant \overline{F}(1) \sqcup \overline{\alpha} = \overline{F}^{\overline{\alpha}}(1) = \overline{F}^{\overline{\alpha}}(x \otimes y)$. from $\overline{F} \in \mathbf{IVFil}(L)$ and (IVF2).

(ii) Assume that $x \otimes y \neq 1$. If $x = y = 1$, it is obvious a contradiction.

If $x = 1, y \neq 1$ or $x \neq 1, y = 1$, let's assume $x = 1, y \neq 1$, then $x \otimes y = 1 \otimes y = y$, and so $\overline{F}^{\overline{\alpha}}(x) \sqcap \overline{F}^{\overline{\alpha}}(y) \sqcap [0.5, 0.5] \leqslant \overline{F}^{\overline{\alpha}}(y) = \overline{F}(y) = \overline{F}(x \otimes y) = \overline{F}^{\overline{\alpha}}(x \otimes y)$. If $x \neq 1$ and $y \neq 1$, it follows that $\overline{F}^{\overline{\alpha}}(x \otimes y) = \overline{F}(x \otimes y) \geqslant \overline{F}(x) \sqcap \overline{F}(y) \sqcap [0.5, 0.5] = \overline{F}^{\overline{\alpha}}(x) \sqcap \overline{F}^{\overline{\alpha}}(y) \sqcap [0.5, 0.5]$ from $\overline{F} \in \mathbf{IVFil}(L)$ and (IVF2).

Summarize these two cases, we conclude that $\overline{F}^{\overline{\alpha}}(x \otimes y) \geqslant \overline{F}^{\overline{\alpha}}(x) \sqcap \overline{F}^{\overline{\alpha}}(y) \sqcap [0.5, 0.5]$, for all $x, y \in L$. i.e., $\overline{F}^{\overline{\alpha}}$ satisfies (IVF2).

Thus it follows that $\overline{F}^{\overline{\alpha}} \in \mathbf{IVFil}(L)$ from Definition 3.

Definition 5. *Let L be an MTL-algebra and $\overline{F} = [F^-, F^+], \overline{G} = [G^-, G^+]$ IV-fuzzy sets on L. IV-fuzzy sets $\overline{F}^{\overline{G}}$ and $\overline{G}^{\overline{F}}$ on L are defined such as for all $x \in L$,*

$$\overline{F}^{\overline{G}}(x) = \begin{cases} \overline{F}(x), & x \neq 1, \\ \overline{F}(1) \sqcup \overline{G}(1), & x = 1, \end{cases} \quad \text{and} \quad \overline{G}^{\overline{F}}(x) = \begin{cases} \overline{G}(x), & x \neq 1, \\ \overline{G}(1) \sqcup \overline{F}(1), & x = 1. \end{cases} \tag{2}$$

Corollary 1. *Let L be an MTL-algebra and $\overline{F} = [F^-, F^+], \overline{G} = [G^-, G^+] \in$ **IVFil**(L). Then $\overline{F}^{\overline{G}} \in$ **IVFil**(L) and $\overline{G}^{\overline{F}} \in$ **IVFil**(L).*

Definition 6. *Let L be an MTL-algebra and $\overline{F} = [F^-, F^+], \overline{G} = [G^-, G^+]$ IV-fuzzy sets on L. An IV-fuzzy set $\overline{F} \uplus \overline{G}$ on L is defined as follows: for all $x, a, b \in L$, $\left(\overline{F} \uplus \overline{G}\right)(x) = \bigsqcup_{a \otimes b \leqslant x} \{\overline{F}(a) \sqcap \overline{G}(b) \sqcap [0.5, 0.5]\}$.*

Theorem 2. *Let L be an MTL-algebra and $\overline{F} = [F^-, F^+], \overline{G} = [G^-, G^+] \in$ **IVFil**(L). Then $\overline{F}^{\overline{G}} \uplus \overline{G}^{\overline{F}} \in$ **IVFil**(L).*

Proof. Firstly, for all $x, y \in L$, let $x \leqslant y$, then $\{a \otimes b | a \otimes b \leqslant x\} \subseteq \{a \otimes b | a \otimes b \leqslant y\}$, so $(\overline{F}^{\overline{G}} \uplus \overline{G}^{\overline{F}})(y) = \bigsqcup_{a \otimes b \leqslant y} \{\overline{F}^{\overline{G}}(a) \sqcap \overline{G}^{\overline{F}}(b) \sqcap [0.5, 0.5]\} \geqslant$

$\bigsqcup_{a\otimes b\leqslant x} \{\overline{F}^{\overline{G}}(a) \sqcap \overline{G}^{\overline{F}}(b) \sqcap [0.5, 0.5]\} = [0.5, 0.5] \sqcap \bigsqcup_{a\otimes b\leqslant x} \{\overline{F}^{\overline{G}}(a) \sqcap \overline{G}^{\overline{F}}(b) \sqcap [0.5, 0.5]\} = (\overline{F}^{\overline{G}} \uplus \overline{G}^{\overline{F}})(x) \sqcap [0.5, 0.5]$. Hence $\overline{F}^{\overline{G}} \uplus \overline{G}^{\overline{F}}$ satisfies (IVF1). Secondly, for all $x, y \in L$, we have that $(\overline{F}^{\overline{G}} \uplus \overline{G}^{\overline{F}})(x \otimes y) = \bigsqcup_{a\otimes b\leqslant x\otimes y} \{\overline{F}^{\overline{G}}(a) \sqcap \overline{G}^{\overline{F}}(b) \sqcap [0.5, 0.5]\} \geqslant \bigsqcup_{a_1\otimes a_2\leqslant x \text{ and } b_1\otimes b_2\leqslant y} \{\overline{F}^{\overline{G}}(a_1 \otimes b_1) \sqcap \overline{G}^{\overline{F}}(a_2 \otimes b_2) \sqcap [0.5, 0.5]\} \geqslant \bigsqcup_{a_1\otimes a_2\leqslant x \text{ and } b_1\otimes b_2\leqslant y} \{\overline{F}^{\overline{G}}(a_1) \sqcap \overline{F}^{\overline{G}}(b_1) \sqcap \overline{G}^{\overline{F}}(a_2) \sqcap \overline{G}^{\overline{F}}(b_2) \sqcap [0.5, 0.5]\} = \bigsqcup_{a_1\otimes a_2\leqslant x} \{\overline{F}^{\overline{G}}(a_1) \sqcap \overline{G}^{\overline{F}}(a_2) \sqcap [0.5, 0.5]\} \sqcap \bigsqcup_{b_1\otimes b_2\leqslant y} \{\overline{F}^{\overline{G}}(b_1) \sqcap \overline{G}^{\overline{F}}(b_2) \sqcap [0.5, 0.5]\} = [0.5, 0.5] \sqcap \bigsqcup_{a_1\otimes a_2\leqslant x} \{\overline{F}^{\overline{G}}(a_1) \sqcap \overline{G}^{\overline{F}}(a_2) \sqcap [0.5, 0.5]\} \sqcap [0.5, 0.5] \sqcap \bigsqcup_{b_1\otimes b_2\leqslant y} \{\overline{F}^{\overline{G}}(b_1) \sqcap \overline{G}^{\overline{F}}(b_2) \sqcap [0.5, 0.5]\} = (\overline{F}^{\overline{G}} \uplus \overline{G}^{\overline{F}})(x) \sqcap (\overline{F}^{\overline{G}} \uplus \overline{G}^{\overline{F}})(y) \sqcap [0.5, 0.5]$, and so $\overline{F}^{\overline{G}} \uplus \overline{G}^{\overline{F}}$ also satisfies (IVF2). Hence $\overline{F}^{\overline{G}} \uplus \overline{G}^{\overline{F}} \in \mathbf{IVFil}(L)$ by Definition 3.

4 The Lattice of Interval Valued $(\in, \in\vee q)$-fuzzy Filters

Definition 7. Let L be an MTL-algebra and $\overline{F} = [F^-, F^+], \overline{G} = [G^-, G^+]$ IV-fuzzy sets on L. The binary relation $\sqsubseteq$ is defined as follows:

$$\overline{F} \sqsubseteq \overline{G} \Longleftrightarrow \overline{F}(x) \sqcap [0.5, 0.5] \leqslant \overline{G}(x), \text{ for all } x \in L, \tag{3}$$

$$\overline{F} = \overline{G} \Longleftrightarrow \overline{F} \sqsubseteq \overline{G} \text{ and } \overline{G} \sqsubseteq \overline{F}. \tag{4}$$

Remark 1. Let L be an MTL-algebra and $\overline{F} = [F^-, F^+], \overline{G} = [G^-, G^+], \overline{H} = [H^-, H^+]$ IV-fuzzy sets on L. Then by Definition 7, we have that

$$\overline{F} \sqsubseteq \overline{G} \text{ implies } \overline{F}(x) \sqcap [0.5, 0.5] \leqslant \overline{G}(x) \sqcap [0.5, 0.5], \text{ for all } x \in L. \tag{5}$$

At the same time, by the definition of $\sqsubseteq$, it is easy to verify that the following assertions are hold: (i) $\overline{F} \sqsubseteq \overline{F}$, (ii) $\overline{F} \sqsubseteq \overline{G}$ and $\overline{G} \sqsubseteq \overline{F}$ imply $\overline{F} = \overline{G}$ and (iii) $\overline{F} \sqsubseteq \overline{G}$ and $\overline{G} \sqsubseteq \overline{H}$ imply $\overline{F} \sqsubseteq \overline{H}$. Hence, we can see that $\sqsubseteq$ is a partial order on the set of all IV-fuzzy sets on L.

Definition 8. *Let L be an MTL-algebra and $\overline{F} = [F^-, F^+]$ an IV-fuzzy set on L. An IV-$(\in, \in\vee q)$-fuzzy filter $\overline{G} = [G^-, G^+]$ of L is called the generated IV-$(\in, \in\vee q)$-fuzzy filter by $\overline{F}$, denoted $\langle\overline{F}\rangle$, if the following conditions are hold:(i) $\overline{F} \sqsubseteq \overline{G}$ and (ii) $\overline{F} \sqsubseteq \overline{H}$ implies $\overline{G} \sqsubseteq \overline{H}$, for all $\overline{H} \in$* **IVFil**(L).

Theorem 3. *Let L be an MTL-algebra and $\overline{F} = [F^-, F^+]$ an IV-fuzzy set on L. An IV-fuzzy set $\overline{G} = [G^-, G^+]$ on L is defined such as $\overline{G}(x) = \bigsqcup_{a_1\otimes\cdots\otimes a_n\leqslant x} \{\overline{F}(a_1) \sqcap \cdots \overline{F}(a_n) \sqcap [0.5, 0.5]\}$ for all $x \in L$. Then $\overline{G} = \langle\overline{F}\rangle$.*

Proof. It is similar to Theorem 3.4 in [8].

Theorem 4. *Let L be an MTL-algebra. Then $(\mathbf{IVFil}(L), \sqsubseteq)$ is a complete lattice.*

Proof. For any $\{\overline{F}_\lambda\}_{\lambda\in\Lambda} \subseteq \mathbf{IVFil}(L)$, where Λ is an indexed set. Define $\Cap_{\lambda\in\Lambda}\overline{F}_\lambda$ and $\Cup_{\lambda\in\Lambda}\overline{F}_\lambda$ as following: for all $x \in L$,

$$\left(\Cap_{\lambda\in\Lambda}\overline{F}_\lambda\right)(x) = \bigsqcap_{\lambda\in\Lambda} \overline{F}_\lambda(x) \text{ and } \left(\Cup_{\lambda\in\Lambda}\overline{F}_\lambda\right)(x) = \bigsqcup_{\lambda\in\Lambda} \overline{F}_\lambda(x).$$

Then it is easy to verify that $\Cap_{\lambda\in\Lambda}\overline{F}_\lambda$ is infimum of $\{\overline{F}_\lambda\}_{\lambda\in\Lambda}$, i. e., $\bigwedge_{\lambda\in\Lambda}\overline{F}_\lambda = \Cap_{\lambda\in\Lambda}\overline{F}_\lambda$. And $\langle\Cup_{\lambda\in\Lambda}\overline{F}_\lambda\rangle$ is supermun of $\{\overline{F}_\lambda\}_{\lambda\in\Lambda}$, where $\langle\Cup_{\lambda\in\Lambda}\overline{F}_\lambda\rangle$ is the IV-$(\in, \in\vee q)$-fuzzy filter generated by $\Cup_{\lambda\in\Lambda}\overline{F}_\lambda$ of L. i. e., $\bigvee_{\lambda\in\Lambda}\overline{F}_\lambda = \langle\Cup_{\lambda\in\Lambda}\overline{F}_\lambda\rangle$. Therefor $(\mathbf{IVFil}(L), \sqsubseteq)$ is a complete lattice. The proof is completed.

Theorem 5. *Let L be an MTL-algebra. Then For all $\overline{F}, \overline{G} \in \mathbf{IVFil}(L)$, $\overline{F}\vee\overline{G} = \langle\overline{F}\Cup\overline{G}\rangle = \overline{F}^{\overline{G}} \uplus \overline{G}^{\overline{F}}$ in the complete lattice $(\mathbf{IVFil}(L), \sqsubseteq)$.*

Proof. For all $\overline{F}, \overline{G} \in \mathbf{IVFil}(L)$, it is obvious that $\overline{F} \sqsubseteq \overline{F}^{\overline{G}} \uplus \overline{G}^{\overline{F}}$ and $\overline{G} \sqsubseteq \overline{F}^{\overline{G}} \uplus \overline{G}^{\overline{F}}$, i.e., $\overline{F}(x)\sqcap[0.5, 0.5] \leqslant (\overline{F}^{\overline{G}}\uplus\overline{G}^{\overline{F}})(x)$ and $\overline{G}(x)\sqcap[0.5, 0.5] \leqslant (\overline{F}^{\overline{G}}\uplus\overline{G}^{\overline{F}})(x)$ for all $x \in L$. Thus $(\overline{F} \Cup \overline{G})(x) \sqcap [0.5, 0.5] = [\overline{F}(x) \sqcup \overline{G}(x)] \sqcap [0.5, 0.5] = (\overline{F}(x) \sqcap [0.5, 0.5]) \sqcup (\overline{G}(x) \sqcap [0.5, 0.5]) \leqslant (\overline{F}^{\overline{G}} \uplus \overline{G}^{\overline{F}})(x)$, that is, $\overline{F} \Cup \overline{G} \sqsubseteq \overline{F}^{\overline{G}} \uplus \overline{G}^{\overline{F}}$, and thus $\langle\overline{F} \Cup \overline{G}\rangle \sqsubseteq \overline{F}^{\overline{G}} \uplus \overline{G}^{\overline{F}} \in \mathbf{IVFil}(L)$ by Theorem 2. Let $\overline{H} \in \mathbf{IVFil}(L)$ such that $\overline{F} \Cup \overline{G} \sqsubseteq \overline{H}$, for all $x \in L$, we consider the following two cases:

(i) If $x = 1$, then $(\overline{F}^{\overline{G}} \uplus \overline{G}^{\overline{F}})(1) \sqcap [0.5, 0.5] = [0.5, 0.5] \sqcap \bigsqcup_{a\otimes b\leqslant 1} \{\overline{F}^{\overline{G}}(a) \sqcap \overline{G}^{\overline{F}}(b) \sqcap [0.5, 0.5]\} = \overline{F}^{\overline{G}}(1) \sqcap \overline{G}^{\overline{F}}(1) \sqcap [0.5, 0.5] = [\overline{F}(1) \sqcup \overline{G}(1)] \sqcap [0.5, 0.5] = (\overline{F} \Cup \overline{G})(1) \sqcap [0.5, 0.5] \leqslant \overline{H}(1)$, thus $\overline{F}^{\overline{G}} \uplus \overline{G}^{\overline{F}} \sqsubseteq \overline{H}$ for this case.

(ii) If $x < 1$, then $(\overline{F}^{\overline{G}}\uplus\overline{G}^{\overline{F}})(x)\sqcap[0.5, 0.5] = [0.5, 0.5]\sqcap \bigsqcup_{a\otimes b\leqslant x} \{\overline{F}^{\overline{G}}(a)\sqcap\overline{G}^{\overline{F}}(b)\sqcap[0.5, 0.5]\} = \bigsqcup_{a\otimes b\leqslant x, a\neq 1, b\neq 1} \{\overline{F}^{\overline{G}}(a)\sqcap\overline{G}^{\overline{F}}(b)\sqcap[0.5, 0.5]\}\sqcup \bigsqcup_{a\leqslant x} \{\overline{F}(a)\sqcap[\overline{F}(1)\sqcup\overline{G}(1)]\sqcap[0.5, 0.5]\} \sqcup \bigsqcup_{b\leqslant x} \{[\overline{F}(1) \sqcup \overline{G}(1)] \sqcap \overline{G}(b) \sqcap [0.5, 0.5]\} = \bigsqcup_{a\otimes b\leqslant x, a\neq 1, b\neq 1} \{\overline{F}(a) \sqcap \overline{G}(b) \sqcap [0.5, 0.5]\} \sqcup \bigsqcup_{a\leqslant x} \{\overline{F}(a)\sqcap [0.5, 0.5]\} \sqcup \bigsqcup_{b\leqslant x} \{\overline{G}(b)\sqcap [0.5, 0.5]\} \leqslant \bigsqcup_{a\otimes b\leqslant x, a\neq 1, b\neq 1} \{\overline{H}(a)\sqcap \overline{H}(b)\sqcap [0.5, 0.5]\} \sqcup \bigsqcup_{a\leqslant x} \{\overline{H}(a)\sqcap [0.5, 0.5]\} \sqcup \bigsqcup_{b\leqslant x} \{\overline{H}(b)\sqcap [0.5, 0.5]\} = \bigsqcup_{a\otimes b\leqslant x} \{\overline{H}(a)\sqcap \overline{H}(b) \sqcap [0.5, 0.5]\} = [0.5, 0.5] \sqcap \bigsqcup_{a\otimes b\leqslant x} \{\overline{H}(a) \sqcap \overline{H}(b) \sqcap [0.5, 0.5]\} \leqslant \bigsqcup_{a\otimes b\leqslant x} \{\overline{H}(a \otimes b) \sqcap [0.5, 0.5]\} \leqslant \overline{H}(x)$, thus $\overline{F}^{\overline{G}} \uplus \overline{G}^{\overline{F}} \sqsubseteq \overline{H}$ for this case too.

By Definition 8 and Theorem 5 we have that $\overline{F} \vee \overline{G} = \langle\overline{F} \Cup \overline{G}\rangle = \overline{F}^{\overline{G}} \uplus \overline{G}^{\overline{F}}$.

Theorem 6. *Let L be an MTL-algebra. Then $(\mathbf{IVFil}(L), \sqsubseteq)$ is a distributive lattice, where, $\overline{F} \wedge \overline{G} = \overline{F} \Cap \overline{G}$ and $\overline{F} \vee \overline{G} = \langle \overline{F} \Cup \overline{G} \rangle$, for all $\overline{F}, \overline{G} \in \mathbf{IVFil}(L)$.*

Proof. To finish the proof, it suffices to show that $\overline{H} \wedge (\overline{F} \vee \overline{G}) = (\overline{H} \wedge \overline{F}) \vee (\overline{H} \wedge \overline{G})$, for all $\overline{F}, \overline{G}, \overline{H} \in \mathbf{IVFil}(L)$. Since the inequality $(\overline{H} \wedge \overline{F}) \vee (\overline{H} \wedge \overline{G}) \sqsubseteq \overline{H} \wedge (\overline{F} \vee \overline{G})$ holds automatically in a lattice, we need only to show the inequality $\overline{H} \wedge (\overline{F} \vee \overline{G}) \sqsubseteq (\overline{H} \wedge \overline{F}) \vee (\overline{H} \wedge \overline{G})$. i.e., we need only to show that $\left(\overline{H} \Cap \left(\overline{F}^{\overline{G}} \uplus \overline{G}^{\overline{H}}\right)\right)(x) \sqcap [0.5, 0.5] \leqslant \left((\overline{H} \Cap \overline{F})^{\overline{H} \Cap \overline{G}} \uplus (\overline{H} \Cap \overline{G})^{\overline{H} \Cap \overline{F}}\right)(x)$, for all $x \in L$. For these, we consider the following two cases:

(i) If $x = 1$, we have $(\overline{H} \Cap (\overline{F}^{\overline{G}} \uplus \overline{G}^{\overline{H}}))(1) \sqcap [0.5, 0.5] = \overline{H}(1) \sqcap (\overline{F}^{\overline{G}} \uplus \overline{G}^{\overline{F}})(1) \sqcap [0.5, 0.5] = \overline{H}(1) \sqcap \bigsqcup_{a \otimes b \leqslant 1} \{\overline{F}^{\overline{G}}(a) \sqcap \overline{G}^{\overline{F}}(b) \sqcap [0.5, 0.5]\} = \overline{H}(1) \sqcap [\overline{F}^{\overline{G}}(1) \sqcap \overline{G}^{\overline{F}}(1) \sqcap [0.5, 0.5]] = \overline{H}(1) \sqcap [\overline{F}(1) \sqcup \overline{G}(1)] \sqcap [0.5, 0.5] = \{[\overline{H}(1) \sqcap \overline{F}(1)] \sqcup [\overline{H}(1) \sqcap \overline{G}(1)]\} \sqcap [0.5, 0.5] = [(\overline{H} \Cap \overline{F})(1) \sqcup (\overline{H} \Cap \overline{G})(1)] \sqcap [0.5, 0.5] = ((\overline{H} \Cap \overline{F})^{\overline{H} \Cap \overline{G}})(1) \sqcap ((\overline{H} \Cap \overline{G})^{\overline{H} \Cap \overline{F}})(1) \sqcap [0.5, 0.5] = \bigsqcup_{a \otimes b \leqslant 1} \{((\overline{H} \Cap \overline{F})^{\overline{H} \Cap \overline{G}})(a) \sqcap ((\overline{H} \Cap \overline{G})^{\overline{H} \Cap \overline{F}})(b) \sqcap [0.5, 0.5]\} = ((\overline{H} \Cap \overline{F})^{\overline{H} \Cap \overline{G}} \uplus (\overline{H} \Cap \overline{G})^{\overline{H} \Cap \overline{F}})(1)$.

(ii) If $x < 1$, we have

$$
\begin{aligned}
&\left(\overline{H} \Cap \left(\overline{F}^{\overline{G}} \uplus \overline{G}^{\overline{H}}\right)\right)(x) \sqcap [0.5, 0.5] = \overline{H}(x) \sqcap \left(\overline{F}^{\overline{G}} \uplus \overline{G}^{\overline{F}}\right)(x) \sqcap [0.5, 0.5] \\
&= \overline{H}(x) \sqcap \bigsqcup_{a \otimes b \leqslant x} \left\{\overline{F}^{\overline{G}}(a) \sqcap \overline{G}^{\overline{F}}(b) \sqcap [0.5, 0.5]\right\} \\
&= \bigsqcup_{a \otimes b \leqslant x} \left\{\overline{H}(x) \sqcap \overline{F}^{\overline{G}}(a) \sqcap \overline{G}^{\overline{F}}(b) \sqcap [0.5, 0.5]\right\} \\
&= \bigsqcup_{a \otimes b \leqslant x, a \neq 1, b \neq 1} \left\{\overline{H}(x) \sqcap \overline{F}^{\overline{G}}(a) \sqcap \overline{G}^{\overline{F}}(b) \sqcap [0.5, 0.5]\right\} \sqcup \\
&\qquad \bigsqcup_{b \leqslant x} \left\{\overline{H}(x) \sqcap \overline{F}^{\overline{G}}(1) \sqcap \overline{G}(b) \sqcap [0.5, 0.5]\right\} \sqcup \\
&\qquad \bigsqcup_{a \leqslant x} \left\{\overline{H}(x) \sqcap \overline{F}(a) \sqcap \overline{G}^{\overline{F}}(1) \sqcap [0.5, 0.5]\right\} \\
&= \bigsqcup_{a \otimes b \leqslant x, a \neq 1, b \neq 1} \left\{[\overline{H}(x) \sqcap \overline{F}(a)] \sqcap [\overline{H}(x) \sqcap \overline{G}(b)] \sqcap [0.5, 0.5]\right\} \sqcup \\
&\qquad \bigsqcup_{b \leqslant x} \left\{\left[\overline{H}(x) \sqcap [0.5, 0.5] \sqcap \overline{F}^{\overline{G}}(1)\right] \sqcap [\overline{H}(x) \sqcap \overline{G}(b)] \sqcap [0.5, 0.5]\right\} \sqcup \\
&\qquad \bigsqcup_{a \leqslant x} \left\{[\overline{H}(x) \sqcap \overline{F}(a)] \sqcap \left[\overline{H}(x) \sqcap [0.5, 0.5] \sqcap \overline{G}^{\overline{F}}(1)\right] \sqcap [0.5, 0.5]\right\}
\end{aligned}
$$

$$\leqslant \bigsqcup_{a\otimes b\leqslant x, a\neq 1, b\neq 1} \left\{\left[\overline{H}(x\vee a)\sqcap \overline{F}(x\vee a)\right]\sqcap\left[\overline{H}(x\vee b)\sqcap \overline{G}(x\vee b)\right]\sqcap[0.5,0.5]\right\}\sqcup$$
$$\bigsqcup_{b\leqslant x}\left\{\left[\overline{H}(1)\sqcap\left(\overline{F}(1)\sqcup\overline{G}(1)\right)\right]\sqcap\left[\overline{H}(x\vee b)\sqcap\overline{G}(x\vee b)\right]\sqcap[0.5,0.5]\right\}\sqcup$$
$$\bigsqcup_{a\leqslant x}\left\{\left[\overline{H}(x\vee a)\sqcap\overline{F}(x\vee a)\right]\sqcap\left[\overline{H}(1)\sqcap\left(\overline{G}(1)\sqcup\overline{F}(1)\right)\right]\sqcap[0.5,0.5]\right\}$$
$$= \bigsqcup_{a\otimes b\leqslant x, a\neq 1, b\neq 1}\left\{\left(\overline{H}\Cap\overline{F}\right)(x\vee a)\sqcap\left(\overline{H}\Cap\overline{G}\right)(x\vee b)\sqcap[0.5,0.5]\right\}\sqcup$$
$$\bigsqcup_{b\leqslant x}\left\{\left[\left(\overline{H}\Cap\overline{F}\right)(1)\sqcup\left(\overline{H}\Cap\overline{G}\right)(1)\right]\sqcap\left(\overline{H}\Cap\overline{G}\right)(x\vee b)\sqcap[0.5,0.5]\right\}\sqcup$$
$$\bigsqcup_{a\leqslant x}\left\{\left(\overline{H}\Cap\overline{F}\right)(x\vee a)\sqcap\left[\left(\overline{H}\Cap\overline{G}\right)(1)\sqcup\left(\overline{H}\Cap\overline{F}\right)(1)\right]\sqcap[0.5,0.5]\right\}$$
$$= \bigsqcup_{a\otimes b\leqslant x, a\neq 1, b\neq 1}\left\{\left(\left(\overline{H}\Cap\overline{F}\right)^{\overline{H}\Cap\overline{G}}\right)(x\vee a)\sqcap\left(\left(\overline{H}\Cap\overline{G}\right)^{\overline{H}\Cap\overline{F}}\right)(x\vee b)\sqcap[0.5,0.5]\right\}$$
$$\sqcup\bigsqcup_{b\leqslant x}\left\{\left(\left(\overline{H}\Cap\overline{F}\right)^{\overline{H}\Cap\overline{G}}\right)(1)\sqcap\left(\left(\overline{H}\Cap\overline{G}\right)^{\overline{H}\Cap\overline{F}}\right)(x\vee b)\sqcap[0.5,0.5]\right\}$$
$$\sqcup\bigsqcup_{a\leqslant x}\left\{\left(\left(\overline{H}\Cap\overline{F}\right)^{\overline{H}\Cap\overline{G}}\right)(x\vee a)\sqcap\left(\left(\overline{H}\Cap\overline{G}\right)^{\overline{H}\Cap\overline{F}}\right)(1)\sqcap[0.5,0.5]\right\}$$
$$= \bigsqcup_{a\otimes b\leqslant x}\left\{\left(\left(\overline{H}\Cap\overline{F}\right)^{\overline{H}\Cap\overline{G}}\right)(x\vee a)\sqcap\left(\left(\overline{H}\Cap\overline{G}\right)^{\overline{H}\Cap\overline{F}}\right)(x\vee b)\sqcap[0.5,0.5]\right\}.$$

Let $a\vee x=c$ and $b\vee x=d$, since $a\otimes b\leqslant x$, using Lemma 1 we get that $c\otimes d=(a\vee x)\otimes(b\vee x)=((a\vee x)\otimes b)\vee((a\vee x)\otimes x)=(a\otimes b)\vee(x\otimes b)\vee(a\otimes x)\vee(x\otimes x)\leqslant(a\otimes b)\vee x\vee x\vee x=(a\otimes b)\vee x\leqslant x\vee x=x$. Hence we can conclude that

$$\left(\overline{H}\Cap\left(\overline{F}^{\overline{G}}\uplus\overline{G}^{\overline{H}}\right)\right)(x)\sqcap[0.5,0.5]$$
$$\leqslant \bigsqcup_{a\otimes b\leqslant x}\left\{\left(\left(\overline{H}\Cap\overline{F}\right)^{\overline{H}\Cap\overline{G}}\right)(x\vee a)\sqcap\left(\left(\overline{H}\Cap\overline{G}\right)^{\overline{H}\Cap\overline{F}}\right)(x\vee b)\sqcap[0.5,0.5]\right\}$$
$$\leqslant \bigsqcup_{c\otimes d\leqslant x}\left\{\left(\left(\overline{H}\Cap\overline{F}\right)^{\overline{H}\Cap\overline{G}}\right)(c)\sqcap\left(\left(\overline{H}\Cap\overline{G}\right)^{\overline{H}\Cap\overline{F}}\right)(d)\sqcap[0.5,0.5]\right\}$$
$$=\left(\left(\overline{H}\Cap\overline{F}\right)^{\overline{H}\Cap\overline{G}}\uplus\left(\overline{H}\Cap\overline{G}\right)^{\overline{H}\Cap\overline{F}}\right)(x).$$

To sum up, we have $(\overline{H}\Cap(\overline{F}^{\overline{G}}\uplus\overline{G}^{\overline{H}}))(x)\sqcap[0.5,0.5]\leqslant((\overline{H}\Cap\overline{F})^{\overline{H}\Cap\overline{G}}\uplus(\overline{H}\Cap\overline{G})^{\overline{H}\Cap\overline{F}})(x)$, for all $x\in L$. The proof is completed.

References

1. Hájek, P.: Meta Mathematics of Fuzzy Logic. Kluwer Academic Publisher, Dordrecht (1998)
2. Esteva, F., Godo, L.: Monoidal t-norm based logic: towards a logic for left-continuous t-norms. Fuzzy Sets Syst. **124**, 271–288 (2001)
3. Borzooei, R.A., Shoar, S.K., Ameri, R.: Some types of filters in MTL-algebras. Fuzzy Sets Syst. **187**, 92–102 (2012)
4. Haveshky, M.: A note on some types of filters in MTL-algebras. Fuzzy Sets Syst. **247**, 135–137 (2014)
5. Zadeh, L.A.: Fuzzy sets. Inf. Control **8**, 338–353 (1965)
6. Kim, K.H., Zhang, Q., Jun, Y.B.: On fuzzy filters of MTL-algebras. J. Fuzzy Math. **10**(4), 981–989 (2002)
7. Liu, C.H.: On fuzzy filters of MTL-algebras. Fuzzy Syst. Math. **29**(2), 98–108 (2015)
8. Jun, Y.B., Xu, Y., Zhang, X.H.: Fuzzy filters of MTL-algebras. Inf. Sci. **175**(1–2), 120–138 (2005)
9. Zadeh, L.A.: The concept of a lingistic variable and its application to approximate reason. Inf. Control **18**, 199–249 (1975)
10. Ma, X.L., Zhan, J.M., Shum, K.P.: Interval valued $(\in, \in\vee q)$-fuzzy filters of MTL-algebras. J. Math. Res. Exposition **30**(2), 265–276 (2010)
11. Belohlavek, R.: Some properties of residuated lattices. Czechoslovak Math. J. **123**, 161–171 (2003)
12. Belohlavek, R.: Fuzzy Relational Systems: Foundations and Principles. Kluwer Academic, New York (2002)

A Ranking Model for Intuitionistic Fuzzy Preference Relation Under Uncertainty for Targeted Poverty

Dongxu Yu[1] and Sidong Xian[2](✉)

[1] School of Automation, Chongqing University of Posts and Telecommunications, Chongqing 400065, People's Republic of China
dxuyu@qq.com

[2] Key Laboratory of Intelligent Analysis and Decision on Complex Systems, Chongqing University of Posts and Telecommunications, Chongqing 400065, People's Republic of China
sidx@163.com

Abstract. Intuitionistic fuzzy preference relation (IFPR) is a suitable tool to present preference information and hesitation for decision maker among poverty alleviation candidates. This paper aims to provide an effective decision method for helping decision maker accurately to obtain priority weights by using IFPRs to represent the proportion of importance from candidates and rank them. Firstly, a fractional interactive minimizing deviation programming model based on multiplicative consistency-based method to derive weight vectors of alternatives from IFPRs based on multiplicative consistency is presented. Specifically, for any IFPR, by minimizing its absolute deviation from the corresponding consistent IFPR, the weight vectors are generated. Then, a new score function based on Euclidean distance is proposed. This defuzzification value is computed by constructing the idea of closest to ideal solution based on distance measure and ranks alternatives. Subsequently, a numerical example about targeted poverty alleviation selection is given to demonstrate the methods effectiveness.

Keywords: Intuitionistic fuzzy preference relation (IFPR) · Priority weights · New score function · Targeted poverty alleviation

1 Introduction

After nearly seventy years of poverty alleviation, China has basically solved the problem of providing food and clothing to the rural poor. However, Chinas poverty alleviation work is still facing challenges, and there is no fundamental change in the backward development of poverty-stricken areas. Hence, targeted poverty alleviation and helping those who really needed help is indispensable. Some study [1,2] have researched about targeted poverty alleviation. Here, we use the decision-making method of preference relation to achieve targeted

Y. Liu et al. (Eds.): ICNC-FSKD 2019, AISC 1074, pp. 704–711, 2020.
https://doi.org/10.1007/978-3-030-32456-8_77

poverty alleviation. In the process of decision making (DM), a decision maker needs to compare a set of alternatives to give his/her preference. The intuitionistic fuzzy preference relation(IFPR) which provides pairwise comparison over the given alternatives are widely used in expressing decision makers preference information in decision problems [3,4]. Meanwhile, the IFPRs which is consistent ensure that a benefit judgment of DM is not paradox or leading to happen logic disorder problem when comparing candidates in pairs. Generally, existing consistency of IFPR can be divided into multiplication consistency [5–7] and addition consistency [8,9]. In order to comprehensively consider the indicators of the needy family to achieve targeted poverty alleviation, the most important thing is that how to find the optimal priority weight vector of alternatives and measure the worth of these weights. Therefore, we developed a new approach which constructs a goal program model for intuitionistic fuzzy decision-making based on IFPR with uncertain priority weights. Then, the score function based on Euclidean distance by closest to ideal solution is proposed to ranks candidates.

2 Preliminaries

2.1 Intuitionistic Fuzzy Set

Definition 1 [10]. *Let a crisp set x be fixed and $A \in X$ be a fixed set, an intuitionistic fuzzy set on x is defined as:*

$$A = \{<x, \mu_A(x), \nu_A(x)> | x \in X\} \tag{1}$$

where the function: $\mu_A(x) \subseteq [0,1]$, $\nu_A(x) \subseteq [0,1]$, $\mu_A(x)$ and $\nu_A(x)$ present degree of membership and degree of non-membership, respectively. And the condition: $0 \leq \mu_A(x) + \nu_A(x) \leq 1$ for $x \in X$ is a constraint condition. The degree of hesitancy denotes: $\pi_A(x) = 1 - \mu_A(x) - \nu_A(x)$. Particularly, if $\pi_A(x) = 0$, the IFS A is degenerated to a fuzzy set.

Definition 2 [5]. *Let $\alpha = (\mu_\alpha, \nu_\alpha)$ be an IFS, the score function is defined as:*

$$s(\alpha) = \mu_\alpha - \nu_\alpha \tag{2}$$

Based on the function Eq. (2), a ranking method for any two IFS α and β is as: (1) If $s(\alpha) < s(\beta)$, then $\alpha < \beta$; (2) If $s(\alpha) = s(\beta)$, then $\alpha = \beta$.

Definition 3 [11]. *Let $\alpha = (\mu_\alpha, \nu_\alpha)$ and $\beta = (\mu_\beta, \nu_\beta)$ be two IFSs, then Euclidean distance measure between α and β is defined as following:*

$$d(\alpha, \beta) = \sqrt{(\mu_\alpha - \mu_\beta)^2 + (\nu_\alpha - \nu_\beta)^2} \tag{3}$$

2.2 Multiplicative Consistency of Intuitionistic Fuzzy Preference Relation

Definition 4 [5]. *An IFPR $\tilde{R}$ on the set $X = \{x_1, x_2, ..., x_n\}$ is represent by an intuitionistic fuzzy judgments matrix $\tilde{R} = (\tilde{r}_{ij})_{n \times n} \subset X \times X$ with $\tilde{r}_{ij} = (\mu_{ij}, \nu_{ij})$, and satisfies the conditions*

$$\mu_{ij} + \nu_{ij} \leq 1, \mu_{ij} = \nu_{ji}, \mu_{ij}, \nu_{ij} \in [0,1], \mu_{ii} = \nu_{ii} = 0.5, i, j = 1, 2, ..., n \tag{4}$$

where μ_{ij} denotes the degree to which x_i is not less than x_j, ν_{ij} is the degree to which x_i is less than x_j, and $\pi_{ij} = 1 - \mu_{ij} - \nu_{ij}$ is interpreted as the hesitancy degree.

Definition 5 [5]. *An intuitionistic fuzzy judgments matrix $\tilde{R} = (\tilde{r}_{ij})_{n \times n}$ with $\tilde{r}_{ij} = (\mu_{ij}, \nu_{ij})$ is called multiplicative consistent if it satisfies the following condition*

$$\mu_{ij}\mu_{jk}\mu_{ki} = \nu_{ij}\nu_{jk}\nu_{ki}, \forall i, j, k = 1, 2, ..., n \tag{5}$$

As $\mu_{ij} = \nu_{ji}$ holds for all IFPR according to Definition 4, (5), we can consider another form: $\mu_{ij}\mu_{jk}\mu_{ki} = \mu_{ik}\mu_{kj}\mu_{ji}$, for $\forall i, j, k = 1, 2, ..., n$ which only contain membership degree.

Definition 6 [5]. *An intuitionistic fuzzy weight vector $\tilde{\omega} = (\tilde{\omega}_1, \tilde{\omega}_2, ..., \tilde{\omega}_n)$ with $\tilde{\omega}_i = (\tilde{\omega}_i^\mu, \tilde{\omega}_i^\nu)$, $\tilde{\omega}_i^\mu, \tilde{\omega}_i^\nu \in [0,1]$ and $\omega_i^\mu + \omega_i^\nu \leq 1$ is said to be normalized if it satisfies the following conditions*

$$\sum_{j=1, j \neq i}^{n} \omega_j^\mu \leq \omega_i^\nu, \sum_{j=1, j \neq i}^{n} \omega_j^\nu \leq \omega_i^\mu + n - 2, \forall i = 1, 2, ..., n \tag{6}$$

With such a multiplicative consistency preference relationship (5) and the intuitionistic fuzzy weight conditions (6), we can construct a multiplicative consistency IFPR that suppose that

$$\tilde{p}_{ij} = (p_{ij}^\mu, p_{ij}^\nu) = \begin{cases} (0.5, 0.5), if i = j \\ \left(\frac{\omega_i^\mu}{\sqrt{\omega_i^\mu(1-\omega_i^\nu)} + \sqrt{\omega_j^\mu(1-\omega_j^\nu)}}, \frac{\omega_j^\mu}{\sqrt{\omega_i^\mu(1-\omega_i^\nu)} + \sqrt{\omega_j^\mu(1-\omega_j^\nu)}}\right), \\ if i \neq j \end{cases} \tag{7}$$

Then we have the following results.

Theorem 1. *Let $P = (\tilde{p}_{ij})_{n \times n}$ be a preference relation matrix defined by Eq. (7), then P is an IFPR.*

Proof. It is obvious that for $\forall i, j = 1, 2, ..., n$, $p_{ij}^\mu = p_{ji}^\nu$ and $p_{ij}^\nu = p_{ji}^\mu$. Since $0 \leq \omega_i^\mu, \omega_i^\nu \leq 1$ and $\omega_i^\mu + \omega_i^\nu \leq 1$, it follows that

$$0 \leq \frac{\omega_i^\mu}{\sqrt{\omega_i^\mu(1-\omega_i^\nu)} + \sqrt{\omega_j^\mu(1-\omega_j^\nu)}} \leq \frac{\omega_i^\mu}{\omega_i^\mu + \omega_j^\mu} \leq 1, 0 \leq \frac{\omega_j^\mu}{\sqrt{\omega_i^\mu(1-\omega_i^\nu)} + \sqrt{\omega_j^\mu(1-\omega_j^\nu)}} \leq \frac{\omega_j^\mu}{\omega_i^\mu + \omega_j^\mu} \leq 1$$

$$0 \leq \frac{\omega_i^\mu}{\sqrt{\omega_i^\mu(1-\omega_i^\nu)} + \sqrt{\omega_j^\mu(1-\omega_j^\nu)}} + \frac{\omega_j^\mu}{\sqrt{\omega_i^\mu(1-\omega_i^\nu)} + \sqrt{\omega_j^\mu(1-\omega_j^\nu)}} \leq \frac{\omega_i^\mu}{\omega_i^\mu + \omega_j^\mu} + \frac{\omega_j^\mu}{\omega_i^\mu + \omega_j^\mu} \leq 1$$

Due to definition 2.4, P is an IFPR.

Theorem 2. *Let* $P = (\tilde{p}_{ij})_{n\times n}$ *be a IFPR matrix defined by Eq. (7), P is multiplicative consistency.*

Proof. From (7), we obtain

$$p^{\mu}_{ij}p^{\mu}_{jk}p^{\mu}_{ki} = \frac{\omega^{\mu}_{i}}{\sqrt{\omega^{\mu}_{i}(1-\omega^{\nu}_{i})}+\sqrt{\omega^{\mu}_{j}(1-\omega^{\nu}_{j})}} \times \frac{\omega^{\mu}_{j}}{\sqrt{\omega^{\mu}_{j}(1-\omega^{\nu}_{j})}+\sqrt{\omega^{\mu}_{k}(1-\omega^{\nu}_{k})}} \times \frac{\omega^{\mu}_{k}}{\sqrt{\omega^{\mu}_{k}(1-\omega^{\nu}_{k})}+\sqrt{\omega^{\mu}_{i}(1-\omega^{\nu}_{i})}}$$

$$p^{\nu}_{ij}p^{\nu}_{jk}p^{\nu}_{ki} = \frac{\omega^{\mu}_{j}}{\sqrt{\omega^{\mu}_{i}(1-\omega^{\nu}_{i})}+\sqrt{\omega^{\mu}_{j}(1-\omega^{\nu}_{j})}} \times \frac{\omega^{\mu}_{k}}{\sqrt{\omega^{\mu}_{j}(1-\omega^{\nu}_{j})}+\sqrt{\omega^{\mu}_{k}(1-\omega^{\nu}_{k})}} \times \frac{\omega^{\mu}_{i}}{\sqrt{\omega^{\mu}_{k}(1-\omega^{\nu}_{k})}+\sqrt{\omega^{\mu}_{i}(1-\omega^{\nu}_{i})}}$$

By Definition 5, it is obviously established multiplicative consistency.

From Eq. (7), if all $\omega^{\mu}_{i} = 1 - \omega^{\nu}_{i}$, it can verify that IFPR P is equivalent to a fuzzy preference relation.

Based on Theorems 1 and 2, the following corollary can be given.

Corollary 1. *For an IFPR* $\tilde{R} = (\tilde{r}_{ij})_{n\times n}$*, if there exists a normalized intuitionistic fuzzy weight vector* $\tilde{\omega} = (\tilde{\omega}_1, \tilde{\omega}_2, ..., \tilde{\omega}_n)$ *such as*

$$\tilde{r}_{ij} = (\mu_{ij}, \nu_{ij}) = \begin{cases} (0.5, 0.5), if\ i = j \\ (\frac{\omega^{\mu}_{i}}{\sqrt{\omega^{\mu}_{i}(1-\omega^{\nu}_{i})}+\sqrt{\omega^{\mu}_{j}(1-\omega^{\nu}_{j})}}, \frac{\omega^{\mu}_{j}}{\sqrt{\omega^{\mu}_{i}(1-\omega^{\nu}_{i})}+\sqrt{\omega^{\mu}_{j}(1-\omega^{\nu}_{j})}}) \\ if\ i \neq j \end{cases} \tag{8}$$

then $\tilde{R}$ *is multiplicative consistency IFPR.*

3 A Decision Model with IFPR

In this section, we develop a new score function based on ideal Euclidean distance with to measure IFSs. Then, a goal programs is proposed based on the multiplicative consistency for deriving intuitionistic fuzzy weights from IFPR. The final ranking results will emerge from these weights with new score function.

3.1 The Score Function Based on Euclidean Distance

For the applications of IFSs, finding an ideal score function plays a key role for the evaluating and ranking of IFS. So we define an intuitionistic fuzzy score function based on Euclidean distance by using the closest to the ideal point. And the idea point must be $A^* = (1, 0)$ which presents the membership degree $\mu = 1$ and the non-membership degree $\nu = 0$. Let $A_i = (\mu_i, \nu_i)$ be an IFS, according to Definition 3, the score function based on Euclidean distance is shown as following:

$$h(A_i) = \sqrt{2} - d(A_i, A^*) \tag{9}$$

where $d(A_i, A^*) = \sqrt{(1 - \mu_i)^2 + (0 - \nu_i)^2}$.

Example 1. Let $A_1 = (0.8, 0.5)$ and $A_2 = (0.6, 0.3)$ be two IFSs, then using (3) and the proposed score function, we have $s(A_1) = 0.4$, $s(A_2) = 0.4$, $s(A_1) = s(A_2)$, but $h(A_1) = 0.50$, $h(A_2) = 0.52$, $h(A_1) < h(A_2)$. It is obvious the results are different, and the proposed score function can be better.

3.2 Generating Intuitionistic Fuzzy Weights Based on IFPR

Considering a decision problem based on IFPR with actions set $X = \{x_1, x_2, ..., x_n\}$, and DM provides an IFPR $\tilde{R} = (\tilde{r}_{ij})_{n\times n}$ to represent the preference on X. Then, unified intuitionistic fuzzy weight vector $\tilde{\omega} = \{\tilde{\omega}_i, \tilde{\omega}_2, ..., \tilde{\omega}_n\}$ can be indicate as Eq. (8).

Every IFPR built by DM are expected that evaluation information is consistent, but it is difficult to achieve this in actual decision environment. Multiplicative consistency by Eq. (5) is to ensure that the DM's judgment is not random, but a valid and logical basis, and in which case some deviation are allowed. Then, priority weights will be derived by minimizing the absolute deviation from a multiplicative consistent IFPR from Eq. (8). Under this case, the deviation variables are shown as following:

$$\varepsilon_{ij} = \frac{\omega_i^\mu}{\sqrt{\omega_i^\mu(1-\omega_i^\nu)} + \sqrt{\omega_j^\mu(1-\omega_j^\nu)}} - u_{ij}^\mu, (i,j = 1,2,...,n, i \neq j). \tag{10}$$

$$\xi_{ij} = \frac{\omega_j^\mu}{\sqrt{\omega_i^\mu(1-\omega_i^\nu)} + \sqrt{\omega_j^\mu(1-\omega_j^\nu)}} - \nu_{ij}^\mu, (i,j = 1,2,...,n, i \neq j). \tag{11}$$

The smaller the absolute difference, the closer the IFPR will be to multiplicative consistent. As $\mu_{ij} = \nu_{ji}$, $\mu_{ji} = \nu_{ij}$, that has $\varepsilon_{ij} = \frac{\omega_i^\mu}{\sqrt{\omega_i^\mu(1-\omega_i^\nu)}+\sqrt{\omega_j^\mu(1-\omega_j^\nu)}} - u_{ij}^\mu = \frac{\omega_j^\mu}{\sqrt{\omega_i^\mu(1-\omega_i^\nu)}+\sqrt{\omega_j^\mu(1-\omega_j^\nu)}} - \nu_{ji}^\mu = \xi_{ji}$, for all $i, j = 1, 2, ..., n, i \neq j$. This leads to a goal programs model such that as Eqs. (12)–(13).

$$\min Z = \sum_{i=1}^{n-1} \sum_{j=2,j>i}^{n} |\varepsilon_{ij}| + |\xi_{ij}|$$
$$\begin{cases} \frac{\omega_i^\mu}{\sqrt{\omega_i^\mu(1-\omega_i^\nu)}+\sqrt{\omega_j^\mu(1-\omega_j^\nu)}} - u_{ij}^\mu - \varepsilon_{ij} = 0, i,j = 1,2,...,n, i \neq j \\ \frac{\omega_j^\mu}{\sqrt{\omega_i^\mu(1-\omega_i^\nu)}+\sqrt{\omega_j^\mu(1-\omega_j^\nu)}} - \nu_{ij}^\mu - \xi_{ij} = 0, i,j = 1,2,...,n, i \neq j \\ \omega_i^\mu \omega_i^\nu \in [0,1], \omega_i^\mu + \omega_i^\nu \leq 1, i = 1,2,...,n \\ \sum_{j=1,j\neq i}^{n} \omega_j^\mu \leq \omega_i^\nu, \omega_i^\mu + n - 2 \geq \sum_{j=1,j\neq i}^{n} \omega_j^\nu, i = 1,2,...,n \end{cases} \tag{12}$$

$$\min Z = \sum_{i=1}^{n-1} \sum_{j=2,j>i}^{n} \varepsilon_{ij}^{+} + \varepsilon_{ij}^{-} + \xi_{ij}^{+} + \xi_{ij}^{-}$$
$$\begin{cases} \frac{\omega_i^{\mu}}{\sqrt{\omega_i^{\mu}(1-\omega_i^{\nu})}+\sqrt{\omega_j^{\mu}(1-\omega_j^{\nu})}} - u_{ij}^{\mu} - \varepsilon_{ij}^{+} + \varepsilon_{ij}^{-} = 0, i,j = 1,2,...,n, i \neq j \\ \frac{\omega_j^{\mu}}{\sqrt{\omega_i^{\mu}(1-\omega_i^{\nu})}+\sqrt{\omega_j^{\mu}(1-\omega_j^{\nu})}} - \nu_{ij}^{\mu} - \xi_{ij}^{+} + \xi_{ij}^{-} = 0, i,j = 1,2,...,n, i \neq j \\ \omega_i^{\mu}\omega_i^{\nu} \in [0,1], \omega_i^{\mu} + \omega_i^{\nu} \leq 1, i = 1,2,...,n-1, \\ \sum_{j=1,j\neq i}^{n} \omega_j^{\mu} \leq \omega_i^{\nu}, \omega_i^{\mu} + n - 2 \geq \sum_{j=1,j\neq i}^{n} \omega_j^{\nu}, i = 1,2,...,n-1 \\ \varepsilon_{ij}^{+}, \varepsilon_{ij}^{-}, \xi_{ij}^{+}, \xi_{ij}^{-} \geq 0, \varepsilon_{ij}^{+}\varepsilon_{ij}^{-} = 0, \xi_{ij}^{+}\xi_{ij}^{-} = 0, i = 1,2,...,n-1, j = i+1,...,n \end{cases} \tag{13}$$

Similar to the treatment in Ref. [5]. Let $\varepsilon_{ij}^{+} = 0.5(|\varepsilon_{ij}|+\varepsilon_{ij})$, $\varepsilon_{ij}^{-} = 0.5(|\varepsilon_{ij}|-\varepsilon_{ij})$ and $\xi_{ij}^{+} = 0.5(|\xi_{ij}| + \xi_{ij})$, $\xi_{ij}^{-} = 0.5(|\xi_{ij}| - \xi_{ij})$. Then, $|\varepsilon_{ij}| = \varepsilon_{ij}^{+} + \varepsilon_{ij}^{-}$, $\varepsilon_{ij} = \varepsilon_{ij}^{+} - \varepsilon_{ij}^{-}$, $|\xi_{ij}| = \xi_{ij}^{+} + \xi_{ij}^{-}$, $\xi_{ij} = \xi_{ij}^{+} - \xi_{ij}^{-}$. Where $\varepsilon_{ij}^{+}, \varepsilon_{ij}^{-}, \xi_{ij}^{+}, \xi_{ij}^{-} > 0$, and $\varepsilon_{ij}^{+} \times \varepsilon_{ij}^{-} = 0$, $\xi_{ij}^{+} \times \xi_{ij}^{-} = 0$. Therefore, the optimization model can be transformed as Eq. (13).

By the tool like MATLAB and lingo, the optimal result can be computed. Then, the optimal weight vector is denoted as $\tilde{\omega} = \{\tilde{\omega}_1, \tilde{\omega}_2, ..., \tilde{\omega}_n\}$. If $Z^* = 0$, we have $\varepsilon_{ij}^{+} = \varepsilon_{ij}^{-} = \xi_{ij}^{+} = \xi_{ij}^{-} = 0$.

Finally, according to intuitionistic fuzzy weight vector $\tilde{\omega}$ by Eq. (13), the order of alternatives will be proposed by using score function by Eq. (9).

4 Numerical Example and Comparison

In this section, a detailed illustration of the decision-making process is given with the following numerical example. A practical example about needy family selection problem for targeted poverty alleviation is utilized to verify the applicability of the proposed approach. There are three optional family A_1, A_2, A_3. In order to pick out the most appropriate alternative from the proposed family, the government invites expert to give his/her objective evaluation of each needy family.

Table 1. Preference relation matrix

	A_1	A_2	A_2
A_1	(0.50,0.50)	(0.065,0.935)	(0.48,0.52)
A_1	(0.065,0.935)	(0.50,0.50)	(0.915,0.085)
A_1	(0.52,0.48)	(0.085,0.915)	(0.50,0.50)

Step 1. For professional analysis and discussion, establish IFPR for preparatory rescue family in Table 1.

Step 2. By solving goal programs Eq. (13), the optimal intuitionistic fuzzy weight $\tilde{\omega} = \{\tilde{\omega}_1, \tilde{\omega}_2, ..., \tilde{\omega}_n\}$ is shown: $\tilde{\omega}_1 = (0.169, 0.779)$, $\tilde{\omega}_2 = (0.619, 0.328)$, $\tilde{\omega}_3 = (0.159, 0.841)$.

Step 3. Ranking the results. Calculated by Eq. (9), we have $h(A_1) = 0.275$, $h(A_2) = 0.911$, $h(A_3) = 0.225$ which gets the final ranking of $A_2 \succ A_1 \succ A_3$, it means the needy family A_1 ranks first.

Wang [8] used the additive consistency-based method to derive a priority weight vector, Lin [6] proposed a absolute deviation programming model to get priority weights based on multiplicative consistency-based method. There are comparisons between the proposed method and previous methods is shown in Table 2. Due to that the proposed method further considers the deviation value in IFPR and constructs a fractional interactive minimizing deviation programming model based on multiplicative consistency-based method, the ranking order is different with other methods.

Table 2. Comparison for priority weights with different methods

Approach	Optimal priority weights	Ranking order
The method in [8]	(0.03,0.97), (0.90,0.10), (0.07,0.93)	$A_2 \succ A_3 \succ A_1$
The method in [6]	(0.039,0.961), (0.893,0.107), (0.068,0.932)	$A_2 \succ A_3 \succ A_1$
The proposed method	(0.169,0.779), (0.619,0.328), (0.159,0.841)	$A_2 \succ A_1 \succ A_3$

5 Conclusion

In this paper, we propose an effective decision-making method when the preference information is IFPRs. The main idea of the method is to further consider the existent deviation value in IFPR and constructs a fractional interactive minimizing deviation programming model based on multiplicative consistency-based method to minimize the deviation of the given fuzzy comprehensive evaluation criterion and the converted consistent fuzzy comprehensive evaluation criterion, then obtains the priority weight of the alternative based on the multiplicative constraint. This method also takes into account the prioritization based on membership and non-membership of two alternatives which measures the importance of the alternative which compares with another in decision-making. By minimizing the deviation factors, the optimal intuitionistic fuzzy weights are determined. Then, a score function by the closest ideal solution based on Euclidean distance is proposed. The closer the positive ideal solution is, the better the alternative is. Furthermore, this method could be effectively integrated with targeted poverty alleviation to evaluate the comprehensive strength of the targeted poverty alleviation, and provide decision support for government. In the future, this IFPR method can be applied to solve other decision problem in other areas.

Acknowledgment. The authors express their gratitude to the anonymous Reviewers for their valuable and constructive comments. And this work was supported by the Chongqing Social Science Planning Project (No. 2018YBSH085), Graduate Teaching Reform Research Program of Chongqing Municipal Education Commission (No.

YJG183074), Major entrustment projects of the Chongqing Bureau of quality and technology supervision (No. CQZJZD2018001), Chongqing research and innovation project of graduate students (No. CYS18252, No. CYS17227) and the Science and Technology Research Project of Chongqing Municipal Education Commission(Grant number: KJQN201800624).

References

1. Wilandari, Y., Mukid, M.A., Megawati, N., et al.: Analisis Klasifikasi Kabupaten di Jawa Tengah Berdasarkan Populasi Ternak Menggunakan Fuzzy Cluster Means. Media Statistika **7**(2), 77–88 (2014)
2. Diaz, Y., Espinoza, F.A., Markaki, Y., et al.: Targeting grenada's most deprived population: a multidimensional living conditions assessment. OPHI Working Papers (2015)
3. Bashir, Z., Rashir, T., Iqbal, M.: Hesitant probabilistic fuzzy preference relations in decision making. Math. Probl. Eng. **2018**, 1–24 (2018)
4. Zhou, W., Xu, Z.: Probability calculation and element optimization of probabilistic hesitant fuzzy preference relations based on expected consistency. IEEE Trans. Fuzzy Syst. **26**(3), 1367–1378 (2018)
5. Liao, H.C., Xu, Z.S.: Priorities of intuitionistic fuzzy preference relation based on multiplicative consistency. IEEE Trans. Fuzzy Syst. **22**(6), 1669–1681 (2014)
6. Lin, Y., Wang, Y.M.: Group decision making with consistency of intuitionistic fuzzy preference relations under uncertainty. IEEE/CAA J. Autom. Sinica **5**(3), 741–748 (2018)
7. Gong, Z.W., Zhang, N., Chiclana, F.: The optimization ordering model for intuitionistic fuzzy preference relations with utility functions. Knowl.-Based Syst. **162**, 174–184 (2018)
8. Wang, Z.J.: Derivation of intuitionistic fuzzy weights based on intuitionistic fuzzy preference relations. Appl. Math. Model. **37**(9), 6377–6388 (2013)
9. Tong, X., Wang, Z.J.: A group decision framework with intuitionistic preference relations and its application to low carbon supplier selection. Int. J. Environ. Res. Public Health **13**(9), 923 (2016)
10. Atanassov, K.T.: Intuitionistic fuzzy sets. Fuzzy Sets Syst. **20**, 87–96 (1986)
11. Das, S., Kar, M.B., Kar, S.: Group multi-criteria decision making using intuitionistic multi-fuzzy sets. J. Uncertainty Anal. Appl. **1**(1), 10 (2013)

A New Distance Measure Between Atanassov's Intuitionistic Fuzzy Sets Based on Mapping

Fang Su(✉), Jing Li, Dongxuan Xiao, and Chengrui Duan

School of Information and Telecommunication Engineering,
Beijing University of Posts and Telecommunication, No. 10, Xitucheng Road,
Haidian District, Beijing 100876, China
sufang@bupt.edu.cn

Abstract. A new trapezoidal approximation of Atanassov's intuitionistic fuzzy sets (AIFSs) and a new distance measure between AIFSs based on the area relations of membership function graphs of their corresponding trapezoidal fuzzy sets are proposed. By the conversion from Atanassov's intuitionistic fuzzy numbers (AIFNs) to trapezoidal fuzzy numbers, the measurement of distance between AIFSs can be transformed into the calculation of area relations of the corresponding trapezoidal fuzzy sets. Then the new distance measurement can integrate continuous information rather than only taking discrete variables into consideration as before. The main innovation of our distance is that it can integrate comprehensive and continuous information in calculation instead of partial discrete variables, which can efficiently avoid information loss. The contrastive tests demonstrate the efficiency and practicality of the approach.

Keywords: Atanassov's intuitionistic fuzzy sets · Distance measure · Trapezoidal fuzzy sets · Pattern recognition · Classification · Similarity measures

1 Introduction

Today, Atanassov's intuitionistic fuzzy sets (AIFSs) [1] have been widely used in multi-attribute decision making (MADM) problems, pattern recognition and other related theoretical and practical issues. In these fields a common but interesting question has been a hot research topic: How to calculate the distance between intuitionistic fuzzy sets (IFSs). At present, there are three main research directions for the solution to this problem: IFSs distance [2–7], IFSs cross entropy [12, 13], and IFSs similarity [21–28].

In the above measurement methods, some methods only consider some attributes, such as only membership degree and non-membership degree; there are also many methods to obtain the difference by directly performing the difference operation on the discrete attribute values in different IFSs. However, many simulations and theoretical derivations have proved that these methods will lose a certain amount of information, resulting in inaccurate results [9, 15].

Therefore, some researchers now try to calculate the difference of IFSs in another way: Instead of directly performing the difference operation on attributes such as membership degree, they first approximate the IFSs and then calculate the distance. For

Y. Liu et al. (Eds.): ICNC-FSKD 2019, AISC 1074, pp. 712–726, 2020.
https://doi.org/10.1007/978-3-030-32456-8_78

example, IFSs are mapped to interval fuzzy numbers, triangular fuzzy numbers, etc. [36, 37], so that the operation of solving the distance between IFSs can be converted into an operation between the membership functions. The advantage of this type of method is that since the membership function is a region function, the operation for discrete points can be converted into operations on consecutive points, thereby effectively reducing the loss of information.

However, in the above mapping, how to define an appropriate mapping relationship is a key issue. When mapping, it can neither be mapped to a fuzzy number function that is too complicated, which will significantly increase the calculation, nor can it be too simple, because it will also cause the loss of information [3, 5, 9, 37].

In order to better solve the above mapping problem, this paper proposes a new mapping relationship: mapping IFSs to Trapezoidal fuzzy numbers (TrFNs), because the interval only contains points on the number axis, which is not comprehensive enough; A triangle can be thought of as a special trapezoid. Next, this paper proposes a new IFSs distance formula based on the mapping relationship, that is, first convert the Intuitionistic fuzzy numbers (IFNs) in the IFS into trapezoidal fuzzy numbers, thus, each IFS can be converted into a Trapezoidal fuzzy set (TrFS). So, the distance between different IFSs is converted to the area relationship of the corresponding membership function of the TrFSs. This new distance method provides a better mapping. The simulation shows that it can reduce the loss of IFSs information more effectively and have better discrimination for different IFSs.

The rest of this paper is organized as follows. A brief review of some representative methods of the uncertainty of AIFSs from aspects of distance and similarity is presented respectively in Sect. 2. Follow on we propose the new distance measure between AIFSs in Sect. 3. After that, Sect. 4 draws specific comparison between our algorithmic method and other related measures with numerical examples. In Sect. 5 we substantiate the practicality of our distance approach by dealing with pattern recognition problems and medical diagnosis. The paper ends with some conclusive remarks in Sect. 6.

2 Review and Contrastive Analysis on the Current Measures of the Uncertainty of AIFSs

There exist many types of measures comparing the discrepancy between AIFSs. In particular, we focused on distance and similarity approaches, owing to their duality and popularity in MADM process, pattern recognition and other related fields. The following made a brief review and comparison of some representative measures.

2.1 Distance Methods Between AIFSs

In order to simplify the distances' definitions, the following notations are used:

$$\Delta_{\mu}(i) = \mu_A(x_i) - \mu_B(x_i)$$

$$\Delta_v(i) = v_A(x_i) - v_B(x_i)$$

$$\Delta_\pi(i) = \pi_A(x_i) - \pi_B(x_i)$$

Where $\mu_A(x_i) \to [0,1]$ and $v_A(x_i) \to [0,1]$ are membership and non-membership functions, respectively. And $\pi_A(x_i)$ defines the hesitation degree.

The above notations describe the *i-th* ($i \in \{1, 2, \ldots, n\}$) difference between the membership, non-membership and hesitancy degree of two AIFSs A, B in $X = \{x_1, \ldots, x_n\}$.

Partial incipient distance measures for AIFSs were extended from measures for FSs. Such as by incorporating the non-membership function $v(\bullet)$, Atanassov pioneered the distance for AIFSs as presented in the sets Eqs. (1)–(4) [2].

$$d_H^1(A,B) = \frac{1}{2}\sum_{i=1}^{n}\left[|\Delta_\mu(i)| + |\Delta_v(i)|\right] \tag{1}$$

$$d_{nH}^1(A,B) = \frac{1}{2n}\sum_{i=1}^{n}\left[|\Delta_\mu(i)| + |\Delta_v(i)|\right] \tag{2}$$

$$d_E^1(A,B) = \sqrt{\frac{1}{2}\sum_{i=1}^{n}\left[(\Delta_\mu(i))^2 + (\Delta_v(i))^2\right]} \tag{3}$$

$$d_{nE}^1(A,B) = \sqrt{\frac{1}{2n}\sum_{i=1}^{n}\left[(\Delta_\mu(i))^2 + (\Delta_v(i))^2\right]} \tag{4}$$

Nevertheless, Szmidt and Kacprzyk [3, 5, 9] successively justified that the omission of the hesitant index $\pi(\bullet)$ from the above definitions may lead to incorrect results in some instances. And Szmidt and Kacprzyk [3] put forward modified forms of the above distances by adding the hesitancy part as.

$$d_H^2(A,B) = \frac{1}{2}\sum_{i=1}^{n}\left[|\Delta_\mu(i)| + |\Delta_v(i)| + |\Delta_\pi(i)|\right] \tag{5}$$

$$d_{nH}^2(A,B) = \frac{1}{2n}\sum_{i=1}^{n}\left[|\Delta_\mu(i)| + |\Delta_v(i)| + |\Delta_\pi(i)|\right] \tag{6}$$

$$d_E^2(A,B) = \sqrt{\frac{1}{2}\sum_{i=1}^{n}\left[(\Delta_\mu(i))^2 + (\Delta_v(i))^2 + (\Delta_\pi(i))^2\right]} \tag{7}$$

$$d_{nE}^2(A,B) = \sqrt{\frac{1}{2n}\sum_{i=1}^{n}\left[(\Delta_\mu(i))^2 + (\Delta_v(i))^2 + (\Delta_\pi(i))^2\right]} \tag{8}$$

The set of formulas has been widely used. But our simulations proved that these methods cause counter intuition results in some cases as presented in Table 2, just as pointed by Papakostas et al. [17], Zhang and Yu [37].

Besides Grzegorzewski [4] introduced Hausdorff-based distance measures for AIFSs as.

$$d_H^h(A,B) = \sum_{i=1}^{n} \max\{|\Delta_\mu(i)|, |\Delta_v(i)|\} \tag{9}$$

$$d_{nH}^h(A,B) = \frac{1}{n}\sum_{i=1}^{n} \max\{|\Delta_\mu(i)|, |\Delta_v(i)|\} \tag{10}$$

$$d_E^h(A,B) = \sqrt{\sum_{i=1}^{n} \max\left\{(\Delta_\mu(i))^2, (\Delta_v(i))^2\right\}} \tag{11}$$

$$d_{nE}^h(A,B) = \sqrt{\frac{1}{n}\sum_{i=1}^{n} \max\left\{(\Delta_\mu(i))^2, (\Delta_v(i))^2\right\}} \tag{12}$$

Once again, Yang and Chiclana [5] verified that the hesitant index should not be omitted and proposed a generalization of the above methods as:

$$d_H^{eh}(A,B) = \sum_{i=1}^{n} \max\{|\Delta_\mu(i)|, |\Delta_v(i)|, |\Delta_\pi(i)|\} \tag{13}$$

$$d_{nH}^{eh}(A,B) = \frac{1}{n}\sum_{i=1}^{n} \max\{|\Delta_\mu(i)|, |\Delta_v(i)|, |\Delta_\pi(i)|\} \tag{14}$$

$$d_E^{eh}(A,B) = \sqrt{\sum_{i=1}^{n} \max\left\{(\Delta_\mu(i))^2, (\Delta_v(i))^2, (\Delta_\pi(i))^2\right\}} \tag{15}$$

$$d_{nE}^{eh}(A,B) = \sqrt{\frac{1}{n}\sum_{i=1}^{n} \max\left\{(\Delta_\mu(i))^2, (\Delta_v(i))^2, (\Delta_\pi(i))^2\right\}} \tag{16}$$

Later on, Wang and Xin [6] made some improvements on some aforementioned measures as:

$$d_1(A,B) = \frac{1}{n}\sum_{i=1}^{n}\left[\frac{|\Delta_\mu(i)| + |\Delta_v(i)|}{4} + \frac{\max\{|\Delta_\mu(i)|, |\Delta_v(i)|\}}{2}\right] \tag{17}$$

$$d_{1\omega}(A,B) = \sum_{i=1}^{n}\omega_i\left[\frac{|\Delta_\mu(i)| + |\Delta_v(i)|}{4} + \frac{\max\{|\Delta_\mu(i)|, |\Delta_v(i)|\}}{2}\right] \Big/ \sum_{i=1}^{n}\omega_i \tag{18}$$

$$d_2^p(A,B) = \frac{1}{\sqrt[p]{n}} \sqrt[p]{\sum_{i=1}^{n} \left(\frac{|\Delta_\mu(i)| + |\Delta_\nu(i)|}{2} \right)^p} \tag{19}$$

In the above formulas, ω_i $(0 \le \omega_i \le 1)$ corresponds to the weight of $x_i \in X = \{x_1, x_2, \ldots, x_n\}$ element, and p is a positive integer.

But our simulations verified that Eqs. (9)–(17) all cause counter intuition results in some cases as shown in Table 2.

2.2 Similarity Methods Between AIFSs

Chen [21] developed the similarity measures for AIFSs (Since Bustince and Burillo [10] concluded that the AIFSs and the vague sets are similar) defined as.

$$S_C^1(A,B) = \frac{1}{n} \sum_{i=1}^{n} \left(1 - \left| \frac{S_A(x_i) - S_B(x_i)}{2} \right| \right) \tag{20}$$

$$S_C^{1\omega}(A,B) = \sum_{i=1}^{n} \omega_i \left(1 - \left| \frac{S_A(x_i) - S_B(x_i)}{2} \right| \right) \Big/ \sum_{i=1}^{n} \omega_i \tag{21}$$

where $S_k(x_i) = \mu_k(x_i) - \nu_k(x_i)$, $k = \{A, B\}$ $0 \le \omega_i \le 1$.

We have to pointed that the Eqs. (20)–(21) cannot distinguish AIFSs which satisfy.

Like two AIFSs $A(0.1, 0.1)$ and $B(0.5, 0.5)$, we will get based on Eqs. (20)–(21). But we can definitely find that the hesitancy degree is different.

Afterwards Chen [22] made changes in formulas as below.

$$S_C^{\omega}(A,B) = \sum_{i=1}^{n} \omega_i \left(1 - \frac{|a * \Delta_\mu(i) + b * \Delta_\nu(i) - c * (\Delta_\mu(i) + \Delta_\nu(i))|}{a - b} \right) \Big/ \sum_{i=1}^{n} \omega_i \tag{22}$$

where $0 \le \omega_i \le 1$, $i \in \{1, 2, \ldots, n\}$, $a \ge c \ge 0 \ge b$.

However, Hong and Kim [23] indicated that Chen's similarity measures shown unreasonable results sometimes and proposed the following $(0 \le \omega_i \le 1$, $i \in \{1, 2, \ldots, n\}$ $a, b, c \ge 0)$.

$$S_H^1(A,B) = \frac{1}{n} \sum_{i=1}^{n} \left(1 - \frac{|\Delta_\mu(i)| + |\Delta_\nu(i)|}{2} \right) \tag{23}$$

$$S_H^{\omega}(A,B) = \sum_{i=1}^{n} \omega_i \left(1 - \frac{a * |\Delta_\mu(i)| + b * |\Delta_\nu(i)| + c * |\Delta_\mu(i) + \Delta_\nu(i)|}{a + b + c} \right) \tag{24}$$

Park et al. [26] put forward a modified version of similarity measures in Li et al. [24] as $(1 \le p < +\infty$, $0 \le \omega_i \le 1$, $\sum_{i=1}^{n} \omega_i = 1)$.

$$S_g^p(A,B) = 1 - \frac{1}{\sqrt[p]{n}} \sqrt[p]{\sum_{i=1}^{n} \left(\frac{|\Delta_\mu(i)| + |\Delta_\nu(i)| + |\Delta_\pi(i)|}{2} \right)^p} \tag{25}$$

$$S_{g\omega}^p(A,B) = 1 - \sqrt[p]{\sum_{i=1}^{n} \omega_i \left(\frac{|\Delta_\mu(i)| + |\Delta_\nu(i)| + |\Delta_\pi(i)|}{2} \right)^p} \tag{26}$$

Moreover, Mitchell [27] provided a new similarity measure in the following form.

$$S_{\text{mod},p}(A,B) = \frac{1}{2}\left(\rho_\mu(A,B) + \rho_f(A,B)\right) \tag{27}$$

$$\rho_\mu(A,B) = S_d^p(\mu_A(x_i), \mu_B(x_i)), \rho_f(A,B) = S_d^p(1 - \nu_A(x_i), 1 - \nu_B(x_i))$$

Hung and Yang [28] proposed another set of similarity measures based on d_{nH}^h as.

$$S_1(A,B) = 1 - d_{nH}^h(A,B) \tag{28}$$

$$S_e(A,B) = \frac{e^{-d_{nH}^h(A,B)} - e^{-1}}{1 - e^{-1}} \tag{29}$$

$$S_C(A,B) = \frac{1 - d_{nH}^h(A,B)}{1 + d_{nH}^h(A,B)} \tag{30}$$

Our simulations proved that Eqs. (23), (25), (27)–(30) may cause incorrect results as demonstrated in Table 2.

In summary partial distance and similarity measures which are merely based on adjacent discrete points may cause counter intuitive results in some cases. So we proposed a novel distance measure between AIFSs.

3 New Distance Measure

An increasing number of researchers attempt to avoid information loss through making appropriate approximation of them. Atanassov and Gargov [36] transformed the AIFNs into the interval-valued fuzzy numbers, Zhang and Yu [37] replaced the AIFNs with the triangular fuzzy numbers. However, the approximation of the AIFSs should be neither too complex (less regular membership functions will lead to more complicated calculations) nor too simple (simple mapping will not attain a good management of uncertainty, like interval and triangular approximation) [30, 31]. Consequently, we define a novel trapezoidal approximation of the AIFSs as follow to conciliate the aforementioned conflict.

For an arbitrary AIFN $\alpha = (\mu_\alpha, \nu_\alpha, \pi_\alpha)$, where $\mu_\alpha, \nu_\alpha, \pi_\alpha$ defines the membership, non-membership and hesitation degree, respectively, its corresponding trapezoidal fuzzy number is $\tilde{\alpha}$, signed as $\tilde{\alpha} = (a, b, c, d)$, which satisfy.

$$a = \mu_\alpha, \quad b = \mu_\alpha + \frac{\pi_\alpha}{4}, \quad c = \mu_\alpha + \frac{3\pi_\alpha}{4}, \quad d = 1 - \nu_\alpha \tag{31}$$

Where a, b, c, d defines the four vertices of trapezoids as shown in Figs. 1, 2, 3, 4, 5 and 6. Obviously, we can draw the conclusion that $a<b<c<d$ using the property $\mu_\alpha + \nu_\alpha + \pi_\alpha = 1$.

Then we give the definition of the membership function of the trapezoidal fuzzy number in Definition 5.

Definition 5

For an arbitrary trapezoidal fuzzy number (TFN) $\tilde{\alpha}$, its membership function, signed as $\tilde{\mu}_\alpha(t)$, is defined as follow.

$$\tilde{\mu}_\alpha(t) = \begin{cases} 4/\pi \cdot (t - \mu_\alpha), & a \le t \le b \\ 1, & b \le t \le c \\ 4/\pi \cdot (\mu_\alpha + \pi_\alpha - t), & c \le t \le d \\ 0, & t \le a \; or \; t \ge d \end{cases} \tag{32}$$

Firstly, the definition of the distance between AIFSs is introduced as below.

For two AIFNs $\alpha_1 = (\mu_1, \nu_1, \pi_1)$ and $\alpha_2 = (\mu_2, \nu_2, \pi_2)$, whose corresponding trapezoidal fuzzy numbers (TFNs) are $\tilde{\alpha}_1$ and $\tilde{\alpha}_2 (\tilde{\alpha}_1 = (a_1, b_1, c_1, d_1), \tilde{\alpha}_2 = (a_2, b_2, c_2, d_2))$, respectively. Then the distance of AIFNs is equal to the distance of their TFNs as denoted by.

$$d(\alpha_1, \alpha_2) = d(\tilde{\alpha}_1, \tilde{\alpha}_2) = S_C - S_I \tag{33}$$

Where S_C describes the total area of two regions: one region is enclosed by the horizontal axis and the outmost lines of $\tilde{\alpha}_1$ and $\tilde{\alpha}_2$, the other region is the area between $\tilde{\alpha}_1$ and $\tilde{\alpha}_2$, which is always a trapezoid. That is, S_C equals with the size of the trapezoid whose endpoints of both upper and bottom lines are denoted as (a, b, c, d) sorted the sequence in descending order $a = \min(a_1, a_2)$, $b = \min(b_1, b_2)$, $c = \max(c_1, c_2)$, $d = \max(d_1, d_2)$; S_I describe the intersection between $\tilde{\alpha}_1$ and $\tilde{\alpha}_2$.

Then the distance between AIFSs $A, B \subseteq AIFS(X)$ is defined as

$$d_\omega(A, B) = d(\tilde{A}, \tilde{B}) = \sum_{i=1}^{n} \omega_i d(\tilde{a}_i, \tilde{b}_i) \tag{34}$$

$A = \{a_1, a_2, \ldots, a_n\}$ $B = \{b_1, b_2, \ldots, b_n\}$ $a_i = (\mu_A(x_i), \nu_A(x_i))$ $b_i = (\mu_B(x_i), \nu_B(x_i))$, $\omega_i \in [0, 1]$ represents the weight of $x_i \in X$ their corresponding trapezoidal fuzzy number is $\tilde{a}_i, \tilde{b}_i$, $i = 1, 2, \ldots n$.

If each attribute has the same weight, namely $\omega_i = 1/n$. Then Eq. (34) can be simplified as.

$$d(A, B) = d(\tilde{A}, \tilde{B}) = \frac{1}{n} \sum_{i=1}^{n} d(\tilde{a}_i, \tilde{b}_i) \tag{35}$$

Then we will give detailed formulas of $d(\tilde{\alpha}_1, \tilde{\alpha}_2)$ according to different position relations of their membership function.

Without generality we assume $a_1 < a_2$.

Case 1 (Positional relation is disjoint as shown in Fig. 1)

$$d(\tilde{\alpha}_1, \tilde{\alpha}_2) = S_C - S_I = S_{ABCD} - 0 = \frac{c_2 - b_1 + d_2 - a_1}{2}$$

Case 2 (Positional relation is intersection, whose graphic is a triangle, as shown in Fig. 2)

$$\begin{aligned} S_I &= S_{LMN} = \int_{a_2}^{d_1} \min\left(\mu_{\tilde{\alpha}_1}, \mu_{\tilde{\alpha}_2}\right) dt \\ S_C &= S_{ABCD} = \tfrac{c_2 - b_1 + d_2 - a_1}{2} \\ d(\tilde{\alpha}_1, \tilde{\alpha}_2) &= S_C - S_I \end{aligned}$$

Case 3 (Positional relation is a quadrilateral intersection as shown in Fig. 3)

$$\begin{aligned} S_I &= S_{EFGH} = \tfrac{d_1 - a_2 + c_1 - b_2}{2} \\ S_C &= S_{ABCD} = \tfrac{d_2 - a_1 + c_2 - b_1}{2} \\ d(\tilde{\alpha}_1, \tilde{\alpha}_2) &= S_C - S_I \end{aligned}$$

Case 4 (Positional relation is another kind of a quadrangular intersection as shown in Fig. 4)

$$\begin{aligned} S_I &= S_{EFGH} = \tfrac{d_2 - a_2 + c_2 - b_2}{2} \\ S_C &= S_{ABCD} = \tfrac{d_1 - a_1 + c_1 - b_1}{2} \\ d(\tilde{\alpha}_1, \tilde{\alpha}_2) &= S_C - S_I \end{aligned}$$

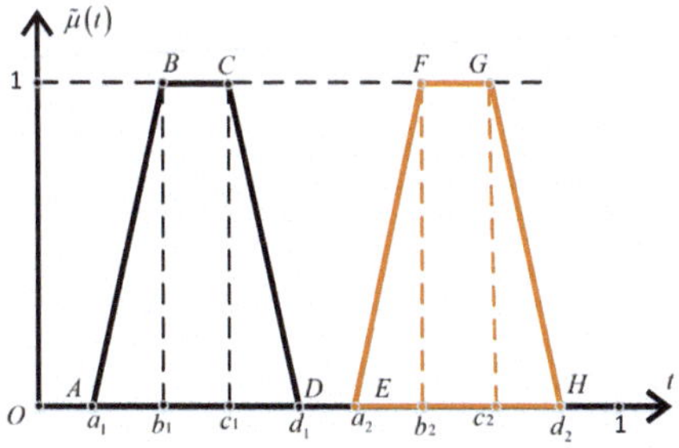

Fig. 1. (Condition: $a_1 < d_1 < a_2$)

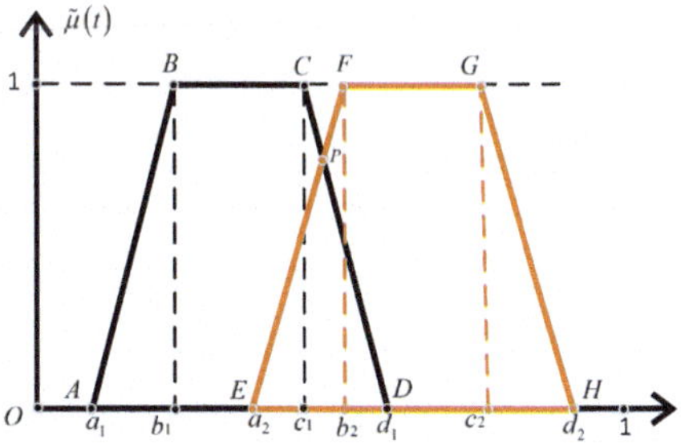

Fig. 2. (Conditions: $a_1 < a_2 < d_1 < d_2$ and $c_1 < b_2$)

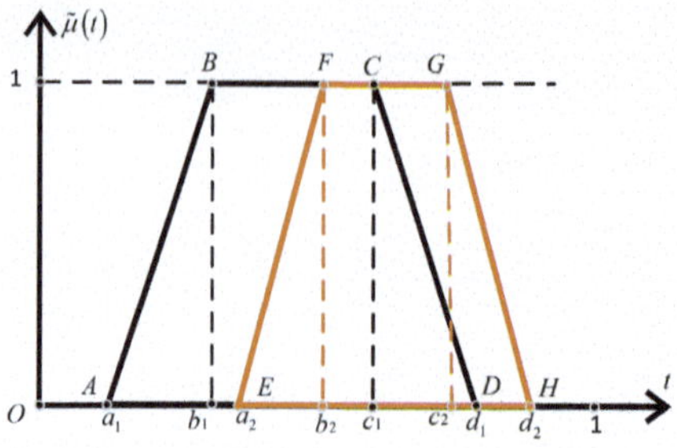

Fig. 3. (Conditions: $a_1 < a_2 < d_1 < d_2$ and $c_1 > b_2$)

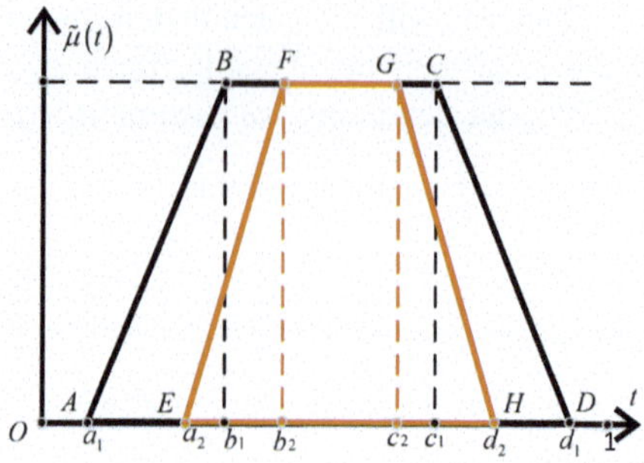

Fig. 4. (Conditions: $a_1 < a_2 < d_2 < d_1$ and $c_1 > c_2$)

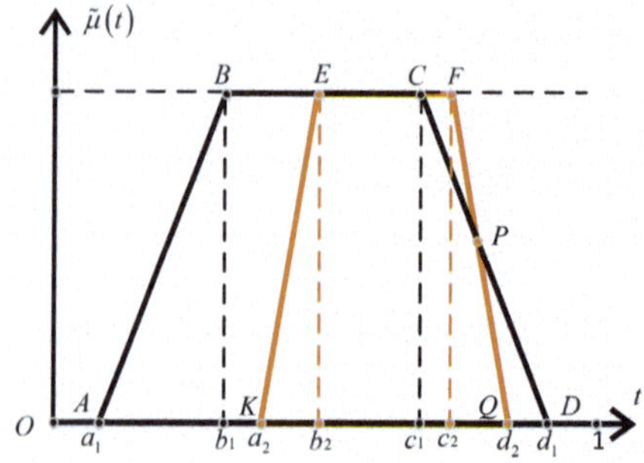

Fig. 5. (Conditions: $a_1 < a_2 < d_2 < d_1$ and $c_1 < c_2$)

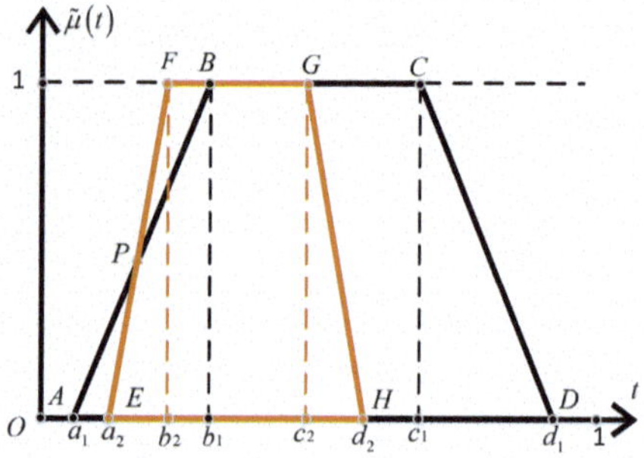

Fig. 6. (Conditions: $a_1 < a_2 < d_2 < d_1$ and $c_1 > c_2$)

Case 5 (Positional relation is intersection, whose graphic is a pentagon in Fig. 5)

$$S_I = S_{KECPQ}$$
$$S_C = S_{ABFDA}$$
$$d(\tilde{\alpha}_1, \tilde{\alpha}_2) = S_C - S_I = S_{ABEK} + S_{CFP} + S_{PQD}$$

Case 6 (Positional relation is intersection, whose graphic is a pentagon in Fig. 6)

$$S_I = S_{EPBGH}$$
$$S_C = S_{APFCDA}$$
$$d(\tilde{\alpha}_1, \tilde{\alpha}_2) = S_C - S_I = S_{GCDH} + S_{PFB} + S_{PAE}$$

Note: all the detailed formulas have been implemented based on MATLAB (version 2013a) software.

4 Comparison and Analysis

In order to study the discrimination ability of the proposed measures, a set of experiments has been arranged. Note that symbol d means results based on our distance measure Eq. (35).

Example1

Those test AIFSs used by Papakostas et al. [17] are depicted in Table 1 and the performance of distances' and similarities' measures for these test AIFSs is presented in Table 2.

Table 1. Test AIFSs

	1	2	3	4	5	6
$A = \{(\mu_A, \nu_A)\}$	{(0.3, 0.3)}	{(0.3, 0.4)}	{(1.0, 0.0)}	{(0.5, 0.5)}	{(0.4, 0.2)}	{(0.4, 0.2)}
$B = \{(\mu_B, \nu_B)\}$	{(0.4, 0.4)}	{(0.4, 0.3)}	{(0.0, 0.0)}	{(0.0, 0.0)}	{(0.5, 0.3)}	{(0.5, 0.2)}

Those counter intuitive cases are highlighted by bold font in Table 2, just as presented in the article of Papakostas [17].

Table 2. Measure for the test AIFSs based on different measures

	1	2	3	4	5	6
d	0.15	0.20	0.875	0.75	0.15	0.10
d_{nH}^1	**0.10**	**0.10**	**0.50**	**0.50**	0.10	**0.50**
d_{nE}^1	**0.10**	**0.10**	0.71	0.50	0.10	**0.70**
d_{nH}^2	0.20	**0.10**	**1.00**	**1.00**	0.20	**0.10**
d_{nE}^2	0.17	**0.10**	1.00	0.87	0.17	**0.10**
d_{nH}^h	**0.10**	**0.10**	1.00	0.50	0.10	**0.10**
d_{nE}^h	**0.10**	**0.10**	1.00	0.50	0.10	**0.10**
d_{nH}^{eh}	0.20	**0.10**	**1.00**	**1.00**	0.20	**0.10**
d_{nE}^{eh}	0.20	**0.10**	**1.00**	**1.00**	0.20	**0.10**
d_1	**0.10**	**0.10**	**0.75**	0.50	0.10	**0.75**
S_C^1	**1.00**	0.90	0.50	**1.00**	**1.00**	0.95
S_H^1	**0.90**	**0.90**	**0.50**	**0.50**	0.90	0.95
$S_g^{p=1}$	0.80	0.90	**0.00**	**0.00**	0.80	**0.90**
$S_g^{p=2}$	0.80	0.90	**0.00**	**0.00**	0.80	**0.90**
$S_{\text{mod},p=1}$	**0.90**	**0.90**	**0.50**	**0.50**	0.90	0.95
$S_{\text{mod},p=2}$	**0.90**	**0.90**	**0.50**	**0.50**	0.90	0.95
S_1	**0.90**	**0.90**	**0.00**	0.50	**0.90**	**0.90**
S_e	**0.85**	**0.85**	**0.00**	0.38	**0.85**	**0.85**
S_C	**0.82**	**0.82**	**0.00**	0.33	**0.82**	**0.82**

For instance, both the distance measures $d^h_{nH} d^h_{nE}$ and the similarity measures S_1, S_e, S_C demonstrate that the fifth and sixth couples of AIFSs have the same difference. However, we and Papakostas [17] both think that the above observations may be counter-intuitive.

In conclusion, we can definitely depict that results based on our method (sighed in Table 2) are more precise and convincing.

5 Applications

Example1: Bacteria Detection

Firstly, we cited the representative example discussed in the work of Khatibi and Montazer [38] and Beliakov et al. [16]. Bacteria are often classified through their shapes. The features used for the bacteria classification are the domical shape, the single microscopic shape, the double microscopic shape, and existence of the flagellum, denoted as x_1–x_4 respectively in Table 3. And the degree of membership and non-membership values for each feature indicating the feature's presence and non-presence in a specific class of bacteria are presented, respectively. Then the experts provided values of the features for different undiagnosed samples as depicted in Table 4.

Table 3. The features' values in the studied bacteria classes (base patterns)

Classes	Domical (x_1) (μ_{x_1}, ν_{x_1})	Single microscopic (x_2) (μ_{x_2}, ν_{x_2})	Double microscopic (x_3) (μ_{x_3}, ν_{x_3})	Flagellum (x_4) (μ_{x_4}, ν_{x_4})
Bacillus coli	(0.85, 0.05)	(0.87, 0.01)	(0.02, 0.97)	(0.92, 0.06)
Shigella	(0.83, 0.08)	(0.92, 0.05)	(0.05, 0.92)	(0.08, 0.91)
Salmonella	(0.79, 0.12)	(0.78, 0.11)	(0.11, 0.85)	(0.87, 0.01)
Klebsiella	(0.82, 0.15)	(0.72, 0.15)	(0.22, 0.75)	(0.12, 0.85)

Table 4. Samples information

	x_1 (μ_{x_1}, ν_{x_1})	x_2 (μ_{x_2}, ν_{x_2})	x_3 (μ_{x_3}, ν_{x_3})	x_4 (μ_{x_4}, ν_{x_4})
$Sample_1$	(0.837, 0.133)	(0.718, 0.159)	(0.064, 0.897)	(0.021, 0.806)
$Sample_2$	(0.911, 0.029)	(0.831, 0.031)	(0.028, 0.894)	(0.952, 0.036)
$Sample_3$	(0.929, 0.037)	(0.812, 0.033)	(0.021, 0.926)	(0.054, 0.922)
$Sample_4$	(0.815, 0.091)	(0.949, 0.048)	(0.020, 0.880)	(0.833, 0.042)
$Sample_5$	(0.864, 0.020)	(0.610, 0.230)	(0.243, 0.624)	(0.0004, 0.964)
$Sample_6$	(0.905, 0.016)	(0.878, 0.015)	(0.072, 0.917)	(0.789, 0.114)

In Table 5 we give the distance between the undiagnosed samples and the four base samples based on our new distance measure (Eq. (35)).

Table 5. Distance measure between samples and base patterns

Distance	Bacillus coli	Shigella	Salmonella	Klebsiella
$Sample_1$	0.3264	0.1053	0.2912	**0.0824**
$Sample_2$	**0.0642**	0.2804	0.1201	0.3500
$Sample_3$	0.2737	**0.0711**	0.3255	0.1610
$Sample_4$	**0.0838**	0.2419	0.0980	0.3258
$Sample_5$	0.4119	0.1999	0.4056	**0.1512**
$Sample_6$	**0.0739**	0.2513	0.1481	0.3272

In Table 5 the minimum distances of each sample are in bold font. The undiagnosed samples should be classified to the base sample which has the minimum distance with it. For example, sample S_1 should be categorized as Klebsiella. The final result is S_1 (Klebsiella), S_2 (Bacillus coli), S_3 (Shigella), S_4 (Bacillus coli), S_5 (Klebsiella), S_6 (Bacillus coli). Our conclusion is in accord with that in the essay of Beliakov et al. [16] and Khatibi and Montazer [38].

Example2: Medical Recognition

Follow on the new distance measure was applied to the colorectal cancer diagnosis, which was a typical medical recognition quoted by Lancu [19], Zhang and Fu [18], etc.

In the example, the patient may fall into any of the following states: metastasis, recurrence, bad (metastasis and recurrence simultaneously) and well. Let be an elements set of a patient and the main five characters (the change of habit and character of stool, bellyache, ictus sileus, chronic sileus, anemia) are used to quantify the element, denoted as x_1, x_2, x_3, x_4, x_5, respectively. Let a colorectal cancer patient whose five characters quantify as B, and A_1, A_2, A_3, A_4 are the character sets of the samples denoted metastasis, recurrence, bad and well. Their values are presented in Table 6.

The distance between B and the other four character sets (in Table 6) based on our distance method (Eq. (35)) are in Table 7.

According to simulations in Table 7, sample B is most similar to A_1, namely, this patient is a metastasis patient. The conclusion is consistent with those obtained in the work of Zhang [18] and Lancu [19].

Table 6. Symptoms characteristic for the patient

	x_1	x_2	x_3	x_4	x_5
	(μ_{x_1}, ν_{x_1})	(μ_{x_2}, ν_{x_2})	(μ_{x_3}, ν_{x_3})	(μ_{x_4}, ν_{x_4})	(μ_{x_5}, ν_{x_5})
B	(0.3, 0.5)	(0.4, 0.4)	(0.6, 0.2)	(0.5, 0.1)	(0.9, 0.0)
A_1	(0.4, 0.4)	(0.3, 0.3)	(0.5, 0.1)	(0.5, 0.2)	(0.6, 0.2)
A_2	(0.2, 0.6)	(0.3, 0.5)	(0.2, 0.3)	(0.7, 0.1)	(0.8, 0.0)
A_3	(0.1, 0.9)	(0.0, 1.0)	(0.2, 0.7)	(0.1, 0.8)	(0.2, 0.8)
A_4	(0.8, 0.2)	(0.9, 0.0)	(1, 0.0)	(0.7, 0.2)	(0.6, 0.4)

Table 7. Distance between B and the other four character sets.

	(A_1, B)	(A_2, B)	(A_3, B)	(A_4, B)
Distance	**0.1925**	0.2368	0.6075	0.4050

6 Conclusions

In this study a new trapezoidal approximation of AIFSs and a novel distance measure between them are proposed. The distance approach transforms the distance between AIFSs into the area relations of the membership function of trapezoidal fuzzy numbers, which integrated more comprehensive and associated information than some current distance, similarity measures. A detailed review and contrastive analysis on some representative distance and similarity methods for AIFSs is presented in the previous sections. And the counter intuitive cases for each measure are extracted by using test AIFSs. Moreover, we demonstrate the performance of our new distance method in pattern recognition application and medical diagnosis. The simulations verified the practical utility of our distance measures.

As future work, we will explore more precise trapezoidal approximation operators as well as the membership function of trapezoidal fuzzy numbers.

References

1. Zadeh, L.A.: Fuzzy sets. Inf. Control **8**(3), 338–353 (1965)
2. Atanassov, K.: Intuitionistic fuzzy sets. Fuzzy Sets Syst. **20**, 87–96 (1986)
3. Szmidt, E., Kacprzyk, J.: Distances between intuitionistic fuzzy sets. Fuzzy Sets Syst. **114**, 505–518 (2000)
4. Grzegorzewski, P.: Distance between intuitionistic fuzzy sets and/or interval-valued fuzzy sets based on the Hausdorff metric. Fuzzy Sets Syst. **148**, 319–328 (2004)
5. Yang, Y.J., Chiclana, F.: Consistency of 2D and 3D distances of intuitionistic fuzzy sets. Expert Syst. Appl. **39**(10), 8665–8670 (2012)
6. Wang, W.Q., Xin, X.L.: Distance measure between intuitionistic fuzzy sets. Pattern Recogn. Lett. **26**(13), 2063–2069 (2005)
7. Song, Y.T., Zhou, X.Z., Zhang, Q.: New properties and measures of distance measure between intuitionistic fuzzy sets. In: 6th International Conference on Fuzzy Systems and Knowledge Discovery, vol. 6, pp. 340–343. IEEE (2009)
8. Hung, W.L., Yang, M.S.: On the J-divergence of intuitionistic fuzzy sets with its application to pattern recognition. Inf. Sci. **178**(6), 1641–1650 (2008)
9. Szmidt, E., Kacprzyk, J.: Distance between intuitionistic fuzzy sets: straightforward approaches may not work in intelligent systems. In: 3rd International IEEE Conference on Intelligent Systems, pp. 716–721. IEEE (2006)
10. Burillo, P., Bustince, H.: Entropy on intuitionistic fuzzy sets and on interval-valued fuzzy sets. Fuzzy Sets Syst. **78**(3), 305–316 (1996)
11. Szmidt, E., Kacprzyk, J.: Entropy for intuitionistic fuzzy sets. Fuzzy Sets Syst. **118**(3), 467–477 (2011)
12. Mao, J.J., Yao, D.B., Wang, C.C.: A novel cross-entropy and entropy measures of IFSs and their applications. Knowl.-Based Syst. **48**, 37–45 (2013)

13. Vlachos, I.K., Sergiadis, G.D.: Intuitionistic fuzzy information-application to pattern recognition. Pattern Recogn. Lett. **28**(2), 197–206 (2007)
14. Ye, J.: Cosine similarity measures for intuitionistic fuzzy sets and their applications. Math. Comput. Model. **53**(1–2), 91–97 (2011)
15. Li, Y.H., Olson, D.L., Qin, Z.: Similarity measures between intuitionistic fuzzy (vague) sets: a comparative analysis. Pattern Recogn. Lett. **28**(2), 278–285 (2007)
16. Beliakov, G., Pagola, M., Wilkin, T.: Vector valued similarity measures for Atanassov's intuitionistic fuzzy sets. Inf. Sci. **280**, 352–367 (2014)
17. Papakostas, G.A., Hatimichailidis, A.G., Kabulasos, V.G.: Distance and similarity measures between intuitionistic fuzzy sets: a comparative analysis from a pattern recognition point of view. Pattern Recogn. Lett. **34**(14), 1609–1622 (2013)
18. Zhang, C.Y., Fu, H.Y.: Similarity measures on three kinds of fuzzy sets. Pattern Recogn. Lett. **27**(12), 1307–1317 (2006)
19. Lancu, I.: Intuitionistic fuzzy similarity measures based on Frank t-norms family. Pattern Recogn. Lett. **42**, 128–136 (2014)
20. Bustince, H., Burillo, P.: Vague sets are intuitionistic fuzzy sets. Fuzzy Sets Syst. **79**(3), 403–405 (1996)
21. Chen, S.M.: Measures of similarity between vague sets. Fuzzy Sets Syst. **74**(2), 217–223 (1995)
22. Chen, S.M.: Similarity measures between vague sets and between elements. IEEE Trans. Syst. **27**(1), 153–158 (1997)
23. Hong, D.H., Kim, C.: A note on similarity measures between vague sets and between elements. Inf. Sci. **115**(1–4), 83–96 (1999)
24. Li, D.F., Cheng, C.T.: New similarity measures of intuitionistic fuzzy sets and application to pattern recognitions. Pattern Recogn. Lett. **23**(1–3), 221–225 (2002)
25. Liang, Z.Z., Shi, P.F.: Similarity measures on intuitionistic fuzzy sets. Pattern Recogn. Lett. **24**(15), 2687–2693 (2003)
26. Park, J.H., Park, J.S., Kwun, Y.C., Lim, K.M.: New similarity measures on intuitionistic fuzzy sets. In: Advances in Soft Computing, pp. 22–30. Springer, Heidelberg (2007)
27. Mitchell, H.B.: On the Dengfeng-Chuntian similarity measure and its application to pattern recognition. Pattern Recogn. Lett. **24**(16), 3101–3104 (2003)
28. Hung, W.L., Yang, M.S.: Similarity measures of intuitionistic fuzzy sets based on Hausdorff distance. Pattern Recogn. Lett. **25**(14), 1603–1611 (2004)
29. Hwang, C.M., Yang, M.S.: Modified cosine similarity measure between intuitionistic fuzzy sets. In: Fourth International Conference, AICI, vol. 7530, pp. 285–293. Springer, Heidelberg (2012)
30. Grzegorzewski, P., Mrowka, E.: Trapezoidal approximations of fuzzy numbers. Fuzzy Sets Syst. **153**(1), 115–135 (2005)
31. Coroianu, L., Gagolewski, M., Grzegorzewski, P.: Nearest piecewise linear approximation of fuzzy numbers. Fuzzy Sets Syst. **233**, 26–51 (2013)
32. Xu, Z.S.: Intuitionistic fuzzy multi-attribute decision making: an interactive method. IEEE Trans. Fuzzy Syst. **20**(3), 514–525 (2012)
33. Park, J.H., Cho, H.J., Kwun, Y.C.: Extension of the VIKOR method to dynamic intuitionistic fuzzy multiple attribute decision making. Comput. Math. Appl. **65**(4), 31–744 (2013)
34. Hung, W.L., Wu, J.W.: Correlation of intuitionistic fuzzy sets by centroid method. Inf. Sci. **144**(1–4), 219–225 (2002)
35. Mitchell, H.B.: A correlation coefficient for intuitionistic fuzzy sets. Int. J. Intell. Syst. **19**(5), 483–490 (2004)

36. Atanassov, K., Gargov, G.: Interval valued intuitionistic fuzzy sets. Fuzzy Sets Syst. **31**(3), 343–349 (1989)
37. Zhang, H.M., Yu, L.Y.: New distance measures between intuitionistic fuzzy sets and interval-valued fuzzy sets. Inf. Sci. **245**, 181–196 (2013)
38. Khatibi, V., Montazer, G.A.: Intuitionistic fuzzy set vs. fuzzy set application in medical pattern recognition. Artif. Intell. Med. **47**(1), 43–52 (2009)
39. Chen, T.Y.: Multiple criteria decision analysis using a likelihood-based outranking method based on interval-valued intuitionistic fuzzy sets. Inf. Sci. **286**, 188–208 (2014)
40. Yu, D.J.: Prioritized information fusion method for triangular intuitionistic fuzzy set and its application to teaching quality evaluation. Int. J. Intell. Syst. **28**(5), 411–435 (2013)

A Constraint Framework for Uncertain Spatiotemporal Data in RDF Graphs

Jinyao Wang, Xiaofeng Di, Jiemin Liu, and Luyi Bai(✉)

School of Computer and Communication Engineering, Northeastern University (Qinhuangdao), Qinhuangdao 066004, China
baily@neuq.edu.cn

Abstract. Researches on spatiotemporal data based on the Resource Description Framework (RDF) have received increasing attention owing to that RDF has a lot of advantages such as extensibility and flexibility. At the same time, the works for uncertain data based on RDF have been extensively studied because of imprecision and uncertainty of data in many real-world applications. Although RDF has been employed to model and handle spatiotemporal data and uncertain data respectively, relatively little work has been carried out to further investigate spatiotemporal data with uncertainty, especially the consistencies of uncertain spatiotemporal data in RDF graphs. In this paper, we first propose an uncertain spatiotemporal data model and define the corresponding constraint framework for the model, then study the types of inconsistencies for uncertain spatiotemporal data in RDF graphs. On this basis, we present the methods for fixing the inconsistencies in the uncertain spatiotemporal RDF graphs caused by updating operations, inserting operations, and deleting operations.

Keywords: Consistency · RDF graph · Uncertain spatiotemporal data

1 Introduction

The Resource Description Framework (RDF), a language proposed by the World Wide Web Consortium (W3C), is a flexible model for representing information about resources on the Web. An RDF statement is a triple presented as (*subject*, *predicate*, *object*), which describes a property value of a subject or the relation between the two entities – the subject and the object. In practice, a huge amount of entities and statements contains spatial and temporal information.

According to the above current status, a number of scholars have began to investigate deeply the spatiotemporal RDF data and been committed to developing a model that works in most spatiotemporal RDF scenarios [1–3]. At the same time, it is common case that the descriptions of objects are accompanied by a degree of uncertainty in real-world applications. Much researches are studied for modeling the uncertainty of data based on RDF [4–6] according to this. Vaneková et al. [4] proposed that each triple is annotated with a degree in [0, 1] while Lv et al. [5] considered that three components of triples were annotated respectively with degrees in [0, 1] to represent the uncertainty of RDF data. In addition, Ma et al. [6] have suggested that only the object component of triples need to be annotated with [0, 1]. Unfortunately, although

Y. Liu et al. (Eds.): ICNC-FSKD 2019, AISC 1074, pp. 727–735, 2020.
https://doi.org/10.1007/978-3-030-32456-8_79

RDF has been employed to model the spatiotemporal data and uncertain data respectively, relatively little research has been carried out to model and handle the data which combines both of them. However, their efforts provide basis of modeling the uncertain spatiotemporal data in RDF graphs for references.

On the other hand, owing to that the inconsistencies existed in the spatiotemporal data, it may have a negative impact on any data analysis and processing. Thus, it is significant to discuss the methods of fixing inconsistencies in the uncertain spatiotemporal RDF graphs. However, almost all the existing work [7–9] on consistencies problems only consider general data, temporal data, spatial data or spatiotemporal data; it falls short in capacity to deal with uncertain spatiotemporal data in RDF graphs. Nevertheless, their efforts play a fundamental role in studying consistencies of spatiotemporal data, and studies on it can be carried out on the basis of their efforts.

In this paper, we develop an uncertain spatiotemporal RDF data model with the constraint. Then we present the types of inconsistency in the uncertain spatiotemporal RDF graphs and methods for fixing inconsistencies of uncertain spatiotemporal data in RDF graphs caused by changing operations. The paper makes the following main contributions:

- We develop an uncertain spatiotemporal data model and then define the constraint framework of uncertain spatiotemporal data in RDF graphs on the basis of the model.
- According to the proposed constraint framework, the possible types of inconsistency in uncertain spatiotemporal RDF graphs are studied
- Furthermore, we present methods for fixing inconsistencies of uncertain spatiotemporal data in RDF graphs caused by these changing operations are proposed.

The rest of the paper is organized as follows. On the basis of uncertain spatiotemporal data model proposed in Sect. 2. Section 3 studies the possible types of consistencies in uncertain spatiotemporal data in RDF graphs. In Sect. 4, we investigate the methods for fixing inconsistencies. And then Sect. 5 concludes the paper.

2 The Uncertain Spatiotemporal Data Model with Constraints

Uncertain spatiotemporal RDF data model is the basis for investigating consistencies of uncertain spatiotemporal data in RDF graphs. In this section, we propose an uncertain spatiotemporal data model and then define the constraint framework on the basis of this so that future consistency discussions can be continued.

Definition 1 (Uncertain spatiotemporal data model). Uncertain spatiotemporal data (denoted as *USP*) is a 6-tuple, $USP = (s, p, o, UL, UT, \mu)$, where

- s, p and o represent subject, predicate, object, respectively.
- *UL and UT* denote location and time interval of the uncertain spatiotemporal data, respectively.
- $\mu \in [0, 1]$ represents the membership degree of the value o for subject s in the location UL to the appointed property p during a time interval UT.

The definition above consists of four parts: (a) the RDF triple of (s, p, o) is an element of $(U \cup B) \times U \times (U \cup L \cup B)$ which U is a set of all Unique Resource Identifiers, B is a set of all blank nodes and L is a set of all literals. It is clear that $U \cap B \cap L_P \cap L_T = \Phi$. (b) the *UL* part denotes the spatial feature (the coordinates) of a 6-tuple which use MBR (minimum bounding rectangle) to represent two-dimensional uncertain spatiotemporal data. (c) the *UT* part whose unit time is set to 1, has the start time T_s, and the end time T_e to denote the valid time interval of a 6-tuple, specially, $T_s = T_e$ if it happens at a time point. (d) $\mu \in [0, 1]$ is determined by a membership function whose formula is $\mu = \mu_{UL} \times \mu_{UT} \times \mu_o$, where μ_{UL} and μ_{UT} are the membership degree of *UL* and *UT* respectively, and μ_o represents the membership degree of the value of the specified property. The three membership degrees need to be obtained through the corresponding membership function.

A set of uncertain spatiotemporal RDF 6-tuples can be represented as a graph. We present a formal definition of uncertain spatiotemporal RDF graph as follows.

Definition 2 (Uncertain spatiotemporal RDF graph). An uncertain spatiotemporal RDF graph is a labeled and directed graph, denoted as $G_F = (N, E, \Sigma_{NE}, \Sigma_{UL}, \Sigma_{UT}, L, \mu)$, where

- N is a finite set of nodes.
- $E \subset N \times N$ is a set of directed edges.
- Σ_{NE} is a set of labels to nodes and edges which contain the temporal and spatial labels in the node besides general labels.
- Σ_{UL} denote the spatial labels of edges where the spatial labels denote the position of the entity (the event).
- Σ_{UT} represent the temporal labels of edges which indicate that the value o of property p at time *UT* is valid.
- L: $N \cup E \rightarrow \Sigma_{NE}$ is a function assigning respectively labels to nodes and edges.
- μ: $E \rightarrow [0, 1]$ is an uncertain subset.

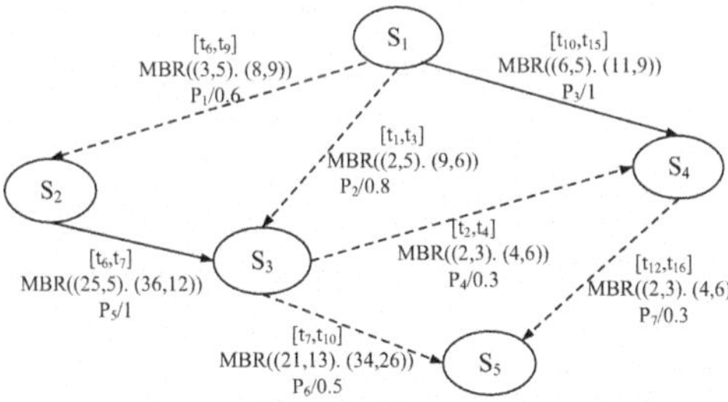

Fig. 1. A sample of an uncertain spatiotemporal RDF graph

In Definition 2, only the temporal and spatial labels on the edges are listed separately (Σ_{UT} and Σ_{UL}), while the temporal and spatial labels in the node are grouped together with general labels (Σ_{NE}). The reason is that in this paper, we only discuss the consistency problem of the temporal and spatial labels on the edges. Figure 1 shows a sample of an uncertain spatiotemporal RDF graph where an edge in the RDF graph connecting the subject vertex and the object vertex with the edge label (the predicate), and *UT* and *UL* are the spatiotemporal information, placed near the edge. In addition, a predicate is followed by a notation "/" and the membership degree μ to represent the uncertainty of a 6-tuple. When $\mu < 1$, the solid line near the edge becomes the dashed line arrow to represent the uncertainty of the 6-tuple in the figure.

We are now ready to define the constraint framework for uncertain spatiotemporal RDF graphs. Firstly, we introduce the notion of lifespan of a node in the RDF graph.

Definition 3 (Lifespan of a node). The lifespan of a node n in the uncertain spatiotemporal RDF graph is denoted as $lifespan(n) = \left[\min(\bigcup_{i=1}^{m} T_i), \max(\bigcup_{i=1}^{m} T_i)\right]$, where T_i denotes uncertain temporal elements of edges incoming and outgoing to the node and m is the number of these edges.

Then we propose a formal definition of the constraint framework for uncertain spatiotemporal RDF graphs in Definition 4.

Definition 4 (The constraints on uncertain spatiotemporal RDF graphs). An uncertain spatiotemporal RDF graph is a directed graph augmented with uncertain spatiotemporal labels. Assume that two nodes n_i and n_j exist in RDF graph, and n_j is the child node of n_i. It satisfies the following conditions:

- There is a directed edge e from n_i to n_j whose lifespan is T. Then $T \subseteq lifespan(n_i) \cap lifespan(n_j)$.
- There is a directed edge e from n_i to n_j whose temporal label, spatial label and the degree of membership are T, L and μ respectively. Then it has the unique and maximum membership degree μ under this condition.
- There is a directed edge e from n_i to n_j whose edge label and temporal label are p and T. Then the edge e has one and only one spatial label L at T when its degree of membership μ is 1.
- There is a directed edge e from n_i to n_j whose temporal label, spatial label and edge label are T, L and p respectively. If its degree of membership μ is 1, there is a one-to-one correspondence between the two nodes of the edge e which has the edge label p.

3 Inconsistency of Uncertain Spatiotemporal RDF Graph

Uncertain spatiotemporal RDF graphs are subject to continuous changing operations, namely update, insert, and delete, which may lead to inconsistencies in uncertain spatiotemporal RDF graphs. Thus, in this section, the types of inconsistency for uncertain spatiotemporal RDF graphs based on the presented constraint framework in Sect. 2, are studied in Definitions 5–7, which are referred to as inconsistency of *type*(*i*), *type*(*ii*) and *type*(*iii*) respectively.

Definition 5 (Inconsistency of *type*(*i*)). Given two nodes n_i and n_j, n_j is the child node of n_i. There is a directed edge from n_i to n_j whose lifespan T $\not\subset$ *lifespan*(n_i) $\cap$ *lifespan*(n_j).

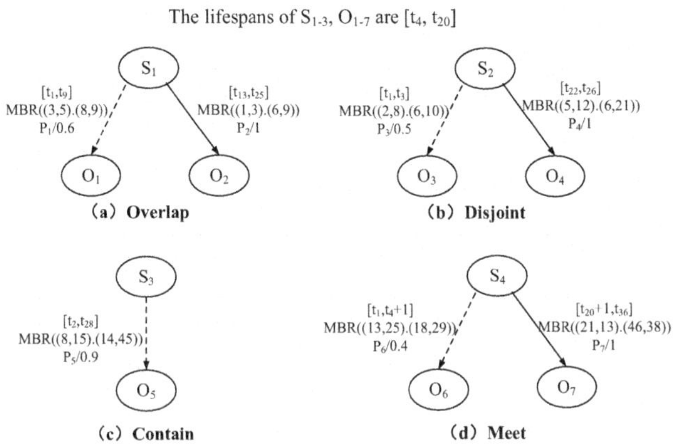

Fig. 2. The inconsistencies of *type*(*i*) in the uncertain spatiotemporal RDF graph

The inconsistency of *type*(*i*) in above definition breaks the first constraint of Definition 5. As shown in Fig. 2, the *type*(*i*) in the RDF graph can be classified into four types: (a) overlap (b) disjoint (c) contain (d) meet. The lifespan of each edge is beyond the lifespan of the corresponding nodes in the figure.

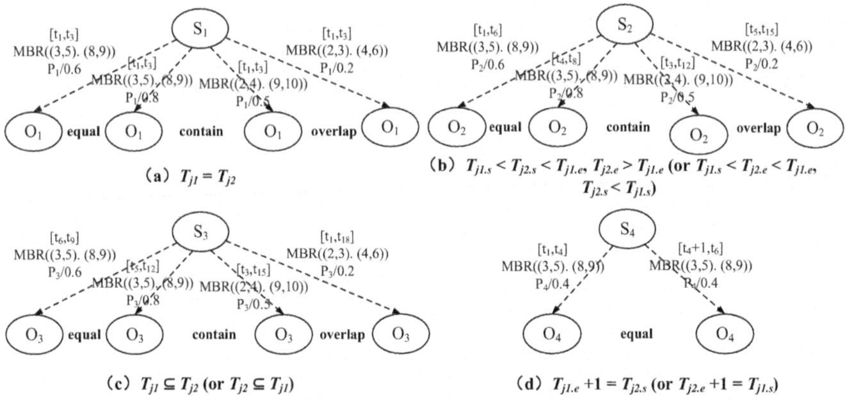

Fig. 3. The inconsistencies of *type*(*ii*) in the uncertain spatiotemporal RDF graph

Definition 6 (Inconsistency of *type*(*ii*)). Given a node n_i, n_{j1} and n_{j2} are the possible operated child nodes of n_i, where μ_1 and μ_2 are the membership degree of n_{j1} and n_{j2}. When the value of n_{j1} is equal to the n_{j2} and the value of (n_i, n_{j1}) is equal to (n_i, n_{j2}),

$\mu_1 \neq \mu_2$ $(0 < \mu_1 < 1,\ \ 0 < \mu_2 < 1)$ if $T_{j1} \cap T_{j2} \neq \Phi$ or $T_{j1.e} + 1 = T_{j2.s}$ (or $T_{j2.e} + 1 = T_{j1.s}$), and $L_{j1} \cap L_{j2} \neq \Phi$.

Similarly, the inconsistency of *type*(*ii*) in Definition 6 breaks the second constraint of Definition 4. Figure 3 shows the four cases of *type*(*ii*) in Definition 6. And then in each case, the spatial topological relations which may cause inconsistencies are labeled. The *type*(*ii*) mainly indicates the possible inconsistencies caused by the insertion of 6-tuples whose *s*, *p* and o are all the same in the model.

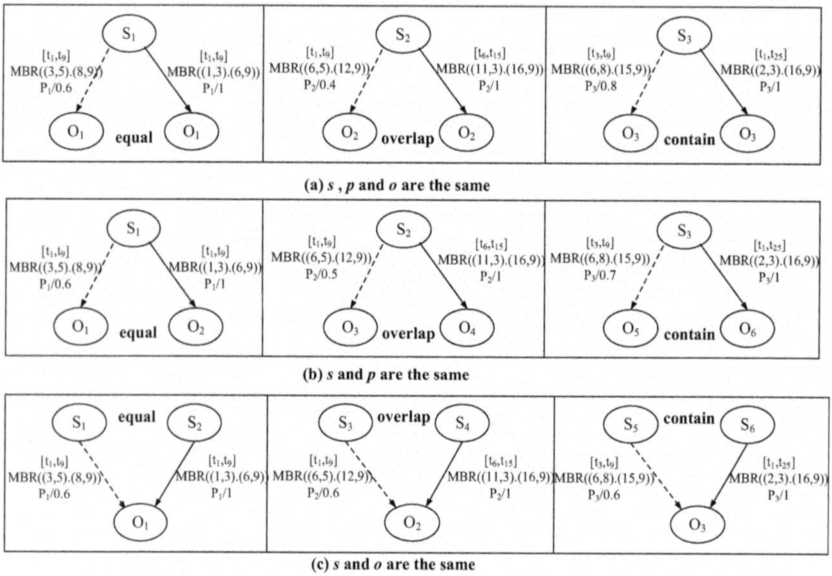

Fig. 4. The inconsistencies of *type*(*iii*) in the uncertain spatiotemporal RDF graph

Definition 7 (Inconsistency of *type*(*iii*)). Given two nodes n_i and n_j, n_j is the child node of n_i, where μ_j is the membership degree, temporal label of n_j and *T*, *L* and *p* are the temporal label, spatial label and the value of (n_i, n_j), respectively. If $\mu_j = 1$, it exists another spatial label L' of (n_i, n_j) which satisfies $L' \cap L \neq \Phi$, or not the a one-to-one correspondence between n_i and n_j under this condition.

In Definition 7, the consistency of *type*(*iii*) violates the last two constraint of Definition 4. In Fig. 4, it shows three cases of *type*(*iii*) where the temporal topological relations which may cause inconsistencies are labeled in each case. Note that in Definition 7, it does not need to consider the effect of spatial topological relationships.

4 Fixing the Inconsistency Caused by Changing Operations

In this section, we investigate the types of inconsistency proposed in Sect. 2 caused by three changing operations, namely update, insert, and delete, based on the presented the constraint framework on uncertain spatiotemporal RDF graphs in Sect. 2.

Then according to this, we present the method for fixing the inconsistencies caused by three changing operations in RDF graphs.

4.1 Insert

When a new RDF 6-tuple is inserted, there are three steps for fixing the inconsistencies if it causes the inconsistencies. The first step is to detects whether the lifespan of inserted 6-tuple is included in the intersected lifespan (if any) of its two nodes. And the second step determines whether the insertion of 6-tuple satisfies any one case for the inconsistency of *type(iii)* in Definition 7. Finally, we need to detect whether (s, p, o) of the inserted 6-tuple exists in the RDF graph and whether its insertion results in inconsistency of *type(ii)* in Definition 6.

Note that it is possible that two nodes that have a parent-child relationship do not both have a lifespan, in which case the temporal label T of the edge to them is specified to be contained in the lifespan of the node that has the lifespan.

Example 1. Assume that a 6-tuple $USP_1 = (s, p, o, L, T, \mu)$ is inserted into an RDF graph and the 6-tuple $USP_2 = (s, p, o, L', T', \mu')$ exists in the RDF graph where $\mu < \mu' = 1$, $T' \subseteq T$ and $L \cap L' \neq \Phi$. According to above fixing steps, modify T of USP_1 as $T \subseteq lifespan(s) \cap lifespan(o)$ first. Then update the T as $[T_s, T'_s]$ and insert the USP_1, and insert a new 6-tuple $USP'_1 = (s, p, o, L, T_1, \mu)$ $(T_1 = [T'_e, T_e])$. Finally, skip the third step because the condition is not met. The other cases for inconsistencies caused by insert operation can be extended in an analogous way according to the Example 1.

4.2 Delete

For deleting an uncertain spatiotemporal RDF 6-tuple, the basis for deletion is that there is data in the graph that can be deleted, so when one or more of s, p and o does not exist, or corresponding location input error, the deletion is not performed.

If the inconsistencies are caused by the delete operation, fix it as follows. Firstly, intersect the lifespan of the 6-tuple in the *delete* command with the lifespan of the 6-tuple to be deleted in the RDF graph, and then delete the intersected lifespan from the lifespan of the 6-tuple to be deleted.

Example 2. Assume that a 6-tuple $USP_1 = (s, p, o, L, T, \mu)$ is deleted from an RDF graph and the 6-tuple in the *delete* command is $USP_2 = (s, p, o, L', T', \mu')$ where $T' \subseteq T$, modify lifespan of USP_1 as $T = [T_s, T'_s - 1]$ and insert a new 6-tuple (s, p, o, L, T_1, μ) which the lifespan T_1 is $[T'_e + 1, T_e]$. Note that the other cases for inconsistencies caused by delete operation can be extended in an analogous way according to the Example 2.

4.3 Update

The update operation can be divided into two parts. Firstly, delete the original data, and then insert the new data. Therefore, the inconsistency analysis involved in the update operation partially coincides with two previous subsections, which will not be explained in detail in this section.

Example 3. Assume that the *update* command is *update* $USP_1 = (s, p, o, L_1, T_1, \mu_1)$ *set* $USP_2 = (s, p, o, L_2, T_2, \mu_2)$ and $USP_3 = (s, p, o, L_3, T_3, \mu_3)$ exists in the RDF graph where $T_1 \subseteq T_3$, $T_{2.s} < T_{3.s}$ and $T_{3.s} < T_{2.e} < T_{1.s} < T_{3.e}$, $L_2 \cap L_3 \neq \Phi$ and $\mu_2 > \mu_3$. The first phase is to *delete* the USP_1: modify lifespan of USP_3 as $T_3 = [T_{3.s}, T_{1.s} - 1]$ and insert a new 6-tuple $USP_4 = (s, p, o, L_3, T_4, \mu_3)$ which the lifespan T_4 is $[T_{1.e} + 1, T_{3.e}]$. The second phase is to *insert* the USP_2: (a) modify T_2 of USP_2 as $T_2 \subseteq$ *lifespan* (*s*) ∩ *lifespan*(*o*) and insert the USP_2 into the RDF graph. (b) skip the second step because the condition is not met. (c) modify T_3 of USP_3 as $T_3 = [T_{2.e}, T_{1.s} - 1]$. The other cases for inconsistencies caused by update operation can be extended in an analogous way according to the Example 3.

5 Conclusion

In this study, we defined an uncertain spatiotemporal data model with the constraint framework. Then, we studied the types of inconsistencies in RDF graph and proposed the repair steps for fixing the consistencies caused by changing operations.

Future work will focus on the algorithms of fixing the inconsistency caused by three primitive changing operations. We also plan to further refine our uncertain spatiotemporal RDF model which can be made to apply in most common cases.

Acknowledgment. This work was supported by the National Natural Science Foundation of China (61402087), the Natural Science Foundation of Hebei Province (F2019501030), the Natural Science Foundation of Liaoning Province (2019-MS-130), and the Fundamental Research Funds for the Central Universities (N172304026).

References

1. Lyell, M., Voyadgis, D., Song, M., et al.: An ontology-based spatiotemporal data model and query language for use in GIS-type applications. In: Proceedings of the 2nd International Conference on Computing for Geospatial Research & Applications, p. 15, DC (2011)
2. Perry, M., Jain, P., Sheth, A.P., et al.: SPARQL-ST: extending SPARQL to support spatiotemporal queries. In: Geospatial Semantics and the Semantic Web, pp. 61–86 (2011)
3. Wang, D., Zou, L., Zhao, D.: g^{st}-store: querying large spatiotemporal RDF graphs. Data Inf. Manag. **1**(2), 84–103 (2017)
4. Vaneková, V., Bella, J., Gurský, P., et al.: Fuzzy RDF in the semantic web: deduction and induction. In: Proceedings of the 6th Workshop on Data Analysis, Abaujszanto, pp. 16–29 (2005)
5. Lv, Y.H., Ma, Z.M., Yan, L.: Fuzzy RDF: a data model to represent fuzzy metadata. In: Proceedings of the 2008 IEEE International Conference on Fuzzy Systems, Hong Kong, China, pp. 1439–1445 (2008)

6. Ma, Z.M., Yan, L.: Modeling fuzzy data with RDF and fuzzy relational database models. Int. J. Intell. Syst. **33**(7), 1534–1554 (2018)
7. Del, M.G., Rodríguez, M.A., Claramunt, C.: Modeling consistency of spatiotemporal graphs. Data Knowl. Eng. **84**, 59–80 (2013)
8. Shuhadah, W.N., Deris, M.M., Noraziah, A.: Database consistency using update-ordering in distributed databases. J. Algorithms Comput. Technol. **1**(1), 17–44 (2007)
9. Sun, J., Xing, Y.J.: An effective image retrieval mechanism using family-based spatial consistency filtration with object region. Int. J. Autom. Comput. **7**(1), 23–30 (2010)

The Probability of Dual Hesitant Fuzzy Event and Its Applications to Bayesian Network Inferences

Xiu Jiu Yuan(✉), Xue Jun Zhao, Jiang Li, and Lin Huang

Department of Basic Sciences, Air Force Engineering University, Xi'an, China
yuanxiujiu@sohu.com

Abstract. The concepts of probabilities and conditional probabilities of dual hesitant fuzzy events have been introduced. The basic properties, the countable additive formula, the law of total probability, the Bayes formula and the continuity of probability of classical probability theory are generalized to that of the probabilities of dual hesitant fuzzy events in frequently used t-norm and s-norm. The correlation coefficient between two dual hesitant fuzzy sets has been defined in random environment. An example of the dual hesitant fuzzy Bayesian network inference is presented finally.

Keywords: Dual hesitant fuzzy set · Dual hesitant fuzzy event · Probability · Bayes formula · Bayesian network

1 Introduction

Hesitant fuzzy sets (HFSs) were originally proposed by Torra (2009, 2010) as an extension of Zadeh's fuzzy sets, permitting membership degree of an element to a set to have a set of possible values. Since HFSs were introduced, many extensions have been given. For example, Chen et al. (2013) extended the notion of hesitant fuzzy set to that of interval-valued hesitant fuzzy set. Qian et al. (2013) introduced generalized hesitant fuzzy sets.

Zhu et al. (2012) introduced the concept of dual hesitant fuzzy sets (DHFSs) which is another generalization of HFSs, by adding the nonmembership hesitancy function to HFSs. Considering such function can provide us with a more exemplary and flexible access to assign values for every element in the domain. DHFSs are more suitable for reflecting people's hesitancy objectively than the other extensions of fuzzy sets, such as, intuitionistic fuzzy sets, fuzzy multisets, type 2 fuzzy sets, etc.

The dual hesitant fuzzy set theory has received much attention from researchers. Farhadinia (2014) has introduced dual interval-valued hesitant fuzzy set (DIHFS) and investigated the correlation measures of DHFS and DIHFS. Ye (2014) introduced correlation coefficients between dual hesitant fuzzy sets and applied it to multiple attribute decision making. Singh (2014, 2017) and Su (2015) presented some similarity measures and distance measures between two dual hesitant fuzzy sets. Yu et al. (2016) proposed a dual hesitant fuzzy group decision-making method and investigated information systems security assessment based on the proposed method. Wang et al.

Y. Liu et al. (Eds.): ICNC-FSKD 2019, AISC 1074, pp. 736–755, 2020.
https://doi.org/10.1007/978-3-030-32456-8_80

(2016) gave some operations on dual hesitant fuzzy based on Archimedean t-conorm and t-norm and researched some dual hesitant fuzzy power aggregation operators. Hao et al. (2017) presented the concept of probabilistic dual hesitant fuzzy set and applied it to the Arctic geopolitical risk evaluation. Zhang et al. (2017) defined dual hesitant fuzzy rough sets by integrating rough set theory with dual hesitant fuzzy set theory.

Most of above studies were performed assumed the elements in reference set are certainty. If the occurrence of elements in the reference set is stochastic, then the dual hesitant fuzzy sets on reference set are the dual hesitant fuzzy events. Fan and Lei (2011) defined the probability of intuitionistic fuzzy events. It is very natural to define the probability of dual hesitant fuzzy event and to develop relative theories, however, there is no research about this issue.

The concept of probabilities of fuzzy events is proposed firstly by Zadeh (1968). After it was introduced, the fuzzy probability theory has attracted more and more scholars' attention. Yager (1979) introduced the concept of fuzzy probabilities of fuzzy events. Klement (1982) modified Yager's definition, which leads the probability to a piecewise continuous fuzzy set. Hesamian and Shams (2017) modified the Yager's definition of fuzzy probability again, to make it to have some desirable properties from mathematical points of view. Chen and Huang (2017) extended the notion of fuzzy probabilities by representing probabilities through the dual hesitant fuzzy numbers instead of intuitionistic fuzzy numbers.

We are mainly focusing on the combination of dual hesitant fuzzy set theory and the probability theory in this paper. The purpose of the present paper is to present the concept of the probabilities of the dual hesitant fuzzy events firstly. Then, we are mainly devoted to studying its various properties.

The rest of this paper is organized as follows. Section 2 reviews the concepts related to dual hesitant fuzzy sets. Section 3 introduces the concept of the probabilities of dual hesitant fuzzy events and its properties. Section 4 researches the condition probabilities of dual hesitant fuzzy events. Section 5 defines the correlation coefficients between two dual hesitant fuzzy sets in stochastic environment. Section 6 presents an example of Bayesian network inference. The last section ends the paper.

2 Preliminaries

2.1 Trigonometric Norms

Definition 2.1. The function $t : [0,1]^2 \rightarrow [0,1]$ is said to be a t-norm, if for any $a, b, c, d \in [0,1]$, it satisfies (1)–(4), that is: (1) it is symmetric: $t(a,b) = t(b,a)$; (2) it is associative: $t(a, t(b,c)) = t(t(a,b),c)$; (3) it is increasing with respect to each component: if $a \leq c$, $b \leq d$, then $t(a,b) \leq t(c,d)$; (4) It satisfies 1-identity: $t(a,1) = a$.

If the (4) is modified to be $t(0,a) = a$, the function t is said to be a anti-triangle model, and denoted as s, s is also called s-norm. If $t(1-a, 1-b) = 1 - s(a,b)$ holds for all $a, b \in [0,1]$, then t-norm and s-norm are said to be dual trigonometric norm. All the t-norm and s-norm used in this paper are dual trigonometric norms.

Denote $t_0(a,b) = a \wedge b$, $s_0(a,b) = a \vee b$; $t_1(a,b) = a \cdot b$, $s_1(a,b) = a + b - a \cdot b$; $t_\infty(a,b) = \max(0, a+b-1)$, $s_\infty(a,b) = \min(1, a+b)$. Literature (Zhang et al. 2005) proved Lemma 2.1.

Lemma 2.1. (1) $t(a,b) \leq t_0(a,b)$; (2) $s(a,b) \geq s_0(a,b)$.

2.2 The Dual Hesitant Fuzzy Sets

Zhu et al. (2012) gave the definitions of dual hesitant fuzzy set as follows.

Definition 2.2. Let X be a nonempty reference set, a dual hesitant fuzzy set D on X is defined as $D = \{<\omega, h(\omega), g(\omega)> | \omega \in X\}$ with the conditions: $0 \leq \gamma^+ + \eta^+ \leq 1$, for all $\omega \in X$, where $h(\omega)$ and $g(\omega)$ are two sets of some values in [0, 1], denoting all possible membership degrees and nonmembership degrees of the element $\omega \in X$ to the set D respectively. $\gamma^+ = \max_{\gamma \in h(\omega)} \{\gamma\}$, $\eta^+ = \max_{\eta \in g(\omega)} \{\eta\}.d(\omega) = (h_D(\omega), g_D(\omega))$ is called a dual hesitant fuzzy element of D.

If $D = \{<\omega, \{1\}, \{0\}> | \omega \in X\}$, D is said to be complete certainty, dented as $D = (\{1\}, \{0\})$, or $D = X$.

If $D = \{<\omega, \{0\}, \{1\}> | \omega \in X\}$, D is said to be complete uncertainty, dented as $D = (\{0\}, \{1\})$ or $D = \Phi$.

If $h(\omega)$ and $g(\omega)$ contain only one element respectively, then the dual hesitant fuzzy set D reduces to an intuitionistic fuzzy set.

Let $l(h_D(\omega))$ be the number of values in $h_D(\omega)$, $l(g_D(\omega))$ be the number of values in $g_D(\omega)$. If $\max_{\omega \in X} l(h_D(\omega)) < +\infty$, $\max_{\omega \in X} l(g_D(\omega)) < +\infty$, D is referred to as a typical dual hesitant fuzzy set on X.

The dual hesitant fuzzy sets mentioned below are typical dual hesitant fuzzy sets. DHFS (X) denotes the set of all dual hesitant fuzzy sets on X.

In this paper, the elements in $h(\omega)$ and $g(\omega)$ are arranged in increasing order, and let $h_i(\omega)$ be the ith largest value in $h(\omega)$, $g_j(\omega)$ be the jth largest value in $g(\omega)$.

For two dual hesitant fuzzy set A and B, in most cases, $l(h_A(\omega)) \neq l(h_B(\omega))$, $l(g_A(\omega)) \neq l(g_B(\omega))$. Denote $l_\omega^1 = \max\{l(h_A(\omega)), l(h_B(\omega))\}$, $l_\omega^2 = \max\{l(g_A(\omega)), l(g_B(\omega))\}$, for each $\omega \in X$. To operate correctly, when $l(h_A(\omega)) < l(h_B(\omega))$, we extend $h_A(\omega)$ by adding the maximum value in it until it has the same number of values in $h_B(\omega)$. When $l(h_A(\omega)) > l(h_B(\omega))$, we extend $h_B(\omega)$ by adding the maximum value in it until it has the same number of values in $h_A(\omega)$. For $g_A(\omega)$ and $g_B(\omega)$, we also manipulate like $h_A(\omega)$ and $h_B(\omega)$. This idea first appeared in literature (Xu and Xa 2011).

2.3 The Operations of Dual Hesitant Fuzzy Sets

Suppose

$$\begin{aligned}
&A = \{<\omega, \{h_1^A(\omega), \cdots, h_{l_\omega^1}^A(\omega)\}, \{g_1^A(\omega), \cdots, g_{l_\omega^2}^A(\omega)\} > |\omega \in X\},\\
&B = \{<\omega, \{h_1^B(\omega), \cdots, h_{l_\omega^1}^B(\omega)\}, \{g_1^B(\omega), \cdots, g_{l_\omega^2}^B(\omega)\} > |\omega \in X\},\\
&h_A(\omega) = \{h_1^A(\omega), \cdots, h_{l_\omega^1}^A(\omega)\}, g_A(\omega) = \{g_1^A(\omega), \cdots, g_{l_\omega^2}^A(\omega)\},\\
&h_B(\omega) = \{h_1^B(\omega), \cdots, h_{l_\omega^1}^B(\omega)\}, g_B(\omega) = \{g_1^B(\omega), \cdots, g_{l_\omega^2}^B(\omega)\},\\
&a(\omega) = (h_A(\omega), g_A(\omega)).
\end{aligned}$$

Zhu et al. (2012) defined the basic operations of dual hesitant fuzzy sets as follows:

(1) $A \cup_Z B = \{<\omega, \{h_i^A(\omega) \cup h_j^B(\omega), i,j = 1,2,\cdots,l_\omega^1\}, \quad \{g_{i'}^A(\omega) \cap g_{j'}^B(\omega), i',j' = 1,2,\cdots,l_\omega^2\} > \}$;

(2) $A \cap_Z B = \{<\omega, \{h_i^A(\omega) \cap h_j^B(\omega), i,j = 1,2,\cdots,l_\omega^1\}, \quad \{g_{i'}^A(\omega) \cup g_{j'}^B(\omega), i',j' = 1,2,\cdots,l_\omega^2\} > \}$.

However, the dimension of the derived DHFS may increase as union or intersection operations are done, which may increase the complexity of the calculations. To overcome this disadvantage, we define some operations on dual hesitant sets A and B as follows: $A \cup_s B = \{<\omega, \{s(h_1^A(\omega), h_1^B(\omega)), \cdots, s(h_{l_\omega^1}^A(\omega), h_{l_\omega^1}^B(\omega))\}, \{t(g_1^A(\omega), g_1^B(\omega)), \cdots, t(g_{l_\omega^2}^A(\omega), g_{l_\omega^2}^B(\omega))\} > |\omega \in X$.

$$A \cap_t B = \{<\omega, \{t(h_1^A(\omega), h_1^B(\omega)), \cdots, t(h_{l_\omega^1}^A(\omega), h_{l_\omega^1}^B(\omega))\},$$

$$\{s(g_1^A(\omega), g_1^B(\omega)), \cdots, s(g_{l_\omega^2}^A(\omega), g_{l_\omega^2}^B(\omega))\} > |\omega \in X\}.$$

$$a^c(\omega) = \begin{cases} (g_A(\omega), h_A(\omega)), & \text{if } h_A(\omega) \neq \Phi, g_A(\omega) \neq \Phi,\\ (\{1 - h_1^A(\omega), \cdots, 1 - h_{l_\omega^1}^A(\omega)\}, \{\Phi\}), & \text{if } h_A(\omega) \neq \Phi, g_A(\omega) = \Phi,\\ (\{\Phi\}, \{1 - g_1^A(\omega), \cdots, 1 - g_{l_\omega^2}^A(\omega)\}), & \text{if } h_A(\omega) = \Phi, g_A(\omega) \neq \Phi. \end{cases}$$

$A^c = \{<\omega, a^c(\omega) > |\omega \in X\}$. By Definition 2.1 (3), $s(h_1^A(\omega), h_1^B(\omega)) \leq s(h_2^A(\omega), h_2^B(\omega)) \leq \ldots \leq s(h_{l_\omega^1}^A(\omega), h_{l_\omega^1}^B(\omega))$, $t(h_1^A(\omega), h_1^B(\omega)) \leq t(h_2^A(\omega), h_2^B(\omega)) \leq \ldots \leq t(h_{l_\omega^2}^A(\omega), h_{l_\omega^2}^B(\omega))$.

Proposition 2.1. $A \cup_s B$, $A \cap_t B$, and A^c are all dual hesitant fuzzy sets.

Proof. Applying Definition 2.1 (3), $\max\limits_{1 \leq k \leq l_1} t(h_k^A(\omega), h_k^B(\omega)) + \max\limits_{1 \leq k' \leq l_2} s(g_{k'}^A(\omega), g_{k'}^B(\omega)) = t(h_{l_1}^A(\omega), h_{l_1}^B(\omega)) + s(g_{l_2}^A(\omega), g_{l_2}^B(\omega)) = t(h_{l_1}^A(\omega), h_{l_1}^B(\omega)) + 1 - t(1 - g_{l_2}^A(\omega), 1 - g_{l_2}^B(\omega))$. Since A and B are two dual hesitant fuzzy sets, so

$h_{l_1}^A(\omega) \leq 1 - g_{l_1}^A(\omega), h_{l_1}^B(\omega) \leq 1 - g_{l_1}^B(\omega)$, thus $1 - t(1 - g_{l_2}^A(\omega), 1 - g_{l_2}^B(\omega)) \leq 1 - t(h_{l_1}^A(\omega), h_{l_1}^B(\omega))$.

Consequently, $\max_{1\le k\le l_1} t(h_k^A(\omega), h_k^B(\omega)) + \max_{1\le k'\le l_2} s(g_{k'}^A(\omega), g_{k'}^B(\omega)) \le t(h_{l_1}^A(\omega), h_{l_1}^B(\omega)) + 1 - t(h_{l_1}^A(\omega), h_{l_1}^B(\omega)) = 1$.

By the definition of t-norm and s-norm, for all $\omega \in X$, $0 \le t(h_k^A(\omega), h_k^B(\omega)) \le 1$, $0 \le s(g_{k'}^A(\omega), g_{k'}^B(\omega)) \le 1$, $k = 1, 2, \cdots, l_2$, $k' = 1, 2, \cdots, l_2$. To sum up, $A \cup_s B$ is a dual hesitant fuzzy set.

In the same way, we can prove that $A \cap_s B$ is a dual hesitant fuzzy set. It is obvious that A^c is a dual hesitant fuzzy set.

When $s(a,b) = s_0(a,b) = a \vee b$, $t(a,b) = t_0(a,b) = a \wedge b$, denote $A \cup_s B = A \cup B$, $A \cap_t B = A \cap B$.

When $s(a,b) = s_1(a,b) = a + b - ab$, $t(a,b) = t_1(a,b) = a \cdot b$, denote $A \cup_s B = A \oplus B$, $A \cap_t B = A \otimes B$.

When $s(a,b) = s_\infty(a,b) = \min\{a+b, 1\}$, $t(a,b) = t_\infty(a,b) = \max\{a+b-1, 0\}$, denote $A \cup_s B = A \hat{+} B$, $A \cap_t B = A \bullet B$.

Definition 2.3. The partial order "$\subseteq$" on DHFS(X) is defined as follows: $A \subseteq B$ if and only if for every $\omega \in X$, $h_A(\omega) \le h_B(\omega)$, $g_A(\omega) \ge g_B(\omega)$, namely $h_1^A(\omega) \le h_1^B(\omega), \cdots, h_{l_\omega^1}^A(\omega) \le h_{l_\omega^1}^B(\omega), g_1^A(\omega) \ge g_1^B(\omega), \cdots, g_{l_\omega^2}^A(\omega) \ge g_{l_\omega^2}^B(\omega)$.

3 The Probability of Dual Hesitant Fuzzy Event

3.1 The Concept of the Probability of Dual Hesitant Fuzzy Event

For a dual hesitant fuzzy set A, notice that the number of elements in $h_A(\omega)$ (or $g_A(\omega)$) may be different for different $\omega \in X$. Hereafter, denote $l_1 = \max_{\omega}\{l(h_A(\omega))\}$, $l_1 = \max_{\omega \in X}\{l(g_A(\omega))\}$. If $l(h_A(\omega)) < l_1$), we extend $h_A(\omega)$ by adding the maximum value in it until its cardinality is l_1. If $l(g_A(\omega)) < l_2$), we extend $g_A(\omega)$ by adding the maximum value in it until its cardinality is l_2.

Definition 3.1. Suppose $X = \{\omega_1, \omega_2, \ldots\}$ is a sample space, a dual hesitant fuzzy A on X is called dual hesitant fuzzy event.

Denoted a dual hesitant fuzzy event A as $A = \{<\omega, h(\omega), g(\omega)> | \omega \in X\}$, where $h(\omega) = \{h_1(\omega), h_2(\omega), \ldots, h_{l_1}(\omega)\}$, $g(\omega) = \{g_1(\omega), g_2(\omega), \ldots, g_{l_2}(\omega)\}$.

The sample spaces mentioned below are countable sample spaces, that is, the random experiment has either a finite or countable infinite number of possible outcomes.

Definition 3.2. Suppose $X = \{\omega_1, \omega_2, \ldots, \omega_n, \ldots\}$ is a sample space, and $P(\omega_i) = p_i$, $i = 1, 2, \ldots, n, \ldots, A$ is a dual hesitant fuzzy event. Let $P(A) = (\{\sum_{i=1}^{\infty} h_1(\omega_i)P(\omega_i), \sum_{i=1}^{\infty} h_2(\omega_i)P(\omega_i), \ldots \sum_{i=1}^{\infty} h_{l_1}(\omega_i)P(\omega_i)\}, \{1 - \sum_{i=1}^{\infty} g_1(\omega_i)P(\omega_i), 1 - \sum_{i=1}^{\infty} g_2(\omega_i)P(\omega_i), \ldots, 1 - \sum_{i=1}^{\infty} g_{l_2}(\omega_i)P(\omega_i)\})$, $P(A)$ is called the probability of the dual hesitant fuzzy event A.

Denote $P_k^h(A) = \sum_{i=1}^{\infty} h_k(\omega_i)P(\omega_i), k = 1, 2, \ldots, l_1$, $P_{k'}^g(A) = \sum_{i=1}^{\infty} g_{k'}(\omega_i)P(\omega_i), k' = 1, 2, \ldots, l_2$. The $P(A)$ can also be represented as:

$P(A) = (\{P_k^h(A), k = 1, 2, \ldots, l_1\}, \{1 - P_{k'}^g(A), k' = 1, 2, \ldots, l_2\})$. $\{P_k^h(A), k = 1, 2, \ldots, l_1\}$ is the possible probabilities of occurrence of the dual hesitant fuzzy event A. $\{P_{k'}^g(A), k' = 1, 2, \ldots, l_2\}$ is the possible probabilities of no occurrence of the dual hesitant fuzzy event A.

3.2 The Basic Properties of the Probability of Dual Hesitant Fuzzy Event

Property 3.1. (1) $0 \le P_k^h(A) \le 1, k = 1, 2, \cdots, l_1$, $0 \le P_{k'}^g(A) \le 1, k' = 1, 2, \cdots, l_2$; (2) $P(X) = (\{1\}, \{1\})$; (3) $P(\{0\}, \{1\}) = (\{0\}, \{0\})$; (4) For all $k = 1, 2, \ldots, l_1$, $k' = 1, 2, \ldots, l_2$, $P_k^h(A) \le 1 - P_{k'}^g(A)$; (5) If for all $\omega \in X$, $h_A(\omega) \ne \Phi, g_A(\omega) \ne \Phi$ then $P(A^c) = (\{P_{k'}^g(A), k' = 1, 2, \ldots, l_2\}, \{1 - P_k^h(A), k = 1, 2, \ldots, l_1\})$.

Proof: We only proof (4). By Definition 2.3, for all $\omega \in X$, $k = 1, 2, \ldots, l_1$, $k' = 1, 2, \ldots, l_2$, $h_k(\omega) \le \max_{1 \le j \le l_1} \{h_i(\omega)\} \le 1 - \max_{1 \le j \le l_2} \{g_i(\omega)\} \le 1 - g_{k'}(\omega)$, so $P_k^h(A) = \sum_{i=1}^{\infty} h_k(\omega_i)P(\omega_i) \le \sum_{i=1}^{\infty} (1 - g_{k'}(\omega_i))P(\omega_i) = 1 - P_{k'}^g(A)$.

Property 3.2. $P(A \cup B) = P(A) + P(B) - p(A \cap B)$.

Proof: Suppose $A = \{<\omega, h_A(\omega), g_A(\omega)> | \omega \in X\}$, $B = \{<\omega, h_B(\omega), g_B(\omega)> | \omega \in X\}$, then

$$
\begin{aligned}
&P(A \cup B) + P(A \cap B) = (\{\sum_{i=1}^{\infty} (h_k^A(\omega_i) \vee h_k^B(\omega_i))P(\omega_i), k = 1, \cdots, l_1\},\\
&\{1 - \sum_{i=1}^{\infty} (g_{k'}^A(\omega_i) \wedge g_{k'}^B(\omega_i))P(\omega_i), k' = 1, 2, \cdots, l_2\})\\
&+ (\{\sum_{i=1}^{\infty} (h_k^A(\omega_i) \wedge h_k^B(\omega_i))P(\omega_i), k = 1, 2, \cdots, l_1\}, \{1 - \sum_{i=1}^{\infty} (g_{k'}^A(\omega_i) \vee g_{k'}^B(\omega_i))P(\omega_i), k' = 1, 2, \cdots, l_2\})\\
&= (\{\sum_{i=1}^{\infty} (h_k^A(\omega_i) \vee h_k^B(\omega_i) + h_k^A(\omega_i) \wedge h_k^B(\omega_i))P(\omega_i), k = 1, 2, \cdots, l_1\},\\
&\{2 - \sum_{i=1}^{\infty} (g_{k'}^A(\omega_i) \wedge g_{k'}^B(\omega_i) + g_{k'}^A(\omega_i) \vee g_{k'}^B(\omega_i))P(\omega_i), k' = 1, 2, \cdots, l_2\})\\
&= (\{\sum_{i=1}^{\infty} (h_k^A(\omega_i) + h_k^B(\omega_i))P(\omega_i), k = 1, 2, \cdots, l_1\}, \{2 - \sum_{i=1}^{\infty} (g_{k'}^A(\omega_i) + g_{k'}^B(\omega_i))P(\omega_i), k' = 1, 2, \cdots, l_2\})\\
&= (\{\sum_{i=1}^{\infty} h_k^A(\omega_i)P(\omega_i), k = 1, 2, \cdots, l_1\}, \{1 - \sum_{i=1}^{\infty} g_{k'}^A(\omega_i)P(\omega_i), k' = 1, 2, \cdots, l_2\})\\
&+ (\{\sum_{i=1}^{\infty} h_k^B(\omega_i)P(\omega_i), k = 1, 2, \cdots, l_1\},\\
&\{1 - \sum_{i=1}^{\infty} g_{k'}^B(\omega_i)P(\omega_i), k' = 1, 2, \cdots, l_2\}) = P(A) + P(B).
\end{aligned}
$$

The following equation can be proved by mathematical induction.

$$P(A_1 \cup A_2 \cup \cdots \cup A_n) = \sum_{i=1}^{n} P(A_i) - \sum_{1 \le i < j \le n} P(A_i \cap A_j) + \sum_{1 \le i < j < k \le n} P(A_i \cap A_j \cap A_k) + \cdots + (-1)^{n+1} P(A_1 \cap A_2 \cap \cdots \cap A_n).$$

Property 3.3. Suppose $A \subseteq B$, then $P(A) \le P(B)$.

Proof: Because $A \subseteq B$, so for all $\omega \in X$, $h_k^A(\omega) \le h_k^B(\omega), k = 1, 2, \cdots, l_1$, $g_{k'}^A(\omega) \ge g_{k'}^B(\omega), k' = 1, 2, \cdots, l_2$, we have $\sum_{i=1}^{\infty} h_k^A(\omega_i)P(\omega_i) \le \sum_{i=1}^{\infty} h_k^B(\omega_i)P(\omega_i), k = 1, 2, \cdots, l_1$, $1 - \sum_{i=1}^{\infty} g_{k'}^A(\omega_i)P(\omega_i) \le 1 - \sum_{i=1}^{\infty} g_{k'}^B(\omega_i)P(\omega_i), k' = 1, 2, \cdots, l_2$, thus $P(A) \le P(B)$.

Property 3.4. Suppose A, B are two dual hesitant fuzzy events, then (1) $P(A \cup_s B) \ge P(A)$, $P(A \cup_s B) \ge P(B)$; (2) $P(A \cup_s B) \le P(A)$, $P(A \cup_s B) \le P(B)$.

Proof: We only show $P(A \cup_s B) \ge P(A)$. $P(A \cup_s B) = (\{\sum_{i=1}^{\infty} s(h_k^A(\omega_i), h_k^B(\omega_i))P(\omega_i), k = 1, 2, \cdots, l_1\}, \{1 - \sum_{i=1}^{\infty} t(g_{k'}^A(\omega_i), g_{k'}^B(\omega_i))P(\omega_i), k' = 1, 2, \cdots, l_2\})$. By Lemma 2.1 $t(x, y) \le x, s(x, y) \ge x$.

$$P(A \cup_s B) \ge (\{\sum_{i=1}^{\infty} (h_k^A(\omega_i)P(\omega_i), k = 1, 2, \cdots, l_1\}, \{1 - \sum_{i=1}^{\infty} t(g_{k'}^A(\omega_i)P(\omega_i), k' = 1, 2, \cdots, l_2\}) = P(A).$$

Property 3.5. If $A_1, A_2, \cdots, A_i, \cdots$ is a dual hesitant fuzzy event sequence with $A_i \cap A_j = (\{0\}, \{1\})$, for $i \ne j$, then $P\left(\bigcup_{i=1}^{\infty} A_i\right) = \sum_{i=1}^{\infty} P(A_i)$.

Proof: Because $A_i \cap A_j = (\{0\}, \{1\})$, for $i \ne j$, hence for arbitrary $\omega \in X$, $1 \le k \le l_1$, $1 \le k' \le l_2$, there are at most one is not 0 in $h_k^{A_1}(\omega), h_k^{A_2}(\omega), \cdots, h_k^{A_1}(\omega), \cdots$, and at most one is not 1 in $g_k^{A_1}(\omega), g_k^{A_2}(\omega), \cdots, g_k^{A_j}(\omega), \cdots$, thus, $\bigvee_{j=1}^{\infty} h_k^{A_j}(\omega) = \sum_{j=1}^{\infty} h_k^{A_j}(\omega)$, $k = 1, 2, \cdots, l_1$, $\bigwedge_{j=1}^{\infty} g_{k'}^{A_j}(\omega) = 1 - \sum_{j=1}^{\infty} \left(1 - g_{k'}^{A_j}(\omega)\right), k' = 1, 2, \cdots, l_2$. Hence

$$\begin{aligned} P\left(\bigcup_{i=1}^{\infty} A_i\right) &= (\{\sum_{i=1}^{\infty} (\bigvee_{j=1}^{\infty} h_k^{A_j}(\omega_i))P(\omega_i), k = 1, 2, \cdots, l_1\}, \{1 - \sum_{i=1}^{\infty} (\bigwedge_{j=1}^{\infty} g_{k'}^{A_j}(\omega_i))P(\omega_i), k' = 1, 2, \cdots, l_2\}) \\ &= (\{\sum_{i=1}^{\infty} (\sum_{j=1}^{\infty} h_k^{A_j}(\omega_i))P(\omega_i), k = 1, 2, \cdots, l_1\}, \{1 - \sum_{i=1}^{\infty} (1 - \sum_{j=1}^{\infty} (1 - g_{k'}^{A_j}(\omega_i))P(\omega_i)), k' = 1, 2, \cdots, l_2\}) \\ &= \sum_{j=1}^{\infty} (\{\sum_{i=1}^{\infty} h_k^{A_j}(\omega_i)P(\omega_i), k = 1, 2, \cdots, l_1\}, \{1 - \sum_{i=1}^{\infty} g_{k'}^{A_j}(\omega_i)P(\omega_i), k' = 1, 2, \cdots, l_2\}) = \sum_{j=1}^{\infty} P(A_j). \end{aligned}$$

Corollary 3.1. Suppose $A_1, A_2, \cdots, A_n$ are dual hesitant fuzzy events and $A_i \cap A_j = (\{0\}, \{1\})$ for $i \neq j$, then $P(A_1 \cup A_2 \cup \cdots \cup A_n) = \sum_{i=1}^{n} P(A_i)$.

Proof: Let $A_{n+1} = A_{n+2} = \cdots = (\{0\}, \{1\})$, then $A_1, A_2, \cdots, A_n, A_{n+1}, \cdots$ is a dual hesitant fuzzy event sequence with $A_i \cap A_j = (\{0\}, \{1\})$ for $i \neq j$, by Property 3.1 (3) $P(A_{n+1}) = P(A_{n+2}) = \cdots = P(\{0\}, \{1\}) = (\{0\}, \{0\})$. By Property 3.5, we have

$$P(A_1 \cup A_2 \cup \cdots \cup A_n) = P(\bigcup_{i=1}^{\infty} A_i) = \sum_{i=1}^{\infty} P(A_i) = \sum_{i=1}^{n} P(A_i).$$

Property 3.6.

$$P(A \oplus B) + P(A \otimes B) = P(A) + P(B).$$

Proof:

$$\begin{aligned}
&P(A \oplus B) + P(A \otimes B) = (\{\sum_{i=1}^{\infty} \left(h_k^A(\omega_i) \oplus h_k^B(\omega_i)\right) P(\omega_i), k = 1, 2, \cdots, l_1\},\\
&\{1 - \sum_{i=1}^{\infty} \left(g_{k'}^A(\omega_i) \otimes h_{k'}^B(\omega_i)\right) P(\omega_i), k' = 1, 2, \cdots, l_2\})\\
&+ (\{\sum_{i=1}^{\infty} \left(h_k^A(\omega_i) \otimes h_k^B(\omega_i)\right) P(\omega_i), k = 1, 2, \cdots, l_1\}, \{1 - \sum_{i=1}^{\infty} \left(g_{k'}^A(\omega_i) \oplus h_{k'}^B(\omega_i)\right)\\
&P(\omega_i), k' = 1, 2, \cdots, l_2\})\\
&= (\{\sum_{i=1}^{\infty} \left(h_k^A(\omega_i) \oplus h_k^B(\omega_i) + h_k^A(\omega_i) \otimes h_k^B(\omega_i)\right) P(\omega_i), k = 1, \cdots, l_1\},\\
&\{2 - \sum_{i=1}^{\infty} \left(g_{k'}^A(\omega_i) \oplus g_{k'}^B(\omega_i) + g_{k'}^A(\omega_i) \otimes g_{k'}^B(\omega_i)\right) P(\omega_i), k' = 1, \cdots, l_2\})\\
&= (\{\sum_{i=1}^{\infty} \left(h_k^A(\omega_i) + h_k^B(\omega_i)\right) P(\omega_i), k = 1, \cdots, l_1\}, \{2 - \sum_{i=1}^{\infty} \left(g_{k'}^A(\omega_i) + g_{k'}^B(\omega_i)\right)\\
&P(\omega_i), k' = 1, 2, \cdots, l_2\}) = P(A) + P(B).
\end{aligned}$$

Property 3.7. Suppose X is a sample space, $A, B \in DHFE(X)$ with $A \bullet B = (\{0\}, \{1\})$, then $P(A \hat{+} B) = P(A) + P(B)$.

Proof: Due to $A \bullet B = (\{0\}, \{1\})$, for every $1 \le k \le l_1$, $1 \le k' \le l_2$, $\omega \in X$, max $\left(h_k^A(\omega) + h_k^B(\omega) - 1, 0\right) = 0$, $\min\left(g_{k'}^A(\omega) + g_{k'}^B(\omega), 1\right) = 1$, thus $0 \le h_k^A(\omega) + h_k^B(\omega) \le 1$, $g_{k'}^A(\omega) + g_{k'}^B(\omega) \ge 1$, therefore $h_k^A(\omega) \hat{+} h_k^B(\omega) = \min\left(h_k^A(\omega) + h_k^B(\omega), 1\right) = h_k^A(\omega) + h_k^B(\omega)$, $g_{k'}^A(\omega) \cdot g_{k'}^B(\omega) = \max\left(g_{k'}^A(\omega) + g_{k'}^B(\omega) - 1, 0\right) = g_{k'}^A(\omega) + g_{k'}^B(\omega) - 1$.

$$P(A \hat{+} B) = (\{\sum_{i=1}^{\infty} (h_k^A(\omega_i) + h_k^B(\omega_i))P(\omega_i), k = 1,2,\cdots,l_1\},$$
$$\{1 - \sum_{i=1}^{\infty} (g_{k'}^A(\omega_i) + g_{k'}^B(\omega_i) - 1)P(\omega_i), k' = 1,2,\cdots,l_2\})$$
$$= (\{\sum_{i=1}^{\infty} h_k^A(\omega_i)P(\omega_i) + \sum_{i=1}^{\infty} h_k^B(\omega_i)P(\omega_i), k = 1,2,\cdots,l_1\},$$
$$\{2 - \sum_{i=1}^{\infty} (g_{k'}^A(\omega_i) + g_{k'}^B(\omega_i))P(\omega_i), k' = 1,2,\cdots,l_2\}) = P(A) + P(B).$$

Denote $\overset{n}{\underset{i=1}{\bullet}} A_i = A_n \bullet (\overset{n-1}{\underset{i=1}{\bullet}} A_i)$, $\overset{n}{\underset{i=1}{\hat{+}}} A_i = A_n \hat{+} (\overset{n-1}{\underset{i=1}{\hat{+}}} A_i)$

Corollary 3.2. If $A_k \bullet (\overset{k-1}{\underset{i=1}{\hat{+}}} A_i) = (\{0\},\{1\}), k = 1,2,\cdots,n,$ then $P(\overset{n}{\underset{i=1}{\hat{+}}} A_i) = \sum_{i=1}^{n} P(A_i)$.

Proof: We give the proof by using mathematical induction. When $n = 2$, by Property 3.7, $P(A \hat{+} B) = P(A) + P(B)$, equality established. Suppose when $n = k$, $P(\overset{k}{\underset{i=1}{\hat{+}}} A_i) = \sum_{i=1}^{k} P(A_i)$, then by theorem's assume conditions, when $n = k + 1$, $P(\overset{k+1}{\underset{i=1}{\hat{+}}}) = P(A_{k+1} \hat{+} (\overset{k}{\underset{i=1}{\hat{+}}} A_i)) = P(A_{k+1}) + P(\overset{k}{\underset{i=1}{\hat{+}}} A_i) = \sum_{i=1}^{k+1} P(A_i)$.

Property 3.8. Suppose X is a sample space, $A, B \in DHFE(X)$, then $P(A \hat{+} B) = P(A) + P(B) - P(A \bullet B)$.

Proof: For any $1 \le k \le l_1, 1 \le k' \le l_2, \omega \in X$, $h_k^A(\omega) \hat{+} h_k^B(\omega) + h_k^A(\omega) \bullet h_k^B(\omega) = \min(h_k^A(\omega) + h_k^B(\omega), 1) + \max(h_k^A(\omega) + h_k^B(\omega) - 1, 0)$. When $h_k^A(\omega) \hat{+} h_k^B(\omega) \ge 1$, $\min(h_k^A(\omega) \hat{+} h_k^B(\omega), 1) = 1$, and $\max(h_k^A(\omega) + h_k^B(\omega) - 1, 0) = h_k^A(\omega) + h_k^B(\omega) - 1$. Thus $h_k^A(\omega) \hat{+} h_k^B(\omega) + h_k^A(\omega) \bullet h_k^B(\omega) = h_k^A(\omega) + h_k^B(\omega)$. When $h_k^A(\omega) \hat{+} h_k^B(\omega) < 1$, $\min(h_k^A(\omega)) \hat{+} h_k^B(\omega), 1) = h_k^A(\omega) + h_k^B(\omega)$, $\max(h_k^A(\omega) \hat{+} h_k^B(\omega) - 1, 0) = 0$, thus $h_k^A(\omega) \hat{+} h_k^B(\omega) + h_k^A(\omega) \bullet h_k^B(\omega) = h_k^A(\omega) + h_k^B(\omega)$. In the same way, we can prove the equality $g_{k'}^A(\omega) \bullet g_{k'}^B(\omega) + g_{k'}^A(\omega) \hat{+} g_{k'}^B(\omega) = g_{k'}^A(\omega) + g_{k'}^B(\omega)$. Therefore

$$\sum_{i=1}^{\infty} (h_k^A(\omega_i) \hat{+} h_k^B(\omega_i))P(\omega_i) + \sum_{i=1}^{\infty} (h_k^A(\omega_i) \bullet h_k^B(\omega_i))P(\omega_i) = \sum_{i=1}^{\infty} h_k^A(\omega_i)P(\omega_i) + \sum_{i=1}^{\infty} h_k^A(\omega_i)P(\omega_i)$$
$$\sum_{i=1}^{\infty} (g_{k'}^A(\omega_i) \bullet h_{k'}^B(\omega_i))P(\omega_i) + \sum_{i=1}^{\infty} (g_{k'}^A(\omega_i) \hat{+} g_{k'}^B(\omega_i))P(\omega_i) = \sum_{i=1}^{\infty} (g_{k'}^A(\omega_i) + g_{k'}^B(\omega_i))P(\omega_i)$$
$= \sum_{i=1}^{\infty} g_{k'}^A(\omega_i)P(\omega_i) + \sum_{i=1}^{\infty} g_{k'}^A(\omega_i)P(\omega_i), k' = 1,2,\cdots,l_2$. Consequently $P(A \hat{+} B) = P(A) + P(B) - P(A \bullet B)$.

3.3 The Continuity of the Probability of Dual Hesitant Fuzzy Event

Lemma 3.1. Suppose $A_1 \subseteq A_2 \subseteq \cdots \subseteq A_n \subseteq \cdots$ (or $A_1 \supseteq A_2 \supseteq \cdots \supseteq A_n \supseteq \cdots$), then $\lim_{n\to\infty} A_n$ exists, and $\lim_{n\to\infty} A_n$ is also a dual hesitant fuzzy set.

Proof: (1) Because $A_1 \subseteq A_2 \subseteq \cdots \subseteq A_n \subseteq \cdots$, so for all $\omega \in X$, and positive integer $1 \le k \le l_1, 1 \le k' \le l_2$, $0 \le h_k^{A_1}(\omega) \le h_k^{A_2}(\omega) \le \cdots \le h_k^{A_n}(\omega) \le \cdots \le 1$, $1 \ge g_{k'}^{A}(\omega) \ge g_{k'}^{A_2}(\omega) \ge \cdots \ge g_{k'}^{A_n}(\omega) \ge \cdots \ge 0$. Namely, $\{h_k^{A}(\omega)\}$ and $\{g_k^{A_i}(\omega)\}$ are the monotone and bounded sequence, therefore $\lim_{n\to\infty} h_k^{A_n}(\omega)$ · and · $\lim_{n\to\infty} g_{k'}^{A_n}(\omega)$ exist.

Due to events $A_n, n = 1, 2, \cdots$ are dual hesitant fuzzy sets, for every $\omega \in X$ $0 \le \max_{1\le k\le 1}\{h_k^{A_n}(\omega)\} + \max_{1\le k'\le l_2}\{g_{k'}^{A_n}(\omega)\} \le 1$, therefore, for arbitrary $1 \le k \le l_1, 1 \le k' \le l_2$, $n = 1, 2, 3, \cdots$, $0 \le h_k^{A_n}(\omega) + g_{k'}^{A_n}(\omega) \le 1$, or $0 \le \lim_{n\to\infty} h_k^{A_n}(\omega) + \lim_{n\to\infty} g_{k'}^{A_n}(\omega) \le 1$ or $0 \le \max_{1\le k\le 4} \lim_{n\to\infty} h_k^{A_n}(\omega) + \max_{1\le k\le l_2} \lim_{n\to\infty} g_{k'}^{A_n}(\omega) \le 1$. Consequently, $\lim_{n\to\infty} A_n$ is a dual hesitant fuzzy set.

Theorem 3.1. Suppose $A_1, A_2, \cdots, A_n, \cdots$ is a dual hesitant fuzzy event sequence, then $\lim_{n\to\infty} P(A_n) = P\left(\lim_{n\to\infty} A_n\right)$ if either of the following conditions holds: (1) $A_1 \subseteq A_2 \subseteq \cdots \subseteq A_n \subseteq \cdots$; (2) $A_1 \supseteq A_2 \supseteq \cdots \supseteq A_n \supseteq \cdots$.

Proof: We will prove part (1), The proof of part (2) is similar to the proof of part (1). By the assumption of the theorem, and Lemma 3.1, for arbitrary $k \in \{1, 2, \cdots, l_1\}, k' \in \{1, 2, \cdots, l_2\}, \omega \in X$, $\lim_{n\to\infty} h_k^{A_n}(\omega)$ and $\lim_{n\to\infty} g_{k'}^{A_k}(\omega)$ exist. Also because of $\sum_{i=1}^{\infty} h_k^{A_n}(\omega)P(\omega_i) \le \sum_{i=1}^{\infty} P(\omega_i) = 1$, $\sum_{i=1}^{\infty} g_k^{A_n}(\omega)P(\omega_i) \le \sum_{i=1}^{\infty} P(\omega_i) = 1$, therefore for any $k \in \{1, 2, \cdots, l_1\}, k' \in \{1, 2, \cdots, l_2\}$, the series $\sum_{i=1}^{\infty} h_k^{A_k}(\omega)P(\omega_i)$ and the series $\sum_{i=1}^{\infty} g_{k'}^{A_n}(\omega)P(\omega_i)$ are uniformly convergent. Consequently

$$
\begin{aligned}
\lim_{n\to\infty} P(A_n) &= \left(\left\{\lim_{n\to\infty}\sum_{i=1}^{\infty} h_1^{A_n}(\omega_i)P(\omega_i), \lim_{n\to\infty}\sum_{i=1}^{\infty} h_2^{A_n}(\omega_i)P(\omega_i), \cdots, \lim_{n\to\infty}\sum_{i=1}^{\infty} h_h^{A_n}(\omega_i)P(\omega_i)\right\},\right. \\
&\left.\left\{1 - \lim_{n\to\infty}\sum_{i=1}^{\infty} g_1^{A_n}(\omega_i)P(\omega_i), 1 - \lim_{n\to\infty}\sum_{i=1}^{\infty} g_2^{A_n}(\omega_i)P(\omega_i), \cdots, 1 - \lim_{n\to\infty}\sum_{i=1}^{\infty} g_{l_2}^{A_n}(\omega_i)P(\omega_i)\right\}\right) \\
&= \left(\left\{\sum_{i=1}^{\infty}\lim_{n\to\infty} h_1^{A_n}(\omega_i), \sum_{i=1}^{\infty}\lim_{n\to\infty} h_2^{A_n}(\omega_i)P(\omega_i), \cdots, \sum_{i=1}^{\infty}\lim_{n\to\infty} h_i^{A_n}(\omega_i)P(\omega_i)\right\},\right. \\
&\left.\left\{1 - \sum_{i=1}^{\infty}\lim_{n\to\infty} g_1^{A_n}(\omega_i)P(\omega_i), 1 - \sum_{i=1}^{\infty}\lim_{n\to\infty} g_2^{A_n}(\omega_i)P(\omega_i), \cdots, 1 - \sum_{i=1}^{\infty}\lim_{n\to\infty} g_{l_2}^{A_n}(\omega_i)P(\omega_i)\right\}\right) = P\left(\lim_{n\to\infty} A_n\right)
\end{aligned}
$$

Lemma 3.2. Suppose t, s are t-norm and s-norm respectively. Denote $\overset{n}{\underset{i=1}{t}}(x_i) = t\left(x_n, \overset{n-1}{\underset{i=1}{t}}(x_i)\right)$, $\overset{n}{\underset{i=1}{s}}(x_i) = S\left(x_n, \overset{n}{\underset{i=1}{s}}(x_i)\right)$, then (1) $\overset{n}{\underset{i=1}{t}}(x_i) \leq \overset{n-1}{\underset{i=1}{t}}(x_i)$; (2) $\overset{n}{\underset{i=1}{s}}(x_i) \geq \overset{n-1}{\underset{i=1}{s}}(x_i)$.

By the Lemma 2.1, the proof is obvious.

Since t-norm and s-norm are commutative and associative. So the value of $\overset{n}{\underset{i=1}{t}}(x_i)$ is independent of the order of $x_1, x_2, \cdots, x_n$.

Proposition 3.1. (1) $\lim_{n\to\infty} \overset{n}{\underset{i=1}{s}}(A_i)$ exists. (2) $\lim_{n\to\infty} \overset{n}{\underset{i=1}{t}}(A_i)$ exists.

Proof: We only prove part (1).

$$\overset{n}{\underset{i=1}{s}}(A_i) = s\left(A_n, \overset{n-1}{\underset{i=1}{s}}(A_i)\right) = \left\{<\omega, \left\{s\left(h_1^{A_n}(\omega), h_1^{\overset{n-1}{\underset{i=1}{s}}(A_i)}(\omega)\right), \cdots, s\left(h_{l_1}^{A_n}(\omega), h_{l_1}^{\overset{n-1}{\underset{i=1}{s}}(A_i)}(\omega)\right)\right\},\right.$$
$$\left.\left\{t\left(g_1^{A_n}(\omega), g_1^{\overset{n-1}{\underset{i=1}{t}}(A_i)}(\omega)\right)\cdots, t\left(g_{l_2}^{A_n}(\omega), g_{l_2}^{\overset{n-1}{\underset{i=1}{t}}(A_i)}(\omega)\right)\right\} > \right\},$$

by Lemma 3.2, $s\left(h_k^{A_n}(\omega), h_k^{\overset{n-1}{\underset{i=1}{s}}(A_i)}(\omega)\right) \geq h_k^{\overset{n-1}{\underset{i=1}{s}}(A_i)}(\omega), k = 1, 2, \cdots, l_1, t(g_{k'}^{A_n}(\omega),$ $g_{k'}^{\overset{n-1}{\underset{i=1}{s}}(A_i)}(\omega)) \leq g_{k'}^{\overset{n-1}{\underset{i=1}{s}}(A_i)}(\omega), k' = 1, 2, \cdots, l_2$. Also because $\overset{n-1}{\underset{i=1}{s}}(A_i) = \left(<\omega, \left\{h_k^{\overset{n-1}{\underset{i=1}{s}}(A_i)}(\omega_i), k = 1, \cdots, l_1\right\}, \left\{g_{k'}^{\overset{n-1}{\underset{i=1}{s}}(A_i)}(\omega_i), k' = 1, \cdots, l_2\right\} > \right)$, thus $\overset{n}{\underset{i=1}{s}}(A_i) \supseteq \overset{n-1}{\underset{i=1}{s}}(A_i), n = 1, 2, \cdots$. By Lemma 3.1, $\lim_{n\to\infty} \overset{n}{\underset{i=1}{s}}(A_i)$ exists.

Denote $\lim_{n\to\infty} \overset{n}{\underset{i=1}{s}}(A_i) = \overset{\infty}{\underset{i=1}{s}}(A_i)$, $\lim_{n\to\infty} \overset{n}{\underset{i=1}{t}}(A_i) = \overset{\infty}{\underset{i=1}{t}}(A_i)$.

Theorem 3.2. Suppose $A_1, A_2, \cdots, A_n, \cdots$ is a dual hesitant fuzzy event sequence, then (1) $\lim_{n\to\infty} P\left(\overset{n}{\underset{i=1}{s}}(A_i)\right) = P\left(\overset{\infty}{\underset{i=1}{s}}(A_i)\right)$; (2) $\lim_{n\to\infty} P\left(\overset{n}{\underset{i=1}{t}}(A_i)\right) = P\left(\overset{\infty}{\underset{i=1}{t}}(A_i)\right)$.

Proof: By the proof process of Proposition 3.1, $\overset{n}{\underset{i=1}{s}}(A_i) \supseteq \sum_{i=1}^{n-1}(A_i), n = 1, 2, \cdots$. By Theorem 3.1 (1), $\lim_{n\to\infty} P\left(\overset{n}{\underset{i=1}{s}}(A_i)\right) = P\left(\overset{\infty}{\underset{i=1}{s}}(A_i)\right)$. It can be proved in the same way that $\lim_{n\to\infty} P\left(\overset{n}{\underset{i=1}{t}}(A_i)\right) = P\left(\overset{\infty}{\underset{i=1}{t}}(A_i)\right)$.

Corollary 3.3. Let $A_1, A_2, \ldots, A_n, \ldots$ be a dual hesitant fuzzy event sequence, then (1) $\lim\limits_{n\to\infty} P(\bigcup\limits_{i=1}^{n} A_i) = P(\bigcup\limits_{i=1}^{\infty} A_i)$, $\lim\limits_{n\to\infty} P(\bigcap\limits_{i=1}^{n} A_i) = P(\bigcap\limits_{i=1}^{\infty} A_i)$; (2) $\lim\limits_{n\to\infty} P(\bigoplus\limits_{i=1}^{n} A_i = P(\bigoplus\limits_{i=1}^{\infty} A_i)$, $\lim\limits_{n\to\infty} P(\bigotimes\limits_{i=1}^{n} A_i) = P(\bigotimes\limits_{i=1}^{\infty} A_i)$; (3) $\lim\limits_{n\to\infty} P(\hat{+}_{i=1}^{n} A_i) = P(\hat{+}_{i=1}^{\infty} A_i)$, $\lim\limits_{n\to\infty} P(\overset{n}{\underset{i=1}{\bullet}} A_i) = P(\overset{\infty}{\underset{i=1}{\bullet}} A_i)$.

Definition 3.3. $\sum\limits_{n=1}^{\infty} P(A_n) < +\infty$ if and only for arbitrary $1 \le k \le l_1$, $1 \le k' \le l_2$, $\sum\limits_{n=1}^{\infty}\sum\limits_{i=1}^{\infty} h_k^{A_n}(\omega_i)P(\omega_i) < +\infty$, $\sum\limits_{n=1}^{\infty} (1 - \sum\limits_{i=1}^{\infty} g_{k'}^{A_n}(\omega_i)P(\omega_i)) < +\infty$.

Lemma 3.3 (Boole's inequality). If $\{A_n\}$ is a dual hesitant fuzzy event sequence, then $P(\bigcup\limits_{n=1}^{\infty} A_n) \le \sum\limits_{n=1}^{\infty} P(A_n)$.

Proof: $\sum\limits_{i=1}^{\infty} (\bigvee\limits_{n=1}^{\infty} h_k^{A_n}(\omega_i))P(\omega_i) \le \sum\limits_{i=1}^{\infty} (\sum\limits_{n=1}^{\infty} h_k^{A_n}(\omega_i))P(\omega_i) = \sum\limits_{n=1}^{\infty} (\sum\limits_{i=1}^{\infty} h_k^{A_n}(\omega_i))\ \ P(\omega_i), k = 1, 2, \cdots, l_1$.

$$1 - \sum_{i=1}^{\infty} \bigwedge_{n=1}^{\infty} g_{k'}^{A_n}(\omega_i)P(\omega_i) = \sum_{i=1}^{\infty} (1 - \bigwedge_{n=1}^{\infty} g_{k'}^{A_n}(\omega_i))P(\omega_i) = \sum_{i=1}^{\infty} \bigvee_{n=1}^{\infty} (1 - g_{k'}^{A_n}(\omega_i))$$

$$P(\omega_i) \le \sum_{i=1}^{\infty}\sum_{n=1}^{\infty} (1 - g_{k'}^{A_n}(\omega_i))P(\omega_i) = \sum_{n=1}^{\infty} (1 - \sum_{i=1}^{\infty} g_{k'}^{A_n}(\omega_i)P(\omega_i)) \qquad 1 \le k' \le l_2,$$

therefore $P(\bigcup\limits_{n=1}^{\infty} A_n) \le \sum\limits_{n=1}^{\infty} P(A_n)$.

Theorem 3.3. If $\sum\limits_{n=1}^{\infty} P(A_n) < +\infty$, then $P(\bigcap\limits_{n=1}^{\infty} (\bigcup\limits_{k=n}^{\infty} A_k)) = (\{0\}, \{0\})$.

Proof: For convenience, set $E_n = \bigcup\limits_{k=n}^{\infty} A_k$. We have $E_1 \supseteq E_2 \supseteq \cdots$ and $\bigcap\limits_{n=1}^{\infty} E_n = \bigcap\limits_{n=1}^{\infty} (\bigcup\limits_{k=n}^{\infty} A_k)$. Applying Corollary 3.3, and Lemma 3.3, we obtain that

$$P(\bigcap_{n=1}^{\infty} (\bigcup_{k=n}^{\infty} A_k)) = P(\bigcap_{n=1}^{\infty} E_n) = \lim_{n\to\infty} P(\bigcap_{k=1}^{n} E_k) = \lim_{n\to\infty} P(E_n) = \lim_{n\to\infty} P(\bigcup_{k=n}^{\infty} A_k)$$
$$\le \lim_{m\to\infty} \sum_{n=m}^{\infty} P(A_n) = \lim_{m\to\infty} (\{\sum_{n=m}^{\infty} (\sum_{i=1}^{\infty} h_k^{A_n}(\omega_i)P(\omega_i), k = 1, 2, \cdots, l_1\},$$
$$\{\sum_{n=m}^{\infty} (1 - \sum_{i=1}^{\infty} g_{k'}^{A_n}(\omega_i)P(\omega_i)), k' = 1, 2, \cdots, l_2\} = \lim_{m\to\infty} \sum_{n=m}^{\infty} P(A_m) = (\{0\}, \{0\}).$$

Where the last equation holds because $\sum\limits_{n=1}^{\infty} P(A_n) < +\infty$.

Theorem 3.4. Suppose X is a sample space, if P is a nonnegative set function with $P(X) = 1$. Zadeh's operators are used in the intersection and union operations of dual hesitant fuzzy sets. Then the set function P is countable additive if and only if (1) The P is the finite additive. (2) The P is lower continuous.

Proof: The necessary conditions of the theorem have been proved. Next the proof of the sufficiency of the theorem is given.

Suppose $A_1, A_2, \ldots, A_n, \ldots$ is a dual hesitant fuzzy event sequence with $A_i \bigcap A_j = (\{0\}, \{1\})$ for $i \neq j$. Because P is the finitely additive, $P(A_1 \bigcup A_2 \bigcup \cdots \bigcup A_n) = \sum_{i=1}^{n} P(A_i)$. Also, $A_1 \subseteq A_1 \bigcup A_2 \subseteq \cdots \subseteq A_1 \bigcup A_2 \bigcup \cdots \bigcup A_n \subseteq \cdots$, thus by the lower continuous of the probability P, $P(\bigcup_{i=1}^{\infty} A_i) = P(\lim_{n\to\infty} \bigcup_{i=1}^{n} A_i) = \lim_{n\to\infty} P(\bigcup_{i=1}^{n} A_i) = \lim_{n\to\infty} \sum_{i=1}^{n} P(A_i) = \sum_{i=1}^{\infty} P(A_i)$.

Theorem 3.5. If $A_k \bullet (\hat{+}_{i=1}^{k-1} A_i) = (\{0\}, \{1\})$, $k = 1, 2, \cdots, n$, then $P(\hat{+}_{i=1}^{\infty} A_i) = \sum_{i=1}^{\infty} P(A_i)$.

Proof: Since $A_k \bullet (\hat{+}_{i=1}^{k-1} A_i) = (\{0\}, \{1\})$, $k = 1, 2, \cdots, n$, thus by Corollary 3.2, $P(\hat{+}_{i=1}^{n} A_i) = \sum_{i=1}^{n} P(A_i)$. Applying Corollary 3.3, $P(\hat{+}_{i=1}^{\infty} A_i) = \lim_{n\to\infty} P(\hat{+}_{i=1}^{n} A_i) = \lim_{n\to\infty} \sum_{i=1}^{n} P(A_i) = \sum_{i=1}^{\infty} P(A_i)$.

4 The Conditional Probability of DHFEs

4.1 The Concept of Conditional Probability of DHFEs

Definition 4.1. Let X is a sample space, A, B are two dual hesitant fuzzy events.

(1) If $P(B) > (\{0\}, \{0\})$, then the conditional probability of dual hesitant fuzzy event A given that dual hesitant fuzzy event B has occurred is defined by

$$P(A|B) = P(t(A,B))/P(B) \text{ or } P(A|B) = (\{\frac{\sum_{i=1}^{\infty} t(h_k^A(\omega_i), h_k^B(\omega_i))P(\omega_i)}{\sum_{i=1}^{\infty} h_k^B(\omega_i)P(\omega_i)}, k = 1, 2, \cdots, l_1\}, \{\frac{1 - \sum_{i=1}^{\infty} s(g_{k'}^A(\omega_i), g_{k'}^B(\omega_i))P(\omega_i)}{1 - \sum_{i=1}^{\infty} g_{k'}^B(\omega_i)P(\omega_i)}, k' = 1, 2, \cdots, l_2\})$$

(2) If $P(t(A,B)) = P(A) \cdot P(B)$, dual hesitant fuzzy events A and B are said to be independent.

Denote $P_k^h(A|B) = \frac{\sum_{i=1}^{\infty} t\left(h_k^A(\omega_i), h_k^B(\omega_i)\right) P(\omega_i)}{\sum_{i=1}^{\infty} h_k^B(\omega_i) P(\omega_i)}, k = 1, 2, \cdots, l_1,$ $P_{k'}^g(A|B) = \frac{1 - \sum_{i=1}^{\infty} s(g_{k'}^A(\omega_i), g_{k'}^B(\omega_i)) P(\omega_i)}{1 - \sum_{i=1}^{\infty} g_{k'}^B(\omega_i) P(\omega_i)}, k' = 1, 2, \cdots, l_2..$

4.2 The Properties of Conditional Probability of DHFEs

Property 4.1. Suppose $P(B) > (\{0\}, \{0\})$, then $0 \leq P_k^h(A|B) \leq 1, k = 1, 2, \cdots, l_1$; (2) $0 \leq P_{k'}^g(A|B) \leq 1, k' = 1, 2, \cdots, l_2$; (3) $P((\{0\}, \{1\}|B) = (\{0\}, \{0\})$; (4) $P((\{1\}, \{0\}|B) = (\{1\}, \{1\})$; (5) If $B \subseteq A, P(A|B) = (\{1\}, \{1\})$.

Most properties of the probability of dual hesitant fuzzy events can be extended to conditional probability of dual hesitant fuzzy event. Property 4.2 lists some such properties.

Property 4.2. Suppose $P(B) > (\{0\}, \{0\})$, then (1) $P(A_1 \bigcup A_2|B) = P(A_1|B) + P(A_2|B) - P(A_1 \bigcap A_2|B)$; (2) If $A_1 \subseteq A_2$, $P(A_1|B) \leq P(A_2|B)$; (3) Suppose $A_1, A_2, \ldots, A_n$ are dual hesitant fuzzy events and $A_i \bigcap A_j = (\{0\}, \{1\})$ for $i \neq j$, then $P(A_1 \bigcup A_2 \bigcup \cdots \bigcup A_n|B) = \sum_{i=1}^{n} P(A_i|B)$.

Proposition 4.1. If $P(B) > (\{0\}, \{0\})$, dual hesitant fuzzy events A and B are independent if and only if $P(A|B) = P(A)$.

Property 4.3. If $A = (\{0\}, \{1\})$ or $A = (\{1\}, \{0\})$, then A is independent with any dual hesitant fuzzy event B.

By the definition of conditional probability (with $P(B) > (\{0\}, \{0\})$), $P(T(A,B)) = P(A|B)P(B)$.

This formula, the multiplication formula for probabilities of DHFEs, can be generalized (by induction) as follows:

Theorem 4.1. Let X be a sample space, $A_1, A_2, \ldots, A_n \in DHFE(X)$, and $P\left(\overset{n}{\underset{t=1}{t}}(A_i)\right) > 0$, then

$$P\left(\overset{n}{\underset{i=1}{t}}(A_i)\right) = P(A_1)P(A_2|A_1)\cdots P\left(A_n|\overset{n-1}{\underset{i=1}{t}}(A_i)\right)$$

Theorem 4.2 (Law of total probability). Let X be a sample space, $A_1, A_2, \ldots, A_n \in DHFE(X)$, and $A_i \bigcap A_j = (\{0\}, \{1\})$, for $i \neq j$, $P(A_j) > (\{0\}, \{1\})$,

$j = 1, 2, \cdots, n$, then for arbitrary dual hesitant fuzzy event $A \subseteq \bigcup_{i=1}^{n} A_i$, $P(A) = \sum_{i=1}^{n} P(A_i)P(A|A_i)$.

Proof: Since $A_i \bigcap A_j = (\{0\}, \{1\})$ for $i \neq j$, so $(A \bigcap A_i) \bigcap (A \bigcap A_j) = (\{0\}, \{1\})$ for $i \neq j$. Because $A \subseteq \bigcup_{i=1}^{n} A_i$, thus $P(A) = P\left(A \bigcap \left(\bigcup_{i=1}^{n} A_i\right)\right) = P\left(\bigcup_{i=1}^{n} (A \bigcap A_i)\right)$, by Corollary 3.1, $P(A) = \sum_{i=1}^{n} P(A \bigcap A_i) = \sum_{i=1}^{n} P(A_i)P(A|A_i)$.

Theorem 4.3 (Bayes formula). Let X be a sample space, $A_1, A_2, \ldots, A_n \in DHFE(X)$, and $A_i \bigcap A_j = (\{0\}, \{1\})$, for $i \neq j$, $P(A_j) > (\{0\}, \{1\}), j = 1, 2, \cdots, n$, $P(A) > (\{0\}, \{1\}), A \subseteq \bigcup_{i=1}^{n} A_i$, then $P(A_i|A) = \frac{P(A|A_i)P(A_i)}{\sum_{j=1}^{n} P(A|A_j)P(A_j)}$.

Proof: By the theorem's assumptions, $P(A) = \sum_{i=1}^{n} P(A_i)P(A|A_i)$. And also

$$P(A_i|A) = P\left(A_i \bigcap A\right)/P(A) = P(A_i)P(A|A_i)/P(A) \text{ so } P(A_i|A) = \frac{P(A|A_i)P(A_i)}{\sum_{j=1}^{n} P(A|A_j)P(A_j)}.$$

4.3 The Comparison Laws of Probabilities of DHFEs

We give the following comparison laws of probabilities of dual hesitant fuzzy events.

Definition 4.2. Suppose the probabilities of dual hesitant fuzzy event A, B are

$$P(A) = \left(\left\{P_k^h(A), k = 1, 2, \cdots, l_1^A\right\}, \left\{1 - P_{k'}^g(A), k' = 1, 2, \cdots, l_2^A\right\}\right).$$
$$P(B) = \left(\left\{P_k^h(B), k = 1, 2, \cdots, l_1^B\right\}, \left\{1 - P_{k'}^g(B), k' = 1, 2, \cdots, l_2^B\right\}\right).$$

The score function of $P(A)$ and $P(B)$ are defined as $H(A) = \frac{1}{2l_1^A} \sum_{k=1}^{l_1^A} P_k^h(A) + \frac{1}{2l_2^A} \sum_{k'=1}^{l_2^A} \left(1 - P_{k'}^g(A)\right)$, $H(B) = \frac{1}{2l_1^B} \sum_{k=1}^{l_1^B} P_k^h(B) + \frac{1}{2l_2^B} \sum_{k'=1}^{l_2^B} \left(1 - P_{k'}^g(B)\right)$. The accuracy functions of $P(A)$ and $P(B)$ are defined as

$$S(A) = \frac{1}{2l_1^A} \sum_{k=1}^{l_1^A} \left(P_k^h(A) - \frac{1}{l_1^A} \sum_{k=1}^{l_1^A} P_k^h(A)\right)^2 + \frac{1}{2l_2^A} \sum_{k'=1}^{l_2^A} \left(P_{k'}^g(A) - \frac{1}{l_2^A} \sum_{k'=1}^{l_2^A} P_{k'}^g(A)\right)^2,$$
$$S(B) = \frac{1}{2l_1^B} \sum_{k=1}^{l_1^B} \left(P_k^h(B) - \frac{1}{l_1^B} \sum_{k=1}^{l_1^B} P_k^h(B)\right)^2 + \frac{1}{2l_2^B} \sum_{k'=1}^{l_2^B} \left(P_{k'}^g(B) - \frac{1}{l_2^B} \sum_{k'=1}^{l_2^B} P_{k'}^g(B)\right)^2.$$

The orders of $P(A)$ and $P(B)$ are defined as:

(1) If $H(A) > H(B)$, then $P(A) > P(B)$; (2) If $H(A) < H(B)$, then $P(A) < P(B)$;

(3) If $H(A) = H(B)$, $S(A) < S(B)$, then $P(A) > P(B)$; If $H(A) = H(B)$, $S(A) > S(B)$, then $P(A) < P(B)$;

If $H(A) = H(B)$, $S(A) = S(B)$, then $P(A) = P(B)$.

The comparison laws of conditional probabilities of dual hesitant fuzzy events can be defined in the same way.

5 Correlation Coefficients Between DHFEs

The correlation and correlation coefficient of dual hesitant fuzzy sets were given by Ye (2014), but them were defined not in random circumstances.

In this paper, we define correlation and correlation coefficients of dual hesitant fuzzy sets in random environments.

Definition 5.1. Let $X = \{\omega_1, \omega_2, \cdots, \omega_i, \cdots\}$ is a sample space, suppose $A = \{<\omega, \{h_k^A(\omega), k = 1, 2, \cdots, l_1\}, \{g_{k'}^A(\omega), k' = 1, 2, \cdots, l_2\} > |\omega \in X\}$ then, the informational energy of A is defined as

$$E(A) = \frac{1}{l_1}\sum_{k=1}^{l_1}\sum_{i=1}^{\infty} h_k^2(\omega_i)P(\omega_i) + \frac{1}{l_2}\sum_{k'=1}^{l_2}\sum_{i=1}^{\infty} g_{k'}^2(\omega_i)P(\omega_i)$$

Definition 5.2. Let $X = \{\omega_1, \omega_2, \cdots, \omega_i, \cdots\}$ is a sample space, suppose $A = \{<\omega, \{h_k^A(\omega), k = 1, 2, \cdots, l_1\}, \{g_{k'}^A(\omega), k' = 1, 2, \cdots, l_2\} > |\omega \in X\}$ $B = \{<\omega, \{h_k^B(\omega), k = 1, 2, \cdots, l_1\}, \{g_{k'}^B(\omega), k' = 1, 2, \cdots, l_2\} > |\omega \in X\}$ then the correlation between A and B is defined by $C(A, B) = 1/(l_1)\sum_{k=1}^{l_1}\sum_{i=1}^{\infty} h_k^A(\omega_i)h_k^B(\omega_i)P(\omega_i) + \frac{1}{l_2}\sum_{k'=1}^{l_2}\sum_{i=1}^{\infty} g_{k'}^A(\omega_i)g_{k'}^B(\omega_i)P(\omega_i)$.

It is obvious that the correlation of two dual hesitant fuzzy sets satisfies the following properties:

(1) $C(A, A) = E(A)$; (2) $C(A, B) = C(B, A)$.

Definition 5.3. Let $X = \{\omega_1, \omega_2, \cdots, \omega_i, \cdots\}$ is a sample space, $A, B \in DHFE(X)$, suppose $A = \{<\omega, \{h_k^A(\omega), k = 1, 2, \cdots, l_1\}, \{g_{k'}^A(\omega), k' = 1, 2, \cdots, l_2\} > |\omega \in X\}$ $B = \{<\omega, \{h_k^B(\omega), k = 1, 2, \cdots, l_1\}, \{g_{k'}^B(\omega), k' = 1, 2, \cdots, l_2\} > |\omega \in X\}$ then the correlation coefficient between A and B is defined as follows:

$$\rho(A, B) = \frac{C(A, B)}{\sqrt{C(A, A)}\sqrt{C(B, B)}}$$

$$= \frac{\frac{1}{l_1}\sum_{k=1}^{l_1}\sum_{i=1}^{\infty} h_k^A(\omega_i)h_k^B(\omega_i)P(\omega_i) + \frac{1}{l_2}\sum_{k'=1}^{l_2}\sum_{i=1}^{\infty} g_{k'}^A(\omega_i)g_k^B(\omega_i)P(\omega_i)}{\sqrt{\frac{1}{l_1}\sum_{k=1}^{l_1}\sum_{i=1}^{\infty}(h_k^A(\omega_i))^2P(\omega_i) + \frac{1}{l_2}\sum_{k=1}^{l_2}\sum_{i=1}^{\infty}(g_k^A(\omega_i))^2P(\omega_i)}\sqrt{\frac{1}{l_1}\sum_{k=1}^{l_1}\sum_{i=1}^{\infty}(h_k^B(\omega_i))^2P(\omega_i) + \frac{1}{l_2}\sum_{k=1}^{l_2}\sum_{i=1}^{\infty}(g_k^B(\omega_i))^2P(\omega_i)}}$$

Theorem 5.1. The correlation coefficient between two dual hesitant fuzzy sets A and B satisfies following properties:

(1) $\rho(A,B) = \rho(B,A)$; (2) $0 \le \rho(A,B) \le 1$; (3) $\rho(A,B) = 1$, if $A = B$.

Proof The correctness of (1) is evident. The proof of (2), (3). Due to the Cauchy-Schwarze inequality.

$$C(A,B) = \left(\frac{1}{l_1}\sum_{k=1}^{l_1}\sum_{i=1}^{\infty} h_k^A(\omega_i)h_k^B(\omega_i)P(\omega_i) + \frac{1}{l_2}\sum_{k'=1}^{l_2}\sum_{i=1}^{\infty} g_{k'}^A(\omega_i)g_{k'}^B(\omega_i)P(\omega_i)\right)^2$$

$$= \lim_{m\to\infty}\left(\sum_{i=1}^{m}\left[\frac{1}{l_1}\sum_{k=1}^{l_1} h_k^A(\omega_i)h_k^B(\omega_i)P(\omega_i) + \frac{1}{l_2}\sum_{k'=1}^{l_2} g_{k'}^A(\omega_i)g_{k'}^B(\omega_i)P(\omega_i)\right]\right)^2 \le \lim_{m\to\infty}$$

$$\left(\sum_{i=1}^{m}\left[\frac{1}{l_1}\sum_{k=1}^{l_1}(h_k^A(\omega_i))^2P(\omega_i) + \frac{1}{l_2}\sum_{k'=1}^{l_2}(g_{k'}^A(\omega_i))^2P(\omega_i)\right]\right)$$

$$\left(\sum_{i=1}^{m}\left[\frac{1}{l_1}\sum_{k=1}^{l_1}(h_k^B(\omega_i))^2P(\omega_i) + \frac{1}{l_2}\sum_{k'=1}^{l_2}(g_{k'}^B(\omega_i))^2P(\omega_i)\right]\right)$$

$$= \lim_{m\to\infty}\left(\frac{1}{l_1}\sum_{k=1}^{l_1}\sum_{i=1}^{\infty}(h_k^A(\omega_i))^2P(\omega_i) + \frac{1}{l_2}\sum_{k'=1}^{l_2}\sum_{i=1}^{\infty}(g_{k'}^A(\omega_i))^2P(\omega_i)\right)$$

$$\left(\frac{1}{l_1}\sum_{k=1}^{l_1}\sum_{i=1}^{\infty}(h_k^B(\omega_i))^2P(\omega_i) + \frac{1}{l_2}\sum_{k'=1}^{l_2}\sum_{i=1}^{\infty}(g_{k'}^B(\omega_i))^2P(\omega_i)\right) = C(A,\,A)\,C(B,\,B).$$

And also when $h_k^A(\omega_i) = h_k^B(\omega_i)$ and $g_{k'}^A(\omega_i) = h_{k'}^B(\omega_i)$ for $k = 1, 2, \cdots, l_1$, $k' = 1, 2, \cdots, l_2$, $i = 1, 2, \cdots$, $C^2(A,B) = C(A,A)C(B,B)$, so if $A = B$, $\rho(A,B) = 1$.

6 Dual Hesitant Fuzzy Bayesian Network Inference

Formally, a Bayesian network is a directed acyclic graph $G = (NG, AG)$ where NG is a set of nodes, AG is a set of arcs. Every node is associated with a random variable. Each random variable has a conditional probability distribution table, it describes the probability distribution of a variable when giving values of its parent node. The probability distribution associated with initial node is called initial distribution. The arcs AG in the graph G represent possible dependences between variables, nodes that are not connected show variables that are conditionally independent of each other.

Due to a Bayesian network can represent a joint probability distribution of all variables, it can be used to computing any interesting probability. For example, the Bayesian network can be used to compute probability distribution of some variables when other variables (the evidence variables) are given. The process of calculating the posterior distribution of variables given other variables is called probabilistic inference. For a Bayesian network, if the probabilities are described as probabilities of dual hesitant fuzzy events or conditional probabilities are described as conditional probabilities of dual hesitant fuzzy events, the Bayesian network is called dual hesitant fuzzy Bayesian network.

Next we give a simple example to show inference task for a dual hesitant fuzzy Bayesian network.

Example 6.1. Suppose that there are only two events which promote student's academic performance: either going to extracurricular tutoring classes or improving the quality of school teaching. Also, suppose that improving the quality of school teaching directly affects the enthusiasm of going to extracurricular tutoring classes (namely that when a school has been improved the quality of teaching, the students may be not to go to the extracurricular classes). Then the situation can be modeled with Bayesian network (shown Fig. 1).

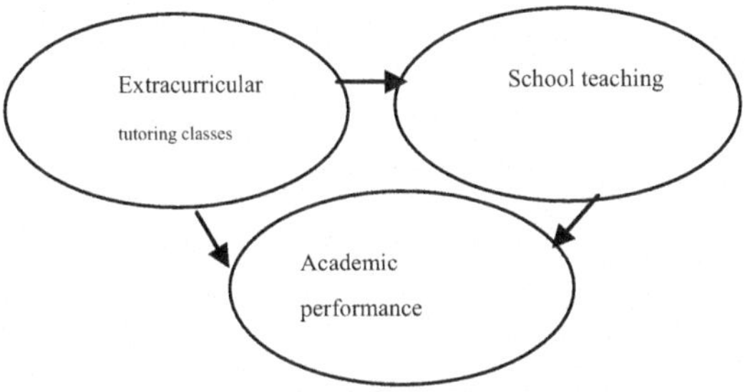

Fig. 1. A simple Bayesian network

Table 1. The conditional distribution of E given

Q	E	
	T	T
F	0.7	0.3
T	0.1	0.9

The variables going to extracurricular tutoring classes and improving the quality of school teaching have two possible values, T (for true) and F (for false). Promoting student's academic performance is regarded as a dual hesitant fuzzy event.

The names of the variables and dual hesitant fuzzy event have been abbreviated to E = going to extracurricular tutoring classes (yes/no). Q = improving the quality of school teaching (yes/no), SA = promote student's academic performance. The conditional distribution (or probability distribution) tables of all variables are shown as Tables 1, 2, and 3.

Table 2. The distribution of variable Q

Q	T	F
	0.2	0.8

Table 3. The condition probability $P(SA|E,Q)$

E	Q	$P(SA\|E,Q)$
F	F	$(\{0.0\}, \{0.0\})$
T	F	$(\{0.75, 0.8\}, \{0.80, 0.86\})$
F	T	$(\{0.7, 0.9\}, \{0.99, 0.992\})$
T	T	$(\{0.98, 0.99\}, \{0.991, 0.993\})$

Please answer the question: given that the student academic performance is promoted, whether the quality of school teaching is improved.

By using definition of the conditional probability of dual hesitant fuzzy events and Law of total probability formula of dual hesitant fuzzy events,

$$\begin{aligned}
&P(SA) = P\{SA|E = F, Q = F\}P\{E = F, Q = F\} + P\{SA|E = T, Q = F\}\\
&P\{E = T, Q = F\} + P\{SA|E = F, Q = T\}P\{E = F, Q = T\}\\
&+ P\{SA|E = T, Q = T\}P\{E = T, Q = T\} = P\{SA|E = F, Q = F\}\\
&P\{E = F|Q = F\}P\{Q = F\} + P\{SA|E = T, Q = F\}P\{E = T|Q = F\}P\{Q = F\}\\
&+ P\{SA|E = F, Q = T\}P\{E = F|Q = T\}P\{Q = T\}\\
&+ P\{SA|E = T, Q = T\}P\{E = T|Q = T\}P\{Q = T\}.
\end{aligned}$$

Substituting the data in Tables 1, 2 and 3 into the last expression, we get $P(SA) = (\{0.5936, 0.6018\}, \{0.6461, 0.6801\})$. $P(SA, Q = T) = P(SA, Q = T, E = T) + P(SA, Q = T, E = F) = P(Q = T)P(E = T|Q = T)P(SA|Q = T, E = T) + P(Q = T)P(E = F|Q = T)P(SA|Q = T, E = F)$. Substituting the data in Tables 1, 2 and 3 into the last expression, we get $P(SA, Q = T) = (\{0.1456, 0.1818\}, \{0.1980, 0.1984\})$. Substituting the values of the $P(SA)$ and $P(SA, Q = T)$ into $P(Q = T|SA) = \frac{P(SA,Q=T)}{P(SA)}$, we get $P(Q = T|SA) = (\{0.1556, 0.3021\}, \{0.3065, 0.2940\})$. Similarly, we can get $P(Q = F|SA) = \frac{P(SA,Q=F)}{P(SA)} = (\{0.2311, 0.2303\}, \{0.2148, 0.2044\})$. The score function of $P(Q = T|SA)$ is 0.2640. The score function of $P(Q = F|SA)$ is 0.2202. Therefore $P(Q = T|SA) > P(Q = F|SA)$.

The school can be regarded to improve the quality of teaching.

7 Conclusion

To deal with the uncertainty about the fuzzy, the hesitancy and random uniformly, the concepts of dual hesitant fuzzy event, its probability and conditional probability have been introduced. The basic properties, the law of total probability, the Bayes formula and the continuity of probability of classical probability theory are generalized to that of the probabilities of dual hesitant fuzzy events. The correlation and correlation coefficient between two dual hesitant fuzzy sets have been defined in random environment. An example of dual hesitant Bayesian network inference is given. The works of this paper lay the foundation for further the dual hesitant fuzzy probability theory research.

References

Torra, V., Narukawa, Y.: On hesitant fuzzy sets and decision. In: The 18th IEEE International Conference on Fuzzy Systems, Jeju Island, Korea, pp. 1378–1382 (2009)

Torre, V.: Hesitant fuzzy sets. Int. J. Intell. Syst. **25**(6), 529–539 (2010)

Chen, N., Xu, Z.S., Xia, M.M.: Interval-valued hesitant preference relations and their applications to group decision making. Knowl.-Based Syst. **37**(2), 528–540 (2013)

Qian, G., Wang, H., Feng, X.: Generalized hesitant fuzzy sets and their application in decision support system. Knowl.-Based Syst. **37**(4), 357–365 (2013)

Zhu, B., Xu, Z.S., Xia, M.M.: Dual hesitant fuzzy sets. J. Appl. Math. **2012**(11), 1–13 (2012)

Xu, Z.S., Xa, M.M.: Distance and similarity measures for hesitant fuzzy sets. Inf. Sci. **181**(11), 2128–2138 (2011)

Farhadinia, B.: Correlation for dual hesitant fuzzy sets and dual interval-valued hesitant fuzzy sets. Int. J. Intell. Syst. **29**(2), 184–205 (2014)

Ye, J.: Correlation coefficient of dual hesitant fuzzy sets and its application to multiple attribute decision making. Appl. Math. Model. **38**(2), 659–666 (2014)

Singh, P.: A new method for solving dual hesitant fuzzy assignment problems with restrictions based on similarity. Appl. Soft Comput. J. **24**(c), 559–6716 (2014)

Singh, P.: Distance and similarity measures for multiple-attribute decision making with dual hesitant fuzzy sets. Comput. Appl. Math. **36**(1), 111–126 (2017)

Su, Z., Xu, Z., et al.: Distance and similarity measures for dual hesitant fuzzy sets and their application in pattern recognition. J. Intell. Fuzzy Syst. **29**(2), 731–745 (2015)

Yu, D.J., Merigó, J.M., et al.: Group decision making in information systems security assessment using dual hesitant fuzzy set. Int. J. Intell. Syst. **31**(6), 786–812 (2016)

Wang, L., Shen, Q.G., Zhu, L.: Dual hesitant fuzzy power aggregation operators based on Archimedean t-conorm and t-norm and their application to multiple attribute group decision making. Appl. Soft Comput. J. **38**©, 23–50 (2016)

Zhang, H.D., Shu, L., et al.: Dual hesitant fuzzy rough set and its application. Soft. Comput. **21** (12), 3287–3305 (2017)

Hao, Z., Xu, Z.S., et al.: Probabilistic dual hesitant fuzzy set and its application in risk evaluation. Knowl.-Based Syst. **127**(c), 16–28 (2017)

Fan, L., Lei, Y.J.: Intuitionistic fuzzy statistic adjudging and decision-making. Control Decis. **26** (3), 357–362 (2011). (in Chinese)

Zadeh, L.A.: Probability measures of fuzzy events. J. Math. Anal. Appl. **23**(3), 421–427 (1968)

Yager, R.R.: A note on probabilities of fuzzy events. Inf. Sci. **18**(2), 113–129 (1979)

Klement, E.: Some remarks on a paper of R.R. Yager. Inf. Sci. 1 **27**, 211–220 (1982)

Hesamian, G., Shams, M.: A note on fuzzy probability of a fuzzy event. Int. J. Intell. Syst. **32**(7), 676–685 (2017)

Chen, J.J., Huang, X.J.: Dual hesitant fuzzy probability. Symmetry **9**(52), 1–11 (2017)

Zhang, W.X., Wang, G.J., et al.: Introduction of Fuzzy Mathematics, 1st edn. Xi'an Jiao Tong University Press, Xi'an (2005). (in Chinese)

Order Relations Between Interval Numbers

Haohao Li(✉)

School of Data Sciences, Zhejiang University of Finance and Economics, Hangzhou 310018, China
hhli@zufe.edu.cn

Abstract. In addition to fuzzy and stochastic theory, interval analysis is a powerful tool for solving uncertain problems. An important problem in interval analysis is the ranking of interval numbers. This paper analyzes the definitions of satisfaction index for comparing interval numbers. Sometimes different definitions of the satisfaction indices lead to confusion. The main result of the paper is to show that some existing definitions of the satisfaction index are equivalent.

Keywords: Interval analysis · Interval optimization · Satisfaction index

1 Introduction

In addition to fuzzy mathematics and stochastic theory, interval mathematics is another effective tool for dealing with uncertainties. In many practical problems, the choice between interval data alternatives requests a ranking of real interval. This topic has been treated by some authors in the literature, by introducing a kind of satisfaction index. There are different versions of satisfaction index. Numerous papers devoted to this fields have appeared in various journals, see [1–10] among others. Sometimes different definitions of the satisfaction index for comprising interval numbers may lead to confusion.

The aim of this paper is to discuss the close affinity among some existing ranking methods. The equivalence relation of 9 satisfaction indices proposed by different authors is established.

2 The Equivalence of Several Satisfaction Indices for Comparing any Two Interval Numbers

2.1 The Basic Interval Arithmetic

An interval matrix is defined as

$$\boldsymbol{A} = [\underline{A}, \bar{A}] = \{A \in R^{m\times n} | \underline{A} \le A \le \bar{A}\},$$

Y. Liu et al. (Eds.): ICNC-FSKD 2019, AISC 1074, pp. 756–763, 2020.
https://doi.org/10.1007/978-3-030-32456-8_81

where $\underline{A} \le \bar{A}$ are fixed matrices; n-dimensional interval vectors can be regarded as a $n \times 1$ interval matrices. Thus, a interval number is defined as

$$\boldsymbol{a} = [\underline{a}, \bar{a}] = \{a \in R | \underline{a} \le a \le \bar{a}\}$$

By

$$m(\boldsymbol{A}) = \frac{1}{2}(\underline{A} + \bar{A}),\ w(\boldsymbol{A}) = \frac{1}{2}(\bar{A} - \underline{A}),\ l_a = \bar{a} - \underline{a}$$

we denote the midpoint, radius of $\boldsymbol{A}$ and the length of the interval number $\boldsymbol{a}$, respectively. We say that $\boldsymbol{a}$ is degenerate if $\bar{a} = \underline{a}$. Such an interval contains a single real number a. By convention, we agree to identify a degenerate interval $[a, a]$ with the real number a.

2.2 Definitions of the Satisfaction Index

Two interval numbers $\boldsymbol{a}$ and $\boldsymbol{b}$, should often be compared one with another. The result of such comparison, unlike it is in case of crisp numbers, is not unique, that is why in the literature many ways of comparing interval numbers are considered.

In 1989, Ishibuchi and Tanaka [1] introduced a kind of satisfaction index (grade of satisfaction) P, for comparing any two interval numbers $\boldsymbol{a}$ and $\boldsymbol{b}$ on the real line.

Definition 1 [1]

$$P(\boldsymbol{a} \le \boldsymbol{b}) = max\left\{0,\ min\left\{\frac{\bar{b} - \underline{a}}{l_a + l_b}, 1\right\}\right\}. \tag{1}$$

It is clear that $0 \le P(\boldsymbol{a} \le \boldsymbol{b}) \le 1$ and the following result hold.

Theorem 1 [1]**.** For any two interval numbers $\boldsymbol{a}$ and $\boldsymbol{b}$, we have

1–1 For any interval number $\boldsymbol{a}$, $P(\boldsymbol{a} \le \boldsymbol{a}) = 0.5$.
1–2 $P(\boldsymbol{a} \le \boldsymbol{b}) = 0.5$ if and only if $a_c = b_c$.
1–3 For any two intervals $\boldsymbol{a}$ and $\boldsymbol{b}$, $P(\boldsymbol{a} \le \boldsymbol{b}) = 1$ if and only if $\bar{a} \le \underline{b}$. (similarly we have $P(\boldsymbol{a} \le \boldsymbol{b}) = 0$ if and only if $\bar{b} \le \underline{a}$.)
1–4 For any two intervals $\boldsymbol{a}$ and $\boldsymbol{b}$, $P(\boldsymbol{a} \le \boldsymbol{b}) > 0$ if and only if $\underline{a} < \bar{b}$. Thus, if $P(\boldsymbol{a} \le \boldsymbol{b}) > 0$, then $a < b$ for some real numbers a and b $(a \in \boldsymbol{a},\ b \in \boldsymbol{b})$.
1–5 For any two intervals $\boldsymbol{a}$ and $\boldsymbol{b}$, $P(\boldsymbol{a} \le \boldsymbol{b}) + P(\boldsymbol{b} \le \boldsymbol{a}) = 1$.

Some other kinds of the satisfaction index have been introduced later by different authors.

Definition 2 [2]

$$P(\boldsymbol{a} \le \boldsymbol{b}) = min\left\{max\left\{\frac{\bar{b} - \underline{a}}{l_a + l_b},\ 0\right\},\ 1\right\}. \tag{2}$$

Definition 3 [3]

$$P(\boldsymbol{a} \le \boldsymbol{b}) = \frac{max\{0,\ l_a + l_b - max\{\bar{a} - \underline{b},\ 0\}\}}{l_a + l_b}. \tag{3}$$

Definition 4 [6]

$$P(\mathbf{a} \leq \mathbf{b}) = \frac{max\{0,\ \bar{b} - \underline{a}\} - max\{0, \underline{b} - \bar{a}\}}{l_a + l_b}. \tag{4}$$

Definition 5 [7]

$$P(\mathbf{a} \leq \mathbf{b}) = \frac{min\{l_a + l_b, max\{\bar{b} - \underline{a}, 0\}\}}{l_a + l_b}. \tag{5}$$

Definition 6 [9]

$$P(\mathbf{a} \leq \mathbf{b}) = min\left\{max\left\{0, \frac{m(\mathbf{b}) - m(\mathbf{a})}{2w(\mathbf{a}) + 2w(\mathbf{b})} + \frac{1}{2}\right\}, 1\right\}. \tag{6}$$

Definition 7 [10]

$$P(\mathbf{a} \leq \mathbf{b}) = \frac{min\{\bar{a} - \underline{b}, 0\} - min\{0,\ \underline{a} - \bar{b}\}}{l_a + l_b}. \tag{7}$$

Definition 8 [10]

$$P(\mathbf{a} \leq \mathbf{b}) = min\left\{\frac{\bar{b} - \underline{a}}{l_a + l_b}, 1\right\} - min\left\{0, \frac{\bar{b} - \underline{a}}{l_a + l_b}\right\}. \tag{8}$$

Definition 9 [10]

$$P(\mathbf{a} \leq \mathbf{b}) = max\left\{0, min\left\{\frac{m(\mathbf{b}) - m(\mathbf{a})}{2w(\mathbf{a}) + 2w(\mathbf{b})} + \frac{1}{2}, 1\right\}\right\}. \tag{9}$$

Clearly, at least one of the interval numbers of **a** and **b** in above definitions must be non-degenerate. In next section, we prove that the satisfaction indices given in Definitions 1–9 are equivalent.

2.3 Equivalence Relations Among Satisfaction Indices

First we state an auxiliary result.

Lemma 1. There holds that

$$max\{0,\ min\{x, 1\}\} = min\{max\{x, 0\}, 1\} \tag{10}$$

for any $x \in R$.

Proof. It is easy to see that

$$max\{0,\ min\{x, 1\}\} = max\left\{0, \tfrac{1}{2}(x + 1) - \tfrac{1}{2}|1 - x|\right\}$$

$$= \tfrac{1}{4}\left(x + 1 - |1 - x| + |x + 1 - |1 - x||\right),$$

and

$$min\{\ max\{x, 0\}, 1\} = min\left\{\tfrac{1}{2}(x + |x|), 1\right\}$$

$$= \tfrac{1}{4}\left(x + 2 + |x| - |x + |x| - 2|\right).$$

Then we have

$$max\{0,\ min\{x,1\}\} - min\{\ max\{x,0\},1\}$$
$$= \frac{1}{4}(-|1-x| + |x+1-|1-x|| - |x| + |x+|x|-2| - 1)$$
$$= 0.$$

for any $x \in R$. Then (10) holds.

Theorem 2. The satisfaction indices given in Definitions 1–9 are equivalent.

Proof. We only need to prove that (1) $\Rightarrow$ (2) $\Rightarrow$ (5) $\Leftrightarrow$ (4) $\Rightarrow$ (7) $\Rightarrow$ (6) $\Rightarrow$ (9) $\Rightarrow$ (3) $\Rightarrow$ (8) $\Rightarrow$ (1).
(1) $\Rightarrow$ (2):

According to Lemma 1, let $x = \frac{\bar{b}-\underline{a}}{l_a+l_b}$, then (1) is equal to (2).
(2) $\Rightarrow$ (5):

$$min\left\{max\left\{\frac{\bar{b}-\underline{a}}{l_a+l_b},\ 0\right\},\ 1\right\}$$
$$= min\left\{\frac{max\{\bar{b}-\underline{a},0\}}{l_a+l_b},\ 1\right\}$$
$$= \frac{min\{l_a+l_b, max\{\bar{b}-\underline{a},0\}\}}{l_a+l_b}.$$

(5) $\Rightarrow$ (4):

$$\frac{min\{l_a+l_b, max\{\bar{b}-\underline{a},0\}\}}{l_a+l_b}$$
$$= \frac{max\{\bar{b}-\underline{a},0\} + min\{l_a+l_b-max\{\bar{b}-\underline{a},0\},0\}}{l_a+l_b}$$
$$= \frac{max\{\bar{b}-\underline{a},0\} + min\{l_a+l_b+min\{\underline{a}-\bar{b},0\},0\}}{l_a+l_b}$$
$$= \frac{max\{\bar{b}-\underline{a},0\} + min\{min\{\bar{a}-\underline{b},l_a+l_b\},0\}}{l_a+l_b}$$
$$= \frac{max\{\bar{b}-\underline{a},0\} + min\{0,\bar{a}-\underline{b},l_a+l_b\}}{l_a+l_b}$$
$$= \frac{max\{\bar{b}-\underline{a},0\} + min\{0,\bar{a}-\underline{b}\}}{l_a+l_b}$$
$$= \frac{max\{\bar{b}-\underline{a},0\} - max\{0,\underline{b}-\bar{a}\}}{l_a+l_b}.$$

(4) ⇒ (7):

$$\frac{max\{\bar{b}-\underline{a},0\}-max\{0,\underline{b}-\bar{a}\}}{l_a+l_b}$$
$$=\frac{-min\{\underline{a}-\bar{b},0\}-max\{0,\underline{b}-\bar{a}\}}{l_a+l_b}$$
$$=\frac{min\{0,\bar{a}-\underline{b}\}-min\{\underline{a}-\bar{b},0\}}{l_a+l_b}.$$

(7) ⇒ (6): It is obvious that

$$\frac{\bar{b}-\underline{a}}{l_a+l_b}=\frac{m(\mathbf{b})-m(\mathbf{a})}{2w(\mathbf{a})+2w(\mathbf{b})}+\frac{1}{2}.$$

Then we have

$$\frac{min\{0,\bar{a}-\underline{b}\}-min\{\underline{a}-\bar{b},0\}}{l_a+l_b}$$
$$=\frac{min\{0,min\{\bar{a}-\underline{b},l_a+l_b\}\}+max\{\bar{b}-\underline{a},0\}}{l_a+l_b}$$
$$=\frac{min\{0,l_a+l_b+min\{0,\underline{a}-\bar{b}\}\}+max\{\bar{b}-\underline{a},0\}}{l_a+l_b}$$
$$=\frac{min\{0,l_a+l_b-max\{\bar{b}-\underline{a},0\}\}+max\{\bar{b}-\underline{a},0\}}{l_a+l_b}$$
$$=\frac{min\{max\{\bar{b}-\underline{a},0\},l_a+l_b\}}{l_a+l_b}$$
$$=min\left\{max\left\{\frac{\bar{b}-\underline{a}}{l_a+l_b},0\right\},1\right\}$$
$$=min\left\{max\left\{0,\frac{m(\mathbf{b})-m(\mathbf{a})}{2w(\mathbf{a})+2w(\mathbf{b})}+\frac{1}{2}\right\},1\right\}.$$

(6) ⇒ (9):

According to Lemma 1, let $x=\frac{m(\mathbf{b})-m(\mathbf{a})}{2w(\mathbf{a})+2w(\mathbf{b})}+\frac{1}{2}$, then (6) is equal to (9).

(9) ⇒ (3):

$$max\left\{0,min\left\{\frac{m(\mathbf{b})-m(\mathbf{a})}{2w(\mathbf{a})+2w(\mathbf{b})}+\frac{1}{2},1\right\}\right\}$$
$$=max\left\{0,min\left\{\frac{\bar{b}-\underline{a}}{l_a+l_b},1\right\}\right\}$$
$$=max\left\{0,\frac{min\{\bar{b}-\underline{a},l_a+l_b\}}{l_a+l_b}\right\}$$

$$= max\left\{0, \frac{l_a + l_b + min\{\bar{a} - \underline{b},\ 0\}}{l_a + l_b}\right\}$$

$$= \frac{max\{0,\ l_a + l_b - max\{\bar{a} - \underline{b},\ 0\}\}}{l_a + l_b}.$$

(3) ⇒ (8):

$$\frac{max\{0,\ l_a + l_b - max\{\bar{a} - \underline{b},\ 0\}\}}{l_a + l_b}$$

$$= \frac{max\{0, min\{\bar{b} - \underline{a}, l_a + l_b\}\}}{l_a + l_b}$$

$$= \frac{min\{\bar{b} - \underline{a}, l_a + l_b\}}{l_a + l_b} + \frac{max\{0, -min\{\bar{b} - \underline{a}, l_a + l_b\}}{l_a + l_b}$$

$$= min\left\{\frac{\bar{b} - \underline{a}}{l_a + l_b}, 1\right\} - \frac{min\{0, min\{\bar{b} - \underline{a}, l_a + l_b\}\}}{l_a + l_b}$$

$$= min\left\{\frac{\bar{b} - \underline{a}}{l_a + l_b}, 1\right\} - \frac{min\{0, \bar{b} - \underline{a}, l_a + l_b\}}{l_a + l_b}$$

$$= min\left\{\frac{\bar{b} - \underline{a}}{l_a + l_b}, 1\right\} - \frac{min\{0, \bar{b} - \underline{a}\}}{l_a + l_b}$$

$$= min\left\{\frac{\bar{b} - \underline{a}}{l_a + l_b}, 1\right\} - min\left\{0, \frac{\bar{b} - \underline{a}}{l_a + l_b}\right\}.$$

(8) ⇒ (1):

$$min\left\{\frac{\bar{b} - \underline{a}}{l_a + l_b}, 1\right\} - min\left\{0, \frac{\bar{b} - \underline{a}}{l_a + l_b}\right\}$$

$$= min\left\{\frac{\bar{b} - \underline{a}}{l_a + l_b}, 1\right\} - min\left\{0, min\left\{\frac{\bar{b} - \underline{a}}{l_a + l_b}, 1\right\}\right\}$$

$$= min\left\{\frac{\bar{b} - \underline{a}}{l_a + l_b}, 1\right\} + max\left\{0, -min\left\{\frac{\bar{b} - \underline{a}}{l_a + l_b}, 1\right\}\right\}$$

$$= max\left\{0, min\left\{\frac{\bar{b} - \underline{a}}{l_a + l_b}, 1\right\}\right\}.$$

□

To sum up, we can see that these 9 definitions of satisfaction indexes are the same in essence, as proved above. Thus we can choose different definitions according to different occasions such that are more convenient for application.

For instance, let intervals $a = [1, 5], b = [3, 7]$. Then

$$\underline{a} = 1, \bar{a} = 5, m(a) = \frac{1}{2}(\underline{a} + \bar{a}) = 3, w(a) = \frac{1}{2}(\bar{a} - \underline{a}) = 2, l_a = \bar{a} - \underline{a} = 4,$$

and

$$\underline{b} = 3, \bar{b} = 7, m(b) = \frac{1}{2}(\underline{b} + \bar{b}) = 5, w(a) = \frac{1}{2}(\bar{b} - \underline{b}) = 2, l_b = \bar{b} - \underline{b} = 4.$$

Thus, by Definition 1,

$$P(\boldsymbol{a} \le \boldsymbol{b}) = max\left\{0,\ min\left\{\frac{7-1}{4+4}, 1\right\}\right\} = \frac{3}{4}.$$

By Definition 2,

$$P(\boldsymbol{a} \le \boldsymbol{b}) = min\left\{max\left\{\frac{7-1}{4+4},\ 0\right\},\ 1\right\} = \frac{3}{4}.$$

By Definition 3,

$$P(\boldsymbol{a} \le \boldsymbol{b}) = \frac{max\{0,\ 4+4-max\{5-3,\ 0\}\}}{4+4} = \frac{3}{4}.$$

And satisfaction indexes $P(\boldsymbol{a} \le \boldsymbol{b})$ under 6 other definitions are all the same.

3 Conclusion

This paper establishes the equivalence relations of some existing works on comparing and ranking any two interval numbers on the real line. Clearly, our approach here can be extended to interval optimization problems. Moreover, it is better to develop an axiomatic definition for ranking interval numbers. These are our interest of further studies.

Acknowledgements. This research was supported by the NNSF of China (Grant No. 11701506).

References

1. Ishibuchi, H., Tanaka, H.: Formulation and analysis of linear programming problem with interval coefficients. J. Jpn. Ind. Manag. Assoc. **40**, 320–329 (1989). (in Japanese)
2. Facchinetti, G., Ricci, R.G., Muzzioli, S.: Note on ranking fuzzy triangular numbers. J. Int. J. Intell. Syst. **13**, 613–622 (1998)
3. Da, Q.L., Liu, X.W.: Interval number linear programming and its satisfactory solution. J. Syst. Eng. Theory Pract. **19**, 3–7 (1999). (in Chinese)
4. Sengupta, A., Pal, T.K.: On comparing interval numbers. J. Eur. J. Oper. Res. **127**, 28–43 (2000)
5. Sengupta, A., Pal, T.K., Chakraborty, D.: Interpretation of inequality constraints involving interval coefficients and a solution to interval linear programming. J. Fuzzy Sets Syst. **119**, 129–138 (2001)
6. Liu, Z.X., Zhang, C., Gang, J.T.: Ranking of fuzzy numbers. J. Liaoning Normal Univ. **25**, 62–365 (2002)
7. Xu, Z.S., Da, Q.L.: Possibility degree method for ranking interval numbers and its application. J. Syst. Eng. **18**, 67–70 (2003). (in Chinese)
8. Lan, J.B., Cao, L.J., Lin, J.: Method for ranking interval numbers based on two-dimensional priority degree. J. Chongqing Inst. Technol. (Nat. Sci. Edn.) **21**, 63–67 (2007)

9. Sun, H.L., Yao, W.X.: Comments on methods for ranking interval numbers. J. Syst. Eng. **25**, 304–312 (2010). (in Chinese)
10. Qin, C.Y., Wang, H.P., Li, W.: Satisfaction indices for comparing any two interval numbers, Technical report, Institute of Operational Research & Cybernetics, Hangzhou Dianzi University (2007)

Some New Distance Measures for Generalized Hesitant Fuzzy Sets

Chen Bin(✉)

School of Mathematical Sciences, University of Jinan, Jinan 250022, China
jnchenbin@126.com

Abstract. Distance and similarity measures are very important topics in fuzzy set theory. In this paper, we propose some new distance measures for generalized hesitant fuzzy sets. We investigate the connections of the aforementioned distance measures and further develop a number of generalized hesitant ordered weighted distance measures.

Keywords: Fuzzy set · Generalized hesitant fuzzy set · Distance measure

1 Introduction

Distance and similarity measures have attached a lot of attention in the last decades due to the fact that they can be applied to many areas such as pattern recognition [7], clustering analysis [4], approximate reasoning [9], image processing [10], medical diagnosis [11] and decision making [1,12]. A lot of distance and similarity measures have been developed for fuzzy sets, intuitionistic fuzzy sets [13–17], linguistic fuzzy sets, hesitant fuzzy sets [1–3], and fuzzy multisets.

The rest of this paper is organized as follows. In Sect. 2, we review the notions of GHFS. In Sect. 3, we present some new geometric distance measures between GHFSs base on geometric distance model.

2 Preliminaries

Gang Qian et al. [6] defined the generalized hesitant fuzzy set in terms of two functions that return two sets of membership values and nonmembership values, respectively, for each element in the domain as follows.

Definition 1. *Let X be a fixed set, then a generalized hesitant fuzzy set (GHF set) G on X is described as:*

$G = \{\frac{x}{h(x), g(x)} | x \in X\}$,

in which $h(x)$ and $g(x)$ are two sets of some values in [0, 1], denoting the possible membership degrees and nonmembership degrees of the element $x \in X$ to the set G, respectively, with the conditions:

$0 \leq \mu_i(x), \nu_i(x) \leq 1, 0 \leq \mu_i(x) + \nu_i(x) \leq 1,\ 1 \leq i \leq N_x = |h(x)| = |g(x)|,$

Y. Liu et al. (Eds.): ICNC-FSKD 2019, AISC 1074, pp. 764–770, 2020.
https://doi.org/10.1007/978-3-030-32456-8_82

where $\mu_i(x) \in h(x)$, $\nu_i(x) \in g(x)$, *for all* $x \in X$. *And* $|h(x)|$ *denote the cardinality of the set* $h(x)$. *For convenience, the pair* $G(x) = (h(x), g(x))$ *is called a generalized hesitant fuzzy element (GHFE) denoted by* $G = (h, g)$.

Definition 2. *Given a GHFE represented by* G, *we define its complement as follows:*

$$G^c(x) = \cup_{\mu_i(x)\in h(x),\nu_i(x)\in g(x)}\{\{\nu_i(x)\}, \{\mu_i(x)\}\}.$$

Definition 3. *Given two generalized hesitant fuzzy sets* $G_1 = (h_1(x), g_1(x))$ *and* $G_2 = (h_2(x), g_2(x))$, *we define their union represented by* $G_1 \cup G_2$ *as* $G_1 \cup G_2(x) = \{(\mu(x), \nu(x)), \mu(x) \in h_1(x) \cup h_2(x))|\mu(x) \geq max(h_1^-(x), h_2^-(x)), \nu(x) \in (g_1(x) \cup g_2(x))|\nu(x) \leq min(g_1^+(x), g_2^+(x))\}$.

Definition 4. *Given two generalized hesitant fuzzy sets* $G_1 = (h_1(x), g_1(x))$ *and* $G_2 = (h_2(x), g_2(x))$, *we define their union represented by* $G_1 \cap G_2$ *as* $G_1 \cap G_2(x) = \{(\mu(x), \nu(x)), \mu(x) \in (h_1(x) \cup h_2(x))|\mu(x) \leq min(h_1^+(x), h_2^+(x)), \nu(x) \in (g_1(x) \cup g_2(x))|\nu(x) \geq max(g_1^-(x), g_2^-(x))\}$.

Example 1. Let $G_1 = \{\{0.1, 0.3, 0.5\}, \{0.8, 0.6, 0.4\}\}$ and $G_2 = \{\{0.2, 0.4\}, \{0.7, 0.5\}\}$ be two GHFEs, then we have.

(1) complement: $G_1^c(x) = \{\{0.8, 0.6, 0.4\}, \{0.1, 0.3, 0.5\}\}$,
(2) union: $G_1 \cup G_2(x) = \{\{0.3, 0.5, 0.2, 0.4\}, \{0.6, 0.4, 0.7, 0.5\}\}$,
(3) intersection: $G_1 \cap G_2(x) = \{\{0.1, 0.3, 0.2, 0.4\}, \{0.8, 0.6, 0.7, 0.5\}\}$.

Definition 5. *Let* M, N *be two GHF set on* X. *Then,* M *is a generalized hesitant fuzzy subset of* N, *if for each* $x \in X$, $1 \leq i \leq l_x$, *we have* $\mu_i^M(x) \leq \mu_i^N(x)$ *and* $\nu_i^M(x) \geq \nu_i^N(x)$. *And denote by* $M \sqsubseteq N$.

Example 2. Let $X = \{x_1, x_2, x_3\}$ be the discourse set, and

$$M = \{\frac{x_1}{(0.4),(0.5)}, \frac{x_2}{(0.4,0.5),(0.5,0.3)}, \frac{x_3}{(0.2,0.3,0.5,0.6),(0.7,0.6,0.4,0.3)}\},$$
$$N = \{\frac{x_1}{(0.5,0.7),(0.4,0.2)}, \frac{x_2}{(0.6),(0.3)}, \frac{x_3}{(0.5,0.6),(0.3,0.2)}\}.$$

be two GHFS sets on X. Then, the GHF sets M, N can be respectively represented as

$$M = \{\frac{x_1}{(0.4,0.4),(0.5,0.5)}, \frac{x_2}{(0.4,0.5),(0.5,0.3)}, \frac{x_3}{(0.2,0.3,0.5,0.6),(0.7,0.6,0.4,0.3)}\},$$
$$N = \{\frac{x_1}{(0.5,0.7),(0.4,0.2)}, \frac{x_2}{(0.6,0.6),(0.3,0.3)}, \frac{x_3}{(0.5,0.6,0.6,0.6),(0.3,0.2,0.2,0.2)}\}.$$

We can find that $\mu_j^M(x_i) \leq \mu_j^N(x_i)$ and $\nu_j^M(x_i) \geq \nu_j^N(x_i)$, for each $x_i \in X$ and each $1 \leq j \leq l_{x_i}$. Then, M is a generalized hesitant fuzzy subset of N and denote by $M \sqsubseteq N$.

3 Some New Distance Between Two Generalized Hesitant Fuzzy Sets

Definition 6. *Let* M, N *be two GHFS sets on* X *and* $d_{max} = max\{d(A, B), A, B \in GHFS\}$. *Then the distance measure between* M, N *is defined as* $d(M, N)$, *which satisfies the following properties:*

(D1) $0 \leq d(M,N) \leq d_{max}$
(D2) $d(M,N) = 0 \Leftrightarrow M = N$
(D3) $d(M,N) = d(N,M)$
(D4) Let P be a GHFS set on X. If $M \sqsubseteq N \sqsubseteq P$. *Then,* $d(M,N) \leq d(M,P)$ *and* $d(N,P) \leq d(M,P)$.

If (D1)′ replaces (D1), the $d(M,N)$ is called a normalized distance measure, where (D1)′ $0 \leq d(M,N) \leq 1$.

Some distance are defined in [18], we generalized them in three parametric.

Definition 7. *GH normalized Hamming distance:*
$d_{GHnh}(M,N) = \frac{1}{2n}\sum_{i=1}^{n}[\frac{1}{l_{x_i}}\sum_{j=1}^{l_{x_i}}|\mu_M^{\sigma(j)}(x_i) - \mu_N^{\sigma(j)}(x_i)| + \frac{1}{l_{x_i}}\sum_{j=1}^{l_{x_i}}|\nu_M^{\sigma(j)}(x_i) - \nu_N^{\sigma(j)}(x_i)| + \frac{1}{l_{x_i}}\sum_{j=1}^{l_{x_i}}|\pi_M^{\sigma(j)}(x_i) - \pi_N^{\sigma(j)}(x_i)|]$

Definition 8. *GH normalized Euclidean distance:*
$d_{GHne}(M,N) = [\frac{1}{2n}\sum_{i=1}^{n}[\frac{1}{l_{x_i}}\sum_{j=1}^{l_{x_i}}|\mu_M^{\sigma(j)}(x_i) - \mu_N^{\sigma(j)}(x_i)|^2 + \frac{1}{l_{x_i}}\sum_{j=1}^{l_{x_i}}|\nu_M^{\sigma(j)}(x_i) - \nu_N^{\sigma(j)}(x_i)|^2 + \frac{1}{l_{x_i}}\sum_{j=1}^{l_{x_i}}|\pi_M^{\sigma(j)}(x_i) - \pi_N^{\sigma(j)}(x_i)|^2]^{\frac{1}{2}}$

where $\mu_M^{\sigma(j)}(x_i)$, $\nu_M^{\sigma(j)}(x_i)$, $\pi_M^{\sigma(j)}(x_i)$, $\mu_N^{\sigma(j)}(x_i)$, $\nu_N^{\sigma(j)}(x_i)$ and $\pi_N^{\sigma(j)}(x_i)$ are the jth values in $\mu_M(x_i)$, $\nu_M(x_i)$, $\pi_M(x_i)$, $\mu_N(x_i)$, $\nu_N(x_i)$ and $\pi_N(x_i)$, respectively, which will be used thereafter.

We can further extend Def 7 and 8 into a generalized GH normalized distance:

Definition 9. *Generalized GH normalized distance:*
$d_{gGHn}(M,N) = [\frac{1}{2n}\sum_{i=1}^{n}[\frac{1}{l_{x_i}}\sum_{j=1}^{l_{x_i}}|\mu_M^{\sigma(j)}(x_i) - \mu_N^{\sigma(j)}(x_i)|^\alpha + \frac{1}{l_{x_i}}\sum_{j=1}^{l_{x_i}}|\nu_M^{\sigma(j)}(x_i) - \nu_N^{\sigma(j)}(x_i)|^\alpha + \frac{1}{l_{x_i}}\sum_{j=1}^{l_{x_i}}|\pi_M^{\sigma(j)}(x_i) - \pi_N^{\sigma(j)}(x_i)|^\alpha]^{\frac{1}{\alpha}}$, *where* $\alpha > 0$.

Especially, if $\alpha = 1$, then the generalized GH normal distance is reduced to the GH normalized Hamming distance; If $\alpha = 2$, then it is reduced to the GH normalized Euclidean distance.

Combining the above equations, we define a hybrid GH normalized Hamming distance, a hybrid GH normalized Euclidean distance, and a generalized hybrid GH normalized distance as follows, respectively:

Definition 10. *Hybrid GH normalized Hamming distance:*
$d_{hGHnh}(M,N) = \frac{1}{4n}\sum_{i=1}^{n}[\frac{1}{l_{x_i}}\sum_{j=1}^{l_{x_i}}|\mu_M^{\sigma(j)}(x_i) - \mu_N^{\sigma(j)}(x_i)| + \frac{1}{l_{x_i}}\sum_{j=1}^{l_{x_i}}|\nu_M^{\sigma(j)}(x_i) - \nu_N^{\sigma(j)}(x_i)| + \frac{1}{l_{x_i}}\sum_{j=1}^{l_{x_i}}|\pi_M^{\sigma(j)}(x_i) - \pi_N^{\sigma(j)}(x_i)|] + \frac{1}{2n}\sum_{i=1}^{n} max_j\{|\mu_M^{\sigma(j)}(x_i) - \mu_N^{\sigma(j)}(x_i)|, |\nu_M^{\sigma(j)}(x_i) - \nu_N^{\sigma(j)}(x_i)|\}$

Definition 11. *Hybrid GH normalized Euclidean distance:*
$d_{hGHne}(M,N) = [\frac{1}{4n}\sum_{i=1}^{n}[\frac{1}{l_{x_i}}\sum_{j=1}^{l_{x_i}}|\mu_M^{\sigma(j)}(x_i) - \mu_N^{\sigma(j)}(x_i)|^2 + \frac{1}{l_{x_i}}\sum_{j=1}^{l_{x_i}}|\nu_M^{\sigma(j)}(x_i) - \nu_N^{\sigma(j)}(x_i)|^2 + \frac{1}{l_{x_i}}\sum_{j=1}^{l_{x_i}}|\pi_M^{\sigma(j)}(x_i) - \pi_N^{\sigma(j)}(x_i)|^2] + \frac{1}{2n}\sum_{i=1}^{n} max_j\{|\mu_M^{\sigma(j)}(x_i) - \mu_N^{\sigma(j)}(x_i)|^2, |\nu_M^{\sigma(j)}(x_i) - \nu_N^{\sigma(j)}(x_i)|^2\}]^{\frac{1}{2}}$

Definition 12. *Generalized hybrid GH normalized distance:*

$d_{ghGHn}(M,N) = [\frac{1}{4n}\sum_{i=1}^{n}[\frac{1}{l_{x_i}}\sum_{j=1}^{l_{x_i}}|\mu_M^{\sigma(j)}(x_i) - \mu_N^{\sigma(j)}(x_i)|^\alpha + \frac{1}{l_{x_i}}\sum_{j=1}^{l_{x_i}}|\nu_M^{\sigma(j)}(x_i) - \nu_N^{\sigma(j)}(x_i)|^\alpha + \frac{1}{l_{x_i}}\sum_{j=1}^{l_{x_i}}|\pi_M^{\sigma(j)}(x_i) - \pi_N^{\sigma(j)}(x_i)|^\alpha] + \frac{1}{2n}\sum_{i=1}^{n} max_j\{|\mu_M^{\sigma(j)}(x_i) - \mu_N^{\sigma(j)}(x_i)|^\alpha, |\nu_M^{\sigma(j)}(x_i) - \nu_N^{\sigma(j)}(x_i)|^\alpha\}]^{\frac{1}{\alpha}}$, *where* $\alpha > 0$.

Assume that the weight of the element $x_i \in X$ is $\omega_i (i = 1, 2, ..., n)$ with $\omega_i \in [0, 1]$ and $\Sigma_{i=1}^{n}\omega_i = 1$, then we get a generalized GH weighted distance:

Definition 13. *Generalized GH weighted distance:*

$d_{gGHw}(M,N) = [\frac{1}{2}\sum_{i=1}^{n}\omega_i(\frac{1}{l_{x_i}}\sum_{j=1}^{l_{x_i}}|\mu_M^{\sigma(j)}(x_i) - \mu_N^{\sigma(j)}(x_i)|^\alpha + \frac{1}{l_{x_i}}\sum_{j=1}^{l_{x_i}}|\nu_M^{\sigma(j)}(x_i) - \nu_N^{\sigma(j)}(x_i)|^\alpha + \frac{1}{l_{x_i}}\sum_{j=1}^{l_{x_i}}|\pi_M^{\sigma(j)}(x_i) - \pi_N^{\sigma(j)}(x_i)|^\alpha)]^{\frac{1}{\alpha}}$, *where* $\alpha > 0$.

Definition 14. *Generalized hybrid GH weighted distance:*

$d_{ghGHw}(M,N) = [\frac{1}{2}\sum_{i=1}^{n}\omega_i(\frac{1}{l_{x_i}}\sum_{j=1}^{l_{x_i}}|\mu_M^{\sigma(j)}(x_i) - \mu_N^{\sigma(j)}(x_i)|^\alpha + \frac{1}{l_{x_i}}\sum_{j=1}^{l_{x_i}}|\nu_M^{\sigma(j)}(x_i) - \nu_N^{\sigma(j)}(x_i)|^\alpha$
$+\frac{1}{l_{x_i}}\sum_{j=1}^{l_{x_i}}|\pi_M^{\sigma(j)}(x_i) - \pi_N^{\sigma(j)}(x_i)|^\alpha) + \sum_{i=1}^{n}\omega_i max_j\{|\mu_M^{\sigma(j)}(x_i) - \mu_N^{\sigma(j)}(x_i)|^\alpha, |\nu_M^{\sigma(j)}(x_i) - \nu_N^{\sigma(j)}(x_i)|^\alpha\}]^{\frac{1}{\alpha}}$, *where* $\alpha > 0$.

In the special cases where $\alpha = 1, 2$, Def 14 is reduced to a hybrid GH weighted Hamming distance and a hybrid GH weighted Euclidean distance as follows, respectively:

$d_{hGHwh}(M,N) = \frac{1}{2}\sum_{i=1}^{n}\omega_i(\frac{1}{l_{x_i}}\sum_{j=1}^{l_{x_i}}|\mu_M^{\sigma(j)}(x_i) - \mu_N^{\sigma(j)}(x_i)| + \frac{1}{l_{x_i}}\sum_{j=1}^{l_{x_i}}|\nu_M^{\sigma(j)}(x_i) - \nu_N^{\sigma(j)}(x_i)|$
$+\frac{1}{l_{x_i}}\sum_{j=1}^{l_{x_i}}|\pi_M^{\sigma(j)}(x_i) - \pi_N^{\sigma(j)}(x_i)|) + \sum_{i=1}^{n}\omega_i max_j\{|\mu_M^{\sigma(j)}(x_i) - \mu_N^{\sigma(j)}(x_i)|, |\nu_M^{\sigma(j)}(x_i) - \nu_N^{\sigma(j)}(x_i)|\}$

$d_{hGHwe}(M,N) = [\frac{1}{2}\sum_{i=1}^{n}\omega_i(\frac{1}{l_{x_i}}\sum_{j=1}^{l_{x_i}}|\mu_M^{\sigma(j)}(x_i) - \mu_N^{\sigma(j)}(x_i)|^2 + \frac{1}{l_{x_i}}\sum_{j=1}^{l_{x_i}}|\nu_M^{\sigma(j)}(x_i) - \nu_N^{\sigma(j)}(x_i)|^2$
$+\frac{1}{l_{x_i}}\sum_{j=1}^{l_{x_i}}|\pi_M^{\sigma(j)}(x_i) - \pi_N^{\sigma(j)}(x_i)|^2) + \sum_{i=1}^{n}\omega_i max_j\{|\mu_M^{\sigma(j)}(x_i) - \mu_N^{\sigma(j)}(x_i)|^2, |\nu_M^{\sigma(j)}(x_i) - \nu_N^{\sigma(j)}(x_i)|^2\}]^{\frac{1}{2}}$.

If both the universe of discourse and the weight of element are continuous, then we define a continuous GH weighted Hamming distance, a continuous GH weighted Euclidean distance and a generalized continuous GH weighted distance as follows, respectively:

Definition 15. *Continuous GH weighted Hamming distance:*

$d_{cGHwh}(M,N) = \frac{1}{2}\int_a^b \omega(x)(\frac{1}{l_x}\sum_{j=1}^{l_x}|\mu_M^{\sigma(j)}(x) - \mu_N^{\sigma(j)}(x)| + \frac{1}{l_x}\sum_{j=1}^{l_x}|\nu_M^{\sigma(j)}(x) - \nu_N^{\sigma(j)}(x)| + \frac{1}{l_x}\sum_{j=1}^{l_x}|\pi_M^{\sigma(j)}(x) - \pi_N^{\sigma(j)}(x)|)dx$

Definition 16. *Continuous GH weighted Euclidean distance:*

$d_{cGHwe}(M,N) = [\frac{1}{2}\int_a^b \omega(x)(\frac{1}{l_x}\sum_{j=1}^{l_x}|\mu_M^{\sigma(j)}(x) - \mu_N^{\sigma(j)}(x)|^2 + \frac{1}{l_x}\sum_{j=1}^{l_x}|\nu_M^{\sigma(j)}(x) - \nu_N^{\sigma(j)}(x)|^2 + \frac{1}{l_x}\sum_{j=1}^{l_x}|\pi_M^{\sigma(j)}(x) - \pi_N^{\sigma(j)}(x)|^2)dx]^{\frac{1}{2}}$

Definition 17. *Generalized continuous GH weighted distance:*
$d_{gcGHw}(M,N) = [\frac{1}{2}\int_a^b \omega(x)(\frac{1}{l_x}\sum_{j=1}^{l_x}|\mu_M^{\sigma(j)}(x) - \mu_N^{\sigma(j)}(x)|^\alpha + \frac{1}{l_x}\sum_{j=1}^{l_x}|\nu_M^{\sigma(j)}(x) - \nu_N^{\sigma(j)}(x)|^\alpha + \frac{1}{l_x}\sum_{j=1}^{l_x}|\pi_M^{\sigma(j)}(x) - \pi_N^{\sigma(j)}(x)|^\alpha)dx]^{\frac{1}{\alpha}}$, *where* $\alpha > 0$.

If $\omega(x) = \frac{1}{b-a}$, for any $x \in [a,b]$, then the continuous GH weighted Hamming distance is reduced to a continuous GH normalized Hamming distance:

Definition 18. *Continuous GH normalized Hamming distance:*
$d_{cGHnh}(M,N) = \frac{1}{2(b-a)}\int_a^b(\frac{1}{l_x}\sum_{j=1}^{l_x}|\mu_M^{\sigma(j)}(x) - \mu_N^{\sigma(j)}(x)| + \frac{1}{l_x}\sum_{j=1}^{l_x}|\nu_M^{\sigma(j)}(x) - \nu_N^{\sigma(j)}(x)| + \frac{1}{l_x}\sum_{j=1}^{l_x}|\pi_M^{\sigma(j)}(x) - \pi_N^{\sigma(j)}(x)|)dx$

while the continuous GH weighted Euclidean distance is reduced to a continuous GH normalized Euclidean distance:

Definition 19. *Continuous GH normalized Euclidean distance*
$d_{cGHne}(M,N) = [\frac{1}{2(b-a)}\int_a^b(\frac{1}{l_x}\sum_{j=1}^{l_x}|\mu_M^{\sigma(j)}(x) - \mu_N^{\sigma(j)}(x)|^2 + \frac{1}{l_x}\sum_{j=1}^{l_x}|\nu_M^{\sigma(j)}(x) - \nu_N^{\sigma(j)}(x)|^2 + \frac{1}{l_x}\sum_{j=1}^{l_x}|\pi_M^{\sigma(j)}(x) - \pi_N^{\sigma(j)}(x)|^2)dx]^{\frac{1}{2}}$

and the generalized continuous GH weighted distance is reduced to a generalized continuous GH normalized distance:

Definition 20. *Generalized continuous GH normalized distance:*
$d_{gcGHn}(M,N) = [\frac{1}{2(b-a)}\int_a^b(\frac{1}{l_x}\sum_{j=1}^{l_x}|\mu_M^{\sigma(j)}(x) - \mu_N^{\sigma(j)}(x)|^\alpha + \frac{1}{l_x}\sum_{j=1}^{l_x}|\nu_M^{\sigma(j)}(x) - \nu_N^{\sigma(j)}(x)|^\alpha + \frac{1}{l_x}\sum_{j=1}^{l_x}|\pi_M^{\sigma(j)}(x) - \pi_N^{\sigma(j)}(x)|^\alpha)dx]^{\frac{1}{\alpha}}$, *where* $\alpha > 0$.

Analogous to the generalized hybrid GH weighted distance, we develop a generalized hybrid continuous GH weighted distance as

Definition 21. *Generalized hybrid continuous GH weighted distance:*
$d_{ghcGHw}(M,N) = [\int_a^b \omega(x)[\frac{1}{4l_x}\sum_{j=1}^{l_x}|\mu_M^{\sigma(j)}(x) - \mu_N^{\sigma(j)}(x)|^\alpha + \frac{1}{4l_x}\sum_{j=1}^{l_x}|\nu_M^{\sigma(j)}(x) - \nu_N^{\sigma(j)}(x)|^\alpha$
$+\frac{1}{4l_x}\sum_{j=1}^{l_x}|\pi_M^{\sigma(j)}(x) - \pi_N^{\sigma(j)}(x)|^\alpha + \frac{1}{2}max_j\{|\mu_M^{\sigma(j)}(x) - \mu_N^{\sigma(j)}(x)|^\alpha, |\nu_M^{\sigma(j)}(x) - \nu_N^{\sigma(j)}(x)|^\alpha\}]dx]^{\frac{1}{\alpha}}$, *where* $\alpha > 0$.

If $\omega(x) = \frac{1}{b-a}$, *for any* $x \in [a,b]$, *then the generalized hybrid continuous GH weighted distance becomes a generalized hybrid continuous GH normalized distance:*

Definition 22. *Generalized hybrid continuous GH normalized distance*
$d_{ghcGHn}(M,N) = [\frac{1}{b-a}\int_a^b[\frac{1}{4l_x}\sum_{j=1}^{l_x}|\mu_M^{\sigma(j)}(x) - \mu_N^{\sigma(j)}(x)|^\alpha + \frac{1}{4l_x}\sum_{j=1}^{l_x}|\nu_M^{\sigma(j)}(x) - \nu_N^{\sigma(j)}(x)|^\alpha$
$+\frac{1}{4l_x}\sum_{j=1}^{l_x}|\pi_M^{\sigma(j)}(x) - \pi_N^{\sigma(j)}(x)|^\alpha + \frac{1}{2}max_j\{|\mu_M^{\sigma(j)}(x) - \mu_N^{\sigma(j)}(x)|^\alpha, |\nu_M^{\sigma(j)}(x) - \nu_N^{\sigma(j)}(x)|^\alpha\}]dx]^{\frac{1}{\alpha}}$, *where* $\alpha > 0$.

Let $\alpha = 1, 2$, then from Definition 21, we get a hybrid continuous GH weighted Hamming distance and a hybrid continuous GH weighted Euclidean distance as:

$$d_{hcGHwh}(M,N) = \int_a^b \omega(x)[\frac{1}{4l_x}\sum_{j=1}^{l_x}|\mu_M^{\sigma(j)}(x) - \mu_N^{\sigma(j)}(x)| + \frac{1}{4l_x}\sum_{j=1}^{l_x}|\nu_M^{\sigma(j)}(x) - \nu_N^{\sigma(j)}(x)| + \frac{1}{4l_x}\sum_{j=1}^{l_x}|\pi_M^{\sigma(j)}(x) - \pi_N^{\sigma(j)}(x)| + \frac{1}{2}max_j\{|\mu_M^{\sigma(j)}(x) - \mu_N^{\sigma(j)}(x)|, |\nu_M^{\sigma(j)}(x) - \nu_N^{\sigma(j)}(x)|\}]dx$$

and

$$d_{hcGHwe}(M,N) = [\int_a^b \omega(x)[\frac{1}{4l_x}\sum_{j=1}^{l_x}|\mu_M^{\sigma(j)}(x) - \mu_N^{\sigma(j)}(x)|^2 + \frac{1}{4l_x}\sum_{j=1}^{l_x}|\nu_M^{\sigma(j)}(x) - \nu_N^{\sigma(j)}(x)|^2 + \frac{1}{4l_x}\sum_{j=1}^{l_x}|\pi_M^{\sigma(j)}(x) - \pi_N^{\sigma(j)}(x)|^2 + \frac{1}{2}max_j\{|\mu_M^{\sigma(j)}(x) - \mu_N^{\sigma(j)}(x)|^2, |\nu_M^{\sigma(j)}(x) - \nu_N^{\sigma(j)}(x)|^2\}]dx]^{\frac{1}{2}}$$

respectively.

Let $\omega(x) = \frac{1}{b-a}$, for any $x \in [a,b]$, then the above two distances are reduced to a hybrid continuous GH normalized Hamming distance:

$$d_{hcGHnh}(M,N) = \frac{1}{b-a}\int_a^b[\frac{1}{4l_x}\sum_{j=1}^{l_x}|\mu_M^{\sigma(j)}(x) - \mu_N^{\sigma(j)}(x)| + \frac{1}{4l_x}\sum_{j=1}^{l_x}|\nu_M^{\sigma(j)}(x) - \nu_N^{\sigma(j)}(x)| + \frac{1}{4l_x}\sum_{j=1}^{l_x}|\pi_M^{\sigma(j)}(x) - \pi_N^{\sigma(j)}(x)| + \frac{1}{2}max_j\{|\mu_M^{\sigma(j)}(x) - \mu_N^{\sigma(j)}(x)|, |\nu_M^{\sigma(j)}(x) - \nu_N^{\sigma(j)}(x)|\}]dx$$

and a hybrid continuous GH normalized Euclidean distance:

$$d_{hcGHne}(M,N) = [\frac{1}{b-a}\int_a^b[\frac{1}{4l_x}\sum_{j=1}^{l_x}|\mu_M^{\sigma(j)}(x) - \mu_N^{\sigma(j)}(x)|^2 + \frac{1}{4l_x}\sum_{j=1}^{l_x}|\nu_M^{\sigma(j)}(x) - \nu_N^{\sigma(j)}(x)|^2 + \frac{1}{4l_x}\sum_{j=1}^{l_x}|\pi_M^{\sigma(j)}(x) - \pi_N^{\sigma(j)}(x)|^2 + \frac{1}{2}max_j\{|\mu_M^{\sigma(j)}(x) - \mu_N^{\sigma(j)}(x)|^2, |\nu_M^{\sigma(j)}(x) - \nu_N^{\sigma(j)}(x)|^2\}]dx]^{\frac{1}{2}}$$

respectively.

Concluding Remarks. In the following paper, we will study entropy, similarity of generalized hesitant fuzzy sets and their applications in decision-making as in [12]. Also, we will investigate the applications of generalized hesitant fuzzy sets with multiopinions in decision-making problems in the following paper.

Acknowledgment. This paper is supported by the Project of Shandong Province Higher Educational Science and Technology Program 2016 (J16LI08).

References

1. Xu, Z.S., Xia, M.M.: Distance and similarity measures for hesitant fuzzy sets. Inf. Sci. **181**(11), 2128–2138 (2011)
2. Xu, Z.S., Xia, M.M.: On distance and correlation measures of hesitant fuzzy information. Int. J. Intell. Syst. **26**(5), 410–425 (2011)
3. Farhadinia, B.: Information measures for hesitant fuzzy sets and interval-valued hesitant fuzzy sets. Inform. Sci. **240**, 129–144 (2013)
4. Chen, N., Xu, Z., Xia, M.: Correlation coefficients of hesitant fuzzy sets and their applications to clustering analysis. Appl. Math. Model. **37**(4), 2197–2211 (2013)

5. Bedregal, B., Reiser, R., Bustince, H., Lopez-Molina, C., Torra, V.: Aggregation functions for typical hesitant fuzzy elements and the action of automorphisms. Inform. Sci. **255**, 82–99 (2014)
6. Qian, G., Wanga, H., Feng, X.: Generalized hesitant fuzzy sets and their application in decision support system. Knowl.-Based Syst. **37**, 357–365 (2013)
7. Li, D.F., Cheng, C.T.: New similarity measures of intuitionistic fuzzy sets and application to pattern recognitions. Pattern Recogn. Lett. **23**, 221–225 (2002)
8. Yang, M.S., Lin, D.C.: On similarity and inclusion measures between type-2 fuzzy sets with an application to clustering. Comput. Math. Appl. **57**, 896–907 (2009)
9. Wang, T.J., Lu, Z.D., Li, F.: Bidirectional approximate reasoning based on weighted similarity measures of vague sets. J. Comput. Eng. Sci. **24**, 96–100 (2002)
10. Pal, S.K., King, R.A.: Image enhancement using smoothing with fuzzy sets. IEEE Trans. Syst. Man Cybern. **11**, 495–501 (1981)
11. Szmidt, E., Kacprzyk, J.: Intuitionistic fuzzy sets in intelligent data analysis for medical diagnosis. In: Alexandrov, V.N., Dongarra, J., Juliano, B.A., Renner, R.S., Tan, C.J.K. (eds.) ICCS 2001. LNCS, vol. 2074, pp. 263–271. Springer, Heidelberg (2001)
12. Xu, Z.S.: A method based on distance measure for interval-valued intuitionistic fuzzy group decision making. Inf. Sci. **180**, 181–190 (2010)
13. Liang, Z.Z., Shi, P.F.: Similarity measures on intuitionistic fuzzy sets. Pattern Recogn. Lett. **24**, 2687–2693 (2003)
14. Mitchell, H.B.: On the Dengfeng-Chuntian similarity measure and its application to pattern recognition. Pattern Recogn. Lett. **24**, 3101–3104 (2003)
15. Szmidt, E., Kacprzyk, J.: Distances between intuitionistic fuzzy sets. Fuzzy Sets Syst. **114**, 505–518 (2000)
16. Wang, W.Q., Xin, X.L.: Distance measure between intuitionistic fuzzy sets. Pattern Recogn. Lett. **26**, 2063–2069 (2005)
17. Hung, W.L., Yang, M.S.: Similarity measures of intuitionistic fuzzy sets based on Hausdorff distance. Pattern Recogn. Lett. **25**, 1603–1611 (2004)
18. Bin, C.: Distance and similarity measures for generalized hesitant fuzzy soft sets. In: Advances in Computational Science and Computing. ISCSC 2018. Advances in Intelligent Systems and Computing, vol. 877, pp. 396–403. Springer (2019)

Determination of Multiple q Values for Tsallis-Entropy-Maximized-FCM

Makoto Yasuda(✉)

National Institute of Technology, Gifu College,
Motosu 2236-2, Gifu 501-0495, Japan
yasuda@gifu-nct.ac.jp

Abstract. Based on the fuzzy c-means clustering method maximized with the Tsallis entropy, we have achieved its extension that assigns the q parameter of Tsallis entropy to each cluster as q_i.

In this method, however, there remains three problems. That is, (1) determination q_i according to the data distribution, (2) occurrence of abnormal bias of q_i, and (3) calculation termination condition of q_i. In this article, we propose a new calculation method and a termination condition of q_i, and show its effects by experiments.

1 Introduction

Statistical mechanics investigates the macroscopic properties of a physical system consisting of several elements. Recently, research activities have become popular that attempt to apply statistical mechanical models or tools to information science.

There exists a strong relationship between the membership functions of fuzzy c-means (FCM) clustering [1] with the maximum entropy or entropy regularization methods [2,3] and the statistical mechanical distribution functions. That is, FCM regularized or maximized with Shannon entropy gives a membership function similar to the Boltzmann (or Gibbs) distribution function [2,4], and FCM regularized or maximized with fuzzy entropy [5] gives a membership function similar to the Fermi-Dirac distribution function [6]. These membership functions are suitable for the annealing methods because they contain a parameter corresponding to a system temperature.

Tsallis [7] achieved nonextensive extension of the Boltzmann-Gibbs statistics. Tsallis postulated a generalization form of entropy with a generalization parameter q, which, in a limit of q to 1, reaches Shannon entropy. Tsallis entropy is applicable to many fields such as physics, chemistry, bioscience, networks and computer science, and proved to be useful [8–10]. In the field of fuzzy clustering, Menard et.al. [11] derived a membership function by regularizing FCM with Tsallis entropy.

We achieved Tsallis entropy maximization within the framework of FCM [12]. One of the important characteristics of this membership function is that its

Y. Liu et al. (Eds.): ICNC-FSKD 2019, AISC 1074, pp. 771–780, 2020.
https://doi.org/10.1007/978-3-030-32456-8_83

extent strongly depends on q as well as a system temperature. This method is combined with the deterministic annealing or DA [4] as Tsallis-DAFCM [13].

Then, we proposed a multi-q extension method of Tsallis entropy based FCM. In this method, qs are assigned to all clusters one by one, that is, q_i to ith cluster. This method is also combined with DA as q_i-Tsallis-DAFCM [14]. The advantege of q_i-Tsallis-DAFCM is that it can improve the accuracy of clustering without consuming a processing time.

However, the following problems are still remaining in q_i-Tsallis-DAFCM.

(1) The determination method of q_i according to the data distribution is not fully established.
(2) Depending on the data distribution, the bias of q_i sometimes becomes very large.
(3) The calculation termination condition reflecting the accuracy of q_i is unknown.

In this article, we improve the conventional q_i determination method and verfy its effects by experiments.

2 Tsallis-DAFCM

Let $X = \{\boldsymbol{x}_1, \ldots, \boldsymbol{x}_n\}(\boldsymbol{x}_k = (x_k^1, \ldots, x_k^p) \in R^p)$ $V = \{\boldsymbol{v}_1, \ldots, \boldsymbol{v}_c\}(\boldsymbol{v}_i = (v_i^1, \ldots, v_i^p))$ be a data set the cluster centers in the p-dimensional real space. Let $u_{ik} \in [0, 1](i = 1, \ldots, c; k = 1, \ldots, n)$ be the membership functions.

In Tsallis-DAFCM, the Tsallis entorpy and the objective function are given as follows:

$$S_q = -\frac{1}{q-1}\left(\sum_{k=1}^{n}\sum_{i=1}^{c} u_{ik}^q - 1\right) = -\sum_{k=1}^{n}\sum_{i=1}^{c} u_{ik}^q \mathtt{ln}_q u_{ik}, \tag{1}$$

$$J_{Tsallis} = \sum_{k=1}^{n}\sum_{i=1}^{c} u_{ik}^q d_{ik}. \tag{2}$$

where $\mathtt{ln}_q$ represents the q-logarithm [10], q is any real number, and $d_{ik} = \|\boldsymbol{x}_k - \boldsymbol{v}_i\|^2$.

By applying the Tsallis entropy maximization method, the membership function and the cluster centers become

$$u_{ik} = \frac{\{1-\beta(1-q)d_{ik}\}^{\frac{1}{1-q}}}{\sum_{j=1}^{c}\{1-\beta(1-q)d_{jk}\}^{\frac{1}{1-q}}}, \tag{3}$$

$$\boldsymbol{v}_i = \frac{\sum_{k=1}^{n} u_{ik}^q \boldsymbol{x}_k}{\sum_{k=1}^{n} u_{ik}^q}. \tag{4}$$

Comparing Eq. (2) with the objective function

$$J = \sum_{k=1}^{n} \sum_{i=1}^{c} u_{ik}^{m} d_{ik} \quad (1 < m) \tag{5}$$

of FCM, it can be seen that the q parameter represents the degree of fuzzification.

3 q_iTsallis-DAFCM

In q_i-Tsallis-DAFCM, to assign q_i to each cluster, the Tsallis entropy and the objective function are replaced by the followings:

$$S_{q_i} = -\sum_{k=1}^{n} \sum_{i=1}^{c} u_{ik}^{q_i} \ln_{q_i} u_{ik}, \tag{6}$$

$$J_{q_i Tsallis} = \sum_{k=1}^{n} \sum_{i=1}^{c} u_{ik}^{q_i} d_{ik}. \tag{7}$$

Then, the membership function and the cluster centers are rewritten as

$$u_{ik} = \frac{q_i^{\frac{1}{1-q_i}} \{1 - \beta(1 - q_i) d_{ik}\}^{\frac{1}{1-q_i}}}{\sum_{j=1}^{c} q_j^{\frac{1}{1-q_j}} \{1 - \beta(1 - q_j) d_{jk}\}^{\frac{1}{1-q_j}}}, \tag{8}$$

$$\boldsymbol{v}_i = \frac{\sum_{k=1}^{n} u_{ik}^{q_i} \boldsymbol{x}_k}{\sum_{k=1}^{n} u_{ik}^{q_i}}. \tag{9}$$

3.1 Determination of q_i

In q_i-Tsallis-DAFCM, q_i is updated according to the following equations when the temperature of DA is updated.

$$a_i = \frac{1}{n'} \sum_{k=1}^{n'} \|\boldsymbol{x}'_k - \boldsymbol{v}_i\|^2 \tag{10}$$

$$q_i = (q_{def} - k) + k \frac{a_i}{\bar{a}_i}. \tag{11}$$

In Eq. (10), a_i, an index of the distribution range of data belonging to a cluster i, is determined by letting $X'_i = \{x'_1, \ldots, x'_n\}$ be the data set with the highest degree of membership to the cluster C_i. In Eq. (11), $\bar{a}_i$ represents the average value of a_1 to a_c, and q_{def} and k are given parameters. In an average cluster with $a_i = \bar{a}_i$, q_i becomes q_{def}. Thus, q_{def} corresponds to the average value of q_i. Furthermore, in the case of $k = 0$, q_i becomes q_{def}, which matches Tsallis-FCM.

3.2 Problems of q_i-Tsallis-FCM

(1) q_i-Tsallis-FCM has the following three problems:
The determination method of q_i according to the data distribution is not fully established.
In [3], the value of k is adjusted based on the experimental results. However, it is difficult to obtain appropriate values for a large amount of data or multidimensional data. Furthermore, since there is no restriction on the possible values of k, there is a possibility that $q_i < 1$ if $q_{def} - 1 < k$. Conversely, if k is too small, there is no difference between q_i.

(2) Depending on the data distribution, the bias of q_i sometimes becomes very large.
According to the definition of subset X_i' in Eq. (10), a data that belongs almost equally to multiple clusters affects only a specific a_i. Therefore, in the case of a data set in which multiple clusters are close to each other, X_i' changes according to a slight change in the degree of membership, and then a_i also changes. Especially, when k is large, the bias of a_i becomes remarkable, and q_i is also greatly skewed. As a result, most data belong to one cluster and clustering fails.

(3) The calculation termination condition reflecting the accuracy of q_i is unknown.
Every time the temperature is updated by annealing, a new q_i reflecting the state of the cluster is calculated. However, since optimization at the next temperature is resumed from the cluster state optimized for q_i obtained at the previous temperature, q_i may continue to change and clustering may not converge. Therefore, the trade-off between complexity and an accuracy q_i has to be solved.

3.3 Proposed Method

To solve the problems mentioned in Sect. 3.2, instead of Eqs. (10) and (11), we propose the following equations:

$$a_i = \sum_{k=1}^{n} u_{ik}^{q_i} d_{ik}, \tag{12}$$

$$q_i = 1.0 + (q_{def} - 1.0)\frac{a_i}{\bar{a}_i}. \tag{13}$$

In Eq. (12), the objective function value of each cluster is used as an index of the distribution range (cluster size) of data belonging to the cluster. The difference from Eq. (10) is that $u_{ik}^{q_i}$ is weighted to emphasize the degree of membership, and all data are used to reduce the sensitivity to changes in the degree of membership of a_i. This solves the problem 2 in Sect. 3.2. Also, this makes the calculation of X_i' unnecessary.

Equation (13) is obtained by replacing k by $q_{def} - 1$ in Eq. (11). The purpose is to determine the k value and satisfy the fuzzification constraint $1 < q_i$ at the same time. a_i satisfies $0 < a_i < 1$ because $u_{ik} \neq 0$. Therefore, the second term is a positive number when $1 < q_{def}$, and the constraint is satisfied.

In the proposed method, in addition to replacement of the update equations, a new calculation termination condition of q_i is introduced. The purpose is to prevent the increase in the computational complexity of problem 3 of Sect. 3.2. Therefore, we update q_i at high temperature by regarding high temperature as initialization of clustering because it is at low temperature that detailed clustering is performed according to the data distribution. Then, calculation is finished at low temperature. To distinguish between high and low temperature, we use the sum of distances between cluster centers. The temperature is defined as high temperature while the sum does not exceed the value at the previous temperature.

4 Clustering Algorithm

The algorithm of the proposed method is shown below. δ_1 and δ_2 are the thresholds of convergence conditions. The number of clusters c is assumed to be known in advance.

1 Set the number of clusters c, the highest temperature T_{high}, the temperature reduction parameter, the thresholds of convergence test δ_1 and δ_2. Also, set $q_i = q_{def} (\forall i)$.
2 Generate initial clusters at random positions. Set current temperature to T_{high}.
3 Calculate the membership function u_{ik} by Eq. (8).
4 Calculate the cluster centers $\boldsymbol{v}_i$ by Eq. (9).
5 Compare the difference between the current centers and the those obtained at the previous iteration $\boldsymbol{v}'_i$. If the convergence condition $max_{1 \leq i \leq c} \|\boldsymbol{v}_i - \boldsymbol{v}'_i\| < \delta_1$ is satisfied, then go to **6**. Otherwise go back to **3**.
6 Compare the difference between the current centers and those obtained at the previous temperature $\boldsymbol{v}''_i$. If the convergence condition $max_{1 \leq i \leq c} \|\boldsymbol{v}_i - \boldsymbol{v}''_i\| < \delta_2$ is satisfied, then stop. Otherwise, if the q_i's update condition is satisfied, update q_i using Eqs. (12) and (13), decrease the temperature, and go back to **3**.

5 Experiments

Experiments were conducted to show the effectiveness of the proposed method. We used two-dimensional artificial data obeying normal distribution and SDSS quasar spectra (SDSS) data [15] for the experiment. Tsallis-DAFCM and q_i-Tsallis-DAFCM were used for comparison.

Very-fast annealing [16] was used as the cooling schedule for DA, and the dimensional parameter c_d was set to 2.0 [12].

5.1 Experiment Using Artificial Data

The artificial data is two-dimensional data consisting of 500 data points and three clusters with different sizes (see Fig. 1). An experiment was conducted on this data under the following conditions [3]:

- Initial temperature $T_{high} = 2.0$.
- $q_{def} = 3.0$.
- $\delta_1 = \delta_2 = 0.01$.
- q_i-Tsallis-FCM was implemented with $k = 0.2, 0.4, 0.6$.

Table 1. Mean numbers of a_i, $\bar{a_i}$, q_i and misclassified data points for the artificial data set of the previous methods and the proposed method.

		a_1	a_2	a_3	$\bar{a_3}$	q_1	q_2	q_3	Missclassified
Tsallis-DAFCM		1.73	1.59	0.90	1.41	3.00	3.00	3.00	28.0
q_i-Tsallis-DAFCM	$k = 0.2$	1.77	1.66	0.64	1.37	3.06	3.04	2.89	19.0
	$k = 0.4$	1.79	1.70	0.57	1.35	3.10	3.12	2.76	14.0
	$k = 0.6$	6.60	0.01	0.00	2.20	4.20	2.40	2.40	294.0
Proposed method		0.18	0.18	0.16	0.17	3.08	3.04	2.88	20.0

Clustering was performed 100 times for each method, and the average of the following values after clustering convergence was evaluated.

- $a_i, \bar{a_i}$.
- q_i.
- A number of misclassified data points.

The results are summarized in Table 1.

While a_i obtained by Tsallis-DAFCM and q_i-Tsallis-DAFCM ($k = 0.2, 0.4$) reflects the cluster size in Fig. 1, a_i obtained by the proposed method, as expected, only represents the magnitude relationship of cluster size.

In q_i-Tsallis-DAFCM, the number of misclassified data points increases rapidly when $k = 0.6$, reflecting the situation of problem 2 in Sect. 3.2. In this case, only a_1 became larger, and clustering failed. On the other hand, although k for the proposed method was set to $q_{def} - 1 = 2$, the bias of a_i became smaller, and q_i is almost equal to the one obtained by q_i-Tsallis-DAFCM ($k = 0.2$). As a result, clustering succeeded.

However, the proposed method has more misclassifications than q_i-Tsallis-DAFCM ($k = 0.4$). The cause is the number of updates of q_i. In q_i-Tsallis-DAFCM, q_i is updated each time the temperature is updated, but in the proposed method, the number of updates of q_i is restricted. Under the conditions of this experiment, the proposed method updates q_i only at the time of the first temperature update, and does not update it thereafter. For this reason, q_i was not optimized sufficiently.

From the above results, it is confirmed that the proposed method can determine q_i according to the data distribution with accuracy close to q_i-Tsallis-DAFCM. In addition, we could avoid the occurrence of large bias of a_i, which was pointed out in the problem 2 of Sect. 3.2.

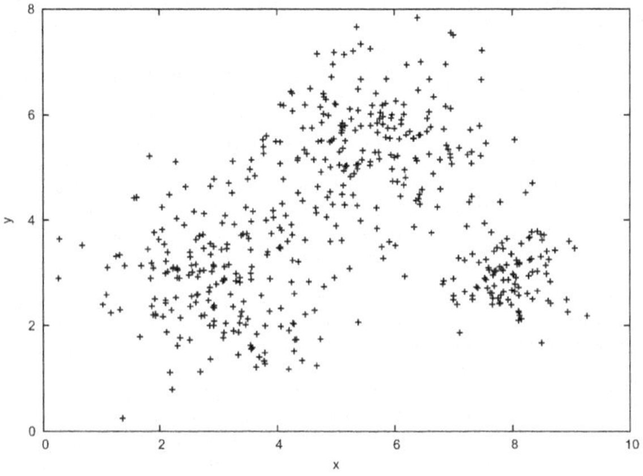

Fig. 1. The artificial data.

5.2 Experiment Using SDSS Data

SDSS Data. SDSS data is observation data of Sloan Digital Sky Survey, which is a large-scale survey of galactic redshifts, and clustering is an important issue of its statistical analysis [17]. In this experiment, the redshifts for spectra of quasars consising of 12-dimensional 9,899 data points were extracted from "Data release 7" including more than 100,000 quasars.

Parallel Clustering Using GPU. In the experiment, calculation of degree of attribution, cluster center and a_i, a'_i were implemented in parallel by CUDA [18] and executed on GPU (GTX 980 Ti). $c \times p$ two-dimensional blocks are used, and the first dimension is parallelized per cluster and the second dimension is parallelized per dimension of the data set. The membership function is updated in each thread, and the cluster center is updated in each block.

Experimental Conditions and Results. The experimental conditions are set as follows:

- $c = 30$.
- $T_{high} = 200.0$.
- $q_{def} = 3.0$.
- $\delta_1 = \delta_2 = 1.0$.
- q_i-Tsallis-DAFCM was implemented with $k = 2^{-4}, 2^{-5}, 2^{-6}$.

These parameters were determined experimentally so that clustering converges. Clustering was performed 100 times for each method, and the maximum, minimum, average and standard deviation of the following values after clustering convergence were evaluated.

Table 2. Maximum, minimum adn mean numbers of $J_{q_i Tsallis}$, a_i, q_i and computational time for the SDSS data set of the previous methods and the proposed method.

		Max	Min	Mean	Std. dev.
$J_{q_i Tsallis} \times 10^6$					
Tsallis-DAFCM		5.93	5.91	5.92	5.24×10^{-3}
q_i-Tsallis-DAFCM	$k = 2^{-6}$	5.94	5.91	5.92	4.57×10^{-3}
	$k = 2^{-5}$	5.95	5.92	5.93	6.64×10^{-3}
	$k = 2^{-4}$	49.23	5.95	44.50	4.10×10^{-3}
Proposed method		6.13	5.89	5.96	4.79×10^{-3}
a_i					
Tsallis-DAFCM		6.41	0.00	2.57	1.55
q_i-Tsallis-DAFCM	$k = 2^{-6}$	8.47	0.00	2.60	1.77
	$k = 2^{-5}$	11.14	0.00	2.67	2.06
	$k = 2^{-4}$	4,666	0.00	47.98	419.89
Proposed method		22.28	18.43	19.93	0.68
q_i					
Tsallis-DAFCM		3.00	3.00	3.00	0
q_i-Tsallis-DAFCM	$k = 2^{-6}$	3.03	2.98	3.00	0.011
	$k = 2^{-5}$	3.09	2.97	3.00	0.026
	$k = 2^{-4}$	4.81	2.94	3.00	0.334
Proposed method		3.23	2.84	3.00	0.062
Computational time [ms]					
Tsallis-DAFCM		1,772	837	1,236	207
q_i-Tsallis-DAFCM	$k = 2^{-6}$	1,707	836	1,238	179
	$k = 2^{-5}$	1,697	866	1,238	179
	$k = 2^{-4}$	6,797	1,262	5,072	790
Proposed method		4,054	1,081	1,016	692

- $J_{q_i Tsallis}$.
- $a_i, \bar{a}_i$.
- q_i.
- Computational time.

The results are summarized in Table 2.

No significant difference of $J_{q_i Tsallis}$ was found in Tsallis-DAFCM, q_i-Tsallis-DAFCM ($k = 2^{-5}, 2^{-6}$) and the proposed method. $J_{q_i Tsallis}$ close to the optimum is considered to be obtained at this time. Principal component analysis of the used SDSS data showed that the data density was high and there was no significant difference in the distribution shape of the data for each cluster. For this reason, the amount of change of q_i became small, and k of q_i-Tsallis-DAFCM also became a minute value.

However, the range of q_i of the proposed method is larger than that of q_i-Tsallis-DAFCM ($k = 2^{-5}, 2^{-6}$). The effect of randomness of the initial cluster positions is more noticeable because k is effectively large in the proposed method.

On the other hand, the average of computational time of the proposed method is about 1.6 times larger than that of Tsallis-DAFCM and q_i-Tsallis-DAFCM ($k = 2^{-5}, 2^{-6}$). This is because the calculation of a_i in q_i-Tsallis-DAFCM converges faster than the update completion of q_i in the proposed method. In addition, the larger range of computational time of the proposed method is are considered to be caused by the random initial cluster positions.

6 Conclusion

The q_i update method in q_i-Tsallis-DAFCM has three problems. In this article, we proposed Eqs. (12) and (13), and the update termination condition for q_i to solve these problems.

The experimental results show that the proposed method can obtain q_i suitable for data distribution. Although the proposed method requires more computational time than the conventional methods, the superiority of the proposed method is clear because it is necessary to find the optimal k in q_i-Tsallis-DAFCM.

On the other hand, the tradeoff between complexity and accuracy of q_i needs to be examined in more detail. In addition, we intend to conduct larger-scale clustering using SDSS data and evaluate its accuracy.

References

1. Bezdek, J.C.: Pattern Recognition with Fuzzy Objective Function Algorithms. Prenum Press, New York (1981)
2. Li, R.P., Mukaidono, M.: A maximum entropy approach to fuzzy clustering. In: Proceedings of the 4th IEEE International Conference on Fuzzy Systems (FUZZ-IEEE/IFES 1995), pp. 2227–2232 (1995)
3. Miyamoto, S., Mukaidono, M.: Fuzzy c-means as a regularization and maximum entropy approach. In: Proceedings of the 7th International Fuzzy Systems Association World Congress, vol. II, pp. 86–92 (1997)

4. Rose, K., Gurewitz, E., Fox, B.C.: A deterministic annealing approach to clustering. Pattern Recogn. Lett. **11**(9), 589–594 (1990)
5. De Luca, A., Termini, S.: A definition of a nonprobabilistic entropy in the setting of fuzzy sets theory. Inf. Control **20**, 301–312 (1972)
6. Yasuda, M., Furuhashi, T., Matsuzaki, M., Okuma, S.: Fuzzy clustering using deterministic annealing method and its statistical mechanical characteristics. In: Proceedings of the 10th IEEE International Conference on Fuzzy Systems, P294 (2001)
7. Tsallis, C.: Possible generalization of Boltzmann-Gibbs statistics. J. Stat. Phys. **52**, 479–487 (1988)
8. Abe, S., Okamoto, Y. (eds.): Nonextensive Statistical Mechanics and Its Applications. Springer (2001)
9. Gell-Mann, M., Tsallis, C. (eds.): Nonextensive Entropy - Interdisciplinary Applications. Oxford University Press, New York (2004)
10. Tsallis, C.: (ed.) Introduction to Nonextensive Statistical Mechanics. Springer (2009)
11. Menard, M., Courboulay, V., Dardignac, P.: Possibilistic and probabilistic fuzzy clustering: unification within the framework of the non-extensive thermostatistics. Pattern Recogn. **36**, 1325–1342 (2003)
12. Yasuda, M.: Entropy maximization and very fast deterministic annealing approach to fuzzy c-means clustering. In: Proceedings of the Joint 5th International Conference on Soft Computing and 11th International Symposium on Intelligent Systems, SU-B1-3, pp. 1515–1520 (2010)
13. Yasuda, M.: Deterministic annealing approach to fuzzy c-means clustering based on entropy maximization. Advances in Fuzzy Systems, 960635 (2011)
14. Yasuda, M., Orito, Y.: Multi-q extension of Tsallis entropy based fuzzy c-means clustering. J. Adv. Comput. Intell. Intell. Inform. **18**(3), 289–296 (2014)
15. Sloan Digital Sky Survey database (2010). http://classic.sdss.org/dr7/
16. Rosen, B.E.: Function optimization based on advanced simulated annealing. In: IEEE Workshop on Physics and Computation, PhysComp 1992, pp. 289–293 (1992)
17. De, T., Burnet, D.F., Chattopadhyay, A.K.: Clustering large number of extragalactic spectra of galaxies and quasars through canopies. Commun. Stat. - Theory Methods **45**(9), 2638–2653 (2016)
18. Sanders, J., Kandrot, E.: CUDA by Example: An Introduction to General-Purpose GPU Programming. Addison-Wesley Professional, Boston (2010)

Very True Operators on Quasi-pseudo-MV Algebras

Guoqing Yang[1], Wenjuan Chen[1(✉)], and Anran Chen[2]

[1] School of Mathematical Sciences, University of Jinan, No. 336, West Road of Nan Xinzhuang, Jinan 250022, Shandong, China
yangguoqing94@163.com, wjchenmath@gmail.com
[2] College of Science and Engineering, University of Minnesota, Twin Cities, USA
chen6265@umn.edu

Abstract. In this paper, we introduce the notion of very true operator on a quasi-pseudo-MV algebra (qpMV-algebra, for short) and investigate the new algebraic structure $qp\mathrm{MV}_{vt}$-algebra which will generalize $ps\mathrm{MV}_{vt}$-algebra defined in [10]. First we discuss some properties of very true operator on a qpMV-algebra. Next we define the dual notion very false operator on a qpMV-algebra and prove that there exists a one-to-one correspondence between very true operators and very false operators on any qpMV-algebra. Finally, some cases of $qp\mathrm{MV}_{vt}$-algebras with truth-depressing hedges are given.

Keywords: Very true operators · Quasi-pseudo-MV algebras · $qp\mathrm{MV}_{vt}$-algebras

1 Introduction

In [3], quasi-pseudo-MV algebras (qpMV-algebras, for short) were introduced as the generalizations of both pseudo-MV algebras [7] and quasi-MV algebras [9]. It is well known that a pseudo-MV algebra is the modal of non-commutative many valued fuzzy logic while a quasi-MV algebra arising from the quantum computational logic is the algebraic model for describing the set of all density operators of the Hilbert space C^2, endowed with a suitable stock of quantum logical gates. Hence qpMV-algebras may play more important role on finding the connection between many valued fuzzy logic and quantum computational logic.

The notion of very true operator on a BL-algebra was introduced by Hajek as a tool for reducing the number of possible logical values in many valued fuzzy logic. Consequently, the concepts of very true operator had been extended to other logical algebras such as pseudo-MV algebras [10], pseudo-BCK algebras [6], equality algebras [14], effect algebras [2], residuated lattices [12] and so on. Thus it is natural to ask whether the concepts of very true operator can be generalized to a qpMV-algebra for the more general results and new applications.

Y. Liu et al. (Eds.): ICNC-FSKD 2019, AISC 1074, pp. 781–790, 2020.
https://doi.org/10.1007/978-3-030-32456-8_84

In this paper, we introduce the notion of very true operator on the qpMV-algebra and investigate the new algebraic structure $qp\text{MV}_{vt}$-algebra. The paper is organized as follows. In Sect. 2, we recall some definitions and results which will be used in what follows. In Sect. 3, we discuss some properties of a very true operator. And then we define the very false operator on a qpMV-algebra and prove that there exists a one-to-one correspondence between very true operators and very false operators on any qpMV-algebra. Some cases of $qp\text{MV}_{vt}$-algebras with truth-depressing hedges are also given.

2 Preliminary

In this section, we list the definition of a qpMV-algebra and some related properties of it. We also present the definitions of very true operators on a BL-algebra and a pseudo-MV algebra.

Definition 1 [3]. *An algebra* $\mathbf{A} = \langle A; \oplus, ^-, ^\sim, 0\rangle$ *of type* $\langle 2, 1, 1, 0\rangle$ *is called a* quasi-pseudo-MV algebra *(qpMV-algebra, for short), if it satisfies the following axioms for any* $x, y, z \in A$*:*

(qpMV1) $x \oplus (y \oplus z) = (x \oplus y) \oplus z$;
(qpMV2) $0^- = 0^\sim$;
(qpMV3) $x \oplus 0 = 0 \oplus x$;
(qpMV4) $x \oplus 0^- = 0^- = 0^- \oplus x$;
(qpMV5) $(x^- \oplus y^-)^\sim = (x^\sim \oplus y^\sim)^-$;
(qpMV6) $x^{-\sim} = x = x^{\sim -}$;
(qpMV7) $y \oplus (x^- \oplus y)^\sim = (y \oplus x^\sim)^- \oplus y = x \oplus (y^- \oplus x)^\sim = (x \oplus y^\sim)^- \oplus x$;
(qpMV8) $(x \oplus 0)^- = x^- \oplus 0$ *and* $(x \oplus 0)^\sim = x^\sim \oplus 0$;
(qpMV9) $x \oplus y \oplus 0 = x \oplus y$.

A qpMV-algebra in which the binary operation $\oplus$ is commutative and the unary operations $^- = ^\sim$, is a quasi-MV algebra (qMV-algebra, for short). On the other hand, a qpMV-algebra satisfying the axiom $x \oplus 0 = x$ is a pseudo-MV algebra (psMV-algebra, for short).

On any qpMV-algebra we can define some operations:

$$\begin{aligned}
x \odot y &= (x^- \oplus y^-)^\sim;\\
x \vee y &= x \oplus (y^- \oplus x)^\sim;\\
x \wedge y &= (x^- \vee y^-)^\sim;\\
x \to y &= x^- \oplus y;\\
x \rightsquigarrow y &= y \oplus x^\sim;\\
x \ominus^L y &= y^- \odot x;\\
x \ominus^R y &= x \odot y^\sim.
\end{aligned}$$

We can also define a relation $x \leq y$ iff $x \vee y = y \oplus 0$, or equivalently, $x \wedge y = x \oplus 0$. This is a quasi-ordering relation [3]. Moreover, if $x \leq y$ and $y \leq x$, then $x \oplus 0 = y \oplus 0$.

Below we list some properties of these operations and the relation. The proofs can be seen in [3,5].

Proposition 1. *Let **A** be a qpMV-algebra. Then for any $x, y, z \in A$:*
(1) $0 \oplus 0 = 0$ *and* $1 \oplus 0 = 1$;
(2) $x \wedge y = (x \to y) \odot x = x \odot (x \rightsquigarrow y)$;
(3) $x \vee y = x \oplus (y \ominus^R x) = (y \ominus^L x) \oplus x$;
(4) $1 \to x = 1 \rightsquigarrow x$;
(5) $x \vee y = (x \vee y) \oplus 0 = (x \oplus 0) \vee y = x \vee (y \oplus 0)$,
$x \wedge y = (x \wedge y) \oplus 0 = (x \oplus 0) \wedge y = x \wedge (y \oplus 0)$;
(6) $x \to y = (x \to y) \oplus 0 = (x \oplus 0) \to y = x \to (y \oplus 0)$,
$x \rightsquigarrow y = (x \rightsquigarrow y) \oplus 0 = (x \oplus 0) \rightsquigarrow y = x \rightsquigarrow (y \oplus 0)$;
(7) $x \leq y$ *iff* $x \to y = 1$ *iff* $x \rightsquigarrow y = 1$;
(8) $0 \leq x \leq 1$;
(9) $x \leq 1 \to x$ *and* $1 \to x \leq x$;
(10) $x \leq y$ *iff* $y^- \leq x^-$ *iff* $y^\sim \leq x^\sim$;
(11) $x \to y \leq (y \to z) \rightsquigarrow (x \to z)$ *and* $x \rightsquigarrow y \leq (y \rightsquigarrow z) \to (x \rightsquigarrow z)$;
(12) $x \to y \leq (z \to x) \to (z \to y)$ *and* $x \rightsquigarrow y \leq (z \rightsquigarrow x) \rightsquigarrow (z \rightsquigarrow y)$;
(13) $x \leq y \to x$ *and* $x \leq y \rightsquigarrow x$;
(14) $x \leq y \to z$ *iff* $y \leq x \rightsquigarrow z$;
(15) *If* $x \leq y$, *then* $y \to z \leq x \to z$ *and* $y \rightsquigarrow z \leq x \rightsquigarrow z$;
(16) *If* $x \leq y$, *then* $z \to x \leq z \to y$ *and* $z \rightsquigarrow x \leq z \rightsquigarrow y$;
(17) $(x \vee y)^- = x^- \wedge y^-$ *and* $(x \vee y)^\sim = x^\sim \wedge y^\sim$,
$(x \wedge y)^- = x^- \vee y^-$ *and* $(x \wedge y)^\sim = x^\sim \vee y^\sim$;
(18) $(x \vee y) \to z = (x \to z) \wedge (y \to z)$ *and* $(x \vee y) \rightsquigarrow z = (x \rightsquigarrow z) \wedge (y \rightsquigarrow z)$,
$z \to (x \vee y) = (z \to x) \vee (z \to y)$ *and* $z \rightsquigarrow (x \vee y) = (z \rightsquigarrow x) \vee (z \rightsquigarrow y)$;
(19) $z \to (x \wedge y) = (z \to x) \wedge (z \to y)$ *and* $z \rightsquigarrow (x \wedge y) = (z \rightsquigarrow x) \wedge (z \rightsquigarrow y)$,
$(x \wedge y) \to z = (x \to z) \vee (y \to z)$ *and* $(x \wedge y) \rightsquigarrow z = (x \rightsquigarrow z) \vee (y \rightsquigarrow z)$;
(20) $(x \to y) \vee (y \to x) = 1$ *and* $(x \rightsquigarrow y) \vee (y \rightsquigarrow x) = 1$.

Let **A** be a *qp*MV-algebra and consider the set $R(A) = \{x \in A | x \oplus 0 = x\}$. Elements of $R(A)$ are called *regular*. Following from Proposition 1, we have that 0 and 1 belong to $R(A)$, so $R(A)$ is a non-empty subset of A. Moreover, $\mathbf{R_A} = \langle R(A); \oplus, ^-, ^\sim, 0\rangle$ is a pseudo-MV subalgebra of **A**. We recall that a *qp*MV-algebra in which $0 = 1$ is called *flat*. Then we can show the following result.

Theorem 1 [3]. *For any qpMV-algebra **A**, there exist a psMV-algebra **M** and a flat qpMV-algebra **F** such that **A** can be embedded into the direct product **M**× **F**.*

Hajek introduced the notion of very true operator on a BL-algebra in [8]. In his paper, a very true operator t is defined as a mapping satisfying $t(1) = 1$, $t(x) \leq x$, $t(x \to y) \leq t(x) \to t(y)$ and $t(x \vee y) \leq t(x) \vee t(y)$ for any x, y. Subsequently, Leustean defined the similar operator on a *ps*MV-algebra [10].

Definition 2 [10]. *A* $psMV_{vt}$-algebra *is a pair* (A, t) *where* A *is a psMV-algebra and* $t : A \to A$ *is a mapping which satisfies the following conditions for any* $x, y \in A$:
(1) $t(1) = 1$;
(2) $t(x \to y) \leq t(x) \to t(y)$;

(3) $t(x) \leq x$;
(4) $t(x \rightarrow y) = t(x \rightsquigarrow y)$;
(5) $t(x \rightarrow y) \vee (y \rightarrow x) = 1$.

In [10], author also proved that the condition (5) is equivalent to (6) $t(x \vee y) \leq t(x) \vee t(y)$ for any $x, y \in A$.

3 Very True Operators and Very False Operators

In this section, we introduce the notion of very true operator and its dual notion very false operator on a *qp*MV-algebra. We investigate the related properties of very true operators and discuss the relationship between very true operators and very false operators on any *qp*MV-algebra.

Definition 3. *Let* $\boldsymbol{A}$ *be a qpMV-algebra and* $t : A \rightarrow A$ *be a mapping. Then* t *is called a* weak very true operator *(wvt-*operator, *for short) on* $\boldsymbol{A}$, *if the following conditions are satisfied for any* $x, y \in A$:
(VT1) $t(1) = 1$;
(VT2) $t(x) \leq x$;
(VT3) $t(x \rightarrow y) \leq t(x) \rightarrow t(y)$ *and* $t(x \rightsquigarrow y) \leq t(x) \rightsquigarrow t(y)$.
If a wvt-operator t *also satisfies* (VT4) $t(x \vee y) \leq t(x) \vee t(y)$, *then it is called a* very true operator *(vt-*operator, *for short) on* $\boldsymbol{A}$. *The pair* $(\boldsymbol{A}, t)$ *is called a* $qpMV_{vt}$-algebra.

For any *wvt*-operator, the axiom (VT1) means TRUE is very true and the axiom (VT2) means very true is really true. The axiom (VT3) means the truth-value of $t(x) \rightarrow t(y)$ (or $t(x) \rightsquigarrow t(y)$) is at least the same as the truth-value of $t(x \rightarrow y)$ (or $t(x \rightsquigarrow y)$). The axiom (VT4) in a *vt*-operator means the truth-value of $t(x) \vee t(y)$ is at least the same as the truth-value of $t(x \vee y)$.

Remark 1. Following from Definition 3, we have that any *ps*MV$_{vt}$-algebra is a *qp*MV$_{vt}$-algebra.

Example 1. Let $\mathbf{F}$ be a flat *qp*MV-algebra and $t : F \rightarrow F$ be a mapping satisfying $t(1) = 1$ and $t(x) \leq x$ for any $x \in F$. Then t is a *vt*-operator on $\mathbf{F}$. In fact, since $1 = 0$ and $x \oplus y = x \oplus y \oplus 0 = x \oplus y \oplus 1 = 1$, we have that all regular elements are equal to 1. Hence (VT3) and (VT4) are satisfied.

Example 2. Let $\mathbf{A}$ be a *qp*MV-algebra. It is easy to see that the identity mapping $\mathbf{Id}_A$ is a *vt*-operator on $\mathbf{A}$. That is to say that any *qp*MV-algebra can be regarded as a *qp*MV$_{vt}$-algebra.

Example 3. Let $\mathbf{A}$ be a linearly ordered *qp*MV-algebra, i.e., for any $x, y \in A$, we have $x \leq y$ or $y \leq x$. We define a mapping $t : A \rightarrow A$ by $t(1) = 1$ and $t(x) = 0$ for any $x \neq 1$. Then t is a *vt*-operator on $\mathbf{A}$.

Example 4. Let $\mathbf{A}$ be a qpMV-algebra in which the operations are defined as follows:

$\oplus$	0	a	b	1
0	0	b	b	1
a	b	1	1	1
b	b	1	1	1
1	1	1	1	1

and

$'$	
0	1
a	a
b	b
1	0

. In fact, it is a qMV-algebra [9] and then $x \rightarrow y = x' \oplus y = x \rightsquigarrow y$. Define $t(1) = 1$, $t(0) = 0$, $t(a) = b$ and $t(b) = a$. Then t is a vt-operator on $\mathbf{A}$.

Example 5. Let $(\mathbf{M}, t_1)$ be a $ps\mathrm{MV}_{vt}$-algebra and $(\mathbf{F}, t_2)$ be a flat $qp\mathrm{MV}_{vt}$-algebra. Then $\mathbf{M} \times \mathbf{F}$ is a qpMV-algebra. We define $t(\langle x, y\rangle) = (t_1(x), t_2(y))$. Then we have that t is a vt-operator on $\mathbf{M} \times \mathbf{F}$.

Proposition 2. *Let $\boldsymbol{A}$ be a qpMV-algebra and t be a wvt-operator on $\boldsymbol{A}$. Then for any $x, y \in A$:*

(1) $t(0) \oplus 0 = 0$;
(2) *If* $t(x) = 1$, *then* $x \oplus 0 = 1$;
(3) *If* $x \leq y$, *then* $t(x) \leq t(y)$;
(4) *If* $t(x) \leq t(y)$, *then* $t(x) \leq y$;
(5) $t(x^-) \leq (t(x))^-$ *and* $t(x^\sim) \leq (t(x))^\sim$;
(6) $t(x) \odot t(y) \leq t(x \odot y)$;
(7) $t(x \oplus 0) \oplus 0 = t(x) \oplus 0$.

In addition, if t is a vt-operator on $\boldsymbol{A}$, then

(8) $t(x) \vee t(y) = t(x \vee y) \oplus 0$;
(9) $t(x) \wedge t(y) = t(x \wedge y) \oplus 0$;
(10) $t(x \rightarrow y) \vee (y \rightarrow x) = 1$.

Proof. (1) By (VT2), we have $t(0) \leq 0$. And $0 \leq t(0)$, it turns out that $t(0) \oplus 0 = 0$.

(2) Since $1 = t(x) \leq x$ and $x \leq 1$, we have $x \oplus 0 = 1$.

(3) Since $x \leq y$, we have $x \rightarrow y = 1$. By (VT1) and (VT3), $1 = t(1) = t(x \rightarrow y) \leq t(x) \rightarrow t(y)$. Note that $t(x) \rightarrow t(y)$ is a regular element and $t(x) \rightarrow t(y) \leq 1$, we have $t(x) \rightarrow t(y) = 1$, so $t(x) \leq t(y)$.

(4) Since $t(x) \leq t(y)$ and $t(y) \leq y$ by (VT2), we have $t(x) \leq y$.

(5) Since $x^- \leq (1 \rightarrow x)^- = (1 \rightarrow x) \rightarrow 0$, we have $t(x^-) \leq t((1 \rightarrow x) \rightarrow 0) \leq t(1 \rightarrow x) \rightarrow t(0) \leq t(x) \rightarrow t(0) = t(x) \rightarrow (t(0) \oplus 0) = t(x) \rightarrow 0 \leq (t(x))^-$ using (VT3) and (1). The other can be proved similarly.

(6) Since $x \odot y \leq x \odot y$, we have $x \leq y \rightarrow (x \odot y)$, it turns out that $t(x) \leq t(y \rightarrow (x \odot y)) \leq t(y) \rightarrow t(x \odot y)$ by (3) and (VT3), so $t(x) \odot t(y) \leq t(x \odot y)$.

(7) Since $x \leq x \oplus 0$ and $x \oplus 0 \leq x$, we have $t(x) \leq t(x \oplus 0)$ and $t(x \oplus 0) \leq t(x)$ by (3), so $t(x) \oplus 0 = t(x \oplus 0) \oplus 0$.

(8) Since $t(x) \leq t(x \vee y)$ and $t(y) \leq t(x \vee y)$ by (3), we have $t(x) \vee t(y) \leq t(x \vee y)$. By (VT4), $t(x \vee y) \leq t(x) \vee t(y)$. Note that $t(x) \vee t(y)$ is a regular element, we have $t(x) \vee t(y) = t(x \vee y) \oplus 0$.

(9) On the one hand, $t(x \wedge y) \leq t(x) \wedge t(y)$. On the other hand, because $(x \rightarrow y) \vee (y \rightarrow x) = 1$, we have $1 = t(1) = t((x \rightarrow y) \vee (y \rightarrow x)) \leq t((x \rightarrow y) \vee (y \rightarrow x)) \oplus 0 = t(x \rightarrow y) \vee t(y \rightarrow x) = t(x \rightarrow (x \wedge y)) \vee t(y \rightarrow (x \wedge y)) \leq$

$(t(x) \to t(x \wedge y)) \vee (t(y) \to t(x \wedge y)) \leq 1$, so $(t(x) \wedge t(y)) \to t(x \wedge y) = (t(x) \to t(x \wedge y)) \vee (t(y) \to t(x \wedge y)) = 1$, which implies that $t(x) \wedge t(y) \leq t(x \wedge y)$, we get $t(x) \wedge t(y) = t(x \wedge y) \oplus 0$.

(10) Since $1 = t(x \to y) \vee t(y \to x) \leq t(x \to y) \vee (y \to x) \leq 1$ by (9) and (VT2), we get the result for the regularity of $t(x \to y) \vee (y \to x)$.

Proposition 3. *Let $\boldsymbol{A}$ be a qpMV-algebra and t be a vt-operator on $\boldsymbol{R_A}$. Then t can be extended to a vt-operator on $\boldsymbol{A}$.*

Proof. Define $\bar{t}(x) = \begin{cases} t(x), & x \in R(A); \\ t(x \oplus 0), & x \in A \setminus R(A). \end{cases}$ Then $\bar{t}$ is a vt-operator on $\mathbf{A}$. Indeed, $\bar{t}(1) = t(1) = 1$, so the condition (VT1) is true. Now we check the conditions (VT2)-(VT4).

(VT2) For any $x \in A$, if $x \in R(A)$, then $\bar{t}(x) = t(x) \leq x$. If $x \notin R(A)$, then $\bar{t}(x) = t(x \oplus 0) \leq x \oplus 0 \leq x$. Hence $\bar{t}(x) \leq x$.

(VT3) We only prove the first one. The other can be proved similarly. If $x, y \in R(A)$, then $\bar{t}(x \to y) = t(x \to y) \leq t(x) \to t(y) = \bar{t}(x) \to \bar{t}(y)$. If $x \in R(A)$ and $y \notin R(A)$, then $\bar{t}(x \to y) = t(x \to y) = t(x \to (y \oplus 0)) \leq t(x) \to t(y \oplus 0) = \bar{t}(x) \to \bar{t}(y)$. If $x \notin R(A)$ and $y \in R(A)$, the proof is similar as above. If $x, y \notin R(A)$, then $\bar{t}(x \to y) = t(x \to y) = t((x \oplus 0) \to (y \oplus 0)) \leq t(x \oplus 0) \to t(y \oplus 0) = \bar{t}(x) \to \bar{t}(y)$.

(VT4) It is similar to (VT3).

Below we deal with the dual notion of very true operator on a qpMV-algebra. In [1], authors showed that there exists the one-to-one correspondence between very true operators and very false operators in CBA fuzzy logic. However, the correspondence may not be one-to-one in general [12]. For any qpMV-algebra, we will show that the one-to-one correspondence exists.

Definition 4. *Let $\mathbf{A}$ be a qpMV-algebra and $f : A \to A$ be a mapping. Then f is called a* weak very false operator *(wvf-*operator*, for short) on $\mathbf{A}$, if the following conditions are satisfied for any $x, y \in A$:*

(VF1) $f(0) = 0$;

(VF2) $x \leq f(x)$;

(VF3) $f(x) \ominus^L f(y) \leq f(x \ominus^L y)$ *and* $f(x) \ominus^R f(y) \leq f(x \ominus^R y)$.

If a wvf-operator f also satisfies (VF4) $f(x \wedge y) \geq f(x) \wedge f(y)$, *then it is called a* very false operator *(vf-*operator*, for short) on $\mathbf{A}$.*

Let $\mathbf{A}$ be a qpMV-algebra and $\varphi : A \to A$ be a mapping. We define two mappings on $\mathbf{A}$ as follows: $\varphi^{\sim}_{-}(x) = (\varphi(x^{\sim}))^{-}$ and $\varphi^{-}_{\sim}(x) = (\varphi(x^{-}))^{\sim}$ for any $x \in A$. Then we have the following results.

Lemma 1. *Let $(\boldsymbol{A}, t)$ be a $qpMV_{vt}$-algebra. Then*

(1) If $x \leq y$, then $t^{\sim}_{-}(x) \leq t^{\sim}_{-}(y)$ and $t^{-}_{\sim}(x) \leq t^{-}_{\sim}(y)$;

(2) $t^{\sim}_{-}(x \to y) \leq t(x) \to t^{\sim}_{-}(y)$ and $t^{-}_{\sim}(x \rightsquigarrow y) \leq t(x) \rightsquigarrow t^{-}_{\sim}(y)$;

(3) If $x \odot y \leq z$, then $t^{\sim}_{-}(x) \odot t(y) \leq t^{\sim}_{-}(z)$ and $t(x) \odot t^{-}_{\sim}(y) \leq t^{-}_{\sim}(z)$;

(4) $t^{\sim}_{-}(x) \odot t(y) \leq t^{\sim}_{-}(x \odot y)$ and $t(x) \odot t^{-}_{\sim}(y) \leq t^{-}_{\sim}(x \odot y)$.

Proof. (1) Since $x \leq y$, we have $y^{\sim} \leq x^{\sim}$ and then $t(y^{\sim}) \leq t(x^{\sim})$, it turns out that $(t(x^{\sim}))^{-} \leq (t(y^{\sim}))^{-}$, so $t_{-}^{\sim}(x) \leq t_{-}^{\sim}(y)$. The other can be proved similarly.

(2) We have $t_{-}^{\sim}(x \to y) = (t((x \to y)^{\sim}))^{-} = (t(x \odot y^{\sim}))^{-} \leq (t(x) \odot t(y^{\sim}))^{-} = (t(x))^{-} \oplus (t(y^{\sim}))^{-} = t(x) \to t_{-}^{\sim}(y)$ by Proposition 1(6). The other can be proved similarly.

(3) If $x \odot y \leq z$, then $x \leq y \to z$, it follows that $t_{-}^{\sim}(x) \leq t_{-}^{\sim}(y \to z) \leq t(y) \to t_{-}^{\sim}(z)$ by Proposition 1(3) and (1), so we have $t_{-}^{\sim}(x) \odot t(y) \leq t_{-}^{\sim}(z)$. Meanwhile, if $x \odot y \leq z$, then $y \leq x \rightsquigarrow z$, it follows that $t_{\sim}^{-}(y) \leq t_{\sim}^{-}(x \rightsquigarrow z) \leq t(x) \rightsquigarrow t_{\sim}^{-}(z)$, so we have $t(x) \odot t_{\sim}^{-}(y) \leq t_{\sim}^{-}(z)$.

(4) Since $x \odot y \leq x \odot y$, the result is obtained by (3).

Proposition 4. *Let* $\mathbf{A}$ *be a qpMV-algebra. Then*

(1) *If* t *is a vt-operator on* $\mathbf{A}$ *and define* $t_{-}^{\sim} = t_{\sim}^{-} = \bar{t}$, *then* $\bar{t}$ *is a vf-operator on* $\mathbf{A}$;

(2) *If* f *is a vf-operator on* $\mathbf{A}$ *and define* $f_{-}^{\sim} = f_{\sim}^{-} = \tilde{f}$, *then* $\tilde{f}$ *is a vt-operator on* $\mathbf{A}$.

Proof. (1) For any $x, y \in A$, we need to check that $\bar{t}$ satisfies the conditions (VF1)-(VF4) in Definition 3. Now, we have $\bar{t}(0) = t_{-}^{\sim}(0) = (t(0^{\sim}))^{-} = (t(1))^{-} = 1^{-} = 0$ and $\bar{t}(x) = t_{-}^{\sim}(x) = (t(x^{\sim}))^{-} \geq x^{\sim -} = x$ by (VT2) and Proposition 1. Moreover, we have $\bar{t}(x) \ominus^{L} \bar{t}(y) = (\bar{t}(y))^{-} \odot \bar{t}(x) = t(y^{-}) \odot \bar{t}(x) \leq \bar{t}(y^{-} \odot x) = \bar{t}(x \ominus^{L} y)$ and $\bar{t}(x) \ominus^{R} \bar{t}(y) = \bar{t}(x) \odot (\bar{t}(y))^{\sim} = \bar{t}(x) \odot t(y^{\sim}) \leq \bar{t}(x \odot y^{\sim}) = \bar{t}(x \ominus^{R} y)$ by Lemma 1. Finally, $\bar{t}(x \wedge y) = t_{-}^{\sim}(x \wedge y) = (t((x \wedge y)^{\sim}))^{-} = (t(x^{\sim} \vee y^{\sim}))^{-} \geq (t(x^{\sim}) \vee t(y^{\sim}))^{-} = t_{-}^{\sim}(x) \wedge t_{-}^{\sim}(y)$.

(2) It is dual to (1).

Remark 2. Let $\mathbf{A}$ be a qpMV-algebra and t be a vt-operator on $\mathbf{A}$. Following from the proof of Proposition 4, we can see that $t_{-}^{\sim}$ satisfies the (VF1), (VF2), (VF4) and the second equation in (VF3), $t_{\sim}^{-}$ satisfies the (VF1), (VF2), (VF4) and the first equation in (VF3).

Suppose that $\mathbf{A}$ is a qpMV-algebra. If for any $x \in A$, we have $x^{-} = x^{\sim}$ (denoted by x'), then it is a pseudo-quasi-MV algebra, which is indeed a non-commutative generalization of qMV-algebra [4,11].

Corollary 1. *Let* $\mathbf{A}$ *be a pseudo-quasi-MV algebra. Then*

(1) *If* t *is a vt-operator on* $\mathbf{A}$, *then* $\bar{t} = (t(x'))'$ *is a vf-operator on* $\mathbf{A}$;

(2) *If* f *is a vf-operator on* $\mathbf{A}$, *then* $\tilde{f} = (f(x'))'$ *is a vt-operator on* $\mathbf{A}$.

Proposition 5. *Let* $\mathbf{A}$ *be a qpMV-algebra. Then there is a one-to-one correspondence between vt-operators and vf-operators on* $\mathbf{A}$.

Proof. If t is a vt-operator on $\mathbf{A}$, then $\bar{t}$ is a vf-operator on $\mathbf{A}$ by Proposition 4(1). Moreover, for any $x \in A$, we have $\tilde{\bar{t}}(x) = \bar{t}_{-}^{\sim}(x) = (\bar{t}(x^{\sim}))^{-} = (t(x^{\sim -}))^{\sim -} = t(x)$. Conversely, if f is a vf-operator on $\mathbf{A}$, then $\tilde{f}$ is a vt-operator on $\mathbf{A}$ Proposition 4(2). Moreover, for any $x \in A$, we have $\bar{\tilde{f}}(x) = \tilde{f}_{-}^{\sim}(x) = (\tilde{f}(x^{\sim}))^{-} = (f(x^{\sim -}))^{\sim -} = f(x)$.

In [13], author introduced a $\mathrm{BL}_{vt,st}$-algebra which extended a BL_{vt}-algebra $(\mathbf{A}, t)$ by adding an unary operation $s : A \to A$ satisfying the following conditions: $s(0) = 0$, $x \leq s(x)$ and $t(x \to y) \leq s(x) \to s(y)$. Similarly, Ciungu defined the pseudo-$\mathrm{BCK}_{vt,st}$-algebra by adding two truth-depressing hedges operators in [6]. Below we show the similar contents within the background of $qp\mathrm{MV}_{vt}$-algebras.

Definition 5. *Let $(\mathbf{A}, t)$ be a $qpMV_{vt}$-algebra and $s_1, s_2 : A \to A$ be two additional unary operators. If they satisfy the following conditions for any $x, y \in A$:*

(ST1) $s_1(0) = s_2(0) = 0$;

(ST2) $x \leq s_1(x)$ and $x \leq s_2(x)$;

(ST3) $t(x \to y) \leq s_1(x) \to s_1(y)$ and $t(x \rightsquigarrow y) \leq s_2(x) \rightsquigarrow s_2(y)$.

*Then $(\mathbf{A}, t, s_1, s_2)$ is called a $qpMV_{vt,st}$-*algebra *and the pair $s = (s_1, s_2)$ is called a* truth-depressing hedge associated with t.

Remark 3. Let $(\mathbf{A}, t, s_1, s_2)$ be a $qp\mathrm{MV}_{vt,st}$-algebra. If $\mathbf{A}$ is a $q\mathrm{MV}$-algebra, then $x \to y = x \rightsquigarrow y$ for any $x, y \in A$. Under this condition, we require $s_1 = s_2 = s$ and denote this algebra $(\mathbf{A}, t, s)$ for short.

Proposition 6. *Let $\mathbf{A}$ be a linearly ordered $qpMV$-algebra and t be the operator defined in Example 3. If $s_1, s_2 : A \to A$ are monotone operators satisfying the conditions (ST1) and (ST2), then $(\mathbf{A}, t, s_1, s_2)$ is a $qpMV_{vt,st}$-algebra.*

Proof. We only check that s_1 and s_2 satisfy the condition (ST3). If $x \to y \neq 1$, then $t(x \to y) = 0$, so we have $t(x \to y) \leq s_1(x) \to s_1(y)$. If $x \to y = 1$, then $x \leq y$, it follows that $s_1(x) \leq s_1(y)$ by the monotony of s_1, so $s_1(x) \to s_1(y) = 1 = t(x \to y)$. The case $t(x \rightsquigarrow y) \leq s_2(x) \rightsquigarrow s_2(y)$ can be proved similarly.

Proposition 7. *Let $(\mathbf{A}, t)$ be a $qpMV_{vt}$-algebra and define $s_1, s_2 : A \to A$ by $s_1(x) = t^{-}_{\sim}(x)$ and $s_2(x) = t^{\sim}_{-}(x)$ for any $x \in A$. Then $(\mathbf{A}, t, s_1, s_2)$ is a $qpMV_{vt,st}$-algebra.*

Proof. By Remark 2, we have that s_1 and s_2 satisfy (ST1) and (ST2). Since $(x \to y) \odot x \leq y$, we have $t(x \to y) \odot s_1(x) \leq s_1(y)$ by Lemma 1, so $t(x \to y) \leq s_1(x) \to s_1(y)$. Similarly, since $x \odot (x \rightsquigarrow y) \leq y$, we have $t(x \rightsquigarrow y) \leq s_2(x) \rightsquigarrow s_2(y)$. Hence $(\mathbf{A}, t, s_1, s_2)$ is a $qp\mathrm{MV}_{vt,st}$-algebra.

Proposition 8. *Let $(\mathbf{A}, t, s_1, s_2)$ be a $qpMV_{vt,st}$-algebra. Suppose that $s_{\min} : A \to A$, $s_{\max} : A \to A$ and $s'_{max} : A \to A$ are mappings defined by $s_{\min}(x) = x$, $s_{\max}(x) = t(x \to 0) \rightsquigarrow 0$ and $s'_{\max}(x) = t(x \rightsquigarrow 0) \to 0$ for any $x \in A$. Then*

(1) $s_{\min}(x) \leq s_1(x) \leq s_{\max}(x)$ and $s_{\min}(x) \leq s_2(x) \leq s'_{\max}(x)$;

(2) $(\mathbf{A}, t, s_{\min}, s_{\min})$ is a $qpMV_{vt,st}$-algebra;

(3) $(\mathbf{A}, t, s_{\max}, s'_{\max})$ is a $qpMV_{vt,st}$-algebra.

Proof. (1) Obviously, $s_{\min}(x) = x \leq s_1(x)$. On the other hand, since $t(x \to 0) \leq s_1(x) \to s_1(0)$, we have $(s_1(x) \to s_1(0)) \rightsquigarrow 0 \leq t(x \to 0) \rightsquigarrow 0$ and $s_1(0) = 0$, we have $(s_1(x) \to 0) \rightsquigarrow 0 \leq s_{\max}(x)$. Note that $s_1(x) \leq s_1(x) \oplus 0 =$

$(s_1(x) \to 0) \rightsquigarrow 0$, it turns out that $s_1(x) \leq s_{\max}(x)$. Similarly, we can prove that $s_{\min}(x) \leq s_2(x) \leq s'_{\max}(x)$.

(2) Obviously, $s_{\min}$ satisfies (ST1), (ST2) and (ST3). Hence $(\mathbf{A}, t, s_{\min}, s_{\min})$ is a $qp\mathrm{MV}_{vt,st}$-algebra.

(3) We have that $s_{\max}(0) = t(0 \to 0) \rightsquigarrow 0 = t(1) \rightsquigarrow 0 = 1 \rightsquigarrow 0 = 0$ and $s'_{\max}(0) = t(0 \rightsquigarrow 0) \to 0 = t(1) \to 0 = 1 \to 0 = 0$. Since $t(x \to 0) \leq x \to 0$, we have $(x \to 0) \rightsquigarrow 0 \leq t(x \to 0) \rightsquigarrow 0$. Note that $x \leq (x \to 0) \rightsquigarrow 0$, so $x \leq s_{\max}(x)$. Similarly, $x \leq s'_{\max}(x)$. Now, since $x \to y \leq (y \to 0) \rightsquigarrow (x \to 0)$, we have $t(x \to y) \leq t((y \to 0) \rightsquigarrow (x \to 0))$, it follows that $t(x \to y) \leq t(y \to 0) \rightsquigarrow t(x \to 0) \leq (t(x \to 0) \rightsquigarrow 0) \to (t(y \to 0) \rightsquigarrow 0) = s_{\max}(x) \to s_{\max}(y)$. Similarly, we can prove that $t(x \rightsquigarrow y) \leq s'_{\max}(x) \rightsquigarrow s'_{\max}(y)$. Hence $(\mathbf{A}, t, s_{\max}, s'_{\max})$ is a $qp\mathrm{MV}_{vt,st}$-algebra.

4 Conclusions

In this paper, we introduce the very true operator on any qpMV-algebra and investigate the algebraic structure $qp\mathrm{MV}_{vt}$-algebra. We discuss the properties of very true operator on a qpMV-algebra and show that there exists a one-to-one correspondence between very true operators and very false operators on any qpMV-algebra. Some cases of $qp\mathrm{MV}_{vt}$-algebras with truth-depressing hedges are also given. In the future, we will focus on studying the hedges on any qpMV-algebra and investigate the relation between hedges and very true operators on any qpMV-algebra.

References

1. Botur, M., Svrcek, F.: Very true on CBA fuzzy logic. Math. Slovaca. **60**(4), 435–446 (2010)
2. Chajda, I., Kolarik, M.: Very true operators in effect algebras. Soft. Comput. **16**(7), 1213–1218 (2012)
3. Chen, W.J., Dudek, W.A.: Quantum computational algebra with a non-commutative generalization. Math. Slovaca. **66**(1), 1–16 (2016)
4. Chen, W.J., Davvaz, B.: Some classes of quasi-pseudo-MV algebras. Logic J. IGPL. **24**(5), 655–673 (2016)
5. Chen, W.J., Dudek, W.A.: Ideals and congruences in quasi-pseudo-MV algebras. Soft. Comput. **22**(12), 3879–3889 (2018)
6. Ciungu, L.C.: Very true pseudo-BCK algebras. Soft Comput. (2019). https://doi.org/10.1007/s00500-019-03859-x
7. Georgescu, G., Iorgulescu, A.: Pseudo MV algebras. Mult.-Valued Log. **6**, 95–135 (2001)
8. Hajek, P.: On very true. Fuzzy Sets Syst. **124**, 329–333 (2001)
9. Ledda, A., Konig, M., Paoli, F., Giuntini, R.: MV algebras and quantum computation. Studia Log. **82**, 245–270 (2006)
10. Leustean, I.: Non-commutative Lukasiewicz propositional logic. Arch. Math. Logic **45**, 191–213 (2006)

11. Liu, J.M., Chen, W.J.: A non-communicative generalization of quasi-MV algebras. In: 2016 IEEE International Conference on Fuzzy Systems, pp. 122–126. FUZZ-IEEE (2016)
12. Rachunek, J., Salounova, D.: Truth values on generalizations of some commutative fuzzy structures. Fuzzy Sets Syst. **157**, 3159–3168 (2006)
13. Vychodil, V.: Truth-depressing hedges and BL-logic. Fuzzy Sets Syst. **157**, 2074–2090 (2006)
14. Wang, J.T., Xin, X.L., Jun, Y.B.: Very true operators on equality algebras. J. Comput. Anal. Appl. **24**(3) (2018)

The Analysis of the Fuzzy Solution to Fully Fuzzy Linear Systems in Two Perturbation Situations

Kun Liu[1(✉)], Wei-peng Li[1], Yong-ling Li[1], and Hong-ying Duan[2]

[1] College of Mathematics and Statistics, Longdong University, Qingyang 745000, Gansu, China
liukunws@163.com

[2] School of Information Engineering, Longdong University, Qingyang 745000, Gansu, China
http://www.sxx.ldxy.edu.cn

Abstract. The fuzzy solution to fully fuzzy linear systems $\tilde{A} \otimes \tilde{x} = \tilde{b}$ (shown as FFLS) in two perturbation situations are discussed in detail in this paper, where $\tilde{A}$ and $\tilde{b}$ are respectively a fuzzy matrix and a fuzzy vector. This paper aims to show how the perturbations of the coefficient matrix or the right hand vector impact the fuzzy approximate solution vector to FFLS, we first transform the original fully fuzzy linear systems into tree crisp linear systems. And then two perturbation situations are studied: (I) the coefficient matrix is slightly perturbed while the right hand side remains unchanged; (II) the coefficient matrix and right hand side are all slightly perturbed. Finally, we deduce the relative error bounds to two perturbation situations based on the distance of LR-type triangular fuzzy vector.

Keywords: LR-type fuzzy numbers · The fully fuzzy linear systems · Perturbation analysis · The relative error bounds

1 Introduction

It is well known the round-off error is inevitable when we use numerical methods to solve fuzzy or fully fuzzy linear systems. If we desire a perfect numerical result, we must measure the error between the computational result and the exact one. In addition, the perturbation analysis can assert whether the problem is well-conditioned or ill-conditioned and then help us find out efficient numerical methods. Therefore, perturbation analysis is very important issue in numerical methods for solving fuzzy or fully fuzzy linear systems. Tang [1] considered the perturbation problems of a fuzzy matrix equation by perturbation methods [2]. Wang et al. [3] analyzed the perturbation of an $n \times n$ fuzzy linear system $Ax = \tilde{b}$. They focused their discussion only on fuzzy linear systems with the right hand side being triangular fuzzy numbers and their results were only valid for numerical approaches based on the embedding method. Tian et al. [4] defined

Y. Liu et al. (Eds.): ICNC-FSKD 2019, AISC 1074, pp. 791–799, 2020.
https://doi.org/10.1007/978-3-030-32456-8_85

a metric in fuzzy vector space and uses the embedding method to transform the original fuzzy linear system $Ax = \tilde{b}$ into a functional system. By solving this functional linear system, they gave the relative error bounds for the perturbed system in terms of the metric of the fuzzy vectors and the spectral norm of the real matrix. However, they focused their discussion only on the fuzzy linear systems of the form $Ax = \tilde{b}$ by the embedding method. Liu et al. [5] studied the perturbation problems of an $n \times n$ fully fuzzy linear system $\tilde{A} \otimes \tilde{x} = \tilde{b}$, which the right hand side is slightly perturbed while the coefficient matrix remains unchanged.

In this paper, we transform first the original fully fuzzy linear systems into tree crisp linear systems, then we give the fuzzy approximate solutions to the fully fuzzy linear systems and the perturbed fully fuzzy linear systems by solving three crisp linear systems, and finally we get the relative error bounds for three cases perturbed fully fuzzy linear system based on the distance of LR-type triangular fuzzy vector.

The outline of this paper is as follows: In Sect. 2, some important fundamental results are listed. In Sect. 3, the fuzzy solution to fully fuzzy linear systems in two perturbation situations are discussed. The conclusion is drawn in Sect. 4.

2 Preliminaries

In this section, we give some definitions and introduce the notation which will be used throughout the paper.

Definition 2.1. [6–8] A fuzzy number M is said to be an LR-type fuzzy number if

$$\mu_M(x) = \begin{cases} L(\frac{m-x}{\alpha}),\ x \leq m,\ \alpha > 0, \\ R(\frac{x-m}{\beta}),\ x \geq m,\ \beta > 0, \end{cases}$$

where m is the mean value of M, and α and β are left and right spreads, respectively, and a function $L(\cdot)$ the left shape function satisfying:

(i) $L(x) = L(-x)$,

(ii) $L(0) = 1$ and $L(1) = 0$,

(iii) $L(x)$ is non-increasing on $[0, \infty)$.

Naturally, a right shape function $R(\cdot)$ is similarly defined as $L(\cdot)$. Using its mean value and left and right spreads, and shape functions, such an LR-type fuzzy number M is symbolically written as $M = (m, \alpha, \beta)_{\mathrm{LR}}$.

The fuzzy arithmetic operations of fuzzy numbers have grown in importance during recent years as an advanced tool in numerical calculation of all kinds of fuzzy problems. Based on the extension principle, Dubois and Prade designed exact formulas for $\oplus$ and $\ominus$ together with the approximate formulas for $\otimes$ and scalar multiplication [6,7] to LR-fuzzy numbers, whose advantage comes from the fact that arithmetic operations with them can be performed in a relatively easy manner.

Let $F_{LR}(R)$ denote the set of all LR-type fuzzy numbers. Consider two triangular fuzzy numbers $M = (m, \alpha, \beta)_{\mathrm{T}}$ and $N = (n, \gamma, \delta)_{\mathrm{T}}$ of the space $F_T(R)$

which represents the collection of all triangular fuzzy numbers, the distance on $F_T(R)$ can be defined as

$d_T^2(M,N) = \frac{1}{3}\{(m-n)^2 + ((m-\frac{1}{2}\alpha)-(n-\frac{1}{2}\gamma))^2 + ((m+\frac{1}{2}\beta)-(n+\frac{1}{2}\delta))^2\}$
$= \frac{1}{3}\{(m-n)^2 + ((m-n)-\frac{1}{2}(\alpha-\gamma))^2 + ((m-n)+\frac{1}{2}(\beta-\delta))^2\}.$

Denotes $F_T^n(R)$ n-dimensional fuzzy vectors space, which consists of all n-dimensional triangular fuzzy number vectors. Let $U, V \in F_T^n(R)$ and

$U = (u_1, u_2, \cdots, u_n)^T, V = (v_1, v_2, \cdots, v_n)^T$. We define the distance of LR-type fuzzy vector U and V as follows:

$d_T^2(U,V) = \frac{1}{3}\sum_{i=1}^{n}\{(u_i-v_i)^2+((u_i-v_i)-\frac{1}{2}(\alpha_i-\gamma_i))^2+((u_i-v_i)+\frac{1}{2}(\beta_i-\delta_i))^2\},$

where $u_i = (u_i, \alpha_i, \beta_i)_{\mathrm{T}}, v_i = (v_i, \gamma_i, \delta_i)_{\mathrm{T}}, i = 1, 2, \cdots, n$.

Up to rest of this paper, we use LR-type triangular fuzzy numbers and multiplication formula, in place of exact $\otimes$. We may represent $n \times n$ fuzzy matrix $\tilde{A} = (\widetilde{a_{ij}})_{n\times n}$, that $\widetilde{a_{ij}} = (a_{ij}, \alpha_{ij}, \beta_{ij})_{\mathrm{T}}$, with new notation $\tilde{A} = (A, M, N)$, where $A = (a_{ij})$, $M = (\alpha_{ij})$ and $N = (\beta_{ij})$ are three crisp $n \times n$ matrices, with the same size of $\tilde{A}$.

Definition 2.2. [9] Consider the $n \times n$ linear system of equations:

$$\begin{cases} \tilde{a}_{11} \otimes \tilde{x}_1 \oplus \tilde{a}_{12} \otimes \tilde{x}_2 \oplus \cdots \oplus \tilde{a}_{1n} \otimes \tilde{x}_n = \tilde{b}_1, \\ \tilde{a}_{21} \otimes \tilde{x}_1 \oplus \tilde{a}_{22} \otimes \tilde{x}_2 \oplus \cdots \oplus \tilde{a}_{2n} \otimes \tilde{x}_n = \tilde{b}_2, \\ \cdots\cdots\cdots\cdots\cdots\cdots\cdots\cdots\cdots \\ \tilde{a}_{n1} \otimes \tilde{x}_1 \oplus \tilde{a}_{n2} \otimes \tilde{x}_2 \oplus \cdots \oplus \tilde{a}_{nn} \otimes \tilde{x}_n = \tilde{b}_n. \end{cases} \tag{1}$$

The matrix form of the above equations is

$$\tilde{A} \otimes \tilde{x} = \tilde{b} \tag{2}$$

or simply $\tilde{A}\tilde{x} = \tilde{b}$, where the coefficient matrix $\tilde{A} = (\tilde{a}_{ij}), 1 \le i, j \le n$ is an $n \times n$ fuzzy matrix and $\tilde{x}_i, \tilde{b}_i \in F_T(R), 1 \le i \le n$. This system is called a fully fuzzy linear systems (in short, FFLS). Also, if each element of $\tilde{A}$ and $\tilde{b}$ is an non-negative LR-type fuzzy number, then the Eq. (2) is called an non-negative FFLS.

Definition 2.3. [9] The fuzzy vector $\tilde{x} = (x, y, z) = ((x_1, y_1, z_1), (x_2, y_2, z_2), \cdots, (x_n, y_n, z_n))^{\mathrm{T}}$ denotes a solution vector to non-negative fully fuzzy linear systems (2), if and only if

$$Ax = b, Ay + Mx = g, Az + N = h,$$

where the membership function of each element of $\{x \mid \mu_{\tilde{x}}(x) > 0\}$ can be defined with the same functions $L(\cdot)$ and $R(\cdot)$ which used in $\tilde{A}$ and $\tilde{b}$. Moreover, if $y \ge 0, z \ge 0$ and $x - y \ge 0$, $\tilde{x} = (x, y, z)$ be called a consistent solution of non-negative fully fuzzy linear systems (2). Otherwise, it will be called a dummy solution.

We will discuss the perturbation analysis of positive fuzzy solution to non-negative fully fuzzy linear systems (2) throughout this paper. When the coefficient matrix $\tilde{A} = (A, M, N)$ is non-singular, we can transform the fully fuzzy linear systems $\tilde{A} \otimes \tilde{x} = \tilde{b}$ into three crisp linear systems $Ax = b, Ay + Mx = g, Az + Nx = h$ by using the Dehghan's method in [9]. If $x = A^{-1}b, y = A^{-1}g - A^{-1}Mx$ and $z = A^{-1}h - A^{-1}Nx$ are respectively the solutions of equations $Ax = b, Ay + Mx = g, Az + Nx = h$, then $\tilde{x} = (A^{-1}b, A^{-1}g - A^{-1}Mx, A^{-1}h - A^{-1}Nx)$ must be a fuzzy solution vector to $\tilde{A} \otimes \tilde{x} = \tilde{b}$.

3 Perturbation Analysis of the Fully Fuzzy Linear Systems

In this section, we will show how the perturbations of the coefficient matrix or the right hand vector impact the fuzzy approximate solution vector to FFLS (2). To this end, we investigate two perturbation situations: (I) the coefficient matrix is slightly perturbed while the right hand side remains unchanged; (II) the coefficient matrix and right hand side are all slightly perturbed. Moreover, we provide the relative error between the fuzzy approximate solution to FFLS (2) and the perturbed fuzzy approximate solution to the perturbed fully fuzzy linear systems by transforming a fully fuzzy linear system into three crisp linear systems.

(I) We assume that the fuzzy coefficient matrix $\tilde{A}$ in the fully fuzzy linear systems $\tilde{A} \otimes \tilde{x} = \tilde{b}$ is slightly perturbed to another fuzzy coefficient matrix $\tilde{A} + \widetilde{\delta A}$ (the perturbed fully fuzzy linear systems is $((\tilde{A} + \widetilde{\delta A}) \otimes \tilde{x^*} = \tilde{b})$ while the fuzzy vector $\tilde{b}$ remain unchanged. Assuming that $A + \delta A$ is an non-singular crisp matrix, using approximate multiplication, $\tilde{A} \otimes \tilde{x} = \tilde{b}$ and $(\tilde{A} + \widetilde{\delta A}) \otimes \tilde{x^*} = \tilde{b}$ can be written as $(Ax, Ay + Mx, Az + Nx) = (b, g, h)$ and $((A + \delta A)x^*, (A + \delta A)y^* + (M + \delta M)x^*, (A + \delta A)z^* + (N + \delta N)x^*) = (b, g, h)$, respectively. By using the Dehghan's method, the fuzzy approximately solution vector to perturbed fuzzy linear systems can be written as

$$\tilde{x^*} = (x^*, y^*, z^*) = ((A + \delta A)^{-1}b, (A + \delta A)^{-1}[g - (M + \delta M)x^*], (A + \delta A)^{-1}[h - (N + \delta N)x^*]). \quad (3)$$

Theorem 3.1. Let $A, A + \delta A$ be two non-singular crisp matrixes. Denote the fuzzy approximate solution vectors to $\tilde{A} \otimes \tilde{x} = \tilde{b}$ and $(\tilde{A} + \widetilde{\delta A}) \otimes \tilde{x^*} = \tilde{b}$ by $\tilde{x} = (\tilde{x}_1, \tilde{x}_2, \cdots, \tilde{x}_n)^T$ and $\tilde{x^*} = (\tilde{x}_1^*, \tilde{x}_2^*, \cdots, \tilde{x}_n^*)^T$, respectively. Then

(a)

$$\frac{d(x^*, x)}{\|x^*\|} \leq (\|A\|_2 \|A^{-1}\|_2) \frac{\|\delta A\|_2}{\|A\|_2};$$

(b)

$$\frac{d(y^*, y)}{\|y^*\|} \leq (\|A\|_2 \|A^{-1}\|_2) \frac{\|\delta A\|_2}{\|A\|_2} \left(\frac{\|M + \delta M\|_2 \|x - x^*\|}{\|g\| + \|M + \delta M\|_2 \|x^*\|}\right);$$

(c)

$$\frac{d(z^*, z)}{\|z^*\|} \leq (\|A\|_2 \|A^{-1}\|_2) \frac{\|\delta A\|_2}{\|A\|_2} \left(\frac{\|N + \delta N\|_2 \|x - x^*\|}{\|h\| + \|N + \delta N\|_2 \|x^*\|}\right).$$

Moreover

$$\frac{d(\tilde{x},\tilde{x^*})}{\|\tilde{x^*}\|} \le (\|A\|_2\|A^{-1}\|_2)\frac{\|\delta A\|_2}{\|A\|_2}\max\{1,$$

$$\frac{\|M+\delta M\|_2\|x-x^*\|}{\|g\|+\|M+\delta M\|_2\|x^*\|}, \frac{\|N+\delta N\|_2\|x-x^*\|}{\|h\|+\|N+\delta N\|_2\|x^*\|}\}.$$

Proof. By using the Dehghan's method, we can transform $\tilde{A}\otimes\tilde{x}=\tilde{b}$ and $(\tilde{A}+\widetilde{\delta A})\otimes\tilde{x^*}=\tilde{b}$ into three crisp linear systems, respectively. Let $x=(x_1,x_2,\cdots,x_n)$ and $x^*=(x_1^*,x_2^*,\cdots,x_n^*)$ are solution vectors to $Ax=b$ and $(A+\delta A)x^*=b$, respectively. It follows that

$$\begin{aligned} x^*-x &= [(A+\delta A)^{-1}-A^{-1}]b \\ &= A^{-1}[A-(A+\delta A)](A+\delta A)^{-1}b \\ &= -A^{-1}\delta A(A+\delta A)^{-1}b \\ &= -A^{-1}\delta Ax^*. \end{aligned}$$

Taking supremum norm, we have

$$\|x^*-x\| \le \|A^{-1}\|_2\|\delta A\|_2\|x^*\|.$$

Dividing both sides by $\|x^*\|$, we have

$$\frac{\|x^*-x\|}{\|x^*\|} \le (\|A\|_2\|A^{-1}\|_2)\frac{\|\delta A\|_2}{\|A\|_2}.$$

According to the definition of vector norm in Sect. 2, we obtain the desired inequality

$$\frac{d(x^*,x)}{\|x^*\|} \le (\|A\|_2\|A^{-1}\|_2)\frac{\|\delta A\|_2}{\|A\|_2}.$$

Hence the inequality (a) holds.

Next we prove (b). Let $y=(y_1,y_2,\cdots,y_n)$ and $y^*=(y_1^*,y_2^*,\cdots,y_n^*)$ are solution vectors to $(A+\delta A)y+(M+\delta M)x=g$ and $(A+\delta A)y^*+(M+\delta M)x^*=g$, respectively. When $(A+\delta A)$ is an non-singular matrix, then

$$y=(A+\delta A)^{-1}(g-(M+\delta M)x),$$

$$y^*=(A+\delta A)^{-1}(g-(M+\delta M)x^*).$$

It follows that

$$\begin{aligned} y^*-y = A^{-1}[A-(A+\delta A)](A+\delta A)^{-1}(g-(M+\delta M)x^* \\ -g+(M+\delta M)x) = -A^{-1}\delta A(A+\delta A)^{-1}(M+\delta M(x-x^*)). \end{aligned}$$

Taking supremum norm, we have

$$\|y^* - y\| \le \|A^{-1}\|_2\|\delta A\|_2\|(A+\delta A)^{-1}\|_2\|(M+\delta M)\|_2\|x - x^*\|.$$

Since

$$\|y^*\| \le \|(A+\delta A)^{-1}\|_2\|g - (M+\delta M)x^*\|,$$

so

$$\frac{\|y^* - y\|}{\|y^*\|} \le (\|A^{-1}\|_2\|A\|_2)\frac{\|\delta A\|_2}{\|A\|_2}\frac{\|M+\delta M\|_2\|x - x^*\|}{(\|g\| + \|M+\delta M\|_2\|x^*\|)}.$$

According to the definition of vector norm in Sect. 2, we obtain the desired inequality

$$\frac{d(y^*, y)}{\|y\|} \le (\|A^{-1}\|_2\|A\|_2)\frac{\|\delta A\|_2}{\|A\|_2}\frac{\|M+\delta M\|_2\|x - x^*\|}{(\|g\| + \|M+\delta M\|_2\|x^*\|)},$$

This completes the proof of (b).

In the same way, we have the inequality (c).

It is obvious that

$$\frac{d(\tilde{x}, \tilde{x^*})}{\|\tilde{x^*}\|} \le (\|A\|_2\|A^{-1}\|_2)\frac{\|\delta A\|_2}{\|A\|_2}\max\{1,$$
$$\frac{\|M+\delta M\|_2\|x - x^*\|}{\|g\| + \|M+\delta M\|_2\|x^*\|}, \frac{\|N+\delta N\|_2\|x - x^*\|}{\|h\| + \|N+\delta N\|_2\|x^*\|}\}.$$

(II) We assume that the coefficient matrix $\tilde{A}$ and the right hand side $\tilde{b}$ in fully fuzzy linear systems $\tilde{A}\otimes\tilde{x} = \tilde{b}$ are all slightly perturbed and the perturbed fully fuzzy linear systems is $(\tilde{A} + \widetilde{\delta A})\otimes\tilde{x^*} = \tilde{b'}$. Assuming that $A + \delta A$ is an non-singular crisp matrix, using approximate multiplication, $\tilde{A}\otimes\tilde{x} = \tilde{b}$ and $(\tilde{A} + \widetilde{\delta A})\otimes\tilde{x^*} = \tilde{b'}$ can be written as $(Ax, Ay + Mx, Az + Nx) = (b, g, h)$ and $((A+\delta A)x^*, (A+\delta A)y^* + (M+\delta M)x^*, (A+\delta A)z^* + (N+\delta N)x^*) = (b', g', h')$, respectively. By using the Dehghan's method, the fuzzy approximately solution vector to perturbed fuzzy linear systems can be written as

$$\begin{aligned}\tilde{x^*} = (x^*, y^*, z^*) = ((A+\delta A)^{-1}b', (A+\delta A)^{-1}[g' - (M+\delta M)x^*],\\ (A+\delta A)^{-1}[h' - (N+\delta N)x^*]).\end{aligned} \tag{4}$$

Theorem 3.2. Let $A, A+\delta A$ be two non-singular crisp matrixes. Denote the fuzzy approximate solution vectors to $\tilde{A}\otimes\tilde{x} = \tilde{b}$ and $(\tilde{A} + \widetilde{\delta A})\otimes\tilde{x^*} = \tilde{b'}$ by $\tilde{x} = (\tilde{x}_1, \tilde{x}_2, \cdots, \tilde{x}_n)^T$ and $\tilde{x^*} = (\tilde{x_1^*}, \tilde{x_2^*}, \cdots, \tilde{x_n^*})^T$, respectively. Also, let $\delta b = b - b', \delta g = g - g', \delta h = h - h'$. Then

(a)

$$\frac{d(x^*, x)}{\|x\|} \le \frac{\|A\|_2\|A^{-1}\|_2}{1 - \|\delta A\|_2\|A^{-1}\|_2}(\frac{\|\delta A\|_2}{\|A\|_2} + \frac{\|\delta b\|}{\|b\|});$$

(b)

$$\frac{d(y^*, y)}{\|y\|} \leq \frac{\|A\|_2\|A^{-1}\|_2}{1 - \|\delta A\|_2\|A^{-1}\|_2}(\frac{\|A + \delta A\|_2}{\|A\|_2} + \frac{\|g' - (M + \delta M)(A + \delta A)^{-1}b'\|}{\|g - MA^{-1}b\|});$$

(c)

$$\frac{d(z^*, z)}{\|z\|} \leq \frac{\|A\|_2\|A^{-1}\|_2}{1 - \|\delta A\|_2\|A^{-1}\|_2}(\frac{\|A + \delta A\|_2}{\|A\|_2} + \frac{\|h' - (N + \delta N)(A + \delta A)^{-1}b'\|}{\|h - NA^{-1}b\|}).$$

Moreover

$$\frac{d(x^*, x)}{\|x^*\|} \leq \frac{\|A\|_2\|A^{-1}\|_2}{1 - \|\delta A\|_2\|A^{-1}\|_2}\max\{\frac{\|\delta A\|_2}{\|A\|_2} + \frac{\|\delta b\|}{\|b\|}, \frac{\|A + \delta A\|_2}{\|A\|_2} + \frac{\|g' - (M + \delta M)(A + \delta A)^{-1}b'\|}{\|g - MA^{-1}b\|}, \frac{\|A + \delta A\|_2}{\|A\|_2} + \frac{\|h' - (N + \delta N)(A + \delta A)^{-1}b'\|}{\|h - NA^{-1}b\|}\}.$$

Proof. By using the Dehghan's method, we can transform $\tilde{A} \otimes \tilde{x} = \tilde{b}$ and $(\tilde{A} + \widetilde{\delta A}) \otimes \tilde{x^*} = \tilde{b'}$ into three crisp linear systems, respectively. Let $x = (x_1, x_2, \cdots, x_n)$ and $x^* = (x_1^*, x_2^*, \cdots, x_n^*)$ are solution vectors to $Ax = b$ and $(A + \delta A)x^* = b'$, respectively. Since

$$A + \delta A = A(I + A^{-1}\delta A),$$

$$\|A^{-1}\|_2\|\delta A\|_2 < 1,$$

and

$$\|(I + A^{-1}\delta A)^{-1}\|_2 \leq \frac{1}{1 - \|A^{-1}\delta A\|_2} \leq \frac{1}{1 - \|A^{-1}\|_2\|\delta A\|_2},$$

so

$$x^* - x = (A + \delta A)^{-1}[b + \delta b - (A + \delta A)x] = (I + A^{-1}\delta A)^{-1}A^{-1}[\delta b - (\delta A)x].$$

Taking supremum norm, we have

$$\|x^* - x\| \leq \frac{\|A^{-1}\|_2}{1 - \|\delta A\|_2\|A^{-1}\|_2}$$

$$(\frac{\|\delta A\|_2}{\|A\|_2}\|A\|_2\|x\| + \frac{\|\delta b\|_2}{\|b\|_2}\|A\|_2\|x\|),$$

thus

$$\frac{\|x^* - x\|}{\|x\|^*} \leq \frac{\|A\|_2\|A^{-1}\|_2}{1 - \|\delta A\|_2\|A^{-1}\|_2}(\frac{\|\delta A\|_2}{\|A\|_2} + \frac{\|\delta b\|_2}{\|b\|_2}).$$

According to the definition of vector norm in Sect. 2, we obtain the desired inequality

$$\frac{d(x^*, x)}{\|x^*\|} \leq \frac{\|A\|_2\|A^{-1}\|_2}{1 - \|\delta A\|_2\|A^{-1}\|_2}\left(\frac{\|\delta A\|_2}{\|A\|_2} + \frac{\|\delta b\|_2}{\|b\|_2}\right).$$

Hence the inequality (a) holds.

Next we prove (b). Let $y = (y_1, y_2, \cdots, y_n)$ and $y^* = (y_1^*, y_2^*, \cdots, y_n^*)$ are solution vectors to $Ay+Mx = g$ and $(A+\delta A)y^*+(M+\delta M)x^* = g'$, respectively. Since $(A + \delta A)$ is an non-singular matrix, so

$$y = A^{-1}(g - Mx),$$
$$y^* = (A + \delta A)^{-1}[g' - (M + \delta M)(A + \delta A)^{-1}b'].$$

Hence

$$\begin{aligned} y^* - y = (A + \delta A)^{-1}[g' - (M + \delta M)(A + \delta A)^{-1}b' \\ -(A + \delta A)A^{-1}(g - MA^{-1}b)] = (A + \delta A)^{-1} \\ [g' - (M + \delta M)(A + \delta A)^{-1}b' - (A + \delta A)y]. \end{aligned}$$

Taking supremum norm, we have

$$\begin{aligned} \|y^* - y\| \leq \frac{\|A^{-1}\|_2\|A\|_2}{1 - \|\delta A\|_2\|A^{-1}\|_2}\Big(\frac{\|A + \delta A\|_2}{\|A\|_2}\|y\| \\ + \frac{\|g' - (M + \delta M)(A + \delta A)^{-1}b'\|}{\|g - MA^{-1}b\|}\|y\|\Big). \end{aligned}$$

So

$$\begin{aligned} \frac{\|y^* - y\|}{\|y\|} \leq \frac{\|A^{-1}\|_2\|A\|_2}{1 - \|\delta A\|_2\|A^{-1}\|_2}\Big(\frac{\|A + \delta A\|_2}{\|A\|_2} \\ + \frac{\|g' - (M + \delta M)(A + \delta A)^{-1}b'\|}{\|g - MA^{-1}b\|}\Big). \end{aligned}$$

According to the definition of vector norm in Sect. 2, we obtain the desired inequality

$$\begin{aligned} \frac{d(y^*, y)}{\|y\|} \leq \frac{\|A\|_2\|A^{-1}\|_2}{1 - \|\delta A\|_2\|A^{-1}\|_2}\Big(\frac{\|A + \delta A\|_2}{\|A\|_2} \\ + \frac{\|g' - (M + \delta M)(A + \delta A)^{-1}b'\|}{\|g - MA^{-1}b\|}\Big). \end{aligned}$$

This completes the proof of (b).

In the same way, we have the inequality (c).

It is obvious that

$$\begin{aligned} \frac{d(x^*, x)}{\|x^*\|} \leq \frac{\|A\|_2\|A^{-1}\|_2}{1 - \|\delta A\|_2\|A^{-1}\|_2}\max\Big\{\frac{\|\delta A\|_2}{\|A\|_2} + \frac{\|\delta b\|}{\|b\|}, \\ \frac{\|A + \delta A\|_2}{\|A\|_2} + \frac{\|g' - (M + \delta M)(A + \delta A)^{-1}b'\|}{\|g - MA^{-1}b\|}, \\ \frac{\|A + \delta A\|_2}{\|A\|_2} + \frac{\|h' - (N + \delta N)(A + \delta A)^{-1}b'\|}{\|h - NA^{-1}b\|}\Big\}. \end{aligned}$$

4 Conclusion

In this paper, the perturbation analysis of the fully fuzzy linear systems $\tilde{A}\otimes\tilde{x} = \tilde{b}$ (shown as FFLS) is investigated. Based on LR-type triangular fuzzy numbers and its approximate arithmetic operators, the original fully fuzzy linear systems are converted to three crisp linear equations. And using the distance of LR-type triangular fuzzy vector and the Dehghan's method, the relative error bounds for two cases perturbed fully fuzzy linear systems are obtained.

Acknowledgment. Thanks to the support by the Provincial Science and Technology Program Foundation of Gansu (18JR3RM238).

References

1. Tang, F.C.: Perturbation teechniques for fuzzy matrix equations. Fuzzy Sets Syst. **109**, 363–369 (2000)
2. Nayfeh, A.H.: Perturbation Methods. Wiley, New York (1973)
3. Wang, K., Chen, G., Wei, Y.: Perturbation analysis for a class of fuzzy linear systems. J. Comput. Appl. Math. **224**, 54–65 (2009)
4. Tian, Z.F., Hu, L.J., Greenhalgh, D.: Perturbation analysis of fuzzy linear systems. Inf. Sci. **180**, 4706–4713 (2009)
5. Liu, K., Li, H., Guo, Y.: Perturbation analysis of the fully fuzzy linear systems. In: Advances in Computational Science and Computing, ISCSC. AISC, vol. 877, pp. 94–101 (2018)
6. Dubois, D., Prade, H.: Operations on fuzzy numbers. Int. J. Syst. Sci. **9**, 613–626 (1978)
7. Dubois, D., Prade, H.: Fuzzy Sets and Systems: Theory and Applications. Academic Press, New York (1980)
8. Gong, Z.T., Zhao, W.C., Liu, K.: A straightforward approach for solving fully fuzzy linear programming problem with LR-type fuzzy numbers. J. Oper. Res. Soc. Japan **61**, 172–185 (2018)
9. Dehghan, M., Hashemi, B., Ghatee, M.: Computational methods for solving fully fuzzy linear systems. Appl. Math. Comput. **179**, 328–343 (2006)

A Novel Approach to Hesitant Fuzzy Soft Set Based Decision Making

Yanping He(✉)

School of Electrical Engineering, Northwest MinZu University, Lanzhou 730030, People's Republic of China
he_yanping@126.com

Abstract. Babitha et al. defined a hybrid model called hesitant fuzzy soft sets by combining the hesitant fuzzy set with the soft set. The aim of this paper is to give deeper insights into decision making based on hesitant fuzzy soft sets. We point out that the algorithm designed by Babitha et al. based decision making is not fit to solve some decision making problems and a counterexample is given. By means of reduct fuzzy soft sets and level hesitant fuzzy soft sets, we present an adjustable approach to hesitant fuzzy soft sets based decision making and some numerical examples are provided to illustrate the developed approach.

Keywords: Soft set · Hesitant fuzzy soft set · Level hesitant fuzzy soft set · Decision making

1 Introduction

As another generalization of fuzzy set [1], hesitant fuzzy set introduced by Torra [2,3] permits the membership of an element to a given set represented by several possible values between 0 and 1, which is powerful to determine the membership degree especially when we have several different values on it. After it was introduced by Torra, the hesitant fuzzy set has attracted more and more scholars' attention [4–6].

Molodtsov [7] initiated a novel concept called soft sets as a new mathematical tool for dealing with uncertainties. The soft set theory is free from many difficulties that have troubled the usual theoretical approaches. In recent years, research works on soft sets are very active and progressing rapidly [8–12]. Very recently, by combining hesitant fuzzy sets and soft sets, Babitha et al. [13] defined a hybrid model called hesitant fuzzy soft sets and investigated some of their basic properties. They also presented an algorithm to solve decision making problems based on hesitant fuzzy soft sets. However, the algorithm designed by Babitha et al. based decision making is not fit to solve some decision making problems. In view of the defect of the algorithm, we also present a novel approach to hesitant fuzzy soft set based decision making problems by using reduct fuzzy soft sets and level hesitant fuzzy soft sets. In general, the final optimal decisions based

Y. Liu et al. (Eds.): ICNC-FSKD 2019, AISC 1074, pp. 800–808, 2020.
https://doi.org/10.1007/978-3-030-32456-8_86

on different level hesitant fuzzy soft sets could be different. Thus the newly proposed approach is actually an adjustable method which efficiently captures the uncertainty and subjectivity of some decision making problems.

The remainder of this paper is organized as follows. We first recall the concepts of hesitant fuzzy sets and hesitant fuzzy soft sets in Sect. 2. Section 3 is then devoted to a deep analysis on the application of hesitant fuzzy soft set to fuzzy decision making. A counterexample is used to show that Babitha's algorithm has the limitation which arises from the improper choice result. Then a novel approach to hesitant fuzzy soft sets based decision making is proposed. Finally, we conclude the paper with a summary and outlook for further research in Sect. 4.

2 Preliminaries

In this section, we briefly review the concepts of hesitant fuzzy sets and hesitant fuzzy soft sets. Further details could be found in [2,3,13]. The pair (U, E) will be called a soft universe. Throughout this paper, unless otherwise stated, U refers to an initial universe, and E is a set of parameters.

Definition 1 ([2,3]). *Let X be a fixed set, a hesitant fuzzy set (HFS, for short) $\tilde{A}$ on X is in terms of a function $h_{\tilde{A}}(x)$ that when applied to X returns a subset of [0,1], that is,*

$\tilde{A} = \{< x, h_{\tilde{A}}(x) > | x \in X\}$,

where $h_{\tilde{A}}(x)$ is a set of some different values in [0,1], representing the possible membership degrees of the element $x \in X$ to $\tilde{A}$.

For convenience, we call $h_{\tilde{A}}(x)$ a hesitant fuzzy element (HFE, for short). The set of all hesitant fuzzy sets on U is denoted by $HF(U)$.

Babitha et al. [13] introduced the notion of hesitant fuzzy soft sets, which is a hybrid model combining hesitant fuzzy sets with soft sets.

Definition 2 ([13]). *Let (U, E) be a soft universe and $A \subseteq E$. A pair $\mathfrak{S} = (\tilde{F}, A)$ is called a hesitant fuzzy soft set over U, where $\tilde{F}$ is a mapping given by $\tilde{F} : A \to HF(U)$.*

A hesitant fuzzy soft set is a parameterized family of hesitant fuzzy subsets of U. Following the standard notations, $\tilde{F}(e)$ can be written as $\tilde{F}(e) = \{< x, \tilde{F}(e)(x) >: x \in U\}$. Sometimes we write $\tilde{F}$ as $(\tilde{F}, E)$. If $A \subseteq E$, we can also have a hesitant fuzzy set $(\tilde{F}, A)$.

Example 1. Let U be a set of four participants performing dance programme, which is denoted by $U = \{x_1, x_2, x_3, x_4\}$. Let E be a parameter set, where $E = \{e_1, e_2, e_3\} = \{$confident; creative; timing$\}$. Suppose that the evaluation of the performance of candidates given by three judges is a hesitant fuzzy soft set $\mathfrak{S} = (\tilde{F}, A)$ which can defined as Table 1.

Next, we introduce the concept of reduct fuzzy soft sets based on hesitant fuzzy soft sets.

Table 1. Hesitant fuzzy soft set $\mathfrak{S} = (\tilde{F}, A)$

U	e_1	e_2	e_3
x_1	{0.70,0.60,0.80}	{0.50,0.60,0.60}	{0.80,0.60,0.90}
x_2	{0.40,0.50,0.70}	{0.60,0.70,0.55}	{0.60,0.80,0.50}
x_3	{0.70,0.80,0.80}	{0.80,0.90,0.85}	{0.78,0.90,0.76}
x_4	{0.80,0.80,0.90}	{1.00,0.80,0.90}	{0.80,0.65,0.80}

Definition 3. *The optimistic reduct fuzzy soft set (ORFS) of a hesitant fuzzy soft set* $(\tilde{F}, A)$ *is defined as a fuzzy soft set* $(\tilde{F}_+, A)$ *over* U *such that for all* $\gamma^{\sigma(k)} \in \tilde{F}(e)(x)$,

$$\tilde{F}_+(e) = \{(x, \tilde{F}_+(e)(x)) : x \in U\} = \{(x, \bigvee_{k=1}^{l} \gamma^{\sigma(k)}) : x \in U\}, \forall e \in A,$$

where $\gamma^{\sigma(k)}$ *is the* k*th largest value in the HFE* $\tilde{F}(e)(x)$, *and* l *stands for the number of values in the HFE* $\tilde{F}(e)(x)$.

Definition 4. *The neutral reduct fuzzy soft set (NRFS) of a hesitant fuzzy soft set* $(\tilde{F}, A)$ *is defined as a fuzzy soft set* $(\tilde{F}_N, A)$ *over* U *such that for all* $\gamma^{\sigma(k)} \in \tilde{F}(e)(x)$,

$$\tilde{F}_N(e) = \{(x, \tilde{F}_N(e)(x)) : x \in U\} = \{(x, \frac{1}{l}\sum_{k=1}^{l} \gamma^{\sigma(k)}) : x \in U\}, \forall e \in A,$$

where $\gamma^{\sigma(k)}$ *is the* k*th largest value in the HFE* $\tilde{F}(e)(x)$, *and* l *stands for the number of values in the HFE* $\tilde{F}(e)(x)$.

3 Decision Making Based on Hesitant Fuzzy Soft Sets

In [13], by integrating hesitant fuzzy sets and soft sets Babitha et al. presented the concept of hesitant fuzzy soft sets, and explored the following algorithm to solve fuzzy decision making problems based on the model.

3.1 Babitha 's Algorithm Based on Hesitant Fuzzy Soft Sets

Algorithm 1 [13]

1. Input the hesitant fuzzy soft set $(\tilde{F}, A)$.
2. Obtain the score matrix $(\tilde{F}_S, A)$ corresponding to $(\tilde{F}, A)$, where the entries h_{ji} of the score matrix $(\tilde{F}_S, A)$ are given by $h_{ji} = \frac{\sum_{\gamma \in \tilde{F}(e_i)(x_j)} \gamma}{l}$.
3. Calculate the average of $\tilde{F}_S(e_i)$ for each x_j and let it be denoted as a_j.
4. Select the optimal alternative x_k if $a_k = \max_j a_j$.
5. If k has more than one value then any one of x_k may be chosen.

The use of Algorithm 1 was illustrated in [13] by a concrete example (see Sect. 4 of [13] for details). Let us simply sketch out that example here. In the example, the tabular representation of the hesitant fuzzy soft set $(\tilde{F}, A)$ is shown in Table 2. By Algorithm 1, we can obtain the score matrix $(\tilde{F_S}, A)$ given in Table 3. Then based on step 3 in Algorithm 1, we calculate the average of $\tilde{F_S}(e_i)$ for each x_j and let it be denoted as a_j. So $a_1 = 0.76497, a_2 = 0.6083, a_3 = 0.688, a_4 = 0.7733$. We see that the optimal alternative is the candidate x_4, since it has the maximum score $a_4 = 0.7733$.

Table 2. Hesitant fuzzy soft set $(\tilde{F}, A)$

U	e_1	e_2	e_3	e_4
x_1	{0.6,0.7,0.8}	{0.7,0.8,0.85}	{0.76,0.82,0.65}	{0.8,0.82,0.88}
x_2	{0.3,0.4,0.45}	{0.6,0.7,0.65}	{0.76,0.7,0.8}	{0.74,0.68,0.52}
x_3	{0.5,0.6,0.7}	{0.7,0.8,0.65}	{0.81,0.66,0.9}	{0.56,0.7,0.68}
x_4	{0.7,0.8,0.8}	{0.6,0.72,0.8}	{0.9,0.8,0.9}	{0.76,0.7,0.8}

Table 3. Score matrix $(\tilde{F_S}, A)$

U	e_1	e_2	e_3	e_4
x_1	0.7	0.7833	0.7433	0.8333
x_2	0.3833	0.65	0.7533	0.64667
x_3	0.6	0.71667	0.79	0.64667
x_4	0.766667	0.70667	0.866667	0.7533

3.2 A Counterexample

The following example will show that Algorithm 1 may not be successfully applied to some fuzzy decision making problems. Suppose that $\Im = (\tilde{G}, B)$ is a hesitant fuzzy soft set and Table 4 gives its tabular representation.

From Table 4, we can obtain the score matrix $(\tilde{G_S}, B)$ with the scores of each person x_j and Table 5 gives its tabular representation. Meanwhile, it follows from Table 5 that the scores of x_1, x_2, and x_3 are the same, namely $a_1 = a_2 = a_3 = 0.525$, whence by Algorithm 1 three of them could be selected as the best choice. Obviously it is unreasonable. In general we should choose the candidate x_2 as the optimal alternative because x_2 relatively satisfies more decision criteria than the other two candidates.

3.3 An Adjustable Approach Using Level Hesitant Fuzzy Soft Sets

In the current subsection, we shall present an adjustable approach to hesitant fuzzy soft set based decision making problems. Next, we will introduce the concept of level hesitant fuzzy soft sets.

Table 4. Hesitant fuzzy soft set $\Im = (\tilde{G}, B)$

U	e_1	e_2	e_3	e_4
x_1	{0.75,0.95,0.7}	{0.6,0.7,0.8}	{0.43,0.57,0.5}	{0.1,0.1,0.1}
x_2	{0.60,0.40,0.2}	{0.65,0.7,0.45}	{0.5,0.6,0.7}	{0.46,0.54,0.5}
x_3	{0.4,0.5,0.6}	{0.1,0.5,0.3}	{0.8,0.9,1.0}	{0.35,0.55,0.3}

Table 5. Score matrix $(\tilde{G_S}, B)$

U	e_1	e_2	e_3	e_4	values
x_1	0.8	0.7	0.5	0.1	0.525
x_2	0.4	0.6	0.6	0.5	0.525
x_3	0.5	0.3	0.9	0.4	0.525

Definition 5. *Let $(\tilde{F}, A)$ be the hesitant fuzzy soft set over U, where $A \subseteq E$ and E is the parameter set. For $t \in [0,1]$, $x \in U$, and $\gamma^{\sigma(k)} \in \tilde{F}(e)(x)$, define*

$$\alpha^{\sigma(k)} = \begin{cases} 1, & \gamma^{\sigma(k)} \geq t, \\ 0, & \gamma^{\sigma(k)} < t. \end{cases}$$

Then the hesitant fuzzy soft set $(\tilde{F}_t, A)$ is called a t-level hesitant fuzzy soft set $L(\mathfrak{S}; t) = (\tilde{F}_t, A)$ defined by

$$\tilde{F}_t(e) = L(\tilde{F}(e); t) = \{< x, \tilde{F}_t(e)(x) >: x \in U\} = \{< x, \{\alpha^{\sigma(k)}\} >: x \in U\},$$

for all $e \in A$.

Remark 1. In Definition 5, t can be viewed as a given threshold on membership values. In practical applications, these thresholds are in advance chosen by decision makers and represent their requirements on "membership levels". It is easy to see that HFEs of level hesitant soft sets consist of 1 and 0, which is very important in real-life applications of hesitant fuzzy soft sets based decision making.

To illustrate Definition 5, let us consider the hesitant fuzzy soft set $(\tilde{F}, A)$ shown in Example 1.

Example 2. (The 0.85-level Hesitant Fuzzy Soft Set of the Hesitant Fuzzy Soft Set $\mathfrak{S}$) Reconsider Example 1. By taking $t = 0.85$, we obtain the 0.85-level hesitant fuzzy soft sets as follows:

$$L(\tilde{F}(e_1); 0.85) = \{<x_1, \{0,0,0\}>, <x_2, \{0,0,0\}>, <x_3, \{0,0,0\}>, <x_4, \{0,0,1\}>\},$$

$$L(\tilde{F}(e_2); 0.85) = \{<x_1, \{0,0,0\}>, <x_2, \{0,0,0\}>, <x_3, \{0,1,1\}>, <x_4, \{1,0,1\}>\},$$

$$L(\tilde{F}(e_3); 0.85) = \{<x_1, \{0,0,1\}>, <x_2, \{0,0,0\}>, <x_3, \{0,1,0\}>, <x_4, \{0,0,0\}>\}.$$

Then the 0.85-level hesitant fuzzy soft set of $\mathfrak{S} = (\tilde{F}, A)$ is a special hesitant fuzzy soft set $L(\mathfrak{S}; 0.85) = (\tilde{F}_{0.85}, A)$, where for $x_j \in U$, HFEs of the hesitant fuzzy sets $\tilde{F}_{0.85}(e_i) = L(\tilde{F}(e_i); 0.85)$ consist of 1 and 0. For example, $\tilde{F}_{0.85}(e_2)(x_3) = L(\tilde{F}(e_2)(x_3); 0.85) = \{0, 1, 1\}$, which means that the second judge and the third judge think the performance of candidate x_3 is creative on the membership degree 0.85.

In Definition 5, the level (or threshold) assigned to each parameter is always the constant value $t \in [0, 1]$. However, in some practical applications, decision makers need to impose different thresholds on different parameters. To address this issue, we replace a constant value by a function as the thresholds on membership values.

Definition 6. *Let $(\tilde{F}, A)$ be the hesitant fuzzy soft set over U, where $A \subseteq E$ and E is the parameter set. Let $\lambda : A \to [0, 1]$ be a fuzzy set in A which is called a threshold fuzzy set. For $x \in U$, and $\gamma^{\sigma(k)} \in \tilde{F}(e)(x)$, define*

$$\alpha^{\sigma(k)} = \begin{cases} 1, & \gamma^{\sigma(k)} \geq \lambda(e), \\ 0, & \gamma^{\sigma(k)} < \lambda(e). \end{cases}$$

Then the hesitant fuzzy soft set $(\tilde{F}_\lambda, A)$ is called a level hesitant fuzzy soft set $L(\mathfrak{S}; \lambda(e)) = (\tilde{F}_\lambda, A)$ with respect to the fuzzy set λ defined by $\tilde{F}_\lambda(e) = L(\tilde{F}(e); \lambda(e)) = \{< x, \tilde{F}_\lambda(e)(x) >: x \in U\} = \{< x, \{\alpha^{\sigma(k)}\} >: x \in U\}$, for all $e \in A$.

In what follows, we shall develop an adjustable approach to hesitant fuzzy soft sets based decision making. This approach involves the following steps:

Algorithm 2

Step 1. Input a hesitant fuzzy soft set $\mathfrak{S} = (\tilde{F}, A)$.

Step 2. Input a threshold fuzzy set $\lambda : A \to [0, 1]$ (or give a threshold value $t \in [0, 1]$; or choose the mid-level decision rule; or choose the top-level decision rule) for decision making.

Step 3. Compute the level hesitant fuzzy soft set $L(\mathfrak{S}; \lambda)$ of $\mathfrak{S}$ with respect to the threshold fuzzy set λ (or the t-level hesitant fuzzy soft set $L(\mathfrak{S}; t)$; or the mid-level hesitant fuzzy soft set $L(\mathfrak{S}; mid)$; or the top-level hesitant fuzzy soft set $L(\mathfrak{S}; max)$.

Step 4. Present the level hesitant fuzzy soft set $L(\mathfrak{S}; \lambda)$ (or $L(\mathfrak{S}; t)$; or $L(\mathfrak{S}; mid)$; or $L(\mathfrak{S}; max)$ in tabular form. For any $x_j \in U$, compute the choice value c_j of x_j.

Step 5. The optimal decision is to select x_k if $c_k = \max_j c_j$.

Step 6. If k has more than one value then any one of x_k may be chosen.

Remark 2. The primary motivation for designing Algorithm 2 is to solve hesitant fuzzy soft set based decision making problem by using level hesitant fuzzy soft sets initiated in this study. This novel idea is completely different from the existing method presented by Babitha et al. in [13].

To illustrate the basic idea of Algorithm 2, we give the following example.

Example 3. Let us reconsider the decision making problem in [13] that involves the hesitant fuzzy soft set $\mathfrak{S} = (\tilde{F}, A)$ with tabular representation as in Table 2. Assume that the opinions of three interviewers must be considered by the company. So we intend to select the object fulfilling the different standards of three interviewers in most aspects as the optimal person. Hence, it is reasonable to use the mid-level decision rule for decision making in such cases. For the neutral reduct fuzzy soft set $(\tilde{F}_N, A)$, we have the mid-threshold fuzzy set $mid_{\mathfrak{S}}$ that can be written as follows:

$$mid_{\mathfrak{S}} = \begin{pmatrix} e_1 & e_2 & e_3 & e_4 \\ 0.6125 & 0.7141 & 0.7883 & 0.7199 \end{pmatrix}.$$

By step 3 of Algorithm 2, we can further obtain the mid-level hesitant fuzzy soft set $L(\mathfrak{S}; mid)$ of $\mathfrak{S}$ with respect to the threshold fuzzy set $mid_{\mathfrak{S}}$ with its tabular representation given by Table 6.

In Table 6, we can calculate choice values as follows:

$c_1 = 1 \times 2 + 1 \times 2 + 1 \times 1 + 1 \times 3 = 8, \quad c_2 = 0 + 0 + 1 + 1 = 2,$
$c_3 = 1 + 1 + 1 \times 2 + 0 = 4, \quad c_4 = 1 \times 3 + 1 \times 2 + 1 \times 3 + 1 \times 2 = 10.$

From the choice value listed in Table 6, it is clear that the optimal decision is to select the candidate x_4 because it has the maximum choice value $c_4 = 10$. Although this result coincides with what was obtained by using Algorithm 1 in [13], it is clear that the calculation is greatly simplified by using Algorithm 2.

In the following, we use Algorithm 2 to handle the decision making problem mentioned in the Subsect. 3.2, which cannot be successfully solved by Algorithm 1. Table 4 gives the tabular representation of the hesitant fuzzy soft set $\mathfrak{I} = (\tilde{G}, B)$. Furthermore we shall use the mid-level decision rule for decision making. On the basic of the neutral reduct fuzzy soft set $(\tilde{G}_N, B)$, the mid-threshold fuzzy set $mid_{\mathfrak{I}}$ can be induced as follows:

$$mid_{\mathfrak{I}} = \begin{pmatrix} e_1 & e_2 & e_3 & e_4 \\ 0.567 & 0.533 & 0.667 & 0.333 \end{pmatrix}.$$

Thus we can obtain the mid-level hesitant fuzzy soft set $L(\mathfrak{I}; mid)$ of $\mathfrak{I}$ with its tabular representation as in Table 7. From Table 7, it follows that the maximum choice value is $c_2 = 7$ and so the optimal decision is to select x_2. The result is identical with our previous subjective judgment in the Subsect. 3.2.

Table 6. Tabular representation of $L(\mathfrak{S}; mid)$ with choice values

U	e_1	e_2	e_3	e_4	Choice value (c_j)
x_1	{0,1,1}	{0,1,1}	{0,1,0}	{1,1,1}	$c_1 = 8$
x_2	{0,0,0}	{0,0,0}	{0,0,1}	{1,0,0}	$c_2 = 2$
x_3	{0,0,1}	{0,1,0}	{1,0,1}	{0,0,0}	$c_3 = 4$
x_4	{1,1,1}	{0,1,1}	{1,1,1}	{1,0,1}	$c_4 = 10$

Table 7. Tabular representation of $L(\Im; mid)$ with choice values

U	e_1	e_2	e_3	e_4	Choice value (c_j)
x_1	{1,1,1}	{1,1,1}	{0,0,0}	{0,0,0}	$c_1 = 6$
x_2	{1,0,0}	{1,1,0}	{0,0,1}	{1,1,1}	$c_2 = 7$
x_3	{0,0,1}	{0,0,0}	{1,1,1}	{1,1,0}	$c_3 = 6$

4 Conclusion

In this study, we have revisited the hesitant fuzzy soft set based decision making problems in [13]. We find that the algorithm designed by Babitha et al. based decision making in [13] is not fit to solve some decision making problems and a counterexample is given by us. So we have proposed an adjustable approach by using level hesitant fuzzy soft sets and illustrated this novel method with some concrete examples. This new proposal proves to be not only more suitable but more feasible for some real-life applications of decision making in an imprecise environment. In the future, we can further use hesitant fuzzy soft sets to address multi-criteria group decision making problems. Moreover, it is important and interesting to further investigate level hesitant fuzzy soft set approach to decision making based on other extensions of hesitant fuzzy soft set theory, such as interval-valued hesitant fuzzy soft set theory.

Acknowledgements. This work is supported by the National Natural Science Foundation of China (No. 61966032), the Fundamental Research Funds for the Central Universities of Northwest MinZu University (No. 31920170010) and the Research Project Funds for Higher Education Institutions of Gansu Province (No. 2016B-005).

References

1. Zadeh, L.A.: Fuzzy sets. Inf. Control **8**, 352–378 (1965)
2. Torra, V., Narukawa, Y.: On hesitant fuzzy sets and decision. In: 18th IEEE International Conference on Fuzzy Systems, pp. 1378–1382. IEEE Press, Jeju Island (2009)
3. Torra, V.: Hesitant fuzzy sets. Int. J. Intell. Syst. **25**, 529–539 (2010)
4. Rodriguez, R.M., Martinez, L., Herrera, F.: Hesitant fuzzy linguistic term sets for decision making. IEEE Trans. Fuzzy Syst. **1**, 109–119 (2012)
5. Xia, M.M., Xu, Z.S.: Hesitant fuzzy information aggregation in decision making. Int. J. Approx. Reason. **52**, 395–407 (2011)
6. Xu, Z.S., Xia, M.M.: Distance and similarity measures for hesitant fuzzy sets. Inf. Sci. **181**, 2128–2138 (2011)
7. Molodtsov, D.: Soft set theory-first results. Comput. Math. Appl. **37**, 19–31 (1999)
8. Jun, Y.B., Park, C.H.: Applications of soft sets in ideal theory of BCK/BCI-algebras. Inf. Sci. **178**, 2466–2475 (2008)
9. Feng, F., Jun, Y.B., Zhao, X.Z.: Soft semirings. Comput. Math. Appl. **56**, 2621–2628 (2008)

10. Roy, A.R., Maji, P.K.: A fuzzy soft set theoretic approach to decision making problems. J. Comput. Appl. Math. **203**, 412–418 (2007)
11. Feng, F., Li, C.X., Davvaz, B., Ali, M.I.: Soft sets combined with fuzzy sets and rough sets: a tentative approach. Soft Computing **14**, 899–911 (2010)
12. Shabir, M., Ali, M.I., Shaheen, T.: Another approach to soft rough sets. Knowl.-Based Syst. **40**, 72–80 (2013)
13. Babitha, K.V., John, S.J.: Hesitant fuzzy soft sets. J. New Results Sci. **3**, 98–107 (2013)

A Novel Probability Evaluation Method Based on Fuzzy Fault Dependent Matrix

Haiyong Dong[1(✉)], Zhengjun Zhai[1], Qingfan Gu[2], Yanhong Lu[1], Guoqing Wang[1,3], and Miao Wang[3]

[1] School of Computer Science and Engineering, Northwestern Polytechnical University, Xi'an, China
donghaiyong@mail.nwpu.edu.cn

[2] China National Aeronautical Radio Electronics Research Institute, Shanghai, China

[3] School of Aeronautics and Astronautics, Shanghai Jiao Tong University, Shanghai, China

Abstract. Safety-critical system is a system in which any design error or failure has the potential to lead to high economic costs or loss of life, and the reliable operation of safety-critical system requires high safety of critical data. Therefore, it is necessary to use fault tree analysis or other traditional methods to analyze the availability and integrity of critical data during the design process. However, traditional methods cannot handle the uncertain fault causal relationships of these safety-critical systems, and availability and integrity cannot be evaluated. Here we propose a fuzzy fault dependent matrix to evaluate the probability of failure condition related to uncertain fault causal relationships based on the thinking of statistics. The qualitative and quantitative analysis methods from the perspective of probability evaluation are given. We take a fault-tolerant display unit an example to illustrate the use of fuzzy fault dependent matrix. A Monte-Carlo simulation is carried out to verify the evaluated results, which proved the correctness of the proposed method.

Keywords: Probability evaluation · Fuzzy fault dependent matrix · Fuzzy cut set

1 Introduction

With the increase of complexity, the safety of safety-critical system has been improved significantly, especially in aviation, aerospace and nuclear industry [1]. The avionics display system is a safety-critical system, and its designation and development should be complied with SAE ARP4754A [2]. The safety evaluation could be carried out based on the safety analysis methods in SAE ARP 4761 [3]. However, there are some uncertain fault causal relationships in a safety-critical system, which are very difficult to analyze by traditional safety analysis methods. For example, if the input voltage of one process unit from a voltage converter is unstable, it is unable to determine whether calculation errors occur or not [4].

Netherlands Aerospace Centre NLR began to maintain a safety method database since 2004, and more than 847 safety methods until 2016 [5]. There are more than 20

Y. Liu et al. (Eds.): ICNC-FSKD 2019, AISC 1074, pp. 809–817, 2020.
https://doi.org/10.1007/978-3-030-32456-8_87

methods that based on fuzzy logical or fuzzy set, including fuzzy fault tree, fuzzy Petri net, etc [6–9]. Fuzzy fault tree analysis describes the event with the fuzzy set and describes the false causal relationship by fuzzy logical. Fuzzy Petri net describes the transition by virtue of the fuzzy logical. However, these fuzzy methods increase the analytical ability while increasing the difficulty of expression. Dong proposed fault dependent matrix based on failure mode and effects affect (FMEA), with comprehensive ability of traditional fault tree analysis and Pet net analysis [10], and it has been used to analyze the availability of deterministic avionics display systems [11].

This paper intends to propose a fuzzy fault dependent matrix for probability evaluation of uncertain fault causal relationships based on the fault dependent matrix. In addition, we take a fault-tolerant display unit model to demonstrate the proposed method. In the last section, future research directions are presented.

2 Fuzzy Fault Dependent Matrix

2.1 Fuzzy Fault Causal Relationship

The traditional safety analysis method analyzes the certain fault causal relationship between the cause and the result of the fault under the defined conditions. Usually, the causes are considered as inputs, and the result is regarded as output. In this paper, we define a fuzzy fault causal relationship, whose inputs have an uncertain effect on the output. Since the inputs may be a combination of events, here we express the influence by fuzzy fault tree as shown in Fig. 1. Probability of the influence of the inputs on the output is defined as the Fuzzy Rate (FRate), which can be calculated by using Bayesian formula or other statistical method. For the traditional safety analysis method, FRate is assigned to be 1. If the FRate is estimated at 0.9, it means that 90% of the input events will have effects on the causal relationship, and the left 10% is exactly the same things as if it did not happen.

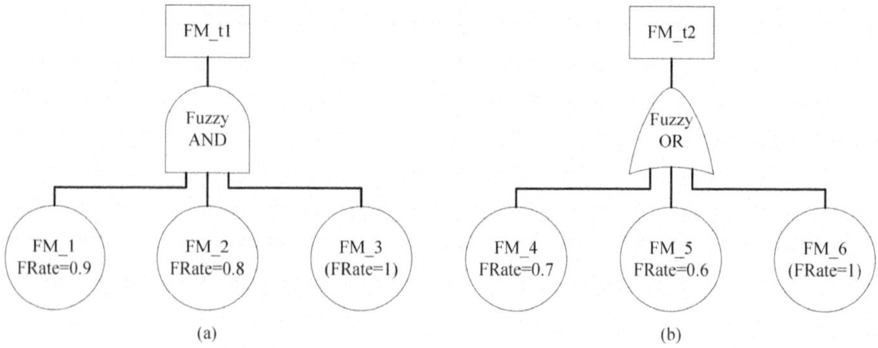

Fig. 1. Fuzzy AND gate and fuzzy OR gate

As shown in Fig. 1, sub graph (a) is a fuzzy AND gate fault tree, which means that statistically, in the case where all three input basic events occur, the probability of FM_t1 occurring is only 72%. Sub graph (b) is a fuzzy OR gate fault tree, indicating that statistically, in the case of the basic event FM_4 occurs, the probability of FM_t2 occurring is 70%, whereas in the case of the basic event FM_6 occurs, the probability becomes 100%.

2.2 Fuzzy Fault Dependent Matrix

The safety analysis method based on fault dependent matrix analyzes the fault causal relationship of the system by several tables, which describe all possible failure modes of the system, and the fault propagation relationships among them. It not only supports the cut set analysis of specific failure conditions, but also be able to analyze the reach ability analysis of the combination of multiple failure modes. In [11], the algorithms of safety analysis based on fault dependent matrix are presented, but the original method cannot express and analyze the uncertain fault causal relationship. Based on the fuzzy fault causal relationship, fuzzy fault dependent matrix can express the fuzzy fault propagation model. For Fig. 1, the fuzzy fault propagation model can be described through Table 1.

Table 1. The fuzzy fault propagation model for model described in Fig. 1.

fpmID	Description	Source	Effect
FPM_1	The occurrence of FM_1 (FRate = 0.9) and FM_2 (FRate = 0.8) and FM_3 may lead to the occurrence of FM_t1	FM_1 (FRate = 0.9) FM_2 (FRate = 0.8) FM_3	FM_t1
FPM_2	The occurrence of FM_4(FRate = 0.7) may lead to the occurrence of FM_t2	FM_4 (FRate = 0.7)	FM_t2
FPM_3	The occurrence of FM_5(FRate = 0.6) may lead to the occurrence of FM_t2	FM_5 (FRate = 0.6)	FM_t2
FPM_4	The occurrence of FM_6 will lead to the occurrence of FM_t2	FM_6	FM_t2

Referring to the incidence matrix of the Petri net, the fuzzy fault propagation model can be expressed by the fuzzy fault dependent matrix as follows.

$$M_{FPM} = \begin{bmatrix} V_{FPM\_1} \\ V_{FPM\_2} \\ V_{FPM\_3} \\ V_{FPM\_4} \end{bmatrix} = \begin{bmatrix} -0.9 & -0.8 & -1 & 0 & 0 & 0 & 1 & 0 \\ 0 & 0 & 0 & -0.7 & 0 & 0 & 0 & 1 \\ 0 & 0 & 0 & 0 & -0.6 & 0 & 0 & 1 \\ 0 & 0 & 0 & 0 & 0 & -1 & 0 & 1 \end{bmatrix} \quad (1)$$

(columns: FM_1, FM_2, FM_3, FM_4, FM_5, FM_6, FM_t1, FM_t2)

In above equation, the negative number indicates an input, and the magnitude of the value means the possibility of affecting fault causal relationship. In addition,1 indicates that the functional fault affected by the fault source, and 0 is meaningless.

Probability evaluationofa failure condition requires obtaining a combination of failure modes with corresponding FRate, which can be derived by the process as below (Fig. 2).

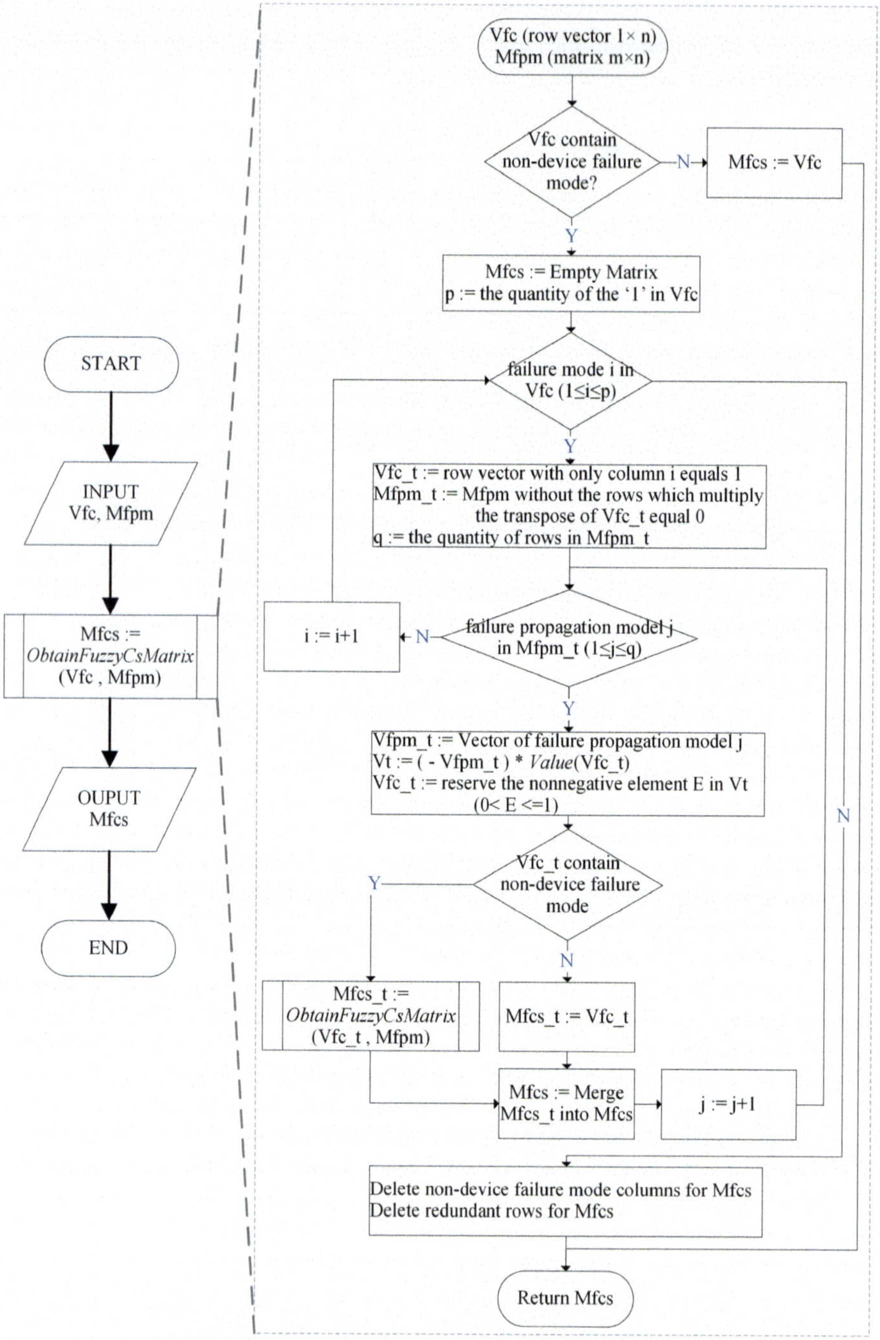

Fig. 2. Process for obtaining the fuzzy cut set matrix

3 Case Study

3.1 Display Unit Model

The fault-tolerant display unit provides data by digital tube and meter. As shown in Fig. 3, the data displayed in digital tube come from the Process Modular (PM), which requires an external 3.3 V DC power supply. Power Supply Modular (PSM) provides power for Power Convert Modular (PCM), which can convert 28 V DC to 12 V DC and convert 12 V DC to 3.3 V DC. The data displayed in the meter come from the standby Sensor (SEN), and it is a backup of the digital tube display. It is known that the display of this data plays a very important role in the operation of the system, and the probability of the data display function lost or deviated too much should be less than 1E-5 per hour.

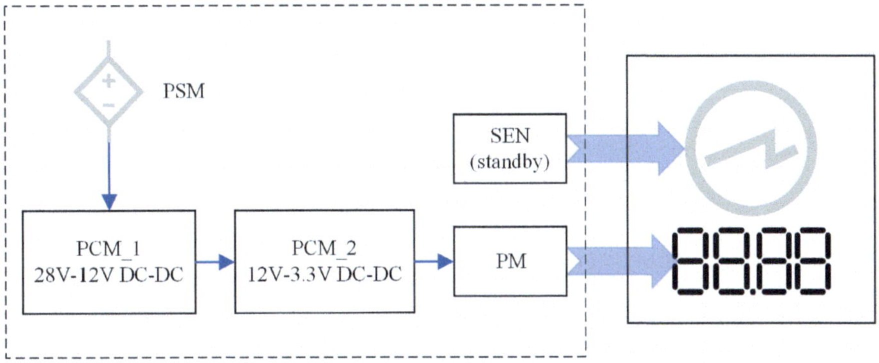

Fig. 3. Fault-tolerance Display Unit model

There are 5 devices in the Display Unit model as shown in Fig. 3. To simplify the model, we define 3 general assumptions [12]: (1) the failure rate of each device is modeled as a statistically independent distributed constant, and all the cables keep working; (2) A fault occurs instantaneously and at most one fault event in a minimum time slice; (3) The system and its devices are irreparable while in use.

3.2 Fuzzy Fault Dependent Matrix of Display Unit

Each type device may malfunction by different failure mode. Refer to the historical experience data, the failure mode information table of the device type can be summarized as shown in Table 2.

Table 2. The description of the failure mode for device type.

Device	Description	Probability
Power Supply Modular (PSM)	PSM cannot provide power	4E-4
Power Convert Modular (PCM)	PCM cannot provide power	5E-4
Power Convert Modular (PCM)	PCM provides unstable power	3E-3
Process Modular (PM)	PM cannot process data to provide information	8E-4
Sensor (SEN)	SEN cannot obtain data to provide information	1E-3
Sensor (SEN)	SEN provides unstable information	3E-3

As there are 2 PCMs, the failure modes for devices of the display unit could be concluded in Table 3.

Table 3. The description of the failure mode for device.

fmID	Device	Description
FM_1	PSM	PSM cannot provide power
FM_2	PCM_1	PCM_1 cannot provide power
FM_3	PCM_1	PCM_1 provides unstable power
FM_4	PCM_2	PCM_2 cannot provide power
FM_5	PCM_2	PCM_2 provides unstable power
FM_6	PM	PM cannot process data to provide information
FM_7	SEN	SEN cannot obtain data to provide information
FM_8	SEN	SEN provides unstable information

Besides the failure mode for device, all functional fault in the display unit need to be defined as failure modes as shown in Table 4.

The failure mode in Tables 3 and 4 is the basis of safety model. Referring to the process of the FMEA, the fault propagation model is presented in Table 5.

Table 4. The description of the failure mode for function.

fmID	Device	Description
FM_9	/	Loss of the ability to provide power for PCM_1 from PSM
FM_10	/	Loss of the ability to provide power for PCM_2 from PCM_1
FM_11	/	Loss of the ability to provide power for PM from PCM_2
FM_12	/	Loss of the ability to provide information for Display Unit from PM
FM_13	/	Undetected PCM_2 cannot work well as input unstable power from PCM_1
FM_14	/	Undetected PM cannot provide correct information as input unstable power from PCM_2
FM_15	/	Digital Display cannot provide correct information
FM_16	/	Analogue Display cannot provide correct information
FM_17	/	Display Unit cannot provide correct information

Table 5. The fault propagation model of the display unit.

fpmID	Description	Source	Effect
FPM_1	The occurrence of FM_1will lead to the occurrence of FM_9	FM_1	FM_9
FPM_2	The occurrence of FM_2 and FM_9will lead to the occurrence of FM_10	FM_2, FM_9	FM_10
FPM_3	The occurrence of FM_4 and FM_10will lead to the occurrence of FM_11	FM_4, FM_10	FM_11
FPM_4	The occurrence of FM_6 and FM_11will lead to the occurrence of FM_12	FM_6, FM_11	FM_12
FPM_5	The occurrence of FM_3(FRate = 0.2) may lead to the occurrence of FM_13	FM_3 (FRate = 0.2)	FM_13
FPM_6	The occurrence of FM_5(FRate = 0.2) and FM_13 (FRate = 0.6) may lead to the occurrence of FM_14	FM_5 (FRate = 0.2) FM_13 (FRate = 0.6)	FM_14
FPM_7	The occurrence of FM_12 and FM_14 will lead to the occurrence of FM_15	FM_12, FM_6	FM_15
FPM_8	The occurrence of FM_7 and FM_8(FRate = 0.05) may lead to the occurrence of FM_16	FM_7, FM_8 (FRate = 0.6)	FM_16
FPM_9	The occurrence of FM_15 and FM_16 will lead to the occurrence of FM_17	FM_15, FM_16	FM_17

Then the fault propagation model can be expressed by the fault dependent matrix as shown in below equation.

$$M_{FPM} = \begin{bmatrix} V_{FPM\_1} \\ V_{FPM\_2} \\ V_{FPM\_3} \\ V_{FPM\_4} \\ V_{FPM\_5} \\ V_{FPM\_6} \\ V_{FPM\_7} \\ V_{FPM\_8} \\ V_{FPM\_9} \end{bmatrix} = \begin{bmatrix} -1 & 0 & 0 & 0 & 0 & 0 & 0 & 0 & 1 & 0 & 0 & 0 & 0 & 0 & 0 & 0 & 0 \\ 0 & -1 & 0 & 0 & 0 & 0 & 0 & 0 & -1 & 1 & 0 & 0 & 0 & 0 & 0 & 0 & 0 \\ 0 & 0 & 0 & -1 & 0 & 0 & 0 & 0 & 0 & -1 & 1 & 0 & 0 & 0 & 0 & 0 & 0 \\ 0 & 0 & 0 & 0 & 0 & -1 & 0 & 0 & 0 & 0 & -1 & 1 & 0 & 0 & 0 & 0 & 0 \\ 0 & 0 & -0.2 & 0 & 0 & 0 & 0 & 0 & 0 & 0 & 0 & 0 & 1 & 0 & 0 & 0 & 0 \\ 0 & 0 & 0 & 0 & -0.2 & 0 & 0 & 0 & 0 & 0 & 0 & 0 & -0.6 & 1 & 0 & 0 & 0 \\ 0 & 0 & 0 & 0 & 0 & 0 & 0 & 0 & 0 & 0 & 0 & -1 & 0 & -1 & 1 & 0 & 0 \\ 0 & 0 & 0 & 0 & 0 & 0 & -1 & -0.05 & 0 & 0 & 0 & 0 & 0 & 0 & 0 & 1 & 0 \\ 0 & 0 & 0 & 0 & 0 & 0 & 0 & 0 & 0 & 0 & 0 & 0 & 0 & 0 & -1 & 0 & 1 \end{bmatrix} \quad (2)$$

For the display unit, the situation that both two types of data display cannot provide correct information is not allowed. Therefore the failure condition is failure mode FM_17, and its failure rate should be less than 1E-5, as shown in Table 6.

Table 6. The failure condition of the display unit

fcID	fmID	Description	Probability
FC_1	FM_17	Display Unit cannot provide correct information	1E-5

3.3 Probability Evaluation

The fuzzy fault cut set of FC_1 can be obtained by utilizing the method in Sect. 2.2, and the probability of FC_1 occurrence is 3.634E-6. Compared with Table 6, it can be verified that the system design has met the safety requirement.

In addition, we also use Monte-Carlo method to simulate the operation of a display unit. After 100,000,000 time simulation for operation time of 1 h, FC_1 occurs only 357 times. The simulation result 3.57E-6 is approximately equal to 3.634E-6. To sum up, the Monte-Carlo simulation verifies the validity of the proposed method.

4 Conclusion

This paper takes statistical values to describe the possibility of uncertainty causal relationships, so as to evaluate the occurring probability of unexpected events. This method simplifies the complex calculation of conventional fuzzy sets and the result is evaluated to the maximum possible value, not a range.

In the future, we plan to use the proposed method to evaluate the availability of avionics display system with multiple fuzzy fault causal relationships. In addition, we also plan to extend fault dependent matrix to express the dynamic fault causal relationship.

References

1. Bahr, N.J.: System Safety Engineering and Risk Assessment: A Practical Approach, 2nd edn. CRC Press, Boca Raton (2018)
2. ARP4754A: Guidelines for Development of Civil Aircraft and Systems. SAE International, Warrendale, PA, USA (2010)
3. ARP4761: Guidelines and Methods for Conducting the Safety Assessment Process on Civil Airborne Systems and Equipment. SAE International, Warrendale, PA, USA (1996)
4. Liang-biao, Z., Li-ye, C., Peng-fei, Y., Yuan, C., Yu-dong, L.: A PHM research for data processing modules in avionics system. In: 2017 Prognostics and System Health Management Conference, pp. 1–6. IEEE, Harbin (2017)
5. Everdij, M.H.C., Blom, H.A.P.: Safety Methods Database, 1.1st edn. Netherlands Aerospace Centre NLR, Netherlands (2016)
6. Singer, D.: A fuzzy set approach to fault tree and reliability analysis. Fuzzy Sets Syst. **34**(2), 145–155 (1990)
7. Dong, Y., Yu, D.: Estimation of failure probability of oil and gas transmission pipelines by fuzzy fault tree analysis. J. Loss Prev. Process Ind. **18**(2), 83–88 (2005)
8. Mahmood, Y.A., Ahmadi, A., Verma, A.K., Srividya, A., Kumar, U.: Fuzzy fault tree analysis: a review of concept and application. Int. J. Syst. Assur. Eng. Manag. **4**(1), 19–32 (2013)
9. Sun, J., Qin, S.Y., Song, Y.H.: Fault diagnosis of electric power systems based on fuzzy Petri nets. IEEE Trans. Power Syst. **19**(4), 2053–2059 (2004)
10. Dong, H., Gu, Q., Wang, G., Zhai, Z., Lu, Y., Wang, M.: A novel safety assessment method based on fault dependent matrix. Int. J. Perform. Eng. **15**(9), 2392–2399 (2019)
11. Dong, H., Zhai, Z., Lu, Y., Gu, Q., Wang, G., Wang, M.: Availability assessment of avionics display system based on fault dependent matrix. In: 9th IEEE International Conference on Electronics Information and Emergency Communication, pp. 561–565. IEEE, Beijing (2019)
12. Dong, H., Gu, Q., Wang, G., Zhai, Z., Lu, Y., Wang, M.: Availability Assessment of IMA System Based on Model-Based Safety Analysis Using AltaRica 3.0. Processes **7**(2), 1–14 (2019)

A Multi-attribute Decision Making Method Based on Interval Pythagorean Fuzzy Language and the PROMETHEE Method

Wenyu Zhang[1,2], Yuting Zhu[1(✉)], Danshu Wang[1], Songmin Zhao[1], and Dadi Dong[1]

[1] Xi'an University of Posts and Telecommunications, Xi'an 710016, China
529403847@qq.com

[2] China Research Institute of Aerospace Systems Science and Engineering, Beijing 100048, China

Abstract. In order to deal with the multi-attribute decision making the problem which is related to the actual decision problem by randomness and uncertainty in the Pythagorean fuzzy environment, this paper proposes a method based on the interval value Pythagorean fuzzy set and the PROMETHEE to solve the multi-attribute decision problem.Firstly, the fuzzy interval Pythagorean decision matrix of each decision maker is constructed, and the weights of decision makers and attributes are determined by the relative proximity coefficient method, the entropy method and the distance of the Pythagorean fuzzy number. Secondly, using the comprehensive weight and IVPFWA operator to aggregate the decision matrix of each decision maker into a comprehensive matrix and Pythagorean fuzzy comprehensive scoring matrix, and then obtain the final sorting result through the PROMETHEE; Finally, the effectiveness and feasibility of the algorithm are verified by the factory location example, and compared with the results of TOPSIS and VIKOR methods.

Keywords: Pythagorean fuzzy · PROMETHEE · Multi-attribute decision

1 Introduction

As the number of decision-making participants increases, the decision-making process becomes complex and slow. Therefore, the multi-attribute group decision-making has developed into an extremely important research field in modern decision-making theory and decision-making science, and has been widely used in engineering, logistics, medicine and military, etc.In the traditional multi-objective decision making method, the evaluation values are completely determined. Due to the complexity of the objective world, it is difficult for policy makers to make accurate numerical evaluations of the program. Zadeh [1]

Y. Liu et al. (Eds.): ICNC-FSKD 2019, AISC 1074, pp. 818–826, 2020.
https://doi.org/10.1007/978-3-030-32456-8_88

proposed the uncertainty and ambiguity of decision information by membership degree. Then Gau and Buehrer [2] defined the Vague set. Torra et al. proposed the concept of the Hesitant Fuzzy Set, which allows membership to exist in the form of multiple sets of possible values to express the degree of hesitation of experts in expressing their target preferences in the decision-making process.To let the sum of membership and non-affiliation is greater than or equal to 1, Yager [3] proposed the Pythagorean Fuzzy Set.

This paper expounds the concept and operation of the interval value Pythagorean fuzzy set, and proposes a method based on a interval value Pythagorean fuzzy set and PROMETHEE to solve the actual multi-attribute decision problem [4]. This method can effectively improve the accuracy of decision making in Pythagorean fuzzy environment.

2 Related Concepts

2.1 Pythagorean Fuzzy Set

Definition 1. Let $\tilde{p} = ([\mu^L \mu^U], [v^L, v^U\ ])$ be the interval value Pythagorean fuzzy number, the score function and precision function are defined as formula (1) and (2).

$$S(\tilde{p}) = \frac{1}{2}\left(\frac{(\mu^L)^2 + (\mu^U)^2}{2} + 1 - \frac{(v^L)^2 + (v^U)^2}{2})\right) \tag{1}$$

$$A(\tilde{p}) = \frac{1}{2}\left(\frac{(\mu^L)^2 + (\mu^U)^2}{2} + \frac{(v^L)^2 + (v^U)^2}{2})\right) \tag{2}$$

Definition 2. Let $\tilde{p}_1 = ([\mu_1^L, \mu_1^U], [v_1^L, v_1^U])$, $\tilde{p}_2 = ([\mu_2^L \mu_2^U], [v_2^L, v_2^U\ ])$ be the interval value Pythagorean fuzzy number, then the Hamming distance $d(\tilde{p}_1, \tilde{p}_2)$ between $\tilde{p}_1$ and $\tilde{p}_2$ is defined as

$$d(\tilde{p}_1, \tilde{p}_2) = \frac{1}{4}(|\mu_1^L - \mu_2^L| + |\mu_1^U - \mu_2^U| + |v_1^L - v_2^L| + |v_1^U - v_2^U|) \tag{3}$$

Definition 3. The Pythagorean weighted average operator (IVPFWA) is defined as:

$$\mathrm{IVPFWA}(\tilde{\mathrm{p}}_1, \tilde{\mathrm{p}}_2, \cdots, \tilde{\mathrm{p}}_\mathrm{n}) = \omega_1\tilde{\mathrm{p}}_1 \oplus \omega_2\tilde{\mathrm{p}}_2 \oplus ... \oplus \omega_\mathrm{n}\tilde{\mathrm{p}}_\mathrm{n} \tag{4}$$

2.2 PROMETHEE Method

The PROMETHEE method [5] introduces a priority function to describe the priority of decision making on the same attribute C_j, to judge the pros and cons of each pair of decisions according to the difference between the values of each decision attribute. Therefore, the PROMETHEE method defines a priority function for each indicator.

Therefore, the method first sets a priority function for each indicator definition, the value range of the function is 0–1, the smaller the function value, the smaller the difference between the schemes and. when the function value is 0, there is no difference between the schemes A_i and A_k. The closer the value is to 1, the higher the degree of scheme A_i is better than the scheme A_k. And when the function value is 1, the scheme A_i is strictly superior to option A_k.

Brans and Vincke summed up the six forms of priority functions based on the actual situation of various problems. In the specific application, the decision-making expert should select the appropriate priority function as the judgment criterion for decision making according to its actual situation.

3 Multi-attribute Decision Making Method Based on Pythagorean Fuzzy Information and PROMETHEE

This section mainly introduces the algorithm steps of the PROMETHEE method based on Pythagorean fuzzy set language [6].

Let $A = \{A_i, i = 1, 2, ..., m\}$ be the program set, $C = \{C_j, j = 1, 2, ..., n\}$ be the indicator set, set a total of t decision experts $E = \{E_k, k = 1, 2, ..., t\}$ to participate in the evaluation, $\tilde{p}_{ij}^{(k)}$ denotes the evaluation value of the kth expert on the plan A_i in the index C_j, where $\tilde{p}_{ij}^{(k)}$ is the interval value Pythagorean fuzzy number [7].

Step1. A number of decision makers give the evaluation value of the program on the index to form a decision matrix is $D^{(k)} = (\tilde{p}_{ij}^{(k)})_{m\times n}$.

Step2. Determine the objective weight of each expert under each indicator. First defined the interval value Pythagorean fuzzy positive ideal matrix $D^+ = (\tilde{p}_{ij}^+)_{m\times n}$ and negative ideal matrix $D_u^- = (\tilde{p}_{ij}^u)_{m\times n}$, $D_d^- = (\tilde{p}_{ij}^d)_{m\times n}$ as follows:

$$D^+ = \begin{bmatrix} \tilde{p}_{11}^+ & \tilde{p}_{12}^+ & \cdots & \tilde{p}_{1n}^+ \\ \tilde{p}_{21}^+ & \tilde{p}_{22}^+ & \cdots & \tilde{p}_{2n}^+ \\ & & \cdot & \\ \cdots & \cdots & \cdot & \cdots \\ & & \cdot & \\ \tilde{p}_{m1}^+ & \tilde{p}_{m2}^+ & \cdots & \tilde{p}_{m3}^+ \end{bmatrix}, \quad D_d^- = \begin{bmatrix} \tilde{p}_{11}^d & \tilde{p}_{12}^d & \cdots & \tilde{p}_{1n}^d \\ \tilde{p}_{21}^d & \tilde{p}_{22}^d & \cdots & \tilde{p}_{2n}^d \\ & & \cdot & \\ \cdots & \cdots & \cdot & \cdots \\ & & \cdot & \\ \tilde{p}_{m1}^d & \tilde{p}_{m2}^d & \cdots & \tilde{p}_{mn}^d \end{bmatrix}, \quad D_u^- = \begin{bmatrix} \tilde{p}_{11}^u & \tilde{p}_{12}^u & \cdots & \tilde{p}_{1n}^u \\ \tilde{p}_{21}^u & \tilde{p}_{22}^u & \cdots & \tilde{p}_{2n}^u \\ & & \cdot & \\ \cdots & \cdots & \cdot & \cdots \\ & & \cdot & \\ \tilde{p}_{m1}^u & \tilde{p}_{m2}^u & \cdots & \tilde{p}_{mn}^u \end{bmatrix}$$

$\tilde{p}_{ij}^+ = \frac{\sum_{k=1}^t \tilde{p}_{ij}^{(k)}}{t} = \left(\left[\mu_{ij}^{L+}, \mu_{ij}^{U+}\right], \left[v_{ij}^{L+}, v_{ij}^{U+}\right]\right)$ is the average value of the evaluation value given by t decision-making experts. The negative ideal evaluation value is divided into two parts, one part is $\tilde{p}_{ij}^u = \max_k \left\{\tilde{p}_{ij}^{(k)}\right\} = \left(\left[\mu_{ij}^{Lu}, \mu_{ij}^{Uu}\right], \left[v_{ij}^{Lu}, v_{ij}^{Uu}\right]\right)$ over the average value, and the other part is $\tilde{p}_{ij}^d = \max_k \left\{\tilde{p}_{ij}^{(k)}\right\} = \left(\left[\mu_{ij}^{Ld}, \mu_{ij}^{Ud}\right],\right.$
$\left.\left[v_{ij}^{Ld}, v_{ij}^{Ud}\right]\right)$ below the average value.

Calculate the closeness of each evaluation value as follows

$$c_{ij}^{(k)} = \frac{d(\tilde{p}_{ij}^{(k)}, \tilde{p}_{ij}^u) + d(\tilde{p}_{ij}^{(k)}, \tilde{p}_{ij}^d)}{d(\tilde{p}_{ij}^{(k)}, \tilde{p}_{ij}^u) + d(\tilde{p}_{ij}^{(k)}, \tilde{p}_{ij}^d) + d(\tilde{p}_{ij}^{(k)}, \tilde{p}_{ij}^+)} \tag{5}$$

Then the weight can be determined as follows

$$\omega_{ij}^{k} = \frac{c_{ij}^{k}}{\sum_{k=1}^{t} c_{ij}^{(k)}}, (i = 1, 2, ..., m,\ j = 1, 2, ..., n,\ k = 1, 2, ..., t) \tag{6}$$

Step3. Integrate the decision matrix into a comprehensive decision matrix. Based on the IVPFWA operator and weight, Construct the Pythagorean fuzzy score matrix $D_S = [S_{ij}]_{m \times n}$ according to the Definition 1, where

$$S_{ij} = \frac{1}{2}\left(\frac{(\mu_{ij}^{L})^2 + (\mu_{ij}^{U})^2}{2} + 1 - \frac{(v_{ij}^{L})^2 + (v_{ij}^{U})^2}{2})\right) \tag{7}$$

Step 4. Calculate the weight ω_j of each indicator $C = \{C_j\}$ by using the entropy method.The steps for calculating the entropy weight are as follows:

Step 4.1 Standardize the indicator data matrix X of the decision plan with the power factor method as $Y_{ij} = \frac{S_{ij} - S_{\min(j)}}{S_{\max(j)} - S_{\min(j)}}\alpha + (1 - \alpha)$.

Step 4.2 Calculate the index entropy value. Assuming that the entropy value of the jth indicator is e_j, the entropy value of the jth index is calculated as $e_j = \frac{1}{(\ln m)\sum_{i=1}^{m} Y_{ij}}\left[(\ln \sum_{i=1}^{m} Y_{ij}) \sum_{i=1}^{m} Y_{ij} - \sum_{i=1}^{m} Y_{ij} \ln Y_{ij}\right]$.

Step 4.3 Calculate the indicator weights. Let the weight of the jth indicator is w_j, then the weight calculation formula is as formula (8):

$$w_j = \frac{1 - e_j}{\sum_{j=1}^{n} (1 - e_j)} \tag{8}$$

Step5. Calculate the priority index [8]. The priority index is defined as formula (9):

$$\Pi(A_i, A_k) = \frac{\sum_{j=1}^{m} \omega_j p_j(A_i, A_k)}{\sum_{j=1}^{m} \omega_j} \tag{9}$$

Step6. Calculate the outflow value is $\varphi^{+}(A_i) = \sum_{x \in X}^{m} \Pi(\mathrm{A_i}, \mathrm{A})$, the inflow value is $\varphi^{-}(A_i) = \sum_{x \in X}^{m} \Pi(\mathrm{A}, \mathrm{A_i})$.The net flow of option A_i calculation method such as formula (10).

$$\varphi(A_i) = \varphi^{+}(A_i) - \varphi^{-}(A_i) \tag{10}$$

Step7. By calculating the net flow of each decision on decision set A, the level relationship between the decisions can be determined [9]. The results are as follows: If $\varphi(A_i) > \varphi(A_k)$, the level A_i is higher than A_k. If $\varphi(A_i) = \varphi(A_k)$, there$A_i$ is no difference with the level A_k. If $\varphi(A_i) < \varphi(A_k)$, the level A_i is lower than A_k [10].

4 Case Analysis

4.1 Analysis of Examples

A company needs to select a new address to build a factory. There are five candidate addresses $A_i(i = 1, 2, ..., 5)$. To evaluate the plant address, we need to examine the following four criteria: C_1-traffic conditions, C_2-topographic conditions, C_2-public facilities, and C_4-economic costs. Four experts E_1E_2, E_3 and E_4 were invited to evaluate the candidate addresses. The candidate addresses are evaluated and ranked by the algorithm of this paper, and the recommended construction address is selected.

Step 1. Four experts E_i according to four criteria C_i give evaluation values of five plant addresses A_i to form the decision matrix $D^{(k)}(1, 2, ..., 4)$. Take the evaluation matrix given by the first expert as an example, as shown in Table 1.

Table 1. Evaluation matrix given by E_1 decision makers

	C_1	C_2	C_3	C_4
A_1	([0.5,0.6],[0.3,0.4])	([0.7,0.8],[0.1,0.2])	([0.3,0.4],[0.4,0.5])	([0.5,0.6],[0.3,0.4])
A_2	([0.7,0.8],[0.2,0.3])	([0.5,0.6],[0.4,0.5])	([0.5,0.6],[0.1,0.2])	([0.4,0.5],[0.3,0.5])
A_3	([0.5,0.6],[0.1,0.2])	([0.2,0.3],[0.5,0.6])	([0.6,0.7],[0.2,0.3])	([0.8,0.9],[0.1,0.2])
A_4	([0.7,0.8],[0.3,0.4])	([0.3,0.4],[0.4,0.5])	([0.5,0.6],[0.2,0.4])	([0.4,0.5],[0.5,0.6])
A_5	([0.4,0.5],[0.6,0.7])	([0.7,0.8],[0.2,0.3])	([0.6,0.8],[0.2,0.3])	([0.5,0.6],[0.1,0.2])

Step 2. Determine the interval value positive and negative ideal matrix D_d^- as shown in Table 2. And determine the objective weight of each expert under each indicator, as shown in Table 3.

Table 2. Interval value Pythagorean positive and negative ideal matrix D_d^-

	C_1	C_2	C_3	C_4
A_1	([0.5,0.6],[0.3,0.4])	([0.5,0.6],[0.2,0.3])	([0.2,0.3],[0.4,0.5])	([0.5,0.6],[0.3,0.4])
A_2	([0.5,0.6],[0.4,0.5])	([0.5,0.6],[0.4,0.5])	([0.7,0.8],[0.3,0.4])	([0.4,0.5],[0.3,0.5])
A_3	([0.5,0.7],[0.4,0.5])	([0.2,0.3],[0.5,0.6])	([0.5,0.6],[0.3,0.4])	([0.7,0.8],[0.3,0.4])
A_4	([0.7,0.8],[0.5,0.6])	([0.3,0.4],[0.4,0.5])	([0.5,0.6],[0.4,0.5])	([0.3,0.4],[0.5,0.6])
A_5	([0.4,0.5],[0.6,0.7])	([0.5,0.6],[0.2,0.3])	([0.3,0.4],[0.6,0.7])	([0.5,0.6],[0.1,0.2])

Table 3. Objective weights of experts under different indicators

	C_1	C_2	C_3	C_4
E_1	0.2012	0.1973	0.193	0.2298
E_2	0.2035	0.1905	0.1859	0.2221
E_3	0.1921	0.1936	0.1835	0.1853
E_4	0.206	0.1878	0.2356	0.1919

Step 3. Construct the Pythagorean fuzzy score synthesis matrix $D_S = [S_{ij}]_{m \times n}$, the results are shown in Table 4.

Table 4. Pythagorean fuzzy score synthesis matrix D_S

	C_1	C_2	C_3	C_4
A_1	0.2362	0.2295	0.1364	0.2695
A_2	0.2313	0.1385	0.3323	0.1951
A_3	0.2292	0.0807	0.2148	0.3594
A_4	0.2955	0.1393	0.2417	0.2175
A_5	0.2244	0.1225	0.2479	0.2286

Step 4. Calculate the weight ω_j of the index $C = \{C_j\}$ by the entropy method, the results are shown in Table 5.

Table 5. Weights of indicators determined by entropy method

	C_1	C_2	C_3	C_4
ω	0.1291	0.3889	0.3324	0.1496

Step 5. Calculate the priority index, the results are shown in Table 6.

Table 6. Priority index

	A_1	A_2	A_3	A_4	A_5
A_1	0	−0.2039	−0.04	0.0014	0.1353
A_2	−0.1913	0	0.3121	0.1996	0.3352
A_3	−0.2138	−0.5371	0	−0.2981	−0.1463
A_4	−0.002	−0.2018	0.2118	0	0.1333
A_5	−0.1353	−0.3352	0.0785	−0.1333	0

Step 6. Calculate the outflow value φ^+, inflow value φ^- and net flow value φ of each scheme. The calculation results are shown in Table 7.

Table 7. Outflow, inflow and net flow values of each scenari

	A_1	A_2	A_3	A_4	A_5
φ^+	−0.1073	0.6556	−1.1953	0.1412	−0.5253
φ^-	−0.5424	−1.278	0.5623	−0.2304	0.4575
φ	**0.4351**	**1.9336**	**-1.7576**	**0.3716**	**-0.9828**

Step 7. The comprehensive evaluation results of the site selection of the five sites are sorted: $A_2 > A_1 > A_4 > A_5 > A_3$, the optimal candidate address is A_2.

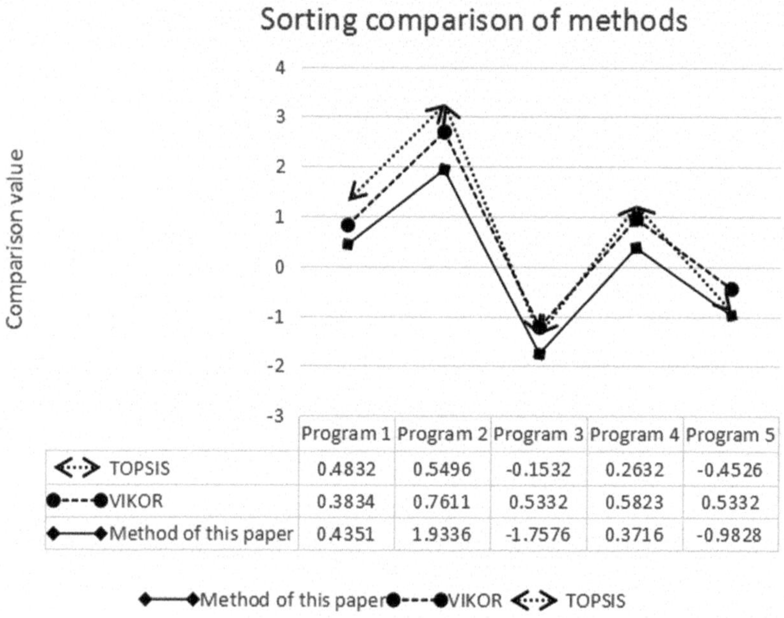

Fig. 1. Sorting comparison of methods

4.2 Analysis of Results

In order to compare the effectiveness of the proposed algorithm, by comparing with other methods based on utility value sorting, the comparison of the theoretical values of each sorting method can be clearly seen from Fig. 1. It can be seen from the figure that the results of different sorting methods are not much different, TOPSIS [11] determines $A_3 > A_5$ because it ignores the poor evaluation of A_5 under the criterion C_3. VIKOR [12] cannot determine the specific relationship between A_3 and A_5, because the compromise theory has limitations in calculating the final value. It can be seen from the above results that the new algorithm given in this paper can solve different types of attribute information problems [13]. For different attribute weights, the obtained decision results are not exactly the same, which is consistent with the actual situation, each candidate address has its own advantages. In the actual decision-making process, each attribute weight can be determined according to the requirements of the company to obtain an optimal candidate address.

5 Conclusion

In the multi-attribute decision problem,for the case where the randomness and the ambiguity uncertainty occur simultaneously and the decision information is biased. This paper is based on a multi-attribute decision making method combining Pythagorean fuzzy information and the PROMETHEE method [14], which providing a new class of research methods for Pythagorean fuzzy numbers.This method can better overcome the subjective influence of decision makers in the Pythagorean fuzzy environment. It can fully process random and fuzzy information, and can effectively improve the validity and accuracy of Pythagorean fuzzy decision-making conclusions. This method is the enrichment and development of multi-criteria group decision making methods.

References

1. Zadeh, L.A.: Fuzzy sets. Inf. Control **8**(3), 338–353 (1965)
2. Gau, W.L., Buehrer, D.J.: Vague sets. IEEE Trans. Syst. Man Cybern. **23**(2), 610–614 (1993)
3. Yager, R.R.: Pythagorean membership grades in multicriteria decision making. IEEE Trans. Fuzzy Syst. **22**(4), 958–965 (2014)
4. Miao, L., Zhang, J., Wu, Y., Wang, Y.: Multi-attribute decision making method based on Pythagorean hesitant fuzzy set similarity measure. J. Wuhan Univ. Technol. (Inf. Manag. Eng.) **40**(02), 219–223 (2018)
5. Ju, P., Chen, Z., Ran, Y., Hu, X.: The improved FMEA method based on PROMETHEE in multi-granular probability language environment. J. Beijing Univ. Aeronaut. Astronaut. 1-10 (2019)
6. Fan, J., Yan, Y., Wu, M.: Multi-criteria decision making method based on cross entropy and TOPSIS in Pythagorean fuzzy environment. Comput. Eng. Appl. **54**(16), 146–151 (2018)
7. Peng, X.: New development of Pythagorean fuzzy set theory and its application. Northwest Normal University (2016)
8. Haddad, M.J., Sanders, D.A.: Selecting a best compromise direction for a powered wheelchair using PROMETHEE II. IEEE Trans. Neural Syst. Rehabil. Eng. **27**, 228–235 (2018)

9. Prasetyo, E.D.W., Handajani, M.: Criteria analysis, weight and priority for handling bridges in Kudus District using AHP and Promethee II methods. J. Phys. Conf. Ser. **1167**(1), 012009 (2019)
10. Wang, Y.: Research on interval hesitant fuzzy decision based on level higher than relationship. Xi'an University of Technology (2017)
11. Zeng, S., Mu, Z.: Study on Pedagoras fuzzy TOPSIS multi-attribute decision making method based on mixed weighted distance. CMS **27**(03), 198–205 (2019)
12. Hu, J., Zhang, X.,Yang, Y., et al.: New doctors ranking system based on VIKOR method. Int. Trans. Oper. Res. (2018)
13. Chen, X., Wang, Y.: Multi-attribute emergency decision-making method considering information source correlation. Syst. Eng. Theor. Pract. **38**(08), 2045–2056 (2018)
14. Zhao, Y., Yi, Y.: Modified PROMETHEEII for venture capital investment selection decision-making towards SMEs. J. Interdiscip. Math. **21**(4), 1–15 (2018)

A Multi-criterion Group Decision Making Method: CRITIC-WASPAS Based on Trapezoidal Intuitionistic Fuzzy Numbers

Wenyu Zhang[1,2], Songmin Zhao[1(✉)], Dadi Dong[1], Yuting Zhu[1], and Danshu Wang[1]

[1] Xi'an University of Posts and Telecommunications, Xi'an 710016, China
1369060478@qq.com

[2] China Research Institute of Aerospace Systems Science and Engineering, Beijing 100048, China

Abstract. In order to solve the multi-criteria group decision problem in the fuzzy environment due to the uncertainty caused by the uncertainty, a new metho CRITIC-WASPAS based on trapezoidal intuitionistic fuzzy number is proposed. Firstly, the intuitionistic scoring matrix of each decision maker is transformed into trapezoidal intuitionistic fuzzy decision matrix. and the objective weight of each decision maker under different criteria is obtained by entropy weight method, and then the matrix of each decision maker is aggregated into a comprehensive matrix by using the trapezoidal intuitionistic fuzzy weighted average operator. secondly, the CRITIC method is used to obtain the objective weights of each criterion, and then the measure values of the WASPAS method based on the trapezoidal intuitionistic fuzzy numbers are obtained, which are clarified and sorted; finally, the feasibility and effectiveness of this algorithm are verified by analyzing the calculation examples selected by third-party logistics service providers.

Keywords: Trapezoidal intuitionistic fuzzy number · Entropy weight method · CRITIC · WASPAS

1 Introduction

Multi-criterion group decision making (MCGDM) refers to the process in which multiple decision makers choose the best alternative by adopting specific methods under different criteria. Fuzzy sets were introduced into multi-criterion group decision problems by Zadeh and Bellman in 1970, which can better deal with fuzzy decision problems [1]. It is difficult for decision makers to express the value of fuzzy information directly. Since then, the proposal of intuitionistic fuzzy number (IFNs) makes it an important direction of MCGDM to express fuzzy decision

Xi'an University of Posts and Telecommunications Graduate Innovation Fund (CXJJWY2018094).

Y. Liu et al. (Eds.): ICNC-FSKD 2019, AISC 1074, pp. 827–835, 2020.
https://doi.org/10.1007/978-3-030-32456-8_89

information based on fuzzy number. Extended forms of IFNs, such as trapezoidal intuitionistic fuzzy numbers (TIFNs) [2], trapezoidal fuzzy numbers [3] are also widely used in multi-criterion decision-making problems. Therefore, the concept of trapezoidal intuitionistic fuzzy numbers was introduced to make the decision-making information clear and improve the accuracy of decision-making in this paper.

At present, many effective methods to solve MCGDM have been proposed by scholars at home and abroad, such as TOPSIS [5], VIKOR [6], DEMATEL [7] and so on. These methods cannot objectively consider the weight of decision-makers, resulting in biased decision results. While WASPAS (Weighted Aggregated Sum Product Assessment Method) is the Weighted Sum model and Weighted linear combination of Product model, compared with other methods, this Method is simple and easy to use, can improve the accuracy of decision target selecting [8]. However, the new method based on trapezoidal intuitionistic fuzzy number has not been applied in multi-criterion group decision problems by scholars.

2 Relevant Concepts

2.1 The Concept of Trapezoidal Intuitionistic Fuzzy Numbers

The trapezoidal intuitionistic fuzzy numbers are an extension of trapezoidal fuzzy numbers. Compared with other fuzzy numbers, it can clearly show membership degree and non-membership degree, and fully reveal the essence of decision information fuzziness and more accurately describe decision information.

Definition 1.1: Let $A = (a, b, c, d)$ be a trapezoidal fuzzy number, and trapezoidal intuitionistic fuzzy number is an extension of trapezoidal fuzzy number, which can be expressed as $\{[u_a, u_b, u_c, u_d], [v_a, v_b, v_c, v_d]\}$, $u_A(x)$ is membership, $v_A(x)$ is non-membership [9].

Definition 1.2: A_1 and A_2 are two trapezoidal intuitionistic fuzzy numbers. The algorithm is as follows:

$$A_1 \oplus A_2 = \left\{ \begin{matrix} \left[\left\{ \begin{matrix} u_{a1} + u_{a2} - u_{a1a2}, u_{b1} + u_{b2} - u_{b1}u_{b2}, \\ u_{c1} + u_{c2} - u_{c1}u_{c2}, u_{d1} + u_{d2} - u_{d1}u_{d2} \end{matrix} \right. \right], \\ [v_{a1}v_{a2}, v_{b1}v_{b2}, v_{c1}v_{c2}, v_{d1}\ v_{d2}] \end{matrix} \right\} \tag{1}$$

$$A_1 \otimes A_2 = \left\{ \begin{matrix} [u_{a1}\ u_{a2}, u_{b1}u_{b2}, u_{c1}u_{c2}, u_{d1}\ u_{d2}], [v_{a1} + v_{a2} - v_{a1}v_{a2}, \\ v_{b1} + v_{b2} - v_{b1}v_{b2}, v_{c1} + v_{c2} - v_{c1}v_{c2}, v_{d1} + v_{d2} - v_{d1}\ v_{d2}] \end{matrix} \right\} \tag{2}$$

$$\beta A_1 = \{[1 - (1 - u_{a1})^\beta, ..., 1 - (1 - u_{d1})^\beta][v_{a1}{}^\beta, ..., v_{d1}^\beta]\}, \beta > 0 \tag{3}$$

$$\frac{A_1}{A_2} = \left\{ \begin{matrix} [\min(u_{a1}, u_{a2}), \min(u_{b1}, u_{b2}), \min(u_{c1}, u_{c2}), \min(u_{d1}, u_{d2})], \\ [\max(v_{a1}, v_{a2}), \max(v_{b1}, v_{b2}), \max(v_{c1}, v_{c2}), \max(v_{d1}, v_{d2})] \end{matrix} \right\} \tag{4}$$

$$\begin{gathered} A_1^\beta = \left\{ \left[u_{a1}^\beta, u_{b1}^\beta, u_{c1}^\beta, u_{d1}^\beta \right], \right. \\ \left. \left[1 - (1 - v_{a1})^\beta, 1 - (1 - v_{a2})^\beta, 1 - (1 - v_{a3})^\beta, 1 - (1 - v_{a4})^\beta \right] \right\} \end{gathered} \tag{5}$$

Definition 1.3: The trapezoidal intuitionistic fuzzy number A on the real number set is defuzzified to obtain the clear value as follows:

$$\& = \sum_{i=a}^{d} \frac{u_i}{4(u_i + \sum\limits_{i=a}^{d} \frac{v_i}{4})} \tag{6}$$

The clear value $\bar{u}$ of A applies to the calculation and ranking of each trapezoidal intuitionistic fuzzy number under different conditions. The larger the value $\bar{u}$, the larger the trapezoidal intuitionistic fuzzy number, and vice versa.

Definition 1.4: According to the operational properties of intuitive trapezoidal fuzzy numbers, For $\bar{A}_i = \{[u_{ai}, u_{bi} , u_{ci}, u_{di}], [v_{ai}, v_{bi}, v_{ci}, v_{di}]\}\, i = 1, 2, ..., n$ be a set of trapezoidal intuitive fuzzy numbers, and then extend then extend the weighted average operator of the intuitionistic trapezoidal fuzzy number to the trapezoidal intuitionistic fuzzy number weighted average operator [10], which is defined as follows:

$$\begin{aligned}\text{TIFNWA}\left(\overline{A}_1, \overline{A}_2 \ldots, \overline{A}_n\right) = \{&[\textstyle\sum_{i=1}^{n} w_i u_{ai}, \sum_{i=1}^{n} w_i u_{bi}, \sum_{i=1}^{n} w_i u_{ci}, \sum_{i=1}^{n} w_i u_{di}], \\ &[\textstyle\sum_{i=1}^{n} w_i v_{ai}, \sum_{i=1}^{n} w_i v_{bi}, \sum_{i=1}^{n} w_i v_{ci}, \sum_{i=1}^{n} w_i v_{di}]\}\end{aligned} \tag{7}$$

2.2 Fuzzy Language Translates into Trapezoidal Intuitionistic Fuzzy Numbers

For the fuzzy language evaluation scale $S = \{s_1, s_2, ..., s_m, ..., s_n\}\, (n\ is\ odd)$, S_1 and S_n are respectively the upper and lower limits of language terms [11]. For any fuzzy language variable value S_i, there is its corresponding trapezoidal fuzzy number.

Definition 1.5: Transform the fuzzy language terms into the corresponding trapezoidal fuzzy number

$$\overset{\approx}{S} = \left\{ \begin{array}{l} s_1\left(0, 0, \frac{1}{n}, \frac{1}{n}\right), s_2\left(\frac{1}{n}, \frac{1}{n}, \frac{2}{n}, \frac{2}{n}\right), \ldots \\ s_m\left(\frac{m-1}{n}, \frac{m-1}{n}, \frac{m}{n}, \frac{m}{n}\right), \ldots, s_n\left(\frac{n-1}{n}, \frac{n-1}{n}, 1, 1\right) \end{array} \right\}$$

$s_i(s_i \in S)$ has complement operator $cpl\,(s_i)$, $cpl\,(s_i) = s_{n-i}\,(0 < i < n)$.

Definition 1.6: For $s = [s_i, s_j]\,(s_i \in \overset{\approx}{S})$, The trapezoidal intuitionistic fuzzy number corresponding to s is recorded a T, $T(s) = \{t\,(s_i)\,t\,(cpl\,(s_j))\} = \{[u_{a_i}, u_{b_i}, u_{c_i} , u_{d_i}], [v_{a_{n-j}} , v_{b_{n-j}}, v_{c_{n-j}}, v_{d_{n-j}}]\}$.

3 CRITIC-WASPAS Method Based on Trapezoidal Intuitionistic Fuzzy numbers

In a fuzzy environment, decision makers are easily influenced by subjective factors such as their own cognition, thus a CRITIC-WASPAS method based on trapezoidal intuitionistic fuzzy number is proposed. The specific decision steps of the CRITIC-WASPAS multi-criteria group decision making method based on trapezoidal intuitionistic fuzzy numbers are as follows:$E = \{E_k, k = 1, \ldots, l\}$ is a set of decision makers; $D = \{D_j, j = 1, \ldots, n\}$ is the set of criteria:

Step 1. $y_{ij}^k = \{s_{ij}^k, S_{ij}^k\} \quad (k = 1, \ldots, l)$ represents the intuitionistic score interval of expert k for alternative j under criterion i, s_{ij}^k is the lowest score of expert k, and S_{ij}^k is the highest score. Therefore, the decision results of a single expert constitute the expert intuition score set as $Y = \left[y_{ij}^k\right]_{m \times n} (i = 1, \ldots, m; j = 1, \ldots, n)$
Step 2. According to Definitions 1.5 and 1.6, the expert intuition score set is processed with trapezoidal fuzzy processing;
Step 3. Fuzzy entropy weight method is used to obtain the objective weight of decision makers under each criterion [12] :$\bar{Y} = (T(y_{ij}^k))_{m \times n}$:

(1) Standardize according to the following formula, and get the standardized decision matrix

$$\overline{Y}_n = \left[\bar{y}_{ij}^k\right]_{m \times n} : \quad \bar{y}_{ij}^k = \begin{cases} \frac{T(y_{ij}^k) - \min_j T(y_{ij}^k)}{\max_j T(y_{ij}^k) - \min_j T(y_{ij}^k)}, i \in benefit\ criteria \\ \frac{\max_j T(y_{ij}^k) - T(y_{ij}^k)}{\max_j T(y_{ij}^k) - \min_j T(y_{ij}^k)}, i \in \text{cost } criteria \end{cases}$$

(2) The fuzzy entropy matrix $\stackrel{\approx}{\Theta}$ calculated, for $\bar{y}_{ij}^k$ is a fuzzy number: $\Theta = 1 - \sqrt{\sum_{i=a}^d (u_i - 0.5)^2 + \sum_{i=a}^d (v_i - 0.5)^2}$
(3) Calculate the index entropy. Assuming that the entropy value of the jth indicator is e_j, the entropy value of the jth indicator is calculated as: $e_j = \frac{1}{(\ln m) \sum_{i=1}^m Y_{ij}} \left[(\ln \sum_{i=1}^m Y_{ij}) \sum_{i=1}^m Y_{ij} - \sum_{i=1}^m Y_{ij} \ln Y_{ij} \right]$
(4) Calculate the objective weight of each decision maker under each index. The weight calculation formula is as follows: $w_j(E_k) = \frac{1 - e_j}{\sum_{j=1}^n (1 - e_j)}$

Step 4. Using the trapezoidal intuitionistic fuzzy number weighted average operator TIFNWA, the TIFN decision matrix of each decision maker is aggregated into a single TIFN matrix according to formulas (1) and (3): $\stackrel{\approx}{Y} = [\stackrel{\approx}{y}_{ij}^k]_{m \times n}$
Step 5. The objective weight of each criterion is calculated using the CRITIC method: (1) Clarify the single ITFN matrix $\stackrel{\approx}{Y}$ according to formula (6), calculate the correlation coefficient $\mathfrak{R}_{jk}$ between criteria, and then construct the correlation coefficient matrix: $R = [\mathfrak{R}_{jk}]_{m \times n}$; (2) the information contained in the jth evaluation criterion: $M_j = \mathfrak{I}_j \sum_{k=1}^m (1 - \mathfrak{R}_{jk})$; (3) The objective weight of the jth evaluation criterion is: $\omega_j(D_j) = \frac{M_j}{\sum_{k=1}^m M_k}$
Step 6. WSM$^{\varphi_i^{(1)}}$, WPM$^{\varphi_i^{(2)}}$ and WASPAS $^{\varphi_i}$ values were obtained respectively:

$$WSM = \varphi_i^{(1)} = \sum_{j=1}^n Y_{ij} \omega_j(D_j) \, WPM = \varphi_i^{(2)} = \prod_{j=1}^n (Y_{ij})^{\omega_j(D_j)}$$

$$WASPAS = \varphi_i = 0.5\varphi_i^{(1)} + 0.5\varphi_i^{(2)} = 0.5\sum_{j=1}^{n} Y_{ij}\omega_j\left(D_j\right) + 0.5\prod_{j=1}^{n}\left(Y_{ij}\right)^{\omega_j(D_j)}$$

Step 7. The values of $\varphi_i^{(1)}, \varphi_i^{(2)}$ and φ_i were defuzzified according to formula (6), and alternative schemes were sorted by using clear values.

4 Case Analysis

Now it is necessary to choose a suitable logistics service provider from the four third-party logistics service providers (C_1, C_2, C_3, C_4) to establisha long-term relationship. Relying on three experienced experts (E_1, E_2, E_3), the assessment of third-party logistics service providers is conducted according to the following three criteria: employment cost (D_1), basic business level (D_2) and operation and management ability (D_3).

Step 1. Three invited experts scored intuitively, and the lowest score and the highest score were respectively given to alternative plan C_i under each criterion C_j. Table 1 shows the intuitive scoring range given by expert 1, and the other experts also scored intuitively:

Table 1. Intuitive score interval of expert 1

	D_1	D_2	D_3
C_1	{[S2,S3]}	{[S3,S4]}	{[S4,S5]}
C_2	{[S5,S6]}	{[S2,S3]}	{[S4,S5]}
C_3	{[S4,S5]}	{[S3,S4]}	{[S2,S3]}
C_4	{[S3,S4]}	{[S4,S5]}	{[S3,S4]}

Step 2. Convert the intuitive scores of experts into the expert trapezoidal intuitionistic fuzzy decision matrix according to the Definition 1.6–1.8, and the results are shown in Table 2.

Table 2. Expert 1 trapezoidal intuitionistic fuzzy decision matrix

	D_1	D_2	D_3
C_1	{[0.14,0.14,029,029] [0.43,043,057,0.57]}	{[0.29,029,043,0.43] [0.29,029,043,0.43]}	{[0.43,0.43,057,0.57] [0.14,0.14,029,029]}
C_2	{[0.57,057,071,0.71] [0.,0,0.14,0.14]}	{[0.14,014,0.29,029] [0.43,043,0.57,057]}	{[0.43,0.43,057,057] [0.14,014,029,029]}
C_3	{[0.43,0.43,057,057] [0.14,014,029,029]}	{[0.29,029,043,0.43] [0.29,029,043,0.43]}	{[0.14,0.14,029,029] [0.43,043,057,0.57]}
C_4	{[0.29,029,043,0.43] [0.29,029,043,0.43]}	{[0.43,0.43,057,0.57] [0.14,0.14,029,029]}	{[0.29,029,043,0.43] [0.29,029,043,0.43]}

Step 3. The entropy weight method is used to obtain the objective weight of decision makers under each criterion (see Table 3)

Table 3. Objective weights of each expert under each criterion

	D_1	D_2	D_3
E_1	0.3211	0.3335	0.3454
E_2	0.3412	0.3329	0.3259
E_3	0.3398	0.3475	0.3127

Step 4. Calculate the aggregate TIFN matrix according to formula (1), formula (3) and formula (7) (see Table 4)

Table 4. Aggregation TIFN decision matrix

	D_1	D_2	D_3
C_1	{[0.26,0.27,0.3,0.27]	{[0.24,0.24,0.41,0.41]	{[0.42,0.36,0.32,0.43]
	[0.4,0.24,0.61,0.36]}	[0.37,0.47,0.58,0.51]}	[0.39,0.42,0.4,0.23]}
C_2	{[0.43,0.47,0.13,0.3]	{[0.3,0.37,0.27,0.1]	{[0.23,0.36,0.53,0.16]
	[0.2,0.34,0.37,0.34]}	[0.3,0.44,0.34,0.27]}	[0.1,0.42,0.29,0.29]}
C_3	{[0.36,0.43,0.23,0.43]	{[0.5,0.37,0.24,0.1]	{[0.5,0.48,0.37,0.43]
	[0.17,0.47,0.34,0.24]}	[0.34,0.27,0.34,0.2]}	[0.45,0.29,0.16,0.23]}
C_4	{[0.2,0.47,0.4,0.4]	{[0.57,0.24,0.44,0.27]	{[0.27,0.39,0.56,0.35]
	[0.4,0.26,0.3,0.23]}	[0.34,0.3,0.3,0.27]}	[0.4,0.2,0.33,0.26]}

Step 5. The weight of each criterion is calculated according to the method: $\omega(D_1) = 0.3129$ $\omega(D_2) = 0.3299$ $\omega(D_3) = 0.3572$
Step 6. WSM, WPM and WASPAS were calculated respectively (see Table 5),

Table 5. Measures of WSM, WPM and WASPAS

	TIFN-WSM	TIFN-WPM	TIFN-WASPAS
C_1	{[0.21,0.19,0.12,0.15]	{[0.33,0.25,0.47,0.18]	{[0.47,0.39,0.53,0.3]
	[0.72,0.61,0.49,0.53]}	[0.51,0.42,0.34,0.49]}	[0.37,0.26,0.17,0.26]}
C_2	{[0.16,0.20,0.11,0.27]	{[0.22,0.35,0.41,0.15]	{[0.34,0.48,0.47,0.38]
	[0.63,0.59,0.41,0.49]}	[0.46,0.57,0.21,0.32]}	[0.29,0.33,0.1,0.16]}
C_3	{[0.52,0.11,0.48,0.32]	{[0.29,0.3,0.16,0.14]	{[0.66,0.38,0.56,0.42]
	[0.31,0.29,0.4,0.35]}	[0.66,0.46,0.29,0.37]}	[0.2,0.13,0.12,0.13]}
C_4	{[0.56,0.47,0.32,0.61]	{[0.61,0.58,0.42,0.31]	{[0.83,0.78,0.61,0.73]
	[0.36,0.27,0.18,0.29]}	[0.33,0.25,0.2,0.17]}	[0.12,0.1,0.1,0.1]}

Step 7. The measurement values of WSM, WPM and WASPAS were clarified (Fig. 1),

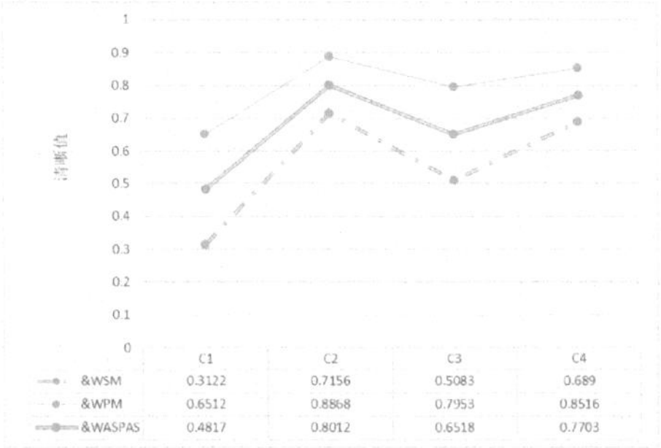

Fig. 1. Clear values of WSM, WPM and WASPAS

According to Fig. 1, the final order of the four third-party logistics service providers is $C_4 > C_3 > C_2 > C_1$, so C_4 is the best partner.

5 Results Analysis

In order to compare the results of the proposed method with other existing methods (see Table 6). The TIFN-CRITIC-WASPAS method proposed in this paper has the following advantages: (1) use trapezoidal intuitionistic fuzzy number to express the information of expert preference, and take into account the degree of membership and the degree of non-membership; (2) weight information is quantitatively considered by using the shadow method to reduce the influence of subjective factors of decision makers; (3) compared with traditional WSM model and WPM model, WASPAS adds them together to determine the optimal alternative, which can improve the accuracy of alternative selection.

Table 6. Applies different methods to sort alternatives

	CRITIC-WASPAS	TFN-MULTIMOORA	TIFN-DEMATAL	TOPSIS
C_1	4	4	4	3
C_2	3	3	3	4
C_3	2	1	2	2
C_4	1	2	1	1

Then, Spearman's rank correlation coefficient r is adopted. When $0.6 < r$, it can be said that there is a high statistical dependence between the results.

except for the correlation coefficient between TOPSIS and TIFNDEMATAL, other correlation coefficients are all greater than or equal to 0.8. Therefore, the relationship between ranking results is very strong. Based on the similarity analysis of four methods, the new trapezoidal intuitionistic fuzzy number method is proved to be more expressive.

6 Conclusion

For in multiple criteria group decision making problems due to the decision-makers of professional background and cognitive aspects of differences lead to the decision problem of information loss, this paper puts forward the extension based on trapezoidal intuitionistic fuzzy number CRITIC-WASPAS can better solve the problem of fuzziness of multicriteria decision making and subjectivity, introduced considering membership and the membership degree of trapezoidal intuitionistic fuzzy number to express decision makers better opinions, can reduce the loss of information, policy makers and the use of entropy method and the CRITIC method obtained rule of objective weight and objective weight, then you can avoid decision bias caused by subjective factors, Finally, WASPAS can improves the accuracy of alternative selection. This method can overcome the subjective influence of decision makers in fuzzy environment, effectively deal with the fuzzy information in multi-criterion decision-making problems

References

1. Bellman, R.E., Zadeh, L.A.: Decision-making in a fuzzy environment. Manag. Sci. **17**, B–141 (1970)
2. Wang, Y., Zhang, L., Jiang, X.: EDW trapezoidal intuitionistic fuzzy numbers with a combination of science and technology awards evaluation. Soft Sci. **28**(10), 121–124 (2014)
3. Wang, Q., Li, G.: Ordered expression of trapezoidal fuzzy number and the center average ranking method. Oper. Manag. **26**(05), 130–136 (2017)
4. Meng, B., Shen, S., Kuang, H., Li, F., Feng, H.: Corporate social responsibility evaluation model based on fuzzy-topsis -taking transportation industry as an example. Manag. Rev. **31**(05), 191–202 (2019)
5. Zhang, N., Zhao, S., Zang, Y.: Method and application of range-wise fuzzy VIKOR decision based on projection. Syst. Eng. **36**(11), 153–158 (2018)
6. Qi, Z., Renjing, L., Jingyu, Y.: Analysis of factors affecting knowledge transfer performance. Sci. Res. **37**(02), 311–319 (2019)
7. Edmundas, K.Z., Romualdas, B., Marius, L.: Sustainable assessment of alternative sites for the construction of a waste incineration plant by applying WASPAS method with single-valued neutrosophic set. Sustainability **7**(12), 15923–15936 (2015)
8. Jiaxiang, Z., Jianfei, C., Xuxun, L.: Trapezoidal intuitionistic fuzzy multicriterion group decision classification method based on CI-TOPSIS. Chin. J. Manag. Sci. **26**(09), 148–157 (2018)
9. Haibin, X.: New ranking index of intuitionistic trapezoidal fuzzy number and its application in multi-attribute decision making. Stat. Decis. **34**(12), 76–79 (2018)

10. Qi, C.: Hybrid multi-attribute group decision making method based on trapezoidal fuzzy MULTICOMORA. Stat. Decis. **35**(05), 41–45 (2019)
11. Wu, C., Wan, X.: Interval intuitionistic fuzzy TOPSIS method based on improved entropy weight method. Oper. Res. Manag. Sci. **23**(05), 42–47 (2014)

An Interval Intuitionistic Fuzzy VIKOR Evaluation Method Based on Unknown Weight

Wenyu Zhang[1,2], Dadi Dong[1(✉)], Songmin Zhao[1], Yuting Zhu[1], and Danshu Wang[1]

[1] Xi'an University of Posts and Telecommunications, Xi'an 710016, China
1362373049@qq.com

[2] China Research Institute of Aerospace Systems Science and Engineering, Beijing 100048, China

Abstract. For multi-attribute group decision-making problems with unknown attribute weights and decision makers' weights, an interval intuitionistic fuzzy VIKOR evaluation method with unknown weights is proposed. Firstly, the interval intuitionistic fuzzy number is used to construct the evaluation matrix of each decision maker. Secondly, the information entropy of interval intuitionistic fuzzy numbers is used to obtain the objective weight of attributes. Then, the scoring weight of each decision-maker is calculated based on the scoring correlation coefficient of the decision-maker and the group decision-maker, and then the above weights are aggregated to obtain a comprehensive evaluation matrix; Thirdly, the VIKOR sorting method is used to sort the comprehensive evaluation matrix to obtain the optimal scheme. Finally, the feasibility and effectiveness of the algorithm are verified by an example of decision-making of the optimal knowledge service platform scheme. The results show that the method is feasible to solve the optimization problem of multi-attribute group decision making with unknown attribute weight and decision maker weight.

Keywords: Unknown weight · Interval intuitionistic fuzziness · VIKOR

1 Introduction

Formulti-attribute group decision-making problems, due to the complexity of the problem and the limitations of decision-makers on problem cognition [1], a certain numerical evaluation is usually too subjective [2]. In addition it is the key to study the problem to reasonably determine the attribute weight, the decision-maker weight and to choose a scientific ranking method to optimize the scheme [3].

Xi'an University of Posts and Telecommunications Graduate Innovation Fund (CXJJWY2018088).

Y. Liu et al. (Eds.): ICNC-FSKD 2019, AISC 1074, pp. 836–844, 2020.
https://doi.org/10.1007/978-3-030-32456-8_90

The introduction of interval-valued intuitionistic fuzzy sets makes information expression more comprehensive and accurate [4,5]. Aiming at how to determine the weights objectively and reasonably in multi-attribute group decision-making problems, most of the existing literatures use the evaluation trend between individual evaluation and group decision-making to determine the weight [6–8]. The above method can only obtain the approximate value of the group decision set node, and there is a certain error.At present, for the decision-making of the scheme, it can be roughly divided into two categories. One is to calculate the final score of each scheme by determining the attribute weight and the expert weight, and use the score value to determine the optimal scheme [9–11]. Since the determination of the weight has certain risks, the direct aggregation of the weights has certain risks for the evaluation of the scheme.The second is to sort the final scoring matrix by TOPSIS, PROMETHEE and VIKOR sorting methods. This method is to sort the schemes objectively after calculating the weights by using the characteristics of each sorting method [12].

In summary, this paper uses the interval intuitionistic fuzzy set to more objectively express the decision maker's evaluation.Using the interval intuitionistic fuzzy information entropy to calculate the objective weight of each attribute, the decision weight of the decision maker is obtained by calculating the score correlation coefficient between the decision individual and the group decision maker. Finally, the VIKOR sorting method is used to select the scheme.

2 The Interval Intuitionistic Fuzzy VIKOR Related Theory

2.1 Interval Intuitionistic Fuzzy Sets

Definition 1: $\widetilde{a} = \{[u_L(x), u_U(x)], [v_L(x), v_U(x)]\}$ is an interval intuitionistic fuzzy number $e(\widetilde{a})$ is called information entropy of interval intuitionistic fuzzy number. $e(\widetilde{a}) = 1 - d(\widetilde{a})$

Among them: $d(\widetilde{a}) = \sqrt{(\mu_L - 0.5)^2 + (\mu_L - 0.5)^2 + (v_L - 0.5)^2 + (v_L - 0.5)^2}$

Definition 2: Let $\widetilde{a} =< [a, b], [c, d] >$be the interval intuitionistic fuzzy number, and its score function is defined as: $h(\widetilde{a}) = \frac{(d-c)^2-(b-a)^2-2(a+b)}{2(a+b+c+d)}$

The precision function is defined as: $\alpha(\widetilde{a}) = b + d - \frac{1}{2}\left[\frac{(b-a)^2}{b} + \frac{(d-c)^2}{d}\right]$

(1) If $h(\widetilde{a}_1) > h(\widetilde{a}_2)$, then $\widetilde{a}_1 > \widetilde{a}_2$
(2) If $h(\widetilde{a}_1) = h(\widetilde{a}_2)$ and $\alpha(\widetilde{a}_1) > \alpha(\widetilde{a}_2)$, then $\widetilde{a}_1 > \widetilde{a}_2$; If $\alpha(\underset{1}{\widetilde{a}}) = \alpha(\underset{2}{\widetilde{a}})$ then $\widetilde{a}_1 \sim \widetilde{a}_2$

2.2 VIKOR Sorting Principles

Definition 3: The Group benefit value S_j and the individual regret value $R_j$are as follows:

$$S_j = \sum_{i=1}^{n} w_i d(\widetilde{x}_i^* - \widetilde{x}_{ij})/d(\widetilde{x}_i^* - \widetilde{x}_i^-) \quad \forall j \tag{1}$$

$$R_j = \max w_i [d(\widetilde{x}_i^* - \widetilde{x}_{ij})/d(\widetilde{x}_i^* - \widetilde{x}_i^-)] \quad \forall j \tag{2}$$

Where w_i is the weight of the attribute, x_i^* and x_i^- are the positive and negative ideal points of the attribute respectively.

Definition 4: Calculation of VIKOR sorting value Q

$$Q_j = \lambda(S_j - S^*)/(S^- - S^*) + (1 - \lambda)(R_j - R^*)/(R^- - R^*) \quad \forall j \tag{3}$$

Where, Q_j represents the Q value of VIKOR in the j-th scheme; S^* is the group's maximum utility of the program evaluation value $S^* = \min S_j$; R^* is the individual minimum regret of the scheme evaluation value $S^* = \min S_j$.

3 The Evaluation Method Based on Interval Intuitionistic Fuzzy VIKOR

For multi-attribute group decision-making problems, let $X = \{x_1, x_2, ..., x_t\}$ be the decision-maker set, $A = \{A_1, A_2, ..., A_n\}$ be the scheme set, and $C = \{c_1, c_2, ..., c_m\}$ be the index set.The specific steps of the interval intuitionistic fuzzy VIKOR ranking method based on unknown weights are as follows:

Step1: Building a decision maker's scoring matrix

After examining each scheme, experts give the score $\widetilde{x}_{ij}^k$ of each scheme expressed by interval fuzzy number under different attributes, $\widetilde{x}_{ij}^k = < [u_{Lij}^k, u_{Uij}^k], [v_{Lij}^k, v_{Uij}^k] >$ $\widetilde{x}_{ij}^k$ represents the evaluation value of the k-th $(k = 1, 2, ..., t)$ expert on the $i(i = 1, 2, ..., n)$ scheme according to index $j = (1, 2, ..., m)$.

Step2: According to the interval intuitionistic information entropy method, the objective weight of attributes is determined.

(1) According to Definition 1, the fuzzy entropy matrix of each decision maker's intuitionistic fuzzy matrix is calculated.
(2) Calculate entropy value of each attribute E_j^k.

$$E_j^k = \frac{1}{n} \sum_{i=1}^{n} e_{ij} \tag{4}$$

E_j^k represents the information entropy value of decision maker k for attribute j.

(3) Calculate that weight of each decision maker to each attribute ω_j^k:

$$\omega_j^k = \frac{1 - E_j^k}{\sum\limits_{j=1}^{m} (1 - E_j^k)} \tag{5}$$

ω_j^k represents the weight of decision maker k on attribute j.

Step3: Calculate the weight of decision makers.

(1) The evaluation value of decision makers after attribute weighting is:

$$p_{ij}^k = \omega_j^k x_{ij}^k \tag{6}$$

(2) The average evaluation value of group decision is

$$p_{ij} = \frac{1}{t} \sum_{k=1}^{t} p_{ij}^k \tag{7}$$

(3) The correlation coefficient between each decision maker and group decision is:

$$\delta(p_{ij}^k, p_{ij}) = \frac{\min d(p_{ij}^k, p_{ij}) + \rho \max d(p_{ij}^k, p_{ij})}{d(p_{ij}^k, p_{ij}) + \rho \max d(p_{ij}^k, p_{ij})} \tag{8}$$

where ρ is the resolution coefficient, $0 < \rho < 1$.

(4) To calculate the degree of correlation:

$$\zeta_k = \frac{1}{m} \sum_{i=1}^{m} \delta(p_{ij}^k, p_{ij}) \tag{9}$$

(5) To establish a scoring weight model for decision makers:

$$\max[\mu \sum_{k=1}^{t} (\lambda_k \zeta_k)^2 - (1 - \mu) \sum_{k=1}^{t} \lambda_k \ln\lambda_k] \quad s.t. \sum_{k=1}^{t} \lambda_k = 1 \tag{10}$$

Finally, the decision-maker weight vector $\lambda_k = (\lambda_1, \cdots, \lambda_t)$ is obtained.
Step4: According to the attribute weight obtained in step2 and the decision maker weight in step3the final comprehensive interval intuitionistic fuzzy number matrix $\tilde{X} = (x_{ij})_{mn}$ is obtained by bringing the attribute weight and the decision maker weight into the scoring matrix.
Step5: Determine positive ideal solution and negative ideal solution. According to Definition 3:

the positive ideal solution is $\tilde{A}_i^+ = [(\max_j x_{ij} | i \in I_1), (\min_j x_{ij} | i \in I_2)] \ \forall i$

Negative ideal solution is $\tilde{A}_i^- = [(\min_j x_{ij} | i \in I_1), (\max_j x_{ij} | i \in I_2)] \ \forall i$

Where x_{ij} represents the evaluation value of scheme $i(i = 1, 2, ..., n)$ under index $j(j = 1, 2, ..., m)$. $\overset{\sim+}{A_i}$ represents the maximum value of the index of the scheme. $\overset{\sim-}{a_i}$ represents the minimum value of the index of the scheme. I_1 represents the benefit index and I_2 represents the cost index.

Step6: Decision makers select different decision-making mechanism coefficients λ according to their acceptance of risks, and calculate the values of S_j, R_j and Q_j of the scheme according to formulas (1), (2) and (3).

Step7: According to the value of Q_j, S_j, R_j each scheme is sorted.

4 Case Study

4.1 Case Study

At present, the comprehensive level of knowledge service of the three knowledge service platforms $A(A_1, A_2, A_3)$ is objectively evaluated, and the service platform with the best comprehensive service level is selected. According to three industry experienced decision makers $k(k_1, k_2, k_3)$, the knowledge service platform is evaluated according to the following four evaluation criteria $C(C_1, C_2, C_3, C_4)$: knowledge retrieval function (C_1), knowledge maintenance function (C_2), knowledge visualization function (C_3) and user management function (C_4).

Step1: Firstly, the interval intuitionistic fuzzy evaluation matrix of each decision maker is obtained, as shown in Table 1, and then the fuzzy entropy matrix H^k of each decision maker's intuitionistic fuzzy matrix is calculated according to Definition 1 as shown in Table 2 (taking a decision maker as an example).

Table 1. The Interval Intuitionistic Fuzzy Evaluation Table of Decision Maker X_1

Programme	C_1	C_2	C_3	C_4
A_1	{[0.4,0.7],[0.2,0.3]}	{[0.3,0.5],[0.2,0.3]}	{[0.2,0.3],[0.4,0.5]}	{0.2,0.4],[0.3,0.5]}
A_2	{[0.7,0.8],[0.2,0.3]}	{[0.5,0.7],[0.2,0.3]}	{[0.4,0.5],[0.3,0.4]}	{[0.3,0.5],[0.4,0.6]}
A_3	{[0.4,0.6],[0.1,0.3]}	{[0.4,0.6],[0.2,0.4]}	{[0.5,0.7],[0.2,0.3]}	{[0.7,0.8],[0.1,0.2]}

Table 2. The Intuitionistic Fuzzy Entropy Evaluation Table of Decision Maker X_1

Programme	C_1	C_2	C_3	C_4
A_1	0.4234	0.4123	0.3742	0.3742
A_2	0.5099	0.4123	0.2499	0.2455
A_3	0.469	0.3464	0.3464	0.4472

Step2: Calculate the weight of decision makers.

The weight of each attribute of each decision maker calculated according to formulas (4) and (5) is: $\omega_j^1 = (0.2160, 0.2473, 0.2751, 0.2651)(j = 1, 2, 3, 4)$, $\omega_j^2 = (0.2519, 0.2293, 0.2459, 0.2729)(j = 1, 2, 3, 4)$, $\omega_j^3 = (0.2678, 0.2592, 0.2422, 0.2308)$ $(j = 1, 2, 3, 4)$.

Step3: Calculate the weight of decision makers

The attribute weights of each decision maker obtained in step2 are brought in, and the decision matrix weighted by each decision maker attribute is calculated according to formulas (6) and (7), and then the average evaluation of the expert group on the scheme is shown in Table 3:

Table 3. Average Evaluation Table of Group Decision Makers

Programme	C1	C2	C3	C4
A1	{[0.1160,0.1638], [0.0485,0.0856]}	{[0.1321,0.1725], [0.0479,0.0646]}	{[0.0834,0.1089], [0.1016,0.0948]}	{[0.1021,0.1405], [0.0824,0.1136]}
A2	{[0.1118,0.1453], [0.0563,0.1072]}	{[0.1162,0.1481], [0.0658,0.0809]}	{[0.1180,0.1435], [0.0357,0.0774]}	{[0.0963,0.1188], [0.0819,0.1041]}
A3	{[0.0836,0.1471], [0.0419,0.0736]}	{[0.1393,0.1721], [0.0414,0.0742]}	{[0.1192,0.1781], [0.0509,0.0763]}	{[0.1359,0.1705], [0.1809,0.0678]}

According to formula (8), the correlation coefficient between each decision-maker scheme and the group scheme is as follows: $\delta^1 = (0.6749, 0.6739, 0.6877)$; $\delta^2 = (0.7130, 0.9398, 0.7488)$; $\delta^3 = (0.5757, 0.6823, 0.7289)$.

According to formula (9), the correlation degree between each decision-maker scheme and the group scheme is as follows:$\zeta_1 = 0.6788$, ζ_2=0.8005, $\zeta_3 = 0.6788$.

According to formula (10), the weight of decision makers is: $\lambda_1 = 0.3788$, $\lambda_2 = 0.2979$, $\lambda_3 = 0.3233$.

Step4: The comprehensive decision matrix obtained by bringing in attribute weight and decision maker weight is shown in Table 4.

Table 4. Comprehensive decision matrix

Programme	C1	C2	C3	C4
A1	{[0.1188,0.1676], [0.0466,0.0813]}	{[0.1376,0.1762], [0.0493,0.0652]}	{[0.0838,0.1092], [0.1034,0.0921]}	{[0.1025,0.1450], [0.0777,0.1086]}
A2	{[0.1146,0.1494], [0.0514,0.1039]}	{[0.1176,0.1488], [0.0660,0.0819]}	{[0.1003,0.1257], [0.0429,0.0848]}	{[0.1039,0.1185], [0.0786,0.1008]}
A3	{[0.0985,0.1478], [0.0423,0.0739]}	{[0.1417,0.1744], [0.0424,0.0751]}	{[0.1178,0.1777], [0.0508,0.0762]}	{[0.1354,0.1689], [0.0422,0.0675]}

Step5: Determining Positive and Negative Ideal Solutions

Positive and negative ideal solutions are obtained according to definition 2: as shown in Table 5

Table 5. Positive and negative ideal solutions

	C1	C2	C3	C4
A^+	{[0.1188,0.1676], [0.0466,0.0813]}	{[0.1417,0.1744], [0.0424,0.0751]}	{[0.1178,0.1777], [0.0508,0.0762]}	{[0.1354,0.1689], [0.0422,0.0675]}
A^-	{[0.0985,0.1478], [0.0423,0.0739]}	{[0.1176,0.1488], [0.0660,0.0819]}	{[0.0838,0.1092], [0.1034,0.0921]}	{[0.1025,0.1450], [0.0777,0.1086]}

Step6: calculate the values ofS_j, R_j and Q_j

According to equations (1) and (2), S_j and R_j are calculated as: $S_1 = 0.5493$, $S_2 = 0.8568$, $S_3 = 0.2678$; $R_1 = 0.2422$, $R_2 = 0.2623$, $R_3 = 0.2678$.

Let the coefficient of the decision-making mechanism be $\lambda = 0.5$ and use equation (3) to obtain Q_j as: $\tilde{Q}_1 = 0.7611$, $\tilde{Q}_2 = 0.1082$, $\tilde{Q}_3 = 0.500$.

Step7: According to the value of $Q_j(j = 1,2,3)$, each scheme is sorted $\tilde{Q}_1 > \tilde{Q}_3 > \tilde{Q}_2$, The best plan is A_1.

4.2 Comparison of the VIKOR Sorting Method with Other Sorting Methods

TOPSIS and PROMETHEE sorting methods were used to sort the schemes and compared with VIKOR sorting method. the results are shown in Table 6.

Table 6. Comparison of sorting methods

Method	Comparative value			
	A1	A2	A3	Sorting
VIKOR	0.7611	0.1082	0.5	$A_1 > A_3 > A_2$
TOPSIS	−0.3319	−3.0231	-0.8134	$A_1 > A_3 > A_2$
PROMETHEE	0.2993	0.0684	0.1313	$A_1 > A_2 > A_3$

From the sorting results, we can see that TOPSIS sorting is consistent with VIKOR(λ=0.5). the sorting results of PROMETHEE and the first two schemes are different in the last two positions, but the optimal scheme is the same, which shows that the method in this paper is effective. In the TOPS method of the existing literature, the schedule function is used to sort the schemes, but sometimes the scheme with the largest close value cannot meet the requirements of being closest to the positive ideal solution and farthest from the negative ideal solution, which is also the disadvantage of TOPS sorting method. In addition, VIKOR ranking method can choose different schemes according to the decision makers' attitudes towards risks, so VIKOR ranking method is more reasonable on the whole.

5 Conclusion

In view of how to objectively determine the weight in multi-attribute group decision-making problems. In this paper, the discrete degree of interval intuitionistic fuzzy information entropy is minimized to determine the weight of each attribute. The scoring weight of each decision-maker is calculated based on the scoring correlation coefficient of decision-makers and group decision-makers. On this basis, the comprehensive decision matrix is sorted according to VIKOR compromise sorting method, and the optimal scheme is obtained. Finally, an example is given to verify the effectiveness of the scheme.

References

1. Garg, H., Kumar, K.: An extended technique for order preference by similarity to ideal solution group decision-making method with linguistic interval-valued intuitionistic fuzzy information. J. Multi-criteria Decis. Anal. **26**(1–2), 16–26 (2019)
2. Zhang, F.W., Huang, W.W., Sun, J.: Generalized fuzzy additive operators on intuitionistic fuzzy sets and interval-valued intuitionistic fuzzy sets and their application. IEEE Access 45734–45743 (2019)
3. Couso, I., Bustince, H.: From fuzzy sets to interval-valued and atanassov intuitionistic fuzzy sets: a unified view of different axiomatic measures. IEEE Trans. Fuzzy Syst. **27**(2), 362–371 (2019)
4. Zheng, T.T., Zhang, M.Y., Zheng, W.R.: A new uncertainty measure of covering-based rough interval-valued intuitionistic fuzzy sets. IEEE Access 53213–53224 (2019)
5. Zhang, Z.G., Sheng, Y., Ou, C.: A method for determining index weight based on FAHP-CEEMDAN. Stat. Decis. Mak. **35**(02), 79–83 (2019)
6. Li, L., Wang, Y.Q.: A new method for determining the weight of multi-attribute group decision experts. Stat. Decis. (02), 14–18 (2017)
7. Yeni, F.B., Özçelik, G.: Interval-valued Atanassov intuitionistic fuzzy CODAS method for multi criteria group decision making problems. Group Decis. Negot. 28, 433-452 (2018)
8. Narayanamoorthy, S., Geetha, S., Rakkiyappan, R., et al.: Interval-valued intuitionistic hesitant fuzzy entropy based VIKOR method for industrial robots selection. Expert Syst. Appl. **121**, 28–37 (2019)
9. Wu, L.P., Wang, J., Gao, H.: Models for competiveness evaluation of tourist destination with some interval-valued intuitionistic fuzzy Hamy mean operators. J. Intell. Fuzzy Syst. 1–17 (2019)
10. Liu, H.C., Jian, X.Y., Chun, Y.D.: An integrated approach for failure mode and effect analysis under interval-valued intuitionistic fuzzy environment. Int. J. Product. Econ. **207**, 163–172 (2019)

11. Li, D.F.: TOPSIS-based nonlinear-programming methodology for multiattribute decision making with interval-valued intuitionistic fuzzy sets. IEEE Trans. Fuzzy Syst. **18**(2), 299–311 (2010)
12. Xu, W.H., Shang, X.P., Wang, J.: A novel approach to multi-attribute group decision-making based on interval-valued intuitionistic fuzzy power Muirhead mean. Symmetry-Basel. 11(3) (2019)

Fuzzy Theory and Algorithms: Soft Computing and Granular Computation

The Residual Power Series Method for Solving the Fractional Fuzzy Delay Differential Equation

Qiujuan Tong(✉), Yongzhen Zang(✉), and Jianke Zhang(✉)

Xi'an University of Posts and Telecommunications, Xi'an 710121, Shaanxi, China
xiaotong0929@163.com, zyzhen_123@163.com, jiankezh@163.com

Abstract. In this paper, the fuzzy delay differential equation is expressed in fractional form, and the Residual Power Series Method (RPSM) is used to solve the equation. The delay differential equation indicates that in a system, the speed of the system is not only related to the current state of the system, but also depends on the historical trajectory until this moment. Due to the existence of systematic errors, fractional fuzzy delay differential equations play an important role in more and more system models in biology, engineering, physics and other sciences. Maple was used to calculate the numerical solution of this equation, and the results are presented in the form of a graph. The results show that it is an effective method when solving the fractional fuzzy delay differential equation. By this method, the approximate solutions of the fractional fuzzy delay differential equation can be obtained with just a few steps.

Keywords: Residual Power Series Method · Fractional fuzzy delay differential equation · Caputo derivative

1 Introduction

Recently, how to solve fuzzy differential equations and fuzzy integral equations has become a new research for many scholars. The concept of fuzzy set was proposed by Zadeh (see [1]) in 1965. In recent years, due to its potential applications, methods for solving fuzzy differential equations have been studied in many published articles. In [2], the Bernstein Neural Networks were used to estimate the solutions of fuzzy differential equations. In [3], the Euler method was used to solve the hybrid fuzzy differential equations. In [4], the fuzzy Laplace transform method was used to approach the solution of n-th order fuzzy problem with initial value. Homotopy Perturbation Method was successfully applied to find the solution of n-th order fuzzy differential equations in [5]. In [6], the Taylor method was used to find the numerical solution of some linear and nonlinear fuzzy differential equations.

In this paper, the RPSM was used to approach the series solution of fractional fuzzy delay differential equation. The method is an effective method for solving

Y. Liu et al. (Eds.): ICNC-FSKD 2019, AISC 1074, pp. 847–855, 2020.
https://doi.org/10.1007/978-3-030-32456-8_91

fuzzy differential equations [7], which gives the solution of the convergent power series, and its analytical approximate solution is a polynomial form, and the components are easy to calculate. In [8], the RPSM was used to solve nonlinear time fractional reaction–diffusion equations. In [9], the RPSM was used to obtain the analytical solutions of fractional foam drainage equation by RPSM, and the result shows that the RPSM is a significant method for exploring nonlinear fractional models.

2 Preliminaries

In this section, the main basic definitions of the fuzzy numbers and fractional calculus theory are illustrated.

Definition 1. *A fuzzy number indicates an indeterminate value, which is different from a real number. A real number refers to a specific value, but a fuzzy number refers to a set of possible values. Every possible value has its own weight from 0 to 1. In this paper the class of all fuzzy subsets of* $\mathbb{R}$ *will be denoted by* $\tilde{E}$ *and satisfy the following properties [10];*

$\mu(t)$ *is normal, i.e* $\exists t_0 \in \mathbb{R}$ *with* $\mu(t_0) = 1$,

1. $\mu(t)$ *is convex fuzzy set, i.e.*
2. $\mu(\lambda t + (1-\lambda)s) \geq \min\{\mu(t), \mu(s)\} \forall t, s \in \mathbb{R}, \lambda \in [1, 0]$,
3. μ *upper semi-continuous on* $\mathbb{R}$, *and* $\{\mathrm{t} \in \mathbb{R}{:}\mu(\mathrm{t}) > 0\}$ *is compact* $\tilde{E}$ *is called the space of fuzzy numbers and* $\mathbb{R}$ *is a proper subset of* $\tilde{E}$.

Define the r-level set $x \in \mathbb{R}$, $[\mu]_r = \{x|\mu(x) \geq r\}$, $0 \leq r \leq 1$ *where* $[\mu]_0 = \{x|\mu(x) > 0\}$ *is compact which is a closed bounded interval and denoted by* $[\mu]_r = (\underline{\mu}(t), \overline{\mu}(t))$. *In the parametric form, a fuzzy number is represented by an ordered pair of functions* $(\underline{\mu}(t), \overline{\mu}(t)), r \in [0, 1]$ *which satisfies [11]:*

1. $\underline{\mu}(t)$ *is a bounded left continuous non-decreasing function over* $[0, 1]$.
2. $\overline{\mu}(t)$ *is a bounded left continuous non-increasing function over* $[0, 1]$.
3. $\underline{\mu}(t) \leq \overline{\mu}(t), \quad r \in [0, 1]$.

A crisp number r is simply represented by $\underline{\mu}(t) = \overline{\mu}(t) = r, \quad r \in [0, 1]$.

Definition 2 *(see [12]). If* $\tilde{E}$ *be the set of all fuzzy numbers we say that* $f(x)$ *is a fuzzy function if* $f : \mathbb{R} \to E$.

Definition 3 *(see [12]). A mapping* $f : T \to \tilde{E}$ *for some interval* $T \subseteq \tilde{E}$ *is called a fuzzy function process and we denote r-level set by:*

$$[\tilde{f}(\mathrm{t})]_r = [\underline{f}(\mathrm{t}; \mathrm{r}), \overline{f}(\mathrm{t}; \mathrm{r})], t \in T, r \in [0, 1] \tag{1}$$

The r-level sets of a fuzzy number are much more effective as representation forms of fuzzy set than the above. Fuzzy sets can be defined by the families of their r-level sets based on the resolution identity theorem.

Definition 4 *(see [13]). The fuzzy integral of fuzzy process* $\tilde{f}(t;r)$, $\int_a^b \tilde{f}(t;r)dt$ *for* $a, b \in T$ *and* $r \in [0,1]$ *is defined by:*

$$\int_a^b \tilde{f}(t;r)dt = \left[\int_a^b \underline{f}(t;r)dt, \int_a^b \overline{f}(t;r)dt\right] \tag{2}$$

Definition 5 *(see [14]). The Riemann-Liouville fractional integral operator of order* $\alpha(\alpha \geq 0)$ *is defined as*

$$\begin{cases} J^\alpha f(x) = \frac{1}{\Gamma(v)}\int_0^x (x-t)^{\alpha-1} f(t)dt, \quad a > 0, x > 0 \\ J^0 f(x) = f(x) \end{cases} \tag{3}$$

Definition 6 *(see [14]). The Caputo fractional derivatives of order* α *are defined as*

$$\begin{aligned} &D^\alpha f(x) = J^{m-\alpha} D^m f(x) = \frac{1}{\Gamma(m-\alpha)}\int_0^x (x-t)^{m-\alpha-1}\frac{d^m}{dt^m} f(t)dt \\ &m-1 < \alpha \leq m, x > 0 \end{aligned} \tag{4}$$

Where D^m *is the classical differential operator of order* m.

For the Caputo derivative we have

$$\begin{cases} D^\alpha x^\beta = 0, \beta < \alpha \\ D^\alpha x^\beta = \frac{\Gamma(\beta+1)}{\Gamma(\beta+1-\alpha)} x^{\beta-\alpha}, \beta \geq \alpha \end{cases} \tag{5}$$

Definition 7 *(see [15]). For* n *to be the smallest integer that exceeds* α, *the Caputo time fractional derivative operator of order* α *of* $u(x,t)$ *is defined as*

$$\begin{cases} D_t^\alpha u(x,t) = \frac{\partial^\alpha u(x,t)}{\partial t^\alpha} = \frac{1}{\Gamma(n-\alpha)}\int_0^t (t-\tau)^{n-\alpha-1}\frac{\partial^n u(x,\tau)}{\partial t^n} d\tau, \\ n-1 < \alpha < n \\ D_t^n u(x,t) = \frac{\partial^n u(x,t)}{\partial t^n}, \quad n \in N \end{cases} \tag{6}$$

Definition 8 *(see [15]). A power series expansion of the form*

$$\begin{aligned} &\sum_{m=0}^{\infty} f_m (t-t_0)^{m\alpha} = f_0 + f_1(t-t_0)^\alpha + f_2(t-t_0)^{2\alpha} + \cdots, \\ &0 \leq m-1 < \alpha \leq m, t \geq t_0 \end{aligned} \tag{7}$$

is named fractional power series at $t = t_0$.

3 Applications for the Residual Power Series Method

In [16], the second order fuzzy pantograph equation has been studied by the Bernoulli wavelet method and it has the following form

$$\tilde{y}''(t) = \frac{3}{4}\tilde{y}(t) + \tilde{y}\left(\frac{t}{2}\right) - t^2 + 2, \quad 0 \leq t \leq 1 \tag{8}$$

subject to

$$\tilde{y}(0) = [r, 2-r], \tilde{y}'(0) = [r, 3-2r], \quad \forall r \in [0,1] \tag{9}$$

In this paper, we study the Fractional Fuzzy Delay differential equations of the form

$$D_t^{2\alpha}\tilde{y}(t) = \frac{3}{4}\tilde{y}(t) + \tilde{y}\left(\frac{t}{2}\right) - t^2 + 2, \quad 0 \le t \le 1, 0.5 \le \alpha \le 1 \tag{10}$$

subject to

$$\tilde{y}(0) = [r, 2-r], D_t^{\alpha}\tilde{y}(0) = [r, 3-2r], \quad \forall r \in [0,1] \tag{11}$$

In order to defuzzify Eq. (10), we denote the fuzzy function $\tilde{y}$ by $\tilde{y} = [\overline{y}, \underline{y}]$ for $t \in [\mathrm{t_0}, \mathrm{T}]$ and $r \in [0,1]$. It means that the r-level set of $\tilde{y}(t)$ can be defined as:

$$\begin{aligned} [\tilde{y}(t)]_r &= [\underline{y}(t;r), \overline{y}(t;r)] \\ [\tilde{y}(t_0)]_r &= [\underline{y}(t_0;r), \overline{y}(t_0;r)] \end{aligned} \tag{12}$$

We write the equation into a fractional form as

$$\begin{cases} D_t^{2\alpha}\underline{y}(t;r) = \frac{3}{4}\underline{y}(t;r) + \underline{y}\left(\frac{t}{2};r\right) - t^2 + 2, \quad 0 \le t \le 1, 0.5 \le \alpha \le 1 \\ \underline{y}(0;r) = r, D_t^{\alpha}\underline{y}(0;r) = r, \quad \forall r \in [0,1] \end{cases} \tag{13}$$

$$\begin{cases} D_t^{2\alpha}\overline{y}(t,r) = \frac{3}{4}\overline{y}(t;r) + \overline{y}\left(\frac{t}{2};r\right) - t^2 + 2, \quad 0 \le t \le 1, 0.5 \le \alpha \le 1 \\ \overline{y}(0;r) = 2-r, D_t^{\alpha}\overline{y}(0;r) = 3-2r, \quad \forall r \in [0,1] \end{cases} \tag{14}$$

We assume that the form of the solution is as follows

$$\tilde{y}(t;r) = \begin{cases} \underline{y}(t;r) = \sum_{n=0}^{\infty} \underline{f}_n(r) \frac{t^{n\alpha}}{\Gamma(1+n\alpha)} \\ \overline{y}(t;r) = \sum_{n=0}^{\infty} \overline{f}_n(r) \frac{t^{n\alpha}}{\Gamma(1+n\alpha)} \end{cases} \tag{15}$$

Therefore, the truncated series solution can be expressed as

$$\tilde{y}_k(t;r) = \begin{cases} \underline{y}_k(t;r) = \sum_{n=0}^{k} \underline{f}_n(r) \frac{t^{n\alpha}}{\Gamma(1+n\alpha)} \\ \overline{y}_k(t;r) = \sum_{n=0}^{k} \overline{f}_n(r) \frac{t^{n\alpha}}{\Gamma(1+n\alpha)} \end{cases} \tag{16}$$

The residual functions $Res_{\underline{y}}$ and $Re\,s_{\overline{y}}$ are defined as

$$Res_{\tilde{y}} = \begin{cases} Res_{\underline{y}} = D_t^{2\alpha}\underline{y}(t;r) - \frac{3}{4}\underline{y}(t;r) - \underline{y}\left(\frac{t}{2};r\right) + t^2 - 2 \\ Res_{\overline{y}} = D_t^{2\alpha}\overline{y}(t;r) - \frac{3}{4}\overline{y}(t;r) - \overline{y}\left(\frac{t}{2};r\right) + t^2 - 2 \end{cases} \tag{17}$$

And the truncated residual function can be write as

$$Res_{\tilde{y}_k} = \begin{cases} Res_{\underline{y}_k} = D_t^{2\alpha}\underline{y}_k(t;r) - \frac{3}{4}\underline{y}_k(t;r) - \underline{y}_k\left(\frac{t}{2};r\right) + t^2 - 2 \\ Res_{\overline{y}_k} = D_t^{2\alpha}\overline{y}_k(t;r) - \frac{3}{4}\overline{y}_k(t;r) - \overline{y}_k\left(\frac{t}{2};r\right) + t^2 - 2 \end{cases} \tag{18}$$

In order to find the coefficient $\underline{f}_n(r)$ and $\overline{f}_n(r)$, we can solve this equation

$$D_t^{(n-2)\alpha} Res_{\tilde{y}_k} = 0, \quad n \in N^* \tag{19}$$

In order to find the coefficient $\underline{f}_2(r)$ and $\overline{f}_2(r)$, we write $k = 2$ in Eq. (18)

$$\begin{cases} Res_{\underline{y}_2} = D_t^{2\alpha}\underline{y}_2(t;r) - \frac{3}{4}\underline{y}_2(t;r) - \underline{y}_2\left(\frac{t}{2};r\right) + t^2 - 2 \\ Res_{\overline{y}_2} = D_t^{2\alpha}\overline{y}_2(t;r) - \frac{3}{4}\overline{y}_2(t;r) - \overline{y}_2\left(\frac{t}{2};r\right) + t^2 - 2 \end{cases} \tag{20}$$

where

$$\begin{cases} \underline{y}_2(t;r) = \underline{f}_0(r) + \underline{f}_1(r)\frac{t^\alpha}{\Gamma(1+\alpha)} + \underline{f}_2(r)\frac{t^{2\alpha}}{\Gamma(1+2\alpha)} \\ \overline{y}_2(t;r) = \overline{f}_0(r) + \overline{f}_1(r)\frac{t^\alpha}{\Gamma(1+\alpha)} + \overline{f}_2(r)\frac{t^{2\alpha}}{\Gamma(1+2\alpha)} \end{cases} \tag{21}$$

for

$$\begin{cases} \underline{f}_0(r) = \underline{y}_0 = \underline{y}(0;r) = r \\ \overline{f}_0(r) = \overline{y}_0 = \overline{y}(0;r) = 2 - r \end{cases} \tag{22}$$

$$\begin{cases} \underline{f}_1(r) = D_t^\alpha \underline{y} = r \\ \overline{f}_1(r) = D_t^\alpha \overline{y} = 3 - 2r \end{cases} \tag{23}$$

Therefore

$$\begin{cases} Res_{\underline{y}_2} = \underline{f}_2(r) - \frac{3}{4}(\underline{f}_0(r) + \underline{f}_1(r)\frac{t^\alpha}{\Gamma(1+\alpha)} + \underline{f}_2(r)\frac{t^{2\alpha}}{\Gamma(1+2\alpha)}) \\ \quad - \left(\underline{f}_0(r) + \underline{f}_1(r)\frac{(\frac{t}{2})^\alpha}{\Gamma(1+\alpha)} + \underline{f}_2(r)\frac{(\frac{t}{2})^{2\alpha}}{\Gamma(1+2\alpha)}\right) + t^2 - 2 \\ Res_{\overline{y}_2} = \overline{f}_2(r) - \frac{3}{4}(\overline{f}_0(r)+\overline{f}_1(r)\frac{t^\alpha}{\Gamma(1+\alpha)} + \overline{f}_2(r)\frac{t^{2\alpha}}{\Gamma(1+2\alpha)}) \\ \quad - (\overline{f}_0(r)+\overline{f}_1(r)\frac{(\frac{t}{2})^\alpha}{\Gamma(1+\alpha)} + \overline{f}_2(r)\frac{(\frac{t}{2})^{2\alpha}}{\Gamma(1+2\alpha)}) + t^2 - 2 \end{cases} \tag{24}$$

From Eq. (19) we deduce that $Res_{\underline{y}_2}$=0, $Res_{\overline{y}_2} = 0\,(t = 0)$ and after the calculation we have

$$\begin{cases} \underline{f}_2(r) = \frac{7}{4}r + 2 \\ \overline{f}_2(r) = -\frac{7}{4}r + \frac{11}{2} \end{cases} \tag{25}$$

Therefore, we get the second residual power series (RPS) approximate solutions

$$\begin{cases} \underline{y}_2(t;r) = r + r\frac{t^\alpha}{\Gamma(1+\alpha)} + \left(\frac{7}{4}r + 2\right)\frac{t^{2\alpha}}{\Gamma(1+2\alpha)} \\ \overline{y}_2(t;r) = 2 - r + (3 - 2r)\frac{t^\alpha}{\Gamma(1+\alpha)} + \left(-\frac{7}{4}r + \frac{11}{2}\right)\frac{t^{2\alpha}}{\Gamma(1+2\alpha)} \end{cases} \tag{26}$$

In order to obtain the coefficient $\underline{f}_3(r)$ and $\overline{f}_3(r)$, we write $k = 3$ in Eq. (18)

$$\begin{cases} Res_{\underline{y}_3} = D_t^{2\alpha}\underline{y}_3(t;r) - \frac{3}{4}\underline{y}_3(t;r) - \underline{y}_3\left(\frac{t}{2};r\right) + t^2 - 2 \\ Res_{\overline{y}_3} = D_t^{2\alpha}\overline{y}_3(t;r) - \frac{3}{4}\overline{y}_3(t;r) - \overline{y}_3\left(\frac{t}{2};r\right) + t^2 - 2 \end{cases} \tag{27}$$

where

$$\begin{cases} \underline{y}_3\,(t;r) = \underline{f}_0\,(r) + \underline{f}_1\,(r)\,\frac{t^\alpha}{\Gamma(1+\alpha)} + \underline{f}_2\,(r)\,\frac{t^{2\alpha}}{\Gamma(1+2\alpha)} \\ \quad + \underline{f}_3\,(r)\,\frac{t3^\alpha}{\Gamma(1+3\alpha)} \\ \overline{y}_3\,(t;r) = \overline{f}_0\,(r) + \overline{f}_1\,(r)\,\frac{t^\alpha}{\Gamma(1+\alpha)} + \overline{f}_2\,(r)\,\frac{t^{2\alpha}}{\Gamma(1+2\alpha)} \\ \quad + \overline{f}_3\,(r)\,\frac{t^{3\alpha}}{\Gamma(1+3\alpha)} \end{cases} \tag{28}$$

Therefore

$$\begin{cases} \mathrm{Res}_{\underline{y}_3} = \underline{f}_2(r) + \underline{f}_3(r)\frac{t^\alpha}{\Gamma(1+\alpha)} - \frac{3}{4}(\underline{f}_0(r) + \underline{f}_1(r)\frac{t^\alpha}{\Gamma(1+\alpha)} + \underline{f}_2(r) \\ \cdot \frac{t^{2\alpha}}{\Gamma(1+2\alpha)} + \underline{f}_3(r)\frac{t^{3\alpha}}{\Gamma(1+3\alpha)}) - (\underline{f}_0(r) + \underline{f}_1(r)\frac{(\frac{t}{2})^\alpha}{\Gamma(1+\alpha)} + \underline{f}_2(r) \\ \cdot \frac{(\frac{t}{2})^{2\alpha}}{\Gamma(1+2\alpha)} + \underline{f}_3(r)\frac{(\frac{t}{2})^{3\alpha}}{\Gamma(1+3\alpha)}) + t^2 - 2 \\ \mathrm{Res}_{\bar{y}_3} = \bar{f}_2(r) + \bar{f}_3(r)\frac{t^\alpha}{\Gamma(1+\alpha)} - \frac{3}{4}(\bar{f}_0(r) + \bar{f}_1(r)\frac{t^\alpha}{\Gamma(1+\alpha)} + \bar{f}_2(r) \\ \cdot \frac{t^{2\alpha}}{\Gamma(1+2\alpha)} + \bar{f}_3(r)\frac{t^{3\alpha}}{\Gamma(1+3\alpha)}) - (\bar{f}_0(r) + \bar{f}_1(r)\frac{(\frac{t}{2})^\alpha}{\Gamma(1+\alpha)} + \bar{f}_2(r)\frac{(\frac{t}{2})^{2\alpha}}{\Gamma(1+2\alpha)} \\ + \bar{f}_3(r)\frac{(\frac{t}{2})^{3\alpha}}{\Gamma(1+3\alpha)}) + t^2 - 2 \end{cases} \tag{29}$$

From Eq. (19) we deduce that $D_t^\alpha Res_{\underline{y}_3} = 0, D_t^\alpha Res_{\bar{y}_3} = 0\,(t = 0)$ and after the calculation we have

$$\begin{cases} \underline{f}_3(r) = \left(\frac{3}{4} + \frac{1}{2^\alpha}\right) r \\ \bar{f}_3(r) = \left(\frac{3}{4} + \frac{1}{2^\alpha}\right)(3 - 2r) \end{cases} \tag{30}$$

Therefore, we get the third RPS approximate solutions

$$\begin{cases} \underline{y}_3(t;r) = r + r\frac{t^\alpha}{\Gamma(1+\alpha)} + \left(\frac{7}{4}r + 2\right)\frac{t^{2\alpha}}{\Gamma(1+2\alpha)} + \left(\frac{3}{4} + \frac{1}{2^\alpha}\right) \\ \cdot r\frac{t^{3\alpha}}{\Gamma(1+3\alpha)} \\ \bar{y}_3(t;r) = 2 - r + (3 - 2r)\frac{t^\alpha}{\Gamma(1+\alpha)} + \left(-\frac{7}{4}r + \frac{11}{2}\right)\frac{t^{2\alpha}}{\Gamma(1+2\alpha)} \\ + \left(\frac{3}{4} + \frac{1}{2^\alpha}\right)(3 - 2r)\frac{t^{3\alpha}}{\Gamma(1+3\alpha)} \end{cases} \tag{31}$$

Similarly, applying the same procedure for $k = 4$ in Eq. (18), after the simple calculation, the following form of $\underline{f}_4(r)$ and $\bar{f}_4(r)$ will be obtained.

$$\begin{cases} \underline{f}_4(r) = \left(\frac{3}{4} + \frac{1}{2^{2\alpha}}\right)\left(\frac{7}{4}r + 2\right) \\ \bar{f}_4(r) = \left(\frac{3}{4} + \frac{1}{2^{2\alpha}}\right)\left(\frac{11}{2} - \frac{7}{4}r\right) \end{cases} \tag{32}$$

Therefore, the fourth RPS approximate solutions are

$$\begin{cases} \underline{y}_4(t;r) = r + r\frac{t^\alpha}{\Gamma(1+\alpha)} + \left(\frac{7}{4}r + 2\right)\frac{t^{2\alpha}}{\Gamma(1+2\alpha)} + \left(\frac{3}{4} + \frac{1}{2^\alpha}\right) \\ \cdot r\frac{t^{3\alpha}}{\Gamma(1+3\alpha)} + \left(\frac{3}{4} + \frac{1}{2^{2\alpha}}\right)\left(\frac{7}{4}r + 2\right)\frac{t^{4\alpha}}{\Gamma(1+4\alpha)} \\ \bar{y}_4(t;r) = 2 - r + (3 - 2r)\frac{t^\alpha}{\Gamma(1+\alpha)} + \left(-\frac{7}{4}r + \frac{11}{2}\right)\frac{t^{2\alpha}}{\Gamma(1+2\alpha)} \\ + \left(\frac{3}{4} + \frac{1}{2^\alpha}\right)(3 - 2r)\frac{t^{3\alpha}}{\Gamma(1+3\alpha)} + \left(\frac{3}{4} + \frac{1}{2^{2\alpha}}\right)\left(\frac{11}{2} - \frac{7}{4}r\right) \\ \cdot \frac{t^{4\alpha}}{\Gamma(1+4\alpha)} \end{cases} \tag{33}$$

4 Result and Analysis

When the power series is expanded to the eighth term, we get the approximate solution as $\tilde{y}_8$, and the image of $\tilde{y}_8$ is as shown in Fig. 1. From Fig. 1(a) to (f), the values of α are 0.5, 0.6, 0.7, 0.8, 0.9, 1.0 respectively. It is clear from Fig. 1(a), (b), (c), (d), (e) and (f) that the value of the solutions is become small when the derivative $0.5 \le \alpha \le 1$ tends to 1.

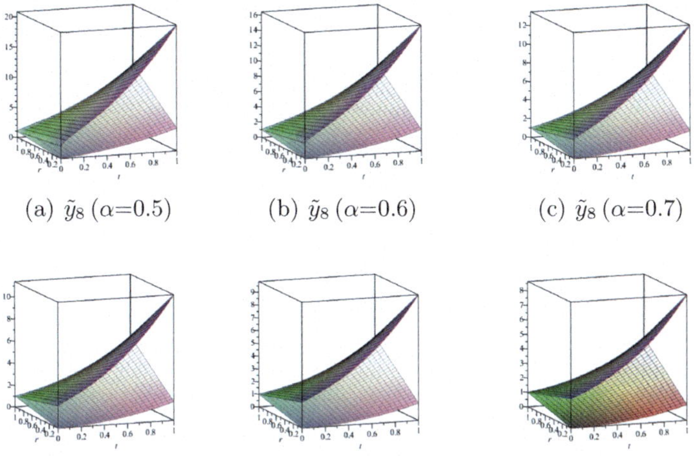

(a) $\tilde{y}_8$ (α=0.5) (b) $\tilde{y}_8$ (α=0.6) (c) $\tilde{y}_8$ (α=0.7)

(d) $\tilde{y}_8$ (α=0.8) (e) $\tilde{y}_8$ (α=0.9) (f) $\tilde{y}_8$ (α=1.0)

Fig. 1. Approximate solution $\tilde{y}_8$ $(\alpha = 0.5, 0.6, 0.7, 0.8, 0.9, 1.0)$

In Table 1, the error of this method is listed, when the values of α are different and $t = 0.1, r = 0.5$. It can be seen from the table that the value of $Res_{\underline{y}_8}$ decreases when the derivative $0.5 \leq \alpha \leq 1$ tends to integer 1, but the value of $Res_{\overline{y}_8}$ increases, and finally they almost reach the same value.

In Table 2, the error of this method is listed, when the values of r are different and $t = 0.1, \alpha = 0.9$. We can see from the table that the value of $Res_{\underline{y}_8}$ is relatively small when the value of r is close to 0 and 1, and the value of $Res_{\overline{y}_8}$ is relatively stable. The error of this result is larger than that in [16], but this method is simple and easy to operate, and better results can be obtained in a few steps.

Table 1. The error of results when $t = 0.1, r = 0.5$

α	$Res_{\underline{y}_8}$	$Res_{\overline{y}_8}$
0.5	0.0298071368	0.009914940
0.6	0.0298071368	0.009914940
0.7	0.0124418419	0.009999759
0.8	0.0108328348	0.009999986
0.9	0.0102783048	0.010000000
1.0	0.0100911815	0.010000000

Table 2. The error of results when $t = 0.1, \alpha = 0.9$

r	$Res_{\underline{y}_8}$	$Res_{\overline{y}_8}$
0	0.0099999999	0.010000000
0.2	0.0101781153	0.010000000
0.4	0.0102671727	0.010000000
0.6	0.0102671730	0.009999998
0.8	0.0101781161	0.010000002
1.0	0.0100000010	0.010000001

5 Conclusion

In this paper, the fuzzy delay differential equation is expressed in fractional form, and the RPSM is used to solve the equation. With this method, we can quickly get an approximate solution to the fractional fuzzy delay differential equation. The results show that the method is an effective method when solving the fractional fuzzy differential equation, and by this method, the approximate solutions can be obtained with just a few steps.

References

1. Zadeh, L.A.: Fuzzy sets. Inf. Control **8**(3), 338–353 (1965)
2. Jafari, R.: Solving fuzzy differential equation with bernstein neural networks. In: 2016 IEEE International Conference on Systems, Man, and Cybernetics • SMC 2016, pp. 001245-001250. IEEE, Budapest (2016)
3. Allahviranloo, T., Salahshour, S.: Euler method for solving hybrid fuzzy differential equation. Soft Comput. - Fusion Found. Methodologies Appl. **15**(7), 1247–1253 (2011)
4. Ahmad, L., Farooq, M., Abdullah, S.: Solving n-th order fuzzy differential equation by fuzzy Laplace transform. Indian J. Pure Appl. Math. (2014)
5. Tapaswini, S., Chakraverty, S.: Numerical solution of n-th order fuzzy linear differential equations by homotopy perturbation method. Int. J. Comput. Appl. **64**(6), 5–10 (2013)
6. Tehran, I.: Numerical solutions of fuzzy differential equations by taylor method. Comput. Methods Appl. Math. **2**(2), 113–124 (2002)
7. Abu, A.O.: Series solution of fuzzy equations under strongly generalized differentiability. J. Adv. Res. Appl. Math. **5**, 31–52 (2013)
8. Tchier, F., Inc, M., Korpinar, Z.S.: Solutions of the time fractional reaction-diffusion equations with residual power series method. Adv. Mech. Eng. **8**(10), 1–10 (2016)
9. Alquran, M.: Analytical solutions of fractional foam drainage equation by residual power series method. Math. Sci. **8**(4), 153–160 (2015)
10. Dubois, D.: Towards fuzzy differential calculus part 3: differentiation. Fuzzy Sets Syst. **8**(3), 225–233 (1982)
11. Goetschel, R., Voxman, W.: Elementary fuzzy calculus. Fuzzy Sets Syst. **18**(1), 31–43 (1986)

12. Seikkala, S.: On the fuzzy initial value problem. Fuzzy Sets Syst. **24**(3), 319–330 (1987)
13. Wu, C., Gong, Z.: On Henstock integral of fuzzy-number-valued functions. Fuzzy Sets Syst. **120**(120), 523–532 (2001)
14. Podlubny, I.: Fractional differential equations: an introduction to fractional derivatives, fractional differential equations, to methods of their solution and some of their applications. In: Mathematics in Science and Engineering , vol. 198, pp. 1–340 (1999)
15. El-Ajou, A., Abu Arqub, O., Al, Z.Z.: New results on fractional power series: theories and applications. Entropy **15**, 5305–5323 (2013)
16. Hamood, M.Y., Ali, F.J.: New analytical method for solving fuzzy delay differential equations. In: AIP Conference Proceedings, vol. 1691, pp. 040028-1–040028-8. AIP Publishing, Malaysia (2015)

Credibility Assessment of Simulation Models Using Hesitant Cloud Linguistic Term Sets

Xiaojun Yang(✉), Zhongfu Xu, Chuan Shi, Hao Lei, and Changwei Yan

Luoyang Electronic Equipment Test Center of China, Luoyang 471003, He'nan, China
yangxiaojun2007@gmail.com, xuzf6904@aliyun.com

Abstract. Credibility assessment of a simulation model is a hard problem as subjectivity is always involved. Linguistic expressions are much more straightforward and friendlier for a subject matter expert (SME) to represent qualitative judgments. Therefore, an assessment method based on hesitant cloud linguistic term sets (HCLTSs) is proposed in this paper. SMEs use context-free grammars and linguistic expressions to express their opinions. Then, these linguistic expressions are transformed into HCLTSs. Finally, aggregation results are calculated by synthetic cloud algorithm and weighted average cloud algorithm. An example under multiple-criteria group decision making is presented to illustrate the proposed method for credibility assessment.

Keywords: Credibility assessment · Simulation model · Hesitant cloud linguistic term sets · Cloud model · Linguistic expressions

1 Introduction

When a system does not yet exist or it is too difficult to directly manipulate it, simulation becomes an indispensable role to observe and study the system. With the development of computer, modeling and simulation technologies, simulation models have been applied widely in military, industrial, economic, environmental and social domains. The widespread use of simulation models in various areas has provoked more and more concerns about their credibility.

Credibility assessment of a simulation model is encountering many challenges, involves the measurements and evaluations of hundreds of qualitative and quantitative indicators and mandates Subject Matter Expert (SME) evaluations [1, 2]. As modeling is an art and credibility assessment is situation dependent, Balci [3] pointed out that subjectivity is and will always be part of the credibility assessment. This indicates that qualitative methods based on SME group and peer assessments are effective in addressing the problem of credibility assessment.

Qualitative approaches use data in the form of words or linguistic expressions rather than numbers. Azadeh *et al.* [4] model linguistic variables using triangular fuzzy numbers. To deal with both intrapersonal uncertainty and interpersonal uncertainty, words are modeled by interval type-2 fuzzy sets [5]. Furthermore, to deal with both fuzziness and randomness of uncertain concepts simultaneously, Li *et al.* [6] presented

Y. Liu et al. (Eds.): ICNC-FSKD 2019, AISC 1074, pp. 856–863, 2020.
https://doi.org/10.1007/978-3-030-32456-8_92

the concept of cloud model based on fuzzy sets and probability theory. Thus, linguistic terms can be modeled by cloud models.

In previous researches, an SME provides a single linguistic term to represent his/her qualitative judgment. However, in practice, an SME may hesitate among several linguistic terms and require richer expressions to express the qualitative judgment more accurately. In this paper, to handle the uncertainty and hesitancy of subjective judgments, an assessment method based on hesitant cloud linguistic term sets (HCLTSs) is proposed.

The rest of the paper is organized as follows. In Sect. 2, the concept of hesitant cloud linguistic term sets are introduced. Section 3 describes a step-by-step instruction for using HCLTSs to assess the credibility of a simulation model. In Sect. 4, an example is presented to illustrate our proposed method. Finally, Sect. 5 draws our conclusions.

2 Hesitant Cloud Linguistic Term Sets

Randomness and fuzziness are the two most important uncertainties inherent in human cognition. To express both fuzziness and randomness simultaneously, Li *et al.* [7] defined the cloud model based on fuzzy sets and probability theory. Cloud models can describe many phenomena with uncertainties in both natural and social sciences.

Thus far, many kinds of cloud models were proposed, including normal cloud models, triangular cloud models, trapezoidal cloud models, half cloud models and combined cloud models etc. The normal cloud model based on normal distribution and Gaussian membership function is the most commonly used one. It has been applied in many fields successfully [8], such as decision making, image processing, data mining, uncertainty reasoning, time series prediction and knowledge representation. In this paper, we focus on the normal cloud model.

Definition 1 (*Normal Cloud Model*) [9, 10]: Let l be a qualitative concept in the universe of discourse U. If $x \in U$ is a random instantiation of the concept l, and the distribution of x satisfies $x \sim N(Ex, En'^2)$, i.e. normal distribution with expectation Ex and variance En'^2, where $En' \sim N(En, He^2)$, and the certainty degree of x belonging to l is determined by the Gaussian function

$$\mu_T(x) = e^{-\frac{(x-Ex)^2}{2(En')^2}} \tag{1}$$

then the distribution of x in the universe U is called a *normal cloud* and each x with the certainty degree $\mu_T(x)$ is defined as a *cloud drop* (quantitative instantiation) of the qualitative concept l.

According to this definition, the forward normal cloud generator algorithm [10] was proposed to transform a normal cloud model from its qualitative representation into its quantitative instantiations.

The normal cloud model reflects randomness by a random occurrence of x from normal distribution and describes fuzziness by assigning x with a Gaussian membership degree. A normal cloud model is characterized by three parameters fully, namely,

Expectation (*Ex*), Entropy (*En*), and Hyper-entropy (*He*). *Ex* is the mathematical expectation of the cloud drops belonging to a concept in the universe. It is the most representative and typical sample of the qualitative concept. *En* represents the uncertainty measurement of a qualitative concept. It is determined by both randomness and fuzziness of the concept. A larger *En* is accompanied by a wider scope and larger certainty degrees. *He* is the entropy of entropy *En*. Thus, it describes the uncertainty of *En*. *He* reflects the dispersion of the cloud drops. Larger *He* leads to larger dispersion (thickness) of cloud drops and larger uncertainty of certainty degrees.

Linguistic terms usually involve both fuzzy type and random type uncertainties. Therefore, a qualitative linguistic concept l can be expressed by a normal cloud model with three parameters, i.e. $l(Ex, En, He)$. When subjective assessment is performed, an SME may hesitate among several linguistic terms or look for a more complex linguistic term, such as "lower than medium" or "between medium and very high", that is not defined in the linguistic term set. To deal with such situations, Huang and Yang [11] introduced the concept of hesitant cloud linguistic term sets (HCLTSs) based on hesitant fuzzy linguistic term sets [12] and normal cloud models.

Definition 2 (*Hesitant Cloud Linguistic Term Sets, HCLTSs*) [11]: Suppose that $L = \{l_0,\ldots,l_n\}$ is a finite and totally ordered discrete linguistic term set, where l_i represents a linguistic term modeled by a normal cloud model. An HCLTS, H_L, is defined as an ordered finite subset of the consecutive linguistic terms of L.

Example 1. Let $L = \{l_0$: *non* (none), l_1: *vl* (very low), l_2: *low*, l_3: *sl* (slightly low), l_4: *med* (medium), l_5: *sh* (slightly high), l_6: *high*, l_7: *vh* (very high), l_8: *max* (maximum)$\}$ be a linguistic term set, two different HCLTSs might be $H_L^1 = \{l_4$: *med*, l_5: *sh*, l_6: *high*$\}$ and $H_L^2 = \{l_6$: *high*, l_7: *vh*$\}$.

3 Credibility Assessment Using HCLTSs

In this section, a step-by-step instruction for using HCLTSs to assess the credibility of a simulation model is provided.

3.1 Selection of Indicators

The credibility of a simulation model is defined as how much confidence one has that the simulation results closely reflect reality with respect to the intended use of the simulation model. A simulation model is developed for an intended purpose. If a simulation model is designed for answering several questions, then the credibility assessment needs to be determined with respect to each question [13].

The first step of credibility assessment is the selection of indicators/criteria. The overall credibility of a simulation model is a complex qualitative concept and cannot be directly measured, involves the SME evaluations of many indicators. Evaluations of diverse characteristics of a simulation model make the credibility assessment become a difficult multiple criteria decision making problem. Administrators and SMEs select indicators for a simulation model in a collaborative manner. The selected indicators should cover all essential elements for credibility assessment of a simulation model.

Liao *et al.* [14] pointed out that a single characteristic is usually not enough to obtain a reliable assessment but multiple characteristics need to be aggregated to a comprehensive combination. Olsen *et al.* [15] calculated the total model credibility by aggregating the assessment results of structure credibility, behavior credibility and data credibility. The structure of a simulation model describes the internal organization and interactions, which is primarily related to the conceptual model. The behaviors of a simulation model describe the external manifestations, which are generally tested by SMEs using the examination approach. The data property of a simulation model regards the data used to operate the model. Therefore, three indicators as structure, behavior and data are selected to determine the credibility of a simulation model.

3.2 Definition of a Linguistic Term Set

Balci [1] defined a linguistic term set involving fifteen nominal terms using interval numbers. However, psychologist Miller [16] found out that people could distinguish subjective information simultaneously involving at most 5–9 grades. With more, they become confused and cannot handle the information. Thus, we define a linguistic term set involving nine linguistic terms with different grades in this paper. The nine linguistic terms are modeled by normal cloud models as listed in Table 1.

Table 1. A linguistic term set involving nine linguistic terms modeled by normal cloud models.

Linguistic terms	Normal cloud models	Linguistic terms	Normal cloud models
none (non)	(0.00, 1.00, 0.20)	*slightly high (sh)*	(71.05, 10.27, 0.41)
very low (vl)	(1.97, 4.57, 0.22)	*high*	(87.43, 12.42, 0.41)
low	(12.46, 10.62, 0.34)	*very high (vh)*	(97.75, 10.54, 0.27)
slightly low (sl)	(30.83, 10.30, 0.33)	*maximum (max)*	(100.00, 1.00, 0.20)
medium (med)	(50.58, 11.09, 0.56)		

The nine linguistic terms are adapted from the 32 words modeled by Yang *et al.* [17]. The main difference is the universe of discourse, as Balci [1] suggested that the scale of credibility assessment is 0–100.

3.3 Context-Free Grammars

An SME can directly select several linguistic terms using the concept of HCLTS for credibility assessment, but such elements are not close to human beings' cognitive models. Therefore, the comparative linguistic methods defined by context-free grammars G_H [12] are proposed to elicit an SME's linguistic expressions.

Definition 3 [12]. Let G_H be a context-free grammar, and $L = \{l_0,\ldots,l_n\}$ be a linguistic term set. The elements of $G_H = (V_N, V_T, I, P)$ are defined as follows:

V_N = {*<primary term>*, *<composite term>*, *<unary relation>*, *<binary relation>*, *<conjunction>*}

V_T = {*lower than*, *greater than*, *at least*, *at most*, *between*, *and*, l_0, $l_1,\ldots,l_n$}

$I \in V_N$

For the context-free grammar G_H, the production rules are as follows:

P = {*I* ::= *<primary term>*|*<composite term>*
<primary term> ::= $l_0|l_1|\ldots|l_n$
<composite term> ::= *<unary relation><primary term>*|*<binary relation><primary term><conjunction><primary term>*
<unary relation> ::= *lower than*|*greater than*|*at least*|*at most*
<binary relation> ::= *between*
<conjunction> ::= *and*}

The comparative linguistic methods defined by G_H are more similar to human beings' expressions.

3.4 Transformation Functions

Linguistic expressions produced by context-free grammars cannot be calculated directly. Some transformation functions were used to transform these linguistic expressions into HCLTSs.

Definition 4 *[11]*. Let f_{G_H} be a function that transforms linguistic expressions generated by G_H into HCLTSs. Then f_{G_H} is defined in different ways according to the meanings of corresponding production rules:

(1) $f_{G_H}(l_i) = \{l_i | l_i \in L\}$;
(2) $f_{G_H}(\text{at most } l_i) = \{l_j | l_j \in L \; and \; l_j \leq l_i\}$;
(3) $f_{G_H}(\text{lower than } l_i) = \{l_j | l_j \in L \; and \; l_j < l_i\}$;
(4) $f_{G_H}(\text{at least } l_i) = \{l_j | l_j \in L \; and \; l_j \geq l_i\}$;
(5) $f_{G_H}(\text{greater than } l_i) = \{l_j | l_j \in L \; and \; l_j > l_i\}$;
(6) $f_{G_H}(\text{between } l_i \text{ and } l_j) = \{l_k | l_k \in L \; and \; l_i \leq l_k \leq l_j\}$.

Therefore, using these transformation functions, all the linguistic expressions produced by G_H can be transformed into HCLTSs for subsequent calculations.

3.5 Synthetic Cloud Model of an HCLTS

An HCLTS $H_L = \{l_i, \ldots, l_{i+k}\}$ defined on $L = \{l_0, \ldots, l_n\}$, where $i \geq 0$, $k \geq 0$, and $i + k \leq n$, can be modeled by a synthetic cloud model $l_s(Ex_s, En_s, He_s)$.

$$\begin{cases} Ex_s = \frac{1}{k} \sum_{j=0}^{k} Ex_{i+j} \\ En_s = \frac{1}{6} \left(\max_{i \leq p \leq i+k} (Ex_p + 3En_p) - \min_{i \leq q \leq i+k} (Ex_q - 3En_q) \right) \\ He_s = \sqrt{\sum_{j=0}^{k} He_{i+j}^2} \end{cases} \quad (2)$$

The synthetic cloud model preserves hesitation information of the SME using parameters En_s and He_s.

3.6 Aggregations

When multiple-criteria group decision making is performed, two types of aggregations should be defined. One is the aggregation of all SMEs' opinions. The other is the aggregation of all indicators. The weighted average cloud is calculated as the aggregation result.

The weighted average cloud $l_{wa}(Ex_{wa}, En_{wa}, He_{wa})$ of n normal cloud models $l_i(Ex_i, En_i, He_i)$, $i = 1, 2, \ldots, n$, is defined as

$$l_{wa} = \sum_{i=1}^{n} w_i l_i \Big/ \sum_{i=1}^{n} w_i \quad (3)$$

where w_i is the weight of l_i.

When multiple SMEs' opinions are aggregated, w_i is calculated as

$$w_i = \frac{1}{En_i + 3 \times He_i} \quad (4)$$

Thus, the weight of an SME is determined by his/her fuzziness, randomness and hesitancy. The SMEs whose opinions have smaller uncertainties are preferred, so are assigned with larger weights.

When multiple indicators are aggregated, w_i is determined by the importance of corresponding indicator.

4 An Illustrative Example

A simple example is presented herein to illustrate the method proposed in this paper.

Three indicators structure, behavior and data are selected to assess the credibility of a simulation model. Four SMEs expressed their opinions using the context-free grammars as listed in Table 2.

Table 2. Linguistic expressions determined by four SMEs on three indicators.

SMEs	Structure	Behavior	Data
SME_1	at least *high*	between *medium* and *high*	at most *slightly high*
SME_2	between *low* and *high*	at least *slightly high*	between *low* and *high*
SME_3	greater than *medium*	at most *high*	greater than *medium*
SME_4	between *medium* and *high*	greater than *high*	between *high* and *very high*

These linguistic expressions are transformed into HCLTSs as follows:

$$\begin{pmatrix} \{l_6, l_7, l_8\} & \{l_4, l_5, l_6\} & \{l_0, l_1, l_2, l_3, l_4, l_5\} \\ \{l_2, l_3, l_4, l_5, l_6\} & \{l_5, l_6, l_7, l_8\} & \{l_2, l_3, l_4, l_5, l_6\} \\ \{l_4, l_5, l_6, l_7, l_8\} & \{l_0, l_1, l_2, l_3, l_4, l_5, l_6\} & \{l_4, l_5, l_6, l_7, l_8\} \\ \{l_4, l_5, l_6\} & \{l_6, l_7, l_8\} & \{l_6, l_7\} \end{pmatrix} \tag{5}$$

The synthetic cloud models of these HCLTSs in Eq. (5) are calculated as

$$\begin{pmatrix} (95.06, 13.20, 0.53) & (69.69, 17.90, 0.81) & (27.82, 20.21, 0.89) \\ (50.47, 24.02, 0.94) & (89.06, 14.86, 0.67) & (50.47, 24.02, 0.94) \\ (81.36, 18.68, 0.87) & (36.33, 24.02, 0.98) & (81.36, 18.68, 0.87) \\ (69.69, 17.90, 0.81) & (95.06, 13.20, 0.53) & (92.59, 13.20, 0.49) \end{pmatrix} \tag{6}$$

On each indicator, the weights of four SMEs are calculated by Eq. (4) as

$$\begin{pmatrix} 0.0676 & 0.0492 & 0.0437 \\ 0.0373 & 0.0593 & 0.0373 \\ 0.0470 & 0.0371 & 0.0470 \\ 0.0492 & 0.0676 & 0.0682 \end{pmatrix} \tag{7}$$

Thus, the aggregations of four SMEs on each indicator are calculated by Eq. (3) as

$$((77.38, 8.82, 0.38), (77.32, 8.31, 0.36), (67.46, 9.07, 0.38)) \tag{8}$$

The weights of the three indicators structure, behavior and data are 0.3, 0.5 and 0.2 respectively [15]. Therefore, the aggregation result of all three indicators is (75.36, 5.25, 0.22). It indicates that the credibility of this simulation model is between *slightly high* to *high*, much closer to *slightly high*.

5 Conclusion

The use of linguistic expressions is much more straightforward and friendlier due to the uncertainties inherent in human judgments. In many cases, an SME might hesitate among several linguistic terms to express his/her qualitative judgment. To deal with such situations, an assessment method based on HCLTSs is proposed in this paper. As far as we know, there is no existing simulation credibility assessment method using HCLTSs. Thus, other methods are difficult to describe fuzziness, randomness and hesitancy of SMEs simultaneously.

What is more, group decision making technique is used to overcome the knowledge limitation of a single SME and hesitation information of the SMEs is preserved and propagated through the entire calculation process. The final assessment result represented by a cloud model not only contains magnitude and grade information but also reflects uncertainty information of SMEs.

Acknowledgment. This work was supported by the Equipment Pre-Research Project of the 'Thirteenth Five-Year-Plan' of China under Grant 6140001010506.

References

1. Balci, O., Adams, R.J., Myers, D.S., Nance, R.E.: A collaborative evaluation environment for credibility assessment of modeling and simulation applications. In: Proceedings of the 2002 Winter Simulation Conference, pp. 214–220. IEEE, San Diego (2002)
2. Yang, Y.N., Kumaraswamy, M.M., Pam, H.J., Mahesh, G.: Integrated qualitative and quantitative methodology to assess validity and credibility of models for bridge maintenance management system development. J. Manag. Eng. **27**(3), 149–158 (2011)
3. Balci, O.: How to assess the acceptability and credibility of simulation results. In: Proceedings of the 1989 Winter Simulation Conference, pp. 62–71. ACM, Washington (1989)
4. Azadeh, A., Abdolhossein Zadeh, S.: An integrated fuzzy analytic hierarchy process and fuzzy multiple-criteria decision-making simulation approach for maintenance policy selection. Simulation **92**(1), 3–18 (2016)
5. Wu, D.R., Mendel, J.M.: Computing with words for hierarchical decision making applied to evaluating a weapon system. IEEE Trans. Fuzzy Syst. **18**(3), 441–460 (2010)
6. Li, D.Y., Meng, H.J., Shi, X.M.: Membership clouds and membership cloud generators. Comput. Res. Dev. **42**(8), 32–41 (1995)
7. Li, D.Y., Han, J.W., Shi, X.M., Chan, M.C.: Knowledge representation and discovery based on linguistic atoms. Knowl.-Based Syst. **10**(7), 431–440 (1998)
8. Yang, X.J., Yan, L.L., Zeng, L.: How to handle uncertainties in AHP: the cloud delphi hierarchical analysis. Inf. Sci. **222**, 384–404 (2013)
9. Li, D.Y., Liu, C.Y., Gan, W.Y.: A new cognitive model: cloud model. Int. J. Intell. Syst. **24**, 357–375 (2009)
10. Li, D.Y., Du, Y.: Artificial Intelligence with Uncertainty. Chapman & Hall/CRC Press, Boca Raton (2007)
11. Huang, H.C., Yang, X.J.: Representation of the pairwise comparisons in AHP using hesitant cloud linguistic term sets. Fundam. Inform. **144**(3–4), 349–362 (2016)
12. Rodríguez, R.M., Martínez, L., Herrera, F.: Hesitant fuzzy linguistic term sets for decision making. IEEE Trans. Fuzzy Syst. **20**(1), 109–119 (2012)
13. Sargent, R.G.: Verification and validation of simulation models. J. Simul. **7**, 12–24 (2013)
14. Liao, W.C., Zhang, J., Zheng, X.P., Zhao, Y.: A generalized validation procedure for pedestrian models. Simul. Model. Pract. Theory **77**, 20–31 (2017)
15. Olsen, M.M., Raunak, M., Setteducati, M.: Enabling quantified validation for model credibility. In: Proceedings of the 50th Computer Simulation Conference, pp. 1–10. Society for Computer Simulation International, Bordeaux, France (2018)
16. Miller, G.A.: The magical number seven, plus or minus two: some limits on our capacity for processing information. Psychol. Rev. **63**, 81–97 (1956)
17. Yang, X.J., Yan, L.L., Peng, H., Gao, X.D.: Encoding words into cloud models from interval-valued data via fuzzy statistics and membership function fitting. Knowl.-Based Syst. **55**, 114–124 (2014)

Fuzzy Entropy Clustering Image Segmentation Algorithm Based on Potential Two-Dimensional Histogram

Changxing Li[1(✉)], Xiaolu Zhang[2(✉)], and Liu Lei[2(✉)]

[1] School of Science, Xi'an University of Posts and Telecommunications, Xi'an 710121, China
shuxueshiyanshi@163.com

[2] School of Communication and Information Engineering, Xi'an University of Posts and Telecommunications, Xi'an 710121, China
1528462447@qq.com, 347320410@qq.com

Abstract. Fuzzy c-means (FCM) algorithm has been widely used in image segmentation. However, the traditional FCM algorithm does not take into account any image spatial information, which makes it very sensitive to noise. In order to make improvements on the basis of FCM, we proposed a fuzzy entropy clustering image segmentation algorithm based on potential two-dimensional histogram. Firstly, the two-dimensional gray histogram is constructed by using the gray value of image pixels and its local spatial uniform gray value, and the potential function is used to map-the mutual influence between image pixels into the data field, so as to describe the interaction between pixels more accurately. Secondly, the kernel function will be used to map the data to a high dimensional space to complete the linear transformation, this makes the data in the original space linearly indivisible become linearly separable or approximately linearly separable, which overcomes the defect that FCM is not suitable for multiple data distributions in some extent. Finally, the iterative formula of image segmentation is obtained by Lagrange multiplier method. Experimental results show that the proposed algorithm has higher correct segmentation rate and stronger robustness.

Keywords: Potential two-dimensional histogram · Fuzzy c-means algorithm · Kernel space

1 Introduction

FCM [1] introduces the fuzziness of each pixel, retains more information in the original image, but does not consider any spatial information, so it's sensitive to noise and get a poor segmentation effect.

Liu Jianzhuang proposed the fuzzy clustering image segmentation (FCM2D) [2] based on two-dimensional histogram algorithm. The method not only considers the gray information of pixels, but also considers the spatial correlation information of pixels and their neighborhoods. FCM algorithm is used for clustering in FCM2D, and a membership threshold was set after FCM algorithm convergence to perform binarization

Y. Liu et al. (Eds.): ICNC-FSKD 2019, AISC 1074, pp. 864–872, 2020.
https://doi.org/10.1007/978-3-030-32456-8_93

segmentation. FCM2D algorithm can better retain image details due to the consideration of the image's neighborhood spatial information, but this threshold method cannot be extended to multiple segmentation cases, and there is a problem of optimal selection of binarization parameters. If the parameters are not selected properly, its segmentation effect will be significantly reduced.

Therefore, a fuzzy entropy clustering for image segmentation based on potential two-dimensional histogram is proposed, whose main idea is to replace one-dimensional pixel grayscale values with two-dimensional histogram of potential values of images, then the fuzzy c-means clustering is carried out on gray histogram of images, which can more completely reflect the relationship between each pixels, and we also add the fuzzy entropy to the objective function, which can save the original information of the image more completely.

2 Related Works

2.1 FCM2D

The FCM2D algorithm [2, 3] proposed by Liu, the advantage of the method is that the gray level information of image pixels and the spatial neighborhood information between pixels are used at the same time. For image I with size $M \times N$, $I(x, y)$ represents the gray value of pixels with position (x, y) on the image, and $G(x, y)$ represents the mean of neighborhood pixels with position (x, y). The binary group (a, b) composed of $I(x, y)$ and $G(x, y)$ is defined as the two-dimensional histogram of the image, the objective function of FCM2D algorithm is as follows

$$J_m(U, V) = \sum_{k=1}^{c} \sum_{a=0}^{L-1} \sum_{b=0}^{L-1} h(a, b) u_{(a,b)k}^{m} d^2((a, b), v_k) \tag{1}$$

Where L is the maximum grayscale value of an image, $h(a, b)$ represents the number of (a, b) at the same position (x, y), $u_{(a,b)k}^{m}$ is the membership degree of the binary group (a, b) belonging to the kth cluster, $m \in [1, \infty)$ is the weighing exponent and usually set as 2, $d^2((a, b), v_k)$ is the Euclidean distance between (a, b) and the cluster centre v_k, it should satisfy $u_{(a,b)k} \in [0, 1]$ and $\sum_{k=1}^{c} u_{(a,b)k} = 1$.

By using Lagrange multiplier method to minimize the objective function, the updating formulas of membership degree and clustering center show in Eq. (2), (3).

After clustering convergence, the algorithm performs binary partition of pixels, Suppose $\mathbf{V}_1$ and $\mathbf{V}_2$ correspond to the clustering center of object and background, if the average gray value of the object is larger than the background, which satisfy the condition: $\|\mathbf{V}_1\| = \sqrt{v_{1,x}^2 + v_{1,y}^2} > \|\mathbf{V}_2\| = \sqrt{v_{2,x}^2 + v_{2,y}^2}$, then the pixel binarization method is defined as Eq. (4).

Where α is a threshold value satisfies $\alpha \in [0.5, 1]$. However when the value of α is not selected properly, the segmentation effect of FCM2D algorithm will be significantly reduced.

$$u_{(a,b)k} = \frac{1}{\sum_{r=1}^{c} \left[\frac{(a-v_{k,x})^2 + (b-v_{k,y})^2}{(a-v_{r,x})^2 + (b-v_{r,y})^2} \right]^{\frac{1}{m-1}}} \tag{2}$$

$$v_{k,x} = \frac{\sum_{a=0}^{L-1} \sum_{b=0}^{L-1} h(a,b) u^m_{(a,b)k} a}{\sum_{a=0}^{L-1} \sum_{b=0}^{L-1} h(a,b) u^m_{(a,b)k}}, v_{k,y} = \frac{\sum_{a=0}^{L-1} \sum_{b=0}^{L-1} h(a,b) u^m_{(a,b)k} b}{\sum_{a=0}^{L-1} \sum_{b=0}^{L-1} h(a,b) u^m_{(a,b)k}} \tag{3}$$

$$I(x,y) = \begin{cases} 255 & u_{1k} \geq \alpha \\ 0 & u_{1k} < \alpha \end{cases} \tag{4}$$

2.2 FEC

The FEC algorithm [4] proposed by Tran and Wagner. Let $\mathbf{X} = \{x_1, x_2, \cdots x_n\}$ represents an image with n pixels, the objective function is given as follows

$$J_{FEC}(U,V) = \sum_{k=1}^{c} \sum_{i=1}^{n} u_{ki} d^2(x_i, v_k) + n \sum_{k=1}^{c} \sum_{i=1}^{n} u_{ki} \log(u_{ki}) \tag{5}$$

Where x_i represents the graycale value of ith pixel, u_{ki} is the membership degree of the ith pixel belonging to the kth cluster, $d^2(x_i, v_k)$ is the Euclidean distance between x_i and the cluster centre v_k, $1/n$ denotes the degree of fuzzy entropy. It should satisfy $u_{ki} \in [0, 1]$ and $\sum_{k=1}^{c} u_{ki} = 1$.

When the algorithm converges, it convert the partition matrix $\mathbf{U} = (u_{ki})_{M \times N}$ to a clear partition. However, FEC does not consider spatial information of images, this will lead to the degradation of the anti-noise performance of the algorithm.

3 Proposed Fuzzy Entropy Clustering Algorithm for Image Segmentation Based on Potential Two-Dimensional Histogram

3.1 Two-Dimensional Histogram Based on Potential

Cognitive physics is a cognitive process from data to information which achieved the mapping from image pixel features to image particle potential space. In a given space $\Omega \subseteq R^p$ includes of n data objects $\{x_1, x_2, \cdots x_n\}$, $x_i = (x_{i1}, x_{i2}, \cdots, x_{ip})'$. Each object corresponds to a point mass in the p-dimensional space, and there is a field around it, any objects in the field will be affected by other objects. Due to that a potential function which is stable and is a single valued function of the position of the field space, we propose to describe the distribution of the field by the potential function [5, 6], which is defined as follows

$$\varphi_x(y) = m_x \times \exp(-(\|x - y\|/\sigma)^k) \tag{6}$$

Where $m_x \geq 0$ is the quality of the data object, it represents the intensity of the field source, $\|x - y\|$ is the Chebyshev distance between the two objects, $\sigma \in (0, +\infty)$ is the influence factor used to control the interaction between objects, k is the distance index. Based on the traditional two-dimensional histogram, we proposed the two-dimensional potential function of the image, it's advantage is that the relationships between each pixels can be reflected more completely through the potential fields, and the two-dimensional potential function is defined as follows

$$\varphi(a, b) = \sum m_i \times \exp(-(\|a - b\|/\sigma)^k) \tag{7}$$

Where $m_i = x_i$ represents the grayscale value of the current pixel i, $b = \bar{x}_i = mean\{x_p, p \in S_i\}$ represents the mean gray value of i's neighborhood pixels. Figure 1(a), (b), (c) show the original image, traditional two-dimensional histogram and the two-dimensional histogram based on potential respectively.

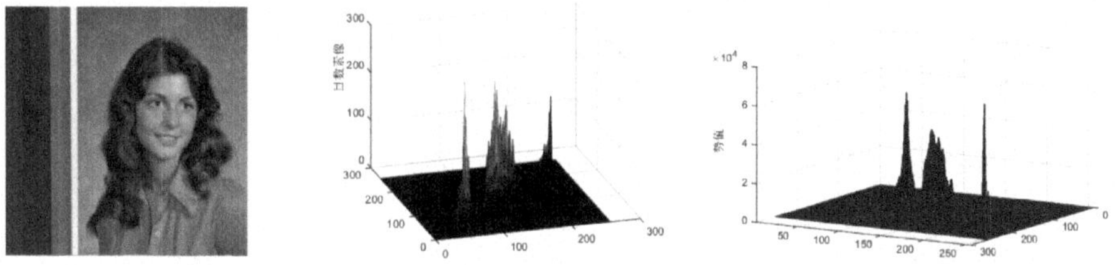

(a) (b) (c)

Fig. 1. (a) Original image. (b) Traditional two-dimensional histogram. (c) Two-dimensional histogram based on potential.

It can be seen clearly that compared with the traditional two-dimensional histogram, two-dimensional histogram based on potential is more able to highlight the peak of two-dimensional histogram, it is due to the interactions of pixels in the potential field.

3.2 Proposed Algorithm

The clustering is carried out on the two-dimensional potential histogram of the image, and the new objective function is defined as Eq. (8).

Where $\varphi(a, b)$ is the potential value of the binary group (a, b). In order to overcome the shortcomings of the clustering algorithm which is not suitable for multiple data distributions, we proposed to map the pixels of the original space through the kernel function to the high-dimensional Hilbert regeneration kernel space [7, 8], so that the pixels that are linearly inseparable in the original space become linear separable or approximately linear separable. Its main expression is as Eq. (9), (10).

$$J(U,V) = \sum_{k=1}^{c}\sum_{a=0}^{L-1}\sum_{b=0}^{L-1} \varphi(a,b)u_{(a,b)k}^{m} d^2((a,b),v_k) + n\sum_{k=1}^{c}\sum_{a=0}^{L-1}\sum_{b=0}^{L-1} u_{(a,b)k}\log u_{(a,b)k} \quad (8)$$

$$d^2((a,b),v_k) = \|\Phi(a,b) - \Phi(v_k)\|^2 \quad (9)$$

$$\|\Phi(a,b) - \Phi(v_k)\|^2 = K((a,b),(a,b)) + K(v_k,v_k) - 2K((a,b),v_k) \quad (10)$$

Gaussian kernel is the most widely used kernel, and its specific form is as follows:

$$K(x,y) = \exp(-\sigma^{-2}\|x-y\|^2) \quad (11)$$

The final objective function is defined as

$$J(U,V) = 2\sum_{k=1}^{c}\sum_{a=0}^{L-1}\sum_{b=0}^{L-1} \varphi(a,b)u_{(a,b)k}^{m}[1 - K((a,b),v_k)] + n\sum_{k=1}^{c}\sum_{a=0}^{L-1}\sum_{b=0}^{L-1} u_{(a,b)k}\log u_{(a,b)k} \quad (12)$$

The iterative updating formulas of membership degree and clustering center obtained by Lagrange multiplier method are as follows

$$u_{(a,b),k} = \sum_{r=1}^{c} \frac{\exp\left\{-\frac{1}{n}\left[2m\varphi(a,b)(1-K((a,b),v_k))u_{(a,b)k}^{m-1}\right]\right\}}{\exp\left\{-\frac{1}{n}\left[2m\varphi(a,b)(1-K((a,b),v_r))u_{(a,b)r}^{m-1}\right]\right\}} \quad (13)$$

$$v_{k,x} = \frac{\sum_{a=0}^{L-1}\sum_{b=0}^{L-1}\varphi(a,b)u_{(a,b)k}^{m}K(a,v_{k,x})a}{\sum_{a=0}^{L-1}\sum_{b=0}^{L-1}\varphi(a,b)u_{(a,b)k}^{m}K(a,v_{k,x})} \quad v_{k,y} = \frac{\sum_{a=0}^{L-1}\sum_{b=0}^{L-1}\varphi(a,b)u_{(a,b)k}^{m}K(b,v_{k,y})b}{\sum_{a=0}^{L-1}\sum_{b=0}^{L-1}\varphi(a,b)u_{(a,b)k}^{m}K(b,v_{k,y})} \quad (14)$$

4 Experimental Results

For the experiments, we used the CMU-Cornell iCoseg Dataset [9] of 643 images total. The other two fuzzy clustering algorithms FCM and FCM2D are chosen as the contrast algorithms to compare the efficiency and robustness.

The parameters are set as follows: the maximum iteration $T = 100$, the threshold value $\varepsilon = 10^{-5}$, the neighborhood search window size $\omega = 3 \times 3$. Besides, two objective evaluation standards, segmentation accuracy (SA) and (PSNR) are used to evaluate the algorithms, they are defined as follows

$$SA = \sum_{i=1}^{c} A_i \cap C_i \Big/ \sum_{j=1}^{c} C_j \tag{15}$$

Where i is the number of classification of the clustering, A_i is the pixels of class i that is segmented by the algorithm, C_i is the pixels of class i that is segmented by the standard image. The closer the value of SA is to 1, the higher the correct segmentation rate of the algorithm is.

$$PSNR = 10 \times \lg\left(\frac{MAX_I^2}{MSE}\right) \quad MSE = \frac{1}{M \times N}\sum_{i=0}^{M-1}\sum_{j=0}^{N-1}\|I(i,j) - K(i,j)\|^2 \tag{16}$$

The larger the value on *PSNR*, the better the effect of this algorithm on noise image segmentation. Figure 2 shows the segmentation results of FCM, FCM2D and the proposed method. The image size of Statue of Liberty is 375 × 500. Figure 2(a) adds a gaussian noise of (0, 0.01). Table 1 shows the value of SA and PSNR of each algorithms.

(a) (b) (c) (d) (e)

Fig. 2. Segmentation results on Statue of Liberty. (a) Add noise image. (b) Ground truth. (c) FCM. (d) FCM2D. (e) Proposed.

As can be seen from Fig. 1, compared with other algorithms, the proposed algorithm can preserve the details of original image relatively intact and is less interfered by noise. As can be seen from Table 1, the value of *SA* and *PSNR* from the proposed algorithm are the highest, indicating that the proposed algorithm is more suitable than other algorithms for the segmentation of noise images.

Table 1. SA and PSNR of each algorithm.

Method	SA	PSNR
FCM	0.9447	12.5780
FCM2D	0.9455	12.6385
Proposed	0.9888	19.5161

The image size of the Pyramid is 500 × 355. Figure 3 shows the segmentation results of the FCM, FCM2D and proposed method, and Fig. 3(a) adds a gaussian noise with (0, 0.01). Table 2 shows the value of SA and PSNR of each algorithms.

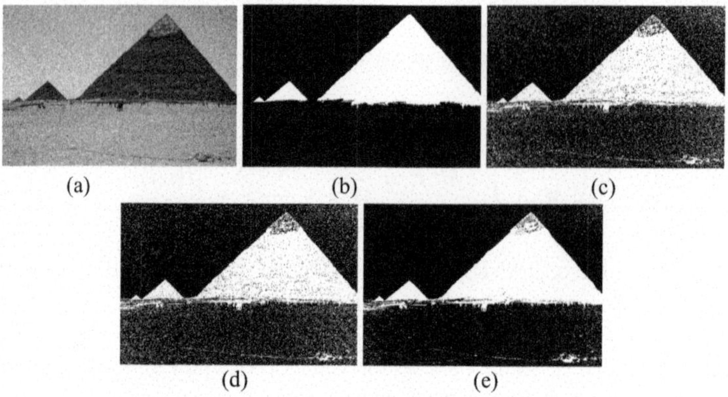

Fig. 3. Segmentation results on Pyramid. (a) Add noise image. (b) Ground truth. (c) FCM. (d) FCM2D. (e) Proposed.

It can be seen from Fig. 2 that the proposed algorithm can preserve image details more completely under the interference of noise, and the segmentation effect is relatively good. As can be seen from Table 2, the *SA* and *PSNR* values corresponding to proposed algorithm is the largest.

Table 2. SA and PSNR of each algorithm.

Method	SA	PSNR
FCM	0.9104	10.4779
FCM2D	0.9023	10.1029
Proposed	0.9718	15.4987

The image size of the Aircraft is 500 × 303. Figure 4 shows the segmentation results of the FCM, FCM2D and proposed method, and Fig. 4(a) adds a gaussian noise with (0, 0.005). Table 2 shows the value of SA and PSNR of each algorithms.

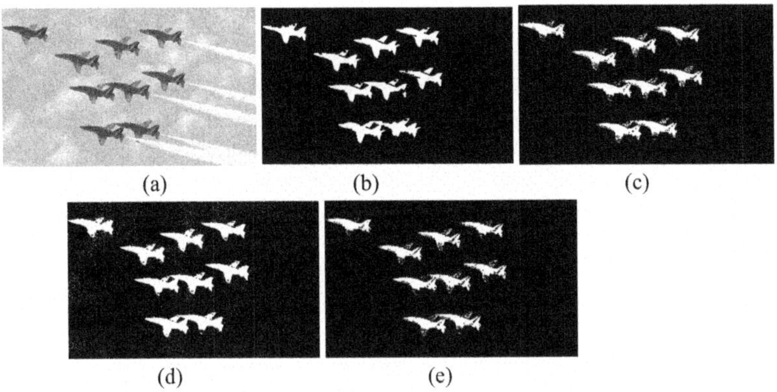

Fig. 4. Segmentation results on Aircraft. (a) Add noise image. (b) Ground truth. (c) FCM. (d) FCM2D. (e) Proposed.

As can be seen from Fig. 4, FCM2D algorithm has poor image integrity retention and poor noise resistance, and the other two algorithms have better effects. Similarly, it can be seen from Table 3 that the SA and PSNR values of FCM algorithm and the proposed algorithm are higher, and objectively, their segmentation results are better.

Table 3. SA and PSNR of each algorithm.

Method	SA	PSNR
FCM	0.9818	17.4298
FCM2D	0.9383	12.0993
Proposed	0.9819	17.4155

5 Conclusion

In order to improve the robustness and validity in image segmentation, a fuzzy entropy clustering for image segmentation based on potential two-dimensional histogram is proposed. The potential function is used to map an image's two-dimensional gray information to the potential field, and its advantage is to reflect the relationship between each pixels more completely and accurately. The potential values are added into the objective function of the clustering algorithm, then a new objective function and iterative formulas are obtained. Besides, the fuzzy entropy is also incorporated into the objective function to preserve more original image information. Compared with other two fuzzy clustering algorithms, the proposed algorithm has higher accuracy and stronger robustness in noise image segmentation.

References

1. Mahata, N., Kahali, S., Adhikari, S.K., et al.: Local contextual information and gaussian function induced fuzzy clustering algorithm for brain mr image segmentation and intensity inhomogeneity estimation. Appl. Soft Comput. **68**, 586–596 (2018)
2. Liu, J.A.: Fuzzy clustering method for image segmentation based on two-dimensional histogram. Acta Electronica Sin. (9), 40–46 (1992)
3. Hou, X., Wu, C.: Adaptive weighted two-dimensional histogram FCM segmentation algorithm. J. Image Graph. **20**(10), 1331–1339 (2015)
4. Tran, D., Wagner, M.: Fuzzy entropy clustering. In: IEEE International Conference on Fuzzy Systems (2000)
5. Tao, W.: Cognitive physics method for image segmentation. China Water & Power Press, GuangDong (2015)
6. Tao, W., Yi-Fu, J., Rui, H., et al.: Cognitive physics-based method for image edge representation and extraction with uncertainty. Acta Phys. Sin.-Chin. Ed. **62**(6), 675 (2013)
7. Wu, Q., Wu., C.: Fast robust kernel-based fuzzy c-means clustering segmentation. J. Image Graph. **23**(12), 1838–1851 (2018)
8. Wu, Z.D., Xie, W.X., Yu, J.P.: Fuzzy c-means clustering algorithm based on kernel method. In: Computational Intelligence and Multimedia Applications (2003)
9. Advanced Multimedia Processing (AMP) Lab. http://chenlab.ece.cornell.edu/downloa-d-s.html. Accessed 6 May 2019

Fuzzy Applications: Fuzzy Pattern Recognition and Diagnostics

Kernel Circulant Object Tracking Based on Illumination Invariant Features

Zhenhai Wang[1], Bo Xu[2], and Xing Zhang[1(✉)]

[1] School of Information Science and Engineering, Linyi University, Linyi, Shandong, China
lywzhh@163.com, zhangxing_1013@163.com
[2] School of Mechanical and Vehicle Engineering, Linyi University, Linyi, Shandong, China

Abstract. In order to improve accuracy and robustness of object tracking, and to resolve the problem that kernel circulant object tracking is easily affected by illumination changes, this paper presents a kernel circulant object tracking method based on illumination invariant features. Firstly, the locality sensitive histograms of input image are calculated and illumination invariant features are extracted. Then, the precise position of tracking object is obtained by the responded confidence image, which can be quickly computed in frequency domain between input image and template image. Tests of many video sequences demonstrate that the proposed method can track the target accurately and effectively. Comparing to the conditional kernel circulant tracking algorithm, the tracking error using our method decreases 24 pixels and the tracking precision increases 28%. As a result, the proposed algorithm has the better robustness under abruptly variant illumination and pose.

Keywords: Object tracking · Locality sensitive histograms · Illumination invariant features · Kernel circulant algorithm

1 Introduction

Object tracking has become one of the most active research topics in the field of intelligent monitoring, intelligent driving, human-computer interaction, precision guidance and other fields.

Because of the diversity of the apparent information, it is always a hot topic in the field of object tracking algorithm [1]. And the discriminant modeling method has become a hot research topic of object tracking algorithm [2]. The discriminant apparent model transforms the object tracking problem into a classification problem of distinguishing the target and the background. By determining whether the pixel belongs to the target or the background, the confidence map is established and the tracking target is determined by finding the peak of the confidence graph [3–5].

Henriques et al. [6] proposed a new discriminant-based object tracking method using the cyclic matrix structure to perform dense sampling. The similarity comparison between the candidate target and the template is carried out by the kernel function. The Henriques team [7] uses the cyclic matrix to diagonalize the nature of the nuclear cycle

Y. Liu et al. (Eds.): ICNC-FSKD 2019, AISC 1074, pp. 875–882, 2020.
https://doi.org/10.1007/978-3-030-32456-8_94

to further track the process of object tracking, the original pixel can be tracked through the Kernel Correlation Filter (KCF) to achieve, at the same time (Hog) with HOG (Histogram of Oriented Gradient) feature is used to perform object tracking on the kernel correlation filter, and good results are obtained.

Because the gray information of the original pixel is susceptible to noise [8], and the gray level changes drastically with the change of ambient light, it will affect the tracking accuracy, but the literature [6] does not consider this situation.

Recently, a Locality Sensitive Histograms (LSH) extraction method is proposed to extract illumination invariant features. In this paper, the object tracking is performed efficiently using gradient direction histogram and compression sensing. Combining the background weight histogram and Kalman filter, the robust object tracking is realized. In this paper, we propose a kernel-based object tracking algorithm based on illumination invariant feature, which makes the target feature more stable, especially when the illumination is changed violently. Experimental results show that the proposed algorithm improves the tracking accuracy and robustness while maintaining high tracking speed.

2 Kernel Circulant Object Tracking

Object tracking can be seen as a minimized optimization problem.

$$\min_{w,b} \sum_i (f(x_i) - y_i)^2 + \lambda ||\omega||^2 \tag{1}$$

Where, $f(x) = <\omega, x> + b$, x_i is a sample, y_i is a regression target, λ is a regularization parameter. $f(x_i)$ is a linear classifier. We can get,

$$\omega = (X^T X + \lambda I)^{-1} X^T y \tag{2}$$

In order to further improve the performance of classifier, the classification of ridge regression in classical Support Vector Machine is adopted. The sample x is mapped to feature space through $\varphi(x)$. The mapping function $\varphi(x)$ can be indirectly expressed by the kernel function $\kappa(x_i, x_j) = <\varphi(x_i), \varphi(x_j)>$.

Equation (2) can be considered as linear component of $\varphi(x)$, that is $\omega = \sum \alpha_i \varphi(x_i)$. Finding optimal α_i and ω through training samples, and making (1) to reach its minimization.

In the process of sampling, Henriques [6] proposed a method of dense sampling by circulant matrix, searching the surrounding areas of objects, which is equal to filter image by circulant matrix. The structure of circulant matrix is P, the product of P and $x = [x_1, x_2, x_3, \ldots, x_{n-1}, x_n]^T$, shifts x by one element, which is equal to cyclic shift element to the right. Then,

$$X = P^u x = C(x) = \begin{bmatrix} x_1 & x_2 & x_3 & \cdots & x_n \\ x_n & x_1 & x_2 & \cdots & x_{n-1} \\ x_{n-1} & x_n & x_1 & \cdots & x_{n-2} \\ \vdots & \vdots & \vdots & \ddots & \vdots \\ x_2 & x_3 & x_4 & \cdots & x_1 \end{bmatrix} \tag{3}$$

Where $u = 0, \ldots, n-1$.

Circulant matrix means it can be computed through the vector convolution. The fact that the convolution of $x \otimes v$ is equivalent to the product between Circulant matrix $C(x)$ and vector v. Usually the convolution can be obtained by Fourier transform. The convolution in time domain transforms to the element-wise product in the Fourier domain, and then the result of convolution is obtained by Fourier inverse transform.

Henriques [6] further provided that the matrix K with element $K_{ij} = \kappa(P^i x, P^j x)$ is circulant matrix if κ is a unitarily invariant kernel. This theorem provides a useful bridge between the kernel matrix and circulant matrix. The computing of kernel matrix can be obtained by calculating circulant matrix using fast Fourier transform.

We can obtain the solution of classification by applying fast Fourier transform,

$$\alpha = F^{-1}\left(\frac{F(y)}{F(k) + \lambda}\right) \tag{4}$$

The algorithm which Henriques proposed is divided into two parts. First, computing α according to (4) through dense sampling. Then, the confidence map between all samples and the template is obtained by (5). The position of the confidence maximum is the result of tracking object.

$$\hat{y} = F^{-1}(F(\bar{k}) \odot F(\alpha)) \tag{5}$$

Gaussian kernel function in that paper is as follows:

$$k^{gauss} = \exp\left(-\frac{1}{\sigma^2}(\|x\|^2 + \|x'\|^2 - 2F^{-1}(F(x) \odot F^*(x')))\right) \tag{6}$$

3 The Kernel Circulant Tracking Algorithm Based on Illumination Invariant Features

3.1 Illumination Invariant Features of Locality Sensitive

Intensity histograms does not contain spacial information of image [10]. Fragment-based tracking divides the object into several regions [12]. Each region is represented by local histograms, which record statistics within a region of a pixel in an image. For the local histogram, pixel inside a local neighborhood has equal contribution. However, for application such as object tracking, pixels further away from the target center should be weighted less as they more likely contain background information or occluding objects. Hence, their contribution to the histogram should be diminished. He [9] proposed a locality sensitive histogram algorithm to address this problem. The locality sensitive histogram computed at pixel p can be written as,

$$H_p^E(d) = \sum_{q=1}^{W} \alpha^{|p-q|} Q(I_q, d), \quad d = 1, \ldots, D \tag{7}$$

Where $\alpha \hat{I}$ (0, 1) is a parameter controlling the decreasing weight as a pixel moves away from the target center. W is the number of pixels in local section. D is the total number of bins, $Q(I_q, d)$ is zero except when intensity value I_q belongs to bin d.

The probability of intensity value I_p belonging to bin d is the probability of four neighborhood of this pixel belonging to bin d multiplies location weight α. As shows in Fig. 1,

$$\begin{aligned} H_p^{(i,j)}(d) = {} & \alpha_y H_p^{(i-1,j)}(d) + \alpha_y H_p^{(i+1,j)}(d) \\ & + \alpha_x H_p^{(i,j-1)}(d) + \alpha_x H_p^{(i,j+1)}(d) \end{aligned} \tag{8}$$

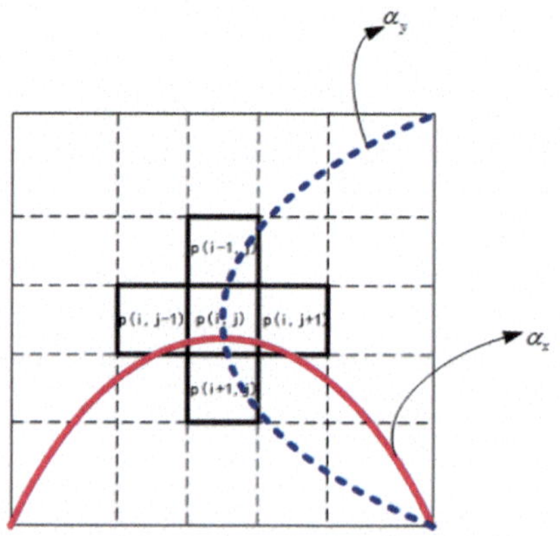

Fig. 1. Illustrating of pixel gray frequency statistics

The statistics of intensity value is computed by iteration respectively in the direction of x and y. In order to overcome the effect of illumination changes, based on LSH, author also proposed a new method for extracting dense illumination invariant features (IIF).

$$L_p = \sum_{d=1}^{D} \exp\left(-\frac{(d - d_p)^2}{2 \max(k, r_p)^2} \right) . H_p^E(d) \tag{9}$$

Where k is a constant, $k = 0.1$, r_p can be approximately represented by kI_p. I_p denotes the intensity values of pixel p. d_p denotes the bin corresponding to the intensity values of pixel p.

3.2 Kernel Circulant Object Tracking Based on Illumination Invariant Features

The model Henriques sets is pixel-level which can distinguish the target from background according to the apparent characteristics trained to some extent. This tracking method based on pixel-level will cause object deviation or drift when the ambient light changes, the target and background are quite similar, the pose varies. Considering the local spatial information, we proposed the kernel circulant object tracking algorithm based on the locality sensitive histogram to solve the object deviation or drift. The algorithm can achieve better tracking performance through calculating image locality sensitive histogram, extracting the illumination invariant features which input the kernel circulant matrix. So the locality sensitive histogram based kernel circulant matrix tracking has the stable tracking performance in conditions of illumination variation dramatically.

The algorithm procedure is as following:

① Initializing the tracking object, selecting the object manually, preprocessing the object with cosine window to eliminate the influence of boundary discontinuity, then calculating the normalized local sensitivity histogram and the illumination invariant features according to (7) and (9) respectively, so generating the initial template, finally computing the weight α of each pixel according to (4).

② Reading a new image and calculating the normalized local sensitivity histogram and the illumination invariant features according to (7) and (9) respectively, then computing the confidence map between the current image and the template and finding the maximum position of the response which is the location of the target.

③ Re-training α in the new position according to (4), and generating the new template and α through the interpolation factor.

④ Going to step ② to process the next frame, up to deal with all frames.

4 Experimental Results and Analysis

4.1 Qualitative Comparison and Analysis

The David videos consisting of 462 frames in which the target moves from the dark region to the bright region. This experiment verifies the adaptability of illumination variation. The results are shown in Fig. 2. The two algorithms can both track the target accurately at the beginning, and the error increases due to the light changing slowly in subsequent frames while adopting the kernel circulant tracking algorithm, the results are shown the 49th and 300th frames in Fig. 2. When tracking the 408th frame, the target drifts and losses while our algorithm can track the target accurately.

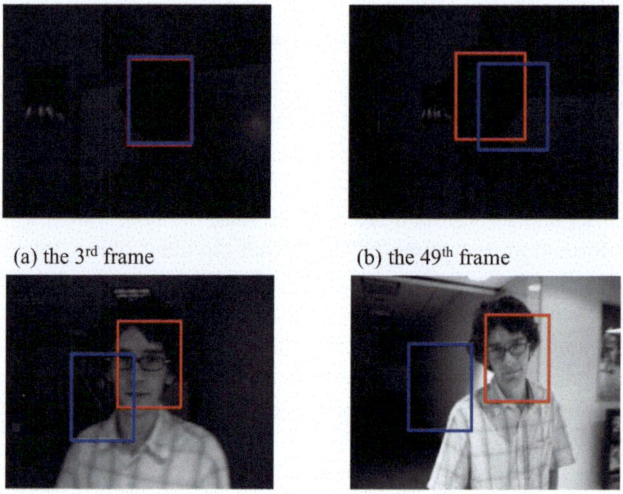

(a) the 3rd frame (b) the 49th frame

(c) the 300th frame (d) the 408th frame

Fig. 2. Tracking results of David Sequences

4.2 Quantitative Comparison and Analysis

Two methods are used to quantify the performance of our algorithm, one is the tracking error which computes the distance between the center point of tracking target and actual target, the other is the tracking precision presented by Babenko et al. which measure the tracking accuracy by the actual position of target hand-drafted. If the distance range within a preset threshold, the tracking result is considered to be a successful tracking, the tracking precision is the percentage of successful tracking frame of video sequence according to the Eq. (10).

$$p = \frac{Num(\text{dis} < \text{threshold})}{Total\_Num} \tag{10}$$

Where, *dis* denotes the distance between the tracking target location and the actual target location hand-drafted; *threshold* denotes the preset threshold which varies from 1 to 50; *Total_Num* denotes the total number of video sequences.

The tracking error curve for David Sequences is shown in Fig. 3(a), the two tracking errors is close before the first 270 frames because of the light changing relatively gentle. After the 270th frame, the errors in the kernel circulant method largen, but the errors in our method keep stable when the lights are gradually increasing. The tracking precision curve for David Sequences is shown in Fig. 3(b), where the horizontal coordinate denotes the threshold that is the distance between the object tracking position and the real target position, and the longitudinal coordinate is accuracy which is the tracking accuracy calculated according to (10). The tracking result is superior to the kernel circulant tracking method proposed by literature [6].

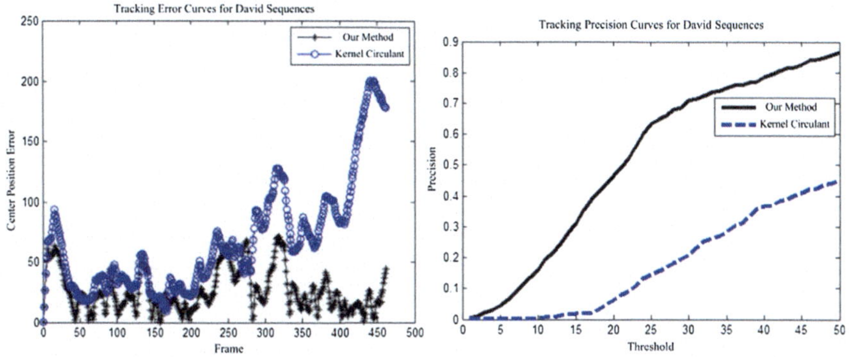

Fig. 3. Comparisons of tracking error and precision for David Sequences

The average tracking error, the average tracking accuracy and the tracking speed of the test video are calculated in Table 1. It can be seen from the table that the average tracking error of our algorithm is much lower than that of the kernel circulant method, and the tracking precision is higher than the kernel circulant method. However, the time consuming increases and the speed decreases because of adopting the local histogram and illumination invariant feature. The track speed arrives the 50 FPS. The tracking accuracy and speed are well balanced to meet the needs of real-time applications.

Table 1. Comparisons of the average tracking performances

Video sequences	Mean error		Mean precision		Mean FPS	
	Our method	Kernel circulant	Our method	Kernel circulant	Our method	Kernel circulant
David	25.69	67.25	51.92%	17.7%	58	141
Girl head	29.2	35.5	64.6%	42.4%	62	172

5 Conclusion

This paper proposes a novel track method combining the illumination invariant features for overcoming the defect of illumination variation sensitivity in the kernel circulant method which only uses the pixel gray feature. Our algorithm not only has the robust to illumination and attitude changing, but also keeps the real-time performance. The object tracking experiments are carried out on David video sequences. The tracking results shows that the average tracking error of our method is about 24 pixels lower than that of the kernel circulant algorithm, and the tracking accuracy is increased by 28%. The key problems in follow-up research are how to solve the problem of long-term occlusion and improve the tracking speed.

Acknowledgments. This research was supported by National Natural Science Foundation of China (No. 61771230).

References

1. Wang, X., Hua, G., Han, T.X.: Discriminative tracking by metric learning. In: Proceedings of European Conference on Computer Vision, Crete, Greece, pp. 200–214 (2010)
2. Kalal, Z., Mikolajczyk, K., Matas, J.: Tracking learning detection. IEEE Trans. Pattern Anal. Mach. Intell. **34**, 1409–1422 (2012)
3. Avida, S.: Ensemble tracking. IEEE Trans. Pattern Anal. Mach. Intell. **29**(2), 261–271 (2007)
4. Zhang, K., Zhang, L., Yang, M.H.: Real-time compressive tracking. In: ECCV, pp. 866–879 (2012)
5. Babenko, B., Yang, M.H., Belongie, S.: Robust object tracking with online multiple instance learning. IEEE Trans. Pattern Anal. Mach. Intell. **33**(8), 1619–1632 (2011)
6. Henriques, J.F., Caseiro, R., Martins, P., et al.: Exploiting the circulant structure of tracking-by-detection with kernels. In: ECCV, pp. 702–715 (2012)
7. Henriques, J.F., Caseiro, R., Martins, P., et al.: High-speed tracking with kernelized correlation filters. IEEE Trans. Pattern Anal. Mach. Intell. **37**(3), 583–596 (2015)
8. Yilmaz, A., Javed, O., Shah, M.: Object tracking: a survey. ACM Comput. Surv. **38**(4), 13–58 (2006)
9. He, S.F., Yang, Q.X., Rynson, W.H.L., et al.: Visual tracking via locality sensitive histograms. In: CVPR, pp. 2427–2434 (2013)
10. Sun, X.Y., Chang, F.L.: Object tracking based on sparse representation of gradient feature. Opt. Precis. Eng. **21**(12), 3191–3197 (2013)
11. Tian, H., Ju, Y.F., Meng, F.K., Li, F.F.: Improved tracking algorithm with background-weighted histogram. J. Image Graph. **20**(1), 0072–0084 (2015)
12. Adam, A., Rivlin, E., Shimshoni, I.: Robust fragments-based tracking using the integral histogram. In: CVPR, pp. 798–805 (2006)

Hybrid Fuzzy CNN Network in the Problem of Medical Images Classification and Diagnostics

Yuriy Zaychenko[1(✉)] and Galib Hamidov[2]

[1] Institute for Applied System Analysis, Igor Sikorsky Kyiv Polytechnic Kyiv, Kiev, Ukraine
zaychenkoyuri@ukr.net
[2] Information Technologies Department, Azershig, Baku, Azerbaijan
galib.hamidov@gmail.com

Abstract. The problem of classification of breast tumors on medical images is considered. For its solution new class of convolutional neural networks-hybrid CNN-FNN network is developed in which convolutional neural network VGG-16 is used as feature extractor while fuzzy neural network NEFClass is used as classifier. Training algorithms of FNN were implemented. The experimental investigations of the suggested hybrid network on the standard data set were carried out and comparison with known results was performed. The problem of data dimensionality reduction is considered and application of PCM method is investigated.

Keywords: Breast tumor classification · FNN NEFclass · hybrid CNN

1 Introduction. State-of-Art Problem Analysis

Nowadays, in practice, at every stage of diagnostics information technologies are utilized. The main goal of medical automated systems are extension of spheres of practical tasks which may be solved with computers aid, raise of level intellectual decision support of doctors in particularly in process of express diagnostics based on processing and analysis of medical images of human tissue obtained by different source (MRT, CT, etc.) [1, 2].

The advantages of medical diagnostics systems are speed, automation and stability of work which make them very comfortable tools for express medical diagnostics. Despite young age of medical informatics which doesn't exceed 30 years information technologies are fast penetrating in various spheres of medicine and health defense (family medicine, insurance medicine, integration in European medical space, etc.)

The latest achievements in images processing technologies and machine learning enable to construct systems of automatic detection and diagnostics that may help pathologists-anatomists to make true diagnosis and accelerate their work.

Classification of images histopathology on different patterns which corresponds to cancer and not-cancer states of tissue is often first rank goal in images analysis systems for automatic cancer diagnostics. Among the different types of cancer breast cancer

Y. Liu et al. (Eds.): ICNC-FSKD 2019, AISC 1074, pp. 883–891, 2020.
https://doi.org/10.1007/978-3-030-32456-8_95

takes the second plag by its occurrence in women. Besides, mortality of it very high as compared with other cancer diseases [4].

Up to date several models and methods were developed for breast cancer detection using various machine learning algorithms. Using such methods and technologies of AI as neuron networks and SVM [4, 5] accuracy of diagnostics from 76% to 94% was attained at data set with 92 images.

The most of last papers refer to the field of breast cancer classification oriented on integer images (WSI) [5–8]. But wide implementation of BIC (breast image classification) and other forms of digital pathology faces with such disturbances as high cost of implementation, insufficient productivity for huge amount of clinic procedures, interior technologic problems and opposition from pathology-anatomists side.

Till now the most of works based on histology breast cancer analysis were performed on not large datasets. Some improvement presents data set with 7909 breast images obtained from 82 patients [9]. In this research the authors estimated various texture descriptors and various classifiers and carried out the experiments with accuracy from 82% to 85%.

The alternative to this approach is the application of Deep learning [2] and convolutional neural networks (CNN) for medical images processing and diagnostics, which is considered and developed in the present research.

It was shown that CNNs are able to overcome the conventional texture descriptors [10, 11]. Besides traditional approach to detection of features based on descriptors demands much efforts and high level knowledge of experts and usually is specific for every task that prevents its direct application for another similar tasks.

Therefore, in our research we suggested and developed hybrid fuzzy CNN medical images classification system in which CNN is utilized to extract informative features of images and FNN NEFClass is applied for classification of detected breast tumors on images in two classes: benign and malignant ones.

The main goal of this work is development and investigation of algorithmic and software tools for fast analysis of breast tissue images, detection of tumors and their classification into two classes: benign or malignant.

This will enable to provide express analysis of images and raise the quality medical diagnostics.

2 Data Set Description

For present investigation data set BreakHis was used, specially created for estimation efficiency of different approaches and methods for medical images of breast tumor diagnostics.

Data set BreakHis [9] contains microscope biopsies from benign and malignant tumors of breast. The images were obtained in clinic research since January 2014 till December 2014. BreakHis consists of 7909 clinically representative microscopic images of breast tumors received from 82 patients with different scale augmentation (40X, 100X, 200X, 400X).

All patients during this period were investigated in R&D medical lab with clinical conclusion of breast cancer were invited to take part in this investigation. All data were anonymized. The patterns are collected by surgery biopsy prepared for histologic research and marked by pathologist-anatomists of R&D lab.

The main goal was to preserve original structure of tissue and molecular composition which allows to observe it with optical microscope.

For investigation all images were split into slides of size 3 mkm. The final conclusion of each case was made by experienced pathologist-anatomist which was confirmed by additional investigation such as immune histology-chemistry (IHC).

The microscope system Olympus BX-50 with augmentation 3.3 connected with digital camera Samsung SCC-131AN, was used for obtaining digitized images of breast tissue. Images were obtained in 3-channels color space True color (24 bits value, 8 bits color channels RGB) with magnification coefficients 40X, 100X, 200X and 400X.

In the Figs. 1, 2, 3 and 4 four images are presented with four magnification coefficients: (a) 40×, (b) 100×, (c) 200×, (d) 400×, obtained from one slide of breast tumor which contains malignant tumor (breast cancer). Separated rectangular (added by hand for illustrative aims) is a region of interest (ROI) which was chosen by pathologist-anatomist. In the Fig. 5 a slide of benign tumor is presented.

Up to date dataset BreakHis consists of 7909 images, divided into benign and malignant tumors.

Table 1 presents the distribution of images [9].

Table 1. Distribution of images by magnification coefficients and class

Magnification	Benign	Malignant	Total
40x	625	1370	1995
100x	644	1437	2081
200x	623	1390	2013
400x	588	1232	1820
Total	2480	5429	7909
Number of patients	24	58	82

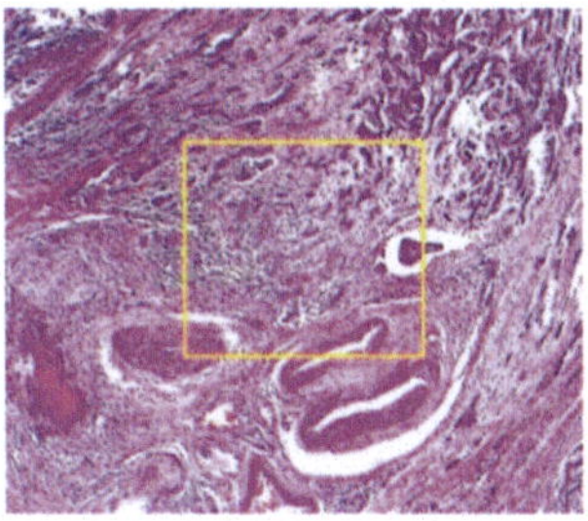

Fig. 1. Slide of malign tumor with tumor magnification 40X

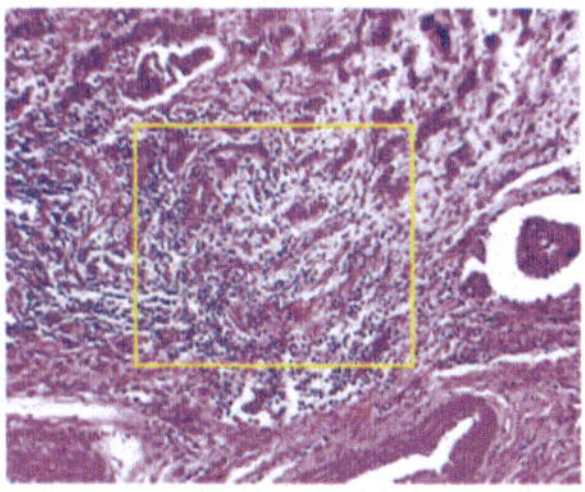

Fig. 2. Slide of malign with magnification 100X.

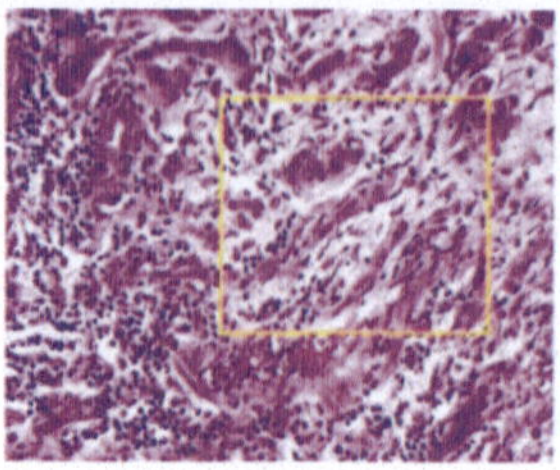

Fig. 3. Slide of malign tumor with with magnification 200X

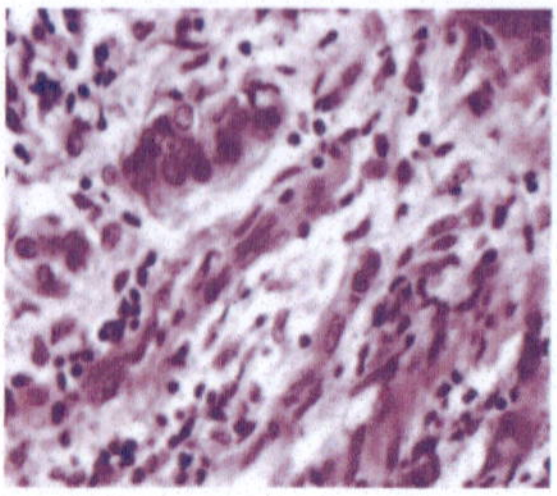

Fig. 4. Slide of malign tumor with magnification 400X

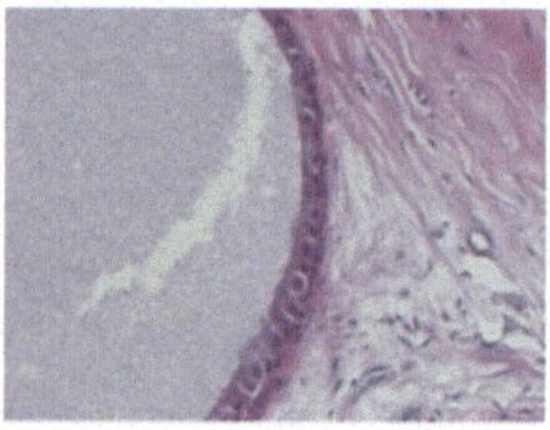

Fig. 5. Slide of benign tumor with magnification 100X

3 Convolutional Neural Networks: Brief Description

A CNN model is a state-of-the-art method that has been largely utilized for image processing. A CNN model has the ability to extract global features in a hierarchical manner that ensures local connectivity as well as the weight-sharing property. It consists of the following layers [10, 11].

Convolutional Layer. The Convolutional layer is considered as the main working ingredient in CNN model and plays a vital determining part of this model. A kernel (filter), which is basically n × n matrix successively goes through all the pixels and extracts the information from them.

Pooling Operation. A CNN model produces a large amount of feature information. To reduce the feature dimensionality a down-sampling method named a pooling operation has been performed. A few pooling operation methods are well known such as Max Pooling Average Pooling, etc. For our analysis, Max Pooling operation was utilized.

Nonlinear Performance. Each layer of the NN produces a linear output and by definition adding two linear functions will also produce another linear output. Due to the linear nature of the output, taking more NN layers will show the same behavior as a single NNlayer. To overcome this drawback, rectifier functions, such as Rectified Linear Unit (ReLU), Leaky ReLU, Tanh, Sigmoid, etc., have been introduced to make the output nonlinear.

Drop-Out. Due to the overtraining of the model, it shows very poor performance on the test data set, which is known as over-fitting. These over-fitting issues have been controlled by removing some of the neurons from the network, known as *Drop-Out*.

Decision Layer. For the classification decision, at the end of a CNN model a decision layer (usually MLP) is introduced. Normally, Softmax layer or SVM layer is introduced for this purpose. This layer contains a normalized exponential function and calculates the loss function for the data classification.

4 CNN Model for Image Classification

In the next Fig. 6 the architecture of CNN VGG-16 is presented which was used in our work as detector of informative features.

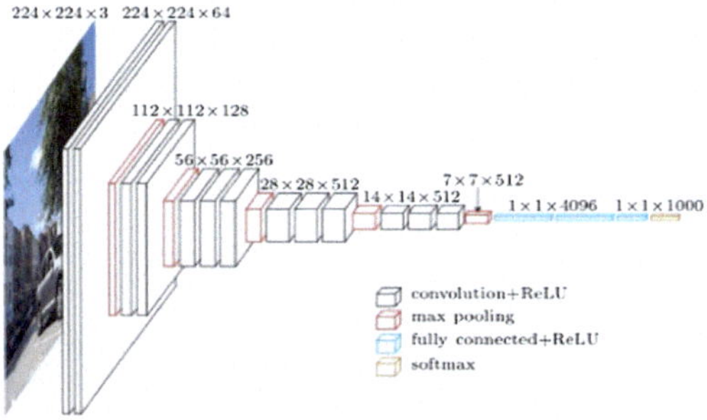

Fig. 6. Convolutional neural network VGG-16

5 Fuzzy Neural Network NEFClass

As the classifier of obtained features it was suggested to use FNN NEFClass. FNN NEFClass was firstly suggested by Nauck and Kruse in [13].

It was modified and developed in [14, 15] (so-called FNN NEFClass M).

The learning algorithms for FNN NEFClass stochastic gradient SG, conjugate gradient descent (CGS) and genetic algorithm were developed and investigated in [15] for the problem of optical images pattern recognition.

FNN NEFCLASS was successfully applied for analysis of medical images of cervix tissue obtained with use of colposcope and diagnostics [16].

The main advantages of FNN NEFClass as a classifier are: possibility to work with incomplete and fuzzy input data; performing fuzzy classification of input patterns (images) using so-called membership functions; speed and high accuracy [14, 15].

The NEFClass model is used for definition of class or category of an input sample (so-called patterns). Patterns are feature vectors $X = (x_1, x_2, \ldots, x_n) \in R^n$ of a certain object, and a class is, respectively, some set in R^n.

A feature of a pattern (sample) is represented by a fuzzy set, and classification is defined by a set of linguistic rules. For each input feature x_i there are q_i fuzzy sets described by membership functions (MF) $\mu_1^i, \ldots, \mu_{q_i}^i$.

Also FNN NEFClass has a **rule base** which contains k fuzzy linguistic rules, denoted as $R_1, \ldots, R_k$. Fuzzy rules which describe data have the following form [16]:

– if x_1 is μ_{1i} and x_2 is μ_{2i} and … and x_n is μ_{ni}, then the sample $(x_1, x_2, \ldots, x_n)$ belongs to a class i, where $\mu_{1i}, \ldots, \mu_{ni}$ are fuzzy sets.

The main task of NEFClass is the definition of these rules and a type of membership functions for fuzzy sets.

FNN NEFClass has three-layer architecture. The first layer is a layer of input neurons which contain the input samples. Activation of neuron of this layer doesn't change input value. The hidden layer contains fuzzy rules neurons R_1, R_2, …, R_k and the third layer consists of output neurons of each class $(C_1, \ldots, C_m)$.

Training algorithm of FNN NEF Class consists of two stages.

In the first stage fuzzy rules base is generated using training sample.

After generating rules base the second training stage starts at which the parameters of fuzzy sets MF are trained. For this goal several training algorithms were developed, incl.: SGD (stochastic gradient descent), conjugate gradient descent; differential evolution; and so-called, baisin hopping algorithm [12].

6 Experimental Investigations and Analysis

The series of experiments were carried out and the results were compared with works of predecessors. In the following Tables 2 and 3 the results of classification (accuracy, %) with different parameters are presented. All sample was divided into training and testing subsamples with ratio 80%/20%.

In the first experiment the number of linguistic variables (terms) and rules were varied that to determine the best parameters values (Table 2).

Table 2. Clasification results of FNN NEFClass

Initial number of fuzzy sets (linguistic terms)/number of rules	40X	100X	200X	400X
2/2	73%	74%	74.2%	73.5%
4/2	75.3%	74.8%	75.7%	75.4%
6/2	78.2%	79%	78.4%	78%
8/2	76%	75.4%	76.5%	75.8%
2,4	75%	74%	73.8%	73%
4/4	78.3%	76.3%	75.7%	75.4%
6/4	82%	83%	82.4%	83.2%
8/4	82.2%	81.5%	81.5%	83.8%
2/6	75.4%	73.8%	74.4%	73.2%
4/6	**90%**	**91%**	**90.5%**	**90%**
6/6	89%	89.7%	90.2%	89.5%
8/6	90.3%	90.5%	92%	91.2%
4/8	89.3%	89.8%	89.7%	89.3%
6/8	89.2%	88%	89.4%	88.4%
8/8	88%	87.2%	87.2%	87%

As it follows from table the best values of parameters for FNN NEFClass are 4 fuzzy sets per variable and 6 rules. For comparison let's present the results of known work obtained with different classifiers for the same problem [9] (see Table 3).

Table 3. Comparison of different classifiers accuracy

Classifier/magnification coefficient	40X	100X	200X	400X
Linear SVM	89%	89%	88%	88%
Polynomial SVM	88%	90%	89%	85%
Random forest	89.18%	88%	87.74%	80%
FNN NEFClass	90%	91%	90.5%	90%

As one can see from the Table 3 FNN NEFClass shows better results than known classifiers: SVM machine and Random forest which were investigated in [9].

In our work for training of FNN NEFClass three algorithms were applied, baisin hopping [12], stochastic gradient descent and differential evolution [13].

Using algorithms baisin hopping and stochastic gradient descent we obtained approximately equal results close to optimal while the training results of differential evolution appeared to be much worse.

It's worth to note that in this problem the number of features extracted by CNN VGG16 was very large – 4096 features. Therefore, it was decided to cut the number of features and reduce dimensionality of classification problem. For this aim principal components method (PCM) was applied. In the Table 4 the results of such reduction are presented.

Table 4. The dependence of total variance on number of components and training time

Number of principal components	Variation	Approximate training time (in hours)
100	0.84058	~2 h
200	0.89736	~3 h
250	0.91232	~4 h
500	0.95486	~9 h

From the Table 4 it follows that the results of reduction with 250 principal components are the most acceptable as the complexity of training increases is approximately proportional to dimension of input data. Due to lack of time the next experiments were performed using data with magnificant factor 100X (2081 images). In the next Table 5 the accuracy of classification using 250 features is presented with various NEFClass parameters.

Table 5. Classification accuracy with 250 features

Number of fuzzy sets/number of rules	100X
4/4	80.64%
4/6	87.24%
4/8	88.18%

In the Table 6 the dependence of classification accuracy versus number of features is presented. One can see from this table that accuracy dropped only by some percent due such features reduction. But by this reduction the training time was cut.

Table 6. Classification accuracy with different number of features

/	100	250	4096
4,4	75.23%	80.64%	86.3%
4/6	83.34%	87.24%	91%
4/8	84.21%	88.18%	89.8%

From this table one can easily see that the accuracy drops with decrease of features number but insignificant only by 3–5% if compare results with 100 and 250 features. For comparison the classification with the full set of features 4096 was performed and it was detected that with decrease features number in 20 times the accuracy falls, in average, only by 3–5%.

7 Conclusions

1. The problem of analysis of breast tumor medical images and classification of detected tumor in two classes: benign and malignant is considered.
2. For pattern recognition of breast tumors new hybrid CNN- FNN network is suggested in which the CNN VGG 16 is used for informative features extraction while FNN NEFClass is used for classification of detected tumors.
3. For training FNN NEFClass algorithms baisin hopping, stochastic gradient descent and differential evolution were suggested and investigated.
4. The experimental investigations of suggested hybrid CNN-FNN network in the problem of classification real images of breast tumors using dataset BreakHis were carried out.
5. The comparison of classification accuracy of the hybrid CNN-FNN network with known results based on use of classification algorithms SVM and Random forest was performed which confirmed the efficiency of the suggested approach.
6. The problem of reducing number of features in medical images classification problem using PCM method was investigated and its efficiency explored.

References

1. Mohan, G., Subashini, M.M.: MRI based medical image analysis: survey on brain tumor grade classification. Biomed. Sig. Process. Control **39**, 139–161 (2018)
2. Ker, J., Wang, L.P., Rao, J., Lim, T.: Deep learning applications in medical image analysis. IEEE Access **6**, 9375–9389 (2017)
3. Boyle, P., Levin, B. (eds.) World Cancer Report 2012. IARC, Lyon (2012). http://www.iarc.fr/en/publications/pdfs-online/wcr/2008/wcr_2012.pdf

4. Lakhani, S.R., Ellis, I.O., Schnitt, S., Tan, P., van de Vijver, M.: WHO classification of tumors of the breast, 4th edn. WHO Press, Lyon (2012)
5. Zhang, Y., Zhang, B., Coenen, F., Lu, W.: Breast cancer diagnosis from biopsy images with highly reliable random subspace classifier ensembles. Mach. Vis. Appl. **24**(7), 1405–1420 (2013)
6. Zhang, Y., Zhang, B., Coenen, F., Xiau, J., Lu, W.: One-class kernel subspace ensemble for medical image classification. EURASIP J. Adv. Sig. Process. **2014**(17), 1–13 (2014)
7. Doyle, S., Agner, S., Madabhushi, A., Feldman, M., Tomaszewski, J.: Automated grading of breast cancer histopathology using spectral clustering with textural and architectural image features. In: Proceedings of the 5th IEEE International Symposium on Biomedical Imaging (ISBI): From Nano to Macro, vol. 61, pp. 496–499. IEEE, May 2008
8. Singh, A., Mansourifar, H., Bilgrami, H., Makkar, N., Shah, T.: Classifying Biological Images Using Pre-trained CNNs. https://docs.google.com/document/d/1H7xVK7nwXcv11CYh7hl5F6pM0m218FQloAXQODP-Hsg/edit?usp=sharing
9. Spanhol, F., Oliveira, L.S., Petitjean, C., Heutte, L.: A dataset for breast cancer histopathological image classification. IEEE Trans. Biomed. Eng. **63**, 1455–1462 (2016)
10. LeCun, Y., Bengio, Y., Hinton, G.: Deep learning. Nature **521**, 436–444 (2015)
11. Krizhevsky, A., Sutskever, I., Hinton, G.E.: Image net classification with deep convolutional neural networks. In: Advances in Neural Information Processing Systems, vol. 25, pp. 1097–1105 (2012)
12. Olson, B., Hashmi, I., Molloy, K., Shehu, A.: Basin hopping as a general and versatile optimization framework for the characterization of biological macromolecules. Adv. Artif. Intell. **2012** (2012). Article ID 674832
13. Nauck, D., Kruse, R.: New learning strategies for NEFCLASS. In: Proceedings of Seventh International Fuzzy Systems Association World Congress, IFSA 1997, vol. IV, pp. 50–55. Academia Prague (1997)
14. Zaychenko, Y.P., Sevaee, F., Matsak, A.V.: Fuzzy neural networks for economic data classification. Vestnik Natl. Tech. Univ. Ukraine "KPI" Sect. "Informatic, Control Comput. Eng. **42**, 121–133 (2004)
15. Zaychenko, Y.P., Petrosyuk I.M., Jaroshenko, M.S.: The investigations of fuzzy neural networks in the problems of electro-optical images recognition. Syst. Res. Inf. Technol. (4), 61–76 (2009)
16. Zaychenko, Y., Huskova, V.: Recognition of objects on optical images in medical diagnostics using fuzzy neural network NEFClass. Int. J. Inf. Models Anal. **4**(1), 13–22 (2015)

Fuzzy Applications: Fuzzy Optimization, Design, and Modeling

The Assessment of Airline Service Performance with Dependent Evaluation Criteria by Generalized QFD and SAW Under Interval-Valued Fuzzy Environment

Yu-Jie Wang[1], Li-Jen Liu[2], and Tzeu-Chen Han[1](✉)

[1] Department of Shipping and Transportation Management, National Penghu University of Science and Technology, Penghu 880, Taiwan, Republic of China
tchan@gms.gmail.com.tw

[2] Dinos International Corporation, Taipei 104, Taiwan, Republic of China

Abstract. For airlines, service is regarded as an essential item in their enterprises, and thus they emphasize service performance on management. Due to varied messages' imprecision and vagueness, the assessment of airline service performance is a fuzzy multi-criteria decision-making (FMCDM) problem for management. In FMCDM problems, classical multi-criteria decision-making (MCDM) methods, including simple additive weighting (SAW), have been extended into FMCDM methods to encompass imprecise and vague messages. The generalizations were first used in FMCDM with independent evaluation criteria, and then FMCDM could be further associated with quality function deployment (QFD) to resolve the tie of the dependent evaluation criteria. Alternative ratings and criteria weights of FMCDM were commonly presented by general (i.e., triangular or trapezoidal) fuzzy numbers. Recently, FMCDM with independent evaluation criteria under an interval-valued fuzzy environment was proposed; however, FMCDM with dependent evaluation criteria under the environment has scarcely been mentioned for high computation difficulty. Moreover, QFD has been generalized under a general fuzzy environment but not an interval-valued fuzzy environment. However, interval-valued fuzzy numbers can present more messages than triangular or trapezoidal fuzzy numbers. Additionally, the assessment of airline service performance using several criteria is not only an FMCDM problem but also a problem with dependent evaluation criteria. In this paper, we generalize QFD and SAW under an interval-valued fuzzy environment for the assessment of airline service performance with dependent evaluation criteria for obtaining more messages. By the association of QFD and SAW, the computation tie of the dependent evaluation criteria corresponding to the interval-valued fuzzy numbers is resolved and more messages are gained for FMCDM.

Keywords: Airline service performance · Dependent evaluation criteria · Interval-valued fuzzy environment

Y. Liu et al. (Eds.): ICNC-FSKD 2019, AISC 1074, pp. 895–903, 2020.
https://doi.org/10.1007/978-3-030-32456-8_96

1 Introduction

Generally, decision-making using evaluation criteria belongs to multi-criteria decision-making (MCDM) [1, 2]. In some MCDM problems [3, 4], alternative ratings and evaluation criteria weights assessed with imprecision or vagueness are fuzzy multi-criteria decision-making (FMCDM) problems [5] because these ratings and criteria weights are expressed by linguistic terms [6, 7] and then converted into fuzzy numbers [8]. Service is a critical and essential item for airlines, and thus they often emphasize service performance. Practically, the assessment of airline service performance using several criteria is an FMCDM problem [9] for management due to varied messages' imprecision and vagueness. Generally, FMCDM problems [10], including related assessment of airline service performance using independent criteria [11], have been widely investigated in the past. In contrast, problems using dependent criteria have been scarcely mentioned due to their computation complexity. Recently, more and more FMCDM problems with dependent criteria have been considered based on decision-making requirements. In order to resolve such complicated computations, quality function deployment (QFD) [12, 13] has been used in the aggregation of experts' and customers' opinions to yield dependent criteria weights. In QFD [14], customer requirements represent customer opinions and technical solutions stand for expert opinions. The customer requirement priorities in QFD [15] are presented in an important level matrix and the relationships between customer requirements and technical solutions in QFD are expressed in a relation matrix. Sometimes, entries in the important level matrix and relation matrix are fuzzy numbers that must be processed in fuzzy quality function deployment (FQFD) [16]. Due to the characteristics of fuzzy numbers, FQFD [17] is complicated, and it is difficult to combine the important level matrix with the relation matrix.

Recently, some approaches have proposed FMCDM with independent evaluation criteria under an interval-valued fuzzy environment [18, 19]. However, FMCDM using dependent evaluation criteria for interval-valued fuzzy numbers [20–22] has scarcely been mentioned due to additional computations about dependency. SAW [23], being a traditional MCDM method, is regarded as the simplest of the MCDM methods. To extend SAW under a fuzzy environment [10], fuzzy operations including fuzzy multiplication, addition, and ranking are crucial. The characteristics of pooled fuzzy numbers [24, 25], which are formed by multiplying two fuzzy numbers cause the complexity and difficulty [17] of related computations for generalized SAW under a fuzzy environment, as described by Wang [10]. In fact, the critical issue of fuzzy generalization for SAW is similar to the tie of extending QFD under a fuzzy environment. In addition, the assessment of airline service performance using several criteria is not only an FMCDM problem but also a problem with dependent evaluation criteria for varied messages of imprecision and vagueness. Due to the similar ties of QFD and SAW generalized under a fuzzy environment composed of interval-valued fuzzy numbers [26, 27], we combine QFD with SAW for the assessment of airline service performance using dependent evaluation criteria to grasp more information based on an extended fuzzy preference relation, which was revised from Lee's [28, 29] extended fuzzy preference relation and was an improvement of fuzzy preference relation [30, 31], thereby satisfying the total ordering relation [32].

For the sake of clarity, related concepts of interval-valued fuzzy numbers are described in Sect. 2. In Sect. 3, generalized QFD and SAW under an interval-valued fuzzy environment for FMCDM with dependent evaluation criteria is proposed to solve decision-making problems. An illustrative example about the assessment of airline service performance on dependent evaluation criteria is expressed in Sect. 4. Finally, conclusions are shown in Sect. 5.

2 Preliminaries

In this section, important mathematical notions [8] are presented as follows.

Definition 2.1. As $A = ((a_l^U, a_l^L), a_h, (a_u^L, a_u^U))$ and $B = ((b_l^U, b_l^L), b_h, (b_u^L, b_u^U))$ are two triangular interval-valued fuzzy numbers, the extended fuzzy preference relation P^* with $\mu_{p*}(A, B)$ is yielded as $\frac{(a_l^U - b_u^U) + (a_l^L - b_u^L) + 4(a_h - b_h) + (a_u^L - b_l^L) + (a_u^U - b_l^U)}{2}$.

Definition 2.2. Let $A= ((a_l^U, a_l^L), a_h, (a_r^L, a_r^U))$ and $B= ((b_l^U, b_l^L), b_h, (b_r^L, b_r^U))$ denote two triangular interval-valued fuzzy numbers. The addition $\oplus$ [33] of A and B is $A \oplus B = ((a_l^U + b_l^U, a_l^L + b_l^L), a_h + b_h, (a_r^L + b_r^L, a_r^U + b_r^U))$, and the multiplication $\otimes$[33] of triangular interval-valued fuzzy number A and real number v is $v \otimes A = ((va_l^U, va_l^L), va_h, (va_r^L, va_r^U))$.

3 Generalized QFD and SAW Under Interval-Valued Fuzzy Environment

Let $C_1, C_2, \ldots, C_n$ be the dependent evaluation criteria and $W_1, W_2, \ldots, W_n$ be the related weights of $C_1, C_2, \ldots, C_n$. WD is then assumed to be a fuzzy important level matrix composed of $WD_1, WD_2, \ldots, WD_t$. By QFD, $W_1, W_2, \ldots, W_n$ are yielded by associating the above important level matrix of customer requirements with a relation matrix. The relation matrix between customer requirements over technical solutions is employed by f experts. Herein, R_{kjv}, assessed by the v th expert, is the fuzzy relationship strength rating for customer requirement D_k over technical solution C_j for $k = 1, 2, \ldots, t$; $j = 1, 2, \ldots, n$; and $v = 1, 2, \ldots, f$. R_{kj} through f experts indicates the relationship strength mean rating for customer requirement D_k over technical solution C_j for $k = 1, 2, \ldots, t$; $j = 1, 2, \ldots, n$. Thus, $R_{kj} = (1/f) \otimes (R_{kj1} \oplus R_{kj2} \oplus \ldots \oplus R_{kjf})$, $k = 1, 2, \ldots, t$; $j = 1, 2, \ldots, n$. R is assumed to be the relation matrix consisting of the relationship strength mean ratings $R_{11}, R_{12}, \ldots, R_{tn}$ for customer requirements over technical solutions. In addition, W is derived by QFD into a weight matrix consisting of criteria weights $W_1, W_2, \ldots, W_n$. Herein, $W_j = (1/t) \otimes (WD_1 \otimes R_{1j} \oplus WD_2 \otimes R_{2j} \oplus \ldots \oplus WD_t \otimes R_{tj})$ for $j = 1, 2, \ldots, n$.

Obviously, multiplying WD_k with R_{kj} forms a pooled fuzzy number instead of a triangular interval-valued fuzzy number. The computation of pooled fuzzy numbers is too difficult to derive. To resolve the computations of pooled fuzzy numbers, the extended fuzzy preference relations of the fuzzy important levels over zero will substitute for the fuzzy important levels of customer requirements. An extended fuzzy preference relation with the membership function $\mu_{P*}(WD_k, 0)$ is yielded to represent the preference degree of WD_k over 0, where $k = 1, 2, \ldots, t$. Let W' be an adjusted fuzzy weight matrix consisting of adjusted fuzzy weights $W'_1, W'_2, \ldots, W'_n$ for dependent criteria, where $W'_j = (1/t) \otimes (\mu_{P*}(WD_1, 0) \otimes R_{1j} \oplus \mu_{P*}(WD_2, 0) \otimes R_{2j} \oplus \ldots \oplus \mu_{P*}$ $(WD_t, 0) \otimes R_{tj})$ for $j = 1, 2, \ldots, n$.

Assume $\mu_{P*}(W'_1, 0)$, $\mu_{P*}(W'_2, 0)$, …, $\mu_{P*}(W'_n, 0)$ to be the representation values of $W_1, W_2, \ldots, W_n$ and can be used in interval-valued FMCDM with dependent evaluation criteria. G_{ijv} is the evaluation rating assessed by the v th expert for A_i on C_j, where $v = 1, 2, \ldots, f$. Let G_{ij} be the evaluation mean rating of A_i on C_j, where $i = 1, 2, \ldots, m$ and $j = 1, 2, \ldots, n$. Therefore, $G_{ij} = (1/f) \otimes (G_{ij1} \oplus G_{ij2} \oplus \ldots \oplus G_{ijf})$, $i = 1, 2, \ldots, m$; $j = 1, 2, \ldots, n$. Additionally, let $\tilde{G}_{ij}$ indicate the normalized value of G_{ij} for $i = 1, 2, \ldots, m; j = 1, 2, \ldots, n$. Herein, $\tilde{G}_{ij} = G_{ij}$ as G_{ij} is yielded from linguistic [6, 7] terms and converted into an interval-valued fuzzy number in the interval [0,1].

Let $\tilde{A}$ be the fuzzy performance matrix composed of fuzzy performance indices $\tilde{A}_1, \tilde{A}_2, \ldots, \tilde{A}_m$. With $\mu_{P*}(W'_j, 0)$ and $\tilde{G}_{ij}$ for $j = 1, 2, \ldots, n$, $\tilde{A}_i$ derived by SAW denotes the fuzzy performance index of alternative A_i for $i = 1, 2, \ldots, m$. Therefore, $\tilde{A}_i = (\tilde{G}_{i1} \otimes \mu_{P^*}(W'_1, 0)) \oplus (\tilde{G}_{i2} \otimes \mu_{P^*}(W'_2, 0)) \oplus \ldots \oplus (\tilde{G}_{in} \otimes \mu_{P^*}(W'_n, 0))$ for $i = 1, 2, \ldots, m$. $\tilde{A}_i$ is a triangular interval-valued fuzzy number because $\mu_{P*}(W'_j, 0)$ is a crisp value and $\tilde{G}_{ij}$ is a triangular interval-valued fuzzy number for $i = 1, 2, \ldots, m$ and $j = 1, 2, \ldots, n$. Through the computation of the extended fuzzy preference relation P^*, these alternatives are ranked according to the representations $\mu_{P^*}(\tilde{A}_1, 0)$, $\mu_{P^*}(\tilde{A}_2, 0)$,…, $\mu_{P^*}(\tilde{A}_m, 0)$ that are respectively the preference relation values of fuzzy performance indices over zero of the alternatives' performance indices. The bigger the representing value of the performance index is, the better the alternative. Therefore, the alternatives' ranking order is determined when their performance indices are yielded. The optimal m alternative is thus found, and the interval-valued FMCDM with dependent criteria is finished.

4 The Assessment of Airline Service Performance Using Dependent Criteria

In an example, three Taiwanese airlines assessed their service performance through their customers using dependent evaluation criteria. Regarding the performance problem, convenience, efficiency, and security stood for three customer requirements, and the WD_1, WD_2, WD_3 evaluated by the customers represented the fuzzy importance levels. These customers assessed importance levels through a linguistic term set (very low (VL), low (L), medium (M), high (H), very high (VH)). These linguistic terms and related fuzzy numbers are displayed in Table 1. Additionally, the linguistic assessments employed by the customers were integrated and the fuzzy importance mean levels of the customer requirements were derived as shown in Table 2.

By the extended fuzzy preference relation P^*, the preference relation values of the fuzzy importance levels for customer requirements over zero according to the entries of Table 2 were computed, as shown in Table 3.

Table 1. Linguistic evaluation terms and corresponding fuzzy numbers

Linguistic terms	Fuzzy numbers
Very low (VL)	((0,0), 0, (0.1,0.2))
Low (L)	((0.1,0.2), 0.3, (0.4,0.5))
Medium (M)	((0.3,0.4), 0.5, (0.6,0.7))
High (H)	((0.5,0.6), 0.7, (0.8, 0.9))
Very high (VH)	((0.8,0.9), 1, (1, 1))

Table 2. Fuzzy importance mean levels for customer requirements

Customer requirements	Convenience	Efficiency	Security
Fuzzy importance levels	((0.597,0.697), 0.797, (0.853,0.910))	((0.433,0.527), 0.620, (0.700,0.780))	((0.550,0.640), 0.730, (0.790,0.850))

Table 3. Preference relation values of fuzzy importance levels over zero

Customer requirements	Convenience	Efficiency	Security
Preference relation values	3.122	2.460	2.875

Six technical solutions, including timeliness, facility, supply, reliability, safety, and personnel were deemed as dependent evaluation criteria, and fuzzy weights $W_1, W_2, \ldots W_6$ were derived in the evaluation problem. Moreover, linguistic relationship strength assessments of the three customer requirements for six technical solutions were integrated and the relationship strength mean ratings were aggregated, as shown in Table 4.

Table 4. Relation matrix of three customer requirements for six technical solutions

	Timeliness	Facility	Supply
Convenience	((0.63,0.73), 0.83, (0.88,0.93))	((0.47,0.57), 0.67, (0.74,0.81))	((0.42,0.51), 0.60, (0.67,0.74))
Efficiency	((0.57,0.67), 0.77, (0.82,0.87))	((0.26,0.34), 0.42, (0.52,0.62))	((0.28,0.37), 0.46, (0.55,0.64))
Security	((0.29,0.36), 0.43, (0.53,0.63))	((0.51,0.60), 0.69, (0.75,0.81))	((0.62,0.72), 0.82, (0.86,0.90))

Through the entries in Tables 3 and 4, an adjusted weight matrix consisting of six adjusted weights for the dependent evaluation criteria was derived, as shown in Table 5.

By P^*, the preference relation values of the adjusted weights over zero according to the entries of Table 5 were yielded, as shown in Table 6, and representations of the dependent evaluation criteria weights were obtained.

Table 5. Adjusted weight matrix

W_1'	W_2'	W_3'
((1.401,1.654), 1.907, (2.096,2.285))	((1.191,1.447), 1.703, (1.915,2.128))	((1.261,1.524), 1.787, (1.972,2.157))
W_4'	W_5'	W_6'
((1.281,1.526), 1.772, (1.964,2.156))	((1.409,1.658), 1.907, (2.069,2.231))	((1.621,1.894), 2.168, (2.327,2.487))

Table 6. Preference relation values of adjusted weights over zero

Technical solutions	Timeliness	Facility	Supply	Reliability	Safety	Personnel
Preference relation values	7.533	6.746	7.032	7.007	7.498	8.501

The fuzzy importance mean levels of the customer requirements and the fuzzy relationship strength mean ratings of the customer requirements for technical solutions were merged into the preference relation values of evaluation criteria weights by QFD, and the fuzzy performance matrix composed of the fuzzy performance mean ratings for all evaluation criteria was as shown in Table 7.

Table 7. Fuzzy performance matrix for airlines on all criteria

Airlines	Timeliness	Facility	Supply
Airline 1	((0.26,0.34),0.42, (0.52,0.62))	((0.53,0.63),0.73, (0.80,0.87))	((0.24,0.32),0.40, (0.48,0.56))
Airline 2	((0.58,0.67),0.76, (0.81,0.86))	((0.56,0.66),0.76, (0.82,0.88))	((0.42,0.51),0.60, (0.67,0.74))
Airline 3	((0.48,0.57),0.66, (0.73,0.80))	((0.49,0.59),0.69, (0.76,0.83))	((0.46,0.55),0.64, (0.71,0.78))
Airlines	Reliability	Safety	Personnel
Airline 1	((0.24,0.32),0.40, (0.50,0.60))	((0.26,0.33),0.40, (0.49,0.58))	((0.35,0.45),0.55, (0.64,0.73))
Airline 2	((0.38,0.47),0.56, (0.65,0.74))	((0.41,0.51),0.61, (0.68,0.75))	((0.55,0.65),0.75, (0.80,0.85))
Airline 3	((0.50,0.60),0.70, (0.78,0.86))	((0.55,0.65),0.75, (0.80,0.85))	((0.41,0.50),0.59, (0.67,0.75))

Based on the entries in Tables 6 and 7, the fuzzy performance indices of the three airlines were yielded by SAW in the following.

Using P^*, the representation values of the three airlines' performance indices through the entries in Table 8 were derived as follows. The representation values of the three airlines' performance indices were: Airline 1 (85.745), Airline 2 (117.736), and Airline 3 (117.043). Obviously, Airline 2 was the optimal of the three airlines for service performance.

Table 8. Fuzzy performance indices of the three airlines

Airlines	Airline 1	Airline 2	Airline 3
Performance indices	((13.828,17.603), 21.379, (25.307,29.236))	((21.513,25.729), 29.945, (32.799,35.653))	((21.269,25.470), 29.671, (32.778,35.886))

Besides, we also compared the applied method with other methods to demonstrate the application feasibility. Lee, Chung, Lee, Gan, and Chou [21] translated interval-valued fuzzy numbers into triangular fuzzy numbers and then derived strength and weakness matrices [28, 29] to solve decision-making problems. Practically, Lee et al.'s [21] method has been utilized in triangular fuzzy numbers because interval-valued fuzzy numbers are converted into triangular fuzzy ones on computations. Additionally, deriving the strength and weakness matrices requires pairwise comparisons of fuzzy numbers. It is obvious that our application is more reasonable than Lee et al.'s [21] method, as the proposed method is directly used in interval-valued fuzzy numbers, whereas Lee et al.'s approach is not. Moreover, Kuo and Linag [33] combined VIKOR [34] with interval-valued fuzzy numbers into interval-valued VIKOR to solve decision-making problems. Interval-valued VIKOR is viewed as extending VIKOR under an interval-valued fuzzy environment. In fact, Kuo and Linag's computations and case study included the normalized Euclidean distance [5], hierarchical structures [35], and VIKOR. The association of several techniques causes the complexity of Kuo and Linag's interval-valued VIKOR, and thus they proposed four related appendices to explain the computations. On the contrary, our interval-valued FMCDM based on QFD and SAW is easily realized. Thus, the proposed application is simpler than Kuo and Linag's interval-valued VIKOR as well. Not only can it be utilized in FMCDM with dependent criteria, as indicated as the illustrative example, the application can also be used in FMCDM with independent criteria if the computation procedure utilizes the derived criteria weights.

5 Conclusions

In this paper, airline service performance with dependent evaluation criteria was assessed by generalized QFD and SAW under an interval-valued fuzzy environment. According to the above computations in the previous section, QFD through an extended fuzzy preference relation was used to yield crucial representation values of dependent criteria weights for the assessment of airline service performance. The representative values of the dependent criteria weights and performance ratings of all criteria for an airline defined by SAW were multiplied and summarized into a performance index. Under the extended fuzzy preference relation, multiplying two interval-valued fuzzy numbers into a pooled fuzzy number is unnecessary for generalized QFD and SAW. The airlines' fuzzy performance indices were converted into airlines' representative values of temselves. Obviously, the assessment of airline service performance with dependent evaluation criteria through generalized QFD and SAW under an interval-valued fuzzy environment was easy and reasonable due to the extended fuzzy preference relation. Additionally, the interval-valued FMCDM with dependent evaluation criteria provided the preference relation values of the evaluation criteria ratings over zero, in addition to the representative values of the airlines' performance indices so that decision-makers, through generalized QFD and SAW under an interval-valued fuzzy environment, may evaluate airline performance problem from varied perspectives.

Acknowledgements. This research work was partially supported by the Ministry of Science and Technology of the Republic of China under Grant No. MOST 106-2410-H-346-002-.

References

1. Chen, S.J., Hwang, C.L.: Fuzzy multiple attribute decision making methods and application. Lecture Notes in Economics and Mathematical Systems. Springer, New York (1992)
2. Hwang, C.L., Yoon, K.: Multiple Attribute Decision Making: Methods and Application. Springer, New York (1981)
3. Bellman, R.E., Zadeh, L.A.: Decision-making in a fuzzy environment. Manag. Sci. **17**(4), 141–164 (1970)
4. Wang, Y.J.: Applying FQFD and utility representative functions under fuzzy environment for FMCDM. Test. Eval. **44**(4), 1776–1790 (2016)
5. Chen, C.T.: Extensions to the TOPSIS for group decision-making under fuzzy environment. Fuzzy Sets Syst. **114**(1), 1–9 (2000)
6. Delgado, M., Verdegay, J.L., Vila, M.A.: Linguistic decision-making models. Int. J. Intell. Syst. **7**(5), 479–492 (1992)
7. Herrera, F., Herrera-Viedma, E., Verdegay, J.L.: A model of consensus in group decision making under linguistic assessments. Fuzzy Sets Syst. **78**(1), 73–87 (1996)
8. Zadeh, L.A.: Fuzzy sets. Inf. Control **8**(3), 338–353 (1965)
9. Liang, G.S., Wang, M.J.: A fuzzy multi-criteria decision-making method for facility site selection. Int. J. Prod. Res. **29**(11), 2313–2330 (1991)
10. Wang, Y.J.: A fuzzy multi-criteria decision-making model based on simple additive weighting method and relative preference relation. Appl. Soft Comput. **30**, 412–420 (2015)
11. Wang, Y.J., Kao, C.S.: A fuzzy multi-criteria group decision-making model for the financial performance evaluation of airlines. In: The 6th International Conference on Fuzzy Systems and Knowledge Discovery (IEEE FSKD 2009), Tianjin, China (2009)
12. Chan, L.K., Wu, M.L.: Quality function deployment: a literature review. Eur. J. Oper. Res. **143**(3), 463–497 (2002)
13. Lowe, A., Ridgway, K., Atkinson, H.: QFD in new production technology evaluation. Int. J. Prod. Econ. **67**(2), 103–112 (2000)

14. Matook, S., Indulska, M.: Improving the quality of process reference models: a quality function deployment-based approach. Decis. Support Syst. **47**(1), 60–71 (2009)
15. Partovi, F.Y., Corredoira, R.A.: Quality function deployment for the good of soccer. Eur. J. Oper. Res. **137**(3), 642–656 (2002)
16. Wang, Y.J.: A criteria weighting approach by combining fuzzy quality function deployment with relative preference relation. Appl. Soft Comput. **14**, 419–430 (2014)
17. Liang, G.S.: Applying fuzzy quality function deployment to identify service management requirements. Qual. Quan. **44**(1), 47–57 (2010)
18. Joshi, D., Kumar, S.: Interval-valued intuitionistic hesitant fuzzy Choquet integral based TOPSIS method for multi-criteria group decision making. Eur. J. Oper. Res. **248**(1), 183–191 (2016)
19. Mokhtarian, M.N.: A note on "Extension of fuzzy TOPSIS method based on interval-valued fuzzy sets". Appl. Soft Comput. **26**, 513–514 (2015)
20. Figueroa-García, J.C., Mehra, A., Chandra, S.: Optimal solutions for group matrix games involving interval-valued Fuzzy Numbers. Fuzzy Sets Syst. **362**(1), 55–70 (2019)
21. Lee, C.S., Chung, C.C., Lee, H.S., Gan, G.Y., Chou, M.T.: An interval-valued fuzzy number approach for supplier selection. Marine Sci. Technol. **24**(3), 384–389 (2016)
22. Sen, S., Patra, K., Mondal, S.K.: Fuzzy risk analysis in familial breast cancer using a similarity measure of interval-valued fuzzy numbers. Pac. Sci. Rev. A: Nat. Sci. Eng. **18**(3), 203–221 (2016)
23. Churchman, C.W., Ackoff, R.J., Amoff, E.L.: Introduction to Operation Research. Wiley, New York (1957)
24. Liang, G.S.: Fuzzy MCDM based on ideal and anti-ideal concepts. Eur. J. Oper. Res. **112**(3), 682–691 (1999)
25. Raj, P.A., Kumar, D.N.: Ranking alternatives with fuzzy weights using maximizing set and minimizing set. Fuzzy Sets Syst. **105**(3), 365–375 (1999)
26. Gorzalczany, M.B.: A method of inference in approximate reasoning based on interval-valued fuzzy sets. Fuzzy Sets Syst. **21**(1), 1–17 (1987)
27. Yao, J.S., Lin, F.T.: Constructing a fuzzy flow-shop sequencing model based on statistical data. Int. J. Approximate Reason. **29**(3), 215–234 (2002)
28. Lee, H.S.: A fuzzy multi-criteria decision making model for the selection of the distribution center. Lecture Notes in Artificial Intelligence, vol. 3612, pp. 1290–1299 (2005)
29. Lee, H.S.: On fuzzy preference relation in group decision making. Int. J. Comput. Math. **82**, 133–140 (2005)
30. Kacprzyk, J., Fedrizzi, M., Nurmi, H.: Group decision making and consensus under fuzzy preferences and fuzzy majority. Fuzzy Sets Syst. **49**(1), 21–31 (1992)
31. Tanino, T.: Fuzzy preference in group decision making. Fuzzy Sets Syst. **12**(2), 117–131 (1984)
32. Epp, S.S.: Discrete Mathematics with Applications, Wadsworth, California (1990)
33. Kuo, M.S., Liang, G.S.: A soft computing method of performance evaluation with MCDM based on interval-valued fuzzy numbers. Appl. Soft Comput. **12**, 476–485 (2012)
34. Sanayei, A., Farid Mousavi, S., Yazdankhah, A.: Group decision making process for supplier selection with VIKOR under fuzzy environment. Expert Syst. Appl. **37**(1), 24–30 (2010)
35. Saaty, T.L.: The Analytic Hierarchy Process. McGraw-Hill, New York (1980)

Construction of Multivariable Fuzzy Time Series Model Based on Multidimensional Information Distribution Technology

Ye Xue(✉), Xiaoxiao Li, and Hengchun Fu

College of Economics and Management, Taiyuan University of Technology, Taiyuan 030024, Shanxi, China
xueye0412@126.com

Abstract. In order to enhance the forecast accuracy of multivariable fuzzy time series model for a small sample, a new multivariable fuzzy time series forecasting model was built based on the multidimensional information distribution technology. Then as an example, three variables (the time series data of total energy consumption, per capita GDP and SO_2 emissions from 2001 to 2017 in China) were selected for the case analysis, which was used to verify the feasibility and to discuss the influence of the variation of fuzziness on the forecast accuracy of the model. Furthermore, a comparative analysis with the Markov model is made. The results show that (1) the suggested model can make up for the defects of small sample; (2) the predictive accuracy increases with the decrease of fuzziness; (3) the proposed model has higher forecast accuracy than the Markov model in forecasting SO_2 emissions.

Keywords: Multidimensional information distribution · Fuzzy approximate reasoning · Multivariable fuzzy time series · Emission of SO_2

1 Introduction

Through the construction and prediction of time series model, the development status of objective phenomena can be described, and the future development trend can also be evaluated. In order to achieve these two goals, many scholars have proposed different models, such as Autoregressive model, Autoregressive Moving Average model [1] and Vector Autoregressive model [2]. At the same time, many research methods have been developed, such as the Box-Jenkins [1], Support Vector Machine [3] and Neural Network [4]. Adapting to the characteristics of real-world data, these models have solved some problems existing in time series especially under the environment of big data. However, in practice, there are still a lot of data not easily obtained, and most of them have the problem of small samples. In addition, due to the increased complexity of the objects studied, the observation sequences are generally multivariable time series. To avoid the model causality determination deviation or forecast errors that might result from the use of traditional models, it is of great theoretical and practical significance to study multivariable fuzzy time series with small samples.

Y. Liu et al. (Eds.): ICNC-FSKD 2019, AISC 1074, pp. 904–912, 2020.
https://doi.org/10.1007/978-3-030-32456-8_97

After Song [5] put forward the concept of fuzzy time series, scholars began to study the theory and application of fuzzy time series. Avazbeigi [6] constructed a multi-high-order fuzzy time series model based on the tabu search algorithm. Xue [7] used the information distribution technology to construct a new one-dimensional fuzzy time series forecasting model. Shuai [8] improved the weighted fuzzy C-means clustering algorithm and analyzed the total number of faults in the 100 engines lifetime collected by the integrated support information system. However, we notice that authors in the study of fuzzy time series mainly focus on univariate or multivariable large sample cases. In view of this, this paper proposes a multivariable fuzzy time series model with the help of the information distribution theory, which can efficiently extract the information of small sample data. The rest of the paper is organized as follows. Section 2 introduces the basic principle of information distribution briefly. Section 3 takes the three variables as an example to present the construction process of the multi-variable fuzzy time series model. In Sect. 4, the proposed model is used to predict SO_2 emissions in China, and the results are compared with those of Markov model under different fuzziness. Finally, conclusion and discussion are presented in Sect. 5.

2 Information Distribution

2.1 One-Dimensional Information Distribution

Definition 1 [9]. Let $X = \{x_1, x_2, \cdots x_i \cdots, x_n\}$ be a one-dimensional random sample, and $U = \{u_1, u_2, \cdots, u_m\}$ be the chosen framework space with $u_{j+1} - u_j \equiv \Delta$, $j = 1, 2, \cdots, m$. For any $x_i \in X$ and $u_j \in U$, the following formula is called a one-dimensional information distribution of X on U:

$$\mu(x_i, u_j) = \begin{cases} 1 - \frac{|x_i - u_j|}{\Delta}, & |x_i - u_j| \leq \Delta, \\ 0, & |x_i - u_j| > \Delta. \end{cases} \tag{1}$$

where Δ is called fuzziness.

Note. Let $X = \{x_1, x_2, \cdots x_i \cdots, x_n\}$ be a crisp time series. For any $x_i \in X$, there is a corresponding one-dimensional information distribution $\tilde{F}_i$ determined by Eq. (1), which is called the fuzzification of the time series X. The fuzzy time series obtained is denoted by $\tilde{F}_X$. Clearly, $\tilde{F}_X$ is not unique since it is dependent on the fuzziness Δ.

2.2 Multi-dimensional Information Distribution

Definition 2 [9]. Let X be a r-dimensional random sample. Then x_i is a r-dimensional vector for any $x_i \in X$, say $x_i = (x_{1i}, x_{2i}, \cdots, x_{ri})$. Denote $K = \{1, 2, \cdots, r\}$ and let $X_k = \{x_{ki} | i = 1, 2, \cdots, n\} (k \in K)$.

For any $k \in K$, assume that the chosen framework space of X_k is $U_{X_k} = \{u_{k(j_k)} | j_k = 1, 2, \cdots, m_k\}$ with $u_{k(j_k+1)} - u_{k(j_k)} \equiv \Delta_{X_k}$. Then $U = U_{X_1} \times U_{X_2} \times \cdots \times U_{X_r} = \prod_{k=1}^{r} U_{X_k}$ is the chosen framework space of X. Based on the Eq. (1),

$$q_{i((j_{X_1})(j_{X_2})\cdots)(j_{X_r})} \triangleq \prod_{k=1}^{r} \mu_k(x_{ki}, u_{k(j_{X_k})}) = \begin{cases} \prod_{k=1}^{r} (1 - \frac{\left|x_{ki} - u_{k(j_{X_k})}\right|}{\Delta_{X_k}}), & \bigwedge_{k=1}^{r} \left|x_{ki} - u_{k(j_{X_k})}\right| \le \Delta_{X_k}, \\ 0, & \text{otherwise}. \end{cases} \tag{2}$$

Where Λ represents and.

Then μ is called a r-dimensional information distribution; in other words, $q_{i(j_{X_1})(j_{X_2})\cdots(j_{X_r})}$ is called distributed information on $(U_{X_1}, U_{X_2}, \cdots, U_{X_r})$ from $(x_{1i}, x_{2i}, \cdots, x_{ri})$.

3 Construction of Multivariable Fuzzy Time Series Model

The steps of constructing a multivariable fuzzy time series model mainly include: (1) Fuzzifying time series based on the one-dimensional information distribution function in the Subsect. 2.1; (2) Constructing fuzzy information matrix via the r-dimensional information distribution technique in the Subsect. 2.2; (3) Transforming the fuzzy information matrix into fuzzy relation matrix using fuzzy concept; (4) Constructing a multivariable fuzzy time series forecasting model based on fuzzy approximate reasoning theory. In the following, we take three variables X_1, X_2, X_3 as an example to introduce the construction process of a multivariable fuzzy time series model in detail. Hence $r = 3$ in Definition 2, and then, $K = \{1, 2, 3\}$, the sample space $X = \{X_1, X_2, X_3\}$, where $X_K = \{x_{K1}, x_{K2}, \cdots, x_{Kn}\}$, and the chosen framework space of X_K are $U_{X_K} = \{u_{K1}, u_{K2}, \cdots, u_{K(j_K)}, \cdots, u_{m_K}\}$ with $\Delta_{X_K} = u_{K(j_K+1)} - u_{K(j_K)}$.

3.1 Fuzzification of Time Series

For the variable X_K, the fuzzy time series $\tilde{F}_{X_K} = \{\tilde{F}_{K1}, \tilde{F}_{K2}, \cdots, \tilde{F}_{Ki}, \cdots, \tilde{F}_{Kn}\}$ is obtained by (1) on the chosen framework space U_{X_K} with Δ_{X_K}, where

$$\tilde{F}_{Ki} = \mu(x_{Ki}, u_{K(j_K)}) = \begin{cases} 1 - \frac{|u_{K(j_K)} - x_{Ki}|}{\Delta_{X_K}}, & |u_{K(j_K)} - x_{Ki}| \le \Delta_{X_K}, \\ 0, & u_{K(j_K)} - x_{Ki}| > \Delta_{X_K}. \end{cases} \tag{3}$$

3.2 Establishment of Fuzzy Information Matrix

The fuzzy relation between X_1, X_2 and X_3 will be received by substituting the fuzzy time series of the Eq. (3) into the Eq. (2). For the sake of convenience, transforming the three-dimensional space $(U_{X_1}, U_{X_2}, U_{X_3})$ into two-dimensional space $(U_{X_1} \cap U_{X_2}, U_{X_3})$. Then the distributed information $q_{i(j_{X_1})(j_{X_2})(j_{X_3})}$ of the i-th sample point on $(U_{X_1}, U_{X_2}) \times U_{X_3}$ obtained by (4) is denoted by the following fuzzy information matrix $Q_{i(j_1)(j_2)(j_3)}$ for the sample X on $(U_{X_1}, U_{X_2}) \times U_{X_3}$:

$$q_{i(j_1j_2)(j_3)} = \prod_{k=1}^{3} \tilde{F}_{ki} = \begin{cases} \left(1 - \frac{|u_{j_1} - x_{1i}|}{\Delta_{X_1}}\right)\left(1 - \frac{|u_{j_2} - x_{2i}|}{\Delta_{X_2}}\right)\left(1 - \frac{|u_{j_3} - x_{3i}|}{\Delta_{X_3}}\right), & \bigwedge_{k=1}^{3} |u_{j_k} - x_{ki}| \le \Delta_{X_k}, \\ 0, & \text{otherwise.} \end{cases} \tag{4}$$

$$Q_{(j_1j_2)(j_3)} = \begin{pmatrix} \sum_{i=1}^{n} q_{i(11)(21),31} & \sum_{i=1}^{n} q_{i(11)(21),32} & \cdots & \sum_{i=1}^{n} q_{i(11)(21),3(m_3)} \\ \sum_{i=1}^{n} q_{i(11)(22),31} & \sum_{i=1}^{n} q_{i(11)(22),32} & \vdots & \sum_{i=1}^{n} q_{i(11)(22),3(m_3)} \\ \sum_{i=1}^{n} q_{i(11)(23),31} & \sum_{i=1}^{n} q_{i(11)(23),32} & \vdots & \sum_{i=1}^{n} q_{i(11)(23),3(m_3)} \\ \vdots & \vdots & \vdots & \vdots \\ \sum_{i=1}^{n} q_{i(1(j_1))(2j_2)),31} & \sum_{i=1}^{n} q_{i(1(j_1))(2j_2)),32} & \cdots & \sum_{i=1}^{n} q_{i(1(j_1))(2j_2)),3(m_3)} \\ \vdots & \vdots & \vdots & \vdots \\ \sum_{i=1}^{n} q_{i(m_1)(m_2),31} & \sum_{i=1}^{n} q_{i(m_1)(m_2),32} & \cdots & \sum_{i=1}^{n} q_{i(m_1)(m_2),3(m_3)} \end{pmatrix}. \tag{5}$$

3.3 Establishment of Fuzzy Relation Matrix

Based on the fuzzy concept, the fuzzy information matrix (5) is transformed into fuzzy relation matrix, which can be done as follows:

Set $S_M = \max_{1 \le j_1j_2 \le m_1m_2} Q_{(j_1j_2)(j_3)}$. Let $r_{(j_1j_2)(j_3)} = \frac{Q_{(j_1j_2)(j_3)}}{S_M}$, $j_1j_2 = 1, 2, \cdots, m_1m_2$.

Thus the fuzzy relation matrix is defined as

$$R = \left\{ r_{(j_1j_2)(j_3)} \right\}_{(m_1m_2)\times m_3}. \tag{6}$$

3.4 Establishment of Multivariable Fuzzy Time Series Forecasting Model

Based on the theory of the fuzzy approximate reasoning, a multivariable fuzzy time series forecast model is constructed:

$$\mu(u_{j_3}) = \left[\mu(x_{1i}, u_{j_1})\mu(x_{2i}, u_{j_2})\right] \circ R.$$

Considering the complexity and precision of the model, the "$\circ$" operation is selected as "$\vee - *$". Then the multivariable fuzzy time series model based on multi-dimensional information distribution technology (for short, MIDTMFTS model) is suggested as follows:

$$\mu_{U_{X_3}}(u_{3_{(j_3)}}) = \max_{j_3=1}^{m_3}\{P_{i((j_1-1)+j_2)} * r_{r_{(j_1 j_2)(j_3)}}\}. \tag{7}$$

4 Case Study - the Forecast of SO_2 Emissions in China

4.1 Data and Their Preprocessing

SO_2 is one of the major air pollutants and many factors could lead to SO_2 emissions. In this paper, two indexes of total energy consumption (*TEC*) and per capita GDP (*PCGDP*) are selected to predict the emissions of SO_2(ESO_2) in China. Among them, the data of per capita GDP and total energy consumption in 2017 are from *Statistical Communiqué of the People's Republic of China on the* 2017 *National Economic and Social Development*,; the data of SO_2 emissions in 2017 is calculated from the data in the *Report on the Work of the Government* published by the State Council of China; and the rest of the data are from *China Statistical Yearbook* 2002–2017.

In order to improve forecast accuracy and reduce the degree of dispersion, the data of all indicators will be logarithmically preprocessed. Namely, $X_1 = \ln(TEC)$, $X_2 = \ln(PCGDP)$ and $X_3 = \ln(ESO_2)$.

4.2 Forecast of SO_2 Emissions Based on the MIDTMFTS Model

Forecast in the Fuzziness Δ_1. Let $\Delta_1 = \{\Delta_{X_1} = 0.120, \Delta_{X_2} = 0.200, \Delta_{X_3} = 0.045\}$. Then the universes of discourse are $U_{X_1} = \{11.850, 11.970, \cdots, 13.050\}$; $U_{X_2} = \{9.000, 9.200, \cdots, 11.000\}$; $U_{X_3} = \{7.450, 7.495, \cdots, 7.900\}$.

According to the Eqs. (4) and (5), the fuzzy information matrix Q is obtained. According to the Eq. (6), the fuzzy relation matrix R is obtained. Substituting R into the Eq. (7) to get the forecasts $\tilde{F}_{3i}$ of SO_2 emissions from 2002 to 2017, as shown in Table 1. To compare with the actual value, calculate fuzzy set gravity GC_i of Table 1. In addition, the logarithmic inverse transformation exponent arithmetic is performed on GC_i, and the predictive results of SO_2 emissions are obtained, which are shown in the third column of Table 2.

$$Q = \begin{array}{c} \\ (11.850, 9.000) \\ \cdots \\ (12.330, 9.000) \\ (12.330, 9.200) \\ (12.330, 9.400) \\ (12.330, 9.600) \\ (12.330, 9.800) \\ (12.330, 10.000) \\ \cdots \\ (13.050, 11.000) \end{array} \begin{array}{c} \begin{array}{ccccccccc} 7.450 & \ldots & 7.585 & 7.630 & 7.675 & 7.720 & 7.765 & \ldots & 7.900 \end{array} \\ \left(\begin{array}{ccccccccc} 0.000 & \ldots & 0.000 & 0.100 & 0.001 & 0.009 & 0.000 & \ldots & 0.000 \\ \ldots & \ldots & \ldots & \ldots & \ldots & \ldots & \ldots & \ldots & \ldots \\ 0.000 & \ldots & 0.000 & 0.000 & 0.000 & 0.808 & 0.091 & \ldots & 0.122 \\ 0.000 & \ldots & 0.000 & 0.000 & 0.000 & 0.029 & 0.085 & \ldots & 0.000 \\ 0.000 & \ldots & 0.000 & 0.000 & 0.000 & 0.000 & 0.000 & \ldots & 0.000 \\ 0.000 & \ldots & 0.000 & 0.000 & 0.000 & 0.000 & 0.000 & \ldots & 0.000 \\ 0.000 & \ldots & 0.000 & 0.000 & 0.000 & 0.000 & 0.000 & \ldots & 0.000 \\ 0.000 & \ldots & 0.000 & 0.000 & 0.000 & 0.003 & 0.209 & \ldots & 0.000 \\ \ldots & \ldots & \ldots & \ldots & \ldots & \ldots & \ldots & \ldots & \ldots \\ 0.332 & \ldots & 0.274 & 0.606 & 0.005 & 0.000 & 0.000 & \ldots & 0.000 \end{array} \right) \end{array}.$$

$$R = \begin{array}{c} \\ (11.850, 9.000) \\ \cdots \\ (12.330, 9.000) \\ (12.330, 9.200) \\ (12.330, 9.400) \\ (12.330, 9.600) \\ (12.330, 9.800) \\ (12.330, 10.000) \\ \cdots \\ (13.050, 11.000) \end{array} \begin{array}{c} \begin{array}{ccccccccc} 7.450 & \cdots & 7.585 & 7.630 & 7.675 & 7.720 & 7.765 & \cdots & 7.900 \end{array} \\ \left(\begin{array}{ccccccccc} 0.000 & \ldots & 0.513 & 0.195 & 0.001 & 0.009 & 0.000 & \ldots & 0.000 \\ \ldots & \ldots & \ldots & \ldots & \ldots & \ldots & \ldots & \ldots & \ldots \\ 0.000 & \ldots & 0.000 & 0.000 & 0.000 & 0.029 & 0.094 & \ldots & 0.122 \\ 0.000 & \ldots & 0.000 & 0.000 & 0.000 & 0.212 & 0.239 & \ldots & 0.000 \\ 0.000 & \ldots & 0.000 & 0.000 & 0.000 & 0.319 & 0.361 & \ldots & 0.000 \\ 0.000 & \ldots & 0.000 & 0.000 & 0.002 & 0.606 & 0.230 & \ldots & 0.000 \\ 0.000 & \ldots & 0.021 & 0.129 & 0.240 & 0.259 & 0.098 & \ldots & 0.000 \\ 0.000 & \ldots & 0.066 & 0.416 & 0.755 & 0.608 & 0.143 & \ldots & 0.000 \\ \ldots & \ldots & \ldots & \ldots & \ldots & \ldots & \ldots & \ldots & \ldots \\ 0.662 & \ldots & 0.213 & 0.014 & 0.005 & 0.000 & 0.000 & \ldots & 0.000 \end{array} \right) \end{array}.$$

Influence of the Variation of Fuzziness Δ on the Predictive Accuracy of the MIDTMFTS Model. In order to illustrate the impact of fuzziness Δ on the accuracy of MIDTMFTS model, increase the fuzziness. Namely, let $\Delta_2=\{\Delta_{X_1} = 0.200, \Delta_{X_2} = 0.335, \Delta_{X_3} = 0.070\}$. The forecasts of SO_2 emissions are derived from the Eqs. (3)–(7), as shown in the fourth column 4 of Table 2.

According to Table 2, with different fuzziness, the MAPEs of forecasts of SO_2 emissions by the MIDTMFTS model are all below 0.035, which indicates that the relative error of the model is small. When $\Delta_1 < \Delta_2$, $\mathrm{MAE}_{\Delta_1} < \mathrm{MAE}_{\Delta_2}$ and $\mathrm{MAPE}_{\Delta_1} < \mathrm{MAPE}_{\Delta_2}$. Therefore, the predictive errors of the model are different with diverse fuzziness, that is to say, fuzziness affects the forecast accuracy of the MIDTMFTS model. With the reduction of the fuzziness, the predictive error of MIDTMFTS model decreases, that is, the forecast accuracy increases. Among them, the predictive error of $\mathrm{MIDTMFTSM}_{\Delta_1}$ for the emissions of SO_2 is the smallest and the predictive accuracy is the highest. In other words, it can reflect the change of the actual value more exactly, as shown in Fig. 1.

Table 1. The predictive results of MIDTMFTSM$_{\Delta_1}$

$\tilde{F}_{3i}$	U_{X_3}								
	7.450	7.495	7.540	7.585	7.630	7.675	7.815	7.855	7.900
$\tilde{F}_{3,2002}$	0.000	0.000	0.496	0.055	0.005	0.082	0.000	0.000	0.000
$\tilde{F}_{3,2003}$	0.000	0.000	0.100	0.011	0.031	0.563	0.000	0.000	0.000
$\tilde{F}_{3,2004}$	0.000	0.000	0.000	0.000	0.000	0.000	0.808	0.040	0.000
$\tilde{F}_{3,2005}$	0.000	0.000	0.000	0.000	0.000	0.000	0.029	0.001	0.000
$\tilde{F}_{3,2006}$	0.000	0.000	0.000	0.000	0.000	0.000	0.000	0.000	0.175
$\tilde{F}_{3,2007}$	0.000	0.000	0.000	0.000	0.000	0.029	0.094	0.423	0.423
$\tilde{F}_{3,2008}$	0.000	0.000	0.000	0.000	0.000	0.212	0.239	0.491	0.491
$\tilde{F}_{3,2009}$	0.000	0.000	0.000	0.000	0.000	0.319	0.361	0.512	0.091
$\tilde{F}_{3,2010}$	0.000	0.000	0.000	0.000	0.002	0.606	0.230	0.031	0.000
$\tilde{F}_{3,2011}$	0.000	0.000	0.021	0.129	0.240	0.259	0.098	0.000	0.000
$\tilde{F}_{3,2012}$	0.000	0.000	0.066	0.416	0.775	0.608	0.143	0.000	0.000
$\tilde{F}_{3,2013}$	0.404	0.298	0.243	0.404	0.508	0.398	0.093	0.000	0.000
$\tilde{F}_{3,2014}$	0.597	0.440	0.359	0.597	0.191	0.109	0.026	0.000	0.000
$\tilde{F}_{3,2015}$	0.573	0.423	0.345	0.573	0.184	0.000	0.000	0.000	0.000
$\tilde{F}_{3,2016}$	0.287	0.251	0.173	0.287	0.092	0.000	0.000	0.000	0.000
$\tilde{F}_{3,2017}$	0.662	0.756	0.213	0.014	0.005	0.000	0.000	0.000	0.000

4.3 Forecast of SO_2 Emissions Based on the *Markov* Model

In order to compare with the MIDTMFTS model, the first-order *Markov* model [10] is selected to predict SO_2 emissions in China from 2001 to 2017, whose results are shown in the fifth and sixth column of Table 2.

Table 2. Results of SO_2 emissions based on the suggested model for different Δ

Year	Actual value	MIDTMFTSM$_{\Delta_1}$	MIDTMFTSM$_{\Delta_1}$	*Markov*$_{\Delta_1}$	*Markov*$_{\Delta_2}$
2002	1927.1	1957.5	2007.0	1837.2	1890.7
2003	2159.3	2121.8	2145.7	2115.8	2096.0
2004	2255.4	2270.6	2336.5	2162.9	2267.9
2005	2549.1	2545.5	2439.1	2570.9	2551.2
2006	2588.8	2579.1	2529.8	2393.5	2430.4
2007	2468.1	2409.8	2394.9	2535.9	2449.2
2008	2321.2	2346.1	2394.7	2142.7	2160.7
2009	2214.4	2282.7	2352.9	2136.9	2138.4
2010	2185.1	2193.8	2205.3	2160.3	2101.7
2011	2217.6	2111.8	2073.1	2087.0	2077.5
2012	2117.6	2094.1	2025.2	2077.7	2042.4
2013	2043.9	1982.5	1998.6	2069.2	2036.9

(*continued*)

Table 2. (*continued*)

Year	Actual value	MIDTMFTSM$_{\Delta_1}$	MIDTMFTSM$_{\Delta_1}$	*Markov*$_{\Delta_1}$	*Markov*$_{\Delta_2}$
2014	1974.4	1912.4	1969.6	2033.7	2037.1
2015	1859.1	1896.9	1930.0	2018.3	2055.8
2016	1803.3	1895.0	1857.9	1848.3	2075.2
2017	1855.6	1826.4	1839.4	1847.2	2074.3
Mean absolute error (MAE)		41.750	67.353	78.705	98.331
Mean absolute percentage error (MAPE)		0.020	0.031	0.036	0.046

4.4 Comparison and Analysis

According to Table 2, when the fuzziness is Δ_1, MAE, 41.750 < 78.705; MAPE, 0.020 < 0.036. Hence, MIDTMFTSM$_{\Delta_1}$ obviously outperforms *Markov*M$_{\Delta_1}$; and the difference between the predictive errors of the two model is large. Figure 1 shows that the distance between the predictive value of MIDTMFTSM$_{\Delta_1}$ and the actual value of SO_2 emissions is smaller. MIDTMFTSM$_{\Delta_1}$ can be more ideal to describe the trend of the actual values. When the fuzziness is Δ_2, by comparing the data in Table 2, the performance of MIDTMFTSM$_{\Delta_2}$ is better than *Markov*M$_{\Delta_2}$. The fact reflected in the line chart (Fig. 1) is that *Markov*M$_{\Delta_2}$ is more distant from the actual value. The following conclusions can be drawn from the case study: (1) With the same fuzziness Δ, the MAE and MAPE of MIDTMFTS model are all smaller than those of the *Markov* model. (2) The forecast accuracy of MIDTMFTS model increases with the decrease of Δ. (3) In the process of forecasting SO_2 emissions, the MIDTMFTS model is more convenient and concise than the *Markov* model.

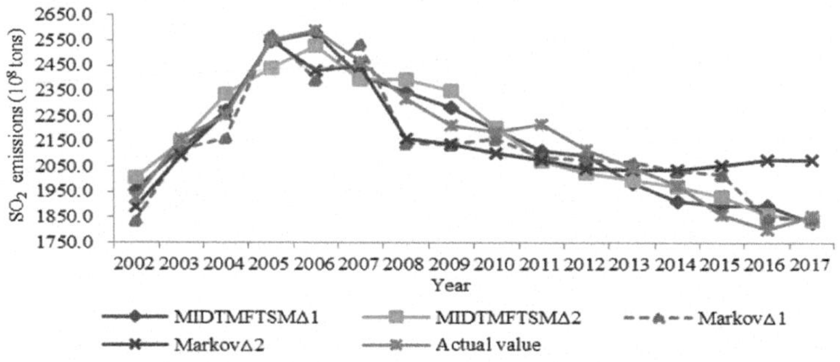

Fig. 1. Actual values and predictive values of the *Markov* model and MIDTMFTS model

5 Conclusions and Discussion

This paper constructs a multivariable fuzzy time series model based on multivariable information distribution technology. It has the following advantages: (1) MIDTMFTS model can make up for the small sample problem by making effective use of the fuzzy transition information in the sample. (2) This model can be used to study multivariable time series. (3) Compared with the *Markov* model, the prediction process of the MIDTMFTS model is simpler and more concise. (4) In the same fuzziness Δ, the MAE and MAPE of MIDTMFTS model are all smaller than those of *Markov* model, so the forecast accuracy of MIDTMFTS model is superior to the *Markov* model. However, it should be pointed out that in this paper, a relatively simple method of ordered isometric partition is used to determine the fuzzy interval length. Although the predictive results are ideal, there is a limitation: forecast results affected by fuzziness Δ. With the decrease of fuzziness, the forecast accuracy increases, otherwise it reduces. Therefore, the method of determining Δ will be studied in depth in the future.

Acknowledgement. This research is financially supported by Program for the soft science of Shanxi Province in China (No. 2017041025-2); and Program for the Philosophy and Social Sciences Research of Higher Learning Institutions of Shanxi (PSSR) in China (No. 2017314).

References

1. Box, G.E.P., Jenkins, G.M., Reinsel, G.C., Ljung, G.M.: Time Series Analysis: Forecasting and Control. Wiley Press, Hoboken (2015)
2. Haldrup, N., Nielsen, F.S., Nielsen, M.Ø.: A vector autoregressive model for electricity prices subject to long memory and regime switching. Energy Econ. **32**(5), 1044–1058 (2010)
3. Mellit, A., Pavan, A.M., Benghanem, M.: Least squares support vector machine for short-term prediction of meteorological time series. Theoret. Appl. Climatol. **111**(1–2), 297–307 (2013)
4. Zecchin, C., Facchinetti, A., Sparacino, G., De Nicolao, G., Cobelli, C.: A new neural network approach for short-term glucose prediction using continuous glucose monitoring time-series and meal information. In: 2011 Annual International Conference of the IEEE Engineering in Medicine and Biology Society, pp. 5653–5656. IEEE Press, New York (2011)
5. Song, Q., Chissom, B.S.: Fuzzy time series and its models. Fuzzy Sets Syst. **54**(3), 269–277 (1993)
6. Avazbeigi, M., Doulabi, S.H.H., Karimi, B.: Choosing the appropriate order in fuzzy time series: a new N-factor fuzzy time series for prediction of the auto industry production. Expert Syst. Appl. **37**(8), 5630–5639 (2010)
7. Xue, Y.: A new fuzzy time series model based on fuzzy information optimization. Stat. Decis. **416**(20), 12–15 (2014)
8. Shuai, Y., Song, T.L., Wang, J.P., Zhan, W.B.: Modified fuzzy time series model interval partitioning algorithm. Comput. Eng. Des. **38**(2), 379–383 + 394 (2017)
9. Huang, C.F., Shi, Y.: Towards Efficient Fuzzy Information Processing Using the Principle of Information Diffusion. Physical-Verlag Press, Heidelberg (2002)
10. Wu, B., Tse, S., Hsu, Y.: A new approach of bivariate fuzzy time series: with applications to the stock index forecasting. Int. J. Uncertainty Fuzziness Knowl. Based Syst. **12**(11), 1793–1811 (2003)

Research on Apple Yield Prediction Model

Zhijun Wang, Xican Li, Yucun Wang, and Shuhan Cheng(✉)

College of Information Science and Engineering,
Shandong Agricultural University, Taian, China
{wzj, shcheng}@sdau.edu.cn

Abstract. This paper has used a fuzzy comprehensive prediction method and taken tree vigor data from 29 orchards as samples. By using 23 of the total samples, the apple yield prediction model base on tree vigor is established. The test precision of the model is 96.36%. With other six orchards to be predicted, this paper calculated fuzzy division and eigenvalues of grade variables and obtained apple yield prediction accuracy which is 91.08%. This model has practical applied value in apple yield prediction.

Keywords: Apple · Yield prediction · Fuzzy comprehensive prediction · Fuzzy partition · Eigenvalues of grade variables

1 Introduction

Orchard is a complex artificial ecosystem, the ecological environment is a sum of all the factors in the living space, including weather conditions, such as temperature, moisture, light, wind speed and so on, soil conditions, terrain conditions, biological factors and human factors, etc. [1]. In the long-term process of growth and development, due to different levels of influence of ecological environment, fruit cultivation technology, plant diseases and pests and other factors, the apple yield showed greater volatility [2]. So studying the changing laws of apple annual production, mastering the dynamic characteristics and predicting it scientifically and accurately can not only provide reliable foundation for social and economic management decisions but also provide important data for related departments to drawing up plan [3, 4].

Aiming at different data samples, this chapter studies the apple yield prediction model based on the tree vigor structure of 49 orchards in Shandong province.

2 Materials and Methods

In order to ensure the requirements for data sequence length of the model and keep some samples to test the model, according to the correlation, choose to keep 29 samples (Table 1) and choose 23 orchards to establish the model and 6 orchards to carry out the tests [4]. Choose the following indicators in each sample: leaves per mu (piece), leaf area per mu (m^2), leafage branch (branch/mu), short branch (branch/mu), medium branch (branch/mu), long branch (branch/mu), developmental branch (branch/mu), yield per mu (kg).

Y. Liu et al. (Eds.): ICNC-FSKD 2019, AISC 1074, pp. 913–920, 2020.
https://doi.org/10.1007/978-3-030-32456-8_98

Table 1. The leaf area, leaf weight, branch group and the yield of apple

Sample no.	Leaves per mu (piece)	Leaf area permu (m^2)	Leafage branch (branch)	Short branch (branch)	Medium branch (branch)	Long branch (branch)	Developmental branch(branch)	Yield per mu (kg)
1	944548	1887.6	48840	34056	21428	11000	3256	4188.8
2	1000736	1680.8	55880	27324	24684	17116	6556	3141.6
3	636648	828.2	44936	23780	15703	9266	1845	2439.5
4	822434	1453.6	48530	24058	21482	14306	5106	3831.8
5	669500	1242.8	26104	12688	14924	16744	14144	4331.6
6	786828	1461.6	52206	30114	16464	17346	11760	3498.6
7	710734	1165.6	55507	37459	19552	16638	5687	3915.1
8	396066	642.6	31314	24888	14382	5457	2091	3034.5
9	976166	1596.2	71070	35144	19596	15732	13064	3284.4
10	819400	1725.0	70500	35450	20300	15350	9650	4165.0
11	628875	1296.0	60885	30825	12960	9360	9540	3213.0
12	538440	948.0	52400	28480	12320	8840	5360	2856.0
13	615105	1012.5	42390	29070	14805	10215	8010	3748.5
14	687150	1075.5	38880	18630	18495	17145	4770	4284.0
15	747890	1501.5	54065	22880	18315	16830	11880	3927.0
16	564150	845.0	50400	30500	16300	11600	3350	3570.0
17	1025568	1833.6	52848	24096	26544	27168	13776	5712.0
18	342678	549.4	25871	18450	12300	5002	1353	2927.4
19	734076	1549.8	47898	31914	20736	15498	4644	4498.2
20	862946	1908.0	49820	29998	17437	16483	10865	3153.5
21	902832	1819.2	61056	25200	21936	16272	14880	3427.2
22	1231148	2263.2	79048	42148	26609	16441	13612	4879.0
23	1239760	2336.0	39160	29200	24240	19440	10600	4760.0
24	556985	1105.5	41030	19800	16775	13860	4290	4581.5
25	949680	1699.2	56928	37536	23616	21216	12864	3998.4
26	503417	813.1	43428	19270	13583	17437	9306	3355.8
27	529975	928.8	43473	30014	10535	10062	7955	3070.2
28	589848	1092.0	34776	19446	16002	7224	4914	2998.8
29	631960	1204.0	42720	20400	20920	16120	6280	3808.0

Correlation analysis of the data in the table above was conducted and then gives the following results (Table 2).

Table 2. Correlation of leaf area, leaf weight, branch group and the apple yield

	Leaves per mu	Leaf area	Leafage branch	Short branch	Medium branch	Long branch	Developmental branch
Correlation coefficient	0.5426	0.5662	0.2952	0.11732	0.53464	0.5956	0.3492

From Table 2, the correlation of leaves per mu, leaf area, long branch, medium branch and apple yield reached significance level.

3 Apple Yield Fuzzy Comprehensive Prediction

Use fuzzy comprehensive prediction model based on fuzzy recognition that proposed by Li [5–8] to predict apple yield, method is given below:

Given sample set composed of n samples to be clustered:

$$X = \{x_1, x_2, \ldots, x_n\} \tag{1}$$

Given m indicators of each sample, the indicators eigenvalues vector of the sample j is:

$$X_j = \{x_{1j}, x_{2j}, \ldots, x_{mj}\}^T \tag{2}$$

The indicators eigenvalues matrix of n samples can be expressed as:

$$X = \begin{bmatrix} x_{11} & \cdots & x_{1n} \\ \vdots & \ddots & \vdots \\ x_{m1} & \cdots & x_{mm} \end{bmatrix} = (x_{ij})_{m\times n} \tag{3}$$

In the function, x_{ij} is the eigenvalue of indicator i of sample j, $i = 1, 2, \ldots, m$; $j = 1, 2, \ldots n$.

By means of normalization treatment of indicators, the influence of dimension and order of the indicators can be overcome. The formulas are as follows:

$$r_{ij} = \frac{x_{ij} - x_{imin}}{x_{imax} - x_{imin}} \tag{4}$$

$$r_{ij} = \frac{x_{imax} - x_{ij}}{x_{imax} - x_{imin}} \tag{5}$$

In the function, r_{ij} represents the normalized value of indicator i, $0 \le r_{ij} \le 1$, $x_{i\max}$ shows the maximum value of indicator i in n models. x_{imin} shows the minimum value of indicator i in n models.

Formulas (4) and (5) are respectively used to do normalization treatment if apple yield is positively or negatively correlated with indicators. Use formula (4) and (5) to translate (3) into numbers between 0 and 1.

$$R = (r_{ij})_{m\times n} \tag{6}$$

A Build Prediction Model

Given vector $Y = (y_1, y_2, \ldots, y_j, \ldots y_n)$ to represents apple yields of n samples, translate the apple yields into numbers between 0 and 1 by using the following formula:

$$x_j = y_j/M \tag{7}$$

In the formula, y_i represents the yield of sample j; M represents the maximum yield of n samples; x_j represents the normalized value of the yield of sample j.

The sample set is divided into c categories by the normalized value based upon the yield of samples. $U = (u_{hj})_{c \times n}$ is the fuzzy dividing matrix. Among the matrix, u_{hj} represents the relative membership degree that sample j attached to class h. $h = 1, 2, \ldots, c$, $\sum_{h=1}^{c} u_{hj} = 1$.

Given vector $S = (s_1, s_2, \ldots, s_h, \ldots, s_c)^T$ represent fuzzy clustering central index of c categories. Among them, s_h represents the central index of category h. The fuzzy dividing matrix and fuzzy cluster center are as follows:

$$\mathrm{u_{hj}} = \frac{1}{\sum_{k=1}^{c} \frac{(x_{j-s_h})^2}{(x_{j-s_k})^2}} \tag{8}$$

$$\mathrm{S_h} = \frac{\sum_{j=1}^{n} u_{hj}^2 x_j}{\sum_{j=1}^{n} u_{hj}^2} \tag{9}$$

Use formulas (8) and (9) and loop iteration to obtain the optimal fuzzy diagnosis matrix U^* and fuzzy clustering center S^* which is divided by yield.

$$\mathrm{U^*} = (\mathrm{u_{hj}^*})_{\mathrm{c \times n}} \tag{10}$$

Calculate the eigenvalues of grade variables of each sample according to calculated fuzzy division,

$$\mathrm{h_j} = (1, 2, \ldots \mathrm{c}) \times (\mathrm{u_{1j}^*}, \mathrm{u_{2j}^*}, \ldots, \mathrm{u_{cj}^*})^{\mathrm{T}} \tag{11}$$

Use the following formula to calculate the correlation coefficient between y and h.

$$\rho = \frac{\sum_{j=1}^{n} (h_j - \bar{h})(y_j - \bar{y})}{\sqrt{\sum_{j=1}^{n} (y_j - \bar{y})^2 \sum_{j=1}^{n} (h_j - \bar{h})^2}} \tag{12}$$

If the correlation coefficient is more than 0.9, equation of linear (nolinear) regression between (predictions) and can be established: $\hat{y} = a + bh$. In the equation: $b = \sum_{j=1}^{n} (h_j - \bar{h})(y_j - \bar{y}) / \sum_{j=1}^{n} (h_j - \bar{h})^2$, $a = \bar{y} - b\bar{h}$. First, according to Table 1, normalization treatment of eigenvalues of 29 sample index should be carried out. Then the samples would be divided into five classes. Based on the given apple yield of sample 1 to 23 (eigenvalue y), by using formulas (8) and (9) loop iteration, the optimal fuzzy diagnosis matrix can be obtained. The results are shown in Table 3.

By using formula (11), eigenvalues of grade variables of each sample can be calculated. Build linear regression model $Y = a + bh$ as the prediction model. The prediction equation can be obtained after calculation.

$$\hat{Y}_J = 725.6622\text{h} + 1879.6620 \tag{13}$$

$\rho = 0.978466$ is the correlation coefficient between y and h which has good correlation. Use formula (13) to make predictions for sample 1 to 23. The results are shown in Table 4.

Table 3. The fuzzy partition of modeling samples

No.	Membership grade					Category
	Class 1	Class 2	Class 3	Class 4	Class 5	
1	0.0151	0.0365	0.8168	0.1196	0.0120	3
2	0.2417	0.7149	0.0303	0.0097	0.0034	2
3	0.7616	0.1544	0.0488	0.0240	0.0112	1
4	0.0253	0.0955	0.8363	0.0358	0.0071	3
5	0.0222	0.0486	0.4664	0.4364	0.0263	3
6	0.0594	0.8142	0.1019	0.0191	0.0054	2
7	0.0066	0.0218	0.9556	0.0136	0.0024	3
8	0.6455	0.3142	0.0271	0.0095	0.0036	1
9	0.0061	0.9908	0.0023	0.0006	0.0002	2
10	0.0125	0.0308	0.8607	0.0867	0.0093	3
11	0.0721	0.9051	0.0164	0.0048	0.0017	2
12	0.9976	0.0019	0.0003	0.0001	0.0001	1
13	0.0516	0.2334	0.6531	0.0507	0.0112	3
14	0.0220	0.0497	0.5939	0.3115	0.0229	3
15	0.0049	0.0159	0.9666	0.0107	0.0019	3
16	0.0769	0.6606	0.2191	0.0343	0.0091	2
17	0.0000	0.0000	0.0000	0.0000	0.9999	5
18	0.9382	0.0514	0.0068	0.0026	0.0010	1
19	0.0090	0.0180	0.1026	0.8532	0.0171	4
20	0.2055	0.7536	0.0286	0.0090	0.0032	2
21	0.0308	0.9279	0.0323	0.0070	0.0021	2
22	0.0092	0.0157	0.0500	0.8692	0.0559	4
23	0.0022	0.0039	0.0140	0.9710	0.0089	4

Table 4. The accuracy test of the prediction model

No.	Eigen value	Measured value (kg)	Predicted value (kg)	Relative error (%)
1	3.077	4188.8	4112.4	−1.823
2	1.818	3141.6	3199.2	1.833
3	1.369	2439.5	2873.0	17.770
4	2.904	3831.8	3986.9	4.049
5	3.396	4331.6	4344.1	0.288
6	2.097	3498.6	3401.1	−2.786
7	2.983	3915.1	4044.6	3.307
8	1.411	3034.5	2903.8	−4.306
9	1.998	3284.4	3329.7	1.378
10	3.049	4165.0	4092.5	−1.740
11	1.959	3213.0	3301.2	2.744
12	1.003	2856.0	2607.6	−8.698
13	2.737	3748.5	3865.5	3.121
14	3.263	4284.0	4247.8	−0.844
15	2.989	3927.0	4048.4	3.091
16	2.238	3570.0	3503.8	−1.855
17	5.000	5712.0	5507.9	−3.573
18	1.077	2927.4	2661.1	−9.097
19	3.851	4498.2	4674.5	3.919
20	1.851	3153.5	3222.7	2.194
21	2.022	3427.2	3346.7	−2.349
22	3.947	4879.0	4743.8	−2.770
23	3.981	4760.0	4768.3	0.175

From Table 4, average error rate between predicted values and measured values is 3.6396%, the result of the prediction is of high precision.

B Calculate Multi-index Fuzzy Clustering Centers and Weights

According to cause analysis, calculate optimum parameter of apple yield prediction model, namely the optimal cluster centers of the samples $S^* = (s_{ih})_{m \times c}$ and weights $W^* = (w_1, w_2, \ldots, w_m)$, $\sum_{i=1}^{m} w_i = 1$. After calculation, fuzzy clustering center matrix S^* and the optimal weights W^* can be obtained. So does the parameters of predication.

$$S^* = \begin{bmatrix} 0.217193 & 0.458843 & 0.479207 & 0.576632 & 0.762984 \\ 0.211729 & 0.455618 & 0.486956 & 0.61643 & 0.72075 \\ 0.258602 & 0.565398 & 0.488403 & 0.389827 & 0.511391 \\ 0.396491 & 0.514104 & 0.566466 & 0.495453 & 0.393191 \\ 0.231269 & 0.424336 & 0.593148 & 0.627936 & 0.99477 \\ 0.150199 & 0.46536 & 0.493265 & 0.504092 & 0.993897 \\ 0.228862 & 0.691754 & 0.50432 & 0.494673 & 0.917072 \end{bmatrix}$$

The weights of seven indicators are:
W* = (0.120282,0.118482,0.131119,0.106067,0.13251,0.305355,0.086185).

C Calculate the Optional Fuzzy Partition of the Samples to Be Tested
Based on calculated fuzzy clustering center matrix S^*, the optimal weights W^* and normalized values of index of samples to be tested, fuzzy division of six samples to be predicted can be calculated. The results are shown in Table 5.

Table 5. Fuzzy partition of prediction samples

No.	Membership degree				
	Class 1	Class 2	Class 3	Class4	Class 5
24	0.1826	0.2691	0.4202	0.0955	0.0326
25	0.0346	0.1680	0.1633	0.4372	0.1969
26	0.1468	0.2925	0.3518	0.1364	0.0725
27	0.6178	0.1498	0.1575	0.0549	0.0200
28	0.8000	0.0725	0.0793	0.0352	0.0131
29	0.0659	0.2677	0.5052	0.1234	0.0377

D Calculate Eigenvalues of Grade Variables of Samples to Be Predicted
By using formula (11) and data from Table 5, eigenvalues of grade variables of samples to be predicted can be obtained, the results are shown in Table 6.

Table 6. Test of the accuracy of the predictive mod

No.	Eigen value	Measured value (kg)	Predicted value (kg)	Relative error (%)
24	2.526	4581.5	3712.9	−18.959
25	3.594	3998.4	4487.6	12.235
26	2.695	3355.8	3835.5	14.295
27	1.709	3070.2	3120.1	1.626
28	1.389	2998.8	2887.5	−3.710
29	2.799	3808.0	3911.0	2.705
MRE				8.92

Plus the eigenvalues of grade variables in Table 6 into the prediction equation $\hat{y}_j = 725.6622h + 1879.6620$, apple yield of the samples can be get. The MRE of apple yield of predicted samples is 8.92%, the prediction accuracy is 91.08%. So this paper apply fuzzy comprehensive prediction model to apple yield prediction and gain higher prediction precision. The results are shown in Table 6.

4 Summary

Taking 29 Red Delicious orchards as samples, this paper uses fuzzy comprehensive prediction method based on fuzzy recognition and chooses 23 orchards to build prediction model based on the structure of tree vigor. Test data from six orchards and the prediction accuracy is high enough. But the factors influencing the apple yield are complicated [9, 10]. The external factors include light intensity, temperature, soil conditions, accumulated temperature, amount of cold, precipitation, irrigation, artificial fertilization and pruning, damage extent of diseases, etc. The internal factors include tree-age, volume of tree crown, the types of trees, branches, leaves, etc. How to combine the main factors to build fruit yield or fruit quality prediction model and improve the prediction accuracy are the future research direction.

References

1. Sun, Z., Sun, Z., Yang, Z., Wang, Y.: Advances on research and application of mathematical simulation in fruit tree ecophysiology. J. Fruit Sci. **4**, 361–366 (2005)
2. Cong, Y., Jun, W.: A comparative study on forecast models of apple output in China. J. Fruit Sci. **5**, 682–684 (2007)
3. Xuelong, Q., Guiping, W., Hua, M.: Research on combination forecasting model of apple output based on genetic algorithm. J. Fruit Sci. **1**, 165–170 (2011)
4. Liu, L., Wang, J., Qu, Z., Gao, F., Chai, Q., Yi, L.: Research on non-linear regression model of meteorological factors for apple yield in Shaanxi Province. Chin. Agric. Sci. Bull. **25**, 248–251 (2012)
5. Li, X., Zong, X., Li, J., Shang, Y., Li, F.: Fuzzy comprehensive prediction model and its application. J. Shandong Agric. Univ. (Nat. Sci.) **2**, 169–173 (2003)
6. Zhu, X., Jiang, Y., Zhao, G., Wang, L., Li, X.: Hyperspectral estimation of kalium content in apple florescence canopy based on fuzzy recognition. Spectroscopy Spectral Anal. **4**, 1023–1027 (2013)
7. Li, X., Xie, M., Xu, D., Duan, J.: On theoretic model of fuzzy classification and fuzzy recognition. Fuzzy Syst. Math. **2**, 58–64 (2002)
8. Li, X., Wang, J., Wang, F., Du, J.: Spectral retrieved deduction of soil properties index based on fuzzy recognition theory. J. Liaoning Tech. Univ. (Nat. Sci.) **2**, 324–327 (2010)
9. Robinson, T.L., Lakso, A.N.: Modifying apple tree canopies for improved production efficiency. Hortsince **8**, 1005–1012 (1991)
10. Lokso, A.N., Grappadelli, L.C.: Implication of pruning and training practices to Carbon partitioning and fruit development in apple. Acta Hort **322**, 231–239 (1992)

A Probability Transition Matrix-Based Recommendation Algorithm for Bipartite Networks

Dongming Chen, Chang Liu, Xinyu Huang, Dongqi Wang(✉), and Jiarui Yan

Software College, Northeastern University, Shenyang 110169, China
wangdq@swc.neu.edu.cn

Abstract. By analyzing the searching and recommendation system in the electronic commerce field, this paper provides a retrieval sorting and recommendation algorithm for e-commerce service. An information transmission matrix is constructed based on the 'customer-product' bipartite networks, and the rankings of customers and products can be obtained by analyzing the network structures. Then we propose a community detection algorithm for bipartite networks, by employing the information transmission matrix. Finally, the recommendation scheme based on both customers and products is obtained. It makes recommendation results more comprehensive and reasonable, which satisfies the various requirements of the customers.

Keywords: Bipartite networks · Sorting · Recommendation · Probability transition

1 Introduction

With the popularization of online shopping, it becomes difficult for customers to find the needed products from huge amounts of items. Being beneficial from the filtering results provided by a search engine, customers can get the products preciously. The working process of search engine is explained as processing the typed keywords, sorting the items and returning the searched results. The quality of the sorting algorithm affects the customers' experience directly. An excellent sorting algorithm can help customers find the products efficiently. However, the searched results sometimes fail to meet the real needs of specific customers, resulted in poor online shopping experience. Considering the interests of the corporations, some people manually modify the ranking of search results rather than according to objective facts, so some irrelevant search results are presented to customers, which affects the objectivity and correctness of search rankings [1].

PageRank [2] was proposed to solve this problem by Sergey Brin and Lawrence Page in 1998 [3], where the web pages are sorted according to their degree centrality. The degree of the web page, i.e., the PageRank value, is determined by the probability that the web page is randomly browsed by the customer. Since the algorithm is built on a random surfing process, the probability of a web page accessing its linked web page

Y. Liu et al. (Eds.): ICNC-FSKD 2019, AISC 1074, pp. 921–929, 2020.
https://doi.org/10.1007/978-3-030-32456-8_99

keeps the same. Thus, the PR value of each web page is divided equally distributed to the web pages it links to, and stability is achieved through iterations. Another important idea in the sorting algorithm is the Markov property of the random process in the Markov chain [4], i.e., the future states relate only to the current state, and independent of past states. The Markov chain holds ergodic property, namely, the transition probability of state does not change with time regardless of the original state after a probability transfer of a sufficiently large number of steps. That is to say, the transition probability of state j after multiple transitions keeps a stable constant, i.e., the system maintains in an equilibrium state. Based on this, we can think that the process of customers' purchasing products and products purchased by customers can constitute a Markov chain process, and the next state can be obtained from the current state space. After a sufficiently large number of purchases, the probability that the customers purchase the products and the products are purchased by the customers tends to be stable.

In addition, the amount of product information in large e-commerce systems is huge, and the traditional retrieval methods only sort by adding time, so that the retrieval results cannot meet the real needs of customers. Aiming at solving this problem, through analyzing the relationships between customers and products in the large online shopping mall system, a bipartite network is constructed based on 'customer-products' relationships [5], and through the analysis of the bipartite network, a bipartite network recommendation algorithm is proposed based on data propagation. The proposed recommendation algorithm provides customers with efficient shopping experience so that the searched results can better meet the needs of customers.

2 Algorithm Description

In the purchasing process, most customers consider the products that are currently needed and regardless of whether the products have been purchased in the past. In the process of a customer purchasing a product and then the product purchased by another customer, a resource propagation network is formed. Based on the special case of customers and products, a bipartite network is finally constructed. By establishing a bipartite network between customers and products, the relationship of probability transfer between them is obtained and then the probability transfer value of each product can be obtained eventually.

When the resource transfer is performed, the traditional probability transfer model equally divides the resource value according to current node degree, which is the average resource value of the current node. Actually, the resource cannot be equally distributed and customers prefer to buy popular or high-weighted products than ordinary products. If a product is more important or there are more customers having bought the product, then more resources are assigned to the product, it is not reasonable to simply equally assign the resource values. Therefore, when performing resources allocation, a weight value should be added, and the amount of resources obtained by the node is proportional to the degree of the node.

Suppose a given bipartite network is composed of two types of nodes, denoted by *User* and *Item*, respectively. Based on the abovementioned ideas, an improved resources allocation model is proposed. The distribution of resources transferred from the *User* nodes to the *Item* nodes is represented as

$$R(B_j) = \sum\nolimits_{i=1}^{m} E_i \times R(A_i) \times \frac{k(B_j)}{\sum\nolimits_{l=1}^{k(A_i)} k(s_l)}, \tag{1}$$

where S represents all nodes connected to the node A_i, B_j denotes the *Item* nodes, A_i denotes the *User* nodes, m is the number of the *User* nodes, k represents the degree of the node, E represents the resource matrix of the bipartite network. When nodes of *Item* acquire resource from nodes of *User*, the larger the node degree of *User*, the more resource the node obtains. Likewise, the distribution of resources transferred from the nodes of *Item* to the nodes of *User*, is formulated as

$$R(A_i) = \sum\nolimits_{j=1}^{n} E_j \times R(B_j) \times \frac{k(A_i)}{\sum\nolimits_{l=1}^{k(B_i)} k(s_l)}, \tag{2}$$

where S represents all nodes connected to node B_j, B_j denotes the Item nodes, A_i denotes the *User* nodes, n is the number of *Item* nodes, k represents the degree of the node, E represents the resource matrix of the bipartite network. After multiple matrix transfers, the final resource value of each node remains stable. Therefore, the final ranking algorithm is written in (3), and the final resource value is a fixed one.

$$P(A)^n = (R(A_i) \times R(B_j))^n \tag{3}$$

The improved resource distribution of the bipartite network is formulated as a matrix. The new resource transfer matrix w_1 in which resources are transferred from *User* nodes to *Item* nodes, in the same way, the matrix w_2 in which resources are transferred from nodes of *Item* to the nodes of *User*. If the initial resources are concentrated on *User* nodes, the initial resource matrix is an identity matrix E_0. The final resource ranking can be written as

$$E_i = E_{i-1} \times w_1 \times w_2. \tag{4}$$

After continuous iteration, the final probability transfer value of each node tends to be stable, and then the sorting relationship between the two types of homogeneous nodes can be obtained respectively. For example, in a '*User-Item*' bipartite network (with m *User* nodes and n *item* nodes), we first assume that all resources are labeled with *User* nodes, then keep iteration in terms of Eq. 4, and a $m \times m$ *User* matrix can be awarded eventually. Furthermore, the values of each column of the matrix keep the same, namely, the ultimate resource value distributed to each customer keeps stable. Similarly, the ranking of *Item* can also be reached. In this way, we can obtain the ranking of the two homogeneous nodes in the bipartite network, and simultaneously we achieve the probability transfer matrix which plays a key role in the recommendation algorithm.

The recommendation algorithm is the core of the recommendation system as well as the main index to measure the quality of a recommendation system. Aiming at the unique structural information to the '*User-Item*' bipartite network and the latent information of customers to customers and products to products, a product recommendation scheme based on the bipartite network is proposed, which makes the recommendation results more accordant with practical circumstances. The detailed process is as follows:

Step 1: By analyzing the bipartite network, recommendations are conducted from customer and products, respectively. A recommendation scheme based on the bipartite network is proposed. Two bipartite network probability transfer matrices w_1 (customer to product) and w_2 (product to customer) are established, and then probability transfer matrices w_3 (customer to customer) and w_4 (product to product) are also derived.

Step 2: Handle w_3 and w_4 separately. Reducing dimension of the matrix mainly performed by finding the eigenvalue. The detailed procedures are as following: calculate the eigenvalues of the matrix, and sort the eigenvalues in descending order, then take out top-*K* eigenvalues and calculate the corresponding eigenvectors, until the similarity matrix is constructed adopting eigenvectors.

Step 3: Conduct fuzzy clustering on the similarity matrix. Calculate the modularity of the community detection by selecting different number of clustering centers [6], until the modularity keeps unreduced, and then the optimal community partition is obtained.

Step 4: Fuzzy clustering was carried out on w_3 and w_4, respectively, to obtain the community partition between products as well as community partition between customers.

Step 5: The final recommendation algorithm is proposed by selecting top-*N* similar customers and products.

Establish a bipartite network adjacency matrix, then generate the probability transfer matrix according to the above-mentioned probability transfer formula. Firstly, the *User* nodes are clustered. The resources of the above bipartite networks are allocated by Eq. 1. We can get two kinds of probability resource matrices as above mentioned w_1 and w_2, respectively.

Equation 5 can be obtained from the above two probability transfer matrices.

$$Q_{ij} = w_1 \times w_2 \tag{5}$$

where Q represents the resource distribution of each node in *User* nodes. The larger the value in matrix Q_{ij}, the more resource value that node i gives to node j. In order to reduce computation, it is necessary to reduce the dimensionality of the matrix Q, and then the PCA algorithm [7] is employed to solve the problem.

After the dimensionality reduction of the resource distribution matrix, the modularity based fuzzy clustering is performed on the vector matrix to conduct community partition. Two main goals for clustering are reached: the minimal distance between internal nodes and the cluster center of a community, and the maximal distance between each cluster center. Customers and products can be separately clustered, and then the community partition among customers and the community partition among products can be obtained, respectively. In order to achieve a better recommendation effect, information mining is performed on perspectives of customers and products respectively, to seek out corresponding recommended products set. The following is an example of recommendation based on similar customers.

In our daily life, two people with a similar lifestyle or background often buy similar products. Based on this, we analyze multiple customers who are in the same community and have a large similarity with the current customer, we can get a set of products that they purchased or were interested in, and these products can be recommended to the current customer. The detailed processes are described as follows:

1. Community detection of all customers is conducted offline to reduce the pressure of the server, and a general understanding of the community partition can be achieved.
2. Find the community in which the current customer is located. The preferences of internal nodes of a community are used to dig out the hidden preferences or required products for the current customer.
3. Process the internal nodes of the community, and calculate the similarity between these nodes and the current node. Set a threshold to get several customers with the maximum similarity.
4. Obtain the recently purchased or favorite products set of each customer, to constitute a recommended alternative product collection.
5. The products in the abovementioned collection are sorted according to the previously proposed algorithm, and top-*N* products are taken out to construct the final recommendation set of products.

In this way, the final set of recommended products is achieved in terms of finding similar customers.

3 Experimental Results and Analyses

3.1 Sorting Algorithm Based on Probabilistic Transfer Model

(1) The South women Dataset [8] indicates a network in which 18 women in the southern United States participate in 14 activities. If a woman participates in an activity, then there is a connection [9] between the activity and the woman. The BPRA algorithm, PageRank algorithm and HITS algorithm [10] were conducted on the South women dataset. The final resource values of each algorithm are obtained and sorted, and the result is plotted and ranked as shown in left panel of Fig. 1, and the time-consuming comparison is shown in the right panel.

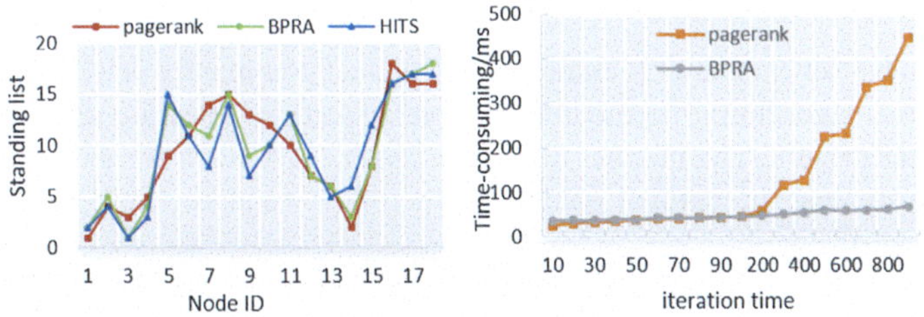

Fig. 1. The sorting Algorithm experimental results based on South women network

We can conclude from the left panel of Fig. 1 that the ranking results of the three algorithms are very close to each other. However, the BPRA algorithm wins more accurate division of the extreme points. As can be seen from right panel of Fig. 1, when the number of iterations increases, the PageRank algorithm takes much more time than the proposed BPRA algorithm.

(2) The Scotland dataset [9] presents the interrelations between some early Scottish companies and shareholders of Scottish chain companies in the early 20th century. There are 108 company nodes and 136 shareholder nodes in the datasets. The BPRA algorithm, PageRank algorithm and HITS algorithm are validated separately on the Scotland dataset, and ranking the company nodes. The experimental results of the three algorithms are shown in left panel of Fig. 2. It obviously demonstrates that the three algorithms achieve appropriate same effect in the Scotland datasets. In addition, the comparison of running time is conducted. By comparing the time spent on different number of iterations, we can get the distinctions of time consumed in the two algorithms. As shown in right panel of Fig. 2, when the number of iterations exceeds 100, the BPRA algorithm takes significantly less time than the PageRank algorithm, and meanwhile the time consumed on the BPRA algorithm does not increase with the number of iterations.

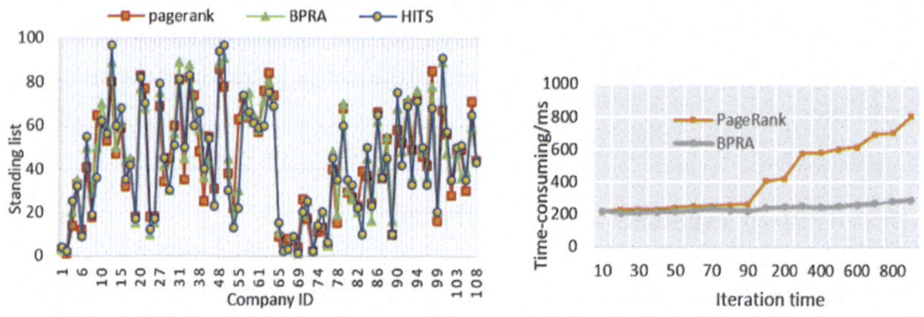

Fig. 2. The sorting Algorithm experimental results based on Scotland network

In summary, the proposed BPRA algorithm can obtain the same ranking results as traditional algorithms for bipartite networks. The proposed algorithm outperforms the traditional algorithms in ranking extreme points, and obtains obvious advantages in time-consuming, especially when the number of iterations increases.

3.2 Recommendation Algorithm Based on Bipartite Network

(1) The south women dataset is employed to conduct community detection to validate the BPRA algorithm. The modularity is calculated by selecting different number of cluster centers. The number of cluster centers is 4 when we gain the maximal modularity, and the optimal community partition result is obtained, as shown in left panel of Fig. 3.

The proposed algorithm is compared with the MP algorithm [11] (matrix decomposition algorithm), BRIM (sensing modularity algorithm) and Davis algorithm on Modularity, as shown in right panel of Fig. 3. The proposed algorithm and the BRIM algorithm hold the highest modularity and are able to achieve a better community structure.

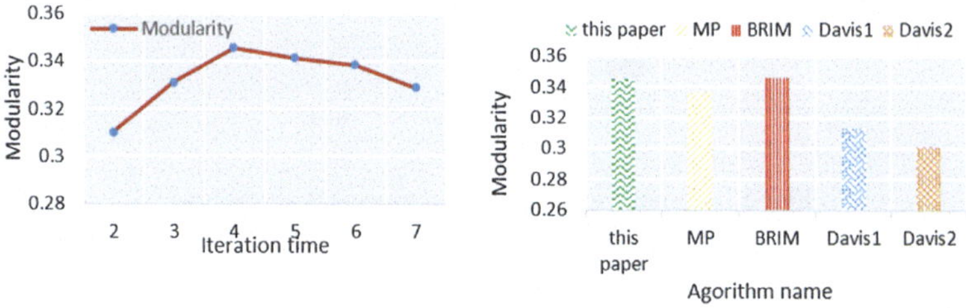

Fig. 3. The recommendation algorithm experimental results based on South women network

(2) In the same way, the community detection is conducted on the Scotland network. The relationship between the number of communities and the modularity is shown in left panel of Fig. 4. The modularity keeps growing when the number of community increases, and the maximal modularity reaches 0.671 when the number of communities turns to 19, and then the optimal community detection result is achieved.
In addition, modularity of the proposed algorithm, the BRIM algorithm and MP algorithm (matrix decomposition algorithm) are also compared as shown in right panel of Fig. 4. It is concluded from right panel of Fig. 4 that the proposed algorithm outperforms the other two with the highest modularity, and absolutely earns a better community detection result.

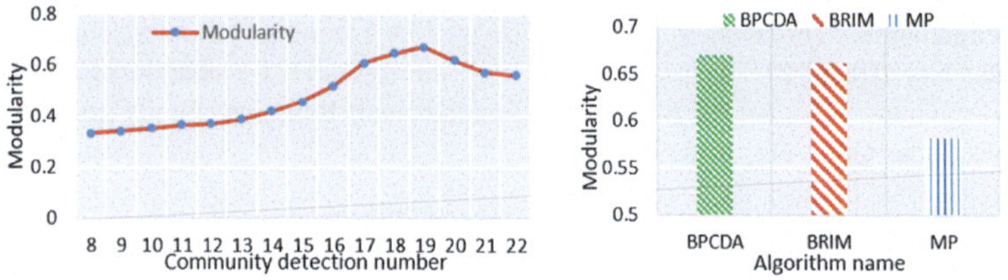

Fig. 4. The recommendation algorithm experimental results based on Scotland network

(3) Comparison of the proposed recommendation algorithm and the customer-based recommendation algorithm. The MSE, recall, and precision of the experiment were compared respectively, and the experimental results are shown in Fig. 5. The MSE of the proposed algorithm is significantly smaller than that of the customer-based collaborative filtering algorithm [12], which shows that the proposed algorithm is closer to the truth. Simultaneously, the precision and recall of the proposed algorithm is higher than that of the customer-based collaborative filtering algorithm. Therefore, based on the statements mentioned above, the effectiveness and efficiency of the proposed algorithm are proved.

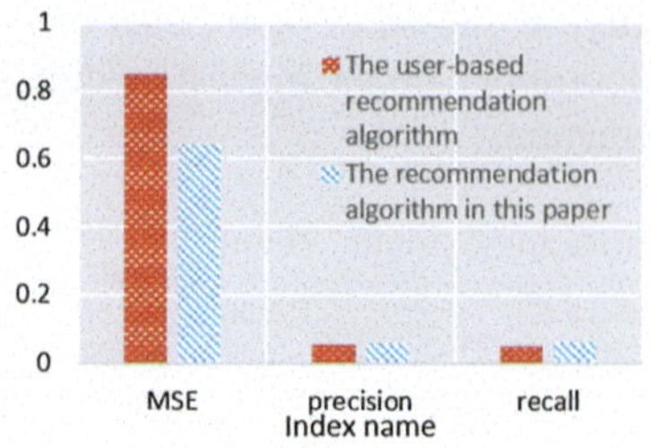

Fig. 5. Comparison of the recommended algorithms

4 Conclusions

Through the experiments conducting on the South women datasets and Scotland datasets, the accuracy and time consumption aspects of the proposed BPRA algorithm, PageRank algorithm and HITS algorithm are compared and analyzed respectively. The experimental results show that the BPRA algorithm outperforms the competitive methods in accuracy and efficiency. Comparison experiments show that the proposed algorithm is more competitive. The recommendation algorithm based on the bipartite network probability transfer model is compared with the customer-based collaborative filtering algorithm, evaluation by the MSE, recall and precision proves the effectiveness of the proposed recommendation algorithm. The proposed ranking algorithm and the recommendation algorithm are capable for customers rapidly and accurately seeking out the needed products. Simultaneously, the hidden information can also be employed to explore the potential customer demands to gain a better customer experience and improve the revenue of e-commerce systems.

Acknowledgements. This work is partially supported by Liaoning Natural Science Foundation under Grant No. 20170540320, the Doctoral Scientific Research Foundation of Liaoning Province under Grant No. 20170520358, the Fundamental Research Funds for the Central Universities under Grant No. N161702001, No. N172410005-2.

References

1. Ye, Q.: URL Rule Based Focused Crawl and its Application. Zhengjiang University (2007)
2. Page, L., Brin, S., et al.: The PageRank citation ranking: bringing order to the web. In: Stanford InfoLab, pp. 1–14 (1998)
3. Huang, D., Qi, H.: PageRank algorithm research. Comput. Eng. **32**(4), 145–146 (2006)
4. Ching, W.-K., Huang, X., Ng, M.K., Siu, T.-K.: Markov Chains Models, Algorithms and Applications. Tsinghua University Press, Beijing (2015)
5. Zhou, T., Ren, J., Medo, M., et al.: Bipartite network projection and personal recommendation. Phys. Rev. E Stat. Nonlinear Soft Matter Phys. **76**(4 Pt 2), 046115 (2007)
6. Newman, M.E.J.: Fast algorithm for detecting community structure in networks. Phys. Rev. E Stat. Nonlinear Soft Matter Phys. **69**(6 Pt 2), 066133 (2003)
7. Wu, X., Yan, D.: Analysis and research on method of data dimensionality reduction. Appl. Res. Comput. **26**(8), 2832–2835 (2009)

8. Yao, F., Chen, L.: Similarity propagation based link prediction in bipartite networks. Comput. Sci. **43**(4), 86–91 (2016)
9. Chen, B.L., Chen, L., Zou, S.R., et al.: Detecting community structure in bipartite networks based on matrix factorisation. Comput. Sci. **6**(6), 599–607 (2013)
10. Chairman-Karloff, H.: Proceedings of the Ninth Annual ACM-SIAM Symposium on Discrete Algorithms, ACM-SIAM Symposium on Discrete Algorithms. Society for Industrial and Applied Mathematics (1998)
11. Yang, B., Cheung, W., Liu, J.: Community mining from signed social networks. IEEE Trans. Knowl. Data Eng. **19**(10), 1333–1348 (2007)
12. Wu, H., Wang, Y.-J., Wang, Z., Wang, X., Du, S.-Z.: Two-phase collaborative filtering algorithm based on co-clustering. J. Softw. **21**(5), 1042–1054 (2010)

CDIA: A Feasible Community Detection Algorithm Based on Influential Nodes in Complex Networks

Xinyu Huang, Dongming Chen(✉), Tao Ren, and Dongqi Wang

Software College, Northeastern University, Shenyang 110169, China
chendm@mail.neu.edu.cn

Abstract. Community detection is a fundamental research in network science, which has attracted researchers all over the world to devoting into this work. However, the existing algorithms can hardly hold performance and efficiency simultaneously. Aiming at addressing the problem, and inspired by the force in physics, this paper defines the node influence from a network perspective. Afterwards, a novel approach to detect communities in terms of influential nodes is proposed. Furthermore, the vital nodes and overlapping nodes can be obtained. Series of experiments on synthetic and real-world networks are conducted, and the experimental results show that the proposed algorithm is capable and effective, which provides a reliable solution for analyzing network structure in-depth.

Keywords: Influential nodes · Nodes gravity · Complex network · Community detection · Overlapping nodes

1 Introduction

Networks are widely-used representation to describe the connectivity of a complex system, such as biological networks, transportation networks, power networks and so on. The basic model of network "graph theory" started from the "Seven Bridge Problem" of Königsberg in 1736. At the end of the 20th century, by employing the graph model, the famous small-world and scale-free features were discovered [1]. In 2002, Girvan and Newman proposed the community structure in networks, which opened up a continuous upsurge of relevant research [2]. The so-called communities are groups of nodes that are more strongly or frequently connected among themselves than with the other nodes of the network [3]. Hence, the community detection is to find the most reasonable clusters based on the network structure. However, scholars have proved that the problem of community detection belongs to the NP problem and cannot be accurately solved [4]. Thus, the modularity [5] proposed by Newman, is widely used as an evaluation for the community results. Applications of community detection include many fields, such as finding potential friends in social media [6], recommending products for customers [7], social opinion analysis [8] and so on.

Y. Liu et al. (Eds.): ICNC-FSKD 2019, AISC 1074, pp. 930–937, 2020.
https://doi.org/10.1007/978-3-030-32456-8_100

In above, community detection is one of the most popular topics in modern network science.

There has been numerous attempts to address this problem through diverse approaches [3,9], such as propagation methods (e.g. LPA), modularity optimizations (e.g. GN algorithm [4] and Louvain algorithm [10]), clustering algorithms (e.g. SCAN algorithm [11]) and so on. Most existing researches are based on non-overlapping communities. However, with the deepening of research, some scholars proposed that there are some nodes usually belong to multiple communities, which are called overlapping nodes [12]. The overlapping nodes play a bridge role in connecting different groups and they are the medium for information diffusion [13]. The algorithms for detecting overlapping communities, such as the CPM algorithm [12], ONES algorithm [14] and NOVER-based algorithm [15], have been criticized for high computational complexity or lower stability. Hence, there are still enormous work to optimize. Some nodes other than overlapping, are also very important in networks. The simplest method to evaluate node importance is by node's degree centrality, which means the greater the node's degree, the more important the node. The disadvantage of this method is that it is easy to miss the nodes that play a role in the bridge between communities. Therefore, the betweenness centrality [16] are proposed. However, it cannot be widely used in large-scale networks for its high computational complexity. In addition to indicators such as clustering coefficients and k-shell, some scholars have also proposed an improved NodeRank based on the famous indicator PageRank to evaluate the importance of nodes in large-scale weighted networks [17]. In above, a simple method to measure the influence of nodes is important for analyzing community structures.

In this paper, we refer to the force in physics, and re-define the node influence and gravity in networks. Afterwards, a community detection algorithm based on this influential nodes (for short CDIA) is proposed. The CDIA algorithm is capable for detecting important nodes and overlapping nodes, which has vital scientific significance and application value.

The rest of this paper is organized as follows: Sect. 2 introduces the related definitions and models, and designs nodes' influence and gravity; Sect. 3 proposes the algorithm and analyses the computational complexity; Sect. 4 conducts CDIA algorithms experiments on several real-world datasets, and compares with other algorithms in performance and efficiency; Sect. 5 summarizes the whole paper.

2 Models and Definitions

This paper employs the network model composed of nodes and edges, can be written as

$$G = \langle V, E, W \rangle, \tag{1}$$

where G represents a network model, V represents a set of nodes, E represents a set of edges and W is the set of weight for edges. For unweighted networks, the model can be simplified as $G = \langle V, E \rangle$.

Node Influence. As we known, the force in physics can be explained as $\Sigma\overrightarrow{F} = m\overrightarrow{a}$, where $\Sigma\overrightarrow{F}$ denotes the resultant force acting on the object. Analogously, a node's influence relies on the neighboring environment, where a node affects its neighbors and is affected by the neighbors likewise. Thus, the influence of node i is defined as

$$I_i = \sum_{j \in \Gamma(i)} (\frac{w_{ij}}{d_j}), \quad (2)$$

where I_i denotes node i's influence, $\Gamma(i)$ depicts the neighbors of node i, d_j is the degree of node j. w_{ij} is the weight of edge e_{ij}. For unweighted networks, $w_{ij} = 1$. By contrast, the node's influence considers more neighboring information than the degree centrality, and has a lower computational complexity than betweenness calculation.

Nodes Gravity. Gravity is the force between objects, which relates to the qualities of the objects and the distance between the two objects. We borrow the idea from the gravity in networks [21] and re-defined the concept as

$$G_{i,j} = \frac{I_i \cdot I_j}{d_{ij}^2}, \quad (3)$$

where $G_{i,j}$ represents the attractive force between node i and j, d_{ij} depicts the shortest path length between node i and node j. The gravity we focus on, in this paper, is utilized to describe the strength of the interactions between the two nodes.

3 Algorithm Description

From the viewpoint of influence and gravity, the nodes with higher influence are locating at the core positions of the local structure, hence, they can attract more nodes around them. The closer the node to a influential node, the greater gravity between them, and the easier to form edges with other surrounding nodes. Thus, we design the CDIA algorithm, and the process is listed as follows:

Step 1: Load the dataset and calculate the influence for each node.

Step 2: Obtain the node with maximum influence and merge it with the neighbors into a community. Then make the nodes of the community as divided.

Step 3: Obtain the node with the second maximum influence in the remainder undivided network and regard it as a new influential node, then merge it with the neighbors into a new community and make these nodes as divided.

Step 4: Repeat step 3 until all the nodes in the network are divided.

After processing with the above steps, we can get a very fine-grained network partitioning results, namely, large quantities communities. However, there are still several optimizations, as given in followings:

(1) Overlapping nodes identification. During the process of detecting overlapping communities, when an overlapping node has already been divided into an community, it is still under consideration in the next traverse. We will

assign the overlapping node to the community by considering on the maximum neighbors with the sum of gravity between them. The rule is represented by

$$C_i = \arg\max_{C_k \in C_i^n} \sum_{j \in \Gamma(i)} G_{i,j} \cdot \delta(j, C_k), \tag{4}$$

where C_i^n represents the neighboring communities of node i, $\delta(j, C_k)$ equals 1 if node j is in community C_k and 0 otherwise.

(2) Modifications for the unreasonable divisions. After the division of the first time, traverse all the nodes and find out the nodes insides a community that have more edges with other communities, put them into the community with maximum connections. If the connections between the competitive communities are the same, the nodes are divided into the community by referring Eq. (4).

(3) Controlling of the community size and amount. Suppose given the parameter of community size (remarked as μ) and total community amount

Algorithm 1: CDIA

Input: G, community size μ (optional), community amount σ (optional)
Output: community C

... // Load nodes V, edges E from G, and calculate the nodes influence.
$l = 0; V^l = V$;
while σ *and getCurrentCommunityAmount()* $< \sigma$ *or* V^l **do**
    $v^l_{max} = \arg\max_{v \in V^l} I_v$ // Marked as influential nodes.
    $v^l_{max}.group = v^l_{max}$
    for i *in* $v^l_{max}.neighbors$ **do**
        i.group += v^l_{max}
    $v^{l+1} = V^l.remove(C_k)$
    if $V^{l+1} == V^l$ **then**
        break;
    $l++$;
for i *in* V **do**
    if *len(i.group)* $!= 1$ *or notReasonable(i.group)* **then**
        if *len(i.group)* > 1 **then**
            i.overlapping = i.group // Marked as overlapping nodes.
        i.group = $\arg\max_{C_k \in C_i^n} \sum_{j \in \Gamma(i)} G_{i,j} \cdot \delta(j, C_k)$
if μ **then**
    for C_k *in* C **do**
        if *len(*C_k*)* $< \mu$ **then**
            for i *in* C_k **do**
                i.group = $\arg\max_{C_k \in C_i^n} \sum_{j \in \Gamma(i)} G_{i,j} \cdot \delta(j, C_k)$
return C

(remarked as σ), if the amount of detected communities is greater than σ, we will select the first σ communities, and then divide the remaindering nodes into the neighboring communities in terms of Eq. (4). Likewise, if the size of a detected community is less than μ, we will re-divide the nodes in this community and put them into the neighboring communities by referring Eq. (4). The pseudo code is shown in Algorithm 1.

Suppose the number of nodes is n, m depicts the edges amount, k is the average degree, l is the iteration times. Initially, the CDIA algorithm traverses the network and calculates the node influence, which requires $O(nk)$ complexity, the calculation of maximum influential nodes requires $O(n)$. Afterwards, it traverses the network, requiring the complexity of $O(2n)$. Generally, $O(nk)$ can be represented as $O(m)$, and the total complexity is less than $O(m + nl + 2n)$. Actually, l is far less than n/k for the scale-free feature of complex networks. Therefore, the total computational complexity can be reduced to $O(m)$, which is close to the complexity of the LPA ($O(n)$) and is superior to Louvain ($O(nlog(n))$). The algorithm stores the nodes, edges and extra information of nodes, which requires the space complexity of $O(n^2 + 2n)$.

4 Experiments

The experimental environment is CentOS Linux release 7.3.1611 (Core), and the system version number is Linux version 3.10.0. The CPU model is Intel(R) Xeon(R) CPU E5-26xx v3 with memory of 1839M. The programming environment of Python 3.6.3, and NetworkX (version 2.0).

The experiments in this paper are represented on NEUSNCP[1]. Statics of datasets and experimental results are given in Table 1. The CDIA algorithm performs well and has acceptable running time. Taking the result on Zachary's Karate Club network as example, the modularity of CDIA algorithm is 0.3718, which is greater than 0.2040 obtained from the real situation, so it is more meaningful. The NMI [18] (i.e, normalized mutual information) of CDIA is 0.6772, which is better than 0.5643 (obtained from Louvain algorithm). The detailed experimental results are shown as Fig. 1.

To verify the performance of the proposed CDIA algorithm, comparison experiments with the competitive algorithms are conducted and the experimental results are shown in Fig. 2.

As shown in Fig. 2, the performance of CDIA algorithm is close to the benchmark algorithms (e.g. Louvain algorithm and GN algorithm), which is better than CPM algorithm, SCAN algorithm and EM algorithm [26]. The comparison on running time is shown in the right panel of Fig. 2. The CDIA algorithm proposed in this paper has a lower time complexity, is close to the liner algorithm (e.g. LPA and SCAN), better than the Louvain algorithm, and much lower than the GN algorithm, EM algorithm. Besides, the optional parameters (μ and σ) improve the feasibility and fineness of the result.

[1] http://neusncp.com/api/cd.

Table 1. Statics of datasets and experimental results of performing with CDIA

Dataset	Nodes	Edges	Communities	Q	Runtime(s)
Southern women activities [19]	32	89	3	0.3294	0.0010
Zachary's karate club [20]	34	78	2	0.3718	0.0022
College students' course registration[a]	47	125	6	0.4827	0.0100
Dolphin social network [22]	62	159	5	0.5088	0.0076
Books about US politics [23]	105	441	5	0.5141	0.0132
American college football team [2]	115	613	10	0.5760	0.0178
LFR synthetic network [24]	500	3527	6	0.7211	0.3900
Coauthorships [25]	1589	2742	320	0.9207	3.6700

[a]The data of students registering course at Software College, Northeastern University is collected by us manually, where the nodes are the students and the edges represent the students have selected the same optional course. Top-10 overlapping nodes are 44, 6, 25, 46, 42, 41, 24, 23, 43 and 47, which are overlapping among over three communities and they are eager to learn in study groups.

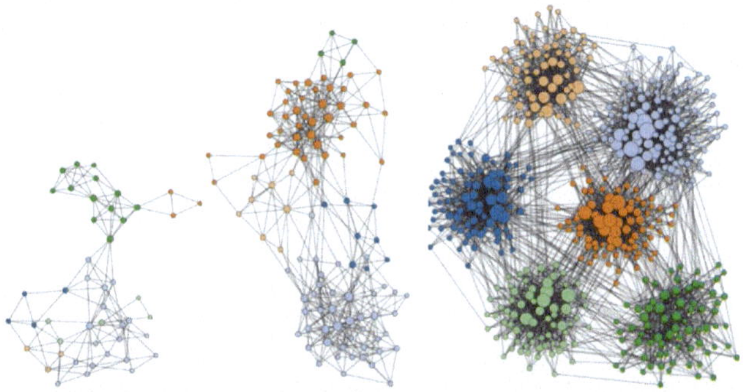

Fig. 1. Experimental results of CDIA conducting on College students' course (a), Politics books (b) and LFR synthetic network (c).

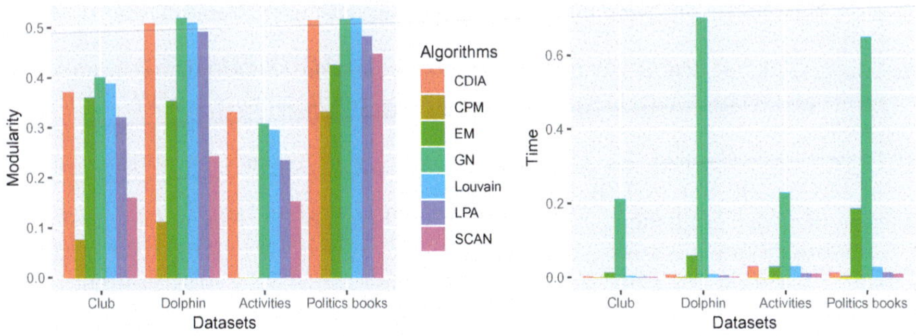

Fig. 2. Comparison on modularity Q and running time. The parameters are setted at the same level, the iteration times of the algorithms is set to 10, and the number of communities is in accordance with the result obtained by other algorithms. The minimum neighbors (μ) in the SCAN algorithm is 3, and the threshold (ε) is 0.6.

5 Conclusion

This paper proposes a community detection algorithm based on the influential nodes, which can obtain a finer-scale community detection result. In order to improve the performance, the optimization is conducted by overlapping nodes identification, unreasonable divisions modification and parameters controlling. The performance of the optimized algorithm is close to state-of-the-art algorithms, but with a lower computational complexity. At the same time, during the process of detecting communities, the overlapping nodes and influential nodes can be obtained, which benefits for analyzing the network structure in-depth. As part of future works, extending the CDIA algorithm on multilayer networks is of great significance.

Acknowledgements. This work is partially supported by Liaoning Natural Science Foundation under Grant No. 20170540320, the Doctoral Scientific Research Foundation of Liaoning Province under Grant No. 20170520358, the National Natural Science Foundation of China under Grant No. 61473073, the Fundamental Research Funds for the Central Universities under Grant No. N161702001, No. N172410005-2.

References

1. Newman, M.: Networks. An introduction. Astron. Nachr. **327**(8), 741–743 (2010)
2. Girvan, M., Newman, M.E.J.: Community structure in social and biological networks. Proc. Natl. Acad. Sci. U.S.A. **99**(12), 7821–7826 (2002)
3. Fortunato, S., Hric, D.: Community detection in networks: a user guide. Phys. Rep. **659**, 1–44 (2016)
4. Fortunato, S.: Community detection in graphs. Phys. Rep. **486**, 75–174 (2010)
5. Newman, M.E.J.: Fast algorithm for detecting community structure in networks. Phys. Rev. E **69**(6), 066133 (2004)
6. Zhu, J., Wang, B., Wu, B., Zhang, W.: Emotional community detection in social network. IEICE Trans. Inf. Syst. **100**(10), 2515–2525 (2017)
7. Chao-Yi, L.I., Zhang, Y.S., Tong, L.L.: A micro-blog personalized recommendation algorithm based on community discovery. Microelectron. Comput. **34**(6), 40–44 (2017)
8. Wang, D., Li, J., Xu, K., Wu, Y.: Sentiment community detection: exploring sentiments and relationships in social networks. Electron. Commer. Res. **17**(1), 103–132 (2017)
9. Rossetti, G., Cazabet, R.: Community discovery in dynamic networks: a survey. ACM Comput. Surv. **51**(2), 35 (2017)
10. Blondel, V.D., Guillaume, J., Lambiotte, R., Lefebvre, E.: Fast unfolding of communities in large networks. J. Stat. Mech: Theory Exp. **2008**(10), 10008 (2008)
11. Xu, X., Yuruk, N., Feng, Z., Schweiger, T.A.J.: SCAN: a structural clustering algorithm for networks. In: KDD, pp. 824–833 (2007)
12. Palla, G., Derenyi, I., Farkas, I.J., Vicsek, T.: Uncovering the overlapping community structure of complex networks in nature and society. Nature **435**(7043), 814–818 (2005)
13. Lei, Z., Pan, H., Su, Y., Zhang, X., Niu, Y.: A mixed representation-based multiobjective evolutionary algorithm for overlapping community detection. IEEE Trans. Cybern. **PP**(99), 1–14 (2017)

14. Chen, D.M., Sima, D.F., Huang, X.Y.: Overlapping community and node discovery algorithm based on edge similarity, no. iceit (2017)
15. Meghanathan, N.: A greedy algorithm for neighborhood overlap-based community detection. Algorithms **9**(1), 8 (2016)
16. Freeman, L.C.: A set of measures of centrality based on betweenness. Sociometry **40**(1), 35 (1977)
17. Pan, Y., Tan, W., Chen, Y.: The analysis of key nodes in complex social networks (2017)
18. Danon, L., Dazguilera, A., Duch, J., Arenas, A.: Comparing community structure identification. J. Stat. Mech. **2005**(09), 09008 (2005)
19. Sehgal, U., Kaur, K., Kumar, P.: Notice of violation of IEEE publication principles - the anatomy of a large-scale hyper textual web search engine. In: 2009 Second International Conference on Computer and Electrical Engineering, Dubai, pp. 491–495 (2009)
20. Zachary, W.W.: An information flow model for conflict and fission in small groups. J. Anthropol. Res. **33**(4), 452–473 (1977)
21. Li, Z., Ren, T., Ma, X., Liu, S., Zhang, Y., Zhou, T.: Identifying influential spreaders by gravity model. Sci. Rep. **9**(1), 8387 (2019)
22. Lusseau, D., Schneider, K., Boisseau, O.J., Haase, P., Slooten, E., Dawson, S.M.: The bottlenose dolphin community of Doubtful Sound features a large proportion of long-lasting associations. Behav. Ecol. Sociobiol. **54**(4), 396–405 (2003)
23. Newman, M.E.J.: Modularity and community structure in networks. Proc. Natl. Acad. Sci. U.S.A. **103**(23), 8577–8582 (2006)
24. Lancichinetti, A., Fortunato, S., Radicchi, F.: Benchmark graphs for testing community detection algorithms. Phys. Rev. E **78**(4), 046110 (2008)
25. Newman, M.E.J.: Finding community structure in networks using the eigenvectors of matrices. Phys. Rev. E **74**(3), 036104 (2006)
26. Newman, M.E.J., Leicht, E.A.: Mixture models and exploratory analysis in networks. Proc. Natl. Acad. Sci. **104**(23), 9564–9569 (2007)

Context-Sensitive Cross- and Auto-correlation Based Supervised Change Detection

Chao Li[1(✉)], Huiying Ru[1], and Xudong Ru[2]

[1] Hebei University of Architecture Engineering, Zhangjiakou, Hebei, China
sammy297@foxmail.com
[2] Beijing Normal University, Beijing, China

Abstract. A supervised method for change detection from remote sensing images based on cross- and auto-correlation is presented. The work is presented as a generic approach to image change detection whereas forest fire bun scars detection just being the specific test used for validation purpose. Using only spectral information of the images for detecting change ignores the within and between image correlation and is not capable of representing change in temporal and spatial domains. The proposed method is designed to improve change detection accuracy by combining spectral information with both local cross- and auto-correlation. A novel way of a simultaneous incorporation of these associations by means of cross- and auto-correlation is put forward and tested here under the logistic regression framework. The proposed method showed the increment of 9% in Kappa accuracy as compared to simple logistic regression ran over spectral information alone. It demonstrated that the use of cross- and auto-correlation leads to substantial increase in burned area detection accuracy.

Keywords: Context-Sensitive · Cross-correlation · Auto-correlation · Supervised · Change detection

1 Introduction

Land use and land cover change detection has been an active area of research and a large number of change detection techniques have been developed [1–6]. New techniques still continue to be developed by efficiently employing spectral shapes [7], balancing misclassification errors [8] and automatic thresholding [9, 10] for improving change detection results.

Detection of a burned area (as a result of fire) is primarily important because of its significant effect on soil, plants, landscape and environment. Visually a burned patch can be detected very easily, but it is challenging for an algorithm to discriminate it automatically due to its diverse spatial and spectral characteristics [11].

The selection of burned area detection methods heavily depends on the characteristics of landscape, the size of the study area and the type of satellite data. The most

Supported by University-Level Research Fund of Hebei University of Architecture (No. B-201803) and Zhangjiakou Science and Technology Research and Development Program (No. 1821142I).

Y. Liu et al. (Eds.): ICNC-FSKD 2019, AISC 1074, pp. 938–946, 2020.
https://doi.org/10.1007/978-3-030-32456-8_101

commonly used methods to estimate burned areas and fire severity from remote sensing images are based on spectral indices [12], spectral mixture analysis [13], fixed and adaptive thresholding algorithms [14] and simulation techniques [15]. Spectral indices are the most widely used technique. Bastarrika et al. [11] used these indices as a first step to only detect severely burned areas with an objective to reduce false alarms. In second step, the results are further improved by using region-growing algorithm reducing missed alarms. If a change has occurred then the same geographic area tends be highly uncorrelated between two dates of imagery. Im et al. [16] exploited this fact using the neighbourhood cross-correlation between two images and reported significant improvement in the change detection results. Instead of using cross-correlation, Myint et al. [17] employed spatial metrics for classification. They were able to support that use of local variability improves classification accuracy. Moreover, reference [18] also used spatial correlation with random forests for land cover classification. They showed that random forests performed significantly better by incorporating spatial neighbourhood information. Kontoes et al. [19] compared two burn scar mapping approaches. Gong et al. [20] combined fixed threshed and adaptive contextual threshold for active fire detection and were able to reduce false alarms.

As mentioned above, it is evident that the use of spatial association [17, 18] or temporal association [16] improve classification accuracy over spectral information alone. However, simultaneous use of spatial and temporal associations still needs to be investigated for detection of burned area. Motivated by this fact, this study is designed to improve change detection capability by utilizing both cross- and auto-correlation with logistic regression for burned area. It is believed that burned and unburned regions may have similar spectral signatures but dissimilar cross- and auto-correlation. It is assumed that pre- and post-fire areas will tend to be uncorrelated. Measuring local cross-correlation between these two areas provides valuable information regarding extent and location of fire. Further more, burned area will exhibit homogeneity represented by high auto-correlation and unburned area will exhibit heterogeneity represented by low auto-correlation, hence serving as surrogate for change. Analyzing the local variability of burned and unburned areas helps in their discrimination. This concept of utilizing temporal and spatial associations with the spectral information for burned area detection has not been applied so far for bi-temporal remote sensing applications.

2 Local Measures of Spatial and Temporal Dependence

There exists two type of dependence in bi-temporal images. One is spatial dependence and other is temporal dependence.

Temporal dependence in change detection framework implies the correlation between two images acquired at same geographical area at two different time instances. The concept of spatial dependence implies that values observed at a given pixel depend on values observed at neighboring pixels. We are only interested in describing the local structure around the pixel under examination and focus on local measures of spatial correlation only.

It is assumed that neighboring pixels are more correlated to each other than those pixels which are at distance. Therefore, a square positive definite weight matrix W is used to control the contiguity structure of the pixels. Figure 1 depicts the layout of a map in which several objects are numbered from 1 to 28. For instance, in order to define the first order contiguity of the object no. 5, it takes value of 1 for its neighbors 2 and 6 in weight matrix as they share a common border with object no. 5 and 0 otherwise. If we consider a second order contiguity for object no. 5 then object no. 1, 3 and 7 besides object no. 2 and 6 will take value of 1 in weight matrix and zero otherwise. The specification of distance to define which elements are nonzero in the spatial weight matrix is a matter of considerable arbitrariness [21]. The negative spatial correlation indicates that neighboring values of ith pixel are mutually dissimilar. On the contrary, the positive spatial correlation is observed when similar values occur in the neighborhood of the ith pixel.

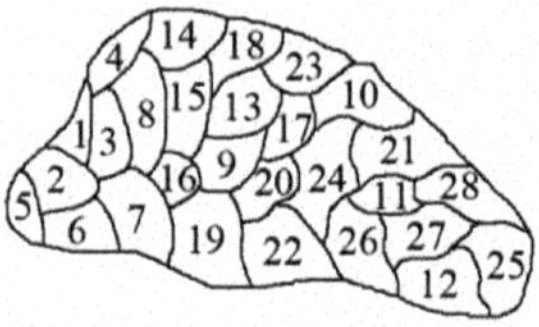

Fig. 1. Contiguity of objects in a map

3 Cross-correlation

Pearson's correlation is an indicator of temporal dependence and widely used as a measure of cross-correlation between two images. The mathematical framework of cross-correlation is presented below. For two coregistered multispectral images I_1 and I_2, size $p \times q \times b$, where p denotes number of lines, q denotes number of pixels per line and b denotes number of bands. For any pixel located at coordinates (r, s) in I_1 and I_2, the following formula calculates cross-correlation $Corr(r, s)$ within its spatial neighbourhood $(l \times l)$ controlled by weights (w_{rs}).

$$Corr(r,s) = \frac{\sum_{r=1}^{p}\sum_{s=1}^{q}\sum_{k=1}^{b} w_{rs}[I_1(r,s,k) - \bar{I}_1][I_2(r,s,k) - \bar{I}_2]}{\left\{\sum_{r=1}^{p}\sum_{s=1}^{q}\sum_{k=1}^{b} w_{rs}[I_1(r,s,k) - \bar{I}_1]^2\right\}^{1/2}\left\{\sum_{r=1}^{p}\sum_{s=1}^{q}\sum_{k=1}^{b} w_{rs}[I_2(r,s,k) - \bar{I}_2]^2\right\}^{1/2}} \tag{1}$$

where

$$\bar{I}_1 = \frac{1}{bl^2}\sum_{r=1}^{p}\sum_{s=1}^{q}\sum_{k=1}^{b} w_{rs} I_1(r,s,k)$$

$$\bar{I}_2 = \frac{1}{bl^2}\sum_{r=1}^{p}\sum_{s=1}^{q}\sum_{k=1}^{b} w_{rs} I_2(r,s,k)$$

l is the size of spatial neighbourhood around $(r,s)^{\text{th}}$ pixel, w_{rs} are the weights specified as $w_{rs} = 1$ if pixels fall in $l \times l$ spatial neighbourhood and $w_{rs} = 0$ if pixels fall outside $l \times l$ spatial neighbourhood. The correlation coefficient value $Corr(r,s)$ ranges from -1 to $+1$, representing perfect negative to perfect positive dependence respectively.

4 Auto-correlation: Getis Statistic

Spatial dependence may be indicated using auto-correlation which is rooted on the idea of reference [22].

The mathematical framework for calculation of the Getis statistic [23] is given below. Since Getis statistic is a measure of local auto-correlation it will be operated on each band of I_1 and I_2 images separately. Let the total number of pixels in each band are $n = p \times q$. For any pixel at location $i(i = 1, \ldots, n)$, the Getis statistic is given by

$$G_i^* = \frac{\sum_{j=1}^{n} w_{ij}(d)x_j - W_i^* \bar{x}}{s[W_i^*(n - W_i^*)/n - 1]^{1/2}} \tag{2}$$

where $w_{ij}(d)$ are the contiguity-based weights which takes value of 1 if pixel j is within distance d from pixel i and takes value of zero otherwise, $\sum_j w_{ij}(d)x_j$ is the sum of all those pixels which are within the distance d from pixel i, w_i^* is the count of pixels within distance d from pixel i, $\bar{x}$ and s are the mean and standard deviation of the entire image, respectively. A large positive G_i^* is observed for a cluster of high pixel values, while a cluster of low pixel value is reflected by negative G_i^* value.

5 Logistic Regression

Logistic regression (LR) is a supervised statistical learning algorithm which is able to model dichotomous response variable with several predictor variables. In this study, response variable Y is a binary variable representing presence or absence of change (fire), whereas predictor variables $X_i(i = 1, 2, \ldots, k)$ are bi-temporal original images supplemented by cross- and auto-correlation. We want to estimate conditional distribution of the response Y, given the input predictor variables, $\Pr(Y|X)$. Let the burned and unburned classes are labelled as 1 and 0, respectively. Assume that $\Pr(Y = 1|X = x) = P(x;\theta)$ for some function P parameterized by θ. In case of binary response variable, logit link function is used to characterize P in LR given by

$$Logit(P) = \beta_0 + \beta_1 X_1 + \beta_2 X_2 + \cdots + \beta_k X_k \tag{3}$$

where $Logit(P) = \log[P/(1 - P)]$, P is the probability of observing fire, β_0 is intercept, X_1 to X_k are predictor variables that help in predicting binary response either 1 (burned) or 0 (unburned) for each pixel, β_0 to β_k are regression coefficients.

To minimize the misclassification rate, we will predict $Y = 1$ when $P \geq 0.5$ and $Y = 0$ when $P \leq 0.5$. It implies that whenever $\beta_0 + \beta_1 X_1 + \beta_2 X_2 + \ldots + \beta_k X_k$ is non-negative, the predicted label is 1 and 0 otherwise. The decision boundary separating the two predicted classes is a solution of $\beta_0 + \beta_1 X_1 + \beta_2 X_2 + \ldots + \beta_k X_k = 0$ and gives us a linear classifier. The distance from decision boundary is $\frac{(\beta_0 + \beta_1 X_1 + \beta_2 X_2 + \ldots + \beta_k X_k)}{\sqrt{\beta_0^2 + \beta_1^2 + \beta_2^2 + \ldots \beta_k^2}}$. LR not only gives where the decision boundary between classes is, but also provides the class probabilities depend on distance from boundaries.

6 Context-Sensitive Cross- and Auto-correlation Based Supervised Change Detection Approach

The flow chart for above mentioned proposed methodology is given in Fig. 2 and can be summarized in the following steps. Consider two multispectral images I_1 and I_2 of size $p \times q \times b$ acquired over same geographical region at two different time instances t_1 and t_2, respectively. It is assumed that change in terms of fire has been occurred between t_1 and t_2.

Step 1. Calculate the cross-correlation between every pixel of I_1 and I_2 images by using Eq. (1).
Step 2. Calculate the auto-correlation of every pixel for each band of I_1 and I_2 images separately by using Eq. (2).
Step 3. Input original images I_1 and I_2, cross-correlation image obtained by step 1 and auto-correlation images obtained by step 2 into LR as predictor variables.
Step 4. Run LR using Eq. (3) to obtain a change map.
Step 5. Evaluate accuracy of change map produced in step 4 by comparing it with reference map.

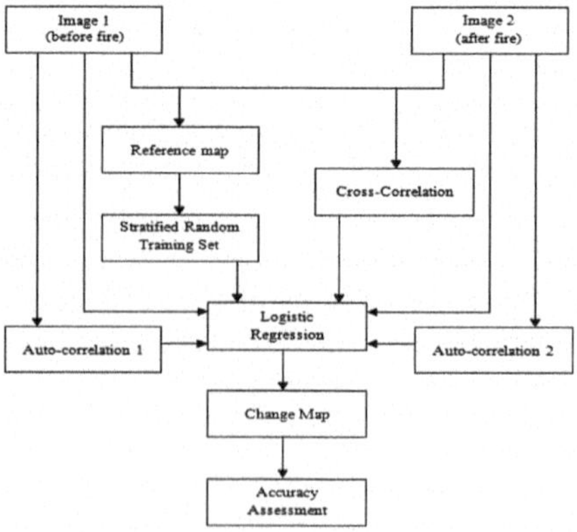

Fig. 2. Flowchart of the proposed methodology.

7 Experiment

7.1 Dataset Related to California 2007

The dataset used in our experiment consists of two multispectral images acquired by Landsat 5 Thematic Mapper (TM), covering some part of Los Angeles, California, on January 12, 2007 and September 9, 2007. These images were obtained from the U.S. Geological Survey (USGS). A subset of 3381 × 2500 pixels has been selected as test data from entire available Landsat scene and only Near- Infrared (0.76 – 0.90 μm) and Mid-Infrared (2.08 – 2.35 μm) bands with a spatial resolution of 30 m were used for analysis. Figure 3(a) and (b) show January and September images respectively. It is apparent that between the aforementioned two dates, a devastating wildfire destroyed a significant portion of vegetation. A reference map (Fig. 4) corresponding to the location of burned area was manually defined by detailed visual analysis of available bi-temporal images and their difference image. Reference map contains 15.17% burned and 84.83% unburned pixels. Experiment was performed, in an automatic manner, to detect this burned area as accurately as possible.

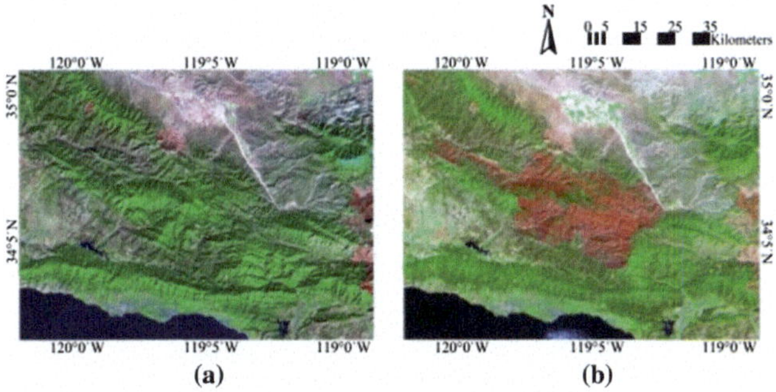

Fig. 3. Location of Study Area (7, 4 and 2 bands as R, G and B) (a) before fire image January 2007 and (b) after fire image September 2007.

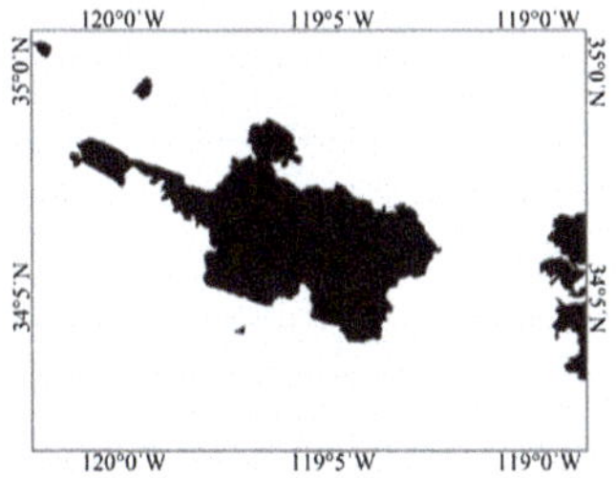

Fig. 4. Reference map corresponding to the location of burned area.

7.2 Results and Discussion

Figure 5 shows the burned area change detection map using simple LR. The resulting change detection map can be easily compared with reference map (Fig. 4). This visual inspection reveals that simple LR is able to capture the global structure of the burned area. However, several false alarms can be identified throughout the change map. Simple LR is also unable to detect large number of burned pixels especially at the centre and middle right of the image, hence resulting in a large number of missed alarms. This large amount of false and missed alarms is due to confusion in spectral signatures of burned and unburned areas, and simple LR is not able to separate them.

Figure 6 shows the burned area detection map obtained by proposed LR with cross- and auto-correlation. In spite of missed and false alarms, the resulting change detection map looks much similar to its reference map (Fig. 4). LR with cross- and auto-correlation is able to detect the burned area at the centre of the image in a much better way as compare to simple LR. Furthermore, it also reduces false alarms at top-middle of the image. LR with cross- and auto-correlation appears to be visually better than simple LR; however quantitative analysis is

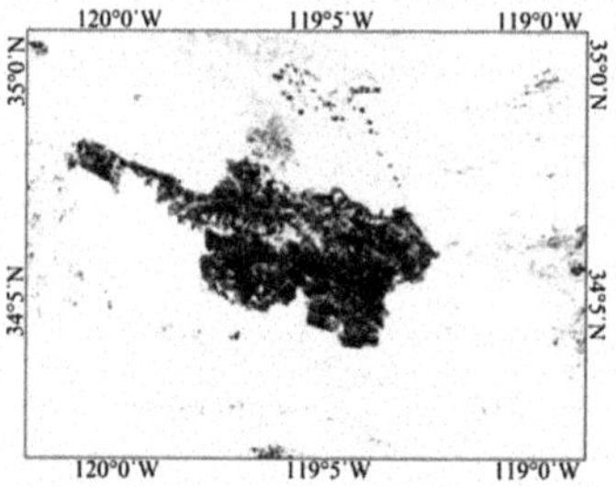

Fig. 5. Burned area detection map using simple LR.

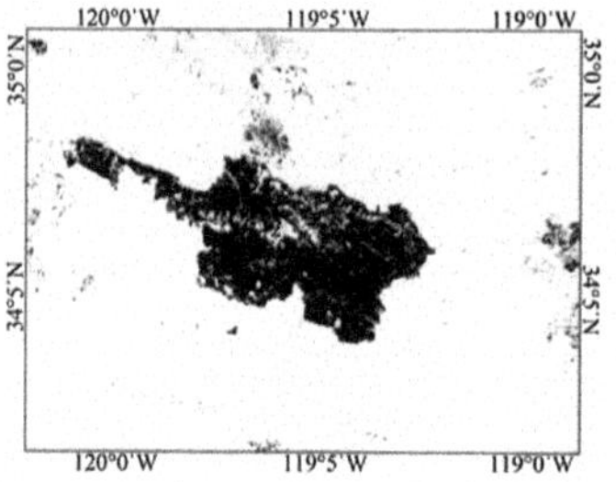

Fig. 6. Burned area detection map using proposed LR with cross- and auto-correlation needed for the validation of the proposed technique.

Quantitative analysis was performed in terms of both accuracies (OA and K) and errors (MA and FA). More significance is attached with MA, as it is better to have less MA because it reflects the actual burned area that the algorithm failed to detect. Furthermore, K should be as large as possible. As Table 1, the simple LR has yielded the OA of 93.59% and K = 0.71, which is significantly lower than LR with cross- and auto-correlation having OA = 95.35% and K = 0.80. In case of LR with cross- and auto-correlation, there is an increase of 1.76% in OA and 9% in K as compared to simple LR. For burned area detection problem when burned area is relatively smaller than unburned area (as in this case 15.17% burned and 84.83% unburned pixels), more importance is attached with K, because it is robust than OA and take care of those changes occurred by chance. Proposed LR with cross- and auto-correlation is also able to reduce both MA and FA, however significant reduction of more than 10% is observed in MA when compared with simple LR. These accuracy assessment indicators suggest that the integration of cross- and auto-correlation improves burned area detection accuracy significantly as compared to use of the spectral information alone.

Table 1. Accruacy assessment of simple LR and proposed LR with cross- and auto-correlation.

Technique	OA(%)	K	MA(%)	FA(%)
Simple LR	93.59	0.7120	37.58	0.8356
LR with cross- and auto-correlation	95.35	0.8002	27.04	0.6446

8 Summary

The simultaneous use of cross- and auto-correlation within proposed logistic regression framework increased the burned area detection accuracy. Experimental results showed that the proposed method improves Kappa coefficient by 9%. The increase in the accuracy of the proposed method can be attributed to the utilization of two different types of local associations, namely temporal and spatial, captured by cross- and auto-correlation respectively. Moreover, their statistical significance showed that incorporating cross- and auto-correlation in proposed LR is meaningful. We are able to support that by simultaneously incorporating both local temporal and spatial associations, we can improve burned area detection capability over spectral information alone. The proposed methodology can be applied to similar images for further validation.

References

1. Radke, R.J., et al.: Image change detection algorithms: a systematic survey. IEEE Trans. Image Process. **14**(3), 294–307 (2005)
2. Singh, A.: Review Article Digital change detection techniques using remotely-sensed data. Int. J. Remote Sens. **10**(6), 989–1003 (1989)
3. Coppin, P., et al.: Review Article Digital change detection methods in ecosystem monitoring: a review. Int. J. Remote Sens. **25**(9), 1565–1596 (2004)
4. Lu, D., et al.: Change detection techniques. Int. J. Remote Sens. **25**(12), 2365–2401 (2004)
5. Bhagat, V.S.: Use of remote sensing techniques for robust digital detection of land: a review. Recent Patents Space Technol. **2**(2), 123–144 (2012)
6. Hussain, M., et al.: Change detection from remotely sensed images: from pixel-based to object-based approaches. ISPRS J. Photogram. Remote Sens. **80**, 91–106 (2013)
7. Chen, J., et al.: A spectral gradient difference based approach for land cover change detection. ISPRS J. Photogram. Remote Sens. **85**, 1–12 (2013)
8. Puertas, O.L., Brenning, A., Meza, F.J.: Balancing misclassification errors of land cover classification maps using support vector machines and Landsat imagery in the Maipo river basin (Central Chile, 1975–2010). Remote Sens. Environ. **137**, 112–123 (2013)

9. Ghaffarian, S., Ghaffarian, S.: Automatic histogram-based fuzzy C-means clustering for remote sensing imagery. ISPRS J. Photogram. Remote Sens. **97**, 46–57 (2014)
10. He, P., et al.: A novel dynamic threshold method for unsupervised change detection from remotely sensed images. Remote Sens. Lett. **5**(4), 396–403 (2014)
11. Bastarrika, A., Chuvieco, E., Martín, M.P.: Mapping burned areas from Landsat TM/ETM + data with a two-phase algorithm: balancing omission and commission errors. Remote Sens. Environ. **115**(4), 1003–1012 (2011)
12. French, N.H.F., et al.: Using landsat data to assess fire and burn severity in the North American boreal forest region: an overview and summary of results. Int. J. Wildland Fire **17**, 443–462 (2008)
13. Quintano, C., Fernández-Manso, A., Roberts, D.A.: Multiple endmember spectral mixture analysis (MESMA) to map burn severity levels from Landsat images in Mediterranean countries. Remote Sens. Environ. **136**, 76–88 (2013)
14. Li, Z., et al.: A review of AVHRR-based fire active fire detection algorithm: principles, limitations, and recommendations. In: Ahern, F., Goldammer, J.G., Justice, C. (eds.) Global and Regional Vegetation Fire Monitoring from Space, Planning and Coordinated International Effort, pp. 199–225 (2001)
15. De Santis, A., Chuvieco, E., Vaughan, P.J.: Short-term assessment of burn severity using the inversion of PROSPECT and GeoSail models. Remote Sens. Environ. **113**(1), 126–136 (2009)
16. Im, J., Jensen, J.R.: A change detection model based on neighborhood correlation image analysis and decision tree classification. Remote Sens. Environ. **99**(3), 326–340 (2005)
17. Myint, S.W., Wentz, E.A., Purkis, S.J.: Employing spatial metrics in urban land-use/land-cover mapping: comparing the Getis and Geary indices. Photogram. Eng. Remote Sens. **73** (12), 1403–1415 (2007)
18. Ghimire, B., Rogan, J., Miller, J.: Contextual land-cover classification: incorporating spatial dependence in land-cover classification models using random forests and the Getis statistic. Remote Sens. Lett. **1**(1), 45–54 (2010)
19. Kontoes, C.C., et al.: A comparative analysis of a fixed thresholding vs. a classification tree approach for operational burn scar detection and mapping. Int. J. Appl. Earth Observ. Geoinf. **11**(5), 299–316 (2009)
20. Gong, P., et al.: An integrated approach to wildland fire mapping of California, USA using NOAA/AVHRR data. Photogram. Eng. Remote Sens. **72**, 139–150 (2006)
21. Anselin, L., Bera, A.: Spatial dependence in linear regression models with an introduction to spatial econometrics. In: Ullah, A., Giles, D.E.A. (eds.) Handbook of Applied Economic Statistics. CRC Press, New York (1998)
22. Tobler, W.R.: A computer movie simulating urban growth in the detroit region. Econ. Geogr. **46**, 234–240 (1970)
23. Getis, A., Ord, J.K.: The analysis of spatial association by use of distance statistics. Geogr. Anal. **24**(3), 189–206 (1992)

Fuzzy Applications: Uncertainty Management

Cultural Ranking of Countries Using the Analytic Hierarchy Process Methodology

Jinling Li[1,2(✉)], Yuying Yang[1], Thomas L. Saaty[3], and Haixiang Guo[1,4,5]

[1] School of Economics and Management, China University of Geosciences, Wuhan 430074, People's Republic of China
{35979923,1458564932}@qq.com, faterdumk0732@sina.com
[2] Wuhan University of Engineering Science, Wuhan 430200, People's Republic of China
[3] University of Pittsburgh, Pittsburgh, PA 15260, USA
saaty@katz.pitt.edu
[4] Key Laboratory of Strategic Studies, Ministry of Land and Resources, China University of Geosciences, Wuhan 430074, People's Republic of China
[5] Mineral Resource Strategy and Policy Research Center, China University of Geosciences, Wuhan 430074, People's Republic of China

Abstract. Different countries have different cultures, but it is difficult to say which culture is best directly. This paper mainly established a hierarchy for the ranking of cultures to prioritize and quantify thinking about and comparing different cultures. It does not only pay attention to the different features of cultures but also focuses on manifestations of cultures. There are 9 criteria and 27 subcriteria in our ranking of cultures. To find some leading cultures in the world, we rank the cultures of different counties by using the Analytic Hierarchy Process (AHP). After we get the priorities of the subcriteria according to judgment matrix obtained by questionnaires, we create different intensity categories for each subcriterion. In the end, we obtain the top 5 countries: the United States, Denmark, Australia, Norway and Canada. The result may be significant for advocating the spread of the better cultures throughout the world. What do they have to offer that others need to emulate? That is what is considered in this paper.

Keywords: Cultural ranking · AHP · Culture globalization

1 Introduction

Humans are hard-wired to adopt their communities' norms, and these norms can include rules for how to treat others - including whether to tolerate diversity or to inhibit outsiders. The difference between tribal societies and the modern world is that now social media make it possible for isolated individuals to adopt the values of a movement halfway around the world. Thus the study of cultures helps us to bring the world together despite climate, color, dressing, food, eating habits, education, environment, genetics, geography, government, language, living style, number of people, race, religion, sexuality, travel and tourism and other factors [1]. Culture, which includes some of these factors, is, in the words of E.B. Tylor, culture is a complex

Y. Liu et al. (Eds.): ICNC-FSKD 2019, AISC 1074, pp. 949–963, 2020.
https://doi.org/10.1007/978-3-030-32456-8_102

whole, including knowledge, beliefs, art, ethics, law, customs, and any skills and habits that people acquire as members of society [2]. According to the Cambridge English Dictionary, culture involves people's lifestyles, especially the customs and beliefs of specific people at specific times [3]. Terror Management Theory holds that culture is a series of activities and worldviews that make human beings value themselves above the mere physical aspects of existence [4]. Culture is the core concept of anthropology, covering many aspects of human society's concepts, behaviors, knowledge systems, customs and habits. The term is used in general as an evolutionary ability to use symbols to classify and express experience, and to act imaginatively and creatively. This ability arose with the evolution of behavioral modernity in humans around 50,000 years ago. This capacity is often thought to be unique to humans, although some other species have demonstrated similar, though much less complex abilities for social learning [5]. Geert Hofstede's short definition of "culture" is: "Culture is the collective programming of the brain, which distinguishes one category or group of people from others" [6]. Thus culture is an abstract concept that is difficult to determine which aspects or criteria characterize it as better or worse.

With globalization, we have become familiar with more and more cultures. Cultural change has gradually become an important development strategy for national competitiveness. Many scholars are occupied today with the study and understanding of different cultures.

Hofstede's Culture's Consequences explores the domain of studying international organizations [7]. He collected data from IBM, a large multinational company, and analyzed data from 49 different countries. Through his empirical data analysis, he concluded that "organizations are culturally-bounded". To analyze work-related cultural values in different countries, he identified four work-related cultural dimensions, including power distance (the power inequality between superiors and subordinates), uncertainty avoidance (people's tolerance of ambiguity), individualism-collectivism (how people value themselves and their groups/organizations), and masculinity (the gender roles in organizations) [8]. Then he proposed a fifth cultural dimension, called Confucian Work Dynamic, which has been renamed as Long-Term Orientation. In addition to identifying work-related cultural dimensions, Hofstede proposed six areas for continued research: (1) non-Anglo cultural dimensions; (2) additional countries; (3) cultural changes over time; (4) subcultures, such as regional, occupational, and organizational cultures; (5) the consequences of cultural dimensions; and (6) foreign organizational and management theories [8].

The research group led by the late Bob House "Global Leadership and Organizational Behavior Effectiveness" (GLOBE) had been working hard to break new ground with regard to extensive cross-cultural investigations. GLOBE researchers used for criteria in their study: uncertainty avoidance, power distance, institutional collectivism, in-group collectivism, assertiveness, future orientation, performance orientation, and humane orientation [9]. Researches on culture in China have mostly focused on cultural competitiveness and cultural differences. Xiong Zhengde put forward the idea that national cultural intangible power is made up of cultural value, cultural productivity, the guiding force of cultural institutions and cultural industry competitiveness. He constructed an evaluation index system of national cultural soft power by Principal Component Analysis [10].

Gannon's book "Understanding Global Cultures" introduces a new concept and method—the cultural metaphor—for understanding and comparing cultures around the globe. He constructed a specific metaphor for each country he studied [11].

Culture is usually a collective phenomenon because it is at least partially shared by people who live in the same social environment now or in the past, and this environment is where people learn culture [12]. The environment affects the culture of a group rather than the genes of the individuals in that group.

In sum, in all previous studies, researchers have focused on the concept of culture and its characteristics rather than on a comprehensive evaluation of the particular performance of that culture. With the deepening and expansion of globalization, cultural exchange activities between different peoples and nations of the world are more frequent, cultural identity and cultural conflict are the growing trends, so that the concept of global culture has spread worldwide. Global cultures in all nations and countries agree that the leading human cultural factors are the backbone and foundation of their development. Culture bears the value of the human community, and the value of the different cultural features "harmony without uniformity" is a significant factor in all these considerations. To promote the development of a worldwide culture, ranking cultures can help us discover the attributes of better cultures [10].

This paper mainly established a hierarchy for the ranking of cultures. It does not only pay attention to the different features of cultures but also focuses on manifestations of cultures. There are 9 criteria and 27 subcriteria in the hierarchy. We selected 42 leading developed and developing countries, obtaining pairwise comparisons by questionnaires of 30 people from diverse countries including China, France, Germany, Greece, India, Indonesia, Iran, Italy, Japan, Malaysia, Poland, Russia, Saudi Arabia and the United States. We then calculated their geometric mean. Subsequently, we obtained the priority weights of all the criteria and subcriteria. We then rated the countries one at time with respect to the subcriteria using a quantitative value for pairwise comparisons of high, medium and low or of excellent, above average, average, below average and poor as appropriate for each subcriterion. Note for example that for obesity, low is more important than high. Finally we evaluated the countries one at a time by assigning it the rating category with its numerical value, then weighted that value by the priority of the corresponding subcriterion and summed over the subcriteria to obtain the final index for a country.

We believe that this paper is the first effort of its kind to prioritize and quantify thinking about and comparing different cultures. We do not claim that the outcome of our sample size of experts provides the last word on our ranking, nor that any size sample would provide the real answer. Still we are confident that the ranks we obtained are not unreasonable in how these countries are generally viewed by knowledgeable people.

Section two is about the methodology where the AHP is introduced and its procedure described along with two examples of its application. In Sect. 3 we construct within the framework of a hierarchy the criteria and subcriteria for cultural ranking. In Sect. 4 we show how to derive the priorities of the criteria and subcriteria. Section 5 deals with rating the different countries we selected. Section 6 is about the conclusions drawn from our analysis.

2 AHP Methodology for Ranking

The Analytic Hierarchy Process (AHP) was first proposed in the 1970s [13] by one author of this paper. It is a concise method for system analysis and evaluation which combine qualitative and quantitative analysis. The analytic hierarchy process first breaks down a complex problem into a hierarchy; each level consists of several manageable elements, each of which is broken down into another set of elements. This process continues down to the most specific elements of the problem, usually the specific course of action considered, which is represented at the lowest level of the hierarchy. Hierarchical construction Any decision problem is an exercise of creative thinking and an effective way to deal with complexity and identify the main components of the problem. Without a single general hierarchy, one of the main features of AHP is that it allows the flexibility of constructing hierarchies to meet the specific needs of decision makers. In addition, metrics are used to prioritize between elements in each layer of the hierarchy. This is achieved by requiring participants to evaluate each set of elements in a paired manner relative to each element of the higher level. This metric provides a framework for ordering alternatives for actions to achieve numerical precedence.

The comparisons of the elements are made between any two criteria. Pairwise comparisons use the "1–9 scale" shown in Table 1. They were derived by using stimulus-response theory.

One uses comparisons or ratings, depending on the circumstances. When the criteria are changeable, as in selecting the best CEO for a company one uses comparisons and its corresponding method of synthesis, called the distributive mode. When criteria

Table 1. The fundamental scale of absolute numbers [13]

Intensity of importance	Definition	Explanation
1	Equal importance	Two activities contribute equally to the objective
3	Moderate importance	Experience and judgment slightly favor one activity over another
5	Strong importance	Experience and judgment strongly favor one activity over another
7	Very strong importance	An activity is favored very strongly over another; its dominance demonstrated in practice
9	Extreme importance	The evidence favoring one activity over another is of the highest possible order of affirmation
2, 4, 6, 8	Intermediate values	Used to represent compromise between the intensities listed above
Reciprocals of above	If activity i has one of the above nonzero numbers assigned to it when compared to activity j, then j has the reciprocal	

are standardized, as in the admission of students to a university, evaluating projects or in evaluating military officers for promotion, we use ratings with its ideal mode. Of course, the ideal itself may change because of adding new alternatives never previously encountered or conceived.

If a_{ij} represents the relative importance of criterion i when compared to criterion j, calculate a_{ji}, using the Eq. (1). A is called the pairwise comparison matrix of the criteria.

$$a_{ji} = \frac{1}{a_{ij}} \tag{1}$$

$$A = \left(a_{ij}\right)_{n*n} = \begin{pmatrix} a_{11} & \cdots & a_{1n} \\ \vdots & \ddots & \vdots \\ a_{n1} & \cdots & a_{nn} \end{pmatrix} \tag{2}$$

Consider n stocks, $A_1, \cdots A_n$ with known worth $w = (w_1, \cdots, w_n)$ respectively, and suppose that a matrix of pairwise is formed whose rows give the ratios of the worth of each stock with respect to all others as follows:

$$\begin{array}{c} \\ A_1 \\ \vdots \\ A_n \end{array} \begin{array}{c} \begin{array}{ccc} A_1 & \ldots & A_n \end{array} \\ \begin{pmatrix} \frac{w_1}{w_1} & \cdots & \frac{w_1}{w_n} \\ \vdots & \ddots & \vdots \\ \frac{w_n}{w_1} & \cdots & \frac{w_n}{w_n} \end{pmatrix} \end{array} \tag{3}$$

We can recover the scale w using the following equation:

$$\begin{pmatrix} \frac{w_1}{w_1} & \cdots & \frac{w_1}{w_n} \\ \vdots & \ddots & \vdots \\ \frac{w_n}{w_1} & \cdots & \frac{w_n}{w_n} \end{pmatrix} \begin{pmatrix} w_1 \\ \vdots \\ w_3 \end{pmatrix} = n \begin{pmatrix} w_1 \\ \vdots \\ w_n \end{pmatrix} \tag{4}$$

One way to obtain the answer is to calculate the eigenvector corresponding to the eigenvalue n in the judgment matrix. When the matrix is consistent one simply does the following [13]:

First step: normalize any column vector and use it for the priorities.

If the matrix is inconsistent raise it to powers as desired for decimal accuracy and take the sum of its rows and normalize these sums.

In order to ensure the consistency of judgment matrix, we use consistency ratio (CR) to test the consistency.

First, calculate $\lambda_{\max}$

$$\lambda_{\max} = \sum_{i=1}^{n} \frac{(AW)_i}{nW_i} \quad (5)$$

Second, calculate the consistency index (CI).

$$CI = (\lambda_{\max} - n)/(n-1) \quad (6)$$

Third, look through the random index table to find the value of RI.

Finally, compute the consistency ratio.

$$CR = CI/RI \quad (7)$$

If $CR < 0.1$, the judgments are acceptable otherwise find the most inconsistent judgment and change it as much as judgment allows and then go to the next most inconsistent judgment and do the same and so on.

3 Criteria and Subcriteria for Cultural Ranking

Culture includes both tangible and intangible criteria. Based on the previous analysis, combined with the research of Hofstede, House, Trompenaars, Hall and others on cultural dimensions and national cultural model, all aspects of culture's influence on human life are also reflected in all aspects of human life. Therefore, this paper selects 9 important criteria to study the cultures of different countries, including diversity, education, food, freedom, health, safety & security, science & technology, values, and wealth [9, 12, 14–18]. There are 4 considerations in the selection of subcriteria: (1) subcriteria should be selected based on the extant literature on its relevance to culture; (2) subcriteria can reflect culture; (3) subcriteria can be converted into comparative forms; and (4) availability of consistent quality data from national data sources. According to the principle of constructing the index system and the data resources, 27 subcriteria are collected for the study, as shown in Table 2.

Table 2. Criteria and subcriteria for cultural ranking

Objective	Criteria	Subcriteria	Description
Cultural ranking A	Diversity B_1	Religions C_1	Number of main religions
		Languages C_2	Number of main languages spoken in the country
		Ethnic groups C_3	Number of main ethnic groups
	Education B_2	Education expenditure C_4	The public expenditure on education as a percent of GDP
		Basic education C_5	The total number of years of schooling (primary to tertiary) that a child can expect to receive
		High-level education C_6	Number of universities listed in Top 500 Global Universities
	Food B_3	Food expenditure C_7	Percent of consumer expenditures spent on food
		Food safety C_8	A condition related to the supply of food, and individuals' access to it
		Food diversification C_9	Main types of food
	Freedom B_4	Economic freedom C_{10}	The fundamental right of every human to control his or her own labor and property
		Political freedom C_{11}	Political rights and civil liberties in the countries around the globe
	Health B_5	Health expenditure C_{12}	The total expenditure on health as a percentage of GDP
		Life expectancy C_{13}	The average number of years to be lived by a group of people born in the same year
		Obesity C_{14}	The percent of a country's population considered to be obese
	Safety & security B_6	Public safety C_{15}	Relevant to security such as domestic violence or traffic deaths, which together with crime rates affect the daily opportunity for citizens to live in safe and secure environment
		Economic security C_{16}	A combination of forms of social and work-related securities in order to flourish and develop
		National security C_{17}	The safekeeping of the nation as a whole
	Science & technology B_7	Internet popularization C_{18}	The percent of a country's population who can access the Internet at home
		High-tech industry C_{19}	Number of companies listed in 500 largest companies worldwide
		Number of patents C_{20}	The total number of patents in recent years
	Values B_8	Aggressiveness C_{21}	The degree to which a collective encourages and rewards group members for performance improvement and excellence
		Collectivism C_{22}	The degree to which individuals express pride, loyalty and cohesiveness
		Gender equality C_{23}	The degree to which a collective minimizes gender inequality
		Well-beings C_{24}	A sense of how people's lives are going, through the interaction between their circumstances, activities and psychological resources or 'mental capital'
	Wealth B_9	GDP per capita C_{25}	GDP on a purchasing power parity basis divided by population
		Poverty C_{26}	The percentage of the population falling below the poverty line
		GINI Index C_{27}	The extent to which the distribution among individuals within an economy deviates from a perfectly equal distribution

4 The Process of Deriving the Priorities

4.1 Constructing the Judgment Matrix

In the concrete operations of evaluation of cultural ranking, constructing the judgment matrices is key. Scoring is based on the "1–9 scale" of absolute numbers in Table 1. Therefore, we used questionnaire survey to get the judgment matrix. The questionnaires asked the following kind of question: "If you are creating a new society what kind of criteria are important for that culture how important is one criterion than another." In order to obtain more scientific and authoritative opinions, through sending questionnaires by email, this study collected effective opinions of 30 experts from famous professors of universities who carry out culture-related research. The judgments in the 30 questionnaires given by people from different countries are in some cases different. The very few with poor consistency were removed from the batch.

We took the geometric mean of the data and formed the integrated questionnaire judgment shown in Tables 3, 4, 5, 6, 7, 8, 9, 10, 11 and 12. The geometric mean is essential to use here because of reciprocal values. Also by using it we can reduce data errors and personal bias. Then, we formed the matrices of the geometric mean ratios of the comparisons received from the questionnaires in Tables 3 4, 5, 6, 7, 8, 9, 10, 11 and 12 which include in the next to last column the derived priorities. Their consistency ratio CR is almost equal to 0 before each value as they are all much less than 10% required by the theory.

Table 3. Judgment matrix of criteria

Cultural ranking	Diversity	Education	Food	Freedom	Health	Safety & security	Science & technology	Values	Wealth	Priority
Diversity	1	1.91	1.11	2.23	1.34	2.20	1.38	1.65	1.22	0.074
Education	1/1.91	1	2.33	1.12	1.22	1.04	1.57	1.25	2.56	0.150
Food	1/1.11	1/2.33	1	2.72	1.57	2.71	1.67	2.10	1.05	0.062
Freedom	1/2.23	1/1.12	1/2.72	1	1.57	1.10	1.30	1.03	2.35	0.153
Health	1/1.34	1/1.22	1/1.57	1/1.57	1	1.75	1.25	1.47	1.59	0.103
Safety & security	1/2.20	1/1.04	1/2.71	1/1.10	1/1.75	1	1.62	1.18	2.45	0.162
Science & technology	1/1.38	1/1.57	1/1.67	1/1.30	1/1.25	1/1.62	1	1.36	1.38	0.098
Values	1/1.65	1/1.25	1/2.10	1/1.03	1/1.47	1/1.18	1/1.36	1	1.60	0.131
Wealth	1/1.22	1/2.56	1/1.05	1/2.35	1/1.59	1/2.45	1/1.38	1/1.60	1	0.066

Table 4. Judgment matrix of diversity

Diversity	Religions	Languages	Ethnic groups	Priority
Religions	1	1.18	1.40	0.309
Languages	1/1.18	1	1.55	0.267
Ethnic groups	1/1.40	1/1.55	1	0.424

Table 5. Judgment matrix of education

Education	Education expenditure	Basic education	High-level education	Priority
Education expenditure	1	1.18	1.17	**0.298**
Basic education	1/1.18	1	1.06	**0.345**
High-level education	1/1.17	1/1.06	1	0.356

Table 6. Judgment matrix of food

Food	Food expenditure	Food safety	Food diversification	Priority
Food expenditure	1	2.66	1.34	0.237
Food safety	1/2.66	1	2.57	0.566
Food diversification	1/1.34	1/3.4994	1	0.197

Table 7. Judgment matrix of freedom

Freedom	Economic freedom	Political freedom	Priority
Economic freedom	1	1.49	0.401
Political freedom	1/1.49	1	0.599

Table 8. Judgment matrix of health

Health	Health expenditure	Life expectancy	Obesity	Priority
Health expenditure	1	1.17	1.98	0.376
Life expectancy	1/1.17	1	2.10	0.427
Obesity	1/1.98	1/2.10	1	0.197

Table 9. Judgment matrix of safety & security

Safety & security	Public safety	Economic security	National security	Priority
Public safety	1	3.20	1.19	0.473
Economic security	1/3.20	1	2.09	0.161
National security	1/1.19	1/2.09	1	0.366

Table 10. Judgment matrix of science & technology

Science & technology	Internet popularization	High-tech industry	Number of patents	Priority
Internet popularization	1	1.03	1.62	0.382
High-tech industry	1/1.03	1	1.97	0.400
Number of patents	1/1.62	1/1.97	1	0.219

Table 11. Judgment matrix of wealth

Wealth	GDP per capita	Poverty	GINI index	Priority
GDP per capita	1	1.69	1.26	0.254
Poverty	1/1.69	1	1.25	0.418
GINI index	1/1.26	1/1.25	1	0.328

Table 12. Judgment matrix of values

Values	Aggressiveness	Collectivism	Gender equality	Well-beings	Priority
Aggressiveness	1	1.16	1.54	2.89	0.151
Collectivism	1/1.16	1	1.30	2.71	0.173
Gender equality	1/1.54	1/1.30	1	1.86	0.232
Well-beings	1/2.89	1/2.71	1/1.86	1	0.445

4.2 Priorities of the Criteria and Subcriteria

Each eigenvector of judgment matrix multiplied by the eigenvector of the corresponding element in the general judgment matrix, we will get the priorities of subcriteria. The subcriteria are shown in Table 13.

Table 13. The priorities of criteria and subcriteria

Index	Diversity	Education	Food	Freedom	Health	Safety & security	Science & technology	Values	Wealth	Global priority
	0.074	0.150	0.062	0.153	0.103	0.162	0.098	0.131	0.066	
Religions	0.309									0.023
Languages	0.267									0.020
Ethnic groups	0.424									0.032
Education expenditure		0.298								0.045
Basic education		0.345								0.052
High-level education		0.356								0.053
Food expenditure			0.237							0.015
Food safety			0.566							0.035
Food diversification			0.197							0.012
Economic freedom				0.401						0.061
Political freedom				0.599						0.092
Health expenditure					0.376					0.039
Life expectancy					0.427					0.044
Obesity					0.197					0.020
Public safety						0.473				0.077
Economic security						0.161				0.026
National security						0.366				0.059
Internet popularization							0.382			0.037
High-tech industry							0.400			0.039
Number of patents							0.219			0.021
Aggressiveness								0.151		0.020
Collectivism								0.173		0.023
Gender equality								0.232		0.030
Well-beings								0.445		0.058
GDP per capita									0.254	0.017
Poverty									0.418	0.028
GINI index									0.328	0.022

According to the last column in Table 19, we can conclude that the priority of C_{11} is the highest, and the priority of C_9 is the lowest one.

4.3 Priority of the Intensities of the Ratings Used for Each Subcriterion

The existing research rarely comprehensively evaluates the specific performance of culture, this paper uses AHP to prioritize and quantify different cultures for the first time. In the general literature, the AHP method is often used to determine the weight of subcriteria, and then the sub-indicator data is combined with the weight. In the present study, after we get the priorities of the subcriteria, we create different intensity categories for each subcriterion of which there are 3 or 5, so that the evaluation results are more accurate. For example, the intensities for education expenditure are excellent, average, and poor; the intensities of food safety are excellent, above average, average, below average and poor. Then we perform pairwise comparisons of the intensities with respect to each subcriterion. Table 14 is an example of how to do pairwise comparison among the intensities. The priority of each intensity for each subcriterion is shown in Tables 15, 16, 17, 18, 19, 20, 21, 22 and 23.

Table 14. Pairwise comparison of the categories

Religion	Excellent	Above average	Average	Below average	Poor
Excellent	1	2	2	3	4
Above average	1/2	1	1	2	2
Average	1/2	1	1	1	2
Below average	1/3	1/2	1	1	1
Poor	1/4	1/2	1/2	1	1

Table 15. Priorities of categories for diversity

Diversity	Excellent	Above average	Average	Below average	Poor
Religions	0.3455	0.2361	0.1732	0.1344	0.1106
Languages	0.3403	0.2435	0.1850	0.1296	0.1016
Ethnic groups	0.2760	0.2297	0.2058	0.1627	0.1257

Table 16. Priorities of categories for education

Education	Excellent	Average	Poor
Education expenditure	0.4356	0.3481	0.2163
Basic education	0.5519	0.3421	0.1060
High-level education	0.6370	0.2583	0.1047

Table 17. Priorities of categories for food

Food	Excellent	Above average	Average	Below average	Poor
Food expenditure	0.0775	0.1200	0.1569	0.2283	0.4173
Food safety	0.4708	0.2290	0.1663	0.0812	0.0528
Food diversification	0.3146	0.2406	0.1945	0.1384	0.1119

Table 18. Priorities of categories for freedom

Freedom	Excellent	Above average	Average	Below average	Poor
Economic freedom	0.3332	0.2530	0.1897	0.1298	0.0943
Political freedom	0.5809	–	0.2791	–	0.1399

Table 19. Priorities of categories for health

Health	Excellent	Above average	Average	Below average	Poor
Health expenditure	0.4403	0.2291	0.1668	0.0954	0.0684
Life expectancy	0.4031	0.2849	0.1450	0.1013	0.0657
Obesity	0.0988	0.1500	0.1903	0.2365	0.3245

Table 20. Priorities of categories for safety & security

Safety & Security	Excellent	Above average	Average	Below average	Poor
Public safety	0.3147	0.2501	0.1872	0.1428	0.1054
Economic security	0.2807	0.2460	0.2069	0.1519	0.1144
National security	0.4195	0.3030	0.1273	0.0858	0.0645

Table 21. Priorities of categories for science & technology

Science & technology	Excellent	Above average	Average	Below average	Poor
Population that uses internet	0.3415	0.2985	0.1832	0.1089	0.0680
High-tech industry	0.3669	0.2367	0.1737	0.1380	0.0847
Number of patents	0.3905	0.2520	0.1587	0.1135	0.0853

Table 22. Priorities of categories for values

Values	Excellent	Above average	Average	Below average	Poor
Aggressiveness	0.3943	0.2785	0.1505	0.1017	0.0751
Collectivism	0.4203	0.2617	0.1404	0.1011	0.0765
Gender equality	0.3194	0.2606	0.1793	0.1366	0.1042
Well-beings	0.3869	0.2587	0.1836	0.1021	0.0688

Table 23. Priorities of categories for wealth

Wealth	Excellent	Above average	Average	Below average	Poor
GDP per capita	0.4023	0.2448	0.1585	0.1045	0.0899
Population below poverty line	0.0835	0.1174	0.1865	0.2534	0.3592
GINI index	0.1043	0.1406	0.1683	0.2501	0.3368

5 Cultural Ranking of Countries

In this paper we chose the following countries to rank. The distribution of the countries is shown in Fig. 1.

Fig. 1. Geographic locations of the sample countries

We give a 3 or a 5 intensity category to each subcriterion, and then we do pairwise comparisons to obtain the priority of each intensity in a category. Let us take the United States as an example for how to obtain the result shown in Table 24. Based on the judgments of each subcriterion for each country, we obtain the final result shown in Table 25. There are final scores and ranking results in it.

Table 24. Calculating the result for the United States

Subcriteria		Weight	Category	Priority	Weighted priority
C1	Religions	0.015	Excellent	0.420	0.006
C2	Languages	0.035	Below average	0.470	0.016
C3	Ethnic groups	0.012	Average	0.110	0.001
C4	Education expenditure	0.061	Average	0.250	0.015
C5	Basic education	0.092	Average	0.580	0.053
C6	High-level education	0.039	Above average	0.440	0.017
C7	Food expenditure	0.044	Poor	0.140	0.006
C8	Food safety	0.020	Excellent	0.100	0.002
C9	Food diversification	0.077	Poor	0.140	0.011
C10	Economic Freedom	0.026	Above average	0.150	0.004
C11	Political Freedom	0.059	Excellent	0.300	0.018
C12	Health expenditure	0.037	Excellent	0.300	0.011
C13	Life expectancy	0.039	Average	0.370	0.014
C14	Obesity	0.021	Excellent	0.390	0.008
C15	Public safety	0.002	Below average	0.280	0.001
C16	Economic security	0.023	Below average	0.080	0.002
C17	National security	0.030	Above average	0.180	0.005
C18	Internet popularization	0.058	Above average	0.260	0.015
C19	High-tech industry	0.017	Excellent	0.240	0.004
C20	Number of patents	0.028	Excellent	0.250	0.007
C21	Aggressiveness	0.022	Above average	0.140	0.003
C22	Collectivism	0.023	Poor	0.350	0.008
C23	Gender equality	0.020	Average	0.130	0.003
C24	Well-beings	0.032	Above average	0.210	0.007
C25	GDP per capita	0.045	Above average	0.350	0.016
C26	Poverty	0.052	Below average	0.340	0.018
C27	GINI Index	0.053	Excellent	0.640	0.034
Final score for the United States					**0.31**

Table 25. The final result of cultural ranking

Countries	Scores	Cultural ranking	Countries	Scores	Cultural ranking	Countries	Scores	Cultural ranking
United States	0.31	1	France	0.26	15	China	0.21	27
Denmark	0.31	1	Belgium	0.26	15	Brazil	0.20	30
Norway	0.30	3	Japan	0.25	17	Malaysia	0.19	31
Australia	0.29	4	South Korea	0.25	17	South Africa	0.19	31
Austria	0.29	4	Spain	0.25	17	Mexico	0.18	33
Switzerland	0.29	4	Singapore	0.24	20	Saudi Arabia	0.18	33
Canada	0.29	4	Poland	0.23	21	Indonesia	0.17	35
United Kingdom	0.29	4	Israel	0.23	21	Thailand	0.17	35
New Zealand	0.29	4	Italy	0.23	21	Turkey	0.16	37
Finland	0.29	4	Portugal	0.23	21	Russia	0.16	37
Sweden	0.28	11	Czech Republic	0.22	25	Iran	0.14	39
Netherlands	0.28	11	Estonia	0.22	25	Egypt	0.14	39
Germany	0.28	11	Hungary	0.21	27	India	0.14	41
Ireland	0.28	11	Greece	0.21	27	Nigeria	0.14	41

6 Conclusions

We have done several things in this analysis. The first, is that we identified from comprehensive search in the literature the criteria and subcriteria as the desired intangibles. Second we developed a quantitative approach to derive priorities for the criteria and subcriteria from pairwise comparisons and then to rate the countries as to how strongly they satisfy the subcriteria.

Our hope is to pursue implementation of the subcriteria in different high ranking cultures in order to globalize the characteristics of these cultures.

Acknowledgement. This research is supported by National Natural Science Foundation of China under Grant No. 71573237, 71874165, New Century Excellent Talents in University of China under Grant No. NCET-13-1012, Research Foundation of Humanities and Social Sciences of Ministry of Education of China No. 15YJA630019, China Institute of Geo-Environment Monitoring No. 0001212016CC60013, Natural Science Foundation of Hubei Province of China No. 2016CFB503, and Open Foundation for Hubei Provincial Ecological Civilization Research Center in China University of Geosciences (Wuhan) under Grant No. STZK2018z09.

References

1. Glowacki, L.: Another View: Tribal violence, then and now. Washington Post, 25 April 2016
2. Tylor, E.B.: Primitive Culture: Researches into the Development of Mythology, Philosophy, Religion, Art, and Custom. Gordon Press, New York (1974). ISBN 978-0-87968-091-6
3. Meaning of culture. Cambridge English Dictionary. Accessed 26 July 2015
4. Greenberg, J., Solomon, S., Pyszczynski, T.: Terror management theory and research: empirical assessments and conceptual refinements. In: Zanna, M.P. (ed.) Advances in Experimental Social Psychology, vol. 29, pp. 61–139. Academic Press, San Diego (1997)
5. http://en.wikipedia.org/wiki/Culture
6. Hofstede, G., Hofstede, G.J., Minkov, M.: Cultures and Organizations: Software of the Mind (2010)
7. Hofstede, G.: Dimensionalizing Cultures: The Hofstede Model in Context. Online readings in Psychology and Culture International Association for Cross-Cultural Psychology (2011)
8. Hofstede, G.: National cultures in four dimensions: a research-based theory of cultural differences among nations. Int. Stud. Manag. Organ. **13**(1/2), 46–74 (1983)
9. Hoppe, M.H.: Culture and leader effectiveness: the GLOBE study. Central Eur. J. Commun. (2007)
10. Xiong, Z.D., Guo, R.F.: Evaluation of national cultural soft power and research on its promotion path. China Ind. Econ. **9**, 18–28 (2011)
11. Gannon, M.J., Pillai, R.: Understanding global cultures: metaphorical journeys through 34 nations, clusters of nations, continents & diversity, London (2015)
12. Wu, M.Y.: Hofstede's cultural dimensions 30 years later: a study of Taiwan and the United States. Intercult. Commun. Stud. **XV**(1), 33–42 (2006)
13. Saaty, T.L.: Relative measurement and its generalization in decision making why pairwise comparisons are central in mathematics for the measurement of intangible factors the analytic hierarchy/network process. RACSAM-Revista de la Real Academia de Ciencias Exactas, Fisicas y Naturales. Serie A. Matematicas **102**(2), 257 (2008)
14. Trompenaars, F., Hampden-Turner, C.: Riding the waves of culture: understanding diversity in global business. Nicholas Brealey International (2011)
15. Hall, E.T., Hall, M.R.: Understanding Cultural Differences: [Germans, French and Americans]. Intercultural Press, Yarmouth (1990)
16. Richerson, P.J., Boyd, R.: Not by Genes Alone: How Culture Transformed Human Evolution. University of Chicago Press, Chicago (2005)
17. Shane, S.A.: Cultural influences on national rates of innovation. J. Bus. Venturing **8**, 59–73 (1993)
18. Vliert, V.E.: Climate, Affluence, and Culture. Cambridge University Press, New York (2009)

Sustainable Performance Evaluation of Urban Mobility Projects Using Multicriteria Group Decision Making Approach

Santoso Wibowo(✉) and Srimannarayana Grandhi

Central Queensland University, Melbourne, VIC 3000, Australia
s.wibowo1@cqu.edu.au

Abstract. This paper presents a multicriteria group decision making approach for effectively evaluating the sustainable performance of urban mobility projects. The subjectiveness and imprecision of the performance evaluation process is adequately modeled by using intuitionistic fuzzy numbers. To avoid the complex and unreliable process of comparing fuzzy numbers usually required in fuzzy multicriteria decision making, an efficient algorithm is developed based on the degree of indeterminacy and the degree of similarities. An example is presented to demonstrate the applicability of the approach.

Keywords: Urban mobility projects · Sustainable · Performance evaluation · Multicriteria · Decision makers · Subjective assessments

1 Introduction

Transport systems have always been considered to be the driving forces of economic and social development throughout the twentieth century [1]. However, these transportation systems have caused negative impacts including traffic congestion, fatalities and injuries, environmental pollution and energy consumption [2]. To overcome this issue, it is critical for government agencies to be active in developing and implementing important measures that encourage sustainable mobility [1].

Sustainable mobility is the ability to meet the needs of society to move freely, gain access, communicate, trade and establish relationships without sacrificing other essential human or ecological values today or in the future [3]. The concept of sustainable mobility includes (a) travel reduction, (b) distance reduction, and (c) effective use of public transport [2]. In order for government agencies to meet their expectations to be economically viable, environmentally friendly, and socially responsible, a structured approach is necessary for evaluating the sustainable performance of the most suitable urban mobility project for development and implementation.

Evaluating the sustainable performance of urban mobility projects is challenging due to (a) the presence of multiple, conflicting criteria, (b) the involvement of multiple decision makers (DMs), and (c) the presence of subjectiveness of the performance evaluation process. As a result, a comprehensive evaluation of the urban mobility project alternatives' overall sustainable performance is necessary.

Y. Liu et al. (Eds.): ICNC-FSKD 2019, AISC 1074, pp. 964–971, 2020.
https://doi.org/10.1007/978-3-030-32456-8_103

Several research has been carried out in evaluating the sustainable performance of urban mobility projects [4–6]. Curiel-Esparza et al. [4] present the integrated approach consisting of the Delphi technique, the analytic hierarchy process (AHP) and the VIKOR approach for evaluating and selecting the optimal urban mobility project. The Delphi technique is used to gather data from a panel of selected experts. The AHP is used to obtain the relative priority of the different criteria in terms of mobility and to assess the alternatives. The VIKOR approach helps to obtain consensus solutions in compromised problems which involve conflicting criteria. Moskolaï Ngossaha et al. [5] present the fuzzy AHP for evaluating the sustainability performance of urban mobility projects. Triangular fuzzy numbers are used for representing the DM's judgments. The concept of fuzzy synthetic extent analysis is applied for deciding the final priority of different criteria. Nassereddine and Eskandari [6] present the Delphi technique, group AHP, and the PROMETHEE for evaluating the performance of urban mobility projects. The Delphi technique is used to determine the most important criteria via the responses of the DMs. The group AHP approach is used to analyze the structure of the public transportation problem and determine the weight of the criteria. Then, the PROMETHEE approach is employed to determine the final ranking of public transportation modes.

These approaches, however, have shortcomings including (a) the inadequacy in modeling the subjectiveness of the evaluation process, (b) the requirement for tedious mathematical computation in the evaluation process, and (c) cognitive demanding on the DMs when the number of alternatives and criteria increases [7, 8].

This paper presents multicriteria group decision making approach for effectively evaluating the sustainable performance of urban mobility projects. The subjectiveness and imprecision of the performance evaluation process is adequately modeled by using intuitionistic fuzzy numbers. To avoid the complex and unreliable process of comparing fuzzy numbers usually required in fuzzy multicriteria decision making, an efficient algorithm is developed based on the degree of indeterminacy and the degree of similarities. An example is presented to show the applicability of the proposed multicriteria group decision making approach.

2 Sustainable Performance Evaluation of Urban Mobility Projects

Much research has been conducted in determining the relevant factors for evaluating the sustainable performance of urban mobility projects in different perspectives [4–6, 9–18]. For example, Banister [4] states the key elements of sustainable urban mobility project should include the reduced need for travel, the modal shift from cars to public transport, the reduction of trip lengths and the promotion of greater efficiency. Curiel-Esparza et al. [5] point out that economy, travel quality, and sustainability are important factors for assessing urban mobility projects. Moskolaï Ngossaha et al. [6] state that average travel time, traffic safety, investment costs, and investment profitability should be taken into account when assessing the sustainable performance of urban mobility projects. Keseru et al. [9] point out that the selection of the most suitable urban mobility project should focus on mixed land uses, high quality public transport networks, and systems of non-motorized transport which enable people to meet most of their needs within short distances. Macedo et al. [10] and believe that social and cultural factors including interpersonal interactions, individuals' values and perceptions of traveling for work, recreation, and school, and their past and present commuting encounters, could influence travel behavior and choices. Hickman et al. [11] and Tsamboulas [12] believe that economic, environmental and socio-political aspects are critical factors to be considered when evaluating the sustainable performance of urban mobility projects. Meanwhile, György et al. [13] reflect on the importance of health, social equity, environmental quality, and viable economy factors for evaluating the sustainable performance of urban mobility projects. Costa et al. [14] state that travel mode which includes work-based trips, travel personal security, emission, and cost of travel are also important factors for assessing the sustainable performance of urban mobility projects. Joumard and Nicolas [15] believe that economic, social, environmental, and the project's adaptability must be considered while evaluating urban mobility projects. Cavalcanti et al. [16] and Mozos-Blanco et al. [17] point out that transport, land-use and spatial planning, environment, economic development, social policy, health, and road safety are critical factors for assessing the sustainable performance of urban mobility projects.

Based on the literature review above, four important criteria for evaluating the sustainable performance of urban mobility projects are determined. The four criteria include (a) Technical (C_1), (b) Social (C_2), (c) Economic (C_3), and (d) Environment (C_4) as shown in Fig. 1.

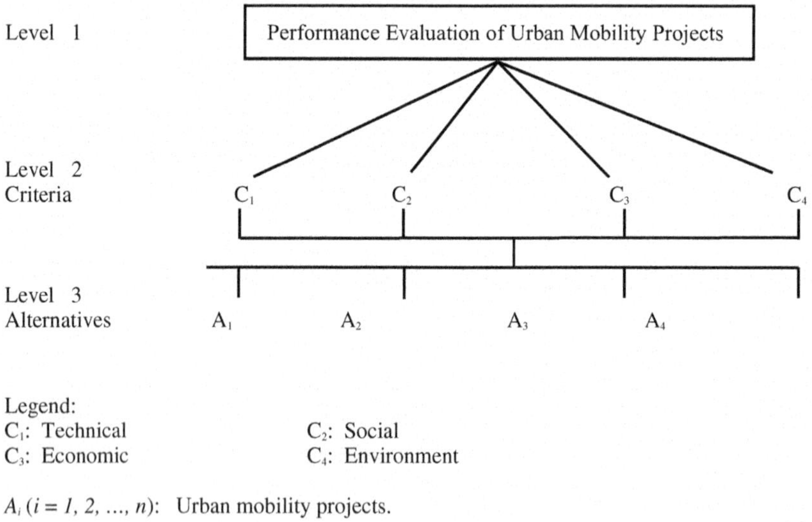

Fig. 1. The hierarchical structure for the urban mobility projects' performance evaluation problem.

Technical (C_1) criteria focus on the technical capabilities of the urban mobility project to meet or exceed the expectations of the users. This is assessed by service reliability, spatial accessibility, service area network, connectivity to multimodal transport, safety, security, suitability to disabled customers, modern and clean facilities, and staff service quality [6, 12]. Social (C_2) criteria consider the improvement of population displacement conditions in urban areas, including proximity to transportation services, comfort, and punctuality. This is measured by the increased number of potential users, social equity, and reduction in city congestion [16]. Economic (C_3) criteria assess the efficiency of the means used for the prioritization of social and environmental goals resulting from the implementation of the urban mobility projects [11]. This is assessed by revenues, investment costs, operating costs, and travel cost [7]. Environment (C_4) plays a key role in assessing the influence of different modes of transportation, such as individual or collective, motorized or non-motorized, and in reducing air pollution [18]. This is measured by the reduction in air pollution caused by local pollutants and greenhouse gas emissions, noise, and fossil fuel consumption [14].

3 A Fuzzy Multicriteria Group Decision Making Approach

The urban mobility project performance evaluation problem can be modeled as a multicriteria group decision making problem as it usually involves in (a) discovering all the alternatives, (b) identifying the evaluation criteria, (c) assessing the alternatives' performance ratings and the criteria weights by individual DMs, (d) aggregating the alternative ratings and criteria weights for producing an overall preference value for each alternative across all the criteria [18].

The sustainable performance evaluation of urban mobility project problem usually considers a number of alternatives A $(a_1, a_2, \ldots, a_m)$ with respect to each criterion C $(c_1, c_2, \ldots, c_n)$. Each DM D_k $(k = 1, 2, \ldots, s)$ assesses each alternative in a form of an intuitionistic preference relation $y_{ij}^k = (\mu_{ij}^k, v_{ij}^k)$, and $0 \leq \mu_{ij}^k + v_{ij}^k \leq 1$, $\mu_{ij}^k = v_{ij}^k$, $v_{ij}^k = \mu_{ij}^k = 0.5$. μ_{ij}^k indicates that the alternative A_i satisfies the criterion C_j, v_{ij}^k indicates that the alternative A_i does not satisfy the criterion C_j. As a result, the decision matrix for all the alternatives can be obtained as in (1).

$$y_{ij}^k = \begin{bmatrix} \mu_{11}^k, v_{11}^k & \mu_{12}^k, v_{12}^k & \cdots & \mu_{1m}^k, v_{1m}^k \\ \mu_{21}^k, v_{21}^k & \mu_{22}^k, v_{22}^k & \cdots & \mu_{2m}^k, v_{2m}^k \\ \cdots & \cdots & \cdots & \cdots \\ \mu_{n1}^k, v_{n1}^k & \mu_{n2}^k, v_{n2}^k & \cdots & \mu_{nm}^k, v_{nm}^k \end{bmatrix} \tag{1}$$

The relative importance of the performance evaluation criteria C_j can be assessed as in (2).

$$w_j^k = (w_1^k, w_2^k, \ldots, w_m^k) \tag{2}$$

This is followed by the determination of the overall intuitionistic fuzzy decision matrix and the overall intuitionistic fuzzy weight vector of each alternative as in (3) and (4) respectively.

$$X = \begin{bmatrix} x_{11} & x_{12} & \ldots & x_{1m} \\ x_{21} & x_{22} & \ldots & x_{2m} \\ \ldots & \ldots & \ldots & \ldots \\ x_{n1} & x_{n2} & \ldots & x_{nm} \end{bmatrix} \tag{3}$$

$$\mathrm{W} = \left(w_1, w_2, \ldots, w_j, \ldots, w_m\right) \tag{4}$$

The collective weighted intuitionistic fuzzy performance matrix is determined by the multiplication of decision matrix in (3) and the criteria weightings in (4).

$$R = \begin{bmatrix} r_{11} & r_{12} & \cdots & r_{1m} \\ r_{21} & r_{22} & \cdots & r_{2m} \\ \cdots & \cdots & \cdots & \cdots \\ r_{n1} & r_{n2} & \cdots & r_{nm} \end{bmatrix} \tag{5}$$

Based on (5), a fuzzy maximum ($M^j_{\max}$) and a fuzzy minimum ($M^j_{\min}$) which represent respectively the best and the worst fuzzy performance ratings among all the alternatives with respect to criterion C_j can be determined as in (6)–(7) respectively.

$$\begin{aligned} \alpha^+ = (\alpha_1^+, \alpha_2^+, \ldots, \alpha_m^+) = \{ \Big\langle & ([(\max_i \mu_{ij}) | j \in J_1, (\min_i v_{ij}) | j \in J_2)], \\ & [(\min_i \mu_{ij}) | j \in J_1, (\max_i v_{ij}) | j \in J_2]) \Big\rangle \, i = 1, 2, \ldots, n\} \end{aligned} \tag{6}$$

$$\begin{aligned} \alpha^- = (\alpha_1^-, \alpha_2^-, \ldots, \alpha_m^-) = \{ \Big\langle & ([(\min_i \mu_{ij}) | j \in J_1, (\max_i v_{ij}) | j \in J_2)], \\ & [(\max_i \mu_{ij}) | j \in J_1, (\min_i v_{ij}) | j \in J_2]) \Big\rangle \, i = 1, 2, \ldots, n\} \end{aligned} \tag{7}$$

The degree of indeterminacy of the relative positive ideal value and the relative negative ideal value can be determined by (8) respectively.

$$\pi_j^+ = 1 - \mu_j^+ - v_j^+, \; \pi_j^- = 1 - \mu_j^- - v_j^- \tag{8}$$

Next, the degree of similarity between alternative A_i and the positive ideal solution and the negative solution can be calculated by (9) and (10) respectively.

$$G_i^+ = 1 - \frac{\left|2(\mu_j^+ - \mu_{ij})(v_j^+ - v_{ij})\right|}{3} \times (1 - \frac{\pi_j^+ + \pi_{ij}}{2}) - \frac{\left|2(v_j^+ - v_{ij})(\mu_j^+ - \mu_{ij})\right|}{3} \times (\frac{\pi_j^+ + \pi_{ij}}{2}) \tag{9}$$

$$G_i^- = 1 - \frac{\left|2(\mu_j^- - \mu_{ij})(v_j^- - v_{ij})\right|}{3} \times (1 - \frac{\pi_j^- + \pi_{ij}}{2}) - \frac{\left|2(v_j^- - v_{ij})(\mu_j^- - \mu_{ij})\right|}{3} \times (\frac{\pi_j^- + \pi_{ij}}{2}) \tag{10}$$

An overall preference value for alternative A_i across all the criteria can be determined by (11). The larger the preference value, the more preferred the alternative.

$$P_i = \frac{G_i^-}{G_i^- + G_i^+} \tag{11}$$

4 An Example

In this section, we apply the multicriteria decision making approach presented above to deal with the sustainable performance evaluation of urban mobility projects.

To start with the sustainable performance evaluation process, the relative performance of all available urban mobility projects in regard to DMs D_1, D_2, and D_3 are determined by making the subjective assessments using intuitionistic based fuzzy numbers. Table 1 shows the results.

Table 1. Performance assessments of urban mobility projects

		C_1	C_2	C_3	C_4
A_1	D_1	(0.7, 0.4)	(0.8, 0.4)	(0.4, 0.2)	(0.5, 0.3)
	D_2	(0.5, 0.6)	(0.7, 0.3)	(0.8, 0.6)	(0.7, 0.4)
	D_3	(0.6, 0.5)	(0.6, 0.3)	(0.5, 0.3)	(0.8, 0.6)
A_2	D_1	(0.7, 0.3)	(0.7, 0.5)	(0.8, 0.2)	(0.7, 0.4)
	D_2	(0.5, 0.2)	(0.5, 0.2)	(0.7, 0.5)	(0.8, 0.3)
	D_3	(0.6, 0.4)	(0.6, 0.3)	(0.4, 0.2)	(0.5, 0.4)
A_3	D_1	(0.8, 0.4)	(0.8, 0.6)	(0.7, 0.4)	(0.4, 0.3)
	D_2	(0.6, 0.4)	(0.5, 0.2)	(0.6, 0.3)	(0.4, 0.3)
	D_3	(0.8, 0.6)	(0.4, 0.3)	(0.7, 0.6)	(0.8, 0.5)

Similarly, the criteria weights for evaluating the performance of urban mobility projects are obtained directly from the DMs D_1, D_2, and D_3 as shown in Table 2.

Table 2. Criteria weights of urban mobility projects

Criteria	Decision makers		
	D_1	D_2	D_3
C_1	(0.6, 0.2)	(0.4, 0.1)	(0.7, 0.2)
C_2	(0.4, 0.3)	(0.7, 0.5)	(0.6, 0.4)
C_3	(0.7, 0.2)	(0.6, 0.3)	(0.6, 0.3)
C_4	(0.6, 0.4)	(0.7, 0.4)	(0.4, 0.1)

Based on (3)–(4), the collective weighted intuitionistic fuzzy performance matrix can be calculated. Table 3 shows the results.

Table 3. The collective weighted intuitionistic fuzzy performance matrix

	C_1	C_2	C_3	C_4
A_1	(0.49, 0.72)	(0.52, 0.67)	(0.48, 0.65)	(0.34, 0.48)
A_2	(0.48, 0.63)	(0.43, 0.58)	(0.38, 0.52)	(0.49, 0.57)
A_3	(0.56, 0.71)	(0.35, 0.71)	(0.61, 0.79)	(0.56, 0.74)
A_4	(0.44, 0.58)	(0.36, 0.57)	(0.43, 0.52)	(0.58, 0.73)

By using (6)–(11), the overall preference value for each alternative can be determined in an effective and efficient manner. Table 4 shows that alternative A_4 has the highest performance index value of 0.79 as compared to other alternatives.

Table 4. The preference value of urban mobility projects and their rankings

Alternatives	Preference value	Ranking
A_1	0.64	3
A_2	0.61	4
A_3	0.73	2
A_4	0.79	1

5 Conclusion

This paper has presented the multicriteria group decision making approach for evaluating the sustainable performance of urban mobility projects. The proposed approach is developed based on the degree of indeterminacy and the degree of similarities. The strengths of the proposed approach include (a) the ability to perform sustainable performance evaluation under qualitative assessments, (b) the ability to consider multiple DMs, and (c) generate the final alternative ranking in a simple manner.

References

1. Awasthi, A., Omrani, H., Gerber, P.: Investigating ideal-solution based multicriteria decision making techniques for sustainability evaluation of urban mobility projects. Transp. Res. Part A: Policy Pract. **116**(1), 247–259 (2018)
2. Moradi, A., Vagnoni, E.: A multi-level perspective analysis of urban mobility system dynamics: what are the future transition pathways? Technol. Forecast. Soc. Chang. **126**(C), 231–243 (2018)
3. Lyons, G., Davidson, C.: Guidance for transport planning and policymaking in the face of an uncertain future. Transp. Res. Part A: Policy Pract. **88**(1), 104–116 (2016)
4. Curiel-Esparza, J., Mazario-Diez, J.L., Canto-Perello, J., Martin-Utrillas, M.: Prioritization by consensus of enhancements for sustainable mobility in urban areas. Environ. Sci. Policy **55**(1), 248–257 (2016)
5. Moskolaï Ngossaha, J., Houé Ngouna, R., Archimède, B., Nlong, J.M.: Sustainability assessment of a transportation system under uncertainty: an integrated multicriteria approach. IFAC-PapersOnLine **50**(1), 7481–7486 (2017)
6. Nassereddine, M., Eskandari, H.: An integrated MCDM approach to evaluate public transportation systems in Tehran. Transp. Res. Part A: Policy Pract. **106**(1), 427–439 (2017)
7. Wibowo, S., Grandhi, S.: Benchmarking knowledge management practices in small and medium enterprises: a fuzzy multicriteria group decision making approach. Benchmarking: Int. J. **24**(5), 1–21 (2017)
8. Wibowo, S.: A fuzzy multiattribute evaluation of green supply chain performance. In: Proceedings of the 8th IEEE Conference on Industrial Electronics and Applications, Melbourne, Australia (2013)
9. Keseru, I., Bulckaen, J., Macharis, C.: Sustainable, participatory and practical: the NISTO evaluation framework for urban and regional mobility projects. Transp. Res. Procedia **13**(1), 134–144 (2016)
10. Macedo, J., Rodrigues, F., Tavares, F.: Urban sustainability mobility assessment: indicators proposal. Energy Procedia **134**(1), 731–740 (2017)
11. Hickman, R., Hall, P., Banister, D.: Planning more for sustainable mobility. J. Transp. Geogr. **33**(1), 210–219 (2013)
12. Tsamboulas, D.A.: A tool for prioritizing multinational transport infrastructure investments. Transp. Policy **14**(1), 11–26 (2007)
13. György, K., Attila, A., Tamás, F.: New framework for monitoring urban mobility in European cities. Transp. Res. Procedia **24**(1), 155–162 (2017)
14. Costa, P.B., Morais Neto, G.C., Bertolde, A.I.: Urban mobility indexes: a brief review of the literature. Transp. Res. Procedia **25**(1), 3645–3655 (2017)
15. Joumard, R., Nicolas, J.P.: Transport project assessment methodology within the framework of sustainable development. Ecol. Ind. **10**(2), 136–142 (2010)
16. Cavalcanti, C.O., Limont, D., Dziedzic, M., Fernandes, V.: Sustainability assessment methodology of urban mobility projects. Land Use Policy **60**(1), 334–342 (2017)
17. Mozos-Blanco, M.A., Pozo-Menéndez, E., Arce-Ruiz, R., Baucells-Aletà, N.: The way to sustainable mobility. A comparative analysis of sustainable mobility plans in Spain. Transp. Policy **72**(1), 45–54 (2018)
18. Wibowo, S., Deng, H.: A consensus support system for supplier selection in group decision making. J. Manag. Sci. Stat. Decis. **6**(1), 52–59 (2009)

Fuzzy Applications: Other Fuzzy Applications

A Parallel Fuzzy Method to Reduce Mixed Gaussian-Impulsive Noise in CT Medical Images

Josep Arnal[1(✉)], Juan B. Pérez[1], and Vicente Vidal[2]

[1] Departamento de Ciencia de la Computación e Inteligencia Artificial, Universidad de Alicante, San Vicente del Raspeig, 03690 Alicante, Spain
arnal@ua.es

[2] Departamento de Sistemas Informáticos y Computación, Universidad Politécnica de Valencia, 46730 Valencia, Spain

Abstract. A fuzzy logic filter to remove mixed Gaussian and impulsive noise from CT medical images is presented. The method performs a weighted average operation where a fuzzy rule-based model is defined to compute the coefficients. In addition, a parallel filter based on this method is presented. Implementation of the parallel algorithm on multi-core platform using OpenMP is presented. Efficiency is measured in terms of execution time and in terms of PSNR over medical CT images. Experimental results show that mixed Gaussian-impulsive noise is reduced efficiently and image details are preserved. The multi-core implementation allows reducing mixed Gaussian-impulse noise in real-time.

Keywords: Parallel computing · Image filter · Fuzzy rules · CT images

1 Introduction

Computed tomography (CT) is a medical essential tool extensively used for diagnosis and image-guided methods. CT images are corrupted by noise mainly due to the image acquisition technique and to attempts to reduce patients exposure to radiation. As the amount of radiation is reduced, noise increases. Therefore, denoising is required for adequate CT image analysis. The additive Gaussian noise is the most common type of noise present in CT images [1,2]. This noise may appear combined with other types of noises what hinders the filtering process. In this paper, the denoising of a combination of Gaussian and impulsive noise is analyzed in CT images. Many algorithms have been proposed to reduce either Gaussian (see e.g. [3–5]) or impulsive noise (see e.g. [6–8]). However, not all methods are suitable when images are contaminated simultaneously with impulsive and Gaussian noise. In this paper, a method based on fuzzy logic is proposed to reduce mixed Gaussian and impulsive noise from CT medical images. Moreover, a parallel method based on this filter is introduced. The filtering process is

Y. Liu et al. (Eds.): ICNC-FSKD 2019, AISC 1074, pp. 975–982, 2020.
https://doi.org/10.1007/978-3-030-32456-8_104

based on a weighted average operation where a fuzzy rule-based model is defined to compute the coefficients. In order to compute the weights a fuzzy rule system is used. Section 2 introduces the proposed fuzzy method. Experimental results are presented in Sect. 3 and conclusions are described in Sect. 3.

2 Fuzzy Method

Let A be a gray-scale image, and let W be a $n \times n$ filtering window with central pixel x. The proposed method will replace the pixel x by a new value $\hat{x}$ computed as a weighted average of selected pixels $x^1, x^2, \ldots, x^q$ from the window W as

$$\hat{x} = \frac{\sum_{j=1}^{q} \omega_j \cdot x^j}{\sum_{j=1}^{q} \omega_j}, \tag{1}$$

where $\omega_j \in [0, 1]$ are obtained by defuzzification using a fuzzy rule-based system.

2.1 Computing Rank-Ordered Absolute Differences (ROAD) Statistic for Impulse Detection

Given a pixel x_i, let consider a $n \times n$ filtering window W with central pixel x_i. The pixels in this window are ordered according to their absolute distance to x_i, $d(x_i, x_j) = |x_i - x^j|$, obtaining a sequence $x^1, x^2, \ldots, x^{n^2}$, such that $d(x_i, x^1) \leq d(x_i, x^2) \leq \ldots \leq d(x_i, x^{n^2})$. In this sequence, clearly $x^1 = x_i$. Then the ROAD_m statistic, defined in [9], is computed by

$$\mathrm{ROAD}_m = \sum_{j=1}^{m} d(x_i, x^j). \tag{2}$$

The ROAD_m statistic gives a measure of how close a pixel is to its m most similar neighbors. Then, a low value for $\mathrm{ROAD}_m(x_i)$ means that those m pixels x^j are similar to x_i and then x_i is expected not to be an impulse. On the other hand, higher values for $\mathrm{ROAD}_m(x_i)$ evidence a higher noise degree for x_i. The adjustment of parameter m is analyzed in Sect. 3.

2.2 Impulsivity Degree Function

We consider the fuzzy set given by the following membership function μ. We define the degree for the pixel x_i to be an impulse by the function

$$\mu(x_i) = \begin{cases} 0, & \mathrm{ROAD}_m(x_i) \leq p_1 \\ \frac{\mathrm{ROAD}_m(x_i) - p_1}{p_2 - p1}, & p_1 < \mathrm{ROAD}_m(x_i) < p_2 \\ 1, & \mathrm{ROAD}_m(x_i) \geq p_2 \end{cases} \tag{3}$$

where p_1 and p_2 are two parameters discussed in Sect. 3. Then, using this certainty degree function, the certainty degree for the pixel x_i not to be an impulse can be defined using the negation operator as $1 - \mu(x_i)$.

2.3 Similarity Degree

To represent the data in the fuzzy system, a variable "Similarity" is considered. This variable can take the values "high", "moderate" and "low". For the value "high" we consider the fuzzy set given by the following membership function. We define the certainty degree for pixel x^j to be highly similar to x_i by

$$\mu_H(x_i, x^j) = \begin{cases} 1, & d(x_i, x^j) \leq p_3 \\ \frac{-d(x_i,x^j)}{3p_3} + \frac{4}{3} & p_3 < d(x_i, x^j) < 4p_3 \\ 0, & d(x_i, x^j) \geq 4p_3. \end{cases} \tag{4}$$

And using the fuzzy negation, we define the membership function corresponding to the value "low" as $\mu_L(x_i, x^j) = 1 - \mu_H(x_i, x^j)$. The certainty degree function $\mu_M(x_i, x^j)$ for the value "moderate" is defined as

$$\mu_M(x_i, x^j) = \begin{cases} \frac{d(x_i,x^j)-p_3}{p_3}, & p_3 < d(x_i, x^j) < 2p_3 \\ 1, & 2p_3 \leq d(x_i, x^j) \leq 3p_3 \\ \frac{4p_3-d(x_i,x^j)}{p_3} & 3p_3 < d(x_i, x^j) < 4p_3 \\ 0, & \text{elsewhere.} \end{cases} \tag{5}$$

2.4 Averaging Fuzzy Coefficients

To represent the data used in the output of the fuzzy system, we consider a linguistic variable to represent the value of the weights ω_j in the averaging process. This variable can take the values "high", "moderate" and "low". The membership functions $\eta_H(\omega_j)$, $\eta_M(\omega_j)$, and $\eta_L(\omega_j)$ corresponding to the value "high", "moderate" and "low" are given by

$$\eta_H(\omega_j) = \begin{cases} \frac{\omega_j-1}{1-p_4} + 1, & p_4 < \omega_j \leq 1 \\ 0, & \text{elsewhere,} \end{cases} \quad \eta_M(\omega_j) = \begin{cases} \frac{2\omega_j-1}{2p_4-1} + 1, & 1-p_4 < \omega_j \leq \frac{1}{2} \\ \frac{1-2\omega_j}{2p_4-1} + 1, & \frac{1}{2} < \omega_j < p_4 \\ 0, & \text{elsewhere,} \end{cases}$$

$$\eta_S(\omega_j) = \begin{cases} \frac{\omega_j}{p_4-1} + 1, & 0 \leq \omega_j \leq 1-p_4 \\ 0, & \text{elsewhere.} \end{cases} \tag{6}$$

The experimental adjustment of the parameter p_4 is discussed in Sect. 3.

2.5 Fuzzy Rule-Based System

In order to calculate the averaging coefficients ω_j in Eq. (1), a fuzzy rule-based system is used. The fuzzy rule-based system is based on three rules.

- Fuzzy Rule 1: IF (x^j is not impulsive AND x_i is impulsive AND the similarity between x_i and x_j is moderate) THEN ω_j is moderate.

- Fuzzy Rule 2: IF (x^j is not impulsive AND x_i is impulsive AND the similarity between x^j and x_i is low) OR (x^j is not impulsive AND x_i is not impulsive AND the similarity between x^j and x_i is high) THEN ω_j is high.
- Fuzzy Rule 3: IF (x^j is impulsive) OR (x^j is not impulsive AND x_i is impulsive AND the similarity between x^j and x_i is high) OR (x^j is not impulsive AND x_i is not impulsive AND the similarity between x^j and x_i is moderate) OR (x^j is not impulsive AND x_i is not impulsive AND the similarity between x^j and x_i is low) THEN ω_j is low.

To evaluate the degree of the fuzzy rules, the operation AND has been calculated using as t-norm the usual product, and the disjunction operation OR has been calculated using as s-norm the probabilistic addition.

2.6 Defuzzification

The numerical value for the weight ω_j corresponding to the pixel x^j is obtained by defuzzification. For this purpose, the center of gravity (COG) [10] was used. Let k_H, k_M, and k_L denote for a pixel x^j the membership degrees of the consequents in the fuzzy rules 1, 2, 3. Taking the surface in each of the triangles determined by η_H, η_M, and η_L and under each one of the three constant functions k_H, k_M, and k_L respectively, three trapeziums are formed. Let L_j be the polygonal line determined by the tops and lateral walls of these trapeziums, then

$$\omega_j = \frac{\int_0^1 x \cdot L_j(x)dx}{\int_0^1 L_j(x)dx}. \tag{7}$$

2.7 Parallel Method

The method is applied iteratively until certain stopping criteria is satisfied. As the number of iterations is increased, the computational time increases. Due to this fact, in order to reduce the computational time, a parallel version of the

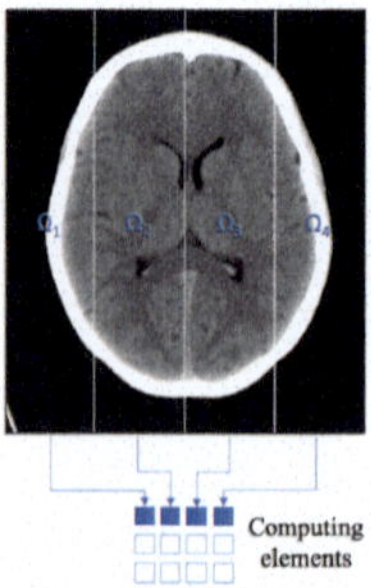

Fig. 1. Image domain decomposition in 4 subdomains.

method is introduced. To distribute the pixels of the CT image among the computing elements in the parallel machine, the image domain Ω is divided in P subdomains $\Omega_i, i = 1, \ldots P$, where P is the number of computing elements. Figure 1 illustrates the subdomains used in the experiments. Algorithm 1 describes the steps of the parallel method. To analyze the parallel performance, the speed-up S_P is defined as $S_P = \frac{T_s}{T_P}$, where T_s is the computational time of the sequential method and T_P is the computational time of the parallel method.

Algorithm 1. Parallel Fuzzy Method.

Require: Image A, domain decomposition $\{A_{\Omega_k}\}_{k=1}^{P}$, parameters $n, q, m, p_1, p_2, p_3, p_4$
Ensure: Filtered image.

Image $A_0 = A$
for *Iteration* $it = 1, \ldots,$ until stoping criterion is satisfied **do**
  Image $A_{it} = A_{it-1}$
  for $k = 1, \ldots, P$, in parallel **do**
    Processor k:
    for x_i pixel in $A_{it\,\Omega_k}$ **do**
      Consider a $n \times n$ filtering window W with central pixel x_i
      Impulsivity degree computation: Compute $\mu(x_i)$ using Eq. (3)
      Similarity degree computation
        Order pixels $x^j \in W$ using $d(x_i, xj)$
        Select q nearest pixels $x^1, \ldots, x^q$ to be used in averaging proces
      for $j = 1, \ldots, q$ **do**
        Compute $\mu_H(x_i, x^j), \mu_L(x_i, x^j), \mu_H(x_i, x^j)$, using Eqs. (4), (5)
      end for
      Computation of averaging weights by defuzzyficattion
      for $j = 1, \ldots, q$ **do**
        Compute certainty degree for antecedents of rules for x^j
        Compute the averaging weight w_j for pixel x^j by using COG
      end for
      Averaging process
        $\hat{x}_i = \frac{\sum_{j=1}^{q} w_j \cdot x^j}{\sum_{j=1}^{q} w_j}$
    end for
  end for
end for

3 Experimental Results

In this section, the proposed filter has been applied to CT medical images. For this purpose, CT medical images (see Fig. 2) from Radiopaedia database (Case Courtesy of A. Prof Frank Gaillard, Radiopaedia.org, rID 35508) have been considered in the study. These images correspond to a normal brain of a 30 years old female. CT images were contaminated with different intensities of Gaussian noise (standard deviation $\sigma \in [5, 30]$) and random impulsive noise (probability $p \in [0.05, 0.3]$). To this end, the random value impulse noise and the classical Gaussian-noise model [3] have been contemplated. Filter performance has been analyzed with the objective measure peak signal-to-noise ratio (PSNR) that measures the noise reduction. For the adjustment of the filter parameters, the PSNR has been analyzed as a function of them. In our experiments the best results were obtained using filtering windows of size 3×3, $q = 7$ in Eq. (1)

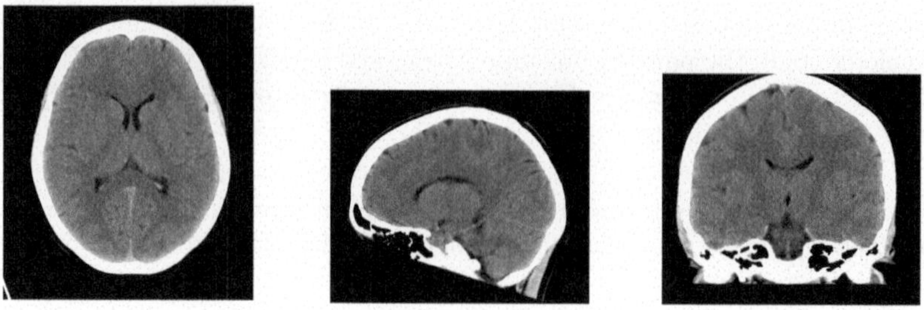

(a) Axial view (b) Sagittal view (c) Coronal view

Fig. 2. CT images used in the experiments (a) Axial: 1024 × 904 pixels, (b) Sagittal: 822 × 1024 pixels, (c) Coronal: 890 × 1024 pixels.

Table 1. PSNR for images corrupted with Gaussian σ and impulsive noise p.

Noise	PSNR					
	Axial view		Sagittal view		Coronal view	
	Noisy	Filtered	Noisy	Filtered	Noisy	Filtered
$\sigma = 5,\ p = 0.05$	19.03	34.97	18.85	34.05	19.21	33.35
$\sigma = 10,\ p = 0.1$	16.06	32.43	15.86	31.75	16.13	30.93
$\sigma = 20,\ p = 0.2$	12.98	28.33	12.82	27.83	13.07	26.93
$\sigma = 30,\ p = 0.3$	11.17	24.93	11.02	24.40	11.25	23.30

and $m = 3$ in Eq. (2). For the parameters p_1, p_2 in Eq. (3), the best results in terms of PSNR were obtained for $p_1 = 90$ and $p_2 = 110$. For the parameter p_3 in Eqs. (4) and (5) we obtained a good performance varying p_3 as a function of the Gaussian noise standard deviation σ, as $p_3 = 0.997\sigma + 1.96$. For the parameter p_4 in Eq. (6) the best results were obtained setting $p_4 = 0.9$. In the experiments the method was applied iteratively until the PSNR reached its maximum value. Figure 3 presents the CT images filtered for sagittal view corrupted with different levels of mixed noise for visual inspection. It can be observed the efficient denoising performance and the good conservation of edges and contours of the images. Table 1 presents the PSNR obtained for the CT images corrupted with different levels of Gaussian noise with standard deviation σ and impulsive noise with appearance probability p. It can be observe that the filter obtains outstanding PSNR values. Figure 4 shows the speed-up obtained for the axial view image corrupted with mixed noise. Similar results were obtained for the other CT images. The experiments were carried out on an Intel Xeon X5660 processor with 12 cores, 48 GB DDR3-1333 RAM memory, under CentOS Linux version 5.6 operative system. Parallel algorithm was implemented using the Open Multi-Processing platform (OpenMP). It can be observed that the level of noise affects the computational time but the speed-up is almost not affected. Considerable efficiencies in the range of 68.4%–98% have been achieved depending on the number of computing elements and the size of the image.

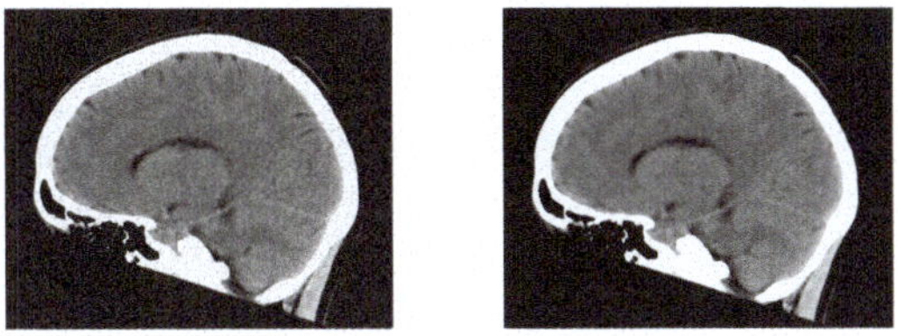

(a) Noisy $\sigma = 5,\ p = 0.05$ (b) Filtered image

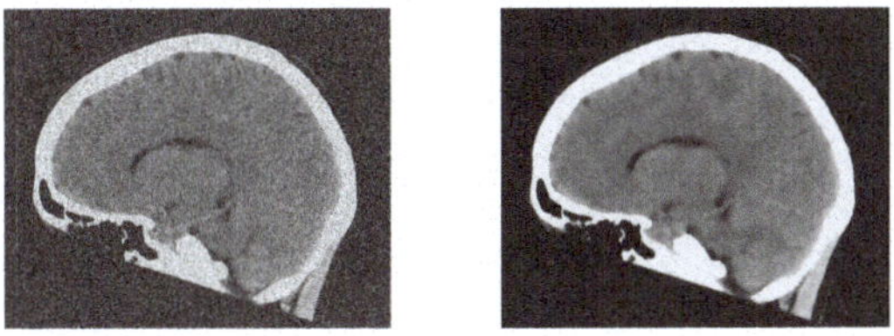

(c) Noisy $\sigma = 30,\ p = 0.30$ (d) Filtered image

Fig. 3. Filter outputs for visual comparison. Sagittal view contaminated with different levels of Gaussian noise σ and impulsive noise p.

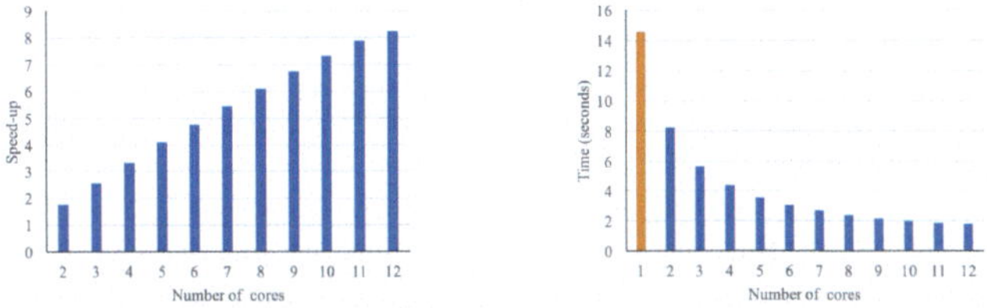

(a) Axial corrupted with $\sigma = 5,\ p = 0.05$

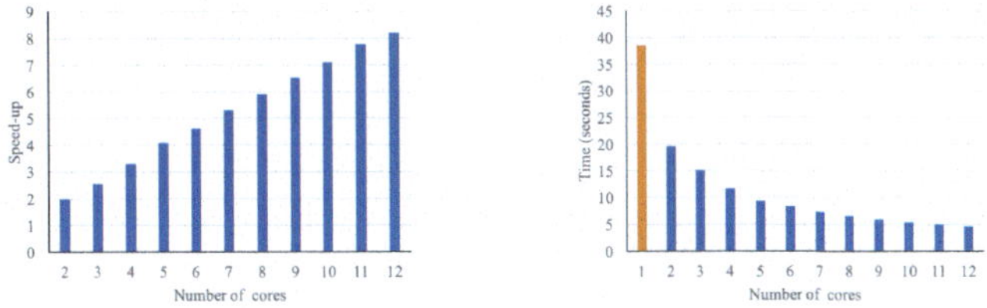

(b) Axial corrupted with $\sigma = 30,\ p = 0.3$

Fig. 4. Speed-up and computational time for CT axial view image corrupted with different densities of Gaussian noise σ and impulsive noise p.

4 Conclusions

A fuzzy filter to reduce mixed impulsive and Gaussian noise from CT medical images has been proposed. The technique performs a weighted averaging, where the weights are determined using a fuzzy rules system. In this process the ROAD statistic is used. Experiments show that the filter efficiently reduces noise and preserves image structures, obtaining competitive values in terms of the MAE and PSNR objective measures. A parallel version of the method is introduced and implemented on shared-memory multiprocessors using OpenMP. Performance in terms of computational time is analyzed on a multi-core machine showing that the new filter is able to reduce mixed Gaussian-impulse noise in real-time.

Acknowledgment. This research was supported by the Spanish Ministry of Science, Innovation and Universities (Grant RTI2018-098156-B-C54) co-financed by FEDER funds.

References

1. Gravel, P., Beaudoin, G., De Guise, J.A.: A method for modeling noise in medical images. IEEE Trans. Med. Imaging **23**(10), 1221–1232 (2004)
2. Lei, T., Sewchand, W.: Statistical approach to X-ray CT imaging and its applications in image analysis. II. A new stochastic model-based image segmentation technique for X-ray CT image. IEEE Trans. Med. Imaging **11**(1), 62–69 (1992)
3. Plataniotis, K.N., Venetsanopoulos, A.N.: Color Image Processing and Applications. Springer, Heidelberg (2013)
4. Tomasi, C., Manduchi, R.: Bilateral filtering for gray and color images. In: Proceedings of the Sixth International Conference on Computer Vision, ICCV 1998, pp. 839–846. IEEE Computer Society, Washington (1998)
5. Li, X.: On modeling interchannel dependency for color image denoising. Int. J. Imaging Syst. Technol. **17**(3), 163–173 (2007)
6. Smolka, B.: Peer group switching filter for impulse noise reduction in color images. Pattern Recogn. Lett. **31**(6), 484–495 (2010)
7. Schulte, S., Witte, V.D., Nachtegael, M., der Weken, D.V., Kerre, E.E.: Fuzzy random impulse noise reduction method. Fuzzy Sets Syst. **158**(3), 270–283 (2007)
8. Melange, T., Nachtegael, M., Kerre, E.E.: Fuzzy random impulse noise removal from color image sequences. IEEE Trans. Image Process. **20**(4), 959–970 (2011)
9. Garnett, R., Huegerich, T., Chui, C., He, W.: A universal noise removal algorithm with an impulse detector. IEEE Trans. Image Process. **14**(11), 1747–1754 (2005)
10. Passino, K.M., Yurkovich, S., Reinfrank, M.: Fuzzy Control, vol. 42. Addison-Wesley, Menlo Park (1998)

Structural Parameters Optimization and Grey Relational Analysis in Honeycomb Spiral Heat Exchangers

Zhengfang Wang[1,2(✉)], Pengfei Han[2], Jia Wang[2], Wenjian Yu[2], and Ming Li[2]

[1] Zibo Vocational Institute, Zibo 255314, China
361185918@qq.com
[2] Shandong University of Technology, Zibo 255012, China

Abstract. Honeycomb structure in spiral baffle heat exchangers can enhance heat transfer, the heat transfer performance will be greatly affected by the change of structural parameters. The orthogonal optimization method was used to design 9 test schemes with 4 factors and 3 levels. For the designed scheme, the mathematical model and entity modeling were established with UG software, and the boundary conditions of mesh division were set and solved numerically with ANSYS Workbench software. In order to optimize the outlet temperature Nusselt number and the surface heat transfer coefficient, the range analysis of the solution results of the test scheme was carried out, the order of the influence factors was determined, and the optimal combination was found out. By setting the ideal scheme sequence, the grey relational degree between each scheme and the ideal scheme is obtained by grey relational analysis based on grey system theory, and the best scheme is the one with the largest relational degree. Finally, it is found that the result of grey correlation analysis is consistent with that of range analysis.

Keywords: Grey relational analysis · Orthogonal optimization method · Honeycomb spiral baffle heat exchanges

1 Introduction

Since 1950's, the spiral plate heat exchanger has been used in many industrial fields in China. After years of research and promotion, not it is widely used in petroleum, chemical, printing and dyeing, pharmaceutical and other industries. After analyzing the advantages and disadvantages of various plate, plate-fin and spiral plate heat exchangers, Song put forward a compact honeycomb spiral plate heat exchanger with the advantages of strong pressure bearing capacity, compact structure and high heat transfer efficiency, as in [1]. The numerical simulation of cylindrical honeycomb spiral plate heat exchanger was carried out by Wang, as in [2]. The honeycomb structure can enhance the heat transfer, and the change of the size parameter of the honeycomb structure will greatly affect the heat transfer performance, as in [3].

With the compression of equipment construction and maintenance cost and the advocacy of energy saving and emission reduction policy, various optimization methods

Y. Liu et al. (Eds.): ICNC-FSKD 2019, AISC 1074, pp. 983–990, 2020.
https://doi.org/10.1007/978-3-030-32456-8_105

have been widely used in machinery. The traditional single-objective optimization method can no longer meet the requirements of design work. There are many performance indexes to measure heat exchanger, such as resistance, heat transfer, structure, economy and so on. In the optimization design, it is often necessary to consider more than one objective [4].

Orthogonal test design is a method of analyzing multi-factor and multi-level test, can help to get better optimization scheme by using a limited number of experiments, can promote efficiency, is a convenient tool for optimization design, as in [5]. GA (Genetic Algorithm) and PSO (Particle Swarm Optimization) are mainly used to optimize algorithms such as function optimization, control theory, artificial algorithm and so on, as in [6, 7]. The entropy production minimization method and the fire dissipation theory are specially used to evaluate the performance of heat exchangers. The evaluation index is limited and the application is limited, as in [8, 9]. At the same time, the orthogonal test design method has strong maneuverability and wide application range. The grey correlation analysis is suitable for the size and regularity of samples, and the calculation is convenient, as in [10, 11]. In view of the above optimization methods, considering the change of structural parameters of rhomboidal honeycomb spiral plate heat exchanger, the optimum combination and primary-secondary relations are determined in many factors and levels, so the orthogonal test design is chosen to complete the parameter optimization. Grey correlation analysis is used to verify the optimal scheme obtained by orthogonal optimization.

2 Orthogonal Optimization Test Scheme

2.1 Selection of Influencing Factors and Design Indexes

For a rhomboidal honeycomb, its basic geometric parameters include transverse diagonal a, longitudinally diagonal b and stack honeycomb height H. Considering the inlet velocity v_{in} effecting on the flow and heat transfer, four parameters are taken and expressed by A, B, C and D respectively. In order to study the effect of these parameters on the heat transfer performance, the outlet temperature, Nusselt number and surface heat transfer coefficient was selected. For the evaluation index of orthogonal optimization design, three levels are selected for each parameter, as shown in Table 1. v_{in} can be found using Eq. (1).

Table 1. Factors and level table of orthogonal design

Factor	Level		
A	1	2	3
B	3	4	5
C	2	3	4
D	0.904	1.206	1.507

$$Re = \frac{\rho v d}{\eta} \tag{1}$$

There Re = 18000, 24000, 30000, ρ = 998.2 kg/m3, η = 0.001003 kg/(m•s), d = 0.02 m. Create $L_9(3^4)$ orthogonal list, refer to Table 2.

Table 2. Table of orthogonal design

Test no.	Factors			
	a/mm	b/mm	H/mm	$v_{in}/(m/s)$
1	1	3	2	0.904
2	1	4	3	1.206
3	1	5	4	1.507
4	2	3	3	1.507
5	2	4	4	0.904
6	2	5	2	1.206
7	3	3	4	1.206
8	3	4	2	1.507
9	3	5	3	0.904

2.2 Model Establishment and Solution

According to the orthogonal test scheme, heat transfer model of diamond honeycomb spiral plate with nine different structural parameters is established using UG software. The overall model is based on the heat transfer model of rhomboidal honeycomb spiral plate, as in [4]. However, the structural parameters of rhomboidal honeycomb are different, and the structural parameters are according to the values in the scheme.

Adopted *Realizable k-ε* turbulent model to calculate, used and standard wall function to simulate the flow near the wall. Considering the influence of boundary layer, adopted Inflation method in all nine models. This method can make the mesh density in the boundary layer more dense, which is helpful to improve the accuracy of the numerical solution of CFD (Computational Fluid Dynamics) problem. The grid numbers of the nine models are 141352, 152146, 243121, 143246, 143242, 243162, 143745, 242479 and 243846.

The medium in the honeycomb spiral channel is treated by water-liquid, in FLUENT software as an incompressible fluid with constant physical properties, and the gravity is ignored. The upper and lower walls are set to Non-slip constant temperature wall, temperature is T_w = 400 K. The front and rear of the model are set to the symmetrical surface; the inlet of the model is set to the velocity inlet; the average velocity is given according to the numerical value in the scheme table; the inlet temperature is T_{in} = 300 K.

The solving process is carried out in FLUENT, the solver is set to double precision, the coupling of pressure and velocity is adopted SIMPLE algorithm, all the terms are adopted second-order upwind scheme, the convergence residual of the energy equation is 10^{-6}.

2.3 Numerical Simulation Results

The internal streamline of the model after numerical simulation test is shown in Fig. 1. The density of the fluidic cluster reflects the velocity variation of each point in the flow field at this time. When the streamline is vortex shape, it means low flow resistance and low heat transfer efficiency, while the streamline arrangement is messy, the random direction is strong, means which is favorable to heat transfer, the disadvantage is that the flow resistance is large. Velocity comparison diagram refer to Fig. 2.

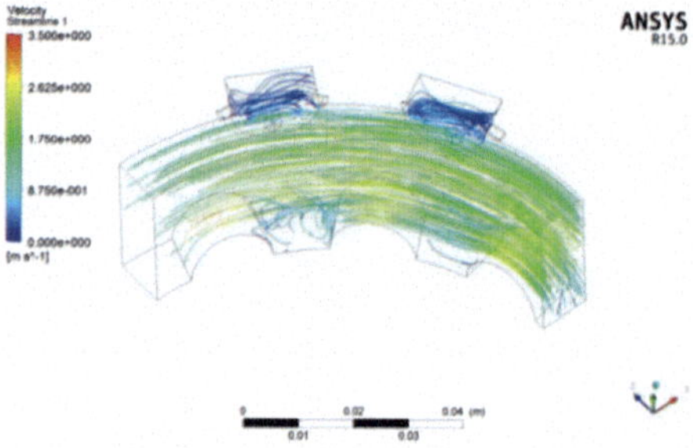

Fig. 1. Streamlines of models

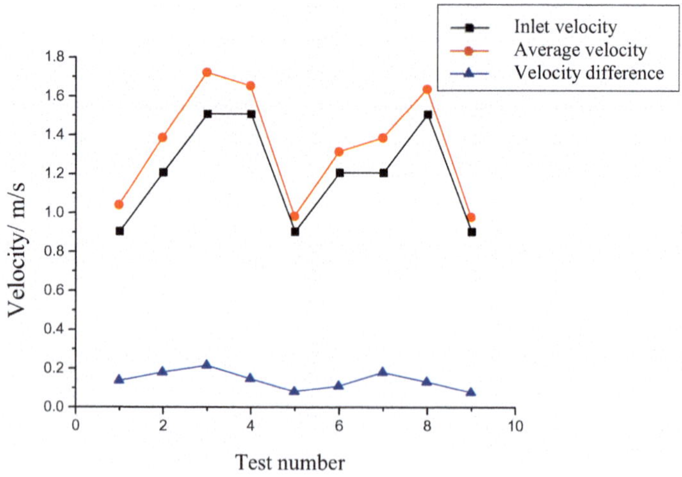

Fig. 2. Comparison of velocity magnitude

3 Range Analysis of Numerical Simulation Results

Through the numerical simulation of nine experimental schemes, the results of three evaluation indexes are obtained. In order to show the relationship between factors and indicators, the trend diagrams of each index and factor are drawn. Table 3 shows the test results of 3 indicators in 9 different schemes.

Table 3. Simulation results

Test no.	Index		
	T_{out}/K	Nu	$h/(W/(m^2 \cdot K))$
1	304.016	12165.420	7299.251
2	303.831	14931.331	8958.798
3	303.653	17423.037	10453.822
4	304.293	20914.277	12554.566
5	305.152	15307.635	9184.581
6	305.455	21914.463	13148.679
7	303.950	14584.455	8750.674
8	305.539	28058.998	16835.411
9	306.060	18080.496	10848.299

Tables 4, 5 and 6 are for each of the three test indicators, and as a result, it is obtained by calculating the data of each test index at each level of each factor.

Table 4. Treatment results of outlet temperature

Index		*a/mm*	*b/mm*	*H/mm*	$v_{in}/(m/s)$
T_{out}/K	K_1	911.500	912.260	915.011	915.229
	K_2	914.902	914.522	914.185	913.238
	K_3	915.549	915.170	912.756	913.485
	$\bar{K}_1$	303.833	304.087	305.004	305.076
	$\bar{K}_2$	304.967	304.841	304.728	304.413
	$\bar{K}_3$	305.183	305.057	304.252	304.495
Range	R	1.350	0.970	0.752	0.663

Table 5. Treatment results of Nusselt number

Index		*a/mm*	*b/mm*	*H/mm*	$v_{in}/(m/s)$
Nu	K_1	44519.787	47664.152	62138.881	45553.551
	K_2	58136.325	58297.963	53926.103	51430.248
	K_3	60723.949	57417.996	47315.127	66396.312
	$\bar{K}_1$	14839.929	15888.051	20712.960	15184.517
	$\bar{K}_2$	19378.775	19432.654	17975.368	17143.416
	$\bar{K}_3$	20241.316	19139.332	15771.709	22132.104
Range	R	5401.387	3544.603	4941.251	6947.587

Table 6. Treatment results of surface heat transfer coefficient

Index		*a/mm*	*b/mm*	*H/mm*	$v_{in}/(m/s)$
$h/(W/(m^2 \cdot K))$	K_1	26711.872	28604.492	37283.331	27332.132
	K_2	34887.826	34978.779	32361.663	30858.151
	K_3	36343.373	34450.800	28389.077	39843.788
	$\bar{K}_1$	8903.957	9534.831	12427.777	9110.711
	$\bar{K}_2$	11629.275	11659.593	10787.221	10286.050
	$\bar{K}_3$	12114.458	11483.600	9463.026	13281.263
Range	R	3210.501	2124.762	2964.751	4170.552

In the range analysis, the order of the influence of each factor on the test index can be judged according to the magnitude of the extreme error. The greater the value of the range, the greater the correlation and importance of the factor to the test index. According to Table 6, it is possible to determine the order of primary and secondary effects of factors under each indicator, as shown in Table 7.

Table 7. Order of primary and secondary affect

Index	Primary and secondary order
Outlet temperature	A > B > C > D
Nusselt number	D > A > C > B
Surface heat transfer coefficient	D > A > C > B

Determine the optimal combination of factors based on the average values of each index at different levels, for outlet temperature the optimized combination is A3B3C1D1, for Nusselt number it is A3B2C1D3, for Surface heat transfer coefficient it is A3B2C1D3. According to the primary and secondary order of each factor, the optimal parameters are determined by the method of comprehensive balance, and the optimized combination is A3B2C1D3, which is the No. 8 scheme.

4 Grey Relational Analysis of Simulation Data

4.1 Find Out Ideal Scheme

Since the outlet temperature, Nusselt number and heat transfer coefficient are both higher and better, the maximum value of each index in the nine schemes is chosen as the ideal value, and the data sequence is obtained, refer to Table 8.

Table 8. Data sequence table

Serial number	Data1	Data2	Data3
X_0	306.060	28058.998	16835.411
X_1	304.016	12165.420	7299.251
X_2	303.831	14931.331	8958.798
X_3	303.653	17423.037	10453.822
X_4	304.293	20914.277	12554.566
X_5	305.152	15307.635	9184.581
X_6	305.455	21914.463	13148.679
X_7	303.950	14584.455	8750.674
X_8	305.539	28058.998	16835.411
X_9	306.060	18080.496	10848.299

Initial value image using Eq. (2)

$$X_i' = X_i/x_i(1) = \left(x_i'(1),\ x_i'(2),\ x_i'(3)\right),\ i = 0,\ 1,\ 2,\ 3,\ 4,\ 5,\ 6,\ 7,\ 8,\ 9 \quad (2)$$

Calculate absolute value sequence using Eq. (3)

$$\Delta_i(k) = \left|x_0'(k) - x_i'(k)\right|;\ i = 1,\ 2,\ 3,\ 4,\ 5,\ 6,\ 7,\ 8,\ 9;\ k = 1,\ 2,\ 3 \quad (3)$$

Seeking extreme error,

$$M = \max_i \max_k \Delta_i(k) = 51.6624,\ m = \min_i \min_k \Delta_i(k) = 0.0000$$

Calculating correlation coefficient using Eq. (4),

$$\gamma_{0i}(k) = \frac{m + \xi M}{\Delta_i(k) + \xi M} = \frac{25.8312}{\Delta_i(k) + 25.8312},\ i = 1, 2, 3, 4, 5, 6, 7, 8, 9;\ K = 1, 2, 3 \quad (4)$$

There $\zeta = 0.5$, $i = 0 \sim 9$, $k = 1 \sim 3$. Calculate of grey correlation degree using Eq. (5)

$$\gamma_{0i} = \frac{1}{n}\sum_{k=1}^{n} \gamma_{0i}(k) \quad (5)$$

There $i = 0 \sim 9$, $k = 1 \sim 3$, then γ_{0i} = (0.5960, 0.6270, 0.6621, 0.7274, 0.6309, 0.7493, 0.6226, 0.9968, 0.6704), γ_{08} is the biggest, so the No. 8 program is the best.

5 Conclusions

Using Orthogonal table of 4 factors 3 levels 9 schemes can complete the scheme design of orthogonal test, building model using UG software, solving problem using ANSYS Workbench, calculating the order of primary and secondary influencing factors is determined, and the optimal combination is obtained by using the method of comprehensive balance. According to outlet temperature, Nusselt number and surface heat transfer coefficient, grey correlation degree between each scheme and the ideal scheme is obtained by grey relational analysis. It is found that the 8 scheme is the best, and the result is consistent with the result obtained by the range analysis.

References

1. Song, H., Li, P.: Honeycomb type spiral plate exchanger. Press. Vessel **27**(04), 13–15 (2010)
2. Wang, Y.: Numerical Simulation and Optimization of New Honeycomb Spiral Plate Heat Exchanger. master's degree thesis, Zhengzhou University, Zhengzhou (2014)
3. Han, P., Wang, Z., Zhang, C., Liu, Z.: Heat transfer research of pile honeycomb structure based on field synergy principle. Coal Mine Mach. **37**(04), 99–101 (2016)
4. Wang, H.: Multi-objective optimization design method of heat exchanger based on genetic algorithm. Beijing: master's degree thesis, North China Electric Power University (Beijing) (2011)
5. Wang, W., Shi, W., Jiang, X.: Optimization design of multistage centrifugal pump impeller by orthogonal experiment and CFD. J. Irrig. Drainage Mach. Eng. **34**(03), 191–197 (2016)
6. Gu, X., Pan, G., Liu, M., Wang, Y.: Optimal design of trefoil-baffle support heat exchanger based on genetic algorithm. Press. Vessel **30**(07), 12–17 (2013)
7. Ren, P., Wu, Z.: Optimization design of plate heat exchanger based on particle swarm optimization algorithm. Energy Technol. **30**(05), 270–271 (2009)
8. Ren, X.: Application of entropy generation minimization method in optimization design of shell-tube heat exchanger. Master's degree thesis, Shandong University, Jinan (2009)
9. Li, M., Guo, J., Xu, M., Cheng, L.: Application of fire dissipation theory to optimal design of tube-shell heat exchanger. J. Eng. Thermophys. **07**, 1189–1192 (2010)
10. Wang, Z., Wang, Y., He, Q., et al.: Effect factors analysis of corrosion of oil atmospheric distillation tower based on grey system theory. J. China Pet. Univ. (Nat. Sci. Ed.) **33**(05), 119–123+129 (2009)
11. Geng, Q., Wei, Z., Du, J.: Cutters' layout optimization method of the full-face rock tunnel boring machine based on grey relational analysis. J. Mech. Eng. **50**(21), 45–53 (2014)

Interval Valued Generalised Fuzzy Soft Expert Set and Its Application

Chen Bin(✉)

School of Mathematical Sciences, University of Jinan, Jinan 250022, China
jnchenbin@126.com

Abstract. The purpose of this paper is to combine the work of interval-valued fuzzy sets and fuzzy soft expert sets, from which we can obtain a new concept: interval-valued generalised fuzzy soft expert sets. We also introduce some of its operations: complement, union, intersection and study their properties. Then, we use interval-valued generalised fuzzy soft expert sets to analyze a decision-making problem.

Keywords: Soft sets · Fuzzy soft sets · Interval-valued generalised fuzzy soft expert sets · Decision making

1 Introduction

Soft set theory [1], proposed by Molodtsov in 1999, is a new general method for dealing with uncertain data. He applied several directions for the applications of soft sets, such as smoothness of functions, game theory, Riemann-integration, and theory of probability. Now, soft set theory and its applications are progressing rapidly in different fields, such as economics, engineering, social science, medical science, etc.

In recent years, research on soft set theory has been done by many researchers in mathematics, computer and information science, including the works of fundamental soft set theory [2], soft set theory in algebra [3,4], soft set theory in topological space [5–8], theory for data analysis, particularly in decision making [9–13]. Yang et al. [14] defined the interval-valued fuzzy soft set which is based on a combination of the interval-valued fuzzy set and soft set. Majumdar and Samanta [15] introduced a concept of generalized fuzzy soft sets and their operations and application of generalised fuzzy soft sets in decision-making problem and medical diagnosis problem. Alkhazaleh and Salleh [16,17] introduced the concept of a soft expert set and fuzzy soft expert set, where the user can know the opinion of all experts in one model without any operations. Even after any operation, the user can know the opinion of all experts. So in this paper we introduce the concepts of the generalised fuzzy soft expert set, which will be more effective and useful which is a combination of a fuzzy soft expert set and interval-valued fuzzy set. We also define its basic operations: complement, union, intersection and study their properties. Finally, we give an application of this concept in decision-making problem.

Y. Liu et al. (Eds.): ICNC-FSKD 2019, AISC 1074, pp. 991–999, 2020.
https://doi.org/10.1007/978-3-030-32456-8_106

2 Preliminaries

Let U be an initial universe set and E_U be a collection of all possible parameters with respect to U, where parameters are the characteristics or properties of objects in U. We will call E_U the universe set of parameters with respect to U. The following definitions appeared in [1,15–19].

Definition 1. *A pair (F, A) is called a soft set over U if $A \subset E_U$ and $F : A \to P(U)$, where $P(U)$ is the set of all subsets of U.*

Definition 2. *A pair (F, A) is called a fuzzy soft set over U if $A \subset E_U$ and $F : A \to I^U$, where I^U denotes the collection of all fuzzy subsets of U.*

Definition 3. *Let $F : A \to I^U$ and μ be a fuzzy subset of U, i.e., $\mu : E \to I = [0, 1]$, where I^U denotes the collection of all fuzzy subsets of U. Let F_μ be a mapping and $F_\mu : E \to I^U \times I$ defined as follows: $F_\mu(e) = (F(e), \mu(e))$, where $F(e) \in I^U$. Then F_μ is called generalised fuzzy soft set over the soft universe (U, E).*

Here for each parameter e_i, $F_\mu(e_i)$ indicates not only the degree of belongingness of the elements of U in $F(e_i)$ but also the degree of possibility of such belongingness which is represented by $\mu(e_i)$.

Definition 4. *An interval-valued fuzzy set F on a universe U is a mapping $F : U \to Int([0, 1])$, where $Int([0, 1])$ stands for the set of all closed subintervals of [0, 1].*

The set of all interval-valued fuzzy sets on U is denoted by $\mathcal{F}(U)$. Suppose that $X \in \mathcal{F}(U)$, for all $h \in U$, $\mu_X(h) = [\mu_X^-(h), \mu_X^+(h)]$ is called the degree of membership of an element h to X. And $\mu_X^-(h)$ and $\mu_X^+(h)$ are referred to as the lower and upper degrees of membership of h to X, where $0 \le \mu_X^-(h) \le \mu_X^+(h) \le 1$.

Definition 5. *Let X and Y be two interval-valued fuzzy sets on universe U. Then the union, intersection, and complement of X and Y are defined as follows:*

$$X \cup Y = \{\frac{h}{[\mu_X^-(h) \vee \mu_Y^-(h), \mu_X^+(h) \vee \mu_Y^+(h)]} : h \in U\} \tag{1}$$

$$X \cap Y = \{\frac{h}{[\mu_X^-(h) \wedge \mu_Y^-(h), \mu_X^+(h) \wedge \mu_Y^+(h)]} : h \in U\} \tag{2}$$

$$X^c = \{\frac{h}{[1 - \mu_X^+(h), 1 - \mu_X^-(h)]} : h \in U\} \tag{3}$$

Let U be an initial universe set and E be the set of parameters. X a set of experts (agent), and O a set of opinions. Let $Z = E \times X \times O$ and $A \subseteq Z$.

Definition 6. *A pair (F, A) is called a soft expert set over U, where F is a mapping given by $F : A \to P^U$, where P^U denote the power set of U.*

Let U be an initial universe set and E be the set of parameters. X a set of experts(agent), and $O = \{1 = agree, 0 = disagree\}$ a set of opinions. Let $Z = E \times X \times O$ and $A \subseteq Z$.

Definition 7. *A pair (F, A) is called a fuzzy soft expert set over U, where F is a mapping given by $F : A \to I^U$, where I^U denote all the fuzzy subsets of U.*

Example 1. Let $U = \{u_1, u_2, u_3, u_4\}$ be a set of universe, and let $E = \{e_1, e_2, e_3\}$ be a set of parameters. Let $Z = E \times X \times O$ and $A \subseteq Z$. Define a function $F : A \to I^U$ as follows:

$$
\begin{aligned}
F(e_1, m, 1) &= \{(u_1, 0.6), (u_2, 0.3), (u_3, 0.2), (u_4, 0.4)\},\\
F(e_1, n, 1) &= \{(u_1, 0.7), (u_2, 0.4), (u_3, 0.1), (u_4, 0.3)\},\\
F(e_1, r, 1) &= \{(u_1, 0.1), (u_2, 0.3), (u_3, 0.8), (u_4, 0.7)\},\\
F(e_2, m, 1) &= \{(u_1, 0.5), (u_2, 0.3), (u_3, 0.2), (u_4, 0.7)\},\\
F(e_2, n, 1) &= \{(u_1, 0.7), (u_2, 0.8), (u_3, 0.6), (u_4, 0.4)\},\\
F(e_2, r, 1) &= \{(u_1, 0.7), (u_2, 0.5), (u_3, 0.6), (u_4, 0.7)\},\\
F(e_3, m, 1) &= \{(u_1, 0.3), (u_2, 0.6), (u_3, 0.8), (u_4, 0.9)\},\\
F(e_3, n, 1) &= \{(u_1, 0.2), (u_2, 0.3), (u_3, 0.7), (u_4, 0.6)\},\\
F(e_3, r, 1) &= \{(u_1, 0.3), (u_2, 0.4), (u_3, 0.6), (u_4, 0.3)\},\\
F(e_1, m, 0) &= \{(u_1, 0.3), (u_2, 0.6), (u_3, 0.7), (u_4, 0.6)\},\\
F(e_1, n, 0) &= \{(u_1, 0.2), (u_2, 0.7), (u_3, 0.8), (u_4, 0.7)\},\\
F(e_1, r, 0) &= \{(u_1, 0.6), (u_2, 0.8), (u_3, 0.4), (u_4, 0.5)\},\\
F(e_2, m, 0) &= \{(u_1, 0.3), (u_2, 0.6), (u_3, 0.8), (u_4, 0.3)\},\\
F(e_2, n, 0) &= \{(u_1, 0.2), (u_2, 0.4), (u_3, 0.5), (u_4, 0.7)\},\\
F(e_2, r, 0) &= \{(u_1, 0.4), (u_2, 0.6), (u_3, 0.5), (u_4, 0.3)\},\\
F(e_3, m, 0) &= \{(u_1, 0.8), (u_2, 0.4), (u_3, 0.3), (u_4, 0.1)\},\\
F(e_3, n, 0) &= \{(u_1, 0.9), (u_2, 0.6), (u_3, 0.5), (u_4, 0.3)\},\\
F(e_3, r, 0) &= \{(u_1, 0.8), (u_2, 0.5), (u_3, 0.4), (u_4, 0.6)\}.
\end{aligned}
$$

Then the family $\{F(e_1, m, 1), F(e_1, n, 1), F(e_1, r, 1), F(e_2, m, 1), F(e_2, n, 1), F(e_2, r, 1), F(e_3, m, 1), F(e_3, n, 1), F(e_3, r, 1), F(e_1, m, 0), F(e_1, n, 0), F(e_1, r, 0), F(e_2, m, 0), F(e_2, n, 0), F(e_2, r, 0), F(e_3, m, 0), F(e_3, n, 0), F(e_3, r, 0)\}$ is a fuzzy soft expert set over U.

3 Interval-Valued Generalised Fuzzy Soft Expert Sets

Obviously, by combining fuzzy soft expert set and the interval-valued fuzzy set, it is natural to define the interval-valued generalised fuzzy soft expert sets model. We first define two types of generalised fuzzy soft expert set as follows. Let U be a universe set, E a set of parameters, X a set of experts (agents), and $O = \{1 = agree, 0 = disagree\}$ a set of opinions. For simplicity we assume in this paper, two-valued opinions only in set O, but multivalued opinions may be assumed as well. Let $Z = E \times X \times O$ and $A \subseteq Z$.

Definition 8. *Let $\tilde{F} : A \to I^U$, and let α be a fuzzy sets of A, that is, $\alpha : A \to [0,1]$. Define a function $\tilde{F}_\alpha : A \to I^U \times [0,1]$ as $\tilde{F}_\alpha(e) = (\tilde{F}(e), \alpha(e))$, where I^U denotes the collection of all fuzzy subsets of U. Then $(\tilde{F}_\alpha, A)$ is called type 1 generalised fuzzy soft expert set over the soft universe (U, E).*

Here for each parameter e, $\tilde{F}_\alpha(e)$ indicates not only the degree of belongingness of elements of U in $\tilde{F}(e)$ but also the degree of preference of such belongingness which is represented by $\alpha(e)$.

Definition 9. *Let $\tilde{F} : A \to I^U$, and let α be an interval-valued fuzzy sets of A, that is, $\alpha : A \to Int([0,1])$, where $Int([0,1])$ stands for the set of all closed subintervals of [0,1]. Define a function $\tilde{F}_\alpha : A \to I^U \times Int([0,1])$ as $\tilde{F}_\alpha(e) = (\tilde{F}(e), \alpha(e))$, where I^U denotes the collection of all fuzzy subsets of U and $\alpha(e) = [\alpha^-(e), \alpha^+(e)]$ is interval values. Then $(\tilde{F}_\alpha, A)$ is called type 2 generalised fuzzy soft expert set over the soft universe (U, E).*

It is clear that if $\alpha^-(e) = \alpha^+(e)$ hold for each $e \in A$, then the type 2 generalised fuzzy soft expert set is just the type 1 generalised fuzzy soft expert set.

In this paper, the type 2 generalised fuzzy soft expert set is denoted by GFSE set for short. To illustrate this idea, let's consider the following example.

Example 2. Let $U = \{u_1, u_2, u_3, u_4\}$ be a set of universe, and let $E = \{e_1, e_2, e_3\}$ be a set of parameters. Let $Z = E \times X \times O$ and $A \subseteq Z$. Let $\tilde{F}_\alpha : A \to I^U \times Int([0,1])$ be a function given as follows:

$$F(e_1, m, 1) = (\{(u_1, 0.6), (u_2, 0.3), (u_3, 0.2), (u_4, 0.4)\}, [0.1, 0.2]),$$
$$F(e_1, n, 1) = (\{(u_1, 0.7), (u_2, 0.4), (u_3, 0.1), (u_4, 0.3)\}, [0.1, 0.2]),$$
$$F(e_1, r, 1) = (\{(u_1, 0.1), (u_2, 0.3), (u_3, 0.8), (u_4, 0.7)\}, [0.2, 0.3]),$$
$$F(e_2, m, 1) = (\{(u_1, 0.5), (u_2, 0.3), (u_3, 0.2), (u_4, 0.7)\}, [0.3, 0.4]),$$
$$F(e_2, n, 1) = (\{(u_1, 0.7), (u_2, 0.8), (u_3, 0.6), (u_4, 0.4)\}, [0.2, 0.3]),$$
$$F(e_2, r, 1) = (\{(u_1, 0.7), (u_2, 0.5), (u_3, 0.6), (u_4, 0.7)\}, [0.25, 0.3]),$$
$$F(e_3, m, 1) = (\{(u_1, 0.3), (u_2, 0.6), (u_3, 0.8), (u_4, 0.9)\}, [0.6, 0.7]),$$
$$F(e_3, n, 1) = (\{(u_1, 0.2), (u_2, 0.3), (u_3, 0.7), (u_4, 0.6)\}, [0.5, 0.6]),$$
$$F(e_3, r, 1) = (\{(u_1, 0.3), (u_2, 0.4), (u_3, 0.6), (u_4, 0.3)\}, [0.8, 0.9]),$$
$$F(e_1, m, 0) = (\{(u_1, 0.3), (u_2, 0.6), (u_3, 0.7), (u_4, 0.6)\}, [0.7, 0.8]),$$
$$F(e_1, n, 0) = (\{(u_1, 0.2), (u_2, 0.7), (u_3, 0.8), (u_4, 0.7)\}, [0.9, 1]),$$
$$F(e_1, r, 0) = (\{(u_1, 0.6), (u_2, 0.8), (u_3, 0.4), (u_4, 0.5)\}, [0.6, 0.7]),$$
$$F(e_2, m, 0) = (\{(u_1, 0.3), (u_2, 0.6), (u_3, 0.8), (u_4, 0.3)\}, [0.6, 0.7]),$$
$$F(e_2, n, 0) = (\{(u_1, 0.2), (u_2, 0.4), (u_3, 0.5), (u_4, 0.7)\}, [0.7, 0.8]),$$
$$F(e_2, r, 0) = (\{(u_1, 0.4), (u_2, 0.6), (u_3, 0.5), (u_4, 0.3)\}, [0.4, 0.5]),$$
$$F(e_3, m, 0) = (\{(u_1, 0.8), (u_2, 0.4), (u_3, 0.3), (u_4, 0.1)\}, [0.3, 0.4]),$$

$$F(e_3, n, 0) = (\{(u_1, 0.9), (u_2, 0.6), (u_3, 0.5), (u_4, 0.3)\}, [0.6, 0.7]),$$
$$F(e_3, r, 0) = (\{(u_1, 0.8), (u_2, 0.5), (u_3, 0.4), (u_4, 0.6)\}, [0.5, 0.6]).$$

Then $(\tilde{F}_\alpha, A) = \{F(e_1, m, 1), F(e_1, n, 1), F(e_1, r, 1), F(e_2, m, 1), F(e_2, n, 1), F(e_2, r, 1), F(e_3, m, 1), F(e_3, n, 1), F(e_3, r, 1), F(e_1, m, 0), F(e_1, n, 0), F(e_1, r, 0), F(e_2, m, 0), F(e_2, n, 0), F(e_2, r, 0), F(e_3, m, 0), F(e_3, n, 0), F(e_3, r, 0)\}$ is a GFSE set.

Definition 10. *Let $(\tilde{F}_\alpha, A)$ and $(\tilde{G}_\beta, B)$ be two GFSE sets over (U, E). Now $(\tilde{F}_\alpha, A)$ is called a GFSE subset of $(\tilde{G}_\beta, B)$ if*

(i) $A \subseteq B$,
(ii) For any $e \in A$, $\tilde{F}(e)$ is a fuzzy soft expert subset of $\tilde{G}(e)$.
(iii) α is an interval-valued fuzzy subset of β, that is, $\alpha^-(e) \leq \beta^-(e)$ and $\alpha^+(e) \leq \beta^+(e)$ for any $e \in A$.

In this case, the above relationship is denoted by $(\tilde{F}_\alpha, A) \tilde{\subseteq} (\tilde{G}_\beta, B)$. And $(\tilde{G}_\beta, B)$ is said to be a GFSE superset of $(\tilde{F}_\alpha, A)$.

Definition 11. *Let $(\tilde{F}_\alpha, A)$ and $(\tilde{G}_\beta, B)$ be two GFSE sets over (U, E). Then $\tilde{F}_\alpha$ and $\tilde{G}_\beta$ are said to be GFSE equal if and only if $\tilde{F}_\alpha \tilde{\subseteq} \tilde{G}_\beta$ and $\tilde{G}_\alpha \tilde{\subseteq} \tilde{F}_\beta$.*

Definition 12. *The complement of a GFSE set $(\tilde{F}_\alpha, A)$ is denoted by $(\tilde{F}_\alpha, A)^C$ and is defined by $(\tilde{F}_\alpha, A)^C = (\tilde{F}_\alpha^C, \neg A)$ where $\tilde{F}_\alpha^C : \neg A \to I^U \times Int([0,1])$ is a mapping given by $\tilde{F}_\alpha^C(e) = 1 - \tilde{F}_\alpha(\neg e), \forall e \in \neg A$ where $\neg A \subset \neg E \times X \times O$.*

Example 3. We consider the GFSE set $(\tilde{F}_\alpha, A)$ given in Example 2 and define a GFSE set $(\tilde{G}_\beta, B)$ as follows:

$$B = \{(e_1, m, 1), (e_2, n, 0), (e_3, r, 1), (e_1, n, 0), (e_2, r, 1), (e_3, m, 0)\}, \text{ Clearly, } B \subseteq A.$$
$$G(e_1, m, 1) = (\{(u_1, 0.4), (u_2, 0.1), (u_3, 0.15), (u_4, 0.2)\}, [0.05, 0.15]),$$
$$G(e_2, n, 0) = (\{(u_1, 0.1), (u_2, 0.3), (u_3, 0.2), (u_4, 0.5)\}, [0.6, 0.7]),$$
$$G(e_3, r, 1) = (\{(u_1, 0.2), (u_2, 0.1), (u_3, 0.4), (u_4, 0.1)\}, [0.7, 0.8]),$$
$$G(e_1, n, 0) = (\{(u_1, 0.1), (u_2, 0.6), (u_3, 0.6), (u_4, 0.5)\}, [0.6, 0.7]),$$
$$G(e_2, r, 1) = (\{(u_1, 0.5), (u_2, 0.4), (u_3, 0.5), (u_4, 0.6)\}, [0.2, 0.3]),$$
$$G(e_3, m, 0) = (\{(u_1, 0.7), (u_2, 0.3), (u_3, 0.2), (u_4, 0.1)\}, [0.2, 0.3]),$$

and for any $e \in A$, $\tilde{G}(e)$ is a fuzzy soft expert subset of $\tilde{F}(e)$, β is an interval-valued fuzzy subset of α.

Therefore, $(\tilde{G}_\beta, B)$ is a GFSE subset of $(\tilde{F}_\alpha, A)$.

Definition 13. *An agree-GFSE set $(\tilde{F}_\alpha, A)_1$ over a common universe U is a GFSE subset of $(\tilde{F}_\alpha, A)$ defined as follows: $(\tilde{F}_\alpha, A)_1 = \{F_1(\alpha) : \alpha \in E \times X \times \{1\}\}$.*

Definition 14. *A disagree-GFSE set* $(\tilde{F}_\alpha, A)_0$ *over a common universe* U *is a GFSE subset of* $(\tilde{F}_\alpha, A)$ *defined as follows:* $(\tilde{F}_\alpha, A)_0 = \{F_0(\alpha) : \alpha \in E\times X\times\{0\}\}$.

Example 4. Consider Example 2. Then the agree-GFSE set of $(\tilde{F}_\alpha, A)$ over U is

$$\begin{aligned}
(\tilde{F}_\alpha, A)_1 = \{&((e_1, m, 1), (\{(u_1, 0.6), (u_2, 0.3), (u_3, 0.2), (u_4, 0.4)\}, [0.1, 0.2])),\\
&((e_1, n, 1), (\{(u_1, 0.7), (u_2, 0.4), (u_3, 0.1), (u_4, 0.3)\}, [0.1, 0.2])),\\
&((e_1, r, 1), (\{(u_1, 0.1), (u_2, 0.3), (u_3, 0.8), (u_4, 0.7)\}, [0.2, 0.3])),\\
&((e_2, m, 1), (\{(u_1, 0.5), (u_2, 0.3), (u_3, 0.2), (u_4, 0.7)\}, [0.3, 0.4])),\\
&((e_2, n, 1), (\{(u_1, 0.7), (u_2, 0.8), (u_3, 0.6), (u_4, 0.4)\}, [0.2, 0.3])),\\
&((e_2, r, 1), (\{(u_1, 0.7), (u_2, 0.5), (u_3, 0.6), (u_4, 0.7)\}, [0.25, 0.3])),\\
&((e_3, m, 1), (\{(u_1, 0.3), (u_2, 0.6), (u_3, 0.8), (u_4, 0.9)\}, [0.6, 0.7])),\\
&((e_3, n, 1), (\{(u_1, 0.2), (u_2, 0.3), (u_3, 0.7), (u_4, 0.6)\}, [0.5, 0.6])),\\
&((e_3, r, 1), (\{(u_1, 0.3), (u_2, 0.4), (u_3, 0.6), (u_4, 0.3)\}, [0.8, 0.9]))\}.
\end{aligned}$$

and the disagree-GFSE set $(\tilde{F}_\alpha, A)_0$ over U is

$$\begin{aligned}
(\tilde{F}_\alpha, A)_0 = \{&((e_1, m, 0), (\{(u_1, 0.3), (u_2, 0.6), (u_3, 0.7), (u_4, 0.6)\}, [0.7, 0.8])),\\
&((e_1, n, 0), (\{(u_1, 0.2), (u_2, 0.7), (u_3, 0.8), (u_4, 0.7)\}, [0.9, 1])),\\
&((e_1, r, 0), (\{(u_1, 0.6), (u_2, 0.8), (u_3, 0.4), (u_4, 0.5)\}, [0.6, 0.7])),\\
&((e_2, m, 0), (\{(u_1, 0.3), (u_2, 0.6), (u_3, 0.8), (u_4, 0.3)\}, [0.6, 0.7])),\\
&((e_2, n, 0), (\{(u_1, 0.2), (u_2, 0.4), (u_3, 0.5), (u_4, 0.7)\}, [0.7, 0.8])),\\
&((e_2, r, 0), (\{(u_1, 0.4), (u_2, 0.6), (u_3, 0.5), (u_4, 0.3)\}, [0.4, 0.5])),\\
&((e_3, m, 0), (\{(u_1, 0.8), (u_2, 0.4), (u_3, 0.3), (u_4, 0.1)\}, [0.3, 0.4])),\\
&((e_3, n, 0), (\{(u_1, 0.9), (u_2, 0.6), (u_3, 0.5), (u_4, 0.3)\}, [0.6, 0.7])),\\
&((e_3, r, 0), (\{(u_1, 0.8), (u_2, 0.5), (u_3, 0.4), (u_4, 0.6)\}, [0.5, 0.6]))\}.
\end{aligned}$$

Definition 15. *The union of two GFSE sets* $(\tilde{F}_\alpha, A)$ *and* $(\tilde{G}_\beta, B)$ *is denoted by* $(\tilde{F}_\alpha, A)\,\tilde{\cup}\,(\tilde{G}_\beta, B)$ *and defined by a GFSE set* $\tilde{H}_\delta : A \cup B \to I^U \times Int([0,1])$ *such that for each* $e \in A \cup B$ *and* $h \in U$,

$$\tilde{H}_\delta(e) = \begin{cases} (\{h, \mu_{\tilde{F}(e)}(h)\}, \alpha(e)), \; if\, e \in A - B; \\ (\{h, \mu_{\tilde{G}(e)}(h)\}, \beta(e)), \; if\, e \in B - A; \\ (\{h, \mu_{\tilde{H}(e)}(h)\}, \delta(e)), \; if\, e \in A \cap B. \end{cases}$$

where $\mu_{\tilde{H}(e)}(h) = S(\mu_{\tilde{F}(e)}(h), \mu_{\tilde{G}(e)}(h))$ *and* $\delta(e) = S(\alpha(e), \beta(e))$.

Example 5. Let us consider the GFSE sets $(\tilde{F}_\alpha, A)$ and $(\tilde{G}_\beta, B)$ as follows. Let us define the t-conorm S as follows: $S(a,b) = max(a,b)$. Then

$A = \{(e_1,m,1),(e_2,n,0),(e_3,r,1),(e_2,m,1),(e_1,n,1),(e_3,n,1),(e_2,r,1),(e_3,m,0)\}$
$B = \{(e_1,m,1),(e_2,n,0),(e_3,r,1),(e_1,n,0),(e_2,r,1),(e_3,m,0)\}$.
$F(e_1,m,1) = (\{(u_1,0.8),(u_2,0.5),(u_3,0.2),(u_4,0.4)\},[0.1,0.2])$,
$F(e_2,n,0) = (\{(u_1,0.2),(u_2,0.4),(u_3,0.9),(u_4,0.8)\},[0.2,0.3])$,
$F(e_3,r,1) = (\{(u_1,0.6),(u_2,0.4),(u_3,0.3),(u_4,0.8)\},[0.4,0.5])$,
$F(e_2,m,1) = (\{(u_1,0.2),(u_2,0.3),(u_3,0.5),(u_4,0.4)\},[0.6,0.7])$,
$F(e_1,n,1) = (\{(u_1,0.8),(u_2,0.6),(u_3,0.7),(u_4,0.8)\},[0.3,0.4])$,
$F(e_3,n,1) = (\{(u_1,0.3),(u_2,0.4),(u_3,0.8),(u_4,0.7)\},[0.5,0.6])$,
$F(e_2,r,1) = (\{(u_1,0.5),(u_2,0.7),(u_3,0.6),(u_4,0.4)\},[0.6,0.7])$,
$F(e_3,m,0) = (\{(u_1,0.4),(u_2,0.6),(u_3,0.5),(u_4,0.3)\},[0.5,0.6])$,
$G(e_1,m,1) = (\{(u_1,0.6),(u_2,0.7),(u_3,0.3),(u_4,0.2)\},[0.3,0.4])$,
$G(e_2,n,0) = (\{(u_1,0.5),(u_2,0.3),(u_3,0.9),(u_4,0.7)\},[0.1,0.2])$,
$G(e_3,r,1) = (\{(u_1,0.6),(u_2,0.5),(u_3,0.1),(u_4,0.7)\},[0.4,0.5])$,
$G(e_1,n,0) = (\{(u_1,0.9),(u_2,0.7),(u_3,1),(u_4,0.9)\},[0.8,0.9])$,
$G(e_2,r,1) = (\{(u_1,0.4),(u_2,0.5),(u_3,0.4),(u_4,0.2)\},[0.4,0.5])$,
$G(e_3,m,0) = (\{(u_1,0.6),(u_2,0.8),(u_3,0.7),(u_4,0.5)\},[0.9,1])$,

Then $(\tilde{F}_\alpha, A)\,\tilde{\cup}\,(\tilde{G}_\beta, B) = (\tilde{H}_\delta, C)$ where

$(\tilde{H}_\delta, C) = \{((e_1,m,1),(\{(u_1,0.8),(u_2,0.7),(u_3,0.3),(u_4,0.4)\},[0.3,0.4]))$,
$((e_2,n,0),(\{(u_1,0.5),(u_2,0.4),(u_3,0.9),(u_4,0.8)\},[0.2,0.3]))$,
$((e_3,r,1),(\{(u_1,0.6),(u_2,0.5),(u_3,0.3),(u_4,0.8)\},[0.4,0.5]))$,
$((e_2,m,1),(\{(u_1,0.2),(u_2,0.3),(u_3,0.5),(u_4,0.4)\},[0.6,0.7]))$,
$((e_1,n,1),(\{(u_1,0.8),(u_2,0.6),(u_3,0.7),(u_4,0.8)\},[0.3,0.4]))$,
$((e_3,n,1),(\{(u_1,0.3),(u_2,0.4),(u_3,0.8),(u_4,0.7)\},[0.5,0.6]))$,
$((e_1,n,0),(\{(u_1,0.9),(u_2,0.7),(u_3,1),(u_4,0.9)\},[0.8,0.9]))$,
$((e_2,r,1),(\{(u_1,0.5),(u_2,0.7),(u_3,0.6),(u_4,0.4)\},[0.6,0.7]))$,
$((e_3,m,0),(\{(u_1,0.6),(u_2,0.8),(u_3,0.7),(u_4,0.5)\},[0.9,1]))\}$

Definition 16. *The intersection of two GFSE sets* $(\tilde{F}_\alpha, A)$ *and* $(\tilde{G}_\beta, B)$ *is denoted by* $(\tilde{F}_\alpha, A)\,\tilde{\cap}\,(\tilde{G}_\beta, B)$ *and defined by a GFSE set* $\tilde{H}_\delta : A \cup B \to I^U \times Int([0,1])$ *such that for each* $e \in A \cup B$ *and* $h \in U$,

$$\tilde{H}_\delta(e) = \begin{cases} (\{h, \mu_{\tilde{F}(e)}(h)\}, \alpha(e)), if e \in A - B; \\ (\{h, \mu_{\tilde{G}(e)}(h)\}, \beta(e)), if e \in B - A; \\ (\{h, \mu_{\tilde{H}(e)}(h)\}, \delta(e)), if e \in A \cap B. \end{cases}$$

where $\mu_{\tilde{H}(e)}(h)) = T(\mu_{\tilde{F}(e)}(h), \mu_{\tilde{G}(e)}(h))$ *and* $\delta(e) = T(\alpha(e), \beta(e))$.

Example 6. Let us consider the GFSE sets $(\tilde{F}_\alpha, A)$ and $(\tilde{G}_\beta, B)$ as follows in Example 2. Let us define the t-norm T as follows: $T(a,b) = min(a,b)$. Then we have$(\tilde{F}_\alpha, A)\tilde{\cap}(\tilde{G}_\beta, B) = (\tilde{H}_\delta, C)$ where

$$
\begin{aligned}
(\tilde{H}_\delta, C) = \{&((e_1, m, 1), (\{(u_1, 0.6), (u_2, 0.5), (u_3, 0.2), (u_4, 0.2)\}, [0.1, 0.2])),\\
&((e_2, n, 0), (\{(u_1, 0.2), (u_2, 0.3), (u_3, 0.9), (u_4, 0.7)\}, [0.1, 0.2])),\\
&((c_3, r, 1), (\{(u_1, 0.6), (u_2, 0.4), (u_3, 0.1), (u_4, 0.7)\}, [0.4, 0.5])),\\
&((e_2, m, 1), (\{(u_1, 0.2), (u_2, 0.3), (u_3, 0.5), (u_4, 0.4)\}, [0.6, 0.7])),\\
&((e_1, n, 1), (\{(u_1, 0.8), (u_2, 0.6), (u_3, 0.7), (u_4, 0.8)\}, [0.3, 0.4])),\\
&((e_3, n, 1), (\{(u_1, 0.3), (u_2, 0.4), (u_3, 0.8), (u_4, 0.7)\}, [0.5, 0.6])),\\
&((e_1, n, 0), (\{(u_1, 0.9), (u_2, 0.7), (u_3, 1), (u_4, 0.9)\}, [0.8, 0.9])),\\
&((e_2, r, 1), (\{(u_1, 0.4), (u_2, 0.5), (u_3, 0.4), (u_4, 0.2)\}, [0.4, 0.5])),\\
&((e_3, m, 0), (\{(u_1, 0.4), (u_2, 0.6), (u_3, 0.5), (u_4, 0.3)\}, [0.5, 0.6]))\}.
\end{aligned}
$$

Theorem 1. *Let $(\tilde{F}_\alpha, A)$, $(\tilde{G}_\beta, B)$ and $(\tilde{H}_\delta, C)$ be any three GIFS sets over (U, E), then the following holds:*

(1) $(\tilde{F}_\alpha, A)\tilde{\cup}(\tilde{G}_\beta, B) = (\tilde{G}_\beta, B)\tilde{\cup}(\tilde{F}_\alpha, A)$.
(2) $(\tilde{F}_\alpha, A)\tilde{\cap}(\tilde{G}_\beta, B) = (\tilde{G}_\beta, B)\tilde{\cap}(\tilde{F}_\alpha, A)$.
(3) $(\tilde{F}_\alpha, A)\tilde{\cup}((\tilde{G}_\beta, B)\tilde{\cup}(\tilde{H}_\delta, C)) = ((\tilde{F}_\alpha, A)\tilde{\cup}(\tilde{G}_\beta, B))\tilde{\cup}(\tilde{H}_\delta, C)$.
(4) $(\tilde{F}_\alpha, A)\tilde{\cap}((\tilde{G}_\beta, B)\tilde{\cap}(\tilde{H}_\delta, C)) = ((\tilde{F}_\alpha, A)\tilde{\cap}(\tilde{G}_\beta, B))\tilde{\cap}(\tilde{H}_\delta, C)$.

Proof. Since the t-norm function and t-conorm functions are commutative and associative, therefore the theorem follows.

Concluding Remarks. In this work we have introduced the concept of interval-valued generalised fuzzy soft expert set and studied some of its properties. The complement, union, and intersection operations have been defined on the interval-valued generalised fuzzy soft expert set. Also, we will investigate the applications of GFSE sets with multiopinions in decision-making problems in the following paper. And the theory of GFSE can be use in the classification for gene expression engineering [21,22].

Acknowledgment. This paper is supported by the Project of Shandong Province Higher Educational Science and Technology Program 2016(J16LI08).

References

1. Molodtsov, D.: Soft set theory-first results. Comput. Math. Appl. **37**, 19–31 (1999)
2. Maji, P.K., Biswas, R., Roy, A.R.: Soft set theory. Comput. Math. Appl. **45**(4–5), 555–562 (2003)
3. Mustafa, H.I.: Soft rough approximation operators on a complete atomic boolean lattice. Math. Probl. Eng **2013**, 11 (2013). Article ID 486321
4. Jun, Y.B., Park, C.H.: Applications of soft sets in ideal theory of BCK/BC Ialgebras. Inf. Sci. **178**, 2466–2475 (2008)

5. Shabir, M., Naz, M.: On soft topological spaces. Comput. Math. Appl. **61**, 1786–1799 (2011)
6. Gunduz, C., Bayramov (Aras), S.: Some results on fuzzy soft topological spaces. Math. Probl. Eng. **2013**, 10 (2013). Article ID 835308
7. Chen, B.: Soft semi-open sets and related properties in soft topological spaces. Appl. Math. Inf. Sci. **7**(1), 287–294 (2013)
8. Chen, B.: Some local properties of soft semi-open sets. Discrete Dyn. Nat. Soc. (2013). Article ID 298032
9. Maji, P.K., Roy, A.R., Biswas, R.: An application of soft sets in a decision making problem. Comput. Math. Appl. **44**, 1077–1083 (2002)
10. Chen, D., Tsang, E.C.C., Yeung, D.S., Wang, X.: The parameterization reduction of soft sets and its applications. Comput. Math. Appl. **49**, 757–763 (2005)
11. Roy, A.R., Maji, P.K.: A fuzzy soft set theoretic approach to decision making problems. J. Comput. Appl. Math. **203**, 412–418 (2007)
12. Zou, Y., Xiao, Z.: Data analysis approaches of soft sets under incomplete information. Knowl. Based Syst. **21**, 941–945 (2008)
13. Kong, Z., Gao, L., Wang, L., Li, S.: The normal parameter reduction of soft sets and its algorithm. Comput. Math. Appl. **56**, 3029–3037 (2008)
14. Yang, X., Lin, T.Y., Yang, J., Li, Y., Yu, D.: Combination of interval-valued fuzzy set and soft set. Comput. Math. Appl. **58**(3), 521–527 (2009)
15. Majumdar, P., Samanta, S.K.: Generalised fuzzy soft sets. Comput. Math. Appl. **59**(4), 1425–1432 (2010)
16. Alkhazaleh, S., Salleh, A.R.: Soft expert sets. Adv. Decis. Sci. **2011**, 12 (2011). Article ID 757868
17. Alkhazaleh, S., Salleh, A.R.: Fuzzy soft expert sets. Ph.D. thesis, Faculty of Science and Technology, University Kebangsaan Malaysia, Selangor, Malaysia
18. Maji, P.K., Biswas, R., Roy, A.R.: Fuzzy soft sets. J. Fuzzy Math. **9**, 589–602 (2001)
19. Gorzalczany, M.B.: A method of inference in approximate reasoning based on interval-valued fuzzy sets. Fuzzy Sets Syst. **21**(1), 1–17 (1987)
20. Schweizer, B., Sklar, A.: Statistical metric space. Pac. J. Mathh. **10**, 314–334 (1960)
21. Kalaiselvi, N., Hannah Inbarani, H.: Fuzzy soft set based classification for gene expression data. http://arxiv.org/abs/1301.1502
22. Handaga, B., Mat, M.: Similarity approach on fuzzy soft set based numerical data classification. In: ICSECS, Part II, CCIS, vol. 180, pp. 575–589 (2011)

Author Index

Y. Liu et al. (Eds.): ICNC-FSKD 2019, AISC 1074, pp. 1001–1004, 2020.
https://doi.org/10.1007/978-3-030-32456-8

The manufacturer's authorised representative in the EU is Springer Nature Customer Service Centre GmbH, Europaplatz 3, 69115 Heidelberg, Germany. If you have any concerns regarding our products, please contact ProductSafety@springernature.com

Printed and bound by CPI Group (UK) Ltd, Croydon, CR0 4YY
30/09/2026
02225232-0017